Fundamentals of Heat and Mass Transfer

FOURTH EDITION

Fundamentals of Heat and Mass Transfer

FRANK P. INCROPERA
DAVID P. DeWITT

School of Mechanical Engineering
Purdue University

JOHN WILEY & SONS

New York • Chichester • Brisbane • Toronto • Singapore

ANTON BASER

ACQUISITIONS EDITOR	Cliff Robichaud
MARKETING MANAGER	Debra Riegert
PRODUCTION MANAGER	Lucille Buonocore
SENIOR PRODUCTION EDITORS	Nancy Prinz, Tracey Kuehn
TEXT DESIGNER	Nancy Field
COVER DESIGNER	Karin Kincheloe
MANUFACTURING MANAGER	Mark Cirillo
ILLUSTRATION EDITOR	Edward Starr

This book was set in Times Roman by General Graphic Services, and printed and bound by R.R. Donnelley, Willard. The cover was printed by Phoenix Color.

Recognizing the importance of preserving what has been written, it is a policy of John Wiley & Sons, Inc., to have books of enduring value published in the United States printed on acid-free paper, and we exert our best efforts to that end.

The paper in this book was manufactured by a mill whose forest management programs include sustained yield harvesting of its timberlands. Sustained yield harvesting principles ensure that the number of trees cut each year does not exceed the amount of new growth.

Library of Congress Cataloging-in-Publication Data:

Incropera, Frank P.
    Fundamentals of heat and mass transfer / Frank P. Incropera, David P. DeWitt. -- 4th ed.
        p.   cm.
    Includes index.
    ISBN 0-471-30460-3 (cloth : alk. paper)
    1. Heat--Transmission.   2. Mass transfer.   I. DeWitt, David P.,
1934–   .   II. Title
QC320.I45   1996
621.402′2--dc20   95-46179
                                                          CIP

Printed in the United States of America

10 9 8

Dedicated to our extended families and their children,

Nicholas DeWitt and Alexandra Joanne Bifano; John Wallace,
Michael Anthony, and Mallory Renee Dant; Patricia Ann and
David Andrew Foley; Michael DeWitt and Sarah Joanne Frederick; and
Brandon Patrick Tafelski

who have brought a new level of love, patience,
and understanding into our lives.

Preface

With the passage of approximately fifteen years since publication of the first edition, this text has clearly become a mature representative of heat transfer pedagogy. Despite this maturation, however, we like to think that, while remaining true to certain basic tenets, our treatment of the subject is constantly evolving.

Preparation of the first edition was motivated by the belief that, above all, a first course in heat transfer should do two things: (i) instill an appreciation for the physical origins of the subject, and (ii) establish the relationship of these origins to the behavior of thermal systems. In so doing, it should develop methodologies which facilitate application of the subject to a broad range of practical problems, and it should nurture the facility to perform the kind of engineering analysis which, if not exact, still provides useful information concerning the design and/or performance of a system or process. Requirements of such an analysis include the ability to discern relevant transport processes and simplifying assumptions, identify pertinent dependent and independent variables, develop appropriate expressions from first principles, and introduce requisite material from the heat transfer knowledge base. In the first edition, achievement of this objective was fostered by couching many of the examples and end-of-chapter problems in terms of actual engineering systems.

The second edition was also driven by the foregoing objectives, as well as by input derived from a questionnaire sent to over 100 colleagues who used, or were otherwise familiar with, the first edition. A major consequence of this input was publication of two versions of the book, *Fundamentals of Heat and Mass Transfer* and *Introduction to Heat Transfer*. As in the first edition, the *Fundamentals* version included mass transfer and provided an integrated treatment of heat, mass and momentum transfer by convection, as well as separate treatments of heat and mass transfer by diffusion. The *Introduction* version of the book was intended for users who embraced the treatment of heat transfer but did not wish to cover mass transfer effects. In both versions, significant improvements were made in the treatments of numerical methods and heat transfer with phase change.

In the third edition, changes were motivated by the desire to expand the scope of applications and to enhance the exposition of physical principles. Coverage of existing material on thermal contact resistance, lumped capacitance analysis, finite-difference methods, and compact heat exchangers was enhanced, while new material on forced convection in submerged jets and free convection in open, parallel plate channels was added. In addition, nearly 300 new problems were provided. In the spirit of past efforts, many of the problems addressed contemporary issues in engineering practice, such as energy conversion and utilization, thermal protection, electronic cooling, manufacturing, and materials processing. It remains our belief that, in addition to reinforcing the student's understanding of principles and applications, the problems serve a motivational role by relating the subject to real engineering needs.

In preparing for the current edition, we were strongly influenced by the intense scrutiny to which engineering education has recently been subjected. On the one hand, we hear that, in placing considerable emphasis on analysis and the engineering sciences, synthesis and systems integration skills commonly required for engineering practice are being neglected. In counterpoint, advocates of post-1950s methods of engineering education argue that a thorough appreciation of basic engineering principles is essential to understanding and improving the operation of existing devices, processes and systems, as well as to the development of new technologies. In our case, we happen to agree with both contentions. We can do a better job preparing our students for engineering practice, and it is important that they understand and be able to apply basic principles. However, we also believe that the two goals are not mutually exclusive, but can be coupled to mutual benefit.

Few educators have not experienced the frustration of seeing students, who have satisfactorily completed the core engineering sciences, flounder in attempting to apply even the most rudimentary principles to design and system level issues. We believe that such difficulties result from a mindset which views each problem as having a unique (the *correct*) solution and only one path to that solution. With the goal of finding the correct path to the correct solution, problem solving runs the risk of becoming a tightly constrained exercise in *pattern recognition.* That is, one's approach to problem solving focuses on a search for existing solutions to similar problems.

At Purdue, as at many other institutions throughout the country, we are using design education as a means of addressing the foregoing deficiencies. An important characteristic of our approach involves *integrating design throughout the curriculum,* including those courses, like heat transfer, which are rooted in the engineering sciences. In such courses *design and open-ended problems* provide fertile ground for relating fundamentals to useful *engineering models* and, in turn, linking these models to design decisions. Although problems may be of limited scope, perhaps requiring no more than a few hours outside the classroom, they can still address real needs and allow for alternative approaches, including *what-if* considerations. In so doing, they provide precisely the context needed for students to develop confidence in applying basic principles to real, open-ended problems and to use such applications as a basis for making design decisions. Through the stimulation they provide, the problems can also heighten interest in and deepen understanding of the basic principles.

In this edition of the book, we have therefore added a significant number of open-ended problems which we believe will enhance student interest in heat transfer, strengthen their ability to apply the subject to real needs, and better prepare them for engineering practice. Because many of these problems involve *exploratory, what-if,* and *parameter sensitivity* considerations, it is recommended that they be addressed by using a computer with an equation-solving software package. Although students can certainly create and solve their models using software with which they may already be familiar, the Windows®-based software developed for this text offers some distinct advantages as a productivity and learning tool. Termed *Interactive Heat Transfer* (IHT) and developed collaboratively with IntelliPro, Inc. of New Brunswick, New Jersey, the software is fully integrated with the text, using the same methodologies and nomenclature.

IHT provides a model-building, problem-solving environment, which includes a *pre-processor,* a *solver,* and a *post-processor.* The pre-processor encompasses a *work space,* into which equations and comments may be entered from preexisting **modules** and/or **tool pads,** as well as from the keyboard. The modules consist of *models,* which cover broader issues, such as energy balances and thermal circuits, while the tool pads provide specific equations for conduction, convection and radiation processes, as well as thermophysical properties for selected substances. The solver provides comprehensive, equation-solving capabilities, while the post-processor includes an *explore option* for parameter sensitivity studies, a *browser* for tabulating results, and a *graphical option* for plotting results. The model-building, problem-solving capabilities of IHT facilitate implementation of the methodologies espoused in the text, as well as the execution of *design* and *what-if* considerations.

Models accessible from the pre-processor are embodied in six different *modules,* each of which contains one or more models. Following the order in which they appear in the text, the **modules** and related *models* are as follows.

1. **First Law:** steady-state energy balances for
   - *isothermal planar, cylindrical and spherical geometries with multimode effects;*
   - *nonisothermal plane wall with multimode effects;*
   - *flow across a tube bank;*
   - *flow within a tube.*

2. **Resistance Networks:** thermal circuit builder and solver for
   - *one-dimensional conduction in plane, cylindrical and spherical walls with convective and/or radiative surface conditions.*

3. **One-dimensional, Steady-state Conduction:** temperature and heat flux distributions with or without uniform energy generation for
   - *one-dimensional conduction in planar, cylindrical and spherical geometries with boundary conditions of the first, second or third kind.*

4. **Extended Surfaces:** models for
   - *temperature and heat rate distributions in a rectangular straight or pin fin;*
   - *performance of a rectangular, triangular or parabolic straight or pin fin and a rectangular circular fin;*
   - *performance of arrays of straight, pin and circular fins.*

5. **Lumped Capacitance:** model builder for
   - *transient response of spacewise isothermal systems with radiation and/or convection surface conditions, with or without energy generation.*

6. **Transient Conduction:** models for one-dimensional, transient conduction in
   - *finite planar, cylindrical, and spherical geometries;*
   - *semi-infinite solids.*

Model-building and problem-solving capabilities are enhanced by features of the following **tool pads** and related *functions*.

1. **Rate Equations:** basic rate equations for
   - *steady-state conduction* (plane, cylindrical and spherical walls);
   - *convection* (plane, cylindrical and spherical surfaces);
   - *radiation* (plane, cylindrical and spherical surfaces).

2. **Thermal Resistances:** expressions for
   - *conduction* (plane, cylindrical and spherical walls);
   - *convection* (plane, cylindrical and spherical surfaces);
   - *radiation* (plane, cylindrical and spherical surfaces).

3. **Finite-Difference Equations:** standard forms of finite-difference equations for
   - *one-dimensional steady and transient systems;*
   - *two-dimensional steady and transient systems.*

4. **Convection Correlations:** correlation equations for
   - *external forced convection* (flat plate, cylinder, sphere, tube bank);
   - *internal forced convection;*
   - *free convection* (vertical and horizontal plates, radial systems);
   - *boiling* (nucleate, film, maximum and minimum heat fluxes);
   - *film condensation* (vertical plate, radial systems).

5. **Heat Exchangers:** effectiveness-NTU relations for design and performance of
   - *concentric tube, shell-and-tube and cross-flow configurations.*

6. **Radiation Exchange:** standard expressions for computing
   - *blackbody functions* (spectral intensity, emissive power and band emission factors);
   - *view factors* (relations and formulas);
   - *radiation exchange in an enclosure.*

7. **Properties:** temperature dependence of thermophysical properties for selected
   - *solids* (aluminum 2024, stainless steel 302, copper, silicon nitride);
   - *liquids* (water, engine oil, ethylene glycol, R12, R113);
   - *gases/vapors* (air, water, helium, R12, R113).

Users of IHT must recognize that it is *not* a collection of pre-solved models to be exercised for different input conditions. Rather, it is a productivity tool which facilitates model construction and solution for the broad range of heat transfer problems embodied in the topical coverage of this text. Construction is facilitated by the ability to *drag* material from any of the modules and tool pads into the work space and, as required for completion of the model, to enter additional equations from the keyboard. For example, if one wished to use the lumped capacitance method (Chapter 5) to determine the transient thermal response of a solid which is cooled by free convection and radiation, the appropriate model could be developed by combining features from module 5 and tool pads 1, 4, and 7. Alternatively, the appropriate energy balance, rate equations, correlation, and properties could be entered from the keyboard. The solver could then be used to compute the desired temperature history, as well as to evaluate and plot the effects of pertinent parameter variations. To facilitate its use, the software also includes a tutorial, worked examples, and options for on-line help.

To minimize frustrations associated with *obtaining incorrect results from an incorrect model,* many of the open-ended problems of this text appear as extensions to single solution problems. In this way students may first develop and test their model under prescribed conditions for which there is only one answer. Having established confidence in the validity of their model, they may then use IHT (or some other solver) to perform parametric calculations from which preferred design or operating conditions may be determined. Such problems are identified by enclosing the exploratory part in a red rectangle, as, for example, (b) , (c) , or (d) . This feature also allows instructors wishing to treat heat transfer without the use of computers to still benefit from the richness of these problems by assigning all but the highlighted portion(s). Problems for which the number itself is highlighted, as, for example, 1.18 , should be solved with a computer.

Relative to use of IHT as a *productivity tool,* it is strongly recommended that students be required to develop their models on paper and to perform limited hand calculations before accessing the software for design/exploratory considerations. Once students have mastered the heat transfer concepts and established familiarity with the software, they will be empowered to address many of the complexities associated with the behavior of real thermal systems. Relative to use of IHT as a *learning tool,* the content and hierarchy of the software reinforce the sequential assimilation and application of heat transfer fundamentals treated in the text.

Preparations for this edition were also influenced by the results of a questionnaire which sought input on four principal issues: (i) is the text too long; (ii) is there a satisfactory balance between treatments of the *science* and *practice* of heat transfer; (iii) should a software package be coupled to the text; and (iv) what is an appropriate balance between end-of-chapter problems which are closed- and open-ended?

Since only eighteen percent of the 310 respondents felt that the text was too long, no attempt was made to reduce its length. A limited amount of new material was added to improve treatments of several topics (the first law; one-dimensional, steady-state conduction with generation; extended surfaces; semi-infinite media), but in each case, with only a small effect on the total length of the text. Although the respondents felt that the book contained a good balance between fundamentals and applications, it was recommended that the new edition include more open-ended, design-oriented problems (approximately 25% of the total) and that simulation software be provided to expedite the solution process. As discussed in preceding paragraphs, we have responded to both of these recommendations.

We are indebted to our many colleagues at Purdue and throughout the world who have provided suggestions and ideas which, in no small way, have contributed to the fabric of this text. We have always strived to remain conscious of student learning needs and difficulties, and we are grateful to the many students, at Purdue and elsewhere, who have provided positive reinforcement for our efforts.

West Lafayette, Indiana

Frank P. Incropera (fpi@ecn.purdue.edu)
David P. DeWitt (dpd@ecn.purdue.edu)

Contents

CHAPTER **7**
External Flow   **345**

CHAPTER **8**
Internal Flow   **419**

Symbols

A	area, m^2
A_c	cross-sectional area, m^2
A_{ff}	free-flow area in compact heat exchanger core (minimum cross-sectional area available for flow through the core), m^2
A_{fr}	heat exchanger frontal area, m^2
A_p	area of prime (unfinned) surface, m^2
A_r	nozzle area ratio
A_s	surface area, m^2
a	acceleration, m/s^2
Bi	Biot number
Bo	Bond number
C	molar concentration, $kmol/m^3$; heat capacity rate, W/K
C_D	drag coefficient
C_f	friction coefficient
C_t	thermal capacitance, J/K
c	specific heat, $J/kg \cdot K$; speed of light, m/s
c_p	specific heat at constant pressure, $J/kg \cdot K$
c_v	specific heat at constant volume, $J/kg \cdot K$
D	diameter, m
D_{AB}	binary mass diffusion coefficient, m^2/s
D_h	hydraulic diameter, m
E	thermal (sensible) internal energy, J; electric potential, V; emissive power, W/m^2
Ec	Eckert number
$\dot{E}_g$	rate of energy generation, W
$\dot{E}_{in}$	rate of energy transfer into a control volume, W
$\dot{E}_{out}$	rate of energy transfer out of control volume, W
$\dot{E}_{st}$	rate of increase of energy stored within a control volume, W
e	thermal internal energy per unit mass, J/kg; surface roughness, m

F	force, N; heat exchanger correction factor; fraction of blackbody radiation in a wavelength band; view factor
Fo	Fourier number
f	friction factor; similarity variable
G	irradiation, W/m^2; mass velocity, $kg/s \cdot m^2$
Gr	Grashof number
Gz	Graetz number
g	gravitational acceleration, m/s^2
g_c	gravitational constant, $1 \text{ kg} \cdot m/N \cdot s^2$ or $32.17 \text{ ft} \cdot lb_m/lb_f \cdot s^2$
H	nozzle height, m
h	convection heat transfer coefficient, $W/m^2 \cdot K$; Planck's constant
h_{fg}	latent heat of vaporization, J/kg
h_m	convection mass transfer coefficient, m/s
h_{rad}	radiation heat transfer coefficient, $W/m^2 \cdot K$
I	electric current, A; radiation intensity, $W/m^2 \cdot sr$
i	electric current density, A/m^2; enthalpy per unit mass, J/kg
J	radiosity, W/m^2
Ja	Jakob number
J_i^*	diffusive molar flux of species i relative to the mixture molar average velocity, $kmol/s \cdot m^2$
j_i	diffusive mass flux of species i relative to the mixture mass average velocity, $kg/s \cdot m^2$
j_H	Colburn j factor for heat transfer
j_m	Colburn j factor for mass transfer
k	thermal conductivity, $W/m \cdot K$; Boltzmann's constant
k_0	zero-order, homogeneous reaction rate constant, $kmol/s \cdot m^3$
k_1	first-order, homogeneous reaction rate constant, s^{-1}

k_1''	first-order, homogeneous reaction rate constant, m/s
L	characteristic length, m
Le	Lewis number
M	mass, kg; number of heat transfer lanes in a flux plot; reciprocal of the Fourier number for finite-difference solutions
$\dot{M}_i$	rate of transfer of mass for species i, kg/s
$\dot{M}_{i,g}$	rate of increase of mass of species i due to chemical reactions, kg/s
$\dot{M}_{\text{in}}$	rate at which mass enters a control volume, kg/s
$\dot{M}_{\text{out}}$	rate at which mass leaves a control volume, kg/s
$\dot{M}_{\text{st}}$	rate of increase of mass stored within a control volume, kg/s
$\mathcal{M}_i$	molecular weight of species i, kg/kmol
m	mass, kg
$\dot{m}$	mass flow rate, kg/s
m_i	mass fraction of species i, ρ_i/ρ
N	number of temperature increments in a flux plot; total number of tubes in a tube bank; number of surfaces in an enclosure
Nu	Nusselt number
NTU	number of transfer units
N_i	molar transfer rate of species i relative to fixed coordinates, kmol/s
N_i''	molar flux of species i relative to fixed coordinates, kmol/s · m²
$\dot{N}_i$	molar rate of increase of species i per unit volume due to chemical reactions, kmol/s · m³
$\dot{N}_i''$	surface reaction rate of species i, kmol/s · m²
n_i''	mass flux of species i relative to fixed coordinates, kg/s · m²
$\dot{n}_i$	mass rate of increase of species i per unit volume due to chemical reactions, kg/s · m³
N_L, N_T	number of tubes in longitudinal and transverse directions
P_L, P_T	dimensionless longitudinal and transverse pitch of a tube bank
P	perimeter, m; general fluid property designation
Pe	Peclet number ($RePr$)
Pr	Prandtl number
p	pressure, N/m²
Q	energy transfer, J
q	heat transfer rate, W
$\dot{q}$	rate of energy generation per unit volume, W/m³
q'	heat transfer rate per unit length, W/m
q''	heat flux, W/m²
R	cylinder radius, m
$\mathcal{R}$	universal gas constant
Ra	Rayleigh number
Re	Reynolds number

R_e	electric resistance, Ω
R_f	fouling factor, m² · K/W
R_m	mass transfer resistance, s/m³
$R_{m,n}$	residual for the m, n nodal point
R_t	thermal resistance, K/W
$R_{t,c}$	thermal contact resistance, K/W
$R_{t,f}$	fin thermal resistance, K/W
$R_{t,o}$	thermal resistance of fin array, K/W
r_o	cylinder or sphere radius, m
r, ϕ, z	cylindrical coordinates
r, θ, ϕ	spherical coordinates
S	solubility, kmol/m³ · atm; shape factor for two-dimensional conduction, m; nozzle pitch, m; plate spacing, m
S_c	solar constant
Sc	Schmidt number
Sh	Sherwood number
St	Stanton number
S_D, S_L, S_T	diagonal, longitudinal, and transverse pitch of a tube bank, m
T	temperature, K
t	time, s
U	overall heat transfer coefficient, W/m² · K; internal energy, J
u, v, w	mass average fluid velocity components, m/s
u^*, v^*, w^*	molar average velocity components, m/s
V	volume, m³; fluid velocity, m/s
v	specific volume, m³/kg
W	width of a slot nozzle, m
$\dot{W}$	rate at which work is performed, W
We	Weber number
X, Y, Z	components of the body force per unit volume, N/m³
x, y, z	rectangular coordinates, m
x_c	critical location for transition to turbulence, m
$x_{\text{fd},c}$	concentration entry length, m
$x_{\text{fd},h}$	hydrodynamic entry length, m
$x_{\text{fd},t}$	thermal entry length, m
x_i	mole fraction of species i, C_i/C

Greek Letters

α	thermal diffusivity, m²/s; heat exchanger surface area per unit volume, m²/m³; absorptivity
β	volumetric thermal expansion coefficient, K⁻¹
Γ	mass flow rate per unit width in film condensation, kg/s · m
δ	hydrodynamic boundary layer thickness, m
δ_c	concentration boundary layer thickness, m
δ_t	thermal boundary layer thickness, m
ε	emissivity; porosity of a packed bed; heat exchanger effectiveness
ε_f	fin effectiveness

ε_H turbulent diffusivity for heat transfer, m²/s

ε_M turbulent diffusivity for momentum transfer, m²/s

ε_m turbulent diffusivity for mass transfer, m²/s

η similarity variable

η_f fin efficiency

η_o overall efficiency of fin array

θ zenith angle, rad; temperature difference, K

κ absorption coefficient, m⁻¹

λ wavelength, μm

μ viscosity, kg/s · m

ν kinematic viscosity, m²/s; frequency of radiation, s⁻¹

ρ mass density, kg/m³; reflectivity

σ Stefan-Boltzmann constant; electrical conductivity, 1/Ω · m; normal viscous stress, N/m²; surface tension, N/m; ratio of heat exchanger minimum cross-sectional area to frontal area

Φ viscous dissipation function, s⁻²

ϕ azimuthal angle, rad

ψ stream function, m²/s

τ shear stress, N/m²; transmissivity

ω solid angle, sr

Subscripts

A, B   species in a binary mixture
abs   absorbed
am   arithmetic mean
b   base of an extended surface; blackbody
c   cross-sectional; concentration; cold fluid
cr   critical insulation thickness
cond   conduction
conv   convection
CF   counterflow
D   diameter; drag
dif   diffusion
e   excess; emission
evap   evaporation

f   fluid properties; fin conditions; saturated liquid conditions
fd   fully developed conditions
g   saturated vapor conditions
H   heat transfer conditions
h   hydrodynamic; hot fluid
i   general species designation; inner surface of an annulus; initial condition; tube inlet condition; incident radiation
L   based on characteristic length
l   saturated liquid conditions
lat   latent energy
lm   log mean condition
M   momentum transfer condition
m   mass transfer condition; mean value over a tube cross section
max   maximum fluid velocity
o   center or midplane condition; tube outlet condition; outer
R   reradiating surface
r, ref   reflected radiation
rad   radiation
S   solar conditions
s   surface conditions; solid properties
sat   saturated conditions
sky   sky conditions
sur   surroundings
t   thermal
tr   transmitted
v saturated vapor conditions
x   local conditions on a surface
λ spectral
∞   free stream conditions

Superscripts

′   fluctuating quantity
*   molar average; dimensionless quantity

Overbar

—   surface average conditions; time mean

CHAPTER 1

Introduction

*F*rom the study of thermodynamics, you have learned that energy can be transferred by interactions of a system with its surroundings. These interactions are called work and heat. However, thermodynamics deals with the end states of the process during which an interaction occurs and provides no information concerning the nature of the interaction or the time rate at which it occurs.

The objective of this text is to extend thermodynamic analysis through study of the *modes* of heat transfer and through development of relations to calculate heat transfer *rates*. In this chapter we lay the foundation for much of the material treated in the text. We do so by raising several questions. *What is heat transfer? How is heat transferred? Why is it important to study it?* In answering these questions, we will begin to appreciate the physical mechanisms that underlie heat transfer processes and the relevance of these processes to our industrial and environmental problems.

1.1
What and How?

A simple, yet general, definition provides sufficient response to the question: What is heat transfer?

> *Heat transfer (or heat) is energy in transit due to a temperature difference.*

Whenever there exists a temperature difference in a medium or between media, heat transfer must occur.

As shown in Figure 1.1, we refer to different types of heat transfer processes as *modes*. When a temperature gradient exists in a stationary medium, which may be a solid or a fluid, we use the term *conduction* to refer to the heat transfer that will occur across the medium. In contrast, the term *convection* refers to heat transfer that will occur between a surface and a moving fluid when they are at different temperatures. The third mode of heat transfer is termed *thermal radiation*. All surfaces of finite temperature emit energy in the form of electromagnetic waves. Hence, in the absence of an intervening medium, there is net heat transfer by radiation between two surfaces at different temperatures.

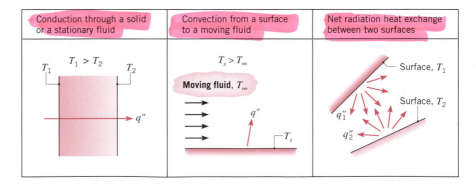

FIGURE 1.1   Conduction, convection, and radiation heat transfer modes.

1.2
Physical Origins and Rate Equations

As engineers it is important that we understand the *physical mechanisms* which underlie the heat transfer modes and that we be able to use the rate equations that quantify the amount of energy being transferred per unit time.

1.2.1   Conduction

At mention of the word *conduction,* we should immediately conjure up concepts of *atomic* and *molecular activity,* for it is processes at these levels that sustain this mode of heat transfer. Conduction may be viewed as the transfer of energy from the more energetic to the less energetic particles of a substance due to interactions between the particles.

The physical mechanism of conduction is most easily explained by considering a gas and using ideas familiar from your thermodynamics background. Consider a gas in which there exists a temperature gradient and assume that there is *no bulk motion.* The gas may occupy the space between two surfaces that are maintained at different temperatures, as shown in Figure 1.2. We associate the temperature at any point with the energy of gas molecules in proximity to the point. This energy is related to the random translational motion, as well as to the internal rotational and vibrational motions, of the molecules.

Higher temperatures are associated with higher molecular energies, and when neighboring molecules collide, as they are constantly doing, a transfer of energy from the more energetic to the less energetic molecules must occur. In the presence of a temperature gradient, energy transfer by conduction must then occur in the direction of decreasing temperature. This transfer is evident from Figure 1.2. The hypothetical plane at x_o is constantly being crossed by molecules from above and below due to their *random* motion. However, molecules from above are associated with a larger temperature than those from below, in which case there must be a *net* transfer of energy in the positive x direction. We may speak of the net transfer of energy by random molecular motion as a *diffusion* of energy.

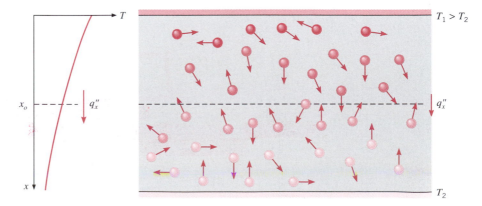

FIGURE 1.2   Association of conduction heat transfer with diffusion of energy due to molecular activity.

The situation is much the same in liquids, although the molecules are more closely spaced and the molecular interactions are stronger and more frequent. Similarly, in a solid, conduction may be attributed to atomic activity in the form of lattice vibrations. The modern view is to ascribe the energy transfer to *lattice waves* induced by atomic motion. In a nonconductor, the energy transfer is exclusively via these lattice waves; in a conductor it is also due to the translational motion of the free electrons. We treat the important properties associated with conduction phenomena in Chapter 2 and in Appendix A.

Examples of conduction heat transfer are legion. The exposed end of a metal spoon suddenly immersed in a cup of hot coffee will eventually be warmed due to the conduction of energy through the spoon. On a winter day there is significant energy loss from a heated room to the outside air. This loss is principally due to conduction heat transfer through the wall that separates the room air from the outside air.

It is possible to quantify heat transfer processes in terms of appropriate *rate equations*. These equations may be used to compute the amount of energy being transferred per unit time. For heat conduction, the rate equation is known as *Fourier's law*. For the one-dimensional plane wall shown in Figure 1.3, having a temperature distribution $T(x)$, the rate equation is expressed as

$$q''_x = -k \frac{dT}{dx} \tag{1.1}$$

The *heat flux* q''_x (W/m²) is the heat transfer rate in the x direction *per* unit area *perpendicular* to the direction of transfer, and it is proportional to the temperature gradient, dT/dx, in this direction. The proportionality constant k is a *transport* property known as the *thermal conductivity* (W/m · K) and is a characteristic of the wall material. The minus sign is a consequence of the fact that heat is transferred in the direction of decreasing temperature. Under the steady-state conditions shown in Figure 1.3, where the temperature distribution is *linear*, the temperature gradient may be expressed as

$$\frac{dT}{dx} = \frac{T_2 - T_1}{L}$$

and the heat flux is then

$$q''_x = -k \frac{T_2 - T_1}{L}$$

or

$$q''_x = k \frac{T_1 - T_2}{L} = k \frac{\Delta T}{L} \tag{1.2}$$

Note that this equation provides a *heat flux,* that is, the rate of heat transfer per unit area. The *heat rate* by conduction, q_x (W), through a plane wall of area A is then the product of the flux and the area, $q_x = q''_x \cdot A$.

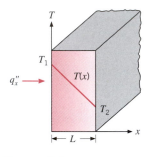

FIGURE 1.3
One-dimensional heat transfer by conduction (diffusion of energy).

EXAMPLE 1.1

The wall of an industrial furnace is constructed from 0.15-m-thick fireclay brick having a thermal conductivity of 1.7 W/m · K. Measurements made during steady-state operation reveal temperatures of 1400 and 1150 K at the inner and outer surfaces, respectively. What is the rate of heat loss through a wall that is 0.5 m by 3 m on a side?

SOLUTION

Known: Steady-state conditions with prescribed wall thickness, area, thermal conductivity, and surface temperatures.

Find: Wall heat loss.

Schematic:

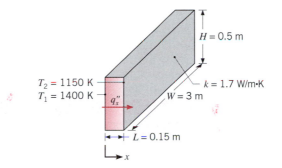

Assumptions:

1. Steady-state conditions.
2. One-dimensional conduction through the wall.
3. Constant thermal conductivity.

Analysis: Since heat transfer through the wall is by conduction, the heat flux may be determined from Fourier's law. Using Equation 1.2, we have

$$q_x'' = k \frac{\Delta T}{L} = 1.7 \text{ W/m} \cdot \text{K} \times \frac{250 \text{ K}}{0.15 \text{ m}} = 2833 \text{ W/m}^2$$

The heat flux represents the rate of heat transfer through a section of unit area. The wall heat loss is then

$$q_x = (HW) \, q_x'' = (0.5 \text{ m} \times 3.0 \text{ m}) \, 2833 \text{ W/m}^2 = 4250 \text{ W} \qquad \triangleleft$$

Comments: Note the direction of heat flow and the distinction between heat flux and heat rate.

1.2.2   Convection

The convection heat transfer *mode* is comprised of *two mechanisms.* In addition to energy transfer due to *random molecular motion (diffusion),* energy is also transferred by the *bulk,* or *macroscopic, motion* of the fluid. This fluid motion is associated with the fact that, at any instant, large numbers of molecules are moving collectively or as aggregates. Such motion, in the presence of a temper-

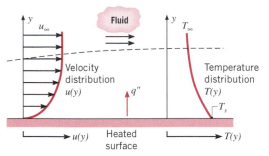

FIGURE 1.4
Boundary layer development in convection heat transfer.

ature gradient, contributes to heat transfer. Because the molecules in the aggregate retain their random motion, the total heat transfer is then due to a superposition of energy transport by the random motion of the molecules and by the bulk motion of the fluid. It is customary to use the term *convection* when referring to this cumulative transport and the term *advection* when referring to transport due to bulk fluid motion.

We are especially interested in convection heat transfer, which occurs between a fluid in motion and a bounding surface when the two are at different temperatures. Consider fluid flow over the heated surface of Figure 1.4. A consequence of the fluid–surface interaction is the development of a region in the fluid through which the velocity varies from zero at the surface to a finite value u_∞ associated with the flow. This region of the fluid is known as the *hydrodynamic,* or *velocity, boundary layer.* Moreover, if the surface and flow temperatures differ, there will be a region of the fluid through which the temperature varies from T_s at $y = 0$ to T_∞ in the outer flow. This region, called the *thermal boundary layer,* may be smaller, larger, or the same size as that through which the velocity varies. In any case, if $T_s > T_\infty$, convection heat transfer will occur between the surface and the outer flow.

The convection heat transfer mode is sustained both by random molecular motion and by the bulk motion of the fluid within the boundary layer. The contribution due to random molecular motion (diffusion) dominates near the surface where the fluid velocity is low. In fact, at the interface between the surface and the fluid ($y = 0$), the fluid velocity is zero and heat is transferred by this mechanism only. The contribution due to bulk fluid motion originates from the fact that the boundary layer *grows* as the flow progresses in the x direction. In effect, the heat that is conducted into this layer is swept downstream and is eventually transferred to the fluid outside the boundary layer. Appreciation of boundary layer phenomena is essential to understanding convection heat transfer. It is for this reason that the discipline of fluid mechanics will play a vital role in our later analysis of convection.

Convection heat transfer may be classified according to the nature of the flow. We speak of *forced convection* when the flow is caused by external means, such as by a fan, a pump, or atmospheric winds. As an example, consider the use of a fan to provide forced convection air cooling of hot electrical components on a stack of printed circuit boards (Figure 1.5a). In contrast, for *free* (or *natural*) *convection* the flow is induced by buoyancy forces, which arise from density differences caused by temperature variations in the fluid. An example is the free convection heat transfer that occurs from hot components on a vertical array of circuit boards in still air (Figure 1.5b). Air that makes

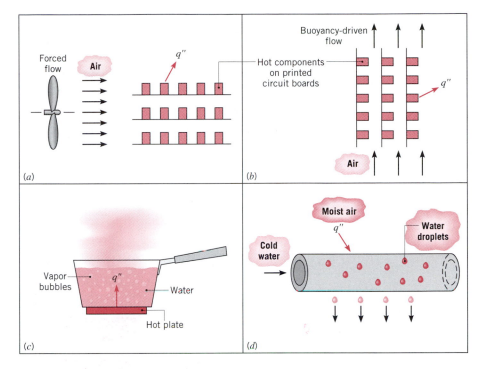

FIGURE 1.5 Convection heat transfer processes. (*a*) Forced convection. (*b*) Natural convection. (*c*) Boiling. (*d*) Condensation.

contact with the components experiences an increase in temperature and hence a reduction in density. Since it is now lighter than the surrounding air, buoyancy forces induce a vertical motion for which warm air ascending from the boards is replaced by an inflow of cooler ambient air.

While we have presumed *pure* forced convection in Figure 1.5*a* and *pure* natural convection in Figure 1.5*b*, conditions corresponding to *mixed (combined) forced* and *natural convection* may exist. For example, if velocities associated with the flow of Figure 1.5*a* are small and/or buoyancy forces are large, a secondary flow that is comparable to the imposed forced flow could be induced. The buoyancy-induced flow would be normal to the forced flow and could have a significant effect on convection heat transfer from the components. In Figure 1.5*b* mixed convection would result if a fan were used to force air upward through the circuit boards, thereby assisting the buoyancy flow, or downward, thereby opposing the buoyancy flow.

We have described the convection heat transfer mode as energy transfer occurring within a fluid due to the combined effects of conduction and bulk fluid motion. Typically, the energy that is being transferred is the *sensible,* or internal thermal, energy of the fluid. However, there are convection processes for which there is, in addition, *latent* heat exchange. This latent heat exchange is generally associated with a phase change between the liquid and vapor states of the fluid. Two special cases of interest in this text are *boiling* and *condensation.* For example, convection heat transfer results from fluid motion induced by vapor bubbles generated at the bottom of a pan of boiling water (Figure 1.5*c*) or by the condensation of water vapor on the outer surface of a cold water pipe (Figure 1.5*d*).

TABLE 1.1   Typical values of the
convection heat transfer coefficient

Process	h (W/m^2 · K)
Free convection	
Gases	2–25
Liquids	50–1000
Forced convection	
Gases	25–250
Liquids	50–20,000
Convection with phase change	
Boiling or condensation	2500–100,000

Regardless of the particular nature of the convection heat transfer process, the appropriate rate equation is of the form

$$q'' = h(T_s - T_\infty) \tag{1.3a}$$

where q'', the convective *heat flux* (W/m^2), is proportional to the difference between the surface and fluid temperatures, T_s and T_∞, respectively. This expression is known as *Newton's law of cooling,* and the proportionality constant h (W/m^2 · K) is termed the *convection heat transfer coefficient.* It depends on conditions in the boundary layer, which are influenced by surface geometry, the nature of the fluid motion, and an assortment of fluid thermodynamic and transport properties.

Any study of convection ultimately reduces to a study of the means by which h may be determined. Although consideration of these means is deferred to Chapter 6, convection heat transfer will frequently appear as a boundary condition in the solution of conduction problems (Chapters 2 to 5). In the solution of such problems we presume h to be known, using typical values given in Table 1.1.

When Equation 1.3a is used, the convection heat flux is presumed to be *positive* if heat is transferred *from* the surface ($T_s > T_\infty$) and *negative* if heat is transferred *to* the surface ($T_\infty > T_s$). However, if $T_\infty > T_s$, there is nothing to preclude us from expressing Newton's law of cooling as

$$q'' = h(T_\infty - T_s) \tag{1.3b}$$

in which case heat transfer is positive if it is to the surface.

1.2.3   Radiation

Thermal radiation is energy *emitted* by matter that is at a finite temperature. Although we will focus on radiation from solid surfaces, emission may also occur from liquids and gases. Regardless of the form of matter, the emission may be attributed to changes in the electron configurations of the constituent atoms or molecules. The energy of the radiation field is transported by electromagnetic waves (or alternatively, photons). While the transfer of energy by conduction or convection requires the presence of a material medium, radiation does not. In fact, radiation transfer occurs most efficiently in a vacuum.

Consider radiation transfer processes for the surface of Figure 1.6*a*. Radiation that is *emitted* by the surface originates from the thermal energy of matter bounded by the surface, and the rate at which energy is released per unit area (W/m²) is termed the surface *emissive power E*. There is an upper limit to the emissive power, which is prescribed by the *Stefan–Boltzmann law*

$$E_b = \sigma T_s^4 \qquad (1.4)$$

where T_s is the *absolute temperature* (K) of the surface and σ is the *Stefan–Boltzmann constant* ($\sigma = 5.67 \times 10^{-8}$ W/m² · K⁴). Such a surface is called an ideal radiator or *blackbody*.

The heat flux emitted by a real surface is less than that of a blackbody at the same temperature and is given by

$$E = \varepsilon \sigma T_s^4 \qquad (1.5)$$

where ε is a radiative property of the surface termed the *emissivity*. With values in the range $0 \le \varepsilon \le 1$, this property provides a measure of how efficiently a surface emits energy relative to a blackbody. It depends strongly on the surface material and finish, and representative values are provided in Table A.11.

Radiation may also be *incident* on a surface from its surroundings. The radiation may originate from a special source, such as the sun, or from other surfaces to which the surface of interest is exposed. Irrespective of the source(s), we designate the rate at which all such radiation is incident on a unit area of the surface as the *irradiation G* (Figure 1.6*a*).

A portion, or all, of the irradiation may be *absorbed* by the surface, thereby increasing the thermal energy of the material. The rate at which radiant energy is absorbed per unit surface area may be evaluated from knowledge of a surface radiative property termed the *absorptivity α*. That is,

$$G_{\text{abs}} = \alpha G \qquad (1.6)$$

where $0 \le \alpha \le 1$. If $\alpha < 1$ and the surface is *opaque*, portions of the irradiation are *reflected*. If the surface is *semitransparent*, portions of the irradiation may

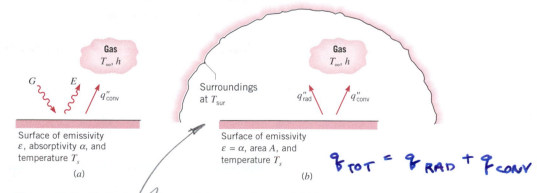

FIGURE 1.6   Radiation exchange: (*a*) at a surface and (*b*) between a surface and large surroundings.

also be *transmitted*. However, while absorbed and emitted radiation increase and reduce, respectively, the thermal energy of matter, reflected and transmitted radiation have no effect on this energy. Note that the value of α depends on the nature of the irradiation, as well as on the surface itself. For example, the absorptivity of a surface to solar radiation may differ from its absorptivity to radiation emitted by the walls of a furnace.

A special case that occurs frequently involves radiation exchange between a small surface at T_s and a much larger, isothermal surface that completely surrounds the smaller one (Figure 1.6b). The *surroundings* could, for example, be the walls of a room or a furnace whose temperature T_{sur} differs from that of an enclosed surface ($T_{sur} \neq T_s$). We will show in Chapter 12 that, for such a condition, the irradiation may be approximated by emission from a blackbody at T_{sur}, in which case $G = \sigma T_{sur}^4$. If the surface is assumed to be one for which $\alpha = \varepsilon$ (a *gray surface*), the *net* rate of radiation heat transfer *from* the surface, expressed per unit area of the surface, is

$$q_{rad}'' = \frac{q}{A} = \varepsilon E_b(T_s) - \alpha G = \varepsilon \sigma (T_s^4 - T_{sur}^4) \tag{1.7}$$

This expression provides the difference between thermal energy that is released due to radiation emission and that which is gained due to radiation absorption.

There are many applications for which it is convenient to express the net radiation heat exchange in the form

$$q_{rad} = h_r A(T_s - T_{sur}) \tag{1.8}$$

where, from Equation 1.7, the *radiation heat transfer coefficient h_r* is

$$h_r \equiv \varepsilon \sigma (T_s + T_{sur})(T_s^2 + T_{sur}^2) \tag{1.9}$$

Here we have modeled the radiation mode in a manner similar to convection. In this sense we have *linearized* the radiation rate equation, making the heat rate proportional to a temperature difference rather than to the difference between two temperatures to the fourth power. Note, however, that h_r depends strongly on temperature, while the temperature dependence of the convection heat transfer coefficient h is generally weak.

The surfaces of Figure 1.6 may also simultaneously transfer heat by convection to an adjoining gas. For the conditions of Figure 1.6b, the total rate of heat transfer *from* the surface is then

$$q = q_{conv} + q_{rad} = hA(T_s - T_\infty) + \varepsilon A \sigma (T_s^4 - T_{sur}^4) \tag{1.10}$$

EXAMPLE 1.2

An uninsulated steam pipe passes through a room in which the air and walls are at 25°C. The outside diameter of the pipe is 70 mm, and its surface temperature and emissivity are 200°C and 0.8, respectively. What are the surface emissive power and irradiation? If the coefficient associated with free convection heat transfer from the surface to the air is 15 W/m² · K, what is the rate of heat loss from the surface per unit length of pipe?

SOLUTION

Known: Uninsulated pipe of prescribed diameter, emissivity, and surface temperature in a room with fixed wall and air temperatures.

Find:

1. Surface emissive power and irradiation.
2. Pipe heat loss per unit length, q'.

Schematic:

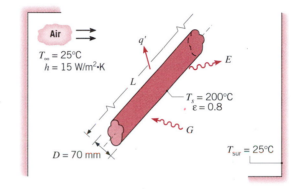

Assumptions:

1. Steady-state conditions exist.
2. Radiation exchange between the pipe and the room is between a small surface in a much larger enclosure.
3. The surface emissivity and absorptivity are equal.

Analysis:

1. The surface emissive power may be evaluated from Equation 1.5, while the irradiation corresponds to $G = \sigma T_{sur}^4$. Hence

$$E = \varepsilon \sigma T_s^4 = 0.8(5.67 \times 10^{-8} \text{ W/m}^2 \cdot \text{K}^4)(473 \text{ K})^4 = 2270 \text{ W/m}^2$$

$$G = \sigma T_{sur}^4 = 5.67 \times 10^{-8} \text{ W/m}^2 \cdot \text{K}^4 (298 \text{ K})^4 = 447 \text{ W/m}^2$$

2. Heat loss from the pipe is by convection to the room air and by radiation exchange with the walls. Hence, from Equation 1.10, with $A = \pi DL$,

$$q = h(\pi DL)(T_s - T_\infty) + \varepsilon(\pi DL)\sigma(T_s^4 - T_{sur}^4)$$

The heat loss per unit length of pipe is then

$$q' = \frac{q}{L} = 15 \text{ W/m}^2 \cdot \text{K } (\pi \times 0.07 \text{ m})(200 - 25)°\text{C}$$

$$+ 0.8(\pi \times 0.07 \text{ m}) 5.67 \times 10^{-8} \text{ W/m}^2 \cdot \text{K}^4 (473^4 - 298^4) \text{ K}^4$$

$$q' = 577 \text{ W/m} + 421 \text{ W/m} = 998 \text{ W/m} \qquad \triangleleft$$

Comments:

1. Note that temperature may be expressed in units of °C or K when evaluating the temperature difference for a convection (or conduction) heat transfer rate. However, temperature must be expressed in kelvins (K) when evaluating a radiation transfer rate.

2. In this situation the radiation and convection heat transfer rates are comparable because T_s is large compared to T_{sur} and the coefficient associated with free convection is small. For more moderate values of T_s and the larger values of h associated with forced convection, the effect of radiation may often be neglected. The radiation heat transfer coefficient may be computed from Equation 1.9, and for the conditions of this problem its value is $h_r = 11 \ \text{W/m}^2 \cdot \text{K}$.

1.2.4   Relationship to Thermodynamics

At this point it is appropriate to note the fundamental differences between heat transfer and thermodynamics. Although thermodynamics is concerned with the heat interaction and the vital role it plays in the first and second laws, it considers neither the mechanisms that provide for heat exchange nor the methods that exist for computing the *rate* of heat exchange. Thermodynamics is concerned with *equilibrium* states of matter, where an equilibrium state necessarily precludes the existence of a temperature gradient. Although thermodynamics may be used to determine the amount of energy required in the form of heat for a system to pass from one equilibrium state to another, it does not acknowledge that *heat transfer is inherently a nonequilibrium process.* For heat transfer to occur, there must be a temperature gradient, hence thermodynamic nonequilibrium. The discipline of heat transfer therefore seeks to do what thermodynamics is inherently unable to do, namely, to quantify the *rate* at which heat transfer occurs in terms of the degree of thermal nonequilibrium. This is done through the rate equations for the three modes, expressed, for example, by Equations 1.2, 1.3, and 1.7.

1.3
The Conservation of Energy Requirement

The subjects of thermodynamics and heat transfer are highly complementary. For example, because it treats the *rate* at which heat is transferred, the subject of heat transfer may be viewed as an extension of thermodynamics. Conversely, for many heat transfer problems, the *first law* of thermodynamics (the *law of conservation of energy*) provides a useful, often essential, tool. In anticipation of such problems, general formulations of the first law are now obtained.

1.3.1   Conservation of Energy for a Control Volume

In applying the first law, we first need to identify the *control volume,* a region of space bounded by a *control surface* through which energy and matter may pass. Once the control volume is identified, an appropriate *time basis* must be specified. There are two options. Since the first law must be satisfied at each and every *instant* of time *t*, one option involves formulating the law on a *rate basis.* That is, at any instant, there must be a balance between all *energy rates,* as measured in joules per second (W). Alternatively, the first law must also be satisfied over any *time interval* Δt. For such an interval, there must be a balance between the *amounts* of all *energy changes,* as measured in joules.

According to the time basis, first law formulations that are well suited for heat transfer analysis may be stated as follows.

At an Instant (*t*)

The rate at which thermal and mechanical energy enters a control volume, plus the rate at which thermal energy is generated within the control volume, minus the rate at which thermal and mechanical energy leaves the control volume must equal the rate of increase of energy stored within the control volume.

Over a Time Interval (Δt)

The amount of thermal and mechanical energy that enters a control volume, plus the amount of thermal energy that is generated within the control volume, minus the amount of thermal and mechanical energy that leaves the control volume must equal the increase in the amount of energy stored in the control volume.

If the inflow and generation of energy exceed the outflow, there will be an increase in the amount of energy stored (accumulated) in the control volume; if the converse is true, there will be a decrease in energy storage. If the inflow and generation of energy equal the outflow, a *steady-state* condition must prevail in which there will be no change in the amount of energy stored in the control volume.

Consider applying energy conservation to the control volume shown in Figure 1.7. The first step is to identify the control surface by drawing a dashed line. The next step is to identify the energy terms. At an instant, these include the rate at which thermal and mechanical energy enter and leave *through* the control surface, $\dot{E}_{in}$ and $\dot{E}_{out}$. Also, thermal energy may be created within the control volume due to conversion from other energy forms. We refer to this process as *energy generation,* and the rate at which it occurs is designated as $\dot{E}_g$. The rate of change of energy stored within the control volume, dE_{st}/dt, is designated as $\dot{E}_{st}$. A general form of the energy conservation requirement may then be expressed on a *rate* basis as

$$\dot{E}_{in} + \dot{E}_g - \dot{E}_{out} = \frac{dE_{st}}{dt} \equiv \dot{E}_{st} \qquad (1.11a)$$

FIGURE 1.7
Conservation of energy for a control volume. Application at an instant.

Equation 1.11a may be applied at any *instant of time.* The alternative form that applies for a *time interval* Δt is obtained by integrating Equation 1.11a over time:

$$E_{in} + E_g - E_{out} = \Delta E_{st} \tag{1.11b}$$

In words this relation says that the amounts of energy inflow and generation act to increase the amount of energy stored within the control volume, whereas outflow acts to decrease the stored energy.

The inflow and outflow terms are *surface phenomena.* That is, they are associated exclusively with processes occurring at the control surface and are proportional to the surface area. A common situation involves energy inflow and outflow due to heat transfer by conduction, convection, and/or radiation. In situations involving fluid flow across the control surface, the terms also include energy advected with matter entering and leaving the control volume. This energy may be composed of internal, kinetic, and potential forms. The inflow and outflow terms may also include work interactions occurring at the system boundaries.

The *energy generation term* is associated with conversion from some other energy form (chemical, electrical, electromagnetic, or nuclear) to thermal energy. It is a *volumetric phenomenon.* That is, it occurs within the control volume and is proportional to the magnitude of this volume. For example, an exothermic chemical reaction may be occurring, converting chemical to thermal energy. The net effect is an increase in the thermal energy of matter within the control volume. Another source of thermal energy is the conversion from electrical energy that occurs due to resistance heating when an electric current is passed through a conductor. That is, if an electric current I passes through a resistance R in the control volume, electrical energy is dissipated at a rate I^2R, which corresponds to the rate at which thermal energy is generated (released) within the volume. Although this process may alternatively be treated as one in which electrical work is done on the system (energy inflow), the net effect is still one of thermal energy creation.

Energy storage is also a *volumetric phenomenon,* and changes within the control volume may be due to changes in the internal, kinetic, and/or potential energies of its contents. Hence, for a time interval, Δt, the storage term of Equation 1.11b, ΔE_{st}, can be equated to the summation, $\Delta U + \Delta KE + \Delta PE$. The change in the *internal energy,* ΔU, consists of a *sensible* or *thermal component,* which accounts for the translational, rotational, and/or vibrational motion of the atoms/molecules comprising the matter; a *latent component,* which relates to intermolecular forces influencing phase change between solid, liquid, and vapor states; a *chemical component,* which accounts for energy stored in the chemical bonds between atoms; and a *nuclear component,* which accounts for binding forces in the nucleus.

In all applications of interest in this text, if chemical or nuclear effects exist, they are treated as sources of thermal energy and hence are included in the generation, rather than the storage, terms of Equations 1.11a,b. Moreover, latent energy effects need only be considered if there is a phase change, as, for example, from solid to liquid (*melting*) or from liquid to vapor (*vaporization, evaporation, boiling*). In such cases, the latent energy increases. Conversely, if the phase change is from vapor to liquid (*condensation*) or from liquid to solid (*solidification, freezing*), the latent energy decreases. Hence, if kinetic and po-

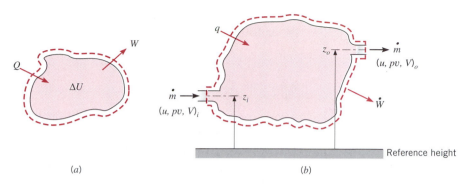

FIGURE 1.8 Conservation of energy: (*a*) application to a closed system over a time interval, and (*b*) application to a steady-flow, open system at an instant of time.

tential energy effects may be neglected, as is almost always the case in heat transfer analysis, changes in energy storage are due only to changes in the internal thermal and/or, in the case of phase change, latent energies ($\Delta E_{st} = \Delta U = \Delta U_t + \Delta U_{lat}$).

Equations 1.11a,b may be used to develop more specific forms of the energy conservation requirement, including those previously considered in your study of thermodynamics. Consider a *closed system* of fixed mass (Figure 1.8*a*), across whose boundaries energy is transferred by heat and work interactions. If, over a time interval Δt, heat is transferred *to* the system in the amount Q (energy inflow), work is done by the system in the amount W (energy outflow), there is no energy conversion occurring within the system ($E_g = 0$), and kinetic and potential energy changes are negligible, Equation 1.11b reduces to

$$Q - W = \Delta U \tag{1.11c}$$

The work term W may be due to displacement of a boundary, a rotating shaft, and/or electromagnetic effects. Alternatively, at an instant of time, the energy conservation requirement is

$$q - \dot{W} = \frac{dU}{dt} \tag{1.11d}$$

The other form of the energy conservation requirement with which you are familiar pertains to an *open system* (Figure 1.8*b*), where mass flow provides for the transport of internal, kinetic, and potential energy into and out of the system. In such cases, it is customary to divide energy exchange in the form of work into two contributions. The first contribution, termed *flow work*, is associated with work done by pressure forces moving fluid through the system boundaries. For a *unit mass*, the amount of work is equivalent to the product of the pressure and the specific volume of the fluid (pv). All other work is presumed to be done by the system and is embodied in the term W. Hence, if heat is presumed to be transferred to the system, there is no energy conversion occurring within the system, and operation is under steady-state conditions ($\dot{E}_{st} = 0$), Equation 1.11a reduces to the following form of the steady-flow energy equation:

$$\dot{m}\left(u + pv + \frac{V^2}{2} + gz\right)_i - \dot{m}\left(u + pv + \frac{V^2}{2} + gz\right)_o + q - \dot{W} = 0 \tag{1.11e}$$

The sum of the internal energy and flow work may, of course, be replaced by the enthalpy, $i = u + pv$.

EXAMPLE 1.3

A long conducting rod of diameter D and electrical resistance per unit length R'_e is initially in thermal equilibrium with the ambient air and its surroundings. This equilibrium is disturbed when an electrical current I is passed through the rod. Develop an equation that could be used to compute the variation of the rod temperature with time during passage of the current.

SOLUTION

Known: Temperature of a rod of prescribed diameter and electrical resistance changes with time due to passage of an electrical current.

Find: Equation that governs temperature change with time for the rod.

Schematic:

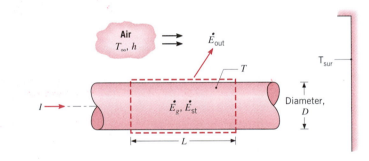

Assumptions:

1. At any time t the temperature of the rod is uniform.
2. Constant properties (ρ, c, $\varepsilon = \alpha$).
3. Radiation exchange between the outer surface of the rod and the surroundings is between a small surface and a large enclosure.

Analysis: The first law of thermodynamics may often be used to determine an unknown temperature. In this case, relevant terms include heat transfer by convection and radiation from the surface, energy generation due to Ohmic heating within the conductor, and a change in thermal energy storage. Since we wish to determine the rate of change of the temperature, the first law should be applied at an instant of time. Hence, applying Equation 1.11a to a control volume of length L about the rod, it follows that

$$\dot{E}_g - \dot{E}_{out} = \dot{E}_{st}$$

where energy generation is due to the electric resistance heating,

$$\dot{E}_g = I^2 R'_e L$$

Heating occurs uniformly within the control volume and could also be expressed in terms of a volumetric heat generation rate $\dot{q}$ (W/m^3). The generation rate for the entire control volume is then $\dot{E}_g = \dot{q}V$, where $\dot{q} = I^2R'_e/(\pi D^2/4)$. Energy outflow is due to convection and net radiation from the surface, Equations 1.3a and 1.7, respectively,

$$\dot{E}_{out} = h(\pi DL)(T - T_\infty) + \varepsilon\sigma(\pi DL)(T^4 - T^4_{sur})$$

and the change in energy storage is due to the temperature change,

$$\dot{E}_{st} = \frac{dU_t}{dt} = \frac{d}{dt}(\rho VcT)$$

The term $\dot{E}_{st}$ is associated with the rate of change in the internal thermal energy of the rod, where ρ and c are the mass density and the specific heat, respectively, of the rod material, and V is the volume of the rod, $V = (\pi D^2/4)L$. Substituting the rate equations into the energy balance, it follows that

$$I^2R'_eL - h(\pi DL)(T - T_\infty) - \varepsilon\sigma(\pi DL)(T^4 - T^4_{sur}) = \rho c\left(\frac{\pi D^2}{4}\right)L\frac{dT}{dt}$$

Hence

$$\frac{dT}{dt} = \frac{I^2R'_e - \pi Dh(T - T_\infty) - \pi D\varepsilon\sigma(T^4 - T^4_{sur})}{\rho c(\pi D^2/4)} \qquad \triangleleft$$

Comments: The above equation could be solved for the time dependence of the rod temperature by integrating numerically. A steady-state condition would eventually be reached for which $dT/dt = 0$. The rod temperature is then determined by an algebraic equation of the form

$$\pi Dh(T - T_\infty) + \pi D\varepsilon\sigma(T^4 - T^4_{sur}) = I^2R'_e$$

For fixed environmental conditions (h, T_∞, T_{sur}), as well as a rod of fixed geometry (D) and properties (ε, R'_e), the temperature depends on the rate of thermal energy generation and hence on the value of the electric current. Consider an uninsulated copper wire ($D = 1$ mm, $\varepsilon = 0.8$, $R'_e = 0.4$ Ω/m) in a relatively large enclosure ($T_{sur} = 300$ K) through which cooling air is circulated ($h = 100$ W/m$^2 \cdot$ K, $T_\infty = 300$ K). Substituting these values into the foregoing equation, the rod temperature has been computed for operating currents in the range $0 \leq I \leq 10$ A and the following results were obtained:

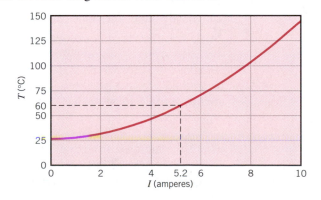

If a maximum operating temperature of $T = 60°C$ is prescribed for safety reasons, the current should not exceed 5.2 A. At this temperature, heat transfer by radiation (0.6 W/m) is much less than heat transfer by convection (10.4 W/m). Hence, if one wished to operate at a larger current while maintaining the rod temperature within the safety limit, the convection coefficient would have to be increased by increasing the velocity of the circulating air. For $h = 250$ W/m² · K, the maximum allowable current could be increased to 8.1 A.

EXAMPLE 1.4

Ice of mass M at the fusion temperature ($T_f = 0°C$) is enclosed in a cubical cavity of width W on a side. The cavity wall is of thickness L and thermal conductivity k. If the outer surface of the wall is at a temperature $T_1 > T_f$, obtain an expression for the time required to completely melt the ice.

SOLUTION

Known: Mass and temperature of ice. Dimensions, thermal conductivity, and outer surface temperature of containing wall.

Find: Expression for time needed to melt the ice.

Schematic:

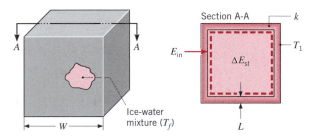

Ice-water
mixture (T_f)

Assumptions:

1. Inner surface of wall is at T_f throughout the process.
2. Constant properties.
3. Steady-state, one-dimensional conduction through each wall.
4. Conduction area of one wall may be approximated as W^2 ($L \ll W$).

Analysis: Since we must determine the melting time t_m, the first law should be applied over the time interval $\Delta t = t_m$. Hence, applying Equation 1.11b to a control volume about the ice–water mixture, it follows that

$$E_{in} = \Delta E_{st} = \Delta U_{lat}$$

where the increase in energy stored within the control volume is due exclusively to the change in latent energy associated with conversion from the solid to liquid state. Heat is transferred to the ice by means of conduction through the container wall, and since the temperature difference across the wall is assumed to remain at $(T_1 - T_f)$ throughout the melting process, the wall conduction rate is a constant

$$q_{\text{cond}} = k(6W^2)\frac{T_1 - T_f}{L}$$

and the amount of energy inflow is

$$E_{\text{in}} = \left[k(6W^2)\frac{T_1 - T_f}{L}\right]t_m$$

The amount of energy required to effect such change per unit mass of solid is termed the *latent heat of fusion* h_{sf}. Hence the increase in energy storage is

$$\Delta E_{\text{st}} = Mh_{sf}$$

By substituting into the first law expression, it follows that

$$t_m = \frac{Mh_{sf}L}{6W^2k(T_1 - T_f)} \qquad \triangleleft$$

Comments: Several complications would arise if the ice were initially subcooled. The storage term would have to include the change in sensible (internal thermal) energy required to take the ice from the subcooled to the fusion temperature. During this process, temperature gradients would develop in the ice.

1.3.2 The Surface Energy Balance

We will frequently have occasion to apply the conservation of energy requirement at the surface of a medium. In this special case the control surface includes no mass or volume and appears as shown in Figure 1.9. Accordingly, the generation and storage terms of the conservation expression, Equation 1.11a, are no longer relevant and it is only necessary to deal with surface phenomena. For this case the conservation requirement becomes

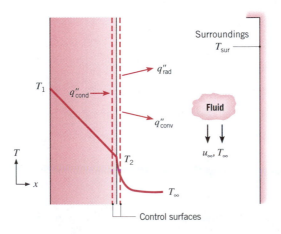

FIGURE 1.9
The energy balance for conservation of energy at the surface of a medium.

$$\dot{E}_{in} - \dot{E}_{out} = 0 \qquad (1.12)$$

Even though thermal energy generation may be occurring in the medium, the process would not affect the energy balance at the control surface. Moreover, this conservation requirement holds for both *steady-state* and *transient* conditions.

In Figure 1.9 three heat transfer terms are shown for the control surface. On a unit area basis they are conduction from the medium *to* the control surface (q''_{cond}), convection *from* the surface to a fluid (q''_{conv}), and net radiation exchange from the surface to the surroundings (q''_{rad}). The energy balance then takes the form

$$q''_{cond} - q''_{conv} - q''_{rad} = 0 \qquad (1.13)$$

and we can express each of the terms using the appropriate rate equations, Equations 1.2, 1.3a, and 1.7.

EXAMPLE 1.5

The hot combustion gases of a furnace are separated from the ambient air and its surroundings, which are at 25°C, by a brick wall 0.15 m thick. The brick has a thermal conductivity of 1.2 W/m · K and a surface emissivity of 0.8. Under steady-state conditions an outer surface temperature of 100°C is measured. Free convection heat transfer to the air adjoining the surface is characterized by a convection coefficient of $h = 20$ W/m² · K. What is the brick inner surface temperature?

SOLUTION

Known: Outer surface temperature of a furnace wall of prescribed thickness, thermal conductivity, and emissivity. Ambient conditions.

Find: Wall inner surface temperature.

Schematic:

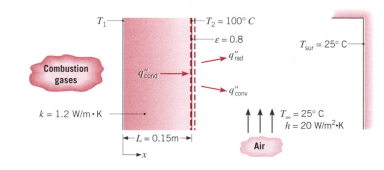

Assumptions:

1. Steady-state conditions.

2. One-dimensional heat transfer by conduction across the wall.

3. Radiation exchange between the outer surface of the wall and the surroundings is between a small surface and a large enclosure.

Analysis: The inside surface temperature may be obtained by performing an energy balance at the outer surface. From Equation 1.12,

$$\dot{E}_{in} - \dot{E}_{out} = 0$$

it follows that, on a unit area basis,

$$q''_{cond} - q''_{conv} - q''_{rad} = 0$$

or, rearranging and substituting from Equations 1.2, 1.3a, and 1.7,

$$k\frac{T_1 - T_2}{L} = h(T_2 - T_\infty) + \varepsilon\sigma(T_2^4 - T_{sur}^4)$$

Therefore, substituting the appropriate numerical values, we find

$$1.2 \text{ W/m} \cdot \text{K} \frac{(T_1 - 373) \text{ K}}{0.15 \text{ m}} = 20 \text{ W/m}^2 \cdot \text{K} (373 - 298) \text{ K}$$

$$+ 0.8(5.67 \times 10^{-8} \text{ W/m}^2 \cdot \text{K}^4)(373^4 - 298^4) \text{ K}^4$$

$$= 1500 \text{ W/m}^2 + 520 \text{ W/m}^2 = 2020 \text{ W/m}^2$$

Solving for T_1,

$$T_1 = 373 \text{ K} + \frac{0.15 \text{ m}}{1.2 \text{ W/m} \cdot \text{K}} (2020 \text{ W/m}^2) = 625 \text{ K} = 352°\text{C} \qquad \triangleleft$$

Comments:

1. Note that the contribution of radiation to heat transfer from the outer surface is significant. The relative contribution would diminish, however, with increasing h and/or decreasing T_2.

2. When using energy balances involving radiation exchange and other modes, it is good practice to express all temperatures in kelvin units. This practice is *necessary* when the unknown temperature appears in the radiation term and in one or more of the other terms.

1.3.3   Application of the Conservation Laws: Methodology

In addition to being familiar with the transport rate equations described in Section 1.2, the heat transfer analyst must be able to work with the energy conservation requirements of Equations 1.11 and 1.12. The application of these balances is simplified if a few basic rules are followed.

1. The appropriate control volume must be defined, with the control surface represented by a dashed line.

2. The appropriate time basis must be identified.

3. The relevant energy processes should be identified. Each process should be shown on the control volume by an appropriately labeled arrow.

4. The conservation equation should then be written, and appropriate rate expressions should be substituted for the terms in the equation.

It is important to note that the energy conservation requirement may be applied to a *finite* control volume or a *differential* (infinitesimal) control volume. In the first case the resulting expression governs overall system behavior. In the second case a differential equation is obtained that can be solved for conditions at each point in the system. Differential control volumes are introduced in Chapter 2, and both types of control volumes are used extensively throughout the text.

1.4
Analysis of Heat Transfer Problems: Methodology

A major objective of this text is to prepare you to solve engineering problems that involve heat transfer processes. To this end numerous problems are provided at the end of each chapter. In working these problems you will gain a deeper appreciation for the fundamentals of the subject, and you will gain confidence in your ability to apply these fundamentals to the solution of engineering problems.

In solving problems, we advocate the use of a systematic procedure characterized by a prescribed format. We consistently employ this procedure in our examples, and we require our students to use it in their problem solutions. It consists of the following steps:

1. *Known:* After carefully reading the problem, state briefly and concisely what is known about the problem. Do not repeat the problem statement.

2. *Find:* State briefly and concisely what must be found.

3. *Schematic:* Draw a schematic of the physical system. If application of the conservation laws is anticipated, represent the required control surface by dashed lines on the schematic. Identify relevant heat transfer processes by appropriately labeled arrows on the schematic.

4. *Assumptions:* List all pertinent simplifying assumptions.

5. *Properties:* Compile property values needed for subsequent calculations and identify the source from which they are obtained.

6. *Analysis:* Begin your analysis by applying appropriate conservation laws, and introduce rate equations as needed. Develop the analysis as completely as possible before substituting numerical values. Perform the calculations needed to obtain the desired results.

7. *Comments:* Discuss your results. Such a discussion may include a summary of key conclusions, a critique of the original assumptions, and an inference of trends obtained by performing additional *what–if* and *parameter sensitivity* calculations.

The importance of following steps 1 through 4 should not be underestimated. They provide a useful guide to thinking about a problem before effecting its solution. In step 7 we hope you will take the initiative to gain additional insights by performing calculations that may be computer-based. The software accompanying this text provides a suitable tool for effecting such calculations.

EXAMPLE 1.6

The coating on a plate is cured by exposure to an infrared lamp providing an irradiation of 2000 W/m². It absorbs 80% of the irradiation and has an emissivity of 0.50. It is also exposed to an air flow and large surroundings for which temperatures are 20°C and 30°C, respectively.

1. If the convection coefficient between the plate and the ambient air is 15 W/m² · K, what is the cure temperature of the plate?
2. Final characteristics of the coating, including wear and durability, are known to depend on the temperature at which curing occurs. An air flow system is able to control the air velocity, and hence the convection coefficient, on the cured surface, but the process engineer needs to know how the temperature depends on the convection coefficient. Provide the desired information by computing and plotting the surface temperature as a function of h for $2 \leq h \leq 200$ W/m² · K. What value of h would provide a cure temperature of 50°C?

SOLUTION

Known: Coating with prescribed radiation properties is cured by irradiation from an infrared lamp. Heat transfer from the coating is by convection to ambient air and radiation exchange with the surroundings.

Find:
1. Cure temperature for $h = 15$ W/m² · K.
2. Effect of air flow on the cure temperature for $2 \leq h \leq 200$ W/m² · K. Value of h for which the cure temperature is 50°C.

Schematic:

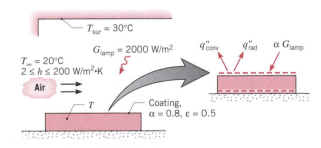

Assumptions:
1. Steady-state conditions.
2. Negligible heat loss from back surface of plate.

3. Plate is small object in large surroundings, and coating has an absorptivity of $\alpha_{sur} = \varepsilon = 0.8$ with respect to irradiation from the surroundings.

Analysis:

1. Since the process corresponds to steady-state conditions and there is no heat transfer at the back surface, the plate must be isothermal ($T_s = T$). Hence the desired temperature may be determined by placing a control surface about the exposed surface and applying Equation 1.12 or by placing the control surface about the entire plate and applying Equation 1.11a. Adopting the latter approach and recognizing that there is no internal energy generation ($\dot{E}_g = 0$), Equation 1.11a reduces to

$$\dot{E}_{in} - \dot{E}_{out} = 0$$

where $\dot{E}_{st} = 0$ for steady-state conditions. With energy inflow due to absorption of the lamp irradiation by the coating and outflow due to convection and net radiation transfer to the surroundings, it follows that

$$(\alpha G)_{lamp} - q''_{conv} - q''_{rad} = 0$$

Substituting from Equations 1.3a and 1.7, we obtain

$$(\alpha G)_{lamp} - h(T - T_\infty) - \varepsilon\sigma(T^4 - T^4_{sur}) = 0$$

Substituting numerical values

$$0.8 \times 2000 \text{ W/m}^2 - 15 \text{ W/m}^2 \cdot \text{K}(T - 293) \text{ K}$$
$$- 0.5 \times 5.67 \times 10^{-8} \text{ W/m}^2 \cdot \text{K}^4(T^4 - 303^4) \text{ K}^4 = 0$$

and solving by trial-and-error, we obtain

$$T = 377 \text{ K} = 104°C \qquad \triangleleft$$

2. Solving the foregoing energy balance for selected values of h in the prescribed range and plotting the results, we obtain

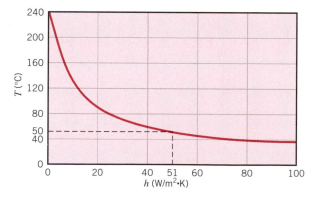

If a cure temperature of 50°C is desired, the air flow must provide a convection coefficient of

$$h(T = 50°C) = 51.0 \text{ W/m}^2 \cdot \text{K} \qquad \triangleleft$$

Comments:

1. The coating (plate) temperature may be reduced by decreasing T_∞ and T_{sur}, as well as by increasing the air velocity and hence the convection coefficient.

2. The relative contributions of convection and radiation to heat transfer from the plate vary greatly with h. For $h = 2$ W/m$^2 \cdot$ K, $T = 477$ K and radiation dominates ($q''_{rad} \approx 1232$ W/m^2, $q''_{conv} \approx 368$ W/m^2). Conversely, for $h = 200$ W/m$^2 \cdot$ K, $T = 301$ K and convection dominates ($q''_{conv} \approx 1606$ W/m^2, $q''_{rad} \approx -6$ W/m^2). In fact, for this condition the plate temperature is slightly less than that of the surroundings and net radiation exchange is *to* the plate.

1.5
Relevance of Heat Transfer

In the vernacular of the time, heat transfer is indeed a relevant subject, not to mention an inherently fascinating part of the engineering sciences. We will devote much time to acquiring an understanding of heat transfer effects and to developing the skills needed to predict heat transfer rates. What is the value of this knowledge, and to what kinds of problems may it be applied?

Heat transfer phenomena play an important role in many industrial and environmental problems. As an example, consider the vital area of energy production and conversion. There is not a single application in this area that does not involve heat transfer effects in some way. In the generation of electrical power, whether it be through nuclear fission or fusion, the combustion of fossil fuels, magnetohydrodynamic processes, or the use of geothermal energy sources, there are numerous heat transfer problems that must be solved. These problems involve conduction, convection, and radiation processes and relate to the design of systems such as boilers, condensers, and turbines. One is often confronted with the need to maximize heat transfer rates and to maintain the integrity of materials in high-temperature environments.

On a smaller scale there are many heat transfer problems related to the development of solar energy conversion systems for space heating, as well as for electric power production. Heat transfer processes also affect the performance of propulsion systems, such as the internal combustion, gas turbine, and rocket engines. Heat transfer problems arise in the design of conventional space and water heating systems, in the design of incinerators and cryogenic storage equipment, in the cooling of electronic equipment, in the design of refrigeration and air conditioning systems, and in many manufacturing processes. Heat transfer is also relevant to air and water pollution and strongly influences local and global climate.

1.6
Units and Dimensions

The physical quantities of heat transfer are specified in terms of *dimensions,* which are measured in terms of units. Four *basic* dimensions are required for the development of heat transfer; they are length (L), mass (M), time (t), and temperature (T). All other physical quantities of interest may be related to these four basic dimensions.

In the United States it has been customary to measure dimensions in terms of an *English system of units,* for which the *base units* are

Dimension		Unit
Length (L)	$\rightarrow$	foot (ft)
Mass (M)	$\rightarrow$	pound mass (lb_m)
Time (t)	$\rightarrow$	second (s)
Temperature (T)	$\rightarrow$	degree Fahrenheit (°F)

The units required to specify other physical quantities may then be inferred from this group. For example, the dimension of force is related to mass through Newton's second law of motion,

$$F = \frac{1}{g_c} Ma \tag{1.14}$$

where the acceleration a has units of feet per square second and g_c is a proportionality constant. If this constant is arbitrarily set equal to unity and made *dimensionless*, the dimensions of force are $(F) = (M) \cdot (L)/(t)^2$ and the unit of force is

$$1 \text{ poundal} = 1 \ lb_m \cdot ft/s^2$$

Alternatively, one could work with a system of basic dimensions that includes both mass and force. However, in this case the proportionality constant must have the dimensions $(M) \cdot (L)/(F) \cdot (t)^2$. Moreover, if one defines the pound force (lb_f) as a unit of force that will accelerate one pound mass by 32.17 ft/s², the proportionality constant must be of the form

$$g_c = 32.17 \ lb_m \cdot ft/lb_f \cdot s^2$$

The units of work may be inferred from its definition as the product of a force times a distance, in which case the units are ft · lb_f. The units of work and energy are, of course, equivalent, although it is customary to use the British thermal unit (Btu) as the unit of thermal energy. One British thermal unit will raise the temperature of 1 lb_m of water at 68°F by 1°F. It is equivalent to 778.16 ft · lb_f, which is termed the *mechanical equivalent of heat.*

In recent years there has been a strong trend toward worldwide usage of a standard set of units. In 1960 the SI (Système International d'Unités) system of units was defined by the Eleventh General Conference on Weights and Measures and recommended as a worldwide standard. In response to this trend, the American Society of Mechanical Engineers (ASME) has required the use of SI units in all of its publications since July 1, 1974. For this reason and because it is operationally more convenient than the English system, the SI system is used for calculations of this text. However, because for some time to come, engineers will also have to work with results expressed in the English system, you should be able to convert from one system to the other. For your convenience conversion factors are provided on the inside back cover of the text.

The SI *base* units required for this text are summarized in Table 1.2. With regard to these units note that 1 mol is the amount of substance that has as

TABLE 1.2   SI base and supplementary units

Quantity and Symbol	Unit and Symbol
Length (L)	meter (m)
Mass (M)	kilogram (kg)
Concentration (C)	mole (mol)
Time (t)	second (s)
Electric current (I)	ampere (A)
Thermodynamic temperature (T)	kelvin (K)
Plane angle[a] (θ)	radian (rad)
Solid angle[a] (ω)	steradian (sr)

[a]Supplementary unit.

many atoms or molecules as there are atoms in 12 g of carbon-12 (^{12}C); this is the gram-mole (mol). Although the mole has been recommended as the unit quantity of matter for the SI system, it is more consistent to work with the kilogram-mol (kmol, kg-mol). One kmol is simply the amount of substance that has as many atoms or molecules as there are atoms in 12 kg of ^{12}C. As long as the use is consistent within a given problem, no difficulties arise in using either mol or kmol. The molecular weight of a substance is the mass associated with a mole or kilogram-mole. For oxygen, as an example, the molecular weight $\mathcal{M}$ is 16 g/mol or 16 kg/kmol.

Although the SI unit of temperature is the kelvin, use of the Celsius temperature scale remains widespread. Zero on the Celsius scale (0°C) is equivalent to 273.15 K on the thermodynamic scale,[1] in which case

$$T\,(\text{K}) = T\,(°\text{C}) + 273.15$$

However, temperature *differences* are equivalent for the two scales and may be denoted as °C or K. Also, although the SI unit of time is the second, other units of time (minute, hour, and day) are so common that their use with the SI system is generally accepted.

The SI units comprise a coherent form of the metric system. That is, all remaining units may be derived from the base units using formulas that do not involve any numerical factors. *Derived* units for selected quantities are listed in Table 1.3. Note that force is measured in newtons, where a 1-N force will accelerate a 1-kg mass at 1 m/s². Hence 1 N = 1 kg · m/s². The unit of pressure

---

[1]The degree symbol is retained for designation of the Celsius temperature (°C) to avoid confusion with use of C for the unit of electrical charge (coulomb).

TABLE 1.3   SI derived units for selected quantities

Quantity	Name and Symbol	Formula	Expression in SI Base Units
Force	newton (N)	m · kg/s²	m · kg/s²
Pressure and stress	pascal (Pa)	N/m²	kg/m · s²
Energy	joule (J)	N · m	m² · kg/s²
Power	watt (W)	J/s	m² · kg/s³

TABLE 1.4   Multiplying prefixes

Prefix	Abbreviation	Multiplier
pico	p	10^{-12}
nano	n	10^{-9}
micro	μ	10^{-6}
milli	m	10^{-3}
centi	c	10^{-2}
hecto	h	10^{2}
kilo	k	10^{3}
mega	M	10^{6}
giga	G	10^{9}
tera	T	10^{12}

(N/m^2) is often referred to as the pascal. In the SI system there is one unit of energy (thermal, mechanical, or electrical), called the joule (J), and 1 J = 1 N · m. The unit for energy rate, or power, is then J/s, where one joule per second is equivalent to one watt (1 J/s = 1 W). Since it is frequently necessary to work with extremely large or small numbers, a set of standard prefixes has been introduced to simplify matters (Table 1.4). For example, 1 megawatt (MW) = 10^6 W, and 1 micrometer (μm) = 10^{-6} m.

1.7
Summary

Although much of the material of this chapter will be discussed in greater detail, you should now have a reasonable overview of heat transfer. You should be aware of the several modes of transfer and their physical origins. Moreover,

TABLE 1.5   Summary of heat transfer processes

Mode	Mechanism(s)	Rate Equation	Equation Number	Transport Property or Coefficient
Conduction	Diffusion of energy due to random molecular motion	$q''_x \, (\text{W/m}^2) = -k\dfrac{dT}{dx}$	(1.1)	k (W/m · K)
Convection	Diffusion of energy due to random molecular motion plus energy transfer due to bulk motion (advection)	$q'' \, (\text{W/m}^2) = h(T_s - T_\infty)$	(1.3a)	h (W/m^2 · K)
Radiation	Energy transfer by electromagnetic waves	$q'' \, (\text{W/m}^2) = \varepsilon\sigma(T_s^4 - T_{\text{sur}}^4)$ or $q \, (\text{W}) = h_r A(T_s - T_{\text{sur}})$	(1.7) (1.8)	ε h_r (W/m^2 · K)

given a physical situation, you should be able to perceive the relevant transport phenomena. The importance of developing this facility must not be underestimated. You will be devoting much time to acquiring the tools needed to calculate heat transfer phenomena. However, before you can begin to use these tools to solve practical problems, you must have the intuition to determine what is happening physically. In short, you must be able to look at a problem and identify the pertinent transport phenomena. The example and problems at the end of this chapter should help you to begin developing this intuition.

You should also appreciate the significance of the rate equations and feel comfortable in using them to compute transport rates. These equations, summarized in Table 1.5, *should be committed to memory*. You must also recognize the importance of the conservation laws and the need to carefully identify control volumes. With the rate equations, the conservation laws may be used to solve numerous heat transfer problems.

EXAMPLE 1.7

A closed container filled with hot coffee is in a room whose air and walls are at a fixed temperature. Identify all heat transfer processes that contribute to cooling of the coffee. Comment on features that would contribute to a superior container design.

SOLUTION

Known: Hot coffee is separated from its cooler surroundings by a plastic flask, an air space, and a plastic cover.

Find: Relevant heat transfer processes.

Schematic:

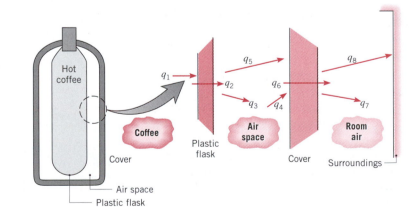

Pathways for energy transfer from the coffee are as follows:

q_1: free convection from the coffee to the flask

q_2: conduction through the flask

q_3: free convection from the flask to the air

q_4:   free convection from the air to the cover

q_5:   net radiation exchange between the outer surface of the flask and the inner surface of the cover

q_6:   conduction through the cover

q_7:   free convection from the cover to the room air

q_8:   net radiation exchange between the outer surface of the cover and the surroundings

Comments: Design improvements are associated with (1) use of aluminized (low emissivity) surfaces for the flask and cover to reduce net radiation, and (2) evacuating the air space or using a filler material to retard free convection.

Problems

Conduction

1.1   A heat rate of 3 kW is conducted through a section of an insulating material of cross-sectional area 10 m² and thickness 2.5 cm. If the inner (hot) surface temperature is 415°C and the thermal conductivity of the material is 0.2 W/m · K, what is the outer surface temperature?

1.2   A concrete wall, which has a surface area of 20 m² and is 0.30 m thick, separates conditioned room air from ambient air. The temperature of the inner surface of the wall is maintained at 25°C, and the thermal conductivity of the concrete is 1 W/m · K.

(a) Determine the heat loss through the wall for ambient temperatures ranging from −15°C to 38°C, which correspond to winter and summer extremes, respectively. Display your results graphically.

(b) On your graph, also plot the heat loss as a function of ambient temperature for wall materials having thermal conductivities of 0.75 and 1.25 W/m · K. Explain the family of curves you have obtained.

1.3   The heat flux through a wood slab 50 mm thick, whose inner and outer surface temperatures are 40 and 20°C, respectively, has been determined to be 40 W/m². What is the thermal conductivity of the wood?

1.4   The inner and outer surface temperatures of a glass window 5 mm thick are 15 and 5°C. What is the heat loss through a window that is 1 m by 3 m on a side? The thermal conductivity of glass is 1.4 W/m · K.

1.5   A freezer compartment consists of a cubical cavity that is 2 m on a side. Assume the bottom to be perfectly insulated. What is the minimum thickness of styrofoam insulation ($k = 0.030$ W/m · K) that must be applied to the top and side walls to ensure a heat load of less than 500 W, when the inner and outer surfaces are −10 and 35°C?

1.6   What is the thickness required of a masonry wall having thermal conductivity 0.75 W/m · K if the heat rate is to be 80% of the heat rate through a composite structural wall having a thermal conductivity of 0.25 W/m · K and a thickness of 100 mm? Both walls are subjected to the same surface temperature difference.

1.7   A square silicon chip ($k = 150$ W/m · K) is of width $w = 5$ mm on a side and of thickness $t = 1$ mm. The chip is mounted in a substrate such that its side and back surfaces are insulated, while the front surface is exposed to a coolant.

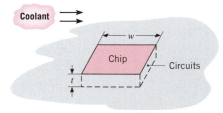

If 4 W are being dissipated in circuits mounted to the back surface of the chip, what is the steady-state temperature difference between back and front surfaces?

1.8 A gage for measuring heat flux to a surface or through a laminated material employs thin-film, chromel/alumel (type K) thermocouples deposited on the upper and lower surfaces of a wafer with a thermal conductivity of 1.4 W/m · K and a thickness of 0.25 mm.

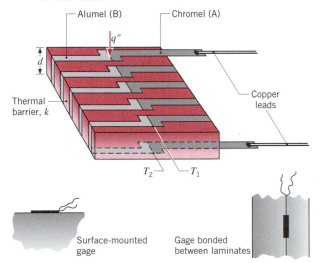

(a) Determine the heat flux q'' through the gage when the voltage output at the copper leads is 350 μV. The Seebeck coefficient of the type-K thermocouple materials is approximately 40 μV/°C.

(b) What precaution should you take in using a gage of this nature to measure heat flow through the laminated structure shown above?

Convection

1.9 You've experienced convection cooling if you've ever extended your hand out the window of a moving vehicle or into a flowing water stream. With the surface of your hand at a temperature of 30°C, determine the convection heat flux for (a) a vehicle speed of 35 km/h in air at −5°C with a convection coefficient of 40 W/m² · K and (b) a velocity of 0.2 m/s in a water stream at 10°C with a convection coefficient of 900 W/m² · K. Which condition would *feel* colder? Contrast these results with a heat loss of approximately 30 W/m² under normal room conditions.

1.10 Air at 40°C flows over a long, 25-mm diameter cylinder with an embedded electrical heater. In a series of tests, measurements were made of the power per unit length, P', required to maintain the cylinder surface temperature at 300°C for different freestream velocities V of the air. The results are as follows:

Air velocity, V (m/s)	1	2	4	8	12
Power, P' (W/m)	450	658	983	1507	1963

(a) Determine the convection coefficient for each velocity, and display your results graphically.

(b) Assuming the dependence of the convection coefficient on the velocity to be of the form $h = CV^n$, determine the parameters C and n from the results of part (a).

1.11 An electric resistance heater is embedded in a long cylinder of diameter 30 mm. When water with a temperature of 25°C and velocity of 1 m/s flows crosswise over the cylinder, the power per unit length required to maintain the surface at a uniform temperature of 90°C is 28 kW/m. When air, also at 25°C, but with a velocity of 10 m/s is flowing, the power per unit length required to maintain the same surface temperature is 400 W/m. Calculate and compare the convection coefficients for the flows of water and air.

1.12 A cartridge electrical heater is shaped as a cylinder of length $L = 200$ mm and outer diameter $D = 20$ mm. Under normal operating conditions the heater dissipates 2 kW, while submerged in a water flow that is at 20°C and provides a convection heat transfer coefficient of $h = 5000$ W/m² · K. Neglecting heat transfer from the ends of the heater, determine its surface temperature T_s. If the water flow is inadvertently terminated while the heater continues to operate, the heater surface is exposed to air that is also at 20°C but for which $h = 50$ W/m² · K. What is the corresponding surface temperature? What are the consequences of such an event?

1.13 A square isothermal chip is of width $w = 5$ mm on a side and is mounted in a substrate such that its side and back surfaces are well insulated, while the front surface is exposed to the flow of a coolant at $T_\infty = 15$°C. From reliability considerations, the chip temperature must not exceed $T = 85$°C.

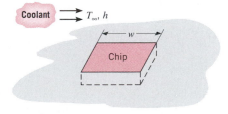

If the coolant is air and the corresponding convection coefficient is $h = 200$ W/m² · K, what is the maximum allowable chip power? If the coolant is a dielectric liquid for which $h = 3000$ W/m$^2 \cdot$ K, what is the maximum allowable power?

1.14 The use of impinging air jets is proposed as a means of effectively cooling high-power logic chips in a computer. However, before the technique can be implemented, the convection coefficient associated with jet impingement on a chip surface must be known. Design an experiment that could be used to determine convection coefficients associated with air jet impingement on a chip measuring approximately 10 mm by 10 mm on a side.

1.15 The temperature controller for a clothes dryer consists of a bimetallic switch mounted on an electrical heater attached to a wall-mounted insulation pad.

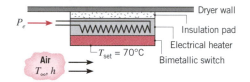

The switch is set to open at 70°C, the maximum dryer air temperature. In order to operate the dryer at a lower air temperature, sufficient power is supplied to the heater such that the switch reaches 70°C (T_{set}) when the air temperature T_∞ is less than T_{set}. If the convection heat transfer coefficient between the air and the exposed switch surface of 30 mm^2 is 25 W/m$^2 \cdot$ K, how much heater power P_e is required when the desired dryer air temperature is $T_\infty = 50$°C?

1.16 The free convection heat transfer coefficient on a thin hot vertical plate suspended in still air can be determined from observations of the change in plate temperature with time as it cools. Assuming the plate is isothermal and radiation exchange with its surroundings is negligible, evaluate the convection coefficient at the instant of time when the plate temperature is 225°C and the change in plate temperature with time (dT/dt) is -0.022 K/s. The ambient air temperature is 25°C and the plate measures 0.3×0.3 m with a mass of 3.75 kg and a specific heat of 2770 J/kg $\cdot$ K.

Radiation

1.17 A spherical interplanetary probe of 0.5-m diameter contains electronics that dissipate 150 W. If the probe surface has an emissivity of 0.8 and the probe does not receive radiation from other surfaces, as, for example, from the sun, what is its surface temperature?

1.18 An instrumentation package has a spherical outer surface of diameter $D = 100$ mm and emissivity $\varepsilon = 0.25$. The package is placed in a large space simulation chamber whose walls are maintained at 77 K. If operation of the electronic components is restricted to the temperature range $40 \leq T \leq 85$°C, what is the range of acceptable power dissipation for the package? Display your results graphically, showing also the effect of variations in the emissivity by considering values of 0.20 and 0.30.

1.19 A surface of area 0.5 m^2, emissivity 0.8, and temperature 150°C is placed in a large, evacuated chamber whose walls are maintained at 25°C. What is the rate at which radiation is emitted by the surface? What is the net rate at which radiation is exchanged between the surface and the chamber walls?

1.20 If $T_s \approx T_{sur}$ in Equation 1.9, the radiation heat transfer coefficient may be approximated as

$$h_{r,a} = 4\varepsilon\sigma\bar{T}^3$$

where $\bar{T} \equiv (T_s + T_{sur})/2$. We wish to assess the validity of this approximation by comparing values of h_r and $h_{r,a}$ for the following conditions. In each case represent your results graphically and comment on the validity of the approximation.

(a) Consider a surface of either polished aluminum ($\varepsilon = 0.05$) or black paint ($\varepsilon = 0.9$), whose temperature may exceed that of the surroundings ($T_{sur} = 25$°C) by 10 to 100°C. Also compare your results with values of the coefficient associated with free convection in air ($T_\infty = T_{sur}$), where h (W/m$^2 \cdot$ K) $= 0.98 \, \Delta T^{1/3}$.

(b) Consider initial conditions associated with placing a workpiece at $T_s = 25$°C in a large furnace whose wall temperature may be varied over the range $100 \leq T_{sur} \leq 1000$°C. According to the surface finish or coating, its emissivity may assume values of 0.05, 0.2, and 0.9. For each emissivity, plot the relative error, $(h_r - h_{r,a})/h_r$, as a function of the furnace temperature.

1.21 Consider the conditions of Problem 1.13. With heat transfer by convection to air, the maximum allowable chip power is found to be 0.35 W. If consideration is also given to net heat transfer by radiation from the chip surface to large surroundings at 15°C, what is the percentage increase in the maximum allowable chip power afforded by this consideration? The chip surface has an emissivity of 0.9.

1.22 A vacuum system, as used in sputtering electrically conducting thin films on microcircuits, is comprised of a baseplate maintained by an electrical heater at

300 K and a shroud within the enclosure maintained at 77 K by a liquid-nitrogen coolant loop. The baseplate, insulated on the lower side, is 0.3 m in diameter and has an emissivity of 0.25.

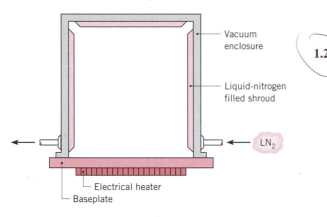

(a) How much electrical power must be provided to the baseplate heater?

(b) At what rate must liquid nitrogen be supplied to the shroud if its heat of vaporization is 125 kJ/kg?

(c) To reduce the liquid-nitrogen consumption, it is proposed to bond a thin sheet of aluminum foil ($\varepsilon = 0.09$) to the baseplate. Will this have the desired effect?

Energy Balance and Multimode Effects

1.23 An electrical resistor is connected to a battery, as shown schematically. After a brief transient, the resistor assumes a nearly uniform, steady-state temperature of 95°C, while the battery and lead wires remain at the ambient temperature of 25°C. Neglect the electrical resistance of the lead wires.

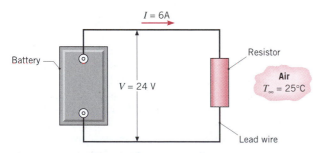

(a) Consider the resistor as a system about which a control surface is placed and Equation 1.11a is applied. Determine the corresponding values of $\dot{E}_{in}$ (W), $\dot{E}_g$ (W), $\dot{E}_{out}$ (W), and $\dot{E}_{st}$ (W). If a control surface is placed about the entire system, what are the values of $\dot{E}_{in}$, $\dot{E}_g$, $\dot{E}_{out}$, and $\dot{E}_{st}$?

(b) If electrical energy is dissipated uniformly within the resistor, which is a cylinder of diameter $D = 60$ mm and length $L = 250$ mm, what is the volumetric heat generation rate, $\dot{q}$ (W/m³)?

(c) Neglecting radiation from the resistor, what is the convection coefficient?

1.24 The variation of temperature with position in a wall is shown below for a specific time, t_1, during a transient (time-varying) process.

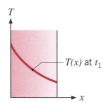

Is the wall being heated or cooled?

1.25 A solid sphere of diameter $D = 1$ m and surface emissivity $\varepsilon = 0.30$ is preheated and then suspended in a large, cryogenically cooled vacuum chamber whose interior surfaces are maintained at 80 K. What is the rate of change of energy stored by the solid when its temperature is 600 K?

1.26 A solid aluminum sphere of emissivity ε is initially at an elevated temperature and is cooled by placing it in a chamber. The walls of the chamber are maintained at a lower temperature, and a cold gas is circulated through the chamber. Obtain an equation that could be used to predict the variation of the aluminum temperature with time during the cooling process. Do not attempt to solve.

1.27 An aluminum plate 4 mm thick is mounted in a horizontal position, and its bottom surface is well insulated. A special, thin coating is applied to the top surface such that it absorbs 80% of any incident solar radiation, while having an emissivity of 0.25. The density ρ and specific heat c of aluminum are known to be 2700 kg/m³ and 900 J/kg · K, respectively.

(a) Consider conditions for which the plate is at a temperature of 25°C and its top surface is suddenly exposed to ambient air at $T_\infty = 20$°C and to solar radiation that provides an incident flux of 900 W/m². The convection heat transfer coefficient between the surface and the air is $h = 20$ W/m² · K. What is the initial rate of change of the plate temperature?

(b) What will be the equilibrium temperature of the plate when steady-state conditions are reached?

(c) The surface radiative properties depend on the specific nature of the applied coating. Compute and plot the steady-state temperature as a function of the emissivity for $0.05 \leq \varepsilon \leq 1$, with all other conditions remaining as prescribed. Repeat your calculations for values of $\alpha_S = 0.5$ and 1.0, and plot the results with those obtained for $\alpha_S = 0.8$. If the intent is to maximize the plate temperature, what is the most desirable combination of the plate emissivity and its absorptivity to solar radiation?

1.28 In an orbiting space station, an electronic package is housed in a compartment having a surface area, $A_s = 1$ m², which is exposed to space. Under normal operating conditions, the electronics dissipate 1 kW, all of which must be transferred from the exposed surface to space. If the surface emissivity is 1.0 and the surface is not exposed to the sun, what is its steady-state temperature? If the surface is exposed to a solar flux of 750 W/m² and its absorptivity to solar radiation is 0.25, what is its steady-state temperature?

1.29 The energy consumption associated with a home water heater has two components: (i) the energy that must be supplied to bring the temperature of groundwater to the heater storage temperature, as it is introduced to replace hot water that has been used, and (ii) the energy needed to compensate for heat losses incurred while the water is stored at the prescribed temperature. In this problem, we will evaluate the first of these components for a family of four, whose daily hot water consumption is approximately 100 gallons. If groundwater is available at 15°C, what is the annual energy consumption associated with heating the water to a storage temperature of 55°C? For a unit electrical power cost of $0.08/kWh, what is the annual cost associated with supplying hot water by means of (a) electric resistance heating or (b) a heat pump having a COP of 3 and a compressor efficiency (conversion from electrical energy to mechanical work) of 85%

1.30 Hot rolling is a process by which steel ingots are successively flattened due to passage through a sequence of compression rollers. Strip (sheet) metal emerges from the last set of rollers and is cooled as it moves along transport rollers before being coiled. Three cooling zones may be identified. Just beyond the last set of rollers and just before the coil, there exist regions in which the strip is exposed to the mill surroundings. Between these regions, there exists an accelerated cooling zone in which planar jets of water impinge on the strip. Water is maintained in a liquid phase through much of the jet impingement region, but the large plate temperatures induce boiling and creation of a vapor blanket in a film boiling region.

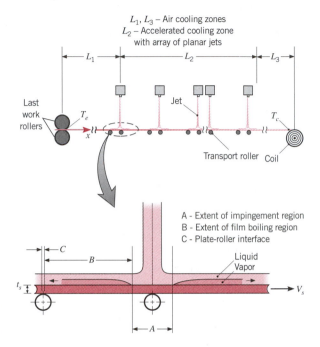

(a) For the production of low chromium strip steel ($\rho = 7840$ kg/m³, $c_p = 970$ J/kg · K), representative operating conditions correspond to a compression roller exit temperature of $T(x = 0) = T_e = 940$°C, a water temperature of 25°C in the impinging jets, and a strip speed, width, and thickness of $V_s = 10$ m/s, $W_s = 2$ m, and $t_s = 4$ mm, respectively. What is the rate at which heat must be extracted from the strip to achieve a strip coiling temperature of $T(x = L_1 + L_2 + L_3) = T_c = 540$°C?

(b) Identify all of the heat transfer processes that contribute to cooling of the plate.

1.31 In one stage of an annealing process, 304 stainless steel sheet is taken from 300 K to 1250 K as it passes through an electrically heated oven at a speed of $V_s = 10$ mm/s. The sheet thickness and width are $t_s = 8$ mm and $W_s = 2$ m, respectively, while the height, width, and length of the oven are $H_o = 2$ m, $W_o = 2.4$ m, and $L_o = 25$ m, respectively. The top and four sides of the oven are exposed to ambient air and large surroundings, each at 300 K, and the corresponding surface temperature, convection coefficient, and emissivity are $T_s = 350$ K, $h = 10$ W/m² · K, and $\varepsilon_s = 0.8$. The bottom surface of the oven is also at 350 K and rests on a 0.5-m thick concrete pad whose base is at 300 K.

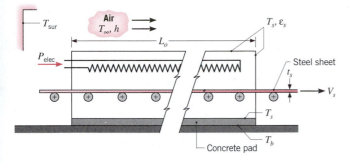

Estimate the required electric power input, P_{elec}, to the oven.

1.32 Radioactive wastes are packed in a long, thin-walled cylindrical container. The wastes generate thermal energy nonuniformly according to the relation $\dot{q} = \dot{q}_o[1 - (r/r_o)^2]$, where $\dot{q}$ is the local rate of energy generation per unit volume, $\dot{q}_o$ is a constant, and r_o is the radius of the container. Steady-state conditions are maintained by submerging the container in a liquid that is at T_∞ and provides a uniform convection coefficient h.

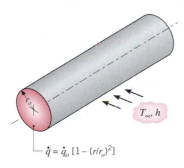

$$\dot{q} = \dot{q}_o\,[1 - (r/r_o)^2]$$

Obtain an expression for the total rate at which energy is generated in a unit length of the container. Use this result to obtain an expression for the temperature T_s of the container wall.

1.33 A spherical, stainless steel (AISI 302) canister is used to store reacting chemicals that provide for a uniform heat flux q_i'' to its inner surface. The canister is suddenly submerged in a liquid bath of temperature $T_\infty < T_i$, where T_i is the initial temperature of the canister wall.

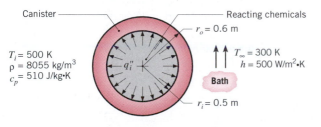

Canister — Reacting chemicals
$r_o = 0.6$ m
$T_i = 500$ K
$\rho = 8055$ kg/m^3
$c_p = 510$ J/kg·K
q_i''
$T_\infty = 300$ K
$h = 500$ W/m^2·K
Bath
$r_i = 0.5$ m

(a) Assuming negligible temperature gradients in the canister wall and a constant heat flux q_i'', develop an equation that governs the variation of the wall temperature with time during the transient process. What is the initial rate of change of the wall temperature if $q_i'' = 10^5$ W/m^2?

(b) What is the steady-state temperature of the wall?

(c) The convection coefficient depends on the velocity associated with fluid flow over the canister and whether or not the wall temperature is large enough to induce boiling in the liquid. Compute and plot the steady-state temperature as a function of h for the range $100 \le h \le 10,000$ W/m$^2 \cdot$ K. Is there a value of h below which operation would be unacceptable?

1.34 Radioactive wastes are packed in a thin-walled spherical container. The wastes generate thermal energy nonuniformly according to the relation $\dot{q} = \dot{q}_o[1 - (r/r_o)^2]$, where $\dot{q}$ is the local rate of energy generation per unit volume, $\dot{q}_o$ is a constant, and r_o is the radius of the container. Steady-state conditions are maintained by submerging the container in a liquid that is at T_∞ and provides a uniform convection coefficient h.

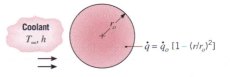

$$\dot{q} = \dot{q}_o\,[1 - (r/r_o)^2]$$

Obtain an expression for the total rate at which thermal energy is generated in the container. Use this result to obtain an expression for the temperature T_s of the container wall.

1.35 Liquid oxygen, which has a boiling point of 90 K and a latent heat of vaporization of 214 kJ/kg, is stored in a spherical container whose outer surface is of 500-mm diameter and at a temperature of $-10°$C. The container is housed in a laboratory whose air and walls are at 25°C.

(a) If the surface emissivity is 0.20 and the heat transfer coefficient associated with free convection at the outer surface of the container is 10 W/m$^2 \cdot$ K, what is the rate, in kg/s, at which oxygen vapor must be vented from the system?

(b) Moisture in the ambient air will result in frost formation on the container, causing the surface emissivity to increase. Assuming the surface temperature and convection coefficient to remain at $-10°$C and 10 W/m$^2 \cdot$ K, respectively, compute the oxygen evaporation rate (kg/s) as a function of surface emissivity over the range 0.2 $\le \varepsilon \le 0.94$.

1.36 A slab of ice in a thin-walled container 10 mm thick and 300 mm on each side is placed on a well-insulated pad. At its top surface, the ice is exposed to ambient air for which $T_\infty = 25°C$ and the convection coefficient is 25 W/m² · K. Neglecting heat transfer from the sides and assuming the ice–water mixture remains at 0°C, how long will it take to completely melt the ice? The density and latent heat of fusion of ice are 920 kg/m³ and 334 kJ/kg, respectively.

1.37 Following the hot vacuum forming of a paper–pulp mixture, the product, an egg carton, is transported on a conveyor for 18 s toward the entrance of a gas-fired oven where it is dried to a desired final water content. To increase the productivity of the line, it is proposed that a bank of infrared radiation heaters, which provide a uniform radiant flux of 5000 W/m², be installed over the conveyor. The carton has an exposed area of 0.0625 m² and a mass of 0.220 kg, 75% of which is water after the forming process.

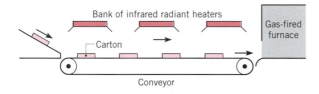

The chief engineer of your plant will approve the purchase of the heaters if the carton water content is reduced from 75% to 65%. Would you recommend the purchase? Assume the heat of vaporization of water is $h_{fg} = 2400$ kJ/kg.

1.38 Electronic power devices are mounted to a heat sink having an exposed surface area of 0.045 m² and an emissivity of 0.80. When the devices dissipate a total power of 20 W and the air and surroundings are at 27°C, the average sink temperature is 42°C. What average temperature will the heat sink reach when the devices dissipate 30 W for the same environmental condition?

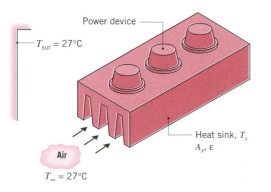

1.39 The roof of a car in a parking lot absorbs a solar radiant flux of 800 W/m², while the underside is perfectly insulated. The convection coefficient between the roof and the ambient air is 12 W/m² · K.

(a) Neglecting radiation exchange with the surroundings, calculate the temperature of the roof under steady-state conditions if the ambient air temperature is 20°C.

(b) For the same ambient air temperature, calculate the temperature of the roof if its surface emissivity is 0.8.

(c) The convection coefficient depends on airflow conditions over the roof, increasing with increasing air speed. Compute and plot the plate temperature as a function of h for $2 \le h \le 200$ W/m² · K.

1.40 The operating temperature of an infrared detector for a space telescope is to be controlled by adjusting the electrical power, q_{elec}, to a thin heater sandwiched between the detector and the "cold finger" whose opposite end is immersed in liquid nitrogen at 77 K. The cold finger rod of 5-mm diameter has a thermal conductivity of 10 W/m · K and extends 50 mm above the level of the liquid nitrogen in the dewar. Assume the detector surface has an emissivity of 0.9 and the vacuum enclosure is maintained at 300 K.

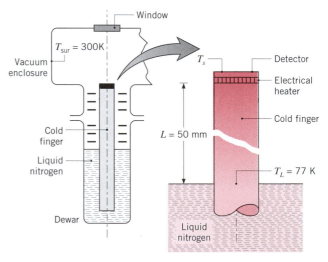

(a) What is the temperature of the detector when no power is supplied to the heater?

(b) How much heater power is required to maintain the detector at 195 K?

(c) Compute and plot the heater power required to maintain a detector temperature of 195 K as a function of the thermal conductivity of the cold finger for $0.1 \leq k \leq 400$ W/m · K. Select a suitable finger material that would allow for maintaining the prescribed detector temperature at a low level of power consumption.

1.41 Consider the physical system depicted in Example 1.5 under conditions for which the combustion gases are at 1300°C and heat transfer by convection from the gases to the inner surface is characterized by a convection coefficient of $h_1 = 50$ W/m² · K. The furnace wall is made from a diatomaceous silica brick for which $k = 0.3$ W/m · K and $\varepsilon = 0.8$, while the ambient air and surroundings remain at 25°C. Radiation exchange between the combustion gases and the inner surface may be neglected. Compute and plot the inner and outer surface temperatures, T_1 and T_2, as a function of the wall thickness ($0.025 \leq L \leq 0.50$ m) for an outside convection coefficient of $h_2 = 10$ W/m² · K and as a function of the convection coefficient ($2 \leq h_2 \leq 50$ W/m² · K) for $L = 0.15$ m. Recommend values of L and h_2 suitable for maintaining T_2 below a maximum allowable value of 100°C.

1.42 The heat flux by diffusion through a plane wall to a surface is known to be 400 W/m². Determine the surface temperature for each of the following conditions:

(a) Convection between the surface and an airstream at 20°C with a heat transfer coefficient $h = 10$ W/m² · K.

(b) The same convection process occurs along with radiative heat transfer between the surface and cold surroundings at −150°C, with a radiative heat transfer coefficient of $h_r = 5$ W/m² · K.

1.43 A surface whose temperature is maintained at 400°C is separated from an airflow by a layer of insulation 25 mm thick for which the thermal conductivity is 0.1 W/m · K. If the air temperature is 35°C and the convection coefficient between the air and the outer surface of the insulation is 500 W/m² · K, what is the temperature of this outer surface?

1.44 The wall of an oven used to cure plastic parts is of thickness $L = 0.05$ m and is exposed to large surroundings and air at its outer surface. The air and the surroundings are at 300 K.

(a) If the temperature of the outer surface is 400 K and its convection coefficient and emissivity are $h = 20$ W/m² · K and $\varepsilon = 0.8$, respectively, what is the temperature of the inner surface if the wall has a thermal conductivity of $k = 0.7$ W/m · K?

(b) Consider conditions for which the temperature of the inner surface is maintained at 600 K, while the air and large surroundings to which the outer surface is exposed are maintained at 300 K. Explore the effects of variations in k, h, and ε on (i) the temperature of the outer surface, (ii) the heat flux through the wall, and (iii) the heat fluxes associated with convection and radiation heat transfer from the outer surface. Specifically, compute and plot the foregoing dependent variables for parametric variations about baseline conditions of $k = 10$ W/m · K, $h = 20$ W/m² · K, and $\varepsilon = 0.5$. Suggested ranges of the independent variables are $0.1 \leq k \leq 400$ W/m · K, $2 \leq h \leq 200$ W/m² · K, and $0.05 \leq \varepsilon \leq 1$. Discuss the physical implications of your results. Under what conditions will the temperature of the outer surface be less than 45°C, which is a reasonable upper limit to avoid burn injuries if contact is made?

1.45 An experiment to determine the convection coefficient associated with airflow over the surface of a thick steel casting involves insertion of thermocouples in the casting at distances of 10 and 20 mm from the surface along a hypothetical line normal to the surface. The steel has a thermal conductivity of 15 W/m · K. If the thermocouples measure temperatures of 50 and 40°C in the steel when the air temperature is 100°C, what is the convection coefficient?

1.46 A thin electrical heating element provides a uniform heat flux q_o'' to the outer surface of a duct through which air flows. The duct wall has a thickness of 10 mm and a thermal conductivity of 20 W/m · K.

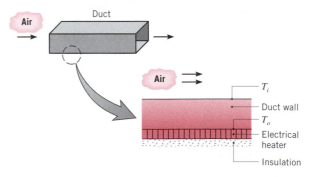

(a) At a particular location, the air temperature is 30°C and the convection heat transfer coefficient between the air and inner surface of the duct is 100 W/m² · K. What heat flux q_o'' is required to maintain the inner surface of the duct at $T_i = 85$°C?

(b) For the conditions of part (a), what is the temperature (T_o) of the duct surface next to the heater?

(c) With $T_i = 85°C$, compute and plot q_o'' and T_o as a function of the air-side convection coefficient h for the range $10 \le h \le 200$ W/m² · K. Briefly discuss your results.

1.47 One surface of a 10-mm-thick wall of stainless steel ($k = 15$ W/m · K) is maintained at 90°C by condensing steam, while the opposite surface is exposed to an airstream for which $T_\infty = 20°C$ and $h = 25$ W/m² · K. What is the temperature of the surface adjoining the air?

1.48 Plate glass at 600°C is cooled by passing air over its surface such that the convection heat transfer coefficient is $h = 5$ W/m² · K. To prevent cracking, it is known that the temperature gradient must not exceed 15°C/mm at any point in the glass during the cooling process. If the thermal conductivity of the glass is 1.4 W/m · K and its surface emissivity is 0.8, what is the lowest temperature of the air that can initially be used for the cooling? Assume that the temperature of the air equals that of the surroundings.

1.49 A solar flux of 700 W/m² is incident on a flat-plate solar collector used to heat water. The area of the collector is 3 m², and 90% of the solar radiation passes through the cover glass and is absorbed by the absorber plate. The remaining 10% is reflected away from the collector. Water flows through the tube passages on the back side of the absorber plate and is heated from an inlet temperature T_i to an outlet temperature T_o. The cover glass, operating at a temperature of 30°C, has an emissivity of 0.94 and experiences radiation exchange with the sky at $-10°C$. The convection coefficient between the cover glass and the ambient air at 25°C is 10 W/m² · K.

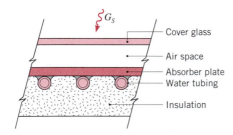

(a) Perform an overall energy balance on the collector to obtain an expression for the rate at which useful heat is collected per unit area of the collector, q_u''. Determine the value of q_u''.

(b) Calculate the temperature rise of the water, $T_o - T_i$, if the flow rate is 0.01 kg/s. Assume the specific heat of the water to be 4179 J/kg · K.

(c) The collector efficiency η is defined as the ratio of the useful heat collected to the rate at which solar energy is incident on the collector. What is the value of η?

1.50 Consider a flat-plate solar collector operating under steady-state conditions. Solar radiation is incident, per unit surface area of the collector, at a rate G_S (W/m²). The cover glass is completely transparent to this radiation, and the fraction of the radiation that is absorbed by the blackened absorber plate is designated as α (the absorptivity). The fraction of the radiation not absorbed by the absorber plate $(1 - \alpha)$ may be assumed to be reflected through the cover plate and back to the atmosphere and space.

Useful energy is obtained from the collector by passing a working fluid through copper tubing that is brazed to the underside of the absorber plate. The tubing forms a serpentine arrangement for which fluid, at a constant flow rate $\dot{m}$ and specific heat c_p, is heated from an inlet temperature T_i to an outlet temperature T_o. Although the bottom of the collector may be assumed to be perfectly insulated (no heat loss), there will be heat loss from the absorber plate due to convection across the air space and radiation exchange with the cover plate. Assuming that the absorber and cover plates have uniform temperatures T_a and T_c, respectively, the parallel convection and radiation heat fluxes may be expressed as $h_a(T_a - T_c)$ and $h_{r, ac}(T_a - T_c)$. The quantity h_a is the convection heat transfer coefficient associated with the air space, while $h_{r, ac}$ is the radiation heat transfer coefficient associated with the absorber plate–cover plate combination. The cover glass also transfers heat by convection to the ambient air, $h_\infty(T_c - T_\infty)$, and exchanges energy in the form of radiation with its surroundings, $h_{r, cs}(T_c - T_{sur})$. The effective temperature T_{sur} of the sky and surrounding surfaces seen by the cover glass is generally less than the ambient air temperature.

(a) Write an equation for the rate at which useful energy q_u (W) is collected by the working fluid, expressing your result in terms of $\dot{m}$, c_p, T_i, and T_o.

(b) Perform an energy balance on the absorber plate. Use this balance to obtain an expression for q_u in terms of G_S, α, T_a, T_c, h_a, $h_{r, ac}$, and A (the surface area of the absorber and cover plates).

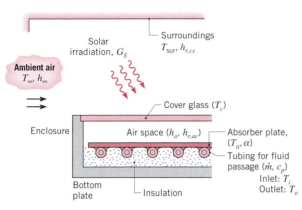

(c) Perform an energy balance on the cover plate.

(d) Perform an overall energy balance on the entire collector, working with a control volume about the collector. Compare your result with that obtained in parts (b) and (c).

(e) The collector efficiency η is defined as the ratio of the useful heat collected to the rate at which solar energy is incident on the collector. Obtain an expression for η.

(f) Comment on what effect the value of $\dot{m}$ will have on T_a, T_o, and η. What would happen to T_a if the cover plate were removed?

1.51 Consider a surface-mount type transistor on a circuit board whose temperature is maintained at 35°C. Air at 20°C flows over the upper surface of dimensions 4 mm by 8 mm with a convection coefficient of 50 W/m² · K. Three wire leads, each of cross section 1 mm by 0.25 mm and length 4 mm, conduct heat from the case to the circuit board. The gap between the case and the board is 0.2 mm.

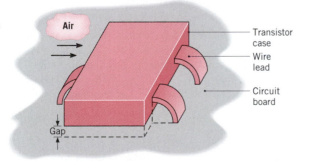

(a) Assuming the case is isothermal and neglecting radiation, estimate the case temperature when 150 mW are dissipated by the transistor and (i) stagnant air or (ii) a conductive paste fills the gap. The thermal conductivities of the wire leads, air, and conductive paste are 25, 0.0263, and 0.12 W/m · K, respectively.

(b) Using the conductive paste to fill the gap, we wish to determine the extent to which increased heat dissipation may be accommodated, subject to the constraint that the case temperature not exceed 40°C. Options include increasing the air speed to achieve a larger convection coefficient h and/or changing the lead wire material to one of larger thermal conductivity. Independently considering leads fabricated from materials with thermal conductivities of 200 and 400 W/m · K, compute and plot the maximum allowable heat dissipation for variations in h over the range 50 ≤ h ≤ 250 W/m² · K.

Process Identification

1.52 In analyzing the performance of a thermal system, the engineer must be able to identify the relevant heat transfer processes. Only then can the system behavior be properly quantified. For the following systems identify the pertinent processes, designating them by appropriately labeled arrows on a sketch of the system. Answer additional questions that appear in the problem statement.

(a) Identify the heat transfer processes that determine the temperature of an asphalt pavement on a summer day. Write an energy balance for the surface of the pavement.

(b) Microwave radiation is known to be transmitted by plastics, glass, and ceramics, but to be absorbed by materials having polar molecules such as water. Water molecules exposed to microwave radiation align and reverse alignment with the microwave radiation at frequencies up to 10^9 s^{-1}, causing heat to be generated. Contrast cooking in a microwave oven with cooking in a conventional radiant or convection oven. In each case what is the physical mechanism responsible for heating the food? Which oven has the greater energy utilization efficiency? Why? Microwave heating is being considered for drying clothes. How would operation of a microwave clothes dryer differ from a conventional dryer? Which is likely to have the greater energy utilization efficiency and why?

(c) Consider an exposed portion of your body (e.g., your forearm with a short-sleeved shirt) while you are sitting in a room. Identify all heat transfer processes that occur at the surface of your skin. In the interest of conserving fuel and funds, and engineer's spouse insists on keeping the thermostat of their home at 15°C (59°F) throughout the winter months. The engineer is able to tolerate this condition if the outside ambient air temperature exceeds $-10°C$ (14°F) but complains of being cold if the ambient temperature falls much below this value. Is the engineer imagining things?

(d) Consider an incandescent light source that consists of a tungsten filament enclosed in a gas-filled glass bulb. Assuming steady-state operation with the filament at a temperature of approximately 2900 K, list all the pertinent heat transfer processes for (i) the filament and (ii) the glass bulb.

(e) There is considerable interest in developing building materials that have improved insulating qualities. The development of such materials would do much to enhance energy conservation by reducing space heating requirements. It has been suggested that superior structural and insulating qualities could be obtained by using the composite shown. The material consists of a honeycomb, with cells of square cross section, sandwiched between solid slabs. The cells are filled with air, and the slabs, as well as the honeycomb matrix, are fabricated from plastics of low thermal conductivity. Identify all heat transfer processes pertinent to the performance of the composite. Suggest ways in which this performance could be enhanced.

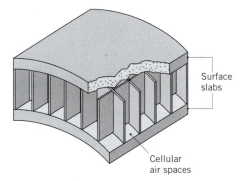

(f) A thermocouple junction is used to measure the temperature of a hot gas stream flowing through a channel by inserting the junction into the mainstream of the gas. The surface of the channel is cooled such that its temperature is well below that of the gas. Identify the heat transfer processes associated with the junction surface. Will the junction sense a temperature that is less than, equal to, or greater than the gas temperature? A radiation shield is a small, open-ended tube that encloses the thermocouple junction, yet allows for passage of the gas through the tube. How does use of such a shield improve the accuracy of the temperature measurement?

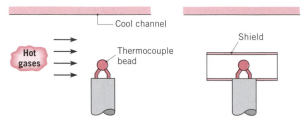

(g) A double-glazed, glass fire screen is inserted between a wood-burning fireplace and the interior of a room. The screen consists of two vertical glass plates that are separated by a space through which room air may flow (the space is open at the top and bottom). Identify the heat transfer processes associated with the fire screen.

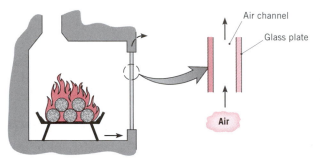

1.53 In considering the following problems involving heat transfer in the natural environment (outdoors), recognize that solar radiation is comprised of long and short wavelength components. If this radiation is incident on a *semitransparent medium,* such as water or glass, two things will happen to the nonreflected portion of the radiation. The long wavelength component will be absorbed at the surface of the medium, whereas the short wavelength component will be transmitted by the surface.

(a) The number of panes in a window can strongly influence the heat loss from a heated room to the outside ambient air. Compare the single- and double-paned units shown by identifying relevant heat transfer processes for each case.

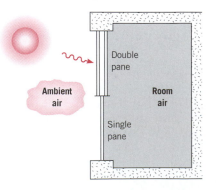

(b) In a typical flat-plate solar collector, energy is collected by a working fluid that is circulated through tubes that are in good contact with the back face of an absorber plate. The back face is insulated from the surroundings, and the absorber plate receives solar radiation on its front face, which is typically covered by one or more transparent plates. Identify the relevant heat transfer processes, first for the absorber plate with no cover plate and then for the absorber plate with a single cover plate.

(c) The solar energy collector design shown below has been used for agricultural applications. Air is blown through a long duct whose cross section is in the form of an equilateral triangle. One side of the triangle is comprised of a double-paned, semitransparent cover, while the other two sides are constructed from aluminum sheets painted flat black on the inside and covered on the outside with a layer of styrofoam insulation. During sunny periods, air entering the system is heated for delivery to either a greenhouse, grain drying unit, or a storage system.

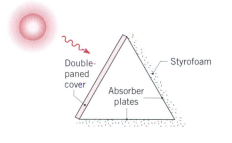

Identify all heat transfer processes associated with the cover plates, the absorber plate(s), and the air.

(d) Evacuated-tube solar collectors are capable of improved performance relative to flat-plate collectors. The design consists of an inner tube enclosed in an outer tube that is transparent to solar radiation. The annular space between the tubes is evacuated. The outer, opaque surface of the inner tube absorbs solar radiation, and a working fluid is passed through the tube to collect the solar energy. The collector design generally consists of a row of such tubes arranged in front of a reflecting panel. Identify all heat transfer processes relevant to the performance of this device.

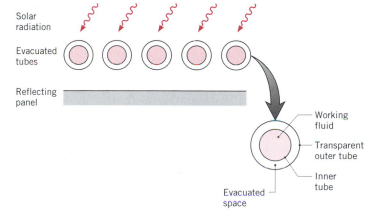

# CHAPTER 2

Introduction to Conduction

Recall that conduction (heat transfer by diffusion) refers to the transport of energy in a medium due to a temperature gradient, and the physical mechanism is that of random atomic or molecular activity. In this chapter we consider in greater detail the conduction rate equation and the relationship of energy conservation to the conduction process. In Chapters 3 to 5 we consider applications involving various geometrical and time-dependent conditions.

2.1
The Conduction Rate Equation

Although the conduction (energy diffusion) rate equation, Fourier's law, was introduced in Section 1.2, it is now appropriate to consider its origin. Fourier's law is phenomenological; that is, it is developed from observed phenomena rather than being derived from first principles. Hence we view the rate equation as a generalization based on much experimental evidence. For example consider the steady-state conduction experiment of Figure 2.1. A cylindrical rod of known material is insulated on its lateral surface, while its end faces are maintained at different temperatures with $T_1 > T_2$. The temperature difference causes conduction heat transfer in the positive x direction. We are able to measure the heat transfer rate q_x, and we seek to determine how q_x depends on the following variables: ΔT, the temperature difference; Δx, the rod length; and A, the cross-sectional area.

We might imagine first holding ΔT and Δx constant and varying A. If we do so, we find that q_x is directly proportional to A. Similarly, holding ΔT and A constant, we observe that q_x varies inversely with Δx. Finally, holding A and Δx constant, we find that q_x is directly proportional to ΔT. The collective effect is then

$$q_x \propto A \frac{\Delta T}{\Delta x}$$

In changing the material (e.g., from a metal to a plastic), we would find that the above proportionality remains valid. However, we would also find that, for equal values of A, Δx, and ΔT, the value of q_x would be smaller for the plastic than for the metal. This suggests that the proportionality may be converted to an equality by introducing a coefficient that is a measure of the material behavior. Hence we write

$$q_x = kA \frac{\Delta T}{\Delta x}$$

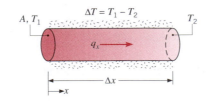

FIGURE 2.1
Steady-state heat conduction experiment.

where k, the *thermal conductivity* (W/m · K), is an important *property* of the material. Evaluating this expression in the limit as $\Delta x \to 0$, we obtain for the heat *rate*

$$q_x = -kA\frac{dT}{dx} \tag{2.1}$$

or for the heat *flux*

$$q''_x = \frac{q_x}{A} = -k\frac{dT}{dx} \tag{2.2}$$

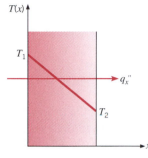

$T(x)$

T_1

q''_x

T_2

x

FIGURE 2.2

The relationship between coordinate system, heat flow direction, and temperature gradient in one dimension.

Recall that the minus sign is necessary because heat is always transferred in the direction of decreasing temperature.

Fourier's law, as written in Equation 2.2, implies that the heat flux is a directional quantity. In particular, the direction of q''_x is *normal* to the cross-sectional area A. Or, more generally, the direction of heat flow will always be normal to a surface of constant temperature, called an *isothermal* surface. Figure 2.2 illustrates the direction of heat flow q''_x in a plane wall for which the *temperature gradient dT/dx* is negative. From Equation 2.2, it follows that q''_x is positive. Note that the isothermal surfaces are planes normal to the x direction.

Recognizing that the heat flux is a vector quantity, we can write a more general statement of the conduction rate equation (*Fourier's law*) as follows:

$$\boldsymbol{q''} = -k\,\boldsymbol{\nabla}T = -k\left(\boldsymbol{i}\frac{\partial T}{\partial x} + \boldsymbol{j}\frac{\partial T}{\partial y} + \boldsymbol{k}\frac{\partial T}{\partial z}\right) \tag{2.3}$$

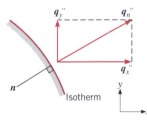

q''_y q''_n

q''_x

n y

Isotherm x

FIGURE 2.3

The heat flux vector normal to an isotherm in a two-dimensional coordinate system.

where $\boldsymbol{\nabla}$ is the three-dimensional del operator and $T(x, y, z)$ is the scalar temperature field. It is implicit in Equation 2.3 that the heat flux vector is in a direction perpendicular to the isothermal surfaces. An alternative form of Fourier's law is therefore

$$q''_n = -k\frac{\partial T}{\partial n} \tag{2.4}$$

where q''_n is the heat flux in a direction n, which is normal to an *isotherm,* as shown for the two-dimensional case in Figure 2.3. The heat transfer is sustained by a temperature gradient along $\boldsymbol{n}$. Note also that the heat flux vector can be resolved into components such that, in Cartesian coordinates, the general expression for $\boldsymbol{q''}$ is

$$\boldsymbol{q''} = \boldsymbol{i}q''_x + \boldsymbol{j}q''_y + \boldsymbol{k}q''_z \tag{2.5}$$

where, from Equation 2.3, it follows that

$$q''_x = -k\frac{\partial T}{\partial x} \qquad q''_y = -k\frac{\partial T}{\partial y} \qquad q''_z = -k\frac{\partial T}{\partial z} \tag{2.6}$$

Each of these expressions relates the heat flux *across a surface* to the temperature gradient in a direction perpendicular to the surface. It is also implicit in Equation 2.3 that the medium in which the conduction occurs is *isotropic*. For such a medium the value of the thermal conductivity is independent of the coordinate direction.

Since Fourier's law is the cornerstone of conduction heat transfer, its key features are summarized as follows. It is *not* an expression that may be derived from first principles; it is instead a generalization based on experimental evidence. It is also an expression that *defines* an important material property, the thermal conductivity. In addition, Fourier's law is a vector expression indicating that the heat flux is normal to an isotherm and in the direction of decreasing temperature. Finally, note that Fourier's law applies for all matter regardless of its state: solid, liquid, or gas.

2.2
The Thermal Properties of Matter

Using Fourier's law necessitates knowledge of the thermal conductivity. This property, which is referred to as a *transport property,* provides an indication of the rate at which energy is transferred by the diffusion process. It depends on the physical structure of matter, atomic and molecular, which is related to the state of the matter. In this section we consider various forms of matter, identifying important aspects of their behavior and presenting typical property values.

2.2.1   Thermal Conductivity

From Fourier's law, Equation 2.6, the thermal conductivity is defined as

$$k \equiv - \frac{q''_x}{(\partial T/\partial x)}$$

It follows that, for a prescribed temperature gradient, the conduction heat flux increases with increasing thermal conductivity. Recalling the physical mechanism associated with conduction (Section 1.2.1), it also follows that, in general, the thermal conductivity of a solid is larger than that of a liquid, which is larger than that of a gas. As illustrated in Figure 2.4, the thermal conductivity of a solid may be more than four orders of magnitude larger than that of a gas. This trend is due largely to differences in intermolecular spacing for the two states.

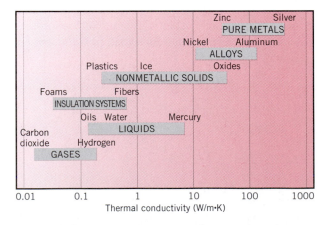

FIGURE 2.4   Range of thermal conductivity for various states of matter at normal temperatures and pressure.

The Solid State   In the modern view of materials, a solid may be comprised of free electrons and of atoms bound in a periodic arrangement called the lattice. Accordingly, transport of thermal energy is due to two effects: the migration of free electrons and lattice vibrational waves. These effects are additive, such that the thermal conductivity k is the sum of the electronic component k_e and the lattice component k_l

$$k = k_e + k_l$$

To a first approximation, k_e is inversely proportional to the electrical resistivity ρ_e. For pure metals, which are of low ρ_e, k_e is much larger than k_l. In contrast, for alloys, which are of substantially larger ρ_e, the contribution of k_l to k is no longer negligible. For nonmetallic solids, k is determined primarily by k_l, which depends on the frequency of interactions between the atoms of the lattice. The regularity of the lattice arrangement has an important effect on k_l, with crystalline (well-ordered) materials like quartz having a higher thermal conductivity than amorphous materials like glass. In fact, for crystalline, nonmetallic solids such as diamond and beryllium oxide, k_l can be quite large, exceeding values of k associated with good conductors, such as aluminum.

The temperature dependence of k is shown in Figure 2.5 for representative metallic and nonmetallic solids. Values for selected materials of technical importance are also provided in Table A.1 (metallic solids) and Tables A.2 and

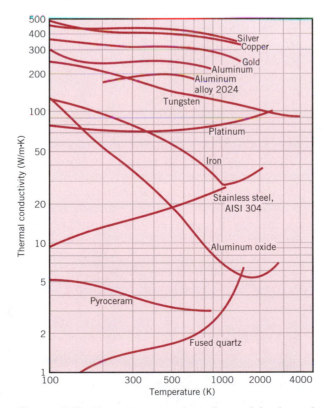

FIGURE 2.5   The temperature dependence of the thermal conductivity of selected solids.

A.3 (nonmetallic solids). More detailed treatments of thermal conductivity are available in the literature [1].

Insulation Systems   Thermal insulations are comprised of low thermal conductivity materials combined to achieve an even lower system thermal conductivity. In *fiber-, powder-,* and *flake*-type insulations, the solid material is finely dispersed throughout an air space. Such systems are characterized by an *effective thermal conductivity,* which depends on the thermal conductivity and surface radiative properties of the solid material, as well as the nature and volumetric fraction of the air or void space. A special parameter of the system is its bulk density (solid mass/total volume), which depends strongly on the manner in which the solid material is interconnected.

If small voids or hollow spaces are formed by bonding or fusing portions of the solid material, a rigid matrix is created. When these spaces are sealed from each other, the system is referred to as a cellular insulation. Examples of such rigid insulations are *foamed* systems, particularly those made from plastic and glass materials. *Reflective* insulations are composed of multilayered, parallel, thin sheets or foils of high reflectivity, which are spaced to reflect radiant heat back to its source. The spacing between the foils is designed to restrict the motion of air, and in high-performance insulations, the space is even evacuated. In all types of insulation, evacuation of the air in the void space will reduce the effective thermal conductivity of the system.

It is important to recognize that heat transfer through any of these insulation systems may include several modes: conduction through the solid materials; conduction or convection through the air in the void spaces; and, if the temperature is sufficiently high, radiation exchange between the surfaces of the solid matrix. The *effective* thermal conductivity accounts for all of these processes, and values for selected insulation systems are summarized in Table A.3. Additional background information and data are available in the literature [2, 3].

The Fluid State   Since the intermolecular spacing is much larger and the motion of the molecules is more random for the fluid state than for the solid state, thermal energy transport is less effective. The thermal conductivity of gases and liquids is therefore generally smaller than that of solids.

The effect of temperature, pressure, and chemical species on the thermal conductivity of a gas may be explained in terms of the kinetic theory of gases [4]. From this theory it is known that the thermal conductivity is directly proportional to the number of particles per unit volume n, the mean molecular speed $\bar{c}$, and the mean free path λ, which is the average distance traveled by a molecule before experiencing a collision. Hence

$$k \propto n\bar{c}\lambda$$

Because $\bar{c}$ increases with increasing temperature and decreasing molecular mass, the thermal conductivity of a gas increases with increasing temperature and decreasing molecular weight. These trends are shown in Figure 2.6. However, because n and λ are directly and indirectly proportional to the gas pressure, respectively, the thermal conductivity is independent of pressure. This assumption is appropriate for the gas pressures of interest in this text. Accord-

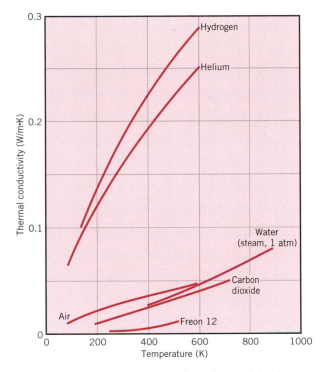

FIGURE 2.6   The temperature dependence of the thermal conductivity of selected gases at normal pressures.

ingly, although the values of k presented in Table A.4 pertain to atmospheric pressure or the saturation pressure corresponding to the prescribed temperature, they may be used over a much wider range.

Molecular conditions associated with the liquid state are more difficult to describe, and physical mechanisms for explaining the thermal conductivity are not well understood [5]. As shown in Figure 2.7, the thermal conductivity of nonmetallic liquids generally decreases with increasing temperature, the notable exceptions being glycerine and water. This property is insensitive to pressure except near the critical point. Also, it generally follows that thermal conductivity decreases with increasing molecular weight. Values of the thermal conductivity are generally tabulated as a function of temperature for the saturated state of the liquid. Tables A.5 and A.6 present such data for several common liquids.

Liquid metals are commonly used in high flux applications, such as occur in nuclear power plants. The thermal conductivity of such liquids is given in Table A.7. Note that the values are much larger than those of the nonmetallic liquids [6].

2.2.2   Other Relevant Properties

In our analysis of heat transfer problems, it will be necessary to use many properties of matter. These properties are generally referred to as *thermophysical*

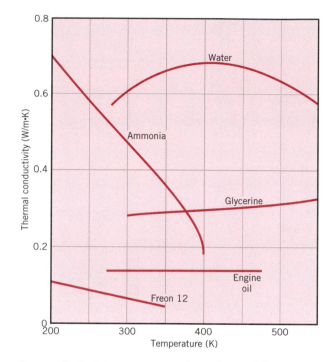

FIGURE 2.7   The temperature dependence of the thermal conductivity of selected nonmetallic liquids under saturated conditions.

properties and include two distinct categories, *transport* and *thermodynamic* properties. The transport properties include the diffusion rate coefficients such as k, the thermal conductivity (for heat transfer), and ν, the kinematic viscosity (for momentum transfer). Thermodynamic properties, on the other hand, pertain to the equilibrium state of a system. Density (ρ) and specific heat (c_p) are two such properties used extensively in thermodynamic analysis. The product $\rho\, c_p$ (J/m^3 · K), commonly termed the *volumetric heat capacity,* measures the ability of a material to store thermal energy. Because substances of large density are typically characterized by small specific heats, many solids and liquids, which are very good energy storage media, have comparable heat capacities ($\rho\, c_p > 1$ MJ/m^3 · K). Because of their very small densities, however, gases are poorly suited for thermal energy storage ($\rho\, c_p \approx 1$ kJ/m^3 · K). Densities and specific heats are provided in the tables of Appendix A for a wide range of solids, liquids, and gases.

In heat transfer analysis, the ratio of the thermal conductivity to the heat capacity is an important property termed the *thermal diffusivity* α, which has units of m^2/s:

$$\alpha = \frac{k}{\rho\, c_p}$$

It measures the ability of a material to conduct thermal energy relative to its ability to store thermal energy. Materials of large α will respond quickly to changes in their thermal environment, while materials of small α will respond more sluggishly, taking longer to reach a new equilibrium condition.

The accuracy of engineering calculations depends on the accuracy with which the thermophysical properties are known [7–9]. Numerous examples could be cited of flaws in equipment and process design or failure to meet performance specifications that were attributable to misinformation associated with the selection of key property values used in the initial system analysis. Selection of reliable property data is an integral part of any careful engineering analysis. The casual use of data from the literature or handbooks, which have not been well characterized or evaluated, is to be avoided. Recommended data values for many thermophysical properties can be obtained from Reference 10. This reference, available in most institutional libraries, was prepared by the Thermophysical Properties Research Center (TPRC) at Purdue University. A continuing program is maintained to provide current, comprehensive coverage on thermophysical properties [11].

EXAMPLE 2.1

The thermal diffusivity α is the controlling transport property for transient conduction. Using appropriate values of k, ρ, and c_p from Appendix A, calculate α for the following materials at the prescribed temperatures: pure aluminum, 300 and 700 K; silicon carbide, 1000 K; paraffin, 300 K.

SOLUTION

Known: Definition of the thermal diffusivity α.

Find: Numerical values of α for selected materials and temperatures.

Properties: Table A.1, pure aluminum (300 K):

$$\left.\begin{array}{l} \rho = 2702 \text{ kg/m}^3 \\ c_p = 903 \text{ J/kg} \cdot \text{K} \\ k = 237 \text{ W/m} \cdot \text{K} \end{array}\right\} \alpha = \frac{k}{\rho\, c_p} = \frac{237 \text{ W/m} \cdot \text{K}}{2702 \text{ kg/m}^3 \times 903 \text{ J/kg} \cdot \text{K}}$$

$$= 97.1 \times 10^{-6} \text{ m}^2\text{/s} \qquad \triangleleft$$

Table A.1, pure aluminum (700 K):

$$\rho = 2702 \text{ kg/m}^3 \qquad \text{at 300 K}$$

$$c_p = 1090 \text{ J/kg} \cdot \text{K} \qquad \text{at 700 K (by linear interpolation)}$$

$$k = 225 \text{ W/m} \cdot \text{K} \qquad \text{at 700 K (by linear interpolation)}$$

Hence

$$\alpha = \frac{k}{\rho\, c_p} = \frac{225 \text{ W/m} \cdot \text{K}}{2702 \text{ kg/m}^3 \times 1090 \text{ J/kg} \cdot \text{K}} = 76 \times 10^{-6} \text{ m}^2\text{/s} \qquad \triangleleft$$

Table A.2, silicon carbide (1000 K):

$$\left.\begin{array}{ll} \rho = 3160 \text{ kg/m}^3 & \text{at 300 K} \\ c_p = 1195 \text{ J/kg} \cdot \text{K} & \text{at 1000 K} \\ k = 87 \text{ W/m} \cdot \text{K} & \text{at 1000 K} \end{array}\right\} \alpha = \frac{87 \text{ W/m} \cdot \text{K}}{3160 \text{ kg/m}^3 \times 1195 \text{ J/kg} \cdot \text{K}}$$

$$= 23 \times 10^{-6} \text{ m}^2/\text{s} \qquad \triangleleft$$

Table A.3, paraffin (300 K):

$$\left.\begin{array}{l} \rho = 900 \text{ kg/m}^3 \\ c_p = 2890 \text{ J/kg} \cdot \text{K} \\ k = 0.024 \text{ W/m} \cdot \text{K} \end{array}\right\} \alpha = \frac{k}{\rho c_p} = \frac{0.024 \text{ W/m} \cdot \text{K}}{900 \text{ kg/m}^3 \times 2890 \text{ J/kg} \cdot \text{K}}$$

$$= 9.2 \times 10^{-9} \text{ m}^2/\text{s} \qquad \triangleleft$$

Comments:

1. Note temperature dependence of the thermophysical properties of aluminum and silicon carbide. For example, for silicon carbide, $\alpha(1000 \text{ K}) \approx 0.1 \times \alpha(300 \text{ K})$; hence properties of this material have a strong temperature dependence.

2. The physical interpretation of α is that it provides a measure of heat transport (k) relative to energy storage (ρc_p). In general, metallic solids have higher α, while nonmetallics (e.g., paraffin) have lower values of α.

3. Linear interpolation of property values is generally acceptable for engineering calculations.

4. Use of the low-temperature (300 K) density at higher temperatures ignores thermal expansion effects but is also acceptable for engineering calculations.

2.3
The Heat Diffusion Equation

A major objective in a conduction analysis is to determine the *temperature field* in a medium resulting from conditions imposed on its boundaries. That is, we wish to know the *temperature distribution,* which represents how temperature varies with position in the medium. Once this distribution is known, the conduction heat flux at any point in the medium or on its surface may be computed from Fourier's law. Other important quantities of interest may also be determined. For a solid, knowledge of the temperature distribution could be used to ascertain structural integrity through determination of thermal stresses, expansions, and deflections. The temperature distribution could also be used to optimize the thickness of an insulating material or to determine the compatibility of special coatings or adhesives used with the material.

We now consider the manner in which the temperature distribution can be determined. The approach follows the methodology described in Section 1.3.3 of applying the energy conservation requirement. That is, we define a differen-

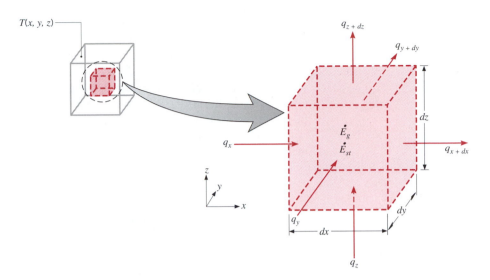

FIGURE 2.8 Differential control volume, $dx\,dy\,dz$, for conduction analysis in Cartesian coordinates.

tial control volume, identify the relevant energy transfer processes, and introduce the appropriate rate equations. The result is a differential equation whose solution, for prescribed boundary conditions, provides the temperature distribution in the medium.

Consider a homogeneous medium within which there is no bulk motion (advection) and the temperature distribution $T(x, y, z)$ is expressed in Cartesian coordinates. Following the methodology of applying conservation of energy (Section 1.3.3), we first define an infinitesimally small (differential) control volume, $dx \cdot dy \cdot dz$, as shown in Figure 2.8. Choosing to formulate the first law at an instant of time, the second step is to consider the energy processes that are relevant to this control volume. If there are temperature gradients, conduction heat transfer will occur across each of the control surfaces. The conduction heat rates perpendicular to each of the control surfaces at the x, y, and z coordinate locations are indicated by the terms q_x, q_y, and q_z, respectively. The conduction heat rates at the opposite surfaces can then be expressed as a Taylor series expansion where, neglecting higher order terms,

$$q_{x+dx} = q_x + \frac{\partial q_x}{\partial x}\,dx \tag{2.7a}$$

$$q_{y+dy} = q_y + \frac{\partial q_y}{\partial y}\,dy \tag{2.7b}$$

$$q_{z+dz} = q_z + \frac{\partial q_z}{\partial z}\,dz \tag{2.7c}$$

In words, Equation 2.7a simply states that the x component of the heat transfer rate at $x + dx$ is equal to the value of this component at x plus the amount by which it changes with respect to x times dx.

Within the medium there may also be an *energy source* term associated with the rate of thermal energy generation. This term is represented as

$$\dot{E}_g = \dot{q}\,dx\,dy\,dz \qquad (2.8)$$

where $\dot{q}$ is the rate at which energy is generated per unit volume of the medium (W/m³). In addition, there may occur changes in the amount of the internal thermal energy stored by the material in the control volume. If the material is not experiencing a change in phase, latent energy effects are not pertinent, and the *energy storage* term may be expressed as

$$\dot{E}_{st} = \rho\,c_p\,\frac{\partial T}{\partial t}\,dx\,dy\,dz \qquad (2.9)$$

where $\rho\,c_p\,\partial T/\partial t$ is the time rate of change of the sensible (thermal) energy of the medium per unit volume.

Once again it is important to note that the terms $\dot{E}_g$ and $\dot{E}_{st}$ represent different physical processes. The energy generation term $\dot{E}_g$ is a manifestation of some energy conversion process involving thermal energy on one hand and chemical, electrical, or nuclear energy on the other. The term is positive (a *source*) if thermal energy is being generated in the material at the expense of some other energy form; it is negative (a *sink*) if thermal energy is being consumed. In contrast, the energy storage term $\dot{E}_{st}$ refers to the rate of change of thermal energy stored by the matter.

The last step in the methodology outlined in Section 1.3.3 is to express conservation of energy using the foregoing rate equations. On a *rate* basis, the general form of the conservation of energy requirement is

$$\dot{E}_{in} + \dot{E}_g - \dot{E}_{out} = \dot{E}_{st} \qquad (1.11a)$$

Hence, recognizing that the conduction rates constitute the energy inflow, $\dot{E}_{in}$, and outflow, $\dot{E}_{out}$, and substituting Equations 2.8 and 2.9, we obtain

$$q_x + q_y + q_z + \dot{q}\,dx\,dy\,dz - q_{x+dx} - q_{y+dy}$$

$$- q_{z+dz} = \rho\,c_p\,\frac{\partial T}{\partial t}\,dx\,dy\,dz \qquad (2.10)$$

Substituting from Equations 2.7, it follows that

$$-\frac{\partial q_x}{\partial x}\,dx - \frac{\partial q_y}{\partial y}\,dy - \frac{\partial q_z}{\partial z}\,dz + \dot{q}\,dx\,dy\,dz = \rho\,c_p\,\frac{\partial T}{\partial t}\,dx\,dy\,dz \quad (2.11)$$

The conduction heat rates may be evaluated from Fourier's law,

$$q_x = -k\,dy\,dz\,\frac{\partial T}{\partial x} \qquad (2.12a)$$

$$q_y = -k\,dx\,dz\,\frac{\partial T}{\partial y} \qquad (2.12b)$$

$$q_z = -k\,dx\,dy\,\frac{\partial T}{\partial z} \qquad (2.12c)$$

where each heat flux component of Equation 2.6 has been multiplied by the appropriate control surface (differential) area to obtain the heat transfer rate. Substituting Equations 2.12 into Equation 2.11 and dividing out the dimensions of the control volume ($dx\,dy\,dz$), we obtain

$$\frac{\partial}{\partial x}\left(k\frac{\partial T}{\partial x}\right) + \frac{\partial}{\partial y}\left(k\frac{\partial T}{\partial y}\right) + \frac{\partial}{\partial z}\left(k\frac{\partial T}{\partial z}\right) + \dot{q} = \rho\,c_p\frac{\partial T}{\partial t} \qquad (2.13)$$

Equation 2.13 is the general form, in Cartesian coordinates, of the *heat diffusion equation*. This equation, usually known as the *heat equation,* provides the basic tool for heat conduction analysis. From its solution, we can obtain the temperature distribution $T(x, y, z)$ as a function of time. The apparent complexity of this expression should not obscure the fact that it describes an important physical condition, that is, conservation of energy. You should have a clear understanding of the physical significance of each term appearing in the equation. For example, the term $\partial(k\,\partial T/\partial x)/\partial x$ is related to the *net* conduction heat flux *into* the control volume for the *x*-coordinate direction. That is, multiplying by dx,

$$\frac{\partial}{\partial x}\left(k\frac{\partial T}{\partial x}\right)dx = q_x'' - q_{x+dx}'' \qquad (2.14)$$

with similar expressions applying for the fluxes in the *y* and *z* directions. In words, the heat equation, Equation 2.13, therefore states that *at any point in the medium the rate of energy transfer by conduction into a unit volume plus the volumetric rate of thermal energy generation must equal the rate of change of thermal energy stored within the volume.*

It is often possible to work with simplified versions of Equation 2.13. For example, if the thermal conductivity is a constant, the heat equation is

$$\frac{\partial^2 T}{\partial x^2} + \frac{\partial^2 T}{\partial y^2} + \frac{\partial^2 T}{\partial z^2} + \frac{\dot{q}}{k} = \frac{1}{\alpha}\frac{\partial T}{\partial t} \qquad (2.15)$$

where $\alpha = k/\rho\,c_p$ is the *thermal diffusivity*. Additional simplifications of the general form of the heat equation are often possible. For example, under *steady-state* conditions, there can be no change in the amount of energy storage; hence Equation 2.13 reduces to

$$\frac{\partial}{\partial x}\left(k\frac{\partial T}{\partial x}\right) + \frac{\partial}{\partial y}\left(k\frac{\partial T}{\partial y}\right) + \frac{\partial}{\partial z}\left(k\frac{\partial T}{\partial z}\right) + \dot{q} = 0 \qquad (2.16)$$

Moreover, if the heat transfer is *one-dimensional* (e.g., in the *x* direction) and there is *no energy generation,* Equation 2.16 reduces to

$$\frac{d}{dx}\left(k\frac{dT}{dx}\right) = 0 \qquad (2.17)$$

The important implication of this result is that *under steady-state, one-dimensional conditions with no energy generation,* the heat flux is a constant in the direction of transfer ($dq_x''/dx = 0$).

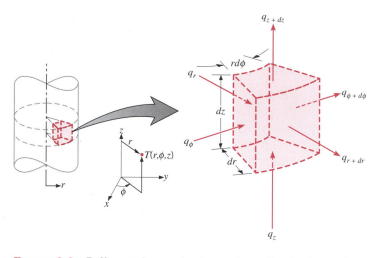

FIGURE 2.9   Differential control volume, $dr \cdot r \, d\phi \cdot dz$, for conduction analysis in cylindrical coordinates (r, ϕ, z).

The heat equation may also be expressed in cylindrical and spherical coordinates. The differential control volumes for these two coordinate systems are shown in Figures 2.9 and 2.10.

Cylindrical Coordinates   When the del operator ∇ of Equation 2.3 is expressed in cylindrical coordinates, the general form of the heat flux vector, and hence of Fourier's law, is

$$\boldsymbol{q}'' = -k \, \boldsymbol{\nabla} T = -k \left(\boldsymbol{i} \frac{\partial T}{\partial r} + \boldsymbol{j} \frac{1}{r} \frac{\partial T}{\partial \phi} + \boldsymbol{k} \frac{\partial T}{\partial z} \right) \tag{2.18}$$

where

$$q_r'' = -k \frac{\partial T}{\partial r} \qquad q_\phi'' = -\frac{k}{r} \frac{\partial T}{\partial \phi} \qquad q_z'' = -k \frac{\partial T}{\partial z} \tag{2.19}$$

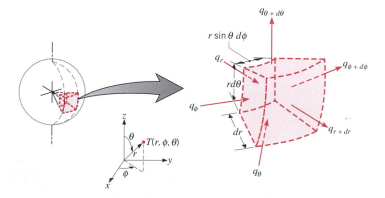

FIGURE 2.10   Differential control volume, $dr \cdot r \sin \theta \, d\phi \cdot r \, d\theta$, for conduction analysis in spherical coordinates (r, ϕ, θ).

are heat flux components in the radial, circumferential, and axial directions, respectively. Applying an energy balance to the differential control volume of Figure 2.9, the following general form of the heat equation is obtained:

$$\frac{1}{r}\frac{\partial}{\partial r}\left(kr\frac{\partial T}{\partial r}\right) + \frac{1}{r^2}\frac{\partial}{\partial \phi}\left(k\frac{\partial T}{\partial \phi}\right)$$

$$+ \frac{\partial}{\partial z}\left(k\frac{\partial T}{\partial z}\right) + \dot{q} = \rho c_p \frac{\partial T}{\partial t} \qquad (2.20)$$

Spherical Coordinates   In spherical coordinates the general form of the heat flux vector and Fourier's law is

$$\mathbf{q''} = -k\,\nabla T = -k\left(\mathbf{i}\,\frac{\partial T}{\partial r} + \mathbf{j}\,\frac{1}{r}\frac{\partial T}{\partial \theta} + \mathbf{k}\,\frac{1}{r\sin\theta}\frac{\partial T}{\partial \phi}\right) \qquad (2.21)$$

where

$$q''_r = -k\frac{\partial T}{\partial r} \qquad q''_\theta = -\frac{k}{r}\frac{\partial T}{\partial \theta} \qquad q''_\phi = -\frac{k}{r\sin\theta}\frac{\partial T}{\partial \phi} \qquad (2.22)$$

are heat flux components in the radial, polar, and azimuthal directions, respectively. Applying an energy balance to the differential control volume of Figure 2.10, the following general form of the heat equation is obtained:

$$\frac{1}{r^2}\frac{\partial}{\partial r}\left(kr^2\frac{\partial T}{\partial r}\right) + \frac{1}{r^2\sin^2\theta}\frac{\partial}{\partial \phi}\left(k\frac{\partial T}{\partial \phi}\right)$$

$$+ \frac{1}{r^2\sin\theta}\frac{\partial}{\partial \theta}\left(k\sin\theta\frac{\partial T}{\partial \theta}\right) + \dot{q} = \rho c_p \frac{\partial T}{\partial t} \qquad (2.23)$$

Since it is important that you be able to apply conservation principles to differential control volumes, you should attempt to derive Equation 2.20 or 2.23 (see Problems 2.32 and 2.33). Note that the temperature gradient in Fourier's law must have units of K/m. Hence, when evaluating the gradient for an angular coordinate, it must be expressed in terms of the differential change in arc *length*. For example, the heat flux component in the circumferential direction of a cylindrical coordinate system is $q''_\phi = -(k/r)(\partial T/\partial \phi)$, and *not* $q''_\phi = -k(\partial T/\partial \phi)$.

EXAMPLE 2.2

The temperature distribution across a wall 1 m thick at a certain instant of time is given as

$$T(x) = a + bx + cx^2$$

where T is in degrees Celsius and x is in meters, while $a = 900°C$, $b = -300°C/m$, and $c = -50°C/m^2$. A uniform heat generation, $\dot{q} = 1000$ W/m^3, is present in the wall of area 10 m^2 having the properties $\rho = 1600$ kg/m^3, $k = 40$ W/m · K, and $c_p = 4$ kJ/kg · K.

1. Determine the rate of heat transfer entering the wall ($x = 0$) and leaving the wall ($x = 1$ m).
2. Determine the rate of change of energy storage in the wall.
3. Determine the time rate of temperature change at $x = 0$, 0.25, and 0.5 m.

SOLUTION

Known: Temperature distribution $T(x)$ at an instant of time t in a one-dimensional wall with uniform heat generation.

Find:

1. Heat rates entering, q_{in} ($x = 0$), and leaving, q_{out} ($x = 1$), the wall.
2. Rate of change of energy storage in the wall, $\dot{E}_{st}$.
3. Time rate of temperature change at $x = 0$, 0.25, and 0.5 m.

Schematic:

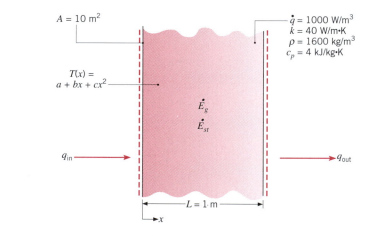

$A = 10 \text{ m}^2$

$\dot{q} = 1000 \text{ W/m}^3$
$k = 40 \text{ W/m·K}$
$\rho = 1600 \text{ kg/m}^3$
$c_p = 4 \text{ kJ/kg·K}$

$T(x) = a + bx + cx^2$

$\dot{E}_g$
$\dot{E}_{st}$

q_{in} q_{out}

$L = 1 \text{ m}$

x

Assumptions:

1. One-dimensional conduction in the x direction.
2. Homogeneous medium with constant properties.
3. Uniform internal heat generation, $\dot{q}$ (W/m^3).

Analysis:

1. Recall that once the temperature distribution is known for a medium, it is a simple matter to determine the conduction heat transfer rate at any point in the medium, or at its surfaces, by using Fourier's law. Hence the desired heat rates may be determined by using the prescribed temperature distribution with Equation 2.1. Accordingly,

$$q_{in} = q_x(0) = -kA \left. \frac{\partial T}{\partial x} \right|_{x=0} = -kA(b + 2cx)_{x=0}$$

$$q_{in} = -bkA = 300°\text{C/m} \times 40 \text{ W/m} \cdot \text{K} \times 10 \text{ m}^2 = 120 \text{ kW} \qquad \triangleleft$$

Similarly,

$$q_{\text{out}} = q_x(L) = -kA \left. \frac{\partial T}{\partial x} \right|_{x=L} = -kA(b + 2cx)_{x=L}$$

$$q_{\text{out}} = -(b + 2cL)kA = -[-300°C/m$$
$$+ 2(-50°C/m^2) \times 1 \text{ m}] \times 40 \text{ W/m} \cdot \text{K} \times 10 \text{ m}^2 = 160 \text{ kW} \quad \triangleleft$$

2. The rate of change of energy storage in the wall $\dot{E}_{\text{st}}$ may be determined by applying an overall energy balance to the wall. Using Equation 1.11a for a control volume about the wall,

$$\dot{E}_{\text{in}} + \dot{E}_g - \dot{E}_{\text{out}} = \dot{E}_{\text{st}}$$

where $\dot{E}_g = \dot{q}AL$, it follows that

$$\dot{E}_{\text{st}} = \dot{E}_{\text{in}} + \dot{E}_g - \dot{E}_{\text{out}} = q_{\text{in}} + \dot{q}AL - q_{\text{out}}$$
$$\dot{E}_{\text{st}} = 120 \text{ kW} + 1000 \text{ W/m}^3 \times 10 \text{ m}^2 \times 1 \text{ m} - 160 \text{ kW}$$
$$\dot{E}_{\text{st}} = -30 \text{ kW} \quad \triangleleft$$

3. The time rate of change of the temperature at any point in the medium may be determined from the heat equation, Equation 2.15, rewritten as

$$\frac{\partial T}{\partial t} = \frac{k}{\rho c_p} \frac{\partial^2 T}{\partial x^2} + \frac{\dot{q}}{\rho c_p}$$

From the prescribed temperature distribution, it follows that

$$\frac{\partial^2 T}{\partial x^2} = \frac{\partial}{\partial x} \left(\frac{\partial T}{\partial x} \right)$$

$$= \frac{\partial}{\partial x}(b + 2cx) = 2c = 2(-50°C/m^2) = -100°C/m^2$$

Note that this derivative is independent of position in the medium. Hence the time rate of temperature change is also independent of position and is given by

$$\frac{\partial T}{\partial t} = \frac{40 \text{ W/m} \cdot \text{K}}{1600 \text{ kg/m}^3 \times 4 \text{ kJ/kg} \cdot \text{K}} \times (-100°C/m^2)$$

$$+ \frac{1000 \text{ W/m}^3}{1600 \text{ kg/m}^3 \times 4 \text{ kJ/kg} \cdot \text{K}}$$

$$\frac{\partial T}{\partial t} = -6.25 \times 10^{-4}°C/s + 1.56 \times 10^{-4}°C/s$$

$$= -4.69 \times 10^{-4}°C/s \quad \triangleleft$$

Comments:

1. From the above result it is evident that the temperature at every point within the wall is decreasing with time.

2. Fourier's law can always be used to compute the conduction heat rate from knowledge of the temperature distribution, even for unsteady conditions with internal heat generation.

2.4
Boundary and Initial Conditions

To determine the temperature distribution in a medium, it is necessary to solve the appropriate form of the heat equation. However, such a solution depends on the physical conditions existing at the *boundaries* of the medium and, if the situation is time dependent, on conditions existing in the medium at some *initial time.* With regard to the *boundary conditions,* there are several common possibilities that are simply expressed in mathematical form. Because the heat equation is second order in the spatial coordinates, two boundary conditions must be expressed for each coordinate needed to describe the system. Because the equation is first order in time, however, only one condition, termed the *initial condition,* must be specified.

The three kinds of boundary conditions commonly encountered in heat transfer are summarized in Table 2.1. The conditions are specified at the surface $x = 0$ for a one-dimensional system. Heat transfer is in the positive x direction with the temperature distribution, which may be time dependent, designated as $T(x, t)$. The first condition corresponds to a situation for which the

TABLE 2.1   Boundary conditions for the heat diffusion equation at the surface ($x = 0$)

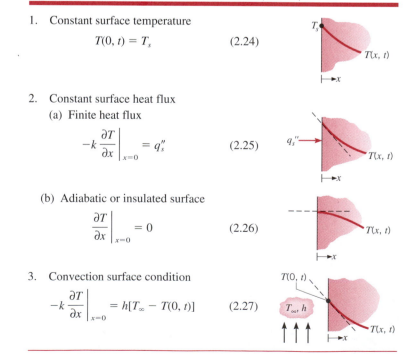

1. Constant surface temperature

$$T(0, t) = T_s \qquad (2.24)$$

2. Constant surface heat flux
    (a) Finite heat flux

$$-k \left. \frac{\partial T}{\partial x} \right|_{x=0} = q_s'' \qquad (2.25)$$

    (b) Adiabatic or insulated surface

$$\left. \frac{\partial T}{\partial x} \right|_{x=0} = 0 \qquad (2.26)$$

3. Convection surface condition

$$-k \left. \frac{\partial T}{\partial x} \right|_{x=0} = h[T_\infty - T(0, t)] \qquad (2.27)$$

surface is maintained at a fixed temperature T_s. It is commonly termed a *Dirichlet condition,* or a boundary condition of the *first kind.* It is closely approximated, for example, when the surface is in contact with a melting solid or a boiling liquid. In both cases there is heat transfer at the surface, while the surface remains at the temperature of the phase change process. The second condition corresponds to the existence of a fixed or constant heat flux q''_s at the surface. This heat flux is related to the temperature gradient at the surface by Fourier's law, Equation 2.6, which may be expressed as

$$q''_x(0) = -k \left. \frac{\partial T}{\partial x} \right|_{x=0}$$

It is termed a *Neumann condition,* or a boundary condition of the *second kind,* and may be realized by bonding a thin film or patch electric heater to the surface. A special case of this condition corresponds to the *perfectly insulated,* or *adiabatic,* surface for which $\partial T/\partial x|_{x=0} = 0$. The boundary condition of the *third kind* corresponds to the existence of convection heating (or cooling) at the surface and is obtained from the surface energy balance discussed in Section 1.3.2.

EXAMPLE 2.3

A long copper bar of rectangular cross section, whose width w is much greater than its thickness L, is maintained in contact with a heat sink at its lower surface, and the temperature throughout the bar is approximately equal to that of the sink, T_o. Suddenly, an electric current is passed through the bar and an airstream of temperature T_∞ is passed over the top surface, while the bottom surface continues to be maintained at T_o. Obtain the differential equation and the boundary and initial conditions that could be solved to determine the temperature as a function of position and time in the bar.

SOLUTION

Known: Copper bar initially in thermal equilibrium with a heat sink is suddenly heated by passage of an electric current.

Find: Differential equation and boundary and initial conditions needed to determine temperature as a function of position and time within the bar.

Schematic:

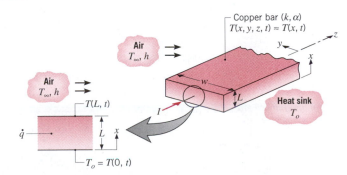

Assumptions:

1. Since $w \gg L$, side effects are negligible and heat transfer within the bar is primarily one dimensional in the x direction.
2. Uniform volumetric heat generation, $\dot{q}$.
3. Constant properties.

Analysis: The temperature distribution is governed by the heat equation (2.13), which, for the one-dimensional and constant property conditions of the present problem, reduces to

$$\frac{\partial^2 T}{\partial x^2} + \frac{\dot{q}}{k} = \frac{1}{\alpha} \frac{\partial T}{\partial t} \qquad (1) \qquad \triangleleft$$

where the temperature is a function of position and time, $T(x, t)$. Since this differential equation is second order in the spatial coordinate x and first order in time t, there must be two boundary conditions for the x direction and one condition, termed the initial condition, for time. The boundary condition at the bottom surface corresponds to case 1 of Table 2.1. In particular, since the temperature of this surface is maintained at a value, T_o, which is fixed with time, it follows that

$$T(0, t) = T_o \qquad (2) \qquad \triangleleft$$

In contrast, the convection surface condition, case 3 of Table 2.1, is appropriate for the top surface. Hence

$$-k \left. \frac{\partial T}{\partial x} \right|_{x=L} = h[T(L, t) - T_\infty] \qquad (3) \qquad \triangleleft$$

The initial condition is inferred from recognition that, before the change in conditions, the bar is at a uniform temperature T_o. Hence

$$T(x, 0) = T_o \qquad (4) \qquad \triangleleft$$

If T_o, T_∞, $\dot{q}$, and h are known, Equations 1 to 4 may be solved to obtain the time-varying temperature distribution $T(x, t)$ following imposition of the electric current.

Comments:

1. The *heat sink* at $x = 0$ could be maintained by exposing the surface to an ice bath or by attaching it to a *cold plate*. A cold plate contains coolant channels machined in a solid of large thermal conductivity (usually copper). By circulating a liquid (usually water) through the channels, the plate, and hence the surface to which it is attached, may be maintained at a nearly uniform temperature.
2. The temperature of the top surface $T(L, t)$ will change with time. This temperature is an unknown and may be obtained after finding $T(x, t)$.
3. How do you expect the temperature to vary with x at different times after the change in conditions? See Problem 2.40.

2.5
Summary

The primary purposes of this chapter were to improve your understanding of the conduction rate equation (Fourier's law) and to familiarize you with the heat equation. You should know the origin and implications of Fourier's law, and you should understand the key thermal properties and how they vary for different substances. You should also know the physical meaning of each term appearing in the heat equation. To what forms does this equation reduce for simplified conditions, and what kinds of boundary conditions may be used for its solution? In short, you should now have a fundamental understanding of the conduction process and its mathematical description. In the three chapters that follow, we undertake conduction analysis for numerous systems and conditions.

References

1. Klemens, P. G., "Theory of the Thermal Conductivity of Solids," in R. P. Tye, Ed., *Thermal Conductivity,* Vol. 1, Academic Press, London, 1969.

2. Mallory, John F., *Thermal Insulation,* Reinhold Book Corp., New York, 1969.

3. American Society of Heating, Refrigeration and Air Conditioning Engineers, *Handbook of Fundamentals,* Chapters 17 and 31, ASHRAE, New York, 1972.

4. Vincenti, W. G., and C. H. Kruger, Jr., *Introduction to Physical Gas Dynamics,* Wiley, New York, 1965.

5. McLaughlin E., "Theory of the Thermal Conductivity of Fluids," in R. P. Tye, Ed., *Thermal Conductivity,* Vol. 2, Academic Press, London, 1969.

6. Foust, O. J., Ed., "Sodium Chemistry and Physical Properties," in *Sodium-NaK Engineering Handbook,* Vol. 1, Gordon & Breach, New York, 1972.

7. Sengers, J. V., and M. Klein, Eds., *The Technical Importance of Accurate Thermophysical Property Infor-mation,* National Bureau of Standards Technical Note No. 590, 1980.

8. Najjar, M. S., K. J. Bell, and R. N. Maddox, *Heat Transfer Eng.,* **2,** 27, 1981.

9. Hanley, H. J. M., and M. E. Baltatu, *Mech. Eng.,* **105,** 68, 1983.

10. Touloukian, Y. S., and C. Y. Ho, Eds., *Thermophysical Properties of Matter, The TPRC Data Series* (13 volumes on thermophysical properties: thermal conductivity, specific heat, thermal radiative, thermal diffusivity, and thermal linear expansion), Plenum Press, New York, 1970 through 1977.

11. Center for Information and Numerical Data Analysis and Synthesis (CINDAS), Purdue University, 2595 Yeager Road, West Lafayette, IN 47906.

Problems

Fourier's Law

2.1 Assume steady-state, one-dimensional heat conduction through the axisymmetric shape shown below.

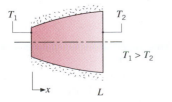

Assuming constant properties and no internal heat generation, sketch the temperature distribution on T–x coordinates. Briefly explain the shape of the curve shown.

2.2 A hot water pipe with outside radius r_1 has a temperature T_1. A thick insulation applied to reduce the heat loss has an outer radius r_2 and temperature T_2. On T–r coordinates, sketch the temperature distribution in the insulation for one-dimensional, steady-state heat transfer with constant properties. Give a brief explanation, justifying the shape of the curve shown.

2.3 A spherical shell with inner radius r_1 and outer radius r_2 has surface temperatures T_1 and T_2, respectively, where $T_1 > T_2$. Sketch the temperature distribution on T–r coordinates assuming steady-state, one-dimensional conduction with constant properties. Briefly justify the shape of the curve shown.

2.4 Assume steady-state, one-dimensional heat conduction through the symmetric shape shown.

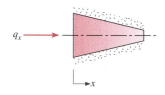

Assuming that there is no internal heat generation, derive an expression for the thermal conductivity $k(x)$ for these conditions: $A(x) = (1 - x)$, $T(x) = 300(1 - 2x - x^3)$, and $q = 6000$ W, where A is in square meters, T in kelvins, and x in meters.

2.5 A solid, truncated cone serves as a support for a system that maintains the top (truncated) face of the cone at a temperature T_1, while the base of the cone is at a temperature $T_2 < T_1$.

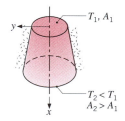

The thermal conductivity of the solid depends on temperature according to the relation $k = k_o - aT$, where a is a positive constant, and the sides of the cone are well insulated. Do the following quantities increase, decrease, or remain the same with increasing x: the heat transfer rate q_x, the heat flux q_x'', the thermal conductivity k, and the temperature gradient dT/dx?

2.6 To determine the effect of the temperature dependence of the thermal conductivity on the temperature distribution in a solid, consider a material for which this dependence may be represented as

$$k = k_o + aT$$

where k_o is a positive constant and a is a coefficient that may be positive or negative. Sketch the steady-state temperature distribution associated with heat transfer in a plane wall for three cases corresponding to $a > 0$, $a = 0$, and $a < 0$.

2.7 One-dimensional, steady-state conduction without heat generation occurs in the system shown. The thermal conductivity is 25 W/m · K and the thickness L is 0.5 m.

Determine the unknown quantities for each case in the accompanying table and sketch the temperature distribution, indicating the direction of the heat flux.

Case	T_1	T_2	dT/dx (K/m)	q_x'' (W/m²)
1	400 K	300 K		
2	100°C		−250	
3	80°C		+200	
4		−5°C		4000
5	30°C			−3000

2.8 Consider steady-state conditions for one-dimensional conduction in a plane wall having a thermal conductivity $k = 50$ W/m · K and a thickness $L = 0.25$ m, with no internal heat generation.

Determine the heat flux and the unknown quantity for each case and sketch the temperature distribution, indicating the direction of the heat flux.

Case	T_1(°C)	T_2(°C)	dT/dx (K/m)
1	50	−20	
2	−30	−10	
3	70		160
4		40	−80
5		30	200

2.9 Consider a plane wall 100 mm thick and of thermal conductivity 100 W/m · K. Steady-state conditions are known to exist with $T_1 = 400$ K and $T_2 = 600$ K. Determine the heat flux q_x'' and the temperature gradient dT/dx for the coordinate systems shown.

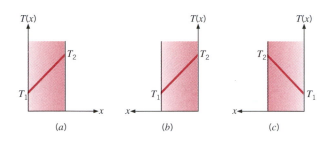

(a)   (b)   (c)

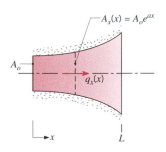

2.10 A cylinder of radius r_o, length L, and thermal conductivity k is immersed in a fluid of convection coefficient h and unknown temperature T_∞. At a certain instant the temperature distribution in the cylinder is $T(r) = a + br^2$, where a and b are constants. Obtain expressions for the heat transfer rate at r_o and the fluid temperature.

2.11 In the two-dimensional body illustrated, the gradient at surface A is found to be $\partial T/\partial y = 30$ K/m. What are $\partial T/\partial y$ and $\partial T/\partial x$ at surface B?

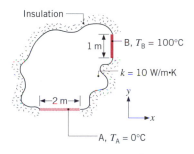

2.12 Sections of the Alaskan pipeline run above the ground and are supported by vertical steel shafts ($k = 25$ W/m · K) that are 1 m long and have a cross-sectional area of 0.005 m². Under normal operating conditions, the temperature variation along the length of a shaft is known to be governed by an expression of the form

$$T = 100 - 150x + 10x^2$$

where T and x have units of °C and meters, respectively. Temperature variations are negligible over the shaft cross section. Evaluate the temperature and conduction heat rate at the shaft–pipeline joint ($x = 0$) and at the shaft–ground interface ($x = 1$ m). Explain the difference in the heat rates.

2.13 Steady-state, one-dimensional conduction occurs in a rod of constant thermal conductivity k and variable cross-sectional area $A_x(x) = A_o e^{ax}$, where A_o and a are constants. The lateral surface of the rod is well insulated.

(a) Write an expression for the conduction heat rate, $q_x(x)$. Use this expression to determine the temperature distribution $T(x)$ and qualitatively sketch the distribution for $T(0) > T(L)$.

(b) Now consider conditions for which thermal energy is generated in the rod at a volumetric rate $\dot{q} = \dot{q}_o \exp(-ax)$, where $\dot{q}_o$ is a constant. Obtain an expression for $q_x(x)$ when the left face ($x = 0$) is well insulated.

Thermophysical Properties

2.14 A solid cylindrical rod of length 0.1 m and diameter 25 mm is well insulated on its side, while its end faces are maintained at temperatures of 100 and 0°C. What is the rate of heat transfer through the rod if it is constructed from (a) pure copper, (b) aluminum alloy 2024-T6, (c) AISI 302 stainless steel, (d) silicon nitride, (e) wood (oak), (f) magnesia, 85%, and (g) Pyrex?

2.15 A one-dimensional system without heat generation has a thickness of 20 mm with surfaces maintained at temperatures of 275 and 325 K. Determine the heat flux through the system if it is constructed from (a) pure aluminum, (b) plain carbon steel, (c) AISI 316 stainless steel, (d) pyroceram, (e) Teflon, and (f) concrete.

2.16 A TV advertisement by a well-known insulation manufacturer states: it isn't the thickness of the insulating material that counts, it's the R-value. The ad shows that to obtain an R-value of 19, you need 18 ft of rock, 15 in. of wood, or just 6 in. of the manufacturer's insulation. Is this advertisement technically reasonable? If you are like most TV viewers, you don't know the R-value is defined as L/k, where L (in.) is the thickness of the insulation and k (Btu · in./hr · ft² · °F) is the thermal conductivity of the material.

2.17 An apparatus for measuring thermal conductivity employs an electrical heater sandwiched between two identical samples of diameter 30 mm and length 60 mm, which are pressed between plates maintained at a uniform temperature $T_o = 77°C$ by a circulating fluid. A conducting grease is placed between all the surfaces to ensure good thermal contact. Differential thermocouples are imbedded in the samples with a spacing of 15 mm. The lateral sides of the samples are insulated to ensure one-dimensional heat transfer through the samples.

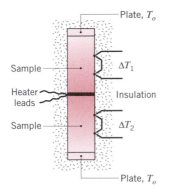

(a) With two samples of SS316 in the apparatus, the heater draws 0.353 A at 100 V and the differential thermocouples indicate $\Delta T_1 = \Delta T_2 = 25.0°C$. What is the thermal conductivity of the stainless steel sample material? What is the average temperature of the samples? Compare your result with the thermal conductivity value reported for this material in Table A.2.

(b) By mistake, an Armco iron sample is placed in the lower position of the apparatus with one of the SS316 samples from part (a) in the upper portion. For this situation, the heater draws 0.601 A at 100 V and the differential thermocouples indicate $\Delta T_1 = \Delta T_2 = 15.0°C$. What are the thermal conductivity and average temperature of the Armco iron sample?

(c) What is the advantage in constructing the apparatus with two identical samples sandwiching the heater rather than with a single heater–sample combination? When would heat leakage out of the lateral surfaces of the samples become significant? Under what conditions would you expect $\Delta T_1 \neq \Delta T_2$?

2.18 A common *comparative* method for measuring the thermal conductivity of metals is illustrated in the sketch. Cylindrical test samples (1 and 2) and a reference sample of equal diameter and length are stacked under pressure and well insulated (not shown on sketch) on their lateral surfaces. The thermal conductivity of the reference material, Armco iron in this case, is assumed known from Table A.2. For the source–sink condition of $T_h = 400$ K and

$T_c = 300$ K, the differential thermocouples imbedded in the samples with a spacing of 10 mm indicate $\Delta T_r = 2.49°C$ and $\Delta T_{t,1} = \Delta T_{t,2} = 3.32°C$ for the reference and test samples, respectively.

(a) What is the thermal conductivity of the test material? What temperature would you assign to this measured value?

(b) Under what conditions would you expect $\Delta T_{t,1}$ not to equal $\Delta T_{t,2}$?

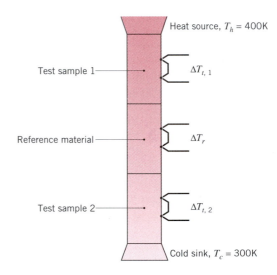

2.19 A method for determining the thermal conductivity k and the specific heat c_p of a material is illustrated in the sketch. Initially the two identical samples of diameter $D = 60$ mm and thickness $L = 10$ mm and the thin heater are at a uniform temperature of $T_i = 23.00°C$, while surrounded by an insulating powder. Suddenly the heater is energized to provide a uniform heat flux q_o'' on each of the sample interfaces, and the heat flux is maintained constant for a period of time, Δt_o. A short time after sudden heating is initiated, the temperature at this interface T_o is related to the heat flux as

$$T_o(t) - T_i = 2q_o'' \left(\frac{t}{\pi \rho\, c_p k} \right)^{1/2}$$

For a particular test run, the electrical heater dissipates 15.0 W for a period of $\Delta t_o = 120$ s and the temperature at the interface is $T_o(30\text{ s}) = 24.57°C$ after 30 s of heating. A long time after the heater is deenergized, $t \gg \Delta t_o$, the samples reach the uniform temperature of $T_o(\infty) = 33.50°C$. The density of the sample materials, determined by measurement of volume and mass, is $\rho = 3965$ kg/m^3.

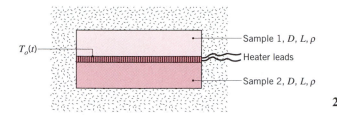

Determine the specific heat and thermal conductivity of the test material. By looking at values of the thermophysical properties in Table A.1 or A.2, identify the test sample material.

The Heat Equation

2.20 At a given instant of time the temperature distribution within an infinite homogeneous body is given by the function

$$T(x, y, z) = x^2 - 2y^2 + z^2 - xy + 2yz$$

Assuming constant properties and no internal heat generation, determine the regions where the temperature changes with time.

2.21 Uniform internal heat generation at $\dot{q}_1 = 5 \times 10^7$ W/m³ is occurring in a cylindrical nuclear reactor fuel rod of 50-mm diameter, and under steady-state conditions the temperature distribution is of the form $T(r) = a + br^2$, where T is in degrees Celsius and r is in meters, while $a = 800°C$ and $b = -4.167 \times 10^5 °C/m^2$. The fuel rod properties are $k = 30$ W/m · K, $\rho = 1100$ kg/m³, and $c_p = 800$ J/kg · K.

(a) What is the rate of heat transfer per unit length of the rod at $r = 0$ (the centerline) and at $r = 25$ mm (the surface)?

(b) If the reactor power level is suddenly increased to $\dot{q}_2 = 10^8$ W/m³, what is the initial time rate of temperature change at $r = 0$ and $r = 25$ mm?

2.22 The steady-state temperature distribution in a one-dimensional wall of thermal conductivity 50 W/m · K and thickness 50 mm is observed to be $T(°C) = a + bx^2$, where $a = 200°C$, $b = -2000°C/m^2$, and x is in meters.

(a) What is the heat generation rate $\dot{q}$ in the wall?

(b) Determine the heat fluxes at the two wall faces. In what manner are these heat fluxes related to the heat generation rate?

2.23 The temperature distribution across a wall 0.3 m thick at a certain instant of time is $T(x) = a + bx + cx^2$, where T is in degrees Celsius and x is in meters, $a = 200°C$, $b = -200°C/m$, and $c = 30°C/m^2$. The wall has a thermal conductivity of 1 W/m · K.

(a) On a unit surface area basis, determine the rate of heat transfer into and out of the wall and the rate of change of energy stored by the wall.

(b) If the cold surface is exposed to a fluid at 100°C, what is the convection coefficient?

2.24 A salt-gradient solar pond is a shallow body of water that consists of three distinct fluid layers and is used to collect solar energy. The upper- and lower-most layers are well mixed and serve to maintain the upper and lower surfaces of the central layer at uniform temperatures T_1 and T_2, where $T_2 > T_1$. Although there is bulk fluid motion in the mixed layers, there is no such motion in the central layer. Consider conditions for which solar radiation absorption in the central layer provides nonuniform heat generation of the form $\dot{q} = Ae^{-ax}$, and the temperature distribution in the central layer is

$$T(x) = -\frac{A}{ka^2} e^{-ax} + Bx + C$$

The quantities A (W/m³), a (1/m), B (K/m), and C (K) are known constants having the prescribed units, and k is the thermal conductivity, which is also constant.

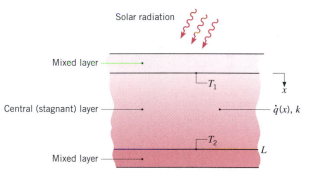

(a) Obtain expressions for the rate at which heat is transferred per unit area from the lower mixed layer to the central layer and from the central layer to the upper mixed layer.

(b) Determine whether conditions are steady or transient.

(c) Obtain an expression for the rate at which thermal energy is generated in the entire central layer, per unit surface area.

2.25 The steady-state temperature distribution in a semi-transparent material of thermal conductivity k and thickness L exposed to laser irradiation is of the form

$$T(x) = -\frac{A}{ka^2}e^{-ax} + Bx + C$$

where A, a, B, and C are known constants. For this situation, radiation absorption in the material is manifested by a distributed heat generation term, $\dot{q}(x)$.

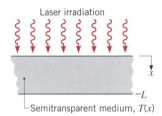

Laser irradiation

x

$-L$

Semitransparent medium, $T(x)$

(a) Obtain expressions for the conduction heat fluxes at the front and rear surfaces.

(b) Derive an expression for $\dot{q}(x)$.

(c) Derive an expression for the rate at which radiation is absorbed in the entire material, per unit surface area. Express your result in terms of the known constants for the temperature distribution, the thermal conductivity of the material, and its thickness.

2.26 The steady-state temperature distribution in a one-dimensional wall of thermal conductivity k and thickness L is of the form $T = ax^3 + bx^2 + cx + d$. Derive expressions for the heat generation rate per unit volume in the wall and the heat fluxes at the two wall faces ($x = 0, L$).

2.27 One-dimensional, steady-state conduction with no internal energy generation is occurring in a plane wall of constant thermal conductivity.

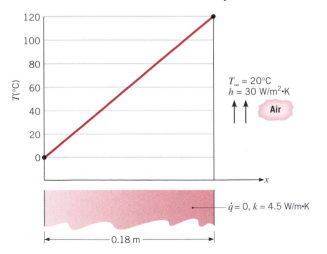

$T_\infty = 20°C$
$h = 30$ W/m²·K

Air

$\dot{q} = 0, k = 4.5$ W/m·K

0.18 m

Is the prescribed temperature distribution possible? Briefly explain your reasoning. With the temperature at $x = 0$ and the fluid temperature fixed at $T(0) = 0°C$ and $T_\infty = 20°C$, respectively, compute and plot the temperature at $x = L$, $T(L)$, as a function of h for $10 \leq h \leq 100$ W/m² · K. Briefly explain your results.

2.28 A plane layer of coal of thickness $L = 1$ m experiences uniform volumetric generation at a rate of $\dot{q} = 20$ W/m³ due to slow oxidation of the coal particles. Averaged over a daily period, the top surface of the layer transfers heat by convection to ambient air for which $h = 5$ W/m² · K and $T_\infty = 25°C$, while receiving solar irradiation in the amount $G_S = 400$ W/m². The solar absorptivity and emissivity of the surface are each $\alpha_S = \varepsilon = 0.95$.

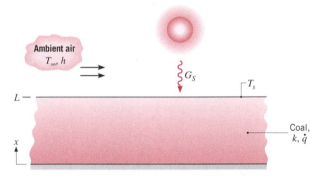

Ambient air
T_∞, h

G_S

T_s

L

x

Coal, $k, \dot{q}$

(a) Write the steady-state form of the heat diffusion equation for the layer of coal. Verify that this equation is satisfied by a temperature distribution of the form

$$T(x) = T_s + \frac{\dot{q}L^2}{2k}\left(1 - \frac{x^2}{L^2}\right)$$

From this distribution, what can you say about conditions at the bottom surface ($x = 0$)? Sketch the temperature distribution and label key features.

(b) Obtain an expression for the rate of heat transfer by conduction per unit area at $x = L$. Applying an energy balance to a control surface about the top surface of the layer, obtain an expression for T_s. Evaluate T_s and $T(0)$ for the prescribed conditions.

(c) Daily average values of G_S and h depend on a number of factors such as time of year, cloud cover, and wind conditions. For $h = 5$ W/m² · K, compute and plot T_s and $T(0)$ as a function of G_S for $50 \leq G_S \leq 500$ W/m². For $G_S = 400$

W/m², compute and plot T_s and $T(0)$ as a function of h for $5 \leq h \leq 50$ W/m² · K.

2.29 The cylindrical system illustrated has negligible variation of temperature in the r and z directions. Assume that $\Delta r = r_o - r_i$ is small compared to r_i and denote the length in the z direction, normal to the page, as L.

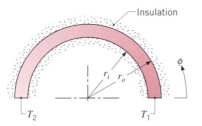

(a) Beginning with a properly defined control volume and considering energy generation and storage effects, derive the differential equation that prescribes the variation in temperature with the angular coordinate ϕ. Compare your result with Equation 2.20.

(b) For steady-state conditions with no internal heat generation and constant properties, determine the temperature distribution $T(\phi)$ in terms of the constants T_1, T_2, r_i, and r_o. Is this distribution linear in ϕ?

(c) For the conditions of part (b) write the expression for the heat rate q_ϕ.

2.30 Beginning with a differential control volume in the form of a cylindrical shell, derive the heat diffusion equation for a one-dimensional, cylindrical, radial coordinate system with internal heat generation. Compare your result with Equation 2.20.

2.31 Beginning with a differential control volume in the form of a spherical shell, derive the heat diffusion equation for a one-dimensional, spherical, radial coordinate system with internal heat generation. Compare your result with Equation 2.23.

2.32 Derive the heat diffusion equation, Equation 2.20, for cylindrical coordinates beginning with the differential control volume shown in Figure 2.9.

2.33 Derive the heat diffusion equation, Equation 2.23, for spherical coordinates beginning with the differential control volume shown in Figure 2.10.

2.34 A steam pipe is wrapped with insulation of inner and outer radii, r_i and r_o, respectively. At a particular instant the temperature distribution in the insulation is known to be of the form

$$T(r) = C_1 \ln\left(\frac{r}{r_o}\right) + C_2$$

Are conditions steady-state or transient? How do the heat flux and heat rate vary with radius?

2.35 For a long circular tube of inner and outer radii r_1 and r_2, respectively, uniform temperatures T_1 and T_2 are maintained at the inner and outer surfaces, while thermal energy generation is occurring within the tube wall ($r_1 < r < r_2$). Consider steady-state conditions for which $T_1 > T_2$. Is it possible to maintain a *linear* radial temperature distribution in the wall? If so, what special conditions must exist?

2.36 Passage of an electric current through a long conducting rod of radius r_i and thermal conductivity k_r results in uniform volumetric heating at a rate of $\dot{q}$. The conducting rod is wrapped in an electrically nonconducting cladding material of outer radius r_o and thermal conductivity k_c, and convection cooling is provided by an adjoining fluid.

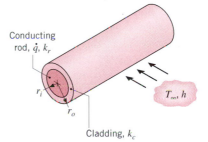

For steady-state conditions, write appropriate forms of the heat equations for the rod and cladding. Express appropriate boundary conditions for the solution of these equations.

2.37 An electric cable of radius r_1 and thermal conductivity k_c is enclosed by an insulating sleeve whose outer surface is of radius r_2 and experiences convection heat transfer and radiation exchange with the adjoining air and large surroundings, respectively. When electric current passes through the cable, thermal energy is generated within the cable at a volumetric rate $\dot{q}$.

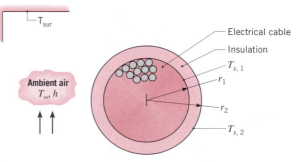

(a) Write the steady-state forms of the heat diffusion equation for the insulation and the cable. Verify that these equations are satisfied by the following temperature distributions:

Insulation: $T(r) = T_{s,2} + (T_{s,1} - T_{s,2}) \dfrac{\ln (r/r_2)}{\ln (r_1/r_2)}$

Cable: $T(r) = T_{s,1} + \dfrac{\dot{q} r^2{}_1}{4 k_c} \left(1 - \dfrac{r^2}{r_1^2} \right)$

Sketch the temperature distribution, $T(r)$, in the cable and the sleeve, labeling key features.

(b) Applying Fourier's law, show that the rate of conduction heat transfer per unit length through the sleeve may be expressed as

$$q'_r = \frac{2\pi k_s (T_{s,1} - T_{s,2})}{\ln (r_2/r_1)}$$

Applying an energy balance to a control surface placed around the cable, obtain an alternative expression for q'_r, expressing your result in terms of $\dot{q}$ and r_1.

(c) Applying an energy balance to a control surface placed around the outer surface of the sleeve, obtain an expression from which $T_{s,2}$ may be determined as a function of $\dot{q}$, r_1, h, T_∞, ε, and T_{sur}.

(d) Consider conditions for which 250 A are passing through a cable having an electric resistance per unit length of $R'_e = 0.005\ \Omega/\mathrm{m}$, a radius of $r_1 = 15$ mm, and a thermal conductivity of $k_c = 200\ \mathrm{W/m \cdot K}$. For $k_s = 0.15\ \mathrm{W/m \cdot K}$, $r_2 = 15.5$ mm, $h = 25\ \mathrm{W/m^2 \cdot K}$, $\varepsilon = 0.9$, $T_\infty = 25°\mathrm{C}$, and $T_{\mathrm{sur}} = 35°\mathrm{C}$, evaluate the surface temperatures, $T_{s,1}$ and $T_{s,2}$, as well as the temperature T_o at the centerline of the cable.

(e) With all other conditions remaining the same, compute and plot T_o, $T_{s,1}$, and $T_{s,2}$ as a function of r_2 for $15.5 \le r_2 \le 20$ mm.

2.38 A spherical shell of inner and outer radii r_i and r_o, respectively, contains heat-dissipating components, and at a particular instant the temperature distribution in the shell is known to be of the form

$$T(r) = \frac{C_1}{r} + C_2$$

Are conditions steady-state or transient? How do the heat flux and heat rate vary with radius?

2.39 A chemically reacting mixture is stored in a thin-walled spherical container of radius $r_1 = 200$ mm, and the exothermic reaction generates heat at a uni-

form, but temperature-dependent volumetric rate of $\dot{q} = \dot{q}_o \exp(-A/T_o)$, where $\dot{q}_o = 5000\ \mathrm{W/m^3}$, $A = 75$ K, and T_o is the mixture temperature in kelvins. The vessel is enclosed by an insulating material of outer radius r_2, thermal conductivity k, and emissivity ε. The outer surface of the insulation experiences convection heat transfer and net radiation exchange with the adjoining air and large surroundings, respectively.

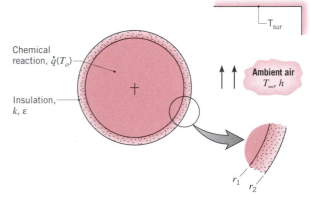

(a) Write the steady-state form of the heat diffusion equation for the insulation. Verify that this equation is satisfied by the temperature distribution

$$T(r) = T_{s,1} - (T_{s,1} - T_{s,2}) \left[\frac{1 - (r_1/r)}{1 - (r_1/r_2)} \right]$$

Sketch the temperature distribution, $T(r)$, labeling key features.

(b) Applying Fourier's law, show that the rate of heat transfer by conduction through the insulation may be expressed as

$$q_r = \frac{4\pi k (T_{s,1} - T_{s,2})}{(1/r_1) - (1/r_2)}$$

Applying an energy balance to a control surface about the container, obtain an alternative expression for q_r, expressing your result in terms of $\dot{q}$ and r_1.

(c) Applying an energy balance to a control surface placed around the outer surface of the insulation, obtain an expression from which $T_{s,2}$ may be determined as a function of $\dot{q}$, r_1, h, T_∞, ε, and T_{sur}.

(d) The process engineer wishes to maintain a reactor temperature of $T_o = T(r_1) = 95°\mathrm{C}$ under conditions for which $k = 0.05\ \mathrm{W/m \cdot K}$, $r_2 = 208$ mm, $h = 5\ \mathrm{W/m^2 \cdot K}$, $\varepsilon = 0.9$, $T_\infty = 25°\mathrm{C}$, and $T_{sur} = 35°C$. What is the outer surface temperature of the insulation, $T_{s,2}$?

(e) Compute and plot the variation of $T_{s,2}$ with r_2 for $201 \leq r_2 \leq 210$ mm. The engineer is concerned about potential burn injuries to personnel who may come into contact with the exposed surface of the insulation. Is increasing the insulation thickness a practical solution to maintaining $T_{s,2} \leq 45°C$? What other parameter could be varied to reduce $T_{s,2}$?

Graphical Representations

2.40 In Example 2.3, we considered a copper bar that was initially at a uniform temperature and suddenly was heated by the passage of an electric current. Assume $T_\infty > T_o$.

(a) On T–x coordinates, sketch the temperature distributions for the following conditions: initial condition ($t \leq 0$), steady-state condition ($t \to \infty$), and for two intermediate times. Assume that the electric current is sufficiently large that the bar outer surface ($x = L$) will eventually be hotter than the air.

(b) On q''_x–t coordinates, sketch the heat flux at the bar faces. That is, show qualitatively how $q''_x(0, t)$ and $q''_x(L, t)$ vary with time.

2.41 The one-dimensional system of mass M with constant properties and no internal heat generation shown in the figure is initially at a uniform temperature T_i. The electrical heater is suddenly energized providing a uniform heat flux q''_o at the surface $x = 0$. The boundaries at $x = L$ and elsewhere are perfectly insulated.

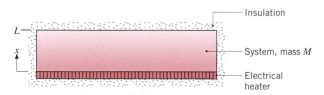

(a) Write the differential equation and identify the boundary and initial conditions that could be used to determine the temperature as a function of position and time in the system.

(b) On T–x coordinates, sketch the temperature distributions for the initial condition ($t \leq 0$) and for several times after the heater is energized. Will a steady-state temperature distribution ever be reached?

(c) On q''_x–t coordinates, sketch the heat flux $q''_x(x, t)$ at the planes $x = 0$, $x = L/2$, and $x = L$ as a function of time.

(d) After a period of time t_e has elapsed, the heater power is switched off. Assuming that the insulation is perfect, the system will eventually reach a final uniform temperature T_f. Derive an expression that can be used to determine T_f as a function of the parameters q''_o, t_e, T_i, and the system characteristics M, c_p, and A_s (the heater surface area).

2.42 The plane wall with constant properties and no internal heat generation shown in the figure is initially at a uniform temperature T_i. Suddenly the surface at $x = L$ is heated by a fluid at T_∞ having a convection heat transfer coefficient h. The boundary at $x = 0$ is perfectly insulated.

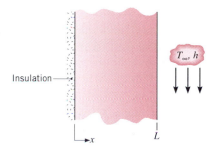

(a) Write the differential equation and identify the boundary and initial conditions that could be used to determine the temperature as a function of position and time in the wall.

(b) On T–x coordinates, sketch the temperature distributions for the following conditions: initial condition ($t \leq 0$), steady-state condition ($t \to \infty$), and two intermediate times.

(c) On q''_x–t coordinates, sketch the heat flux at the locations $x = 0$ and $x = L$. That is, show qualitatively how $q''_x(0, t)$ and $q''_x(L, t)$ vary with time.

(d) Write an expression for the total energy transferred to the wall per unit volume of the wall (J/m³).

2.43 A plane wall has constant properties, no internal heat generation, and is initially at a uniform temperature T_i. Suddenly, the surface at $x = L$ is heated by a fluid at T_∞ having a convection coefficient h. At the same instant, the electrical heater is energized providing a constant heat flux q''_o at $x = 0$.

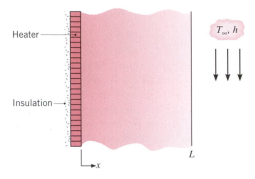

(a) On T–x coordinates, sketch the temperature distributions for the following conditions: initial condition ($t \leq 0$), steady-state condition ($t \rightarrow \infty$), and for two intermediate times.

(b) On q_x''–x coordinates, sketch the heat flux corresponding to the four temperature distributions of part (a).

(c) On q_x''–t coordinates, sketch the heat flux at the locations $x = 0$ and $x = L$. That is, show qualitatively how $q_x''(0, t)$ and $q_x''(L, t)$ vary with time.

(d) Derive an expression for the steady-state temperature at the heater surface, $T(0, \infty)$, in terms of q_o'', T_∞, k, h, and L.

2.44 A plane wall with constant properties is initially at a uniform temperature T_o. Suddenly, the surface at $x = L$ is exposed to a convection process with a fluid at T_∞ ($> T_o$) having a convection coefficient h. Also, suddenly the wall experiences a uniform internal volumetric heating $\dot{q}$ that is sufficiently large to induce a maximum steady-state temperature within the wall, which exceeds that of the fluid. The boundary at $x = 0$ remains at T_o.

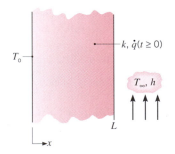

(a) On T–x coordinates, sketch the temperature distributions for the following conditions: initial condition ($t \leq 0$), steady-state condition ($t \rightarrow \infty$), and for two intermediate times. Show also the distribution for the special condition when there is no heat flow at the $x = L$ boundary.

(b) On q_x''–t coordinates, sketch the heat flux for the locations $x = 0$ and $x = L$, that is, $q_x''(0, t)$ and $q_x''(L, t)$, respectively.

2.45 A very thin electrically conducting foil is sandwiched between two electrically nonconducting plane walls of equivalent thickness L and thermal conductivity k. If an electric current is passed through the foil, heat is generated within the foil, creating a uniform heat flux at the interface between the walls. Consider conditions for which the walls are initially at a uniform temperature T_i and Ohmic heating maintains a uniform heat flux q_o'' at the interface for $t \geq 0$. Concurrently, the exposed surfaces are maintained at a fixed temperature T_o that exceeds T_i.

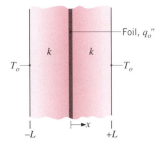

(a) On T–x coordinates, sketch the temperature distribution $T(x)$ in the walls ($-L \leq x \leq +L$) for the initial condition ($t = 0$), for the final steady-state condition ($t \rightarrow \infty$), and for two intermediate times.

(b) On q_x''–t coordinates, sketch the variation of the local heat flux for the locations $x = 0$ and $x = L$, that is, $q_x''(0, t)$ and $q_x''(L, t)$, respectively.

2.46 A plane wall that is insulated on one side ($x = 0$) is initially at a uniform temperature T_i, when its exposed surface at $x = L$ is suddenly raised to a temperature T_s.

(a) Verify that the following equation correctly characterizes the subsequent variation of the wall temperature, $T(x, t)$, with position and time:

$$\frac{T(x, t) - T_s}{T_i - T_s} = C_1 \exp\left(-\frac{\pi^2}{4} \frac{\alpha t}{L^2}\right) \cos\left(\frac{\pi}{2} \frac{x}{L}\right)$$

where C_1 is a constant and α is the thermal diffusivity.

(b) Obtain expressions for the heat flux at $x = 0$ and $x = L$.

(c) Sketch the temperature distribution $T(x)$ at $t = 0$, at $t \rightarrow \infty$, and at an intermediate time. Sketch the variation with time of the heat flux at $x = L$, $q_L''(t)$.

(d) What effect does α have on the thermal response of the material to a change in surface temperature?

One-Dimensional, Steady-State Conduction

*I*n this chapter we treat situations for which heat is transferred by diffusion under *one-dimensional, steady-state* conditions. The term "one-dimensional" refers to the fact that only one coordinate is needed to describe the spatial variation of the dependent variables. Hence, in a *one-dimensional* system, temperature gradients exist along only a single coordinate direction, and heat transfer occurs exclusively in that direction. The system is characterized by *steady-state* conditions if the temperature at each point is independent of time. Despite their inherent simplicity, one-dimensional, steady-state models may be used to accurately represent numerous engineering systems.

We begin our consideration of one-dimensional, steady-state conduction by discussing heat transfer with no internal generation (Sections 3.1 to 3.3). The objective is to determine expressions for the temperature distribution and heat transfer rate in common geometries. The concept of thermal resistance (analogous to electrical resistance) is introduced as an aid to solving conduction heat transfer problems. The effect of internal heat generation on the temperature distribution and heat rate is then treated (Section 3.4). Finally, conduction analysis is used to describe the performance of extended surfaces or fins, wherein the role of convection at the external boundary must be considered (Section 3.5).

3.1
The Plane Wall

For one-dimensional conduction in a plane wall, temperature is a function of the x coordinate only and heat is transferred exclusively in this direction. In Figure 3.1*a*, a plane wall separates two fluids of different temperatures. Heat transfer occurs by convection from the hot fluid at $T_{\infty,1}$ to one surface of the wall at $T_{s,1}$, by conduction through the wall, and by convection from the other surface of the wall at $T_{s,2}$ to the cold fluid at $T_{\infty,2}$.

We begin by considering conditions *within* the wall. We first determine the temperature distribution, from which we can then obtain the conduction heat transfer rate.

3.1.1   Temperature Distribution

The temperature distribution in the wall can be determined by solving the heat equation with the proper boundary conditions. For steady-state conditions with no distributed source or sink of energy within the wall, the appropriate form of the heat equation, Equation 2.17, is

$$\frac{d}{dx}\left(k\,\frac{dT}{dx} \right) = 0 \tag{3.1}$$

Hence, from Equation 2.2, it follows that, for *one-dimensional, steady-state conduction in a plane wall with no heat generation, the heat flux is a constant,*

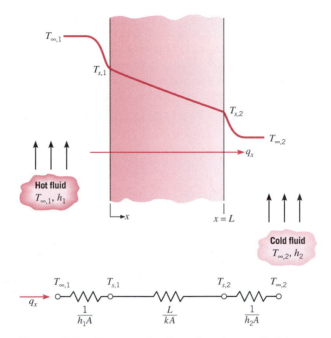

FIGURE 3.1 Heat transfer through a plane wall. (*a*) Temperature distribution. (*b*) Equivalent thermal circuit.

independent of x. If the thermal conductivity of the wall material is assumed to be constant, the equation may be integrated twice to obtain the *general solution*

$$T(x) = C_1 x + C_2 \tag{3.2}$$

To obtain the constants of integration, C_1 and C_2, boundary conditions must be introduced. We choose to apply conditions of the first kind at $x = 0$ and $x = L$, in which case

$$T(0) = T_{s,1} \qquad \text{and} \qquad T(L) = T_{s,2}$$

Applying the condition at $x = 0$ to the general solution, it follows that

$$T_{s,1} = C_2$$

Similarly, at $x = L$,

$$T_{s,2} = C_1 L + C_2 = C_1 L + T_{s,1}$$

in which case

$$\frac{T_{s,2} - T_{s,1}}{L} = C_1$$

Substituting into the general solution, the temperature distribution is then

$$T(x) = (T_{s,2} - T_{s,1}) \frac{x}{L} + T_{s,1} \tag{3.3}$$

From this result it is evident that, *for one-dimensional, steady-state conduction in a plane wall with no heat generation and constant thermal conductivity, the temperature varies linearly with x.*

Now that we have the temperature distribution, we may use Fourier's law, Equation 2.1, to determine the conduction heat transfer rate. That is,

$$q_x = -kA \frac{dT}{dx} = \frac{kA}{L}(T_{s,1} - T_{s,2}) \tag{3.4}$$

Note that A is the area of the wall *normal* to the direction of heat transfer and, for the plane wall, it is a constant independent of x. The heat flux is then

$$q_x'' = \frac{q_x}{A} = \frac{k}{L}(T_{s,1} - T_{s,2}) \tag{3.5}$$

Equations 3.4 and 3.5 indicate that both the heat rate q_x and heat flux q_x'' are constants, independent of x.

In the foregoing paragraphs we have used the *standard approach* to solving conduction problems. That is, the general solution for the temperature distribution is first obtained by solving the appropriate form of the heat equation. The boundary conditions are then applied to obtain the particular solution, which is used with Fourier's law to determine the heat transfer rate. Note that we have opted to prescribe surface temperatures at $x = 0$ and $x = L$ as boundary conditions, even though it is the fluid temperatures, and not the surface temperatures, that are typically known. However, since adjoining fluid and surface temperatures are easily related through a surface energy balance (see Section 1.3.2), it is a simple matter to express Equations 3.3 to 3.5 in terms of fluid, rather than surface, temperatures. Alternatively, equivalent results could be obtained directly by using the surface energy balances as boundary conditions of the third kind in evaluating the constants of Equation 3.2 (see Problem 3.1).

3.1.2 Thermal Resistance

At this point we note that a very important concept is suggested by Equation 3.4. In particular, there exists an analogy between the diffusion of heat and electrical charge. Just as an electrical resistance is associated with the conduction of electricity, a thermal resistance may be associated with the conduction of heat. Defining resistance as the ratio of a driving potential to the corresponding transfer rate, it follows from Equation 3.4 that the *thermal resistance for conduction* is

$$R_{t,\,cond} \equiv \frac{T_{s,1} - T_{s,2}}{q_x} = \frac{L}{kA} \tag{3.6}$$

Similarly, for electrical conduction in the same system, Ohm's law provides an electrical resistance of the form

$$R_e = \frac{E_{s,1} - E_{s,2}}{I} = \frac{L}{\sigma A} \tag{3.7}$$

The analogy between Equations 3.6 and 3.7 is obvious. A thermal resistance may also be associated with heat transfer by convection at a surface. From Newton's law of cooling,

$$q = hA(T_s - T_\infty) \tag{3.8}$$

the *thermal resistance for convection* is then

$$R_{t,\,conv} \equiv \frac{T_s - T_\infty}{q} = \frac{1}{hA} \tag{3.9}$$

Circuit representations provide a useful tool for both conceptualizing and quantifying heat transfer problems. The *equivalent thermal circuit* for the plane wall with convection surface conditions is shown in Figure 3.1b. The heat transfer rate may be determined from separate consideration of each element in the network. Since q_x is constant throughout the network, it follows that

$$q_x = \frac{T_{\infty,\,1} - T_{s,\,1}}{1/h_1 A} = \frac{T_{s,\,1} - T_{s,\,2}}{L/kA} = \frac{T_{s,\,2} - T_{\infty,\,2}}{1/h_2 A} \tag{3.10}$$

In terms of the *overall temperature difference*, $T_{\infty,\,1} - T_{\infty,\,2}$, and the *total thermal resistance*, R_{tot}, the heat transfer rate may also be expressed as

$$q_x = \frac{T_{\infty,\,1} - T_{\infty,\,2}}{R_{tot}} \tag{3.11}$$

Because the conduction and convection resistances are in series and may be summed, it follows that

$$R_{tot} = \frac{1}{h_1 A} + \frac{L}{kA} + \frac{1}{h_2 A} \tag{3.12}$$

Yet another resistance may be pertinent if a surface is separated from *large surroundings* by a gas (Section 1.2.3). In particular, radiation exchange between the surface and its surroundings may be important, and the rate may be determined from Equation 1.8. It follows that a *thermal resistance for radiation* may be defined as

$$R_{t,\,rad} = \frac{T_s - T_{sur}}{q_{rad}} = \frac{1}{h_r A} \tag{3.13}$$

where h_r is determined from Equation 1.9. Surface radiation and convection resistances act in parallel, and if $T_\infty = T_{sur}$, they may be combined to obtain a single, effective surface resistance.

3.1.3 The Composite Wall

Equivalent thermal circuits may also be used for more complex systems, such as *composite walls*. Such walls may involve any number of series and parallel thermal resistances due to layers of different materials. Consider the series composite wall of Figure 3.2. The one-dimensional heat transfer rate for this system may be expressed as

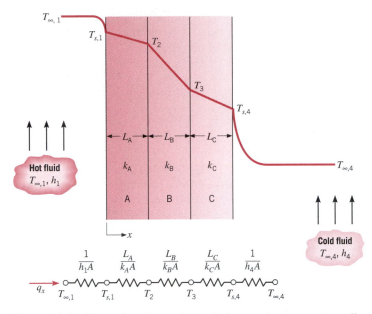

FIGURE 3.2   Equivalent thermal circuit for a series composite wall.

$$q_x = \frac{T_{\infty,1} - T_{\infty,4}}{\sum R_t} \tag{3.14}$$

where $T_{\infty,1} - T_{\infty,4}$ is the *overall* temperature difference and the summation includes all thermal resistances. Hence

$$q_x = \frac{T_{\infty,1} - T_{\infty,4}}{[(1/h_1 A) + (L_A/k_A A) + (L_B/k_B A) + (L_C/k_C A) + (1/h_4 A)]} \tag{3.15}$$

Alternatively, the heat transfer rate can be related to the temperature difference and resistance associated with each element. For example,

$$q_x = \frac{T_{\infty,1} - T_{s,1}}{(1/h_1 A)} = \frac{T_{s,1} - T_2}{(L_A/k_A A)} = \frac{T_2 - T_3}{(L_B/k_B A)} = \cdots \tag{3.16}$$

With composite systems it is often convenient to work with an *overall heat transfer coefficient, U,* which is defined by an expression analogous to Newton's law of cooling. Accordingly,

$$q_x \equiv UA\,\Delta T \tag{3.17}$$

where ΔT is the overall temperature difference. The overall heat transfer coefficient is related to the total thermal resistance, and from Equations 3.14 and 3.17 we see that $UA = 1/R_{tot}$. Hence, for the composite wall of Figure 3.2,

$$U = \frac{1}{R_{tot} A} = \frac{1}{[(1/h_1) + (L_A/k_A) + (L_B/k_B) + (L_C/k_C) + (1/h_4)]} \tag{3.18}$$

In general, we may write

$$R_{tot} = \sum R_t = \frac{\Delta T}{q} = \frac{1}{UA} \tag{3.19}$$

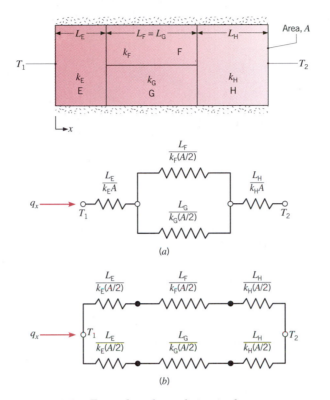

FIGURE 3.3 Equivalent thermal circuits for a series–parallel composite wall.

Composite walls may also be characterized by series–parallel configurations, such as that shown in Figure 3.3. Although the heat flow is now two-dimensional, it is often reasonable to assume one-dimensional conditions. Subject to this assumption, two different thermal circuits may be used. For case (*a*) it is presumed that surfaces normal to the *x* direction are isothermal, while for case (*b*) it is assumed that surfaces parallel to the *x* direction are adiabatic. Different results are obtained for R_{tot}, and the corresponding values of *q* bracket the actual heat transfer rate. These differences increase with increasing $|k_F - k_G|$, as two-dimensional effects become more significant.

3.1.4 Contact Resistance

Although neglected until now, it is important to recognize that, in composite systems, the temperature drop across the interface between materials may be appreciable. This temperature change is attributed to what is known as the *thermal contact resistance*, $R_{t, c}$. The effect is shown in Figure 3.4, and for a unit area of the interface, the resistance is defined as

$$R''_{t, c} = \frac{T_A - T_B}{q''_x} \tag{3.20}$$

The existence of a finite contact resistance is due principally to surface roughness effects. Contact spots are interspersed with gaps that are, in most in-

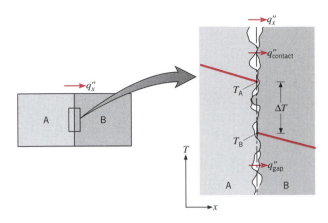

FIGURE 3.4   Temperature drop due to thermal contact resistance.

stances, air filled. Heat transfer is therefore due to conduction across the actual contact area and to conduction and/or radiation across the gaps. The contact resistance may be viewed as two parallel resistances: that due to the contact spots and that due to the gaps. The contact area is typically small, and especially for rough surfaces, the major contribution to the resistance is made by the gaps.

For solids whose thermal conductivities exceed that of the interfacial fluid, the contact resistance may be reduced by increasing the area of the contact spots. Such an increase may be effected by increasing the joint pressure and/or by reducing the roughness of the mating surfaces. The contact resistance may also be reduced by selecting an interfacial fluid of large thermal conductivity. In this respect, no fluid (an evacuated interface) eliminates conduction across the gap, thereby increasing the contact resistance.

Although theories have been developed for predicting $R''_{t,c}$, the most reliable results are those that have been obtained experimentally. The effect of loading on metallic interfaces can be seen in Table 3.1*a*, which presents an approximate range of thermal resistances under vacuum conditions. The effect of interfacial fluid on the thermal resistance of an aluminum interface is shown in Table 3.1*b*.

TABLE 3.1   Thermal contact resistance for (*a*) metallic interfaces under vacuum conditions and (*b*) aluminum interface (10-μm surface roughness, 10^5 N/m^2) with different interfacial fluids [1]

Thermal Resistance, $R''_{t,c} \times 10^4$ (m$^2 \cdot$ K/W)

(a) Vacuum Interface			(b) Interfacial Fluid	
Contact pressure	100 kN/m^2	10,000 kN/m^2	Air	2.75
Stainless steel	6–25	0.7–4.0	Helium	1.05
Copper	1–10	0.1–0.5	Hydrogen	0.720
Magnesium	1.5–3.5	0.2–0.4	Silicone oil	0.525
Aluminum	1.5–5.0	0.2–0.4	Glycerine	0.265

TABLE 3.2   Thermal resistance of representative solid/solid interfaces

Interface	$R''_{t,c} \times 10^4$ (m$^2 \cdot$ K/W)	Source
Silicon chip/lapped aluminum in air (27–500 kN/m^2)	0.3–0.6	[2]
Aluminum/aluminum with indium foil filler ($\sim$100 kN/m^2)	$\sim$0.07	[1, 3]
Stainless/stainless with indium foil filler ($\sim$3500 kN/m^2)	$\sim$0.04	[1, 3]
Aluminum/aluminum with metallic (Pb) coating	0.01–0.1	[4]
Aluminum/aluminum with Dow Corning 340 grease ($\sim$100 kN/m^2)	$\sim$0.07	[1, 3]
Stainless/stainless with Dow Corning 340 grease ($\sim$3500 kN/m^2)	$\sim$0.04	[1, 3]
Silicon chip/aluminum with 0.02-mm epoxy	0.2–0.9	[5]
Brass/brass with 15-μm tin solder	0.025–0.14	[6]

Contrary to the results of Table 3.1, many applications involve contact between dissimilar solids and/or a wide range of possible interstitial (filler) materials (Table 3.2). Any interstitial substance that fills the gap between contacting surfaces and whose thermal conductivity exceeds that of air will decrease the contact resistance. Two classes of materials that are well suited for this purpose are soft metals and thermal greases. The metals, which include indium, lead, tin, and silver, may be inserted as a thin foil or applied as a thin coating to one of the parent materials. Silicon-based thermal greases are attractive on the basis of their ability to completely fill the interstices with a material whose thermal conductivity is as much as 50 times that of air.

Unlike the foregoing interfaces, which are not permanent, many interfaces involve permanently bonded joints. The joint could be formed from an epoxy, a soft solder rich in lead, or a hard solder such as a gold/tin alloy. Due to interface resistances between the parent and bonding materials, the actual thermal resistance of the joint exceeds the theoretical value (L/k) computed from the thickness L and thermal conductivity k of the joint material. The thermal resistance of epoxied and soldered joints is also adversely affected by voids and cracks, which may form during manufacture or as a result of thermal cycling during normal operation.

Comprehensive reviews of thermal contact resistance results and models are provided by Snaith et al. [3], Madhusudana and Fletcher [7], and Yovanovich [8].

EXAMPLE 3.1

A leading manufacturer of household appliances is proposing a self-cleaning oven design that involves use of a composite window separating the oven cavity from the room air. The composite is to consist of two high-temperature plastics (A and B) of thicknesses $L_A = 2L_B$ and thermal conductivities $k_A = 0.15$ W/m $\cdot$ K and $k_B = 0.08$ W/m $\cdot$ K. During the self-cleaning process, the oven wall and air temperatures, T_w and T_a, are 400°C, while the room air temperature

T_∞ is 25°C. The inside convection and radiation heat transfer coefficients h_i and h_r, as well as the outside convection coefficient h_o, are each approximately 25 W/m² · K. What is the minimum window thickness, $L = L_A + L_B$, needed to ensure a temperature that is 50°C or less at the outer surface of the window? This temperature must not be exceeded for safety reasons.

SOLUTION

Known: The properties and relative dimensions of plastic materials used for a composite oven window, and conditions associated with self-cleaning operation.

Find: Composite thickness $L_A + L_B$ needed to ensure safe operation.

Schematic:

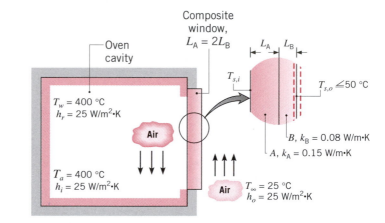

Assumptions:

1. Steady-state conditions exist.

2. Conduction through the window is one-dimensional.

3. Contact resistance is negligible.

4. Radiation absorption *within* the window is negligible; hence no internal heat generation (radiation exchange between window and oven walls occurs at the window inner surface).

5. Radiation exchange between window outer surface and surroundings is negligible.

6. Each plastic is homogeneous with constant properties.

Analysis: The thermal circuit can be constructed by recognizing that resistance to heat flow is associated with convection at the outer surface, conduction in the plastics, and convection and radiation at the inner surface. Accordingly, the circuit and the resistances are of the following form:

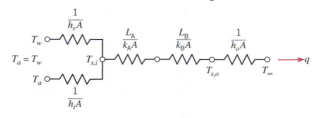

Since the outer surface temperature of the window, $T_{s,o}$, is prescribed, the required window thickness may be obtained by applying an energy balance at this surface. That is, from Equation 1.12

$$\dot{E}_{in} = \dot{E}_{out}$$

where, from Equation 3.19, with $T_w = T_a$,

$$\dot{E}_{in} = q = \frac{T_a - T_{s,o}}{\Sigma R_t}$$

and from Equation 3.8

$$\dot{E}_{out} = q = h_o A(T_{s,o} - T_\infty)$$

The total thermal resistance between the oven cavity and the outer surface of the window includes an effective resistance associated with convection and radiation, which act in parallel at the inner surface of the window, and the conduction resistances of the window materials. Hence

$$\Sigma R_t = \left(\frac{1}{1/h_i A} + \frac{1}{1/h_r A} \right)^{-1} + \frac{L_A}{k_A A} + \frac{L_B}{k_B A}$$

or

$$\Sigma R_t = \frac{1}{A} \left(\frac{1}{h_i + h_r} + \frac{L_A}{k_A} + \frac{L_A}{2k_B} \right)$$

Substituting into the energy balance, it follows that

$$\frac{T_a - T_{s,o}}{(h_i + h_r)^{-1} + (L_A/k_A) + (L_A/2k_B)} = h_o(T_{s,o} - T_\infty)$$

Hence, solving for L_A,

$$L_A = \frac{(1/h_o)(T_a - T_{s,o})/(T_{s,o} - T_\infty) - (h_i + h_r)^{-1}}{(1/k_A + 1/2k_B)}$$

$$L_A = \frac{0.04 \text{ m}^2 \cdot \text{K/W} \left(\dfrac{400 - 50}{50 - 25} \right) - 0.02 \text{ m}^2 \cdot \text{K/W}}{(1/0.15 + 1/0.16) \text{ m} \cdot \text{K/W}} = 0.0418 \text{ m}$$

Since $L_B = L_A/2 = 0.0209$ m,

$$L = L_A + L_B = 0.0627 \text{ m} = 62.7 \text{ mm} \qquad \triangleleft$$

Comments:

1. The self-cleaning operation is a transient process, as far as the thermal response of the window is concerned, and steady-state conditions may not be reached in the time required for cleaning. However, the steady-state condition provides the maximum possible value of $T_{s,o}$ and hence is well suited for the design calculation.

2. Radiation exchange between the oven walls and the composite window actually depends on the inner surface temperature $T_{s,i}$, and although it has been neglected, there is radiation exchange between the window and the surroundings, which depends on $T_{s,o}$. A more complete analysis may be

made to concurrently determine $T_{s,i}$ and $T_{s,o}$. Approximating the oven cavity as a large enclosure relative to the window and applying an energy balance, Equation 1.12, at the inner surface, it follows that

$$q''_{rad,i} + q''_{conv,i} = q''_{cond}$$

or

$$\varepsilon\sigma(T^4_{w,i} - T^4_{s,i}) + h_i(T_a - T_{s,i}) = \frac{T_{s,i} - T_{s,o}}{(L_A/k_A) + (L_B/k_B)} \tag{1}$$

Approximating the kitchen walls as a large isothermal enclosure relative to the window, with $T_{w,o} = T_\infty$, and this time applying an energy balance at the outer surface, it follows that

$$q''_{cond} = q''_{rad,o} + q''_{conv,o}$$

or

$$\frac{T_{s,i} - T_{s,o}}{(L_A/k_A) + (L_B/k_B)} = \varepsilon\sigma(T^4_{s,o} - T^4_{w,o}) + h_o(T_{s,o} - T_\infty) \tag{2}$$

If all other quantities are known, Equations 1 and 2 may be solved for $T_{s,i}$ and $T_{s,o}$.

We wish to explore the effect on $T_{s,o}$ of varying the velocity, and hence the convection coefficient, associated with airflow over the outer surface. With $\varepsilon = 0.9$ and all other conditions remaining the same, Equations 1 and 2 have been solved for values of h_o in the range $0 \leq h_o \leq 100$ W/m² · K, and the results are represented graphically.

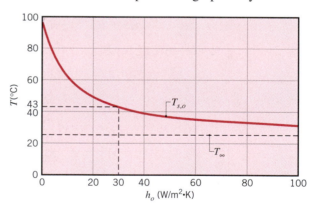

Increasing h_o reduces the corresponding convection resistance, and a value of $h_o = 30$ W/m² · K would yield a safe-to-touch temperature of $T_{s,o} = 43°C$. Because the conduction resistance is so large, the change in h_o has a negligible effect on $T_{s,i}$. However, it does influence the outer surface temperature, and as $h_o \to \infty$, $T_{s,o} \to T_\infty$.

EXAMPLE 3.2

A thin silicon chip and an 8-mm-thick aluminum substrate are separated by a 0.02-mm-thick epoxy joint. The chip and substrate are each 10 mm on a side, and their exposed surfaces are cooled by air, which is at a temperature of 25°C and provides a convection coefficient of 100 W/m² · K. If the chip dissipates 10^4 W/m² under normal conditions, will it operate below a maximum allowable temperature of 85°C?

SOLUTION

Known: Dimensions, heat dissipation, and maximum allowable temperature of a silicon chip. Thickness of aluminum substrate and epoxy joint. Convection conditions at exposed chip and substrate surfaces.

Find: Whether maximum allowable temperature is exceeded.

Schematic:

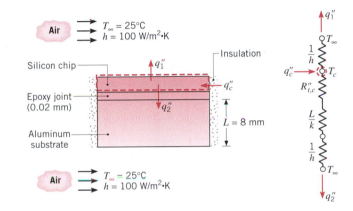

Assumptions:

1. Steady-state conditions.
2. One-dimensional conduction (negligible heat transfer from sides of composite).
3. Negligible chip thermal resistance (an isothermal chip).
4. Constant properties.
5. Negligible radiation exchange with surroundings.

Properties: Table A.1, pure aluminum ($T \sim 350$ K): $k = 238$ W/m · K.

Analysis: Heat dissipated in the chip is transferred to the air directly from the exposed surface and indirectly through the joint and substrate. Performing an energy balance on a control surface about the chip, it follows that, on the basis of a unit surface area,

$$q''_c = q''_1 + q''_2$$

or

$$q''_c = \frac{T_c - T_\infty}{(1/h)} + \frac{T_c - T_\infty}{R''_{t,c} + (L/k) + (1/h)}$$

To conservatively estimate T_c, the maximum possible value of $R''_{t,c} = 0.9 \times 10^{-4}$ m² · K/W is obtained from Table 3.2. Hence

$$T_c = T_\infty + q_c'' \left[h + \frac{1}{R_{t,c}'' + (L/k) + (1/h)} \right]^{-1}$$

or

$$T_c = 25°C + 10^4 \text{ W/m}^2$$

$$\times \left[100 + \frac{1}{(0.9 + 0.34 + 100) \times 10^{-4}} \right]^{-1} \text{ m}^2 \cdot \text{K/W}$$

$$T_c = 25°C + 50.3°C = 75.3°C \qquad \triangleleft$$

Hence the chip will operate below its maximum allowable temperature.

Comments:

1. The joint and substrate thermal resistances are much less than the convection resistance. The joint resistance would have to increase to the unrealistically large value of 50×10^{-4} m² · K/W, before the maximum allowable chip temperature would be exceeded.

2. The allowable power dissipation may be increased by increasing the convection coefficients, either by increasing the air velocity and/or by replacing the air with a more effective heat transfer fluid. Exploring this option for $100 \le h \le 2000$ W/m² · K, the following results are obtained.

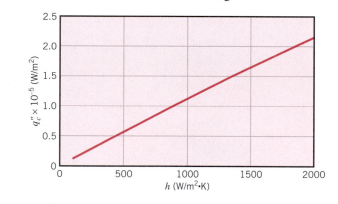

As $h \to \infty$, $q_2'' \to 0$ and virtually all of the chip power is transferred directly to the fluid stream.

3.2
An Alternative Conduction Analysis

The conduction analysis of Section 3.1 was performed using the *standard approach*. That is, the heat equation was solved to obtain the temperature distribution, Equation 3.3, and Fourier's law was then applied to obtain the heat transfer rate, Equation 3.4. However, an alternative approach may be used for the

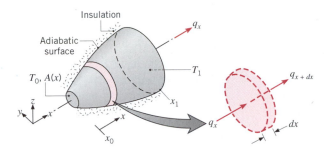

FIGURE 3.5   System with a constant conduction heat transfer rate.

conditions presently of interest. Considering conduction in the system of Figure 3.5, we recognize that, for *steady-state conditions* with *no heat generation* and *no heat loss from the sides,* the heat transfer rate q_x must be a constant independent of x. That is, for any differential element dx, $q_x = q_{x+dx}$. This condition is, of course, a consequence of the energy conservation requirement, and it must apply even if the area varies with position $A(x)$ and the thermal conductivity varies with temperature $k(T)$. Moreover, even though the temperature distribution may be two dimensional, varying with x and y, it is often reasonable to neglect the y variation and to assume a *one-dimensional* distribution in x.

For the above conditions it is possible to work exclusively with Fourier's law when performing a conduction analysis. In particular, since the conduction rate is a *constant,* the rate equation may be *integrated,* even though neither the rate nor the temperature distribution is known. Consider Fourier's law, Equation 2.1, which may be applied to the system of Figure 3.5. Although we may have no knowledge of the value of q_x or the form of $T(x)$, we do know that q_x is a constant. Hence we may express Fourier's law in the integral form

$$q_x \int_{x_0}^{x} \frac{dx}{A(x)} = -\int_{T_0}^{T} k(T)\, dT \tag{3.21}$$

The cross-sectional area may be a known function of x, and the material thermal conductivity may vary with temperature in a known manner. If the integration is performed from a point x_0 at which the temperature T_0 is known, the resulting equation provides the functional form of $T(x)$. Moreover, if the temperature $T = T_1$ at some $x = x_1$ is also known, integration between x_0 and x_1 provides an expression from which q_x may be computed. Note that, if the area A is uniform and k is independent of temperature, Equation 3.21 reduces to

$$\frac{q_x \, \Delta x}{A} = -k\, \Delta T \tag{3.22}$$

where $\Delta x = x_1 - x_0$ and $\Delta T = T_1 - T_0$.

We frequently elect to solve diffusion problems by working with integrated forms of the diffusion rate equations. However, the limiting conditions for which this may be done should be firmly fixed in our minds: *steady-state* and *one-dimensional* transfer with *no heat generation.*

EXAMPLE 3.3

The diagram shows a conical section fabricated from pyroceram. It is of circular cross section with the diameter $D = ax$, where $a = 0.25$. The small end is at $x_1 = 50$ mm and the large end at $x_2 = 250$ mm. The end temperatures are $T_1 = 400$ K and $T_2 = 600$ K, while the lateral surface is well insulated.

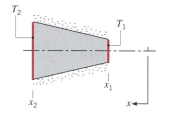

1. Derive an expression for the temperature distribution $T(x)$ in symbolic form, assuming one-dimensional conditions. Sketch the temperature distribution.
2. Calculate the heat rate q_x through the cone.

SOLUTION

Known: Conduction in a circular conical section having a diameter $D = ax$, where $a = 0.25$.

Find:
1. Temperature distribution $T(x)$.
2. Heat transfer rate q_x.

Schematic:

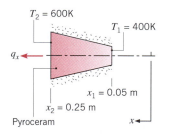

Assumptions:
1. Steady-state conditions.
2. One-dimensional conduction in the x direction.
3. No internal heat generation.
4. Constant properties.

Properties: Table A.2, pyroceram (500 K): $k = 3.46$ W/m · K.

Analysis:
1. Since heat conduction occurs under steady-state, one-dimensional conditions with no internal heat generation, the heat transfer rate q_x is a constant

independent of x. Accordingly, Fourier's law, Equation 2.1, may be used to determine the temperature distribution

$$q_x = -kA \frac{dT}{dx}$$

With $A = \pi D^2/4 = \pi a^2 x^2/4$ and separating variables

$$\frac{4q_x\, dx}{\pi a^2 x^2} = -k\, dT$$

Integrating from x_1 to any x within the cone, and recalling that q_x and k are constants, it follows that

$$\frac{4q_x}{\pi a^2} \int_{x_1}^{x} \frac{dx}{x^2} = -k \int_{T_1}^{T} dT$$

Hence

$$\frac{4q_x}{\pi a^2} \left(-\frac{1}{x} + \frac{1}{x_1} \right) = -k(T - T_1)$$

or solving for T

$$T(x) = T_1 - \frac{4q_x}{\pi a^2 k} \left(\frac{1}{x_1} - \frac{1}{x} \right)$$

Although q_x is a constant, it is as yet an unknown. However, it may be determined by evaluating the above expression at $x = x_2$, where $T(x_2) = T_2$. Hence

$$T_2 = T_1 - \frac{4q_x}{\pi a^2 k} \left(\frac{1}{x_1} - \frac{1}{x_2} \right)$$

and solving for q_x

$$q_x = \frac{\pi a^2 k (T_1 - T_2)}{4[(1/x_1) - (1/x_2)]}$$

Substituting for q_x into the expression for $T(x)$, the temperature distribution becomes

$$T(x) = T_1 + (T_1 - T_2) \left[\frac{(1/x) - (1/x_1)}{(1/x_1) - (1/x_2)} \right] \qquad \triangleleft$$

From this result, temperature may be calculated as a function of x and the distribution is as shown.

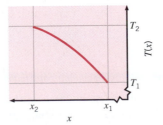

Note that, since $dT/dx = -4q_x/k\pi a^2 x^2$ from Fourier's law, it follows that the temperature gradient and heat flux decrease with increasing x.

2. Substituting numerical values into the foregoing result for the heat transfer rate, it follows that

$$q_x = \frac{\pi(0.25)^2 \times 3.46 \text{ W/m} \cdot \text{K} (400 - 600) \text{ K}}{4\left(\dfrac{1}{0.05 \text{ m}} - \dfrac{1}{0.25 \text{ m}}\right)} = -2.12 \text{ W} \qquad \triangleleft$$

Comments: When the parameter a increases, the one-dimensional assumption becomes less appropriate. That is, the assumption worsens when the cross-sectional area change with distance is more pronounced.

3.3
Radial Systems

Cylindrical and spherical systems often experience temperature gradients in the radial direction only and may therefore be treated as one dimensional. Moreover, under steady-state conditions with no heat generation, such systems may be analyzed by using the *standard* method, which begins with the appropriate form of the heat equation, or the *alternative* method, which begins with the appropriate form of Fourier's law. In this section, the cylindrical system is analyzed by means of the standard method and the spherical system by means of the alternative method.

3.3.1 The Cylinder

A common example is the hollow cylinder, whose inner and outer surfaces are exposed to fluids at different temperatures (Figure 3.6). For steady-state conditions with no heat generation, the appropriate form of the heat equation, Equation 2.20, is

$$\frac{1}{r}\frac{d}{dr}\left(kr\frac{dT}{dr}\right) = 0 \tag{3.23}$$

where, for the moment, k is treated as a variable. The physical significance of this result becomes evident if we also consider the appropriate form of Fourier's law. The rate at which energy is conducted across any cylindrical surface in the solid may be expressed as

$$q_r = -kA\frac{dT}{dr} = -k(2\pi rL)\frac{dT}{dr} \tag{3.24}$$

where $A = 2\pi rL$ is the area normal to the direction of heat transfer. Since Equation 3.23 dictates that the quantity $kr(dT/dr)$ is independent of r, it follows

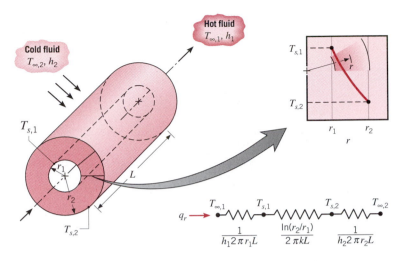

FIGURE 3.6   Hollow cylinder with convective surface conditions.

from Equation 3.24 that the conduction *heat transfer rate* q_r (*not* the heat flux q_r'') is a *constant in the radial direction.*

    We may determine the temperature distribution in the cylinder by solving Equation 3.23 and applying appropriate boundary conditions. Assuming the value of k to be constant, Equation 3.23 may be integrated twice to obtain the general solution

$$T(r) = C_1 \ln r + C_2 \tag{3.25}$$

To obtain the constants of integration C_1 and C_2, we introduce the following boundary conditions:

$$T(r_1) = T_{s,1} \quad \text{and} \quad T(r_2) = T_{s,2}$$

Applying these conditions to the general solution, we then obtain

$$T_{s,1} = C_1 \ln r_1 + C_2 \quad \text{and} \quad T_{s,2} = C_1 \ln r_2 + C_2$$

Solving for C_1 and C_2 and substituting into the general solution, we then obtain

$$T(r) = \frac{T_{s,1} - T_{s,2}}{\ln (r_1/r_2)} \ln \left(\frac{r}{r_2}\right) + T_{s,2} \tag{3.26}$$

Note that the temperature distribution associated with radial conduction through a cylindrical wall is logarithmic, not linear, as it is for the plane wall under the same conditions. The logarithmic distribution is sketched in the insert of Figure 3.6.

    If the temperature distribution, Equation 3.26, is now used with Fourier's law, Equation 3.24, we obtain the following expression for the heat transfer rate:

$$q_r = \frac{2\pi L k(T_{s,1} - T_{s,2})}{\ln (r_2/r_1)} \tag{3.27}$$

From this result it is evident that, for radial conduction in a cylindrical wall, the thermal resistance is of the form

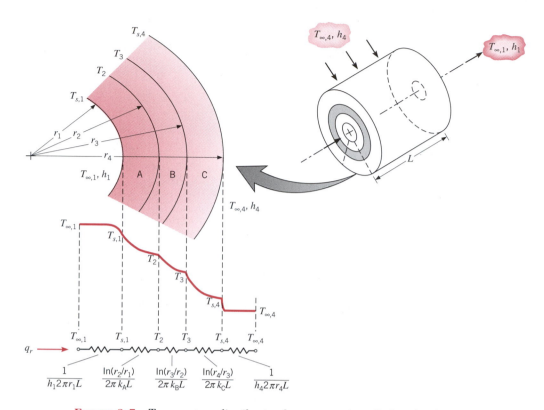

FIGURE 3.7 Temperature distribution for a composite cylindrical wall.

$$R_{t,\,cond} = \frac{\ln (r_2/r_1)}{2\pi Lk} \tag{3.28}$$

This resistance is shown in the series circuit of Figure 3.6. Note that since the value of q_r is independent of r, the foregoing result could have been obtained by using the alternative method, that is, by integrating Equation 3.24.

Consider now the composite system of Figure 3.7. Recalling how we treated the composite plane wall and neglecting the interfacial contact resistances, the heat transfer rate may be expressed as

$$q_r = \frac{T_{\infty,\,1} - T_{\infty,\,4}}{\dfrac{1}{2\pi r_1 L h_1} + \dfrac{\ln (r_2/r_1)}{2\pi k_A L} + \dfrac{\ln (r_3/r_2)}{2\pi k_B L} + \dfrac{\ln (r_4/r_3)}{2\pi k_C L} + \dfrac{1}{2\pi r_4 L h_4}} \tag{3.29}$$

The foregoing result may also be expressed in terms of an overall heat transfer coefficient. That is,

$$q_r = \frac{T_{\infty,\,1} - T_{\infty,\,4}}{R_{tot}} = UA(T_{\infty,\,1} - T_{\infty,\,4}) \tag{3.30}$$

If U is defined in terms of the inside area, $A_1 = 2\pi r_1 L$, Equations 3.29 and 3.30 may be equated to yield

$$U_1 = \cfrac{1}{\cfrac{1}{h_1} + \cfrac{r_1}{k_A}\ln\cfrac{r_2}{r_1} + \cfrac{r_1}{k_B}\ln\cfrac{r_3}{r_2} + \cfrac{r_1}{k_C}\ln\cfrac{r_4}{r_3} + \cfrac{r_1}{r_4}\cfrac{1}{h_4}} \qquad (3.31)$$

This definition is *arbitrary*, and the overall coefficient may also be defined in terms of A_4 or any of the intermediate areas. Note that

$$U_1A_1 = U_2A_2 = U_3A_3 = U_4A_4 = (\Sigma R_t)^{-1} \qquad (3.32)$$

and the specific forms of U_2, U_3, and U_4 may be inferred from Equations 3.29 and 3.30.

EXAMPLE 3.4

The possible existence of an optimum insulation thickness for radial systems is suggested by the presence of competing effects associated with an increase in this thickness. In particular, although the conduction resistance increases with the addition of insulation, the convection resistance decreases due to increasing outer surface area. Hence there may exist an insulation thickness that minimizes heat loss by maximizing the total resistance to heat transfer. Resolve this issue by considering the following system.

1. A thin-walled copper tube of radius r_i is used to transport a low-temperature refrigerant and is at a temperature T_i that is less than that of the ambient air at T_∞ around the tube. Is there an optimum thickness associated with application of insulation to the tube?

2. Confirm the above result by computing the total thermal resistance per unit length of tube for a 10-mm-diameter tube having the following insulation thicknesses: 0, 2, 5, 10, 20, and 40 mm. The insulation is composed of cellular glass, and the outer surface convection coefficient is 5 W/m² · K.

SOLUTION

Known: Radius r_i and temperature T_i of a thin-walled copper tube to be insulated from the ambient air.

Find:
1. Whether there exists an optimum insulation thickness that minimizes the heat transfer rate.

2. Thermal resistance associated with using cellular glass insulation of varying thickness.

Schematic:

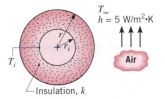

Assumptions:

1. Steady-state conditions.
2. One-dimensional heat transfer in the radial (cylindrical) direction.
3. Negligible tube wall thermal resistance.
4. Constant properties for insulation.
5. Negligible radiation exchange between insulation outer surface and surroundings.

Properties: Table A.3, cellular glass (285 K, assumed): $k = 0.055$ W/m · K.

Analysis:

1. The resistance to heat transfer between the refrigerant and the air is dominated by conduction in the insulation and convection in the air. The thermal circuit is therefore

where the conduction and convection resistances per unit length follow from Equations 3.28 and 3.9, respectively. The total thermal resistance per unit length of tube is then

$$R'_{\text{tot}} = \frac{\ln (r/r_i)}{2\pi k} + \frac{1}{2\pi r h}$$

where the rate of heat transfer per unit length of tube is

$$q' = \frac{T_\infty - T_i}{R'_{\text{tot}}}$$

An optimum insulation thickness would be associated with the value of r that minimized q' or maximized R'_{tot}. Such a value could be obtained from the requirement that

$$\frac{dR'_{\text{tot}}}{dr} = 0$$

Hence

$$\frac{1}{2\pi kr} - \frac{1}{2\pi r^2 h} = 0$$

or

$$r = \frac{k}{h}$$

To determine whether the foregoing result maximizes or minimizes the total resistance, the second derivative must be evaluated. Hence

$$\frac{d^2 R'_{\text{tot}}}{dr^2} = -\frac{1}{2\pi kr^2} + \frac{1}{\pi r^3 h}$$

or, at $r = k/h$,

$$\frac{d^2 R'_{tot}}{dr^2} = \frac{1}{\pi (k/h)^2} \left(\frac{1}{k} - \frac{1}{2k} \right) = \frac{1}{2\pi k^3 / h^2} > 0$$

Since this result is always positive, it follows that $r = k/h$ is the insulation radius for which the total resistance is a minimum, not a maximum. Hence an *optimum* insulation thickness *does not exist*.

From the above result it makes more sense to think in terms of a *critical insulation radius*

$$r_{cr} \equiv \frac{k}{h}$$

below which q' increases with increasing r and above which q' decreases with increasing r.

2. With $h = 5 \text{ W/m}^2 \cdot \text{K}$ and $k = 0.055 \text{ W/m} \cdot \text{K}$, the critical radius is

$$r_{cr} = \frac{0.055 \text{ W/m} \cdot \text{K}}{5 \text{ W/m}^2 \cdot \text{K}} = 0.011 \text{ m}$$

Hence $r_{cr} > r_i$ and heat transfer will increase with the addition of insulation up to a thickness of

$$r_{cr} - r_i = (0.011 - 0.005) \text{ m} = 0.006 \text{ m}$$

The thermal resistances corresponding to the prescribed insulation thicknesses may be calculated and are plotted as follows:

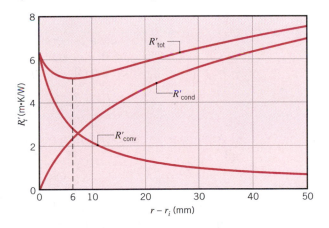

Comments:

1. The effect of the critical radius is revealed by the fact that, even for 20 mm of insulation, the total resistance is not as large as the value for no insulation.

2. If $r_i < r_{cr}$, as it is in this case, the total resistance decreases and the heat rate therefore increases with the addition of insulation. This trend continues until the outer radius of the insulation corresponds to the critical radius. The trend is desirable for electrical current flow through a wire, since the addition of electrical insulation would aid in transferring heat dissipated in

the wire to the surroundings. Conversely, if $r_i > r_{cr}$, any addition of insulation would increase the total resistance and therefore decrease the heat loss. This behavior would be desirable for steam flow through a pipe, where insulation is added to reduce heat loss to the surroundings.

3. For radial systems, the problem of reducing the total resistance through the application of insulation exists only for small diameter wires or tubes and for small convection coefficients, such that $r_{cr} > r_i$. For a typical insulation ($k \approx 0.03$ W/m · K) and free convection in air ($h \approx 10$ W/m² · K), $r_{cr} = (k/h) \approx 0.003$ m. Such a small value tells us that, normally, $r_i > r_{cr}$ and we need not be concerned with the effects of a critical radius.

4. The existence of a critical radius requires that the heat transfer area change in the direction of transfer, as for radial conduction in a cylinder (or a sphere). In a plane wall the area perpendicular to the direction of heat flow is constant and there is no critical insulation thickness (the total resistance always increases with increasing insulation thickness).

3.3.2 The Sphere

Now consider applying the alternative method to analyzing conduction in the hollow sphere of Figure 3.8. For the differential control volume of the figure, energy conservation requires that $q_r = q_{r+dr}$ for steady-state, one-dimensional conditions with no heat generation. The appropriate form of Fourier's law is

$$q_r = -kA \frac{dT}{dr} = -k(4\pi r^2) \frac{dT}{dr} \tag{3.33}$$

where $A = 4\pi r^2$ is the area normal to the direction of heat transfer.

Acknowledging that q_r is a constant, independent of r, Equation 3.33 may be expressed in the integral form

$$\frac{q_r}{4\pi} \int_{r_1}^{r_2} \frac{dr}{r^2} = - \int_{T_{s,1}}^{T_{s,2}} k(T) \, dT \tag{3.34}$$

Assuming constant k, we then obtain

$$q_r = \frac{4\pi k(T_{s,1} - T_{s,2})}{(1/r_1) - (1/r_2)} \tag{3.35}$$

Remembering that the thermal resistance is defined as the temperature difference divided by the heat transfer rate, we obtain

$$R_{t,\text{cond}} = \frac{1}{4\pi k} \left(\frac{1}{r_1} - \frac{1}{r_2} \right) \tag{3.36}$$

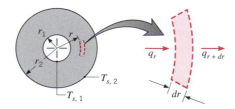

FIGURE 3.8
Conduction in a spherical shell.

Note that the temperature distribution and Equations 3.35 and 3.36 could have been obtained by using the standard approach, which begins with the appropriate form of the heat equation.

Spherical composites may be treated in much the same way as composite walls and cylinders, where appropriate forms of the total resistance and overall heat transfer coefficient may be determined.

EXAMPLE 3.5

A spherical, thin-walled metallic container is used to store liquid nitrogen at 77 K. The container has a diameter of 0.5 m and is covered with an evacuated, reflective insulation composed of silica powder. The insulation is 25 mm thick, and its outer surface is exposed to ambient air at 300 K. The convection coefficient is known to be 20 W/m^2 · K. The latent heat of vaporization and the density of liquid nitrogen are 2×10^5 J/kg and 804 kg/m^3, respectively.

1. What is the rate of heat transfer to the liquid nitrogen?
2. What is the rate of liquid boil-off?

SOLUTION

Known: Liquid nitrogen is stored in a spherical container that is insulated and exposed to ambient air.

Find:

1. The rate of heat transfer to the nitrogen.
2. The mass rate of nitrogen boil-off.

Schematic:

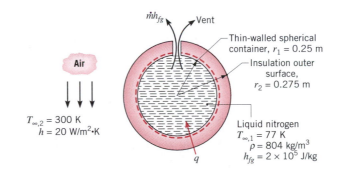

Assumptions:

1. Steady-state conditions.
2. One-dimensional transfer in the radial direction.
3. Negligible resistance to heat transfer through the container wall and from the container to the nitrogen.
4. Constant properties.
5. Negligible radiation exchange between outer surface of insulation and surroundings.

Properties: Table A.3, evacuated silica powder (300 K): $k = 0.0017$ W/m · K.

Analysis:

1. The thermal circuit involves a conduction and convection resistance in series and is of the form

where, from Equation 3.36,

$$R_{t,\,cond} = \frac{1}{4\pi k}\left(\frac{1}{r_1} - \frac{1}{r_2}\right)$$

and from Equation 3.9

$$R_{t,\,conv} = \frac{1}{h4\pi r_2^2}$$

The rate of heat transfer to the liquid nitrogen is then

$$q = \frac{T_{\infty,2} - T_{\infty,1}}{(1/4\pi k)[(1/r_1) - (1/r_2)] + (1/h4\pi r_2^2)}$$

Hence

$$q = [(300 - 77)\,\text{K}]$$

$$\div \left[\frac{1}{4\pi(0.0017\,\text{W/m}\cdot\text{K})}\left(\frac{1}{0.25\,\text{m}} - \frac{1}{0.275\,\text{m}}\right)\right.$$

$$\left. + \frac{1}{(20\,\text{W/m}^2\cdot\text{K})4\pi(0.275\,\text{m})^2}\right]$$

$$q = \frac{223}{17.02 + 0.05}\,\text{W} = 13.06\,\text{W} \qquad \triangleleft$$

2. Performing an energy balance for a control surface about the nitrogen, it follows from Equation 1.12 that

$$\dot{E}_{in} - \dot{E}_{out} = 0$$

where $\dot{E}_{in} = q$ and $\dot{E}_{out} = \dot{m}h_{fg}$ is associated with the loss of latent energy due to boiling. Hence

$$q - \dot{m}h_{fg} = 0$$

and the boil-off $\dot{m}$ is

$$\dot{m} = \frac{q}{h_{fg}}$$

$$\dot{m} = \frac{13.06\,\text{J/s}}{2\times 10^5\,\text{J/kg}} = 6.53\times 10^{-5}\,\text{kg/s}$$

The loss per day is

$$\dot{m} = 6.53 \times 10^{-5} \text{ kg/s} \times 3600 \text{ s/h} \times 24 \text{ h/day}$$

$$\dot{m} = 5.64 \text{ kg/day}$$

or on a volumetric basis

$$\dot{V} = \frac{\dot{m}}{\rho} = \frac{5.64 \text{ kg/day}}{804 \text{ kg/m}^3} = 0.007 \text{ m}^3/\text{day} = 7 \text{ liters/day}$$

Comments:

1. $R_{t, \text{conv}} \ll R_{t, \text{cond}}$.
2. With a container volume of $(4/3)(\pi r_1^3) = 0.065 \text{ m}^3 = 65$ liters, the daily loss amounts to $(7 \text{ liters}/65 \text{ liters}) 100\% = 10.8\%$ of capacity.

3.4
Summary of One-Dimensional Conduction Results

Many important problems are characterized by one-dimensional, steady-state conduction in plane, cylindrical, or spherical walls without thermal energy generation. Key results for these three geometries are summarized in Table 3.3, where ΔT refers to the temperature difference, $T_{s, 1} - T_{s, 2}$, between the inner and outer surfaces identified in Figures 3.1, 3.6, and 3.8. In each case, beginning with the heat equation, you should be able to derive the corresponding expressions for the temperature distribution, heat flux, heat rate, and thermal resistance.

TABLE 3.3 One-dimensional, steady-state solutions to the heat equation with no generation

	Plane Wall	**Cylindrical Wall**[a]	**Spherical Wall**[a]
Heat equation	$\dfrac{d^2T}{dx^2} = 0$	$\dfrac{1}{r}\dfrac{d}{dr}\left(r\dfrac{dT}{dr}\right) = 0$	$\dfrac{1}{r^2}\dfrac{d}{dr}\left(r^2\dfrac{dT}{dr}\right) = 0$
Temperature distribution	$T_{s,1} - \Delta T\dfrac{x}{L}$	$T_{s,2} + \Delta T\dfrac{\ln(r/r_2)}{\ln(r_1/r_2)}$	$T_{s,1} - \Delta T\left[\dfrac{1 - (r_1/r)}{1 - (r_1/r_2)}\right]$
Heat flux (q'')	$k\dfrac{\Delta T}{L}$	$\dfrac{k\,\Delta T}{r\ln(r_2/r_1)}$	$\dfrac{k\,\Delta T}{r^2[(1/r_1) - (1/r_2)]}$
Heat rate (q)	$kA\dfrac{\Delta T}{L}$	$\dfrac{2\pi L k\,\Delta T}{\ln(r_2/r_1)}$	$\dfrac{4\pi k\,\Delta T}{(1/r_1) - (1/r_2)}$
Thermal resistance ($R_{t,\text{cond}}$)	$\dfrac{L}{kA}$	$\dfrac{\ln(r_2/r_1)}{2\pi L k}$	$\dfrac{(1/r_1) - (1/r_2)}{4\pi k}$

[a]The critical radius of insulation is $r_{\text{cr}} = k/h$ for the cylinder and $r_{\text{cr}} = 2k/h$ for the sphere.

3.5
Conduction with Thermal Energy Generation

In the preceding section we considered conduction problems for which the temperature distribution in a medium was determined solely by conditions at the boundaries of the medium. We now want to consider the additional effect on the temperature distribution of processes that may be occurring *within* the medium. In particular, we wish to consider situations for which thermal energy is being *generated* due to *conversion* from some other energy form.

A common thermal energy generation process involves the conversion from *electrical to thermal energy* in a current-carrying medium (*Ohmic* or *resistance heating*). The rate at which energy is generated by passing a current I through a medium of electrical resistance R_e is

$$\dot{E}_g = I^2 R_e \tag{3.37}$$

If this power generation (W) occurs uniformly throughout the medium of volume V, the volumetric generation rate (W/m³) is then

$$\dot{q} \equiv \frac{\dot{E}_g}{V} = \frac{I^2 R_e}{V} \tag{3.38}$$

Energy generation may also occur as a result of the deceleration and absorption of neutrons in the fuel element of a nuclear reactor or exothermic chemical reactions occurring within a medium. Endothermic reactions would, of course, have the inverse effect (a thermal energy sink) of converting thermal energy to chemical bonding energy. Finally, a conversion from electromagnetic to thermal energy may occur due to the absorption of radiation within the medium. The process may occur, for example, because gamma rays are absorbed in external nuclear reactor components (cladding, thermal shields, pressure vessels, etc.) or because visible radiation is absorbed in a semitransparent medium. Remember not to confuse energy generation with energy storage (Section 1.3.1).

3.5.1   The Plane Wall

Consider the plane wall of Figure 3.9*a*, in which there is *uniform* energy generation per unit volume ($\dot{q}$ is constant) and the surfaces are maintained at $T_{s,1}$ and $T_{s,2}$. For constant thermal conductivity k, the appropriate form of the heat equation, Equation 2.16, is

$$\frac{d^2 T}{dx^2} + \frac{\dot{q}}{k} = 0 \tag{3.39}$$

The general solution is

$$T = -\frac{\dot{q}}{2k} x^2 + C_1 x + C_2 \tag{3.40}$$

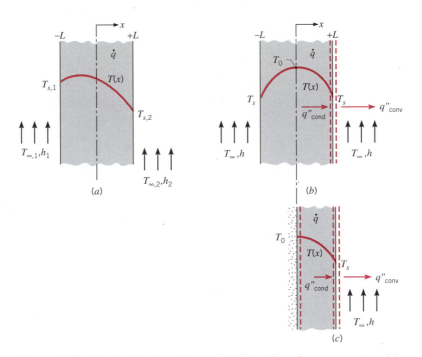

FIGURE 3.9   Conduction in a plane wall with uniform heat generation. (*a*) Asymmetrical boundary conditions. (*b*) Symmetrical boundary conditions. (*c*) Adiabatic surface at midplane.

where C_1 and C_2 are the constants of integration. For the prescribed boundary conditions,

$$T(-L) = T_{s,1} \quad \text{and} \quad T(L) = T_{s,2}$$

The constants may be evaluated and are of the form

$$C_1 = \frac{T_{s,2} - T_{s,1}}{2L} \quad \text{and} \quad C_2 = \frac{\dot{q}}{2k}L^2 + \frac{T_{s,1} + T_{s,2}}{2}$$

in which case the temperature distribution is

$$T(x) = \frac{\dot{q}L^2}{2k}\left(1 - \frac{x^2}{L^2}\right) + \frac{T_{s,2} - T_{s,1}}{2}\frac{x}{L} + \frac{T_{s,1} + T_{s,2}}{2} \tag{3.41}$$

The heat flux at any point in the wall may, of course, be determined by using Equation 3.41 with Fourier's law. Note, however, that *with generation the heat flux is no longer independent of x.*

The preceding result simplifies when both surfaces are maintained at a common temperature, $T_{s,1} = T_{s,2} \equiv T_s$. The temperature distribution is then *symmetrical* about the midplane, Figure 3.9*b*, and is given by

$$T(x) = \frac{\dot{q}L^2}{2k}\left(1 - \frac{x^2}{L^2}\right) + T_s \tag{3.42}$$

The maximum temperature exists at the midplane

$$T(0) \equiv T_0 = \frac{\dot{q}L^2}{2k} + T_s \tag{3.43}$$

in which case the temperature distribution, Equation 3.42, may be expressed as

$$\frac{T(x) - T_0}{T_s - T_0} = \left(\frac{x}{L}\right)^2 \tag{3.44}$$

It is important to note that at the plane of symmetry in Figure 3.9b, the temperature gradient is zero, $(dT/dx)_{x=0} = 0$. Accordingly, there is no heat transfer across this plane, and it may be represented by the *adiabatic* surface shown in Figure 3.9c. One implication of this result is that Equation 3.42 also applies to plane walls that are perfectly insulated on one side ($x = 0$) and maintained at a fixed temperature T_s on the other side ($x = L$).

To use the foregoing results the surface temperature(s) T_s must be known. However, a common situation is one for which it is the temperature of an adjoining fluid, T_∞, and not T_s, which is known. It then becomes necessary to relate T_s to T_∞. This relation may be developed by applying a surface energy balance. Consider the surface at $x = L$ for the symmetrical plane wall (Figure 3.9b) or the insulated plane wall (Figure 3.9c). Neglecting radiation and substituting the appropriate rate equations, the energy balance given by Equation 1.12 reduces to

$$-k\frac{dT}{dx}\bigg|_{x=L} = h(T_s - T_\infty) \tag{3.45}$$

Substituting from Equation 3.42 to obtain the temperature gradient at $x = L$, it follows that

$$T_s = T_\infty + \frac{\dot{q}L}{h} \tag{3.46}$$

Hence T_s may be computed from knowledge of T_∞, $\dot{q}$, L, and h.

Equation 3.46 may also be obtained by applying an *overall* energy balance to the plane wall of Figure 3.9b or 3.9c. For example, relative to a control surface about the wall of Figure 3.9c, the rate at which energy is generated within the wall must be balanced by the rate at which energy leaves via convection at the boundary. Equation 1.11a reduces to

$$\dot{E}_g = \dot{E}_{out} \tag{3.47}$$

or, for a unit surface area,

$$\dot{q}L = h(T_s - T_\infty) \tag{3.48}$$

Solving for T_s, Equation 3.46 is obtained.

Equation 3.46 may be combined with Equation 3.42 to eliminate T_s from the temperature distribution, which is then expressed in terms of the known quantities $\dot{q}$, L, k, h, and T_∞. The same result may be obtained directly by using Equation 3.45 as a boundary condition to evaluate the constants of integration appearing in Equation 3.40.

EXAMPLE 3.6

A plane wall is a composite of two materials, A and B. The wall of material A has uniform heat generation $\dot{q} = 1.5 \times 10^6$ W/m^3, $k_A = 75$ W/m · K, and thickness $L_A = 50$ mm. The wall material B has no generation with $k_B = 150$ W/m · K and thickness $L_B = 20$ mm. The inner surface of material A is well insulated, while the outer surface of material B is cooled by a water stream with $T_\infty = 30°$C and $h = 1000$ W/m^2 · K.

1. Sketch the temperature distribution that exists in the composite under steady-state conditions.

2. Determine the temperature T_0 of the insulated surface and the temperature T_2 of the cooled surface.

SOLUTION

Known: Plane wall of material A with internal heat generation is insulated on one side and bounded by a second wall of material B, which is without heat generation and is subjected to convection cooling.

Find:

1. Sketch of steady-state temperature distribution in the composite.

2. Inner and outer surface temperatures of the composite.

Schematic:

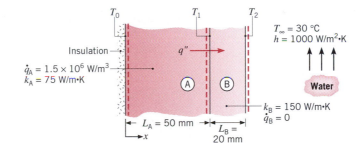

Assumptions:

1. Steady-state conditions.

2. One-dimensional conduction in x direction.

3. Negligible contact resistance between walls.

4. Inner surface of A adiabatic.

5. Constant properties for materials A and B.

Analysis:

1. From the prescribed physical conditions, the temperature distribution in the composite is known to have the following features, as shown:
   (a) Parabolic in material A.
   (b) Zero slope at insulated boundary.

(c) Linear in material B.
(d) Slope change = $k_B/k_A = 2$ at interface.
The temperature distribution in the water is characterized by
(e) Large gradients near the surface.

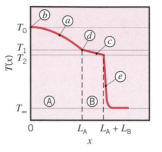

2. The outer surface temperature T_2 may be obtained by performing an energy balance on a control volume about material B. Since there is no generation in this material, it follows that, for steady-state conditions and a unit surface area, the heat flux into the material at $x = L_A$ must equal the heat flux from the material due to convection at $x = L_A + L_B$. Hence

$$q'' = h(T_2 - T_\infty) \tag{1}$$

The heat flux q'' may be determined by performing a second energy balance on a control volume about material A. In particular, since the surface at $x = 0$ is adiabatic, there is no inflow and the rate at which energy is generated must equal the outflow. Accordingly, for a unit surface area,

$$\dot{q}L_A = q'' \tag{2}$$

Combining Equations 1 and 2, the outer surface temperature is

$$T_2 = T_\infty + \frac{\dot{q}L_A}{h}$$

$$T_2 = 30°C + \frac{1.5 \times 10^6 \text{ W/m}^3 \times 0.05 \text{ m}}{1000 \text{ W/m}^2 \cdot \text{K}} = 105°C \qquad \triangleleft$$

From Equation 3.43 the temperature at the insulated surface is

$$T_0 = \frac{\dot{q}L_A^2}{2k_A} + T_1 \tag{3}$$

where T_1 may be obtained from the following thermal circuit:

That is,

$$T_1 = T_\infty + (R''_{\text{cond, B}} + R''_{\text{conv}})q''$$

where the resistances for a unit surface area are

$$R''_{\text{cond, B}} = \frac{L_B}{k_B} \qquad R''_{\text{conv}} = \frac{1}{h}$$

Hence

$$T_1 = 30°C + \left(\frac{0.02 \text{ m}}{150 \text{ W/m} \cdot \text{K}} + \frac{1}{1000 \text{ W/m}^2 \cdot \text{K}} \right)$$

$$\times 1.5 \times 10^6 \text{ W/m}^3 \times 0.05 \text{ m}$$

$$T_1 = 30°C + 85°C = 115°C$$

Substituting into Equation 3,

$$T_0 = \frac{1.5 \times 10^6 \text{ W/m}^3 (0.05 \text{ m})^2}{2 \times 75 \text{ W/m} \cdot \text{K}} + 115°C$$

$$T_0 = 25°C + 115°C = 140°C \qquad \triangleleft$$

Comments:

1. Material A, having heat generation, cannot be represented by a thermal circuit element.

2. Since the resistance to heat transfer by convection is significantly larger than that due to conduction in material B, $R''_{conv}/R''_{cond} = 7.5$, the surface-to-fluid temperature difference is much larger than the temperature drop across material B, $(T_2 - T_\infty)/(T_1 - T_2) = 7.5$. This result is consistent with the temperature distribution plotted in part (1).

3. The surface and interface temperatures (T_0, T_1 and T_2) depend on the generation rate $\dot{q}$, the thermal conductivities k_A and k_B, and the convection coefficient h. Each material will have a maximum allowable operating temperature, which must not be exceeded if thermal failure of the system is to be avoided. We explore the effect of one of these parameters by computing and plotting temperature distributions for values of $h = 200$ and 1000 W/m$^2 \cdot$ K, which would be representative of air and liquid cooling, respectively.

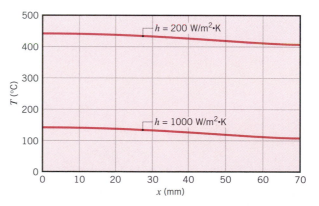

For $h = 200$ W/m$^2 \cdot$ K, there is a significant increase in temperature throughout the system and, depending on the selection of materials, thermal failure could be a problem. Note the slight discontinuity in the temperature gradient, dT/dx, at $x = 50$ mm. What is the physical basis for this discontinuity? We have assumed negligible contact resistance at this location. What would be the effect of such a resistance on the temperature distribution throughout the system? Sketch a representative distribution. What would be the effect on the temperature distribution of an increase in $\dot{q}$, k_A, or k_B? Qualitatively sketch the effect of such changes on the temperature distribution.

3.5.2 Radial Systems

Heat generation may occur in a variety of radial geometries. Consider the long, solid cylinder of Figure 3.10, which could represent a current-carrying wire or a fuel element in a nuclear reactor. For steady-state conditions the rate at which heat is generated within the cylinder must equal the rate at which heat is convected from the surface of the cylinder to a moving fluid. This condition allows the surface temperature to be maintained at a fixed value of T_s.

To determine the temperature distribution in the cylinder, we begin with the appropriate form of the heat equation. For constant thermal conductivity k, Equation 2.20 reduces to

$$\frac{1}{r}\frac{d}{dr}\left(r\frac{dT}{dr}\right) + \frac{\dot{q}}{k} = 0 \tag{3.49}$$

Separating variables and assuming uniform generation, this expression may be integrated to obtain

$$r\frac{dT}{dr} = -\frac{\dot{q}}{2k}r^2 + C_1 \tag{3.50}$$

Repeating the procedure, the general solution for the temperature distribution becomes

$$T(r) = -\frac{\dot{q}}{4k}r^2 + C_1\ln r + C_2 \tag{3.51}$$

To obtain the constants of integration C_1 and C_2, we apply the boundary conditions

$$\left.\frac{dT}{dr}\right|_{r=0} = 0 \qquad \text{and} \qquad T(r_o) = T_s$$

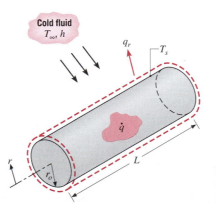

FIGURE 3.10
Conduction in a solid cylinder with uniform heat generation.

The first condition results from the symmetry of the situation. That is, for the solid cylinder the centerline is a line of symmetry for the temperature distribution and the temperature gradient must be zero. Recall that similar conditions existed at the midplane of a wall having symmetrical boundary conditions (Figure 3.9*b*). From the symmetry condition at $r = 0$ and Equation 3.50, it is evident that $C_1 = 0$. Using the surface boundary condition at $r = r_o$ with Equation 3.51, we then obtain

$$C_2 = T_s + \frac{\dot{q}}{4k} r_o^2 \tag{3.52}$$

The temperature distribution is therefore

$$T(r) = \frac{\dot{q}r_o^2}{4k}\left(1 - \frac{r^2}{r_o^2}\right) + T_s \tag{3.53}$$

Evaluating Equation 3.53 at the centerline and dividing the result into Equation 3.53, we obtain the temperature distribution in nondimensional form,

$$\frac{T(r) - T_s}{T_o - T_s} = 1 - \left(\frac{r}{r_o}\right)^2 \tag{3.54}$$

where T_o is the centerline temperature. The heat rate at any radius in the cylinder may, of course, be evaluated by using Equation 3.53 with Fourier's law.

To relate the surface temperature, T_s, to the temperature of the cold fluid, T_∞, either a surface energy balance or an overall energy balance may be used. Choosing the second approach, we obtain

$$\dot{q}(\pi r_o^2 L) = h(2\pi r_o L)(T_s - T_\infty)$$

or

$$T_s = T_\infty + \frac{\dot{q}r}{2h} \tag{3.55}$$

A convenient and systematic procedure for treating the different combinations of surface conditions, which may be applied to one-dimensional cylindrical (and planar) geometries with uniform thermal energy generation, is provided in Appendix C.

EXAMPLE 3.7

Consider a long solid tube, insulated at the outer radius r_2 and cooled at the inner radius r_1, with uniform heat generation $\dot{q}$ (W/m³) within the solid.

1. Obtain the general solution for the temperature distribution in the tube.
2. In a practical application a limit would be placed on the maximum temperature that is permissible at the insulated surface ($r = r_2$). Specifying this limit as $T_{s,2}$, identify appropriate boundary conditions that could be used to determine the arbitrary constants appearing in the general solution. Determine these constants and the corresponding form of the temperature distribution.

3. Determine the heat removal rate per unit length of tube.

4. If the coolant is available at a temperature T_∞, obtain an expression for the convection coefficient that would have to be maintained at the inner surface to allow for operation at prescribed values of $T_{s,2}$ and $\dot{q}$.

SOLUTION

Known: Solid tube with uniform heat generation is insulated at the outer surface and cooled at the inner surface.

Find:

1. General solution for the temperature distribution $T(r)$.

2. Appropriate boundary conditions and the corresponding form of the temperature distribution.

3. Heat removal rate.

4. Convection coefficient at the inner surface.

Schematic:

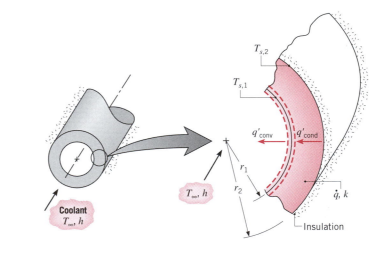

Assumptions:

1. Steady-state conditions.

2. One-dimensional radial conduction.

3. Constant properties.

4. Uniform volumetric heat generation.

5. Outer surface adiabatic.

Analysis:

1. To determine $T(r)$, the appropriate form of the heat equation, Equation 2.20, must be solved. For the prescribed conditions, this expression reduces to Equation 3.49, and the general solution is given by Equation 3.51. Hence this solution applies in a cylindrical shell, as well as in a solid cylinder (Figure 3.10).

2. Two boundary conditions are needed to evaluate C_1 and C_2, and in this problem it is appropriate to specify both conditions at r_2. Invoking the prescribed temperature limit,

$$T(r_2) = T_{s,2} \tag{1}$$

and applying Fourier's law, Equation 3.24, at the adiabatic outer surface

$$\left. \frac{dT}{dr} \right|_{r_2} = 0 \tag{2}$$

Using Equations 3.51 and 1, it follows that

$$T_{s,2} = -\frac{\dot{q}}{4k} r_2^2 + C_1 \ln r_2 + C_2 \tag{3}$$

Similarly, from Equations 3.50 and 2

$$0 = -\frac{\dot{q}}{2k} r_2^2 + C_1 \tag{4}$$

Hence, from Equation 4,

$$C_1 = \frac{\dot{q}}{2k} r_2^2 \tag{5}$$

and from Equation 3

$$C_2 = T_{s,2} + \frac{\dot{q}}{4k} r_2^2 - \frac{\dot{q}}{2k} r_2^2 \ln r_2 \tag{6}$$

Substituting Equations 5 and 6 into the general solution, Equation 3.51, it follows that

$$T(r) = T_{s,2} + \frac{\dot{q}}{4k} (r_2^2 - r^2) - \frac{\dot{q}}{2k} r_2^2 \ln \frac{r_2}{r} \tag{7}$$

3. The heat removal rate may be determined by obtaining the conduction rate at r_1 or by evaluating the total generation rate for the tube. From Fourier's law

$$q_r' = -k2\pi r \frac{dT}{dr}$$

Hence substituting from Equation 7 and evaluating the result at r_1,

$$q_r'(r_1) = -k2\pi r_1 \left(-\frac{\dot{q}}{2k} r_1 + \frac{\dot{q}}{2k} \frac{r_2^2}{r_1} \right) = -\pi \dot{q}(r_2^2 - r_1^2) \tag{8}$$

Alternatively, because the tube is insulated at r_2, the rate at which heat is generated in the tube must equal the rate of removal at r_1. That is, for a control volume about the tube, the energy conservation requirement, Equation 1.11a, reduces to $\dot{E}_g - \dot{E}_{out} = 0$, where $\dot{E}_g = \dot{q}\pi(r_2^2 - r_1^2)L$ and $\dot{E}_{out} = q_{cond}' L = -q_r'(r_1)L$. Hence

$$q_r'(r_1) = -\pi \dot{q}(r_2^2 - r_1^2) \tag{9}$$

4. Applying the energy conservation requirement, Equation 1.12, to the inner surface, it follows that

$$q'_{cond} = q'_{conv}$$

or

$$\pi \dot{q}(r_2^2 - r_1^2) = h2\pi r_1(T_{s,1} - T_\infty)$$

Hence

$$h = \frac{\dot{q}(r_2^2 - r_1^2)}{2r_1(T_{s,1} - T_\infty)} \tag{10}$$

where $T_{s,1}$ may be obtained by evaluating Equation 7 at $r = r_1$.

Comments:

1. Note that, through application of Fourier's law in part 3, the sign on $q'_r(r_1)$ was found to be negative, Equation 8, implying that heat flow is in the negative r direction. However, in applying the energy balance, we acknowledged that heat flow was *out* of the wall. Hence we expressed q'_{cond} as $-q'_r(r_1)$ and we expressed q'_{conv} in terms of $(T_{s,1} - T_\infty)$, rather than $(T_\infty - T_{s,1})$.

2. The temperature distribution, Equation 7, may also be obtained by using the results of Appendix C. Applying a surface energy balance at $r = r_1$, with $q(r_1) = -\dot{q}\pi(r_2^2 - r_1^2)L$, $(T_{s,2} - T_{s,1})$ may be determined from Equation C.8 and the result substituted into Equation C.2 to eliminate $T_{s,1}$ and obtain the desired expression.

3.5.3 Application of Resistance Concepts

We conclude our discussion of heat generation effects with a word of caution. In particular, when such effects are present, the heat transfer rate is not a constant, independent of the spatial coordinate. Accordingly, it would be *incorrect* to use the conduction resistance concepts and the related heat rate equations developed in Sections 3.1 and 3.3.

3.6
Heat Transfer from Extended Surfaces

The term *extended surface* is commonly used in reference to a solid that experiences energy transfer by conduction within its boundaries, as well as energy transfer by convection (and/or radiation) between its boundaries and the surroundings. Such a system is shown schematically in Figure 3.11. A strut is used to provide mechanical support to two walls that are at different temperatures. A temperature gradient in the x direction sustains heat transfer by conduction internally, at the same time that there is energy transfer by convection from the surface.

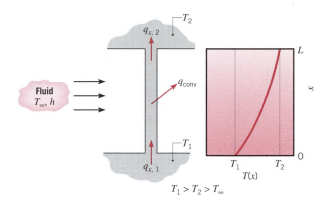

FIGURE 3.11   Combined conduction and convection in a structural element.

Although there are many different situations that involve combined conduction–convection effects, the most frequent application is one in which an extended surface is used specifically to *enhance* the heat transfer rate between a solid and an adjoining fluid. Such an extended surface is termed a *fin*.

Consider the plane wall of Figure 3.12*a*. If T_s is fixed, there are two ways in which the heat transfer rate may be increased. The convection coefficient h could be increased by increasing the fluid velocity, and/or the fluid temperature T_∞ could be reduced. However, many situations would be encountered in which increasing h to the maximum possible value is either insufficient to obtain the desired heat transfer rate or the associated costs are prohibitive. Such costs are related to the blower or pump power requirements needed to increase h through increased fluid motion. Moreover, the second option of reducing T_∞ is often impractical. Examining Figure 3.12*b*, however, we see that there exists a third option. That is, the heat transfer rate may be increased by increasing the surface area across which the convection occurs. This may be done by employing *fins* that *extend* from the wall into the surrounding fluid. The thermal conductivity of the fin material has a strong effect on the temperature distribution along the fin and therefore influences the degree to which the heat transfer rate is en-

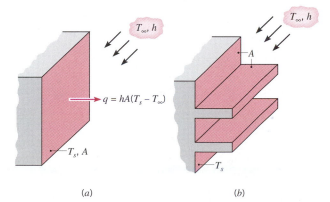

FIGURE 3.12   Use of fins to enhance heat transfer from a plane wall. (*a*) Bare surface. (*b*) Finned surface.

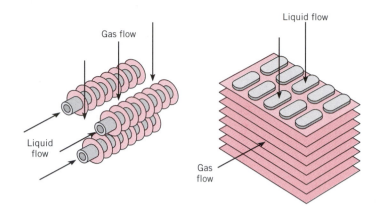

FIGURE 3.13   Schematic of typical finned-tube heat exchangers.

hanced. Ideally, the fin material should have a large thermal conductivity to minimize temperature variations from its base to its tip. In the limit of infinite thermal conductivity, the entire fin would be at the temperature of the base surface, thereby providing the maximum possible heat transfer enhancement.

You are already familiar with several fin applications. Consider the arrangement for cooling engine heads on motorcycles and lawnmowers or for cooling electric power transformers. Consider also the tubes with attached fins used to promote heat exchange between air and the working fluid of an air conditioner. Two common finned-tube arrangements are shown in Figure 3.13.

Different fin configurations are illustrated in Figure 3.14. A *straight fin* is any extended surface that is attached to a *plane wall*. It may be of uniform cross-sectional area, or its cross-sectional area may vary with the distance x from the wall. An *annular fin* is one that is circumferentially attached to a cylinder, and its cross section varies with radius from the centerline of the cylinder. The foregoing fin types have rectangular cross sections, whose area may be expressed as a product of the fin thickness t and the width w for straight fins or the circumference $2\pi r$ for annular fins. In contrast a *pin fin,* or *spine,* is an extended surface of circular cross section. Pin fins may also be of uniform or nonuniform cross section. In any application, selection of a particular fin con-

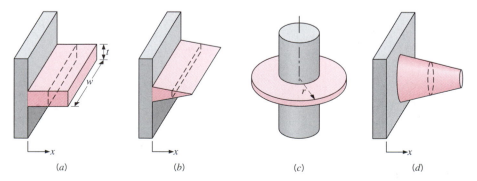

FIGURE 3.14   Fin configurations. (*a*) Straight fin of uniform cross section. (*b*) Straight fin of nonuniform cross section. (*c*) Annular fin. (*d*) Pin fin.

figuration may depend on space, weight, manufacturing, and cost considerations, as well as on the extent to which the fins reduce the surface convection coefficient and increase the pressure drop associated with flow over the fins.

3.6.1   A General Conduction Analysis

As engineers we are primarily interested in knowing the extent to which particular extended surfaces or fin arrangements could improve heat transfer from a surface to the surrounding fluid. To determine the heat transfer rate associated with a fin, we must first obtain the temperature distribution along the fin. As we have done for previous systems, we begin by performing an energy balance on an appropriate differential element. Consider the extended surface of Figure 3.15. The analysis is simplified if certain assumptions are made. We choose to assume one-dimensional conditions in the longitudinal (x) direction, even though conduction within the fin is actually two dimensional. The rate at which energy is convected to the fluid from any point on the fin surface must be balanced by the rate at which energy reaches that point due to conduction in the transverse (y, z) direction. However, in practice the fin is thin and temperature changes in the longitudinal direction are much larger than those in the transverse direction. Hence we may assume one-dimensional conduction in the x direction. We will consider steady-state conditions and also assume that the thermal conductivity is constant, that radiation from the surface is negligible, that heat generation effects are absent, and that the convection heat transfer coefficient h is uniform over the surface.

Applying the conservation of energy requirement, Equation 1.11a, to the differential element of Figure 3.15, we obtain

$$q_x = q_{x+dx} + dq_{conv} \tag{3.56}$$

From Fourier's law we know that

$$q_x = -kA_c \frac{dT}{dx} \tag{3.57}$$

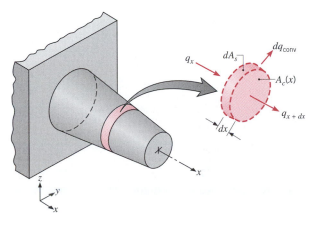

FIGURE 3.15   Energy balance for an extended surface.

where A_c is the *cross-sectional* area, which may vary with x. Since the conduction heat rate at $x + dx$ may be expressed as

$$q_{x+dx} = q_x + \frac{dq_x}{dx}\, dx \tag{3.58}$$

it follows that

$$q_{x+dx} = -kA_c \frac{dT}{dx} - k \frac{d}{dx}\left(A_c \frac{dT}{dx} \right) dx \tag{3.59}$$

The convection heat transfer rate may be expressed as

$$dq_{\text{conv}} = h\, dA_s(T - T_\infty) \tag{3.60}$$

where dA_s is the *surface* area of the differential element. Substituting the foregoing rate equations into the energy balance, Equation 3.56, we obtain

$$\frac{d}{dx}\left(A_c \frac{dT}{dx} \right) - \frac{h}{k} \frac{dA_s}{dx}(T - T_\infty) = 0$$

or

$$\frac{d^2T}{dx^2} + \left(\frac{1}{A_c} \frac{dA_c}{dx} \right)\frac{dT}{dx} - \left(\frac{1}{A_c} \frac{h}{k} \frac{dA_s}{dx} \right)(T - T_\infty) = 0 \tag{3.61}$$

This result provides a general form of the energy equation for one-dimensional conditions in an extended surface. Its solution for appropriate boundary conditions would provide the temperature distribution, which could then be used with Equation 3.57 to calculate the conduction rate at any x.

3.6.2   Fins of Uniform Cross-Sectional Area

To solve Equation 3.61 it is necessary to be more specific about the geometry. We begin with the simplest case of straight rectangular and pin fins of uniform cross section (Figure 3.16). Each fin is attached to a base surface of temperature $T(0) = T_b$ and extends into a fluid of temperature T_∞.

For the prescribed fins, A_c is a constant and $A_s = Px$, where A_s is the surface area measured from the base to x and P is the fin perimeter. Accordingly, with $dA_c/dx = 0$ and $dA_s/dx = P$, Equation 3.61 reduces to

$$\frac{d^2T}{dx^2} - \frac{hP}{kA_c}(T - T_\infty) = 0 \tag{3.62}$$

To simplify the form of this equation, we transform the dependent variable by defining an *excess temperature* θ as

$$\theta(x) \equiv T(x) - T_\infty \tag{3.63}$$

where, since T_∞ is a constant, $d\theta/dx = dT/dx$. Substituting Equation 3.63 into Equation 3.62, we then obtain

$$\frac{d^2\theta}{dx^2} - m^2\theta = 0 \tag{3.64}$$

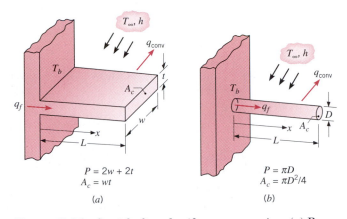

FIGURE 3.16 Straight fins of uniform cross section. (*a*) Rectangular fin. (*b*) Pin fin.

where

$$m^2 \equiv \frac{hP}{kA_c} \qquad (3.65)$$

Equation 3.64 is a linear, homogeneous, second-order differential equation with constant coefficients. Its general solution is of the form

$$\theta(x) = C_1 e^{mx} + C_2 e^{-mx} \qquad (3.66)$$

By substitution it may readily be verified that Equation 3.66 is indeed a solution to Equation 3.64.

To evaluate the constants C_1 and C_2 of Equation 3.66, it is necessary to specify appropriate boundary conditions. One such condition may be specified in terms of the temperature at the *base* of the fin ($x = 0$)

$$\theta(0) = T_b - T_\infty \equiv \theta_b \qquad (3.67)$$

The second condition, specified at the fin tip ($x = L$), may correspond to any one of four different physical situations.

The first condition, case A, considers convection heat transfer from the fin tip. Applying an energy balance to a control surface about this tip (Figure 3.17), we obtain

$$hA_c[T(L) - T_\infty] = -kA_c \left. \frac{dT}{dx} \right|_{x=L}$$

or

$$h\theta(L) = -k \left. \frac{d\theta}{dx} \right|_{x=L} \qquad (3.68)$$

That is, the rate at which energy is transferred to the fluid by convection from the tip must equal the rate at which energy reaches the tip by conduction through the fin. Substituting Equation 3.66 into Equations 3.67 and 3.68, we obtain, respectively,

$$\theta_b = C_1 + C_2 \qquad (3.69)$$

and

$$h(C_1 e^{mL} + C_2 e^{-mL}) = km(C_2 e^{-mL} - C_1 e^{mL})$$

Solving for C_1 and C_2, it may be shown, after some manipulation, that

$$\frac{\theta}{\theta_b} = \frac{\cosh m(L - x) + (h/mk) \sinh m(L - x)}{\cosh mL + (h/mk) \sinh mL} \tag{3.70}$$

The form of this temperature distribution is shown schematically in Figure 3.17. Note that the magnitude of the temperature gradient decreases with increasing x. This trend is a consequence of the reduction in the conduction heat transfer $q_x(x)$ with increasing x due to continuous convection losses from the fin surface.

We are also interested in the total heat transferred by the fin. From Figure 3.17 it is evident that the fin heat transfer rate q_f may be evaluated in two alternative ways, both of which involve use of the temperature distribution. The simpler procedure, and the one that we will use, involves applying Fourier's law at the fin base. That is,

$$q_f = q_b = -kA_c \left.\frac{dT}{dx}\right|_{x=0} = -kA_c \left.\frac{d\theta}{dx}\right|_{x=0} \tag{3.71}$$

Hence, knowing the temperature distribution, $\theta(x)$, q_f may be evaluated, giving

$$q_f = \sqrt{hPkA_c}\,\theta_b \frac{\sinh mL + (h/mk) \cosh mL}{\cosh mL + (h/mk) \sinh mL} \tag{3.72}$$

However, conservation of energy dictates that the rate at which heat is transferred by convection from the fin must equal the rate at which it is conducted

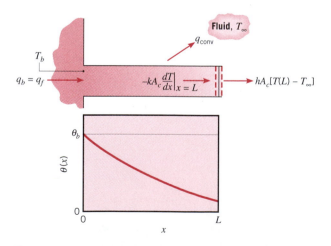

FIGURE 3.17   Conduction and convection in a fin of uniform cross section.

through the base of the fin. Accordingly, the alternative formulation for q_f is

$$q_f = \int_{A_f} h[T(x) - T_\infty]\, dA_s$$

$$q_f = \int_{A_f} h\theta(x)\, dA_s \tag{3.73}$$

where A_f is the *total*, including the tip, *fin surface area*. Substitution of Equation 3.70 into Equation 3.73 would yield Equation 3.72.

The second tip condition, case B, corresponds to the assumption that the convective heat loss from the fin tip is negligible, in which case the tip may be treated as adiabatic and

$$\left.\frac{d\theta}{dx}\right|_{x=L} = 0 \tag{3.74}$$

Substituting from Equation 3.66 and dividing by m, we then obtain

$$C_1 e^{mL} - C_2 e^{-mL} = 0$$

Using this expression with Equation 3.69 to solve for C_1 and C_2 and substituting the results into Equation 3.66, we obtain

$$\frac{\theta}{\theta_b} = \frac{\cosh m(L - x)}{\cosh mL} \tag{3.75}$$

Using this temperature distribution with Equation 3.71, the fin heat transfer rate is then

$$q_f = \sqrt{hPkA_c}\,\theta_b \tanh mL \tag{3.76}$$

In the same manner, we can obtain the fin temperature distribution and heat transfer rate for case C, where the temperature is prescribed at the fin tip. That is, the second boundary condition is $\theta(L) = \theta_L$, and the resulting expressions are of the form

$$\frac{\theta}{\theta_b} = \frac{(\theta_L/\theta_b)\sinh mx + \sinh m(L - x)}{\sinh mL} \tag{3.77}$$

$$q_f = \sqrt{hPkA_c}\,\theta_b \frac{\cosh mL - \theta_L/\theta_b}{\sinh mL} \tag{3.78}$$

The *very long fin*, case D, is an interesting extension of these results. In particular, as $L \to \infty$, $\theta_L \to 0$ and it is easily verified that

$$\frac{\theta}{\theta_b} = e^{-mx} \tag{3.79}$$

$$q_f = \sqrt{hPkA_c}\,\theta_b \tag{3.80}$$

The foregoing results are summarized in Table 3.4. A table of hyperbolic functions is provided in Appendix B.1.

TABLE 3.4 Temperature distribution and heat loss for fins of uniform cross section

Case	Tip Condition $(x = L)$	Temperature Distribution θ/θ_b	Fin Heat Transfer Rate q_f
A	Convection heat transfer: $h\theta(L) = -kd\theta/dx\|_{x=L}$	$\dfrac{\cosh m(L-x) + (h/mk)\sinh m(L-x)}{\cosh mL + (h/mk)\sinh mL}$ (3.70)	$M\dfrac{\sinh mL + (h/mk)\cosh mL}{\cosh mL + (h/mk)\sinh mL}$ (3.72)
B	Adiabatic $d\theta/dx\|_{x=L} = 0$	$\dfrac{\cosh m(L-x)}{\cosh mL}$ (3.75)	$M\tanh mL$ (3.76)
C	Prescribed temperature: $\theta(L) = \theta_L$	$\dfrac{(\theta_L/\theta_b)\sinh mx + \sinh m(L-x)}{\sinh mL}$ (3.77)	$M\dfrac{(\cosh mL - \theta_L/\theta_b)}{\sinh mL}$ (3.78)
D	Infinite fin $(L \to \infty)$: $\theta(L) = 0$	e^{-mx} (3.79)	M (3.80)

$$\theta \equiv T - T_\infty \qquad m^2 \equiv hP/kA_c$$
$$\theta_b = \theta(0) = T_b - T_\infty \qquad M \equiv \sqrt{hPkA_c}\,\theta_b$$

EXAMPLE 3.8

A very long rod 5 mm in diameter has one end maintained at 100°C. The surface of the rod is exposed to ambient air at 25°C with a convection heat transfer coefficient of 100 W/m² · K.

1. Determine the temperature distributions along rods constructed from pure copper, 2024 aluminum alloy, and type AISI 316 stainless steel. What are the corresponding heat losses from the rods?

2. Estimate how long the rods must be for the assumption of *infinite length* to yield an accurate estimate of the heat loss.

SOLUTION

Known: A long circular rod exposed to ambient air.

Find:

1. Temperature distribution and heat loss when rod is fabricated from copper, an aluminum alloy, or stainless steel.

2. How long rods must be to assume infinite length.

Schematic:

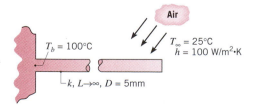

Assumptions:

1. Steady-state conditions.
2. One-dimensional conduction along the rod.
3. Constant properties.
4. Negligible radiation exchange with surroundings.
5. Uniform heat transfer coefficient.
6. Infinitely long rod.

Properties: Table A.1, copper [$T = (T_b + T_\infty)/2 = 62.5°C \approx 335$ K]: $k = 398$ W/m · K. Table A.1, 2024 aluminum (335 K): $k = 180$ W/m · K. Table A.1, stainless steel, AISI 316 (335 K): $k = 14$ W/m · K.

Analysis:

1. Subject to the assumption of an infinitely long fin, the temperature distributions are determined from Equation 3.79, which may be expressed as

$$T = T_\infty + (T_b - T_\infty)e^{-mx}$$

where $m = (hP/kA_c)^{1/2} = (4h/kD)^{1/2}$. Substituting for h and D, as well as for the thermal conductivities of copper, the aluminum alloy, and the stainless steel, respectively, the values of m are 14.2, 21.2, and 75.6 m^{-1}. The temperature distributions may then be computed and plotted as follows.

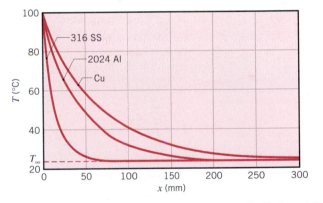

From these distributions, it is evident that there is little additional heat transfer associated with extending the length of the rod much beyond 50, 200, and 300 mm, respectively, for the stainless steel, the aluminum alloy, and the copper.

From Equation 3.80 the heat loss is

$$q_f = \sqrt{hPkA_c}\,\theta_b$$

Hence for copper,

$$q_f = \left[100 \text{ W/m}^2 \cdot \text{K} \times \pi \times 0.005 \text{ m} \right.$$

$$\left. \times 398 \text{ W/m} \cdot \text{K} \times \frac{\pi}{4} (0.005 \text{ m})^2 \right]^{1/2} (100 - 25)°C$$

$$= 8.3 \text{ W}$$ ◁

Similarly, for the aluminum alloy and stainless steel, respectively, the heat rates are $q_f = 5.6$ W and 1.6 W.

2. Since there is no heat loss from the tip of an infinitely long rod, an estimate of the validity of this approximation may be made by comparing Equations 3.76 and 3.80. To a satisfactory approximation, the expressions provide equivalent results if $\tanh mL \geq 0.99$ or $mL \geq 2.65$. Hence a rod may be assumed to be infinitely long if

$$L \geq L_\infty \equiv \frac{2.65}{m} = 2.65 \left(\frac{kA_c}{hP} \right)^{1/2}$$

For copper,

$$L_\infty = 2.65 \left[\frac{398 \text{ W/m} \cdot \text{K} \times (\pi/4)(0.005 \text{ m})^2}{100 \text{ W/m}^2 \cdot \text{K} \times \pi(0.005 \text{ m})} \right]^{1/2} = 0.19 \text{ m} \qquad \triangleleft$$

Results for the aluminum alloy and stainless steel are $L_\infty = 0.13$ m and $L_\infty = 0.04$ m, respectively.

Comments: The foregoing results suggest that the fin heat transfer rate may accurately be predicted from the infinite fin approximation if $mL \gtrsim 2.65$. However, if the infinite fin approximation is to accurately predict the temperature distribution $T(x)$, a larger value of mL would be required. This value may be inferred from Equation 3.79 and the requirement that the tip temperature be very close to the fluid temperature. Hence, if we require that $\theta(L)/\theta_b = \exp(-mL) < 0.01$, it follows that $mL > 4.6$, in which case $L_\infty \approx 0.33$, 0.23, and 0.07 m for the copper, aluminum alloy, and stainless steel, respectively. These results are consistent with the distributions plotted in part 1.

3.6.3   Fin Performance

Recall that fins are used to increase the heat transfer from a surface by increasing the effective surface area. However, the fin itself represents a conduction resistance to heat transfer from the original surface. For this reason, there is no assurance that the heat transfer rate will be increased through the use of fins. An assessment of this matter may be made by evaluating the *fin effectiveness* ε_f. It is defined as the *ratio of the fin heat transfer rate to the heat transfer rate that would exist without the fin.* Therefore

$$\varepsilon_f = \frac{q_f}{hA_{c,b}\theta_b} \tag{3.81}$$

where $A_{c,b}$ is the fin cross-sectional area at the base. In any rational design the value of ε_f should be as large as possible, and in general, the use of fins may rarely be justified unless $\varepsilon_f \gtrsim 2$.

Subject to any one of the four tip conditions that have been considered, the effectiveness for a fin of uniform cross section may be obtained by dividing the appropriate expression for q_f in Table 3.4 by $hA_{c,b}\theta_b$. Although the installation of fins will alter the surface convection coefficient, this effect is commonly neglected. Hence, assuming the convection coefficient of the finned surface to be equivalent to that of the unfinned base, it follows that, for the infinite fin approximation (case D), the result is

$$\varepsilon_f = \left(\frac{kP}{hA_c} \right)^{1/2} \tag{3.82}$$

Several important trends may be inferred from this result. Obviously, fin effectiveness is enhanced by the choice of a material of high thermal conductivity. Aluminum alloys and copper come to mind. However, although copper is superior from the standpoint of thermal conductivity, aluminum alloys are the more common choice because of additional benefits related to lower cost and weight. Fin effectiveness is also enhanced by increasing the ratio of the perimeter to the cross-sectional area. For this reason the use of *thin,* but closely spaced, fins is preferred, with the proviso that the fin gap not be reduced to a value for which flow between the fins is severely impeded, thereby reducing the convection coefficient.

Equation 3.82 also suggests that the use of fins can better be justified under conditions for which the convection coefficient h is small. Hence from Table 1.1 it is evident that the need for fins is stronger when the fluid is a gas rather than a liquid and particularly when the surface heat transfer is by *free* convection. If fins are to be used on a surface separating a gas and a liquid, they are generally placed on the gas side, which is the side of lower convection coefficient. A common example is the tubing in an automobile radiator. Fins are applied to the outer tube surface, over which there is flow of ambient air (small h), and not to the inner surface, through which there is flow of water (large h). Note that, if $\varepsilon_f > 2$ is used as a criterion to justify the implementation of fins, Equation 3.82 yields the requirement that $(kP/hA_c) > 4$.

Equation 3.82 provides an upper limit to ε_f, which is reached as L approaches infinity. However, it is certainly not necessary to use very long fins to achieve near maximum heat transfer enhancement. When an adiabatic tip condition is considered, Equation 3.76 and Table B.1 tell us that 98% of the maximum possible fin heat transfer rate is achieved for $mL = 2.3$. Hence it would make little sense to extend the fins beyond $L = 2.3/m$.

Fin performance may also be quantified in terms of a thermal resistance. Treating the difference between the base and fluid temperatures as the driving potential, a *fin resistance* may be defined as

$$R_{t,f} = \frac{\theta_b}{q_f} \tag{3.83}$$

This result is extremely useful, particularly when representing a finned surface by a thermal circuit. Note that, according to the fin tip condition, an appropriate expression for q_f may be obtained from Table 3.4.

Dividing Equation 3.83 into the expression for the thermal resistance due to convection at the exposed base,

$$R_{t,b} = \frac{1}{hA_{c,b}} \tag{3.84}$$

and substituting from Equation 3.81, it follows that

$$\varepsilon_f = \frac{R_{t,b}}{R_{t,f}} \tag{3.85}$$

Hence the fin effectiveness may be interpreted as a ratio of thermal resistances, and to increase ε_f it is necessary to reduce the conduction/convection resistance of the fin. If the fin is to enhance heat transfer, its resistance must not exceed that of the exposed base.

Another measure of fin thermal performance is provided by the *fin efficiency* η_f. The maximum driving potential for convection is the temperature difference between the base ($x = 0$) and the fluid, $\theta_b = T_b - T_\infty$. Hence the maximum rate at which a fin could dissipate energy is the rate that would exist *if* the entire fin surface were at the base temperature. However, since any fin is characterized by a finite conduction resistance, a temperature gradient must exist along the fin and the above condition is an idealization. A logical definition of fin efficiency is therefore

$$\eta_f \equiv \frac{q_f}{q_{\max}} = \frac{q_f}{hA_f\theta_b} \tag{3.86}$$

where A_f is the surface area of the fin. For a straight fin of uniform cross section and an adiabatic tip, Equations 3.76 and 3.86 yield

$$\eta_f = \frac{M \tanh mL}{hPL\theta_b} = \frac{\tanh mL}{mL} \tag{3.87}$$

Referring to Table B.1, this result tells us that η_f approaches its maximum and minimum values of 1 and 0, respectively, as L approaches 0 and ∞.

In lieu of the somewhat cumbersome expression for heat transfer from a straight rectangular fin with an active tip, Equation 3.72, it has been shown that approximate, yet accurate, predictions may be obtained by using the adiabatic tip result, Equation 3.76, with a corrected fin length of the form $L_c = L + (t/2)$ for a rectangular fin and $L_c = L + (D/4)$ for a pin fin [9]. The correction is based on assuming equivalence between heat transfer from the actual fin with tip convection and heat transfer from a longer, hypothetical fin with an adiabatic tip. Hence, with tip convection, the fin heat rate may be approximated as

$$q_f = M \tanh mL_c \tag{3.88}$$

and the corresponding efficiency as

$$\eta_f = \frac{\tanh mL_c}{mL_c} \tag{3.89}$$

Errors associated with the approximation are negligible if (ht/k) or $(hD/2k) \lesssim 0.0625$ [10].

If the width of a rectangular fin is much larger than its thickness, $w \gg t$, the perimeter may be approximated as $P = 2w$, and

$$mL_c = \left(\frac{hP}{kA_c}\right)^{1/2} L_c = \left(\frac{2h}{kt}\right)^{1/2} L_c$$

Multiplying numerator and denominator by $L_c^{1/2}$ and introducing a corrected fin

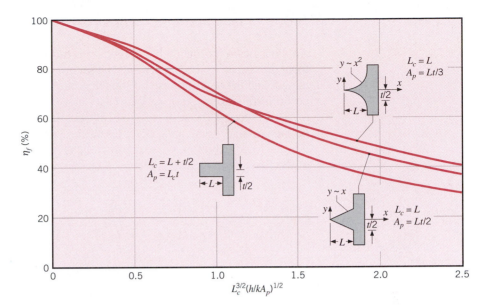

FIGURE 3.18 Efficiency of straight fins (rectangular, triangular, and parabolic profiles).

profile area, $A_p = L_c t$, it follows that

$$mL_c = \left(\frac{2h}{kA_p}\right)^{1/2} L_c^{3/2} \tag{3.90}$$

Hence, as shown in Figures 3.18 and 3.19, the efficiency of a rectangular fin with tip convection may be represented as a function of $L_c^{3/2}(h/kA_p)^{1/2}$.

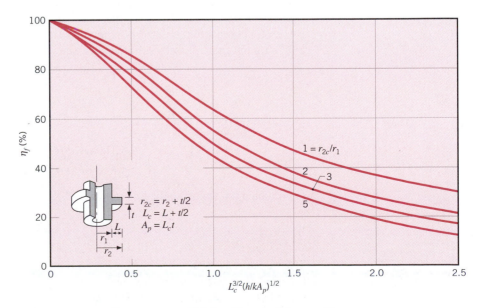

FIGURE 3.19 Efficiency of annular fins of rectangular profile.

3.6.4   Fins of Nonuniform Cross-Sectional Area

Analysis of fin thermal behavior becomes more complex if the fin is of nonuniform cross section. For such cases the second term of Equation 3.61 must be retained, and the solutions are no longer in the form of simple exponential or hyperbolic functions. As a special case, consider the annular fin shown in the inset of Figure 3.19. Although the fin thickness is uniform (t is independent of r), the cross-sectional area, $A_c = 2\pi r t$, varies with r. Replacing x by r in Equation 3.61 and expressing the surface area as $A_s = 2\pi(r^2 - r_1^2)$, the general form of the fin equation reduces to

$$\frac{d^2T}{dr^2} + \frac{1}{r}\frac{dT}{dr} - \frac{2h}{kt}(T - T_\infty) = 0$$

or, with $m^2 \equiv 2h/kt$ and $\theta \equiv T - T_\infty$,

$$\frac{d^2\theta}{dr^2} + \frac{1}{r}\frac{d\theta}{dr} - m^2\theta = 0$$

The foregoing expression is a *modified Bessel equation* of order zero, and its general solution is of the form

$$\theta(r) = C_1 I_0(mr) + C_2 K_0(mr)$$

where I_0 and K_0 are modified, zero-order Bessel functions of the first and second kinds, respectively. If the temperature at the base of the fin is prescribed, $\theta(r_1) = \theta_b$, and an adiabatic tip is presumed, $d\theta/dr|_{r_2} = 0$, C_1 and C_2 may be evaluated to yield a temperature distribution of the form

$$\frac{\theta}{\theta_b} = \frac{I_0(mr)K_1(mr_2) + K_0(mr)I_1(mr_2)}{I_0(mr_1)K_1(mr_2) + K_0(mr_1)I_1(mr_2)}$$

where $I_1(mr) = d[I_0(mr)]/d(mr)$ and $K_1(mr) = -d[K_0(mr)]/d(mr)$ are modified, first-order Bessel functions of the first and second kinds, respectively. The Bessel functions are tabulated in Appendix B.

With the fin heat transfer rate expressed as

$$q_f = -kA_{c,b}\frac{dT}{dr}\bigg|_{r=r_1} = -k(2\pi r_1 t)\frac{d\theta}{dr}\bigg|_{r=r_1}$$

it follows that

$$q_f = 2\pi k r_1 t \theta_b m \frac{K_1(mr_1)I_1(mr_2) - I_1(mr_1)K_1(mr_2)}{K_0(mr_1)I_1(mr_2) + I_0(mr_1)K_1(mr_2)}$$

from which the fin efficiency becomes

$$\eta_f = \frac{q_f}{h2\pi(r_2^2 - r_1^2)\theta_b} = \frac{2r_1}{m(r_2^2 - r_1^2)}\frac{K_1(mr_1)I_1(mr_2) - I_1(mr_1)K_1(mr_2)}{K_0(mr_1)I_1(mr_2) + I_0(mr_1)K_1(mr_2)} \tag{3.91}$$

This result may be applied for an active (convecting) tip, if the tip radius r_2 is replaced by a corrected radius of the form $r_{2c} = r_2 + (t/2)$. Results are represented graphically in Figure 3.19.

Knowledge of the thermal efficiency of a fin may be used to evaluate the fin resistance, where, from Equations 3.83 and 3.86, it follows that

$$R_{t,f} = \frac{1}{hA_f\eta_f} \tag{3.92}$$

Expressions for the efficiency and surface area of several common fin geometries are summarized in Table 3.5. Although results for the fins of uniform thickness or diameter were obtained by assuming an adiabatic tip, the effects of convection may be treated by using a corrected length (Equations 3.89 and 3.95) or radius (Equation 3.91). The triangular and parabolic fins are of nonuniform thickness, which reduces to zero at the fin tip.

TABLE 3.5 Efficiency of common fin shapes

Straight Fins

Rectangular[a]

$A_f = 2wL_c$

$L_c = L + (t/2)$

$$\eta_f = \frac{\tanh mL_c}{mL_c} \tag{3.89}$$

Triangular[a]

$A_f = 2w[L^2 + (t/2)^2]^{1/2}$

$$\eta_f = \frac{1}{mL}\frac{I_1(2mL)}{I_0(2mL)} \tag{3.93}$$

Parabolic[a]

$A_f = w[C_1L + (L^2/t)\ln(t/L + C_1)]$

$C_1 = [1 + (t/L)^2]^{1/2}$

$y = (t/2)(1 - x/L)^2$

$$\eta_f = \frac{2}{[4(mL)^2 + 1]^{1/2} + 1} \tag{3.94}$$

Circular Fin

Rectangular[a]

$A_f = 2\pi(r_{2c}^2 - r_1^2)$

$r_{2c} = r_2 + (t/2)$

$$\eta_f = C_2\frac{K_1(mr_1)I_1(mr_{2c}) - I_1(mr_1)K_1(mr_{2c})}{I_0(mr_1)K_1(mr_{2c}) + K_0(mr_1)I_1(mr_{2c})} \tag{3.91}$$

$$C_2 = \frac{(2r_1/m)}{(r_{2c}^2 - r_1^2)}$$

Pin Fins

Rectangular[b]

$A_f = \pi DL_c$

$L_c = L + (D/4)$

$$\eta_f = \frac{\tanh mL_c}{mL_c} \tag{3.95}$$

Continued on next page.

TABLE 3.5   *Continued*

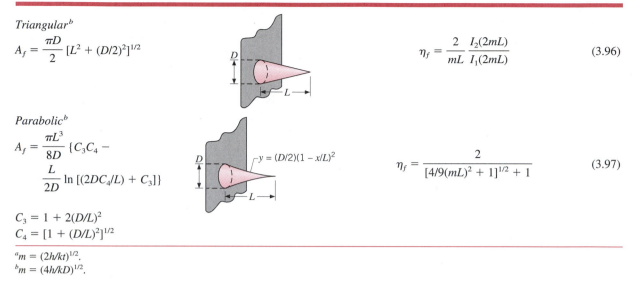

Triangular[b]

$$A_f = \frac{\pi D}{2} [L^2 + (D/2)^2]^{1/2}$$

$$\eta_f = \frac{2}{mL} \frac{I_2(2mL)}{I_1(2mL)} \qquad (3.96)$$

Parabolic[b]

$$A_f = \frac{\pi L^3}{8D} \{C_3 C_4 - \frac{L}{2D} \ln [(2DC_4/L) + C_3]\}$$

$$y = (D/2)(1 - x/L)^2$$

$$\eta_f = \frac{2}{[4/9(mL)^2 + 1]^{1/2} + 1} \qquad (3.97)$$

$$C_3 = 1 + 2(D/L)^2$$
$$C_4 = [1 + (D/L)^2]^{1/2}$$

[a] $m = (2h/kt)^{1/2}$.
[b] $m = (4h/kD)^{1/2}$.

A straight *triangular* fin is attractive because, for equivalent heat transfer, it requires much less volume (fin material) than a rectangular profile. In this regard, heat dissipation per unit volume, $(q/V)_f$, is largest for a *parabolic* profile. However, since $(q/V)_f$ for the parabolic profile is only slightly larger than that for a triangular profile, its use can rarely be justified in view of its larger manufacturing costs. The *annular* fin of rectangular profile is commonly used to enhance heat transfer to or from circular tubes.

3.6.5 Overall Surface Efficiency

In contrast to the fin efficiency η_f, which characterizes the performance of a single fin, the *overall surface efficiency* η_o characterizes an *array* of fins and the base surface to which they are attached. Representative arrays are shown in Figure 3.20, where S designates the fin pitch. In each case the overall efficiency is defined as

$$\eta_o = \frac{q_t}{q_{\max}} = \frac{q_t}{hA_t\theta_b} \qquad (3.98)$$

where q_t is the total heat rate from the surface area A_t associated with both the fins and the exposed portion of the base (often termed the *prime* surface). If there are N fins in the array, each of surface area A_f, and the area of the prime surface is designated as A_b, the total surface area is

$$A_t = NA_f + A_b \qquad (3.99)$$

The maximum possible heat rate would result if the entire fin surface, as well as the exposed base, were maintained at T_b.

The total rate of heat transfer by convection from the fins and the prime (unfinned) surface may be expressed as

$$q_t = N\eta_f hA_f\theta_b + hA_b\theta_b \qquad (3.100)$$

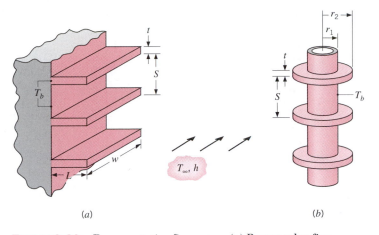

FIGURE 3.20 Representative fin arrays. (*a*) Rectangular fins. (*b*) Annular fins.

where the convection coefficient h is assumed to be equivalent for the finned and prime surfaces and η_f is the efficiency of a single fin. Hence

$$q_t = h[N\eta_f A_f + (A_t - NA_f)]\theta_b = hA_t \left[1 - \frac{NA_f}{A_t}(1 - \eta_f)\right]\theta_b \quad (3.101)$$

Substituting Equation (3.101) into (3.98), it follows that

$$\eta_o = 1 - \frac{NA_f}{A_t}(1 - \eta_f) \quad (3.102)$$

From knowledge of η_o, Equation 3.98 may be used to calculate the total heat rate for a fin array.

Recalling the definition of the fin thermal resistance, Equation 3.83, Equation 3.98 may be used to infer an expression for the thermal resistance of a fin array. That is,

$$R_{t,o} = \frac{\theta_b}{q_t} = \frac{1}{\eta_o hA_t} \quad (3.103)$$

where $R_{t,o}$ is an effective resistance that accounts for parallel heat flow paths by conduction/convection in the fins and by convection from the prime surface. Figure 3.21*a* illustrates the thermal circuits corresponding to the parallel paths and their representation in terms of an effective resistance.

If fins are machined as an integral part of the wall from which they extend, there is no contact resistance at their base. However, more commonly, fins are manufactured separately and are attached to the wall by a metallurgical or adhesive joint. Alternatively, the attachment may involve a *press fit,* for which the fins are forced into slots machined on the wall material. In such cases (Figure 3.21*b*), there is a thermal contact resistance, $R_{t,c}$, which may adversely influence overall thermal performance. An effective circuit resistance may again be obtained, where, with the contact resistance,

$$R_{t,o(c)} = \frac{\theta_b}{q_t} = \frac{1}{\eta_{o(c)} hA_t} \quad (3.104)$$

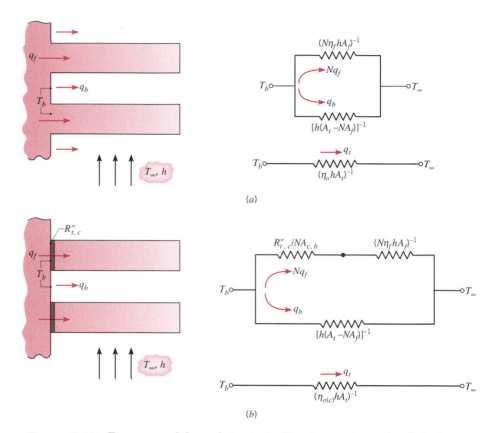

FIGURE 3.21 Fin array and thermal circuit. (*a*) Fins that are integral with the base. (*b*) Fins that are attached to the base.

It is readily shown that the corresponding overall surface efficiency is

$$\eta_{o(c)} = 1 - \frac{NA_f}{A_t}\left(1 - \frac{\eta_f}{C_1}\right) \qquad (3.105a)$$

where

$$C_1 = 1 + \eta_f h A_f (R''_{t,c}/A_{c,b}) \qquad (3.105b)$$

In manufacturing, care must be taken to render $R_{t,c} \ll R_{t,f}$.

EXAMPLE 3.9

The engine cylinder of a motorcycle is constructed of 2024-T6 aluminum alloy and is of height $H = 0.15$ m and outside diameter $D = 50$ mm. Under typical operating conditions the outer surface of the cylinder is at a temperature of 500 K and is exposed to ambient air at 300 K, with a convection coefficient of 50 W/m² · K. Annular fins are integrally cast with the cylinder to increase heat transfer to the surroundings. Consider five such fins, which are of thickness $t = 6$ mm, length $L = 20$ mm, and equally spaced. What is the increase in heat transfer due to use of the fins?

SOLUTION

Known: Operating conditions of a finned motorcycle cylinder.

Find: Increase in heat transfer associated with using fins.

Schematic:

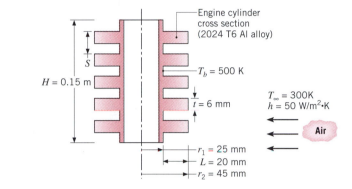

Assumptions:

1. Steady-state conditions.
2. One-dimensional radial conduction in fins.
3. Constant properties.
4. Negligible radiation exchange with surroundings.
5. Uniform convection coefficient over outer surface (with or without fins).

Properties: Table A.1, 2024-T6 aluminum ($T = 400$ K): $k = 186$ W/m · K.

Analysis: With the fins in place, the heat transfer rate is given by Equation 3.101

$$q_t = hA_t \left[1 - \frac{NA_f}{A_t} (1 - \eta_f) \right] \theta_b$$

where $A_f = 2\pi (r_{2c}^2 - r_1^2) = 2\pi [(0.048 \text{ m})^2 - (0.025 \text{ m})^2] = 0.0105 \text{ m}^2$ and, from Equation 3.99, $A_t = NA_f + 2\pi r_1(H - Nt) = 0.0527 \text{ m}^2 + 2\pi(0.025$ m)$[0.15 \text{ m} - 0.03 \text{ m}] = 0.0716 \text{ m}^2$. With $r_{2c}/r_1 = 1.92$, $L_c = 0.023$ m, $A_p = 1.380 \times 10^{-4} \text{ m}^2$, we obtain $L_c^{3/2}(h/kA_p)^{1/2} = 0.15$. Hence, from Figure 3.19, the fin efficiency is $\eta_f \approx 0.95$. With the fins, the total heat transfer rate is then

$$q_t = 50 \text{ W/m}^2 \cdot \text{K} \times 0.0716 \text{ m}^2 \left[1 - \frac{0.0527 \text{ m}^2}{0.0716 \text{ m}^2} (0.05) \right] 200 \text{ K} = 690 \text{ W}$$

Without the fins, the convection heat transfer rate would be

$$q_{\text{wo}} = h(2\pi r_1 H)\theta_b = 50 \text{ W/m}^2 \cdot \text{K}(2\pi \times 0.025 \text{ m} \times 0.15 \text{ m})200 \text{ K} = 236 \text{ W}$$

Hence

$$\Delta q = q_t - q_{\text{wo}} = 454 \text{ W} \qquad \triangleleft$$

Comments: Although the fins significantly increase heat transfer from the cylinder, considerable improvement could still be obtained by increasing the number of fins. We assess this possibility by computing q_t as a function of N, first by fixing the fin thickness at $t = 6$ mm and increasing the number of fins by reducing the spacing between fins. Prescribing a fin clearance of 2 mm at each end of the array and a minimum fin gap of 4 mm, the maximum allowable number of fins is $N = H/S = 0.15$ m/$(0.004 + 0.006)$ m $= 15$. The parametric calculations yield the following variation of q_t with N:

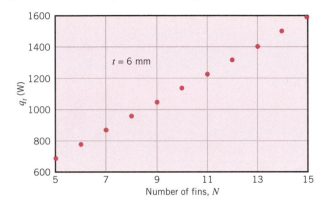

The number of fins could also be increased by reducing the fin thickness. If the fin gap is fixed at $(S - t) = 4$ mm and manufacturing constraints dictate a minimum allowable fin thickness of 2 mm, up to $N = 25$ fins may be accommodated. In this case the parametric calculations yield

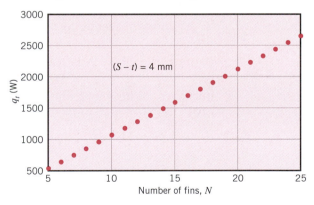

The foregoing calculations are based on the assumption that h is not affected by a reduction in the fin gap. The assumption is reasonable as long as there is no interaction between boundary layers that develop on the opposing surfaces of adjoining fins. Note that, since $NA_f \gg 2\pi r_1(H - Nt)$ for the prescribed conditions, q_t increases nearly linearly with increasing N.

EXAMPLE 3.10

Heat transfer from a transistor may be enhanced by inserting it in an aluminum sleeve ($k = 200$ W/m · K) having 12 integrally machined longitudinal fins on its outer surface. The transistor radius and height are $r_1 = 2$ mm and $H =$

6 mm, respectively, while the fins are of length $L = r_3 - r_2 = 10$ mm and uniform thickness $t = 0.7$ mm. The thickness of the sleeve base is $r_2 - r_1 = 1$ mm, and the contact resistance of the sleeve–transistor interface is $R''_{t,c} = 10^{-3}$ m$^2 \cdot$ K/W. Air at $T_\infty = 20°$C flows over the fin surface, providing an approximately uniform convection coefficient of $h = 25$ W/m$^2 \cdot$ K.

1. Assuming one-dimensional transfer in the radial direction, sketch the equivalent thermal circuit for heat transfer from the transistor case ($r = r_1$) to the air. Clearly label each resistance.
2. Evaluate each of the resistances in the foregoing circuit. If the temperature of the transistor case is $T_1 = 80°$C, what is the rate of heat transfer from the sleeve?

SOLUTION

Known: Dimensions of finned aluminum sleeve inserted over a transistor. Contact resistance between sleeve and transistor. Surface convection conditions and temperature of transistor case.

Find:
1. Equivalent thermal circuit.
2. Rate of heat transfer from sleeve.

Schematic:

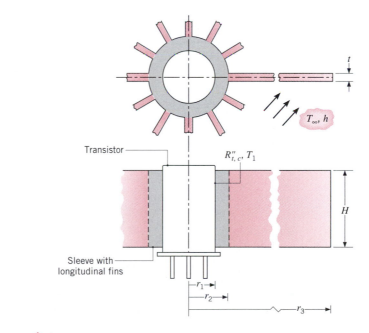

Assumptions:
1. Steady-state conditions.
2. Negligible heat transfer from the top and bottom surfaces of the transistor.
3. One-dimensional radial conduction.

4. Constant properties.

5. Negligible radiation.

Analysis:

1. The circuit must account for the contact resistance, conduction in the sleeve, convection from the exposed base, and conduction/convection from the fins.

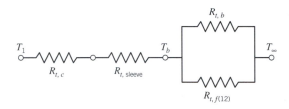

2. Thermal resistances for the contact joint and sleeve are

$$R_{t,c} = \frac{R''_{t,c}}{2\pi r_1 H} = \frac{10^{-3}\ \text{m}^2 \cdot \text{K/W}}{2\pi(0.002\ \text{m})(0.006\ \text{m})} = 13.3\ \text{K/W}$$

$$R_{t,\text{sleeve}} = \frac{\ln(r_2/r_1)}{2\pi k H} = \frac{\ln(3/2)}{2\pi(200\ \text{W/m} \cdot \text{K})(0.006\ \text{m})} = 0.054\ \text{K/W}$$

For a single fin, $R_{t,f} = \theta_b/q_f$, where from Table 3.4,

$$q_f = (hPkA_c)^{1/2}\theta_b\, \frac{\sinh mL + (h/mk)\cosh mL}{\cosh mL + (h/mk)\sinh mL}$$

With $P = 2(H + t) = 13.4\ \text{mm} = 0.0134\ \text{m}$ and $A_c = t \times H = 4.2 \times 10^{-6}$ m^2,

$$m = \frac{hP}{kA_c} = \left(\frac{25\ \text{W/m}^2 \cdot \text{K} \times 0.0134\ \text{m}}{200\ \text{W/m} \cdot \text{K} \times 4.2 \times 10^{-6}\ \text{m}^2}\right)^{1/2} = 20.0\ \text{m}^{-1}$$

$$mL = 20\ \text{m}^{-1} \times 0.01\ \text{m} = 0.20$$

$$\frac{h}{mk} = \frac{25\ \text{W/m}^2 \cdot \text{K}}{20\ \text{m}^{-1} \times 200\ \text{W/m} \cdot \text{K}} = 0.00625$$

and

$$(hPkA_c)^{1/2} = (25\ \text{W/m}^2 \cdot \text{K} \times 0.0134\ \text{m} \times 200\ \text{W/m} \cdot \text{K}$$

$$\times\ 4.2 \times 10^{-6}\ \text{m}^2)^{1/2} = 0.0168\ \text{W/K}$$

use of Table B.1 yields, for a *single fin,*

$$R_{t,f} = \frac{1.020 + 0.00625 \times 0.201}{0.0168\ \text{W/K}\ (0.201 + 0.00625 \times 1.020)} = 293\ \text{K/W}$$

Hence, for 12 fins,

$$R_{t,f(12)} = \frac{R_{t,f}}{12} = 24.4\ \text{K/W}$$

For the exposed base,

$$R_{t,b} = \frac{1}{h(2\pi r_2 - 12t)H}$$

$$= \frac{1}{25 \text{ W/m}^2 \cdot \text{K} (2\pi \times 0.003 - 12 \times 0.0007) \text{ m} \times 0.006 \text{ m}}$$

$$R_{t,b} = 638 \text{ K/W}$$

With

$$R_{\text{equiv}} = [(24.4)^{-1} + (638)^{-1}]^{-1} = 23.5 \text{ K/W}$$

it follows that

$$R_{\text{tot}} = (13.3 + 0.054 + 23.5) \text{ K/W} = 36.9 \text{ K/W}$$

and

$$q_t = \frac{T_1 - T_\infty}{R_{\text{tot}}} = \frac{(80 - 20)°\text{C}}{36.9 \text{ K/W}} = 1.63 \text{ W}$$

Comments:

1. Without the finned sleeve, the convection resistance of the transistor case is $R_{\text{tran}} = (2\pi r_1 Hh)^{-1} = 531$ K/W. Hence there is considerable advantage to using the fins.

2. If an adiabatic fin tip is assumed, tanh $mL = 0.197$ and $R_{t,f} = 302$. Hence the fin resistance is within 3% of that obtained for the actual convecting tip.

3. With $\eta_f = (hA_f R_{t,f})^{-1} = 0.988$, Equation 3.102 yields $\eta_o = 0.988$, from which it follows that $R_{t,o} = (\eta_o hA_t)^{-1} = 23.5$ K/W. This result is, of course, identical to that obtained in the foregoing determination of R_{equiv}.

4. The prescribed fin design and operating conditions are by no means optimized. If it became necessary to dissipate more than 1.63 W, while keeping the base temperature at 80°C, what measures would you take to improve the system's thermal performance? You may, for example, want to consider the effects of doubling h, halving $R''_{t,c}$, increasing L, and/or increasing N.

3.7
Summary

Despite the inherent mathematical simplicity, one-dimensional, steady-state heat transfer occurs in numerous engineering applications. Although one-dimensional, steady-state conditions may not apply exactly, the assumptions may often be made to obtain results of reasonable accuracy. You should therefore be thoroughly familiar with the means by which such problems are treated. In particular, you should be comfortable with the use of equivalent thermal circuits and with the expressions for the conduction resistances that pertain to each of the three common geometries. You should also be familiar with how the heat equation and Fourier's law may be used to obtain temperature distributions and the corresponding fluxes. The implications of an internally distributed source of energy should also be clearly understood. Finally, you should appreciate the important role that extended surfaces can play in the design of thermal systems and should have the facility to effect design and performance calculations for such surfaces.

References

1. Fried, E., "Thermal Conduction Contribution to Heat Transfer at Contacts," in R. P. Tye, Ed., *Thermal Conductivity,* Vol. 2, Academic Press, London, 1969.

2. Eid, J. C., and V. W. Antonetti, "Small Scale Thermal Contact Resistance of Aluminum against Silicon," in C. L. Tien, V. P. Carey, and J. K. Ferrel, Eds., *Heat Transfer—1986,* Vol. 2, Hemisphere, New York, 1986, pp. 659–664.

3. Snaith, B., P. W. O'Callaghan, and S. D. Probert, "Interstitial Materials for Controlling Thermal Conductances across Pressed Metallic Contacts," *Appl. Energy,* **16,** 175, 1984.

4. Yovanovich, M. M., "Theory and Application of Constriction and Spreading Resistance Concepts for Microelectronic Thermal Management," Presented at the International Symposium on Cooling Technology for Electronic Equipment, Honolulu, 1987.

5. Peterson, G. P., and L. S. Fletcher, "Thermal Contact Resistance of Silicon Chip Bonding Materials," Proceedings of the International Symposium on Cooling Technology for Electronic Equipment, Honolulu, 1987, pp. 438–448.

6. Yovanovich, M. M., and M. Tuarze, "Experimental Evidence of Thermal Resistance at Soldered Joints," *AIAA J. Spacecraft Rockets,* **6,** 1013, 1969.

7. Madhusudana, C. V., and L. S. Fletcher, "Contact Heat Transfer—The Last Decade," *AIAA J.,* **24,** 510, 1986.

8. Yovanovich, M. M., "Recent Developments in Thermal Contact, Gap and Joint Conductance Theories and Experiment," in C. L. Tien, V. P. Carey, and J. K. Ferrel, Eds., *Heat Transfer—1986,* Vol. 1, Hemisphere, New York, 1986, pp. 35–45.

9. Harper, D. R., and W. B. Brown, "Mathematical Equations for Heat Conduction in the Fins of Air Cooled Engines," NACA Report No. 158, 1922.

10. Schneider, P. J., *Conduction Heat Transfer,* Addison-Wesley, Reading, MA, 1955.

Problems

Plane Wall

3.1 Consider the plane wall of Figure 3.1, separating hot and cold fluids at temperatures $T_{\infty,1}$ and $T_{\infty,2}$, respectively. Using surface energy balances as boundary conditions at $x = 0$ and $x = L$ (see Equation 2.27), obtain the temperature distribution within the wall and the heat flux in terms of $T_{\infty,1}$, $T_{\infty,2}$, h_1, h_2, k, and L.

3.2 The rear window of an automobile is defogged by passing warm air over its inner surface.

(a) If the warm air is at $T_{\infty,i} = 40°C$ and the corresponding convection coefficient is $h_i = 30$ W/m² · K, what are the inner and outer surface temperatures of 4-mm-thick window glass, if the outside ambient air temperature is $T_{\infty,o} = -10°C$ and the associated convection coefficient is $h_o = 65$ W/m² · K?

(b) In practice $T_{\infty,o}$ and h_o vary according to weather conditions and car speed. For values of $h_o = 2, 65,$ and 100 W/m² · K, compute and plot the inner and outer surface temperatures as a function of $T_{\infty,o}$ for $-30 \le T_{\infty,o} \le 0°C$.

3.3 The rear window of an automobile is defogged by attaching a thin, transparent, film-type heating element to its inner surface. By electrically heating this element, a uniform heat flux may be established at the inner surface.

(a) For 4-mm-thick window glass, determine the electrical power required per unit window area to maintain an inner surface temperature of 15°C when the interior air temperature and convection coefficient are $T_{\infty,i} = 25°C$ and $h_i = 10 \text{ W/m}^2 \cdot \text{K}$, while the exterior (ambient) air temperature and convection coefficient are $T_{\infty,o} = -10°C$ and $h_o = 65 \text{ W/m}^2 \cdot \text{K}$.

(b) In practice $T_{\infty,o}$ and h_o vary according to weather conditions and car speed. For values of $h_o = 2, 20, 65$, and $100 \text{ W/m}^2 \cdot \text{K}$, determine and plot the electrical power requirement as a function of $T_{\infty,o}$ for $-30 \leq T_{\infty,o} \leq 0°C$. From your results, what can you conclude about the need for heater operation at low values of h_o? How is this conclusion affected by the value of $T_{\infty,o}$? If $h \propto V^n$, where V is the vehicle speed and n is a positive exponent, how does the vehicle speed affect the need for heater operation?

3.4 In a manufacturing process, a transparent film is being bonded to a substrate as shown in the sketch. To cure the bond at a temperature T_0, a radiant source is used to provide a heat flux q_0'' (W/m²), all of which is absorbed at the bonded surface. The back of the substrate is maintained at T_1 while the free surface of the film is exposed to air at T_∞ and a convection heat transfer coefficient h.

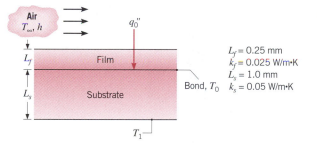

$L_f = 0.25$ mm
$k_f = 0.025$ W/m·K
$L_s = 1.0$ mm
$k_s = 0.05$ W/m·K

(a) Show the thermal circuit representing the steady-state heat transfer situation. Be sure to label *all* elements, nodes, and heat rates. Leave in symbolic form.

(b) Assume the following conditions: $T_\infty = 20°C$, $h = 50 \text{ W/m}^2 \cdot \text{K}$, and $T_1 = 30°C$. Calculate the heat flux q_0'' that is required to maintain the bonded surface at $T_0 = 60°C$.

(c) Compute and plot the required heat flux as a function of the film thickness for $0 \leq L_f \leq 1$ mm.

(d) If the film is not transparent and all of the radiant heat flux is absorbed at its upper surface, determine the heat flux required to achieve bonding. Plot your results as a function of L_f for $0 \leq L_f \leq 1$ mm.

3.5 Both copper and stainless steel (AISI 304) are being considered as wall material for a liquid-cooled rocket nozzle. The cooled exterior of the wall is maintained at 150°C, while combustion gases within the nozzle are at 2750°C. The gas-side heat transfer coefficient is $h_i = 2 \times 10^4 \text{ W/m}^2 \cdot \text{K}$, and the radius of the nozzle is much larger than the wall thickness. Thermal limitations dictate that the temperature of the copper and the steel not exceed 540°C and 980°C, respectively. What is the maximum wall thickness that could be employed for each of the two materials? If the nozzle is constructed with the maximum wall thickness, which material would be preferred?

3.6 A technique for measuring convection heat transfer coefficients involves adhering one surface of a thin metallic foil to an insulating material and exposing the other surface to the fluid flow conditions of interest.

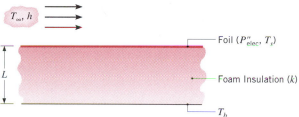

By passing an electric current through the foil, heat is dissipated uniformly within the foil and the corresponding flux, P_{elec}'', may be inferred from related voltage and current measurements. If the insulation thickness L and thermal conductivity k are known and the fluid, foil, and insulation temperatures (T_∞, T_s, T_b) are measured, the convection coefficient may be determined. Consider conditions for which $T_\infty = T_b = 25°C$, $P_{elec}'' = 2000 \text{ W/m}^2$, $L = 10$ mm, and $k = 0.040 \text{ W/m} \cdot \text{K}$.

(a) With water flow over the surface, the foil temperature measurement yields $T_s = 27°C$. Determine the convection coefficient. What error would be incurred by assuming all of the dissipated power to be transferred to the water by convection?

(b) If, instead, air flows over the surface and the temperature measurement yields $T_s = 125°C$, what is the convection coefficient? The foil has an emissivity of 0.15 and is exposed to large surroundings at 25°C. What error would be incurred by assuming all of the dissipated power to be transferred to the air by convection?

(c) Typically, heat flux gages are operated at a fixed temperature (T_s), in which case the power dissipation provides a direct measure of the convection coefficient. For $T_s = 27°C$, plot P''_{elec} as a function of h_o for $10 \leq h_o \leq 1000$ W/m² · K. What effect does h_o have on the error associated with neglecting conduction through the insulation?

3.7 The *wind chill*, which is experienced on a cold, windy day, is related to increased heat transfer from exposed human skin to the surrounding atmosphere. Consider a layer of fatty tissue that is 3 mm thick and whose interior surface is maintained at a temperature of 36°C. On a calm day the convection heat transfer coefficient at the outer surface is 25 W/m² · K, but with 30 km/h winds it reaches 65 W/m² · K. In both cases the ambient air temperature is −15°C.

(a) What is the ratio of the heat loss per unit area from the skin for the calm day to that for the windy day?

(b) What will be the skin outer surface temperature for the calm day? For the windy day?

(c) What temperature would the air have to assume on the calm day to produce the same heat loss occurring with the air temperature at −15°C on the windy day?

3.8 Consider the surface-mount type transistor illustrated in Problem 1.51. Construct the thermal circuit, write an expression for the case temperature T_c, and evaluate T_c for the two situations where stagnant air and a conductive paste fill the gap.

3.9 A 1-m-long steel plate ($k = 50$ W/m · K) is well insulated on its sides, while the top surface is at 100°C and the bottom surface is convectively cooled by a fluid at 20°C. Under steady-state conditions with no generation, a thermocouple at the midpoint of the plate reveals a temperature of 85°C.

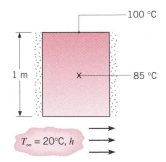

What is the value of the convection heat transfer coefficient at the bottom surface?

3.10 A thermopane window consists of two pieces of glass 7 mm thick that enclose an air space 7 mm thick. The window separates room air at 20°C from outside ambient air at −10°C. The convection coefficient associated with the inner (room-side) surface is 10 W/m² · K.

(a) If the convection coefficient associated with the outer (ambient) air is $h_o = 80$ W/m² · K, what is the heat loss through a window that is 0.8 m long by 0.5 m wide? Neglect radiation, and assume the air enclosed between the panes to be stagnant.

(b) Compute and plot the effect of h_o on the heat loss for $10 \leq h_o \leq 100$ W/m² · K. Repeat this calculation for a triple-pane construction in which a third pane and a second air space of equivalent thickness are added.

3.11 The wall of a passive solar collector consists of a phase change material (PCM) of thickness L enclosed within two structural supporting surfaces.

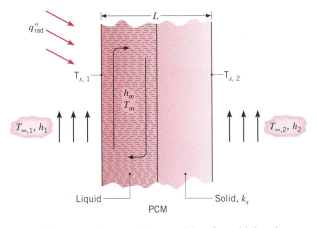

Assume a steady-state condition for which solar radiation absorption at one surface maintains its temperature ($T_{s,1}$) above the melting temperature of the PCM. The liquid and solid portion of the PCM are divided by a sharp vertical interface. The liquid has a core temperature of T_m and is characterized by a buoyancy-driven recirculating flow that maintains the same convection coefficient (h_m) at its interfaces with the surface (s, 1) and the solid. Consider conditions for which the net radiation flux is $q''_{rad} = 1000$ W/m², the ambient temperatures and convection coefficients are $T_{\infty,1} = T_{\infty,2} = 20°C$ and $h_1 = h_2 = 20$ W/m² · K, the temperature and convection coefficient of the liquid PCM are $T_m = 50°C$ and $h_m = 10$ W/m² · K, and the thermal conductivity of the solid PCM is $k_s = 0.5$ W/m · K. Evaluate the surface temperature, $T_{s,1}$. If the total thickness of the PCM is $L = 0.10$ m, what is the thickness of the liquid layer? Compute the surface temperature $T_{s,2}$.

3.12 The wall of a building is a composite consisting of a 100-mm layer of common brick, a 100-mm layer of glass fiber (paper faced, 28 kg/m³), a 10-mm layer of gypsum plaster (vermiculite), and a 6-mm layer of pine panel. If the inside convection coefficient is 10 W/m² · K and the outside convection coefficient is 70 W/m² · K, what are the total resistance and the overall coefficient for heat transfer?

3.13 The composite wall of an oven consists of three materials, two of which are of known thermal conductivity, $k_A = 20$ W/m · K and $k_C = 50$ W/m · K, and known thickness, $L_A = 0.30$ m and $L_C = 0.15$ m. The third material, B, which is sandwiched between materials A and C, is of known thickness, $L_B = 0.15$ m, but unknown thermal conductivity k_B.

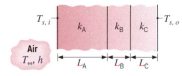

Under steady-state operating conditions, measurements reveal an outer surface temperature of $T_{s,o} = 20°C$, an inner surface temperature of $T_{s,i} = 600°C$, and an oven air temperature of $T_\infty = 800°C$. The inside convection coefficient h is known to be 25 W/m² · K. What is the value of k_B?

3.14 The exterior walls of a building are a composite consisting of 10-mm-thick plaster board, 50-mm-thick urethane foam, and a 10-mm-thick soft wood. On a typical winter day the outside and inside air temperatures are $-15°C$ and $20°C$, respectively, with outer and inner convection coefficients of 15 W/m² · K and 5 W/m² · K, respectively.

(a) What is the heating load for a 1-m² section of the wall?

(b) What is the heating load if the composite wall is replaced by a 3-mm-thick glass pane window?

(c) What is the heating load if the composite wall is replaced by a double-glazed window consisting of two 3-mm-thick glass panes separated by a 5-mm-thick stagnant air gap?

3.15 A house has a composite wall of wood, fiberglass insulation, and plaster board, as indicated in the sketch. On a cold winter day the convection heat transfer coefficients are $h_o = 60$ W/m² · K and $h_i = 30$ W/m² · K. The total wall surface area is 350 m².

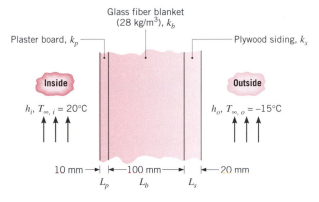

(a) Determine a symbolic expression for the total thermal resistance of the wall, including inside and outside convection effects for the prescribed conditions.

(b) Determine the total heat loss through the wall.

(c) If the wind were blowing violently, raising h_o to 300 W/m² · K, determine the percentage increase in the heat loss.

(d) What is the controlling resistance that determines the amount of heat flow through the wall?

3.16 Consider the composite wall of Problem 3.15 under conditions for which the inside air is still characterized by $T_{\infty,i} = 20°C$ and $h_i = 30$ W/m² · K. However, use the more realistic conditions for which the outside air is characterized by a diurnal (time) varying temperature of the form

$$T_{\infty,o}(K) = 273 + 5 \sin\left(\frac{2\pi}{24}t\right) \qquad 0 \le t \le 12 \text{ h}$$

$$T_{\infty,o}(K) = 273 + 11 \sin\left(\frac{2\pi}{24}t\right) \qquad 12 \le t \le 24 \text{ h}$$

with $h_o = 60$ W/m² · K. Assuming quasisteady conditions for which changes in energy storage within the wall may be neglected, estimate the daily heat loss through the wall if its total surface area is 200 m².

3.17 Consider a composite wall that includes an 8-mm-thick hardwood siding, 40-mm by 130-mm hardwood studs on 0.65-m centers with glass fiber insulation (paper faced, 28 kg/m³), and a 12-mm layer of gypsum (vermiculite) wall board.

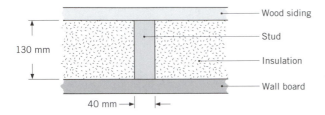

130 mm

40 mm →| |←

Wood siding
Stud
Insulation
Wall board

What is the thermal resistance associated with a wall that is 2.5 m high by 6.5 m wide (having 10 studs, each 2.5 m high)?

3.18 The thermal characteristics of a small, dormitory refrigerator are determined by performing two separate experiments, each with the door closed and the refrigerator placed in ambient air at $T_\infty = 25°C$. In one case, an electric heater is suspended in the refrigerator cavity, while the refrigerator is unplugged. With the heater dissipating 20 W, a steady-state temperature of 90°C is recorded within the cavity. With the heater removed and the refrigerator now in operation, the second experiment involves maintaining a steady-state cavity temperature of 5°C for a fixed time interval and recording the electrical energy required to operate the refrigerator. In such an experiment for which steady operation is maintained over a 12-hour period, the input electrical energy is 125,000 J. Determine the refrigerator's coefficient of performance (COP).

3.19 In the design of buildings, energy conservation requirements dictate that the exterior surface area, A_s, be minimized. This requirement implies that, for a desired floor space, there may be optimum values associated with the number of floors and horizontal dimensions of the building. Consider a design for which the floor space, A_f, and the vertical distance between floors, H_f, are prescribed.

(a) If the building has a square cross section of width W on a side, obtain an expression for the value of W that would minimize heat loss to the surroundings. Heat loss may be assumed to occur from the four vertical side walls and from a flat roof. Express your result in terms of A_f and H_f.

(b) If $A_f = 32,768$ m^2 and $H_f = 4$ m, for what values of W and N_f (the number of floors) is the heat loss minimized? If the average overall heat transfer coefficient is $U = 1$ W/m$^2 \cdot$ K and the difference between the inside and ambient air temperatures is 25°C, what is the corresponding heat loss? What is the percent reduction in heat loss compared with a building for $N_f = 2$?

Contact Resistance

3.20 A composite wall separates combustion gases at 2600°C from a liquid coolant at 100°C, with gas- and liquid-side convection coefficients of 50 and 1000 W/m$^2 \cdot$ K. The wall is composed of a 10-mm-thick layer of beryllium oxide on the gas side and a 20-mm-thick slab of stainless steel (AISI 304) on the liquid side. The contact resistance between the oxide and the steel is 0.05 m$^2 \cdot$ K/W. What is the heat loss per unit surface area of the composite? Sketch the temperature distribution from the gas to the liquid.

3.21 Two stainless steel plates 10 mm thick are subjected to a contact pressure of 1 bar under vacuum conditions for which there is an overall temperature drop of 100°C across the plates. What is the heat flux through the plates? What is the temperature drop across the contact plane?

3.22 Consider a plane composite wall that is composed of two materials of thermal conductivities $k_A = 0.1$ W/m $\cdot$ K and $k_B = 0.04$ W/m $\cdot$ K and thicknesses $L_A = 10$ mm and $L_B = 20$ mm. The contact resistance at the interface between the two materials is known to be 0.30 m$^2 \cdot$ K/W. Material A adjoins a fluid at 200°C for which $h = 10$ W/m$^2 \cdot$ K, and material B adjoins a fluid at 40°C for which $h = 20$ W/m$^2 \cdot$ K.

(a) What is the rate of heat transfer through a wall that is 2 m high by 2.5 m wide?

(b) Sketch the temperature distribution.

3.23 The performance of gas turbine engines may be improved by increasing the tolerance of the turbine blades to hot gases emerging from the combustor. One approach to achieving high operating temperatures involves application of a *thermal barrier coating* (TBC) to the exterior surface of a blade, while passing cooling air through the blade. Typically, the blade is made from a high-temperature superalloy, such as Inconel ($k \approx 25$ W/m $\cdot$ K), while a ceramic, such as zirconia ($k \approx 1.3$ W/m $\cdot$ K), is used as a TBC.

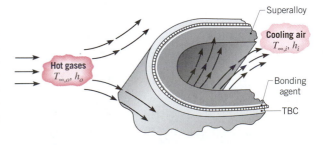

Superalloy
Cooling air
$T_{\infty,i}, h_i$
Hot gases
$T_{\infty,o}, h_o$
Bonding agent
TBC

Consider conditions for which hot gases at $T_{\infty,o} = 1700$ K and cooling air at $T_{\infty,i} = 400$ K provide outer and inner surface convection coefficients of $h_o = 1000$ W/m² · K and $h_i = 500$ W/m² · K, respectively. If a 0.5-mm-thick zirconia TBC is attached to a 5-mm-thick Inconel blade wall by means of a metallic bonding agent, which provides an interfacial thermal resistance of $R''_{t,c} = 10^{-4}$ m² · K/W, can the Inconel be maintained at a temperature that is below its maximum allowable value of 1250 K? Radiation effects may be neglected, and the turbine blade may be approximated as a plane wall. Plot the temperature distribution with and without the TBC. Are there any limits to the thickness of the TBC?

3.24 A silicon chip is encapsulated such that, under steady-state conditions, all of the power it dissipates is transferred by convection to a fluid stream for which $h = 1000$ W/m² · K and $T_{\infty} = 25°C$. The chip is separated from the fluid by a 2-mm-thick aluminum cover plate, and the contact resistance of the chip/aluminum interface is 0.5×10^{-4} m² · K/W.

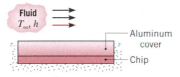

If the chip surface area is 100 mm² and its maximum allowable temperature is 85°C, what is the maximum allowable power dissipation in the chip?

3.25 Approximately 10^6 discrete electrical components can be placed on a single integrated circuit (chip), with electrical heat dissipation as high as 30,000 W/m². The chip, which is very thin, is exposed to a dielectric liquid at its outer surface, with $h_o = 1000$ W/m² · K and $T_{\infty,o} = 20°C$, and is joined to a circuit board at its inner surface. The thermal contact resistance between the chip and the board is 10^{-4} m² · K/W, and the board thickness and thermal conductivity are $L_b = 5$ mm and $k_b = 1$ W/m · K, respectively. The other surface of the board is exposed to ambient air for which $h_i = 40$ W/m² · K and $T_{\infty,i} = 20°C$.

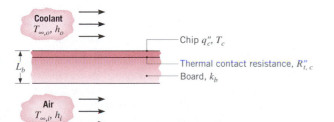

(a) Sketch the equivalent thermal circuit corresponding to steady-state conditions. In variable form, label appropriate resistances, temperatures, and heat fluxes.

(b) Under steady-state conditions for which the chip heat dissipation is $q''_c = 30,000$ W/m², what is the chip temperature?

(c) The maximum allowable heat flux, $q''_{c,m}$, is determined by the constraint that the chip temperature must not exceed 85°C. Determine $q''_{c,m}$ for the foregoing conditions. If air is used in lieu of the dielectric liquid, the convection coefficient is reduced by approximately an order of magnitude. What is the value of $q''_{c,m}$ for $h_o = 100$ W/m² · K? With air cooling, can significant improvements be realized by using an aluminum oxide circuit board and/or by using a conductive paste at the chip/board interface for which $R''_{t,c} = 10^{-5}$ m² · K/W?

Alternative Conduction Analysis

3.26 The diagram shows a conical section fabricated from pure aluminum. It is of circular cross section having diameter $D = ax^{1/2}$, where $a = 0.5$ m$^{1/2}$. The small end is located at $x_1 = 25$ mm and the large end at $x_2 = 125$ mm. The end temperatures are $T_1 = 600$ K and $T_2 = 400$ K, while the lateral surface is well insulated.

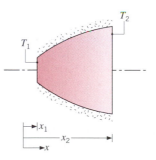

(a) Derive an expression for the temperature distribution $T(x)$ in symbolic form, assuming one-dimensional conditions. Sketch the temperature distribution.

(b) Calculate the heat rate q_x.

3.27 A truncated solid cone is of circular cross section, and its diameter is related to the axial coordinate by an expression of the form $D = ax^{3/2}$, where $a = 1.0$ m$^{-1/2}$.

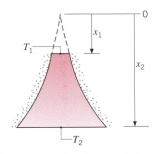

The sides are well insulated, while the top surface of the cone at x_1 is maintained at T_1 and the bottom surface at x_2 is maintained at T_2.

(a) Obtain an expression for the temperature distribution $T(x)$.

(b) What is the rate of heat transfer across the cone if it is constructed of pure aluminum with $x_1 = 0.075$ m, $T_1 = 100°C$, $x_2 = 0.225$ m, and $T_2 = 20°C$.

3.28 From Figure 2.5 it is evident that, over a wide temperature range, the temperature dependence of the thermal conductivity of many solids may be approximated by a linear expression of the form $k = k_o + aT$, where k_o is a positive constant and a is a coefficient that may be positive or negative. Obtain an expression for the heat flux across a plane wall whose inner and outer surfaces are maintained at T_0 and T_1, respectively. Sketch the forms of the temperature distribution corresponding to $a > 0$, $a = 0$, and $a < 0$.

3.29 Consider a tube wall of inner and outer radii r_i and r_o, whose temperatures are maintained at T_i and T_o, respectively. The thermal conductivity of the cylinder is temperature dependent and may be represented by an expression of the form $k = k_o(1 + aT)$, where k_o and a are constants. Obtain an expression for the heat transfer per unit length of the tube. What is the thermal resistance of the tube wall?

3.30 Measurements show that steady-state conduction through a plane wall without heat generation produced a convex temperature distribution such that the midpoint temperature was ΔT_o higher than expected for a linear temperature distribution.

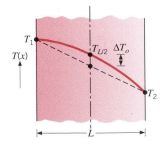

Assuming that the thermal conductivity has a linear dependence on temperature, $k = k_o(1 + \alpha T)$, where α is a constant, develop a relationship to evaluate α in terms of ΔT_o, T_1, and T_2.

3.31 Use the alternative conduction analysis method to derive the expression for the thermal resistance of a hollow cylinder of thermal conductivity k, inner and outer radii r_i and r_o, respectively, and length L.

Cylindrical Wall

3.32 A steam pipe of 0.12-m outside diameter is insulated with a layer of calcium silicate.

(a) If the insulation is 20 mm thick and its inner and outer surfaces are maintained at $T_{s,1} = 800$ K and $T_{s,2} = 490$ K, respectively, what is the heat loss per unit length (q') of the pipe?

(b) We wish to explore the effect of insulation thickness on the heat loss q' and outer surface temperature $T_{s,2}$, with the inner surface temperature fixed at $T_{s,1} = 800$ K. The outer surface is exposed to an airflow ($T_\infty = 25°C$) that maintains a convection coefficient of $h = 25$ W/m² · K and to large surroundings for which $T_{sur} = T_\infty = 25°C$. The surface emissivity of calcium silicate is approximately 0.8. Compute and plot the temperature distribution in the insulation as a function of the dimensionless radial coordinate, $(r - r_1)/(r_2 - r_1)$, where $r_1 = 0.06$ m and r_2 is a variable (0.06 < $r_2 \leq 0.20$ m). Compute and plot the heat loss as a function of the insulation thickness for $0 \leq (r_2 - r_1) \leq 0.14$ m.

3.33 Consider the water heater described in Problem 1.29. We now wish to determine the energy needed to compensate for heat losses incurred while the water is stored at the prescribed temperature of 55°C. The cylindrical storage tank (with flat ends) has a capacity of 100 gallons, and foamed urethane is used to insulate the side and end walls from ambient air at an annual average temperature of 20°C. The resistance to heat transfer is dominated by conduction in the insulation and by free convection in the air, for which $h \approx 2$ W/m² · K. If electric resistance heating is used to compensate for the losses and the cost of electric power is \$0.08/kWh, specify tank and insulation dimensions for which the annual cost associated with the heat losses is less than \$50.

3.34 A thin electrical heater is wrapped around the outer surface of a long cylindrical tube whose inner surface is maintained at a temperature of 5°C. The tube wall has inner and outer radii of 25 and 75 mm, respectively, and a thermal conductivity of 10 W/m · K. The thermal contact resistance between the heater and the outer surface of the tube (per unit length of the tube) is $R'_{t,c} = 0.01$ m · K/W. The outer surface of the heater is exposed to a fluid with $T_\infty = -10°C$ and a convection coefficient of $h = 100$ W/m² · K. Determine the heater power per unit length of tube required to maintain the heater at $T_o = 25°C$.

3.35 In the foregoing problem, the electrical power required to maintain the heater at $T_o = 25°C$ depends on the thermal conductivity of the wall material k, the thermal contact resistance $R'_{t,c}$, and the convection coefficient h. Compute and plot the separate effect of changes in k ($1 \le k \le 200$ W/m · K), $R'_{t,c}$ ($0 \le R'_{t,c} \le 0.1$ m · K/W), and h ($10 \le h \le 1000$ W/m² · K) on the total heater power requirement, as well as the rate of heat transfer to the inner surface of the tube and to the fluid.

3.36 Urethane ($k = 0.026$ W/m · K) is used to insulate the sidewall and the top and bottom of a cylindrical hot water tank. The insulation is 40 mm thick and is sandwiched between sheet metal of thin-wall construction. The height and inside diameter of the tank are 2 m and 0.80 m, respectively, and the tank is in ambient air for which $T_\infty = 10°C$ and $h = 10$ W/m² · K. If the hot water maintains the inner surface at 55°C and energy costs amount to $0.15/kWh, what is the daily cost to maintain the water in storage?

3.37 A thin electrical heater is inserted between a long circular rod and a concentric tube with inner and outer radii of 20 and 40 mm. The rod (A) has a thermal conductivity of $k_A = 0.15$ W/m · K, while the tube (B) has a thermal conductivity of $k_B = 1.5$ W/m · K and its outer surface is subjected to convection with a fluid of temperature $T_\infty = -15°C$ and heat transfer coefficient 50 W/m² · K. The thermal contact resistance between the cylinder surfaces and the heater is negligible.

(a) Determine the electrical power per unit length of the cylinders (W/m) that is required to maintain the outer surface of cylinder B at 5°C.

(b) What is the temperature at the center of cylinder A?

3.38 A long cylindrical rod of 100-mm radius consists of a nuclear reacting material ($k = 0.5$ W/m · K) generating 24,000 W/m³ uniformly throughout its volume. This rod is encapsulated within a tube having an outer radius of 200 mm and a thermal conductivity of 4 W/m · K. The outer surface is surrounded by a fluid at 100°C, and the convection

coefficient between the surface and the fluid is 20 W/m² · K. Find the temperatures at the interface between the two cylinders and at the outer surface.

3.39 A special coating, which is applied to the inner surface of a plastic tube, is cured by placing a cylindrical radiation heat source within the tube. The space between the tube and the source is evacuated, and the source delivers a uniform heat flux q''_1, which is absorbed at the inner surface of the tube. The outer surface of the tube is maintained at a uniform temperature, $T_{s,2}$.

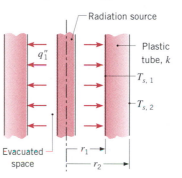

Develop an expression for the temperature distribution $T(r)$ in the tube wall in terms of q''_1, $T_{s,2}$, r_1, r_2, and k. If the inner and outer tube radii are $r_1 = 25$ mm and $r_2 = 38$ mm, what is the power required per unit length of the radiation source to maintain the inner surface at $T_{s,1} = 150°C$, while the outer surface is fixed at $T_{s,2} = 25°C$? The tube wall thermal conductivity is $k = 10$ W/m · K.

3.40 Consider a long, hollow cylinder of thermal conductivity k with inner and outer radii of r_i and r_o, respectively. The inner surface temperature is maintained at T_i while the outer surface experiences a uniform heat flux q''_o.

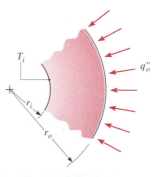

(a) Beginning with the appropriate form of the heat diffusion equation, derive an expression for the temperature distribution, $T(r)$, in terms of r_i, r_o, k, T_i, and q''_o.

(b) Sketch the temperature distribution on T–r coordinates.

(c) Write an expression for the heat rate per unit length of the cylinder at the inner surface, $q_r'(r_i)$, in terms of q_o'' and cylinder geometrical parameters.

3.41 The evaporator section of a refrigeration unit consists of thin-walled, 10-mm-diameter tubes through which refrigerant passes at a temperature of $-18°C$. Air is cooled as it flows over the tubes, maintaining a surface convection coefficient of 100 W/m² · K, and is subsequently routed to the refrigerator compartment.

(a) For the foregoing conditions and an air temperature of $-3°C$, what is the rate at which heat is extracted from the air per unit tube length?

(b) If the refrigerator's defrost unit malfunctions, frost will slowly accumulate on the outer tube surface. Assess the effect of frost formation on the cooling capacity of a tube for frost layer thicknesses in the range $0 \leq \delta \leq 4$ mm. Frost may be assumed to have a thermal conductivity of 0.4 W/m · K.

(c) The refrigerator is disconnected after the defrost unit malfunctions and a 2-mm-thick layer of frost has formed. If the tubes are in ambient air for which $T_\infty = 20°C$ and natural convection maintains a convection coefficient of 2 W/m² · K, how long will it take for the frost to melt? The frost may be assumed to have a mass density of 700 kg/m³ and a latent heat of fusion of 334 kJ/kg.

3.42 A composite cylindrical wall is composed of two materials of thermal conductivity k_A and k_B, which are separated by a very thin, electric resistance heater for which interfacial contact resistances are negligible.

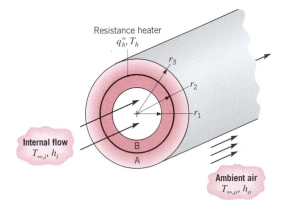

Liquid pumped through the tube is at a temperature $T_{\infty,i}$ and provides a convection coefficient h_i at the inner surface of the composite. The outer surface is exposed to ambient air, which is at $T_{\infty,o}$ and provides a convection coefficient of h_o. Under steady-state conditions, a uniform heat flux of q_h'' is dissipated by the heater.

(a) Sketch the equivalent thermal circuit of the system and express all resistances in terms of relevant variables.

(b) Obtain an expression that may be used to determine the heater temperature, T_h.

(c) Obtain an expression for the ratio of heat flows to the outer and inner fluids, q_o'/q_i'. How might the variables of the problem be adjusted to minimize this ratio?

3.43 An electric wire having a radius of $r_i = 5$ mm and a resistance per unit length of 10^{-4} Ω/m is coated with a plastic insulation of thermal conductivity $k = 0.20$ W/m · K. The insulation is exposed to ambient air for which $T_\infty = 300$ K and $h = 10$ W/m² · K. If the insulation has a maximum allowable temperature of 450 K, what is the maximum possible current that may be passed by the wire?

3.44 An electrical current of 700 A flows through a stainless steel cable having a diameter of 5 mm and an electrical resistance of 6×10^{-4} Ω/m (i.e., per meter of cable length). The cable is in an environment having a temperature of 30°C, and the total coefficient associated with convection and radiation between the cable and the environment is approximately 25 W/m² · K.

(a) If the cable is bare, what is its surface temperature?

(b) If a very thin coating of electrical insulation is applied to the cable, with a contact resistance of 0.02 m² · K/W, what are the insulation and cable surface temperatures?

(c) There is some concern about the ability of the insulation to withstand elevated temperatures. What thickness of this insulation ($k = 0.5$ W/m · K) will yield the lowest value of the maximum insulation temperature? What is the value of the maximum temperature when the thickness is used?

3.45 A 0.20-m diameter, thin-walled steel pipe is used to transport saturated steam at a pressure of 20 bars in a room for which the air temperature is 25°C and the convection heat transfer coefficient at the outer surface of the pipe is 20 W/m² · K.

(a) What is the heat loss per unit length from the bare pipe (no insulation)? Estimate the heat loss per unit length if a 50-mm-thick layer of insulation (magnesia, 85%) is added. The steel and magnesia may each be assumed to have an emissivity of 0.8, and the steam-side convection resistance may be neglected.

(b) The costs associated with generating the steam and installing the insulation are known to be $4/10^9$ J and $100/m of pipe length, respectively. If the steam line is to operate 7500 h/yr, how many years are needed to pay back the initial investment in insulation?

3.46 Steam at a temperature of 250°C flows through a steel pipe (AISI 1010) of 60-mm inside diameter and 75-mm outside diameter. The convection coefficient between the steam and the inner surface of the pipe is 500 W/m² · K, while that between the outer surface of the pipe and the surroundings is 25 W/m² · K. The pipe emissivity is 0.8, and the temperature of the air and the surroundings is 20°C. What is the heat loss per unit length of pipe?

3.47 We wish to determine the effect of adding a layer of magnesia insulation to the steam pipe of the foregoing problem. The convection coefficient at the outer surface of the insulation may be assumed to remain at 25 W/m² · K, and the emissivity is $\varepsilon = 0.8$. Determine and plot the heat loss per unit length of pipe and the outer surface temperature as a function of insulation thickness. If the cost of generating the steam is $4/10^9$ J and the steam line operates 7000 h/yr, recommend an insulation thickness and determine the corresponding annual savings in energy costs. Plot the temperature distribution in the insulation for the recommended thickness.

3.48 An uninsulated, thin-walled pipe of 100-mm diameter is used to transport water to equipment that operates outdoors and uses the water as a coolant. During particularly harsh winter conditions, the pipe wall achieves a temperature of −15°C and a cylindrical layer of ice forms on the inner surface of the wall. If the mean water temperature is 3°C and a convection coefficient of 2000 W/m² · K is maintained at the inner surface of the ice, which is at 0°C, what is the thickness of the ice layer?

3.49 Steam flowing through a long, thin-walled pipe maintains the pipe wall at a uniform temperature of 500 K. The pipe is covered with an insulation blanket comprised of two different materials, A and B.

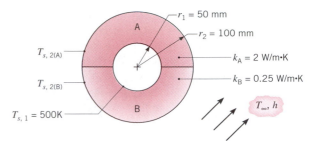

The interface between the two materials may be assumed to have an infinite contact resistance, and the entire outer surface is exposed to air for which $T_\infty = 300$ K and $h = 25$ W/m² · K.

(a) Sketch the thermal circuit of the system. Label (using the above symbols) all pertinent nodes and resistances.

(b) For the prescribed conditions, what is the total heat loss from the pipe? What are the outer surface temperatures $T_{s, 2(A)}$ and $T_{s, 2(B)}$?

3.50 A Bakelite coating is to be used with a 10-mm-diameter conducting rod, whose surface is maintained at 200°C by passage of an electrical current. The rod is in a fluid at 25°C, and the convection coefficient is 140 W/m² · K. What is the critical radius associated with the coating? What is the heat transfer rate per unit length for the bare rod and for the rod with a coating of Bakelite that corresponds to the critical radius? How much Bakelite should be added to reduce the heat transfer associated with the bare rod by 25%?

Spherical Wall

3.51 A storage tank consists of a cylindrical section that has a length and inner diameter of $L = 2$ m and $D_i = 1$ m, respectively, and two hemispherical end sections. The tank is constructed from 20-mm-thick glass (Pyrex) and is exposed to ambient air for which the temperature is 300 K and the convection coefficient is 10 W/m² · K. The tank is used to store heated oil, which maintains the inner surface at a temperature of 400 K. Determine the electrical power that must be supplied to a heater submerged in the oil if the prescribed conditions are to be maintained. Radiation effects may be neglected, and the Pyrex may be assumed to have a thermal conductivity of 1.4 W/m · K.

3.52 Consider the liquid oxygen storage system and the laboratory environmental conditions of Problem 1.35. To reduce oxygen loss due to vaporization, an insulating layer should be applied to the outer surface of the container. Consider using a laminated aluminum foil/glass mat insulation, for which the thermal conductivity and surface emissivity are $k = 0.00016$ W/m · K and $\varepsilon = 0.20$, respectively.

(a) If the container is covered with a 10-mm-thick layer of insulation, what is the percentage reduction in oxygen loss relative to the uncovered container?

(b) Compute and plot the oxygen evaporation rate (kg/s) as a function of the insulation thickness t for $0 \leq t \leq 50$ mm.

3.53 In Example 3.4, an expression was derived for the critical insulation radius of an insulated, cylindrical tube. Derive the expression that would be appropriate for an insulated sphere.

3.54 A hollow aluminum sphere, with an electrical heater in the center, is used in tests to determine the thermal conductivity of insulating materials. The inner and outer radii of the sphere are 0.15 and 0.18 m, respectively, and testing is done under steady-state conditions with the inner surface of the aluminum maintained at 250°C. In a particular test, a spherical shell of insulation is cast on the outer surface of the sphere to a thickness of 0.12 m. The system is in a room for which the air temperature is 20°C and the convection coefficient at the outer surface of the insulation is 30 W/m² · K. If 80 W are dissipated by the heater under steady-state conditions, what is the thermal conductivity of the insulation?

3.55 A spherical tank for storing liquid oxygen on the space shuttle is to be made from stainless steel of 0.80-m outer diameter and 5-mm wall thickness. The boiling point and latent heat of fusion of liquid oxygen are 90 K and 213 kJ/kg, respectively. The tank is to be installed in a large compartment whose temperature is to be maintained at 240 K. Design a thermal insulation system that will maintain oxygen losses due to boiling below 1 kg/day.

3.56 A spherical, cryosurgical probe may be imbedded in diseased tissue for the purpose of freezing, and thereby destroying, the tissue. Consider a probe of 3-mm diameter whose surface is maintained at −30°C when imbedded in tissue that is at 37°C. A spherical layer of frozen tissue forms around the probe, with a temperature of 0°C existing at the phase front (interface) between the frozen and normal tissue. If the thermal conductivity of frozen tissue is approximately 1.5 W/m · K and heat transfer at the phase front may be characterized by an effective convection coefficient of 50 W/m² · K, what is the thickness of the layer of frozen tissue?

3.57 A composite spherical shell of inner radius $r_1 = 0.25$ m is constructed from lead of outer radius $r_2 = 0.30$ m and AISI 302 stainless steel of outer radius $r_3 = 0.31$ m. The cavity is filled with radioactive wastes that generate heat at a rate of $\dot{q} = 5 \times 10^5$ W/m³. It is proposed to submerge the container in oceanic waters that are at a temperature of $T_\infty = 10$°C and provide a uniform convection coefficient of $h = 500$ W/m² · K at the outer surface of the container. Are there any problems associated with this proposal?

3.58 As an alternative to storing radioactive materials in oceanic waters, it is proposed that the system of Problem 3.57 be placed in a large tank for which the flow of water, and hence the convection coefficient h, can be controlled. Compute and plot the maximum temperature of the lead, $T(r_1)$, as a function of h for $100 \leq h \leq 1000$ W/m² · K. If the temperature of the lead is not to exceed 500 K, what is the minimum allowable value of h? To improve system reliability, it is desirable to increase the thickness of the stainless steel shell. For $h = 300$, 500, and 1000 W/m² · K, compute and plot the maximum lead temperature as a function of shell thickness for $r_3 \geq 0.30$ m. What are the corresponding values of the maximum allowable thickness?

3.59 The energy transferred from the anterior chamber of the eye through the cornea varies considerably depending on whether a contact lens is worn. Treat the eye as a spherical system and assume the system to be at steady state. The convection coefficient h_o is unchanged with and without the contact lens in place. The cornea and the lens cover one-third of the spherical surface area.

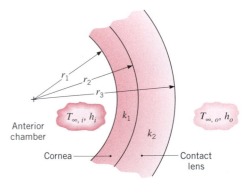

Values of the parameters representing this situation are as follows:

$r_1 = 10.2$ mm $r_2 = 12.7$ mm

$r_3 = 16.5$ mm

$T_{\infty, i} = 37°C$ $T_{\infty, o} = 21°C$

$k_1 = 0.35 \text{ W/m} \cdot \text{K}$ $k_2 = 0.80 \text{ W/m} \cdot \text{K}$

$h_i = 12 \text{ W/m}^2 \cdot \text{K}$ $h_o = 6 \text{ W/m}^2 \cdot \text{K}$

(a) Construct the thermal circuits, labeling all potentials and flows for the systems excluding the contact lens and including the contact lens. Write resistance elements in terms of appropriate parameters.

(b) Determine the heat loss from the anterior chamber with and without the contact lens in place.

(c) Discuss the implication of your results.

3.60 The outer surface of a hollow sphere of radius r_2 is subjected to a uniform heat flux q_2''. The inner surface at r_1 is held at a constant temperature $T_{s, 1}$.

(a) Develop an expression for the temperature distribution $T(r)$ in the sphere wall in terms of q_2'', $T_{s, 1}$, r_1, r_2, and the thermal conductivity of the wall material k.

(b) If the inner and outer tube radii are $r_1 = 50$ mm and $r_2 = 100$ mm, what heat flux q_2'' is required to maintain the outer surface at $T_{s, 2} = 50°C$, while the inner surface is at $T_{s, 1} = 20°C$? The thermal conductivity of the wall material is $k = 10 \text{ W/m} \cdot \text{K}$.

3.61 A spherical shell of inner and outer radii r_i and r_o, respectively, is filled with a heat-generating material that provides for a uniform volumetric generation rate (W/m³) of $\dot{q}$. The outer surface of the shell is exposed to a fluid having a temperature T_∞ and a convection coefficient h. Obtain an expression for the steady-state temperature distribution $T(r)$ in the shell, expressing your result in terms of r_i, r_o, $\dot{q}$, h, T_∞, and the thermal conductivity k of the shell material.

3.62 A transistor, which may be approximated as a hemispherical heat source of radius $r_o = 0.1$ mm, is embedded in a large silicon substrate ($k = 125$ W/m · K) and dissipates heat at a rate q. All boundaries of the silicon are maintained at an ambient temperature of $T_\infty = 27°C$, except for a plane surface that is well insulated.

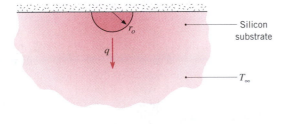

Obtain a general expression for the substrate temperature distribution and evaluate the surface temperature of the heat source for $q = 4$ W.

3.63 One modality for destroying malignant tissue involves imbedding a small spherical heat source of radius r_o within the tissue and maintaining local temperatures above a critical value T_c for an extended period. Tissue that is well removed from the source may be assumed to remain at normal body temperature ($T_b = 37°C$). Obtain a general expression for the radial temperature distribution in the tissue under steady-state conditions for which heat is dissipated at a rate q. If $r_o = 0.5$ mm, what heat rate must be supplied to maintain a tissue temperature of $T \geq T_c = 42°C$ in the domain $0.5 \leq r \leq 5$ mm? The tissue thermal conductivity is approximately 0.5 W/m · K.

Conduction with Thermal Energy Generation

3.64 Consider cylindrical and spherical shells with inner and outer surfaces at r_1 and r_2 maintained at uniform temperatures $T_{s, 1}$ and $T_{s, 2}$, respectively. If there is uniform heat generation within the shells, obtain expressions for the steady-state, one-dimensional radial distributions of the temperature, heat flux, and heat rate. Contrast your results with those summarized in Appendix C.

3.65 The steady-state temperature distribution in a composite plane wall of three different materials, each of constant thermal conductivity, is shown below.

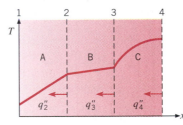

(a) Comment on the relative magnitudes of q_2'' and q_3'' and of q_3'' and q_4''.

(b) Comment on the relative magnitudes of k_A and k_B and of k_B and k_C.

(c) Sketch the heat flux as a function of x.

3.66 A plane wall of thickness 0.1 m and thermal conductivity 25 W/m · K having uniform volumetric heat generation of 0.3 MW/m³ is insulated on one side, while the other side is exposed to a fluid at 92°C. The convection heat transfer coefficient between the wall and the fluid is 500 W/m² · K. Determine the maximum temperature in the wall.

3.67 Consider one-dimensional conduction in a plane composite wall. The outer surfaces are exposed to a fluid at 25°C and a convection heat transfer coefficient of 1000 W/m² · K. The middle wall B experiences uniform heat generation $\dot{q}_B$, while there is no generation in walls A and C. The temperatures at the interfaces are $T_1 = 261°C$ and $T_2 = 211°C$.

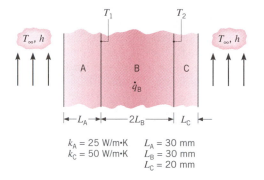

$k_A = 25$ W/m·K $L_A = 30$ mm
$k_C = 50$ W/m·K $L_B = 30$ mm
                   $L_C = 20$ mm

(a) Assuming negligible contact resistance at the interfaces, determine the volumetric heat generation $\dot{q}_B$ and the thermal conductivity k_B.

(b) Plot the temperature distribution, showing its important features.

(c) Consider conditions corresponding to a *loss of coolant* at the exposed surface of material A ($h = 0$). Determine T_1 and T_2 and plot the temperature distribution throughout the system.

3.68 Consider the composite plane wall of Problem 3.67 subjected to the same convection conditions. The middle wall has a thermal conductivity of $k_B = 15$ W/m · K and experiences a uniform heat generation of $\dot{q}_B = 4 \times 10^6$ W/m³, while the outer walls have no generation.

(a) Neglecting interfacial contact resistances, determine T_1 and T_2, as well as the heat fluxes through walls A and C.

(b) Consider conditions for which contact resistances of 0.0025 and 0.001 m² · K/W exist at the A/B and B/C interfaces, respectively. Determine T_1 and T_2 and plot the temperature distribution.

3.69 When passing an electrical current I, a copper bus bar of rectangular cross section (6 mm × 150 mm) experiences uniform heat generation at a rate $\dot{q}$ (W/m³) given by $\dot{q} = aI^2$, where $a = 0.015$ W/m³ · A². If the bar is in ambient air with $h = 5$ W/m² · K and its maximum temperature must not exceed that of the air by more than 30°C, what is the allowable current capacity for the bus bar?

3.70 A semiconductor material of thermal conductivity $k = 2$ W/m · K and electrical resistivity $\rho_e = 2 \times$

10^{-5} Ω · m is used to fabricate a cylindrical rod 10 mm in diameter and 40 mm long. The longitudinal surface of the rod is well insulated, while the ends are maintained at temperatures of 100 and 0°C. If the rod carries a current of 10 A, what is its midpoint temperature? What is the heat rate at each of the ends?

3.71 A rear window defroster in an automobile consists of uniformly distributed high-resistance wires molded into the glass. When power is applied to the wires, uniform heat generation may be assumed to occur within the window. During operation, heat that is generated is transferred by convection from both the interior and exterior surfaces of the window. However, due to the effects of vehicle speed and atmospheric winds, the convection coefficient of the warmer, interior side h_i is less than that on the exterior side h_o. On the same coordinate system, sketch the steady-state temperature distributions that would exist in the glass *before* the defroster has been turned on and *after* the defroster has been on for some time.

3.72 A nuclear fuel element of thickness $2L$ is covered with a steel cladding of thickness b. Heat generated within the nuclear fuel at a rate $\dot{q}$ is removed by a fluid at T_∞, which adjoins one surface and is characterized by a convection coefficient h. The other surface is well insulated, and the fuel and steel have thermal conductivities of k_f and k_s, respectively.

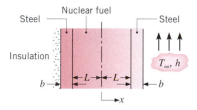

(a) Obtain an equation for the temperature distribution $T(x)$ in the nuclear fuel. Express your results in terms of $\dot{q}$, k_f, L, b, k_s, h, and T_∞.

(b) Sketch the temperature distribution $T(x)$ for the entire system.

3.73 The air *inside* a chamber at $T_{\infty,i} = 50°C$ is heated convectively with $h_i = 20$ W/m² · K by a 200-mm-thick wall having a thermal conductivity of 4 W/m · K and a uniform heat generation of 1000 W/m³. To prevent any heat generated within the wall from being lost to the *outside* of the chamber at $T_{\infty,o} = 25°C$ with $h_o = 5$ W/m² · K, a very thin electrical strip heater is placed on the outer wall to provide a uniform heat flux, q_o''.

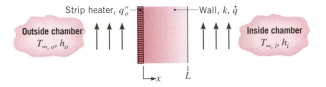

(a) Sketch the temperature distribution in the wall on T–x coordinates for the condition where no heat generated within the wall is lost to the *outside* of the chamber.

(b) What are the temperatures at the wall boundaries, $T(0)$ and $T(L)$, for the conditions of part (a)?

(c) Determine the value of q_o'' that must be supplied by the strip heater so that all heat generated within the wall is transferred to the *inside* of the chamber.

(d) If the heat generation in the wall were switched off while the heat flux to the strip heater remained constant, what would be the steady-state temperature, $T(0)$, of the outer wall surface?

3.74 In the previous problem, the strip heater acts to *guard* against heat losses from the wall to the outside, and the required heat flux q_o'' depends on chamber operating conditions such as $\dot{q}$ and $T_{\infty, i}$. As a first step in designing a controller for the guard heater, compute and plot q_o'' and $T(0)$ as a function of $\dot{q}$ for $200 \leq \dot{q} \leq 2000$ W/m^3 and $T_{\infty, i} = 30, 50,$ and $70°$C.

3.75 An electric current I is passed through a thin metallic wire of diameter D and thermal and electrical conductivities k and σ, respectively. Air at T_∞ flows over the wire maintaining a convection heat transfer coefficient h.

(a) Beginning with a differential control volume, derive the differential equation that governs the temperature distribution $T(x)$ in the wire.

(b) Beginning with an appropriate transformation of the dependent variable, show that the general solution is of the form

$$T(x) = C_1 e^{mx} + C_2 e^{-mx} + T_\infty + \frac{4I^2}{\pi^2 \sigma h D^3}$$

where $m = (4h/kD)^{1/2}$.

(c) Consider conditions for which the wire is connected to two electrodes that are separated by a distance L and are each maintained at the temperature T_E. What is the corresponding temperature distribution?

(d) It is possible to control the current I such that no heat is transferred from the wire to the elec-

trodes. Obtain an expression for this current in terms of σ, D, h, T_E, and T_∞.

3.76 The exposed surface ($x = 0$) of a plane wall of thermal conductivity k is subjected to microwave radiation that causes volumetric heating to vary as

$$\dot{q}(x) = \dot{q}_o \left(1 - \frac{x}{L} \right)$$

where $\dot{q}_o$ (W/m^3) is a constant. The boundary at $x = L$ is perfectly insulated, while the exposed surface is maintained at a constant temperature T_o. Determine the temperature distribution $T(x)$ in terms of x, L, k, $\dot{q}_o$, and T_o.

3.77 Consider a plane wall of thickness L, which is to act as shielding for a nuclear reactor. The inner surface ($x = 0$) receives gamma radiation that is partially absorbed within the shielding and has the effect of an internally distributed heat source. In particular, heat is generated per unit volume within the shielding according to the relation

$$\dot{q}(x) = q_o'' \alpha e^{-\alpha x}$$

where q_o'' is the incident radiation flux and α is a property (the absorption coefficient) of the shielding material.

(a) If the inner ($x = 0$) and outer ($x = L$) surfaces of the shielding are maintained at temperatures of T_1 and T_2, respectively, what is the form of the temperature distribution in the shielding?

(b) Obtain an expression that could be used to determine the x location in the shield at which the temperature is a maximum.

3.78 A quartz window of thickness L serves as a viewing port in a furnace used for annealing steel. The inner surface ($x = 0$) of the window is irradiated with a uniform heat flux q_o'' due to emission from hot gases in the furnace. A fraction, β, of this radiation may be assumed to be absorbed at the inner surface, while the remaining radiation is partially absorbed as it passes through the quartz. The volumetric heat generation due to this absorption may be described by an expression of the form

$$\dot{q}(x) = (1 - \beta)q_o'' \alpha e^{-\alpha x}$$

where α is the absorption coefficient of the quartz. Convection heat transfer occurs from the outer surface ($x = L$) of the window to ambient air at T_∞ and is characterized by the convection coefficient h. Convection and radiation emission from the inner surface may be neglected, along with radiation emission from the outer surface. Determine the temperature distribution in the quartz, expressing your result in terms of the foregoing parameters.

3.79 A copper cable of 30-mm diameter has an electrical resistance of 5×10^{-3} Ω/m and is used to carry an electrical current of 250 A. The cable is exposed to ambient air at 20°C, and the associated convection coefficient is 25 W/m² · K. What are the surface and centerline temperatures of the copper?

3.80 For the conditions described in Problem 1.32, determine the temperature distribution, $T(r)$, in the container, expressing your result in terms of $\dot{q}_o$, r_o, T_∞, h, and the thermal conductivity k of the radioactive wastes.

3.81 A cylindrical shell of inner and outer radii, r_i and r_o, respectively, is filled with a heat-generating material that provides a uniform volumetric generation rate (W/m³) of $\dot{q}$. The inner surface is insulated, while the outer surface of the shell is exposed to a fluid at T_∞ and a convection coefficient h.

(a) Obtain an expression for the steady-state temperature distribution, $T(r)$, in the shell, expressing your result in terms of r_i, r_o, $\dot{q}$, h, T_∞, and the thermal conductivity k of the shell material.

(b) Determine an expression for the heat rate, $q'(r_o)$, at the outer radius of the shell in terms of $\dot{q}$ and shell dimensions.

3.82 The cross section of a long cylindrical fuel element in a nuclear reactor is shown. Energy generation occurs uniformly in the thorium fuel rod, which is of diameter $D = 25$ mm and is wrapped in a thin aluminum cladding.

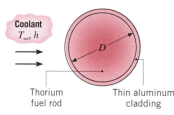

(a) It is proposed that, under steady-state conditions, the system operates with a generation rate of $\dot{q} = 7 \times 10^8$ W/m³ and cooling system characteristics of $T_\infty = 95°C$ and $h = 7000$ W/m² · K. Is this proposal satisfactory?

(b) Explore the effect of variations in $\dot{q}$ and h by plotting temperature distributions, $T(r)$, for a range of parameter values. Suggest an envelope of acceptable operating conditions.

3.83 A nuclear reactor fuel element consists of a solid cylindrical pin of radius r_1 and thermal conductivity k_f. The fuel pin is in good contact with a cladding material of outer radius r_2 and thermal con-

ductivity k_c. Consider steady-state conditions for which uniform heat generation occurs within the fuel at a volumetric rate $\dot{q}$ and the outer surface of the cladding is exposed to a coolant that is characterized by a temperature T_∞ and a convection coefficient h.

(a) Obtain equations for the temperature distributions $T_f(r)$ and $T_c(r)$ in the fuel and cladding, respectively. Express your results exclusively in terms of the foregoing variables.

(b) Consider a uranium oxide fuel pin for which $k_f = 2$ W/m · K and $r_1 = 6$ mm and cladding for which $k_c = 25$ W/m · K and $r_2 = 9$ mm. If $\dot{q} = 2 \times 10^8$ W/m³, $h = 2000$ W/m² · K, and $T_\infty = 300$ K, what is the maximum temperature in the fuel element?

(c) Compute and plot the temperature distribution, $T(r)$, for values of $h = 2000$, 5000, and 10,000 W/m² · K. If the operator wishes to maintain the centerline temperature of the fuel element below 1000 K, can she do so by adjusting the coolant flow and hence the value of h?

3.84 Consider the configuration of Example 3.7, where uniform volumetric heating within a stainless steel tube is induced by an electric current and heat is transferred by convection to air flowing through the tube. The tube wall has inner and outer radii of $r_1 = 25$ mm and $r_2 = 35$ mm, a thermal conductivity of $k = 15$ W/m · K, an electrical resistivity of $\rho_e = 0.7 \times 10^{-6}$ Ω · m, and a maximum allowable operating temperature of 1400 K.

(a) Assuming the outer tube surface to be perfectly insulated and the air flow to be characterized by a temperature and convection coefficient of $T_{\infty,1} = 400$ K and $h_1 = 100$ W/m² · K, determine the maximum allowable electric current I.

(b) Compute and plot the radial temperature distribution in the tube wall for the electric current of part (a) and three values of h_1 (100, 500, and 1000 W/m² · K). For each value of h_1, determine the rate of heat transfer to the air per unit length of tube.

(c) In practice, even the best of insulating materials would be unable to maintain adiabatic conditions at the outer tube surface. Consider use of a refractory insulating material of thermal conductivity $k = 1.0$ W/m · K and neglect radiation exchange at its outer surface. For $h_1 = 100$ W/m² · K and the maximum allowable current determined in part (a), compute and plot the temperature distribution in the *composite* wall for two values of the insulation thickness ($\delta = 25$ and 50 mm). The outer surface of the insulation is exposed to room air for which $T_{\infty, 2} = 300$ K and $h_2 = 25$ W/m² · K. For each insulation thickness, determine the rate of heat transfer per unit tube length to the inner air flow and the ambient air.

3.85 A homeowner, whose water pipes have frozen during a period of cold weather, decides to melt the ice by passing an electric current I through the pipe wall. The inner and outer radii of the wall are designated as r_1 and r_2, and its electrical resistance per unit length is designated as R'_e (Ω/m). The pipe is well insulated on the outside, and during melting the ice (and water) in the pipe remain at a constant temperature T_m associated with the melting process.

    (a) Assuming that steady-state conditions are reached shortly after application of the current, determine the form of the steady-state temperature distribution $T(r)$ in the pipe wall during the melting process.

    (b) Develop an expression for the time t_m required to completely melt the ice. Calculate this time for $I = 100$ A, $R'_e = 0.30$ Ω/m, and $r_1 = 50$ mm.

3.86 A high-temperature, gas-cooled nuclear reactor consists of a composite cylindrical wall for which a thorium fuel element ($k \approx 57$ W/m · K) is encased in graphite ($k \approx 3$ W/m · K) and gaseous helium flows through an annular coolant channel. Consider conditions for which the helium temperature is $T_{\infty} = 600$ K and the convection coefficient at the outer surface of the graphite is $h = 2000$ W/m² · K.

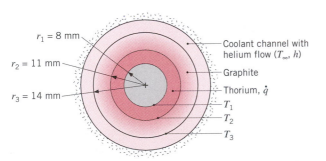

$r_1 = 8$ mm

$r_2 = 11$ mm

$r_3 = 14$ mm

Coolant channel with helium flow (T_{∞}, h)

Graphite

Thorium, $\dot{q}$

T_1

T_2

T_3

    (a) If thermal energy is uniformly generated in the fuel element at a rate $\dot{q} = 10^8$ W/m³, what are the temperatures T_1 and T_2 at the inner and outer surfaces, respectively, of the fuel element?

    (b) Compute and plot the temperature distribution in the composite wall for selected values of $\dot{q}$. What is the maximum allowable value of $\dot{q}$?

3.87 A long cylindrical rod of diameter 200 mm with thermal conductivity of 0.5 W/m · K experiences uniform volumetric heat generation of 24,000 W/m³. The rod is encapsulated by a circular sleeve having an outer diameter of 400 mm and a thermal conductivity of 4 W/m · K. The outer surface of the sleeve is exposed to cross flow of air at 27°C with a convection coefficient of 25 W/m² · K.

    (a) Find the temperature at the interface between the rod and sleeve and on the outer surface.

    (b) What is the temperature at the center of the rod?

3.88 A radioactive material of thermal conductivity k is cast as a solid sphere of radius r_o and placed in a liquid bath for which the temperature T_{∞} and convection coefficient h are known. Heat is uniformly generated within the solid at a volumetric rate of $\dot{q}$. Obtain the steady-state radial temperature distribution in the solid, expressing your result in terms of r_o, $\dot{q}$, k, h, and T_{∞}.

3.89 For the conditions described in Problem 1.34, determine the temperature distribution, $T(r)$, in the container. Express your result in terms of $\dot{q}_o$, r_o, T_{∞}, h, and the thermal conductivity k of the radioactive wastes.

3.90 Radioactive wastes ($k_{rw} = 20$ W/m · K) are stored in a spherical, stainless steel ($k_{ss} = 15$ W/m · K) container of inner and outer radii equal to $r_i = 0.5$ m and $r_o = 0.6$ m. Heat is generated volumetrically within the wastes at a uniform rate of $\dot{q} = 10^5$ W/m³, and the outer surface of the container is exposed to a water flow for which $h = 1000$ W/m² · K and $T_{\infty} = 25$°C.

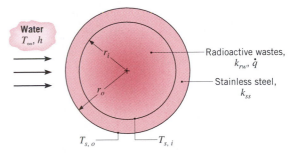

Water
T_{∞}, h

r_i

r_o

Radioactive wastes, k_{rw}, $\dot{q}$

Stainless steel, k_{ss}

$T_{s, o}$

$T_{s, i}$

    (a) Evaluate the steady-state outer surface temperature, $T_{s, o}$.

    (b) Evaluate the steady-state inner surface temperature, $T_{s, i}$.

    (c) Obtain an expression for the temperature distribution, $T(r)$, in the radioactive wastes. Express your result in terms of r_i, $T_{s, i}$, k_{rw}, and $\dot{q}$. Evaluate the temperature at $r = 0$.

(d) A proposed extension of the foregoing design involves storing waste materials having the same thermal conductivity but twice the heat generation ($\dot{q} = 2 \times 10^5$ W/m³) in a stainless steel container of equivalent inner radius ($r_i = 0.5$ m). Safety considerations dictate that the maximum system temperature not exceed 475°C and that the container wall thickness be no less than $t = 0.04$ m and preferably at or close to the original design ($t = 0.1$ m). Assess the effect of varying the outside convection coefficient to a maximum achievable value of $h = 5000$ W/m² · K (by increasing the water velocity) and the container wall thickness. Is the proposed extension feasible? If so, recommend suitable operating and design conditions for h and t, respectively.

3.91 Unique characteristics of biologically active materials such as fruits, vegetables, and other products require special care in handling. Following harvest and separation from producing plants, glucose is catabolized to produce carbon dioxide, water vapor, and heat, with attendant internal energy generation. Consider a carton of apples, each of 80-mm diameter, which is ventilated with air at 5°C and a velocity of 0.5 m/s. The corresponding value of the heat transfer coefficient is 7.5 W/m² · K. Within each apple thermal energy is uniformly generated at a total rate of 4000 J/kg · day. The density and thermal conductivity of the apple are 840 kg/m³ and 0.5 W/m · K, respectively.

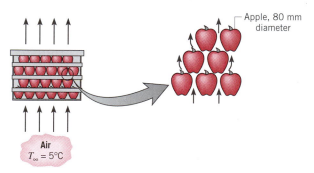

Apple, 80 mm diameter

Air
$T_\infty = 5°C$

(a) Determine the apple center and surface temperatures.

(b) For the stacked arrangement of apples within the crate, the convection coefficient depends on the velocity as $h = C_1 V^{0.425}$, where $C_1 = 10.1$ W/m² · K · (m/s)$^{0.425}$. Compute and plot the center and surface temperatures as a function of the air velocity for $0.1 \leq V \leq 1$ m/s.

Extended Surfaces

3.92 The radiation heat gage shown in the diagram is made from constantan metal foil, which is coated black and is in the form of a circular disk of radius R and thickness t. The gage is located in an evacuated enclosure. The incident radiation flux absorbed by the foil, q_i'', diffuses toward the outer circumference and into the larger copper ring, which acts as a heat sink at the constant temperature $T(R)$. Two copper lead wires are attached to the center of the foil and to the ring to complete a thermocouple circuit that allows for measurement of the temperature difference between the foil center and the foil edge, $\Delta T = T(0) - T(R)$.

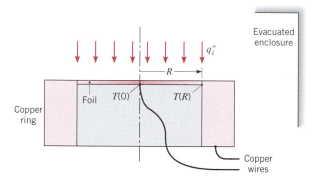

Obtain the differential equation that determines $T(r)$, the temperature distribution in the foil, under steady-state conditions. Solve this equation to obtain an expression relating ΔT to q_i''. You may neglect radiation exchange between the foil and its surroundings.

3.93 Copper tubing is joined to the absorber of a flat-plate solar collector as shown.

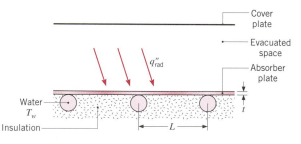

The aluminum alloy (2024-T6) absorber plate is 6 mm thick and well insulated on its bottom. The top surface of the plate is separated from a transparent cover plate by an evacuated space. The tubes are spaced a distance L of 0.20 m from each other, and water is circulated through the tubes to remove the collected energy. The water may be assumed to be at a uniform temperature of $T_w = 60°C$. Under steady-state operating conditions for which the *net* radiation heat flux to the surface is $q''_{rad} = 800$ W/m², what is the maximum temperature on the plate and the heat transfer rate per unit length of tube? Note that q''_{rad} represents the net effect of solar radiation absorption by the absorber plate and radiation exchange between the absorber and cover plates. You may assume the temperature of the absorber plate directly above a tube to be equal to that of the water.

3.94 Copper tubing is joined to a solar collector plate of thickness t, and the working fluid maintains the temperature of the plate above the tubes at T_o. There is a uniform net radiation heat flux q''_{rad} to the top surface of the plate, while the bottom surface is well insulated. The top surface is also exposed to a fluid at T_∞ that provides for a uniform convection coefficient h.

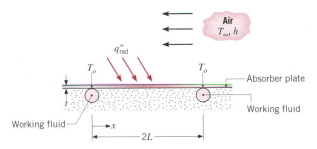

(a) Derive the differential equation that governs the temperature distribution $T(x)$ in the plate.

(b) Obtain a solution to the differential equation for appropriate boundary conditions.

3.95 A thin flat plate of length L, thickness t, and width $W \gg L$ is thermally joined to two large heat sinks that are maintained at a temperature T_o. The bottom of the plate is well insulated, while the net heat flux to the top surface of the plate is known to have a uniform value of q''_o.

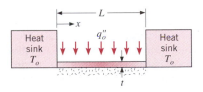

(a) Derive the differential equation that determines the steady-state temperature distribution $T(x)$ in the plate.

(b) Solve the foregoing equation for the temperature distribution, and obtain an expression for the rate of heat transfer from the plate to the heat sinks.

3.96 Consider the flat plate of Problem 3.95, but with the heat sinks at different temperatures, $T(0) = T_o$ and $T(L) = T_L$, and with the bottom surface no longer insulated. Convection heat transfer is now allowed to occur between this surface and a fluid at T_∞, with a convection coefficient h.

(a) Derive the differential equation that determines the steady-state temperature distribution $T(x)$ in the plate.

(b) Solve the foregoing equation for the temperature distribution, and obtain an expression for the rate of heat transfer from the plate to the heat sinks.

(c) For $q''_o = 20,000$ W/m², $T_o = 100°C$, $T_L = 35°C$, $T_\infty = 25°C$, $k = 25$ W/m·K, $h = 50$ W/m²·K, $L = 100$ mm, $t = 5$ mm, and a plate width of $W = 30$ mm, plot the temperature distribution and determine the sink heat rates, $q_x(0)$ and $q_x(L)$. On the same graph, plot three additional temperature distributions corresponding to changes in the following parameters, with the remaining parameters unchanged: (i) $q''_o = 30,000$ W/m², (ii) $h = 200$ W/m²·K, and (iii) the value of q''_o for which $q_x(0) = 0$ when $h = 200$ W/m²·K.

3.97 A bonding operation utilizes a laser to provide a constant heat flux, q''_o, across the top surface of a thin adhesive-backed, plastic film to be affixed to a metal strip as shown in the sketch. The metal strip has a thickness $d = 1.25$ mm and its width is large relative to that of the film. The thermophysical properties of the strip are $\rho = 7850$ kg/m³, $c_p = 435$ J/kg·K, and $k = 60$ W/m·K. The thermal resistance of the plastic film of width $w_1 = 40$ mm is negligible. The upper and lower surfaces of the strip (including the plastic film) experience convection with air at 25°C and a convection coefficient of 10 W/m²·K. The strip and film are very long in the direction normal to the page. Assume the edges of the metal strip are at the air temperature (T_∞).

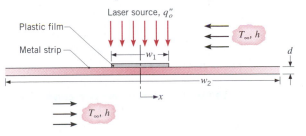

(a) Derive an expression for the temperature distribution in the portion of the steel strip with the plastic film ($-w_1/2 \leq x \leq +w_1/2$).

(b) If the heat flux provided by the laser is 10,000 W/m^2, determine the temperature of the plastic film at the center ($x = 0$) and its edges ($x = \pm w_1/2$).

(c) Plot the temperature distribution for the entire strip and point out its special features.

3.98 A thin metallic wire of thermal conductivity k, diameter D, and length $2L$ is annealed by passing an electrical current through the wire to induce a uniform volumetric heat generation $\dot{q}$. The ambient air around the wire is at a temperature T_∞, while the ends of the wire at $x = \pm L$ are also maintained at T_∞. Heat transfer from the wire to the air is characterized by the convection coefficient h. Obtain an expression for the steady-state temperature distribution $T(x)$ along the wire.

3.99 A motor draws electric power P_{elec} from a supply line and delivers mechanical power P_{mech} to a pump through a rotating copper shaft of thermal conductivity k_s, length L, and diameter D. The motor is mounted on a square pad of width W, thickness t, and thermal conductivity k_p. The surface of the housing exposed to ambient air at T_∞ is of area A_h, and the corresponding convection coefficient is h_h. Opposite ends of the shaft are at temperatures of T_h and T_∞, and heat transfer from the shaft to the ambient air is characterized by the convection coefficient h_s. The base of the pad is at T_∞.

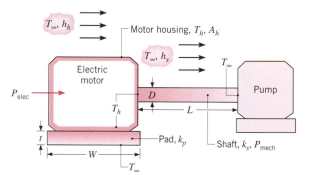

(a) Expressing your result in terms of P_{elec}, P_{mech}, k_s, L, D, W, t, k_p, A_h, h_h, and h_s, obtain an expression for ($T_h - T_\infty$).

(b) What is the value of T_h if $P_{elec} = 25$ kW, $P_{mech} = 15$ kW, $k_s = 400$ W/m · K, $L = 0.5$ m, $D = 0.05$ m, $W = 0.7$ m, $t = 0.05$ m, $k_p = 0.5$ W/m · K, $A_h = 2$ m^2, $h_h = 10$ W/m^2 · K, $h_s = 300$ W/m^2 · K, and $T_\infty = 25°C$?

3.100 A long rod passes through the opening in an oven having an air temperature of 400°C and is pressed firmly onto the surface of a billet. Thermocouples imbedded in the rod at locations 25 and 120 mm from the billet register temperatures of 325 and 375°C, respectively. What is the temperature of the billet?

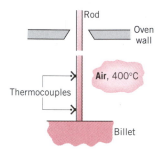

3.101 A probe of overall length $L = 200$ mm and diameter $D = 12.5$ mm is inserted through a duct wall such that a portion of its length, referred to as the immersion length L_i, is in contact with the water stream whose temperature, $T_{\infty,i}$, is to be determined. The convection coefficients over the immersion and ambient-exposed lengths are $h_i = 1100$ W/m^2 · K and $h_o = 10$ W/m^2 · K, respectively. The probe has a thermal conductivity of 177 W/m · K and is in poor thermal contact with the duct wall.

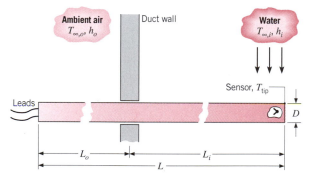

(a) Derive an expression for evaluating the measurement error, $\Delta T_{err} = T_{tip} - T_{\infty,i}$, which is the difference between the tip temperature, T_{tip}, and the water temperature, $T_{\infty,i}$. *Hint:* Define a coordinate system with the origin at the duct wall and treat the probe as two fins extending inward and outward from the duct, but having the same base temperature. Use case A results from Table 3.2.

(b) With the water and ambient air temperatures at 80 and 20°C, respectively, calculate the measurement error, ΔT_{err}, as a function of immersion length for the conditions $L_i/L = 0.225$, 0.425, and 0.625.

(c) Compute and plot the effects of probe thermal conductivity and water velocity (h_i) on the measurement error.

3.102 A rod of diameter $D = 25$ mm and thermal conductivity $k = 60$ W/m · K protrudes normally from a furnace wall that is at $T_w = 200°C$ and is covered by insulation of thickness $L_{ins} = 200$ mm. The rod is welded to the furnace wall and is used as a hanger for supporting instrumentation cables. To avoid damaging the cables, the temperature of the rod at its exposed surface, T_o, must be maintained below a specified operating limit of $T_{max} = 100°C$. The ambient air temperature is $T_\infty = 25°C$, and the convection coefficient is $h = 15$ W/m² · K.

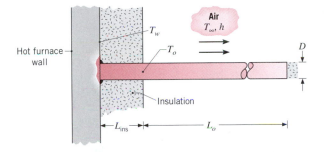

(a) Derive an expression for the exposed surface temperature T_o as a function of the prescribed thermal and geometrical parameters. The rod has an exposed length L_o, and its tip is well insulated.

(b) Will a rod with $L_o = 200$ mm meet the specified operating limit? If not, what design parameters would you change? Consider another material, increasing the thickness of the insulation, and increasing the rod length. Also, consider how you might attach the base of the rod to the furnace wall as a means to reduce T_o.

3.103 From Problem 1.51, consider the wire leads connecting the transistor to the circuit board. The leads are of thermal conductivity k, thickness t, width w, and length L. One end of a lead is maintained at a temperature T_c corresponding to the transistor case, while the other end assumes the temperature T_b of the circuit board. During steady-state operation, current flow through the leads provides for uniform volumetric heating in the amount $\dot{q}$, while there is convection cooling to air that is at T_∞ and maintains a convection coefficient h.

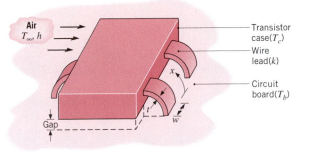

(a) Derive an equation from which the temperature distribution in a wire lead may be determined. List all pertinent assumptions.

(b) Determine the temperature distribution in a wire lead, expressing your results in terms of the prescribed variables.

3.104 Turbine blades mounted to a rotating disc in a gas turbine engine are exposed to a gas stream that is at $T_\infty = 1200°C$ and maintains a convection coefficient of $h = 250$ W/m² · K over the blade.

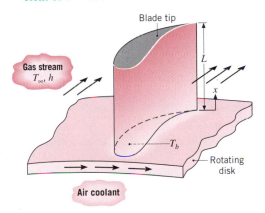

The blades, which are fabricated from Inconel, $k \approx 20$ W/m · K, have a length of $L = 50$ mm. The blade profile has a uniform cross-sectional area of $A_c = 6 \times 10^{-4}$ m² and a perimeter of $P = 110$ mm. A proposed blade-cooling scheme, which involves routing air through the supporting disc, is able to maintain the base of each blade at a temperature of $T_b = 300°C$.

(a) If the maximum allowable blade temperature is 1050°C and the blade tip may be assumed to be adiabatic, is the proposed cooling scheme satisfactory?

(b) For the proposed cooling scheme, what is the rate at which heat is transferred from each blade to the coolant?

3.105 In a test to determine the friction coefficient, μ, associated with a disc brake, one disc and its shaft are rotated at a constant angular velocity ω, while an equivalent disc/shaft assembly is stationary. Each disc has an outer radius of $r_2 = 180$ mm, a shaft radius of $r_1 = 20$ mm, a thickness of $t = 12$ mm, and a thermal conductivity of $k = 15$ W/m · K. A known force F is applied to the system, and the corresponding torque τ required to maintain rotation is measured. The disc contact pressure may be assumed to be uniform (i.e., independent of location on the interface), and the discs may be assumed to be well insulated from the surroundings.

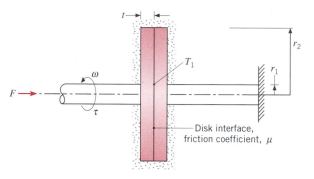

(a) Obtain an expression that may be used to evaluate μ from known quantities.

(b) For the region $r_1 \leq r \leq r_2$, determine the radial temperature distribution, $T(r)$, in the disc, where $T(r_1) = T_1$ is presumed to be known.

(c) Consider test conditions for which $F = 200$ N, $\omega = 40$ rad/s, $\tau = 8$ N · m, and $T_1 = 80°C$. Evaluate the friction coefficient and the maximum disc temperature.

Simple Fins

3.106 A long, circular aluminum rod is attached at one end to a heated wall and transfers heat by convection to a cold fluid.

(a) If the diameter of the rod is tripled, by how much would the rate of heat removal change?

(b) If a copper rod of the same diameter is used in place of the aluminum, by how much would the rate of heat removal change?

3.107 A brass rod 100 mm long and 5 mm in diameter extends horizontally from a casting at 200°C. The rod is in an air environment with $T_\infty = 20°C$ and $h = 30$ W/m² · K. What is the temperature of the rod 25, 50, and 100 mm from the casting?

3.108 Two long copper rods of diameter $D = 10$ mm are soldered together end to end, with solder having a melting point of 650°C. The rods are in air at 25°C with a convection coefficient of 10 W/m² · K. What is the minimum power input needed to effect the soldering?

3.109 Circular copper rods of diameter $D = 1$ mm and length $L = 25$ mm are used to enhance heat transfer from a surface that is maintained at $T_{s,1} = 100°C$. One end of the rod is attached to this surface (at $x = 0$), while the other end ($x = 25$ mm) is joined to a second surface, which is maintained at $T_{s,2} = 0°C$. Air flowing between the surfaces (and over the rods) is also at a temperature of $T_\infty = 0°C$, and a convection coefficient of $h = 100$ W/m² · K is maintained.

(a) What is the rate of heat transfer by convection from a single copper rod to the air?

(b) What is the total rate of heat transfer from a 1-m by 1-m section of the surface at 100°C, if a bundle of the rods is installed on 4-mm centers?

3.110 Pin fins are widely used in electronic systems to provide cooling as well as to support devices. Consider the pin fin of uniform diameter D, length L, and thermal conductivity k connecting two identical devices of length L_g and surface area A_g. The devices are characterized by a uniform volumetric generation of thermal energy $\dot{q}$ and a thermal conductivity k_g. Assume that the exposed surfaces of the devices are at a uniform temperature corresponding to that of the pin base, T_b, and that heat is transferred by convection from the exposed surfaces to an adjoining fluid. The back and sides of the devices are perfectly insulated.

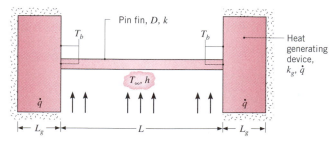

Derive an expression for the base temperature T_b in terms of the device parameters (k_g, $\dot{q}$, L_g, A_g), the convection parameters (T_∞, h), and the fin parameters (k, D, L).

3.111 Consider two long, slender rods of the same diameter but different materials. One end of each rod is attached to a base surface maintained at 100°C, while the surface of the rods are exposed to ambient air at 20°C. By traversing the length of each rod with a thermocouple, it was observed that the temperatures of the rods were equal at the positions $x_A = 0.15$ m and $x_B = 0.075$ m, where x is measured from the base surface. If the thermal conductivity of rod A is known to be $k_A = 70$ W/m · K, determine the value of k_B for rod B.

3.112 Consider a slender rod of length L, which is exposed to convection cooling (T_∞, h) and has both ends maintained at $T_o > T_\infty$. For each of the three cases described below, sketch the temperature distribution on T–x coordinates and identify features of the distribution. Assume the end temperatures and convection heat transfer coefficient are the same for all cases.

(a) The rod has a thermal conductivity k_A.

(b) The rod has a thermal conductivity k_B, where $k_B < k_A$.

(c) The rod is a composite rod with k_A for $0 \le x \le L/2$ and k_B for $L/2 \le x \le L$.

3.113 An experimental arrangement for measuring the thermal conductivity of solid materials involves the use of two long rods that are equivalent in every respect, except that one is fabricated from a standard material of known thermal conductivity k_A while the other is fabricated from the material whose thermal conductivity k_B is desired. Both rods are attached at one end to a heat source of fixed temperature T_b, are exposed to a fluid of temperature T_∞, and are instrumented with thermocouples to measure the temperature at a fixed distance x_1 from the heat source. If the standard material is aluminum, with $k_A = 200$ W/m · K, and measurements reveal values of $T_A = 75°C$ and $T_B = 60°C$ at x_1 for $T_b = 100°C$ and $T_\infty = 25°C$, what is the thermal conductivity k_B of the test material?

Fin Systems and Arrays

3.114 Finned passages are frequently formed between parallel plates to enhance convection heat transfer in compact heat exchanger cores. An important application is in electronic equipment cooling, where one or more air-cooled stacks are placed between heat-dissipating electrical components. Consider a single stack of rectangular fins of length L and thickness t, with convection conditions corresponding to h and T_∞.

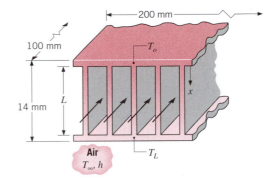

(a) Obtain expressions for the fin heat transfer rates, $q_{f,o}$ and $q_{f,L}$, in terms of the base temperatures, T_o and T_L.

(b) In a specific application, a stack that is 200 mm wide and 100 mm deep contains 50 fins, each of length $L = 12$ mm. The entire stack is made from aluminum, which is everywhere 1.0 mm thick. If temperature limitations associated with electrical components joined to opposite plates dictate maximum allowable plate temperatures of $T_o = 400$ K and $T_L = 350$ K, what are the corresponding maximum power dissipations if $h = 150$ W/m^2 · K and $T_\infty = 300$ K?

3.115 The fin array of Problem 3.114 is commonly found in *compact heat exchangers,* whose function is to provide a large surface area per unit volume in transferring heat from one fluid to another. In such applications, it is desirable to minimize the thermal resistance $R_{t,o}$ of the fin array. Consider a unit heat exchanger core that is 1 m long in the direction of air flow and 1 m wide in a direction normal to both the air flow and the fin surfaces. The length of the fin passages between adjoining parallel plates is $L = 8$ mm, while the fin thermal conductivity and convection coefficient are $k = 200$ W/m · K (aluminum) and $h = 150$ W/m^2 · K, respectively.

(a) If the fin thickness and pitch are $t = 1$ mm and $S = 4$ mm, respectively, what is the value of $R_{t,o}$?

(b) Subject to the constraints that the fin thickness and pitch may not be less than 0.5 and 3 mm, respectively, assess the effect of changes in t and S.

3.116 A disk-shaped transistor, which is mounted in an insulating medium, dissipates 0.25 W during steady-state operation. To reduce the temperature of the transistor, it is proposed that a hollow copper tube be attached to the transistor as shown.

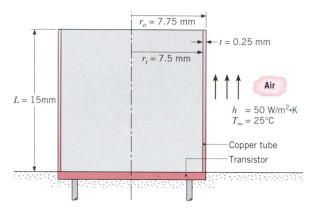

$r_o = 7.75$ mm

$t = 0.25$ mm

$r_i = 7.5$ mm

$L = 15$mm

Air

$h = 50$ W/m²·K
$T_\infty = 25°C$

Copper tube
Transistor

The outer surface of the tube is exposed to ambient air at $T_\infty = 25°C$ with a convection coefficient of $h = 50$ W/m² · K. As a first approximation, heat transfer from the interior surface of the tube and from the exposed surface of the transistor may be neglected. What is the temperature of the transistor with the fin? What is the temperature of the transistor without the fin if h and T_∞ remain the same?

3.117 As more and more components are placed on a single integrated circuit (chip), the amount of heat that is dissipated continues to increase. However, this increase is limited by the maximum allowable chip operating temperature, which is approximately 75°C. To maximize heat dissipation, it is proposed that a 4 × 4 array of copper pin fins be metallurgically joined to the outer surface of a square chip that is 12.7 mm on a side.

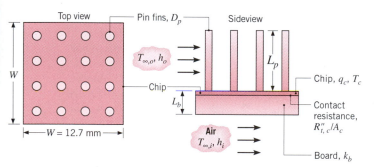

Top view — Pin fins, D_p — Sideview

$T_{\infty,o}, h_o$

Chip

L_b

Air
$T_{\infty,i}, h_i$

W

$W = 12.7$ mm

L_p

Chip, q_c, T_c

Contact resistance, $R''_{t,c}/A_c$

Board, k_b

(a) Sketch the equivalent thermal circuit for the pin–chip–board assembly, assuming one-dimensional, steady-state conditions and negligible contact resistance between the pins and the chip. In variable form, label appropriate resistances, temperatures, and heat rates.

(b) For the conditions prescribed in Problem 3.25, what is the maximum rate at which heat can be dissipated in the chip when the pins are in place? That is, what is the value of q_c for $T_c = 75°C$? The pin diameter and length are $D_p = 1.5$ mm and $L_p = 15$ mm.

3.118 In Problem 3.117, the prescribed value of $h_o = 1000$ W/m² · K is large and characteristic of liquid cooling. In practice it would be far more preferable to use air cooling, for which a reasonable upper limit to the convection coefficient would be $h_o = 250$ W/m² · K. Assess the effect of changes in the pin fin geometry on the chip heat rate if the remaining conditions of Problem 3.117, including a maximum allowable chip temperature of 75°C, remain in effect. Parametric variations that may be considered include the total number of pins, N, in the square array, the pin diameter D_p, and the pin length L_p. However, the product $N^{1/2}D_p$ should not exceed 9 mm to ensure adequate air flow passage through the array. Recommend a design that enhances chip cooling.

3.119 As a means of enhancing heat transfer from high-performance logic chips, it is common to attach a *heat sink* to the chip surface in order to increase the surface area available for convection heat transfer. Because of the ease with which it may be manufactured (by taking orthogonal sawcuts in a block of material), an attractive option is to use a heat sink consisting of an array of square fins of width w on a side. The spacing between adjoining fins would be determined by the width of the sawblade, with the sum of this spacing and the fin width designated as the fin pitch S. The method by which the heat sink is joined to the chip would determine the interfacial contact resistance, $R''_{t,c}$.

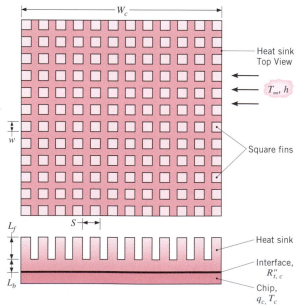

W_c

Heat sink
Top View

T_∞, h

w

Square fins

L_f

S

Heat sink

Interface, $R''_{t,c}$

L_b

Chip, q_c, T_c

Consider a chip of width $W_c = 16$ mm and conditions for which cooling is provided by a dielectric liquid with $T_\infty = 25°C$ and $h = 1500$ W/m² · K.

The heat sink is fabricated from copper ($k = 400$ W/m · K), and its characteristic dimensions are $w = 0.25$ mm, $S = 0.50$ mm, $L_f = 6$ mm, and $L_b = 3$ mm. The prescribed values of w and S represent minima imposed by manufacturing constraints and the need to maintain adequate flow in the passages between fins.

(a) If a metallurgical joint provides a contact resistance of $R''_{t,c} = 5 \times 10^{-6}$ m² · K/W and the maximum allowable chip temperature is 85°C, what is the maximum allowable chip power dissipation q_c? Assume all of the heat to be transferred through the heat sink.

(b) It may be possible to increase the heat dissipation by increasing w, subject to the constraint that $(S - w) \geq 0.25$ mm, and/or increasing L_f (subject to manufacturing constraints that $L_f \leq 10$ mm). Assess the effect of such changes.

3.120 Because of the large number of devices in today's PC chips, finned heat sinks are often used to maintain the chip at an acceptable operating temperature. Two fin designs are to be evaluated, both of which have base (unfinned) area dimensions of 53 mm × 57 mm. The fins are of square cross section and fabricated from an extruded aluminum alloy with a thermal conductivity of 175 W/m · K. Cooling air may be supplied at 25°C, and the maximum allowable chip temperature is 75°C. Other features of the design and operating conditions are tabulated.

Fin Dimensions

Design	Cross Section $w \times w$ (mm)	Length L (mm)	Number of Fins in Array	Convection Coefficient (W/m² · K)
A	3×3	30	6×9	125
B	1×1	7	14×17	375

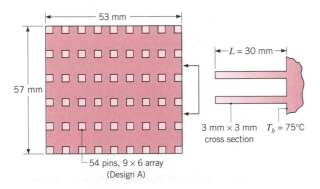

54 pins, 9 × 6 array
(Design A)

Determine which fin arrangement is superior. In your analysis, calculate the heat rate, efficiency, and effectiveness of a single fin, as well as the total heat rate and overall efficiency of the array. Since

real estate inside the computer enclosure is important, compare the total heat rate per unit volume for the two designs.

3.121 One wall of an electrical enclosure is made from copper plate ($k = 400$ W/m · K), 160 mm × 160 mm wide and 5 mm thick. To enhance heat transfer across the plate, 400 copper pin fins, each of 4-mm diameter and 20-mm length, are integrally machined on both sides of the plate in a square pitch on 8-mm centers. Warm air in the enclosure is at a temperature of 65°C, and natural circulation provides an average convection coefficient of 5 W/m² · K on the inner surface of the plate. A forced flow of ambient air at 20°C provides an average convection coefficient of 100 W/m² · K on the outer surface of the plate.

(a) Estimate the rate of heat transfer across the plate. Assuming the same convection coefficients without the fins, determine the amount of heat transfer enhancement afforded by the pins.

(b) It is recommended that manufacturing costs be reduced by silver soldering the pins to the plate, rather than using an expensive process such as electric discharge machining to achieve a continuous plate/pin construction. If the corresponding contact resistance is 5×10^{-6} m² · K/W, what is the rate of heat transfer across the plate?

3.122 A long rod of 20-mm diameter and a thermal conductivity of 1.5 W/m · K has a uniform internal volumetric thermal energy generation of 10^6 W/m³. The rod is covered with an electrically insulating sleeve of 2-mm thickness and thermal conductivity of 0.5 W/m · K. A spider with 12 ribs and dimensions as shown in the sketch has a thermal conductivity of 175 W/m · K, and is used to support the rod and to maintain concentricity with an 80-mm diameter tube. Air at the same temperature as the tube surface, $T_s = T_\infty = 25$°C, passes over the spider surface and the convection coefficient is 20 W/m² · K.

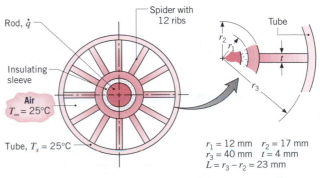

(a) Develop a thermal circuit that can be used to determine the temperature at the outer surface of the rod. Evaluate this temperature.

(b) What is the temperature at the center of the rod?

3.123 Consider the physical system and conditions of Problem 3.122, but now assume that the outer surface of the tube is well insulated. We wish to increase volumetric heating within the rod, while not allowing its centerline temperature to exceed 100°C. Determine the impact of the following changes, which may be effected independently or concurrently: (i) increasing the air speed and hence the convection coefficient; (ii) changing the number and/or thickness of the ribs; and (iii) using an electrically nonconducting sleeve material of larger thermal conductivity (e.g., amorphous carbon or quartz). Recommend a realistic configuration that yields a significant increase in $\dot{q}$.

3.124 An air heater consists of a steel tube ($k = 20$ W/m · K), with inner and outer radii of $r_1 = 13$ mm and $r_2 = 16$ mm, respectively, and eight integrally machined longitudinal fins, each of thickness $t = 3$ mm. The fins extend to a concentric tube, which is of radius $r_3 = 40$ mm and insulated on its outer surface. Water at a temperature $T_{\infty,i} = 90$°C flows through the inner tube, while air at $T_{\infty,o} = 25$°C flows through the annular region formed by the larger concentric tube.

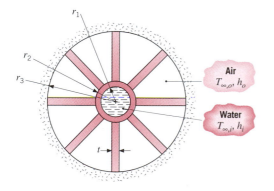

(a) Sketch the equivalent thermal circuit of the heater and relate each thermal resistance to appropriate system parameters.

(b) If $h_i = 5000$ W/m² · K and $h_o = 200$ W/m² · K, what is the heat rate per unit length?

(c) Assess the effect of increasing the number of fins N and/or the fin thickness t on the heat rate, subject to the constraint that $Nt < 50$ mm.

3.125 Determine the percentage increase in heat transfer associated with attaching aluminum fins of rectan-

gular profile to a plane wall. The fins are 50 mm long, 0.5 mm thick, and are equally spaced at a distance of 4 mm (250 fins/m). The convection coefficient associated with the bare wall is 40 W/m² · K, while that resulting from attachment of the fins is 30 W/m² · K.

3.126 Consider the use of straight, stainless steel (304) fins of rectangular and triangular profiles on a plane wall whose temperature is 100°C. The adjoining fluid is at 20°C, and the associated convection coefficient is 75 W/m² · K. Each fin is 6 mm thick and 20 mm long. Compare the efficiency, the effectiveness, and the heat loss per unit width associated with the two kinds of fins.

3.127 Aluminum fins of triangular profile are attached to a plane wall whose surface temperature is 250°C. The fin base thickness is 2 mm, and its length is 6 mm. The system is in ambient air at a temperature of 20°C, and the surface convection coefficient is 40 W/m² · K.

(a) What are the fin efficiency and effectiveness?

(b) What is the heat dissipated per unit width by a single fin?

3.128 An annular aluminum fin of rectangular profile is attached to a circular tube having an outside diameter of 25 mm and a surface temperature of 250°C. The fin is 1 mm thick and 10 mm long, and the temperature and the convection coefficient associated with the adjoining fluid are 25°C and 25 W/m² · K, respectively.

(a) What is the heat loss per fin?

(b) If 200 such fins are spaced at 5-mm increments along the tube length, what is the heat loss per meter of tube length?

3.129 Annular aluminum fins of rectangular profile are attached to a circular tube having an outside diameter of 50 mm and an outer surface temperature of 200°C. The fins are 4 mm thick and 15 mm long. The system is in ambient air at a temperature of 20°C, and the surface convection coefficient is 40 W/m² · K.

(a) What are the fin efficiency and effectiveness?

(b) If there are 125 such fins per meter of tube length, what is the rate of heat transfer per unit length of tube?

3.130 Annular aluminum fins that are 2 mm thick and 15 mm long are installed on an aluminum tube of 30-mm diameter. The thermal contact resistance between a fin and the tube is known to be 2×10^{-4} m² · K/W. If the tube wall is at 100°C and the adjoining fluid is at 25°C, with a convection coefficient of 75 W/m² · K, what is the rate of heat transfer from a single fin? What would be the rate of heat transfer if the contact resistance could be eliminated?

3.131 It is proposed to air-cool the cylinders of a combustion chamber by joining an aluminum casing with annular fins ($k = 240$ W/m · K) to the cylinder wall ($k = 50$ W/m · K).

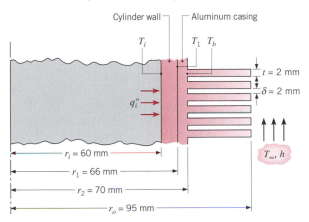

Cylinder wall ⌐ Aluminum casing

The air is at 320 K and the corresponding convection coefficient is 100 W/m² · K. Although heating at the inner surface is periodic, it is reasonable to assume steady-state conditions with a time-averaged heat flux of $q_i'' = 10^5$ W/m². Assuming negligible contact resistance between the wall and the casing, determine the wall inner temperature T_i, the interface temperature T_1, and the fin base temperature T_b. Determine these temperatures if the interface contact resistance is $R_{t,c}'' = 10^{-4}$ m² · K/W.

3.132 Consider the air-cooled combustion cylinder of Problem 3.131, but instead of imposing a uniform heat flux at the inner surface, consider conditions for which the time-averaged temperature of the combustion gases is $T_g = 1100$ K and the corresponding convection coefficient is $h_g = 150$ W/m² · K. All other conditions, including the cylinder/casing contact resistance, remain the same. Determine the heat rate per unit length of cylinder (W/m), as well as the cylinder inner temperature T_i, the interface temperatures $T_{1,i}$ and $T_{1,o}$, and the fin base temperature T_b. Subject to the constraint that the fin gap is fixed at $\delta = 2$ mm, assess the effect of increasing the fin thickness at the expense of reducing the number of fins.

3.133 In Example 3.10, we considered a heat sink design and operating conditions that maintained a transistor case temperature of 80°C, while 1.63 W were dissipated by the transistor. Identify all of the measures that could be taken to improve design and/or

operating conditions, such that heat dissipation may be increased while still maintaining a case temperature of 80°C. In words, assess the relative merits of each measure. Choose what you believe to be the three most promising measures, and numerically assess the effect of corresponding changes in design and/or operating conditions on thermal performance.

3.134 Water is heated by submerging 50-mm diameter, thin-walled copper tubes in a tank and passing hot combustion gases ($T_g = 750$ K) through the tubes. To enhance heat transfer to the water, four straight fins of uniform cross section, which form a cross, are inserted in each tube. The fins are 5 mm thick and are also made of copper ($k = 400$ W/m · K).

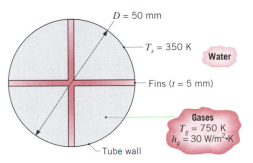

If the tube surface temperature is $T_s = 350$ K and the gas-side convection coefficient is $h_g = 30$ W/m² · K, what is the rate of heat transfer to the water per meter of pipe length?

3.135 Consider the conditions of Problem 3.134, but now allow for a tube wall thickness of 5 mm (inner and outer diameters of 50 and 60 mm), a fin-to-tube thermal contact resistance of 10^{-4} m² · K/W, and the fact that it is the water temperature, $T_w = 350$ K, rather than the tube surface temperature, that is known. The water-side convection coefficient is $h_w = 2000$ W/m² · K. Determine the rate of heat transfer per unit tube length (W/m) to the water. What would be the separate effect of each of the following design changes on the heat rate: (i) elimination of the contact resistance; (ii) increasing the number of fins from four to eight; and (iii) changing the tube wall and fin material from copper to AISI 304 stainless steel ($k = 20$ W/m · K)?

3.136 A scheme for concurrently heating separate water and air streams involves passing them through and over an array of tubes, respectively, while the tube wall is heated electrically. To enhance gas-side heat transfer, annular fins of rectangular profile are attached to the outer tube surface. Attachment is facilitated with a dielectric adhesive that electrically isolates the fins from the current-carrying tube wall.

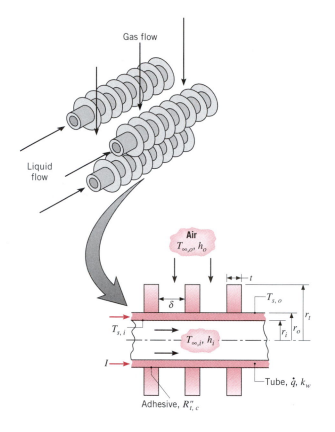

(a) Assuming uniform volumetric heat generation within the tube wall, obtain expressions for the heat rate per unit tube length (W/m) at the inner (r_i) and outer (r_o) surfaces of the wall. Express your results in terms of the tube inner and outer surface temperatures, $T_{s,i}$ and $T_{s,o}$, and other pertinent parameters.

(b) Obtain expressions that could be used to determine $T_{s,i}$ and $T_{s,o}$ in terms of parameters associated with the water- and air-side conditions.

(c) Consider conditions for which the water and air are at $T_{\infty,i} = T_{\infty,o} = 300$ K, with corresponding convection coefficients of $h_i = 2000$ W/m² · K and $h_o = 100$ W/m² · K. Heat is uniformly dissipated in a stainless steel tube ($k_w = 15$ W/m · K), having inner and outer radii of $r_i = 25$ mm and $r_o = 30$ mm, and aluminum fins ($t = \delta = 2$ mm, $r_t = 55$ mm) are attached to the outer surface, with $R''_{t,c} = 10^{-4}$ m² · K/W. Determine the heat rates and temperatures at the inner and outer surfaces as a function of the rate of volumetric heating $\dot{q}$. The upper limit to $\dot{q}$ will be determined by the constraints that $T_{s,i}$ not exceed the boiling point of water (100°C) and $T_{s,o}$ not exceed the decomposition temperature of the adhesive (250°C).

Two-Dimensional, Steady-State Conduction

*T*o this point we have restricted our attention to conduction problems in which the temperature gradient is significant for only one coordinate direction. However, in many cases such problems are grossly oversimplified if a one-dimensional treatment is used, and it is necessary to account for multidimensional effects. In this chapter, we consider several techniques for treating two-dimensional systems under steady-state conditions.

4.1
Alternative Approaches

Consider a long, prismatic solid in which two-dimensional conduction effects are important (Figure 4.1). With two surfaces insulated and the other surfaces maintained at different temperatures, $T_1 > T_2$, heat transfer by conduction will occur from surface 1 to 2. According to Fourier's law, Equation 2.3 or 2.4, the local heat flux in the solid is a vector that is everywhere perpendicular to lines of constant temperature (*isotherms*). The directions of the heat flux vector are represented by the *heat flow lines* of Figure 4.1, and the vector itself is the resultant of heat flux components in the *x* and *y* directions. These components are determined by Equation 2.6.

Recall that, in any conduction analysis, there exist two major objectives. The first objective is to determine the temperature distribution in the medium, which, for the present problem, necessitates determining $T(x, y)$. This objective is generally achieved by solving the appropriate form of the heat equation. For two-dimensional, steady-state conditions with no generation and constant thermal conductivity, this form is, from Equation 2.16,

$$\frac{\partial^2 T}{\partial x^2} + \frac{\partial^2 T}{\partial y^2} = 0 \qquad (4.1)$$

If Equation 4.1 can be solved for $T(x, y)$, it is then a simple matter to satisfy the second major objective, which is to determine the heat flux components q_x'' and q_y'' by applying the rate equations (2.6). Methods for solving Equation 4.1 include the use of *analytical, graphical,* and *numerical (finite-difference, finite-element,* or *boundary-element*) approaches.

The analytical method involves effecting an exact mathematical solution to Equation 4.1. The problem is more difficult than those considered in Chapter 3,

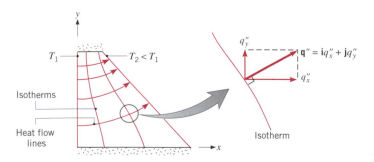

FIGURE 4.1   Two-dimensional conduction.

since it now involves a partial, rather than an ordinary, differential equation. Although several techniques are available for solving such equations, the solutions typically involve complicated mathematical series and functions and may be obtained for only a restricted set of simple geometries and boundary conditions [1–5]. Nevertheless, the solutions are of considerable value, since the dependent variable T is determined as a continuous function of the independent variables (x, y). Hence the solution could be used to compute the temperature at *any* point of interest in the medium. To illustrate the nature and importance of analytical techniques, an exact solution to Equation 4.1 is obtained in Section 4.2 by the method of *separation of variables.*

In contrast to the analytical methods, which provide *exact* results at *any* point, the graphical and numerical methods can provide only *approximate* results at *discrete* points. However, because the methods can accommodate complex geometries and boundary conditions, they often offer the only means by which multidimensional conduction problems can be solved. The graphical, or flux-plotting, method (Section 4.3) may be used to obtain a rough estimate of the temperature field, while the numerical method (Sections 4.4 and 4.5) may be used to obtain extremely accurate results for complex geometries.

4.2
The Method of Separation of Variables

To appreciate how the method of separation of variables may be used to solve two-dimensional conduction problems, we consider the system of Figure 4.2. Three sides of the rectangular plate are maintained at a constant temperature T_1, while the fourth side is maintained at a constant temperature $T_2 \neq T_1$. We are interested in the temperature distribution $T(x, y)$, but to simplify the solution we introduce the transformation

$$\theta \equiv \frac{T - T_1}{T_2 - T_1} \qquad (4.2)$$

Substituting Equation 4.2 into Equation 4.1, the transformed differential equation is then

$$\frac{\partial^2 \theta}{\partial x^2} + \frac{\partial^2 \theta}{\partial y^2} = 0 \qquad (4.3)$$

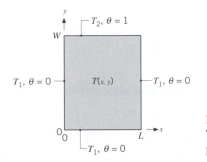

FIGURE 4.2
Two-dimensional conduction in a rectangular plate.

Since the equation is second order in both x and y, two boundary conditions are needed for each of the coordinates. They are

$$\theta(0, y) = 0 \qquad \text{and} \qquad \theta(x, 0) = 0$$

$$\theta(L, y) = 0 \qquad \text{and} \qquad \theta(x, W) = 1$$

Note that, through the transformation of Equation 4.2, three of the four boundary conditions are now homogeneous and the value of θ is restricted to the range from 0 to 1.

We now apply the separation of variables technique by assuming that the desired solution can be expressed as the product of two functions, one of which depends only on x while the other depends only on y. That is, we assume the existence of a solution of the form

$$\theta(x, y) = X(x) \cdot Y(y) \tag{4.4}$$

Substituting into Equation 4.3 and dividing by XY, we obtain

$$-\frac{1}{X}\frac{d^2X}{dx^2} = \frac{1}{Y}\frac{d^2Y}{dy^2} \tag{4.5}$$

and it is evident that the differential equation is, in fact, separable. That is, the left-hand side of the equation depends only on x and the right-hand side depends only on y. Hence the equality can apply in general (for any x or y) only if both sides are equal to the same constant. Identifying this, as yet unknown, *separation constant* as λ^2, we then have

$$\frac{d^2X}{dx^2} + \lambda^2 X = 0 \tag{4.6}$$

$$\frac{d^2Y}{dy^2} - \lambda^2 Y = 0 \tag{4.7}$$

and the partial differential equation has been reduced to two ordinary differential equations. Note that the designation of λ^2 as a positive constant was not arbitrary. If a negative value were selected or a value of $\lambda^2 = 0$ chosen, it is readily shown (Problem 4.1) that it would be impossible to obtain a solution that satisfies the prescribed boundary conditions.

The general solutions to Equations 4.6 and 4.7 are, respectively,

$$X = C_1 \cos \lambda x + C_2 \sin \lambda x$$

$$Y = C_3 e^{-\lambda y} + C_4 e^{+\lambda y}$$

in which case the general form of the two-dimensional solution is

$$\theta = (C_1 \cos \lambda x + C_2 \sin \lambda x)(C_3 e^{-\lambda y} + C_4 e^{\lambda y}) \tag{4.8}$$

Applying the condition that $\theta(0, y) = 0$, it is evident that $C_1 = 0$. In addition from the requirement that $\theta(x, 0) = 0$, we obtain

$$C_2 \sin \lambda x (C_3 + C_4) = 0$$

which may only be satisfied if $C_3 = -C_4$. Although the requirement could also be satisfied by having $C_2 = 0$, this equality would totally eliminate the x dependence and hence would provide an unacceptable solution. If we now invoke the requirement that $\theta(L, y) = 0$, we obtain

$$C_2 C_4 \sin \lambda L (e^{\lambda y} - e^{-\lambda y}) = 0$$

The only way in which this condition may be satisfied (and still have an acceptable solution) is by requiring that λ assume discrete values for which $\sin \lambda L = 0$. These values must then be of the form

$$\lambda = \frac{n\pi}{L} \qquad n = 1, 2, 3, \ldots \tag{4.9}$$

where the integer $n = 0$ is precluded, since it provides an unacceptable solution. The desired solution may now be expressed as

$$\theta = C_2 C_4 \sin \frac{n\pi x}{L} (e^{n\pi y/L} - e^{-n\pi y/L}) \tag{4.10}$$

Combining constants and acknowledging that the new constant may depend on n, we obtain

$$\theta(x, y) = C_n \sin \frac{n\pi x}{L} \sinh \frac{n\pi y}{L}$$

where we have also used the fact that $(e^{n\pi y/L} - e^{-n\pi y/L}) = 2 \sinh (n\pi y/L)$. In the above form we have really obtained an infinite number of solutions that satisfy the original differential equation and the boundary conditions. However, since the problem is linear, a more general solution may be obtained from a superposition of the form

$$\theta(x, y) = \sum_{n=1}^{\infty} C_n \sin \frac{n\pi x}{L} \sinh \frac{n\pi y}{L} \tag{4.11}$$

To determine C_n we now apply the remaining boundary condition, which is of the form

$$\theta(x, W) = 1 = \sum_{n=1}^{\infty} C_n \sin \frac{n\pi x}{L} \sinh \frac{n\pi W}{L} \tag{4.12}$$

Although Equation 4.12 would seem to be an extremely complicated relation for evaluating C_n, a standard method is available. It involves writing an analogous infinite series expansion in terms of *orthogonal functions*. An infinite set of functions $g_1(x), g_2(x), \ldots, g_n(x), \ldots$ is said to be orthogonal in the domain $a \leq x \leq b$ if

$$\int_a^b g_m(x) g_n(x) \, dx = 0 \qquad m \neq n \tag{4.13}$$

Many functions exhibit orthogonality, including the trigonmetric functions sin $(n\pi x/L)$ and cos $(n\pi x/L)$ for $0 \leq x \leq L$. Their utility in the present problem rests with the fact that any function $f(x)$ may be expressed in terms of an infinite series of orthogonal functions

$$f(x) = \sum_{n=1}^{\infty} A_n g_n(x) \tag{4.14}$$

The form of the coefficients A_n in this series may be determined by multiplying each side of the equation by $g_n(x)$ and integrating between the limits a and b.

$$\int_a^b f(x)g_n(x)\, dx = \int_a^b g_n(x) \sum_{n=1}^{\infty} A_n g_n(x)\, dx \tag{4.15}$$

However, from Equation 4.13 it is evident that all but one of the terms on the right-hand side of Equation 4.15 must be zero, leaving us with

$$\int_a^b f(x)g_n(x)\, dx = A_n \int_a^b g_n^2(x)\, dx$$

Hence

$$A_n = \frac{\int_a^b f(x)g_n(x)\, dx}{\int_a^b g_n^2(x)\, dx} \tag{4.16}$$

The properties of orthogonal functions may be used to solve Equation 4.12 for C_n by formulating an *analogous* infinite series for the appropriate form of $f(x)$. From Equation 4.12 it is evident that we should choose $f(x) = 1$ and the orthogonal function $g_n(x) = \sin(n\pi x/L)$. Substituting into Equation 4.16 we obtain

$$A_n = \frac{\int_0^L \sin \dfrac{n\pi x}{L}\, dx}{\int_0^L \sin^2 \dfrac{n\pi x}{L}\, dx} = \frac{2}{\pi}\frac{(-1)^{n+1}+1}{n}$$

Hence from Equation 4.14, we have

$$1 = \sum_{n=1}^{\infty} \frac{2}{\pi}\frac{(-1)^{n+1}+1}{n} \sin\frac{n\pi x}{L} \tag{4.17}$$

which is simply the expansion of unity in a Fourier series. Comparing Equations 4.12 and 4.17 we obtain

$$C_n = \frac{2[(-1)^{n+1}+1]}{n\pi \sinh(n\pi W/L)} \qquad n = 1,2,3,\ldots \tag{4.18}$$

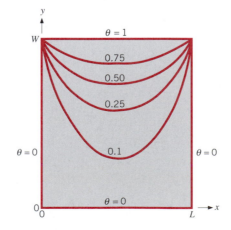

FIGURE 4.3

Isotherms for two-dimensional conduction in a rectangular plate.

Substituting Equation 4.18 into Equation 4.11, we then obtain for the final solution

$$\theta(x, y) = \frac{2}{\pi} \sum_{n=1}^{\infty} \frac{(-1)^{n+1} + 1}{n} \sin \frac{n\pi x}{L} \frac{\sinh (n\pi y/L)}{\sinh (n\pi W/L)} \qquad (4.19)$$

Equation 4.19 is a convergent series, from which the value of θ may be computed for any x and y. Representative results are shown in the form of isotherms for a schematic of the rectangular plate (Figure 4.3). The temperature T corresponding to a value of θ may be obtained for Equation 4.2. Exact solutions for other geometries and boundary conditions are provided in the literature [1–5].

4.3
The Graphical Method

The graphical method may be employed for two-dimensional problems involving adiabatic and isothermal boundaries. The approach demands some patience and an artistic flair (not to mention the use of durable paper and a good eraser) and has been superseded largely by computer solutions based on numerical procedures. Despite its limitations, however, the method may still be used to obtain a first estimate of the temperature distribution and to develop a physical appreciation for the nature of the temperature field and heat flow in a system.

4.3.1 Methodology of Constructing a Flux Plot

The rationale for the graphical method comes from the fact that lines of constant temperature must be perpendicular to lines that indicate the direction of heat flow (Figure 4.1). The objective of the graphical method is to systematically construct such a network of isotherms and heat flow lines. This network, commonly termed a *flux plot,* is used to infer the temperature distribution and the rate of heat flow through the system.

Consider a square, two-dimensional channel whose inner and outer surfaces are maintained at T_1 and T_2, respectively. A cross section of the channel is shown in Figure 4.4a. A procedure for constructing the flux plot, a portion of which is shown in Figure 4.4b, is enumerated as follows.

1. The first step in any flux plot should be to *identify all relevant lines of symmetry.* Such lines are determined by thermal, as well as geometrical, conditions. For the square channel of Figure 4.4a, such lines include the designated vertical, horizontal, and diagonal lines. For this system it is therefore possible to consider only one-eighth of the configuration, as shown in Figure 4.4b.

2. *Lines of symmetry are adiabatic* in the sense that there can be no heat transfer in a direction perpendicular to the lines. They are therefore heat flow lines and should be treated as such. Since there is no heat flow in a direction perpendicular to a heat flow line, such a line can be termed an *adiabat.*

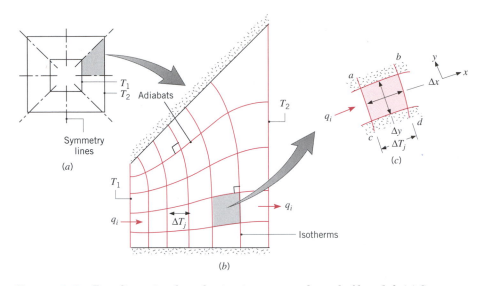

FIGURE 4.4 Two-dimensional conduction in a square channel of length *l*. (*a*) Symmetry planes. (*b*) Flux plot. (*c*) Typical curvilinear square.

3. After all known lines of constant temperature associated with the system boundaries have been identified, an attempt should be made to sketch lines of constant temperature within the system. Note that *isotherms should always be perpendicular to adiabats.*

4. The heat flow lines should then be drawn with an eye toward creating a network of *curvilinear squares.* This is done by having the *heat flow lines and isotherms intersect at right angles* and by requiring that *all sides of each square be of approximately the same length.* It is often impossible to satisfy this second requirement exactly, and it is more realistic to strive for equivalence between the sums of the opposite sides of each square, as shown in Figure 4.4*c*. Assigning the *x* coordinate to the direction of heat flow and the *y* coordinate to the direction normal to this flow, the requirement may be expressed as

$$\Delta x \equiv \frac{ab + cd}{2} \approx \Delta y \equiv \frac{ac + bd}{2} \tag{4.20}$$

It is difficult to create a satisfactory network of curvilinear squares in the first attempt, and numerous iterations must often be made. This trial-and-error process involves adjusting the isotherms and adiabats until satisfactory curvilinear squares are obtained for most of the network.[1] Once the flux plot has been obtained, it may be used to infer the temperature distribution in the medium. From a simple analysis, the heat transfer rate may then be obtained.

---

[1]In certain regions, such as corners, it may be impossible to approach the curvilinear square requirements. However, such difficulties generally have little effect on the overall accuracy of the results obtained from the flux plot.

4.3.2 Determination of the Heat Transfer Rate

The rate at which energy is conducted through a *lane,* which is the region between adjoining adiabats, is designated as q_i. If the flux plot is properly constructed, the value of q_i will be the same for all lanes and the total heat transfer rate may be expressed as

$$q = \sum_{i=1}^{M} q_i = Mq_i \tag{4.21}$$

where M is the *number of lanes* associated with the plot. From the curvilinear square of Figure 4.4c and the application of Fourier's law, q_i may be expressed as

$$q_i \approx kA_i \frac{\Delta T_j}{\Delta x} \approx k(\Delta y \cdot l) \frac{\Delta T_j}{\Delta x} \tag{4.22}$$

where ΔT_j is the temperature difference between successive isotherms, A_i is the conduction heat transfer area for the lane, and l is the length of the channel normal to the page. However, if the flux plot is properly constructed, the temperature increment is the same for all adjoining isotherms, and the overall temperature difference between boundaries, ΔT_{1-2}, may be expressed as

$$\Delta T_{1-2} = \sum_{j=1}^{N} \Delta T_j = N \, \Delta T_j \tag{4.23}$$

where N is the total number of temperature increments. Combining Equations 4.21 to 4.23 and recognizing that $\Delta x \approx \Delta y$ for curvilinear squares, we obtain

$$q \approx \frac{Ml}{N} k \, \Delta T_{1-2} \tag{4.24}$$

The manner in which a flux plot may be used to obtain the heat transfer rate for a two-dimensional system is evident from Equation 4.24. The ratio of the number of heat flow lanes to the number of temperature increments (the value of M/N) may be obtained from the plot. Recall that specification of N is based on step 3 of the foregoing procedure, and the value, which is an integer, may be made large or small depending on the desired accuracy. The value of M is then a consequence of following step 4. Note that M is not necessarily an integer, since a fractional lane may be needed to arrive at a satisfactory network of curvilinear squares. For the network of Figure 4.4b, $N = 6$ and $M = 5$. Of course, as the network, or *mesh,* of curvilinear squares is made finer, N and M increase and the estimate of M/N becomes more accurate.

4.3.3 The Conduction Shape Factor

Equation 4.24 may be used to define the *shape factor, S,* of a two-dimensional system. That is, the heat transfer rate may be expressed as

$$q = Sk \, \Delta T_{1-2} \tag{4.25}$$

where, for a flux plot,

$$S \equiv \frac{Ml}{N} \tag{4.26}$$

From Equation 4.25, it also follows that a *two-dimensional conduction resistance* may be expressed as

$$R_{t,\,cond(2D)} = \frac{1}{Sk} \tag{4.27}$$

Shape factors have been obtained for numerous two-dimensional systems, and results are summarized in Table 4.1 for some common configurations. In each case, two-dimensional conduction is presumed to occur between boundaries that are maintained at uniform temperatures, with $\Delta T_{1-2} \equiv T_1 - T_2$. Shape factors may also be defined for one-dimensional geometries, and from the results of Table 3.3, it follows that for plane, cylindrical, and spherical walls, respectively, the shape factors are A/L, $2\pi L/\ln(r_2/r_1)$, and $4\pi r_1 r_2/(r_2 - r_1)$. Results are available for many other configurations [6–9].

TABLE 4.1 Conduction shape factors for selected two-dimensional systems $[q = Sk(T_1 - T_2)]$

System	Schematic	Restrictions	Shape Factor
Case 1. Isothermal sphere buried in a semi-infinite medium		$z > D/2$	$\dfrac{2\pi D}{1 - D/4z}$
Case 2. Horizontal isothermal cylinder of length L buried in a semi-infinite medium		$L \gg D$	$\dfrac{2\pi L}{\cosh^{-1}(2z/D)}$
		$L \gg D$ $z > 3D/2$	$\dfrac{2\pi L}{\ln(4z/D)}$
Case 3. Vertical cylinder in a semi-infinite medium		$L \gg D$	$\dfrac{2\pi L}{\ln(4L/D)}$
Case 4. Conduction between two cylinders of length L in infinite medium		$L \gg D_1, D_2$ $L \gg w$	$\dfrac{2\pi L}{\cosh^{-1}\left(\dfrac{4w^2 - D_1^2 - D_2^2}{2D_1 D_2}\right)}$
Case 5. Horizontal circular cylinder of length L midway between parallel planes of equal length and infinite width		$z \gg D/2$ $L \gg z$	$\dfrac{2\pi L}{\ln(8z/\pi D)}$

TABLE 4.1 *Continued*

System	Schematic	Restrictions	Shape Factor
Case 6. Circular cylinder of length L centered in a square solid of equal length		$w > D$ $L \gg w$	$\dfrac{2\pi L}{\ln\,(1.08\,w/D)}$
Case 7. Eccentric circular cylinder of length L in a cylinder of equal length		$D > d$ $L \gg D$	$\dfrac{2\pi L}{\cosh^{-1}\left(\dfrac{D^2 + d^2 - 4z^2}{2Dd}\right)}$
Case 8. Conduction through the edge of adjoining walls		$D > L/5$	$0.54D$
Case 9. Conduction through corner of three walls with a temperature difference ΔT_{1-2} across the walls		$L \ll$ length and width of wall	$0.15L$
Case 10. Disk of diameter D and T_1 on a semi-infinite medium of thermal conductivity k and T_2		None	$2D$

EXAMPLE 4.1

A hole of diameter $D = 0.25$ m is drilled through the center of a solid block of square cross section with $w = 1$ m on a side. The hole is drilled along the length, $l = 2$ m, of the block, which has a thermal conductivity of $k = 150$ W/m · K. A warm fluid passing through the hole maintains an inner surface temperature of $T_1 = 75°C$, while the outer surface of the block is kept at $T_2 = 25°C$.

1. Using the flux plot method, determine the shape factor for the system.
2. What is the rate of heat transfer through the block?

SOLUTION

Known: Dimensions and thermal conductivity of a block with a circular hole drilled along its length.

Find:

1. Shape factor.
2. Heat transfer rate for prescribed surface temperatures.

Schematic:

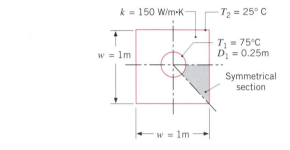

Assumptions:

1. Steady-state conditions.
2. Two-dimensional conduction.
3. Constant properties.
4. Ends of block are well insulated.

Analysis:

1. The flux plot may be simplified by identifying lines of symmetry and reducing the system to the one-eighth section shown in the schematic. Using a fairly coarse grid involving $N = 6$ temperature increments, the flux plot was generated. The resulting network of curvilinear squares is as follows.

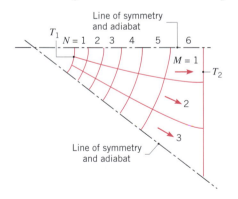

With the number of heat flow lanes for the section corresponding to $M = 3$, it follows from Equation 4.26 that the shape factor for the entire block is

$$S = 8 \frac{Ml}{N} = 8 \frac{3 \times 2 \text{ m}}{6} = 8 \text{ m} \qquad \triangleleft$$

where the factor of 8 results from the number of symmetrical sections. The accuracy of this result may be determined by referring to Table 4.1, where, for the prescribed system, it follows that

$$S = \frac{2\pi L}{\ln(1.08 \, w/D)} = \frac{2\pi \times 2 \text{ m}}{\ln(1.08 \times 1 \text{ m}/0.25 \text{ m})} = 8.59 \text{ m}$$

Hence the result of the flux plot underpredicts the shape factor by approximately 7%. Note that, although the requirement $l \gg w$ is not satisfied for this problem, the shape factor result from Table 4.1 remains valid if there is negligible axial conduction in the block. This condition is satisified if the ends are insulated.

2. Using $S = 8.59$ m with Equation 4.25, the heat rate is

$$q = Sk(T_1 - T_2)$$

$$q = 8.59 \text{ m} \times 150 \text{ W/m} \cdot \text{K} \, (75 - 25)°\text{C} = 64.4 \text{ kW} \qquad \triangleleft$$

Comments: The accuracy of the flux plot may be improved by using a finer grid (increasing the value of N). How would the symmetry and heat flow lines change if the vertical sides were insulated? If one vertical and one horizontal side were insulated? If both vertical and one horizontal side were insulated?

4.4
Finite-Difference Equations

As discussed in Sections 4.1 and 4.2, analytical methods may be used, in certain cases, to effect exact mathematical solutions to steady, two-dimensional conduction problems. These solutions have been generated for an assortment of simple geometries and boundary conditions and are well documented in the literature [1–5]. However, more often than not, two-dimensional problems involve geometries and/or boundary conditions that preclude such solutions. In these cases, the best alternative is often one that uses a *numerical* technique such as the *finite-difference, finite-element,* or *boundary-element* method. Because of its ease of application, the finite-difference method is well suited for an introductory treatment of numerical techniques.

4.4.1   The Nodal Network

In contrast to an analytical solution, which allows for temperature determination at *any* point of interest in a medium, a numerical solution enables determination of the temperature at only *discrete* points. The first step in any numerical analysis must therefore be to select these points. Referring to Figure 4.5, this is done by subdividing the medium of interest into a number of small regions and assigning to each a reference point that is at its center. The reference point is frequently termed a *nodal point* (or simply a *node*), and the aggregate of points is termed a *nodal network, grid,* or *mesh.* The nodal points are designated by a numbering scheme that, for a two-dimensional system, may take the form shown in Figure 4.5*a*. The x and y locations are designated by the m and n indices, respectively.

Each node represents a certain region, and its temperature is a measure of the *average* temperature of the region. For example, the temperature of the

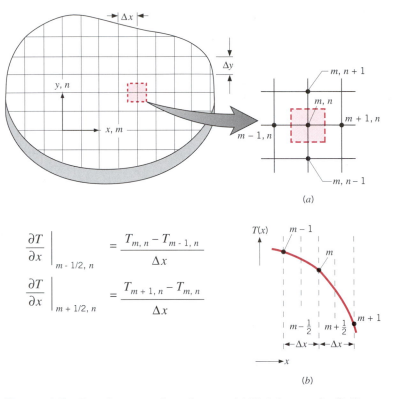

FIGURE 4.5   Two-dimensional conduction. (*a*) Nodal network. (*b*) Finite-difference approximation.

node *m, n* of Figure 4.5*a* may be viewed as the average temperature of the surrounding shaded area. The selection of nodal points is rarely arbitrary, depending often on matters such as geometric convenience and the desired accuracy. The numerical accuracy of the calculations depends strongly on the number of designated nodal points. If this number is large (a *fine mesh*), extremely accurate solutions can be obtained.

4.4.2   Finite-Difference Form of the Heat Equation

Determination of the temperature distribution numerically dictates that an appropriate conservation equation be written for *each* of the nodal points of unknown temperature. The resulting set of equations may then be solved simultaneously for the temperature at each node. For *any interior* node of a two-dimensional system with no generation and uniform thermal conductivity, the *exact* form of the energy conservation requirement is given by the heat equation, Equation 4.1. However, if the system is characterized in terms of a nodal network, it is necessary to work with an *approximate,* or *finite-difference,* form of this equation.

A finite-difference equation that is suitable for the interior nodes of a two-dimensional system may be inferred directly from Equation 4.1. Consider the second derivative, $\partial^2 T/\partial x^2$. From Figure 4.5*b*, the value of this derivative at the *m, n* nodal point may be approximated as

$$\frac{\partial^2 T}{\partial x^2}\bigg|_{m,n} \approx \frac{\partial T/\partial x|_{m+1/2,\,n} - \partial T/\partial x|_{m-1/2,\,n}}{\Delta x} \qquad (4.28)$$

The temperature gradients may in turn be expressed as a function of the nodal temperatures. That is,

$$\left.\frac{\partial T}{\partial x}\right|_{m+1/2,\,n} \approx \frac{T_{m+1,\,n} - T_{m,\,n}}{\Delta x} \tag{4.29}$$

$$\left.\frac{\partial T}{\partial x}\right|_{m-1/2,\,n} \approx \frac{T_{m,\,n} - T_{m-1,\,n}}{\Delta x} \tag{4.30}$$

Substituting Equations 4.29 and 4.30 into 4.28, we obtain

$$\left.\frac{\partial^2 T}{\partial x^2}\right|_{m,\,n} \approx \frac{T_{m+1,\,n} + T_{m-1,\,n} - 2T_{m,\,n}}{(\Delta x)^2} \tag{4.31}$$

Proceeding in a similar fashion, it is readily shown that

$$\left.\frac{\partial^2 T}{\partial y^2}\right|_{m,\,n} \approx \frac{\partial T/\partial y|_{m,\,n+1/2} - \partial T/\partial y|_{m,\,n-1/2}}{\Delta y}$$

$$\approx \frac{T_{m,\,n+1} + T_{m,\,n-1} - 2T_{m,\,n}}{(\Delta y)^2} \tag{4.32}$$

Using a network for which $\Delta x = \Delta y$ and substituting Equations 4.31 and 4.32 into Equation 4.1, we obtain

$$T_{m,\,n+1} + T_{m,\,n-1} + T_{m+1,\,n} + T_{m-1,\,n} - 4T_{m,\,n} = 0 \tag{4.33}$$

Hence for the m, n node the heat equation, which is an *exact differential equation,* is reduced to an *approximate algebraic equation.* This approximate, *finite-difference form of the heat equation* may be applied to any interior node that is equidistant from its four neighboring nodes. It simply requires that the sum of the temperatures associated with the neighboring nodes be four times the temperature of the node of interest.

4.4.3 The Energy Balance Method

The finite-difference equation for a node may also be obtained by applying conservation of energy to a control volume about the nodal region. Since the actual direction of heat flow (into or out of the node) is often unknown, it is convenient to formulate the energy balance by *assuming* that *all* the heat flow is *into the node.* Such a condition is, of course, impossible, but if the rate equations are expressed in a manner consistent with this assumption, the correct form of the finite-difference equation is obtained. For steady-state conditions with generation, the appropriate form of Equation 1.11a is then

$$\dot{E}_{\text{in}} + \dot{E}_{g} = 0 \tag{4.34}$$

Consider applying Equation 4.34 to a control volume about the interior node m, n of Figure 4.6. For two-dimensional conditions, energy exchange is influenced by conduction between m, n and its four adjoining nodes, as well as by generation. Hence Equation 4.34 reduces to

$$\sum_{i=1}^{4} q_{(i) \rightarrow (m,\,n)} + \dot{q}(\Delta x \cdot \Delta y \cdot 1) = 0$$

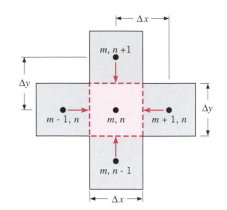

FIGURE 4.6

Conduction to an interior node from its adjoining nodes.

where i refers to the neighboring nodes, $q_{(i) \to (m, n)}$ is the conduction rate between nodes, and unit depth is assumed. To evaluate the conduction rate terms, we *assume* that conduction transfer occurs exclusively through *lanes* that are oriented in either the x or y direction. Simplified forms of Fourier's law may therefore be used. For example, the rate at which energy is transferred by conduction from node $m - 1, n$ to m, n may be expressed as

$$q_{(m-1, n) \to (m, n)} = k(\Delta y \cdot 1) \frac{T_{m-1, n} - T_{m, n}}{\Delta x} \quad (4.35)$$

The quantity $(\Delta y \cdot 1)$ is the heat transfer area, and the term $(T_{m-1, n} - T_{m, n})/\Delta x$ is the finite-difference approximation to the temperature gradient at the boundary between the two nodes. The remaining conduction rates may then be expressed as

$$q_{(m+1, n) \to (m, n)} = k(\Delta y \cdot 1) \frac{T_{m+1, n} - T_{m, n}}{\Delta x} \quad (4.36)$$

$$q_{(m, n+1) \to (m, n)} = k(\Delta x \cdot 1) \frac{T_{m, n+1} - T_{m, n}}{\Delta y} \quad (4.37)$$

$$q_{(m, n-1) \to (m, n)} = k(\Delta x \cdot 1) \frac{T_{m, n-1} - T_{m, n}}{\Delta y} \quad (4.38)$$

Note that in evaluating each conduction rate, we have subtracted the temperature of the m, n node from the temperature of its adjoining node. This convention is necessitated by the assumption of heat flow into m, n and is consistent with the direction of the arrows shown in Figure 4.6. Substituting Equations 4.35 through 4.38 into the energy balance and remembering that $\Delta x = \Delta y$, it follows that the finite-difference equation for an interior node with generation is

$$T_{m, n+1} + T_{m, n-1} + T_{m+1, n} + T_{m-1, n} + \frac{\dot{q}(\Delta x)^2}{k} - 4T_{m, n} = 0 \quad (4.39)$$

If there is no internally distributed source of energy ($\dot{q} = 0$), this expression reduces to Equation 4.33.

It is important to note that a finite-difference equation is needed for each nodal point at which the temperature is unknown. However, it is not always possible to classify all such points as interior and hence to use Equation 4.33 or 4.39. For example, the temperature may be unknown at an insulated surface or at a surface that is exposed to convective conditions. For points on such surfaces, the finite-difference equation must be obtained by applying the energy balance method.

To illustrate this method, consider the node corresponding to the internal corner of Figure 4.7. This node represents the three-quarter shaded section and exchanges energy by convection with an adjoining fluid at T_∞. Conduction to the nodal region (m, n) will occur along four different lanes from neighboring nodes in the solid. The conduction heat rates q_{cond} may be expressed as

$$q_{(m-1, n) \rightarrow (m, n)} = k(\Delta y \cdot 1) \frac{T_{m-1, n} - T_{m, n}}{\Delta x} \tag{4.40}$$

$$q_{(m, n+1) \rightarrow (m, n)} = k(\Delta x \cdot 1) \frac{T_{m, n+1} - T_{m, n}}{\Delta y} \tag{4.41}$$

$$q_{(m+1, n) \rightarrow (m, n)} = k\left(\frac{\Delta y}{2} \cdot 1\right) \frac{T_{m+1, n} - T_{m, n}}{\Delta x} \tag{4.42}$$

$$q_{(m, n-1) \rightarrow (m, n)} = k\left(\frac{\Delta x}{2} \cdot 1\right) \frac{T_{m, n-1} - T_{m, n}}{\Delta y} \tag{4.43}$$

Note that the areas for conduction from nodal regions $(m - 1, n)$ and $(m, n + 1)$ are proportional to Δy and Δx, respectively, whereas conduction from $(m + 1, n)$ and $(m, n - 1)$ occurs along lanes of half-width $\Delta y/2$ and $\Delta x/2$, respectively.

Conditions in the nodal region m, n are also influenced by convective exchange with the fluid, and this exchange may be viewed as occurring along half-lanes in the x and y directions. The total convection rate q_{conv} may be expressed as

$$q_{(\infty) \rightarrow (m, n)} = h\left(\frac{\Delta x}{2} \cdot 1\right)(T_\infty - T_{m, n}) + h\left(\frac{\Delta y}{2} \cdot 1\right)(T_\infty - T_{m, n}) \tag{4.44}$$

Implicit in this expression is the assumption that the exposed surfaces of the corner are at a uniform temperature corresponding to the nodal temperature $T_{m, n}$. This assumption is consistent with the concept that the entire nodal region is characterized by a single temperature, which represents an average of the ac-

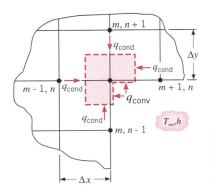

FIGURE 4.7
Formulation of the finite-difference equation for an internal corner of a solid with surface convection.

TABLE 4.2 Summary of nodal finite-difference equations

Configuration	Finite-Difference Equation for $\Delta x = \Delta y$

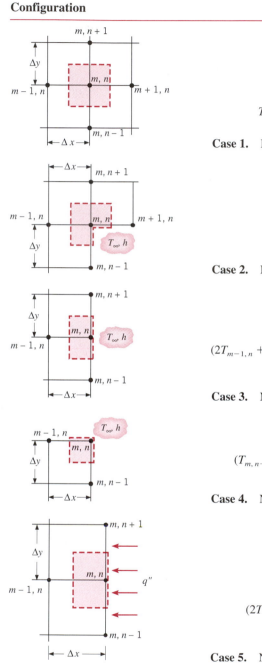

$$T_{m,n+1} + T_{m,n-1} + T_{m+1,n} + T_{m-1,n} - 4T_{m,n} = 0 \tag{4.33}$$

Case 1. Interior node

$$2(T_{m-1,n} + T_{m,n+1}) + (T_{m+1,n} + T_{m,n-1})$$
$$+ 2\frac{h\,\Delta x}{k}T_\infty - 2\left(3 + \frac{h\,\Delta x}{k}\right)T_{m,n} = 0 \tag{4.45}$$

Case 2. Node at an internal corner with convection

$$(2T_{m-1,n} + T_{m,n+1} + T_{m,n-1}) + \frac{2h\,\Delta x}{k}T_\infty - 2\left(\frac{h\,\Delta x}{k} + 2\right)T_{m,n} = 0 \tag{4.46}^{a}$$

Case 3. Node at a plane surface with convection

$$(T_{m,n-1} + T_{m-1,n}) + 2\frac{h\,\Delta x}{k}T_\infty - 2\left(\frac{h\,\Delta x}{k} + 1\right)T_{m,n} = 0 \tag{4.47}$$

Case 4. Node at an external corner with convection

$$(2T_{m-1,n} + T_{m,n+1} + T_{m,n-1}) + \frac{2q''\,\Delta x}{k} - 4T_{m,n} = 0 \tag{4.48}^{b}$$

Case 5. Node at a plane surface with uniform heat flux

[a,b] To obtain the finite-difference equation for an adiabatic surface (or surface of symmetry), simply set h or q'' equal to zero.

tual temperature distribution in the region. In the absence of transient, three-dimensional, and generation effects, conservation of energy, Equation 4.34, requires that the sum of Equations 4.40 to 4.44 be zero. Summing these equations and rearranging, we therefore obtain

$$T_{m-1,n} + T_{m,n+1} + \tfrac{1}{2}(T_{m+1,n} + T_{m,n-1}) + \frac{h\,\Delta x}{k}\,T_{\infty} - \left(3 + \frac{h\,\Delta x}{k}\right)T_{m,n} = 0$$

(4.45)

where again the mesh is such that $\Delta x = \Delta y$.

Nodal energy balance equations pertinent to several common geometries are presented in Table 4.2.

EXAMPLE 4.2

Using the energy balance method, derive the finite-difference equation for the m, n nodal point located on a plane, insulated surface of a medium with uniform heat generation.

SOLUTION

Known: Network of nodal points adjoining an insulated surface.

Find: Finite-difference equation for the surface nodal point.

Schematic:

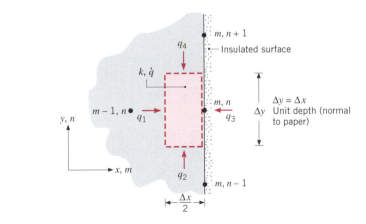

Assumptions:

1. Steady-state conditions.
2. Two-dimensional conduction.
3. Constant properties.
4. Uniform internal heat generation.

Analysis: Applying the energy conservation requirement, Equation 4.34, to the control surface about the region $(\Delta x/2 \cdot \Delta y \cdot 1)$ associated with the m, n node, it follows that, with volumetric heat generation at a rate $\dot{q}$,

$$q_1 + q_2 + q_3 + q_4 + \dot{q}\left(\frac{\Delta x}{2} \cdot \Delta y \cdot 1\right) = 0$$

where

$$q_1 = k(\Delta y \cdot 1)\frac{T_{m-1,n} - T_{m,n}}{\Delta x}$$

$$q_2 = k\left(\frac{\Delta x}{2} \cdot 1\right)\frac{T_{m,n-1} - T_{m,n}}{\Delta y}$$

$$q_3 = 0$$

$$q_4 = k\left(\frac{\Delta x}{2} \cdot 1\right)\frac{T_{m,n+1} - T_{m,n}}{\Delta y}$$

Substituting into the energy balance and dividing by $k/2$, it follows that

$$2T_{m-1,n} + T_{m,n-1} + T_{m,n+1} - 4T_{m,n} + \frac{\dot{q}(\Delta x \cdot \Delta y)}{k} = 0 \qquad \triangleleft$$

Comments: The same result could be obtained by using the symmetry condition, $T_{m+1,n} = T_{m-1,n}$, with the finite-difference equation (Equation 4.39) for an interior nodal point. If $\dot{q} = 0$, the desired result could also be obtained by setting $h = 0$ in Equation 4.46 (Table 4.2).

It is useful to note that heat rates between adjoining nodes may also be formulated in terms of the corresponding thermal resistances. Referring, for example, to Figure 4.7, the rate of heat transfer by conduction from node $(m - 1, n)$ to (m, n) may be expressed as

$$q_{(m-1,n) \rightarrow (m,n)} = \frac{T_{m-1,n} - T_{m,n}}{R_{t,\,\text{cond}}} = \frac{T_{m-1,n} - T_{m,n}}{\Delta x/k\,(\Delta y \cdot 1)}$$

yielding a result that is equivalent to that of Equation 4.40. Similarly, the rate of heat transfer by convection to (m, n) may be expressed as

$$q_{(\infty) \rightarrow (m,n)} = \frac{T_\infty - T_{m,n}}{R_{t,\,\text{conv}}} = \frac{T_\infty - T_{m,n}}{\{h[(\Delta x/2) \cdot 1 + (\Delta y/2) \cdot 1]\}^{-1}}$$

which is equivalent to Equation 4.44.

As an example of the utility of resistance concepts, consider an interface that separates two dissimilar materials and is characterized by a thermal contact resistance $R''_{t,\,c}$ (Figure 4.8). The rate of heat transfer from node (m, n) to $(m, n - 1)$ may be expressed as

$$q_{(m,n) \rightarrow (m,n-1)} = \frac{T_{m,n} - T_{m,n-1}}{R_{\text{tot}}} \qquad (4.49)$$

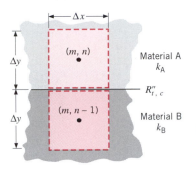

FIGURE 4.8
Conduction between adjoining, dissimilar materials
with an interface contact resistance.

where, for a unit depth,

$$R_{\text{tot}} = \frac{\Delta y/2}{k_A(\Delta x \cdot 1)} + \frac{R''_{t,c}}{\Delta x \cdot 1} + \frac{\Delta y/2}{k_B(\Delta x \cdot 1)} \tag{4.50}$$

4.5
Finite-Difference Solutions

Once the nodal network has been established and an appropriate finite-difference equation has been written for each node, the temperature distribution may be determined. The problem reduces to one of solving a system of linear, algebraic equations. Numerous methods are available for this purpose and may be classified according to whether they are *direct* or *iterative*. Direct methods involve a fixed and predetermined number of arithmetic operations and are suitable for use when the number of equations (unknown nodal temperatures) is small. However, such methods are subject to excessive computer memory and time requirements, and it is often more efficient to use an iterative technique. Although the required number of arithmetic operations cannot be predetermined, iterative methods are characterized by reduced computer requirements and are especially appropriate when the number of equations is large.

In this section we consider matrix inversion and Gauss–Seidel iteration as examples of the direct and iterative methods, respectively. More detailed descriptions of such procedures, as well as related algorithms, may be found in the literature [10, 11].

4.5.1   The Matrix Inversion Method

Consider a system of N finite-difference equations corresponding to N unknown temperatures. Identifying the nodes by a single integer subscript, rather than by the double subscript (m, n), the procedure for performing a matrix inversion begins by expressing the equations as

$$a_{11}T_1 + a_{12}T_2 + a_{13}T_3 + \cdots + a_{1N}T_N = C_1$$

$$a_{21}T_1 + a_{22}T_2 + a_{23}T_3 + \cdots + a_{2N}T_N = C_2$$

$$\vdots \qquad \vdots \qquad \vdots \qquad \vdots \qquad \vdots \qquad \vdots$$

$$a_{N1}T_1 + a_{N2}T_2 + a_{N3}T_3 + \cdots + a_{NN}T_N = C_N \tag{4.51}$$

where the quantities $a_{11}, a_{12}, \ldots, C_1, \ldots$ are known coefficients and constants involving quantities such as Δx, k, h, and T_∞. Using matrix notation, these equations may be expressed as

$$[A][T] = [C] \tag{4.52}$$

where

$$A \equiv \begin{bmatrix} a_{11} & a_{12} & \cdots & a_{1N} \\ a_{21} & a_{22} & \cdots & a_{2N} \\ \vdots & \vdots & & \vdots \\ a_{N1} & a_{N2} & \cdots & a_{NN} \end{bmatrix}, \qquad T \equiv \begin{bmatrix} T_1 \\ T_2 \\ \vdots \\ T_n \end{bmatrix}, \qquad C \equiv \begin{bmatrix} C_1 \\ C_2 \\ \vdots \\ C_N \end{bmatrix}$$

The *coefficient matrix* $[A]$ is square ($N \times N$), and its *elements* are designated by a double subscript notation, for which the first and second subscripts refer to rows and columns, respectively. The matrices $[T]$ and $[C]$ have a single column and are known as *column vectors*. Typically, they are termed the *solution* and *right-hand side vectors*, respectively. If the matrix multiplication implied by the left-hand side of Equation 4.52 is performed, Equations 4.51 are obtained.

The solution vector may now be expressed as

$$[T] = [A]^{-1}[C] \tag{4.53}$$

where $[A]^{-1}$ is the inverse of $[A]$ and is defined as

$$[A]^{-1} \equiv \begin{bmatrix} b_{11} & b_{12} & \cdots & b_{1N} \\ b_{21} & b_{22} & \cdots & b_{2N} \\ \vdots & \vdots & & \vdots \\ b_{N1} & b_{N2} & \cdots & b_{NN} \end{bmatrix}$$

Evaluating the right-hand side of Equation 4.53, it follows that

$$T_1 = b_{11}C_1 + b_{12}C_2 + \cdots + b_{1N}C_N$$

$$T_2 = b_{21}C_1 + b_{22}C_2 + \cdots + b_{2N}C_N$$

$$\vdots \qquad \vdots \qquad \vdots \qquad \vdots \qquad \vdots$$

$$T_N = b_{N1}C_1 + b_{N2}C_2 + \cdots + b_{NN}C_N \tag{4.54}$$

and the problem reduces to one of determining $[A]^{-1}$. That is, if $[A]$ is inverted, its elements $b_{11}, b_{12}, \ldots$ may be determined and the unknown temperatures may be computed from the above expressions.

Matrix inversion may readily be performed on a programmable calculator or a personal computer, depending on the size of the matrix. Hence the method provides a *convenient* means of solving two-dimensional conduction problems. Despite this convenience, however, matrix inversion is rarely numerically efficient, and it is often preferable to use an iterative numerical procedure.

4.5.2   Gauss–Seidel Iteration

The Gauss–Seidel method is a powerful and extremely popular iterative technique. Application to the system of equations represented by Equation 4.51 is facilitated by the following procedure.

1. To whatever extent possible, the equations should be reordered to provide diagonal elements whose magnitudes are larger than those of other elements in the same row. That is, it is desirable to sequence the equations such that $|a_{11}| > |a_{12}|, |a_{13}|, \ldots, |a_{1N}|; |a_{22}| > |a_{21}|, |a_{23}|, \ldots, |a_{2N}|$; and so on.

2. After reordering, each of the N equations should be written in explicit form for the temperature associated with its diagonal element. Each temperature in the solution vector would then be of the form

$$T_i^{(k)} = \frac{C_i}{a_{ii}} - \sum_{j=1}^{i-1} \frac{a_{ij}}{a_{ii}} T_j^{(k)} - \sum_{j=i+1}^{N} \frac{a_{ij}}{a_{ii}} T_j^{(k-1)} \qquad (4.55)$$

where $i = 1, 2, \ldots, N$. The superscript k refers to the level of the iteration.

3. An initial ($k = 0$) value is assumed for each temperature T_i. Subsequent computations may be reduced by selecting values based on rational estimates.

4. New values of T_i are then calculated by substituting assumed ($k = 0$) or new ($k = 1$) values of T_j into the right-hand side of Equation 4.55. This step is the first iteration ($k = 1$).

5. Using Equation 4.55, the iteration procedure is continued by calculating new values of $T_i^{(k)}$ from the $T_j^{(k)}$ values of the current iteration, where $1 \leq j \leq i - 1$, and the $T_j^{(k-1)}$ values of the previous iteration, where $i + 1 \leq j \leq N$.

6. The iteration is terminated when a prescribed *convergence criterion* is satisfied. The criterion may be expressed as

$$\left| T_i^{(k)} - T_i^{(k-1)} \right| \leq \varepsilon \qquad (4.56)$$

where ε represents an error in the temperature, which is considered to be acceptable.

If step 1 can be accomplished for each equation, the resulting system is said to be *diagonally dominant,* and the rate of convergence is maximized (the number of required iterations is minimized). However, convergence may also be achieved in many situations for which diagonal dominance cannot be obtained, although the rate of convergence is slowed. The manner in which new values of T_i are computed (steps 4 and 5) should also be noted. Because the T_i for a particular iteration are calculated sequentially, each value can be computed by using the *most recent estimates* of the other T_i. This feature is implicit in Equation 4.55, where the value of each unknown is updated as soon as possible, that is, for $1 \leq j \leq i - 1$.

EXAMPLE 4.3

A large industrial furnace is supported on a long column of fireclay brick, which is 1 m by 1 m on a side. During steady-state operation, installation is such that three surfaces of the column are maintained at 500 K while the remaining surface is exposed to an airstream for which $T_\infty = 300$ K and $h = 10$ W/m² · K. Using a grid of $\Delta x = \Delta y = 0.25$ m, determine the two-dimensional temperature distribution in the column and the heat rate to the airstream per unit length of column.

SOLUTION

Known: Dimensions and surface conditions of a support column.

Find: Temperature distribution and heat rate per unit length.

Schematic:

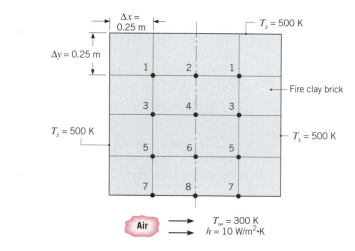

Assumptions:

1. Steady-state conditions.
2. Two-dimensional conduction.
3. Constant properties.
4. No internal heat generation.

Properties: Table A.3, fireclay brick ($T \approx 478$ K): $k = 1$ W/m · K.

Analysis: The prescribed grid consists of 12 nodal points at which the temperature is unknown. However, the number of unknowns is reduced to 8 through symmetry, in which case the temperature of nodal points to the left of the symmetry line must equal the temperature of those to the right.

Nodes 1, 3, and 5 are interior points for which the finite-difference equations may be inferred from Equation 4.33. Hence

$$\text{Node 1:} \quad T_2 + T_3 + 1000 - 4T_1 = 0$$
$$\text{Node 3:} \quad T_1 + T_4 + T_5 + 500 - 4T_3 = 0$$
$$\text{Node 5:} \quad T_3 + T_6 + T_7 + 500 - 4T_5 = 0$$

Equations for points 2, 4, and 6 may be obtained in a like manner or, since they lie on a symmetry adiabat, by using Equation 4.46 with $h = 0$. Hence

$$\text{Node 2:} \quad 2T_1 + T_4 + 500 - 4T_2 = 0$$
$$\text{Node 4:} \quad T_2 + 2T_3 + T_6 - 4T_4 = 0$$
$$\text{Node 6:} \quad T_4 + 2T_5 + T_8 - 4T_6 = 0$$

From Equation 4.46 and the fact that $h\,\Delta x/k = 2.5$, it also follows that

Node 7: $2T_5 + T_8 + 2000 - 9T_7 = 0$
Node 8: $2T_6 + 2T_7 + 1500 - 9T_8 = 0$

Having the required finite-difference equations, a matrix inversion solution may be obtained by rearranging them as follows:

$$
\begin{aligned}
-4T_1 + T_2 + T_3 + 0 + 0 + 0 + 0 + 0 &= -1000 \\
2T_1 - 4T_2 + 0 + T_4 + 0 + 0 + 0 + 0 &= -500 \\
T_1 + 0 - 4T_3 + T_4 + T_5 + 0 + 0 + 0 &= -500 \\
0 + T_2 + 2T_3 - 4T_4 + 0 + T_6 + 0 + 0 &= 0 \\
0 + 0 + T_3 + 0 - 4T_5 + T_6 + T_7 + 0 &= -500 \\
0 + 0 + 0 + T_4 + 2T_5 - 4T_6 + 0 + T_8 &= 0 \\
0 + 0 + 0 + 0 + 2T_5 + 0 - 9T_7 + T_8 &= -2000 \\
0 + 0 + 0 + 0 + 0 + 2T_6 + 2T_7 - 9T_8 &= -1500
\end{aligned}
$$

In matrix notation, following Equation 4.52, these equations are of the form $[A][T] = [C]$, where

$$
[A] = \begin{bmatrix}
-4 & 1 & 1 & 0 & 0 & 0 & 0 & 0 \\
2 & -4 & 0 & 1 & 0 & 0 & 0 & 0 \\
1 & 0 & -4 & 1 & 1 & 0 & 0 & 0 \\
0 & 1 & 2 & -4 & 0 & 1 & 0 & 0 \\
0 & 0 & 1 & 0 & -4 & 1 & 1 & 0 \\
0 & 0 & 0 & 1 & 2 & -4 & 0 & 1 \\
0 & 0 & 0 & 0 & 2 & 0 & -9 & 1 \\
0 & 0 & 0 & 0 & 0 & 2 & 2 & -9
\end{bmatrix}
\qquad
[C] = \begin{bmatrix}
-1000 \\
-500 \\
-500 \\
0 \\
-500 \\
0 \\
-2000 \\
-1500
\end{bmatrix}
$$

Using a standard matrix inversion routine, it is a simple matter to find the inverse of $[A]$, $[A]^{-1}$, giving

$$[T] = [A]^{-1}[C]$$

where

$$
[T] = \begin{bmatrix}
T_1 \\
T_2 \\
T_3 \\
T_4 \\
T_5 \\
T_6 \\
T_7 \\
T_8
\end{bmatrix}
=
\begin{bmatrix}
489.30 \\
485.15 \\
472.07 \\
462.01 \\
436.95 \\
418.74 \\
356.99 \\
339.05
\end{bmatrix}
\text{K}
$$

The heat rate from the column to the airstream may be computed from the expression

$$\left(\frac{q}{L}\right) = 2h\left[\left(\frac{\Delta x}{2}\right)(T_s - T_\infty) + \Delta x(T_7 - T_\infty) + \left(\frac{\Delta x}{2}\right)(T_8 - T_\infty)\right]$$

where the factor of 2 outside the brackets originates from the symmetry condition. Hence

$$\left(\frac{q}{L}\right) = 2 \times 10 \text{ W/m}^2 \cdot \text{K} \, [(0.125 \text{ m } (200 \text{ K})$$

$$+ \, 0.25 \text{ m } (56.99 \text{ K}) + 0.125 \text{ m } (39.05 \text{ K})] = 883 \text{ W/m} \quad \triangleleft$$

Comments:

1. To ensure that no errors have been made in formulating the finite-difference equations or in effecting their solution, a check should be made to verify that the results satisfy conservation of energy for the nodal network. For steady-state conditions, the requirement dictates that the rate of energy inflow be balanced by the rate of outflow *for a control surface about those nodal regions whose temperatures have been evaluated.*

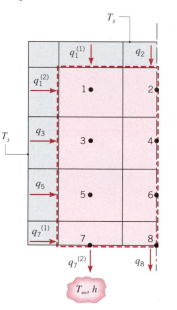

For the one-half symmetrical section shown schematically, it follows that conduction into the nodal regions must be balanced by convection from the regions. Hence

$$q_1^{(1)} + q_1^{(2)} + q_2 + q_3 + q_5 + q_7^{(1)} = q_7^{(2)} + q_8$$

The cumulative conduction rate is then

$$\frac{q_{\text{cond}}}{L} = k \left[\Delta x \, \frac{(T_s - T_1)}{\Delta y} + \Delta y \, \frac{(T_s - T_1)}{\Delta x} + \frac{\Delta x}{2} \, \frac{(T_s - T_2)}{\Delta y} \right.$$

$$\left. + \, \Delta y \, \frac{(T_s - T_3)}{\Delta x} + \Delta y \, \frac{(T_s - T_5)}{\Delta x} + \frac{\Delta y}{2} \, \frac{(T_s - T_7)}{\Delta x} \right]$$

$$= 191.31 \text{ W/m}$$

and the convection rate is

$$\frac{q_{conv}}{L} = h\left[\Delta x(T_7 - T_\infty) + \frac{\Delta x}{2}(T_8 - T_\infty)\right] = 191.29 \text{ W/m}$$

Agreement between the conduction and convection rates is excellent (within round-off error), confirming that mistakes have not been made in formulating and solving the finite-difference equations. Note that convection transfer from the entire bottom surface (883 W/m) is obtained by adding transfer from the edge node at 500 K (250 W/m) to that from the interior nodes (191.3 W/m) and multiplying by 2 from symmetry.

2. Although the computed temperatures satisfy the finite-difference equations, they do not provide us with the exact temperature field. Remember that the equations are approximations whose accuracy may be improved by reducing the grid size (increasing the number of nodal points).

3. The temperature distribution may also be determined by using the Gauss–Seidel iteration method. Referring to the arrangement of finite-difference equations, it is evident that the order is already characterized by diagonal dominance. This behavior is typical of finite-difference solutions to conduction problems. We therefore begin with step 2 and express the equations in explicit form

$$T_1^{(k)} = 0.25T_2^{(k-1)} + 0.25T_3^{(k-1)} + 250$$

$$T_2^{(k)} = 0.50T_1^{(k)} + 0.25T_4^{(k-1)} + 125$$

$$T_3^{(k)} = 0.25T_1^{(k)} + 0.25T_4^{(k-1)} + 0.25T_5^{(k-1)} + 125$$

$$T_4^{(k)} = 0.25T_2^{(k)} + 0.50T_3^{(k)} + 0.25T_6^{(k-1)}$$

$$T_5^{(k)} = 0.25T_3^{(k)} + 0.25T_6^{(k-1)} + 0.25T_7^{(k-1)} + 125$$

$$T_6^{(k)} = 0.25T_4^{(k)} + 0.50T_5^{(k)} + 0.25T_8^{(k-1)}$$

$$T_7^{(k)} = 0.2222T_5^{(k)} + 0.1111T_8^{(k-1)} + 222.22$$

$$T_8^{(k)} = 0.2222T_6^{(k)} + 0.2222T_7^{(k)} + 166.67$$

Having the finite-difference equations in the required form, the iteration procedure may be implemented by using a table that has one column for the iteration (step) number and a column for each of the nodes labeled as T_i. The calculations proceed as follows:

1. For each node, the initial temperature estimate is entered on the row for $k = 0$. Values are selected rationally to reduce the number of required iterations.

2. Using the N finite-difference equations and values of T_i from the first and second rows, the new values of T_i are calculated for the first iteration ($k = 1$). These new values are entered on the second row.

3. This procedure is repeated to calculate $T_i^{(k)}$ from the previous values of $T_i^{(k-1)}$ and the current values of $T_i^{(k)}$, until the temperature difference between iterations meets the prescribed criterion, $\varepsilon \leq 0.2$ K, at every nodal point.

k	T_1	T_2	T_3	T_4	T_5	T_6	T_7	T_8
0	480	470	440	430	400	390	370	350
1	477.5	471.3	451.9	441.3	428.0	411.8	356.2	337.3
2	480.8	475.7	462.5	453.1	436.6	413.9	355.8	337.7
3	484.6	480.6	467.6	457.4	434.3	415.9	356.2	338.2
4	487.1	482.9	469.7	459.6	435.5	417.2	356.6	338.6
5	488.1	484.0	470.8	460.7	436.1	417.9	356.7	338.8
6	488.7	484.5	471.4	461.3	436.5	418.3	356.9	338.9
7	489.0	484.8	471.7	461.6	436.7	418.5	356.9	339.0
8	489.1	485.0	471.9	461.8	436.8	418.6	356.9	339.0

The results given in row 8 are in excellent agreement with those obtained by matrix inversion, although better agreement could be obtained by reducing the value of ε. However, given the approximate nature of the finite-difference equations, the results still represent approximations to the actual temperatures. The accuracy of the approximation may be improved by using a finer grid (increasing the number of nodes).

4.5.3 Some Precautions

As previously noted, it is good practice to verify that a numerical solution has been correctly formulated by performing an energy balance on a control surface about the nodal regions whose temperatures have been evaluated. The temperatures should be substituted into the energy balance equation, and if the balance is not satisfied to a high degree of precision, the finite-difference equations should be checked for errors.

Even when the finite-difference equations have been properly formulated and solved, the results may still represent a coarse approximation to the actual temperature field. This behavior is a consequence of the finite spacings (Δx, Δy) between nodes and of finite-difference approximations, such as $k(\Delta y \cdot 1)(T_{m-1, n} - T_{m, n})/\Delta x$, to Fourier's law of conduction, $-k(dy \cdot 1)dT/dx$. We have previously indicated that the finite-difference approximations become more accurate as the nodal network is refined (Δx and Δy are reduced). Hence, if accurate results are desired, grid studies should be performed, whereby results obtained for a fine grid are compared with those obtained for a coarse grid. One could, for example, reduce Δx and Δy by a factor of 2, thereby increasing the number of nodes and finite-difference equations by a factor of 4. If the agreement is unsatisfactory, further grid refinements could be made until the computed temperatures no longer depend on the choice of Δx and Δy. Such *grid-independent* results would provide an accurate solution to the physical problem.

Another option for validating a numerical solution involves comparing results with those obtained from an exact solution. For example, a finite-difference solution of the physical problem described in Figure 4.2 could be compared with the exact solution given by Equation 4.19. However, this option is limited by the fact that we seldom seek numerical solutions to problems for which there exist exact solutions. Nevertheless, if we seek a numerical solution to a complex problem for which there is no exact solution, it is often useful to test our finite-difference procedures by applying them to a simpler version of the problem.

EXAMPLE 4.4

A major objective in advancing gas turbine engine technologies is to increase the temperature limit associated with operation of the gas turbine blades. This limit determines the permissible turbine gas inlet temperature, which, in turn, strongly influences overall system performance. In addition to fabricating turbine blades from special, high-temperature, high-strength superalloys, it is common to use internal cooling by machining flow channels within the blades and routing air through the channels. We wish to assess the effect of such a scheme by approximating the blade as a rectangular solid in which rectangular channels are machined. The blade, which has a thermal conductivity of $k = 25$ W/m $\cdot$ K, is 6 mm thick, and each channel has a 2 mm $\times$ 6 mm rectangular cross section, with a 4-mm spacing between adjoining channels.

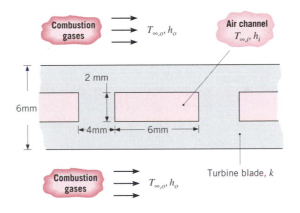

Under operating conditions for which $h_o = 1000$ W/m^2 $\cdot$ K, $T_{\infty, o} = 1700$ K, $h_i = 200$ W/m^2 $\cdot$ K, and $T_{\infty, i} = 400$ K, determine the temperature field in the turbine blade and the rate of heat transfer per unit length to the channel. At what location is the temperature a maximum?

SOLUTION

Known: Dimensions and operating conditions for a gas turbine blade with embedded channels.

Find: Temperature field in the blade, including a location of maximum temperature. Rate of heat transfer per unit length to the channel.

Schematic:

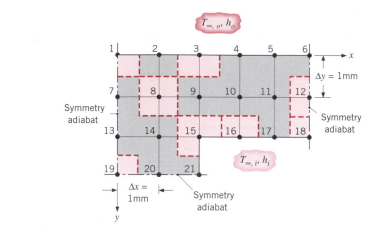

Assumptions:

1. Steady-state, two-dimensional conduction.
2. Constant properties.

Analysis: Adopting a grid space of $\Delta x = \Delta y = 1$ mm and identifying the three lines of symmetry, the foregoing nodal network is constructed. The corresponding finite-difference equations may be obtained by applying the energy balance method to nodes 1, 6, 18, 19, and 21 and by using the results of Table 4.2 for the remaining nodes.

Heat transfer to node 1 may occur by conduction from nodes 2 and 7, as well as by convection from the outer fluid. Since there is no heat transfer from the region beyond the symmetry adiabat, application of an energy balance to the one-quarter section associated with node 1 yields a finite-difference equation of the form

$$\text{Node 1:} \quad T_2 + T_7 - \left(2 + \frac{h_o \Delta x}{k}\right) T_1 = -\frac{h_o \Delta x}{k} T_{\infty,o}$$

A similar result may be obtained for nodal region 6, which is characterized by equivalent surface conditions (2 conduction, 1 convection, 1 adiabatic). Nodes 2 to 5 correspond to case 3 of Table 4.2, and choosing node 3 as an example, it follows that

$$\text{Node 3:} \quad T_2 + T_4 + 2T_9 - 2\left(\frac{h_o \Delta x}{k} + 2\right) T_3 = -\frac{2h_o \Delta x}{k} T_{\infty,o}$$

Nodes 7, 12, 13, and 20 correspond to case 5 of Table 4.2, with $q'' = 0$, and choosing node 12 as an example, it follows that

$$\text{Node 12:} \quad T_6 + 2T_{11} + T_{18} - 4T_{12} = 0$$

Nodes 8 to 11 and 14 are interior nodes (case 1), in which case the finite-difference equation for node 8 is

$$\text{Node 8:} \quad T_2 + T_7 + T_9 + T_{14} - 4T_8 = 0$$

Node 15 is an internal corner (case 2) for which

$$\text{Node 15:} \quad 2T_9 + 2T_{14} + T_{16} + T_{21} - 2\left(3 + \frac{h_i \Delta x}{k}\right) T_{15} = -2\frac{h_i \Delta x}{k} T_{\infty, i}$$

while nodes 16 and 17 are situated on a plane surface with convection (case 3):

$$\text{Node 16:} \quad 2T_{10} + T_{15} + T_{17} - 2\left(\frac{h_i \Delta x}{k} + 2\right) T_{16} = -\frac{2h_i \Delta x}{k} T_{\infty, i}$$

In each case heat transfer to nodal regions 18 and 21 is characterized by conduction from two adjoining nodes and convection from the internal flow, with no heat transfer occurring from an adjoining adiabat. Performing an energy balance for nodal region 18, it follows that

$$\text{Node 18:} \quad T_{12} + T_{17} - T_{18}\left(2 + \frac{h_i \Delta x}{k}\right) = -\frac{h_i \Delta x}{k} T_{\infty, i}$$

The last special case corresponds to nodal region 19, which has two adiabatic surfaces and experiences heat transfer by conduction across the other two surfaces.

$$\text{Node 19:} \quad T_{13} + T_{20} - 2T_{19} = 0$$

The 21 finite-difference equations may be solved for the unknown temperatures, and for the prescribed conditions, the following results are obtained:

T_1	T_2	T_3	T_4	T_5	T_6
1526.0K	1525.3 K	1523.6K	1521.9K	1520.8K	1520.5K

T_7	T_8	T_9	T_{10}	T_{11}	T_{12}
1519.7 K	1518.8K	1516.5 K	1514.5 K	1513.3 K	1512.9 K

T_{13}	T_{14}	T_{15}	T_{16}	T_{17}	T_{18}
1515.1 K	1513.7 K	1509.2 K	1506.4 K	1505.0 K	1504.5 K

T_{19}	T_{20}	T_{21}
1513.4 K	1511.7 K	1506.0 K

The temperature field may also be represented in the form of isotherms, and four such lines of constant temperature are shown schematically.

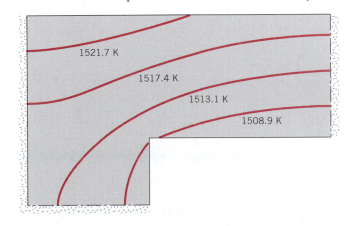

As expected, the maximum temperature exists at the location furthest removed from the coolant, which corresponds to node 1. Temperatures along the surface of the turbine blade exposed to the combustion gases are of particular interest, and using an interpolation scheme with the finite-difference predictions, the following distribution is obtained:

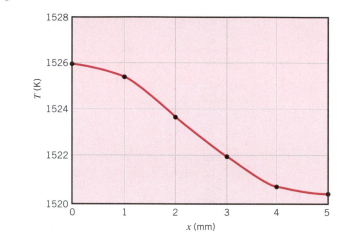

The rate of heat transfer per unit length of channel may be expressed as

$$q' = 4h_i[(\Delta y/2)(T_{21} - T_{\infty,i}) + (\Delta y/2 + \Delta x/2)(T_{15} - T_{\infty,i})$$
$$+ (\Delta x)(T_{16} - T_{\infty,i}) + \Delta x(T_{17} - T_{\infty,i}) + (\Delta x/2)(T_{18} - T_{\infty,i})]$$

or, alternatively, as

$$q' = 4h_o[(\Delta x/2)(T_{\infty,o} - T_1) + (\Delta x)(T_{\infty,o} - T_2) + (\Delta x)(T_{\infty,o} - T_3)$$
$$+ (\Delta x)(T_{\infty,o} - T_4) + (\Delta x)(T_{\infty,o} - T_5) + (\Delta x/2)(T_{\infty,o} - T_6)]$$

where the factor of 4 originates from the symmetry conditions. In any case, we obtain

$$q' = 3540.6 \text{ W/m} \qquad \qquad \triangleleft$$

Comments:

1. The accuracy of the finite-difference solution may be improved by refining the grid. If, for example, we halve the grid spacing ($\Delta x = \Delta y = 0.5$ mm), thereby increasing the number of unknown nodal temperatures to 65, we obtain the following results for selected temperatures and the heat rate:

$$T_1 = 1525.9 \text{ K}, \qquad T_6 = 1520.5 \text{ K}, \qquad T_{15} = 1509.2 \text{ K},$$
$$T_{18} = 1504.5 \text{ K}, \qquad T_{19} = 1513.5 \text{K}, \qquad T_{21} = 1505.7 \text{ K},$$
$$q' = 3539.9 \text{ W/m}$$

Agreement between the two sets of results is excellent. Of course, use of the finer mesh increases set up and computation time, and in many cases results obtained from a coarse grid are satisfactory. Selection of the appropriate grid is a judgment that must be made by the engineer.

2. In the gas turbine industry, there is great interest in adopting measures that reduce blade temperatures. Such measures could include use of a different alloy of larger thermal conductivity and/or increasing coolant flow through

the channel, thereby increasing h_i. Using the finite-difference solution with $\Delta x = \Delta y = 1$ mm, the following results are obtained for parametric variations of k and h_i:

k (W/m · K)	h_i (W/m² · K)	T_1 (K)	q' (W/m)
25	200	1526.0	3540.6
50	200	1523.4	3563.3
25	1000	1154.5	11,095.5
50	1000	1138.9	11,320.7

Why do increases in k and h_i reduce temperature in the blade? Why is the effect of the change in h_i more significant than that of k?

3. Note that, because the exterior surface of the blade is at an extremely high temperature, radiation losses to its surroundings may be significant. In the finite-difference analysis, such effects could be considered by linearizing the radiation rate equation (see Equations 1.8 and 1.9) and treating radiation in the same manner as convection. However, because the radiation coefficient h_r depends on the surface temperature, an iterative finite-difference solution would be necessary to ensure that the resulting surface temperatures correspond to the temperatures at which h_r is evaluated at each nodal point.

4.6
Summary

You should now have an appreciation for the nature of a two-dimensional conduction problem and the methods that are available for its solution. When confronted with a two-dimensional problem, you should first determine whether an exact solution is known. This may be done by examining one or more of the many excellent references in which exact solutions to the heat equation are obtained [1–5]. You may also want to determine whether the shape factor is known for the system of interest [6–9]. However, conditions often are such that the use of a shape factor or an exact solution is not possible, and it is necessary to use a finite-difference solution. You should appreciate the inherent nature of the *discretization process,* and you should know how to formulate the finite-difference equations for the discrete points of a nodal network. Although you may find it convenient to solve these equations using hand calculations for a coarse mesh, you should be able to treat fine meshes using standard computer algorithms involving direct or iterative techniques.

References

1. Schneider, P. J., *Conduction Heat Transfer,* Addison-Wesley, Reading, MA, 1955.

2. Arpaci, V. S., *Conduction Heat Transfer,* Addison-Wesley, Reading, MA, 1966.

3. Özisik, M. N., *Boundary Value Problems of Heat Conduction,* International Textbook, Scranton, PA, 1968.

4. Kakac, S., and Y. Yener, *Heat Conduction,* Hemisphere Publishing, New York, 1985.

5. Poulikakos, D., *Conduction Heat Transfer,* Prentice-Hall, Englewood Cliffs, NJ, 1994.

6. Sunderland, J. E., and K. R. Johnson, "Shape Factors for Heat Conduction Through Bodies with Isothermal or Convective Boundary Conditions," *Trans. ASHRAE,* **10,** 237–241, 1964.

7. Kutateladze, S. S., *Fundamentals of Heat Transfer,* Academic Press, New York, 1963.

8. General Electric Co. (Corporate Research and Development), *Heat Transfer Data Book,* Section 502, General Electric Company, Schenectady, NY, 1973.

9. Hahne, E., and U. Grigull, *Int. J. Heat Mass Transfer,* **18,** 751–767, 1975.

10. Gerald, C. F., *Applied Numerical Analysis,* Addison-Wesley, Reading, MA, 1978.

11. Hoffman, J. D., *Numerical Methods for Engineers and Scientists,* McGraw-Hill, New York, 1992.

Problems

Exact Solutions

4.1 In the method of separation of variables (Section 4.2) for two-dimensional, steady-state conduction, the separation constant λ^2 in Equations 4.6 and 4.7 must be a positive constant. Show that a negative or zero value of λ^2 will result in solutions that cannot satisfy the prescribed boundary conditions.

4.2 A two-dimensional rectangular plate is subjected to prescribed boundary conditions. Using the results of the exact solution for the heat equation presented in Section 4.2, calculate the temperature at the midpoint $(1, 0.5)$ by considering the first five nonzero terms of the infinite series that must be evaluated. Assess the error resulting from using only the first three terms of the infinite series. Plot the temperature distributions $T(x, 0.5)$ and $T(1.0, y)$.

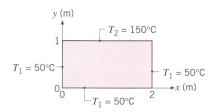

4.3 Consider the two-dimensional rectangular plate of Problem 4.2 having a thermal conductivity of 50 W/m · K. Beginning with the exact solution for the temperature distribution, derive an expression for the heat transfer rate per unit thickness from the plate along the lower surface $(0 \le x \le 2, y = 0)$. Evaluate the heat rate considering the first five nonzero terms of the infinite series.

4.4 A two-dimensional rectangular plate is subjected to the boundary conditions shown. Derive an expression for the steady-state temperature distribution $T(x, y)$.

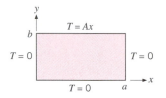

Flux Plotting

4.5 An infinitely long square bar has one of its surfaces maintained at 100°C while the other three are maintained at 0°C. Without performing a flux plot, sketch the 25 and 50°C isotherms. Explain how you arrive at their shapes and positions.

4.6 A long furnace, constructed from refractory brick with a thermal conductivity of 1.2 W/m · K, has the cross section shown with inner and outer surface temperatures of 600 and 60°C, respectively. Determine the shape factor and the heat transfer rate per unit length using the flux plot method.

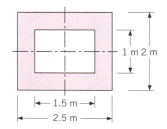

4.7 A hot pipe is embedded eccentrically as shown in a material of thermal conductivity 0.5 W/m · K. Using the flux plot method, determine the shape factor and the heat transfer per unit length when the pipe and outer surface temperatures are 150 and 35°C, respectively.

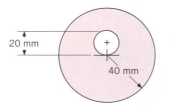

4.8 A supporting strut fabricated from a material with a thermal conductivity of 75 W/m · K has the cross section shown. The end faces are at different temperatures $T_1 = 100°C$ and $T_2 = 0°C$, while the remaining sides are insulated.

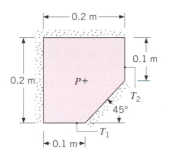

(a) Estimate the temperature at the location P.

(b) Using the flux plot method, estimate the shape factor and the heat transfer rate through the strut per unit length.

(c) Sketch the 25, 50, and 75°C isotherms.

(d) Consider the same geometry, but now with the 0.1-m wide surfaces insulated, the 45° surface maintained at $T_1 = 100°C$, and the 0.2-m wide surfaces maintained at $T_2 = 0°C$. Using the flux plot method, estimate the corresponding shape factor and the heat rate per unit length. Sketch the 25, 50, and 75°C isotherms.

4.9 A hot liquid flows along a V-groove in a solid whose top and side surfaces are well insulated and whose bottom surface is in contact with a coolant.

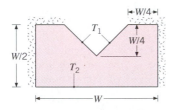

Accordingly, the V-groove surface is at a temperature T_1, which exceeds that of the bottom surface, T_2. Construct an appropriate flux plot and determine the shape factor of the system.

4.10 A very long conduit of inner circular cross section and a thermal conductivity of 1 W/m · K passes a hot fluid, which maintains the inner surface at $T_1 = 50°C$. The outer surfaces of square cross section are insulated or maintained at a uniform temperature of $T_2 = 20°C$, depending on the application. Find the shape factor and the heat rate for each case.

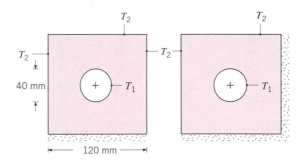

4.11 A long support column of trapezoidal cross section is well insulated on its sides, and temperatures of 100 and 0°C are maintained at its top and bottom surfaces, respectively. The column is fabricated from AISI 1010 steel, and its widths at the top and bottom surfaces are 0.3 and 0.6 m, respectively.

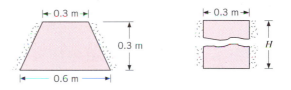

(a) Using the flux plot method, determine the heat transfer rate per unit length of the column.

(b) If the trapezoidal column is replaced by a bar of rectangular cross section 0.3 m wide and the same material, what height H must the bar be to provide an equivalent thermal resistance?

4.12 Hollow prismatic bars fabricated from plain carbon steel are 1 m long with top and bottom surfaces, as well as both ends, well insulated. For each bar, find the shape factor and the heat rate per unit length of the bar when $T_1 = 500$ K and $T_2 = 300$ K.

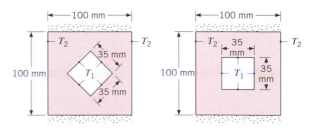

4.13 The two-dimensional, square shapes are maintained at uniform temperatures on portions of their boundaries and are well insulated elsewhere. Use the flux plot method to estimate the shape factors, as well as temperatures at the centers of the geometries.

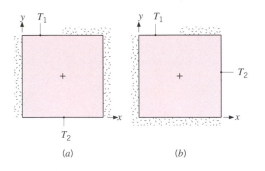

(a)   (b)

4.14 The two-dimensional, square shapes, 1 m to a side, are maintained at uniform temperatures, $T_1 = 100°C$ and $T_2 = 0°C$, on portions of their boundaries and are well insulated elsewhere.

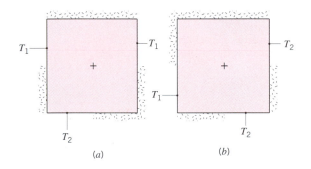

(a)   (b)

Use the flux plot method to estimate the heat rate per unit length normal to the page if the thermal conductivity is 50 W/m · K.

4.15 The two-dimensional circular shapes shown are maintained at uniform temperatures on portions of their boundaries. Use the flux plot method to estimate the shape factors.

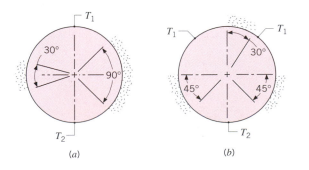

(a)   (b)

Shape Factors

4.16 Using the thermal resistance relations developed in Chapter 3, determine shape factor expressions for the following geometries:

(a) Plane wall, cylindrical shell, and spherical shell.

(b) Isothermal sphere of diameter D buried in an infinite medium.

4.17 Radioactive wastes are temporarily stored in a spherical container, the center of which is buried a distance of 10 m below the earth's surface. The outside diameter of the container is 2 m, and 500 W of heat are released as a result of the radioactive decay process. If the soil surface temperature is 20°C, what is the outside surface temperature of the container under steady-state conditions? On a sketch of the soil–container system drawn to scale, show representative isotherms and heat flow lines in the soil.

4.18 A pipeline, used for the transport of crude oil, is buried in the earth such that its centerline is a distance of 1.5 m below the surface. The pipe has an outer diameter of 0.5 m and is insulated with a layer of cellular glass 100 mm thick. What is the heat loss per unit length of pipe under conditions for which heated oil at 120°C flows through the pipe and the surface of the earth is at a temperature of 0°C?

4.19 A very long electrical conductor is buried in a large trench filled with sand ($k = 0.03$ W/m · K) to a centerline depth of 0.5 m. The conductor has an outer diameter of 25 mm, and the current flow and resistance of the cable cause a dissipation of 1 W per meter of length. The conductor is covered with an insulating sleeve of thickness 3 mm and thermal conductivity 0.01 W/m · K. Estimate the temperature at the interface between the conductor and the insulating sleeve when the temperature at the surface of the sand is 20°C.

4.20 A long power transmission cable is buried at a depth (ground to cable centerline distance) of 2 m. The cable is encased in a thin-walled pipe of 0.1-m diameter, and to render the cable *superconducting* (essentially zero power dissipation), the space between the cable and pipe is filled with liquid nitrogen at 77 K. If the pipe is covered with a superinsulator ($k_i = 0.005$ W/m · K) of 0.05-m thickness and the surface of the earth ($k_g = 1.2$ W/m · K) is at 300 K, what is the cooling load in W/m that must be maintained by a cryogenic refrigerator per unit pipe length?

4.21 An electrical heater 100 mm long and 5 mm in diameter is inserted into a hole drilled normal to the surface of a large block of material having a thermal conductivity of 5 W/m · K. Estimate the temperature reached by the heater when dissipating 50 W with the surface of the block at a temperature of 25°C.

4.22 Two parallel pipelines spaced 0.5 m apart are buried in soil having a thermal conductivity of 0.5 W/m · K. The pipes have outer diameters of 100 and 75 mm with surface temperatures of 175 and 5°C, respectively. Estimate the heat transfer rate per unit length between the two pipelines.

4.23 A tube of diameter 50 mm having a surface temperature of 85°C is embedded in the center plane of a concrete slab 0.1 m thick with upper and lower surfaces at 20°C.

   (a) Using the appropriate tabulated relation for this configuration, find the shape factor. Determine the heat transfer rate per unit lengh of the tube.

   (b) Using the flux plot method, estimate the shape factor and compare with the result of part (a).

4.24 Pressurized steam at 450 K flows through a long, thin-walled pipe of 0.5-m diameter. The pipe is enclosed in a concrete casing that is of square cross section and 1.5 m on a side. The axis of the pipe is centered in the casing, and the outer surfaces of the casing are maintained at 300 K. What is the heat loss per unit length of pipe?

4.25 Hot water at 85°C flows through a thin-walled copper tube of 30 mm diameter. The tube is enclosed by an eccentric cylindrical shell that is maintained at 35°C and has a diameter of 120 mm. The eccentricity, defined as the separation between the centers of the tube and shell, is 20 mm. The space between the tube and shell is filled with an insulating material having a thermal conductivity of 0.05 W/m · K. Calculate the heat loss per unit length of the tube and compare the result with the heat loss for a concentric arrangement.

4.26 A furnace of cubical shape, with external dimensions of 0.35 m, is constructed from a refractory brick (fireclay). If the wall thickness is 50 mm, the inner surface temperature is 600°C, and the outer surface temperature is 75°C, calculate the heat loss from the furnace.

Shape Factors with Thermal Circuits

4.27 A cubical glass melting furnace has exterior dimensions of width $W = 5$ m on a side and is constructed

from refractory brick of thickness $L = 0.35$ m and thermal conductivity $k = 1.4$ W/m · K. The sides and top of the furnace are exposed to ambient air at 25°C, with free convection characterized by an average coefficient of $h = 5$ W/m² · K. The bottom of the furnace rests on a framed platform for which much of the surface is exposed to the ambient air, and a convection coefficient of $h = 5$ W/m² · K may be assumed as a first approximation. Under operating conditions for which combustion gases maintain the inner surfaces of the furnace at 1100°C, what is the heat loss from the furnace?

4.28 A hot fluid passes through circular channels of a cast iron platen (A) of thickness $L_A = 30$ mm which is in poor contact with the cover plates (B) of thickness $L_B = 7.5$ mm. The channels are of diameter $D = 15$ mm with a centerline spacing of $L_o = 60$ mm. The thermal conductivities of the materials are $k_A = 20$ W/m · K and $k_B = 75$ W/m · K, while the contact resistance between the two materials is $R''_{t,c} = 2.0 \times 10^{-4}$ m² · K/W. The hot fluid is at $T_i = 150°C$, and the convection coefficient is 1000 W/m² · K. The cover plate is exposed to ambient air at $T_\infty = 25°C$ with a convection coefficient of 200 W/m² · K.

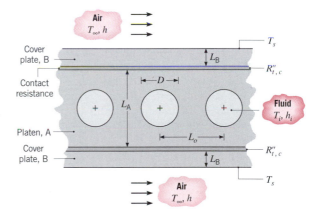

   (a) Determine the heat rate from a single channel per unit length of the platen normal to the page, q'_i.

   (b) Determine the outer surface temperature of the cover plate, T_s.

   (c) Comment on the effects that changing the centerline spacing will have on q'_i and T_s. How would insulating the lower surface affect q'_i and T_s?

4.29 A long constantan wire of 1-mm diameter is butt welded to the surface of a large copper block, forming a thermocouple junction. The wire behaves as a fin, permitting heat to flow from the surface, thereby depressing the sensing junction temperature T_j below that of the block T_o.

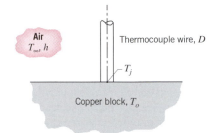

(a) If the wire is in air at 25°C with a convection coefficient of 10 W/m² · K, estimate the measurement error $(T_j - T_o)$ for the thermocouple when the block is at 125°C.

(b) For convection coefficients of 5, 10, and 25 W/m² · K, plot the measurement error as a function of the thermal conductivity of the block material over the range 15 to 400 W/m · K. Under what circumstances is it advantageous to use smaller diameter wire?

4.30 A more realistic set of conditions for Example 4.1 would involve prescription of temperatures and convection coefficients associated with fluids adjoining the inner and outer surfaces, rather than specification of the surface temperatures. Consider conditions for which the outer surfaces are exposed to ambient air, with $T_{\infty, 2} = 25°C$ and $h_2 = 4$ W/m² · K, while hot oil flowing through the hole is characterized by $T_{\infty, 1} = 300°C$ and $h_1 = 50$ W/m² · K. Determine the corresponding heat rate and surface temperatures.

4.31 In Chapter 3 we assumed that, whenever fins are attached to a base material, the base temperature is unchanged. What in fact happens is that, if the temperature of the base material exceeds the fluid temperature, attachment of a fin depresses the junction temperature T_j below the original temperature of the base, and heat flow from the base material to the fin is two-dimensional.

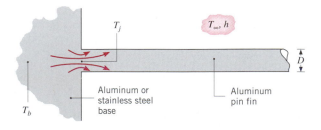

Consider conditions for which a long aluminum pin fin of diameter $D = 5$ mm is attached to a base material whose temperature far from the junction is maintained at $T_b = 100°C$. Fin convection conditions correspond to $h = 50$ W/m² · K and $T_{\infty} = 25°C$.

(a) What are the fin heat rate and junction temperature when the base material is (i) aluminum ($k = 240$ W/m · K) and (ii) stainless steel ($k = 15$ W/m · K)?

(b) Repeat the foregoing calculations if a thermal contact resistance of $R''_{t,j} = 3 \times 10^{-5}$ m² · K/W is associated with the method of joining the pin fin to the base material.

(c) Considering the thermal contact resistance, plot the heat rate as a function of the convection coefficient over the range $10 \leq h \leq 100$ W/m² · K for each of the two materials.

4.32 An igloo is built in the shape of a hemisphere, with an inner radius of 1.8 m and walls of compacted snow that are 0.5 m thick. On the inside of the igloo the surface heat transfer coefficient is 6 W/m² · K; on the outside, under normal wind conditions, it is 15 W/m² · K. The thermal conductivity of compacted snow is 0.15 W/m · K. The temperature of the ice cap on which the igloo sits is −20°C and has the same thermal conductivity as the compacted snow.

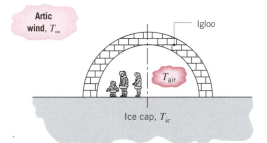

(a) Assuming that the occupants' body heat provides a continuous source of 320 W within the igloo, calculate the inside air temperature when the outside air temperature is $T_{\infty} = -40°C$. Be sure to consider heat losses through the floor of the igloo.

(b) Using the thermal circuit of part (a), perform a parameter sensitivity analysis to determine which variables have a significant effect on the inside air temperature. For instance, for very high wind conditions, the outside convection coefficient could double or even triple. Does it make sense to construct the igloo with walls half or twice as thick?

4.33 A thin, power-dissipating electronic component has a diameter of $D = 10$ mm, and one surface is epoxied to a large block of aluminum ($k = 237$ W/m · K). The thermal resistance for a unit area of the epoxy joint is $R''_{t,c} = 0.5 \times 10^{-4}$ m² · K/W, and at points well removed from the component, the block is maintained at a temperature of $T_b = 25°C$. The other surface is exposed to an airstream for which $h = 25$ W/m² · K and $T_\infty = 25°C$.

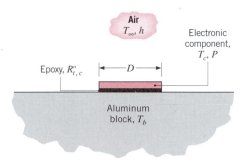

(a) Sketch the thermal circuit of the system and label thermal resistances, the heat flow direction(s), and the temperatures T_b and T_∞.

(b) If the component temperature may not exceed $T_c = 100°C$, what is the maximum allowable operating power P?

4.34 An electronic device, in the form of a disk 20 mm in diameter, dissipates 100 W when mounted flush on a large aluminum alloy (2024) block whose temperature is maintained at 27°C. The mounting arrangement is such that a contact resistance of $R''_{t,c} = 5 \times 10^{-5}$ m² · K/W exists at the interface between the device and the block.

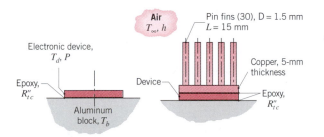

(a) Calculate the temperature the device will reach, assuming that all the power generated by the device must be transferred by conduction to the block.

(b) In order to operate the device at a higher power level, a circuit designer proposes to attach a finned heat sink to the top of the device. The pin fins and base material are fabricated from copper ($k = 400$ W/m · K) and are exposed to an airstream at 27°C for which the convection coefficient is 1000 W/m² · K. For the device temperature computed in part (a), what is the permissible operating power?

Finite-Difference Equations: Derivations

4.35 Consider nodal configuration 2 of Table 4.2. Derive the finite-difference equations under steady-state conditions for the following situations.

(a) The horizontal boundary of the internal corner is perfectly insulated and the vertical boundary is subjected to the convection process (T_∞, h).

(b) Both boundaries of the internal corner are perfectly insulated. How does this result compare with Equation 4.45?

4.36 Consider nodal configuration 3 of Table 4.2. Derive the finite-difference equations under steady-state conditions for the following situations.

(a) The boundary is insulated. Explain how Equation 4.46 can be modified to agree with your result.

(b) The boundary is subjected to a constant heat flux.

4.37 Consider nodal configuration 4 of Table 4.2. Derive the finite-difference equations under steady-state conditions for the following situations.

(a) The upper boundary of the external corner is perfectly insulated and the side boundary is subjected to the convection process (T_∞, h).

(b) Both boundaries of the external corner are perfectly insulated. How does this result compare with Equation 4.47?

4.38 Consider heat transfer in a one-dimensional (radial) cylindrical coordinate system under steady-state conditions with volumetric heat generation.

(a) Derive the finite-difference equation for any interior node m.

(b) Derive the finite-difference equation for the node n located at the external boundary subjected to the convection process (T_∞, h).

4.39 Derive the finite-difference equations required of Problem 4.38, but for a one-dimensional (radial) spherical coordinate system.

4.40 In a two-dimensional cylindrical configuration, the radial (Δr) and angular ($\Delta \phi$) spacings of the nodes are uniform. The boundary at $r = r_i$ is of uniform temperature T_i. The boundaries in the radial direction are adiabatic (insulated) and exposed to surface convection (T_∞, h), as illustrated. Derive the finite-difference equations for (a) node 2, (b) node 3, and (c) node 1.

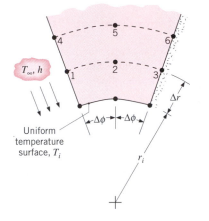

Uniform temperature surface, T_i

4.41 Upper and lower surfaces of a bus bar are convectively cooled by air at T_∞, with $h_u \neq h_l$. The sides are cooled by maintaining contact with heat sinks at T_o, through a thermal contact resistance of $R''_{t,c}$. The bar is of thermal conductivity k, and its width is twice its thickness L.

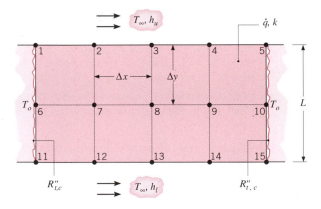

Consider steady-state conditions for which heat is uniformly generated at a volumetric rate $\dot{q}$ due to passage of an electric current. Using the energy balance method, derive finite-difference equations for nodes 1 and 13.

4.42 Derive the nodal finite-difference equations for the following configurations.

(a) Node m, n on a diagonal boundary subjected to convection with a fluid at T_∞ and a heat transfer coefficient h. Assume $\Delta x = \Delta y$.

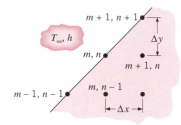

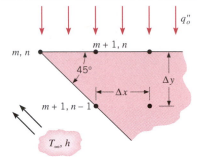

(b) Node m, n at the tip of a cutting tool with the upper surface exposed to a constant heat flux q''_o, and the diagonal surface exposed to a convection cooling process with the fluid at T_∞ and a heat transfer coefficient h. Assume $\Delta x = \Delta y$.

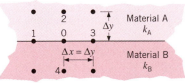

4.43 Consider the nodal point 0 located on the boundary between materials of thermal conductivity k_A and k_B.

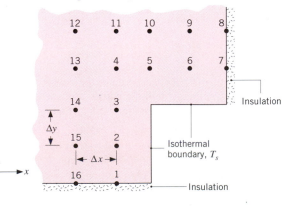

Derive the finite-difference equation, assuming no internal generation.

4.44 Consider the two-dimensional grid ($\Delta x = \Delta y$) representing steady-state conditions with no internal volumetric generation for a system with thermal conductivity k. One of the boundaries is maintained at a constant temperature T_s while the others are adiabatic.

Derive an expression for the heat rate per unit length normal to the page crossing the isothermal boundary (T_s).

4.45 Consider a one-dimensional fin of uniform cross-sectional area, insulated at its tip, $x = L$. (See Table 3.4, case B). The temperature at the base of the fin T_b and of the adjoining fluid T_∞, as well as the heat transfer coefficient h and the thermal conductivity k, are known.

(a) Derive the finite-difference equation for any interior node m.

(b) Derive the finite-difference equation for a node n located at the insulated tip.

Finite-Difference Equations: Analysis

4.46 Consider the network for a two-dimensional system without internal volumetric generation having nodal temperatures shown below. If the grid space is 125 mm and the thermal conductivity of the material is 50 W/m · K, calculate the heat rate per unit length normal to the page from the isothermal surface (T_s).

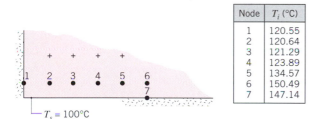

Node	T_i (°C)
1	120.55
2	120.64
3	121.29
4	123.89
5	134.57
6	150.49
7	147.14

$T_s = 100°C$

4.47 Consider the square channel shown in the sketch operating under steady-state conditions. The inner surface of the channel is at a uniform temperature of 600 K, while the outer surface is exposed to convection with a fluid at 300 K and a convection coefficient of 50 W/m² · K. From a symmetrical element of the channel, a two-dimensional grid has been constructed and the nodes labeled. The temperatures for nodes 1, 3, 6, 8, and 9 are identified.

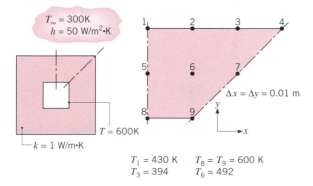

$T_1 = 430$ K $\qquad T_8 = T_9 = 600$ K
$T_3 = 394 \qquad\quad T_6 = 492$

(a) Beginning with properly defined control volumes, derive the finite-difference equations for nodes 2, 4, and 7 and determine the temperatures T_2, T_4, and T_7 (K).

(b) Calculate the heat loss per unit length from the channel.

4.48 Steady-state temperatures (K) at three nodal points of a long rectangular rod are as shown. The rod experiences a uniform volumetric generation rate of 5 × 10⁷ W/m³ and has a thermal conductivity of 20 W/m · K. Two of its sides are maintained at a constant temperature of 300 K, while the others are insulated.

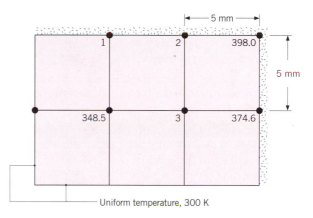

Uniform temperature, 300 K

(a) Determine the temperatures at nodes 1, 2, and 3.

(b) Calculate the heat transfer rate per unit length (W/m) from the rod using the nodal temperatures. Compare this result with the heat rate calculated from knowledge of the volumetric generation rate and the rod dimensions.

4.49 The temperatures (K) at the nodal points of a two-dimensional system are as shown. Surface B is held at a uniform temperature, while surface A is subjected to a convection boundary condition. Calculate the heat rate leaving surface A per unit thickness normal to the page. Estimate the thermal conductivity of the material.

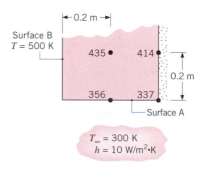

4.50 The steady-state temperatures (°C) associated with selected nodal points of a two-dimensional system having a thermal conductivity of 1.5 W/m · K are shown on the accompanying grid.

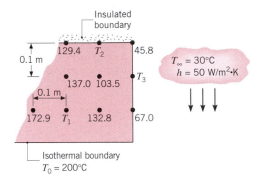

(a) Determine the temperatures at nodes 1, 2, and 3.

(b) Calculate the heat transfer rate per unit thickness normal to the page from the system to the fluid.

4.51 A steady-state, finite-difference analysis has been performed on a cylindrical fin with a diameter of 12 mm and a thermal conductivity of 15 W/m · K. The convection process is characterized by a fluid temperature of 25°C and a heat transfer coefficient of 25 W/m² · K.

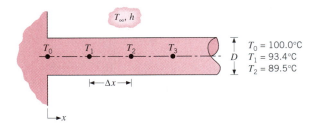

(a) The temperatures for the first three nodes, separated by a spatial increment of $x = 10$ mm, are given in the sketch. Determine the fin heat rate.

(b) Determine the temperature at node 3, T_3.

Finite-Difference Solutions

4.52 A long bar of rectangular cross section is 60 mm by 90 mm on a side and has a thermal conductivity of 1 W/m · K. One surface is exposed to a convection process with air at 100°C and a convection coefficient of 100 W/m² · K, while the remaining surfaces are maintained at 50°C.

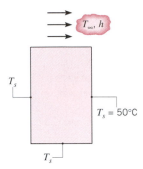

(a) Using a grid spacing of 30 mm and the Gauss–Seidel iteration method, determine the nodal temperatures and the heat rate per unit length normal to the page into the bar from the air.

(b) Determine the effect of grid spacing on the temperature field and heat rate. Specifically, consider a grid spacing of 15 mm. For this grid, explore the effect of changes in h on the temperature field.

4.53 Consider two-dimensional, steady-state conduction in a square cross section with prescribed surface temperatures.

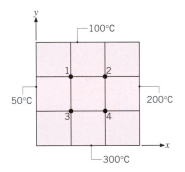

(a) Determine the temperatures at nodes 1, 2, 3, and 4. Estimate the midpoint temperature.

(b) Reducing the mesh size by a factor of 2, determine the corresponding nodal temperatures. Compare your results with those from the coarser grid.

(c) From the results for the finer grid, plot the 75, 150, and 250°C isotherms.

4.54 Consider a long bar of square cross section (0.8 m to the side) and of thermal conductivity 2 W/m · K. Three of these sides are maintained at a uniform temperature of 300°C. The fourth side is exposed to a fluid at 100°C for which the convection heat transfer coefficient is 10 W/m² · K.

(a) Using an appropriate numerical technique with a grid spacing of 0.2 m, determine the midpoint temperature and heat transfer rate between the bar and the fluid per unit length of the bar.

(b) Reducing the grid spacing by a factor of 2, determine the midpoint temperature and heat transfer rate. Plot the corresponding temperature distribution across the surface exposed to the fluid. Also, plot the 200 and 250°C isotherms.

4.55 A long conducting rod of rectangular cross section (20 mm × 30 mm) and thermal conductivity $k = 20$ W/m · K experiences uniform heat generation at a rate $\dot{q} = 5 \times 10^7$ W/m³, while its surfaces are maintained at 27°C.

(a) Using a finite-difference method with a grid spacing of 5 mm, determine the temperature distribution in the rod.

(b) With the boundary conditions unchanged, what heat generation rate will cause the midpoint temperature to reach 600°C?

4.56 A flue passing hot exhaust gases has a square cross section, 300 mm to a side. The walls are constructed of refractory brick 150 mm thick with a thermal conductivity of 0.85 W/m · K. Calculate the heat loss from the flue per unit length when the interior and exterior surfaces are maintained at 350 and 25°C, respectively. Use a grid spacing of 75 mm.

4.57 Consider the system of Problem 4.56. The interior surface is exposed to hot gases at 350°C with a convection coefficient of 100 W/m² · K, while the exterior surface experiences convection with air at 25°C and a convection coefficient of 5 W/m² · K.

(a) Using a grid spacing of 75 mm, calculate the temperature field within the system and determine the heat loss per unit length by convection from the outer surface of the flue to the air. Compare this result with the heat gained by convection from the hot gases to the air.

(b) Determine the effect of grid spacing on the temperature field and heat loss per unit length to the air. Specifically, consider a grid spacing of 25 mm. For $\Delta x = \Delta y = 25$ mm, explore the effect of changes in the convection coefficients on the temperature field, and heat loss.

4.58 The perimeter of a plate of thickness 4 mm, thermal conductivity 2 W/m · K, and square cross section (100 mm to the side) is maintained at 600 K.

One surface of the plate is insulated, while the other surface has an emissivity of 0.9 and is exposed to large surroundings at a temperature of 300 K. Neglecting temperature gradients in the direction of the plate thickness and using the Gauss–Seidel method with a grid spacing of 25 mm, determine the radiation heat exchange between the plate and its surroundings. *Hint:* Linearize the radiation rate equation according to Equation 1.8; after each iteration, update the value of the radiation heat transfer coefficient h_r.

4.59 A common arrangement for heating a large surface area is to move warm air through rectangular ducts below the surface. The ducts are square and located midway between the top and bottom surfaces that are exposed to room air and insulated, respectively.

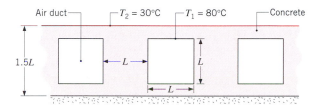

For the condition when the floor and duct temperatures are 30 and 80°C, respectively, and the thermal conductivity of concrete is 1.4 W/m · K, calculate the heat rate from each duct, per unit length of duct. Use a grid spacing with $\Delta x = 2\,\Delta y$, where $\Delta y = 0.125L$ and $L = 150$ mm.

4.60 Consider the gas turbine cooling scheme of Example 4.4. In Problem 3.23, advantages associated with applying a *thermal barrier coating* (TBC) to the exterior surface of a turbine blade are described. If a 0.5-mm-thick zirconia coating ($k = 1.3$ W/m · K, $R''_{t,c} = 10^{-4}$ m² · K/W) is applied to the outer surface of the air-cooled blade, determine the temperature field in the blade for the operating conditions of Example 4.4.

4.61 A long bar of rectangular cross section, 0.4 m × 0.6 m on a side and having a thermal conductivity of 1.5 W/m · K, is subjected to the boundary conditions shown on the next page. Two of the sides are maintained at a uniform temperature of 200°C. One of the sides is adiabatic, and the remaining side is subjected to a convection process with $T_\infty = 30$°C and $h = 50$ W/m² · K. Using an appropriate numerical technique with a grid spacing of 0.1 m, determine the temperature distribution in the bar and the heat transfer rate between the bar and the fluid per unit length of the bar.

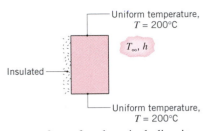

Uniform temperature, $T = 200°C$

T_∞, h

Insulated

Uniform temperature, $T = 200°C$

4.62 The top surface of a plate, including its grooves, is maintained at a uniform temperature of $T_1 = 200°C$. The lower surface is at $T_2 = 20°C$, the thermal conductivity is 15 W/m · K, and the groove spacing is 0.16 m.

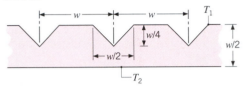

(a) Using a finite-difference method with a mesh size of $\Delta x = \Delta y = 40$ mm, calculate the unknown nodal temperatures and the heat transfer rate per width of groove spacing (w) and per unit length normal to the page.

(b) With a mesh size of $\Delta x = \Delta y = 10$ mm, repeat the foregoing calculations, determining the temperature field and the heat rate. Also, consider conditions for which the bottom surface is not at a uniform temperature T_2 but is exposed to a fluid at $T_\infty = 20°C$. With $\Delta x = \Delta y = 10$ mm, determine the temperature field and heat rate for values of $h = 5, 200$, and 1000 W/m² · K, as well as for $h \rightarrow \infty$.

4.63 Refer to the two-dimensional rectangular plate of Problem 4.2. Using an appropriate numerical method with $\Delta x = \Delta y = 0.25$ m, determine the temperature at the midpoint (1, 0.5).

4.64 A long trapezoidal bar is subjected to uniform temperatures on two surfaces, while the remaining surfaces are well insulated. If the thermal conductivity of the material is 20 W/m · K, estimate the heat transfer rate per unit length of the bar using a finite-difference method. Use the Gauss–Seidel method of solution with a space increment of 10 mm.

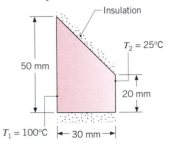

Insulation

$T_2 = 25°C$

50 mm

20 mm

$T_1 = 100°C$ ← 30 mm →

4.65 The shape factor for conduction through the edge of adjoining walls for which $D > L/5$, where D and L are the wall depth and thickness, respectively, is shown in Table 4.1. The two-dimensional symmetrical element of the edge, which is represented by inset (*a*), is bounded by the diagonal symmetry adiabat and a section of the wall thickness over which the temperature distribution is assumed to be linear between T_1 and T_2.

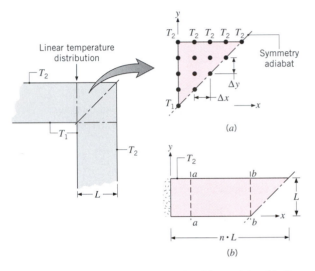

Linear temperature distribution

T_2

T_2 T_2 T_2 T_2 T_2

Symmetry adiabat

Δy

T_1 Δx x

T_1

T_2

L

(a)

T_2

a b

L

a b x

$n \cdot L$

(b)

(a) Using the nodal network of inset (*a*) with $L = 40$ mm, determine the temperature distribution in the element for $T_1 = 100°C$ and $T_2 = 0°C$. Evaluate the heat rate per unit depth ($D = 1$ m) if $k = 1$ W/m · K. Determine the corresponding shape factor for the edge and compare your result with that from Table 4.1.

(b) Choosing a value of $n = 1$ or $n = 1.5$, establish a nodal network for the trapezoid of inset (*b*) and determine the corresponding temperature field. Assess the validity of assuming linear temperature distributions across sections *a–a* and *b–b*.

4.66 The diagonal of a long triangular bar is well insulated, while sides of equivalent length are maintained at uniform temperatures T_a and T_b.

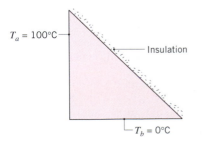

$T_a = 100°C$

Insulation

$T_b = 0°C$

(a) Establish a nodal network consisting of five nodes along each of the sides. For one of the nodes on the diagonal surface, define a suitable control volume and derive the corresponding finite-difference equation. Using this form for the diagonal nodes and appropriate equations for the interior nodes, find the temperature distribution for the bar. On a scale drawing of the shape, show the 25, 50, and 75°C isotherms.

(b) An alternate and simpler procedure to obtain the finite-difference equations for the diagonal nodes follows from recognizing that the insulated diagonal surface is a symmetry plane. Consider a square 5 × 5 nodal network and represent its diagonal as a symmetry line. Recognize which nodes on either side of the diagonal have identical temperatures. If you have done this properly, you can treat the diagonal nodes as "interior" nodes and write the finite-difference equations by inspection.

Special Applications

4.67 A straight fin of uniform cross section is fabricated from a material of thermal conductivity 50 W/m · K, thickness $w = 6$ mm, and length $L = 48$ mm, and is very long in the direction normal to the page. The convection heat transfer coefficient is 500 W/m² · K with an ambient air temperature of $T_\infty = 30$°C. The base of the fin is maintained at $T_b = 100$°C, while the fin tip is well insulated.

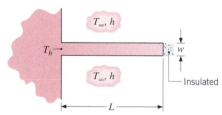

(a) Using a finite-difference method with a space increment of 4 mm, estimate the temperature distribution within the fin. Is the assumption of one-dimensional heat transfer reasonable for this fin?

(b) Estimate the fin heat transfer rate per unit length normal to the page. Compare your result with the one-dimensional, analytical solution, Equation 3.76.

(c) Using the finite-difference mesh of part (a), compute and plot the fin temperature distribution for values of $h = 10, 100, 500,$ and 1000 W/m² · K. Determine and plot the fin heat transfer rate as a function of h.

4.68 A straight fin has a triangular profile given by the function $y = 0.2(L - x)$ and is very long in the direction normal to the page. Both top and bottom surfaces experience convection with $T_\infty = 15$°C and $h = 100$ W/m² · K. The base of the fin is maintained at $T_b = 115$°C and the fin material has a thermal conductivity of 25 W/m · K. Assuming one-dimensional heat transfer and using a finite-difference method with a space increment of 10 mm, determine the fin heat rate and efficiency.

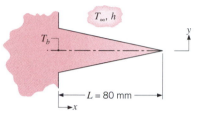

4.69 A rod of 10-mm diameter and 250-mm length has one end maintained at 100°C. The surface of the rod experiences free convection with the ambient air at 25°C and a convection coefficient that depends on the difference between the temperature of the surface and the ambient air. Specifically, the coefficient is prescribed by a correlation of the form, $h_{fc} = 2.89[0.6 + 0.624(T - T_\infty)^{1/6}]^2$, where the units are h_{fc} (W/m² · K) and T (K). The surface of the rod has an emissivity $\varepsilon = 0.2$ and experiences radiation exchange with the surroundings at $T_{sur} = 25$°C. The fin tip also experiences free convection and radiation exchange.

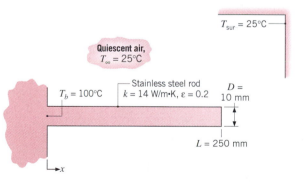

Assuming one-dimensional conduction and using a finite-difference method representing the fin by five nodes, estimate the temperature distribution for the fin. Determine also the fin heat rate and the relative contributions of free convection and radiation exchange. *Hint:* For each node requiring an energy balance, use the linearized form of the radiation rate equation, Equation 1.8, with the radiation coefficient h_r, Equation 1.9, evaluated for each node. Similarly, for the convection rate equation associated with each node, the free convection coefficient h_{fc} must be evaluated for each node.

4.70 A thin metallic foil of thickness 0.25 mm with a pattern of extremely small holes serves as an acceleration grid to control the electrical potential of an ion beam. Such a grid is used in a chemical vapor deposition (CVD) process for the fabrication of semiconductors. The top surface of the grid is exposed to a uniform heat flux caused by absorption of the ion beam, $q''_s = 600 \text{ W/m}^2$. The edges of the foil are thermally coupled to water-cooled sinks maintained at 300 K. The upper and lower surfaces of the foil experience radiation exchange with the vacuum enclosure walls maintained at 300 K. The effective thermal conductivity of the foil material is 40 W/m · K and its emissivity is 0.45.

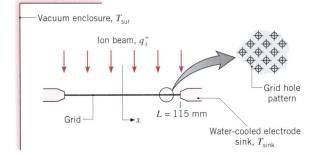

Assuming one-dimensional conduction and using a finite-difference method representing the grid by ten nodes in the x direction, estimate the temperature distribution for the grid. *Hint:* For each node requiring an energy balance, use the linearized form of the radiation rate equation, Equation 1.8, with the radiation coefficient h_r, Equation 1.9, evaluated for each node.

4.71 Small diameter electrical heating elements dissipating 50 W/m (length normal to the sketch) are used to heat a ceramic plate of thermal conductivity 2 W/m · K. The upper surface of the plate is exposed to ambient air at 30°C with a convection coefficient of 100 W/m² · K, while the lower surface is well insulated.

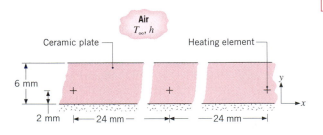

(a) Using the Gauss–Seidel method with a grid spacing of $\Delta x = 6$ mm and $\Delta y = 2$ mm, obtain the temperature distribution within the plate.

(b) Using the calculated nodal temperatures, sketch four isotherms to illustrate the temperature distribution in the plate.

(c) Calculate the heat loss by convection from the plate to the fluid. Compare this value with the element dissipation rate.

(d) What advantage, if any, is there in not making $\Delta x = \Delta y$ for this situation?

(e) With $\Delta x = \Delta y = 2$ mm, calculate the temperature field within the plate and the rate of heat transfer from the plate. Under no circumstances may the temperature at any location in the plate exceed 400°C. Would this limit be exceeded if the air flow were terminated and heat transfer to the air was by natural convection with $h = 10$ W/m² · K?

4.72 A long bar of rectangular cross section is fabricated from two materials with thermal conductivities $k_A = 15$ W/m · K and $k_B = 1$ W/m · K and with thicknesses $L_A = 50$ mm and $L_B = 100$ mm, respectively. Its width is $w = 300$ mm. Three sides of the bar are subjected to convection conditions with $T_\infty = 100°C$ and $h = 30$ W/m² · K. Calculate the rate at which heat must be removed per unit length of the bar by the cooling coil to maintain the bottom surface of the bar at $T_0 = 0°C$.

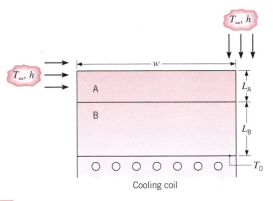

4.73 A simplified representation for cooling in very large-scale integration (VLSI) of microelectronics is shown in the sketch. A silicon chip is mounted in a dielectric substrate, and one surface of the system is convectively cooled, while the remaining surfaces are well insulated from the surroundings. The problem is rendered two-dimensional by assuming the system to be very long in the direction perpendicular to the paper. Under steady-state operation, electric power dissipation in the chip provides for uniform volumetric heating at a rate of $\dot{q}$. However, the heating rate is limited by restrictions on the maximum temperature that the chip is allowed to assume.

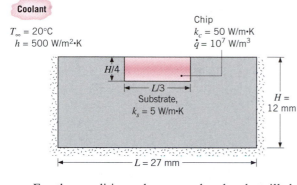

For the conditions shown on the sketch, will the maximum temperature in the chip exceed 85°C, the maximum allowable operating temperature set by industry standards? A grid spacing of 3 mm is suggested.

4.74 Electronic devices dissipating electrical power can be cooled by conduction to a heat sink. The lower surface of the sink is cooled, and the spacing of the devices w_s, the width of the device w_d, and the thickness L and thermal conductivity k of the heat sink material each affect the thermal resistance between the device and the cooled surface. The function of the heat sink is to *spread* the heat dissipated in the device throughout the sink material.

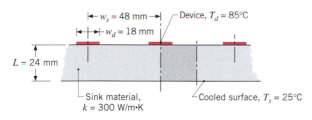

(a) Beginning with the shaded symmetrical element, use the flux plot method to estimate the thermal resistance per unit depth between the device and lower surface of the sink, $R'_{t,\,d-s}$ (m · K/W). How does this value compare with thermal resistances based on the assumption of one-dimensional conduction in rectangular domains of (i) width w_d and length L and (ii) width w_s and length L?

(b) Using a coarse (5×5) nodal network, calculate the thermal resistance $R'_{t,\,d-s}$ (m · K/W).

(c) Using nodal networks with grid spacings three and five times smaller than that in part (b), de-

termine the effect of grid size on the precision of the thermal resistance calculation.

(d) Using the finer nodal network developed for part (c), determine the effect of device width on the thermal resistance. Specifically, keeping w_s and L fixed, find the thermal resistance for values of $w_d/w_s = 0.175, 0.275, 0.375,$ and 0.475.

4.75 A major problem in packaging very large-scale integrated (VLSI) circuits concerns cooling of the circuit elements. The problem results from increasing levels of power dissipation within a chip, as well as from packing chips closer together in a module. A novel technique for cooling multichip modules has been developed by IBM. Termed the thermal conduction module (TCM), the chips are soldered to a multilayer ceramic substrate, and heat dissipated in each chip is conducted through a spring-loaded aluminum piston to a water-cooled *cold plate*.

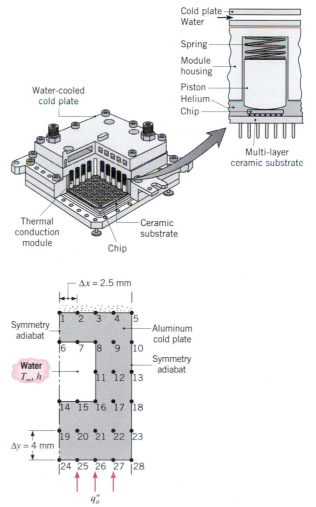

(a) Consider a cold plate fabricated from aluminum ($k = 190$ W/m · K) with regularly spaced rectangular channels through which the water is passed. Under normal operation, power dissipation within the chips results in a uniform heat flux of $q_o'' = 10^5$ W/m² at the base of the cold plate, while water flow provides a temperature of $T_\infty = 15°C$ and a convection coefficient of $h = 5000$ W/m² · K within the channels. We are interested in obtaining the steady-state temperature distribution within the cold plate, and from symmetry considerations, we may confine our attention to the prescribed nodal network. Assuming the top surface of the cold plate to be well insulated, determine the nodal temperatures.

(b) Although there is interest in operating at higher power levels, system reliability considerations dictate that the maximum cold plate temperature must not exceed 40°C. Using the prescribed cold plate geometry and nodal network, assess the effect of changes in operating or design conditions intended to increase the operating heat flux q_o''. Estimate the upper limit for the heat flux.

4.76 A heat sink for cooling computer chips is fabricated from copper ($k_s = 400$ W/m · K), with machined microchannels passing a cooling fluid for which $T_\infty = 25°C$ and $h = 30{,}000$ W/m² · K. The lower side of the sink experiences no heat removal, and a preliminary heat sink design calls for dimensions of $a = b = w_s = w_f = 200$ μm. A symmetrical element of the heat path from the chip to the fluid is shown in the inset.

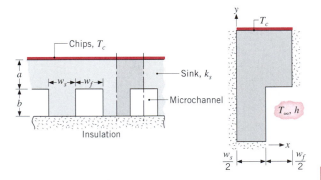

(a) Using the symmetrical element with a square nodal network of $\Delta x = \Delta y = 100$ μm, determine the corresponding temperature field and the heat rate q' to the coolant per unit channel length (W/m) for a maximum allowable chip temperature $T_{c,max} = 75°C$. Estimate the corre-

sponding thermal resistance between the chip surface and the fluid, $R_{t,c-f}'$ (m · K/W). What is the maximum allowable heat dissipation for a chip that measures 10 mm × 10 mm on a side?

(b) The grid spacing used in the foregoing finite-difference solution is coarse, resulting in poor precision for the temperature distribution and heat removal rate. Investigate the effect of grid spacing by considering spatial increments of 50 and 25 μm.

(c) Consistent with the requirement that $a + b = 400$ μm, can the heat sink dimensions be altered in a manner that reduces the overall thermal resistance?

4.77 A test cell for measuring free-convection coefficients between two parallel plates consists of two large copper plates separated by partitions. The cross section of an end partition and the plates is shown below, along with representative thermal conditions.

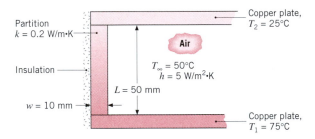

In the experiment, the power required to maintain the lower plate at T_1 is measured, and if there is negligible heat transfer through the partition to the air and the upper plate, all of the power is assumed to be transported by free convection to the upper plate.

(a) Assuming heat transfer through the partition is one-dimensional, calculate the rate of heat transfer per unit depth (W/m) through the partition to the air and to the upper plate. *Hint:* Consider the partition as a fin of uniform cross section with a prescribed end temperature.

(b) Calculate the heat rate by considering two-dimensional conduction within the partition. Compare the results with those from part (a).

4.78 A plate ($k = 10$ W/m·K) is stiffened by a series of longitudinal ribs having a rectangular cross section with length $L = 8$ mm and width $w = 4$ mm. The base of the plate is maintained at a uniform temperature $T_b = 45°C$, while the rib surfaces are exposed to air at a temperature of $T_\infty = 25°C$ and a convection coefficient of $h = 600$ W/m² · K.

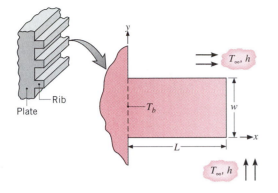

Plate, Rib, T_b, T_∞, h, w, L, T_∞, h, y, x

web is well insulated, while the flange surfaces experience convection with hot gases at $T_\infty = 400°C$ and a convection coefficient of $h = 150$ W/m² · K. Consider the symmetrical element of the flange region (inset *a*), assuming that the temperature distribution across the web is uniform at $T_w = 100°C$. The beam thermal conductivity is 10 W/m · K, and its dimensions are $w_f = 80$ mm, $w_w = 30$ mm, and $L = 30$ mm.

(a) Using a finite-difference method with $\Delta x = \Delta y = 2$ mm and a total of 5 × 3 nodal points and regions, estimate the temperature distribution and the heat rate from the base. Compare these results with those obtained by assuming that heat transfer in the rib is one-dimensional, thereby approximating the behavior of a fin.

(b) The grid spacing used in the foregoing finite-difference solution is coarse, resulting in poor precision for estimates of temperatures and the heat rate. Investigate the effect of grid refinement by reducing the nodal spacing to $\Delta x = \Delta y = 1$ mm (a 9 × 3 grid).

(c) Investigate the nature of two-dimensional conduction in the rib and determine a criterion for which the one-dimensional approximation is reasonable. Do so by extending your finite-difference analysis to determine the heat rate from the base as a function of the length of the rib for the range $1.5 \leq L/w \leq 10$, keeping the length l constant. Compare your results with those determined by approximating the rib as a fin.

4.79 The bottom-half of an I-beam providing support for a furnace roof extends into the heating zone. The

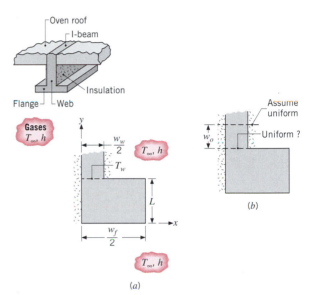

Oven roof, I-beam, Insulation, Flange, Web, Gases T_∞, h, $\dfrac{w_w}{2}$, T_∞, h, T_w, y, L, $\dfrac{w_f}{2}$, x, T_∞, h

(a)

Assume uniform, Uniform?, w_o

(b)

(a) Calculate the heat transfer rate per unit length to the beam using a 5 × 4 nodal network.

(b) Is it reasonable to assume that the temperature distribution across the web–flange interface is uniform? Consider the L-shaped domain of inset (*b*) and use a fine grid to obtain the temperature distribution across the web–flange interface. Make the distance $w_o \geq w_w/2$.

Transient Conduction

*I*n our treatment of conduction we have gradually considered more complicated conditions. We began with the simple case of one-dimensional, steady-
state conduction with no internal generation, and we subsequently considered
complications due to multidimensional and generation effects. However, we
have not yet considered situations for which conditions change with time.

We now recognize that many heat transfer problems are time dependent.
Such *unsteady,* or *transient,* problems typically arise when the boundary conditions of a system are changed. For example, if the surface temperature of a system is altered, the temperature at each point in the system will also begin to
change. The changes will continue to occur until a *steady-state* temperature distribution is reached. Consider a hot metal billet that is removed from a furnace
and exposed to a cool airstream. Energy is transferred by convection and radiation from its surface to the surroundings. Energy transfer by conduction also
occurs from the interior of the metal to the surface, and the temperature at each
point in the billet decreases until a steady-state condition is reached. Such time-
dependent effects occur in many industrial heating and cooling processes.

To determine the time dependence of the temperature distribution within a
solid during a transient process, we could begin by solving the appropriate form
of the heat equation, for example, Equation 2.13. Some cases for which solutions have been obtained are discussed in Sections 5.4 to 5.8. However, under
conditions for which temperature gradients within the solid are small, a simpler
approach, termed the *lumped capacitance method,* may be used.

5.1
The Lumped Capacitance Method

A simple, yet common, transient conduction problem is one in which a solid
experiences a sudden change in its thermal environment. Consider a hot metal
forging that is initially at a uniform temperature T_i and is quenched by immersing it in a liquid of lower temperature $T_\infty < T_i$ (Figure 5.1). If the quenching is
said to begin at time $t = 0$, the temperature of the solid will decrease for time
$t > 0$, until it eventually reaches T_∞. This reduction is due to convection heat
transfer at the solid–liquid interface. The essence of the lumped capacitance
method is the assumption that the temperature of the solid is *spatially uniform*

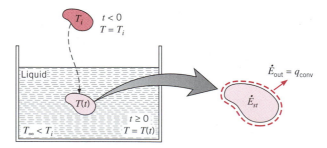

FIGURE 5.1   Cooling of a hot metal forging.

at any instant during the transient process. This assumption implies that temperature gradients within the solid are negligible.

From Fourier's law, heat conduction in the absence of a temperature gradient implies the existence of infinite thermal conductivity. Such a condition is clearly impossible. However, although the condition is never satisfied exactly, it is closely approximated if the resistance to conduction within the solid is small compared with the resistance to heat transfer between the solid and its surroundings. For now we assume that this is, in fact, the case.

In neglecting temperature gradients within the solid, we can no longer consider the problem from within the framework of the heat equation. Instead, the transient temperature response is determined by formulating an overall energy balance on the solid. This balance must relate the rate of heat loss at the surface to the rate of change of the internal energy. Applying Equation 1.11a to the control volume of Figure 5.1, this requirement takes the form

$$-\dot{E}_{out} = \dot{E}_{st} \tag{5.1}$$

or

$$-hA_s(T - T_\infty) = \rho Vc \frac{dT}{dt} \tag{5.2}$$

Introducing the temperature difference

$$\theta \equiv T - T_\infty \tag{5.3}$$

and recognizing that $(d\theta/dt) = (dT/dt)$, it follows that

$$\frac{\rho Vc}{hA_s} \frac{d\theta}{dt} = -\theta$$

Separating variables and integrating from the initial condition, for which $t = 0$ and $T(0) = T_i$, we then obtain

$$\frac{\rho Vc}{hA_s} \int_{\theta_i}^{\theta} \frac{d\theta}{\theta} = -\int_0^t dt$$

where

$$\theta_i \equiv T_i - T_\infty \tag{5.4}$$

Evaluating the integrals it follows that

$$\frac{\rho Vc}{hA_s} \ln \frac{\theta_i}{\theta} = t \tag{5.5}$$

or

$$\frac{\theta}{\theta_i} = \frac{T - T_\infty}{T_i - T_\infty} = \exp\left[-\left(\frac{hA_s}{\rho Vc}\right)t\right] \tag{5.6}$$

Equation 5.5 may be used to determine the time required for the solid to reach some temperature T, or, conversely, Equation 5.6 may be used to compute the temperature reached by the solid at some time t.

The foregoing results indicate that the difference between the solid and fluid temperatures must decay exponentially to zero as t approaches infinity.

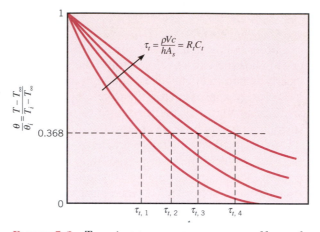

FIGURE 5.2 Transient temperature response of lumped capacitance solids corresponding to different thermal time constants τ_t.

This behavior is shown in Figure 5.2. From Equation 5.6 it is also evident that the quantity ($\rho Vc/hA_s$) may be interpreted as a *thermal time constant.* This time constant may be expressed as

$$\tau_t = \left(\frac{1}{hA_s}\right)(\rho Vc) = R_t C_t \tag{5.7}$$

where R_t is the resistance to convection heat transfer and C_t is the *lumped thermal capacitance* of the solid. Any increase in R_t or C_t will cause a solid to respond more slowly to changes in its thermal environment and will increase the time required to reach thermal equilibrium ($\theta = 0$). This behavior is analogous to the voltage decay that occurs when a capacitor is discharged through a resistor in an electrical *RC* circuit.

To determine the total energy transfer Q occurring up to some time t, we simply write

$$Q = \int_0^t q \, dt = hA_s \int_0^t \theta \, dt$$

Substituting for θ from Equation 5.6 and integrating, we obtain

$$Q = (\rho Vc)\theta_i \left[1 - \exp\left(-\frac{t}{\tau_t}\right)\right] \tag{5.8a}$$

The quantity Q is, of course, related to the change in the internal energy of the solid, and from Equation 1.11b

$$-Q = \Delta E_{\text{st}} \tag{5.8b}$$

For quenching Q is positive and the solid experiences a decrease in energy. Equations 5.5, 5.6, and 5.8a also apply to situations where the solid is heated ($\theta < 0$), in which case Q is negative and the internal energy of the solid increases.

5.2
Validity of the Lumped Capacitance Method

From the foregoing results it is easy to see why there is a strong preference for using the lumped capacitance method. It is certainly the simplest and most convenient method that can be used to solve transient conduction problems. Hence it is important to determine under what conditions it may be used with reasonable accuracy.

To develop a suitable criterion consider steady-state conduction through the plane wall of area A (Figure 5.3). Although we are assuming steady-state conditions, this criterion is readily extended to transient processes. One surface is maintained at a temperature $T_{s,1}$ and the other surface is exposed to a fluid of temperature $T_\infty < T_{s,1}$. The temperature of this surface will be some intermediate value, $T_{s,2}$, for which $T_\infty < T_{s,2} < T_{s,1}$. Hence under steady-state conditions the surface energy balance, Equation 1.12, reduces to

$$\frac{kA}{L}(T_{s,1} - T_{s,2}) = hA(T_{s,2} - T_\infty)$$

where k is the thermal conductivity of the solid. Rearranging, we then obtain

$$\frac{T_{s,1} - T_{s,2}}{T_{s,2} - T_\infty} = \frac{(L/kA)}{(1/hA)} = \frac{R_{\text{cond}}}{R_{\text{conv}}} = \frac{hL}{k} \equiv Bi \tag{5.9}$$

The quantity (hL/k) appearing in Equation 5.9 is a *dimensionless parameter.* It is termed the *Biot number,* and it plays a fundamental role in conduction problems that involve surface convection effects. According to Equation 5.9 and as illustrated in Figure 5.3, the Biot number provides a measure of the temperature drop in the solid relative to the temperature difference between the surface and the fluid. Note especially the conditions corresponding to $Bi \ll 1$. The results suggest that, for these conditions, it is reasonable to *assume* a uniform temperature distribution across a solid at any time during a transient process. This result may also be associated with interpretation of the Biot number as a ratio of thermal resistances, Equation 5.9. If $Bi \ll 1$, *the resistance to conduction within the solid is much less than the resistance to convection across the fluid boundary layer.* Hence the assumption of a uniform temperature distribution is reasonable.

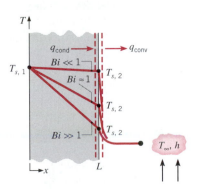

FIGURE 5.3
Effect of Biot number on steady-state temperature distribution in a plane wall with surface convection.

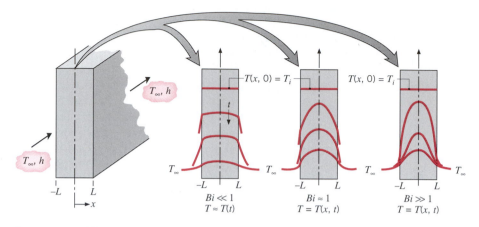

FIGURE 5.4 Transient temperature distribution for different Biot numbers in a plane wall symmetrically cooled by convection.

We have introduced the Biot number because of its significance to transient conduction problems. Consider the plane wall of Figure 5.4, which is initially at a uniform temperature T_i and experiences convection cooling when it is immersed in a fluid of $T_\infty < T_i$. The problem may be treated as one-dimensional in x, and we are interested in the temperature variation with position and time, $T(x, t)$. This variation is a strong function of the Biot number, and three conditions are shown in Figure 5.4. For $Bi \ll 1$ the temperature gradient in the solid is small and $T(x, t) \approx T(t)$. Virtually all the temperature difference is between the solid and the fluid, and the solid temperature remains nearly uniform as it decreases to T_∞. For moderate to large values of the Biot number, however, the temperature gradients within the solid are significant. Hence $T = T(x, t)$. Note that for $Bi \gg 1$, the temperature difference across the solid is now much larger than that between the surface and the fluid.

We conclude this section by emphasizing the importance of the lumped capacitance method. Its inherent simplicity renders it the preferred method for solving transient conduction problems. Hence, when confronted with such a problem, *the very first thing that one should do is calculate the Biot number.* If the following condition is satisfied

$$Bi = \frac{hL_c}{k} < 0.1 \tag{5.10}$$

the error associated with using the lumped capacitance method is small. For convenience, it is customary to define the *characteristic length* of Equation 5.10 as the ratio of the solid's volume to surface area, $L_c \equiv V/A_s$. Such a definition facilitates calculation of L_c for solids of complicated shape and reduces to the half-thickness L for a plane wall of thickness $2L$ (Figure 5.4), to $r_o/2$ for a long cylinder, and to $r_o/3$ for a sphere. However, if one wishes to implement the criterion in a conservative fashion, L_c should be associated with the length scale corresponding to the maximum spatial temperature difference. Accordingly, for a symmetrically heated (or cooled) plane wall of thickness $2L$, L_c would remain equal to the half-thickness L. However, for a long cylinder or sphere, L_c would equal the actual radius r_o, rather than $r_o/2$ or $r_o/3$.

Finally, we note that, with $L_c \equiv V/A_s$, the exponent of Equation 5.6 may be expressed as

$$\frac{hA_s t}{\rho V c} = \frac{ht}{\rho c L_c} = \frac{hL_c}{k} \frac{k}{\rho c} \frac{t}{L_c^2} = \frac{hL_c}{k} \frac{\alpha t}{L_c^2}$$

or

$$\frac{hA_s t}{\rho V c} = Bi \cdot Fo \qquad (5.11)$$

where

$$Fo \equiv \frac{\alpha t}{L_c^2} \qquad (5.12)$$

is termed the Fourier number. It is a *dimensionless time*, which, with the Biot number, characterizes transient conduction problems. Substituting Equation 5.11 into 5.6, we obtain

$$\frac{\theta}{\theta_i} = \frac{T - T_\infty}{T_i - T_\infty} = \exp(-Bi \cdot Fo) \qquad (5.13)$$

EXAMPLE 5.1

A thermocouple junction, which may be approximated as a sphere, is to be used for temperature measurement in a gas stream. The convection coefficient between the junction surface and the gas is known to be $h = 400 \text{ W/m}^2 \cdot \text{K}$, and the junction thermophysical properties are $k = 20 \text{ W/m} \cdot \text{K}$, $c = 400 \text{ J/kg} \cdot \text{K}$, and $\rho = 8500 \text{ kg/m}^3$. Determine the junction diameter needed for the thermocouple to have a time constant of 1 s. If the junction is at 25°C and is placed in a gas stream that is at 200°C, how long will it take for the junction to reach 199°C?

SOLUTION

Known: Thermophysical properties of thermocouple junction used to measure temperature of a gas stream.

Find:

1. Junction diameter needed for a time constant of 1 s.
2. Time required to reach 199°C in gas stream at 200°C.

Schematic:

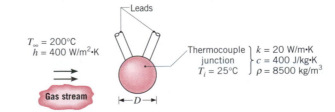

Leads

$T_\infty = 200°C$
$h = 400 \text{ W/m}^2 \cdot \text{K}$

Thermocouple junction
$T_i = 25°C$

$k = 20 \text{ W/m} \cdot \text{K}$
$c = 400 \text{ J/kg} \cdot \text{K}$
$\rho = 8500 \text{ kg/m}^3$

Gas stream

$\leftarrow D \rightarrow$

Assumptions:

1. Temperature of junction is uniform at any instant.
2. Radiation exchange with the surroundings is negligible.
3. Losses by conduction through the leads are negligible.
4. Constant properties.

Analysis:

1. Because the junction diameter is unknown, it is not possible to begin the solution by determining whether the criterion for using the lumped capacitance method, Equation 5.10, is satisfied. However, a reasonable approach is to use the method to find the diameter and to then determine whether the criterion is satisfied. From Equation 5.7 and the fact that $A_s = \pi D^2$ and $V = \pi D^3/6$ for a sphere, it follows that

$$\tau_t = \frac{1}{h\pi D^2} \times \frac{\rho\pi D^3}{6} c$$

Rearranging and substituting numerical values,

$$D = \frac{6h\tau_t}{\rho c} = \frac{6 \times 400 \text{ W/m}^2 \cdot \text{K} \times 1 \text{ s}}{8500 \text{ kg/m}^3 \times 400 \text{ J/kg} \cdot \text{K}} = 7.06 \times 10^{-4} \text{ m} \qquad \triangleleft$$

With $L_c = r_o/3$ it then follows from Equation 5.10 that

$$Bi = \frac{h(r_o/3)}{k} = \frac{400 \text{ W/m}^2 \cdot \text{K} \times 3.53 \times 10^{-4} \text{ m}}{3 \times 20 \text{ W/m} \cdot \text{K}} = 2.35 \times 10^{-4}$$

Accordingly, Equation 5.10 is satisfied (for $L_c = r_o$, as well as for $L_c = r_o/3$) and the lumped capacitance method may be used to an excellent approximation.

2. From Equation 5.5 the time required for the junction to reach $T = 199°C$ is

$$t = \frac{\rho(\pi D^3/6)c}{h(\pi D^2)} \ln \frac{T_i - T_\infty}{T - T_\infty} = \frac{\rho Dc}{6h} \ln \frac{T_i - T_\infty}{T - T_\infty}$$

$$t = \frac{8500 \text{ kg/m}^3 \times 7.06 \times 10^{-4} \text{ m} \times 400 \text{ J/kg} \cdot \text{K}}{6 \times 400 \text{ W/m}^2 \cdot \text{K}} \ln \frac{25 - 200}{199 - 200}$$

$$t = 5.2 \text{ s} \approx 5\tau_t \qquad \triangleleft$$

Comments: Heat transfer due to radiation exchange between the junction and the surroundings and conduction through the leads would affect the time response of the junction and would, in fact, yield an equilibrium temperature that differs from T_∞.

5.3
General Lumped Capacitance Analysis

Although transient conduction in a solid is commonly initiated by convection heat transfer to or from an adjoining fluid, other processes may induce transient thermal conditions within the solid. For example, a solid may be separated from

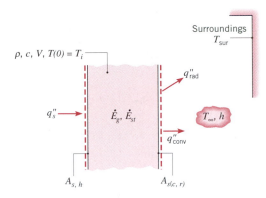

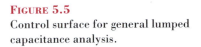

FIGURE 5.5
Control surface for general lumped capacitance analysis.

large surroundings by a gas or vacuum. If the temperatures of the solid and surroundings differ, radiation exchange could cause the internal thermal energy, and hence the temperature, of the solid to change. Temperature changes could also be induced by applying a heat flux at a portion, or all, of the surface and/or by initiating thermal energy generation within the solid. Surface heating could, for example, be applied by attaching a film or sheet electrical heater to the surface, while thermal energy could be generated by passing an electrical current through the solid.

Figure 5.5 depicts a situation for which thermal conditions within a solid may be influenced simultaneously by convection, radiation, an applied surface heat flux, and internal energy generation. It is presumed that, initially ($t = 0$), the temperature of the solid (T_i) differs from that of the fluid, T_∞, and the surroundings, T_{sur}, and that both surface and volumetric heating (q_s'' and $\dot{q}$) are initiated. The imposed heat flux q_s'' and the convection–radiation heat transfer occur at mutually exclusive portions of the surface, $A_{s(h)}$ and $A_{s(c, r)}$, respectively, and convection–radiation transfer is presumed to be *from* the surface. Moreover, although convection and radiation have been prescribed for the same surface, the surfaces may, in fact, differ ($A_{s, c} \neq A_{s, r}$). Applying conservation of energy at any instant t, it follows from Equation 1.11a that

$$q_s'' A_{s, h} + \dot{E}_g - (q_{conv}'' + q_{rad}'') A_{s(c, r)} = \rho V c \frac{dT}{dt} \tag{5.14}$$

or, from Equations 1.3a and 1.7,

$$q_s'' A_{s, h} + \dot{E}_g - [h(T - T_\infty) + \varepsilon \sigma (T^4 - T_{sur}^4)] A_{s(c, r)} = \rho V c \frac{dT}{dt} \tag{5.15}$$

Unfortunately, Equation 5.15 is a nonlinear, first-order, nonhomogenous, ordinary differential equation that cannot be integrated to obtain an exact solution.[1] However, exact solutions may be obtained for simplified versions of the equation. For example, if there is no imposed heat flux or generation and convection is either nonexistent (a vacuum) or negligible relative to radiation, Equation 5.15 reduces to

$$\rho V c \frac{dT}{dt} = -\varepsilon A_{s, r} \sigma (T^4 - T_{sur}^4) \tag{5.16}$$

---

[1] An approximate, finite-difference solution may be obtained by *discretizing* the time derivative (Section 5.9) and *marching* the solution out in time.

Separating variables and integrating from the initial condition to any time t, it follows that

$$\frac{\varepsilon A_{s,r}\sigma}{\rho Vc}\int_0^t dt = \int_{T_i}^{T}\frac{dT}{T_{sur}^4 - T^4} \tag{5.17}$$

Evaluating both integrals and rearranging, the time required to reach the temperature T becomes

$$t = \frac{\rho Vc}{4\varepsilon A_{s,r}\sigma T_{sur}^3}\left\{\ln\left|\frac{T_{sur} + T}{T_{sur} - T}\right| - \ln\left|\frac{T_{sur} + T_i}{T_{sur} - T_i}\right|\right.$$
$$\left. + 2\left[\tan^{-1}\left(\frac{T}{T_{sur}}\right) - \tan^{-1}\left(\frac{T_i}{T_{sur}}\right)\right]\right\} \tag{5.18}$$

This expression cannot be used to evaluate T explicitly in terms of t, T_i, and T_{sur}, nor does it readily reduce to the limiting result for $T_{sur} = 0$ (radiation to deep space). Returning to Equation 5.17, it is readily shown that, for $T_{sur} = 0$,

$$t = \frac{\rho Vc}{3\varepsilon A_{s,r}\sigma}\left(\frac{1}{T^3} - \frac{1}{T_i^3}\right) \tag{5.19}$$

An exact solution to Equation 5.15 may also be obtained if radiation may be neglected and h is independent of time. Introducing a reduced temperature, $\theta \equiv T - T_\infty$, where $d\theta/dt = dT/dt$, Equation 5.15 reduces to a linear, first-order, nonhomogenous differential equation of the form

$$\frac{d\theta}{dt} + a\theta - b = 0 \tag{5.20}$$

where $a \equiv (hA_{s,c}/\rho Vc)$ and $b \equiv [(q_s''A_{s,h} + \dot{E}_g)/\rho Vc]$. Although Equation 5.20 may be solved by summing its homogeneous and particular solutions, an alternative approach is to eliminate the nonhomogeneity by introducing the transformation

$$\theta' \equiv \theta - \frac{b}{a} \tag{5.21}$$

Recognizing that $d\theta'/dt = d\theta/dt$, Equation 5.21 may be substituted into (5.20) to yield

$$\frac{d\theta'}{dt} + a\theta' = 0 \tag{5.22}$$

Separating variables and integrating from 0 to t (θ_i' to θ'), it follows that

$$\frac{\theta'}{\theta_i'} = \exp(-at) \tag{5.23}$$

or substituting for θ' and θ,

$$\frac{T - T_\infty - (b/a)}{T_i - T_\infty - (b/a)} = \exp(-at) \tag{5.24}$$

Hence

$$\frac{T - T_\infty}{T_i - T_\infty} = \exp(-at) + \frac{b/a}{T_i - T_\infty}[1 - \exp(-at)] \tag{5.25}$$

As it must, Equation 5.25 reduces to (5.6) when $b = 0$ and yields $T = T_i$ at $t = 0$. As $t \rightarrow \infty$, Equation 5.25 reduces to $(T - T_\infty) = (b/a)$, which could also be obtained by performing an energy balance on the control surface of Figure 5.5 for steady-state conditions.

EXAMPLE 5.2

A 3-mm-thick panel of aluminum alloy ($k = 177$ W/m · K, $c = 875$ J/kg · K, and $\rho = 2770$ kg/m³) is finished on both sides with an epoxy coating that must be cured at or above $T_c = 150°$C for at least 5 min. The production line for the curing operation involves two steps: (1) heating in a large oven with air at $T_{\infty,o} = 175°$C and a convection coefficient of $h_o = 40$ W/m² · K, and (2) cooling in a large chamber with air at $T_{\infty,c} = 25°$C and a convection coefficient of $h_c = 10$ W/m² · K. The heating portion of the process is conducted over a time interval t_e, which exceeds the time t_c required to reach 150°C by 5 min ($t_e = t_c + 300$ s). The coating has an emissivity of $\varepsilon = 0.8$, and the temperatures of the oven and chamber walls are 175°C and 25°C, respectively. If the panel is placed in the oven at an initial temperature of 25°C and removed from the chamber at a *safe-to-touch* temperature of 37°C, what is the total elapsed time for the two-step curing operation?

SOLUTION

Known: Operating conditions for a two-step heating/cooling process in which a coated aluminum panel is maintained at or above a temperature of 150°C for at least 5 min.

Find: Total time t_t required for the two-step process.

Schematic:

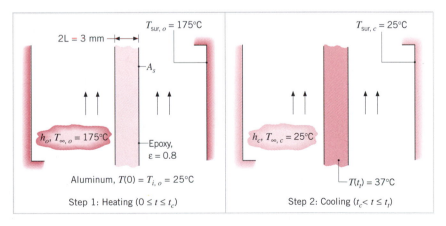

Assumptions:

1. Panel temperature is uniform at any instant.

2. Thermal resistance of epoxy is negligible.

3. Constant properties.

Analysis: To assess the validity of the lumped capacitance approximation, we begin by calculating Biot numbers for the heating and cooling processes.

$$Bi_h = \frac{h_o L}{k} = \frac{(40 \text{ W/m}^2 \cdot \text{K})(0.0015 \text{ m})}{177 \text{ W/m} \cdot \text{K}} = 3.4 \times 10^{-4}$$

$$Bi_c = \frac{h_c L}{k} = \frac{(10 \text{ W/m}^2 \cdot \text{K})(0.0015 \text{ m})}{177 \text{ W/m} \cdot \text{K}} = 8.5 \times 10^{-5}$$

Hence the lumped capacitance approximation is excellent.

To determine whether radiation exchange between the panel and its surroundings should be considered, the radiation heat transfer coefficient is determined from Equation 1.9. The representative value of h_r for the heating process is associated with the cure condition, in which case

$$
\begin{aligned}
h_{r,o} &= \varepsilon \sigma (T_c + T_{\text{sur},o})(T_c^2 + T_{\text{sur},o}^2) \\
&= 0.8 \times 5.67 \times 10^{-8} \text{ W/m}^2 \cdot \text{K}^4 (423 + 448)\text{K}(423^2 + 448^2)\text{K}^2 \\
&= 15 \text{ W/m}^2 \cdot \text{K}
\end{aligned}
$$

Using $T_c = 150°C$ with $T_{\text{sur},c} = 25°C$ for the cooling process, we also obtain $h_{r,c} = 5.1 \text{ W/m}^2 \cdot \text{K}$. Since the values of $h_{r,o}$ and $h_{r,c}$ are comparable to those of h_o and h_c, respectively, radiation effects must be considered.

With $V = 2LA_s$ and $A_{s,c} = A_{s,r} = 2A_s$, Equation 5.15 may be expressed as

$$\int_{T_i}^{T} dT = T(t) - T_i = -\frac{1}{\rho c L} \int_0^t [h(T - T_\infty) + \varepsilon \sigma (T^4 - T_{\text{sur}}^4)]dt$$

Selecting a suitable time increment Δt, the right-hand side of this equation may be evaluated numerically to obtain the panel temperature at $t = \Delta t, 2\Delta t, 3\Delta t$, and so on. At each new step of the calculation, the value of T computed from the previous time step is used in the integrand. Selecting $\Delta t = 10$ s, calculations for the heating process are extended to $t_e = t_c + 300$ s, which is 5 min beyond the time required for the panel to reach $T_c = 150°C$. At t_e the cooling process is initiated and continued until the panel temperature reaches $37°C$ at $t = t_r$. The integration was performed using a fourth-order Runge-Kutta scheme, and results of the calculations are plotted as follows:

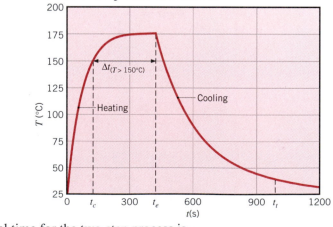

The total time for the two-step process is

$$t_t = 989 \text{ s} \qquad \triangleleft$$

with intermediate times of $t_c = 124$ s and $t_e = 424$ s.

Comments:

1. Generally, the accuracy of a numerical integration improves with decreasing Δt, but at the expense of increased computation time. In this case, however, results obtained for $\Delta t = 1$ s are virtually identical to those obtained for $\Delta t = 10$ s, indicating that the larger time interval is sufficient to accurately depict the temperature history.

2. The duration of the two-step process may be reduced by increasing the convection coefficients and/or by reducing the period of extended heating. The second option is made possible by the fact that, during a portion of the cooling period, the panel temperature remains above 150°C. Hence, to satisfy the cure requirement, it is not necessary to extend heating for as much as 5 min from $t = t_c$. If the convection coefficients are increased to $h_o = h_c = 100$ W/m$^2 \cdot$ K and an extended heating period of 300 s is maintained, the numerical integration yields $t_c = 58$ s and $t_t = 445$ s. The corresponding time interval over which the panel temperature exceeds 150°C is $\Delta t_{(T>150°C)} = 306$ s (58 s $\leq t \leq 364$ s). If the extended heating period is reduced to 294 s, the numerical integration yields $t_c = 58$ s, $t_t = 439$ s, and $\Delta t_{(T>150°C)} = 300$ s. Hence the total process time is reduced, while the curing requirement is still satisfied.

5.4
Spatial Effects

Situations frequently arise for which the lumped capacitance method is inappropriate, and alternative approaches must be used. Regardless of the particular method, we must now cope with the fact that gradients within the medium are no longer negligible.

In their most general form, transient conduction problems are described by the heat equation, Equation 2.13 for rectangular coordinates or Equations 2.20 and 2.23, respectively, for cylindrical and spherical coordinates. The solution to these partial differential equations provides the variation of temperature with both time and the spatial coordinates. However, in many problems, such as the plane wall of Figure 5.4, only one spatial coordinate is needed to describe the internal temperature distribution. With no internal generation and the assumption of constant thermal conductivity, Equation 2.13 then reduces to

$$\frac{\partial^2 T}{\partial x^2} = \frac{1}{\alpha} \frac{\partial T}{\partial t} \tag{5.26}$$

To solve Equation 5.26 for the temperature distribution $T(x, t)$, it is necessary to specify an *initial* condition and two *boundary conditions*. For the typical transient conduction problem of Figure 5.4, the initial condition is

$$T(x, 0) = T_i \tag{5.27}$$

and the boundary conditions are

$$\left. \frac{\partial T}{\partial x} \right|_{x=0} = 0 \tag{5.28}$$

and

$$-k\frac{\partial T}{\partial x}\bigg|_{x=L} = h[T(L, t) - T_\infty] \tag{5.29}$$

Equation 5.27 presumes a uniform temperature distribution at time $t = 0$; Equation 5.28 reflects the *symmetry requirement* for the midplane of the wall; and Equation 5.29 describes the surface condition experienced for time $t > 0$. From Equations 5.26 to 5.29, it is evident that, in addition to depending on x and t, temperatures in the wall also depend on a number of physical parameters. In particular

$$T = T(x, t, T_i, T_\infty, L, k, \alpha, h) \tag{5.30}$$

The foregoing problem may be solved analytically or numerically. These methods will be considered in subsequent sections, but first it is important to note the advantages that may be obtained by *nondimensionalizing* the governing equations. This may be done by arranging the relevant variables into suitable *groups*. Consider the dependent variable T. If the temperature difference $\theta \equiv T - T_\infty$ is divided by the *maximum possible temperature difference* $\theta_i \equiv T_i - T_\infty$, a dimensionless form of the dependent variable may be defined as

$$\theta^* \equiv \frac{\theta}{\theta_i} = \frac{T - T_\infty}{T_i - T_\infty} \tag{5.31}$$

Accordingly, θ^* must lie in the range $0 \le \theta^* \le 1$. A dimensionless spatial coordinate may be defined as

$$x^* \equiv \frac{x}{L} \tag{5.32}$$

where L is the half-thickness of the plane wall, and a dimensionless time may be defined as

$$t^* \equiv \frac{\alpha t}{L^2} \equiv Fo \tag{5.33}$$

where t^* is equivalent to the dimensionless *Fourier number*, Equation 5.12.

Substituting the definitions of Equations 5.31 to 5.33 into Equations 5.26 to 5.29, the heat equation becomes

$$\frac{\partial^2 \theta^*}{\partial x^{*2}} = \frac{\partial \theta^*}{\partial Fo} \tag{5.34}$$

and the initial and boundary conditions become

$$\theta^*(x^*, 0) = 1 \tag{5.35}$$

$$\frac{\partial \theta^*}{\partial x^*}\bigg|_{x^*=0} = 0 \tag{5.36}$$

and

$$\frac{\partial \theta^*}{\partial x^*}\bigg|_{x^*=1} = -Bi\,\theta^*(1, t^*) \tag{5.37}$$

where the Biot number is $Bi \equiv hL/k$. In dimensionless form the functional dependence may now be expressed as

$$\theta^* = f(x^*, Fo, Bi) \tag{5.38}$$

Recall that a similar functional dependence, without the x^* variation, was obtained for the lumped capacitance method, as shown in Equation 5.13.

Comparing Equations 5.30 and 5.38, the considerable advantage associated with casting the problem in dimensionless form becomes apparent. Equation 5.38 implies that *for a prescribed geometry, the transient temperature distribution is a universal function of x^*, Fo, and Bi. That is, the dimensionless solution* assumes a prescribed form that does not depend on the particular value of T_i, T_∞, L, k, α, or h. Since this generalization greatly simplifies the presentation and utilization of transient solutions, the dimensionless variables are used extensively in subsequent sections.

5.5
The Plane Wall with Convection

Exact, analytical solutions to transient conduction problems have been obtained for many simplified geometries and boundary conditions and are well documented [1–4]. Several mathematical techniques, including the method of separation of variables (Section 4.2), may be used for this purpose, and typically the solution for the dimensionless temperature distribution, Equation 5.38, is in the form of an infinite series. However, except for very small values of the Fourier number, this series may be approximated by a single term and the results may be represented in a convenient graphical form.

5.5.1 Exact Solution

Consider the *plane wall* of thickness $2L$ (Figure 5.6a). If the thickness is small relative to the width and height of the wall, it is reasonable to assume that conduction occurs exclusively in the x direction. If the wall is initially at a uniform

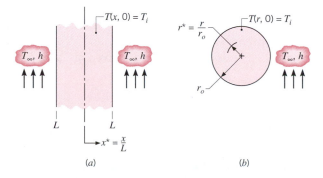

(a)   (b)

FIGURE 5.6   One-dimensional systems with an initial uniform temperature subjected to sudden convection conditions. (*a*) Plane wall. (*b*) Infinite cylinder or sphere.

temperature, $T(x, 0) = T_i$, and is suddenly immersed in a fluid of $T_\infty \neq T_i$, the resulting temperatures may be obtained by solving Equation 5.34 subject to the conditions of Equations 5.35 to 5.37. Since the convection conditions for the surfaces at $x^* = \pm 1$ are the same, the temperature distribution at any instant must be symmetrical about the midplane ($x^* = 0$). An exact solution to this problem has been obtained and is of the form [2]

$$\theta^* = \sum_{n=1}^{\infty} C_n \exp\left(-\zeta_n^2 Fo\right) \cos\left(\zeta_n x^*\right) \tag{5.39a}$$

where $Fo = \alpha t / L^2$ and the coefficient C_n is

$$C_n = \frac{4 \sin \zeta_n}{2\zeta_n + \sin(2\zeta_n)} \tag{5.39b}$$

and the discrete values (*eigenvalues*) of ζ_n are positive roots of the transcendental equation

$$\zeta_n \tan \zeta_n = Bi \tag{5.39c}$$

The first four roots of this equation are given in Appendix B.3.

5.5.2 Approximate Solution

It can be shown (Problem 5.27) that for values of $Fo > 0.2$, the infinite series solution, Equation 5.39a, can be approximated by the first term of the series. Invoking this approximation, the dimensionless form of the temperature distribution becomes

$$\theta^* = C_1 \exp\left(-\zeta_1^2 Fo\right) \cos\left(\zeta_1 x^*\right) \tag{5.40a}$$

or

$$\theta^* = \theta_o^* \cos\left(\zeta_1 x^*\right) \tag{5.40b}$$

where $\theta_o^* \equiv (T_o - T_\infty)/(T_i - T_\infty)$ represents the midplane ($x^* = 0$) temperature

$$\theta_o^* = C_1 \exp\left(-\zeta_1^2 Fo\right) \tag{5.41}$$

An important implication of Equation 5.40b is that *the time dependence of the temperature at any location within the wall is the same as that of the midplane temperature.* The coefficients C_1 and ζ_1 are evaluated from Equations 5.39b and 5.39c, respectively, and are given in Table 5.1 for a range of Biot numbers.

5.5.3 Total Energy Transfer

In many situations it is useful to know the total energy that has left the wall up to any time t in the transient process. The conservation of energy requirement, Equation 1.11b, may be applied for the time interval bounded by the initial condition ($t = 0$) and any time $t > 0$

$$E_{\text{in}} - E_{\text{out}} = \Delta E_{\text{st}} \tag{5.42}$$

TABLE 5.1 Coefficients used in the one-term approximation to the series solutions for transient one-dimensional conduction

Bi^a	Plane Wall		Infinite Cylinder		Sphere	
	ζ_1 (rad)	C_1	ζ_1 (rad)	C_1	ζ_1 (rad)	C_1
0.01	0.0998	1.0017	0.1412	1.0025	0.1730	1.0030
0.02	0.1410	1.0033	0.1995	1.0050	0.2445	1.0060
0.03	0.1732	1.0049	0.2439	1.0075	0.2989	1.0090
0.04	0.1987	1.0066	0.2814	1.0099	0.3450	1.0120
0.05	0.2217	1.0082	0.3142	1.0124	0.3852	1.0149
0.06	0.2425	1.0098	0.3438	1.0148	0.4217	1.0179
0.07	0.2615	1.0114	0.3708	1.0173	0.4550	1.0209
0.08	0.2791	1.0130	0.3960	1.0197	0.4860	1.0239
0.09	0.2956	1.0145	0.4195	1.0222	0.5150	1.0268
0.10	0.3111	1.0160	0.4417	1.0246	0.5423	1.0298
0.15	0.3779	1.0237	0.5376	1.0365	0.6608	1.0445
0.20	0.4328	1.0311	0.6170	1.0483	0.7593	1.0592
0.25	0.4801	1.0382	0.6856	1.0598	0.8448	1.0737
0.30	0.5218	1.0450	0.7465	1.0712	0.9208	1.0880
0.4	0.5932	1.0580	0.8516	1.0932	1.0528	1.1164
0.5	0.6533	1.0701	0.9408	1.1143	1.1656	1.1441
0.6	0.7051	1.0814	1.0185	1.1346	1.2644	1.1713
0.7	0.7506	1.0919	1.0873	1.1539	1.3525	1.1978
0.8	0.7910	1.1016	1.1490	1.1725	1.4320	1.2236
0.9	0.8274	1.1107	1.2048	1.1902	1.5044	1.2488
1.0	0.8603	1.1191	1.2558	1.2071	1.5708	1.2732
2.0	1.0769	1.1795	1.5995	1.3384	2.0288	1.4793
3.0	1.1925	1.2102	1.7887	1.4191	2.2889	1.6227
4.0	1.2646	1.2287	1.9081	1.4698	2.4556	1.7201
5.0	1.3138	1.2402	1.9898	1.5029	2.5704	1.7870
6.0	1.3496	1.2479	2.0490	1.5253	2.6537	1.8338
7.0	1.3766	1.2532	2.0937	1.5411	2.7165	1.8674
8.0	1.3978	1.2570	2.1286	1.5526	2.7654	1.8921
9.0	1.4149	1.2598	2.1566	1.5611	2.8044	1.9106
10.0	1.4289	1.2620	2.1795	1.5677	2.8363	1.9249
20.0	1.4961	1.2699	2.2881	1.5919	2.9857	1.9781
30.0	1.5202	1.2717	2.3261	1.5973	3.0372	1.9898
40.0	1.5325	1.2723	2.3455	1.5993	3.0632	1.9942
50.0	1.5400	1.2727	2.3572	1.6002	3.0788	1.9962
100.0	1.5552	1.2731	2.3809	1.6015	3.1102	1.9990
∞	1.5707	1.2733	2.4050	1.6018	3.1415	2.0000

$^a Bi = hL/k$ for the plane wall and hr_o/k for the infinite cylinder and sphere. See Figure 5.6.

Equating the energy transferred from the wall Q to E_{out} and setting $E_{in} = 0$ and $\Delta E_{st} = E(t) - E(0)$, it follows that

$$Q = -[E(t) - E(0)] \tag{5.43a}$$

or

$$Q = -\int \rho c[T(r, t) - T_i]\, dV \tag{5.43b}$$

where the integration is performed over the volume of the wall. It is convenient to nondimensionalize this result by introducing the quantity

$$Q_o = \rho c V(T_i - T_\infty) \tag{5.44}$$

which may be interpreted as the initial internal energy of the wall relative to the fluid temperature. It is also the *maximum* amount of energy transfer that could occur if the process were continued to time $t = \infty$. Hence, assuming constant properties, the ratio of the total energy transferred from the wall over the time interval t to the maximum possible transfer is

$$\frac{Q}{Q_o} = \int \frac{-[T(x, t) - T_i]}{T_i - T_\infty}\frac{dV}{V} = \frac{1}{V}\int (1 - \theta^*)\, dV \tag{5.45}$$

Employing the approximate form of the temperature distribution for the plane wall, Equation 5.40b, the integration prescribed by Equation 5.45 can be performed to obtain

$$\frac{Q}{Q_o} = 1 - \frac{\sin \zeta_1}{\zeta_1}\theta_o^* \tag{5.46}$$

where θ_o^* can be determined from Equation 5.41, using Table 5.1 for values of the coefficients C_1 and ζ_1.

5.5.4 Additional Considerations

Because the mathematical problem is precisely the same, the foregoing results may also be applied to a plane wall of thickness L, which is insulated on one side ($x^* = 0$) and experiences convective transport on the other side ($x^* = +1$). This equivalence is a consequence of the fact that, regardless of whether a symmetrical or an adiabatic requirement is prescribed at $x^* = 0$, the boundary condition is of the form $\partial\theta^*/\partial x^* = 0$.

It should also be noted that the foregoing results may be used to determine the transient response of a plane wall to a sudden change in *surface* temperature. The process is equivalent to having an infinite convection coefficient, in which case the Biot number is infinite ($Bi = \infty$) and the fluid temperature T_∞ is replaced by the prescribed surface temperature T_s.

Finally, we note that graphical representations of the one-term approximations have been developed [5, 6] and are presented in Appendix D. Although the associated charts provide a convenient means of solving one-dimensional transient conduction problems for $Fo > 0.2$, better accuracy may be obtained by using Equations 5.40 and 5.46.

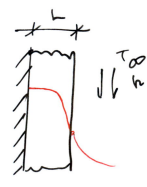

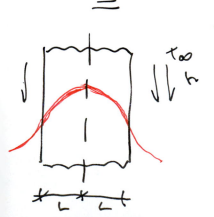

5.6
Radial Systems with Convection

For an infinite cylinder or sphere of radius r_o (Figure 5.6b), which is at an initial uniform temperature and experiences a change in convective conditions, results similar to those of Section 5.5 may be developed. That is, an exact series solution may be obtained for the time dependence of the radial temperature distribution, and a one-term approximation may be used for most conditions. The infinite cylinder is an idealization that permits the assumption of one-dimensional conduction in the radial direction. It is a reasonable approximation for cylinders having $L/r_o \gtrsim 10$.

5.6.1   Exact Solutions

Exact solutions to the transient, one-dimensional form of the heat equation have been developed for the infinite cylinder and for the sphere. For a uniform initial temperature and convective boundary conditions, the solutions [2] are as follows.

Infinite Cylinder   In dimensionless form, the temperature is

$$\theta^* = \sum_{n=1}^{\infty} C_n \exp\left(-\zeta_n^2 Fo\right) J_0(\zeta_n r^*) \tag{5.47a}$$

where $Fo = \alpha t/r_o^2$,

$$C_n = \frac{2}{\zeta_n} \frac{J_1(\zeta_n)}{J_0^2(\zeta_n) + J_1^2(\zeta_n)} \tag{5.47b}$$

and the discrete values of ζ_n are positive roots of the transcendental equation

$$\zeta_n \frac{J_1(\zeta_n)}{J_0(\zeta_n)} = Bi \tag{5.47c}$$

The quantities J_1 and J_0 are Bessel functions of the first kind and their values are tabulated in Appendix B.4. Roots of the transcendental equation (5.47c) are tabulated by Schneider [2].

Sphere   Similarly, for the sphere

$$\theta^* = \sum_{n=1}^{\infty} C_n \exp\left(-\zeta_n^2 Fo\right) \frac{1}{\zeta_n r^*} \sin\left(\zeta_n r^*\right) \tag{5.48a}$$

where $Fo = \alpha t/r_o^2$,

$$C_n = \frac{4[\sin\left(\zeta_n\right) - \zeta_n \cos\left(\zeta_n\right)]}{2\zeta_n - \sin\left(2\zeta_n\right)} \tag{5.48b}$$

and the discrete values of ζ_n are positive roots of the transcendental equation

$$1 - \zeta_n \cot \zeta_n = Bi \tag{5.48c}$$

Roots of the transcendental equation are tabulated by Schneider [2].

5.6.2   Approximate Solutions

For the infinite cylinder and sphere, the foregoing series solutions can again be approximated by a single term for $Fo > 0.2$. Hence, as for the case of the plane wall, the time dependence of the temperature at any location within the radial system is the same as that of the centerline or centerpoint.

Infinite Cylinder   The one-term approximation to Equation 5.47 is

$$\theta^* = C_1 \exp{(-\zeta_1^2 Fo)} J_0(\zeta_1 r^*) \tag{5.49a}$$

or

$$\theta^* = \theta_o^* J_0(\zeta_1 r^*) \tag{5.49b}$$

where θ_o^* represents the centerline temperature and is of the form

$$\theta_o^* = C_1 \exp{(-\zeta_1^2 Fo)} \tag{5.49c}$$

Values of the coefficients C_1 and ζ_1 have been determined and are listed in Table 5.1 for a range of Biot numbers.

Sphere   From equation 5.48a, the one-term approximation is

$$\theta^* = C_1 \exp{(-\zeta_1^2 Fo)} \frac{1}{\zeta_1 r^*} \sin{(\zeta_1 r^*)} \tag{5.50a}$$

or

$$\theta^* = \theta_o^* \frac{1}{\zeta_1 r^*} \sin{(\zeta_1 r^*)} \tag{5.50b}$$

where θ_o^* represents the center temperature and is of the form

$$\theta_o^* = C_1 \exp{(-\zeta_1^2 Fo)} \tag{5.50c}$$

Values of the coefficients C_1 and ζ_1 have been determined and are listed in Table 5.1 for a range of Biot numbers.

5.6.3   Total Energy Transfer

As in Section 5.5.3, an energy balance may be performed to determine the total energy transfer from the infinite cylinder or sphere over the time interval $\Delta t = t$. Substituting from the approximate solutions, Equations 5.49b and 5.50b, and introducing Q_o from Equation 5.44, the results are as follows.

Infinite Cylinder

$$\frac{Q}{Q_o} = 1 - \frac{2\theta_o^*}{\zeta_1} J_1(\zeta_1) \tag{5.51}$$

Sphere

$$\frac{Q}{Q_o} = 1 - \frac{3\theta_o^*}{\zeta_1^3} [\sin{(\zeta_1)} - \zeta_1 \cos{(\zeta_1)}] \tag{5.52}$$

Values of the center temperature θ_o^* are determined from Equation 5.49c or 5.50c, using the coefficients of Table 5.1 for the appropriate system.

5.6.4 Additional Considerations

As for the plane wall, the foregoing results may be used to predict the transient response of long cylinders and spheres subjected to a sudden change in *surface* temperature. Namely, an infinite Biot number is prescribed, and the fluid temperature T_∞ is replaced by the constant surface temperature T_s.

Graphical representations of the one-term approximations are presented in Appendix D.

EXAMPLE 5.3

Consider a steel pipeline (AISI 1010) that is 1 m in diameter and has a wall thickness of 40 mm. The pipe is heavily insulated on the outside, and before the initiation of flow, the walls of the pipe are at a uniform temperature of $-20°C$. With the initiation of flow, hot oil at 60°C is pumped through the pipe creating a convective surface condition corresponding to $h = 500 \text{ W/m}^2 \cdot \text{K}$ at the inner surface of the pipe.

1. What are the appropriate Biot and Fourier numbers 8 min after the initiation of flow?

2. At $t = 8$ min, what is the temperature of the exterior pipe surface covered by the insulation?

3. What is the heat flux q'' (W/m²) to the pipe from the oil at $t = 8$ min?

4. How much energy per meter of pipe length has been transferred from the oil to the pipe at $t = 8$ min?

SOLUTION

Known: Wall subjected to sudden change in convective surface condition.

Find:

1. Biot and Fourier numbers after 8 min.
2. Temperature of exterior pipe surface after 8 min.
3. Heat flux to the wall at 8 min.
4. Energy transferred to pipe per unit length after 8 min.

Schematic:

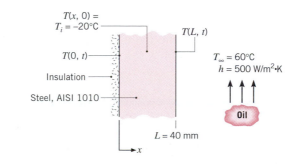

Assumptions:

1. Pipe wall can be approximated as plane wall, since thickness is much less than diameter.
2. Constant properties.
3. Outer surface of pipe is adiabatic.

Properties: Table A.1, steel type AISI 1010 [$T = (-20 + 60)°C/2 \approx 300$ K]: $\rho = 7823$ kg/m³, $c = 434$ J/kg · K, $k = 63.9$ W/m · K, $\alpha = 18.8 \times 10^{-6}$ m²/s.

Analysis:

1. At $t = 8$ min, the Biot and Fourier numbers are computed from Equations 5.10 and 5.12, respectively, with $L_c = L$. Hence

$$Bi = \frac{hL}{k} = \frac{500 \text{ W/m}^2 \cdot \text{K} \times 0.04 \text{ m}}{63.9 \text{ W/m} \cdot \text{K}} = 0.313 \qquad \triangleleft$$

$$Fo = \frac{\alpha t}{L^2} = \frac{18.8 \times 10^{-6} \text{ m}^2/\text{s} \times 8 \text{ min} \times 60 \text{ s/min}}{(0.04 \text{ m})^2} = 5.64 \qquad \triangleleft$$

2. With $Bi = 0.313$, use of the lumped capacitance method is inappropriate. However, since $Fo > 0.2$ and transient conditions in the insulated pipe wall of thickness L correspond to those in a plane wall of thickness $2L$ experiencing the same surface condition, the desired results may be obtained from the one-term approximation for a plane wall. The midplane temperature can be determined from Equation 5.41

$$\theta_o^* = \frac{T_o - T_\infty}{T_i - T_\infty} = C_1 \exp\left(-\zeta_1^2 Fo\right)$$

where, with $Bi = 0.313$, $C_1 = 1.047$ and $\zeta_1 = 0.531$ rad from Table 5.1. With $Fo = 5.64$,

$$\theta_o^* = 1.047 \exp\left[-(0.531 \text{ rad})^2 \times 5.64\right] = 0.214$$

Hence after 8 min, the temperature of the exterior pipe surface, which corresponds to the midplane temperature of a plane wall, is

$$T(0, 8 \text{ min}) = T_\infty + \theta_o^*(T_i - T_\infty) = 60°C + 0.214(-20 - 60)°C = 42.9°C \qquad \triangleleft$$

3. Heat transfer to the inner surface at $x = L$ is by convection, and at any time t the heat flux may be obtained from Newton's law of cooling. Hence at $t = 480$ s,

$$q_x''(L, 480 \text{ s}) \equiv q_L'' = h[T(L, 480 \text{ s}) - T_\infty]$$

Using the one-term approximation for the surface temperature, Equation 5.40b with $x^* = 1$ has the form

$$\theta^* = \theta_o^* \cos(\zeta_1)$$

$$T(L, t) = T_\infty + (T_i - T_\infty)\theta_o^* \cos(\zeta_1)$$

$$T(L, 8 \text{ min}) = 60°C + (-20 - 60)°C \times 0.214 \times \cos(0.531 \text{ rad})$$

$$T(L, 8 \text{ min}) = 45.2°C$$

The heat flux at $t = 8$ min is then

$$q_L'' = 500 \text{ W/m}^2 \cdot \text{K } (45.2 - 60)°\text{C} = -7400 \text{ W/m}^2 \qquad \triangleleft$$

4. The energy transfer to the pipe wall over the 8-min interval may be obtained from Equations 5.44 and 5.46. With

$$\frac{Q}{Q_o} = 1 - \frac{\sin(\zeta_1)}{\zeta_1} \theta_o^*$$

$$\frac{Q}{Q_o} = 1 - \frac{\sin(0.531 \text{ rad})}{0.531 \text{ rad}} \times 0.214 = 0.80$$

it follows that

$$Q = 0.80 \, \rho c V(T_i - T_\infty)$$

or with a volume per unit pipe length of $V' = \pi DL$,

$$Q' = 0.80 \, \rho c \pi DL(T_i - T_\infty)$$

$$Q' = 0.80 \times 7823 \text{ kg/m}^3 \times 434 \text{ J/kg} \cdot \text{K}$$

$$\times \pi \times 1 \text{ m} \times 0.04 \text{ m } (-20 - 60)°\text{C}$$

$$Q' = -2.73 \times 10^7 \text{ J/m} \qquad \triangleleft$$

Comments:

1. The minus sign associated with q'' and Q' simply implies that the direction of heat transfer is from the oil to the pipe (into the pipe wall).

2. The foregoing results could also be obtained by applying the Heisler and Gröber charts of Appendix D. For example, using Figure D.1 with $Bi^{-1} = 3.2$, it follows that $\theta_o^* \approx 0.22$ and the corresponding value of the midplane temperature is $T_o \approx 42°\text{C}$. For $x^* = 1$ and $Bi^{-1} = 3.2$, Figure D.2 yields $\theta(L, 8 \text{ min})/\theta_o(8 \text{ min}) \approx 0.86$, from which it follows that $T(L, 8 \text{ min}) \approx T_\infty + 0.86[T_o(8 \text{ min}) - T_\infty] \approx 45°\text{C}$ and $q_L'' = -7500 \text{ W/m}^2$. With $Bi = 0.313$ and $Bi^2 Fo = 0.55$, Figure D.3 yields $Q/Q_o \approx 0.78$. Substituting from Equation 5.44, it follows that

$$Q' \approx 0.78 \rho c \pi DL(T_i - T_\infty) = -2.7 \times 10^7 \text{ J/m}$$

The foregoing results are in good agreement with those obtained directly from the one-term approximations.

EXAMPLE 5.4

A new process for treatment of a special material is to be evaluated. The material, a sphere of radius $r_o = 5$ mm, is initially in equilibrium at 400°C in a furnace. It is suddenly removed from the furnace and subjected to a two-step cooling process.

Step 1 Cooling in air at 20°C for a period of time t_a until the center temperature reaches a critical value, $T_a(0, t_a) = 335°\text{C}$. For this situation, the convective heat transfer coefficient is $h_a = 10 \text{ W/m}^2 \cdot \text{K}$.

After the sphere has reached this critical temperature, the second step is initiated.

Step 2 Cooling in a well-stirred water bath at 20°C, with a convective heat transfer coefficient of $h_w = 6000$ W/m² · K.

The thermophysical properties of the material are $\rho = 3000$ kg/m³, $k = 20$ W/m · K, $c = 1000$ J/kg · K, and $\alpha = 6.66 \times 10^{-6}$ m²/s.

1. Calculate the time t_a required for step 1 of the cooling process to be completed.
2. Calculate the time t_w required during step 2 of the process for the center of the sphere to cool from 335°C (the condition at the completion of step 1) to 50°C.

SOLUTION

Known: Temperature requirements for cooling a sphere.

Find:

1. Time t_a required to accomplish desired cooling in air.
2. Time t_w required to complete cooling in water bath.

Schematic:

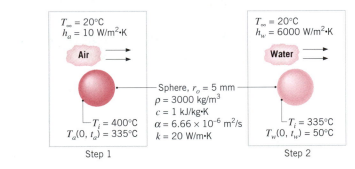

Assumptions:

1. One-dimensional conduction in r.
2. Constant properties.

Analysis:

1. To determine whether the lumped capacitance method can be used, the Biot number is calculated. From Equation 5.10, with $L_c = r_o/3$,

$$Bi = \frac{h_a r_o}{3k} = \frac{10 \text{ W/m}^2 \cdot \text{K} \times 0.005 \text{ m}}{3 \times 20 \text{ W/m} \cdot \text{K}} = 8.33 \times 10^{-4}$$

Accordingly, the lumped capacitance method may be used, and the temperature is nearly uniform throughout the sphere. From Equation 5.5 it follows that

$$t_a = \frac{\rho V c}{h_a A_s} \ln \frac{\theta_i}{\theta_a} = \frac{\rho r_o c}{3 h_a} \ln \frac{T_i - T_\infty}{T_a - T_\infty}$$

where $V = (4/3)\pi r_o^3$ and $A_s = 4\pi r_o^2$. Hence

$$t_a = \frac{3000 \text{ kg/m}^3 \times 0.005 \text{ m} \times 1000 \text{ J/kg} \cdot \text{K}}{3 \times 10 \text{ W/m}^2 \cdot \text{K}} \ln \frac{400 - 20}{335 - 20} = 94 \text{ s} \quad \triangleleft$$

2. To determine whether the lumped capacitance method may also be used for the second step of the cooling process, the Biot number is again calculated. In this case

$$Bi = \frac{h_w r_o}{3k} = \frac{6000 \text{ W/m}^2 \cdot \text{K} \times 0.005 \text{ m}}{3 \times 20 \text{ W/m} \cdot \text{K}} = 0.50$$

[handwritten: $L_c = \frac{r_o}{3}$ (general form)]

and the lumped capacitance method is not appropriate. However, to an excellent approximation, the temperature of the sphere is uniform at $t = t_a$ and the one-term approximation may be used for the calculations from $t = t_a$ to $t = t_a + t_w$. The time t_w at which the center temperature reaches 50°C, that is, $T(0, t_w) = 50$°C, can be obtained by rearranging Equation 5.50c

$$Fo = -\frac{1}{\zeta_1^2} \ln \left[\frac{\theta_o^*}{C_1} \right] = -\frac{1}{\zeta_1^2} \ln \left[\frac{1}{C_1} \times \frac{T(0, t_w) - T_\infty}{T_i - T_\infty} \right]$$

where $t_w = Fo \, r_o^2 / \alpha$. With the Biot number now defined as

$$Bi = \frac{h_w r_o}{k} = \frac{6000 \text{ W/m}^2 \cdot \text{K} \times 0.005 \text{ m}}{20 \text{ W/m} \cdot \text{K}} = 1.50$$

[handwritten: $L_c = r_o$ (depends on the table)]

Table 5.1 yields $C_1 = 1.376$ and $\zeta_1 = 1.800$ rad. It follows that

$$Fo = -\frac{1}{(1.800 \text{ rad})^2} \ln \left[\frac{1}{1.376} \times \frac{(50 - 20)\text{°C}}{(335 - 20)\text{°C}} \right] = 0.82$$

and

$$t_w = Fo \frac{r_o^2}{\alpha} = 0.82 \frac{(0.005 \text{ m})^2}{6.66 \times 10^{-6} \text{ m}^2/\text{s}} = 3.1 \text{ s} \quad \triangleleft$$

Note that, with $Fo = 0.82$, use of the one-term approximation is justified.

Comments:

1. If the temperature distribution in the sphere at the conclusion of step 1 were not uniform, the one-term approximation could not be used for the calculations of step 2.

2. The surface temperature of the sphere at the conclusion of step 2 may be obtained from Equation 5.50b. With $\theta_o^* = 0.095$ and $r^* = 1$,

$$\theta^*(r_o) = \frac{T(r_o) - T_\infty}{T_i - T_\infty} = \frac{0.095}{1.800 \text{ rad}} \sin(1.800 \text{ rad}) = 0.0514$$

and

$$T(r_o) = 20\text{°C} + 0.0514(335 - 20)\text{°C} = 36\text{°C}$$

The infinite series, Equation 5.48a, and its one-term approximation, Equation 5.50b, may be used to compute the temperature at any location in the sphere and at any time $t > t_a$. For $(t - t_a) < 0.2(0.005 \text{ m})^2/6.66 \times 10^{-6}$ m²/s = 0.75 s, a sufficient number of terms must be retained to ensure convergence of the series. For $(t - t_a) > 0.75$ s, satisfactory convergence is provided by the one-term approximation. Computing and plotting the temperatures histories for $r = 0$ and $r = r_o$, we obtain the following results for $0 \leq (t - t_a) \leq 5$ s:

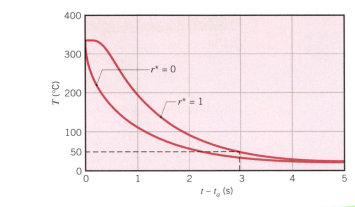

3. The Heisler charts of Appendix D could also be used to analyze the step 2 process. With $Bi^{-1} = 0.67$ and $\theta_o^* = 0.095$, Figure D.7 yields $Fo \approx 0.8$, in which case $t_w \approx 3.0$ s. From Figure D.8, with $r^* = 1$, $\theta(r_o)/\theta_o \approx 0.52$, in which case $T(r_o) \approx 20°C + 0.52(50 - 20)°C \approx 36°C$.

5.7
The Semi-Infinite Solid

Another simple geometry for which analytical solutions may be obtained is the *semi-infinite solid*. Since such a solid extends to infinity in all but one direction, it is characterized by a single identifiable surface (Figure 5.7). If a sudden change of conditions is imposed at this surface, transient, one-dimensional conduction will occur within the solid. The semi-infinite solid provides a *useful idealization* for many practical problems. It may be used to determine transient heat transfer near the surface of the earth or to approximate the transient response of a finite solid, such as a thick slab. For this second situation the approximation would be reasonable for the early portion of the transient, during which temperatures in the slab interior (well removed from the surface) are uninfluenced by the change in surface conditions.

The heat equation for transient conduction in a semi-infinite solid is given by Equation 5.26. The initial condition is prescribed by Equation 5.27, and the interior boundary condition is of the form

$$T(x \to \infty, t) = T_i \tag{5.53}$$

Closed-form solutions have been obtained for three important surface conditions, instantaneously applied at $t = 0$ [1, 2]. These conditions are shown in Figure 5.7. They include application of a constant surface temperature $T_s \neq T_i$, application of a constant surface heat flux q_o'', and exposure of the surface to a fluid characterized by $T_\infty \neq T_i$ and the convection coefficient h.

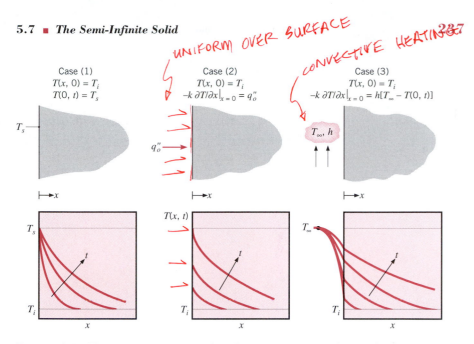

FIGURE 5.7   Transient temperature distributions in a semi-infinite solid for three surface conditions: constant surface temperature, constant surface heat flux, and surface convection.

The solution for case 1 may be obtained by recognizing the existence of a *similarity variable* η, through which the heat equation may be transformed from a partial differential equation, involving two independent variables (x and t), to an ordinary differential equation expressed in terms of the single similarity variable. To confirm that such a requirement is satisfied by $\eta \equiv x/(4\alpha t)^{1/2}$, we first transform the pertinent differential operators, such that

$$\frac{\partial T}{\partial x} = \frac{dT}{d\eta} \frac{\partial \eta}{\partial x} = \frac{1}{(4\alpha t)^{1/2}} \frac{dT}{d\eta}$$

$$\frac{\partial^2 T}{\partial x^2} = \frac{d}{d\eta}\left[\frac{\partial T}{\partial x}\right] \frac{\partial \eta}{\partial x} = \frac{1}{4\alpha t} \frac{d^2 T}{d\eta^2}$$

$$\frac{\partial T}{\partial t} = \frac{dT}{d\eta} \frac{\partial \eta}{\partial t} = -\frac{x}{2t(4\alpha t)^{1/2}} \frac{dT}{d\eta}$$

Substituting into Equation 5.26, the heat equation becomes

$$\frac{d^2 T}{d\eta^2} = -2\eta \frac{dT}{d\eta} \tag{5.54}$$

With $x = 0$ corresponding to $\eta = 0$, the surface condition may be expressed as

$$T(\eta = 0) = T_s \tag{5.55}$$

and with $x \to \infty$, as well as $t = 0$, corresponding to $\eta \to \infty$, both the initial condition and the interior boundary condition correspond to the single requirement that

$$T(\eta \to \infty) = T_i \tag{5.56}$$

Since the transformed heat equation and the initial/boundary conditions are independent of x and t, $\eta \equiv x/(4\alpha t)^{1/2}$ is, indeed, a similarity variable. Its existence implies that the *shape* of the temperature distribution in the medium, $T(x)$, is independent of time and that, irrespective of the values of x and t, the temperature may be represented as a unique function of η.

The specific form of the temperature dependence, $t(\eta)$, may be obtained by separating variables in Equation 5.54, such that

$$\frac{d(dT/d\eta)}{(dT/d\eta)} = -2\eta \, d\eta$$

Integrating, it follows that

$$\ln(dT/d\eta) = -\eta^2 + C_1'$$

or

$$\frac{dT}{d\eta} = C_1 \exp(-\eta^2)$$

Integrating a second time, we obtain

$$T = C_1 \int_0^\eta \exp(-u^2) \, du + C_2$$

where u is a dummy variable. Applying the boundary condition at $\eta = 0$, Equation 5.55, it follows that $C_2 = T_s$ and

$$T = C_1 \int_0^\eta \exp(-u^2) \, du + T_s$$

From the second boundary condition, Equation 5.56, we obtain

$$T_i = C_1 \int_0^\infty \exp(-u^2) \, du + T_s$$

or, evaluating the definite integral,

$$C_1 = \frac{2(T_i - T_s)}{\pi^{1/2}}$$

Hence the temperature distribution may be expressed as

$$\frac{T - T_s}{T_i - T_s} = (2/\pi^{1/2}) \int_0^\eta \exp(-u^2) \, du \equiv \operatorname{erf} \eta \tag{5.57}$$

where the *Gaussian error function,* erf η, is a standard mathematical function that is tabulated in Appendix B. The surface heat flux may be obtained by applying Fourier's law at $x = 0$, in which case

$$q_s'' = -k \left. \frac{\partial T}{\partial x} \right|_{x=0} = -k(T_i - T_s) \left. \frac{d(\operatorname{erf} \eta)}{d\eta} \frac{\partial \eta}{\partial x} \right|_{\eta=0}$$

$$q_s'' = k(T_s - T_i)(2/\pi^{1/2}) \exp(-\eta^2)(4\alpha t)^{-1/2} \big|_{\eta=0}$$

$$q_s'' = \frac{k(T_s - T_i)}{(\pi \alpha t)^{1/2}} \tag{5.58}$$

Analytical solutions may also be obtained for the case 2 and case 3 surface conditions, and results for all three cases are summarized as follows.

Case 1   Constant Surface Temperature: $T(0, t) = T_s$

$$\frac{T(x, t) - T_s}{T_i - T_s} = \text{erf}\left(\frac{x}{2\sqrt{\alpha t}}\right) \qquad (5.57)$$

$$q_s''(t) = \frac{k(T_s - T_i)}{\sqrt{\pi \alpha t}} \qquad (5.58)$$

Case 2   Constant Surface Heat Flux: $q_s'' = q_o''$

$$T(x, t) - T_i = \frac{2q_o''(\alpha t/\pi)^{1/2}}{k} \exp\left(\frac{-x^2}{4\alpha t}\right) - \frac{q_o'' x}{k} \text{erfc}\left(\frac{x}{2\sqrt{\alpha t}}\right) \qquad (5.59)$$

Case 3   Surface Convection: $-k \left.\dfrac{\partial T}{\partial x}\right|_{x=0} = h[T_\infty - T(0, t)]$

$$\frac{T(x, t) - T_i}{T_\infty - T_i} = \text{erfc}\left(\frac{x}{2\sqrt{\alpha t}}\right)$$

$$- \left[\exp\left(\frac{hx}{k} + \frac{h^2 \alpha t}{k^2}\right)\right]\left[\text{erfc}\left(\frac{x}{2\sqrt{\alpha t}} + \frac{h\sqrt{\alpha t}}{k}\right)\right] \qquad (5.60)$$

The *complementary error function,* erfc w, is defined as erfc $w \equiv 1 - \text{erf } w$.

Temperature histories for the three cases are shown in Figure 5.7, and distinguishing features should be noted. With a step change in the surface temperature, case 1, temperatures within the medium monotonically approach T_s with increasing t, while the magnitude of the surface temperature gradient, and hence the surface heat flux, decreases as $t^{-1/2}$. In contrast, for a fixed surface heat flux (case 2), Equation 5.59 reveals that $T(0, t) = T_s(t)$ increases monotonically as $t^{1/2}$. For surface convection (case 3), the surface temperature and temperatures within the medium approach the fluid temperature T_∞ with increasing time. As T_s approaches T_∞, there is, of course, a reduction in the surface heat flux, $q_s''(t) = h[T_s(t) - T_\infty]$. Specific temperature histories computed from Equation 5.60 are plotted in Figure 5.8. The result corresponding to $h = \infty$ is equivalent to that associated with a sudden change in surface temperature, case 1.

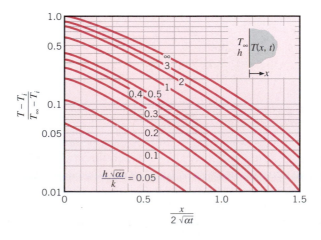

FIGURE 5.8
Temperature histories in a semi-infinite solid with surface convection [2].
Adapted with permission.

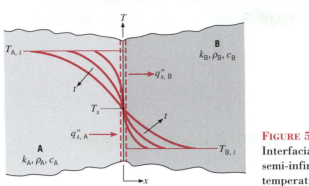

FIGURE 5.9

Interfacial contact between two semi-infinite solids at different initial temperatures.

That is, for $h = \infty$, the surface instantaneously achieves the imposed fluid temperature ($T_s = T_\infty$), and with the second term on the right-hand side of Equation 5.60 reducing to zero, the result is equivalent to Equation 5.57.

An interesting permutation of case 1 results when two semi-infinite solids, initially at uniform temperatures $T_{A,i}$ and $T_{B,i}$, are placed in contact at their free surfaces (Figure 5.9). If the contact resistance is negligible, the requirement of thermal equilibrium dictates that, at the instant of contact ($t = 0$), both surfaces must assume the same temperature T_s, for which $T_{B,i} < T_s < T_{A,i}$. Since T_s does not change with increasing time, it follows that the transient thermal response and the surface heat flux of each of the solids is determined by Equations 5.57 and 5.58, respectively.

The equilibrium surface temperature of Figure 5.9 may be determined from a surface energy balance, which requires that

$$q''_{s,A} = q''_{s,B} \tag{5.61}$$

Substituting from Equation 5.58 for $q''_{s,A}$ and $q''_{s,B}$ and recognizing that the x coordinate of Figure 5.9 requires a sign change for $q''_{s,A}$, it follows that

$$\frac{-k_A(T_s - T_{A,i})}{(\pi\alpha_A t)^{1/2}} = \frac{k_B(T_s - T_{B,i})}{(\pi\alpha_B t)^{1/2}} \tag{5.62}$$

or, solving for T_s,

$$T_s = \frac{(k\rho c)_A^{1/2} T_{A,i} + (k\rho c)_B^{1/2} T_{B,i}}{(k\rho c)_A^{1/2} + (k\rho c)_B^{1/2}} \tag{5.63}$$

Hence the quantity $m \equiv (k\rho c)^{1/2}$ is a weighting factor that determines whether T_s will more closely approach $T_{A,i}(m_A > m_B)$ or $T_{B,i}(m_B > m_A)$.

EXAMPLE 5.5

In laying water mains, utilities must be concerned with the possibility of freezing during cold periods. Although the problem of determining the temperature in soil as a function of time is complicated by changing surface conditions, reasonable estimates can be based on the assumption of a constant surface temperature over a prolonged period of cold weather. What minimum burial depth x_m would you recommend to avoid freezing under conditions for which soil, initially at a uniform temperature of 20°C, is subjected to a constant surface temperature of −15°C for 60 days?

SOLUTION

Known: Temperature imposed at the surface of soil initially at 20°C.

Find: The depth x_m to which the soil has frozen after 60 days.

Schematic:

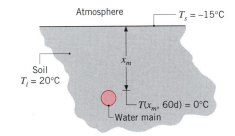

Assumptions:

1. One-dimensional conduction in x.
2. Soil is a semi-infinite medium.
3. Constant properties.

Properties: Table A.3, soil (300 K): $\rho = 2050$ kg/m³, $k = 0.52$ W/m · K, $c = 1840$ J/kg · K, $\alpha = (k/\rho c) = 0.138 \times 10^{-6}$ m²/s.

Analysis: The prescribed conditions correspond to those of case 1 of Figure 5.7, and the transient temperature response of the soil is governed by Equation 5.57. Hence at the time $t = 60$ days after the surface temperature change,

$$\frac{T(x_m, t) - T_s}{T_i - T_s} = \mathrm{erf}\left(\frac{x_m}{2\sqrt{\alpha t}}\right)$$

or

$$\frac{0 - (-15)}{20 - (-15)} = 0.429 = \mathrm{erf}\left(\frac{x_m}{2\sqrt{\alpha t}}\right)$$

Hence from Appendix B.1

$$\frac{x_m}{2\sqrt{\alpha t}} = 0.40$$

and

$$x_m = 0.80\alpha t = 0.80(0.138 \times 10^{-6} \text{ m}^2/\text{s} \times 60 \text{ days} \times 24 \text{ h/day}$$
$$\times 3600 \text{ s/h})^{1/2} = 0.68 \text{ m} \qquad \triangleleft$$

Comments: The properties of soil are highly variable, depending on the nature of the soil and its moisture content, and a representative range of thermal diffusivities is $1 \times 10^{-7} < \alpha < 3 \times 10^{-7}$ m²/s. To assess the effect of soil properties on freezing conditions, we use Equation 5.57 to compute temperature histories at $x_m = 0.68$ m for $\alpha \times 10^7 = 1.0$, 1.38 and 3.0 m²/s.

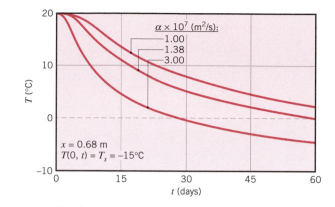

If $\alpha > 1.38 \times 10^{-7}$ m²/s, the design criterion is not achieved at $x_m = 0.68$ m and freezing would occur. It is also instructional to examine temperature distributions in the soil at representative times during the cooling period. Using Equation 5.57 with $\alpha = 1.38 \times 10^{-7}$ m²/s, the following results are obtained:

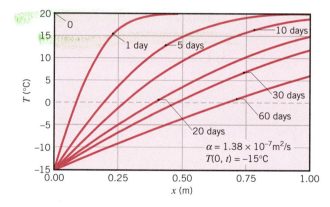

As thermal penetration increases with increasing time, the temperature gradient at the surface, $\partial T/\partial x|_{x=0}$, and hence the rate of heat extraction from the soil, decreases.

5.8
Multidimensional Effects

Transient problems are frequently encountered for which two- and even three-dimensional effects are significant. Solution to a class of such problems can be obtained from the one-dimensional results of Sections 5.5 to 5.7.

Consider immersing the *short* cylinder of Figure 5.10, which is initially at a uniform temperature T_i, in a fluid of temperature $T_\infty \neq T_i$. Because the length and diameter are comparable, the subsequent transfer of energy by conduction will be significant for both the r and x coordinate directions. The temperature within the cylinder will therefore depend on r, x, and t.

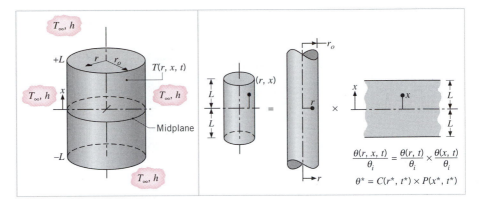

Figure 5.10   Two-dimensional, transient conduction in a short cylinder. (*a*) Geometry. (*b*) Form of the product solution.

Assuming constant properties and no generation, the appropriate form of the heat equation is, from Equation 2.20,

$$\frac{1}{r}\frac{\partial}{\partial r}\left(r\frac{\partial T}{\partial r}\right) + \frac{\partial^2 T}{\partial x^2} = \frac{1}{\alpha}\frac{\partial T}{\partial t}$$

where x has been used in place of z to designate the axial coordinate. A closed-form solution to this equation may be obtained by the separation of variables method. Although we will not consider the details of this solution, it is important to note that the end result may be expressed in the following form:

$$\frac{T(r, x, t) - T_\infty}{T_i - T_\infty} = \left.\frac{T(x, t) - T_\infty}{T_i - T_\infty}\right|_{\substack{\text{Plane}\\\text{wall}}} \cdot \left.\frac{T(r, t) - T_\infty}{T_i - T_\infty}\right|_{\substack{\text{Infinite}\\\text{cylinder}}}$$

That is, the two-dimensional solution may be expressed as a *product* of one-dimensional solutions that correspond to those for a plane wall of thickness $2L$ and an infinite cylinder of radius r_o. For $Fo > 0.2$, these solutions are provided by the one-term approximations of Equations 5.40 and 5.49, as well as by Figures D.1 and D.2 for the plane wall and Figures D.4 and D.5 for the infinite cylinder.

Results for other multidimensional geometries are summarized in Figure 5.11. In each case the multidimensional solution is prescribed in terms of a product involving one or more of the following one-dimensional solutions:

$$S(x, t) \equiv \left.\frac{T(x, t) - T_\infty}{T_i - T_\infty}\right|_{\substack{\text{Semi-infinite}\\\text{solid}}} \tag{5.64}$$

$$P(x, t) \equiv \left.\frac{T(x, t) - T_\infty}{T_i - T_\infty}\right|_{\substack{\text{Plane}\\\text{wall}}} \tag{5.65}$$

$$C(r, t) \equiv \left.\frac{T(r, t) - T_\infty}{T_i - T_\infty}\right|_{\substack{\text{Infinite}\\\text{cylinder}}} \tag{5.66}$$

The x coordinate for the semi-infinite solid is measured from the surface, whereas for the plane wall it is measured from the midplane. In using Figure 5.11 the coordinate origins should carefully be noted. The transient, three-di-

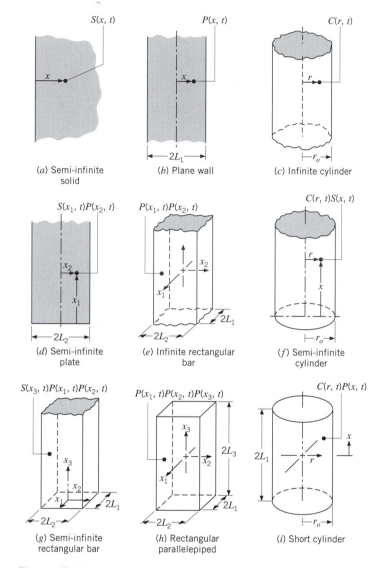

$S(x, t)$ $P(x, t)$ $C(r, t)$

(a) Semi-infinite solid   (b) Plane wall   (c) Infinite cylinder

$S(x_1, t)P(x_2, t)$ $P(x_1, t)P(x_2, t)$ $C(r, t)S(x, t)$

(d) Semi-infinite plate   (e) Infinite rectangular bar   (f) Semi-infinite cylinder

$S(x_3, t)P(x_1, t)P(x_2, t)$ $P(x_1, t)P(x_2, t)P(x_3, t)$ $C(r, t)P(x, t)$

(g) Semi-infinite rectangular bar   (h) Rectangular parallelepiped   (i) Short cylinder

FIGURE 5.11 Solutions for multidimensional systems expressed as products of one-dimensional results.

mensional temperature distribution in a rectangular parallelepiped, Figure 5.11*h*, is then, for example, the product of three one-dimensional solutions for plane walls of thicknesses $2L_1$, $2L_2$, and $2L_3$. That is,

$$\frac{T(x_1, x_2, x_3, t) - T_\infty}{T_i - T_\infty} = P(x_1, t) \cdot P(x_2, t) \cdot P(x_3, t)$$

The distances x_1, x_2, and x_3 are all measured with respect to a rectangular coordinate system whose origin is at the center of the parallelepiped.

The amount of energy Q transferred to or from a solid during a multidimensional transient conduction process may also be determined by combining one-dimensional results, as shown by Langston [7].

EXAMPLE 5.6

In a manufacturing process stainless steel cylinders (AISI 304) initially at 600 K are quenched by submersion in an oil bath maintained at 300 K with $h = 500$ W/m$^2 \cdot$ K. Each cylinder is of length $2L = 60$ mm and diameter $D = 80$ mm. Consider a time 3 min into the cooling process and determine temperatures at the center of the cylinder, at the center of a circular face, and at the midheight of the side.

SOLUTION

Known: Initial temperature and dimensions of cylinder and temperature and convection conditions of an oil bath.

Find: Temperatures $T(r, x, t)$ after 3 min at the cylinder center, $T(0, 0, 3$ min), at the center of a circular face, $T(0, L, 3$ min), and at the midheight of the side, $T(r_o, 0, 3$ min).

Schematic:

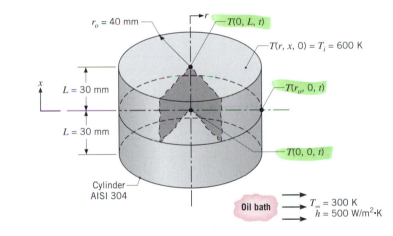

Assumptions:

1. Two-dimensional conduction in r and x.

2. Constant properties.

Properties: Table A.1, stainless steel, AISI 304 [$T = (600 + 300)/2 = 450$ K]: $\rho = 7900$ kg/m^3, $c = 526$ J/kg $\cdot$ K, $k = 17.4$ W/m $\cdot$ K, $\alpha = k/\rho c = 4.19 \times 10^{-6}$ m^2/s.

Analysis: The solid steel cylinder corresponds to case (*i*) of Figure 5.11, and the temperature at any point in the cylinder may be expressed as the following product of one-dimensional solutions.

$$\frac{T(r, x, t) - T_\infty}{T_i - T_\infty} = P(x, t)C(r, t)$$

where $P(x, t)$ and $C(r, t)$ are defined by Equations 5.65 and 5.66, respectively. Accordingly, for the center of the cylinder,

$$\frac{T(0, 0, 3 \text{ min}) - T_\infty}{T_i - T_\infty} = \frac{T(0, 3 \text{ min}) - T_\infty}{T_i - T_\infty}\bigg|_{\substack{\text{Plane}\\\text{wall}}} \cdot \frac{T(0, 3 \text{ min}) - T_\infty}{T_i - T_\infty}\bigg|_{\substack{\text{Infinite}\\\text{cylinder}}}$$

Hence, for the plane wall, with

$$Bi^{-1} = \frac{k}{hL} = \frac{17.4 \text{ W/m} \cdot \text{K}}{500 \text{ W/m}^2 \cdot \text{K} \times 0.03 \text{ m}} = 1.16$$

$$Fo = \frac{\alpha t}{L^2} = \frac{4.19 \times 10^{-6} \text{ m}^2/\text{s} \times 180 \text{ s}}{(0.03 \text{ m})^2} = 0.84$$

it follows from Equation 5.41 that

$$\theta_o^* = \frac{\theta_o}{\theta_i} = C_1 \exp\left(-\zeta_1^2 Fo\right)$$

where, with $Bi = 0.862$, $C_1 = 1.109$ and $\zeta_1 = 0.814$ rad from Table 5.1. With $Fo = 0.84$,

$$\frac{\theta_o}{\theta_i} = \frac{T(0, 3 \text{ min}) - T_\infty}{T_i - T_\infty}\bigg|_{\substack{\text{Plane}\\\text{wall}}} = 1.109 \exp\left[-(0.814 \text{ rad})^2 \times 0.84\right] = 0.636$$

Similarly, for the infinite cylinder, with

$$Bi^{-1} = \frac{k}{hr_o} = \frac{17.4 \text{ W/m} \cdot \text{K}}{500 \text{ W/m}^2 \cdot \text{K} \times 0.04 \text{ m}} = 0.87$$

$$Fo = \frac{\alpha t}{r_o^2} = \frac{4.19 \times 10^{-6} \text{ m}^2/\text{s} \times 180 \text{ s}}{(0.04 \text{ m})^2} = 0.47$$

it follows from Equation 5.49c that

$$\theta_o^* = \frac{\theta_o}{\theta_i} = C_1 \exp\left(-\zeta_1^2 Fo\right)$$

where, with $Bi = 1.15$, $C_1 = 1.227$ and $\zeta_1 = 1.307$ from Table 5.1. With $Fo = 0.47$,

$$\frac{\theta_o}{\theta_i}\bigg|_{\substack{\text{Infinite}\\\text{cylinder}}} = 1.109 \exp\left[-(1.307 \text{ rad})^2 \times 0.47\right] = 0.550$$

Hence, for the center of the cylinder,

$$\frac{T(0, 0, 3 \text{ min}) - T_\infty}{T_i - T_\infty} = 0.636 \times 0.550 = 0.350$$

$$T(0, 0, 3 \text{ min}) = 300 \text{ K} + 0.350(600 - 300) \text{ K} = 405 \text{ K} \qquad \triangleleft$$

The temperature at the center of a circular face may be obtained from the requirement that

$$\frac{T(0, L, 3 \text{ min}) - T_\infty}{T_i - T_\infty} = \frac{T(L, 3 \text{ min}) - T_\infty}{T_i - T_\infty}\bigg|_{\substack{\text{Plane}\\\text{wall}}} \cdot \frac{T(0, 3 \text{ min}) - T_\infty}{T_i - T_\infty}\bigg|_{\substack{\text{Infinite}\\\text{cylinder}}}$$

where, from Equation 5.40b,

$$\frac{\theta^*}{\theta_o^*} = \frac{\theta}{\theta_o} = \cos(\zeta_1 x^*)$$

Hence, with $x^* = 1$, we have

$$\frac{\theta(L)}{\theta_o} = \frac{T(L, 3\ \text{min}) - T_\infty}{T(0, 3\ \text{min}) - T_\infty}\bigg|_{\substack{\text{Plane} \\ \text{wall}}} = \cos(0.814\ \text{rad} \times 1) = 0.687$$

Hence

$$\frac{T(L, 3\ \text{min}) - T_\infty}{T_i - T_\infty}\bigg|_{\substack{\text{Plane} \\ \text{wall}}} = \frac{T(L, 3\ \text{min}) - T_\infty}{T(0, 3\ \text{min}) - T_\infty}\bigg|_{\substack{\text{Plane} \\ \text{wall}}} \cdot \frac{T(0, 3\ \text{min}) - T_\infty}{T_i - T_\infty}\bigg|_{\substack{\text{Plane} \\ \text{wall}}}$$

$$\frac{T(L, 3\ \text{min}) - T_\infty}{T_i - T_\infty}\bigg|_{\substack{\text{Plane} \\ \text{wall}}} = 0.687 \times 0.636 = 0.437$$

Hence

$$\frac{T(0, L, 3\ \text{min}) - T_\infty}{T_i - T_\infty} = 0.437 \times 0.550 = 0.240$$

$$T(0, L, 3\ \text{min}) = 300\ \text{K} + 0.24(600 - 300)\ \text{K} = 372\ \text{K} \qquad \triangleleft$$

The temperature at the midheight of the side may be obtained from the requirement that

$$\frac{T(r_o, 0, 3\ \text{min}) - T_\infty}{T_i - T_\infty} = \frac{T(0, 3\ \text{min}) - T_\infty}{T_i - T_\infty}\bigg|_{\substack{\text{Plane} \\ \text{wall}}} \cdot \frac{T(r_o, 3\ \text{min}) - T_\infty}{T_i - T_\infty}\bigg|_{\substack{\text{Infinite} \\ \text{cylinder}}}$$

where, from Equation 5.49b,

$$\frac{\theta^*}{\theta_o^*} = \frac{\theta}{\theta_o} = J_0(\zeta_1 r^*)$$

With $r^* = 1$ and the value of the Bessel function determined from Table B.4,

$$\frac{\theta(r_o)}{\theta_o} = \frac{T(r_o, 3\ \text{min}) - T_\infty}{T(0, 3\ \text{min}) - T_\infty}\bigg|_{\substack{\text{Infinite} \\ \text{cylinder}}} = J_0(1.307\ \text{rad} \times 1) = 0.616$$

Hence

$$\frac{T(r_o, 3\ \text{min}) - T_\infty}{T_i - T_\infty}\bigg|_{\substack{\text{Infinite} \\ \text{cylinder}}} = \frac{T(r_o, 3\ \text{min}) - T_\infty}{T(0, 3\ \text{min}) - T_\infty}\bigg|_{\substack{\text{Infinite} \\ \text{cylinder}}}$$

$$\cdot \frac{T(0, 3\ \text{min}) - T_\infty}{T_i - T_\infty}\bigg|_{\substack{\text{Infinite} \\ \text{cylinder}}}$$

$$\frac{T(r_o, 3\ \text{min}) - T_\infty}{T_i - T_\infty}\bigg|_{\substack{\text{Infinite} \\ \text{cylinder}}} = 0.616 \times 0.550 = 0.339$$

Hence

$$\frac{T(r_o, 0, 3\ \text{min}) - T_\infty}{T_i - T_\infty} = 0.636 \times 0.339 = 0.216$$

$$T(r_o, 0, 3\ \text{min}) = 300\ \text{K} + 0.216(600 - 300)\ \text{K} = 365\ \text{K} \qquad \triangleleft$$

Comments:

1. Verify that the temperature at the edge of the cylinder is $T(r_o, L, 3 \text{ min}) = 344 \text{ K}$.

2. The Heisler charts of Appendix D could also be used to obtain the desired results. Accessing these charts, one would obtain $\theta_o/\theta_i|_{\text{Plane wall}} \approx 0.64$, $\theta_o/\theta_i|_{\text{Infinite cylinder}} \approx 0.55$, $\theta(L)/\theta_o|_{\text{Plane wall}} \approx 0.68$, and $\theta(r_o)/\theta_o \approx 0.61$, which are in good agreement with results obtained from the one-term approximations.

5.9
Finite-Difference Methods

Analytical solutions to transient problems are restricted to simple geometries and boundary conditions, such as those considered in the preceding sections. However, in many cases the geometry and/or boundary conditions preclude the use of analytical techniques, and recourse must be made to *finite-difference* methods. Such methods, introduced in Section 4.4 for steady-state conditions, are readily extended to transient problems. In this section we consider *explicit* and *implicit* forms of finite-difference solutions to transient conduction problems.

5.9.1   Discretization of the Heat Equation: The Explicit Method

Once again consider the two-dimensional system of Figure 4.5. Under transient conditions with constant properties and no internal generation, the appropriate form of the heat equation, Equation 2.15, is

$$\frac{1}{\alpha} \frac{\partial T}{\partial t} = \frac{\partial^2 T}{\partial x^2} + \frac{\partial^2 T}{\partial y^2} \tag{5.67}$$

To obtain the finite-difference form of this equation, we may use the *central-difference* approximations to the spatial derivatives prescribed by Equations 4.31 and 4.32. Once again the m and n subscripts may be used to designate the x and y locations of *discrete nodal points*. However, in addition to being discretized in space, the problem must be discretized in time. The integer p is introduced for this purpose, where

$$t = p \, \Delta t \tag{5.68}$$

and the finite-difference approximation to the time derivative in Equation 5.67 is expressed as

$$\frac{\partial T}{\partial t}\bigg|_{m,n} \approx \frac{T_{m,n}^{p+1} - T_{m,n}^{p}}{\Delta t} \tag{5.69}$$

The superscript p is used to denote the time dependence of T, and the time derivative is expressed in terms of the difference in temperatures associated with the *new* $(p + 1)$ and *previous* (p) times. Hence calculations must be performed at successive times separated by the interval Δt, and just as a finite-difference solution restricts temperature determination to discrete points in space, it also restricts it to discrete points in time.

If Equation 5.69 is substituted into Equation 5.67, the nature of the finite-difference solution will depend on the specific time at which temperatures are evaluated in the finite-difference approximations to the spatial derivatives. In the *explicit method* of solution, these temperatures are evaluated at the *previous* (p) time. Hence Equation 5.69 is considered to be a *forward-difference* approximation to the time derivative. Evaluating terms on the right-hand side of Equations 4.31 and 4.32 at p and substituting into Equation 5.67, the explicit form of the finite-difference equation for the interior node m, n is

$$\frac{1}{\alpha} \frac{T_{m,n}^{p+1} - T_{m,n}^{p}}{\Delta t} = \frac{T_{m+1,n}^{p} + T_{m-1,n}^{p} - 2T_{m,n}^{p}}{(\Delta x)^2}$$

$$+ \frac{T_{m,n+1}^{p} + T_{m,n-1}^{p} - 2T_{m,n}^{p}}{(\Delta y)^2} \tag{5.70}$$

Solving for the nodal temperature at the new $(p + 1)$ time and assuming that $\Delta x = \Delta y$, it follows that

$$T_{m,n}^{p+1} = Fo(T_{m+1,n}^{p} + T_{m-1,n}^{p} + T_{m,n+1}^{p} + T_{m,n-1}^{p})$$

$$+ (1 - 4Fo)T_{m,n}^{p} \tag{5.71}$$

where Fo is a finite-difference form of the Fourier number

$$Fo = \frac{\alpha \, \Delta t}{(\Delta x)^2} \tag{5.72}$$

If the system is one-dimensional in x, the explicit form of the finite-difference equation for an interior node m reduces to

$$T_{m}^{p+1} = Fo(T_{m+1}^{p} + T_{m-1}^{p}) + (1 - 2Fo)T_{m}^{p} \tag{5.73}$$

Equations 5.71 and 5.73 are *explicit* because *unknown* nodal temperatures for the new time are determined exclusively by *known* nodal temperatures at the previous time. Hence calculation of the unknown temperatures is straightforward. Since the temperature of each interior node is known at $t = 0$ $(p = 0)$ from prescribed initial conditions, the calculations begin at $t = \Delta t$ $(p = 1)$, where Equation 5.71 or 5.73 is applied to each interior node to determine its temperature. With temperatures known for $t = \Delta t$, the appropriate finite-difference equation is then applied at each node to determine its temperature at $t = 2 \Delta t$ $(p = 2)$. In this way, the transient temperature distribution is obtained by *marching out in time,* using intervals of Δt.

The accuracy of the finite-difference solution may be improved by decreasing the values of Δx and Δt. Of course, the number of interior nodal points that must be considered increases with decreasing Δx, and the number of time intervals required to carry the solution to a prescribed final time increases with decreasing Δt. Hence the computation time increases with decreasing Δx and Δt.

The choice of Δx is typically based on a compromise between accuracy and computational requirements. Once this selection has been made, however, the value of Δt may not be chosen independently. It is, instead, determined by *stability* requirements.

An undesirable feature of the explicit method is that it is not unconditionally *stable*. In a transient problem, the solution for the nodal temperatures should continuously approach final (steady-state) values with increasing time. However, with the explicit method, this solution may be characterized by numerically induced oscillations, which are physically impossible. The oscillations may become *unstable*, causing the solution to diverge from the actual steady-state conditions. To prevent such erroneous results, the prescribed value of Δt must be maintained below a certain limit, which depends on Δx and other parameters of the system. This dependence is termed a *stability criterion*, which may be obtained mathematically or demonstrated from a thermodynamic argument (see Problem 5.78). For the problems of interest in this text, *the criterion is determined by requiring that the coefficient associated with the node of interest at the previous time is greater than or equal to zero*. In general, this is done by collecting all terms involving $T_{m,n}^p$ to obtain the form of the coefficient. This result is then used to obtain a limiting relation involving Fo, from which the maximum allowable value of Δt may be determined. For example, with Equations 5.71 and 5.73 already expressed in the desired form, it follows that the stability criterion for a one-dimensional interior node is $(1 - 2Fo) \geq 0$, or

$$Fo \leq \tfrac{1}{2} \tag{5.74}$$

and for a two-dimensional node, it is $(1 - 4Fo) \geq 0$, or

$$Fo \leq \tfrac{1}{4} \tag{5.75}$$

For prescribed values of Δx and α, these criteria may be used to determine upper limits to the value of Δt.

Equations 5.71 and 5.73 may also be derived by applying the energy balance method of Section 4.4.3 to a control volume about the interior node. Accounting for changes in thermal energy storage, a general form of the energy balance equation may be expressed as

$$\dot{E}_{\text{in}} + \dot{E}_g = \dot{E}_{\text{st}} \tag{5.76}$$

In the interest of adopting a consistent methodology, it is again assumed that all heat flow is *into* the node.

To illustrate application of Equation 5.76, consider the surface node of the one-dimensional system shown in Figure 5.12. To more accurately determine thermal conditions near the surface, this node has been assigned a thickness that is one-half that of the interior nodes. Assuming convection transfer from an adjoining fluid and no generation, it follows from Equation 5.76 that

$$hA(T_{\infty} - T_0^p) + \frac{kA}{\Delta x}(T_1^p - T_0^p) = \rho c A \frac{\Delta x}{2} \frac{T_0^{p+1} - T_0^p}{\Delta t}$$

or, solving for the surface temperature at $t + \Delta t$,

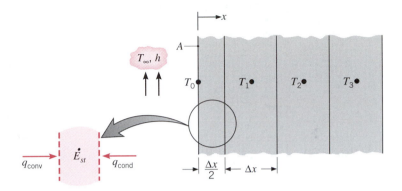

FIGURE 5.12 Surface node with convection and one-dimensional transient conduction.

$$T_0^{p+1} = \frac{2h\,\Delta t}{\rho c\,\Delta x}\,(T_\infty - T_0^p) + \frac{2\alpha\,\Delta t}{\Delta x^2}\,(T_1^p - T_0^p) + T_0^p$$

Recognizing that $(2h\,\Delta t/\rho c\,\Delta x) = 2(h\,\Delta x/k)(\alpha\,\Delta t/\Delta x^2) = 2\,Bi\,Fo$ and grouping terms involving T_0^p, it follows that

$$T_0^{p+1} = 2Fo(T_1^p + Bi\,T_\infty) + (1 - 2Fo - 2Bi\,Fo)T_0^p \qquad (5.77)$$

The finite-difference form of the Biot number is

$$Bi = \frac{h\,\Delta x}{k} \qquad (5.78)$$

Recalling the procedure for determining the stability criterion, we require that the coefficient for T_0^p be greater than or equal to zero. Hence

$$1 - 2Fo - 2Bi\,Fo \geq 0$$

or

$$Fo(1 + Bi) \leq \tfrac{1}{2} \qquad (5.79)$$

Since the complete finite-difference solution requires the use of Equation 5.73 for the interior nodes, as well as Equation 5.77 for the surface node. Equation 5.79 must be contrasted with Equation 5.74 to determine which requirement is the more stringent. Since $Bi \geq 0$, it is apparent that the limiting value of Fo for Equation 5.79 is less than that for Equation 5.74. To ensure stability for all nodes, Equation 5.79 should therefore be used to select the maximum allowable value of Fo, and hence Δt, to be used in the calculations.

Forms of the explicit finite-difference equation for several common geometries are presented in Table 5.2. Each equation may be derived by applying the energy balance method to a control volume about the corresponding node. To develop confidence in your ability to apply this method, you should attempt to verify at least one of these equations.

TABLE 5.2 Summary of transient, two-dimensional finite-difference equations ($\Delta x = \Delta y$)

Configuration	Explicit Method		Implicit Method
	Finite-Difference Equation	**Stability Criterion**	

1. Interior node

Explicit:
$$T_{m,n}^{p+1} = Fo(T_{m+1,n}^p + T_{m-1,n}^p + T_{m,n+1}^p + T_{m,n-1}^p) + (1 - 4Fo)T_{m,n}^p \tag{5.71}$$

Stability: $Fo \leq \frac{1}{4}$ (5.75)

Implicit:
$$(1 + 4Fo)T_{m,n}^{p+1} - Fo(T_{m+1,n}^{p+1} + T_{m-1,n}^{p+1} + T_{m,n+1}^{p+1} + T_{m,n-1}^{p+1}) = T_{m,n}^p \tag{5.87}$$

2. Node at interior corner with convection

Explicit:
$$T_{m,n}^{p+1} = \tfrac{2}{3}Fo(T_{m+1,n}^p + 2T_{m-1,n}^p + 2T_{m,n+1}^p + T_{m,n-1}^p + 2Bi\,T_\infty) + (1 - 4Fo - \tfrac{4}{3}Bi\,Fo)T_{m,n}^p \tag{5.80}$$

Stability: $Fo(3 + Bi) \leq \frac{3}{4}$ (5.81)

Implicit:
$$(1 + 4Fo(1 + \tfrac{1}{3}Bi))T_{m,n}^{p+1} - \tfrac{2}{3}Fo \cdot (T_{m+1,n}^{p+1} + 2T_{m-1,n}^{p+1} + 2T_{m,n+1}^{p+1} + T_{m,n-1}^{p+1}) = T_{m,n}^p + \tfrac{4}{3}Bi\,Fo\,T_\infty \tag{5.90}$$

3. Node at plane surface with convection[a]

Explicit:
$$T_{m,n}^{p+1} = Fo(2T_{m-1,n}^p + T_{m,n+1}^p + T_{m,n-1}^p + 2Bi\,T_\infty) + (1 - 4Fo - 2Bi\,Fo)T_{m,n}^p \tag{5.82}$$

Stability: $Fo(2 + Bi) \leq \frac{1}{2}$ (5.83)

Implicit:
$$(1 + 2Fo(2 + Bi))T_{m,n}^{p+1} - Fo(2T_{m-1,n}^{p+1} + T_{m,n+1}^{p+1} + T_{m,n-1}^{p+1}) = T_{m,n}^p + 2Bi\,Fo\,T_\infty \tag{5.91}$$

4. Node at exterior corner with convection

Explicit:
$$T_{m,n}^{p+1} = 2Fo(T_{m-1,n}^p + T_{m,n-1}^p + 2Bi\,T_\infty) + (1 - 4Fo - 4Bi\,Fo)T_{m,n}^p \tag{5.84}$$

Stability: $Fo(1 + Bi) \leq \frac{1}{4}$ (5.85)

Implicit:
$$(1 + 4Fo(1 + Bi))T_{m,n}^{p+1} - 2Fo(T_{m-1,n}^{p+1} + T_{m,n-1}^{p+1}) = T_{m,n}^p + 4Bi\,Fo\,T_\infty \tag{5.92}$$

[a]To obtain the finite-difference equation and/or stability criterion for an adiabatic surface (or surface of symmetry), simply set Bi equal to zero.

EXAMPLE 5.7

A fuel element of a nuclear reactor is in the shape of a plane wall of thickness $2L = 20$ mm and is convectively cooled at both surfaces, with $h = 1100$ W/m² · K and $T_\infty = 250°C$. At normal operating power, heat is generated uniformly within the element at a volumetric rate of $\dot{q}_1 = 10^7$ W/m³. A departure from the steady-state conditions associated with normal operation will occur if there is a change in the generation rate. Consider a sudden change to $\dot{q}_2 = 2 \times 10^7$ W/m³, and use the explicit finite-difference method to determine the fuel element temperature distribution after 1.5 s. The fuel element thermal properties are $k = 30$ W/m · K and $\alpha = 5 \times 10^{-6}$ m²/s.

SOLUTION

Known: Conditions associated with heat generation in a rectangular fuel element with surface cooling.

Find: Temperature distribution 1.5 s after a change in operating power.

Schematic:

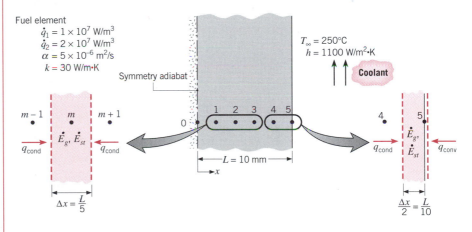

Assumptions:

1. One-dimensional conduction in x.
2. Uniform generation.
3. Constant properties.

Analysis: A numerical solution will be obtained using a space increment of $\Delta x = 2$ mm. Since there is symmetry about the midplane, the nodal network yields six unknown nodal temperatures. Using the energy balance method, Equation 5.76, an explicit finite-difference equation may be derived for any interior node m.

$$kA \frac{T^p_{m-1} - T^p_m}{\Delta x} + kA \frac{T^p_{m+1} - T^p_m}{\Delta x} + \dot{q}A \Delta x = \rho A \Delta x \, c \frac{T^{p+1}_m - T^p_m}{\Delta t}$$

Solving for T_m^{p+1} and rearranging,

$$T_m^{p+1} = Fo\left[T_{m-1}^p + T_{m+1}^p + \frac{\dot{q}(\Delta x)^2}{k}\right] + (1 - 2Fo)T_m^p \qquad (1)$$

This equation may be used for node 0, with $T_{m-1}^p = T_{m+1}^p$, as well as for nodes 1, 2, 3, and 4. Applying energy conservation to a control volume about node 5,

$$hA(T_\infty - T_5^p) + kA\frac{T_4^p - T_5^p}{\Delta x} + \dot{q}A\frac{\Delta x}{2} = \rho A\frac{\Delta x}{2}c\frac{T_5^{p+1} - T_5^p}{\Delta t}$$

or

$$T_5^{p+1} = 2Fo\left[T_4^p + Bi\,T_\infty + \frac{\dot{q}(\Delta x)^2}{2k}\right] + (1 - 2Fo - 2Bi\,Fo)T_5^p \qquad (2)$$

Since the most restrictive stability criterion is associated with Equation 2, we select Fo from the requirement that

$$Fo(1 + Bi) \leq \tfrac{1}{2}$$

Hence, with

$$Bi = \frac{h\,\Delta x}{k} = \frac{1100\ \text{W/m}^2 \cdot \text{K}\,(0.002\ \text{m})}{30\ \text{W/m} \cdot \text{K}} = 0.0733$$

it follows that

$$Fo \leq 0.466$$

or

$$\Delta t = \frac{Fo(\Delta x)^2}{\alpha} \leq \frac{0.466(2 \times 10^{-3}\ \text{m})^2}{5 \times 10^{-6}\ \text{m}^2/\text{s}} \leq 0.373\ \text{s}$$

To be well within the stability limit, we select $\Delta t = 0.3$ s, which corresponds to

$$Fo = \frac{5 \times 10^{-6}\ \text{m}^2/\text{s}(0.3\ \text{s})}{(2 \times 10^{-3}\ \text{m})^2} = 0.375$$

Substituting numerical values, including $\dot{q} = \dot{q}_2 = 2 \times 10^7$ W/m³, the nodal equations become

$$T_0^{p+1} = 0.375(2T_1^p + 2.67) + 0.250T_0^p$$
$$T_1^{p+1} = 0.375(T_0^p + T_2^p + 2.67) + 0.250T_1^p$$
$$T_2^{p+1} = 0.375(T_1^p + T_3^p + 2.67) + 0.250T_2^p$$
$$T_3^{p+1} = 0.375(T_2^p + T_4^p + 2.67) + 0.250T_3^p$$
$$T_4^{p+1} = 0.375(T_3^p + T_5^p + 2.67) + 0.250T_4^p$$
$$T_5^{p+1} = 0.750(T_4^p + 19.67) + 0.195T_5^p$$

To begin the marching solution, the initial temperature distribution must be known. This distribution is given by Equation 3.42, with $\dot{q} = \dot{q}_1$. Obtaining $T_s = T_5$ from Equation 3.46,

$$T_5 = T_\infty + \frac{\dot{q}L}{h} = 250°\text{C} + \frac{10^7\ \text{W/m}^3 \times 0.01\ \text{m}}{1100\ \text{W/m}^2 \cdot \text{K}} = 340.91°\text{C}$$

it follows that

$$T(x) = 16.67\left(1 - \frac{x^2}{L^2}\right) + 340.91°C$$

Computed temperatures for the nodal points of interest are shown in the first row of the accompanying table.

Using the finite-difference equations, the nodal temperatures may be sequentially calculated with a time increment of 0.3 s until the desired final time is reached. The results are illustrated in rows 2 through 6 of the table and may be contrasted with the new steady-state condition (row 7), which was obtained by using Equations 3.42 and 3.46 with $\dot{q} = \dot{q}_2$:

Tabulated nodal temperatures

p	t(s)	T_0	T_1	T_2	T_3	T_4	T_5
0	0	357.58	356.91	354.91	351.58	346.91	340.91
1	0.3	358.08	357.41	355.41	352.08	347.41	341.41
2	0.6	358.58	357.91	355.91	352.58	347.91	341.88
3	0.9	359.08	358.41	356.41	353.08	348.41	342.35
4	1.2	359.58	358.91	356.91	353.58	348.89	342.82
5	1.5	360.08	359.41	357.41	354.07	349.37	343.27
∞	∞	465.15	463.82	459.82	453.15	443.82	431.82

Comments: It is evident that at 1.5 s, the wall is in the early stages of the transient process and that many additional calculations would have to be made to reach steady-state conditions with the finite-difference solution. The computation time could be reduced slightly by using the maximum allowable time increment ($\Delta t = 0.373$ s), but with some loss of accuracy. In the interest of maximizing accuracy, the time interval should be reduced until the computed results become independent of further reductions in Δt.

Extending the finite-difference solution, the time required to achieve the new steady-state condition may be determined, with temperature histories computed for the midplane (0) and surface (5) nodes having the following forms:

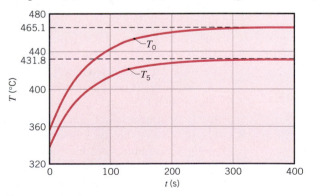

With steady-state temperatures of $T_0 = 465.15°C$ and $T_5 = 431.82°C$, it is evident that the new equilibrium condition is reached within 250 s of the step change in operating power.

5.9.2 Discretization of the Heat Equation: The Implicit Method

In the *explicit* finite-difference scheme, the temperature of any node at $t + \Delta t$ may be calculated from knowledge of temperatures at the same and neighboring nodes for the *preceding time t*. Hence determination of a nodal temperature at some time is *independent* of temperatures at other nodes for the *same time*. Although the method offers computational convenience, it suffers from limitations on the selection of Δt. For a given space increment, the time interval must be compatible with stability requirements. Frequently, this dictates the use of extremely small values of Δt, and a very large number of time intervals may be necessary to obtain a solution.

A reduction in the amount of computation time may often be realized by employing an *implicit,* rather than explicit, finite-difference scheme. The implicit form of a finite-difference equation may be derived by using Equation 5.69 to approximate the time derivative, while evaluating all other temperatures at the *new* ($p + 1$) time, instead of the previous (p) time. Equation 5.69 is then considered to provide a *backward-difference* approximation to the time derivative. In contrast to Equation 5.70, the implicit form of the finite-difference equation for the interior node of a two-dimensional system is then

$$\frac{1}{\alpha} \frac{T_{m,n}^{p+1} - T_{m,n}^{p}}{\Delta t} = \frac{T_{m+1,n}^{p+1} + T_{m-1,n}^{p+1} - 2T_{m,n}^{p+1}}{(\Delta x)^2}$$

$$+ \frac{T_{m,n+1}^{p+1} + T_{m,n-1}^{p+1} - 2T_{m,n}^{p+1}}{(\Delta y)^2} \qquad (5.86)$$

Rearranging and assuming $\Delta x = \Delta y$, it follows that

$$(1 + 4Fo)T_{m,n}^{p+1} - Fo(T_{m+1,n}^{p+1} + T_{m-1,n}^{p+1} + T_{m,n+1}^{p+1} + T_{m,n-1}^{p+1}) = T_{m,n}^{p} \qquad (5.87)$$

From Equation 5.87 it is evident that the *new* temperature of the m, n node depends on the *new* temperatures of its adjoining nodes, which are, in general, unknown. Hence, to determine the unknown nodal temperatures at $t + \Delta t$, the corresponding nodal equations must be solved *simultaneously*. Such a solution may be effected by using Gauss–Seidel iteration or matrix inversion, as discussed in Section 4.5. The *marching solution* would then involve simultaneously solving the nodal equations at each time $t = \Delta t, 2\Delta t, \ldots$, until the desired final time was reached.

Relative to the explicit method, the implicit formulation has the important advantage of being *unconditionally stable*. That is, the solution remains stable for all space and time intervals, in which case there are no restrictions on Δx and Δt. Since larger values of Δt may therefore be used with an implicit method, computation times may often be reduced, with little loss of accuracy. Nevertheless, to maximize accuracy, Δt should be sufficiently small to ensure that the results are independent of further reductions in its value.

The implicit form of a finite-difference equation may also be derived from the energy balance method. For the surface node of Figure 5.12, it is readily shown that

$$(1 + 2Fo + 2Fo\,Bi)T_0^{p+1} - 2Fo\,T_1^{p+1} = 2Fo\,Bi\,T_\infty + T_0^{p} \qquad (5.88)$$

For any interior node of Figure 5.12, it may also be shown that

$$(1 + 2Fo)T_m^{p+1} - Fo(T_{m-1}^{p+1} + T_{m+1}^{p+1}) = T_m^p \tag{5.89}$$

Forms of the implicit finite-difference equation for other common geometries are presented in Table 5.2. Each equation may be derived by applying the energy balance method.

EXAMPLE 5.8

A thick slab of copper initially at a uniform temperature of 20°C is suddenly exposed to radiation at one surface such that the net heat flux is maintained at a constant value of 3×10^5 W/m². Using the explicit and implicit finite-difference techniques with a space increment of $\Delta x = 75$ mm, determine the temperature at the irradiated surface and at an interior point that is 150 mm from the surface after 2 min have elapsed. Compare the results with those obtained from an appropriate analytical solution.

SOLUTION

Known: Thick slab of copper, initially at a uniform temperature, is subjected to a constant net heat flux at one surface.

Find:

1. Using the explicit finite-difference method, determine temperatures at the surface and 150 mm from the surface after an elapsed time of 2 min.

2. Repeat the calculations using the implicit finite-difference method.

3. Determine the same temperatures analytically.

Schematic:

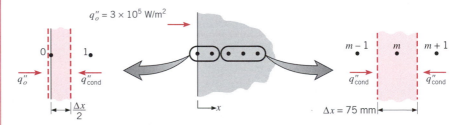

Assumptions:

1. One-dimensional conduction in x.

2. Thick slab may be approximated as a semi-infinite medium with constant surface heat flux.

3. Constant properties.

Properties: Table A.1, copper (300 K): $k = 401$ W/m · K, $\alpha = 117 \times 10^{-6}$ m²/s.

Analysis:

1. An explicit form of the finite-difference equation for the surface node may be obtained by applying an energy balance to a control volume about the node.

$$q_o''A + kA \frac{T_1^p - T_0^p}{\Delta x} = \rho A \frac{\Delta x}{2} c \frac{T_0^{p+1} - T_0^p}{\Delta t}$$

or

$$T_0^{p+1} = 2Fo\left(\frac{q_o'' \Delta x}{k} + T_1^p\right) + (1 - 2Fo)T_0^p$$

The finite-difference equation for any interior node is given by Equation 5.73. Both the surface and interior nodes are governed by the stability criterion

$$Fo \leq \tfrac{1}{2}$$

Noting that the finite-difference equations are simplified by choosing the maximum allowable value of Fo, we select $Fo = \tfrac{1}{2}$. Hence

$$\Delta t = Fo \frac{(\Delta x)^2}{\alpha} = \frac{1}{2} \frac{(0.075 \text{ m})^2}{117 \times 10^{-6} \text{ m}^2/\text{s}} = 24 \text{ s}$$

With

$$\frac{q_o'' \Delta x}{k} = \frac{3 \times 10^5 \text{ W/m}^2 (0.075 \text{ m})}{401 \text{ W/m} \cdot \text{K}} = 56.1°\text{C}$$

the finite-difference equations become

$$T_0^{p+1} = 56.1°\text{C} + T_1^p \qquad \text{and} \qquad T_m^{p+1} = \frac{T_{m+1}^p + T_{m-1}^p}{2}$$

for the surface and interior nodes, respectively. Performing the calculations, the results are tabulated as follows:

Explicit finite-difference solution for $Fo = \tfrac{1}{2}$

p	$t(s)$	T_0	T_1	T_2	T_3	T_4
0	0	20	20	20	20	20
1	24	76.1	20	20	20	20
2	48	76.1	48.1	20	20	20
3	72	104.2	48.1	34.1	20	20
4	96	104.2	69.1	34.1	27.1	20
5	120	125.3	69.1	48.1	27.1	20

After 2 min, the surface temperature and the desired interior temperature are $T_0 = 125.3°\text{C}$ and $T_2 = 48.1°\text{C}$.

Note that calculation of identical temperatures at successive times for the same node is an idiosyncrasy of using the maximum allowable value of Fo with the explicit finite-difference technique. The actual physical condition is, of course, one in which the temperature changes continuously with time. The idiosyncrasy is eliminated and the accuracy of the calculations is improved by reducing the value of Fo.

To determine the extent to which the accuracy may be improved by reducing Fo, let us redo the calculations for $Fo = \frac{1}{4}(\Delta t = 12 \text{ s})$. The finite-difference equations are then of the form

$$T_0^{p+1} = \tfrac{1}{2}(56.1°C + T_1^p) + \tfrac{1}{2}T_0^p$$

$$T_m^{p+1} = \tfrac{1}{4}(T_{m+1}^p + T_{m-1}^p) + \tfrac{1}{2}T_m^p$$

and the results of the calculations are tabulated as follows:

Explicit finite-difference solution for $Fo = \frac{1}{4}$

p	$t(s)$	T_0	T_1	T_2	T_3	T_4	T_5	T_6	T_7	T_8
0	0	20	20	20	20	20	20	20	20	20
1	12	48.1	20	20	20	20	20	20	20	20
2	24	62.1	27.0	20	20	20	20	20	20	20
3	36	72.6	34.0	21.8	20	20	20	20	20	20
4	48	81.4	40.6	24.4	20.4	20	20	20	20	20
5	60	89.0	46.7	27.5	21.3	20.1	20	20	20	20
6	72	95.9	52.5	30.7	22.6	20.4	20.0	20	20	20
7	84	102.3	57.9	34.1	24.1	20.8	20.1	20.0	20	20
8	96	108.1	63.1	37.6	25.8	21.5	20.3	20.0	20.0	20
9	108	113.7	68.0	41.0	27.6	22.2	20.5	20.1	20.0	20.0
10	120	118.9	72.6	44.4	29.6	23.2	20.8	20.2	20.0	20.0

After 2 min, the desired temperatures are $T_0 = 118.9°C$ and $T_2 = 44.4°C$. Comparing the above results with those obtained for $Fo = \frac{1}{2}$, it is clear that by reducing Fo we have eliminated the problem of recurring temperatures. We have also predicted greater thermal penetration (to node 6 instead of node 3). An assessment of the improvement in accuracy must await a comparison with results based on an exact solution.

2. Performing an energy balance on a control volume about the surface node, the implicit form of the finite-difference equation is

$$q_o'' + k\,\frac{T_1^{p+1} - T_0^{p+1}}{\Delta x} = \rho\,\frac{\Delta x}{2}\,c\,\frac{T_0^{p+1} - T_0^p}{\Delta t}$$

or

$$(1 + 2Fo)T_0^{p+1} - 2FoT_1^{p+1} = \frac{2\alpha q_o''\,\Delta t}{k\,\Delta x} + T_0^p$$

Arbitrarily choosing $Fo = \frac{1}{2}(\Delta t = 24 \text{ s})$, it follows that

$$2T_0^{p+1} - T_1^{p+1} = 56.1 + T_0^p$$

From equation 5.89, the finite-difference equation for any interior node is then of the form

$$-T_{m-1}^{p+1} + 4T_m^{p+1} - T_{m+1}^{p+1} = 2T_m^p$$

Since we are dealing with a semi-infinite solid, the number of nodes is, in principle, infinite. In practice, however, the number may be limited to the nodes that are affected by the change in the boundary condition for the time period of interest. From the results of the explicit method, it is evident that we are safe in choosing nine nodes corresponding to $T_0, T_1, \ldots, T_8$. We are thereby assuming that, at $t = 120$ s, there has been no change in T_8.

We now have a set of nine equations that must be solved simultaneously for each time increment. Using the matrix inversion method, we express the equations in the form $[A][T] = [C]$, where

$$[A] = \begin{bmatrix} 2 & -1 & 0 & 0 & 0 & 0 & 0 & 0 & 0 \\ -1 & 4 & -1 & 0 & 0 & 0 & 0 & 0 & 0 \\ 0 & -1 & 4 & -1 & 0 & 0 & 0 & 0 & 0 \\ 0 & 0 & -1 & 4 & -1 & 0 & 0 & 0 & 0 \\ 0 & 0 & 0 & -1 & 4 & -1 & 0 & 0 & 0 \\ 0 & 0 & 0 & 0 & -1 & 4 & -1 & 0 & 0 \\ 0 & 0 & 0 & 0 & 0 & -1 & 4 & -1 & 0 \\ 0 & 0 & 0 & 0 & 0 & 0 & -1 & 4 & -1 \\ 0 & 0 & 0 & 0 & 0 & 0 & 0 & -1 & 4 \end{bmatrix}$$

$$[C] = \begin{bmatrix} 56.1 + T_0^p \\ 2T_1^p \\ 2T_2^p \\ 2T_3^p \\ 2T_4^p \\ 2T_5^p \\ 2T_6^p \\ 2T_7^p \\ 2T_8^p + T_9^{p+1} \end{bmatrix}$$

Note that numerical values for the components of $[C]$ are determined from previous values of the nodal temperatures. Note also how the finite-difference equation for node 8 appears in matrices $[A]$ and $[C]$.

A table of nodal temperatures may be compiled, beginning with the first row ($p = 0$) corresponding to the prescribed initial condition. To obtain nodal temperatures for subsequent times, the inverse of the coefficient matrix $[A]^{-1}$ must first be found. At each time $p + 1$, it is then multiplied by the column vector $[C]$, which is evaluated at p, to obtain the temperatures $T_0^{p+1}, T_1^{p+1}, \ldots, T_8^{p+1}$. For example, multiplying $[A]^{-1}$ by the column vector corresponding to $p = 0$,

$$[C]_{p=0} = \begin{bmatrix} 76.1 \\ 40 \\ 40 \\ 40 \\ 40 \\ 40 \\ 40 \\ 40 \\ 60 \end{bmatrix}$$

the second row of the table is obtained. Updating $[C]$, the process is repeated four more times to determine the nodal temperatures at 120 s. The desired temperatures are $T_0 = 114.7°C$ and $T_2 = 44.2°C$.

Implicit finite-difference solution for $Fo = \frac{1}{2}$

p	$t(s)$	T_0	T_1	T_2	T_3	T_4	T_5	T_6	T_7	T_8
0	0	20.0	20.0	20.0	20.0	20.0	20.0	20.0	20.0	20.0
1	24	52.4	28.7	22.3	20.6	20.2	20.0	20.0	20.0	20.0
2	48	74.0	39.5	26.6	22.1	20.7	20.2	20.1	20.0	20.0
3	72	90.2	50.3	32.0	24.4	21.6	20.6	20.2	20.1	20.0
4	96	103.4	60.5	38.0	27.4	22.9	21.1	20.4	20.2	20.1
5	120	114.7	70.0	44.2	30.9	24.7	21.9	20.8	20.3	20.1

3. Approximating the slab as a semi-infinite medium, the appropriate analytical expression is given by Equation 5.59, which may be applied to any point in the slab.

$$T(x, t) - T_i = \frac{2q_o''(\alpha t/\pi)^{1/2}}{k} \exp\left(-\frac{x^2}{4\alpha t}\right) - \frac{q_o'' x}{k} \operatorname{erfc}\left(\frac{x}{2\sqrt{\alpha t}}\right)$$

At the surface, this expression yields

$$T(0, 120 \text{ s}) - 20°\text{C} = \frac{2 \times 3 \times 10^5 \text{ W/m}^2}{401 \text{ W/m} \cdot \text{K}} (117 \times 10^{-6} \text{ m}^2/\text{s} \times 120 \text{ s}/\pi)^{1/2}$$

or

$$T(0, 120 \text{ s}) = 120.0°\text{C} \qquad \triangleleft$$

At the interior point ($x = 0.15$ m)

$$T(0.15 \text{ m}, 120 \text{ s}) - 20°\text{C} = \frac{2 \times 3 \times 10^5 \text{ W/m}^2}{401 \text{ W/m} \cdot \text{K}}$$

$$\times (117 \times 10^{-6} \text{ m}^2/\text{s} \times 120 \text{ s}/\pi)^{1/2}$$

$$\times \exp\left[-\frac{(0.15 \text{ m})^2}{4 \times 117 \times 10^{-6} \text{ m}^2/\text{s} \times 120 \text{ s}}\right] - \frac{3 \times 10^5 \text{ W/m}^2 \times 0.15 \text{ m}}{401 \text{ W/m} \cdot \text{K}}$$

$$\times \left[1 - \operatorname{erf}\left(\frac{0.15 \text{ m}}{2\sqrt{117 \times 10^{-6} \text{ m}^2/\text{s} \times 120 \text{ s}}}\right)\right] = 45.4°\text{C} \qquad \triangleleft$$

Comments:

1. Comparing the exact results with those obtained from the three approximate solutions, it is clear that the explicit method with $Fo = \frac{1}{4}$ provides the most accurate predictions.

Method	$T_0 = T(0, 120 \text{ s})$	$T_2 = T(0.15 \text{ m}, 120 \text{ s})$
Explicit ($Fo = \frac{1}{2}$)	125.3	48.1
Explicit ($Fo = \frac{1}{4}$)	118.9	44.4
Implicit ($Fo = \frac{1}{2}$)	114.7	44.2
Exact	120.0	45.4

This is not unexpected, since the corresponding value of Δt is 50% smaller than that used in the other two methods. Although computations are simplified by using the maximum allowable value of *Fo* in the explicit method, the accuracy of the results is seldom satisfactory.

2. The accuracy of the foregoing calculations is adversely affected by the coarse grid ($\Delta x = 75$ mm), as well as by the large time steps ($\Delta t = 24$ s, 12 s). Applying the implicit method with $\Delta x = 18.75$ mm and $\Delta t = 6$ s ($Fo = 2.0$), the solution yields $T_0 = T(0, 120\text{ s}) = 119.2°\text{C}$ and $T_2 = T(0.15\text{ m}, 120\text{ s}) = 45.3°\text{C}$, both of which are in good agreement with the exact solution. Complete temperature distributions may be plotted at any of the discrete times, and results obtained at $t = 60$ and 120 s are as follows:

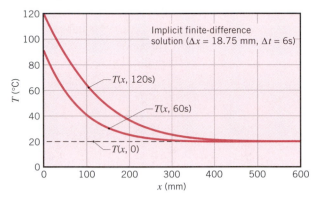

Note that, at $t = 120$ s, the assumption of a semi-infinite medium would remain valid if the thickness of the slab exceeded approximately 500 mm.

3. Note that the coefficient matrix $[A]$ is *tridiagonal*. That is, all elements are zero except those that are on, or to either side of, the main diagonal. Tridiagonal matrices are associated with one-dimensional conduction problems.

4. A more general radiative heating condition would be one in which the surface is suddenly exposed to large surroundings at an elevated temperature T_{sur} (Problem 5.91). The net rate at which radiation is transferred to the surface may then be calculated from Equation 1.7. Allowing for convection heat transfer to the surface, application of conservation of energy to the surface node yields an explicit finite-difference equation of the form

$$\varepsilon\sigma[T_{\text{sur}}^4 - (T_0^p)^4] + h(T_\infty - T_0^p) + k\,\frac{T_1^p - T_0^p}{\Delta x} = \rho\,\frac{\Delta x}{2}\,c\,\frac{T_0^{p+1} - T_0^p}{\Delta t}$$

Use of this finite-difference equation in a numerical solution is complicated by the fact that it is *nonlinear*. However, the equation may be *linearized* by introducing the radiation heat transfer coefficient h_r defined by Equation 1.9, and the finite-difference equation is

$$h_r^p(T_{\text{sur}} - T_0^p) + h(T_\infty - T_0^p) + k\,\frac{T_1^p - T_0^p}{\Delta x} = \rho\,\frac{\Delta x}{2}\,c\,\frac{T_0^{p+1} - T_0^p}{\Delta t}$$

The solution may proceed in the usual manner, although the effect of a radiative Biot number ($Bi_r \equiv h_r\,\Delta x/k$) must be included in the stability criterion and the value of h_r must be updated at each step in the calculations. If the implicit method is used, h_r is calculated at $p + 1$, in which case an iterative calculation must be made at each time step.

5.10
Summary

Transient conduction occurs in numerous engineering applications and may be treated using different methods. There is certainly much to be said for simplicity, in which case, when confronted with a transient problem, the first thing you should do is calculate the Biot number. If this number is much less than unity, you may use the lumped capacitance method to obtain accurate results with minimal computational requirements. However, if the Biot number is not much less than unity, spatial effects must be considered, and some other method must be used. Analytical results are available in convenient graphical and equation form for the plane wall, the infinite cylinder, the sphere, and the semi-infinite solid. You should know when and how to use these results. If geometrical complexities and/or the form of the boundary conditions preclude their use, recourse must be made to an approximate numerical technique, such as the finite-difference method.

References

1. Carslaw, H. S., and J. C. Jaeger, *Conduction of Heat in Solids,* 2nd ed., Oxford University Press, London, 1959.

2. Schneider, P. J., *Conduction Heat Transfer,* Addison-Wesley, Reading, MA, 1955.

3. Kakac, S., and Y. Yener, *Heat Conduction,* Hemisphere Publishing, New York, 1985.

4. Poulikakos, D., *Conduction Heat Transfer,* Prentice-Hall, Englewood Cliffs, NJ, 1994.

5. Heisler, M. P., "Temperature Charts for Induction and Constant Temperature Heating," *Trans. ASME,* **69,** 227–236, 1947.

6. Gröber, H., S. Erk, and U. Grigull, *Fundamentals of Heat Transfer,* McGraw-Hill, New York, 1961.

7. Langston, L. S., "Heat Transfer from Multidimensional Objects Using One-Dimensional Solutions for Heat Loss," *Int. J. Heat Mass Transfer,* **25,** 149–150, 1982.

Problems

Qualitative Considerations

5.1 Consider a thin electrical heater attached to a plate and backed by insulation. Initially, the heater and plate are at the temperature of the ambient air, T_∞. Suddenly, the power to the heater is activated, yielding a constant heat flux q''_o (W/m²) at the inner surface of the plate.

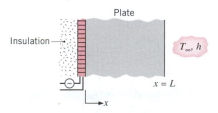

(a) Sketch and label, on T–x coordinates, the temperature distributions: initial, steady-state, and at two intermediate times.

(b) Sketch the heat flux at the outer surface $q''_x(L, t)$ as a function of time.

5.2 The inner surface of a plane wall is insulated while the outer surface is exposed to an airstream at T_∞. The wall is at a uniform temperature corresponding to that of the airstream. Suddenly, a radiation heat source is switched on applying a uniform flux q''_o to the outer surface.

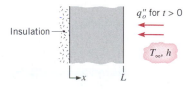

(a) Sketch and label, on T–x coordinates, the temperature distributions: initial, steady-state, and at two intermediate times.

(b) Sketch the heat flux at the outer surface $q''_x(L, t)$ as a function of time.

5.3 A microwave oven operates on the principle that application of a high-frequency field causes electrically polarized molecules in food to oscillate. The net effect is a nearly *uniform generation* of thermal energy within the food. Consider the process of cooking a slab of beef of thickness $2L$ in a microwave oven and compare it with cooking in a conventional oven, where *each side* of the slab is *heated by radiation.* In each case the meat is to be heated from 0°C to a *minimum* temperature of 90°C. Base your comparison on a sketch of the temperature distribution at selected times for each of the cooking processes. In particular, consider the time t_0 at which heating is initiated, a time t_1 during the heating process, the time t_2 corresponding to the conclusion of heating, and a time t_3 well into the subsequent cooling process.

5.4 A plate of thickness $2L$, surface area A_s, mass M, and specific heat c_p, initially at a uniform temperature T_i, is suddenly heated on both surfaces by a convection process (T_∞, h) for a period of time t_o, following which the plate is insulated. Assume that the midplane temperature does not reach T_∞ within this period of time.

(a) Assuming $Bi \gg 1$ for the heating process, sketch and label, on T–x coordinates, the following temperature distributions: initial, steady-state $(t \to \infty)$, $T(x, t_o)$, and at two intermediate times between $t = t_o$ and $t \to \infty$.

(b) Sketch and label, on T–t coordinates, the midplane and exposed surface temperature distributions.

(c) Repeat parts (a) and (b) assuming $Bi \ll 1$ for the plate.

(d) Derive an expression for the steady-state temperature $T(x, \infty) = T_f$, leaving your result in terms of plate parameters (M, c_p), thermal conditions (T_i, T_∞, h), the surface temperature $T(L, t)$, and the heating time t_o.

Lumped Capacitance Method

5.5 Steel balls 12 mm in diameter are annealed by heating to 1150 K and then slowly cooling to 400 K in an air environment for which $T_\infty = 325$ K and $h = 20$ W/m$^2 \cdot$ K. Assuming the properties of the steel to be $k = 40$ W/m $\cdot$ K, $\rho = 7800$ kg/m^3, and $c = 600$ J/kg $\cdot$ K, estimate the time required for the cooling process.

5.6 The heat transfer coefficient for air flowing over a sphere is to be determined by observing the temper-ature–time history of a sphere fabricated from pure copper. The sphere, which is 12.7 mm in diameter, is at 66°C before it is inserted into an airstream having a temperature of 27°C. A thermocouple on the outer surface of the sphere indicates 55°C 69 s after the sphere is inserted in the airstream. Assume, and then justify, that the sphere behaves as a spacewise isothermal object and calculate the heat transfer coefficient.

5.7 A solid steel sphere (AISI 1010), 300 mm in diameter, is coated with a dielectric material layer of thickness 2 mm and thermal conductivity 0.04 W/m $\cdot$ K. The coated sphere is initially at a uniform temperature of 500°C and is suddenly quenched in a large oil bath for which $T_\infty = 100$°C and $h = 3300$ W/m$^2 \cdot$ K. Estimate the time required for the coated sphere temperature to reach 140°C. *Hint:* Neglect the effect of energy storage in the dielectric material, since its thermal capacitance $(\rho c V)$ is small compared to that of the steel sphere.

5.8 A spherical lead bullet of 6-mm diameter is moving at a Mach number of approximately 3. The resulting shock wave heats the air around the bullet to 700 K, and the average convection coefficient for heat transfer between the air and the bullet is 500 W/m$^2 \cdot$ K. If the bullet leaves the barrel at 300 K and the time of flight is 0.4 s, what is its surface temperature on impact?

5.9 Carbon steel (AISI 1010) shafts of 0.1-m diameter are heat treated in a gas-fired furnace whose gases are at 1200 K and provide a convection coefficient of 100 W/m$^2 \cdot$ K. If the shafts enter the furnace at 300 K, how long must they remain in the furnace to achieve a centerline temperature of 800 K?

5.10 A thermal energy storage unit consists of a large rectangular channel, which is well insulated on its outer surface and encloses alternating layers of the storage material and the flow passage.

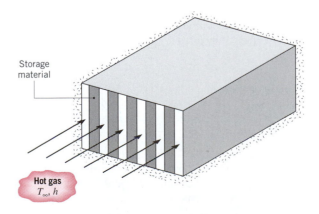

Storage material

Hot gas
T_∞, h

Each layer of the storage material is an aluminum slab of width $W = 0.05$ m, which is at an initial temperature of 25°C. Consider conditions for which the storage unit is charged by passing a hot gas through the passages, with the gas temperature and the convection coefficient assumed to have constant values of $T_\infty = 600°C$ and $h = 100$ W/m² · K throughout the channel. How long will it take to achieve 75% of the maximum possible energy storage? What is the temperature of the aluminum at this time?

5.11 A leaf spring of dimensions 32 mm by 10 mm by 1.1 m is sprayed with a thin anticorrosion coating, which is heat treated by suspending the spring vertically in the lengthwise direction and passing it through a conveyor oven maintained at an air temperature of 175°C. Satisfactory coatings have been obtained on springs, initially at 25°C, with an oven residence time of 35 min. The coating supplier has specified that the coating should be treated for 10 min above a temperature of 140°C. How long should a spring of dimensions 76 mm by 35 mm by 1.6 m remain in the oven in order to properly heat treat the coating? The thermophysical properties of the spring material are $\rho = 8131$ kg/m³, $c_p = 473$ J/kg · K, and $k = 42$ W/m · K.

5.12 A tool used for fabricating semiconductor devices consists of a chuck (thick metallic, cylindrical disk) onto which a very thin silicon wafer ($\rho = 2700$ kg/m³, $c = 875$ J/kg · K, $k = 177$ W/m · K) is placed by a robotic arm. Once in position, an electric field in the chuck is energized, creating an electrostatic force that holds the wafer firmly to the chuck. To ensure a reproducible thermal contact resistance between the chuck and the wafer from cycle-to-cycle, pressurized helium gas is introduced at the center of the chuck and flows (very slowly) radially outward between the asperites of the interface region.

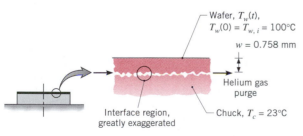

Wafer, $T_w(t)$,
$T_w(0) = T_{w, i} = 100°C$

$w = 0.758$ mm

Helium gas purge

Interface region, greatly exaggerated

Chuck, $T_c = 23°C$

An experiment has been performed under conditions for which the wafer, initially at a uniform temperature $T_{w, i} = 100°C$, is suddenly placed on the chuck, which is at a uniform and constant temperature $T_c = 23°C$. With the wafer in place, the

electrostatic force and the helium gas flow are applied. After 15 seconds, the temperature of the wafer is determined to be 33°C. What is the thermal contact resistance $R''_{t, c}$ (m² · K/W) between the wafer and chuck? Will the value of $R''_{t, c}$ increase, decrease, or remain the same if air, instead of helium, is used as the purge gas?

5.13 Strip steel emerges from the last set of rollers in a hot rolling mill and is cooled (from both surfaces) by convection and radiation heat transfer to the ambient air and large surroundings, respectively, where $T_\infty = T_{sur} = 300$ K.

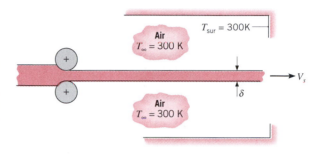

Air
$T_\infty = 300$ K
$T_{sur} = 300K$
V_s
δ
Air
$T_\infty = 300$ K

The strip thickness is $\delta = 5$ mm, and its density, specific heat, thermal conductivity, and emissivity are $\rho = 7900$ kg/m³, $c_p = 640$ J/kg · K, $k = 30$ W/m · K, and $\varepsilon = 0.7$, respectively.

(a) Assuming a uniform convection coefficient of $h = 25$ W/m² · K, determine the time required to cool the strip from an initial temperature of 940°C to 540°C, at which point it may be coiled for shipment. If the strip is moving at 10 m/s, how long must the cooling section be? What can you conclude about the effectiveness of this cooling method?

(b) Determine the time required to cool the strip, first assuming heat transfer only by radiation, and then assuming heat transfer only by convection. For each of the three cases (convection and radiation, radiation only, and convection only), plot the strip temperature as a function of time over the range 540°C $\leq T \leq$ 940°C. Over this range, also plot the radiation heat transfer coefficient, h_r, as a function of time.

5.14 A plane wall of a furnace is fabricated from plain carbon steel ($k = 60$ W/m · K, $\rho = 7850$ kg/m³, $c = 430$ J/kg · K) and is of thickness $L = 10$ mm. To protect it from the corrosive effects of the furnace combustion gases, one surface of the wall is coated with a thin ceramic film that, for a unit surface area, has a thermal resistance of $R''_{t, f} = 0.01$ m² · K/W. The opposite surface is well insulated from the surroundings.

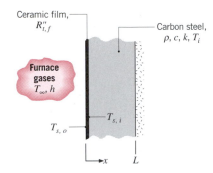

At furnace start-up the wall is at an initial temperature of $T_i = 300$ K, and combustion gases at $T_\infty = 1300$ K enter the furnace, providing a convection coefficient of $h = 25$ W/m² · K at the ceramic film. Assuming the film to have negligible thermal capacitance, how long will it take for the inner surface of the steel to achieve a temperature of $T_{s,i} = 1200$ K? What is the temperature $T_{s,o}$ of the exposed surface of the ceramic film at this time?

5.15 In an industrial process requiring high dc currents, water-jacketed copper rods, 20 mm in diameter, are used to carry the current. The water, which flows continuously between the jacket and the rod, maintains the rod temperature at 75°C during normal operation at 500 A. The electrical resistance of the rod is known to be 0.15 Ω/m. Problems would arise if the coolant water ceased to be available (e.g., because of a valve malfunction). In such a situation heat transfer from the rod surface would diminish greatly, and the rod would eventually melt.

(a) Assuming no heat transfer from the rod following loss of coolant, estimate how long it would take for the rod to melt.

(b) To protect against thermal failure due to loss of coolant, installation of a back-up cooling system is proposed. From consideration of the flow dynamics, it is determined that a back-up system could be activated within 5 seconds. Again assuming no heat transfer from the rod following loss of coolant, determine its temperature after 5 s. Two back-up cooling systems are being considered, one involving water and the other compressed air, which would each be at a temperature of 15°C and would provide convection coefficients of 10,000 W/m² · K and 1000 W/m² · K, respectively. Determine the transient thermal response of the rod following activation of each of the back-up systems. In both cases, neglect radiation transfer. Plot the temperature histories, $T(t)$, for the

two cases and select the most suitable back-up system.

5.16 A steel strip of thickness $\delta = 12$ mm is annealed by passing it through a large furnace whose walls are maintained at a temperature T_w corresponding to that of combustion gases flowing through the furnace ($T_w = T_\infty$). The strip, whose density, specific heat, thermal conductivity, and emissivity are $\rho = 7900$ kg/m³, $c_p = 640$ J/kg · K, $k = 30$ W/m · K, and $\varepsilon = 0.7$, respectively, is to be heated from 300°C to 600°C.

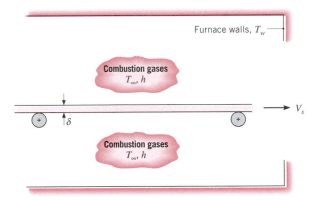

(a) For a uniform convection coefficient of $h = 100$ W/m² · K and $T_w = T_\infty = 700$°C, determine the time required to heat the strip. If the strip is moving at 0.5 m/s, how long must the furnace be?

(b) The annealing process may be accelerated (the strip speed increased) by increasing the environmental temperatures. For the furnace length obtained in part (a), determine the strip speed for $T_w = T_\infty = 850$°C and $T_w = T_\infty = 1000$°C. For each set of environmental temperatures (700, 850, and 1000°C), plot the strip temperature as a function of time over the range 25°C ≤ T ≤ 600°C. Over this range, also plot the radiation heat transfer coefficient, h_r, as a function of time.

5.17 A long wire of diameter $D = 1$ mm is submerged in an oil bath of temperature $T_\infty = 25$°C. The wire has an electrical resistance per unit length of $R'_e = 0.01$ Ω/m. If a current of $I = 100$ A flows through the wire and the convection coefficient is $h = 500$ W/m² · K, what is the steady-state temperature of the wire? From the time the current is applied, how long does it take for the wire to reach a temperature that is within 1°C of the steady-state value? The properties of the wire are $\rho = 8000$ kg/m³, $c = 500$ J/kg · K, and $k = 20$ W/m · K.

5.18 Consider the system of Problem 5.1 where the temperature of the plate is spacewise isothermal during the transient process.

(a) Obtain an expression for the temperature of the plate as a function of time $T(t)$ in terms of q_o'', T_∞, h, L, and the plate properties ρ and c.

(b) Determine the thermal time constant and the steady-state temperature for a 12-mm-thick plate of pure copper when $T_\infty = 27°C$, $h = 50$ W/m² · K, and $q_o'' = 5000$ W/m². Estimate the time required to reach steady-state conditions.

(c) For the conditions of part (b), as well as for $h = 100$ and 200 W/m² · K, compute and plot the corresponding temperature histories of the plate for $0 \leq t \leq 2500$ s.

5.19 An electronic device, such as a power transistor mounted on a finned heat sink, can be modeled as a spatially isothermal object with internal heat generation and an external convection resistance.

(a) Consider such a system of mass M, specific heat c, and surface area A_s, which is initially in equilibrium with the environment at T_∞. Suddenly, the electronic device is energized such that a constant heat generation $\dot{E}_g$ (W) occurs. Show that the temperature response of the device is

$$\frac{\theta}{\theta_i} = \exp\left(-\frac{t}{RC}\right)$$

where $\theta \equiv T - T(\infty)$ and $T(\infty)$ is the steady-state temperature corresponding to $t \rightarrow \infty$; $\theta_i = T_i - T(\infty)$; T_i = initial temperature of device; R = thermal resistance $1/\bar{h}A_s$; and C = thermal capacitance Mc.

(b) An electronic device, which generates 60 W of heat, is mounted on an aluminum heat sink weighing 0.31 kg and reaches a temperature of 100°C in ambient air at 20°C under steady-state conditions. If the device is initially at 20°C, what temperature will it reach 5 min after the power is switched on?

5.20 Before being injected into a furnace, pulverized coal is preheated by passing it through a cylindrical tube whose surface is maintained at $T_{sur} = 1000°C$. The coal pellets are suspended in an airflow and are known to move with a speed of 3 m/s. If the pellets may be approximated as spheres of 1-mm diameter and it may be assumed that they are heated by radiation transfer from the tube surface, how long must the tube be to heat coal entering at 25°C to a temperature of 600°C? Is the use of the lumped capacitance method justified?

5.21 A metal sphere of diameter D, which is at a uniform temperature T_i, is suddenly removed from a furnace and suspended from a fine wire in a large room with air at a uniform temperature T_∞ and the surrounding walls at a temperature T_{sur}.

(a) Neglecting heat transfer by radiation, obtain an expression for the time required to cool the sphere to some temperature T.

(b) Neglecting heat transfer by convection, obtain an expression for the time required to cool the sphere to the temperature T.

(c) How would you go about determining the time required for the sphere to cool to the temperature T if both convection and radiation are of the same order of magnitude?

(d) Consider an anodized aluminum sphere ($\varepsilon = 0.75$) 50 mm in diameter, which is at an initial temperature of $T_i = 800$ K. Both the air and surroundings are at 300 K, and the convection coefficient is 10 W/m² · K. For the conditions of parts (a), (b), and (c), determine the time required for the sphere to cool to 400 K. Plot the corresponding temperature histories. Repeat the calculations for a polished aluminum sphere ($\varepsilon = 0.1$).

5.22 As permanent space stations increase in size, there is an attendant increase in the amount of electrical power they dissipate. To keep station compartment temperatures from exceeding prescribed limits, it is necessary to transfer the dissipated heat to space. A novel heat rejection scheme that has been proposed for this purpose is termed a Liquid Droplet Radiator (LDR). The heat is first transferred to a high vacuum oil, which is then injected into outer space as a stream of small droplets. The stream is allowed to traverse a distance L, over which it cools by radiating energy to outer space at absolute zero temperature. The droplets are then collected and routed back to the space station.

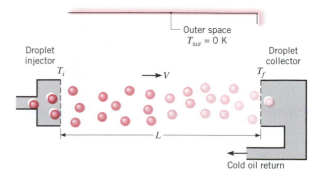

Consider conditions for which droplets of emissivity $\varepsilon = 0.95$ and diameter $D = 0.5$ mm are injected at a temperature of $T_i = 500$ K and a velocity of $V = 0.1$ m/s. Properties of the oil are $\rho = 885$ kg/m^3, $c = 1900$ J/kg · K, and $k = 0.145$ W/m · K. Assuming each drop to radiate to deep space at $T_{sur} = 0$ K, determine the distance L required for the droplets to impact the collector at a final temperature of $T_f = 300$ K. What is the amount of thermal energy rejected by each droplet?

5.23 Plasma spray-coating processes are often used to provide surface protection for materials exposed to hostile environments, which induce degradation through factors such as wear, corrosion, or outright thermal failure. *Ceramic* coatings are commonly used for this purpose. By injecting ceramic powder through the nozzle (anode) of a plasma torch, the particles are entrained by the plasma jet, within which they are then accelerated and heated.

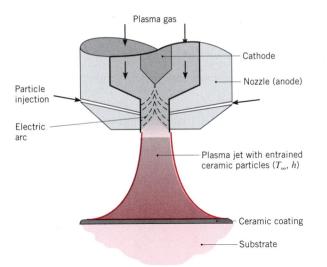

Plasma gas

Cathode

Nozzle (anode)

Particle injection

Electric arc

Plasma jet with entrained ceramic particles (T_∞, h)

Ceramic coating

Substrate

During their *time-in-flight*, the ceramic particles must be heated to their melting point and experience complete conversion to the liquid state. The coating is formed as the molten droplets impinge (*splat*) on the substrate material and experience rapid solidification. Consider conditions for which spherical alumina (Al$_2$O$_3$) particles of diameter $D_p = 50$ μm, density $\rho_p = 3970$ kg/m^3, thermal conductivity $k_p = 10.5$ W/m · K, and specific heat $c_p = 1560$ J/kg · K are injected into an arc plasma, which is at $T_\infty = 10{,}000$ K and provides a coefficient of $h = 30{,}000$ W/m^2 · K for convective heating of the particles. The melting point and latent heat of fusion of alumina are $T_{mp} = 2318$ K and $h_{sf} = 3577$ kJ/kg, respectively.

(a) Neglecting radiation, obtain an expression for the time-in-flight, t_{i-f}, required to heat a particle from its initial temperature T_i to its melting point T_{mp}, and, once at the melting point, for the particle to experience complete melting. Evaluate t_{i-f} for $T_i = 300$ K and the prescribed heating conditions.

(b) Assuming alumina to have an emissivity of $\varepsilon_p = 0.4$ and the particles to exchange radiation with large surroundings at $T_{sur} = 300$ K, assess the validity of neglecting radiation.

5.24 Long metallic rods of circular cross section are heat treated by passing an electric current through the rods to provide uniform volumetric heat generation at a rate $\dot{q}$ (W/m^3). The rods are of diameter D and are placed in a large chamber whose walls are maintained at the same temperature T_∞ as the enclosed air. Convection from the surface of the rods to the air is characterized by the coefficient h.

(a) Obtain an expression that could be used to determine the steady-state temperature of the rod.

(b) Neglecting radiation and prescribing an initial ($t = 0$) rod temperature of $T_i = T_\infty$, obtain the transient temperature response of the rod.

5.25 A chip that is of length $L = 5$ mm on a side and thickness $t = 1$ mm is encased in a ceramic substrate, and its exposed surface is convectively cooled by a dielectric liquid for which $h = 150$ W/m^2 · K and $T_\infty = 20°$C.

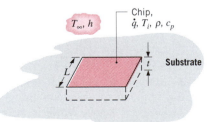

Chip, $\dot{q}$, T_i, ρ, c_p

T_∞, h

Substrate

In the off-mode the chip is in thermal equilibrium with the coolant ($T_i = T_\infty$). When the chip is energized, however, its temperature increases until a new steady-state is established. For purposes of analysis, the energized chip is characterized by uniform volumetric heating with $\dot{q} = 9 \times 10^6$ W/m^3. Assuming an infinite contact resistance between the chip and substrate and negligible conduction resistance within the chip, determine the steady-state chip temperature T_f. Following activation of the chip, how long does it take to come within 1°C of this temperature? The chip density and specific heat are $\rho = 2000$ kg/m^3 and $c = 700$ J/kg · K, respectively.

5.26 Consider the conditions of Problem 5.25. In addition to treating heat transfer by convection directly from the chip to the coolant, a more realistic analysis would account for indirect transfer from the chip to the substrate and then from the substrate to the coolant. The total thermal resistance associated with this indirect route includes contributions due to the chip–substrate interface (a contact resistance), multidimensional conduction in the substrate, and convection from the surface of the substrate to the coolant. If this total thermal resistance is $R_t = 200$ K/W, what is the steady-state chip temperature T_f? Following activation of the chip, how long does it take to come within 1°C of this temperature?

One-Dimensional Conduction: The Plane Wall

5.27 Consider the series solution, Equation 5.39, for the plane wall with convection. Calculate midplane ($x^* = 0$) and surface ($x^* = 1$) temperatures θ^* for $Fo = 0.1$ and 1, using $Bi = 0.1$, 1, and 10. Consider only the first four eigenvalues. Based on these results discuss the validity of the approximate solutions, Equations 5.40 and 5.41.

5.28 Consider the one-dimensional wall shown in the sketch, which is initially at a uniform temperature T_i and is suddenly subjected to the convection boundary condition with a fluid at T_∞.

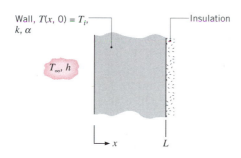

Wall, $T(x, 0) = T_i$, k, α — Insulation — T_∞, h — x — L

For a particular wall, case 1, the temperature at $x = L_1$ after $t_1 = 100$ s is $T_1(L_1, t_1) = 315$°C. Another wall, case 2, has different thickness and thermal conditions as shown below.

How long will it take for the second wall to reach 28.5°C at the position $x = L_2$? Use as the basis for analysis, the dimensionless functional dependence for the transient temperature distribution expressed in Equation 5.38.

5.29 Referring to the semiconductor processing tool of Problem 5.12, it is desired at some point in the manufacturing cycle to cool the chuck, which is made of aluminum alloy 2024. The proposed cooling scheme passes air at 15°C between the air-supply head and the chuck surface.

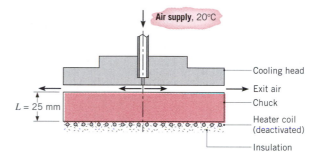

Air supply, 20°C — Cooling head — Exit air — Chuck — $L = 25$ mm — Heater coil (deactivated) — Insulation

(a) If the chuck is initially at a uniform temperature of 100°C, calculate the time required for its lower surface to reach 25°C, assuming a uniform convection coefficient of 50 W/m² · K at the head–chuck interface.

(b) Generate a plot of the time-to-cool as a function of the convection coefficient for the range $10 \leq h \leq 2000$ W/m² · K. If the lower limit represents a free convection condition without any head present, comment on the effectiveness of the head design as a method for cooling the chuck.

5.30 After a long, hard week on the books, you and your friend are ready to relax. You take a steak 50 mm thick from the freezer. How long do you have to let the good times roll before the steak has thawed? Assume that the steak is initially at −6°C, that it thaws when the midplane temperature reaches 4°C, and that the room temperature is 23°C with a convection heat transfer coefficient of 10 W/m² · K. Treat the steak as a slab having the properties of liquid water at 0°C. Neglect the heat of fusion associated with the melting phase change.

Case	L (m)	α (m²/s)	k (W/m · K)	T_i (°C)	T_∞ (°C)	h (W/m² · K)
1	0.10	15×10^{-6}	50	300	400	200
2	0.40	25×10^{-6}	100	30	20	100

5.31 A one-dimensional plane wall with a thickness of 0.1 m initially at a uniform temperature of 250°C is suddenly immersed in an oil bath at 30°C. Assuming the convection heat transfer coefficient for the wall in the bath is 500 W/m² · K, calculate the surface temperature of the wall 9 min after immersion. The properties of the wall are $k = 50$ W/m · K, $\rho = 7835$ kg/m³, and $c = 465$ J/kg · K.

5.32 Consider the thermal energy storage unit of Problem 5.10, but with a masonry material of $\rho = 1900$ kg/m³, $c = 800$ J/kg · K, and $k = 0.70$ W/m · K used in place of the aluminum. How long will it take to achieve 75% of the maximum possible energy storage? What are the maximum and minimum temperatures of the masonry at this time?

5.33 The wall of a rocket nozzle is of thickness $L = 25$ mm and is made from a high alloy steel for which $\rho = 8000$ kg/m³, $c = 500$ J/kg · K, and $k = 25$ W/m · K. During a test firing, the wall is initially at $T_i = 25°C$ and its inner surface is exposed to hot combustion gases for which $h = 500$ W/m² · K and $T_\infty = 1750°C$. The outer surface is well insulated.

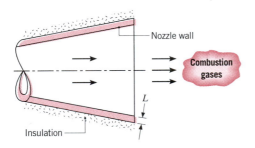

(a) If the wall must be maintained at least 100°C below its melting point of $T_{mp} = 1600°C$, what is the maximum allowable firing time t_f? The diameter of the nozzle is much larger than its thickness.

(b) To increase t_f, consideration is being given to changing the wall thickness L. Should L be increased or decreased? Why? For wall thicknesses of $L = 10$, 25, and 50 mm, compute and plot the inner and outer surface temperature histories over the period $0 \leq t \leq 600$ s. The value of t_f could also be increased by selecting a wall material with different thermophysical properties. Should materials of larger or smaller values of ρ, c, and k be chosen?

5.34 During transient operation, the steel nozzle of a rocket engine must not exceed a maximum allowable operating temperature of 1500 K when exposed to combustion gases characterized by a temperature of 2300 K and a convection coefficient of 5000 W/m² · K. To extend the duration of engine operation, it is proposed that a ceramic *thermal barrier coating* ($k = 10$ W/m · K, $\alpha = 6 \times 10^{-6}$ m²/s) be applied to the interior surface of the nozzle.

(a) If the ceramic coating is 10 mm thick and at an initial temperature of 300 K, obtain a conservative estimate of the maximum allowable duration of engine operation. The nozzle radius is much larger than the combined wall and coating thickness.

(b) Compute and plot the inner and outer surface temperatures of the coating as a function of time for $0 \leq t \leq 150$ s. Repeat the calculations for a coating thickness of 40 mm.

5.35 In a tempering process, glass plate, which is initially at a uniform temperature T_i, is cooled by suddenly reducing the temperature of both surfaces to T_s. The plate is 20 mm thick, and the glass has a thermal diffusivity of 6×10^{-7} m²/s.

(a) How long will it take for the midplane temperature to achieve 50% of its maximum possible temperature reduction?

(b) If $(T_i - T_s) = 300°C$, what is the maximum temperature gradient in the glass at the above time?

5.36 The strength and stability of tires may be enhanced by heating both sides of the rubber ($k = 0.14$ W/m · K, $\alpha = 6.35 \times 10^{-8}$ m²/s) in a steam chamber for which $T_\infty = 200°C$. In the heating process, a 20-mm-thick rubber wall (assumed to be unthreaded) is taken from an initial temperature of 25°C to a midplane temperature of 150°C.

(a) If steam flow over the tire surfaces maintains a convection coefficient of $h = 200$ W/m² · K, how long will it take to achieve the desired midplane temperature?

(b) To accelerate the heating process, it is recommended that the steam flow be made sufficiently vigorous to maintain the tire surfaces at 200°C throughout the process. Compute and plot the midplane and surface temperatures for this case, as well as for the conditions of part (a).

5.37 Copper-coated, epoxy-filled fiberglass circuit boards are treated by heating a stack of them under high pressure as shown in the sketch. The purpose of the pressing–heating operation is to cure the epoxy that bonds the fiberglass sheets, imparting stiffness to the boards. The stack, referred to as a *book,* is comprised of 10 *boards* and 11 *pressing* *plates,* which prevent epoxy from flowing between the boards and impart a smooth finish to the cured boards. In order to perform simplified thermal analyses, it is reasonable to approximate the book as having an effective thermal conductivity (k) and an effective thermal capacitance (ρc_p). Calculate the effective properties if each of the boards and plates has a thickness of 2.36 mm and the following thermophysical properties: board (b) $\rho_b = 1000$ kg/m^3, $c_{p,b} = 1500$ J/kg · K, $k_b = 0.30$ W/m · K; plate (p) $\rho_p = 8000$ kg/m^3, $c_{p,p} = 480$ J/kg · K, $k_p = 12$ W/m · K.

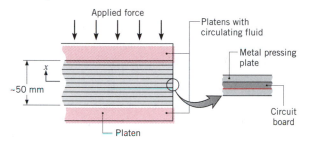

5.38 Circuit boards are treated by heating a stack of them under high pressure as illustrated in Problem 5.37. The platens at the top and bottom of the stack are maintained at a uniform temperature by a circulating fluid. The purpose of the pressing–heating operation is to cure the epoxy, which bonds the fiberglass sheets, and impart stiffness to the boards. The cure condition is achieved when the epoxy has been maintained at or above 170°C for at least 5 min. The effective thermophysical properties of the stack or *book* (boards and metal pressing plates) are $k = 0.613$ W/m · K and $\rho c_p = 2.73 \times 10^6$ J/m^3 · K.

(a) If the book is initially at 15°C and, following application of pressure, the platens are suddenly brought to a uniform temperature of 190°C, calculate the elapsed time t_e required for the midplane of the book to reach the cure temperature of 170°C.

(b) If, at this instant of time, $t = t_e$, the platen temperature were reduced suddenly to 15°C, how much energy would have to be removed from the book by the coolant circulating in the platen, in order to return the stack to its initial uniform temperature?

5.39 An ice layer forms on a 5-mm-thick windshield of a car while parked during a cold night for which the ambient temperature is −20°C. Upon start-up, using a new defrost system, the interior surface is suddenly exposed to an airstream at 30°C. Assuming that the ice behaves as an insulating layer on the exterior surface, what interior convection coef-

ficient would allow the exterior surface to reach 0°C in 60 s? The windshield thermophysical properties are $\rho = 2200$ kg/m^3, $c_p = 830$ J/kg · K, and $k = 1.2$ W/m · K.

One-Dimensional Conduction: The Long Cylinder

5.40 Cylindrical steel rods (AISI 1010), 50 mm in diameter, are heat treated by drawing them through an oven 5 m long in which air is maintained at 750°C. The rods enter at 50°C and achieve a centerline temperature of 600°C before leaving. For a convection coefficient of 125 W/m^2 · K, estimate the speed at which the rods must be drawn through the oven.

5.41 Estimate the time required to cook a hot dog in boiling water. Assume that the hot dog is initially at 6°C, that the convection heat transfer coefficient is 100 W/m^2 · K, and that the final temperature is 80°C at the centerline. Treat the hot dog as a long cylinder of 20-mm diameter having the properties: $\rho = 880$ kg/m^3, $c = 3350$ J/kg · K, and $k = 0.52$ W/m · K.

5.42 A long rod of 60-mm diameter and thermophysical properties $\rho = 8000$ kg/m^3, $c = 500$ J/kg · K, and $k = 50$ W/m · K is initially at a uniform temperature and is heated in a forced convection furnace maintained at 750 K. The convection coefficient is estimated to be 1000 W/m^2 · K.

(a) What is the centerline temperature of the rod when the surface temperature is 550 K?

(b) In a heat-treating process, the centerline temperature of the rod must be increased from $T_i = 300$ K to $T = 500$ K. Compute and plot the centerline temperature histories for $h = 100$, 500, and 1000 W/m^2 · K. In each case the calculation may be terminated when $T = 500$ K.

5.43 A long cylinder of 30-mm diameter, initially at a uniform temperature of 1000 K, is suddenly quenched in a large, constant-temperature oil bath at 350 K. The cylinder properties are $k = 1.7$ W/m · K, $c = 1600$ J/kg · K, and $\rho = 400$ kg/m^3, while the convection coefficient is 50 W/m^2 · K.

(a) Calculate the time required for the surface of the cylinder to reach 500 K.

(b) Compute and plot the surface temperature history for $0 \leq t \leq 300$ s. If the oil were agitated, providing a convection coefficient of 250 W/m^2 · K, how would the temperature history change?

5.44 A long pyroceram rod of diameter 20 mm is clad with a very thin metallic tube for mechanical protection. The bonding between the rod and the tube has a thermal contact resistance of $R'_{t,c} = 0.12$ m · K/W.

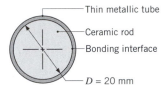

- Thin metallic tube
- Ceramic rod
- Bonding interface
- $D = 20$ mm

(a) If the rod is initially at a uniform temperature of 900 K and is suddenly cooled by exposure to an airstream for which $T_\infty = 300$ K and $h = 100$ W/m² · K, at which time will the centerline reach 600 K?

(b) Cooling may be accelerated by increasing the airspeed and hence the convection coefficient. For values of $h = 100$, 500, and 1000 W/m² · K, compute and plot the centerline and surface temperatures of the pyroceram as a function of time for $0 \le t \le 300$ s. Comment on the implications of achieving enhanced cooling solely by increasing h.

5.45 A long rod 40 mm in diameter, fabricated from sapphire (aluminum oxide) and initially at a uniform temperature of 800 K, is suddenly cooled by a fluid at 300 K having a heat transfer coefficient of 1600 W/m² · K. After 35 s, the rod is wrapped in insulation and experiences no heat losses. What will be the temperature of the rod after a long period of time?

5.46 A long bar of 70-mm diameter and initially at 90°C is cooled by immersing it in a water bath that is at 40°C and provides a convection coefficient of 20 W/m² · K. The thermophysical properties of the bar are $\rho = 2600$ kg/m³, $c = 1030$ J/kg · K, and $k = 3.50$ W/m · K.

(a) How long should the bar remain in the bath in order that, when it is removed and allowed to equilibrate while isolated from any surroundings, it achieves a uniform temperature of 55°C?

(b) What is the surface temperature of the bar when it is removed from the bath?

5.47 A long plastic rod of 30-mm diameter ($k = 0.3$ W/m · K and $\rho c_p = 1040$ kJ/m³ · K) is uniformly heated in an oven as preparation for a pressing operation. For best results, the temperature in the rod should not be less than 200°C. To what uniform temperature should the rod be heated in the oven if,

for the worst case, the rod sits on a conveyor for 3 min while exposed to convection cooling with ambient air at 25°C and with a convection coefficient of 8 W/m² · K? A further condition for good results is a maximum–minimum temperature difference of less than 10°C. Is this condition satisfied and, if not, what could you do to satisfy it?

One-Dimensional Conduction: The Sphere

5.48 In heat treating to harden steel ball bearings ($c = 500$ J/kg · K, $\rho = 7800$ kg/m³, $k = 50$ W/m · K), it is desirable to increase the surface temperature for a short time without significantly warming the interior of the ball. This type of heating can be accomplished by sudden immersion of the ball in a molten salt bath with $T_\infty = 1300$ K and $h = 5000$ W/m² · K. Assume that any location within the ball whose temperature exceeds 1000 K will be hardened. Estimate the time required to harden the outer millimeter of a ball of diameter 20 mm, if its initial temperature is 300 K.

5.49 A sphere of 80-mm diameter ($k = 50$ W/m · K and $\alpha = 1.5 \times 10^{-6}$ m²/s) is initially at a uniform, elevated temperature and is quenched in an oil bath maintained at 50°C. The convection coefficient for the cooling process is 1000 W/m² · K. At a certain time, the surface temperature of the sphere is measured to be 150°C. What is the corresponding center temperature of the sphere?

5.50 A cold air chamber is proposed for quenching steel ball bearings of diameter $D = 0.2$ m and initial temperature $T_i = 400$°C. Air in the chamber is maintained at -15°C by a refrigeration system, and the steel balls pass through the chamber on a conveyor belt. Optimum bearing production requires that 70% of the initial thermal energy content of the ball above -15°C be removed. Radiation effects may be neglected, and the convection heat transfer coefficient within the chamber is 1000 W/m² · K. Estimate the residence time of the balls within the chamber, and recommend a drive velocity of the conveyor. The following properties may be used for the steel: $k = 50$ W/m · K, $\alpha = 2 \times 10^{-5}$ m²/s, and $c = 450$ J/kg · K.

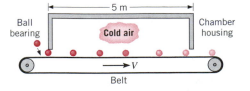

Ball bearing — 5 m — Chamber housing — Cold air — V — Belt

5.51 Stainless steel (AISI 304) ball bearings, which have uniformly been heated to 850°C, are hardened by quenching them in an oil bath that is maintained at 40°C. The ball diameter is 20 mm, and the convection coefficient associated with the oil bath is 1000 W/m² · K.

(a) If quenching is to occur until the surface temperature of the balls reaches 100°C, how long must the balls be kept in the oil? What is the center temperature at the conclusion of the cooling period?

(b) If 10,000 balls are to be quenched per hour, what is the rate at which energy must be removed by the oil bath cooling system in order to maintain its temperature at 40°C?

5.52 A spherical hailstone that is 5 mm in diameter is formed in a high-altitude cloud at −30°C. If the stone begins to fall through warmer air at 5°C, how long will it take before the outer surface begins to melt? What is the temperature of the stone's center at this point in time, and how much energy (J) has been transferred to the stone? A convection heat transfer coefficient of 250 W/m² · K may be assumed, and the properties of the hailstone may be taken to be those of ice.

5.53 A sphere 30 mm in diameter initially at 800 K is quenched in a large bath having a constant temperature of 320 K with a convection heat transfer coefficient of 75 W/m² · K. The thermophysical properties of the sphere material are: $\rho = 400$ kg/m³, $c = 1600$ J/kg · K, and $k = 1.7$ W/m · K.

(a) Show, in a qualitative manner on T–t coordinates, the temperatures at the center and at the surface of the sphere as a function of time.

(b) Calculate the time required for the surface of the sphere to reach 415 K.

(c) Determine the heat flux (W/m²) at the outer surface of the sphere at the time determined in part (b).

(d) Determine the energy (J) that has been lost by the sphere during the process of cooling to the surface temperature of 415 K.

(e) At the time determined by part (b), the sphere is quickly removed from the bath and covered with perfect insulation, such that there is no heat loss from the surface of the sphere. What will be the temperature of the sphere after a long period of time has elapsed?

(f) Compute and plot the center and surface temperature histories over the period $0 \leq t \leq 150$ s. What effect does an increase in the convec-

tion coefficient to $h = 200$ W/m² · K have on the foregoing temperature histories? For $h = 75$ and 200 W/m² · K, compute and plot the surface heat flux as a function of time for $0 \leq t \leq 180$ s.

5.54 Spheres A and B are initially at 800 K, and they are simultaneously quenched in large constant temperature baths, each having a temperature of 320 K. The following parameters are associated with each of the spheres and their cooling processes.

	Sphere A	Sphere B
Diameter (mm)	300	30
Density (kg/m³)	1600	400
Specific heat (kJ/kg · K)	0.400	1.60
Thermal conductivity (W/m · K)	170	1.70
Convection coefficient (W/m² · K)	5	50

(a) Show in a qualitative manner, on T versus t coordinates, the temperatures at the center and at the surface for each sphere as a function of time. Briefly explain the reasoning by which you determine the relative positions of the curves.

(b) Calculate the time required for the surface of each sphere to reach 415 K.

(c) Determine the energy that has been gained by each of the baths during the process of the spheres cooling to 415 K.

5.55 The convection coefficient for flow over a solid sphere may be determined by submerging the sphere, which is initially at 25°C, into the flow, which is at 75°C, and measuring its surface temperature at some time during the transient heating process.

(a) If the sphere has a diameter of 0.1 m, a thermal conductivity of 15 W/m · K, and a thermal diffusivity of 10^{-5} m²/s, at what time will a surface temperature of 60°C be recorded if the convection coefficient is 300 W/m² · K?

(b) Assess the effect of thermal diffusivity on the thermal response of the material by computing center and surface temperature histories for $\alpha = 10^{-6}$, 10^{-5}, and 10^{-4} m²/s. Plot your results for the period $0 \leq t \leq 300$ s. In a similar manner, assess the effect of thermal conductivity by considering values of $k = 1.5$, 15, and 150 W/m · K.

5.56 In a process to manufacture glass beads ($k = 1.4$ W/m · K, $\rho = 2200$ kg/m³, $c_p = 800$ J/kg · K) of

3-mm diameter, the beads are suspended in an up-wardly directed airstream that is at $T_\infty = 15°C$ and maintains a convection coefficient of $h = 400$ W/m² · K.

(a) If the beads are at an initial temperature of $T_i = 477°C$, how long must they be suspended to achieve a center temperature of 80°C? What is the corresponding surface temperature?

(b) Compute and plot the center and surface tem-peratures as a function of time for $0 \le t \le 20$ s and $h = 100, 400,$ and 1000 W/m² · K.

Semi-infinite Media

5.57 Two large blocks of different materials, such as copper and concrete, have been sitting in a room (23°C) for a very long time. Which of the two blocks, if either, will feel colder to the touch? As-sume the blocks to be semi-infinite solids and your hand to be at a temperature of 37°C.

5.58 Asphalt pavement may achieve temperatures as high as 50°C on a hot summer day. Assume that such a temperature exists throughout the pavement, when suddenly a rainstorm reduces the surface temperature to 20°C. Calculate the total amount of energy (J/m²) that will be transferred from the as-phalt over a 30-min period in which the surface is maintained at 20°C.

5.59 A furnace wall is fabricated from fireclay brick ($\alpha = 7.1 \times 10^{-7}$ m²/s), and its inner surface is maintained at 1100 K during furnace operation. The wall is designed according to the criterion that, for an initial temperature of 300 K, its midpoint temperature will not exceed 325 K after 4 h of fur-nace operation. What is the minimum allowable wall thickness?

5.60 A block of material of thickness 20 mm with known thermophysical properties ($k = 15$ W/m · K and $\alpha = 2.0 \times 10^{-5}$ m²/s) is imbedded in the wall of a channel that is initially at 25°C and is sud-denly subjected to a convection process with gases at 325°C. A thermocouple (TC) is installed 2 mm below the surface of the channel wall for the pur-pose of sensing the temperature–time history (fol-lowing start-up of the hot gas flow) and thereby determining the transient heat flux. At an elapsed time of 10 s, the thermocouple indicates a tempera-ture of 167°C.

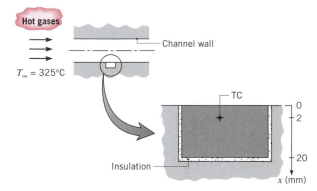

Calculate the corresponding surface convective heat flux assuming the block behaves as a semi-in-finite solid. Compare this result with that obtained from the one-term approximation for a plane wall.

5.61 A tile-iron consists of a massive plate maintained at 150°C by an imbedded electrical heater. The iron is placed in contact with a tile to soften the ad-hesive, allowing the tile to be easily lifted from the subflooring. The adhesive will soften sufficiently if heated above 50°C for at least 2 min, but its tem-perature should not exceed 120°C to avoid deterio-ration of the adhesive. Assume the tile and sub-floor to have an initial temperature of 25°C and to have equivalent thermophysical properties of $k = 0.15$ W/m · K and $\rho c_p = 1.5 \times 10^6$ J/m³ · K.

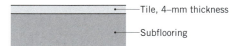

(a) How long will it take a worker using the tile-iron to lift a tile? Will the adhesive temperature exceed 120°C?

(b) If the tile-iron has a square surface area 254 mm to the side, how much energy has been re-moved from it during the time it has taken to lift the tile?

5.62 The manufacturer of a heat flux gage like that il-lustrated in Problem 1.8 claims the time constant for a 63.2% response to be $\tau = (4d^2\rho c_p)/\pi^2 k$, where ρ, c_p, and k are the thermophysical proper-ties of the gage material and d is its thickness. Not knowing the origin of this relation, your task is to model the gage considering the two extreme cases illustrated below. In both cases, the gage, initially at a uniform temperature T_i, is exposed to a sudden change in surface temperature, $T(0, t) = T_s$. For case (a) the backside of the gage is insulated, and for case (b) the gage is imbedded in a semi-infinite solid having the same thermophysical properties as those of the gage.

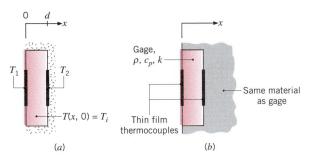

(a)

(b)

Develop relationships for predicting the time constant of the gage for the two cases and compare them to the manufacturer's relation. What conclusion can you draw from this analysis regarding the transient response of gages for different applications?

5.63 A simple procedure for measuring surface convection heat transfer coefficients involves coating the surface with a thin layer of material having a precise melting point temperature. The surface is then heated and, by determining the time required for melting to occur, the convection coefficient is determined. The following experimental arrangement uses the procedure to determine the convection coefficient for gas flow normal to a surface. Specifically, a long copper rod is encased in a super insulator of very low thermal conductivity, and a very thin coating is applied to its exposed surface.

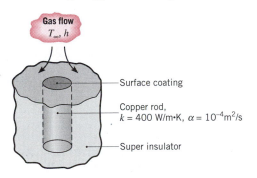

If the rod is initially at 25°C and gas flow for which $h = 200 \ \text{W/m}^2 \cdot \text{K}$ and $T_\infty = 300°C$ is initiated, what is the melting point temperature of the coating if melting is observed to occur at $t = 400$ s?

5.64 An insurance company has hired you as a consultant to improve their understanding of burn injuries. They are especially interested in injuries in-

duced when a portion of a worker's body comes into contact with machinery that is at elevated temperatures in the range of 50 to 100°C. Their medical consultant informs them that irreversible thermal injury (cell death) will occur in any living tissue that is maintained at $T \geq 48°C$ for a duration $\Delta t \geq 10$ s. They want information concerning the extent of irreversible tissue damage (as measured by distance from the skin surface) as a function of the machinery temperature and the time during which contact is made between the skin and the machinery. Assume that living tissue has a normal temperature of 37°C, is isotropic, and has constant properties equivalent to those of liquid water.

(a) To assess the seriousness of the problem, compute locations in the tissue at which the temperature will reach 48°C after 10 s of exposure to machinery at 50°C and 100°C.

(b) For a machinery temperature of 100°C and $0 \leq t \leq 30$ s, compute and plot temperature histories at tissue locations of 0.5, 1, 2, and 5 mm from the skin.

5.65 A procedure for determining the thermal conductivity of a solid material involves embedding a thermocouple in a thick slab of the solid and measuring the response to a prescribed change in temperature at one surface. Consider an arrangement for which the thermocouple is embedded 10 mm from a surface that is suddenly brought to a temperature of 100°C by exposure to boiling water. If the initial temperature of the slab was 30°C and the thermocouple measures a temperature of 65°C, 2 min after the surface is brought to 100°C, what is its thermal conductivity? The density and specific heat of the solid are known to be 2200 kg/m³ and 700 J/kg · K.

5.66 An electric heater in the form of a sheet is placed in good contact with the surface of a thick slab of Bakelite having a uniform temperature of 300 K. Determine the temperature of the slab at the surface and at a depth of 25 mm, 10 min after the heater has been energized and is providing a constant heat flux to the surface of 2500 W/m².

5.67 A very thick slab with thermal diffusivity 5.6×10^{-6} m²/s and thermal conductivity 20 W/m · K is initially at a uniform temperature of 325°C. Suddenly, the surface is exposed to a coolant at 15°C for which the convection heat transfer coefficient is 100 W/m² · K.

(a) Determine temperatures at the surface and at a depth of 45 mm after 3 min have elapsed.

(b) Compute and plot temperature histories ($0 \leq t \leq 300$ s) at $x = 0$ and $x = 45$ mm for the following parametric variations: (i) $\alpha = 5.6 \times 10^{-7}$, 5.6×10^{-6}, and 5.6×10^{-5} m²/s; and (ii) $k = 2$, 20, and 200 W/m · K.

5.68 A thick oak wall, initially at 25°C, is suddenly exposed to combustion products for which $T_\infty = 800°C$ and $h = 20$ W/m² · K.

(a) Determine the time of exposure required for the surface to reach the ignition temperature of 400°C.

(b) Plot the temperature distribution $T(x)$ in the medium at $t = 325$ s. The distribution should extend to a location for which $T \approx 25°C$.

5.69 It is well known that, although two materials are at the same temperature, one may feel cooler to the touch than the other. Consider thick plates of copper and glass, each at an initial temperature of 300 K. Assuming your finger to be at an initial temperature of 310 K and to have thermophysical properties of $\rho = 1000$ kg/m³, $c = 4180$ J/kg · K, and $k = 0.625$ W/m · K, determine whether the copper or the glass will feel cooler to the touch.

5.70 Two stainless steel plates ($\rho = 8000$ kg/m³, $c = 500$ J/kg · K, $k = 15$ W/m · K), each 20 mm thick and insulated on one surface, are initially at 400 and 300 K when they are pressed together at their uninsulated surfaces. What is the temperature of the insulated surface of the hot plate after 1 min has elapsed?

5.71 Special coatings are often formed by depositing thin layers of a molten material on a solid substrate. Solidification begins at the substrate surface and proceeds until the thickness S of the solid layer becomes equal to the thickness δ of the deposit.

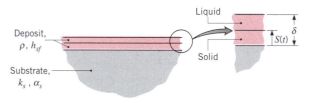

(a) Consider conditions for which molten material at its fusion temperature T_f is deposited on a *large* substrate that is at an initial uniform temperature T_i. With $S = 0$ at $t = 0$, develop an expression for estimating the time t_d required to completely solidify the deposit if it remains at T_f throughout the solidification process. Express your result in terms of the substrate thermal conductivity and thermal diffusivity (k_s,

α_s), the density and latent heat of fusion of the deposit (ρ, h_{sf}), the deposit thickness δ, and the relevant temperatures (T_f, T_i).

(b) The plasma spray deposition process of Problem 5.23 is used to apply a thin ($\delta = 2$ mm) alumina coating on a thick tungsten substrate. The substrate has a uniform initial temperature of $T_i = 300$ K, and its thermal conductivity and thermal diffusivity may be approximated as $k_s = 120$ W/m · K and $\alpha_s = 4.0 \times 10^{-5}$ m²/s, respectively. The density and latent heat of fusion of the alumina are $\rho = 3970$ kg/m³ and $h_{sf} = 3577$ kJ/kg, respectively, and the alumina solidifies at its fusion temperature ($T_f = 2318$ K). Assuming that the molten layer is instantaneously deposited on the substrate, estimate the time required for the deposit to solidify.

5.72 When a molten metal is cast in a mold that is a poor conductor, the dominant resistance to heat flow is within the mold wall. Consider conditions for which a liquid metal is solidifying in a thick-walled mold of thermal conductivity k_w and thermal diffusivity α_w. The density and latent heat of fusion of the metal are designated as ρ and h_{sf}, respectively, and in both its molten and solid states, the thermal conductivity of the metal is very much larger than that of the mold.

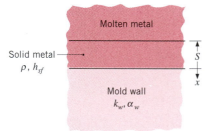

Just before the start of solidification ($S = 0$), the mold wall is everywhere at an initial uniform temperature T_i and the molten metal is everywhere at its fusion (melting point) temperature of T_f. Following the start of solidification, there is conduction heat transfer into the mold wall and the thickness of the solidified metal, S, increases with time t.

(a) Sketch the one-dimensional temperature distribution, $T(x)$, in the mold wall and the metal at $t = 0$ and at two subsequent times during the solidification. Clearly indicate any underlying assumptions.

(b) Obtain a relation for the variation of the solid layer thickness S with time t, expressing your result in terms of appropriate parameters of the system.

Multidimensional Conduction

5.73 A long steel (plain carbon) billet of square cross section 0.3 m by 0.3 m, initially at a uniform temperature of 30°C, is placed in a soaking oven having a temperature of 750°C. If the convection heat transfer coefficient for the heating process is 100 W/m² · K, how long must the billet remain in the oven before its center temperature reaches 600°C?

5.74 Fireclay brick of dimensions 0.06 m × 0.09 m × 0.20 m is removed from a kiln at 1600 K and cooled in air at 40°C with $h = 50$ W/m² · K. What is the temperature at the center and at the corners of the brick after 50 min of cooling?

5.75 A cylindrical copper pin 100 mm long and 50 mm in diameter is initially at a uniform temperature of 20°C. The end faces are suddenly subjected to an intense heating rate that raises them to a temperature of 500°C. At the same time, the cylindrical surface is subjected to heating by gas flow with a temperature of 500°C and a heat transfer coefficient of 100 W/m² · K.

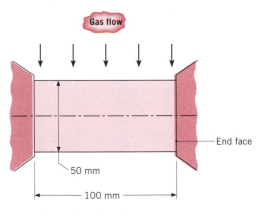

Gas flow

End face

50 mm

100 mm

(a) Determine the temperature at the center point of the cylinder 8 s after sudden application of the heat.

(b) Considering the parameters governing the temperature distribution in transient heat diffusion problems, can any simplifying assumptions be justified in analyzing this particular problem? Explain briefly.

5.76 Recalling that your mother once said that meat should be cooked until every portion has attained a temperature of 80°C, how long will it take to cook a 2.25-kg roast? Assume that the meat is initially at 6°C and that the oven temperature is 175°C with a convection heat transfer coefficient of 15 W/m² · K. Treat the roast as a cylinder with properties of liquid water, having a diameter equal to its length.

5.77 A long rod 20 mm in diameter is fabricated from alumina (polycrystalline aluminum oxide) and is initially at a uniform temperature of 850 K. The rod is suddenly exposed to fluid at 350 K with $h = 500$ W/m² · K. Estimate the centerline temperature of the rod after 30 s at an exposed end and at an axial distance of 6 mm from the end.

Finite-Difference Equations: Derivations

5.78 The stability criterion for the explicit method requires that the coefficient of the T_m^p term of the one-dimensional, finite-difference equation be zero or positive. Consider the situation for which the temperatures at the two neighboring nodes (T_{m-1}^p, T_{m+1}^p) are 100°C while the center node (T_m^p) is at 50°C. Show that for values of $Fo > \frac{1}{2}$, the finite-difference equation will predict a value of T_m^{p+1} that violates the second law of thermodynamics.

5.79 A thin rod of diameter D is initially in equilibrium with its surroundings, a large vacuum enclosure at temperature T_{sur}. Suddenly an electrical current I (A) is passed through the rod having an electrical resistivity ρ_e and emissivity ε. Other pertinent thermophysical properties are identified in the sketch. Derive the transient, finite-difference equation for node m.

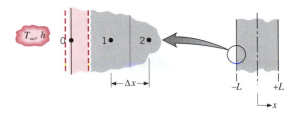

5.80 A one-dimensional slab of thickness $2L$ is initially at a uniform temperature T_i. Suddenly, electric current is passed through the slab causing a uniform volumetric heating $\dot{q}$ (W/m³). At the same time, both outer surfaces ($x = \pm L$) are subjected to a convection process at T_∞ with a heat transfer coefficient h.

Write the finite-difference equation expressing conservation of energy for node 0 located on the outer surface at $x = -L$. Rearrange your equation and identify any important dimensionless coefficients.

5.81 A plane wall ($\rho = 4000$ kg/m³, $c_p = 500$ J/kg · K, $k = 10$ W/m · K) of thickness $L = 20$ mm initially has a linear, steady-state temperature distribution with boundaries maintained at $T_1 = 0°$C and $T_2 = 100°$C. Suddenly, an electric current is passed through the wall, causing uniform energy generation at a rate $\dot{q} = 2 \times 10^7$ W/m³. The boundary conditions T_1 and T_2 remain fixed.

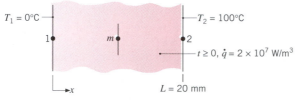

(a) On T–x coordinates, sketch temperature distributions for the following cases: (i) initial condition ($t \leq 0$); (ii) steady-state conditions ($t \rightarrow \infty$), assuming that the maximum temperature in the wall exceeds T_2; and (iii) for two intermediate times. Label all important features of the distributions.

(b) For the system of three nodal points shown schematically (1, m, 2), define an appropriate control volume for node m and, identifying all relevant processes, derive the corresponding finite-difference equation using either the *explicit* or *implicit* method.

(c) With a time increment of $\Delta t = 5$ s, use the finite-difference method to obtain values of T_m for the first 45 s of elapsed time. Determine the corresponding heat fluxes at the boundaries, that is, q_x'' (0, 45 s) and q_x'' (20 mm, 45 s).

(d) To determine the effect of mesh size, repeat your analysis using grids of 5 and 11 nodal points ($\Delta x = 5.0$ and 2.0 mm, respectively).

5.82 A round solid cylinder made of a plastic material ($\alpha = 6 \times 10^{-7}$ m²/s) is initially at a uniform temperature of 20°C and is well insulated along its lateral surface and at one end. At time $t = 0$, heat is applied to the left boundary causing T_0 to increase linearly with time at a rate of 1°C/s.

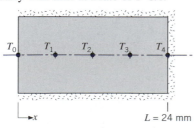

(a) Using the explicit method with $Fo = \frac{1}{2}$, derive the finite-difference equations for nodes 1, 2, 3, and 4.

(b) Format a table with headings of p, t (s), and the nodal temperatures T_0 to T_4. Determine the surface temperature T_0 when $T_4 = 35°$C.

Finite-Difference Solutions: One-Dimensional Systems

5.83 A wall 0.12 m thick having a thermal diffusivity of 1.5×10^{-6} m²/s is initially at a uniform temperature of 85°C. Suddenly one face is lowered to a temperature of 20°C, while the other face is perfectly insulated.

(a) Using the explicit finite-difference technique with space and time increments of 30 mm and 300 s, respectively, determine the temperature distribution at $t = 45$ min.

(b) With $\Delta x = 30$ mm and $\Delta t = 300$ s, compute $T(x, t)$ for $0 \leq t \leq t_{ss}$, where t_{ss} is the time required for the temperature at each nodal point to reach a value that is within 1°C of the steady-state temperature. Repeat the foregoing calculations for $\Delta t = 75$ s. For each value of Δt, plot temperature histories for each face and the midplane.

5.84 The plane wall of Problem 2.43 ($k = 1.5$ W/m · K, $\alpha = 7.5 \times 10^{-6}$ m²/s) has a thickness of $L = 50$ mm and an initial uniform temperature of 25°C. Suddenly, the boundary at $x = L$ experiences heating by a fluid for which $T_\infty = 50°$C and $h = 75$ W/m² · K, while the other boundary at $x = 0$ experiences an applied heat flux of $q_o'' = 2000$ W/m².

(a) With $\Delta x = 5$ mm and $\Delta t = 20$ s, compute and plot temperature distributions in the wall for (i) the initial condition, (ii) the steady-state condition, and (iii) two intermediate times.

(b) On q_x''–x coordinates, plot the heat flux distributions corresponding to the four temperature distributions represented in part (a).

(c) On q_x''–t coordinates, plot the heat flux at $x = 0$ and $x = L$.

5.85 The plane wall of Problem 2.44 ($k = 50$ W/m · K, $\alpha = 1.5 \times 10^{-6}$ m²/s) has a thickness of $L = 40$ mm and an initial uniform temperature of $T_o = 25°$C. Suddenly, the boundary at $x = L$ experiences heating by a fluid for which $T_\infty = 50°$C and $h = 1000$ W/m² · K, while heat is uniformly generated within the wall at $\dot{q} = 1 \times 10^7$ W/m³. The boundary at $x = 0$ remains at T_o.

(a) With $\Delta x = 4$ mm and $\Delta t = 1$ s, plot temperature distributions in the wall for (i) the initial condition, (ii) the steady-state condition, and (iii) two intermediate times.

(b) On q_x''–t coordinates, plot the heat flux at $x = 0$ and $x = L$. At what elapsed time is there zero heat flux at $x = L$?

5.86 Consider the fuel element of Example 5.7. Initially, the element is at a uniform temperature of 250°C with no heat generation. Suddenly, the element is inserted into the reactor core causing a uniform volumetric heat generation rate of $\dot{q} = 10^8$ W/m³. The surfaces are convectively cooled with $T_\infty = 250°C$ and $h = 1100$ W/m² · K. Using the explicit method with a space increment of 2 mm, determine the temperature distribution 1.5 s after the element is inserted into the core.

5.87 A plane wall of thickness 100 mm with a uniform volumetric heat generation of $\dot{q} = 1.5 \times 10^6$ W/m³ is exposed to convection conditions of $T_\infty = 30°C$ and $h = 1000$ W/m² · K on both surfaces. The wall is maintained under steady-state conditions when, suddenly, the heat generation level ($\dot{q}$) is reduced to zero. The thermal diffusivity and thermal conductivity of the wall material are 1.6×10^{-6} m²/s and 75 W/m · K. A space increment of 10 mm is suggested.

(a) Estimate the midplane temperature 3 min after the generation has been switched off.

(b) Plot on T–x coordinates the temperature distribution obtained in part (a). Show also the initial and steady-state temperature distributions for the wall.

5.88 A large plastic casting with thermal diffusivity 6.0×10^{-7} m²/s is removed from its mold at a uniform temperature of 150°C. The casting is then exposed to a high-velocity airstream such that the surface experiences a sudden change in temperature to 20°C. Assuming the casting approximates a semi-infinite medium and using a finite-difference method with a space increment of 6 mm, estimate the temperature at a distance 18 mm from the surface after 3 min have elapsed. Verify your result by comparison with the appropriate analytical solution.

5.89 The cross section of an oven wall is composed of 30-mm-thick insulation sandwiched between two thin (1.5-mm-thick) stainless steel sheets. Under steady-state conditions, the oven is operating with an inside air temperature of $T_{\infty,i} = 150°C$ and an ambient air temperature of $T_{\infty,o} = 20°C$ with $h_i = 100$ W/m² · K and $h_o = 10$ W/m² · K. When the

oven heater level is changed and the fan speed changed to substantially increase air circulation within the oven, the inside surface of the oven experiences a sudden temperature change to 100°C. The insulation has a thermal conductivity of 0.03 W/m · K and a thermal diffusivity of 7.5×10^{-7} m²/s. In your finite-difference solution, use a space increment of 6 mm. Assume that the effect of the stainless steel sheets is negligible and that the outside convection heat transfer coefficient h_o remains unchanged. Estimate the time required for the oven wall to approximate steady-state conditions after the inner wall temperature is changed to 100°C.

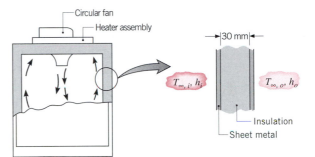

5.90 A very thick plate with thermal diffusivity 5.6×10^{-6} m²/s and thermal conductivity 20 W/m · K is initially at a uniform temperature of 325°C. Suddenly, the surface is exposed to a coolant at 15°C for which the convection heat transfer coefficient is 100 W/m² · K. Using the finite-difference method with a space increment of $\Delta x = 15$ mm and a time increment of 18 s, determine temperatures at the surface and at a depth of 45 mm after 3 min have elapsed.

5.91 Referring to Example 5.8, Comment 4, consider a sudden exposure of the surface to large surroundings at an elevated temperature (T_{sur}) and to convection (T_∞, h).

(a) Derive the explicit, finite-difference equation for the surface node in terms of Fo, Bi, and Bi_r.

(b) Obtain the stability criterion for the surface node. Does this criterion change with time? Is the criterion more restrictive than that for an interior node?

(c) A thick slab of material ($k = 1.5$ W/m · K, $\alpha = 7 \times 10^{-7}$ m²/s, $\varepsilon = 0.9$), initially at a uniform temperature of 27°C, is suddenly exposed to large surroundings at 1000 K. Neglecting convection and using a space increment of 10 mm, determine temperatures at the surface and 30 mm from the surface after an elapsed time of 1 min.

5.92 Consider the bonding operation described in Problem 3.97, which was analyzed under steady-state conditions. In this case, however, the laser will be used to heat the film for a prescribed period of time, creating the transient heating situation shown in the sketch.

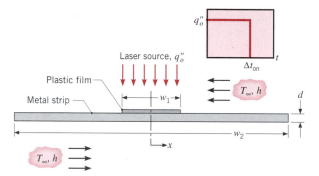

The strip is initially at 25°C and the laser provides a uniform flux of 85,000 W/m² over a time interval of $\Delta t_{on} = 10$ s. The system dimensions and thermophysical properties remain the same, but the convection coefficient to the ambient air at 25°C is now 100 W/m² · K.

(a) Using an implicit finite-difference method with $\Delta x = 4$ mm and $\Delta t = 1$ s, obtain temperature histories for $0 \le t \le 30$ s at the center and film edge, $T(0, t)$ and $T(w_1/2, t)$, respectively, to determine if the adhesive is satisfactorily cured above 90°C for 10 s and if its degradation temperature of 200°C is exceeded.

(b) Validate your program code by comparing it against the steady-state results of Problem 3.97. What type of analytical solution would you seek in order to test the proper transient behavior of your code?

5.93 One end of a stainless steel (AISI 316) rod of diameter 10 mm and length 0.16 m is inserted into a fixture maintained at 200°C. The rod, covered with an insulating sleeve, reaches a uniform temperature throughout its length. When the sleeve is removed, the rod is subjected to ambient air at 25°C such that the convection heat transfer coefficient is 30 W/m² · K.

(a) Using the explicit finite-difference technique with a space increment of $\Delta x = 0.016$ m, estimate the time required for the midlength of the rod to reach 100°C.

(b) With $\Delta x = 0.016$ m and $\Delta t = 10$ s, compute $T(x, t)$ for $0 \le t \le t_1$, where t_1 is the time required for the midlength of the rod to reach 50°C. Plot the temperature distribution for $t = 0, 200$ s, 400 s, and t_1.

5.94 A tantalum rod of diameter 3 mm and length 120 mm is supported by two electrodes within a large vacuum enclosure. Initially the rod is in equilibrium with the electrodes and its surroundings, which are maintained at 300 K. Suddenly, an electrical current, $I = 80$ A, is passed through the rod. Assume the emissivity of the rod is 0.1 and the electrical resistivity is $95 \times 10^{-8} \, \Omega \cdot$ m. Use Table A.1 to obtain the other thermophysical properties required in your solution. Use a finite-difference method with a space increment of 10 mm.

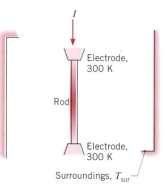

(a) Estimate the time required for the midlength of the rod to reach 1000 K.

(b) Determine the steady-state temperature distribution and estimate approximately how long it will take to reach this condition.

5.95 A support rod ($k = 15$ W/m · K, $\alpha = 4.0 \times 10^{-6}$ m²/s) of diameter $D = 15$ mm and length $L = 100$ mm spans a channel whose walls are maintained at a temperature of $T_b = 300$ K. Suddenly, the rod is exposed to a cross flow of hot gases for which $T_\infty = 600$ K and $h = 75$ W/m² · K. The channel walls are cooled and remain at 300 K.

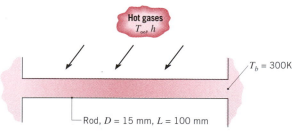

(a) Using an appropriate numerical technique, determine the thermal response of the rod to the convective heating. Plot the midspan temperature as a function of elapsed time. Using an appropriate analytical model of the rod, determine the steady-state temperature distribution and compare the result with that obtained numerically for very long elapsed times.

(b) After the rod has reached steady-state conditions, the flow of hot gases is suddenly terminated, and the rod cools by free convection to ambient air at $T_\infty = 300$ K and by radiation exchange with large surroundings at $T_{sur} = 300$ K. The free convection coefficient can be expressed as h (W/m$^2 \cdot$ K) $= C \Delta T^n$, where $C = 4.4$ W/m$^2 \cdot$ K$^{1.188}$ and $n = 0.188$. The emissivity of the rod is 0.5. Determine the subsequent thermal response of the rod. Plot the midspan temperature as a function of cooling time, and determine the time required for the rod to reach a *safe-to-touch temperature* of 315 K.

5.96 Consider the acceleration-grid foil ($k = 40$ W/m $\cdot$ K, $\alpha = 3 \times 10^{-5}$ m^2/s, $\varepsilon = 0.45$) of Problem 4.70. Develop an implicit, finite-difference model of the foil, which can be used for the following purposes.

(a) Assuming the foil to be at a uniform temperature of 300 K when the ion beam source is activated, obtain a plot of the midspan temperature–time history. At what elapsed time does this point on the foil reach a temperature within 1 K of the steady-state value?

(b) The foil is operating under steady-state conditions when, suddenly, the ion beam is deactivated. Obtain a plot of the subsequent midspan temperature–time history. How long does it take for the hottest point on the foil to cool to 315 K, a safe-to-touch condition?

5.97 Circuit boards are treated by heating a stack of them under high pressure as illustrated in Problem 5.37 and described further in Problem 5.38. A finite-difference method of solution is sought with two additional considerations. First, the book is to be treated as having distributed, rather than lumped, characteristics, by using a grid spacing of $\Delta x = 2.36$ mm with nodes at the center of the individual circuit board or plate. Second, rather than bringing the platens to 190°C in one sudden change, the heating schedule $T_p(t)$ shown below is to be used in order to minimize excessive thermal stresses induced by rapidly changing thermal gradients in the vicinity of the platens.

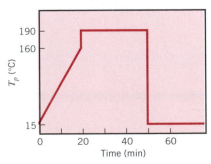

(a) Using a time increment of $\Delta t = 60$ s and the implicit method, find the temperature history of the midplane of the book and determine whether curing will occur (170°C for 5 min).

(b) Following the reduction of the platen temperatures to 15°C ($t = 50$ min), how long will it take for the midplane of the book to reach 37°C, a safe temperature at which the operator can begin unloading the press?

(c) Validate your program code by using the heating schedule of a sudden change of platen temperature from 15 to 190°C and compare results with those from an appropriate analytical solution (see Problem 5.38).

Finite-Difference Equations: Cylindrical Coordinates

5.98 A thin circular disk is subjected to induction heating from a coil, the effect of which is to provide a uniform heat generation within a ring section as shown. Convection occurs at the upper surface, while the lower surface is well insulated.

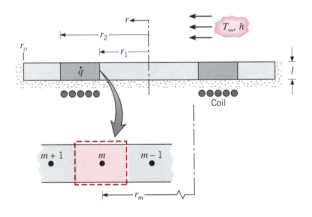

(a) Derive the transient, finite-difference equation for node m, which is within the region subjected to induction heating.

(b) On T–r coordinates sketch, in a qualitative manner, the steady-state temperature distribution, identifying important features.

5.99 An electrical cable, experiencing a uniform volumetric generation $\dot{q}$, is half buried in an insulating material while the upper surface is exposed to a convection process (T_∞, h).

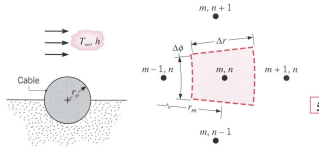

(a) Derive the explicit, finite-difference equations for an interior node (m, n), the center node $(m = 0)$, and the outer surface nodes (M, n) for the convection and insulated boundaries.

(b) Obtain the stability criterion for each of the finite-difference equations. Identify the most restrictive criterion.

Finite-Difference Solutions: Two-Dimensional Systems

5.100 Two very long (in the direction normal to the page) bars having the prescribed initial temperature distributions are to be soldered together. At time $t = 0$, the $m = 3$ face of the copper (pure) bar contacts the $m = 4$ face of the steel (AISI 1010) bar. The solder and flux act as an interfacial layer of negligible thickness and effective contact resistance $R''_{t, c} = 2 \times 10^{-5} \, \text{m}^2 \cdot \text{K/W}$.

Initial Temperatures (K)

n/m	1	2	3	4	5	6
1	700	700	700	1000	900	800
2	700	700	700	1000	900	800
3	700	700	700	1000	900	800

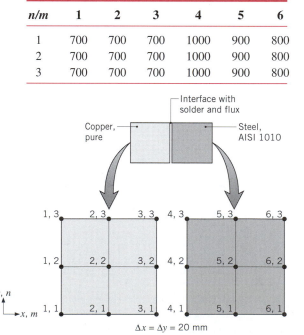

$\Delta x = \Delta y = 20 \, \text{mm}$

(a) Derive the explicit, finite-difference equation in terms of Fo and $Bi_c = \Delta x / k R''_{t, c}$ for $T_{4, 2}$ and determine the corresponding stability criterion.

(b) Using $Fo = 0.01$, determine $T_{4, 2}$ one time step after contact is made. What is Δt? Is the stability criterion satisfied?

5.101 Consider the system of Problem 4.57. Initially with no flue gases flowing, the walls ($\alpha = 5.5 \times 10^{-6}$ m²/s) are at a uniform temperature of 25°C. Using the implicit, finite-difference method with a time increment of 1 h, find the temperature distribution in the wall 1, 2, 5, and 20 h after introduction of the flue gases.

5.102 Consider the system of Problem 4.71. Initially, the ceramic plate ($\alpha = 1.5 \times 10^{-6}$ m²/s) is at a uniform temperature of 30°C, and suddenly the electrical heating elements are energized. Using the implicit, finite-difference method, estimate the time required for the difference between the surface and initial temperatures to reach 95% of the difference for steady-state conditions. Use a time increment of 2 s.

5.103 Consider the thermal conduction module and operating conditions of Problem 4.75. To evaluate the transient response of the cold plate, which has a thermal diffusivity of $\alpha = 75 \times 10^{-6}$ m²/s, assume that, when the module is activated at $t = 0$, the initial temperature of the cold plate is $T_i = 15°C$ and a uniform heat flux of $q''_o = 10^5$ W/m² is applied at its base. Using the implicit finite-difference method and a time increment of $\Delta t = 0.1$ s, compute the designated nodal temperatures as a function of time. From the temperatures computed at a particular time, evaluate the ratio of the rate of heat transfer by convection to the water to the heat input at the base. Terminate the calculations when this ratio reaches 0.99. Print the temperature field at 5-s intervals and at the time for which the calculations are terminated.

CHAPTER 6

Introduction to Convection

*T*hus far we have focused on heat transfer by conduction and have considered convection only to the extent that it provides a possible boundary condition for conduction problems. In Section 1.2.2 we used the term "convection" to describe energy transfer between a surface and a fluid moving over the surface. Although the mechanism of diffusion (random motion of fluid molecules) contributes to this transfer, the dominant contribution is generally made by the bulk or gross motion of fluid particles.

In our treatment of convection, we have two major objectives. In addition to obtaining an understanding of the physical mechanisms that underlie convection transfer, we wish to develop the means to perform convection transfer calculations. This chapter is devoted primarily to achieving the former objective. In particular an effort has been made to concentrate many of the fundamentals in one place. Physical origins are discussed, and relevant dimensionless parameters, as well as important analogies, are developed.

A unique feature of this chapter is the manner in which convection mass transfer effects are introduced by analogy to those of convection heat transfer. In mass transfer by convection, gross fluid motion combines with diffusion to promote the transport of a species for which there exists a concentration gradient. In this text, we focus on convection mass transfer that occurs at the surface of a volatile solid or a liquid due to motion of a gas over the surface.

With conceptual foundations established, subsequent chapters are used to develop useful tools for quantifying convection effects. Chapters 7 and 8 present methods for computing the coefficients associated with *forced convection* in *external* and *internal flow* configurations, respectively. Chapter 9 describes methods for determining these coefficients in *free convection,* and Chapter 10 considers the problem of *convection with phase change* (*boiling* and *condensation*). Chapter 11 develops methods for designing and evaluating the performance of *heat exchangers,* devices that are widely used in engineering practice to effect heat transfer between fluids.

Accordingly, we begin by developing our understanding of the nature of convection.

6.1
The Convection Transfer Problem

Consider the flow condition of Figure 6.1*a*. A fluid of velocity V and temperature T_∞ flows over a surface of arbitrary shape and of area A_s. The surface is presumed to be at a uniform temperature, T_s, and if $T_s \neq T_\infty$, we know that convection heat transfer will occur. The *local heat flux* q'' may be expressed as

$$q'' = h(T_s - T_\infty) \tag{6.1}$$

where h is the *local convection coefficient.* Because flow conditions vary from point to point on the surface, both q'' and h also vary along the surface. The *total heat transfer rate q* may be obtained by integrating the local flux over the entire surface. That is,

$$q = \int_{A_s} q'' \, dA_s \tag{6.2}$$

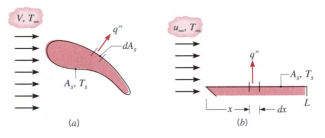

FIGURE 6.1 Local and total convection heat transfer effects. (*a*) Surface of arbitrary shape. (*b*) Flat plate.

or, from Equation 6.1,

$$q = (T_s - T_\infty) \int_{A_s} h \, dA_s \qquad (6.3)$$

Defining an *average convection coefficient* $\bar{h}$ for the entire surface, the total heat transfer rate may also be expressed as

$$q = \bar{h} A_s (T_s - T_\infty) \qquad (6.4)$$

Equating Equations 6.3 and 6.4, it follows that the average and local convection coefficients are related by an expression of the form

$$\bar{h} = \frac{1}{A_s} \int_{A_s} h \, dA_s \qquad (6.5)$$

Note that for the special case of flow over a flat plate (Figure 6.1*b*), *h* varies with the distance *x* from the leading edge and Equation 6.5 reduces to

$$\bar{h} = \frac{1}{L} \int_0^L h \, dx \qquad (6.6)$$

Similar results may be obtained for convection mass transfer. If a fluid of species molar concentration $C_{A,\infty}$ flows over a surface at which the species concentration is maintained at some uniform value $C_{A,s} \neq C_{A,\infty}$ (Figure 6.2*a*), transfer of the species by convection will occur. Species A is typically a vapor that is transferred into a gas stream due to evaporation or sublimation at a liquid or solid surface, respectively, and we are interested in determining the rate at which this transfer occurs. As for the case of heat transfer, such a calculation may be based on the use of a convection coefficient. In particular, we may re-

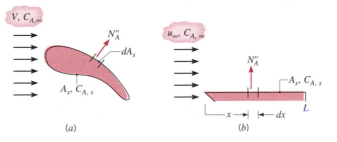

FIGURE 6.2 Local and total convection species transfer effects. (*a*) Surface of arbitrary shape. (*b*) Flat plate.

late the molar flux of species A to the product of a transfer coefficient and a concentration difference. Referring to Figure 6.2, the molar flux of species A, N_A'' (kmol/s · m^2), may be expressed as

$$N_A'' = h_m(C_{A,s} - C_{A,\infty}) \qquad (6.7)$$

where h_m (m/s) is the *convection mass transfer coefficient*. The molar concentrations $C_{A,s}$ and $C_{A,\infty}$ have units of kilogram-moles per cubic meter (kmol/m^3). The total molar transfer rate for an entire surface, N_A (kmol/s), may then be expressed as

$$N_A = \bar{h}_m A_s(C_{A,s} - C_{A,\infty}) \qquad (6.8)$$

where the average and local mass transfer convection coefficients are related by an equation of the form

$$\bar{h}_m = \frac{1}{A_s} \int_{A_s} h_m \, dA_s \qquad (6.9)$$

For the flat plate of Figure 6.2b, it follows that

$$\bar{h}_m = \frac{1}{L} \int_0^L h_m \, dx \qquad (6.10)$$

Species transfer may also be expressed as a mass flux, n_A'' (kg/s · m^2), or as a mass transfer rate, n_A (kg/s), simply by multiplying both sides of Equations 6.7 and 6.8, respectively, by the molecular weight $\mathcal{M}_A$ (kg/kmol) of species A. Accordingly,

$$n_A'' = h_m(\rho_{A,s} - \rho_{A,\infty}) \qquad (6.11)$$

and

$$n_A = \bar{h}_m A_s(\rho_{A,s} - \rho_{A,\infty}) \qquad (6.12)$$

where ρ_A (kg/m^3) is the mass density of species A.[1]

To perform a convection mass transfer calculation, it is necessary to determine the value of $C_{A,s}$ or $\rho_{A,s}$. Such a determination may readily be made by noting that thermodynamic equilibrium exists at the interface between the gas and the liquid or solid phase. One implication of this equilibrium condition is that the temperature of the vapor at the interface is equal to the surface temperature T_s. A second implication is that the vapor is in a *saturated state*, in which case thermodynamic tables, such as Table A.6 for water, may be used to obtain its density from knowledge of T_s. To a good approximation, the molar concentration of the vapor at the surface may also be determined from the vapor pressure through application of the equation of state for an ideal gas. That is,

$$C_{A,s} = \frac{p_{sat}(T_s)}{\mathcal{R} T_s} \qquad (6.13)$$

---

[1]Although the foregoing nomenclature is well suited for characterizing mass transfer processes of interest in this text, there is, by no means, a standard nomenclature, and it is often difficult to reconcile results from different publications. A review of the different ways in which driving potentials, fluxes, and convection coefficients may be formulated is provided by Webb [1].

where $\mathscr{R}$ is the universal gas constant and $p_{sat}(T_s)$ is the vapor pressure corresponding to saturation at T_s. Note that the vapor mass density and molar concentration are related by $\rho_A = \mathscr{M}_A C_A$.

The local flux and/or the total transfer rate are of paramount importance in any convection problem. These quantities may be determined from the rate equations, Equations 6.1, 6.4, 6.7, and 6.8, which depend on knowledge of the local and average convection coefficients. It is for this reason that determination of these coefficients is viewed as *the problem of convection*. However, the problem is not a simple one, for in addition to depending on numerous *fluid properties* such as density, viscosity, thermal conductivity, and specific heat, the coefficients depend on the *surface geometry* and the *flow conditions*. This multiplicity of independent variables results because convection transfer is determined by the *boundary layers* that develop on the surface.

EXAMPLE 6.1

Experimental results for the local heat transfer coefficient h_x for flow over a flat plate with an extremely rough surface were found to fit the relation

$$h_x(x) = ax^{-0.1}$$

where a is a coefficient (W/m$^{1.9}$ · K) and x (m) is the distance from the leading edge of the plate.

1. Develop an expression for the ratio of the average heat transfer coefficient $\bar{h}_x$ for a plate of length x to the local heat transfer coefficient h_x at x.
2. Show, in a qualitative manner, the variation of h_x and $\bar{h}_x$ as a function of x.

SOLUTION

Known: Variation of the local heat transfer coefficient, $h_x(x)$.

Find:

1. The ratio of the average heat transfer coefficient $\bar{h}(x)$ to the local value $h_x(x)$.
2. Sketch of the variation of h_x and $\bar{h}_x$ with x.

Schematic:

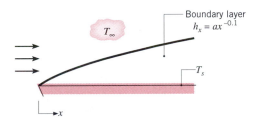

Analysis:

1. From Equation 6.6 the average value of the convection heat transfer coefficient over the region from 0 to x is

$$\bar{h}_x = \bar{h}_x(x) = \frac{1}{x}\int_0^x h_x(x)\,dx$$

Substituting the expression for the local heat transfer coefficient

$$h_x(x) = ax^{-0.1}$$

and integrating, we obtain

$$\bar{h}_x = \frac{1}{x} \int_0^x ax^{-0.1}\, dx = \frac{a}{x} \int_0^x x^{-0.1}\, dx = \frac{a}{x}\left(\frac{x^{+0.9}}{0.9}\right) = 1.11ax^{-0.1}$$

or

$$\bar{h}_x = 1.11h_x \qquad\qquad \triangleleft$$

2. The variation of h_x and $\bar{h}_x$ with x is as follows:

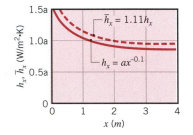

Comments: Boundary layer development causes both the local and average coefficients to decrease with increasing distance from the leading edge. The average coefficient up to x must therefore exceed the local value at x.

EXAMPLE 6.2

A long circular cylinder 20 mm in diameter is fabricated from solid naphthalene, a common moth repellant, and is exposed to an airstream that provides for a convection mass transfer coefficient of $\bar{h}_m = 0.05$ m/s. The molar concentration of naphthalene vapor at the cylinder surface is 5×10^{-6} kmol/m³, and its molecular weight is 128 kg/kmol. What is the mass sublimation rate per unit length of cylinder?

SOLUTION

Known: Saturated vapor concentration of naphthalene.

Find: Sublimation rate per unit length, n'_A (kg/s · m).

Schematic:

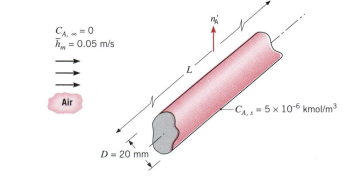

Assumptions:

1. Steady-state conditions.

2. Negligible concentration of naphthalene in free stream of air.

Analysis: Naphthalene is transported to the air by convection, and from Equation 6.8, the molar transfer rate for the cylinder is

$$N_A = \bar{h}_m \pi DL(C_{A,s} - C_{A,\infty})$$

With $C_{A,\infty} = 0$ and $N'_A = N_A/L$, it follows that

$$N'_A = (\pi D)\bar{h}_m C_{A,s} = \pi \times 0.02 \text{ m} \times 0.05 \text{ m/s} \times 5 \times 10^{-6} \text{ kmol/m}^3$$

$$N'_A = 1.57 \times 10^{-8} \text{ kmol/s} \cdot \text{m}$$

The mass sublimation rate is then

$$n'_A = \mathcal{M}_A N'_A = 128 \text{ kg/kmol} \times 1.57 \times 10^{-8} \text{ kmol/s} \cdot \text{m}$$

$$n'_A = 2.01 \times 10^{-6} \text{ kg/s} \cdot \text{m} \qquad \triangleleft$$

6.2
The Convection Boundary Layers

6.2.1 The Velocity Boundary Layer

To introduce the concept of a boundary layer, consider flow over the flat plate of Figure 6.3. When fluid particles make contact with the surface, they assume zero velocity. These particles then act to retard the motion of particles in the adjoining fluid layer, which act to retard the motion of particles in the next layer, and so on until, at a distance $y = \delta$ from the surface, the effect becomes negligible. This retardation of fluid motion is associated with *shear stresses* τ acting in planes that are parallel to the fluid velocity (Figure 6.3). With increasing distance y from the surface, the x velocity component of the fluid, u, must then increase until it approaches the free stream value u_∞. The subscript ∞ is used to designate conditions in the *free stream* outside the boundary layer.

The quantity δ is termed the *boundary layer thickness*, and it is typically defined as the value of y for which $u = 0.99u_\infty$. The *boundary layer velocity profile* refers to the manner in which u varies with y through the boundary layer. Accordingly, the fluid flow is characterized by two distinct regions, a thin fluid layer (the boundary layer) in which velocity gradients and shear stresses are large and a region outside the boundary layer in which velocity gra-

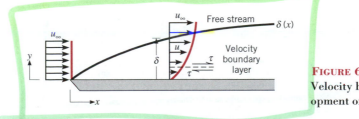

FIGURE 6.3
Velocity boundary layer development on a flat plate.

dients and shear stresses are negligible. With increasing distance from the leading edge, the effects of viscosity penetrate further into the free stream and the boundary layer grows (δ increases with x).

Because it pertains to the fluid velocity, the foregoing boundary layer may be referred to more specifically as the *velocity boundary layer*. It develops whenever there is fluid flow over a surface, and it is of fundamental importance to problems involving convection transport. In fluid mechanics its significance to the engineer stems from its relation to the surface shear stress τ_s, and hence to surface frictional effects. For external flows it provides the basis for determining the local *friction coefficient*

$$C_f \equiv \frac{\tau_s}{\rho u_\infty^2/2} \tag{6.14}$$

a key dimensionless parameter from which the surface frictional drag may be determined. Assuming a *Newtonian fluid,* the surface shear stress may be evaluated from knowledge of the velocity gradient at the surface

$$\tau_s = \mu \left. \frac{\partial u}{\partial y} \right|_{y=0} \tag{6.15}$$

where μ is a fluid property known as the *dynamic viscosity*.

6.2.2   The Thermal Boundary Layer

Just as a velocity boundary layer develops when there is fluid flow over a surface, a *thermal boundary layer* must develop if the fluid free stream and surface temperatures differ. Consider flow over an isothermal flat plate (Figure 6.4). At the leading edge the *temperature profile* is uniform, with $T(y) = T_\infty$. However, fluid particles that come into contact with the plate achieve thermal equilibrium at the plate's surface temperature. In turn these particles exchange energy with those in the adjoining fluid layer, and temperature gradients develop in the fluid. The region of the fluid in which these temperature gradients exist is the thermal boundary layer, and its thickness δ_t is typically defined as the value of y for which the ratio $[(T_s - T)/(T_s - T_\infty)] = 0.99$. With increasing distance from the leading edge, the effects of heat transfer penetrate further into the free stream and the thermal boundary layer grows.

The relation between conditions in this boundary layer and the convection heat transfer coefficient may readily be demonstrated. At any distance x from the leading edge, the *local* heat flux may be obtained by applying Fourier's law to the *fluid* at $y = 0$. That is,

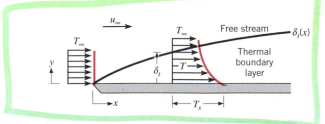

FIGURE 6.4
Thermal boundary layer development on an isothermal flat plate.

$$q''_s = -k_f \frac{\partial T}{\partial y}\bigg|_{y=0} \tag{6.16}$$

This expression is appropriate because, *at the surface, there is no fluid motion and energy transfer occurs only by conduction.* By combining Equation 6.16 with Newton's law of cooling, Equation 6.1, we then obtain

$$h = \frac{-k_f \,\partial T/\partial y|_{y=0}}{T_s - T_\infty} \tag{6.17}$$

Hence conditions in the thermal boundary layer, which strongly influence the wall temperature gradient $\partial T/\partial y|_{y=0}$, determine the rate of heat transfer across the boundary layer. Since $(T_s - T_\infty)$ is a constant, independent of x, while δ_t increases with increasing x, temperature gradients in the boundary layer must decrease with increasing x. Accordingly, the magnitude of $\partial T/\partial y|_{y=0}$ decreases with increasing x, and it follows that q''_s and h decrease with increasing x.

6.2.3 The Concentration Boundary Layer

Just as the velocity and thermal boundary layers determine wall friction and convection heat transfer, the *concentration boundary layer* determines convection mass transfer. If a binary mixture of chemical species A and B flows over a surface and the concentration of species A at the surface, $C_{A,s}$, differs from that in the free stream, $C_{A,\infty}$ (Figure 6.5), a concentration boundary layer will develop. It is the region of the fluid in which concentration gradients exist, and its thickness δ_c is typically defined as the value of y for which $[(C_{A,s} - C_A)/(C_{A,s} - C_{A,\infty})] = 0.99$. Species transfer by convection between the surface and the free stream fluid is determined by conditions in this boundary layer.

The relation between species transfer by convection and the concentration boundary layer may be demonstrated by first recognizing that the molar flux associated with species transfer by diffusion is determined by an expression that is analogous to Fourier's law. For the conditions of interest in this text, the expression, which is termed *Fick's law,* has the form

$$N''_A = -D_{AB} \frac{\partial C_A}{\partial y} \tag{6.18}^2$$

---

[2]This expression results from a more general form of Fick's law of diffusion (Section 14.1.2) when the total molar concentration of the mixture, $C = C_A + C_B$, is a constant.

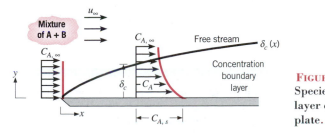

FIGURE 6.5
Species concentration boundary layer development on a flat plate.

where D_{AB} is a property of the binary mixture known as the *binary diffusion coefficient.* At any point corresponding to $y > 0$ in the concentration boundary layer of Figure 6.5, species transfer is due to both bulk fluid motion and diffusion. However, at $y = 0$ there is no fluid motion, and species transfer is by diffusion only. Applying Fick's law at $y = 0$, the species flux at any distance from the leading edge is then

$$N_A'' = -D_{AB} \frac{\partial C_A}{\partial y}\bigg|_{y=0} \tag{6.19}$$

Combining Equations 6.7 and 6.19, it follows that

$$h_m = \frac{-D_{AB}\, \partial C_A/\partial y|_{y=0}}{C_{A,s} - C_{A,\infty}} \tag{6.20}$$

Therefore conditions in the concentration boundary layer, which strongly influence the surface concentration gradient $\partial C_A/\partial y|_{y=0}$, will also influence the convection mass transfer coefficient, and hence the rate of species transfer into the boundary layer.

The foregoing results may also be expressed on a mass, rather than a molar, basis. Multiplying both sides of Equation 6.18 by the species molecular weight $\mathcal{M}_A$, the species mass flux due to diffusion is

$$n_A'' = -D_{AB} \frac{\partial \rho_A}{\partial y} \tag{6.21}$$

Applying this equation at $y = 0$ and combining with Equation 6.11, we obtain

$$h_m = \frac{-D_{AB}\, \partial \rho_A/\partial y|_{y=0}}{\rho_{A,s} - \rho_{A,\infty}} \tag{6.22}$$

EXAMPLE 6.3

At some location on the surface of a pan of water, measurements of the partial pressure of water vapor p_A (atm) are made as a function of the distance y from the surface, and the results are as follows:

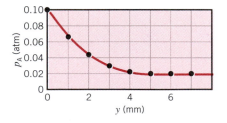

Determine the convection mass transfer coefficient $h_{m,x}$ at this location.

SOLUTION

Known: Partial pressure p_A of water vapor as a function of distance y at a particular location on the surface of a water layer.

Find: Convection mass transfer coefficient at the prescribed location.

Schematic:

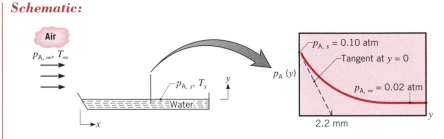

Assumptions:

1. Water vapor may be approximated as an ideal gas.
2. Conditions are isothermal.

Properties: Table A.6, saturated vapor (0.1 atm = 0.101 bar): T_s = 319 K. Table A.8, water vapor–air (319 K): D_{AB} (319 K) = D_{AB} (298 K) × (319 K/298 K)$^{3/2}$ = 0.288 × 10^{-4} m^2/s.

Analysis: From Equation 6.22 the local convection mass transfer coefficient is

$$h_{m,\,x} = \frac{-D_{AB}\,\partial\rho_A/\partial y\big|_{y=0}}{\rho_{A,\,s} - \rho_{A,\,\infty}}$$

or, approximating the vapor as an ideal gas

$$p_A = \rho_A RT$$

with constant T (isothermal conditions),

$$h_{m,\,x} = \frac{-D_{AB}\,\partial p_A/\partial y\big|_{y=0}}{p_{A,\,s} - p_{A,\,\infty}}$$

From the measured vapor pressure distribution

$$\frac{\partial p_A}{\partial y}\bigg|_{y=0} = \frac{(0 - 0.1)\ \text{atm}}{(0.0022 - 0)\ \text{m}} = -45.5\ \text{atm/m}$$

Hence

$$h_{m,\,x} = \frac{-0.288 \times 10^{-4}\ \text{m}^2/\text{s}\ (-45.5\ \text{atm/m})}{(0.1 - 0.02)\ \text{atm}} = 0.0164\ \text{m/s}$$

Comments: From thermodynamic equilibrium at the liquid–vapor interface, the interfacial temperature was determined from Table A.6.

6.2.4 Significance of the Boundary Layers

In summary, the velocity boundary layer is of extent $\delta(x)$ and is characterized by the presence of velocity gradients and shear stresses. The thermal boundary layer is of extent $\delta_t(x)$ and is characterized by temperature gradients and heat transfer. Finally, the concentration boundary layer is of extent $\delta_c(x)$ and is characterized by concentration gradients and species transfer. For the engineer the principal manifestations of the three boundary layers are, respectively, *surface friction, convection heat transfer,* and *convection mass transfer.* The key boundary layer parameters are then the *friction coefficient C_f* and the *heat* and *mass transfer convection coefficients* h and h_m, respectively.

For flow over any surface, there will always exist a velocity boundary layer, and hence surface friction. However, a thermal boundary layer, and hence convection heat transfer, exists only if the surface and free stream temperatures differ. Similarly, a concentration boundary layer and convection mass transfer exist only if the surface concentration of a species differs from its free stream concentration. Situations can arise in which all three boundary layers are present. In such cases, the boundary layers rarely grow at the same rate, and the values of δ, δ_t, and δ_c at a given x location are not the same.

6.3
Laminar and Turbulent Flow

An essential first step in the treatment of any convection problem is to determine whether the boundary layer is *laminar* or *turbulent.* Surface friction and the convection transfer rates depend strongly on which of these conditions exists.

As shown in Figure 6.6, there are sharp differences between laminar and turbulent flow conditions. In the laminar boundary layer, fluid motion is highly ordered and it is possible to identify streamlines along which particles move. Fluid motion along a streamline is characterized by velocity components in both the x and y directions. Since the velocity component v is in the direction normal to the surface, it can contribute significantly to the transfer of momentum, energy, or species through the boundary layer. Fluid motion normal to the surface is necessitated by boundary layer growth in the x direction.

In contrast, fluid motion in the turbulent boundary layer is highly irregular and is characterized by velocity fluctuations. These fluctuations enhance the

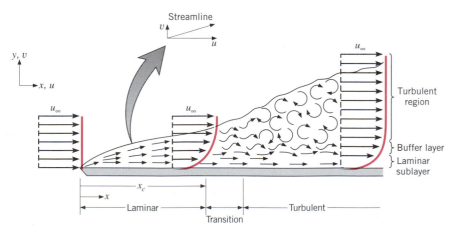

FIGURE 6.6   **Velocity boundary layer development on a flat plate.**

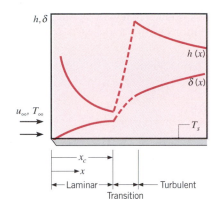

FIGURE 6.7
Variation of velocity boundary layer thickness δ and the local heat transfer coefficient h for flow over an isothermal flat plate.

transfer of momentum, energy, and species, and hence increase surface friction as well as convection transfer rates. Fluid mixing resulting from the fluctuations makes turbulent boundary layer thicknesses larger and boundary layer profiles (velocity, temperature, and concentration) flatter than in laminar flow.

The foregoing conditions are shown schematically in Figure 6.6 for velocity boundary layer development on a flat plate. The boundary layer is initially laminar, but at some distance from the leading edge, small disturbances are amplified and transition to turbulent flow begins to occur. Fluid fluctuations begin to develop in the *transition region,* and the boundary layer eventually becomes completely turbulent. In the fully turbulent region, conditions are characterized by a highly random, three-dimensional motion of relatively large parcels of fluid, and it is not surprising that the transition to turbulence is accompanied by significant increases in the boundary layer thicknesses, the wall shear stress, and the convection coefficients. These effects are illustrated in Figure 6.7 for the velocity boundary layer thickness δ and the local convection heat transfer coefficient h. In the turbulent boundary layer, three different regions may be delineated. We may speak of a *laminar sublayer* in which transport is dominated by diffusion and the velocity profile is nearly linear. There is an adjoining *buffer layer* in which diffusion and turbulent mixing are comparable, and there is a *turbulent zone* in which transport is dominated by turbulent mixing.

In calculating boundary layer behavior, it is frequently reasonable to assume that transition begins at some location x_c. This location is determined by a dimensionless grouping of variables called the *Reynolds number,*

$$Re_x \equiv \frac{\rho u_\infty x}{\mu} \tag{6.23}$$

where the characteristic length x is the distance from the leading edge. The *critical Reynolds number* is the value of Re_x for which transition begins, and for flow over a flat plate, it is known to vary from 10^5 to 3×10^6, depending on surface roughness and the turbulence level of the free stream. A representative value of

$$Re_{x,c} = \frac{\rho u_\infty x_c}{\mu} = 5 \times 10^5 \tag{6.24}$$

is often assumed for boundary layer calculations and, unless otherwise noted, is used for the calculations of this text.

6.4
The Convection Transfer Equations

We can improve our understanding of the physical effects that determine boundary layer behavior and further illustrate its relevance to convection transport by developing the equations that govern boundary layer conditions. Consider the simultaneous development of velocity, thermal, and concentration boundary layers over the surface of Figure 6.8. The fluid is considered to be a binary mixture of species A and B, and the species A concentration boundary layer originates from a difference between the free stream and surface concentrations ($C_{A,\infty} \neq C_{A,s}$). Selection of the relative thicknesses ($\delta_t > \delta_c > \delta$) is arbitrary, for the moment, and the factors that influence relative boundary layer development are discussed later in this chapter. To simplify the development we assume two-dimensional, steady flow conditions for which x is in the direction along the surface and y is normal to the surface. For each of the boundary layers, we will identify the relevant physical processes and apply the appropriate conservation laws to control volumes of infinitesimal size. Extension of this development to three-dimensional flows may readily be made [2, 3].

6.4.1   The Velocity Boundary Layer

One conservation law that is pertinent to the velocity boundary layer is that matter may neither be created nor destroyed. Stated in the context of the differential control volume of Figure 6.9, this law requires that, for steady flow, *the net rate at which mass enters the control volume* (inflow − outflow) *must equal zero.* Mass enters and leaves the control volume exclusively through

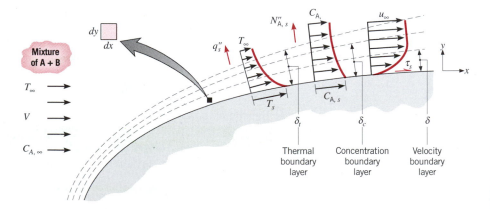

FIGURE 6.8 Development of the velocity, thermal, and concentration boundary layers for an arbitrary surface.

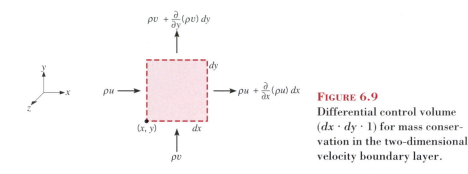

FIGURE 6.9
Differential control volume
(*dx · dy ·* 1) for mass conservation in the two-dimensional
velocity boundary layer.

gross fluid motion. Transport due to such motion is often referred to as *advec-tion.* If one corner of the control volume is located at (x, y), the rate at which mass enters the control volume through the surface perpendicular to x may be expressed as $(\rho u)\, dy$, where ρ is the total mass density ($\rho = \rho_A + \rho_B$) and u is the x component of the *mass average velocity.* The control volume is of unit depth in the z direction. Since ρ and u may vary with x, the rate at which mass leaves the surface at $x + dx$ may be expressed by a Taylor series expansion of the form

$$\left[(\rho u) + \frac{\partial(\rho u)}{\partial x}\, dx \right] dy$$

Using a similar result for the y direction, the conservation of mass requirement becomes

$$(\rho u)\, dy + (\rho v)\, dx - \left[\rho u + \frac{\partial(\rho u)}{\partial x}\, dx \right] dy - \left[\rho v + \frac{\partial(\rho v)}{\partial y}\, dy \right] dx = 0$$

Canceling terms and dividing by $dx\, dy$, we obtain

$$\frac{\partial(\rho u)}{\partial x} + \frac{\partial(\rho v)}{\partial y} = 0 \tag{6.25}$$

Equation 6.25, the *continuity equation,* is a general expression of the *over-all* mass conservation requirement, and it must be satisfied at every point in the velocity boundary layer. The equation applies for a single species fluid, as well as for mixtures in which species diffusion and chemical reactions may be oc-curring.

The second fundamental law that is pertinent to the velocity boundary layer is *Newton's second law of motion.* For a differential control volume in the ve-locity boundary layer, this requirement states that the sum of all forces acting on the control volume must equal the net rate at which momentum leaves the control volume (outflow − inflow).

Two kinds of forces may act on the fluid in the boundary layer: *body forces,* which are proportional to the volume, and *surface forces,* which are pro-portional to area. Gravitational, centrifugal, magnetic, and/or electric fields may contribute to the total body force, and we designate the x and y compo-nents of this force per unit volume of fluid as X and Y, respectively. The surface forces F_s are due to the fluid static pressure as well as to *viscous stresses.* At

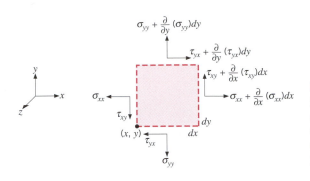

FIGURE 6.10
Normal and shear viscous
stresses for a differential control volume ($dx \cdot dy \cdot 1$) in the
two-dimensional velocity
boundary layer.

any point in the boundary layer, the viscous stress (a force per unit area) may
be resolved into two perpendicular components, which include a *normal stress*
σ_{ii} and a *shear stress* τ_{ij} (Figure 6.10).

A double subscript notation is used to specify the stress components. The
first subscript indicates the surface orientation by providing the direction of its
outward normal, and the second subscript indicates the direction of the force
component. Accordingly, for the x surface of Figure 6.10, the normal stress σ_{xx}
corresponds to a force component normal to the surface, and the shear stress τ_{xy}
corresponds to a force in the y direction along the surface. All the stress components shown are positive in the sense that *both* the surface normal and the force
component are in the same direction. That is, they are both in either the positive
coordinate direction or the negative coordinate direction. By this convention
the normal viscous stresses are *tensile* stresses. In contrast the static pressure
originates from an external force acting on the fluid in the control volume and
is therefore a *compressive* stress.

Several features of the viscous stress should be noted. The associated force
is between adjoining fluid elements and is a natural consequence of the fluid
motion and viscosity. The surface forces of Figure 6.10 are therefore presumed
to act on the fluid within the control volume and are attributed to its interaction
with the surrounding fluid. These stresses would vanish if the fluid velocity, or
the velocity gradient, went to zero. In this respect the normal viscous stresses
(σ_{xx} and σ_{yy}) must not be confused with the static pressure, which does not vanish for zero velocity.

Each of the stresses may change continuously in each of the coordinate directions. Using a Taylor series expansion for the stresses, the *net* surface force
for each of the two directions may be expressed as

$$F_{s,x} = \left(\frac{\partial \sigma_{xx}}{\partial x} - \frac{\partial p}{\partial x} + \frac{\partial \tau_{yx}}{\partial y} \right) dx \, dy \tag{6.26}$$

$$F_{s,y} = \left(\frac{\partial \tau_{xy}}{\partial x} + \frac{\partial \sigma_{yy}}{\partial y} - \frac{\partial p}{\partial y} \right) dx \, dy \tag{6.27}$$

To use Newton's second law, the fluid momentum fluxes for the control
volume must also be evaluated. If we focus on the x direction, the relevant
fluxes are as shown in Figure 6.11. A contribution to the total x-momentum
flux is made by the mass flow in each of the two directions. For example, the
mass flux through the x surface (in the y–z plane) is (ρu), and the corresponding

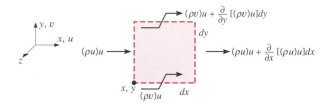

FIGURE 6.11 Momentum fluxes for a differential control volume $(dx \cdot dy \cdot 1)$ in the two-dimensional velocity boundary layer.

x-momentum flux is $(\rho u)u$. Similarly, the x-momentum flux due to mass flow through the y surface (in the x–z plane) is $(\rho v)u$. These fluxes may change in each of the coordinate directions, and the *net* rate at which x momentum leaves the control volume is

$$\frac{\partial[(\rho u)u]}{\partial x} dx(dy) + \frac{\partial[(\rho v)u]}{\partial y} dy(dx)$$

Equating the rate of change in the x momentum of the fluid to the sum of the forces in the x direction, we then obtain

$$\frac{\partial[(\rho u)u]}{\partial x} + \frac{\partial[(\rho v)u]}{\partial y} = \frac{\partial \sigma_{xx}}{\partial x} - \frac{\partial p}{\partial x} + \frac{\partial \tau_{yx}}{\partial y} + X \qquad (6.28)$$

This expression may be put in a more convenient form by expanding the derivatives on the left-hand side and substituting from the continuity equation, Equation 6.25, giving

$$\rho \left(u \frac{\partial u}{\partial x} + v \frac{\partial u}{\partial y} \right) = \frac{\partial}{\partial x} (\sigma_{xx} - p) + \frac{\partial \tau_{yx}}{\partial y} + X \qquad (6.29)$$

A similar expression may be obtained for the y direction and is of the form

$$\rho \left(u \frac{\partial v}{\partial x} + v \frac{\partial v}{\partial y} \right) = \frac{\partial \tau_{xy}}{\partial x} + \frac{\partial}{\partial y} (\sigma_{yy} - p) + Y \qquad (6.30)$$

We should not lose sight of the physics represented by Equations 6.29 and 6.30. The two terms on the left-hand side of each equation represent the *net* rate of momentum flow from the control volume. The terms on the right-hand side account for net viscous and pressure forces, as well as the body force. These equations must be satisfied at each point in the boundary layer, and with Equation 6.25 they may be solved for the velocity field.

Before a solution to the foregoing equations can be obtained, it is necessary to relate the viscous stresses to other flow variables. These stresses are associated with the deformation of the fluid and are a function of the fluid viscosity and velocity gradients. From Figure 6.12 it is evident that a *normal stress* must produce a *linear deformation* of the fluid, whereas a *shear stress* produces an *angular deformation*. Moreover, the magnitude of a stress is proportional to the *rate* at which the deformation occurs. The deformation rate is, in turn, related to

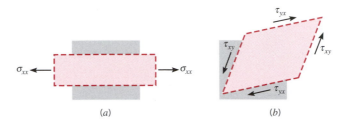

FIGURE 6.12   Deformations of a fluid element due to viscous stresses. (*a*) Linear deformation due to a normal stress. (*b*) Angular deformation due to shear stresses.

the fluid viscosity and to the velocity gradients in the flow. For a *Newtonian fluid*[3] the stresses are proportional to the velocity gradients, where the proportionality constant is the fluid viscosity. Because of its complexity, however, development of the specific relations is left to the literature [2], and we limit ourselves to a presentation of the results. In particular, it has been shown that

$$\sigma_{xx} = 2\mu \frac{\partial u}{\partial x} - \frac{2}{3}\mu \left(\frac{\partial u}{\partial x} + \frac{\partial v}{\partial y} \right) \tag{6.31}$$

$$\sigma_{yy} = 2\mu \frac{\partial v}{\partial y} - \frac{2}{3}\mu \left(\frac{\partial u}{\partial x} + \frac{\partial v}{\partial y} \right) \tag{6.32}$$

$$\tau_{xy} = \tau_{yx} = \mu \left(\frac{\partial u}{\partial y} + \frac{\partial v}{\partial x} \right) \tag{6.33}$$

Substituting Equations 6.31 to 6.33 into Equations 6.29 and 6.30, the *x*- and *y*-momentum equations become

$$\rho \left(u \frac{\partial u}{\partial x} + v \frac{\partial u}{\partial y} \right) = -\frac{\partial p}{\partial x} + \frac{\partial}{\partial x} \left\{ \mu \left[2 \frac{\partial u}{\partial x} - \frac{2}{3} \left(\frac{\partial u}{\partial x} + \frac{\partial v}{\partial y} \right) \right] \right\}$$

$$+ \frac{\partial}{\partial y} \left[\mu \left(\frac{\partial u}{\partial y} + \frac{\partial v}{\partial x} \right) \right] + X \tag{6.34}$$

$$\rho \left(u \frac{\partial v}{\partial x} + v \frac{\partial v}{\partial y} \right) = -\frac{\partial p}{\partial y} + \frac{\partial}{\partial y} \left\{ \mu \left[2 \frac{\partial v}{\partial y} - \frac{2}{3} \left(\frac{\partial u}{\partial x} + \frac{\partial v}{\partial y} \right) \right] \right\}$$

$$+ \frac{\partial}{\partial x} \left[\mu \left(\frac{\partial u}{\partial y} + \frac{\partial v}{\partial x} \right) \right] + Y \tag{6.35}$$

Equations 6.25, 6.34, and 6.35 provide a complete representation of conditions in a two-dimensional velocity boundary layer, and the velocity field in the boundary layer may be determined by solving these equations. Once the veloc-

---

[3]A Newtonian fluid is one for which the shear stress is linearly proportional to the rate of angular deformation. All fluids of interest in this text are Newtonian.

ity field is known, it is a simple matter to obtain the wall shear stress τ_s from Equation 6.15.

6.4.2 The Thermal Boundary Layer

To apply the energy conservation requirement (Equation 1.11a) to a differential control volume in the thermal boundary layer (Figure 6.13), it is necessary to first delineate the relevant physical processes. The energy per unit mass of the fluid includes the thermal internal energy e and the kinetic energy $V^2/2$, where $V^2 \equiv u^2 + v^2$. Accordingly, thermal and kinetic energy are *advected* with the *bulk fluid* motion across the control surfaces, and for the x direction, the *net* rate at which this energy *enters* the control volume is

$$\dot{E}_{adv, x} - \dot{E}_{adv, x+dx} \equiv \rho u \left(e + \frac{V^2}{2} \right) dy - \left\{ \rho u \left(e + \frac{V^2}{2} \right) \right.$$
$$\left. + \frac{\partial}{\partial x} \left[\rho u \left(e + \frac{V^2}{2} \right) \right] dx \right\} dy$$
$$= -\frac{\partial}{\partial x} \left[\rho u \left(e + \frac{V^2}{2} \right) \right] dx \, dy \qquad (6.36)$$

Energy is also transferred across the control surface by *molecular processes*. There may be two contributions: that due to *conduction* and energy transfer due to the *diffusion* of *species* A and B. However, it is only in chemically reacting boundary layers that species diffusion strongly influences thermal conditions. Hence the effect is neglected in this development. For the conduction process, the *net* transfer of energy into the control volume is

$$\dot{E}_{cond, x} - \dot{E}_{cond, x+dx} = -\left(k \frac{\partial T}{\partial x} \right) dy - \left[-k \frac{\partial T}{\partial x} - \frac{\partial}{\partial x} \left(k \frac{\partial T}{\partial x} \right) dx \right] dy$$
$$= \frac{\partial}{\partial x} \left(k \frac{\partial T}{\partial x} \right) dx \, dy \qquad (6.37)$$

Energy may also be transferred to and from the fluid in the control volume by *work* interactions involving the *body* and *surface forces*. The *net* rate

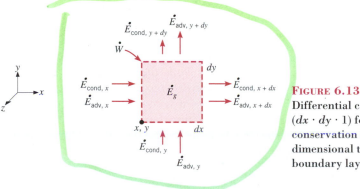

FIGURE 6.13 Differential control volume ($dx \cdot dy \cdot 1$) for energy conservation in the two-dimensional thermal boundary layer.

at which work is done *on* the fluid by forces in the x direction may be expressed as

$$\dot{W}_{\text{net}, x} = (Xu)\, dx\, dy + \frac{\partial}{\partial x}[(\sigma_{xx} - p)u]\, dx\, dy + \frac{\partial}{\partial y}(\tau_{yx}u)\, dx\, dy \quad (6.38)$$

The first term on the right-hand side of Equation 6.38 represents the work done by the body force, and the remaining terms account for the *net* work done by the pressure and viscous forces.

Using Equations 6.36 to 6.38, as well as analogous equations for the y direction, the energy conservation requirement (Equation 1.11a) may be expressed as

$$-\frac{\partial}{\partial x}\left[\rho u\left(e + \frac{V^2}{2}\right)\right] - \frac{\partial}{\partial y}\left[\rho v\left(e + \frac{V^2}{2}\right)\right]$$

$$+ \frac{\partial}{\partial x}\left(k\frac{\partial T}{\partial x}\right) + \frac{\partial}{\partial y}\left(k\frac{\partial T}{\partial y}\right) + (Xu + Yv) - \frac{\partial}{\partial x}(pu) - \frac{\partial}{\partial y}(pv)$$

$$+ \frac{\partial}{\partial x}(\sigma_{xx}u + \tau_{xy}v) + \frac{\partial}{\partial y}(\tau_{yx}u + \sigma_{yy}v) + \dot{q} = 0 \quad (6.39)$$

where $\dot{q}$ is the rate at which energy is generated per unit volume. This expression provides a general form of the energy conservation requirement for the thermal boundary layer.

Because Equation 6.39 represents conservation of *kinetic* and *thermal internal* energy, it is rarely used in solving heat transfer problems. Instead, a more convenient form, which is termed the *thermal energy equation,* is obtained by multiplying Equations 6.29 and 6.30 by u and v, respectively, and subtracting the results from Equation 6.39. After considerable manipulation, it follows that [3]

$$\rho u\frac{\partial e}{\partial x} + \rho v\frac{\partial e}{\partial y} = \frac{\partial}{\partial x}\left(k\frac{\partial T}{\partial x}\right) + \frac{\partial}{\partial y}\left(k\frac{\partial T}{\partial y}\right) - p\left(\frac{\partial u}{\partial x} + \frac{\partial v}{\partial y}\right) + \mu\Phi + \dot{q}$$

$$(6.40)$$

where the term $p(\partial u/\partial x + \partial v/\partial y)$ represents a reversible conversion between kinetic and thermal energy, and $\mu\Phi$, the *viscous dissipation,* is defined as

$$\mu\Phi \equiv \mu\left\{\left(\frac{\partial u}{\partial y} + \frac{\partial v}{\partial x}\right)^2 + 2\left[\left(\frac{\partial u}{\partial x}\right)^2 + \left(\frac{\partial v}{\partial y}\right)^2\right] - \frac{2}{3}\left(\frac{\partial u}{\partial x} + \frac{\partial v}{\partial y}\right)^2\right\} \quad (6.41)$$

The first term on the right-hand side of Equation 6.41 originates from the viscous shear stresses, and the remaining terms arise from the viscous normal stresses. Collectively, the terms account for the rate at which *kinetic energy is irreversibly converted to thermal energy due to viscous effects in the fluid.*

On occasion it is more convenient to work with a formulation of the thermal energy equation that is based on the fluid enthalpy i, rather than its internal energy e. Introducing the definition of the enthalpy

$$i = e + \frac{p}{\rho} \quad (6.42)$$

and substituting from Equation 6.25, Equation 6.40 may be rearranged to yield

$$
\rho u \frac{\partial i}{\partial x} + \rho v \frac{\partial i}{\partial y} = \frac{\partial}{\partial x}\left(k\frac{\partial T}{\partial x}\right) + \frac{\partial}{\partial y}\left(k\frac{\partial T}{\partial y}\right)
$$

$$
+ \left(u\frac{\partial p}{\partial x} + v\frac{\partial p}{\partial y}\right) + \mu\Phi + \dot{q} \tag{6.43}
$$

To express the left-hand side of the thermal energy equation in terms of the temperature, it is necessary to specify the nature of the substance. If, for example, the substance is an *ideal gas*, $di = c_p\,dT$ and Equation 6.43 becomes

$$
\rho c_p\left(u\frac{\partial T}{\partial x} + v\frac{\partial T}{\partial y}\right) = \frac{\partial}{\partial x}\left(k\frac{\partial T}{\partial x}\right) + \frac{\partial}{\partial y}\left(k\frac{\partial T}{\partial y}\right)
$$

$$
+ \left(u\frac{\partial p}{\partial x} + v\frac{\partial p}{\partial y}\right) + \mu\Phi + \dot{q} \tag{6.44}
$$

Alternatively, if the substance is *incompressible*, $c_v = c_p$ and Equation 6.25 reduces to

$$
\frac{\partial u}{\partial x} + \frac{\partial v}{\partial y} = 0 \tag{6.45}
$$

With $de = c_v\,dT = c_p\,dT$, Equation 6.40 then reduces to

$$
\rho c_p\left(u\frac{\partial T}{\partial x} + v\frac{\partial T}{\partial y}\right) = \frac{\partial}{\partial x}\left(k\frac{\partial T}{\partial x}\right) + \frac{\partial}{\partial y}\left(k\frac{\partial T}{\partial y}\right)
$$

$$
+ \mu\Phi + \dot{q} \tag{6.46}
$$

6.4.3   The Concentration Boundary Layer

Since we are considering a binary mixture in which there are species concentration gradients (Figure 6.8), there will be *relative* transport of the species, and *species conservation* must be satisfied at each point in the concentration boundary layer. The pertinent form of the conservation equation may be obtained by identifying the processes that affect the *transport* and *generation* of species A for a differential control volume in the boundary layer.

Consider the control volume of Figure 6.14. Species A may be transported by *advection* (with the mean velocity of the mixture) and by *diffusion* (relative to the mean motion) in each of the coordinate directions. The concentration

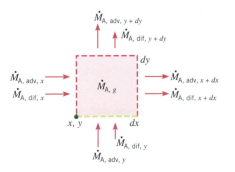

FIGURE 6.14
Differential control volume $(dx \cdot dy \cdot 1)$ for species conservation in the two-dimensional concentration boundary layer.

may also be affected by chemical reactions, and we designate the rate at which the mass of species A is generated per unit volume due to such reactions as $\dot{n}_A$.

The *net* rate at which species A *enters* the control volume due to *advection* in the x direction is

$$\dot{M}_{A,\,adv,\,x} - \dot{M}_{A,\,adv,\,x+dx} = (\rho_A u)\,dy - \left[(\rho_A u) + \frac{\partial(\rho_A u)}{\partial x}dx\right]dy$$

$$= -\frac{\partial(\rho_A u)}{\partial x}dx\,dy \qquad (6.47)$$

Similarly, assuming an incompressible fluid (constant ρ) and using Fick's law (Equation 6.21) to evaluate the diffusion flux, the *net* rate at which species A *enters* the control volume due to *diffusion* in the x direction is

$$\dot{M}_{A,\,dif,\,x} - \dot{M}_{A,\,dif,\,x+dx} = \left(-D_{AB}\frac{\partial \rho_A}{\partial x}\right)dy - \left[\left(-D_{AB}\frac{\partial \rho_A}{\partial x}\right)\right.$$

$$\left.+ \frac{\partial}{\partial x}\left(-D_{AB}\frac{\partial \rho_A}{\partial x}\right)dx\right]dy = \frac{\partial}{\partial x}\left(D_{AB}\frac{\partial \rho_A}{\partial x}\right)dx\,dy \quad (6.48)$$

Expressions similar to Equations 6.47 and 6.48 may be formulated for the y direction.

Referring to Figure 6.14, the species conservation requirement is

$$\dot{M}_{A,\,adv,\,x} - \dot{M}_{A,\,adv,\,x+dx} + \dot{M}_{A,\,adv,\,y} - \dot{M}_{A,\,adv,\,y+dy}$$

$$+ \dot{M}_{A,\,dif,\,x} - \dot{M}_{A,\,dif,\,x+dx} + \dot{M}_{A,\,dif,\,y} - \dot{M}_{A,\,dif,\,y+dy} + \dot{M}_{A,\,g} = 0 \quad (6.49)$$

Substituting from Equations 6.47 and 6.48, as well as from similar forms for the y direction, it follows that

$$\frac{\partial(\rho_A u)}{\partial x} + \frac{\partial(\rho_A v)}{\partial y} = \frac{\partial}{\partial x}\left(D_{AB}\frac{\partial \rho_A}{\partial x}\right) + \frac{\partial}{\partial y}\left(D_{AB}\frac{\partial \rho_A}{\partial y}\right) + \dot{n}_A \qquad (6.50)$$

A more useful form of this equation may be obtained by expanding the terms on the left-hand side and substituting from the overall continuity equation (6.25). With the total mass density ρ assumed to be constant, Equation 6.50 reduces to

$$u\frac{\partial \rho_A}{\partial x} + v\frac{\partial \rho_A}{\partial y} = \frac{\partial}{\partial x}\left(D_{AB}\frac{\partial \rho_A}{\partial x}\right) + \frac{\partial}{\partial y}\left(D_{AB}\frac{\partial \rho_A}{\partial y}\right) + \dot{n}_A \qquad (6.51)$$

or in molar form

$$u\frac{\partial C_A}{\partial x} + v\frac{\partial C_A}{\partial y} = \frac{\partial}{\partial x}\left(D_{AB}\frac{\partial C_A}{\partial x}\right) + \frac{\partial}{\partial y}\left(D_{AB}\frac{\partial C_A}{\partial y}\right) + \dot{N}_A \qquad (6.52)$$

EXAMPLE 6.4

One of the few situations for which *exact* solutions to the convection transfer equations may be obtained involves what is termed *parallel flow*. In this case fluid motion is only in one direction. Consider a special case of parallel flow involving stationary and moving plates of infinite extent separated by a distance L, with the intervening space filled by an incompressible fluid. This situation is referred to as Couette flow and occurs, for example, in a journal bearing.

1. What is the appropriate form of the continuity equation (Equation 6.25)?
2. Beginning with the momentum equation (Equation 6.34), determine the velocity distribution between the plates.
3. Beginning with the energy equation (Equation 6.46), determine the temperature distribution between the plates.
4. Consider conditions for which the fluid is engine oil with $L = 3$ mm. The speed of the moving plate is $U = 10$ m/s, and the temperatures of the stationary and moving plates are $T_0 = 10°C$ and $T_L = 30°C$, respectively. Calculate the heat flux to each of the plates and determine the maximum temperature in the oil.

SOLUTION

Known: Couette flow with heat transfer.

Find:

1. Form of the continuity equation.
2. Velocity distribution.
3. Temperature distribution.
4. Surface heat fluxes and maximum temperature for prescribed conditions.

Schematic:

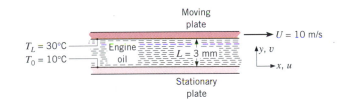

Assumptions:

1. Steady-state conditions.
2. Two-dimensional flow (no variations in z).
3. Incompressible fluid with constant properties.
4. No body forces.
5. No internal energy generation.

Properties: Table A.8, engine oil (20°C): $\rho = 888.2$ kg/m^3, $k = 0.145$ W/m · K, $\nu = 900 \times 10^{-6}$ m^2/s, $\mu = \nu\rho = 0.799$ N · s/m^2.

Analysis:

1. For an incompressible fluid (constant ρ) and parallel flow ($v = 0$), Equation 6.25 reduces to

$$\frac{\partial u}{\partial x} = 0$$

◁

The important implication of this result is that, although depending on y, the x velocity component u is independent of x. It may then be said that the velocity field is *fully developed*.

2. For two-dimensional, steady-state conditions with $v = 0$, $(\partial u/\partial x) = 0$ and $X = 0$, Equation 6.34 reduces to

$$0 = -\frac{\partial p}{\partial x} + \frac{\partial}{\partial y}\left(\mu \frac{\partial u}{\partial y}\right)$$

However, in Couette flow, motion of the fluid is sustained not by the pressure gradient, $\partial p/\partial x$, but by an external force that provides for motion of the top plate relative to the bottom plate. Hence $(\partial p/\partial x) = 0$. Accordingly, with constant viscosity, the x-momentum equation reduces to

$$\frac{\partial^2 u}{\partial y^2} = 0$$

The desired velocity distribution may be obtained by solving this equation. Integrating twice, we obtain

$$u(y) = C_1 y + C_2$$

where C_1 and C_2 are the constants of integration. Applying the boundary conditions

$$u(0) = 0 \qquad u(L) = U$$

it follows that $C_2 = 0$ and $C_1 = U/L$. The velocity distribution is then

$$u(y) = \frac{y}{L} U \qquad\qquad \triangleleft$$

3. The energy equation (6.46) may be simplified for the prescribed conditions. In particular for two-dimensional, steady-state conditions with $v = 0$, $(\partial u/\partial x) = 0$, and $\dot{q} = 0$, it follows that

$$\rho c_p u \frac{\partial T}{\partial x} = \frac{\partial}{\partial x}\left(k\frac{\partial T}{\partial x}\right) + \frac{\partial}{\partial y}\left(k\frac{\partial T}{\partial y}\right) + \mu\left(\frac{\partial u}{\partial y}\right)^2$$

However, because the top and bottom plates are at uniform temperatures, the temperature field must also be fully developed, in which case $(\partial T/\partial x) = 0$. For constant thermal conductivity the appropriate form of the energy equation is then

$$0 = k\frac{\partial^2 T}{\partial y^2} + \mu\left(\frac{\partial u}{\partial y}\right)^2$$

The desired temperature distribution may be obtained by solving this equation. Rearranging and substituting for the velocity distribution,

$$k\frac{d^2 T}{dy^2} = -\mu\left(\frac{du}{dy}\right)^2 = -\mu\left(\frac{U}{L}\right)^2$$

Integrating twice, we obtain

$$T(y) = -\frac{\mu}{2k}\left(\frac{U}{L}\right)^2 y^2 + C_3 y + C_4$$

The constants of integration may be obtained from the boundary conditions

$$T(0) = T_0 \qquad T(L) = T_L$$

in which case

$$C_4 = T_0 \qquad \text{and} \qquad C_3 = \frac{T_L - T_0}{L} + \frac{\mu}{2k}\frac{U^2}{L}$$

and

$$T(y) = T_0 + \frac{\mu}{2k}U^2\left[\frac{y}{L} - \left(\frac{y}{L}\right)^2\right] + (T_L - T_0)\frac{y}{L} \qquad \triangleleft$$

4. Knowing the temperature distribution, the surface heat fluxes may be obtained by applying Fourier's law. Hence

$$q_y'' = -k\frac{dT}{dy} = -k\left[\frac{\mu}{2k}U^2\left(\frac{1}{L} - \frac{2y}{L^2}\right) + \frac{T_L - T_0}{L}\right]$$

At the bottom and top surfaces, respectively, it follows that

$$q_0'' = -\frac{\mu U^2}{2L} - \frac{k}{L}(T_L - T_0) \qquad \text{and} \qquad q_L'' = +\frac{\mu U^2}{2L} - \frac{k}{L}(T_L - T_0)$$

Hence, for the prescribed numerical values,

$$q_0'' = -\frac{0.799 \text{ N} \cdot \text{s/m}^2 \times 100 \text{ m}^2/\text{s}^2}{2 \times 3 \times 10^{-3} \text{ m}} - \frac{0.145 \text{ W/m} \cdot \text{K}}{3 \times 10^{-3} \text{ m}}(30 - 10)°\text{C}$$

$$q_0'' = -13{,}315 \text{ W/m}^2 - 967 \text{ W/m}^2 = -14.3 \text{ kW/m}^2 \qquad \triangleleft$$

$$q_L'' = +13{,}315 \text{ W/m}^2 - 967 \text{ W/m}^2 = 12.3 \text{ kW/m}^2 \qquad \triangleleft$$

The location of the maximum temperature in the oil may be found from the requirement that

$$\frac{dT}{dy} = \frac{\mu}{2k}U^2\left(\frac{1}{L} - \frac{2y}{L^2}\right) + \frac{T_L - T_0}{L} = 0$$

Solving for y it follows that

$$y_{\text{max}} = \left[\frac{k}{\mu U^2}(T_L - T_0) + \frac{1}{2}\right]L$$

or for the prescribed conditions

$$y_{\text{max}} = \left[\frac{0.145 \text{ W/m} \cdot \text{K}}{0.799 \text{ N} \cdot \text{s/m}^2 \times 100 \text{ m}^2/\text{s}^2}(30 - 10)°\text{C} + \frac{1}{2}\right]L = 0.536L$$

Substituting the value of y_{max} into the expression for $T(y)$, it follows that

$$T_{\text{max}} = 89.3°\text{C} \qquad \triangleleft$$

Comments:

1. Given the strong effect of viscous dissipation for the prescribed conditions, the maximum temperature occurs in the oil and there is heat transfer to the

hot, as well as to the cold, plate. The temperature distribution is a function of the velocity of the moving plate, and the effect is shown schematically below.

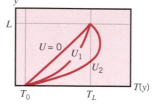

For velocities less than U_1 the maximum temperature corresponds to that of the hot plate. For $U = 0$ there is no viscous dissipation, and the temperature distribution is linear.

2. Recognize that the properties were evaluated at $\bar{T} = (T_L + T_0)/2 = 20°C$, which is *not* a good measure of the average oil temperature. For more precise calculations, the properties should be evaluated at a more appropriate value of the average temperature (e.g., $\bar{T} \approx 55°C$), and the calculations should be repeated.

6.5
Approximations and Special Conditions

The equations of the preceding section provide a complete account of the physical processes that influence conditions in steady, two-dimensional velocity, thermal, and concentration boundary layers. However, it is a rare situation when all of the terms need to be considered, and it is customary to work with simplified forms of the equations. The usual situation is one in which the boundary layer can be characterized as: *incompressible* (ρ is constant), having *constant properties* (k, μ, etc.) and *negligible body forces* ($X = Y = 0$), *nonreacting* ($\dot{n}_A = 0$), and *without energy generation* ($\dot{q} = 0$).

Additional simplifications may be made by invoking what are known as the *boundary layer approximations*. Because the boundary layer thicknesses are typically very small, the following inequalities are known to apply:

$$\left. \begin{array}{l} u \gg v \\[4pt] \dfrac{\partial u}{\partial y} \gg \dfrac{\partial u}{\partial x}, \dfrac{\partial v}{\partial y}, \dfrac{\partial v}{\partial x} \end{array} \right] \quad \begin{array}{l} \text{Velocity} \\ \text{boundary layer} \end{array}$$

$$\left. \dfrac{\partial T}{\partial y} \gg \dfrac{\partial T}{\partial x} \right] \quad \begin{array}{l} \text{Thermal} \\ \text{boundary layer} \end{array}$$

$$\left. \dfrac{\partial C_A}{\partial y} \gg \dfrac{\partial C_A}{\partial x} \right] \quad \begin{array}{l} \text{Concentration} \\ \text{boundary layer} \end{array}$$

That is, the velocity component in the direction along the surface is much larger than that normal to the surface, and gradients normal to the surface are much larger than those along the surface. The normal stresses given by Equations 6.31 and 6.32 are then negligible, and the single relevant shear stress component of Equation 6.33 reduces to

$$\tau_{xy} = \tau_{yx} = \mu\left(\frac{\partial u}{\partial y}\right) \tag{6.53}$$

Moreover, conduction and species diffusion rates for the y direction are much larger than those for the x direction.

Special attention needs to be given to the effect of species transfer on the velocity boundary layer. Recall that velocity boundary layer development is generally characterized by the existence of zero fluid velocity *at the surface*. This condition pertains to the velocity component v normal to the surface, as well as to the velocity component u along the surface. However, if there is simultaneous mass transfer to or from the surface, it is evident that v can no longer be zero at the surface. Nevertheless, for the mass transfer problems of interest in this text, it will be reasonable to assume that $v = 0$, which is equivalent to assuming that mass transfer has a negligible effect on the velocity boundary layer. The assumption is reasonable for problems involving evaporation or sublimation from gas–liquid or gas–solid interfaces, respectively. It is not reasonable, however, for *mass transfer cooling* problems that involve large surface mass transfer rates [4]. In addition we note that with mass transfer, the boundary layer fluid is a binary mixture of species A and B, and its properties should be those of the mixture. However, in all problems of interest, $C_A \ll C_B$ and it is reasonable to assume that the boundary layer properties (such as k, μ, c_p, etc.) are those of species B.

With the foregoing simplifications and approximations, the overall continuity equation (6.25) and the x-momentum equation (6.34) reduce to

$$\frac{\partial u}{\partial x} + \frac{\partial v}{\partial y} = 0 \tag{6.54}$$

$$u\frac{\partial u}{\partial x} + v\frac{\partial u}{\partial y} = -\frac{1}{\rho}\frac{\partial p}{\partial x} + \nu\frac{\partial^2 u}{\partial y^2} \tag{6.55}$$

In addition, from an order-of-magnitude analysis that uses the velocity boundary layer approximations [2], it may be shown that the y-momentum equation (6.35) reduces to

$$\frac{\partial p}{\partial y} = 0 \tag{6.56}$$

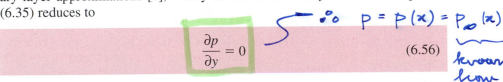
∴ $p = p(x) = p_\infty(x)$

known from the potential flow theory for a given geometry

That is, the *pressure does not vary in the direction normal to the surface.* Hence the pressure in the boundary layer depends only on x and is equal to the pressure in the free stream outside the boundary layer. The form of $p(x)$, which depends on the surface geometry, may then be obtained from a separate consideration of flow conditions in the free stream. Hence, as far as Equation 6.55 is concerned, $(\partial p/\partial x) = (dp/dx)$, and the pressure gradient may be treated as a known quantity.

With the foregoing simplifications, the energy equation (6.46) reduces to

$$u\frac{\partial T}{\partial x} + v\frac{\partial T}{\partial y} = \alpha\frac{\partial^2 T}{\partial y^2} + \frac{\nu}{c_p}\left(\frac{\partial u}{\partial y}\right)^2 \tag{6.57}$$

and the species continuity equation (6.52) becomes

$$u\frac{\partial C_A}{\partial x} + v\frac{\partial C_A}{\partial y} = D_{AB}\frac{\partial^2 C_A}{\partial y^2} \tag{6.58}$$

Note that the last term on the right-hand side of Equation 6.57 is what remains of the viscous dissipation, Equation 6.41. In most situations this term may be neglected relative to those that account for advection (the left-hand side of the equation) and conduction (the first term on the right-hand side). In fact it is only for sonic flows or the high-speed motion of lubricating oils that viscous dissipation may not be neglected.

Equations 6.54, 6.55, 6.57, and 6.58 may be solved to determine the spatial variations of u, v, T, and C_A in the different boundary layers. For incompressible, constant property flow, Equations 6.54 and 6.55 are *uncoupled* from 6.57 and 6.58. That is, Equations 6.54 and 6.55 may be solved for the *velocity field*, $u(x, y)$ and $v(x, y)$, to the exclusion of Equations 6.57 and 6.58. From knowledge of $u(x, y)$, the velocity gradient $(\partial u/\partial y)_{y=0}$ could then be evaluated, and the wall shear stress could be obtained from Equation 6.15. In contrast, through the appearance of u and v in Equations 6.57 and 6.58, the temperature and species concentration are *coupled* to the velocity field. Hence $u(x, y)$ and $v(x, y)$ must be known before Equations 6.57 and 6.58 may be solved for $T(x, y)$ and $C_A(x, y)$. Once $T(x, y)$ and $C_A(x, y)$ have been obtained from such solutions, the convection heat and mass transfer coefficients may be determined from Equations 6.17 and 6.20, respectively. It follows that these coefficients depend strongly on the velocity field.

Because boundary layer solutions generally involve mathematics beyond the scope of this text, our treatment of such solutions will be restricted to the analysis of parallel flow over an isothermal flat plate (Section 7.2 and Appendix E). However, other analytical solutions are discussed in advanced texts on convection [5–7], and detailed boundary layer solutions may be obtained by using numerical (finite-difference or finite-element) techniques [8].

We have not developed the boundary layer equations solely for the purpose of obtaining solutions. In fact, we have been motivated primarily by two other considerations. One major motivation has been to cultivate an appreciation for the different physical processes that occur in the boundary layers. These processes will, of course, affect wall friction, as well as energy and species transfer into the boundary layers. A second motivation arises from the fact that the equations may be used to identify key *boundary layer similarity parameters*, as well as important *analogies* between *momentum*, *heat*, and *mass* transfer.

6.6
Boundary Layer Similarity: The Normalized Convection Transfer Equations

If we examine Equations 6.55, 6.57, and 6.58 more carefully, we note a strong similarity. In fact, if the pressure gradient appearing in Equation 6.55 and the viscous dissipation term of Equation 6.57 are negligible, the three equations are of the same form. *Each equation is characterized by advection terms on the left-hand side and a diffusion term on the right-hand side.* This situation describes *low-speed, forced convection flows,* which are found in many engineering applications and will occupy much of our attention in this text. Implications of this similarity may be developed in a rational manner by first *nondimensionalizing* the governing equations.

6.6.1 Boundary Layer Similarity Parameters

The boundary layer equations are normalized by first defining dimensionless independent variables of the forms

$$x^* \equiv \frac{x}{L} \quad \text{and} \quad y^* \equiv \frac{y}{L} \tag{6.59}$$

where *L* is some *characteristic length* for the surface of interest (e.g., the length of a flat plate). Moreover, dependent dimensionless variables may also be defined as

$$u^* \equiv \frac{u}{V} \quad \text{and} \quad v^* \equiv \frac{v}{V} \tag{6.60}$$

where *V* is the velocity upstream of the surface (Figure 6.8), and as

$$T^* \equiv \frac{T - T_s}{T_\infty - T_s} \tag{6.61}$$

$$C_A^* \equiv \frac{C_A - C_{A,s}}{C_{A,\infty} - C_{A,s}} \tag{6.62}$$

Equations 6.59 to 6.62 may be substituted into Equations 6.55, 6.57, and 6.58 to obtain the dimensionless forms of the conservation equations shown in Table 6.1. Note that viscous dissipation has been neglected and that $p^* \equiv (p/\rho V^2)$ is a dimensionless pressure. The boundary conditions required to solve the equations are also shown in the table.

From the form of Equations 6.63 to 6.65, three *similarity parameters* may be inferred. Similarity parameters are important because they allow us to apply results obtained for a surface experiencing one set of conditions to *geometrically similar* surfaces experiencing entirely different conditions. These conditions may vary, for example, with the nature of the fluid, the fluid velocity, and/or with the size of the surface (as determined by *L*).

TABLE 6.1 The convection transfer equations and their boundary conditions in nondimensional form

Boundary Layer	Conservation Equation	Boundary Conditions Wall	Boundary Conditions Free Stream	Similarity Parameter(s)
Velocity	$u^*\dfrac{\partial u^*}{\partial x^*} + v^*\dfrac{\partial u^*}{\partial y^*}$ $= -\dfrac{dp^*}{dx^*} + \dfrac{\nu}{VL}\dfrac{\partial^2 u^*}{\partial y^{*2}}$ (6.63)	$u^*(x^*, 0) = 0$ $v^*(x^*, 0) = 0$	$u^*(x^*, \infty) = \dfrac{u_\infty(x^*)}{V}$ (6.66)	Re_L
Thermal	$u^*\dfrac{\partial T^*}{\partial x^*} + v^*\dfrac{\partial T^*}{\partial y^*} = \dfrac{\alpha}{VL}\dfrac{\partial^2 T^*}{\partial y^{*2}}$ (6.64)	$T^*(x^*, 0) = 0$	$T^*(x^*, \infty) = 1$ (6.67)	Re_L, Pr
Concentration	$u^*\dfrac{\partial C_A^*}{\partial x^*} + v^*\dfrac{\partial C_A^*}{\partial y^*} = \dfrac{D_{AB}}{VL}\dfrac{\partial^2 C_A^*}{\partial y^{*2}}$ (6.65)	$C_A^*(x^*, 0) = 0$	$C_A^*(x^*, \infty) = 1$ (6.68)	Re_L, Sc

Beginning with Equation 6.63, we note that the quantity ν/VL is a *dimensionless group* whose reciprocal is termed the Reynolds number.

Reynolds Number:

$$Re_L \equiv \frac{VL}{\nu} \tag{6.69}$$

From Equation 6.64 we also note that the term α/VL is a dimensionless group that may be expressed as $(\nu/VL)(\alpha/\nu) = (Re_L)^{-1}(\alpha/\nu)$. The ratio of properties, α/ν, is also dimensionless and its reciprocal is referred to as the Prandtl number.

Prandtl Number:

$$Pr \equiv \frac{\nu}{\alpha} \tag{6.70}$$

Finally, from the species continuity equation, Equation 6.65, we note that the term D_{AB}/VL is equivalent to $(\nu/VL)(D_{AB}/\nu) = (Re_L)^{-1}(D_{AB}/\nu)$. The ratio D_{AB}/ν is dimensionless and its reciprocal is termed the Schmidt number.

Schmidt Number:

$$Sc \equiv \frac{\nu}{D_{AB}} \tag{6.71}$$

Using Equations 6.69 through 6.71 with the boundary layer equations, Equations 6.63 to 6.65, and including the dimensionless form of the continuity equation (6.54), the complete set of boundary layer equations becomes

$$\frac{\partial u^*}{\partial x^*} + \frac{\partial v^*}{\partial y^*} = 0 \tag{6.72}$$

$$u^* \frac{\partial u^*}{\partial x^*} + v^* \frac{\partial u^*}{\partial y^*} = -\frac{dp^*}{dx^*} + \frac{1}{Re_L} \frac{\partial^2 u^*}{\partial y^{*2}} \qquad (6.73)$$

$$u^* \frac{\partial T^*}{\partial x^*} + v^* \frac{\partial T^*}{\partial y^*} = \frac{1}{Re_L \, Pr} \frac{\partial^2 T^*}{\partial y^{*2}} \qquad (6.74)$$

$$u^* \frac{\partial C_A^*}{\partial x^*} + v^* \frac{\partial C_A^*}{\partial y^*} = \frac{1}{Re_L \, Sc} \frac{\partial^2 C_A^*}{\partial y^{*2}} \qquad (6.75)$$

6.6.2 Functional Form of the Solutions

The preceding equations are extremely useful from the standpoint of suggesting how important boundary layer results may be simplified and generalized. The momentum equation (6.73) suggests that, although conditions in the velocity boundary layer depend on the fluid properties ρ and μ, the velocity V and the length scale L, this dependence may be simplified by grouping these variables in the form of the Reynolds number. We therefore anticipate that the solution to Equation 6.73 will be of the functional form

$$u^* = f_1 \left(x^*, y^*, Re_L, \frac{dp^*}{dx^*} \right) \qquad (6.76)$$

Note that the pressure distribution $p^*(x^*)$ depends on the surface geometry and may be obtained independently by considering flow conditions in the free stream. Hence the appearance of dp^*/dx^* in Equation 6.76 represents the influence of geometry on the velocity distribution.

From Equation 6.15, the shear stress at the surface, $y^* = 0$, may be expressed as

$$\tau_s = \mu \frac{\partial u}{\partial y} \bigg|_{y=0} = \left(\frac{\mu V}{L} \right) \frac{\partial u^*}{\partial y^*} \bigg|_{y^*=0}$$

and from Equations 6.14 and 6.69 it follows that the friction coefficient is

$$C_f = \frac{\tau_s}{\rho V^2 / 2} = \frac{2}{Re_L} \frac{\partial u^*}{\partial y^*} \bigg|_{y^*=0} \qquad (6.77)$$

From Equation 6.76 we also know that

$$\frac{\partial u^*}{\partial y^*} \bigg|_{y^*=0} = f_2 \left(x^*, Re_L, \frac{dp^*}{dx^*} \right)$$

Hence *for a prescribed geometry* Equation 6.77 may be expressed as

$$C_f = \frac{2}{Re_L} f_2(x^*, Re_L) \qquad (6.78)$$

The significance of this result should not be overlooked. Equation 6.78 states that the friction coefficient, a dimensionless parameter of considerable importance to the engineer, may be expressed exclusively in terms of a dimensionless space coordinate and the Reynolds number. Hence for a prescribed geometry we expect the function that relates C_f to x^* and Re_L to be *universally* applicable. That is, we expect it to apply to different fluids and over a wide range of values for V and L.

Similar results may be obtained for the convection coefficients of heat and mass transfer. Intuitively, we might anticipate that h depends on the fluid properties (k, c_p, μ, and ρ), the fluid velocity V, the length scale L, and the surface geometry. However, Equation 6.74 suggests the manner in which this dependence may be simplified. In particular, the solution to this equation may be expressed in the form

$$T^* = f_3\left(x^*, y^*, Re_L, Pr, \frac{dp^*}{dx^*}\right) \tag{6.79}$$

where the dependence on dp^*/dx^* originates from the influence of the fluid motion (u^* and v^*) on the thermal conditions. Once again the term dp^*/dx^* represents the effect of surface geometry. From the definition of the convection coefficient, Equation 6.17, and the dimensionless variables, Equations 6.59 and 6.61, we also obtain

$$h = -\frac{k_f}{L}\frac{(T_\infty - T_s)}{(T_s - T_\infty)}\frac{\partial T^*}{\partial y^*}\bigg|_{y^*=0} = +\frac{k_f}{L}\frac{\partial T^*}{\partial y^*}\bigg|_{y^*=0}$$

This expression suggests defining a dependent dimensionless parameter termed the Nusselt number.

Nusselt Number:

$$Nu \equiv \frac{hL}{k_f} = +\frac{\partial T^*}{\partial y^*}\bigg|_{y^*=0} \tag{6.80}$$

This parameter is equal to the dimensionless temperature gradient at the surface, and it provides a measure of the convection heat transfer occurring at the surface. From Equation 6.79 it follows that, *for a prescribed geometry,*

$$Nu = f_4(x^*, Re_L, Pr) \tag{6.81}$$

The Nusselt number is to the thermal boundary layer what the friction coefficient is to the velocity boundary layer. Equation 6.81 implies that for a given geometry, the Nusselt number must be some *universal function* of x^*, Re_L, and Pr. If this function were known, it could be used to compute the value of Nu for different fluids and for different values of V and L. From knowledge of Nu, the local convection coefficient h may be found and the *local* heat flux may then be computed from Equation 6.1. Moreover, since the *average* heat transfer coefficient is obtained by integrating over the surface of the body, it must be independent of the spatial variable x^*. Hence the functional dependence of the *average* Nusselt number is

$$\overline{Nu} = \frac{\overline{h}L}{k_f} = f_5(Re_L, Pr) \tag{6.82}$$

Similarly, it may be argued that, for mass transfer in a gas flow over an evaporating liquid or a sublimating solid, the convection mass transfer coefficient h_m depends on the properties D_{AB}, ρ, and μ, the velocity V, and the characteristic length L. However, Equation 6.75 suggests that this dependence may be simplified. The solution to this equation must be of the form

$$C_A^* = f_6\left(x^*, y^*, Re_L, Sc, \frac{dp^*}{dx^*}\right) \tag{6.83}$$

where the dependence on dp^*/dx^* again originates from the influence of the fluid motion. From the definition of the convection coefficient, Equation 6.20, and the dimensionless variables, Equation 6.59 and 6.62, we know that

$$h_m = -\frac{D_{AB}}{L}\frac{(C_{A,\infty} - C_{A,s})}{(C_{A,s} - C_{A,\infty})}\frac{\partial C_A^*}{\partial y^*}\bigg|_{y^*=0} = +\frac{D_{AB}}{L}\frac{\partial C_A^*}{\partial y^*}\bigg|_{y^*=0}$$

Hence we may define a dependent dimensionless parameter termed the Sherwood number (Sh).

Sherwood Number:

$$Sh \equiv \frac{h_m L}{D_{AB}} = +\frac{\partial C_A^*}{\partial y^*}\bigg|_{y^*=0} \tag{6.84}$$

This parameter is equal to the dimensionless concentration gradient at the surface, and it provides a measure of the convection mass transfer occurring at the surface. From Equation 6.83 it follows that, *for a prescribed geometry,*

$$Sh = f_7(x^*, Re_L, Sc) \tag{6.85}$$

The Sherwood number is to the concentration boundary layer what the Nusselt number is to the thermal boundary layer, and Equation 6.85 implies that it must be a universal function of x^*, Re_L, and Sc. As for the Nusselt number, it is also possible to work with an *average* Sherwood number that depends on only Re_L and Sc.

$$\overline{Sh} = \frac{\bar{h}_m L}{D_{AB}} = f_8(Re_L, Sc) \tag{6.86}$$

From the foregoing development we have obtained the relevant dimensionless parameters for low-speed, forced-convection boundary layers. We have done so by nondimensionalizing the differential equations that describe the physical processes occurring within the boundary layers. An alternative approach could have involved the use of dimensional analysis in the form of the Buckingham pi theorem [9]. However, the success of this method depends on one's ability to select, largely from intuition, the various parameters that influence a problem. For example, knowing beforehand that $\bar{h} = f(k, c_p, \rho, \mu, V, L)$, one could use the Buckingham pi theorem to obtain Equation 6.82. However, having begun with the differential form of the conservation equations, we have eliminated the guesswork and have established the similarity parameters in a rigorous fashion.

The value of an expression such as Equation 6.82 should be fully appreciated. It states that convection heat transfer results, whether obtained theoretically or experimentally, can be represented in terms of three dimensionless groups, instead of the original seven parameters. The convenience afforded by such simplifications is evident. Moreover, once the form of the functional dependence of Equation 6.82 has been obtained for a particular surface geometry, let us say from laboratory measurements, it is known to be *universally* applicable. By this we mean that it may be applied for different fluids, velocities, and length scales, as long as the assumptions implicit in the originating boundary layer equations remain valid (e.g., negligible viscous dissipation and body forces).

EXAMPLE 6.5

Experimental tests on a portion of the turbine blade shown indicate a heat flux to the blade of $q'' = 95{,}000 \text{ W/m}^2$. To maintain a steady-state surface temperature of 800°C, heat transferred to the blade is removed by circulating a coolant inside the blade.

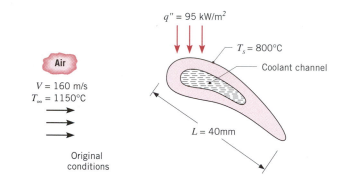

1. Determine the heat flux to the blade if its temperature is reduced to 700°C by increasing the coolant flow.

2. Determine the heat flux at the same dimensionless location for a similar turbine blade having a chord length of $L = 80$ mm, when the blade operates in an airflow at $T_\infty = 1150°C$ and $V = 80$ m/s, with $T_s = 800°C$.

SOLUTION

Known: Operating conditions of an internally cooled turbine blade.

Find:

1. Heat flux to the blade when the surface temperature is reduced.

2. Heat flux to a larger turbine blade of the same shape with reduced air velocity.

Schematic:

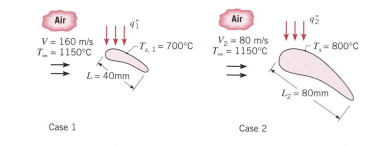

Assumptions:

1. Steady-state conditions.
2. Constant air properties.

Analysis:

1. From Equation 6.81 it follows that, for the prescribed geometry,

$$Nu = \frac{hL}{k} = f_4(x^*, Re_L, Pr)$$

Hence, since there is no change in x^*, Re_L, or Pr associated with a change in T_s for constant properties, the local Nusselt number is unchanged. Moreover, since L and k are also unchanged, the local convection coefficient remains the same. The desired heat flux for case 1 may then be obtained from Newton's law of cooling

$$q_1'' = h_1(T_\infty - T_s)_1$$

where

$$h_1 = h = \frac{q''}{(T_\infty - T_s)}$$

Hence

$$q_1'' = \frac{q''(T_\infty - T_s)_1}{(T_\infty - T_s)} = 95{,}000 \text{ W/m}^2 \frac{(1150 - 700)°\text{C}}{(1150 - 800)°\text{C}} = 122{,}000 \text{ W/m}^2 \quad \triangleleft$$

2. To determine the heat flux associated with the larger blade and the reduced airflow (case 2), we first note that, although L has increased by a factor of 2, the velocity has decreased by the same factor and the Reynolds number has not changed. That is,

$$Re_{L,2} = \frac{V_2 L_2}{\nu} = \frac{VL}{\nu} = Re_L$$

Accordingly, since x^* and Pr are also unchanged, the local Nusselt number remains the same.

$$Nu_2 = Nu$$

Because the characteristic length is different, however, the convection coefficient changes, where

$$\frac{h_2 L_2}{k} = \frac{hL}{k} \qquad \text{or} \qquad h_2 = h\frac{L}{L_2} = \frac{q''}{(T_\infty - T_s)}\frac{L}{L_2}$$

The heat flux is then

$$q_2'' = h_2(T_\infty - T_s) = q''\frac{(T_\infty - T_s)}{(T_\infty - T_s)}\frac{L}{L_2}$$

$$q_2'' = 95{,}000 \text{ W/m}^2 \times \frac{0.04 \text{ m}}{0.08 \text{ m}} = 47{,}500 \text{ W/m}^2 \quad \triangleleft$$

Comments: If the Reynolds number for the two situations of part 2 differed, that is, $Re_{L,2} \neq Re_L$, the heat flux q_2'' could only be obtained if the particular form of the function f_4 were known. Such forms are provided for many different shapes in subsequent chapters.

6.7
Physical Significance of the Dimensionless Parameters

All of the foregoing dimensionless parameters have physical interpretations that relate to conditions in the boundary layers. Consider the *Reynolds number Re* (Equation 6.69), which may be interpreted as the *ratio of inertia to viscous forces* in the velocity boundary layer. For a differential control volume in this boundary layer, inertia forces are associated with an increase in the momentum flux of fluid moving through the control volume. From Equation 6.28, it is evident that these forces are of the form $\partial[(\rho u)u]/\partial x$, in which case an order-of-magnitude approximation gives $F_I \approx \rho V^2/L$. Similarly, the net shear force is of the form $\partial \tau_{yx}/\partial y = \partial[\mu(\partial u/\partial y)]/\partial y$ and may be approximated as $F_s \approx \mu V/L^2$. Therefore the ratio of forces is

$$\frac{F_I}{F_s} \approx \frac{\rho V^2/L}{\mu V/L^2} = \frac{\rho V L}{\mu} = Re_L$$

We therefore expect inertia forces to dominate for large values of Re and viscous forces to dominate for small Re.

There are several important implications of this result. Recall that the Reynolds number determines the existence of laminar or turbulent flow. In any flow there exist small disturbances that can be amplified to produce turbulent conditions. For small Re, however, viscous forces are sufficiently large relative to inertia forces to prevent this amplification. Hence laminar flow is maintained. But, with increasing Re, viscous effects become progressively less important relative to inertia effects, and small disturbances may be amplified to a point where transition occurs. We should also expect the magnitude of the Reynolds number to influence the velocity boundary layer thickness δ. With increasing Re at a fixed location on a surface, we expect viscous forces to become less influential relative to inertia forces. Hence the effects of viscosity do not penetrate as far into the free stream, and the value of δ diminishes.

The physical interpretation of the *Prandtl number* follows from its definition as a ratio of the momentum diffusivity ν to the thermal diffusivity α. The Prandtl number provides a *measure of the relative effectiveness of momentum and energy transport by diffusion in the velocity and thermal boundary layers,* respectively. From Table A.4 we see that the Prandtl number of gases is near unity, in which case energy and momentum transfer by diffusion are comparable. In a liquid metal (Table A.7), $Pr \ll 1$ and the energy diffusion rate greatly exceeds the momentum diffusion rate. The opposite is true for oils (Table A.5), for which $Pr \gg 1$. From this interpretation it follows that the value of Pr strongly influences the relative growth of the velocity and thermal boundary layers. In fact for laminar boundary layers (in which transport by diffusion is *not* overshadowed by turbulent mixing) it is reasonable to expect that

$$\frac{\delta}{\delta_t} \approx Pr^n \qquad \text{ASSUME } n = 1/3 \qquad (6.87)$$

where n is a positive exponent. Hence for a gas $\delta_t \approx \delta$; for a liquid metal $\delta_t \gg \delta$; for an oil $\delta_t \ll \delta$.

Similarly, the *Schmidt number,* which is defined by Equation 6.71, provides *a measure of the relative effectiveness of momentum and mass transport by diffusion in the velocity and concentration boundary layers,* respectively. For convection mass transfer in laminar flows, it therefore determines the relative velocity and concentration boundary layer thicknesses, where

$$\frac{\delta}{\delta_c} \approx Sc^n \qquad n = 1/3 \qquad (6.88)$$

Another parameter, which is related to Pr and Sc, is the *Lewis number (Le).* It is defined as

$$Le = \frac{\alpha}{D_{AB}} = \frac{Sc}{Pr} \qquad (6.89)$$

and is relevant to any situation involving simultaneous heat and mass transfer by convection. From Equations 6.87 to 6.89 it then follows that

$$\frac{\delta_t}{\delta_c} \approx Le^n \qquad n = 1/3 \qquad (6.90)$$

The Lewis number is therefore a measure of the relative thermal and concentration boundary layer thicknesses. For most applications it is reasonable to assume a value of $n = \frac{1}{3}$ in Equations 6.87, 6.88, and 6.90.

Table 6.2 lists the dimensionless groups that appear frequently in the heat and mass transfer literature. The list includes groups already considered, as well as those yet to be introduced for special conditions. As a new group is confronted, its definition and interpretation should be committed to memory. Note that the *Grashof number* provides a measure of the ratio of buoyancy forces to viscous forces in the velocity boundary layer. Its role in free convection (Chapter 9) is much the same as that of the Reynolds number in forced convection. The *Eckert number* provides a measure of the kinetic energy of the flow relative to the enthalpy difference across the thermal boundary layer. It plays an important role in high-speed flows for which viscous dissipation is significant. Note also that, although similar in form, the Nusselt and Biot numbers differ in both definition and interpretation. Whereas the Nusselt number is defined in terms of the thermal conductivity of the fluid, the Biot number is based on the solid thermal conductivity, Equation 5.9.

TABLE 6.2   Selected dimensionless groups of heat and mass transfer

Group	Definition	Interpretation
Biot number (Bi)	$\dfrac{hL}{k_s}$	Ratio of the internal thermal resistance of a solid to the boundary layer thermal resistance.
Mass transfer Biot number (Bi_m)	$\dfrac{h_m L}{D_{AB}}$	Ratio of the internal species transfer resistance to the boundary layer species transfer resistance.
Bond number (Bo)	$\dfrac{g(\rho_l - \rho_v)L^2}{\sigma}$	Ratio of gravitational and surface tension forces.
Coefficient of friction (C_f)	$\dfrac{\tau_s}{\rho V^2/2}$	Dimensionless surface shear stress.
Eckert number (Ec)	$\dfrac{V^2}{c_p(T_s - T_\infty)}$	Kinetic energy of the flow relative to the boundary layer enthalpy difference.
Fourier number (Fo)	$\dfrac{\alpha t}{L^2}$	Ratio of the heat conduction rate to the rate of thermal energy storage in a solid. Dimensionless time.
Mass transfer Fourier number (Fo_m)	$\dfrac{D_{AB} t}{L^2}$	Ratio of the species diffusion rate to the rate of species storage. Dimensionless time.
Friction factor (f)	$\dfrac{\Delta p}{(L/D)(\rho u_m^2/2)}$	Dimensionless pressure drop for internal flow.
Grashof number (Gr_L)	$\dfrac{g\beta(T_s - T_\infty)L^3}{\nu^2}$	Ratio of buoyancy to viscous forces.
Colburn j factor (j_H)	$St\,Pr^{2/3}$	Dimensionless heat transfer coefficient.
Colburn j factor (j_m)	$St_m\,Sc^{2/3}$	Dimensionless mass transfer coefficient.
Jakob number (Ja)	$\dfrac{c_p(T_s - T_{sat})}{h_{fg}}$	Ratio of sensible to latent energy absorbed during liquid–vapor phase change.
Lewis number (Le)	$\dfrac{\alpha}{D_{AB}}$	Ratio of the thermal and mass diffusivities.
Nusselt number (Nu_L)	$\dfrac{hL}{k_f}$	Dimensionless temperature gradient at the surface.
Peclet number (Pe_L)	$\dfrac{VL}{\alpha} = Re_L\,Pr$	Dimensionless independent heat transfer parameter.
Prandtl number (Pr)	$\dfrac{c_p \mu}{k} = \dfrac{\nu}{\alpha}$	Ratio of the momentum and thermal diffusivities.
Reynolds number (Re_L)	$\dfrac{VL}{\nu}$	Ratio of the inertia and viscous forces.
Schmidt number (Sc)	$\dfrac{\nu}{D_{AB}}$	Ratio of the momentum and mass diffusivities.
Sherwood number (Sh_L)	$\dfrac{h_m L}{D_{AB}}$	Dimensionless concentration gradient at the surface.
Stanton number (St)	$\dfrac{h}{\rho V c_p} = \dfrac{Nu_L}{Re_L\,Pr}$	Modified Nusselt number.

TABLE 6.2 *Continued*

Group	Definition	Interpretation
Mass transfer Stanton number (St_m)	$\dfrac{h_m}{V} = \dfrac{Sh_L}{Re_L \, Sc}$	Modified Sherwood number.
Weber number (We)	$\dfrac{\rho V^2 L}{\sigma}$	Ratio of inertia to surface tension forces.

6.8
Boundary Layer Analogies

As engineers, our interest in boundary layer behavior is directed principally toward the dimensionless parameters C_f, Nu, and Sh. From knowledge of these parameters, we may compute the wall shear stress and the convection heat and mass transfer rates. It is therefore understandable that expressions that relate C_f, Nu, and Sh to each other can be useful tools in convection analysis. Such expressions are available in the form of *boundary layer analogies*.

6.8.1 The Heat and Mass Transfer Analogy

If two or more processes are governed by dimensionless equations of the same form, the processes are said to be *analogous*. Clearly then, from Equations 6.64 and 6.65 and the boundary conditions, Equations 6.67 and 6.68, of Table 6.1, convection heat and mass transfer are analogous. Each of the differential equations is composed of advection and diffusion terms of the same form. Moreover, as shown in Equations 6.74 and 6.75, each equation is related to the velocity field through Re_L, and the parameters Pr and Sc assume analogous roles. One implication of this analogy is that dimensionless relations that govern thermal boundary layer behavior must be the same as those that govern the concentration boundary layer. Hence the boundary layer temperature and concentration profiles must be of the same functional form.

Recalling the discussion of Section 6.6.2, features of which are summarized in Table 6.3, an important result of the heat and mass transfer analogy may be obtained. From the foregoing paragraph, it follows that f_3 of Equation 6.79 must be of the same form as f_6 of Equation 6.83. From Equations 6.80 and 6.84 it then follows that the dimensionless temperature and concentration gradients evaluated at the surface, and therefore the values of Nu and Sh, are analogous. That is, f_4 of Equation 6.81 is of the same form as f_7 of Equation 6.85. Similarly, expressions for the average Nusselt and Sherwood numbers, which involve the functions f_5 and f_8 of Equations 6.82 and 6.86, respectively, are also of the same form. *Accordingly, heat and mass transfer relations for a particular geometry are interchangeable.* If, for example, one has performed a set of heat transfer experiments to determine the form of f_4 for a particular surface geometry, the results may be used for convection mass transfer involving the same geometry, simply by replacing Nu with Sh and Pr with Sc.

TABLE 6.3   Functional relations pertinent to the boundary layer analogies

Fluid Flow	Heat Transfer	Mass
$u^* = f_1\left(x^*, y^*, Re_L, \dfrac{dp^*}{dx^*}\right)$ (6.76)	$T^* = f_3\left(x^*, y^*, Re_L, Pr, \dfrac{dp^*}{dx^*}\right)$ (6.79)	$C_A^* = f_6\left(x^*, y^*, Re_L, Sc, \dfrac{dp^*}{dx^*}\right)$ (6.83)
$C_f = \dfrac{2}{Re_L}\dfrac{\partial u^*}{\partial y^*}\Big|_{y^*=0}$ (6.77)	$Nu = \dfrac{hL}{k} = +\dfrac{\partial T^*}{\partial y^*}\Big|_{y^*=0}$ (6.80)	$Sh = \dfrac{h_m L}{D_{AB}} = +\dfrac{\partial C_A^*}{\partial y^*}\Big|_{y^*=0}$ (6.84)
$C_f = \dfrac{2}{Re_L} f_2(x^*, Re_L)$ (6.78)	$Nu = f_4(x^*, Re_L, Pr)$ (6.81)	$Sh = f_7(x^*, Re_L, Sc)$ (6.85)
	$\overline{Nu} = f_5(Re_L, Pr)$ (6.82)	$\overline{Sh} = f_8(Re_L, Sc)$ (6.86)

The analogy may also be used to directly relate the two convection coefficients. In subsequent chapters we will find that Nu and Sh are generally proportional to Pr^n and Sc^n, respectively, where n is a positive exponent less than 1. Anticipating this dependence, we use Equations 6.81 and 6.85 to obtain

$$Nu = f_4'(x^*, Re_L)Pr^n \qquad \text{and} \qquad Sh = f_7'(x^*, Re_L)Sc^n$$

in which case

$$\frac{Nu}{Pr^n} = f_4'(x^*, Re_L) = f_7'(x^*, Re_L) = \frac{Sh}{Sc^n} \tag{6.91}$$

Substituting from Equations 6.80 and 6.84 we then obtain

$$\frac{hL/k}{Pr^n} = \frac{h_m L/D_{AB}}{Sc^n}$$

or, from Equation 6.89,

$$\frac{h}{h_m} = \frac{k}{D_{AB}Le^n} = \rho c_p Le^{1-n} \tag{6.92}$$

This result may often be used in determining one convection coefficient, for example, h_m, from knowledge of the other coefficient. The same relation may be applied to the average coefficients $\bar{h}$ and $\bar{h}_m$, and it may be used in turbulent, as well as laminar, flow. For most applications it is reasonable to assume a value of $n = \frac{1}{3}$.

EXAMPLE 6.6

A solid of arbitrary shape is suspended in atmospheric air having a free stream temperature and velocity of 20°C and 100 m/s, respectively. The solid has a characteristic length of 1 m, and its surface is maintained at 80°C. Under these conditions measurements of the heat flux at a particular point (x^*) on the surface and of the temperature in the boundary layer above this point (x^*, y^*) reveal values of 10^4 W/m^2 and 60°C, respectively. A mass transfer operation is to be effected for a second solid having the same shape but a characteristic length of 2 m. In particular, a thin film of water on the solid is to be evaporated in dry atmospheric air having a free stream velocity of 50 m/s, with the air and the solid both at a temperature of 50°C. What are the molar concentration and the species molar flux of the water vapor at a location (x^*, y^*) corresponding to the point at which the temperature and heat flux measurements were made in the first case?

SOLUTION

Known: A boundary layer temperature and heat flux at a location on a solid in an airstream of prescribed temperature and velocity.

Find: Water vapor concentration and flux associated with the same location on a larger surface of the same shape.

Schematic:

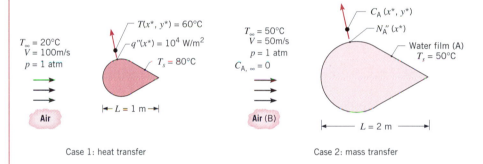

Case 1: heat transfer

Case 2: mass transfer

Assumptions:

1. Steady-state, two-dimensional, incompressible boundary layer behavior; constant properties.
2. Boundary layer approximations are valid.
3. Negligible viscous dissipation.
4. Mole fraction of water vapor in concentration boundary layer is much less than unity.

Properties: Table A.4, air (50°C): $\nu = 18.2 \times 10^{-6}$ m^2/s, $k = 28 \times 10^{-3}$ W/m · K, $Pr = 0.70$. Table A.6, saturated water vapor (50°C): $\rho_{A,\,\text{sat}} = v_g^{-1} = 0.082$ kg/m^3. Table A.8, water vapor–air (50°C): $D_{AB} \approx 0.26 \times 10^{-4}$ m^2/s.

Analysis: The desired molar concentration and flux may be determined by invoking the analogy between heat and mass transfer. From Equations 6.79 and 6.83, we know that

$$T^* \equiv \frac{T - T_s}{T_\infty - T_s} = f_3\left(x^*, y^*, Re_L, Pr, \frac{dp^*}{dx^*}\right)$$

and

$$C_A^* \equiv \frac{C_A - C_{A,s}}{C_{A,\infty} - C_{A,s}} = f_6\left(x^*, y^*, Re_L, Sc, \frac{dp^*}{dx^*}\right)$$

However, for case 1

$$Re_{L,1} = \frac{V_1 L_1}{\nu} = \frac{100 \text{ m/s} \times 1 \text{ m}}{18.2 \times 10^{-6} \text{ m}^2/\text{s}} = 5.5 \times 10^6, \qquad Pr = 0.70$$

while for case 2

$$Re_{L,2} = \frac{V_2 L_2}{\nu} = \frac{50 \text{ m/s} \times 2 \text{ m}}{18.2 \times 10^{-6} \text{ m}^2/\text{s}} = 5.5 \times 10^6$$

$$Sc = \frac{\nu}{D_{AB}} = \frac{18.2 \times 10^{-6} \text{ m}^2/\text{s}}{26 \times 10^{-6} \text{ m}^2/\text{s}} = 0.70$$

Since $Re_{L,1} = Re_{L,2}$, $Pr = Sc$, $x_1^* = x_2^*$, $y_1^* = y_2^*$, and the surface geometries are the same, it follows that $f_3 = f_6$. Hence

$$\frac{C_A(x^*, y^*) - C_{A,s}}{C_{A,\infty} - C_{A,s}} = \frac{T(x^*, y^*) - T_s}{T_\infty - T_s} = \frac{60 - 80}{20 - 80} = 0.33$$

or, with $C_{A,\infty} = 0$,

$$C_A(x^*, y^*) = C_{A,s}(1 - 0.33) = 0.67 C_{A,s}$$

With

$$C_{A,s} = C_{A,\text{sat}}(50°C) = \frac{\rho_{A,\text{sat}}}{\mathcal{M}_A} = \frac{0.082 \text{ kg/m}^3}{18 \text{ kg/kmol}} = 0.0046 \text{ kmol/m}^3$$

it follows that

$$C_A(x^*, y^*) = 0.67 \,(0.0046 \text{ kmol/m}^3) = 0.0031 \text{ kmol/m}^3 \qquad \triangleleft$$

The molar flux may be obtained from Equation 6.7

$$N_A''(x^*) = h_m(C_{A,s} - C_{A,\infty})$$

with h_m evaluated from the analogy. From Equations 6.81 and 6.85 we know that, since $x_1^* = x_2^*$, $Re_{L,1} = Re_{L,2}$, and $Pr = Sc$, it follows that $f_4 = f_7$. Hence

$$Sh = \frac{h_m L_2}{D_{AB}} = Nu = \frac{h L_1}{k}$$

With $h = q''/(T_s - T_\infty)$ from Newton's law of cooling,

$$h_m = \frac{L_1}{L_2} \times \frac{D_{AB}}{k} \times \frac{q''}{(T_s - T_\infty)} = \frac{1}{2} \times \frac{0.26 \times 10^{-4} \text{ m}^2/\text{s}}{0.028 \text{ W/m} \cdot \text{K}} \times \frac{10^4 \text{ W/m}^2}{(80 - 20)°C}$$

$$h_m = 0.077 \text{ m/s}$$

Hence

$$N_A''(x^*) = 0.077 \text{ m/s} \,(0.0046 - 0.0) \text{ kmol/m}^3$$

or

$$N_A''(x^*) = 3.54 \times 10^{-4} \text{ kmol/s} \cdot \text{m}^2 \qquad \triangleleft$$

Comments: Recognize that, since the mole fraction of water vapor in the concentration boundary layer is small, the kinematic viscosity of air (ν_B) may be used to evaluate $Re_{L,2}$.

6.8.2 Evaporative Cooling

An important application of the heat and mass transfer analogy is to the process of *evaporative cooling*, which occurs whenever a gas flows over a liquid (Figure 6.15). Evaporation must occur from the liquid surface, and the energy associated with the phase change is the latent heat of vaporization of the liquid. Evaporation occurs when liquid molecules near the surface experience collisions that increase their energy above that needed to overcome the surface binding energy. The energy required to sustain the evaporation must come from the internal energy of the liquid, which then must experience a reduction in temperature (the cooling effect). However, if steady-state conditions are to be maintained, the latent energy lost by the liquid because of evaporation must be replenished by energy transfer to the liquid from its surroundings. Neglecting radiation effects, this transfer may be due to the convection of sensible energy from the gas or to heat addition by other means, as, for example, by an electrical heater submerged in the liquid. Applying conservation of energy to a control surface about the liquid (Equation 1.11a), it follows that, for a unit surface area,

$$q''_{conv} + q''_{add} = q''_{evap} \tag{6.93}$$

where q''_{evap} may be approximated as the product of the evaporative mass flux and the latent heat of vaporization

$$q''_{evap} = n''_A h_{fg} \tag{6.94}$$

If there is no heat addition by other means, Equation 6.93 reduces to a balance between convection heat transfer from the gas and the evaporative heat loss from the liquid. Substituting from Equations 6.1, 6.11, and 6.94, Equation 6.93 may then be expressed as

$$h(T_\infty - T_s) = h_{fg} h_m [\rho_{A,\,sat}(T_s) - \rho_{A,\,\infty}] \tag{6.95}$$

where the vapor density at the surface is that associated with saturated conditions at T_s. Hence the magnitude of the cooling effect may be expressed as

$$T_\infty - T_s = h_{fg}\left(\frac{h_m}{h}\right)[\rho_{A,\,sat}(T_s) - \rho_{A,\,\infty}] \tag{6.96}$$

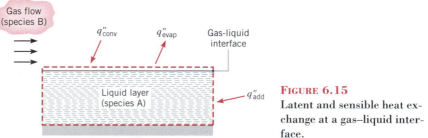

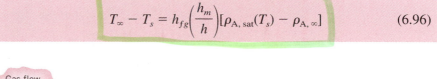

FIGURE 6.15
Latent and sensible heat exchange at a gas–liquid interface.

Substituting for (h_m/h) from Equation 6.92 and for the vapor densities from the ideal gas law, the cooling effect may also be expressed as

$$(T_\infty - T_s) = \frac{\mathcal{M}_A h_{fg}}{\mathcal{R}\rho c_p \, Le^{2/3}}\left[\frac{p_{A,\,sat}(T_s)}{T_s} - \frac{p_{A,\,\infty}}{T_\infty}\right] \tag{6.97}$$

In the interest of accuracy, the gas (species B) properties ρ, c_p, and Le should be evaluated at the arithmetic mean temperature of the thermal boundary layer, $T_{am} = (T_s + T_\infty)/2$. A representative value of $n = \frac{1}{3}$ has been assumed for the Pr and Sc exponent of Equation 6.92.

Equation 6.97 may generally be applied to a good approximation. A somewhat less accurate, but more convenient, form may be obtained by assuming that T_s and T_∞ are approximately equal to T_{am}. Accordingly,

$$(T_\infty - T_s) \approx \frac{\mathcal{M}_A h_{fg}}{\mathcal{R}c_p \, Le^{2/3}\, \rho T_{am}}[p_{A,\,sat}(T_s) - p_{A,\,\infty}]$$

or, recognizing that $m_A \ll m_B$, we may introduce the expression $(\rho T_{am}) = p/(\mathcal{R}/\mathcal{M}_B)$ from the equation of state for an ideal gas to obtain

$$(T_\infty - T_s) \approx \frac{(\mathcal{M}_A/\mathcal{M}_B)h_{fg}}{c_p \, Le^{2/3}}\left[\frac{p_{A,\,sat}(T_s)}{p} - \frac{p_{A,\,\infty}}{p}\right] \tag{6.98}$$

Numerous environmental and industrial applications of the foregoing results arise for situations in which the gas is *air* and the liquid is *water*.

EXAMPLE 6.7

A container, which is wrapped in a fabric that is continually moistened with a highly volatile liquid, may be used to keep beverages cool in hot arid regions. Suppose that the container is placed in dry ambient air at 40°C, with heat and mass transfer between the wetting agent and the air occurring by forced convection. The wetting agent is known to have a molecular weight of 200 kg/kmol and a latent heat of vaporization of 100 kJ/kg. Its saturated vapor pressure for the prescribed conditions is approximately 5000 N/m², and the diffusion coefficient of the vapor in air is 0.2×10^{-4} m²/s. What is the steady-state temperature of the beverage?

SOLUTION

Known: Properties of wetting agent used to evaporatively cool a beverage container.

Find: Steady-state temperature of beverage.

Schematic:

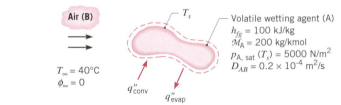

Assumptions:

1. Heat and mass transfer analogy is applicable.
2. Vapor displays ideal gas behavior.
3. Radiation effects are negligible.
4. Air properties may be evaluated at a mean boundary layer temperature assumed to be 300 K.

Properties: Table A.4, air (300 K): $\rho = 1.16$ kg/m^3, $c_p = 1.007$ kJ/kg · K, $\alpha = 22.5 \times 10^{-6}$ m^2/s.

Analysis: Subject to the foregoing assumptions, the evaporative cooling effect is given by Equation 6.97.

$$(T_\infty - T_s) = \frac{\mathcal{M}_A h_{fg}}{\mathcal{R}\rho c_p \, Le^{2/3}}\left[\frac{p_{A, \text{sat}}(T_s)}{T_s} - \frac{p_{A, \infty}}{T_\infty}\right]$$

Setting $p_{A, \infty} = 0$ and rearranging, it follows that

$$T_s^2 - T_\infty T_s + B = 0$$

where the coefficient B is

$$B = \frac{\mathcal{M}_A h_{fg} p_{A, \text{sat}}}{\mathcal{R}\rho c_p \, Le^{2/3}}$$

or

$$B = [200 \text{ kg/kmol} \times 100 \text{ kJ/kg} \times 5000 \text{ N/m}^2 \times 10^{-3} \text{ kJ/N · m}]$$

$$\div \left[8.315 \text{ kJ/kmol · K} \times 1.16 \text{ kg/m}^3 \times 1.007 \text{ kJ/kg · K}\right.$$

$$\left.\times \left(\frac{22.5 \times 10^{-6} \text{ m}^2/\text{s}}{20 \times 10^{-6} \text{ m}^2/\text{s}}\right)^{2/3}\right] = 9514 \text{ K}^2$$

Hence

$$T_s = \frac{T_\infty \pm \sqrt{T_\infty^2 - 4B}}{2} = \frac{313 \text{ K} \pm \sqrt{(313)^2 - 4(9514)} \text{ K}}{2}$$

Rejecting the minus sign on physical grounds (T_s must equal T_∞ if there is no evaporation, in which case $p_{A, \text{sat}} = 0$ and $B = 0$), it follows that

$$T_s = 278.9 \text{ K} = 5.9°\text{C} \qquad \triangleleft$$

Comments: The result is independent of the shape of the container as long as the heat and mass transfer analogy may be used.

6.8.3 The Reynolds Analogy

A second boundary layer analogy may be obtained by noting from Table 6.1 that, for $dp^*/dx^* = 0$ and $Pr = Sc = 1$, the conservation equations, Equations 6.63 to 6.65, are of precisely the same form. Moreover, since $u_\infty = V$ if $dp^*/dx^* = 0$, the boundary conditions, Equations 6.66 to 6.68, also have the same form.

Hence the solutions for u^*, T^*, and C_A^* must be equivalent. That is, from Equations 6.76, 6.79, and 6.83 of Table 6.3, $f_1 = f_3 = f_6$. Moreover, the friction coefficient, Nusselt number, and Sherwood number are related by the requirement that $f_2 = f_4 = f_7$, and, from Equation 6.78, 6.81, and 6.85, we conclude that

$$C_f \frac{Re_L}{2} = Nu = Sh \tag{6.99}$$

Replacing *Nu* and *Sh* by the *Stanton number* (*St*) and the *mass transfer Stanton number* (St_m), respectively,

$$St \equiv \frac{h}{\rho V c_p} = \frac{Nu}{Re\ Pr} \tag{6.100}$$

$$St_m \equiv \frac{h_m}{V} = \frac{Sh}{Re\ Sc} \tag{6.101}$$

Equation 6.99 may also be expressed in the form

$$\frac{C_f}{2} = St = St_m \tag{6.102}$$

Equation 6.102 is known as the *Reynolds analogy,* and it relates the key engineering parameters of the velocity, thermal, and concentration boundary layers. If the velocity parameter is known, the analogy may be used to obtain the other parameters, and vice versa. However, there are numerous restrictions associated with using this result. In addition to relying on the validity of the boundary layer approximations, the accuracy of Equation 6.102 depends on having *Pr* and *Sc* ≈ 1 and dp^*/dx^* ≈ 0. However, it has been shown that the analogy may be applied over a wide range of *Pr* and *Sc*, if certain corrections are added. In particular the *modified Reynolds,* or *Chilton–Colburn, analogies* [10, 11] have the form

$$\frac{C_f}{2} = St\ Pr^{2/3} \equiv j_H \qquad 0.6 < Pr < 60 \tag{6.103}$$

$$\frac{C_f}{2} = St_m\ Sc^{2/3} \equiv j_m \qquad 0.6 < Sc < 3000 \tag{6.104}$$

where j_H and j_m are the *Colburn j factors* for heat and mass transfer, respectively. For laminar flow Equations 6.103 and 6.104 are only appropriate when dp^*/dx^* ≈ 0, but in turbulent flow, conditions are less sensitive to the effect of pressure gradients and these equations remain approximately valid. If the analogy is applicable at every point on a surface, it may be applied to the surface average coefficients.

6.9
The Effects of Turbulence

At this point we recognize that turbulent conditions characterize numerous flows of practical interest. In fact the practicing engineer deals much more often with turbulent flows than with laminar flows. It is well known that small

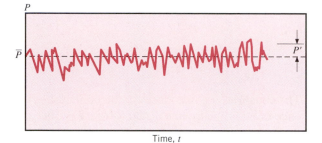

FIGURE 6.16 Property variation with time at some point in a turbulent boundary layer.

disturbances associated with distortions in the fluid streamlines of a laminar flow can eventually lead to turbulent conditions. These disturbances may originate from the free stream, or they may be induced by surface roughness. The onset of turbulence depends on whether these disturbances are amplified or attenuated in the direction of fluid flow, which in turn depends on the ratio of the inertia to viscous forces (the Reynolds number). Recall that if the Reynolds number is small, inertia forces are small relative to viscous forces. The naturally occurring disturbances are then dissipated, and the flow remains laminar. For a large Reynolds number, however, the inertia forces are sufficiently large to amplify the disturbances, and a transition to turbulence occurs. In Section 6.3 we observed that the critical Reynolds number, Re_c, required for transition is approximately 5×10^5 for flow over a flat plate.

Turbulence is associated with the existence of *random fluctuations* in the fluid, and at least on a small scale, the flow is inherently *unsteady*. This behavior is shown in Figure 6.16, where the variation of an arbitrary flow property P is plotted as a function of time at some location in a turbulent boundary layer. The property P could be a velocity component, the fluid temperature, or a species concentration, and at any instant it may be represented as the sum of a *time-mean* value $\bar{P}$ and a fluctuating component P'. The average is taken over a time that is large compared with the period of a typical fluctuation, and if $\bar{P}$ is independent of time, the time-mean flow is said to be *steady*.

The existence of turbulent flow can be advantageous in the sense of providing increased heat and mass transfer rates. However, the motion is extremely complicated and difficult to describe theoretically. Although the boundary layer conservation equations developed in preceding sections remain applicable, the dependent variables (u, v, T, C_A) must be interpreted as *instantaneous values*, and it is impossible to predict their exact variation with time. In the practical sense, however, this inability to determine the variation of the instantaneous properties P with time is not a serious restriction, since the engineer is generally concerned only with the time-mean properties $\bar{P}$. Equations of the form $P = \bar{P} + P'$ may be substituted for each of the flow variables into the boundary layer equations. For steady, incompressible, constant property flow, using well-established, time-averaging procedures [2, 12], the following forms of the x-momentum, energy, and species conservation equations may be obtained:

$$\rho\left(\bar{u}\frac{\partial \bar{u}}{\partial x} + \bar{v}\frac{\partial \bar{u}}{\partial y}\right) = -\frac{dp}{dx} + \frac{\partial}{\partial y}\left(\mu\frac{\partial \bar{u}}{\partial y} - \overline{\rho u'v'}\right) \qquad (6.105)$$

$$\rho c_p \left(\bar{u}\frac{\partial \bar{T}}{\partial x} + \bar{v}\frac{\partial \bar{T}}{\partial y} \right) = \frac{\partial}{\partial y}\left(k\frac{\partial \bar{T}}{\partial y} - \rho c_p \overline{v'T'} \right) \tag{6.106}$$

$$\left(\bar{u}\frac{\partial \bar{C}_A}{\partial x} + \bar{v}\frac{\partial \bar{C}_A}{\partial y} \right) = \frac{\partial}{\partial y}\left(D_{AB}\frac{\partial \bar{C}_A}{\partial y} - \overline{v'C'_A} \right) \tag{6.107}$$

The equations are like those for the laminar boundary layer, except for the presence of additional terms of the form $\overline{a'b'}$. These terms account for the effect of the turbulent fluctuations on momentum, energy, and species transport.

On the basis of the foregoing results, it is customary to speak of a *total* shear stress and *total* fluxes, which are defined as

$$\tau_{tot} = \left(\mu\frac{\partial \bar{u}}{\partial y} - \rho\overline{u'v'} \right) \tag{6.108}$$

$$q''_{tot} = -\left(k\frac{\partial \bar{T}}{\partial y} - \rho c_p\overline{v'T'} \right) \tag{6.109}$$

$$N''_{A,\,tot} = -\left(D_{AB}\frac{\partial \bar{C}_A}{\partial y} - \overline{v'C'_A} \right) \tag{6.110}$$

and consist of contributions due to molecular diffusion and turbulent mixing. From the form of these equations we see how momentum, energy, and species transfer rates are enhanced by the existence of turbulence. The term $\rho\overline{u'v'}$ appearing in Equation 6.108 represents the momentum flux due to the turbulent fluctuations, and it is often termed the *Reynolds stress*.

A simple conceptual model attributes the transport of momentum, heat, and mass in a turbulent boundary layer to the motion of *eddies,* small portions of fluid in the boundary layer that move about for a short time before losing their identity. Because of this motion, the transport of momentum, energy, and species is greatly enhanced. The notion of transport by eddies has prompted the introduction of a transport coefficient defined as the *eddy diffusivity for momentum transfer* ε_M, which has the form

$$\rho\varepsilon_M\frac{\partial \bar{u}}{\partial y} \equiv -\rho\overline{u'v'} \tag{6.111}$$

Hence the total shear stress may be expressed as

$$\tau_{tot} = \rho(\nu + \varepsilon_M)\frac{\partial \bar{u}}{\partial y} \tag{6.112}$$

Similarly, *eddy diffusivities for heat and mass transfer* ε_H and ε_m may be defined by the relations

$$\varepsilon_H\frac{\partial \bar{T}}{\partial y} \equiv -\overline{v'T'} \tag{6.113}$$

$$\varepsilon_m\frac{\partial \bar{C}_A}{\partial y} \equiv -\overline{v'C'_A} \tag{6.114}$$

in which case

$$q''_{tot} = -\rho c_p(\alpha + \varepsilon_H)\frac{\partial \bar{T}}{\partial y} \tag{6.115}$$

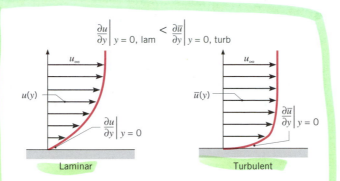

FIGURE 6.17 Comparison of laminar and turbulent velocity boundary layer profiles for the same free stream velocity.

$$N''_{A,\,tot} = -(D_{AB} + \varepsilon_m)\frac{\partial \bar{C}_A}{\partial y} \qquad (6.116)$$

In the region of a turbulent boundary layer removed from the surface (the core region), the eddy diffusivities are much larger than the molecular diffusivities. The enhanced mixing associated with this condition has the effect of making velocity, temperature, and concentration profiles more uniform in the core. This behavior is shown in Figure 6.17 for laminar and turbulent velocity boundary layers corresponding to the same free stream velocity. Accordingly, the velocity gradient at the surface, and therefore the surface shear stress, is much larger for the turbulent boundary layer than for the laminar boundary layer. In like manner, it may be argued that the surface temperature or concentration gradient, and therefore the heat or mass transfer rates, are much larger for turbulent than for laminar flow. Because of this enhancement of convection heat and mass transfer rates, it is desirable to have turbulent flow conditions in many engineering applications. However, the increase in wall shear stress will always have the adverse effect of increasing pump or fan power requirements. A fundamental problem in performing a turbulent boundary layer analysis involves determining the eddy diffusivities as a function of the mean properties of the flow. Unlike the molecular diffusivities, which are strictly fluid properties, the eddy diffusivities depend strongly on the nature of the flow and vary from point to point in a boundary layer. The problem is one that continues to attract many researchers in fluid mechanics.

6.10
The Convection Coefficients

In this chapter we have attempted to develop the fundamentals of convection transport phenomena. In the process, however, it is hoped that you have not lost sight of what remains *the problem of convection*. Our primary objective is still that of developing means to determine the convection coefficients h and h_m. Although these coefficients may be obtained by solving the boundary layer equations, it is only for simple flow situations that such solutions are readily effected. The more practical approach frequently involves calculating h and h_m from empirical relations of the form given by Equations 6.81 and 6.85. The particular form of these equations is obtained by *correlating* measured convection heat and mass transfer results in terms of appropriate dimensionless groups. It is this approach that is emphasized in the chapters that follow.

6.11 Summary

In this chapter attempts have been made to develop, in a logical fashion, the mathematical and physical bases of convection transport. To test your comprehension of the material, you should challenge yourself with appropriate questions. What are the velocity, thermal, and concentration boundary layers? Under what conditions do they develop, and why are they of interest to the engineer? How do laminar and turbulent boundary layers differ, and how does one determine whether a particular boundary layer is laminar or turbulent? There are numerous processes that affect momentum, energy, and species transfer in a boundary layer. What are they? How are they represented mathematically? What are the boundary layer approximations, and in what way do they alter the conservation equations? What are the relevant dimensionless groups for the various boundary layers? How may they be physically interpreted? How will the use of these groups facilitate convection calculations? How are velocity, thermal, and concentration boundary layer behaviors analogous? How may the effects of turbulence be treated in a boundary layer analysis? Finally, what is the central problem of convection?

References

1. Webb, R. L., *Int. Comm. Heat Mass Trans.,* **17,** 529, 1990.
2. Schlichting, H., *Boundary Layer Theory,* 7th ed., McGraw-Hill, New York, 1979.
3. Bird, R. B., W. E. Stewart, and E. N. Lightfoot, *Transport Phenomena,* Chapters 10 and 18, Wiley, New York, 1966.
4. Hartnett, J. P., "Mass Transfer Cooling," in W. M. Rohsenow and J. P. Hartnett, Eds., *Handbook of Heat Transfer,* McGraw-Hill, New York, 1973.
5. Kays, W. M., and M. E. Crawford, *Convective Heat and Mass Transfer,* McGraw-Hill, New York, 1980.
6. Burmeister, L. C., *Convective Heat Transfer,* Wiley-Interscience, New York, 1983.
7. Kaviany, M., *Principles of Convective Heat Transfer,* Springer-Verlag, New York, 1994.
8. Patankar, S. V., *Numerical Heat Transfer and Fluid Flow,* Hemisphere Publishing, New York, 1980.
9. Fox, R. W., and A. T. McDonald, *Introduction to Fluid Mechanics,* Wiley, New York, 1993.
10. Colburn, A. P., *Trans. Am. Inst. Chem. Eng.,* **29,** 174, 1933.
11. Chilton, T. H., and A. P. Colburn, *Ind. Eng. Chem.,* **26,** 1183, 1934.
12. Hinze, J. O., *Turbulence,* 2nd ed., McGraw-Hill, New York, 1975.

Problems

Heat Transfer Coefficients

6.1 For laminar flow over a flat plate, the local heat transfer coefficient h_x is known to vary as $x^{-1/2}$, where x is the distance from the leading edge ($x = 0$) of the plate. What is the ratio of the average coefficient between the leading edge and some location x on the plate to the local coefficient at x?

6.2 For laminar free convection from a heated vertical surface, the local convection coefficient may be expressed as $h_x = Cx^{-1/4}$, where h_x is the coefficient at a distance x from the leading edge of the surface and the quantity C, which depends on the fluid properties, is independent of x. Obtain an expression for the ratio $\bar{h}_x/h_x$, where $\bar{h}_x$ is the average coefficient between the leading edge ($x = 0$) and the x location. Sketch the variation of h_x and $\bar{h}_x$ with x.

6.3 A circular, hot gas jet at T_∞ is directed normal to a circular plate that has radius r_o and is maintained at a uniform temperature T_s. Gas flow over the plate is axisymmetric, causing the local convection coefficient to have a radial dependence of the form $h(r) = a + br^n$, where a, b, and n are constants. Determine the rate of heat transfer to the plate, expressing your result in terms of T_∞, T_s, r_o, a, b, and n.

6.4 Parallel flow of atmospheric air over a flat plate of length $L = 3$ m is disrupted by an array of stationary rods placed in the flow path over the plate.

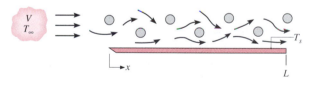

Laboratory measurements of the local convection coefficient at the surface of the plate are made for a prescribed value of V and $T_s > T_\infty$. The results are correlated by an expression of the form $h_x = 0.7 + 13.6x - 3.4x^2$, where h_x has units of W/m$^2 \cdot$ K and x is in meters. Evaluate the average convection coefficient $\bar{h}_L$ for the entire plate and the ratio $\bar{h}_L/h_L$ at the trailing edge.

6.5 Air at a free stream temperature of $T_\infty = 20°$C is in parallel flow over a flat plate of length $L = 5$ m and temperature $T_s = 90°$C. However, obstacles placed in the flow intensify mixing with increasing distance x from the leading edge, and the spatial variation of temperatures measured in the boundary layer is correlated by an expression of the form $T(°\text{C}) = 20 + 70 \exp(-600xy)$, where x and y are in meters. Determine and plot the manner in which the local convection coefficient h varies with x. Evaluate the average convection coefficient $\bar{h}$ for the plate.

6.6 The heat transfer rate per unit width (normal to the page) from a longitudinal section, $x_2 - x_1$, can be expressed as $q'_{12} = \bar{h}_{12}(x_2 - x_1)(T_s - T_\infty)$, where $\bar{h}_{12}$ is the average coefficient for the section of length $(x_2 - x_1)$. Consider laminar flow over a flat plate with a uniform temperature T_s. The spatial variation of the local convection coefficient is of the form $h_x = Cx^{-1/2}$, where C is a constant.

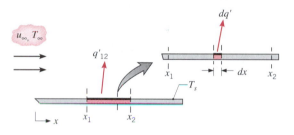

(a) Beginning with the convection rate equation in the form $dq' = h_x \, dx(T_s - T_\infty)$, derive an expression for $\bar{h}_{12}$ in terms of C, x_1, and x_2.

(b) Derive an expression for $\bar{h}_{12}$ in terms of x_1, x_2, and the average coefficients $\bar{h}_1$ and $\bar{h}_2$, corresponding to lengths x_1 and x_2, respectively.

6.7 Experiments to determine the local convection heat transfer coefficient for uniform flow normal to a heated circular disk have yielded a radial Nusselt number distribution of the form

$$Nu_D = \frac{h(r)D}{k} = Nu_o\left[1 + a\left(\frac{r}{r_o}\right)^n\right]$$

where both n and a are positive. The Nusselt number at the stagnation point is correlated in terms of the Reynolds ($Re_D = VD/\nu$) and Prandtl numbers

$$Nu_o = \frac{h(r = 0)D}{k} = 0.814Re_D^{1/2}\,Pr^{0.36}$$

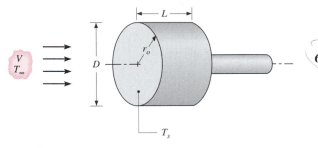

Obtain an expression for the average Nusselt number, $\overline{Nu}_D = \bar{h}D/k$, corresponding to heat transfer from an isothermal disk. Typically, boundary layer development from a stagnation point yields a decaying convection coefficient with increasing distance from the stagnation point. Provide a plausible explanation for why the opposite trend is observed for the disk.

6.8 An experimental procedure for validating results of the foregoing problem involves preheating a copper disk to an initial elevated temperature T_i and recording its temperature history $T(t)$ as it is subsequently cooled by the impinging flow to a final temperature T_f. The measured temperature decay may then be compared with predictions based on the correlation for $\overline{Nu}_D$. Assume that values of $a = 0.30$ and $n = 2$ are associated with the correlation.

Consider experimental conditions for which a disk of diameter $D = 50$ mm and length $L = 25$ mm is preheated to $T_i = 1000$ K and cooled to $T_f = 400$ K by an impinging airflow at $T_\infty = 300$ K. The cooled surface of the disk has an emissivity of $\varepsilon = 0.8$ and is exposed to large, isothermal surroundings for which $T_{sur} = T_\infty$. The remaining surfaces of the disk are well insulated, and heat transfer through the supporting rod may be neglected. Using results from the preceding problem, compute and plot temperature histories corresponding to air velocities of $V = 4, 20,$ and 50 m/s. Constant properties may be assumed for the copper ($\rho = 8933$ kg/m^3, $c_p = 425$ J/kg · K, $k = 386$ W/m · K) and air ($\nu = 38.8 \times 10^{-6}$ m^2/s, $k = 0.0407$ W/m · K, $Pr = 0.684$).

Boundary Layer Profiles

6.9 In flow over a surface, velocity and temperature profiles are of the forms

$$u(y) = Ay + By^2 - Cy^3 \quad \text{and}$$

$$T(y) = D + Ey + Fy^2 - Gy^3$$

where the coefficients A through G are constants. Obtain expressions for the friction coefficient C_f and the convection coefficient h in terms of $u_\infty, T_\infty,$ and appropriate profile coefficients and fluid properties.

6.10 Water at a temperature of $T_\infty = 25°C$ flows over one of the surfaces of a steel wall (AISI 1010) whose temperature is $T_{s,1} = 40°C$. The wall is 0.35 m thick, and its other surface temperature is $T_{s,2} = 100°C$. For steady-state conditions what is the convection coefficient associated with the water flow? What is the temperature gradient in the wall and in the water that is in contact with the wall? Sketch the temperature distribution in the wall and in the adjoining water.

6.11 In a particular application involving airflow over a heated surface, the boundary layer temperature distribution may be approximated as

$$\frac{T - T_s}{T_\infty - T_s} = 1 - \exp\left(-Pr\frac{u_\infty y}{\nu}\right)$$

where y is the distance normal to the surface and the Prandtl number, $Pr = c_p\mu/k = 0.7$, is a dimensionless fluid property. If $T_\infty = 400$ K, $T_s = 300$ K, and $u_\infty/\nu = 5000$ m^{-1}, what is the surface heat flux?

Boundary Layer Transition

6.12 Consider airflow over a flat plate of length $L = 1$ m under conditions for which transition occurs at $x_c = 0.5$ m based on the critical Reynolds number, $Re_{x,c} = 5 \times 10^5$. In the laminar and turbulent regions, the local convection coefficients are, respectively,

$$h_{lam}(x) = C_{lam}x^{-0.5} \quad \text{and} \quad h_{turb} = C_{turb}x^{-0.2}$$

where $C_{lam} = 8.845$ W/m^2 · K$^{0.5}$, $C_{turb} = 49.75$ W/m^2 · K$^{0.8}$, and x has units of m.

(a) Evaluating the thermophysical properties of air at 350 K, determine the velocity of the airflow.

(b) Develop an expression for the average convection coefficient $\bar{h}_{lam}(x)$, as a function of distance from the leading edge, x, for the laminar region, $0 \le x \le x_c$.

(c) Develop an expression for the average convection coefficient, $\bar{h}_{turb}(x)$, as a function of distance from the leading edge, x, for the turbulent region, $x_c \le x \le L$.

(d) On the same coordinates, plot the local and average convection coefficients, h_x and $\bar{h}_x$, respectively, as a function of x for $0 \le x \le L$.

6.13 A fan that can provide air speeds up to 50 m/s is to be used in a low-speed wind tunnel with atmospheric air at 25°C. If one wishes to use the wind tunnel to study flat-plate boundary layer behavior up to Reynolds numbers of $Re_x = 10^8$, what is the minimum plate length that should be used? At what distance from the leading edge would transition occur if the critical Reynolds number were $Re_{x,c} = 5 \times 10^5$?

6.14 Assuming a transition Reynolds number of 5×10^5, determine the distance from the leading edge of a flat plate at which transition will occur for each of the following fluids when $u_\infty = 1$ m/s: atmospheric air, water, engine oil, and mercury. In each case the fluid temperature is 27°C.

Conservation Equations and Solutions

6.15 Consider the control volume shown for the special case of steady-state conditions with $v = 0$, $T = T(y)$, and $\rho = $ const.

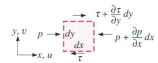

(a) Prove that $u = u(y)$ if $v = 0$ everywhere.

(b) Derive the x-momentum equation and simplify it as much as possible.

(c) Derive the energy equation and simplify it as much as possible.

6.16 Consider a lightly loaded journal bearing using oil having the constant properties $\mu = 10^{-2}$ kg/s · m and $k = 0.15$ W/m · K. If the journal and the bearing are each maintained at a temperature of 40°C, what is the maximum temperature in the oil when the journal is rotating at 10 m/s?

6.17 Consider a lightly loaded journal bearing using oil having the constant properties $\rho = 800$ kg/m³, $\nu = 10^{-5}$ m²/s, and $k = 0.13$ W/m · K. The journal diameter is 75 mm; the clearance is 0.25 mm; and the bearing operates at 3600 rpm.

(a) Determine the temperature distribution in the oil film assuming that there is no heat transfer into the journal and that the bearing surface is maintained at 75°C.

(b) What is the rate of heat transfer from the bearing, and how much power is needed to rotate the journal?

6.18 Consider two large (infinite) parallel plates, 5 mm apart. One plate is stationary, while the other plate is moving at a speed of 200 m/s. Both plates are maintained at 27°C. Consider two cases, one for which

the plates are separated by water and the other for which the plates are separated by air.

(a) For each of the two fluids, what is the force per unit surface area required to maintain the above condition? What is the corresponding power requirement?

(b) What is the viscous dissipation associated with each of the two fluids?

(c) What is the maximum temperature in each of the two fluids?

6.19 A judgment concerning the influence of viscous dissipation in forced convection heat transfer may be made by calculating the quantity $Pr\ Ec$, where the Prandtl number $Pr = c_p \mu / k$ and the Eckert number $Ec = U^2 / c_p\ \Delta T$ are *dimensionless groups*. The characteristic velocity and temperature difference of the problem are designated as U and ΔT, respectively. If $Pr\ Ec \ll 1$, dissipation effects may be neglected. Consider Couette flow for which one plate moves at 10 m/s and a temperature difference of 25°C is maintained between the plates. Evaluating properties at 27°C, determine the value of $Pr\ Ec$ for air, water, and engine oil. What is the value of $Pr\ Ec$ for air if the plate is moving at the sonic velocity?

6.20 Consider Couette flow for which the moving plate is maintained at a uniform temperature and the stationary plate is insulated. Determine the temperature of the insulated plate, expressing your result in terms of fluid properties and the temperature and speed of the moving plate. Obtain an expression for the heat flux at the moving plate.

6.21 Consider Couette flow with heat transfer for which the lower plate (mp) moves with a speed of $U = 5$ m/s and is perfectly insulated. The upper plate (sp) is stationary and is made of a material with thermal conductivity $k_{sp} = 1.5$ W/m · K and thickness $L_{sp} = 3$ mm. Its outer surface is maintained at $T_{sp} = 40$°C. The plates are separated by a distance $L_o = 5$ mm, which is filled with an engine oil of viscosity $\mu = 0.799$ N · s/m² and thermal conductivity $k_o = 0.145$ W/m · K.

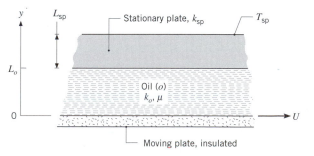

(a) On $T(y)$–y coordinates, sketch the temperature distribution in the oil film and the moving plate.

(b) Obtain an expression for the temperature at the lower surface of the oil film, $T(0) = T_o$, in terms of the plate speed U, the stationary plate parameters (T_{sp}, k_{sp}, L_{sp}) and the oil parameters (μ, k_o, L_o). Calculate this temperature for the prescribed conditions.

6.22 A shaft with a diameter of 100 mm rotates at 9000 rpm in a journal bearing that is 70 mm long. A uniform lubricant gap of 1 mm separates the shaft and the bearing. The lubricant properties are $\mu = 0.03$ N · s/m² and $k = 0.15$ W/m · K, while the bearing material has a thermal conductivity of $k_b = 45$ W/m · K.

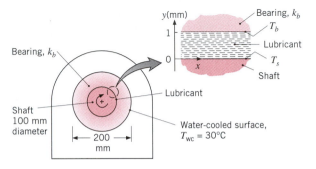

(a) Determine the viscous dissipation, $\mu\Phi$ (W/m³), in the lubricant.

(b) Determine the rate of heat transfer (W) from the lubricant, assuming that no heat is lost through the shaft.

(c) If the bearing housing is water-cooled, such that the outer surface of the bearing is maintained at 30°C, determine the temperatures of the bearing and shaft, T_b and T_s.

6.23 Consider Couette flow with heat transfer as described in Example 6.4.

(a) Rearrange the temperature distribution to obtain the dimensionless form

$$\theta(\eta) = \eta[1 + \tfrac{1}{2}Pr\,Ec(1 - \eta)]$$

where $\theta \equiv [T(y) - T_0]/[T_L - T_0]$ and $\eta = y/L$. The *dimensionless groups* are the Prandtl number $Pr = \mu c_p/k$ and the Eckert number $Ec = U^2/c_p(T_L - T_0)$.

(b) Derive an expression that prescribes the conditions under which there will be no heat transfer to the upper plate.

(c) Derive an expression for the heat transfer rate to the lower plate for the conditions identified in part (b).

(d) Generate a plot of θ versus η for $0 \le \eta \le 1$ and values of $Pr\,Ec = 0, 1, 2, 4, 10$. Explain key features of the temperature distributions.

6.24 Consider the problem of steady, incompressible laminar flow between two stationary, infinite parallel plates maintained at different temperatures.

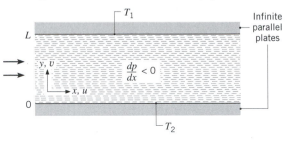

Referred to as *Poiseuille flow* with heat transfer, this special case of parallel flow is one for which the x velocity component is finite, but the y and z components (v and w) are zero.

(a) What is the form of the continuity equation for this case? In what way is the flow *fully developed*?

(b) What forms do the x- and y-momentum equations take? What is the form of the velocity profile? Note that, unlike Couette flow, fluid motion between the plates is now sustained by a finite pressure gradient. How is this pressure gradient related to the maximum fluid velocity?

(c) Assuming viscous dissipation to be significant and recognizing that conditions must be thermally fully developed, what is the appropriate form of the energy equation? Solve this equation for the temperature distribution. What is the heat flux at the upper ($y = L$) surface?

6.25 Consider the conservation equations (6.25, 6.29, and 6.43).

(a) Identify each equation and briefly describe the physical significance of each term.

(b) Identify the approximations and special conditions that are made to reduce these expressions to the boundary layer equations (6.54, 6.55, and 6.57).

(c) Comparing Equations 6.55 and 6.57, identify conditions for which the equations have the same form. Comment on the existence of a heat and momentum transfer analogy.

Similarity and Dimensionless Parameters

6.26 To a good approximation, the dynamic viscosity μ, the thermal conductivity k, and the specific heat c_p are independent of pressure. In what manner do the kinematic viscosity ν and thermal diffusivity α vary with pressure for an incompressible liquid and an ideal gas? Determine ν and α of air at 350 K for pressures of 1 and 10 atm.

6.27 An object of irregular shape has a characteristic length of $L = 1$ m and is maintained at a uniform surface temperature of $T_s = 400$ K. When placed in atmospheric air at a temperature of $T_\infty = 300$ K and moving with a velocity of $V = 100$ m/s, the average heat flux from the surface to the air is 20,000 W/m². If a second object of the same shape, but with a characteristic length of $L = 5$ m, is maintained at a surface temperature of $T_s = 400$ K and is placed in atmospheric air at $T_\infty = 300$ K, what will the value of the average convection coefficient be if the air velocity is $V = 20$ m/s?

6.28 Experiments have shown that, for airflow at $T_\infty = 35°C$ and $V_1 = 100$ m/s, the rate of heat transfer from a turbine blade of characteristic length $L_1 = 0.15$ m and surface temperature $T_{s,1} = 300°C$ is $q_1 = 1500$ W. What would be the heat transfer rate from a second turbine blade of characteristic length $L_2 = 0.3$ m operating at $T_{s,2} = 400°C$ in airflow of $T_\infty = 35°C$ and $V_2 = 50$ m/s? The surface area of the blade may be assumed to be directly proportional to its characteristic length.

6.29 Experimental measurements of the convection heat transfer coefficient for a square bar in cross flow yielded the following values:

$$\bar{h}_1 = 50 \text{ W/m}^2 \cdot \text{K} \quad \text{when} \quad V_1 = 20 \text{ m/s}$$

$$\bar{h}_2 = 40 \text{ W/m}^2 \cdot \text{K} \quad \text{when} \quad V_2 = 15 \text{ m/s}$$

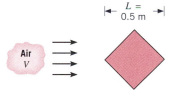

Assume that the functional form of the Nusselt number is $\overline{Nu} = C\,Re^m\,Pr^n$, where C, m, and n are constants.

(a) What will be the convection heat transfer coefficient for a similar bar with $L = 1$ m when $V = 15$ m/s?

(b) What will be the convection heat transfer coefficient for a similar bar with $L = 1$ m when $V = 30$ m/s?

(c) Would your results be the same if the side of the bar, rather than its diagonal, were used as the characteristic length?

6.30 Experimental results for heat transfer over a flat plate with an extremely rough surface were found to be correlated by an expression of the form

$$Nu_x = 0.04Re_x^{0.9}\,Pr^{1/3}$$

where Nu_x is the local value of the Nusselt number at a position x measured from the leading edge of the plate. Obtain an expression for the ratio of the average heat transfer coefficient $\bar{h}_x$ to the local coefficient h_x.

6.31 Consider conditions for which a fluid with a free stream velocity of $V = 1$ m/s flows over a surface with a characteristic length of $L = 1$ m, providing an average convection heat transfer coefficient of $\bar{h} = 100$ W/m² · K. Calculate the dimensionless parameters $\overline{Nu}_L$, Re_L, Pr, and $\bar{j}_H$ for the following fluids: air, engine oil, mercury, and water. Assume the fluids to be at 300 K.

6.32 For flow over a flat plate of length L, the local heat transfer coefficient h_x is known to vary as $x^{-1/2}$, where x is the distance from the leading edge of the plate. What is the ratio of the average Nusselt number for the entire plate $(\overline{Nu}_L)$ to the local Nusselt number at $x = L\ (Nu_L)$?

6.33 For laminar boundary layer flow over a flat plate with air at 20°C and 1 atm, the thermal boundary layer thickness δ_t is approximately 13% larger than the velocity boundary layer thickness δ. Determine the ratio δ/δ_t if the fluid is ethylene glycol under the same flow conditions.

6.34 Sketch the variation of the velocity and thermal boundary layer thicknesses with distance from the leading edge of a flat plate for the laminar flow of air, water, engine oil, and mercury. For each case assume a mean fluid temperature of 300 K.

6.35 Forced air at $T_\infty = 25°C$ and $V = 10$ m/s is used to cool electronic elements on a circuit board. One such element is a chip, 4 mm by 4 mm, located 120 mm from the leading edge of the board. Experiments have revealed that flow over the board is disturbed by the elements and that convection heat transfer is correlated by an expression of the form

$$Nu_x = 0.04Re_x^{0.85}\,Pr^{1/3}$$

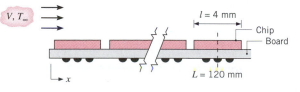

Estimate the surface temperature of the chip if it is dissipating 30 mW.

6.36 Consider the chip on the circuit board of the foregoing problem. To ensure reliable operation over extended periods, the chip temperature should not exceed 85°C. Assuming the availability of forced air at $T_\infty = 25°C$ and applicability of the prescribed heat transfer correlation, compute and plot the maximum allowable chip power dissipation P_c as a function of air velocity for $1 \leq V \leq 25$ m/s. If the chip surface has an emissivity of 0.80 and the board is mounted in a large enclosure whose walls are at 25°C, what is the effect of radiation on the P_c–V plot?

Reynolds Analogy

6.37 A thin, flat plate that is 0.2 m by 0.2 m on a side is oriented parallel to an atmospheric airstream having a velocity of 40 m/s. The air is at a temperature of $T_\infty = 20°C$, while the plate is maintained at $T_s = 120°C$. The air flows over the top and bottom surfaces of the plate, and measurement of the drag force reveals a value of 0.075 N. What is the rate of heat transfer from both sides of the plate to the air?

6.38 Atmospheric air is in parallel flow ($u_\infty = 15$ m/s, $T_\infty = 15°C$) over a flat heater surface that is to be maintained at a temperature of 140°C. The heater surface area is 0.25 m², and the airflow is known to induce a drag force of 0.25 N on the heater. What is the electrical power needed to maintain the prescribed surface temperature?

6.39 For flow over a flat plate with an extremely rough surface, convection heat transfer effects are known to be correlated by the expression of Problem 6.30. For airflow at 50 m/s, what is the surface shear stress at $x = 1$ m from the leading edge of the plate? Assume the air to be at a temperature of 300 K.

Mass Transfer Coefficients

6.40 On a summer day the air temperature is 27°C and the relative humidity is 30%. Water evaporates from the surface of a lake at a rate of 0.10 kg/h per square meter of water surface area. The temperature of the water is also 27°C. Determine the value of the convection mass transfer coefficient.

6.41 It is observed that a 230-mm-diameter pan of water at 23°C has a mass loss rate of 1.5×10^{-5} kg/s when the ambient air is dry and at 23°C.

(a) Determine the convection mass transfer coefficient for this situation.

(b) Estimate the evaporation mass loss rate when the ambient air has a relative humidity of 50%.

(c) Estimate the evaporation mass loss rate when the water and ambient air temperatures are 47°C, assuming that the convection mass transfer coefficient remains unchanged and the ambient air is dry.

6.42 The rate at which water is lost because of evaporation from the surface of a body of water may be determined by measuring the surface recession rate. Consider a summer day for which the temperature of both the water and the ambient air is 305 K and the relative humidity of the air is 40%. If the surface recession rate is known to be 0.1 mm/h, what is the rate at which mass is lost because of evaporation per unit surface area? What is the convection mass transfer coefficient?

6.43 Photosynthesis, as it occurs in the leaves of a green plant, involves transport of carbon dioxide (CO_2) from the atmosphere to the chloroplasts of the leaves, and the rate of photosynthesis may be quantified in terms of the rate of CO_2 assimilation by the chloroplasts. This assimilation is strongly influenced by CO_2 transfer through the boundary layer that develops on the leaf surface. Under conditions for which the density of CO_2 is 6×10^{-4} kg/m³ in the air and 5×10^{-4} kg/m³ at the leaf surface and the convection mass transfer coefficient is 10^{-2} m/s, what is the rate of photosynthesis in terms of kilograms of CO_2 assimilated per unit time and area of leaf surface?

6.44 Species A is evaporating from a flat surface into species B. Assume that the concentration profile for species A in the concentration boundary layer is of the form $C_A(y) = Dy^2 + Ey + F$, where D, E, and F are constants at any x location and y is measured along a normal from the surface. Develop an expression for the mass transfer convection coefficient h_m in terms of these constants, the concentration of A in the free stream $C_{A,\infty}$ and the mass diffusivity D_{AB}. Write an expression for the molar flux of mass transfer by convection for species A.

Species Conservation Equation and Solution

6.45 Consider Problem 6.24, when the fluid is a binary mixture with different molar concentrations $C_{A,1}$ and $C_{A,2}$ at the top and bottom surfaces, respectively. For the region between the plates, what is the appropriate form of the species A continuity equation? Obtain expressions for the species concentration distribution and the species flux at the upper surface.

6.46 A simple scheme for desalination involves maintaining a thin film of saltwater on the lower surface of two large (infinite) parallel plates that are slightly inclined and separated by a distance L.

A slow, incompressible, laminar airflow exists between the plates, such that the x velocity component is finite while the y and z components are zero. Evaporation occurs from the liquid film on the lower surface, which is maintained at an elevated temperature T_0, while condensation occurs at the upper surface, which is maintained at a reduced temperature T_L. The corresponding molar concentrations of water vapor at the lower and upper surfaces are designated as $C_{A,0}$ and $C_{A,L}$, respectively. The species concentration and temperature may be assumed to be independent of x and z.

(a) Obtain an expression for the distribution of the water vapor molar concentration $C_A(y)$ in the air. What is the mass rate of pure water production per unit surface area? Express your results in terms of $C_{A,0}$, $C_{A,L}$, L, and the vapor–air diffusion coefficient D_{AB}.

(b) Obtain an expression for the rate at which heat must be supplied per unit area to maintain the lower surface at T_0. Express your result in terms of $C_{A,0}$, $C_{A,L}$, T_0, T_L, L, D_{AB}, h_{fg} (the latent heat of vaporization of water), and the thermal conductivity k.

6.47 Consider the conservation equations (6.43) and (6.52).

(a) Describe the physical significance of each term.

(b) Identify the approximations and special conditions needed to reduce these expressions to the boundary layer equations (6.57 and 6.58). Comparing these equations, identify the conditions under which they have the same form. Comment on the existence of a heat and mass transfer analogy.

6.48 The *falling film* is widely used in chemical processing for the removal of gaseous species. It involves the flow of a liquid along a surface that may be inclined at some angle $\phi \geq 0$.

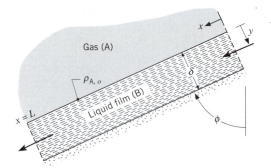

The flow is sustained by gravity, and the gas species A outside the film is absorbed at the liquid–gas interface. The film is in fully developed laminar flow over the entire plate, such that its velocity components in the y and z directions are zero. The mass density of A at $y = 0$ in the liquid is a constant $\rho_{A,o}$ independent of x.

(a) Write the appropriate form of the x-momentum equation for the film. Solve this equation for the distribution of the x velocity component, $u(y)$, in the film. Express your result in terms of δ, g, ϕ, and the liquid properties μ and ρ. Write an expression for the maximum velocity u_{max}.

(b) Obtain an appropriate form of the A species conservation equation for conditions within the film. If it is further assumed that the transport of species A across the gas–liquid interface does not penetrate very far into the film, the position $y = \delta$ may, for all practical purposes, be viewed as $y = \infty$. This condition implies that to a good approximation, $u = u_{max}$ in the region of penetration. Subject to these assumptions, determine an expression for $\rho_A(x, y)$ that applies in the film. *Hint:* This problem is analogous to conduction in a semi-infinite medium with a sudden change in surface temperature.

(c) If a local mass transfer convection coefficient is defined as

$$h_{m,x} \equiv \frac{n''_{A,x}}{\rho_{A,o}}$$

where $n''_{A,x}$ is the local mass flux at the gas–liquid interface, develop a suitable correlation for Sh_x as a function of Re_x and Sc.

(d) Develop an expression for the total gas absorption rate per unit width for a film of length L (kg/s · m).

(e) A water film that is 1 mm thick runs down the inside surface of a vertical tube that is 2 m long and has an inside diameter of 50 mm. An airstream containing ammonia (NH_3) moves through the tube, such that the mass density of NH_3 at the gas–liquid interface (but in the liquid) is 25 kg/m³. A dilute solution of ammonia in water is formed, and the diffusion coefficient is 2×10^{-9} m²/s. What is the mass rate of NH_3 removal by absorption?

Similarity and Heat–Mass Transfer Analogy

6.49 Consider cross flow of gas X over an object having a characteristic length of $L = 0.1$ m. For a Reynolds number of 1×10^4, the average heat transfer coefficient is 25 W/m² · K. The same object is then impregnated with liquid Y and subjected to the same flow conditions. Given the following thermophysical properties, what is the average convection mass transfer coefficient?

	ν (m²/s)	k (W/m · K)	α (m²/s)
Gas X	21×10^{-6}	0.030	29×10^{-6}
Liquid Y	3.75×10^{-7}	0.665	1.65×10^{-7}
Vapor Y	4.25×10^{-5}	0.023	4.55×10^{-5}
Mixture of gas X–vapor Y:	$Sc = 0.72$		

6.50 Consider conditions for which a fluid with a free stream velocity of $V = 1$ m/s flows over an evaporating or subliming surface with a characteristic length of $L = 1$ m, providing an average mass transfer convection coefficient of $\bar{h}_m = 10^{-2}$ m/s. Calculate the dimensionless parameters $\overline{Sh}_L$, Re_L, Sc, and $\bar{j}_m$ for the following combinations: airflow over water, airflow over naphthalene, and warm glycerol over ice. Assume a fluid temperature of 300 K and a pressure of 1 atm.

6.51 Sketch the variation of the velocity and concentration boundary layer thicknesses with distance from the leading edge of a flat plate for the following laminar flow conditions: airflow over a water film, airflow over a layer of dry ice, airflow over a layer of naphthalene, and the flow of warm glycerol over a layer of ice, which melts and dissolves in the glycerol. For each case assume a mean fluid temperature of 300 K.

6.52 An object of irregular shape has a characteristic length of $L = 1$ m and is maintained at a uniform surface temperature of $T_s = 325$ K. It is suspended in an airstream that is at atmospheric pressure ($p = 1$ atm) and has a velocity of $V = 100$ m/s and a temperature of $T_\infty = 275$ K. The average heat flux from the surface to the air is 12,000 W/m². Referring to the foregoing situation as case 1, consider the fol-

lowing cases and determine whether conditions are analogous to those of case 1. Each case involves an object of the same shape, which is suspended in an airstream in the same manner. Where analogous behavior does exist, determine the corresponding value of the average convection coefficient.

(a) The values of T_s, T_∞, and p remain the same, but $L = 2$ m and $V = 50$ m/s.

(b) The values of T_s and T_∞ remain the same, but $L = 2$ m, $V = 50$ m/s, and $p = 0.2$ atm.

(c) The surface is coated with a liquid film that evaporates into the air. The entire system is at 300 K, and the diffusion coefficient for the air–vapor mixture is $D_{AB} = 1.12 \times 10^{-4}$ m²/s. Also, $L = 2$ m, $V = 50$ m/s, and $p = 1$ atm.

(d) The surface is coated with another liquid film for which $D_{AB} = 1.12 \times 10^{-4}$ m²/s, and the system is at 300 K. In this case $L = 2$ m, $V = 250$ m/s, and $p = 0.2$ atm.

6.53 On a cool day in April a scantily clothed runner is known to lose heat at a rate of 500 W because of convection to the surrounding air at $T_\infty = 10°C$. The runner's skin remains dry and at a temperature of $T_s = 30°C$. Three months later, the runner is moving at the same speed, but the day is warm and humid with a temperature of $T_\infty = 30°C$ and a relative humidity of $\phi_\infty = 60\%$. The runner is now drenched in sweat and has a uniform surface temperature of 35°C. Under both conditions constant air properties may be assumed with $\nu = 1.6 \times 10^{-5}$ m²/s, $k = 0.026$ W/m · K, $Pr = 0.70$, and D_{AB} (water vapor–air) $= 2.3 \times 10^{-5}$ m²/s.

(a) What is the rate of water loss due to evaporation on the summer day?

(b) What is the total convective heat loss on the summer day?

6.54 An object of irregular shape 1 m long maintained at a constant temperature of 100°C is suspended in an airstream having a free stream temperature of 0°C, a pressure of 1 atm, and a velocity of 120 m/s. The air temperature measured at a point near the object in the airstream is 80°C. A second object having the same shape is 2 m long and is suspended in an airstream in the same manner. The air free stream velocity is 60 m/s. Both the air and the object are at 50°C, and the total pressure is 1 atm. A plastic coating on the surface of the object is being dried by this process. The molecular weight of the vapor is 82 and the saturation pressure at 50°C for the plastic material is 0.0323 atm. The mass diffusivity for the vapor in air at 50°C is 2.60×10^{-5} m²/s.

(a) For the second object, at a location corresponding to the point of measurement in the first object, determine the vapor concentration and partial pressure.

(b) If the average heat flux q'' is 2000 W/m² for the first object, determine the average mass flux n_A'' (kg/s · m²) for the second object.

6.55 An industrial process involves the evaporation of water from a liquid film that forms on a contoured surface. Dry air is passed over the surface, and from laboratory measurements the convection heat transfer correlation is of the form

$$\overline{Nu}_L = 0.43Re_L^{0.58}\,Pr^{0.4}$$

(a) For an air temperature and velocity of 27°C and 10 m/s, respectively, what is the rate of evaporation from a surface of 1-m² area and characteristic length $L = 1$ m? Approximate the density of saturated vapor as $\rho_{A,\,sat} = 0.0077$ kg/m³.

(b) What is the steady-state temperature of the liquid film?

6.56 A streamlined strut supporting a bearing housing is exposed to a hot airflow from an engine exhaust. It is necessary to run experiments to determine the average convection heat transfer coefficient $\overline{h}$ from the air to the strut in order to be able to cool the strut to the desired surface temperature T_s. It is decided to run mass transfer experiments on an object of the same shape and to obtain the desired heat transfer results by using the heat and mass transfer analogy.

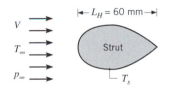

The mass transfer experiments were conducted using a half-size model strut constructed from naphthalene exposed to an airstream at 27°C. Mass transfer measurements yielded these results:

$\overline{Sh}_L$	Re_L
282	60,000
491	120,000
568	144,000
989	288,000

(a) Using the mass transfer experimental results, determine the coefficients C and m for a correlation of the form $\overline{Sh}_L = C\,Re_L^m\,Sc^{1/3}$.

(b) Determine the average convection heat transfer coefficient $\overline{h}$ for the full-sized strut, $L_H = 60$ mm, when exposed to a free stream airflow with $V = 60$ m/s, $T_\infty = 184°C$, and $p_\infty = 1$ atm when $T_s = 70°C$.

(c) The surface area of the strut can be expressed as $A_s = 2.2L_H \cdot l$, where l is the length normal to the page. For the conditions of part (b), what is the change in the rate of heat transfer to the strut if the characteristic length L_H is doubled?

6.57 Warm glycerol flows over a layer of ice whose shape is such that convection heat transfer effects can be correlated by an equation of the form $\overline{Nu}_L = 0.25Re_L^{0.7}\,Pr^{1/3}$. The ice melts and dissolves in the glycerol. Under conditions for which the average convection heat transfer coefficient is $\overline{h}_L = 100$ W/m² · K, what is the average convection mass transfer coefficient?

6.58 Consider the conditions of Problem 6.12, but with a thin film of water on the surface. If the air is dry and the Schmidt number Sc is 0.6, what is the evaporative mass flux? Is there net energy transfer to or from the water?

6.59 Consider the conditions of Problem 6.4, for which a heat transfer experiment yielded the prescribed distribution of the local convection coefficient, $h_x(x)$. The experiment was performed for surface and free stream temperatures of 310 and 290 K, respectively. Now consider repeating the experiment under conditions for which the surface is coated with a thin layer of naphthalene and both the surface and air are at 300 K. What is the corresponding value of the average convection mass transfer coefficient, $\overline{h}_{m,\,L}$?

6.60 Using the naphthalene sublimation technique, the radial distribution of the local convection mass transfer coefficient for uniform flow normal to a circular disk has been correlated by an expression of the form

$$Sh_D \equiv \frac{h_m(r)D}{D_{AB}} = Sh_o\left[1 + a\left(\frac{r}{r_o}\right)^n\right]$$

The stagnation point Sherwood number (Sh_o) depends on the Reynolds ($Re_D = VD/\nu$) and Schmidt ($Sc = \nu/D_{AB}$) numbers, and data have been correlated by the following expression:

$$Sh_o = \frac{h_m(r = 0)D}{D_{AB}} = 0.814Re_D^{1/2}\,Sc^{0.36}$$

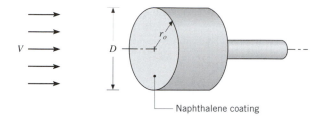

Naphthalene coating

Obtain an expression for the average Nusselt number ($\overline{Nu}_D = \overline{h}D/k$) corresponding to heat transfer from an isothermal disk exposed to the foregoing flow conditions. If $a = 1.2$ and $n = 5.5$, what is the rate of heat transfer from a disk of diameter $D = 20$ mm and surface temperature $T_s = 125°C$ to an airstream for which $Re_D = 5 \times 10^4$ and $T_\infty = 25°C$? Typically, boundary layer development from a stagnation point yields a decaying convection coefficient with increasing distance from the stagnation point. Provide a plausible explanation for why the opposite trend is observed for the disk.

6.61 To reduce the threat of predators, the sand grouse, a bird of Kenya, will lay its eggs in locations well removed from sources of groundwater. To bring water to its chicks, the grouse will then fly to the nearest source and, by submerging the lower part of its body, will entrain water within its plumage. The grouse will then return to its nest, and the chicks will imbibe water from the plumage. Of course, if the time of flight is too long, evaporative losses could cause a significant reduction in the water content of the plumage and the chicks could succumb to dehydration.

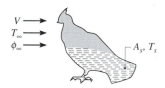

To gain a better understanding of convective transfer during flight, wind tunnel studies were performed using molded models of the grouse. By heating that portion of the model which corresponds to the water-encapsulating plumage, an average convection heat transfer coefficient was determined. Results for different air speeds and model sizes were then used to develop an empirical correlation of the form

$$\overline{Nu}_L = 0.034 Re_L^{4/5} Pr^{1/3}$$

The effective surface area of the water-encapsulating portion of the plumage is designated as A_s, and the characteristic length is defined as $L = (A_s)^{1/2}$.

Consider conditions for which a grouse has entrained 0.05 kg of water within plumage of $A_s = 0.04$ m² and is returning to its nest at a constant speed of $V = 30$ m/s. The ambient air is stagnant and at a temperature and relative humidity of $T_\infty = 37°C$ and $\phi_\infty = 25\%$, respectively. If, throughout the flight, the surface A_s is covered with a liquid water film at $T_s = 32°C$, what is the maximum allowable distance of the nest from the water source, if the bird must return with at least 50% of its initial water supply? Properties of the air and the air–vapor mixture may be taken to be $\nu = 16.7 \times 10^{-6}$ m²/s and $D_{AB} = 26.0 \times 10^{-6}$ m²/s.

6.62 A laboratory experiment involves simultaneous heat and mass transfer from a water-soaked towel experiencing irradiation from a bank of radiant lamps and parallel flow of air over its surface. Using a convection correlation to be introduced in Chapter 7, the average heat transfer convection coefficient is estimated to be $\overline{h} = 28.7$ W/m² · K. Assume that the radiative properties of the towel are those of water, for which $\alpha = \varepsilon = 0.96$, and that the surroundings are at 300 K.

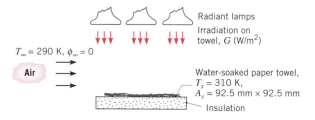

(a) Determine the rate at which water evaporates from the towel, n_A (kg/s).

(b) Perform an energy balance on the towel to determine the net rate of radiation transfer, q_{rad} (W), to the towel. Determine the irradiation G (W/m²).

6.63 In the spring, concrete surfaces such as sidewalks and driveways are sometimes very wet in the morning, even when it has not rained during the night. Typical nighttime conditions are shown in the sketch.

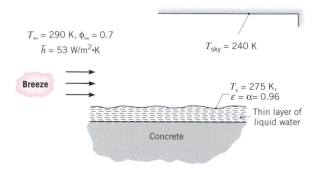

(a) Determine the heat fluxes associated with convection, q''_{conv}, evaporation, q''_{evap}, and radiation exchange with the sky, q''_{rad}.

(b) Do your calculations suggest why the concrete is wet instead of dry? Explain briefly.

(c) Is heat flowing from the liquid layer to the concrete? Or from the concrete to the liquid layer? Determine the heat flux by conduction into or out of the concrete.

6.64 Dry air at 32°C flows over a wetted plate of length 200 mm and width 1 m (case A). An imbedded electrical heater supplies 432 W and the surface temperature is 27°C.

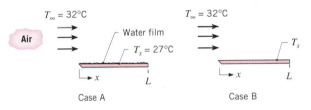

Case A   Case B

(a) What is the evaporation rate of water from the plate (kg/h)?

(b) After a long period of operation, all the water is evaporated from the plate and its surface is dry (case B). For the same free stream conditions and the same heater power as case A, estimate the temperature of the plate, T_s.

Evaporative Cooling

6.65 A successful California engineer has installed a circular hot tub in his back yard and finds that for the typical operating conditions prescribed below, water must be added at a rate of 0.001 kg/s in order to maintain a fixed water level in the tub.

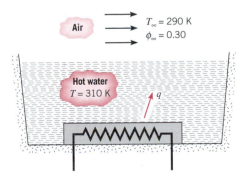

If the tub is well insulated on its sides and bottom and the temperature of the makeup water is equal to that of the tub water, at what rate must electrical heaters supply energy to maintain the tub water at 310 K?

6.66 It is known that on clear nights the air temperature need not drop below 0°C before a thin layer of water on the ground will freeze. Consider such a layer of water on a clear night for which the effective sky temperature is −30°C and the convection heat transfer coefficient due to wind motion is $h = 25$ W/m² · K. The water may be assumed to have an emissivity of 1.0 and to be insulated from the ground as far as conduction is concerned.

(a) Neglecting evaporation, determine the lowest temperature the air can have without the water freezing.

(b) For the conditions given, estimate the mass transfer coefficient for water evaporation h_m (m/s).

(c) Accounting now for the effect of evaporation, what is the lowest temperature the air can have without the water freezing? Assume the air to be dry.

6.67 An expression for the actual water vapor partial pressure in terms of wet-bulb and dry-bulb temperatures, referred to as the Carrier equation, is given as

$$p_v = p_{gw} - \frac{(p - p_{gw})(T_{db} - T_{wb})}{1810 - T_{wb}}$$

where p_v, p_{gw}, and p are the actual partial pressure, the saturation pressure at the wet-bulb temperature, and the total pressure (all in bars), while T_{db} and T_{wb} are the dry- and wet-bulb temperatures in kelvins. Consider air at 1 atm and 37.8°C flowing over a wet-bulb thermometer that indicates 21.1°C.

(a) Using Carrier's equation, calculate the partial pressure of the water vapor in the free stream. What is the relative humidity?

(b) Refer to a psychrometric chart and obtain the relative humidity directly for the conditions indicated. Compare the result with part (a).

(c) Use the value of vapor pressure obtained in part (a) with the evaporative cooling relation, Equation 6.98, to calculate a temperature difference between the bulk air and the wet bulb. Compare this temperature difference with the actual temperature difference. What is the percentage difference between the two values?

6.68 A wet-bulb thermometer consists of a mercury-in-glass thermometer covered with a wetted (water) fabric. When suspended in a stream of air, the steady-state thermometer reading indicates the wet-bulb temperature T_{wb}. Obtain an expression for determining the relative humidity of the air from knowledge of the air temperature (T_∞), the wet-bulb temperature, and appropriate air and water vapor properties. If $T_\infty = 45°C$ and $T_{wb} = 25°C$, what is the relative humidity of the airstream?

6.69 An industrial process involves evaporation of a thin water film from a contoured surface by heating it from below and forcing air across it. Laboratory measurements for this surface have provided the following heat transfer correlation:

$$\overline{Nu_L} = 0.43 Re_L^{0.58} Pr^{0.4}$$

The air flowing over the surface has a temperature of 290 K, a velocity of 10 m/s, and is completely dry ($\phi_\infty = 0$). The surface has a length of 1 m and a surface area of 1 m². Just enough energy is supplied to maintain its steady-state temperature at 310 K.

(a) Determine the heat transfer coefficient and the rate at which the surface loses heat by convection.

(b) Determine the mass transfer coefficient and the evaporation rate (kg/h) of the water on the surface.

(c) Determine the rate at which heat must be supplied to the surface for these conditions.

6.70 A disk of 20-mm diameter is covered with a water film. Under steady-state conditions, a heater power of 200 mW is required to maintain the disk–water film at 305 K in dry air at 295 K and the observed evaporation rate is 2.55×10^{-4} kg/h.

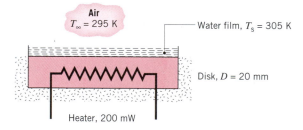

(a) Calculate the average mass transfer convection coefficient, $\overline{h}_m$, for the evaporation process.

(b) Calculate the average heat transfer convection coefficient, $\overline{h}$.

(c) Do the values of $\overline{h}_m$ and $\overline{h}$ satisfy the heat–mass analogy?

(d) If the relative humidity of the ambient air at 295 K were increased from 0 (dry) to 0.50, but the power supplied to the heater was maintained at 200 mW, would the evaporation rate increase or decrease? Would the disk temperature increase or decrease?

6.71 An experiment is conducted to determine the average mass transfer convection coefficient of a small droplet using a heater controlled to operate at a constant temperature. The power history required to completely evaporate the droplet at a temperature of 37°C is shown in the sketch. It was observed that, as the droplet dried, its wetted diameter on the heater surface remained nearly constant at a value of 4 mm.

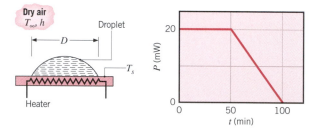

(a) Calculate the average mass transfer convection coefficient based on the wetted area during the evaporation process when the droplet, heater, and the *dry* ambient air are at 37°C.

(b) How much energy will be required to evaporate the droplet if the *dry* ambient air temperature is 27°C, while the droplet–heater temperature remains at 37°C?

6.72 It is desired to develop a simple model for predicting the temperature–time history of a plate during the drying cycle in a dishwasher. Following the wash cycle the plate is at $T_p(t) = T_p(0) = 65°C$ and the air in the dishwasher is completely saturated ($\phi_\infty = 1.0$) at $T_\infty = 55°C$. The values of the plate surface area A_s, mass M, and specific heat c are such that $Mc/A_s = 1600$ J/m² · K.

(a) Assuming the plate is completely covered by a thin film of water and neglecting the thermal resistances of the film and plate, derive a differential equation for predicting the plate temperature as a function of time.

(b) For the initial conditions ($t = 0$) estimate the change in plate temperature with time, dT/dt (°C/s), assuming that the average heat transfer coefficient on the plate is 3.5 W/m² · K.

CHAPTER

External Flow

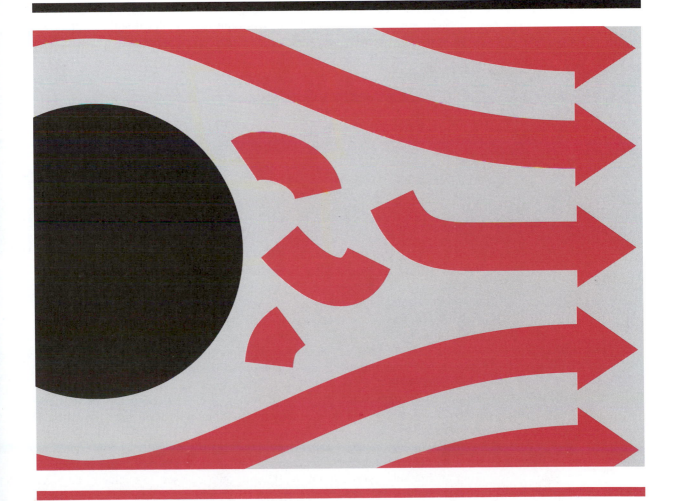

*I*n this chapter we focus on the problem of computing heat and mass transfer rates to or from a surface in *external flow*. In such a flow boundary layers develop freely, without constraints imposed by adjacent surfaces. Accordingly, there will always exist a region of the flow outside the boundary layer in which velocity, temperature, and/or concentration gradients are negligible. Examples include fluid motion over a flat plate (inclined or parallel to the free stream velocity) and flow over curved surfaces such as a sphere, cylinder, airfoil, or turbine blade.

For the moment we confine our attention to problems of *low speed, forced convection* with *no phase change* occurring within the fluid. In *forced* convection, the relative motion between the fluid and the surface is maintained by external means, such as a fan or a pump, and not by buoyancy forces due to temperature gradients in the fluid (*natural* convection). *Internal flows, natural convection,* and *convection with phase change* are treated in Chapters 8, 9, and 10, respectively.

Our primary objective is to determine convection coefficients for different flow geometries. In particular, we wish to obtain specific forms of the functions that represent these coefficients. By nondimensionalizing the boundary layer conservation equations in Chapter 6, we found that the local and average convection coefficients may be correlated by equations of the form

Heat Transfer:

$$Nu_x = f_4(x^*, Re_x, Pr) \tag{6.81}$$

$$\overline{Nu}_x = f_5(Re_x, Pr) \tag{6.82}$$

Mass Transfer:

$$Sh_x = f_7(x^*, Re_x, Sc) \tag{6.85}$$

$$\overline{Sh}_x = f_8(Re_x, Sc) \tag{6.86}$$

The subscript x has been added to emphasize our interest in conditions at a particular location on the surface. The overbar indicates an average from $x^* = 0$, where the boundary layer begins to develop, to the location of interest. Recall that *the problem of convection* is one of obtaining these functions. There are two approaches that we could take, one theoretical and the other experimental.

The *experimental* or *empirical approach* involves performing heat and mass transfer measurements under controlled laboratory conditions and correlating the data in terms of appropriate dimensionless parameters. A general discussion of the approach is provided in Section 7.1. It has been applied to many different geometries and flow conditions, and important results are presented in Sections 7.2 to 7.8.

The *theoretical approach* involves solving the boundary layer equations for a particular geometry. For example, obtaining the temperature profile T^* from such a solution, Equation 6.80 could then be used to evaluate the local Nusselt number Nu_x, and therefore the local convection coefficient h_x. With knowledge of how h_x varies over the surface, Equation 6.5 may then be used to determine the average convection coefficient $\overline{h}_x$, and therefore the Nusselt number $\overline{Nu}_x$. In Section 7.2.1 this approach is illustrated by using the *similarity method* to obtain an *exact solution* of the boundary layer equations for a flat plate in parallel, laminar flow [1–3]. An *approximate solution* to the same problem is obtained in Appendix E by using the *integral method* [4].

7.1
The Empirical Method

The manner in which a convection heat transfer correlation may be obtained experimentally is illustrated in Figure 7.1. If a prescribed geometry, such as the flat plate in parallel flow, is heated electrically to maintain $T_s > T_\infty$, convection heat transfer occurs from the surface to the fluid. It would be a simple matter to measure T_s and T_∞, as well as the electrical power, $E \cdot I$, which is equal to the total heat transfer rate q. The convection coefficient $\overline{h}_L$, which is an average associated with the entire plate, could then be computed from Newton's law of cooling, Equation 6.4. Moreover, from knowledge of the characteristic length L and the fluid properties, the Nusselt, Reynolds, and Prandtl numbers could be computed from their definitions, Equations 6.82, 6.69, and 6.70, respectively.

The foregoing procedure could be repeated for a variety of test conditions. We could vary the velocity u_∞ and the plate length L, as well as the nature of the fluid, using, for example, air, water, and engine oil, which have substantially different Prandtl numbers. We would then be left with many different values of the Nusselt number corresponding to a wide range of Reynolds and Prandtl numbers, and the results could be plotted on a *log–log* scale, as shown in Figure 7.2a. Each symbol represents a unique set of test conditions. As is often the case, the results associated with a given fluid, and hence a fixed Prandtl number, fall close to a straight line, which may be represented by an algebraic expression of the form

$$\overline{Nu}_L = C \, Re_L^m \, Pr^n \tag{7.1}$$

Since the values of C, m, and n are often independent of the nature of the fluid, the family of straight lines corresponding to different Prandtl numbers can be collapsed to a single line by plotting the results in terms of the ratio, $\overline{Nu}_L/Pr^n$, as shown in Figure 7.2b.

Because Equation 7.1 is inferred from experimental measurements, it is termed an *empirical correlation*. However, the specific values of the coefficient

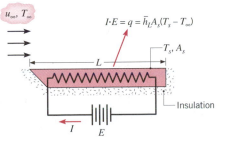

FIGURE 7.1
Experiment for measuring the average convection heat transfer coefficient $\overline{h}_L$.

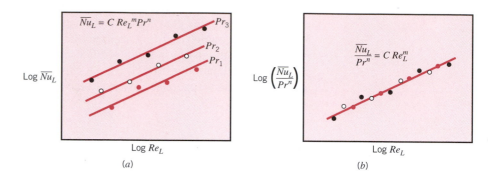

FIGURE 7.2   Dimensionless representation of convection heat transfer measurements.

C and the exponents m and n vary with the nature of the surface geometry and the type of flow.

We will use expressions of the form given by Equation 7.1 for many special cases, and it is important to note that the assumption of *constant fluid properties* is often implicit in the results. However, we know that the fluid properties vary with temperature across the boundary layer and that this variation can certainly influence the heat transfer rate. This influence may be handled in one of two ways. In one method, Equation 7.1 is used with all properties evaluated at a mean boundary layer temperature T_f, termed the *film temperature.*

$$T_f \equiv \frac{T_s + T_\infty}{2} \tag{7.2}$$

The alternate method is to evaluate all properties at T_∞ and to multiply the right-hand side of Equation 7.1 by an additional parameter to account for the property variations. The parameter is commonly of the form $(Pr_\infty/Pr_s)^r$ or $(\mu_\infty/\mu_s)^r$, where the subscripts ∞ and s designate evaluation of the properties at the free stream and surface temperatures, respectively. Both methods are used in the results that follow.

Finally, we note that experiments may also be performed to obtain convection mass transfer correlations. However, under conditions for which the heat and mass transfer analogy (Section 6.8.1) may be applied, the mass transfer correlation assumes the same form as the corresponding heat transfer correlation. Accordingly, we anticipate correlations of the form

$$\overline{Sh}_L = C\, Re_L^m\, Sc^n \tag{7.3}$$

where, for a given geometry and flow condition, the values of C, m, and n are the same as those appearing in Equation 7.1.

7.2
The Flat Plate in Parallel Flow

Despite its simplicity, parallel flow over a flat plate (Figure 7.3) occurs in numerous engineering applications. As discussed in Section 6.3, laminar bound-

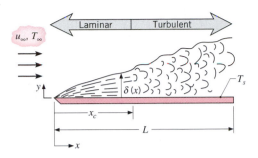

FIGURE 7.3
The flat plate in parallel flow.

ary layer development begins at the leading edge ($x = 0$) and transition to turbulence may occur at a downstream location (x_c) for which a critical Reynolds number $Re_{x,c}$ is achieved. We begin by considering conditions in the laminar boundary layer.

7.2.1 Laminar Flow: A Similarity Solution

The major convection parameters may be obtained by solving the appropriate form of the boundary layer equations. Assuming *steady, incompressible, laminar* flow with *constant fluid properties* and *negligible viscous dissipation* and recognizing that $dp/dx = 0$, the boundary layer equations (6.54, 6.55, 6.57, and 6.58) reduce to

Continuity:

$$\frac{\partial u}{\partial x} + \frac{\partial v}{\partial y} = 0 \tag{7.4}$$

Momentum:

$$u\frac{\partial u}{\partial x} + v\frac{\partial u}{\partial y} = \nu\frac{\partial^2 u}{\partial y^2} \tag{7.5}$$

Energy:

$$u\frac{\partial T}{\partial x} + v\frac{\partial T}{\partial y} = \alpha\frac{\partial^2 T}{\partial y^2} \tag{7.6}$$

Species:

$$u\frac{\partial \rho_A}{\partial x} + v\frac{\partial \rho_A}{\partial y} = D_{AB}\frac{\partial^2 \rho_A}{\partial y^2} \tag{7.7}$$

Solution of these equations is simplified by the fact that for constant properties, conditions in the velocity (hydrodynamic) boundary layer are independent of temperature and species concentration. Hence we may begin by solving the hydrodynamic problem, Equations 7.4 and 7.5, to the exclusion of Equations 7.6

and 7.7. Once the hydrodynamic problem has been solved, solutions to Equations 7.6 and 7.7, which depend on u and v, may be obtained.

The hydrodynamic solution follows the method of Blasius [1, 2]. The velocity components are defined in terms of a stream function $\psi(x, y)$,

$$u \equiv \frac{\partial \psi}{\partial y} \quad \text{and} \quad v \equiv -\frac{\partial \psi}{\partial x} \tag{7.8}$$

such that Equation 7.4 is automatically satisfied and hence is no longer needed. New dependent and independent variables, f and η, respectively, are then defined such that

$$f(\eta) \equiv \frac{\psi}{u_\infty \sqrt{vx/u_\infty}} \tag{7.9}$$

$$\eta \equiv y\sqrt{u_\infty/vx} \tag{7.10}$$

As we will find, use of these variables simplifies matters by reducing the partial differential equation, Equation 7.5, to an ordinary differential equation.

The Blasius solution is termed a *similarity solution,* and η is a *similarity variable.* This terminology is used because, despite growth of the boundary layer with distance x from the leading edge, the velocity profile u/u_∞ remains *geometrically similar.* This similarity is of the functional form

$$\frac{u}{u_\infty} = \phi\left(\frac{y}{\delta}\right)$$

where δ is the boundary layer thickness. Assuming this thickness to vary as $(vx/u_\infty)^{1/2}$, it follows that

$$\frac{u}{u_\infty} = \phi(\eta) \tag{7.11}$$

Hence the velocity profile is assumed to be uniquely determined by the similarity variable η, which depends on both x and y.

From Equations 7.8 through 7.10 we obtain

$$u = \frac{\partial \psi}{\partial y} = \frac{\partial \psi}{\partial \eta}\frac{\partial \eta}{\partial y} = u_\infty \sqrt{\frac{vx}{u_\infty}}\frac{df}{d\eta}\sqrt{\frac{u_\infty}{vx}} = u_\infty\frac{df}{d\eta} \tag{7.12}$$

and

$$v = -\frac{\partial \psi}{\partial x} = -\left(u_\infty\sqrt{\frac{vx}{u_\infty}}\frac{\partial f}{\partial x} + \frac{u_\infty}{2}\sqrt{\frac{v}{u_\infty x}}\,f\right)$$

$$v = \frac{1}{2}\sqrt{\frac{vu_\infty}{x}}\left(\eta\frac{df}{d\eta} - f\right) \tag{7.13}$$

By differentiating the velocity components, it may also be shown that

$$\frac{\partial u}{\partial x} = -\frac{u_\infty}{2x}\eta\frac{d^2f}{d\eta^2} \tag{7.14}$$

$$\frac{\partial u}{\partial y} = u_\infty\sqrt{\frac{u_\infty}{vx}}\frac{d^2f}{d\eta^2} \tag{7.15}$$

$$\frac{\partial^2 u}{\partial y^2} = \frac{u_\infty^2}{\nu x}\frac{d^3 f}{d\eta^3} \tag{7.16}$$

Substituting these expressions into Equation 7.5, we then obtain

$$2\frac{d^3 f}{d\eta^3} + f\frac{d^2 f}{d\eta^2} = 0 \tag{7.17}$$

Hence the hydrodynamic boundary layer problem is reduced to one of solving a nonlinear, third-order ordinary differential equation. The appropriate boundary conditions are

$$u(x, 0) = v(x, 0) = 0 \quad \text{and} \quad u(x, \infty) = u_\infty$$

or, in terms of the similarity variables,

$$\left.\frac{df}{d\eta}\right|_{\eta=0} = f(0) = 0 \quad \text{and} \quad \left.\frac{df}{d\eta}\right|_{\eta=\infty} = 1 \tag{7.18}$$

The solution to Equation 7.17, subject to the conditions of Equation 7.18, may be obtained by a series expansion [2] or by numerical integration [3]. Selected results are presented in Table 7.1, from which useful information may be extracted. We first note that, to a good approximation, $(u/u_\infty) = 0.99$ for $\eta = 5.0$. Defining the boundary layer thickness δ as that value of y for which $(u/u_\infty) = 0.99$, it follows from Equation 7.10 that

$$\delta = \frac{5.0}{\sqrt{u_\infty/\nu x}} = \frac{5x}{\sqrt{Re_x}} \tag{7.19}$$

From Equation 7.19 it is clear that δ increases with increasing x and ν but decreases with increasing u_∞ (the larger the free stream velocity, the *thinner* the boundary layer). In addition, from Equation 7.15 the wall shear stress may be expressed as

$$\tau_s = \mu\left.\frac{\partial u}{\partial y}\right|_{y=0} = \mu u_\infty\sqrt{u_\infty/\nu x}\left.\frac{d^2 f}{d\eta^2}\right|_{\eta=0}$$

Hence from Table 7.1

$$\tau_s = 0.332u_\infty\sqrt{\rho\mu u_\infty/x}$$

The *local* friction coefficient is then

$$C_{f,x} \equiv \frac{\tau_{s,x}}{\rho u_\infty^2/2} = 0.664Re_x^{-1/2} \tag{7.20}$$

From knowledge of conditions in the velocity boundary layer, the energy and species continuity equations may now be solved. To solve Equation 7.6 we introduce the dimensionless temperature $T^* \equiv [(T - T_s)/(T_\infty - T_s)]$ and assume a similarity solution of the form $T^* = T^*(\eta)$. Making the necessary substitutions, Equation 7.6 reduces to

$$\frac{d^2 T^*}{d\eta^2} + \frac{Pr}{2}f\frac{dT^*}{d\eta} = 0 \tag{7.21}$$

TABLE 7.1   Flat plate laminar boundary layer functions [3]

$\eta = y\sqrt{\dfrac{u_\infty}{\nu x}}$	f	$\dfrac{df}{d\eta} = \dfrac{u}{u_\infty}$	$\dfrac{d^2f}{d\eta^2}$
0	0	0	0.332
0.4	0.027	0.133	0.331
0.8	0.106	0.265	0.327
1.2	0.238	0.394	0.317
1.6	0.420	0.517	0.297
2.0	0.650	0.630	0.267
2.4	0.922	0.729	0.228
2.8	1.231	0.812	0.184
3.2	1.569	0.876	0.139
3.6	1.930	0.923	0.098
4.0	2.306	0.956	0.064
4.4	2.692	0.976	0.039
4.8	3.085	0.988	0.022
5.2	3.482	0.994	0.011
5.6	3.880	0.997	0.005
6.0	4.280	0.999	0.002
6.4	4.679	1.000	0.001
6.8	5.079	1.000	0.000

Note the dependence of the thermal solution on hydrodynamic conditions through appearance of the variable f in Equation 7.21. The appropriate boundary conditions are

$$T^*(0) = 0 \quad \text{and} \quad T^*(\infty) = 1 \tag{7.22}$$

Subject to the conditions of Equation 7.22, Equation 7.21 may be solved by numerical integration for different values of the Prandtl number. One important consequence of this solution is that, for $Pr \gtrsim 0.6$, results for the surface temperature gradient $dT^*/d\eta|_{\eta=0}$ may be correlated by the following relation:

$$\left. \frac{dT^*}{d\eta} \right|_{\eta=0} = 0.332 Pr^{1/3}$$

Expressing the local convection coefficient as

$$h_x = \frac{q_s''}{T_s - T_\infty} = -\frac{T_\infty - T_s}{T_s - T_\infty} k \left. \frac{\partial T^*}{\partial y} \right|_{y=0}$$

$$h_x = k \left(\frac{u_\infty}{\nu x} \right)^{1/2} \left. \frac{dT^*}{d\eta} \right|_{\eta=0}$$

it follows that the *local* Nusselt number is of the form

$$Nu_x \equiv \frac{h_x x}{k} = 0.332 Re_x^{1/2} Pr^{1/3} \qquad Pr \gtrsim 0.6 \tag{7.23}$$

From the solution to Equation 7.21, it also follows that the ratio of the velocity to thermal boundary layer thickness is

$$\frac{\delta}{\delta_t} \approx Pr^{1/3} \tag{7.24}$$

where δ is given by Equation 7.19.

The concentration boundary layer solution is obtained in a similar manner. Introducing a normalized species density $\rho_A^* \equiv [(\rho_A - \rho_{A,s})/(\rho_{A,\infty} - \rho_{A,s})]$ and assuming a similarity solution, Equation 7.7 becomes

$$\frac{d^2\rho_A^*}{d\eta^2} + \frac{Sc}{2} f \frac{d\rho_A^*}{d\eta} = 0 \tag{7.25}$$

with

$$\rho_A^*(0) = 0 \quad \text{and} \quad \rho_A^*(\infty) = 1 \tag{7.26}$$

For $Sc \gtrsim 0.6$, the corresponding solution yields

$$\frac{d\rho_A^*}{d\eta}\bigg|_{\eta=0} = 0.332 Sc^{1/3}$$

in which case

$$h_{m,x} \equiv \frac{n_A''}{\rho_{A,s} - \rho_{A,\infty}} = D_{AB} \frac{\partial \rho_A^*}{\partial y}\bigg|_{y=0} = D_{AB} \sqrt{u_\infty/\nu x} \frac{d\rho_A^*}{d\eta}\bigg|_{\eta=0}$$

and

$$Sh_x \equiv \frac{h_{m,x} x}{D_{AB}} = 0.332 Re_x^{1/2} Sc^{1/3} \qquad Sc \gtrsim 0.6 \tag{7.27}$$

From the solution to Equation 7.25, it also follows that the ratio of boundary layer thicknesses is

$$\frac{\delta}{\delta_c} \approx Sc^{1/3} \tag{7.28}$$

The foregoing results may be used to compute important *laminar* boundary layer parameters for any $0 < x < x_c$, where x_c is the distance from the leading edge at which transition begins. Equations 7.20, 7.23, and 7.27 imply that $\tau_{s,x}$, h_x, and $h_{m,x}$ are, in principle, infinite at the leading edge and decrease as $x^{-1/2}$ in the flow direction. Equations 7.24 and 7.28 also imply that, for values of Pr and Sc close to unity, which is the case for most gases, the three boundary layers experience nearly identical growth. Equations 7.23 and 7.27, as well as Equations 7.24 and 7.28, are of precisely the same form, confirming the heat and mass transfer analogy.

From the foregoing local results, average boundary layer parameters may be determined. With the average friction coefficient defined as

$$\overline{C}_{f,x} \equiv \frac{\overline{\tau}_{s,x}}{\rho u_\infty^2/2} \tag{7.29}$$

where

$$\overline{\tau}_{s,x} \equiv \frac{1}{x} \int_0^x \tau_{s,x}\, dx$$

the form of $\tau_{s,x}$ may be substituted from Equation 7.20 and the integration performed to obtain

$$\overline{C}_{f,x} = 1.328Re_x^{-1/2} \tag{7.30}$$

Moreover, from Equations 6.6 and 7.23 the *average* heat transfer coefficient for laminar flow is

$$\overline{h}_x = \frac{1}{x}\int_0^x h_x\, dx = 0.332\left(\frac{k}{x}\right)Pr^{1/3}\left(\frac{u_\infty}{\nu}\right)^{1/2}\int_0^x \frac{dx}{x^{1/2}}$$

Integrating and substituting from Equation 7.23, it follows that $\overline{h}_x = 2h_x$. Hence

$$\overline{Nu}_x \equiv \frac{\overline{h}_x x}{k} = 0.664Re_x^{1/2}\,Pr^{1/3} \qquad Pr \gtrsim 0.6 \tag{7.31}$$

In a similar fashion, it may be shown that

$$\overline{Sh}_x \equiv \frac{\overline{h}_{m,x} x}{D_{AB}} = 0.664Re_x^{1/2}\,Sc^{1/3} \qquad Sc \gtrsim 0.6 \tag{7.32}$$

If the flow is laminar over the entire surface, the subscript x may be replaced by L, and Equations 7.30 to 7.32 may be used to predict average conditions for the entire surface.

From the foregoing expressions we see that, for laminar flow over a flat plate, the *average* friction and convection coefficients from the leading edge to a point x on the surface are *twice* the *local* coefficients at that point. We also note that, in using these expressions, the effect of variable properties can be treated by evaluating all properties at the *film temperature,* Equation 7.2.

For fluids of small Prandtl number, namely, *liquid metals,* Equation 7.23 does not apply. However, for this case the thermal boundary layer development is much more rapid than that of the velocity boundary layer ($\delta_t \gg \delta$), and it is reasonable to assume uniform velocity ($u = u_\infty$) throughout the thermal boundary layer. From a solution to the thermal boundary layer equation based on this assumption [5], it may then be shown that

$$Nu_x = 0.565Pe_x^{1/2} \qquad Pr \lesssim 0.05, \quad Pe_x \gtrsim 100 \tag{7.33}$$

where $Pe_x \equiv Re_x\,Pr$ is the *Peclet number* (Table 6.2). Despite the corrosive and reactive nature of liquid metals, their unique properties (low melting point and vapor pressure, as well as high thermal capacity and conductivity) render them attractive as coolants in applications requiring high heat transfer rates.

A single correlating equation, which applies for all Prandtl numbers, has been recommended by Churchill and Ozoe [6]. For laminar flow over an isothermal plate, the local convection coefficient may be obtained from

$$Nu_x = \frac{0.3387Re_x^{1/2}\,Pr^{1/3}}{[1 + (0.0468/Pr)^{2/3}]^{1/4}} \qquad Pe_x \gtrsim 100 \tag{7.34}$$

with $\overline{Nu}_x = 2Nu_x$.

7.2.2 Turbulent Flow

From experiment [7] it is known that, for turbulent flows with Reynolds numbers up to approximately 10^7, the *local* friction coefficient is well correlated by an expression of the form

$$C_{f,x} = 0.0592 Re_x^{-1/5} \qquad Re_x \lesssim 10^7 \qquad (7.35)$$

The expression may also be used to within 15% accuracy for values of Re_x up to 10^8. Moreover, it is known that, to a reasonable approximation, the velocity boundary layer thickness may be expressed as

$$\delta = 0.37x\, Re_x^{-1/5} \qquad (7.36)$$

Comparing these results with those for the laminar boundary layer, Equations 7.19 and 7.20, we see that turbulent boundary layer growth is much more rapid (δ varies as $x^{4/5}$ in contrast to $x^{1/2}$ for laminar flow) and that the decay in the friction coefficient is more gradual ($x^{-1/5}$ versus $x^{-1/2}$). For turbulent flow, boundary layer development is influenced strongly by random fluctuations in the fluid and not by molecular diffusion. Hence relative boundary layer growth does not depend on the value of Pr or Sc, and Equation 7.36 may be used to obtain the thermal and concentration, as well as the velocity, boundary layer thicknesses. That is, for turbulent flow, $\delta \approx \delta_t \approx \delta_c$.

Using Equation 7.35 with the modified Reynolds, or Chilton–Colburn, analogy, Equations 6.103 and 6.104, the *local* Nusselt number for turbulent flow is

$$Nu_x = St\, Re_x\, Pr = 0.0296 Re_x^{4/5}\, Pr^{1/3} \qquad 0.6 < Pr < 60 \qquad (7.37)$$

and the *local* Sherwood number is

$$Sh_x = St_m\, Re_x\, Sc = 0.0296 Re_x^{4/5}\, Sc^{1/3} \qquad 0.6 < Sc < 3000 \qquad (7.38)$$

Enhanced mixing causes the turbulent boundary layer to grow more rapidly than the laminar boundary layer and to have larger friction and convection coefficients.

Expressions for the average coefficients may now be determined using the procedures of Section 7.2.1. However, since the turbulent boundary layer is generally preceded by a laminar boundary layer, we first consider *mixed* boundary layer conditions.

7.2.3 Mixed Boundary Layer Conditions

For laminar flow over the entire plate, Equations 7.30 to 7.32 may be used to compute the average coefficients. Moreover, if transition occurs toward the rear of the plate, for example, in the range $0.95 \lesssim (x_c/L) \lesssim 1$, these equations may be used to compute the average coefficients to a reasonable approximation. However, when transition occurs sufficiently upstream of the trailing edge, $(x_c/L) \lesssim 0.95$, the surface average coefficients will be influenced by conditions in both the laminar and turbulent boundary layers.

In the mixed boundary layer situation (Figure 7.3), Equation 6.6 may be used to obtain the average convection heat transfer coefficient for the entire plate. Integrating over the laminar region ($0 \leq x \leq x_c$) and then over the turbulent region ($x_c < x \leq L$), this equation may be expressed as

$$\overline{h}_L = \frac{1}{L}\left(\int_0^{x_c} h_{lam}\, dx + \int_{x_c}^{L} h_{turb}\, dx\right)$$

where it is assumed that transition occurs abruptly at $x = x_c$. Substituting from Equations 7.23 and 7.37 for h_{lam} and h_{turb}, respectively, we obtain

$$\overline{h}_L = \left(\frac{k}{L}\right)\left[0.332\left(\frac{u_\infty}{\nu}\right)^{1/2}\int_0^{x_c}\frac{dx}{x^{1/2}} + 0.0296\left(\frac{u_\infty}{\nu}\right)^{4/5}\int_{x_c}^{L}\frac{dx}{x^{1/5}}\right]Pr^{1/3}$$

Integrating, we then obtain

$$\overline{Nu}_L = \left[0.664Re_{x,c}^{1/2} + 0.037\left(Re_L^{4/5} - Re_{x,c}^{4/5}\right)\right]Pr^{1/3}$$
$$\overline{Nu}_L = \left(0.037Re_L^{4/5} - A\right)Pr^{1/3} \tag{7.39}$$

where the constant A is determined by the value of the critical Reynolds number $Re_{x,c}$. That is,

$$A = 0.037Re_{x,c}^{4/5} - 0.664Re_{x,c}^{1/2} \tag{7.40}$$

If a representative transition Reynolds number of $Re_{x,c} = 5 \times 10^5$ is assumed, Equation 7.39 reduces to

$$\overline{Nu}_L = (0.037Re_L^{4/5} - 871)Pr^{1/3} \tag{7.41}$$

$$\left[\begin{array}{l} 0.6 < Pr < 60 \\ 5 \times 10^5 < Re_L \lesssim 10^8 \\ Re_{x,c} = 5 \times 10^5 \end{array}\right]$$

where the bracketed relations indicate the range of applicability. In a similar manner it may be shown that, for convection mass transfer,

$$\overline{Sh}_L = (0.037Re_L^{4/5} - 871)Sc^{1/3} \tag{7.42}$$

$$\left[\begin{array}{l} 0.6 < Sc < 3000 \\ 5 \times 10^5 < Re_L \lesssim 10^8 \\ Re_{x,c} = 5 \times 10^5 \end{array}\right]$$

and for the friction coefficient

$$\overline{C}_{f,L} = \frac{0.074}{Re_L^{1/5}} - \frac{1742}{Re_L} \tag{7.43}$$

$$\left[\begin{array}{l} 5 \times 10^5 < Re_L \lesssim 10^8 \\ Re_{x,c} = 5 \times 10^5 \end{array}\right]$$

In situations for which $L \gg x_c (Re_L \gg Re_{x,c})$, $A \ll 0.037Re_L^{4/5}$ and to a reasonable approximation Equations 7.41 and 7.42 reduce to

$$\overline{Nu}_L = 0.037Re_L^{4/5}\, Pr^{1/3} \tag{7.44}$$

$$\overline{Sh}_L = 0.037Re_L^{4/5}\, Sc^{1/3} \tag{7.45}$$

Similarly, it follows that

$$\overline{C}_{f,L} = 0.074 Re_L^{-1/5} \tag{7.46}$$

Use of the above results is also appropriate when a turbulent boundary layer exists over the entire plate. Such a condition may be realized by *tripping* the boundary layer at the leading edge, using a fine wire or some other turbulence promoter.

All of the foregoing correlations require evaluation of the fluid properties at the film temperature, Equation 7.2. An alternative approach for treating the temperature dependence of the fluid properties has been suggested by Whitaker [8] on the basis of measurements by Zhukauskas [9]. Churchill [6, 10] has critically reviewed available theoretical and experimental results and has suggested a single correlating equation that applies over the complete range of *Re* and *Pr*. Different constants are suggested for use with the equation, where specific values depend on the existence of uniform surface temperature or heat flux and whether one is interested in local or average conditions.

7.2.4 Special Cases

All the foregoing Nusselt number expressions are restricted to situations for which the surface temperature T_s is uniform. A common exception involves existence of an *unheated starting length* ($T_s = T_\infty$) upstream of a heated section ($T_s \neq T_\infty$). As shown in Figure 7.4, velocity boundary layer growth begins at $x = 0$, while thermal boundary layer development begins at $x = \xi$. Hence there is no heat transfer for $0 \leq x \leq \xi$. Through use of an integral boundary layer solution [5], it is known that, for laminar flow,

$$Nu_x = \frac{Nu_x|_{\xi=0}}{[1 - (\xi/x)^{3/4}]^{1/3}} \tag{7.47}$$

where $Nu_x|_{\xi=0}$ is given by Equation 7.23. In both Nu_x and $Nu_x|_{\xi=0}$, the characteristic length x is measured from the leading edge of the unheated starting length. It has also been found that, for turbulent flow,

$$Nu_x = \frac{Nu_x|_{\xi=0}}{[1 - (\xi/x)^{9/10}]^{1/9}} \tag{7.48}$$

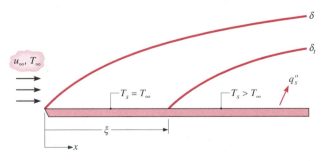

FIGURE 7.4
Flat plate in parallel flow with unheated starting length.

where $Nu_x|_{\xi=0}$ is given by Equation 7.37. Equations 7.47 and 7.48 apply for $x > \xi$, and to obtain average Nusselt numbers for $\xi < x < L$, they would have to be integrated numerically. Analogous mass transfer results are obtained by replacing (Nu_x, Pr) with (Sh_x, Sc).

It is also possible to have a uniform surface heat flux, rather than a uniform temperature, imposed at the plate. For laminar flow, it may be shown that [5]

$$Nu_x = 0.453Re_x^{1/2}\, Pr^{1/3} \qquad Pr \gtrsim 0.6 \qquad (7.49)$$

while for turbulent flow

$$Nu_x = 0.0308Re_x^{4/5}\, Pr^{1/3} \qquad 0.6 \lesssim Pr \lesssim 60 \qquad (7.50)$$

Hence the Nusselt number is 36% and 4% larger than the constant surface temperature result for laminar and turbulent flow, respectively. Correction for the effect of an unheated starting length may be made by using Equations 7.49 and 7.50 with Equations 7.47 and 7.48, respectively. If the heat flux is known, the convection coefficient may be used to determine the local surface temperature

$$T_s(x) = T_\infty + \frac{q_s''}{h_x} \qquad (7.51)$$

Since the total heat rate is readily determined from the product of the uniform flux and the surface area, $q = q_s'' A_s$, it is inappropriate to introduce an average convection coefficient for the purpose of determining q. However, one may still wish to determine an *average surface temperature* from an expression of the form

$$\overline{(T_s - T_\infty)} = \frac{1}{L}\int_0^L (T_s - T_\infty)\, dx = \frac{q_s''}{L}\int_0^L \frac{x}{k\, Nu_x}\, dx \qquad (7.52)$$

where Nu_x is obtained from the appropriate convection correlation. Substituting from Equation 7.49, it follows that

$$\overline{(T_s - T_\infty)} = \frac{q_s''L}{k\, \overline{Nu_L}} \qquad (7.53a)$$

where

$$\overline{Nu_L} = 0.680Re_L^{1/2}\, Pr^{1/3} \qquad (7.53b)$$

This result is only 2% larger than that obtained by evaluating Equation 7.31 at $x = L$. Differences are even smaller for turbulent flow, suggesting that any of the Nu_L results obtained for a uniform surface temperature may be used with Equation 7.53a to evaluate $\overline{(T_s - T_\infty)}$.

Although the equations of this section are suitable for most engineering calculations, in practice they rarely provide exact values for the convection coefficients. Conditions vary according to free stream turbulence and surface roughness, and errors as large as 25% may be incurred by using the expressions. A detailed description of free stream turbulence effects is provided by Blair [11].

7.3
Methodology for a Convection Calculation

Although we have only discussed correlations for parallel flow over a flat plate, selection and application of a convection correlation for *any flow situation* are facilitated by following a few simple rules.

1. *Become immediately cognizant of the flow geometry.* Does the problem involve flow over a flat plate, a sphere, or a cylinder? The specific form of the convection correlation depends, of course, on the geometry.

2. *Specify the appropriate reference temperature and evaluate the pertinent fluid properties at that temperature.* For moderate boundary layer temperature differences, the film temperature, Equation 7.2, may be used for this purpose. However, we will consider correlations that require property evaluation at the free stream temperature and include a property ratio to account for the nonconstant property effect.

$$T_F \equiv \frac{t_s + t_\infty}{2}$$

3. *In mass transfer problems the pertinent fluid properties are those of species B.* In our treatment of convection mass transfer, we are only concerned with *dilute, binary mixtures.* That is, the problems involve transport of some species A for which $x_A \ll 1$. To a good approximation the properties of the mixture may then be assumed to be the properties of species B. The Schmidt number, for example, would be $Sc = \nu_B/D_{AB}$ and the Reynolds number would be $Re_L = (VL/\nu_B)$.

4. *Calculate the Reynolds number.* Boundary layer conditions are strongly influenced by this parameter. If the geometry is a flat plate in parallel flow, determine whether the flow is laminar or turbulent.

5. *Decide whether a local or surface average coefficient is required.* Recall that for constant surface temperature or vapor density, the local coefficient is used to determine the flux at a particular point on the surface, whereas the average coefficient determines the transfer rate for the entire surface.

6. *Select the appropriate correlation.*

EXAMPLE 7.1

Air at a pressure of 6 kN/m² and a temperature of 300°C flows with a velocity of 10 m/s over a flat plate 0.5 m long. Estimate the cooling rate per unit width of the plate needed to maintain it at a surface temperature of 27°C.

SOLUTION

Known: Airflow over an isothermal flat plate.

Find: Cooling rate per unit width of the plate, q' (W/m).

Schematic:

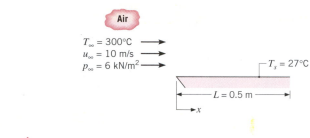

$T_\infty = 300°C$
$u_\infty = 10$ m/s
$p_\infty = 6$ kN/m²

$T_s = 27°C$

$L = 0.5$ m

x

Assumptions:

1. Steady-state conditions.
2. Negligible radiation effects.

Properties: Table A.4, air ($T_f = 437$ K, $p = 1$ atm): $\nu = 30.84 \times 10^{-6}$ m²/s, $k = 36.4 \times 10^{-3}$ W/m · K, $Pr = 0.687$. Properties such as k, Pr, and μ may be assumed to be independent of pressure to an excellent approximation. However, for a gas, the kinematic viscosity $\nu = \mu/\rho$ will vary with pressure through its dependence on density. From the ideal gas law, $\rho = p/RT$, it follows that the ratio of kinematic viscosities for a gas at the same temperature but at different pressures, p_1 and p_2, is $(\nu_1/\nu_2) = (p_2/p_1)$. Hence the kinematic viscosity of air at 437 K and $p_\infty = 6 \times 10^3$ N/m² is

$$\nu = 30.84 \times 10^{-6} \text{ m}^2/\text{s} \times \frac{1.0133 \times 10^5 \text{ N/m}^2}{6 \times 10^3 \text{ N/m}^2} = 5.21 \times 10^{-4} \text{ m}^2/\text{s}$$

Analysis: For a plate of unit width, it follows from Newton's law of cooling that the rate of convection heat transfer *to* the plate is

$$q' = \overline{h}L(T_\infty - T_s)$$

To determine the appropriate convection correlation for computing $\overline{h}$, the Reynolds number must first be determined

$$Re_L = \frac{u_\infty L}{\nu} = \frac{10 \text{ m/s} \times 0.5 \text{ m}}{5.21 \times 10^{-4} \text{ m}^2/\text{s}} = 9597$$

Hence the flow is laminar over the entire plate, and the appropriate correlation is given by Equation 7.31.

$$\overline{Nu}_L = 0.664 Re_L^{1/2} Pr^{1/3} = 0.664(9597)^{1/2}(0.687)^{1/3} = 57.4$$

The average convection coefficient is then

$$\overline{h} = \frac{\overline{Nu}_L k}{L} = \frac{57.4 \times 0.0364 \text{ W/m · K}}{0.5 \text{ m}} = 4.18 \text{ W/m}^2 · \text{K}$$

and the required cooling rate per unit width of plate is

$$q' = 4.18 \text{ W/m}^2 · \text{K} \times 0.5 \text{ m} (300 - 27)°C = 570 \text{ W/m} \qquad \triangleleft$$

Comments: The results of Table A.4 apply to gases at atmospheric pressure. Except for the kinematic viscosity, mass density, and thermal diffusivity, they may generally be used at other pressures without correction. The kinematic viscosity and thermal diffusivity for pressures other than 1 atm may be obtained by dividing the tabulated value by the pressure (atm).

EXAMPLE 7.2

A flat plate of width $w = 1$ m is maintained at a uniform surface temperature, $T_s = 230°C$, by using independently controlled, electrical strip heaters, each of which is 50 mm long. If atmospheric air at 25°C flows over the plate at a velocity of 60 m/s, at what heater is the electrical input a maximum? What is the value of this input?

SOLUTION

Known: Airflow over a flat plate with segmented heaters.

Find: Maximum heater power requirement.

Schematic:

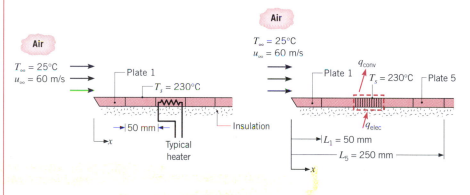

Assumptions:

1. Steady-state conditions.
2. Negligible radiation effects.
3. Bottom surface of plate adiabatic.

Properties: Table A.4, air ($T_f = 400$ K, $p = 1$ atm): $\nu = 26.41 \times 10^{-6}$ m²/s, $k = 0.0338$ W/m · K, $Pr = 0.690$.

Analysis: The location of the heater requiring the maximum electrical power may be determined by first finding the point of boundary layer transition. The Reynolds number based on the length L_1 of the first heater is

$$Re_1 = \frac{u_\infty L_1}{\nu} = \frac{60 \text{ m/s} \times 0.05 \text{ m}}{26.41 \times 10^{-6} \text{ m}^2/\text{s}} = 1.14 \times 10^5$$

If the transition Reynolds number is assumed to be $Re_{x,c} = 5 \times 10^5$, it follows that transition will occur on the fifth heater, or more specifically at

$$x_c = \frac{\nu}{u_\infty} Re_{x,c} = \frac{26.41 \times 10^{-6} \text{ m}^2/\text{s}}{60 \text{ m/s}} 5 \times 10^5 = 0.22 \text{ m}$$

The heater requiring the maximum electrical power is that for which the average convection coefficient is largest. Knowing how the local convection coefficient varies with distance from the leading edge, we conclude that there are three possibilities:

1. Heater 1, since it corresponds to the largest local, laminar convection coefficient.
2. Heater 5, since it corresponds to the largest local, turbulent convection coefficient.
3. Heater 6, since turbulent conditions exist over the entire heater.

For each of these heaters, conservation of energy requires that

$$q_{elec} = q_{conv}$$

For the first heater,

$$q_{conv,1} = \overline{h}_1 L_1 w (T_s - T_\infty)$$

where $\overline{h}_1$ is determined from Equation 7.31,

$$\overline{Nu}_1 = 0.664 Re_1^{1/2} Pr^{1/3} = 0.664(1.14 \times 10^5)^{1/2}(0.69)^{1/3} = 198$$

Hence

$$\overline{h}_1 = \frac{\overline{Nu}_1 k}{L_1} = \frac{198 \times 0.0338 \text{ W/m} \cdot \text{K}}{0.05 \text{ m}} = 134 \text{ W/m}^2 \cdot \text{K}$$

and

$$q_{conv,1} = 134 \text{ W/m}^2 \cdot \text{K} (0.05 \times 1) \text{ m}^2 (230 - 25)°\text{C} = 1370 \text{ W}$$

The power requirement for the fifth heater may be obtained by subtracting the total heat loss associated with the first four heaters from that associated with the first five heaters. Accordingly,

$$q_{conv,5} = \overline{h}_{1-5} L_5 w (T_s - T_\infty) - \overline{h}_{1-4} L_4 w (T_s - T_\infty)$$

$$q_{conv,5} = (\overline{h}_{1-5} L_5 - \overline{h}_{1-4} L_4) w (T_s - T_\infty)$$

The value of $\overline{h}_{1-4}$ may be obtained from Equation 7.31, where

$$\overline{Nu}_4 = 0.664 Re_4^{1/2} Pr^{1/3}$$

With $Re_4 = 4Re_1 = 4.56 \times 10^5$,

$$\overline{Nu}_4 = 0.664(4.56 \times 10^5)^{1/2}(0.69)^{1/3} = 396$$

Hence

$$\overline{h}_{1-4} = \frac{\overline{Nu}_4 k}{L_4} = \frac{396 \times 0.0338 \text{ W/m} \cdot \text{K}}{0.2 \text{ m}} = 67 \text{ W/m}^2 \cdot \text{K}$$

In contrast, the fifth heater is characterized by mixed boundary layer conditions, and $\overline{h}_{1-5}$ must be obtained from Equation 7.41. With $Re_5 = 5Re_1 = 5.70 \times 10^5$,

$$\overline{Nu}_5 = (0.037 Re_5^{4/5} - 871) \, Pr^{1/3}$$

$$\overline{Nu}_5 = [0.037(5.70 \times 10^5)^{4/5} - 871](0.69)^{1/3} = 546$$

Hence

$$\overline{h}_{1-5} = \frac{\overline{Nu}_5 \, k}{L_5} = \frac{546 \times 0.0338 \text{ W/m} \cdot \text{K}}{0.25 \text{ m}} = 74 \text{ W/m}^2 \cdot \text{K}$$

The rate of heat transfer from the fifth heater is then

$$q_{\text{conv}, 5} = (74 \text{ W/m}^2 \cdot \text{K} \times 0.25 \text{ m} - 67 \text{ W/m}^2 \cdot \text{K} \times 0.20 \text{ m})$$
$$\times 1 \text{ m} (230 - 25)°\text{C}$$

$$q_{\text{conv}, 5} = 1050 \text{ W}$$

Similarly, the power requirement for the sixth heater may be obtained by subtracting the total heat loss associated with the first five heaters from that associated with the first six heaters. Hence

$$q_{\text{conv}, 6} = (\overline{h}_{1-6} L_6 - \overline{h}_{1-5} L_5) w (T_s - T_\infty)$$

where $\overline{h}_{1-6}$ may be obtained from Equation 7.41. With $Re_6 = 6Re_1 = 6.84 \times 10^5$,

$$\overline{Nu}_6 = [0.037(6.84 \times 10^5)^{4/5} - 871](0.69)^{1/3} = 753$$

Hence

$$\overline{h}_{1-6} = \frac{\overline{Nu}_6 \, k}{L_6} = \frac{753 \times 0.0338 \text{ W/m} \cdot \text{K}}{0.30 \text{ m}} = 85 \text{ W/m}^2 \cdot \text{K}$$

and

$$q_{\text{conv}, 6} = (85 \text{ W/m}^2 \cdot \text{K} \times 0.30 \text{ m} - 74 \text{ W/m}^2 \cdot \text{K} \times 0.25 \text{ m})$$
$$\times 1 \text{ m} (230 - 25)°\text{C}$$

$$q_{\text{conv}, 6} = 1440 \text{ W} \qquad \triangleleft$$

Hence $q_{\text{conv}, 6} > q_{\text{conv}, 1} > q_{\text{conv}, 5}$, and the sixth plate has the largest power requirement.

Comments:

1. An alternative method of finding the convection heat transfer rate from a particular plate involves estimating an average local convection coefficient for the surface. For example, Equation 7.37 could be used to evaluate the local convection coefficient at the midpoint of the sixth plate. With $x = 0.275$ m, $Re_x = 6.27 \times 10^5$, $Nu_x = 1136$, and $h_x = 140 \text{ W/m}^2 \cdot \text{K}$, the convection heat transfer rate from the sixth plate is

$$q_{\text{conv}, 6} = h_x (L_6 - L_5) w (T_s - T_\infty)$$

$$q_{\text{conv}, 6} = 140 \text{ W/m}^2 \cdot \text{K} (0.30 - 0.25) \text{ m} \times 1 \text{ m} (230 - 25)°\text{C} = 1440 \text{ W}$$

This procedure may be used only when the variation of the local convection coefficient with distance is gradual, such as in turbulent flow. It could lead to significant error when used for a surface that experiences transition.

2. The variation of the local convection coefficient along the flat plate may be determined from Equations 7.21 and 7.30 for laminar and turbulent flow, respectively, and the results are represented by the solid curves of the following schematic:

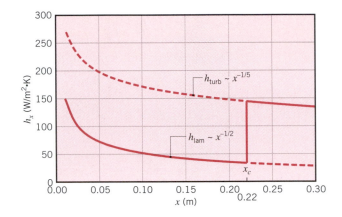

The $x^{-1/2}$ decay of the laminar convection coefficient is presumed to conclude abruptly at $x_c = 0.22$ m, where transition yields more than a fourfold increase in the local convection coefficient. For $x > x_c$, the decay in the convection coefficient is more gradual ($x^{-1/5}$). The dashed lines represent extensions of the distributions, which would apply if the value of x_c were shifted. For example, if the free stream turbulence were to increase and/or the surface were to be roughened, $Re_{x, c}$ would decrease. The smaller value of x_c would cause the laminar and turbulent distributions, respectively, to extend over smaller and larger portions of the plate. A similar effect may be achieved by increasing u_∞. In this case larger values of h_x would be associated with the laminar and turbulent distributions ($h_{lam} \sim u_\infty^{1/2}$, $h_{turb} \sim u_\infty^{4/5}$).

EXAMPLE 7.3

Drought conditions in the southwestern region of the country have prompted officials to question whether the operation of residential swimming pools should be permitted. As the chief engineer of a city that has a large number of pools, you have been asked to estimate the *daily* water loss due to pool evaporation. As representative conditions, you may assume water and ambient air temperatures of 25°C, an ambient relative humidity of 50%, pool surface dimensions of 6 m by 12 m, and a wind speed of 2 m/s in the direction of the long side of the pool. You may assume the free stream turbulence of the air to be negligible and the surface of the water to be smooth and level with the pool deck. What is the water loss for the pool in kilograms per day?

SOLUTION

Known: Ambient air conditions above a swimming pool.

Find: Daily evaporative water loss.

Schematic:

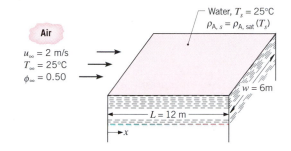

Assumptions:

1. Steady-state conditions.
2. Smooth water surface and negligible free stream turbulence.
3. Heat and mass transfer analogy applicable.
4. Transition Reynolds number 5×10^5.
5. Ideal gas behavior for water vapor in free stream.

Properties: Table A.4, air (25°C): $\nu = 15.7 \times 10^{-6}$ m²/s. Table A.8, water vapor–air (25°C): $D_{AB} = 0.26 \times 10^{-4}$ m²/s, $Sc = \nu/D_{AB} = 0.60$. Table A.6, saturated water vapor (25°C): $\rho_{A,sat} = v_g^{-1} = 0.0226$ kg/m³.

Analysis: With a Reynolds number of

$$Re_L = \frac{u_\infty L}{\nu} = \frac{2 \text{ m/s} \times 12 \text{ m}}{15.7 \times 10^{-6} \text{ m}^2/\text{s}} = 1.53 \times 10^6$$

transition occurs at $x_c = (5 \times 10^5/1.53 \times 10^6)12 = 3.9$ m. Hence a mixed boundary layer condition exists, and Equation 7.42 gives

$$\overline{Sh}_L = (0.037Re_L^{4/5} - 871)Sc^{1/3}$$

$$\overline{Sh}_L = [0.037(1.53 \times 10^6)^{4/5} - 871](0.60)^{1/3} = 2032$$

It follows that

$$\overline{h}_{m,L} = \overline{Sh}_L \left(\frac{D_{AB}}{L}\right) = 2032 \frac{0.26 \times 10^{-4} \text{ m}^2/\text{s}}{12 \text{ m}} = 4.4 \times 10^{-3} \text{ m/s}$$

The evaporation rate for the pool is then

$$n_A = \overline{h}_m A(\rho_{A,s} - \rho_{A,\infty})$$

or, with the free stream water vapor assumed to be an ideal gas

$$\phi_\infty = \frac{\rho_{A,\infty}}{\rho_{A,sat}(T_\infty)}$$

and with $\rho_{A,s} = \rho_{A,sat}(T_s)$,

$$n_A = \bar{h}_m A[\rho_{A,sat}(T_s) - \phi_\infty \rho_{A,sat}(T_\infty)]$$

Since $T_s = T_\infty = 25°C$, it follows that

$$n_A = \bar{h}_m A \rho_{A,sat}(25°C)[1 - \phi_\infty]$$

Hence

$$n_A = 4.4 \times 10^{-3} \text{ m/s} \times 72 \text{ m}^2 \times 0.0226 \text{ kg/m}^3 \times 0.5 \times 86,400 \text{ s/day}$$

$$n_A = 309 \text{ kg/day} \qquad\qquad\qquad\qquad\qquad \triangleleft$$

Comments: The water surface temperature is likely to be slightly less than the air temperature because of the evaporative cooling effect.

7.4
The Cylinder in Cross Flow

7.4.1   Flow Considerations

Another common external flow involves fluid motion normal to the axis of a circular cylinder. As shown in Figure 7.5, the free stream fluid is brought to rest at the *forward stagnation point,* with an accompanying rise in pressure. From this point, the pressure decreases with increasing x, the streamline coordinate, and the boundary layer develops under the influence of a *favorable pressure gradient* ($dp/dx < 0$). However, the pressure must eventually reach a minimum, and toward the rear of the cylinder further boundary layer development occurs in the presence of an *adverse pressure gradient* ($dp/dx > 0$).

In Figure 7.5 the distinction between the upstream velocity V and the free stream velocity u_∞ should be noted. Unlike conditions for the flat plate in parallel flow, these velocities differ, with u_∞ now depending on the distance x from the stagnation point. From Euler's equation for an inviscid flow [12], $u_\infty(x)$ must exhibit behavior opposite to that of $p(x)$. That is, from $u_\infty = 0$ at the stagnation point, the fluid accelerates because of the favorable pressure gradient

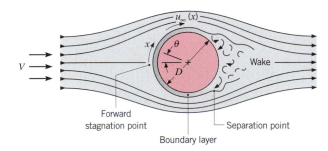

FIGURE 7.5   Boundary layer formation and separation on a circular cylinder in cross flow.

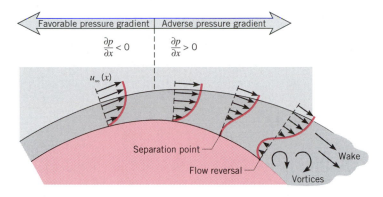

FIGURE 7.6 Velocity profile associated with separation on a circular cylinder in cross flow.

($du_\infty/dx > 0$ when $dp/dx < 0$), reaches a maximum velocity when $dp/dx = 0$, and decelerates because of the adverse pressure gradient ($du_\infty/dx < 0$ when $dp/dx > 0$). As the fluid decelerates, the velocity gradient at the surface, $\partial u/\partial y|_{y=0}$, eventually becomes zero (Figure 7.6). At this location, termed the *separation point,* fluid near the surface lacks sufficient momentum to overcome the pressure gradient, and continued downstream movement is impossible. Since the oncoming fluid also precludes flow back upstream, *boundary layer separation* must occur. This is a condition for which the boundary layer detaches from the surface, and a *wake* is formed in the downstream region. Flow in this region is characterized by vortex formation and is highly irregular. The *separation point* is the location for which $(\partial u/\partial y)_s = 0$.

The occurrence of *boundary layer transition,* which depends on the Reynolds number, strongly influences the position of the separation point. For the circular cylinder the characteristic length is the diameter, and the Reynolds number is defined as

$$Re_D \equiv \frac{\rho V D}{\mu} = \frac{V D}{\nu}$$

Since the momentum of fluid in a turbulent boundary layer is larger than in the laminar boundary layer, it is reasonable to expect transition to delay the occurrence of separation. If $Re_D \lesssim 2 \times 10^5$, the boundary layer remains laminar, and separation occurs at $\theta \approx 80°$ (Figure 7.7). However, if $Re_D \gtrsim 2 \times 10^5$, boundary layer transition occurs, and separation is delayed to $\theta \approx 140°$.

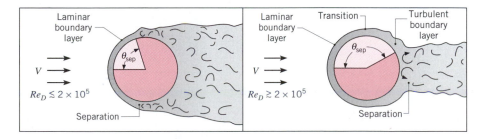

FIGURE 7.7 The effect of turbulence on separation.

The foregoing processes strongly influence the drag force, F_D, acting on the cylinder. This force has two components, one of which is due to the boundary layer surface shear stress (*friction drag*). The other component is due to a pressure differential in the flow direction resulting from formation of the wake (*form, or pressure, drag*). A dimensionless *drag coefficient* C_D may be defined as

$$C_D \equiv \frac{F_D}{A_f(\rho V^2/2)} \tag{7.54}$$

where A_f is the cylinder frontal area (the area projected perpendicular to the free stream velocity). The drag coefficient is a function of Reynolds number and results are presented in Figure 7.8. For $Re_D < 2$ separation effects are negligible, and conditions are dominated by friction drag. However, with increasing Reynolds number, the effect of separation, and therefore form drag, becomes more important. The large reduction in C_D that occurs for $Re_D > 2 \times 10^5$ is due to boundary layer transition, which delays separation, thereby reducing the extent of the wake region and the magnitude of the form drag.

7.4.2 Convection Heat and Mass Transfer

Experimental results for the variation of the local Nusselt number with θ are shown in Figure 7.9 for the cylinder in a cross flow of air. Not unexpectedly, the results are strongly influenced by the nature of boundary layer development on the surface. Consider conditions for $Re_D \lesssim 10^5$. Starting at the stagnation point, Nu_θ decreases with increasing θ as a result of laminar boundary layer development. However, a minimum is reached at $\theta \approx 80°$, where separation occurs and Nu_θ increases with θ due to mixing associated with vortex formation in the wake. In contrast, for $Re_D \gtrsim 10^5$ the variation of Nu_θ with θ is character-

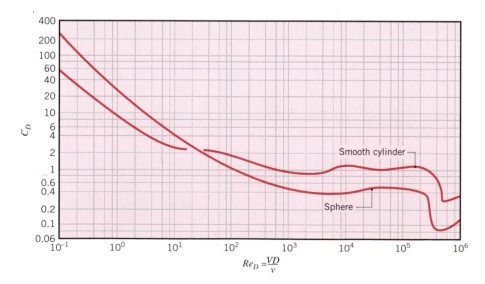

FIGURE 7.8   Drag coefficients for smooth circular cylinder in cross flow and for a sphere [7]. Adapted with permission.

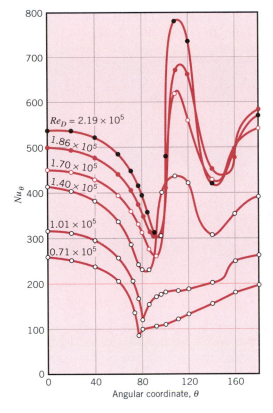

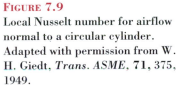

FIGURE 7.9
Local Nusselt number for airflow normal to a circular cylinder. Adapted with permission from W. H. Giedt, *Trans. ASME*, **71**, 375, 1949.

ized by two minima. The decline in Nu_θ from the value at the stagnation point is again due to laminar boundary layer development, but the sharp increase that occurs between 80° and 100° is now due to boundary layer transition to turbulence. With further development of the turbulent boundary layer, Nu_θ again begins to decline. Eventually separation occurs ($\theta \approx 140°$), and Nu_θ increases as a result of mixing in the wake region. The increase in Nu_θ with increasing Re_D is due to a corresponding reduction in the boundary layer thickness.

Correlations may be obtained for the local Nusselt number, and at the forward stagnation point for $Pr \geq 0.6$, boundary layer analysis [5] yields an expression of the form

$$Nu_D(\theta = 0) = 1.15 Re_D^{1/2} Pr^{1/3} \qquad (7.55a)$$

However, from the standpoint of engineering calculations, we are more interested in overall average conditions. The empirical correlation due to Hilpert [13]

$$\overline{Nu_D} \equiv \frac{\overline{h}D}{k} = C\,Re_D^m\,Pr^{1/3} \qquad (7.55b)$$

is widely used, where the constants C and m are listed in Table 7.2. Equation 7.55 may also be used for *gas flow* over cylinders of noncircular cross section, with the characteristic length D and the constants obtained from Table 7.3. In working with Equations 7.55 all properties are evaluated at the film temperature.

TABLE 7.2   Constants of Equation 7.55b for the circular cylinder in cross flow [13, 14]

Re_D	C	m
0.4–4	0.989	0.330
4–40	0.911	0.385
40–4000	0.683	0.466
4000–40,000	0.193	0.618
40,000–400,000	0.027	0.805

Other correlations have been suggested for the circular cylinder in cross flow [8, 16, 17]. The correlation due to Zhukauskas [16] is of the form

$$\overline{Nu}_D = C\,Re_D^m\,Pr^n\left(\frac{Pr}{Pr_s}\right)^{1/4} \tag{7.56}$$

$$\begin{bmatrix} 0.7 < Pr < 500 \\ 1 < Re_D < 10^6 \end{bmatrix}$$

where all properties are evaluated at T_∞, except Pr_s, which is evaluated at T_s. Values of C and m are listed in Table 7.4. If $Pr \leq 10$, $n = 0.37$; if $Pr > 10$, $n = 0.36$. Churchill and Bernstein [17] have proposed a single comprehensive equation that covers the entire range of Re_D for which data are available, as well as a wide range of Pr. The equation is recommended for all $Re_D\,Pr > 0.2$ and has the form

$$\overline{Nu}_D = 0.3 + \frac{0.62Re_D^{1/2}\,Pr^{1/3}}{[1 + (0.4/Pr)^{2/3}]^{1/4}}\left[1 + \left(\frac{Re_D}{282,000}\right)^{5/8}\right]^{4/5} \tag{7.57}$$

where all properties are evaluated at the film temperature.

TABLE 7.3   Constants of Equation 7.55b for noncircular cylinders in cross flow of a gas [15]

Geometry		Re_D	C	m
Square				
$V \rightarrow$ ◇	D	5×10^3–10^5	0.246	0.588
$V \rightarrow$ ▢	D	5×10^3–10^5	0.102	0.675
Hexagon				
$V \rightarrow$ ⬡	D	5×10^3–1.95×10^4	0.160	0.638
		1.95×10^4–10^5	0.0385	0.782
$V \rightarrow$ ⬡	D	5×10^3–10^5	0.153	0.638
Vertical plate				
$V \rightarrow$ ▯	D	4×10^3–1.5×10^4	0.228	0.731

TABLE 7.4 Constants of Equation 7.56 for the circular cylinder in cross flow [16]

Re_D	C	m
1–40	0.75	0.4
40–1000	0.51	0.5
10^3–2×10^5	0.26	0.6
2×10^5–10^6	0.076	0.7

Again we caution the reader not to view any of the foregoing correlations as sacrosanct. Each correlation is reasonable over a certain range of conditions, but for most engineering calculations one should not expect accuracy to much better than 20%. Because they are based on more recent results encompassing a wide range of conditions, Equations 7.56 and 7.57 are used for the calculations of this text. A detailed review of the many correlations that have been developed for the circular cylinder is provided by Morgan [18].

Finally, we note that by invoking the heat and mass transfer analogy, Equations 7.55 to 7.57 may be applied to problems involving convection mass transfer from a cylinder in cross flow. It is simply a matter of replacing $\overline{Nu}_D$ by $\overline{Sh}_D$ and Pr by Sc. In mass transfer problems, boundary layer property variations are typically small. Hence, when using the mass transfer analog of Equation 7.56, the property ratio, which accounts for nonconstant property effects, may be neglected.

EXAMPLE 7.4

Experiments have been conducted on a metallic cylinder 12.7 mm in diameter and 94 mm long. The cylinder is heated internally by an electrical heater and is subjected to a cross flow of air in a low-speed wind tunnel. Under a specific set of operating conditions for which the upstream air velocity and temperature were maintained at $V = 10$ m/s and 26.2°C, respectively, the heater power dissipation was measured to be $P = 46$ W, while the average cylinder surface temperature was determined to be $T_s = 128.4$°C. It is estimated that 15% of the power dissipation is lost through the cumulative effect of surface radiation and conduction through the endpieces.

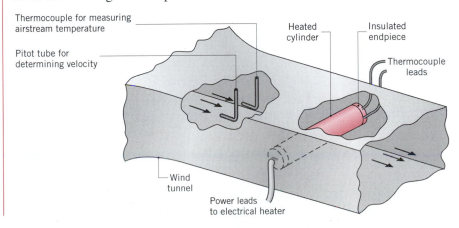

Thermocouple for measuring airstream temperature

Pitot tube for determining velocity

Heated cylinder

Insulated endpiece

Thermocouple leads

Wind tunnel

Power leads to electrical heater

1. Determine the convection heat transfer coefficient from the experimental observations.

2. Compare the experimental result with the convection coefficient computed from an appropriate correlation.

SOLUTION

Known: Operating conditions for a heated cylinder.

Find:

1. Convection coefficient associated with the operating conditions.

2. Convection coefficient from an appropriate correlation.

Schematic:

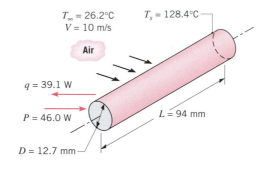

Assumptions:

1. Steady-state conditions.

2. Uniform cylinder surface temperature.

Properties: Table A.4, air ($T_\infty = 26.2°C \approx 300$ K): $\nu = 15.89 \times 10^{-6}$ m²/s, $k = 26.3 \times 10^{-3}$ W/m · K, $Pr = 0.707$. Table A.4, air ($T_f \approx 350$ K): $\nu = 20.92 \times 10^{-6}$ m²/s, $k = 30 \times 10^{-3}$ W/m · K, $Pr = 0.700$. Table A.4, air ($T_s = 128.4°C = 401$ K): $Pr = 0.690$.

Analysis:

1. The convection heat transfer coefficient may be determined from the data by using Newton's law of cooling. That is,

$$\bar{h} = \frac{q}{A(T_s - T_\infty)}$$

With $q = 0.85P$ and $A = \pi DL$, it follows that

$$\bar{h} = \frac{0.85 \times 46 \text{ W}}{\pi \times 0.0127 \text{ m} \times 0.094 \text{ m } (128.4 - 26.2)°C} = 102 \text{ W/m}^2 \cdot \text{K} \quad \triangleleft$$

2. Working with the Zhukauskas relation, Equation 7.56,

$$\overline{Nu}_D = C\, Re_D^m\, Pr^n \left(\frac{Pr}{Pr_s} \right)^{1/4}$$

all properties, except Pr_s, are evaluated at T_∞. Accordingly,

$$Re_D = \frac{VD}{\nu} = \frac{10 \text{ m/s} \times 0.0127 \text{ m}}{15.89 \times 10^{-6} \text{ m}^2/\text{s}} = 7992$$

Hence, from Table 7.4, $C = 0.26$ and $m = 0.6$. Also, since $Pr < 10$, $n = 0.37$. It follows that

$$\overline{Nu}_D = 0.26(7992)^{0.6}(0.707)^{0.37}\left(\frac{0.707}{0.690}\right)^{0.25} = 50.5$$

$$\bar{h} = \overline{Nu}_D \frac{k}{D} = 50.5 \frac{0.0263 \text{ W/m} \cdot \text{K}}{0.0127 \text{ m}} = 105 \text{ W/m}^2 \cdot \text{K} \qquad \triangleleft$$

Comments:

1. Using the Churchill relation, Equation 7.57,

$$\overline{Nu}_D = 0.3 + \frac{0.62Re_D^{1/2} Pr^{1/3}}{[1 + (0.4/Pr)^{2/3}]^{1/4}}\left[1 + \left(\frac{Re_D}{282,000}\right)^{5/8}\right]^{4/5}$$

With all properties evaluated at T_f, $Pr = 0.70$ and

$$Re_D = \frac{VD}{\nu} = \frac{10 \text{ m/s} \times 0.0127 \text{ m}}{20.92 \times 10^{-6} \text{ m}^2/\text{s}} = 6071$$

Hence the Nusselt number and the convection coefficient are

$$\overline{Nu}_D = 0.3 + \frac{0.62(6071)^{1/2}(0.70)^{1/3}}{[1 + (0.4/0.70)^{2/3}]^{1/4}}\left[1 + \left(\frac{6071}{282,000}\right)^{5/8}\right]^{4/5} = 40.6$$

$$\bar{h} = \overline{Nu}_D \frac{k}{D} = 40.6 \frac{0.030 \text{ W/m} \cdot \text{K}}{0.0127 \text{ m}} = 96.0 \text{ W/m}^2 \cdot \text{K}$$

Alternatively, from the Hilpert correlation, Equation 7.55b,

$$\overline{Nu}_D = C Re_D^m Pr^{1/3}$$

With all properties evaluated at the film temperature, $Re_D = 6071$ and $Pr = 0.70$. Hence, from Table 7.2, $C = 0.193$ and $m = 0.618$. The Nusselt number and the convection coefficient are then

$$\overline{Nu}_D = 0.193(6071)^{0.618}(0.700)^{0.333} = 37.3$$

$$\bar{h} = \overline{Nu}_D \frac{k}{D} = 37.3 \frac{0.030 \text{ W/m} \cdot \text{K}}{0.0127 \text{ m}} = 88 \text{ W/m}^2 \cdot \text{K}$$

2. Uncertainties associated with measuring the air velocity, estimating the heat loss from cylinder ends, and averaging the cylinder surface temperature, which varies axially and circumferentially, render the experimental result accurate to no better than 15%. Accordingly, calculations based on each of the three correlations are within the experimental uncertainty of the measured result.

3. Recognize the importance of using the proper temperature when evaluating fluid properties.

7.5
The Sphere

Boundary layer effects associated with flow over a sphere are much like those for the circular cylinder, with transition and separation playing prominent roles. Results for the drag coefficient, which is defined by Equation 7.54, are presented in Figure 7.8. In the limit of very small Reynolds numbers (*creeping flow*), the coefficient is inversely proportional to the Reynolds number and the specific relation is termed *Stokes' law*

$$C_D = \frac{24}{Re_D} \qquad Re_D < 0.5 \tag{7.58}$$

Numerous heat transfer correlations have been proposed, and Whitaker [8] recommends an expression of the form

$$\overline{Nu}_D = 2 + (0.4Re_D^{1/2} + 0.06Re_D^{2/3})\, Pr^{0.4} \left(\frac{\mu}{\mu_s}\right)^{1/4} \tag{7.59}$$

$$\begin{bmatrix} 0.71 < Pr < 380 \\ 3.5 < Re_D < 7.6 \times 10^4 \\ 1.0 < \left(\dfrac{\mu}{\mu_s}\right) < 3.2 \end{bmatrix}$$

All properties except μ_s are evaluated at T_∞, and the result may be applied to mass transfer problems simply by replacing $\overline{Nu}_D$ and Pr with $\overline{Sh}_D$ and Sc, respectively. A special case of convection heat and mass transfer from spheres relates to transport from freely falling liquid drops, and the correlation of Ranz and Marshall [19] is often used

$$\overline{Nu}_D = 2 + 0.6Re_D^{1/2}\, Pr^{1/3} \tag{7.60}$$

In the limit $Re_D \rightarrow 0$, Equations 7.59 and 7.60 reduce to $\overline{Nu}_D = 2$, which corresponds to heat transfer by conduction from a spherical surface to a stationary, infinite medium around the surface.

EXAMPLE 7.5

The decorative plastic film on a copper sphere of 10-mm diameter is cured in an oven at 75°C. Upon removal from the oven, the sphere is subjected to an airstream at 1 atm and 23°C having a velocity of 10 m/s. Estimate how long it will take to cool the sphere to 35°C.

SOLUTION

Known: Sphere cooling in an airstream.

Find: Time t required to cool from $T_i = 75°C$ to $T(t) = 35°C$.

Schematic:

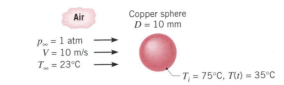

Air

Copper sphere
$D = 10$ mm

$p_\infty = 1$ atm
$V = 10$ m/s
$T_\infty = 23°C$

$T_i = 75°C$, $T(t) = 35°C$

Assumptions:

1. Negligible thermal resistance and capacitance for the plastic film.
2. Spatially isothermal sphere.
3. Negligible radiation effects.

Properties: Table A.1, copper ($T \approx 328$ K): $\rho = 8933$ kg/m^3, $k = 399$ W/m · K, $c_p = 387$ J/kg · K. Table A.4, air ($T_\infty = 296$ K): $\mu = 181.6 \times 10^{-7}$ N · s/m^2, $\nu = 15.36 \times 10^{-6}$ m^2/s, $k = 0.0258$ W/m · K, $Pr = 0.709$. Table A.4, air ($T_s \approx 328$ K): $\mu = 197.8 \times 10^{-7}$ N · s/m^2.

Analysis: The time required to complete the cooling process may be obtained from results for a lumped capacitance. In particular, from Equations 5.4 and 5.5

$$t = \frac{\rho V c_p}{\overline{h} A_s} \ln \frac{T_i - T_\infty}{T - T_\infty}$$

or, with $V = \pi D^3/6$ and $A_s = \pi D^2$,

$$t = \frac{\rho c_p D}{6\overline{h}} \ln \frac{T_i - T_\infty}{T - T_\infty}$$

From Equation 7.59

$$\overline{Nu}_D = 2 + (0.4 Re_D^{1/2} + 0.06 Re_D^{2/3}) \, Pr^{0.4} \left(\frac{\mu}{\mu_s}\right)^{1/4}$$

where

$$Re_D = \frac{VD}{\nu} = \frac{10 \text{ m/s} \times 0.01 \text{ m}}{15.36 \times 10^{-6} \text{ m}^2/\text{s}} = 6510$$

Hence the Nusselt number and the convection coefficient are

$$\overline{Nu}_D = 2 + [0.4(6510)^{1/2} + 0.06(6510)^{2/3}](0.709)^{0.4}$$

$$\times \left(\frac{181.6 \times 10^{-7} \text{ N} \cdot \text{s/m}^2}{197.8 \times 10^{-7} \text{ N} \cdot \text{s/m}^2}\right)^{1/4} = 47.4$$

$$\overline{h} = \overline{Nu}_D \frac{k}{D} = 47.4 \frac{0.0258 \text{ W/m} \cdot \text{K}}{0.01 \text{ m}} = 122 \text{ W/m}^2 \cdot \text{K}$$

The time required for cooling is then

$$t = \frac{8933 \text{ kg/m}^3 \times 387 \text{ J/kg} \cdot \text{K} \times 0.01 \text{ m}}{6 \times 122 \text{ W/m}^2 \cdot \text{K}} \ln\left(\frac{75 - 23}{35 - 23}\right) = 69.2 \text{ s} \quad \triangleleft$$

Comments:

1. The validity of the lumped capacitance method may be determined by calculating the Biot number. From Equation 5.10

$$Bi = \frac{\overline{h}L_c}{k_s} = \frac{\overline{h}(r_o/3)}{k_s} = \frac{122 \text{ W/m}^2 \cdot \text{K} \times 0.005 \text{ m/3}}{399 \text{ W/m} \cdot \text{K}} = 5.1 \times 10^{-4}$$

and the criterion is satisfied.

2. Although their definitions are similar, the Nusselt number is defined in terms of the thermal conductivity of the fluid, whereas the Biot number is defined in terms of the thermal conductivity of the solid.

3. Options for enhancing production rates include accelerating the cooling process by increasing the fluid velocity and/or using a different fluid. Applying the foregoing procedures, the cooling time is computed and plotted for air and helium over the range of velocities, $5 \leq V \leq 25$ m/s.

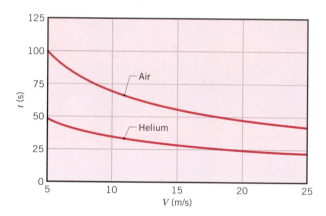

Although Reynolds numbers for He are much smaller than those for air, the thermal conductivity is much larger and, as shown below, convection heat transfer is enhanced.

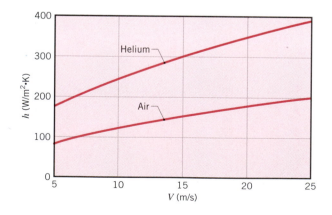

Hence production rates could be increased by substituting helium for air, albeit with a significant increase in cost.

7.6
Flow Across Banks of Tubes

Heat transfer to or from a bank (or bundle) of tubes in cross flow is relevant to numerous industrial applications, such as steam generation in a boiler or air cooling in the coil of an air conditioner. The geometric arrangement is shown schematically in Figure 7.10. Typically, one fluid moves over the tubes, while a second fluid at a different temperature passes through the tubes. In this section we are specifically interested in the convection heat transfer associated with cross flow over the tubes.

The tube rows of a bank are either *staggered* or *aligned* in the direction of the fluid velocity V (Figure 7.11). The configuration is characterized by the tube diameter D and by the *transverse pitch* S_T and *longitudinal pitch* S_L measured between tube centers. Flow conditions within the bank are dominated by boundary layer separation effects and by wake interactions, which in turn influence convection heat transfer.

The heat transfer coefficient associated with a tube is determined by its position in the bank. The coefficient for a tube in the first row is approximately equal to that for a single tube in cross flow, whereas larger heat transfer coefficients are associated with tubes of the inner rows. The tubes of the first few rows act as a turbulence grid, which increases the heat transfer coefficient for tubes in the following rows. In most configurations, however, heat transfer conditions stabilize, such that little change occurs in the convection coefficient for a tube beyond the fourth or fifth row.

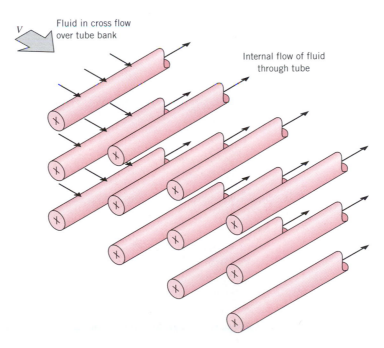

FIGURE 7.10   Schematic of a tube bank in cross flow.

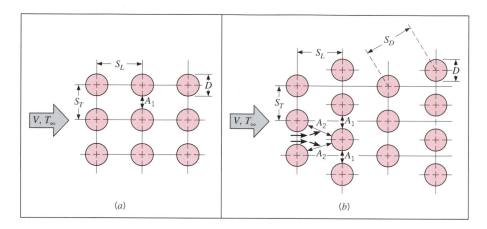

Figure 7.11 Tube arrangements in a bank. (*a*) Aligned. (*b*) Staggered.

Generally, we wish to know the *average* heat transfer coefficient for the *entire* tube bundle. For airflow across tube bundles composed of *10 or more rows* ($N_L \geq 10$), Grimison [20] has obtained a correlation of the form

$$\overline{Nu}_D = C_1 \, Re_{D,\,max}^m \qquad \begin{bmatrix} N_L \geq 10 \\ 2000 < Re_{D,\,max} < 40{,}000 \\ Pr = 0.7 \end{bmatrix} \qquad (7.61)$$

where C_1 and m are listed in Table 7.5 and

$$Re_{D,\,max} \equiv \frac{\rho V_{max} D}{\mu} \qquad (7.62)$$

It has become common practice to extend this result to other fluids through insertion of the factor $1.13 Pr^{1/3}$, in which case

$$\overline{Nu}_D = 1.13 C_1 \, Re_{D,\,max}^m \, Pr^{1/3} \qquad (7.63)$$

$$\begin{bmatrix} N_L \geq 10 \\ 2000 < Re_{D,\,max} < 40{,}000 \\ Pr \geq 0.7 \end{bmatrix}$$

All properties appearing in the above equations are evaluated at the film temperature. If $N_L < 10$, a correction factor may be applied such that

$$\overline{Nu}_D\big|_{(N_L<10)} = C_2 \, \overline{Nu}_D\big|_{(N_L \geq 10)} \qquad (7.64)$$

where C_2 is given in Table 7.6.

The Reynolds number $Re_{D,\,max}$ for the foregoing correlations is based on the *maximum fluid velocity* occurring within the tube bank. For the aligned arrangement, V_{max} occurs at the transverse plane A_1 of Figure 7.11*a*, and from the mass conservation requirement for an incompressible fluid

$$V_{max} = \frac{S_T}{S_T - D} V \qquad (7.65)$$

TABLE 7.5 Constants of Equations 7.61 and 7.63 for airflow over a tube bank of 10 or more rows [20]

| | S_T/D | | | | | | | |
| | 1.25 | | 1.5 | | 2.0 | | 3.0 | |
S_L/D	C_1	m	C_1	m	C_1	m	C_1	m
Aligned								
1.25	0.348	0.592	0.275	0.608	0.100	0.704	0.0633	0.752
1.50	0.367	0.586	0.250	0.620	0.101	0.702	0.0678	0.744
2.00	0.418	0.570	0.299	0.602	0.229	0.632	0.198	0.648
3.00	0.290	0.601	0.357	0.584	0.374	0.581	0.286	0.608
Staggered								
0.600	—	—	—	—	—	—	0.213	0.636
0.900	—	—	—	—	0.446	0.571	0.401	0.581
1.000	—	—	0.497	0.558	—	—	—	—
1.125	—	—	—	—	0.478	0.565	0.518	0.560
1.250	0.518	0.556	0.505	0.554	0.519	0.556	0.522	0.562
1.500	0.451	0.568	0.460	0.562	0.452	0.568	0.488	0.568
2.000	0.404	0.572	0.416	0.568	0.482	0.556	0.449	0.570
3.000	0.310	0.592	0.356	0.580	0.440	0.562	0.428	0.574

For the staggered configuration, the maximum velocity may occur at either the transverse plane A_1 or the diagonal plane A_2 of Figure 7.11b. It will occur at A_2 if the rows are spaced such that

$$2(S_D - D) < (S_T - D)$$

The factor of 2 results from the bifurcation experienced by the fluid moving from the A_1 to the A_2 planes. Hence V_{max} occurs at A_2 if

$$S_D = \left[S_L^2 + \left(\frac{S_T}{2} \right)^2 \right]^{1/2} < \frac{S_T + D}{2}$$

in which case it is given by

$$V_{max} = \frac{S_T}{2(S_D - D)} V \tag{7.66}$$

If V_{max} occurs at A_1 for the staggered configuration, it may again be computed from Equation 7.65.

TABLE 7.6 Correction factor C_2 of Equation 7.64 for $N_L < 10$ [21]

N_L	1	2	3	4	5	6	7	8	9
Aligned	0.64	0.80	0.87	0.90	0.92	0.94	0.96	0.98	0.99
Staggered	0.68	0.75	0.83	0.89	0.92	0.95	0.97	0.98	0.99

TABLE 7.7 Constants of Equation 7.67 for
the tube bank in cross flow [16]

Configuration	$Re_{D,\,max}$	C	m
Aligned	$10–10^2$	0.80	0.40
Staggered	$10–10^2$	0.90	0.40
Aligned	$10^2–10^3$	Approximate as a single	
Staggered	$10^2–10^3$	(isolated) cylinder	
Aligned $(S_T/S_L > 0.7)^a$	$10^3–2 \times 10^5$	0.27	0.63
Staggered $(S_T/S_L < 2)$	$10^3–2 \times 10^5$	$0.35(S_T/S_L)^{1/5}$	0.60
Staggered $(S_T/S_L > 2)$	$10^3–2 \times 10^5$	0.40	0.60
Aligned	$2 \times 10^5–2 \times 10^6$	0.021	0.84
Staggered	$2 \times 10^5–2 \times 10^6$	0.022	0.84

^{a}For $S_T/S_L < 0.7$, heat transfer is inefficient and aligned tubes should not be used.

More recent results have been obtained [8, 16], and Zhukauskas [16] has proposed a correlation of the form

$$\overline{Nu}_D = C\,Re_{D,\,max}^m\,Pr^{0.36}\left(\frac{Pr}{Pr_s}\right)^{1/4} \qquad (7.67)$$

$$\begin{bmatrix} N_L \geq 20 \\ 0.7 < Pr < 500 \\ 1000 < Re_{D,\,max} < 2 \times 10^6 \end{bmatrix}$$

where all properties except Pr_s are evaluated at the arithmetic mean of the fluid inlet and outlet temperatures and the constants C and m are listed in Table 7.7. The need to evaluate fluid properties at the arithmetic mean of the inlet ($T_i = T_\infty$) and outlet (T_o) temperatures is dictated by the fact that the fluid temperature will decrease or increase, respectively, due to heat transfer to or from the tubes. If the fluid temperature change, $|T_i - T_o|$, is large, significant error could result from evaluation of the properties at the inlet temperature. If $N_L < 20$, a correction factor may be applied such that

$$\overline{Nu}_D\big|_{(N_L<20)} = C_2\,\overline{Nu}_D\big|_{(N_L\geq20)} \qquad (7.68)$$

where C_2 is given in Table 7.8.

TABLE 7.8 Correction factor C_2 of Equation 7.68
for $N_L < 20$ $(Re_D > 10^3)$ [16]

N_L	1	2	3	4	5	7	10	13	16
Aligned	0.70	0.80	0.86	0.90	0.92	0.95	0.97	0.98	0.99
Staggered	0.64	0.76	0.84	0.89	0.92	0.95	0.97	0.98	0.99

Flow around tubes in the first row of a tube bank corresponds to that for a single (isolated) cylinder in cross flow. However, for subsequent rows, flow depends strongly on the tube bank arrangement (Figure 7.12). Aligned tubes beyond the first row are in the turbulent wakes of upstream tubes, and for moderate values of S_L convection coefficients associated with downstream rows are enhanced by turbulation of the flow. Typically, the convection coefficient of a row increases with increasing row number until approximately the fifth row, after which there is little change in the turbulence and hence in the convection coefficient. However, for small values of S_T/S_L, upstream rows, in effect, shield downstream rows from much of the flow, and heat transfer is adversely affected. That is, the preferred flow path is in lanes between the tubes and much of the tube surface is not exposed to the main flow. For this reason, operation of aligned tube banks with $S_T/S_L < 0.7$ (Table 7.7) is undesirable. For the staggered array, however, the path of the main flow is more tortuous and a greater portion of the surface area of downstream tubes remains in this path. In general, heat transfer enhancement is favored by the more tortuous flow of a staggered arrangement, particularly for small Reynolds number ($Re_D < 100$).

Since the fluid may experience a large change in temperature as it moves through the tube bank, the heat transfer rate could be significantly overpre-

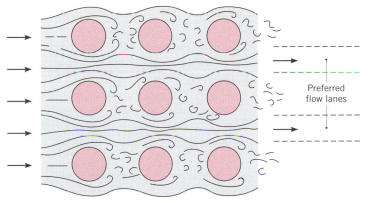

Preferred flow lanes

(a)

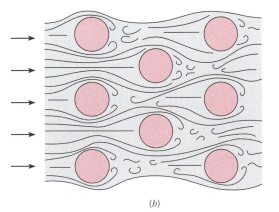

(b)

FIGURE 7.12 Flow conditions for (*a*) aligned and (*b*) staggered tubes.

dicted by using $\Delta T = T_s - T_\infty$ as the temperature difference in Newton's law of cooling. As the fluid moves through the bank, its temperature approaches T_s and $|\Delta T|$ decreases. In Chapter 11 the appropriate form of ΔT is shown to be a *log-mean temperature difference*,

$$\Delta T_{\text{lm}} = \frac{(T_s - T_i) - (T_s - T_o)}{\ln\left(\dfrac{T_s - T_i}{T_s - T_o}\right)} \tag{7.69}$$

where T_i and T_o are temperatures of the fluid as it enters and leaves the bank, respectively. The outlet temperature, which is needed to determine ΔT_{lm}, may be estimated from

$$\frac{T_s - T_o}{T_s - T_i} = \exp\left(-\frac{\pi D N \bar{h}}{\rho V N_T S_T c_p}\right) \tag{7.70}$$

where N is the total number of tubes in the bank and N_T is the number of tubes in the transverse plane. Once ΔT_{lm} is known, the heat transfer rate per unit length of the tubes may be computed from

$$q' = N(\bar{h}\pi D \, \Delta T_{\text{lm}}) \tag{7.71}$$

The foregoing results may be used to determine mass transfer rates associated with evaporation or sublimation from the surfaces of a bank of cylinders in cross flow. Once again it is only necessary to replace $\overline{Nu}_D$ and Pr by $\overline{Sh}_D$ and Sc, respectively.

We close by recognizing that there is generally as much interest in the pressure drop associated with flow across a tube bank as in the overall heat transfer rate. The power required to move the fluid across the bank is often a major operating expense and is directly proportional to the pressure drop,

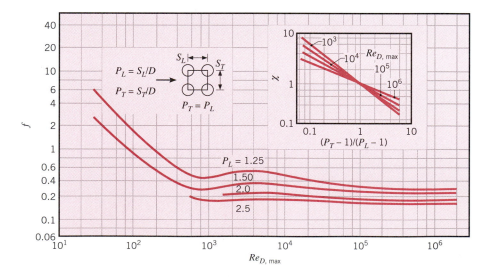

FIGURE 7.13   Friction factor f and correction factor χ for Equation 7.72. In-line tube bundle arrangement [16]. Used with permission.

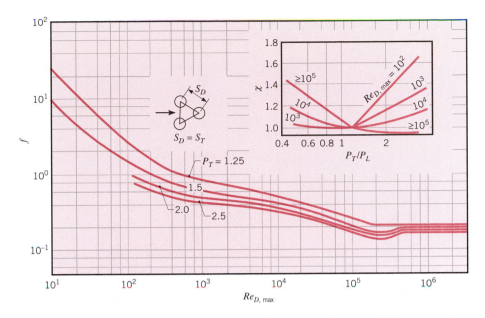

FIGURE 7.14 Friction factor f and correction factor χ for Equation 7.72. Staggered tube bundle arrangement [16]. Used with permission.

which may be expressed as [16]

$$\Delta p = N_L \chi \left(\frac{\rho V_{max}^2}{2} \right) f \qquad (7.72)$$

The friction factor f and the correction factor χ are plotted in Figures 7.13 and 7.14. Figure 7.13 pertains to a square, in-line tube arrangement for which the dimensionless longitudinal and transverse pitches, $P_L \equiv S_L/D$ and $P_T \equiv S_T/D$, respectively, are equal. The correction factor χ, plotted in the inset, is used to apply the results to other in-line arrangements. Similarly, Figure 7.14 applies to a staggered arrangement of tubes in the form of an equilateral triangle ($S_T = S_D$), and the correction factor enables extension of the results to other staggered arrangements. Note that the Reynolds number appearing in Figures 7.13 and 7.14 is based on the maximum fluid velocity V_{max}.

EXAMPLE 7.6

Pressurized water is often available at elevated temperatures and may be used for space heating or industrial process applications. In such cases it is customary to use a tube bundle in which the water is passed through the tubes, while air is passed in cross flow over the tubes. Consider a staggered arrangement for which the tube outside diameter is 16.4 mm and the longitudinal and transverse pitches are $S_L = 34.3$ mm and $S_T = 31.3$ mm. There are seven rows of tubes in the airflow direction and eight tubes per row. Under typical operating conditions the cylinder surface temperature is at 70°C, while the air upstream temperature and velocity are 15°C and 6 m/s, respectively. Determine the air-side convection coefficient and the rate of heat transfer for the tube bundle. What is the air-side pressure drop?

SOLUTION

Known: Geometry and operating conditions of a tube bank.

Find:

1. Air-side convection coefficient and heat rate.
2. Pressure drop.

Schematic:

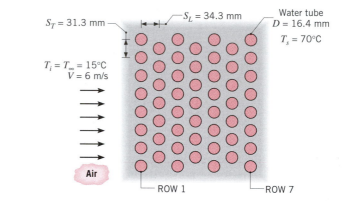

Assumptions:

1. Steady-state conditions.
2. Negligible radiation effects.
3. Negligible effect of change in air temperature across tube bank on air properties.

Properties: Table A.4, air ($T_\infty = 15°C$): $\rho = 1.217$ kg/m³, $c_p = 1007$ J/kg · K, $\nu = 14.82 \times 10^{-6}$ m²/s, $k = 0.0253$ W/m · K, $Pr = 0.710$. Table A.4, air ($T_s = 70°C$): $Pr = 0.701$. Table A.4, air ($T_f = 43°C$): $\nu = 17.4 \times 10^{-6}$ m²/s, $k = 0.0274$ W/m · K, $Pr = 0.705$.

Analysis:

1. From Equations 7.67 and 7.68, the air-side Nusselt number is

$$\overline{Nu}_D = C_2 C\, Re_{D,\,\text{max}}^m\, Pr^{0.36} \left(\frac{Pr}{Pr_s} \right)^{1/4}$$

Since $S_D = [S_L^2 + (S_T/2)^2]^{1/2} = 37.7$ mm is greater than $(S_T + D)/2$, the maximum velocity occurs on the transverse plane, A_1, of Figure 7.11. Hence from Equation 7.65

$$V_{\text{max}} = \frac{S_T}{S_T - D} V = \frac{31.3 \text{ mm}}{(31.3 - 16.4) \text{ mm}} 6 \text{ m/s} = 12.6 \text{ m/s}$$

With

$$Re_{D, max} = \frac{V_{max}D}{\nu} = \frac{12.6 \text{ m/s} \times 0.0164 \text{ m}}{14.82 \times 10^{-6} \text{ m}^2/\text{s}} = 13{,}943$$

and

$$\frac{S_T}{S_L} = \frac{31.3 \text{ mm}}{34.3 \text{ mm}} = 0.91 < 2$$

it follows from Tables 7.7 and 7.8 that

$$C = 0.35 \left(\frac{S_T}{S_L}\right)^{1/5} = 0.34, \qquad m = 0.60, \qquad \text{and} \qquad C_2 = 0.95$$

Hence

$$\overline{Nu}_D = 0.95 \times 0.34(13{,}943)^{0.60}(0.71)^{0.36} \left(\frac{0.710}{0.701}\right)^{0.25} = 87.9$$

and

$$\overline{h} = \overline{Nu}_D \frac{k}{D} = 87.9 \times \frac{0.0253 \text{ W/m} \cdot \text{K}}{0.0164 \text{ m}} = 135.6 \text{ W/m}^2 \cdot \text{K} \qquad \triangleleft$$

From Equation 7.70

$$T_s - T_o = (T_s - T_i) \exp\left(-\frac{\pi D N \overline{h}}{\rho V N_T S_T c_p}\right)$$

$$T_s - T_o = (55°C) \exp\left(-\frac{\pi(0.0164 \text{ m}) \, 56 \, (135.6 \text{ W/m}^2 \cdot \text{K})}{1.217 \text{ kg/m}^3 \, (6 \text{ m/s}) \, 8 \, (0.0313 \text{ m}) \, 1007 \text{ J/kg} \cdot \text{K}}\right)$$

$$T_s - T_o = 44.5°C$$

Hence from Equations 7.69 and 7.71

$$\Delta T_{lm} = \frac{(T_s - T_i) - (T_s - T_o)}{\ln\left(\dfrac{T_s - T_i}{T_s - T_o}\right)} = \frac{(55 - 44.5)°C}{\ln\left(\dfrac{55}{44.5}\right)} = 49.6°C$$

and

$$q' = N(\overline{h}\pi D \, \Delta T_{lm}) = 56\pi \times 135.6 \text{ W/m}^2 \cdot \text{K} \times 0.0164 \text{ m} \times 49.6°C$$

$$q' = 19.4 \text{ kW/m} \qquad \triangleleft$$

2. The pressure drop may be obtained from Equation 7.72.

$$\Delta p = N_L \chi \left(\frac{\rho V_{max}^2}{2}\right) f$$

With $Re_{D, max} = 13{,}943$, $P_T = (S_T/D) = 1.91$, and $(P_T/P_L) = 0.91$, it follows from Figure 7.14 that $\chi \approx 1.04$ and $f \approx 0.35$. Hence with $N_L = 7$

$$\Delta p = 7 \times 1.04 \left[\frac{1.217 \text{ kg/m}^3(12.6 \text{ m/s})^2}{2}\right] 0.35$$

$$\Delta p = 246 \text{ N/m}^2 = 2.46 \times 10^{-3} \text{ bars} \qquad \triangleleft$$

Comments:

1. With properties evaluated at T_f, $Re_{D,\,\mathrm{max}} = 11{,}876$. With $S_T/D \approx 2$ and $S_L/D \approx 2$, it follows from Tables 7.5 and 7.6 that $C_1 = 0.482$, $m = 0.556$, and $C_2 = 0.97$. From Equations 7.63 and 7.64, the Nusselt number is then $\overline{Nu}_D = 86.7$, and $\bar{h} = 144.8$ W/m² · K. Values of $\bar{h}$ obtained from Equations 7.63 and 7.67 therefore agree to within 7%, which is well within their uncertainties.

2. Had $\Delta T_i \equiv T_s - T_i$ been used in lieu of ΔT_{lm} in Equation 7.71, the heat rate would have been overpredicted by 11%.

3. Since the air temperature is predicted to increase by only 10.5°C, evaluation of the air properties at $T_i = 15°C$ is a reasonable approximation. However, if improved accuracy is desired, the calculations could be repeated with the properties reevaluated at $(T_i + T_o)/2 = 20.25°C$.

4. The air outlet temperature and heat rate may be increased by increasing the number of tube rows, and for a fixed number of rows, they may be varied by adjusting the air velocity. For $5 \leq N_L \leq 25$ and $V = 6$ m/s, parametric calculations based on Equations 7.67 to 7.71 yield the following results:

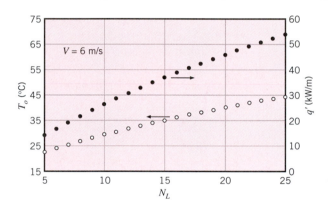

The air outlet temperature would asymptotically approach the surface temperature with increasing N_L, at which point the heat rate approaches a constant value and there is no advantage to adding more tube rows. Note that Δp increases linearly with increasing N_L. For $N_L = 25$ and $1 \leq V \leq 20$ m/s, we obtain

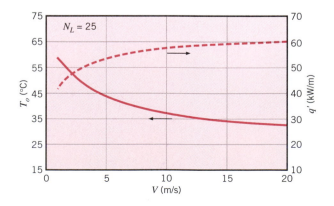

Although the heat rate increases with increasing V, the air outlet temperature decreases, approaching T_i as $V \rightarrow \infty$.

7.7
Impinging Jets

A single gas jet or an array of such jets, impinging normally on a surface, may be used to achieve enhanced coefficients for convective heating, cooling, or drying. Applications include tempering of glass plate, annealing of metal sheets, drying of textile and paper products, cooling of heated components in gas turbine engines, and deicing of aircraft systems.

7.7.1 Hydrodynamic and Geometric Considerations

As shown in Figure 7.15, gas jets are typically discharged into a quiescent ambient from a round nozzle of diameter D or a slot (rectangular) nozzle of width W. Typically, the jet is turbulent and, at the nozzle exit, is characterized by a uniform velocity profile. However, with increasing distance from the exit, momentum exchange between the jet and the ambient causes the free boundary of the jet to broaden and the *potential core,* within which the uniform exit velocity is retained, to contract. Downstream of the potential core the velocity profile is nonuniform over the entire jet cross section and the maximum (center) velocity

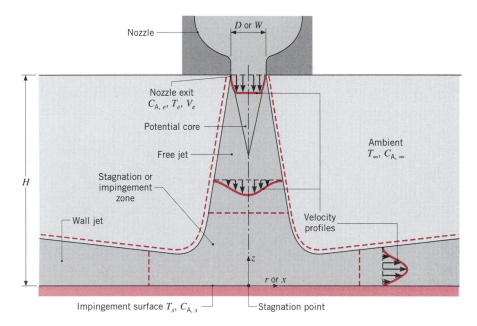

FIGURE 7.15 Surface impingement of a single round or slot gas jet.

decreases with increasing distance from the nozzle exit. The region of the flow over which conditions are unaffected by the impingement (target) surface is termed the *free jet.*

Within the *stagnation* or *impingement zone,* flow is influenced by the target surface and is decelerated and accelerated in the normal (z) and transverse (r or x) directions, respectively. However, since the flow continues to entrain zero momentum fluid from the ambient, horizontal acceleration cannot continue indefinitely and accelerating flow in the stagnation zone is transformed to a decelerating *wall jet.* Hence, with increasing r or x, velocity components parallel to the surface increase from a value of zero to some maximum and subsequently decay to zero. Velocity profiles within the wall jet are characterized by zero velocity at both the impingement and free surfaces. If $T_s \neq T_e$ and/or $C_{A,s} \neq C_{A,e}$, convection heat and/or mass transfer occurs in both the stagnation and wall jet regions.

Many impingement heat (mass) transfer schemes involve an array of jets, as, for example, the array of slot jets shown in Figure 7.16. In addition to flow from each nozzle exhibiting free jet, stagnation, and wall jet regions, secondary stagnation zones result from the interaction of adjoining wall jets. In many such schemes the jets are discharged into a restricted volume bounded by the target surface and the nozzle plate from which the jets originate. The overall rate of heat (mass) transfer depends strongly on the manner in which *spent gas,* whose temperature (species concentration) is between values associated with the nozzle exit and the impingement surface, is vented from the system. For the configuration of Figure 7.16, spent gas cannot flow upward between the nozzles but must instead flow symmetrically in the $\pm y$ directions. As the temperature (surface cooling) or species concentration (surface evaporation) of the gas increases with increasing $|y|$, the local surface-to-gas temperature or concentration difference decreases, causing a reduction in local convection fluxes. A preferable situation is one for which the space between adjoining nozzles is open to the ambient, thereby permitting continuous upflow and direct discharge of the spent gas.

Plan (top) views of single round and slot nozzles, as well as regular arrays of round and slot nozzles, are shown in Figure 7.17. For the isolated nozzles (Figures 7.17a, d), local and average convection coefficients are associated with any $r > 0$ and $x > 0$. For the arrays, symmetry dictates equivalent local and average values for each of the unit cells delineated by dashed lines. For a large number of square-in-line (Figure 7.17b) or equilaterally staggered (Figure

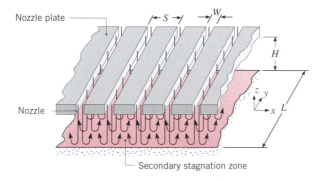

FIGURE 7.16   Surface impingement of an array of slot jets.

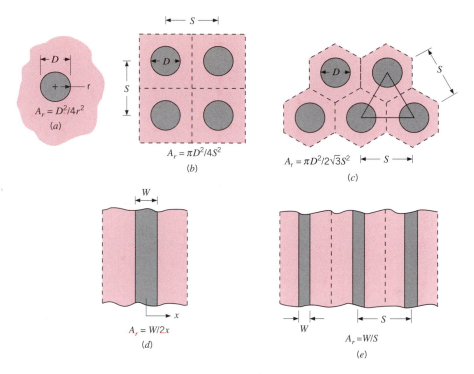

FIGURE 7.17 Plan view of pertinent geometrical features for (*a*) single round jet, (*b*) in-line array of round jets, (*c*) staggered array of round jets, (*d*) single slot jet, and (*e*) array of slot jets.

7.17*c*) round jets, the unit cells correspond to a square or hexagon, respectively. A pertinent geometric parameter is the relative nozzle area, which is defined as the ratio of the nozzle exit cross-sectional area to the surface area of the cell ($A_r \equiv A_{c,e}/A_{cell}$). In each case, S represents the pitch of the array.

7.7.2 Convection Heat and Mass Transfer

In the results that follow, it is presumed that the gas jet exits its nozzle with a uniform velocity V_e, temperature T_e, and species concentration $C_{A,e}$. Thermal and compositional equilibrium with the ambient are presumed ($T_e = T_\infty$, $C_{A,e} = C_{A,\infty}$), while convection heat and/or mass transfer may occur at an impingement surface of uniform temperature and/or species composition ($T_s \neq T_e$, $C_{A,s} \neq C_{A,e}$). Newton's law of cooling and its mass transfer analog are then

$$q'' = h(T_s - T_e) \tag{7.73}$$

$$N''_A = h_m(C_{A,s} - C_{A,e}) \tag{7.74}$$

Conditions are presumed to be uninfluenced by the level of turbulence at the nozzle exit, and the surface is presumed to be stationary. However, this requirement may be relaxed for surface velocities which are much less than the jet impact velocity.

An extensive review of available convection coefficient data for impinging gas jets has been performed by Martin [22], and for a single round or slot nozzle, distributions of the *local* Nusselt number have the characteristic forms shown in Figure 7.18. The characteristic length is the *hydraulic diameter* of the nozzle, which is defined as four times its cross-sectional area divided by its wetted perimeter ($D_h \equiv 4A_{c,e}/P$). Hence the characteristic length is the diameter of a round nozzle, and assuming $L \gg W$, it is twice the width of a slot nozzle. It follows that $Nu = hD/k$ for a round nozzle and $Nu = h(2W/k)$ for a slot nozzle. For large nozzle-to-plate separations, Figure 7.18*a*, the distribution is characterized by a bell-shaped curve for which Nu monotonically decays from a maximum value at the *stagnation point, r/D(x/2W) = 0*.

For small separations, Figure 7.18*b*, the distribution is characterized by a second maximum, whose value increases with increasing jet Reynolds number and may exceed that of the first maximum. The threshold separation of $H/D \approx 5$, below which there is a second maximum, is loosely associated with the length of the potential core (Figure 7.15). Appearance of the second maximum is attributed to a sharp rise in the turbulence level which accompanies the transition from an accelerating stagnation region flow to a decelerating wall jet [22]. Additional maxima have been observed and attributed to the formation of vortices in the stagnation zone, as well as transition to a turbulent wall jet [23].

Secondary maxima in Nu are also associated with the interaction of adjoining wall jets for an array [22, 24]. However, distributions are two-dimensional, exhibiting, for example, variations with both *x* and *y* for the slot jet array of Figure 7.16. Variations with *x* could be expected to yield maxima at the jet centerline and halfway between adjoining jets, while constraint of the exhaust flow to the $\pm y$ direction would induce acceleration with increasing $|y|$ and hence a monotonically increasing Nu with $|y|$. However, variations with *y* decrease with increasing cross-sectional area of the outflow and may be neglected if $S \times H \gtrsim W \times L$ [22].

Average Nusselt (Sherwood) numbers may be obtained by integrating local results over the appropriate surface area. For single nozzles, the corresponding heat transfer correlations are expected to be of the form

$$\overline{Nu} = f(Re, Pr, r(\text{or } x)/D_h, H/D_h) \tag{7.75}$$

where

$$\overline{Nu} \equiv \frac{\overline{h}D_h}{k} \tag{7.76}$$

$$Re = \frac{V_e D_h}{\nu} \tag{7.77}$$

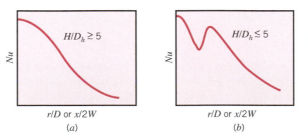

FIGURE 7.18
Distribution of local Nusselt number associated with a single round or slot nozzle for (*a*) large and (*b*) small relative nozzle-to-plate spacings.

and

$$D_h = D \quad \text{(Round nozzle)} \qquad \text{or} \qquad D_h = 2W \quad \text{(Slot nozzle)} \quad (7.78)$$

Having assessed data from several sources, Martin [22] recommends the following correlation for a _single round nozzle_

$$\frac{\overline{Nu}}{Pr^{0.42}} = G\left(\frac{r}{D}, \frac{H}{D}\right) F_1(Re) \qquad (7.79)$$

where

$$F_1 = 2Re^{1/2}(1 + 0.005Re^{0.55})^{1/2} \qquad (7.80)$$

and

$$G = \frac{D}{r} \frac{1 - 1.1D/r}{1 + 0.1(H/D - 6)D/r} \qquad (7.81a)$$

or, replacing D/r by $2A_r^{1/2}$,

$$G = 2A_r^{1/2} \frac{1 - 2.2A_r^{1/2}}{1 + 0.2(H/D - 6)A_r^{1/2}} \qquad (7.81b)$$

The ranges of validity are

$$\begin{bmatrix} 2000 \leq Re \leq 400{,}000 \\[2mm] 2 \leq \dfrac{H}{D} \leq 12 \\[2mm] 2.5 \leq \dfrac{r}{D} \leq 7.5 \\[2mm] \text{or} \\[2mm] 0.04 \geq A_r \geq 0.004 \end{bmatrix}$$

For $r < 2.5D$ ($A_r > 0.04$), results for $\overline{Nu}$ are available in graphical form [22].

For a _single slot nozzle_, the recommended correlation is of the form

$$\frac{\overline{Nu}}{Pr^{0.42}} = \frac{3.06}{x/W + H/W + 2.78} Re^m \qquad (7.82)$$

where

$$m = 0.695 - \left[\left(\frac{x}{2W}\right) + \left(\frac{H}{2W}\right)^{1.33} + 3.06\right]^{-1} \qquad (7.83)$$

and the ranges of validity are

$$\begin{bmatrix} 3000 \leq Re \leq 90{,}000 \\[2mm] 2 \leq \dfrac{H}{W} \leq 10 \\[2mm] 4 \leq \dfrac{x}{W} \leq 20 \end{bmatrix}$$

As a *first approximation,* Equation 7.82 may also be used for $x/W < 4$, yielding predictions for the stagnation point ($x = 0$) that are within 40% of measured results.

For an *array of round nozzles,* Martin [22] recommends a correlation of the form

$$\frac{\overline{Nu}}{Pr^{0.42}} = K\left(A_r, \frac{H}{D}\right) G\left(A_r, \frac{H}{D}\right) F_2(Re) \tag{7.84}$$

where

$$K = \left[1 + \left(\frac{H/D}{0.6/A_r^{1/2}}\right)^6\right]^{-0.05} \tag{7.85}$$

$$F_2 = 0.5\, Re^{2/3} \tag{7.86}$$

and G is the single nozzle function given by Equation 7.81b. The function K accounts for the fact that, for $H/D \gtrsim 0.6/A_r^{1/2}$, the average Nusselt number for the array decays more rapidly with increasing H/D than that for the single nozzle. The correlation is valid over the ranges

$$\begin{bmatrix} 2000 \le Re \le 100{,}000 \\ 2 \le \dfrac{H}{D} \le 12 \\ 0.004 \le A_r \le 0.04 \end{bmatrix}$$

For an array of slot nozzles, the recommended correlation is of the form

$$\frac{\overline{Nu}}{Pr^{0.42}} = \frac{2}{3} A_{r,o}^{3/4} \left(\frac{2Re}{A_r/A_{r,o} + A_{r,o}/A_r}\right)^{2/3} \tag{7.87}$$

where

$$A_{r,o} = \left[60 + 4\left(\frac{H}{2W} - 2\right)^2\right]^{-1/2} \tag{7.88}$$

The correlation pertains to conditions for which the outflow of spent gas is restricted to the $\pm y$ directions of Figure 7.16 and the outflow area is large enough to satisfy the requirement that $(S \times H)/(W \times L) \gtrsim 1$. Additional restrictions are that

$$\begin{bmatrix} 1500 \le Re \le 40{,}000 \\ 2 \le \dfrac{H}{W} \le 80 \\ 0.008 \le A_r \le 2.5 A_{r,o} \end{bmatrix}$$

An *optimal* arrangement of nozzles would be one for which the values of H, S, and D_h yielded the largest value of $\overline{Nu}$ for a prescribed total gas flow rate per unit surface area of the target. For fixed H and for arrays of both round and slot nozzles, optimal values of D_h and S have been found to be [22]

$$D_{h,\,op} \approx 0.2H \tag{7.89}$$

$$S_{op} \approx 1.4H \tag{7.90}$$

The optimum value of $(D_h/H)^{-1} \approx 5$ coincides approximately with the length of the potential core. Beyond the potential core, the midline jet velocity decays causing an attendant reduction in convection coefficients.

Invoking the heat and mass transfer analogy by substituting $\overline{Sh}/Sc^{0.42}$ for $\overline{Nu}/Pr^{0.42}$, the foregoing correlations may also be applied to convection mass transfer. However, for both heat and mass transfer, application of the equations should be restricted to conditions for which they were developed. For example, in their present form, the correlations may not be used if the jets emanate from sharp-edged orifices instead of bell-shaped nozzles. The orifice jet is strongly affected by a flow contraction phenomenon that alters convection heat or mass transfer [22, 23]. In the case of convection heat transfer, conditions are also influenced by differences between the jet exit and ambient temperatures ($T_e \neq T_\infty$). The exit temperature is then an inappropriate temperature in Newton's law of cooling, Equation 7.73, and should be replaced by what is commonly termed the recovery, or adiabatic wall, temperature [25, 26].

7.8
Packed Beds

Gas flow through a *packed bed* of solid particles (Figure 7.19) is relevant to many industrial processes, which include the transfer and storage of thermal energy, heterogeneous catalytic reactions and drying. The term *packed bed* refers to a condition for which the position of the particles is *fixed*. In contrast, a *fluidized bed* is one for which the particles are in motion due to advection with the fluid.

For a packed bed a large amount of heat or mass transfer surface area can be obtained in a small volume, and the irregular flow that exists in the voids of the bed enhances transport through turbulent mixing. Many correlations that have been developed for different particle shapes, sizes, and packing densities are described in the literature [27–30]. One such correlation, which has been recommended for gas flow in a bed of spheres, is of the form

$$\varepsilon \bar{j}_H = \varepsilon \bar{j}_m = 2.06 Re_D^{-0.575} \quad \begin{bmatrix} Pr \approx 0.7 \\ 90 \leq Re_D \leq 4000 \end{bmatrix} \tag{7.91}$$

where $\bar{j}_H$ and $\bar{j}_m$ are the Colburn j factors defined by Equations 6.103 and 6.104. The Reynolds number $Re_D = VD/\nu$ is defined in terms of the sphere diameter and the upstream velocity V that would exist in the empty channel without the packing. The quantity ε is the *porosity,* or *void fraction,* of the bed (volume of

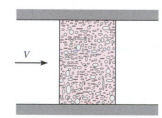

FIGURE 7.19
Gas flow through a packed bed of solid particles.

void space per unit volume of bed), and its value typically ranges from 0.30 to 0.50. The correlation may be applied to packing materials other than spheres by multiplying the right-hand side by an appropriate correction factor. For a bed of uniformly sized cylinders, with length-to-diameter ratio of 1, the factor is 0.79; for a bed of cubes it is 0.71.

In using Equation 7.91, properties should be evaluated at the arithmetic mean of the fluid temperatures entering and leaving the bed. If the particles are at a uniform temperature T_s, the heat transfer rate for the bed may be computed from

$$q = \bar{h} A_{p,t} \Delta T_{\text{lm}} \tag{7.92}$$

where $A_{p,t}$ is the total surface area of the particles and ΔT_{lm} is the log-mean temperature difference defined by Equation 7.69. The outlet temperature, which is needed to compute ΔT_{lm}, may be estimated from

$$\frac{T_s - T_o}{T_s - T_i} = \exp\left(-\frac{\bar{h} A_{p,t}}{\rho V A_{c,b} c_p}\right) \tag{7.93}$$

where ρ and V are the inlet density and velocity, respectively, and $A_{c,b}$ is the bed (channel) cross-sectional area.

7.9
Summary

In this chapter we have compiled convection correlations that may be used to estimate convection transfer rates for a variety of external flow conditions. For simple surface geometries these results may be derived from a boundary layer analysis, but in most cases they are obtained from generalizations based on experiment. You should know when and how to use the various expressions, and you should be familiar with the general methodology of a convection calculation. To facilitate their use, the correlations are summarized in Table 7.9.

TABLE 7.9   **Summary of convection heat transfer correlations for external flow**[a, b]

Correlation		Geometry	Conditions
$\delta = 5x\,Re_x^{-1/2}$	(7.19)	Flat plate	Laminar, T_f
$C_{f,x} = 0.664Re_x^{-1/2}$	(7.20)	Flat plate	Laminar, local, T_f
$Nu_x = 0.332Re_x^{1/2}\,Pr^{1/3}$	(7.23)	Flat plate	Laminar, local, T_f, $0.6 \leq Pr \leq 50$
$\delta_t = \delta Pr^{-1/3}$	(7.24)	Flat plate	Laminar, T_f
$\overline{C}_{f,x} = 1.328Re_x^{-1/2}$	(7.30)	Flat plate	Laminar, average, T_f
$\overline{Nu}_x = 0.664Re_x^{1/2}\,Pr^{1/3}$	(7.31)	Flat plate	Laminar, average, T_f, $0.6 \leq Pr \leq 50$
$Nu_x = 0.565Pe_x^{1/2}$	(7.33)	Flat plate	Laminar, local, T_f, $Pr \leq 0.05$

TABLE 7.9 *(Continued)*

Correlation		Geometry	Conditions
$C_{f,x} = 0.0592 Re_x^{-1/5}$	(7.35)	Flat plate	Turbulent, local, T_f, $Re_x \lesssim 10^8$
$\delta = 0.37x\, Re_x^{-1/5}$	(7.36)	Flat plate	Turbulent, local, T_f, $Re_x \lesssim 10^8$
$Nu_x = 0.0296 Re_x^{4/5}\, Pr^{1/3}$	(7.37)	Flat plate	Turbulent, local, T_f, $Re_x \lesssim 10^8$, $0.6 \lesssim Pr \lesssim 60$
$\overline{C}_{f,L} = 0.074 Re_L^{-1/5} - 1742 Re_L^{-1}$	(7.43)	Flat plate	Mixed, average, T_f, $Re_{x,c} = 5 \times 10^5$, $Re_L \lesssim 10^8$
$\overline{Nu}_L = (0.037 Re_L^{4/5} - 871)Pr^{1/3}$	(7.41)	Flat plate	Mixed, average, T_f, $Re_{x,c} = 5 \times 10^5$, $Re_L \lesssim 10^8$, $0.6 < Pr < 60$
$\overline{Nu}_D = C\, Re_D^m\, Pr^{1/3}$ (Table 7.2)	(7.55b)	Cylinder	Average, T_f, $0.4 < Re_D < 4 \times 10^5$, $Pr \gtrsim 0.7$
$\overline{Nu}_D = C\, Re_D^m\, Pr^n (Pr/Pr_s)^{1/4}$ (Table 7.4)	(7.56)	Cylinder	Average, T_∞, $1 < Re_D < 10^6$, $0.7 < Pr < 500$
$\overline{Nu}_D = 0.3 + [0.62 Re_D^{1/2}\, Pr^{1/3}$ $\times\ [1 + (0.4/Pr)^{2/3}]^{-1/4}]$ $\times\ [1 + (Re_D/282{,}000)^{5/8}]^{4/5}$	(7.57)	Cylinder	Average, T_f, $Re_D\, Pr > 0.2$
$\overline{Nu}_D = 2 + (0.4 Re_D^{1/2}$ $+ 0.06 Re_D^{2/3})Pr^{0.4}$ $\times\ (\mu/\mu_s)^{1/4}$	(7.59)	Sphere	Average, T_∞, $3.5 < Re_D < 7.6 \times 10^4$, $0.71 < Pr < 380$, $1.0 < (\mu/\mu_s) < 3.2$
$\overline{Nu}_D = 2 + 0.6 Re_D^{1/2}\, Pr^{1/3}$	(7.60)	Falling drop	Average, T_∞
$\overline{Nu}_D = 1.13 C_1 Re_{D,\max}^m\, Pr^{1/3}$ (Tables 7.5, 7.6)	(7.63)	Tube bank[c]	Average, $\overline{T}_f$, $2000 < Re_{D,\max}$ $< 4 \times 10^4$, $Pr \gtrsim 0.7$
$\overline{Nu}_D = C\, Re_{D,\max}^m\, Pr^{0.36}(Pr/Pr_s)^{1/4}$ (Tables 7.7, 7.8)	(7.67)	Tube bank[c]	Average, $\overline{T}$, $1000 < Re_D < 2 \times 10^6$, $0.7 < Pr < 500$
Single round nozzle	(7.79)	Impinging jet	Average, T_f, $2000 < Re < 4 \times 10^5$, $2 < (H/D) < 12$, $2.5 < (r/D) < 7.5$
Single slot nozzle	(7.82)	Impinging jet	Average, T_f, $3000 < Re < 9 \times 10^4$, $2 < (H/W) < 10$, $4 < (x/W) < 20$
Array of round nozzles	(7.84)	Impinging jet	Average, T_f, $2000 < Re < 10^5$, $2 < (H/D) < 12$, $0.004 < A_r < 0.04$
Array of slot nozzles	(7.87)	Impinging jet	Average, T_f, $1500 < Re < 4 \times 10^4$, $2 < (H/W) < 80$, $0.008 < A_r < 2.5 A_{r,o}$
$\varepsilon \overline{j}_H = \varepsilon \overline{j}_m = 2.06 Re_D^{-0.575}$	(7.91)	Packed bed of spheres[c]	Average, $\overline{T}$, $90 \leq Re_D \leq 4000$, $Pr \approx 0.7$

[a]Correlations in this table pertain to isothermal surfaces; for special cases involving an unheated starting length or a uniform surface heat flux, see Section 7.2.4.

[b]When the heat and mass transfer analogy is applicable, the corresponding mass transfer correlations may be obtained by replacing Nu and Pr by Sh and Sc, respectively.

[c]For tube banks and packed beds, properties are evaluated at the average fluid temperature, $\overline{T} = (T_i + T_o)/2$, or the average film temperature, $\overline{T}_f = (T_s + \overline{T})/2$.

References

1. Blasius, H., *Z. Math. Phys.,* **56,** 1, 1908. English translation in National Advisory Committee for Aeronautics Technical Memo No. 1256.

2. Schlichting, H., *Boundary Layer Theory,* 4th ed., McGraw-Hill, New York, 1960.

3. Howarth, L., *Proc. R. Soc. Lond., Ser. A,* **164,** 547, 1938.

4. Pohlhausen, E., *Z. Angew. Math. Mech.,* **1,** 115, 1921.

5. Kays, W. M., and M. E. Crawford, *Convective Heat and Mass Transfer,* McGraw-Hill, New York, 1980.

6. Churchill, S. W., and H. Ozoe, *J. Heat Transfer,* **95,** 78, 1973.

7. Schlichting, H., *Boundary Layer Theory,* 6th ed., McGraw-Hill, New York, 1968.

8. Whitaker, S., *AIChE J.,* **18,** 361, 1972.

9. Zhukauskas, A., and A. B. Ambrazyavichyus, *Int. J. Heat Mass Transfer,* **3,** 305, 1961.

10. Churchill, S. W., *AIChE J.,* **22,** 264, 1976.

11. Blair, M. F., *J. Heat Transfer,* **105,** 33 and 41, 1983.

12. Fox, R. W., and A. T. McDonald, *Introduction to Fluid Mechanics,* 3rd ed., Wiley, New York, 1985.

13. Hilpert, R., *Forsch. Geb. Ingenieurwes.,* **4,** 215, 1933.

14. Knudsen, J. D., and D. L. Katz, *Fluid Dynamics and Heat Transfer,* McGraw-Hill, New York, 1958.

15. Jakob, M., *Heat Transfer,* Vol. 1, Wiley, New York, 1949.

16. Zhukauskas, A., "Heat Transfer from Tubes in Cross Flow," in J. P. Hartnett and T. F. Irvine, Jr., Eds., *Advances in Heat Transfer,* Vol. 8, Academic Press, New York, 1972.

17. Churchill, S. W., and M. Bernstein, *J. Heat Transfer,* **99,** 300, 1977.

18. Morgan, V. T., "The Overall Convective Heat Transfer from Smooth Circular Cylinders," in T. F. Irvine, Jr. and J. P. Hartnett, Eds., *Advances in Heat Transfer,* Vol. 11, Academic Press, New York, 1975.

19. Ranz, W., and W. Marshall, *Chem. Eng. Prog.,* **48,** 141, 1952.

20. Grimison, E. D., *Trans. ASME,* **59,** 583, 1937.

21. Kays, W. M. and R. K. Lo, Stanford University Technical Report No. 15, 1952.

22. Martin, H., "Heat and Mass Transfer between Impinging Gas Jets and Solid Surfaces," in J. P. Hartnett and T. F. Irvine, Jr., Eds., *Advances in Heat Transfer,* Vol. 13, Academic Press, New York, 1977.

23. Popiel, Cz. O., and L. Bogusiawski, "Mass or Heat Transfer in Impinging Single Round Jets Emitted by a Bell-Shaped Nozzle and Sharp-Ended Orifice," in C. L. Tien, V. P. Carey, and J. K. Ferrell, Eds., *Heat Transfer 1986,* Vol. 3, Hemisphere Publishing, New York, 1986.

24. Goldstein, R. J., and J. F. Timmers, *Int. J. Heat Mass Transfer,* **25,** 1857, 1982.

25. Hollworth, B. R., and L. R. Gero, *J. Heat Transfer,* **107,** 910, 1985.

26. Goldstein, R. J., A. I. Behbahani, and K. K. Heppelman, *Int. J. Heat Mass Transfer,* **29,** 1227, 1986.

27. Bird, R. B., W. E. Stewart, and E. N. Lightfoot, *Transport Phenomena,* Wiley, New York, 1966.

28. Jakob, M., *Heat Transfer,* Vol. 2, Wiley, New York, 1957.

29. Geankopplis, C. J., *Mass Transport Phenomena,* Holt, Rinehart & Winston, New York, 1972.

30. Sherwood, T. K., R. L. Pigford, and C. R. Wilkie, *Mass Transfer,* McGraw-Hill, New York, 1975.

Problems

Flat Plate in Parallel Flow

7.1   Consider the following fluids at a film temperature of 300 K in parallel flow over a flat plate with velocity of 1 m/s: atmospheric air, water, engine oil, and mercury.

     (a) For each fluid, determine the velocity and thermal boundary layer thicknesses at a distance of 40 mm from the leading edge.

     (b) For each of the prescribed fluids and on the same coordinates, plot the boundary layer thicknesses as a function of distance from the leading edge to a plate length of 40 mm.

7.2   Consider atmospheric air at 25°C in parallel flow at 5 m/s over both surfaces of a 1-m long flat plate maintained at 75°C.

     (a) Determine the velocity boundary layer thickness, the surface shear stress, and the heat flux at the trailing edge.

     (b) Determine the drag force on the plate and the total heat transfer from the plate, each per unit width of the plate.

(c) Plot each of the parameters of part (a) as a function of distance from the leading edge of the plate.

7.3 Engine oil at 100°C and a velocity of 0.1 m/s flows over both surfaces of a 1-mm long flat plate maintained at 20°C. Determine:

(a) The velocity and thermal boundary layer thicknesses at the trailing edge.

(b) The local heat flux and surface shear stress at the trailing edge.

(c) The total drag force and heat transfer per unit width of the plate.

(d) Plot the boundary layer thicknesses and local values of the surface shear stress, convection coefficient, and heat flux as a function of x for $0 \leq x \leq 1$ m.

7.4 Consider steady, parallel flow of atmospheric air over a flat plate. The air has a temperature and free stream velocity of 300 K and 25 m/s.

(a) Evaluate the boundary layer thickness at distances of $x = 1$, 10, and 100 mm from the leading edge. If a second plate were installed parallel to and at a distance of 3 mm from the first plate, what is the distance from the leading edge at which boundary layer merger would occur?

(b) Evaluate the surface shear stress and the y-velocity component at the outer edge of the boundary layer for the single plate at $x = 1$, 10, and 100 mm.

(c) Comment on the validity of the boundary layer approximations.

7.5 A common flow geometry, which includes the flat plate in parallel flow as a special case, is the *wedge*.

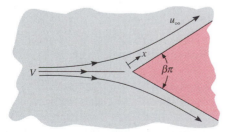

From a potential flow solution, it is known that the velocity at the edge of the boundary layer increases with x according to the relation $u_\infty = Vx^m$, where $m = \beta/(2 - \beta)$ and $\beta\pi$ is the wedge angle. From a laminar boundary layer solution, it is also known that the local Nusselt number may be expressed as

$Nu_x = C_1 Re_x^{1/2}$, where $Re_x \equiv (u_\infty x/\nu)$ and C_1 is a known function of Pr and β, as tabulated below.

				Pr		
		0.7	0.8	1.0	5.0	10.0
β	m			C_1		
0	0	0.292	0.307	0.332	0.585	0.730
0.2	0.111	0.331	0.348	0.378	0.669	0.851
0.5	0.333	0.384	0.403	0.440	0.792	1.013
1.0	1.000	0.496	0.523	0.570	1.043	1.344

(a) Comment on the nature of flow conditions at $x = 0$ for $\beta > 0$. How does u_∞ vary with x for $\beta = 1$?

(b) Obtain the ratio of the average convection coefficient $\bar{h}_x$ to the local coefficient h_x for $\beta = 0.5$ and for $\beta = 1.0$.

(c) It is common practice to approximate wedge flow heat transfer by using convection correlations associated with the flat plate in parallel flow. Comment on the accuracy of such an approximation by computing the ratio $(\bar{h}_{x,\,\beta>0}/\bar{h}_{x,\,\beta=0})$ at $x = 1$ m for air flow over wedges of $\beta = 0.5$ and $\beta = 1.0$.

7.6 Consider a liquid metal ($Pr \ll 1$), with free stream conditions u_∞ and T_∞, in parallel flow over an isothermal flat plate at T_s. Assuming that $u = u_\infty$ throughout the thermal boundary layer, write the corresponding form of the boundary layer energy equation. Applying appropriate initial ($x = 0$) and boundary conditions, solve this equation for the boundary layer temperature field, $T(x, y)$. Use the result to obtain an expression for the local Nusselt number Nu_x. (*Hint:* This problem is analogous to one-dimensional heat transfer in a semi-infinite medium with a sudden change in surface temperature.)

7.7 Consider the velocity boundary layer profile for flow over a flat plate to be of the form $u = C_1 + C_2y$. Applying appropriate boundary conditions, obtain an expression for the velocity profile in terms of the boundary layer thickness δ and the free stream velocity u_∞. Using the integral form of the boundary layer momentum equation (Appendix E), obtain expressions for the boundary layer thickness and the local friction coefficient, expressing your result in terms of the local Reynolds number. Compare your results with those obtained from the exact solution (Section 7.2.1) and the integral solution with a cubic profile (Appendix E).

7.8   Do Problem 7.7 for a velocity boundary layer profile of the form $u = C_1 + C_2 \sin (C_3 y)$.

7.9   Consider a steady, turbulent boundary layer on an isothermal flat plate of temperature T_s. The boundary layer is "tripped" at the leading edge $x = 0$ by a fine wire. Assume constant physical properties and velocity and temperature profiles of the form

$$\frac{u}{u_\infty} = \left(\frac{y}{\delta}\right)^{1/7} \quad \text{and} \quad \frac{T - T_\infty}{T_s - T_\infty} = 1 - \left(\frac{y}{\delta_t}\right)^{1/7}$$

(a) From experiment it is known that the surface shear stress is related to the boundary layer thickness by an expression of the form

$$\tau_s = 0.0228 \, \rho u_\infty^2 \left(\frac{u_\infty \delta}{\nu}\right)^{-1/4}$$

Beginning with the momentum integral equation (Appendix E), show that $\delta/x = 0.376Re_x^{-1/5}$. Determine the average friction coefficient $\overline{C}_{f,x}$.

(b) Beginning with the energy integral equation, obtain an expression for the local Nusselt number Nu_x and use this result to evaluate the average Nusselt number $\overline{Nu}_x$.

7.10 Consider flow over a flat plate for which it is desired to determine the average heat transfer coefficient over the short span x_1 to x_2, $\overline{h}_{1-2}$, where $(x_2 - x_1) \ll L$.

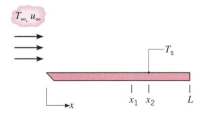

Provide three different expressions that can be used to evaluate $\overline{h}_{1-2}$ in terms of (a) the local coefficient at $x = (x_1 + x_2)/2$, (b) the local coefficients at x_1 and x_2, and (c) the average coefficients at x_1 and x_2. Indicate which of the expressions is approximate. Considering whether the flow is laminar, turbulent, or mixed, indicate when it is appropriate or inappropriate to use each of the equations.

7.11 A flat plate of width 1 m is maintained at a uniform surface temperature of $T_s = 150°C$ by using independently controlled, heat-generating rectangular modules of thickness $a = 10$ mm and length $b = 50$ mm. Each module is insulated from its neighbors, as well as on its back side. Atmospheric air at 25°C flows over the plate at a velocity of 30

m/s. The thermophysical properties of the module are $k = 5.2$ W/m · K, $c_p = 320$ J/kg · K, and $\rho = 2300$ kg/m³.

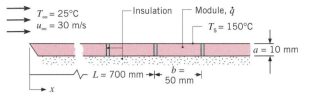

(a) Find the required power generation, $\dot{q}$ (W/m³), in a module positioned at a distance 700 mm from the leading edge.

(b) Find the maximum temperature T_{max} in the heat-generating module.

7.12 An electric air heater consists of a horizontal array of thin metal strips that are each 10 mm long in the direction of an airstream that is in parallel flow over the top of the strips. Each strip is 0.2 m wide, and 25 strips are arranged side by side, forming a continuous and smooth surface over which the air flows at 2 m/s. During operation each strip is maintained at 500°C and the air is at 25°C.

(a) What is the rate of convection heat transfer from the first strip? The fifth strip? The tenth strip? All the strips?

(b) For air velocities of 2, 5, and 10 m/s, determine the convection heat rates for all the locations of part (a). Represent your results in tabular or bar graph form.

(c) Repeat part (b), but under conditions for which the flow is fully turbulent over the entire array of strips.

7.13 Consider atmospheric air at 25°C and a velocity of 25 m/s flowing over both surfaces of a 1-m long flat plate that is maintained at 125°C. Determine the rate of heat transfer per unit width from the plate for values of the critical Reynolds number corresponding to 10^5, 5×10^5, and 10^6.

7.14 Consider air at 27°C and 1 atm in parallel flow over an isothermal, 1-m long flat plate with a velocity of 10 m/s.

(a) Plot the variation of the local heat transfer coefficient, $h_x(x)$, with distance along the plate for three flow conditions corresponding to transition Reynolds numbers of (i) 5×10^5, (ii) 2.5×10^5, and (iii) 0 (the flow is fully turbulent).

(b) Plot the variation of the average heat transfer coefficient, $\overline{h}_x(x)$, with distance for the three flow conditions of part (a).

(c) What are the average heat transfer coefficients for the entire plate, $\bar{h}_L$, for the three flow conditions of part (a)?

7.15 Consider water at 27°C in parallel flow over an isothermal, 1-m long flat plate with a velocity of 2 m/s.

(a) Plot the variation of the local heat transfer coefficient, $h_x(x)$, with distance along the plate for three flow conditions corresponding to transition Reynolds numbers of (i) 5×10^5, (ii) 3×10^5, and (iii) 0 (the flow is fully turbulent).

(b) Plot the variation of the average heat transfer coefficient, $\bar{h}_x(x)$, with distance for the three flow conditions of part (a).

(c) What are the average heat transfer coefficients for the entire plate, $\bar{h}_L$, for the three flow conditions of part (a)?

7.16 Air at a pressure of 1 atm and a temperature of 15°C is in parallel flow at a velocity of 10 m/s over a 3-m long flat plate that is heated to a uniform temperature of 140°C.

(a) What is the average heat transfer coefficient for the plate?

(b) What is the local heat transfer coefficient at the midpoint of the plate?

(c) Plot the variation of the heat flux with distance along the length of the plate.

7.17 Explain under what conditions the total rate of heat transfer from an isothermal flat plate of dimensions L by $2L$ would be the same, independent of whether parallel flow over the plate is directed along the side of length L or $2L$. With a critical Reynolds number of 5×10^5, for what values of Re_L would the total heat transfer be independent of orientation?

7.18 The surface of a 1.5-m long flat plate is maintained at 40°C, and water at a temperature of 4°C and a velocity of 0.6 m/s flows over the surface.

(a) Using the film temperature T_f for evaluation of the properties, calculate the heat transfer rate per unit width of the plate, q' (W/m).

(b) Calculate the error in q' that would be incurred in part (a) if the thermophysical properties of the water were evaluated at the free stream temperature and the same empirical correlation were used.

(c) In part (a), if a wire were placed near the leading edge of the plate to induce turbulence over its entire length, what would be the heat transfer rate?

7.19 Air at a pressure of 1 atm and a temperature of 50°C is in parallel flow over the top surface of a flat plate that is heated to a uniform temperature of 100°C. The plate has a length of 0.20 m (in the flow direction) and a width of 0.10 m. The Reynolds number based on the plate length is 40,000. What is the rate of heat transfer from the plate to the air? If the free stream velocity of the air is doubled and the pressure is increased to 10 atm, what is the rate of heat transfer?

7.20 A thin, flat plate of length $L = 1$ m separates two airstreams that are in parallel flow over opposite surfaces of the plate. One airstream has a temperature of $T_{\infty,1} = 200°C$ and a velocity of $u_{\infty,1} = 60$ m/s, while the other airstream has a temperature of $T_{\infty,2} = 25°C$ and a velocity of $u_{\infty,2} = 10$ m/s. What is the heat flux between the two streams at the midpoint of the plate?

7.21 Consider a rectangular fin that is used to cool a motorcycle engine. The fin is 0.15 m long and at a temperature of 250°C, while the motorcycle is moving at 80 km/h in air at 27°C. The air is in parallel flow over both surfaces of the fin, and turbulent flow conditions may be assumed to exist throughout.

(a) What is the rate of heat removal per unit width of the fin?

(b) Generate a plot of the heat removal rate per unit width of the fin for motorcycle speeds ranging from 10 to 100 km/h.

7.22 Consider the convective cooling conditions described for your hand in Problem 1.9. Use standard correlations to estimate the convection coefficient for each of the two cases. Which condition would *feel* colder? Contrast these results with a heat loss of approximately 30 W/m² under normal conditions.

7.23 Consider the wing of an aircraft as a flat plate of 2.5-m length in the flow direction. The plane is moving at 100 m/s in air that is at a pressure of 0.7 bar and a temperature of −10°C. The top surface of the wing absorbs solar radiation at a rate of 800 W/m². Assume the wing to be of solid construction and to have a single, uniform temperature.

(a) Estimate the steady-state temperature of the wing.

(b) Generate a plot of the steady-state temperature for plane speeds ranging from 100 to 250 m/s.

7.24 The top surface of a heated compartment consists of very smooth (A) and highly roughened (B) portions, and the surface is placed in an atmospheric airstream. In the interest of minimizing total convection heat transfer from the surface, which orientation, (1) or (2), is preferred? If $T_s = 100°C$, $T_\infty = 20°C$, and $u_\infty = 20$ m/s, what is the convection heat transfer from the entire surface for this orientation?

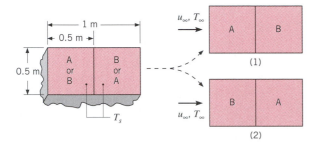

7.25 Initially the top surface of an oven measuring 0.5 m by 0.5 m is at a uniform temperature of 47°C under quiescent room air conditions. The inside air temperature of the oven is 150°C, the room air temperature is 17°C, and the heat transfer from the surface is 40 W. In order to reduce the surface temperature and meet safety requirements, room air is blown across the top surface with a velocity of 20 m/s in a direction parallel to an edge.

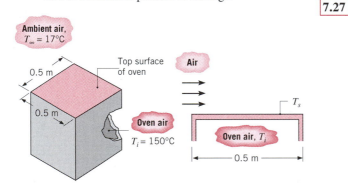

(a) Assuming *internal* convection conditions to remain unchanged, determine the heat loss from the top surface under the forced convection condition.

(b) Estimate the surface temperature achieved with the forced convection condition.

(c) Generate a plot of the surface temperature as a function of room air velocity for $5 \leq u_\infty \leq 30$ m/s.

7.26 Consider weather conditions for which the prevailing wind blows past the penthouse tower on a tall

building. The tower length in the wind direction is 10 m and there are 10 window panels.

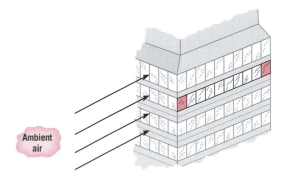

(a) Calculate the average convection coefficient for the first, third, and tenth window panels when the wind speed is 5 m/s. Use a film temperature of 300 K to evaluate the thermophysical properties required of the correlation. Would this be a suitable value of the film temperature for ambient air temperatures in the range $-15 \leq T_\infty \leq 38°C$?

(b) For the first, third, and tenth windows, on one graph, plot the variation of the average convection coefficient with wind speed for the range $5 \leq u_\infty \leq 100$ km/h. Explain the major features of each curve and their relative magnitudes.

7.27 The proposed design for an anemometer to determine the velocity of an airstream in a wind tunnel is comprised of a thin metallic strip whose ends are supported by stiff rods serving as electrodes for passage of current used to heat the strip. A fine-wire thermocouple is attached to the trailing edge of the strip and serves as the sensor for a system that controls the power to maintain the strip at a constant operating temperature for variable airstream velocities. Design conditions pertain to an airstream at $T_\infty = 25°C$ and $1 \leq u_\infty \leq 50$ m/s, with a strip temperature of $T_s = 35°C$.

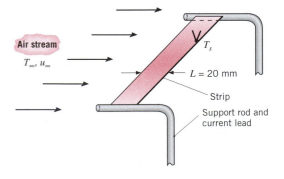

(a) Determine the relationship between the electrical power dissipation per unit width of the strip in the transverse direction, P' (mW/mm), and the airstream velocity. Show this relationship graphically for the specified range of u_∞.

(b) If the accuracy with which the temperature of the operating strip can be measured and maintained constant is $\pm 0.2°C$, what is the uncertainty in the airstream velocity?

(c) The proposed design operates in a strip constant-temperature mode for which the airstream velocity is related to the measured power. Consider now an alternative mode wherein the strip is provided with a constant power, say, 30 mW/mm, and the airstream velocity is related to the measured strip temperature T_s. For this mode of operation, show the graphical relationship between the strip temperature and airstream velocity. If the temperature can be measured with an uncertainty of $\pm 0.2°C$, what is the uncertainty in the airstream velocity?

(d) Compare the features associated with each of the anemometer operating modes.

7.28 Two rooms are arranged on the side of a building. The wall of room A is 13 m long and 3.5 m high, while that of room B is 7 m long and 3.5 m high. Both walls are 0.25 m thick and have an effective thermal conductivity of 1 W/m · K.

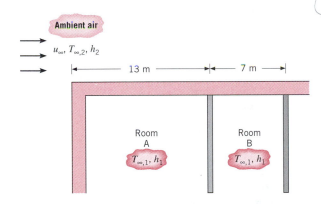

Consider conditions for which the ambient air is in parallel flow over the outer surface of the wall with $u_\infty = 7$ m/s and $T_{\infty,2} = -20°C$. The inside air temperature is maintained at $T_{\infty,1} = 20°C$, and the inside convection coefficient is $h_1 = 5$ W/m² · K. Turbulent boundary layer conditions may be assumed for the ambient airflow over the entire wall.

(a) Estimate the rate at which heat is lost through the 13-m long and 7-m long walls of rooms A and B, respectively.

(b) Investigate the influence of the following parameters on the heat losses: (i) variation of the ambient air velocity from 2 to 15 m/s, (ii) the thermal conductivity of the wall material from 0.5 to 1.25 W/m · K, and (iii) the inside convection coefficient from 3 to 10 W/m² · K.

7.29 As a means of supplying fresh water to arid regions of the world, it has been advocated that icebergs be towed from polar regions. Icebergs that are considered to be best suited for towing are those that are relatively broad and flat. Consider an iceberg that is 1 km long by 0.5 km wide and of depth $D = 0.25$ km. It is proposed that this iceberg be towed at 1 km/h in the direction of its length for 6000 km through water whose average temperature (over the trip) is 10°C. As a first approximation, the interaction of the iceberg with its surroundings may be assumed to be dominated by conditions at the bottom (1 km × 0.5 km) surface. The latent heat of fusion of ice is 3.34×10^5 J/kg.

(a) What is the average recession (melting) rate dD/dt at the bottom surface?

(b) What is the power required to move the iceberg at the designated speed?

(c) If towing costs amount to \$1/kW ·h of power requirement, what is the minimum cost of fresh water at the destination?

7.30 A steel strip emerges from the hot roll section of a steel mill at a speed of 20 m/s and a temperature of 1200 K. Its length and thickness are $L = 100$ m and $\delta = 0.003$ m, respectively, and its density and specific heat are 7900 kg/m³ and 640 J/kg · K, respectively.

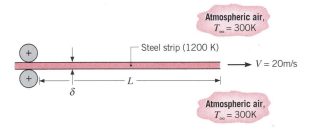

Accounting for heat transfer from the top and bottom surfaces and neglecting radiation and strip conduction effects, determine the initial time rate of change of the strip temperature at a distance of 1 m from the leading edge and at the trailing edge. Determine the distance from the leading edge at which the minimum cooling rate is achieved.

7.31 A heat sink constructed from 2024 aluminum alloy is used to cool a power diode dissipating 5 W. The internal resistance between the diode junction and the case is 0.8°C/W, while the thermal contact resistance between the case and the heat sink is 10^{-5} m² · °C/W. Convection at the fin surface may be approximated as that corresponding to a flat plate in parallel flow.

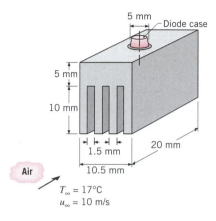

(a) Assuming that all the diode power is transferred to the ambient air through the rectangular fins, estimate the operating temperature of the diode.

(b) Explore options for reducing the diode temperature, subject to the constraints that the air velocity and fin length may not exceed 25 m/s and 20 mm, respectively, while the fin thickness may not be less than 0.5 mm. All other conditions, including the spacing between fins, remain as prescribed.

7.32 An array of heat-dissipating electrical components is mounted on the bottom side of a 1.2 m by 1.2 m horizontal aluminum plate, while the top side is cooled by an airstream for which $u_\infty = 15$ m/s and $T_\infty = 300$ K. The plate is attached to a well-insulated enclosure such that all the dissipated heat must be transferred to the air. Also, the aluminum is sufficiently thick to ensure a nearly uniform plate temperature.

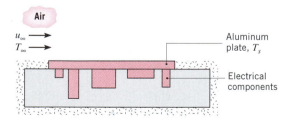

(a) If the temperature of the aluminum is not to exceed 350 K, what is the maximum allowable heat dissipation?

(b) Determine the maximum allowable heat dissipation as a function of air velocity in the range, $5 \leq u_\infty \leq 25$ m/s. With $u_\infty = 25$ m/s, the maximum allowable power may be increased further by using an aluminum plate with longitudinal fins. What is the maximum allowable power if the fin length, thickness, and spacing are 25 mm, 5 mm, and 10 mm, respectively?

7.33 One-hundred electrical components, each dissipating 25 W, are attached to one surface of a square (0.2 m × 0.2 m) copper plate, and all the dissipated energy is transferred to water in parallel flow over the opposite surface. A protuberance at the leading edge of the plate acts to *trip* the boundary layer, and the plate itself may be assumed to be isothermal. The water velocity and temperature are $u_\infty = 2$ m/s and $T_\infty = 17$°C, and the water's thermophysical properties may be approximated as $\nu = 0.96 \times 10^{-6}$ m²/s, $k = 0.620$ W/m · K, and $Pr = 5.2$.

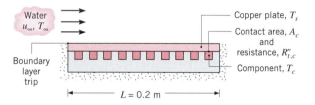

(a) What is the temperature of the copper plate?

(b) If each component has a plate contact surface area of 1 cm² and the corresponding contact resistance is 2×10^{-4} m² · K/W, what is the component temperature? Neglect the temperature variation across the thickness of the copper plate.

7.34 Air at 27°C with a free stream velocity of 10 m/s is used to cool electronic devices mounted on a printed circuit board. Each device, 4 mm by 4 mm, dissipates 40 mW, which is removed from the top surface. A turbulator is located at the leading edge of the board, causing the boundary layer to be turbulent.

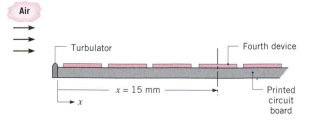

(a) Estimate the surface temperature of the fourth device located 15 mm from the leading edge of the board.

(b) Generate a plot of the surface temperature of the first four devices as a function of the free stream velocity for $5 \leq u_\infty \leq 15$ m/s.

(c) What is the minimum free stream velocity if the surface temperature of the hottest device is not to exceed 80°C?

7.35 Forced air at 25°C and 10 m/s is used to cool electronic elements mounted on a circuit board. Consider a chip of length 4 mm and width 4 mm located 120 mm from the leading edge. Because the board surface is irregular, the flow is disturbed and the appropriate convection correlation is of the form $Nu_x = 0.04Re_x^{0.85} Pr^{0.33}$.

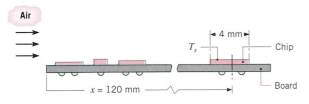

Estimate the surface temperature of the chip, T_s, if its heat dissipation rate is 30 mW.

7.36 Consider atmospheric air at $u_\infty = 2$ m/s and $T_\infty = 300$ K in parallel flow over an isothermal flat plate of length $L = 1$ m and temperature $T_s = 350$ K.

(a) Compute the local convection coefficient at the leading and trailing edges of the heated plate with and without an unheated starting length of $\xi = 1$ m.

(b) Compute the average convection coefficient for the plate for the same conditions as part (a).

(c) Plot the variation of the local convection coefficient over the plate with and without an unheated starting length.

7.37 The cover plate of a flat-plate solar collector is at 15°C, while ambient air at 10°C is in parallel flow over the plate, with $u_\infty = 2$ m/s.

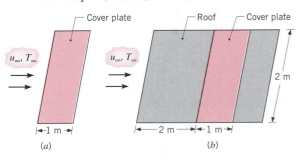

(a) What is the rate of convective heat loss from the plate?

(b) If the plate is installed 2 m from the leading edge of a roof and flush with the roof surface, what is the rate of convective heat loss?

7.38 An array of 10 silicon chips, each of length $L = 10$ mm on a side, is insulated on one surface and cooled on the opposite surface by atmospheric air in parallel flow with $T_\infty = 24$°C and $u_\infty = 40$ m/s. When in use, the same electrical power is dissipated in each chip, maintaining a uniform heat flux over the entire cooled surface.

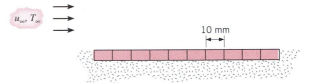

If the temperature of each chip may not exceed 80°C, what is the maximum allowable power per chip? What is the maximum allowable power if a turbulence promoter is used to trip the boundary layer at the leading edge? Would it be preferable to orient the array normal, instead of parallel, to the airflow?

7.39 A square (10 mm × 10 mm) silicon chip is insulated on one side and cooled on the opposite side by atmospheric air in parallel flow at $u_\infty = 20$ m/s and $T_\infty = 24$°C. When in use, electrical power dissipation within the chip maintains a uniform heat flux at the cooled surface. If the chip temperature may not exceed 80°C at any point on its surface, what is the maximum allowable power? What is the maximum allowable power if the chip is flush mounted in a substrate that provides for an unheated starting length of 20 mm?

7.40 Working in groups of two, our students design and perform experiments on forced convection phenomena using the general arrangement shown schematically. The air box consists of two muffin fans, a plenum chamber, and flow straighteners discharging a nearly uniform airstream over the flat *test-plate*. The objectives of one experiment were to measure the heat transfer coefficient and to compare the results with standard convection correlations. The velocity of the airstream was measured using a thermistor-based anemometer, and thermocouples were used to determine the temperatures of the airstream and the *test-plate*.

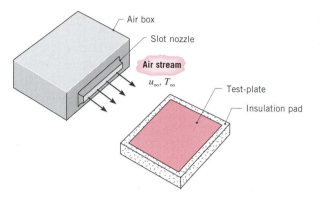

- Air box
- Slot nozzle
- **Air stream** u_∞, T_∞
- Test-plate
- Insulation pad

With the airstream from the box fully stabilized at $T_\infty = 20°C$, an aluminum plate was preheated in a convection oven and quickly mounted in the test-plate holder. The subsequent temperature history of the plate was determined from thermocouple measurements, and histories obtained for airstream velocities of 3 and 9 m/s were fitted by the following polynomial:

$$T(t) = a + bt + ct^2 + dt^3 + et^4$$

The temperature T and time t have units of °C and s, respectively, and values of the coefficients appropriate for the time interval of the experiments are tabulated as follows:

Velocity (m/s)	3	9
Elapsed Time (s)	300	160
a (°C)	56.87	57.00
b (°C/s)	−0.1472	−0.2641
c (°C/s²)	3×10^{-4}	9×10^{-4}
d (°C/s³)	-4×10^{-7}	-2×10^{-6}
e (°C/s⁴)	2×10^{-10}	1×10^{-9}

The plate is square, 133 mm to a side, with a thickness of 3.2 mm, and is made from a highly polished aluminum alloy ($\rho = 2770$ kg/m³, $c = 875$ J/kg · K, $k = 177$ W/m · K).

(a) Determine the heat transfer coefficients for the two cases, assuming the plate behaves as a spacewise isothermal object.

(b) Evaluate the coefficients C and m for a correlation of the form

$$\overline{Nu_L} = C\,Re^m\,Pr^{-1/3}$$

Compare this result with a standard flat-plate correlation. Comment on the *goodness* of the comparison and explain any differences.

Cylinder in Cross Flow

7.41 Consider the following fluids, each with a velocity of $V = 5$ m/s and a temperature of $T_\infty = 20°C$, in cross flow over a 10-mm diameter cylinder maintained at 50°C: atmospheric air, saturated water, and engine oil.

(a) Calculate the rate of heat transfer per unit length, q', using the Churchill–Bernstein correlation.

(b) Generate a plot of q' as a function of fluid velocity for $0.5 \leq V \leq 10$ m/s.

7.42 A circular pipe of 25-mm outside diameter is placed in an airstream at 25°C and 1-atm pressure. The air moves in cross flow over the pipe at 15 m/s, while the outer surface of the pipe is maintained at 100°C. What is the drag force exerted on the pipe per unit length? What is the rate of heat transfer from the pipe per unit length?

7.43 A circular cylinder of 25-mm diameter is initially at 150°C and is quenched by immersion in a 80°C oil bath, which moves at a velocity of 2 m/s in cross flow over the cylinder. What is the initial rate of heat loss per unit length of the cylinder?

7.44 Consider atmospheric air at 25°C and flowing at a velocity of 15 m/s. What is the rate of heat transfer per unit length from the following surfaces, each at 75°C, when the air is in cross flow over the surface?

(a) A circular cylinder of 10-mm diameter.

(b) A square cylinder of 10-mm length on a side.

(c) A vertical plate of 10-mm height.

7.45 A long, cylindrical, electrical heating element of diameter $D = 10$ mm, thermal conductivity $k = 240$ W/m · K, density $\rho = 2700$ kg/m³, and specific heat $c_p = 900$ J/kg · K is installed in a duct for which air moves in cross flow over the heater at a temperature and velocity of 27°C and 10 m/s, respectively.

(a) Neglecting radiation, estimate the steady-state surface temperature when, per unit length of the heater, electrical energy is being dissipated at a rate of 1000 W/m.

(b) If the heater is activated from an initial temperature of 27°C, estimate the time required for the surface temperature to come within 10°C of its steady-state value.

7.46 Consider the conditions of Problem 7.45, but now allow for radiation exchange between the surface of the heating element ($\varepsilon = 0.8$) and the walls of the duct, which form a large enclosure at 27°C.

(a) Evaluate the steady-state surface temperature.

(b) If the heater is activated from an initial temperature of 27°C, estimate the time required for the surface temperature to come within 10°C of the steady-state value.

(c) To guard against overheating due to unanticipated excursions in the blower output, the heater controller is designed to maintain a fixed surface temperature of 275°C. Determine the power dissipation required to maintain this temperature for air velocities in the range $5 \leq V \leq 10$ m/s.

7.47 Air at 40°C flows over a long, 25-mm diameter cylinder with an embedded electrical heater. Measurements of the effect of the free stream velocity, V, on the power per unit length, P', required to maintain the cylinder surface temperature at 300°C yield the following results:

V(m/s)	1	2	4	8	12
P' (W/m)	450	658	983	1507	1963

(a) Determine the convection coefficient for each of the foregoing test conditions. Display your results graphically.

(b) For the corresponding Reynolds number range, determine suitable constants C and m for use with an empirical correlation of the form $\overline{Nu} = C \, Re_D^m \, Pr^{1/3}$.

(c) Compare your experimental correlation to that of Hilpert, Equation 7.55b.

7.48 A pin fin of 10-mm diameter dissipates 30 W by forced convection to air in cross flow with a Reynolds number of 4000. If the diameter of the fin is doubled and all other conditions remain the same, estimate the fin heat rate. Assume the pin to be infinitely long.

7.49 Air at 27°C and a velocity of 5 m/s passes over the small region A_s (20 mm × 20 mm) on a large surface, which is maintained at $T_s = 127°C$. For these conditions, 0.5 W is removed from the surface A_s. In order to increase the heat removal rate, a stainless steel (AISI 304) pin fin of diameter 5 mm is affixed to A_s, which is assumed to remain at $T_s = 127°C$.

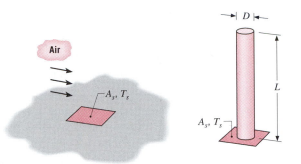

(a) Determine the maximum possible heat removal rate through the fin.

(b) What length of the pin fin will provide for the maximum rate found in part (a)?

(c) Determine the fin effectiveness, ε_f.

(d) What is the percentage increase in the heat rate from A_s due to installation of the fin?

7.50 To enhance heat transfer from a silicon chip of width $W = 4$ mm on a side, a copper pin fin is brazed to the surface of the chip. The pin length and diameter are $L = 12$ mm and $D = 2$ mm, respectively, and atmospheric air at $V = 10$ m/s and $T_\infty = 300$ K is in cross flow over the pin. The surface of the chip, and hence the base of the pin, are maintained at a temperature of $T_b = 350$ K.

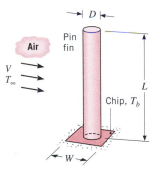

(a) Assuming the chip to have a negligible effect on flow over the pin, what is the average convection coefficient for the surface of the pin?

(b) Neglecting radiation and assuming the convection coefficient at the pin tip to equal that calculated in part (a), determine the pin heat transfer rate.

(c) Neglecting radiation and assuming the convection coefficient at the exposed chip surface to equal that calculated in part (a), determine the total rate of heat transfer from the chip.

(d) Independently determine and plot the effect of increasing velocity ($10 \leq V \leq 40$ m/s) and pin diameter ($2 \leq D \leq 4$ mm) on the total rate of heat transfer from the chip. What is the heat rate for $V = 40$ m/s and $D = 4$ mm?

7.51 A fine wire of diameter D is positioned across a passage to determine flow velocity from heat transfer characteristics. Current is passed through the wire to heat it, and the heat is dissipated to the flowing fluid by convection. The resistance of the wire is determined from electrical measurements, and the temperature is known from the resistance.

(a) For a fluid of arbitrary Prandtl number, develop an expression for its velocity in terms of the difference between the temperature of the wire and the free stream temperature of the fluid.

(b) What is the velocity of an airstream at 1 atm and 25°C, if a wire of 0.5-mm diameter achieves a temperature of 40°C while dissipating 35 W/m?

7.52 To determine air velocity changes, it is proposed to measure the electric current required to maintain a platinum wire of 0.5-mm diameter at a constant temperature of 77°C in a stream of air at 27°C.

(a) Assuming Reynolds numbers in the range 40 $< Re_D <$ 1000, develop a relationship between the wire current and the velocity of the air that is in cross flow over the wire. Use this result to establish a relation between fractional changes in the current, $\Delta I/I$, and the air velocity, $\Delta V/V$.

(b) Calculate the current required when the air velocity is 10 m/s and the electrical resistivity of the platinum wire is 17.1×10^{-5} Ω/m.

7.53 A computer code is being developed to analyze a temperature sensor of 12.5-mm diameter experiencing cross flow of water with a free stream temperature of 80°C and variable velocity. Derive an expression for the convection heat transfer coefficient as a function of the sensor surface temperature T_s for the range $20 < T_s < 80$°C and for velocities V in the range $0.005 < V < 0.20$ m/s. Use the Zhukauskas correlation for the range 40 $< Re_D <$ 1000 and assume, where appropriate, that the thermophysical properties of water have a linear temperature dependence.

7.54 A 25-mm diameter, high-tension line has an electrical resistance of 10^{-4} Ω/m and is transmitting a current of 1000 A.

(a) If ambient air at 10°C and 5 m/s is in cross flow over the line, what is its surface temperature?

(b) If the line may be approximated as a solid copper rod, what is its centerline temperature?

(c) Generate a plot that depicts the variation of the surface temperature with air velocity for $1 \leq V \leq 10$ m/s.

7.55 A horizontal copper rod 10 mm in diameter and 100 mm long is inserted in the airspace between surfaces of an electronic device to enhance heat dissipation. The ends of the rod are at 90°C, while air at 25°C is in cross flow over the cylinder with a velocity of 25 m/s. What is the temperature at the midplane of the rod? What is the rate of heat transfer from the rod?

7.56 An uninsulated steam pipe is used to transport high-temperature steam from one building to another. The pipe is of 0.5-m diameter, has a surface temperature of 150°C, and is exposed to ambient air at −10°C. The air moves in cross flow over the pipe with a velocity of 5 m/s.

(a) What is the heat loss per unit length of pipe?

(b) Consider the effect of insulating the pipe with a rigid urethane foam ($k = 0.026$ W/m · K). Evaluate and plot the heat loss as a function of the thickness δ of the insulation layer for $0 \leq \delta \leq 50$ mm.

7.57 Assume that a person can be approximated as a cylinder of 0.3-m diameter and 1.8-m height with a surface temperature of 24°C. Calculate the body heat loss while this person is subjected to a 15-m/s wind whose temperature is −5°C.

7.58 The walls of a large radiant heating oven are maintained at an elevated surface temperature T_s, while a gas at a lower temperature T_∞ flows through the oven. A product placed in the oven is then heated to a steady-state temperature T_p, which is less than T_s and larger than T_∞. Consider the product to be in the form of a solid cylinder whose length is much larger than its diameter, $D = 50$ mm. The surface emissivity of the solid is $\varepsilon = 0.8$.

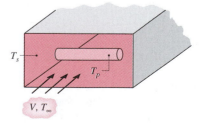

(a) If the product is placed in an oven whose surfaces are maintained at $T_s = 650$ K and atmospheric nitrogen is in cross flow over the product with $T_\infty = 350$ K and $V = 3$ m/s, what is the steady-state temperature of the product?

(b) Suppose you'd like to increase the product temperature 25°C above the result found in part (a). What change would you make to the nitrogen gas velocity? Or to the oven surface temperature?

7.59 A thermocouple is inserted into a hot air duct to measure the air temperature. The thermocouple (T_1) is soldered to the tip of a steel *thermocouple well* of length $L = 0.15$ m and inner and outer diameters of $D_i = 5$ mm and $D_o = 10$ mm. A second thermocouple (T_2) is used to measure the duct wall temperature.

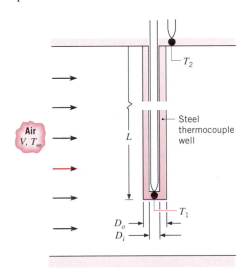

Consider conditions for which the air velocity in the duct is $V = 3$ m/s and the two thermocouples register temperatures of $T_1 = 450$ K and $T_2 = 375$ K. Neglecting radiation, determine the air temperature T_∞. Assume that, for steel, $k = 35$ W/m · K, and, for air, $\rho = 0.774$ kg/m³, $\mu = 251 \times 10^{-7}$ N · s/m², $k = 0.0373$ W/m · K, and $Pr = 0.686$.

7.60 Consider conditions for which a mercury-in-glass thermometer of 4-mm diameter is inserted to a length L through the wall of a duct in which air at 77°C is flowing. If the stem of the thermometer at the duct wall is at the wall temperature $T_w = 15$°C, conduction heat transfer through the glass causes the bulb temperature to be lower than that of the airstream.

(a) Develop a relationship for the *immersion* error, $\Delta T_i = T(L) - T_\infty$, as a function of air velocity, thermometer diameter, and insertion length L.

(b) To what length L must the thermometer be inserted if the immersion error is not to exceed 0.25°C when the air velocity is 10 m/s?

(c) Using the insertion length determined in part (b), calculate and plot the immersion error as a function of air velocity for the range 2 to 20 m/s.

(d) For a given insertion length, will the immersion error increase or decrease if the diameter of the thermometer is increased? Is the immersion error more sensitive to the diameter or air velocity?

7.61 Fluid velocities can be measured using hot-film sensors, and a common design is one for which the sensing element forms a thin film about the circumference of a quartz rod. Electrical power is supplied to the film to maintain its surface temperature T_s at a constant value.

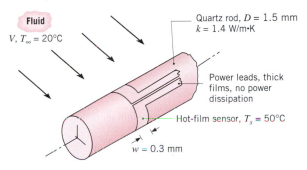

Proper operation is assured only if the heat generated in the film is transferred to the fluid, rather than conducted from the film into the quartz rod. Thermally, the film should therefore be strongly coupled to the fluid and weakly coupled to the quartz rod. This condition is satisfied if the Biot number is very large, $Bi = hD/2k \gg 1$, where h is the convection coefficient between the fluid and the film and k is the thermal conductivity of the rod.

(a) For the following fluids and velocities, calculate and plot the convection coefficient as a function of velocity: (i) water, $0.5 \le V \le 5$ m/s; (ii) air, $1 \le V \le 20$ m/s.

(b) Comment on the suitability of using this hot-film sensor for the foregoing conditions.

7.62 In a manufacturing process, a long coated plastic rod ($\rho = 2200$ kg/m³, $c = 800$ J/kg · K, $k = 1$ W/m · K) of diameter $D = 20$ mm is initially at a uniform temperature of 25°C and is suddenly exposed to a cross flow of air at $T_\infty = 350$°C and $V = 50$ m/s.

(a) How long will it take for the surface of the rod to reach 175°C, the temperature above which the special coating will cure?

(b) Generate a plot of the time-to-reach 175°C as a function of air velocity for $5 \leq V \leq 50$ m/s.

7.63 The objective of an experiment performed by our students is to determine the effect of pin fins on the thermal resistance between a flat plate and an airstream. A 25.9-mm square polished aluminum plate is subjected to an airstream in parallel flow at $T_\infty = 20°C$ and $u_\infty = 6$ m/s. An electrical heating patch is attached to the backside of the plate and dissipates 15.5 W under all conditions. Pin fins of diameter $D = 4.8$ mm and length $L = 25.4$ mm are fabricated from brass and can be firmly attached to the plate at various locations over its surface. Thermocouples are attached to the plate surface and the tip of one of the fins.

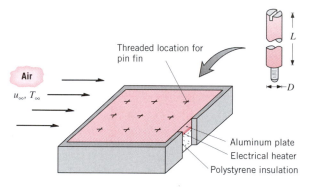

Measured temperatures for five pin-fin configurations are tabulated.

Number of Pin Fins	Temperature (°C)	
	Fin Tip	**Plate Base**
0	—	70.2
1	40.6	67.4
2	39.5	64.7
5	36.4	57.4
8	34.2	52.1

(a) Using the experimental observations and neglecting the effect of flow interactions between pins, determine the thermal resistance between the plate and the airstream for the five configurations.

(b) Develop a model of the plate–pin fin system and using appropriate convection correlations, predict the thermal resistances for the five con-

figurations. Compare your predictions with the observations and explain any differences.

(c) Use your model to predict the thermal resistances when the airstream velocity is doubled.

Spheres

7.64 Water at 20°C flows over a 20-mm diameter sphere with a velocity of 5 m/s. The surface of the sphere is at 60°C. What is the drag force on the sphere? What is the rate of heat transfer from the sphere?

7.65 Air at 25°C flows over a 10-mm diameter sphere with a velocity of 25 m/s, while the surface of the sphere is maintained at 75°C.

(a) What is the drag force on the sphere?

(b) What is the rate of heat transfer from the sphere?

(c) Generate a plot of the heat transfer from the sphere as a function of the air velocity for the range 1 to 25 m/s.

7.66 Atmospheric air at 25°C and a velocity of 0.5 m/s flows over a 50-W incandescent bulb whose surface temperature is at 140°C. The bulb may be approximated as a sphere of 50-mm diameter. What is the rate of heat loss by convection to the air?

7.67 In manufacturing lead shot, molten lead droplets fall from a tower through cool air into a tank of water. The lead is supposed to solidify before hitting the water. Consider each pellet to be a sphere of 3-mm diameter, moving at the terminal velocity (for which the drag force equals the gravitational force) throughout the descent. If pellets at the melting point are to be converted from the molten to the solid state in air at 15°C, what is the necessary height of the tower above the tank? The latent heat of fusion of lead is 2.45×10^4 J/kg.

7.68 Copper spheres of 20-mm diameter are quenched by being dropped into a tank of water that is maintained at 280 K. The spheres may be assumed to reach the terminal velocity on impact and to drop freely through the water. Estimate the terminal velocity by equating the drag and gravitational forces acting on the sphere. What is the approximate height of the water tank needed to cool the spheres from an initial temperature of 360 K to a center temperature of 320 K?

7.69 For the conditions of Problem 7.68, what are the terminal velocity and the tank height if engine oil at 300 K, rather than water, is used as the coolant.

7.70 Consider the plasma spray coating process of Problem 5.23. In addition to the prescribed conditions, the argon plasma jet is known to have a mean velocity of $V = 400$ m/s, while the *initial* velocity of the injected alumina particles may be approximated as zero. The nozzle exit and the substrate are separated by a distance of $L = 100$ mm, and pertinent properties of the argon plasma may be approximated as $k = 0.671$ W/m · K, $c_p = 1480$ J/kg · K, $\mu = 2.70 \times 10^{-4}$ kg/s · m, and $\nu = 5.6 \times 10^{-3}$ m²/s.

(a) Assuming the motion of particles entrained by the plasma jet to be governed by Stokes' law, derive expressions for the particle velocity, $V_p(t)$, and its distance of travel from the nozzle exit, $x_p(t)$, as a function of time, t, where $t = 0$ corresponds to particle injection. Evaluate the time-in-flight required for a particle to traverse the separation distance, $x_p = L$, and the velocity V_p at this time.

(b) Assuming an average relative velocity of $(V - V_p) = 315$ m/s during the time-of-flight, estimate the convection coefficient associated with heat transfer from the plasma to the particle. Using this coefficient and assuming an initial particle temperature of $T_i = 300$ K, estimate the time-in-flight required to heat a particle to its melting point, T_{mp}, and, once at T_{mp}, for the particle to experience complete melting. Is the prescribed value of L sufficient to ensure complete particle melting before surface impact?

7.71 Consider the manufacturing process of Problem 5.56.

(a) Determine the velocity of the airstream required to suspend the beads.

(b) Accounting for radiation with a glass emissivity of $\varepsilon_g = 0.8$, determine the time required for the beads to cool from $T_i = 477°C$ to $80°C$.

7.72 A spherical thermocouple junction 1.0 mm in diameter is inserted in a combustion chamber to measure the temperature T_∞ of the products of combustion. The hot gases have a velocity of $V = 5$ m/s.

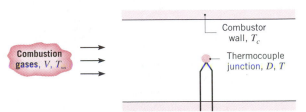

(a) If the thermocouple is at room temperature, T_i, when it is inserted in the chamber, estimate the time required for the temperature difference, $T_\infty - T$, to reach 2% of the initial temperature difference, $T_\infty - T_i$. Neglect radiation and conduction through the leads. Properties of the thermocouple junction are approximated as $k = 100$ W/m · K, $c = 385$ J/kg · K, and $\rho = 8920$ kg/m³, while those of the combustion gases may be approximated as $k = 0.05$ W/m · K, $\nu = 50 \times 10^{-6}$ m²/s, and $Pr = 0.69$.

(b) If the thermocouple junction has an emissivity of 0.5 and the cooled walls of the combustor are at $T_c = 400$ K, what is the steady-state temperature of the thermocouple junction if the combustion gases are at 1000 K? Conduction through the lead wires may be neglected.

(c) To determine the influence of the gas velocity on the thermocouple measurement error, compute the steady-state temperature of the thermocouple junction for velocities in the range $1 \leq V \leq 25$ m/s. The emissivity of the junction can be controlled through application of a thin coating. To reduce the measurement error, should the emissivity be increased or decreased? For $V = 5$ m/s, compute the steady-state junction temperature for emissivities in the range $0.1 \leq \varepsilon \leq 1.0$.

7.73 A thermocouple junction is inserted in a large duct to measure the temperature of hot gases flowing through the duct.

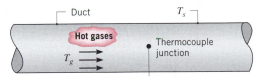

(a) If the duct surface temperature T_s is less than the gas temperature T_g, will the thermocouple sense a temperature that is less than, equal to, or greater than T_g? Justify your answer on the basis of a simple analysis.

(b) A thermocouple junction in the shape of a 2-mm diameter sphere with a surface emissivity of 0.60 is placed in a gas stream moving at 3 m/s. If the thermocouple senses a temperature of 320°C when the duct surface temperature is 175°C, what is the actual gas temperature? The gas may be assumed to have the properties of air at atmospheric pressure.

(c) How would changes in velocity and emissivity affect the temperature measurement error? Determine the measurement error for velocities in the range $1 \leq V \leq 25$ m/s ($\varepsilon = 0.6$) and for emissivities in the range $0.1 \leq \varepsilon \leq 1.0$ ($V = 3$ m/s).

7.74 Consider temperature measurement in a gas stream using the thermocouple junction described in Problem 7.73 ($D = 2$ mm, $\varepsilon = 0.60$). If the gas velocity and temperature are 3 m/s and 500°C, respectively, what temperature will be indicated by the thermocouple if the duct surface temperature is 200°C? The gas may be assumed to have the properties of atmospheric air. What temperature will be indicated by the thermocouple if the gas pressure is doubled and all other conditions remain the same?

7.75 A high-temperature gas reactor (HTGR) consists of *spherical*, uranium oxide fuel elements in which there is uniform volumetric heating ($\dot{q}$). Each fuel element is embedded in a graphite spherical shell, which is cooled by a helium gas flow at 1 atm.

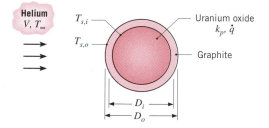

Consider steady-state conditions for which radiation effects may be neglected, the gas velocity and temperature are $V = 20$ m/s and $T_\infty = 500$ K, the pellet and shell diameters are $D_i = 10$ mm and $D_o = 12$ mm, and the shell surface temperature is $T_{s,o} = 1300$ K. The uranium oxide and graphite each have a thermal conductivity of $k_p = k_g = 2$ W/m · K.

(a) What is the rate of heat transfer from a single pellet to the gas stream?

(b) What is the volumetric rate of heat generation in the pellet and what is the temperature at the pellet–graphite interface ($T_{s,i}$)?

(c) Obtain an expression for the radial temperature distribution, $T(r)$, in the pellet, expressing your result in terms of the temperature at the center of the pellet, $T(0)$. Evaluate $T(0)$ for the prescribed conditions.

(d) Determine $T_{s,o}$, $T_{s,i}$ and $T(0)$ as a function of the gas velocity for $5 \leq V \leq 20$ m/s and $\dot{q} = 1.50 \times 10^8$ W/m³.

Tube Banks

7.76 Repeat Example 7.6 for a more compact tube bank in which the longitudinal and transverse pitches are

$S_L = S_T = 20.5$ mm. All other conditions remain the same.

7.77 A preheater involves the use of condensing steam at 100°C on the inside of a bank of tubes to heat air that enters at 1 atm and 25°C. The air moves at 5 m/s in cross flow over the tubes. Each tube is 1 m long and has an outside diameter of 10 mm. The bank consists of 196 tubes in a square, aligned array for which $S_T = S_L = 15$ mm. What is the total rate of heat transfer to the air? What is the pressure drop associated with the airflow?

7.78 A tube bank uses an aligned arrangement of 10-mm diameter tubes with $S_T = S_L = 20$ mm. There are 10 rows of tubes with 50 tubes in each row. Consider an application for which cold water flows through the tubes, maintaining the outer surface temperature at 27°C, while flue gases at 427°C and a velocity of 5 m/s are in cross flow over the tubes. The properties of the flue gas may be approximated as those of atmospheric air at 427°C. What is the total rate of heat transfer per unit length of the tubes in the bank?

7.79 A tube bank uses an aligned arrangement of 30-mm diameter tubes with $S_T = S_L = 60$ mm and a tube length of 1 m. There are 10 tube rows in the flow direction ($N_L = 10$) and 7 tubes per row ($N_T = 7$). Air with upstream conditions of $T_\infty = 27$°C and $V = 15$ m/s is in cross flow over the tubes, while a tube wall temperature of 100°C is maintained by steam condensation inside the tubes. Determine the temperature of air leaving the tube bank, the pressure drop across the bank, and the fan power requirement.

7.80 Electrical components mounted to each of two isothermal plates are cooled by passing atmospheric air between the plates, and an in-line array of aluminum pin fins is used to enhance heat transfer to the air.

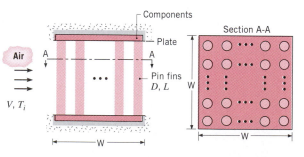

The pins are of diameter $D = 2$ mm, length $L = 100$ mm, and thermal conductivity $k = 240$ W/m · K. The longitudinal and transverse pitches are $S_L = S_T = 4$ mm, with a square array of 625 pins ($N_T = $

$N_L = 25$) mounted to square plates that are each of width $W = 100$ mm on a side. Air enters the pin array with a velocity of 10 m/s and a temperature of 300 K.

(a) Evaluating air properties at 300 K, estimate the average convection coefficient for the array of pin fins.

(b) Assuming a uniform convection coefficient over all heat transfer surfaces (plates and pins), use the result of part (a) to determine the air outlet temperature and total heat rate when the plates are maintained at 350 K. *Hint:* The air outlet temperature is governed by an exponential relation of the form $[(T_s - T_o)/(T_s - T_i)] = \exp[-(\bar{h}A_t\eta_o)/\dot{m}c_p]$, where $\dot{m} = \rho V L N_T S_T$ is the mass flow rate of air passing through the array, A_t is the total heat transfer surface area (plates and fins), and η_o is the overall surface efficiency defined by Equation 3.102.

7.81 Consider the chip cooling scheme of Problem 3.117, but with an insulated top wall placed at the pin tips to force airflow across the pin array. Air enters the array at 20°C and with a velocity V that may be varied but cannot exceed 10 m/s due to pressure drop considerations. The pin fin geometry, which includes the number of pins in the $N \times N$ square array, as well as the pin diameter D_p and length L_p, may also be varied, subject to the constraint that the product ND_p not exceed 9 mm. Neglecting heat transfer through the board, assess the effect of changes in air velocity, and hence h_o, as well as pin fin geometry, on the air outlet temperature and the chip heat rate, if the remaining conditions of Problems 3.117 and 3.25, including a maximum allowable chip temperature of 75°C, remain in effect. Recommend design and operating conditions for which chip cooling is enhanced. *Hint:* The air outlet temperature is governed by a relation of the form $[(T_s - T_o)/(T_s - T_i)] = \exp[-(\bar{h}A_t\eta_o)/\dot{m}c_p]$, where $\dot{m}$ is the mass flow rate of air passing through the array, A_t is the total heat transfer surface area (chip and pins), and η_o is the overall surface efficiency defined by Equation 3.102.

7.82 An air-cooled steam condenser is operated with air in cross flow over a square, in-line array of 400 tubes ($N_L = N_T = 20$), with an outside tube diameter of 20 mm and longitudinal and transverse pitches of $S_L = 60$ mm and $S_T = 30$ mm, respectively. Saturated steam at a pressure of 2.455 bars enters the tubes, and a uniform tube outer surface temperature of $T_s = 390$ K may be assumed to be maintained as condensation occurs within the tubes.

(a) If the temperature and velocity of the air upstream of the array are $T_i = 300$ K and $V = 4$ m/s, what is the temperature T_o of the air that leaves the array? As a first approximation, evaluate the properties of air at 300 K.

(b) If the tubes are 2 m long, what is the total heat transfer rate for the array? What is the rate at which steam is condensed in kg/s?

(c) Assess the effect of increasing N_L by a factor of 2, while reducing S_L to 30 mm. For this configuration, explore the effect of changes in the air velocity.

Impinging Jets

7.83 A circular transistor of 10-mm diameter is cooled by impingement of an air jet exiting a 2-mm diameter round nozzle with a velocity of 20 m/s and a temperature of 15°C. The jet exit and the exposed surface of the transistor are separated by a distance of 10 mm.

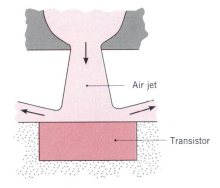

If the transistor is well insulated at all but its exposed surface and the surface temperature is not to exceed 85°C, what is the transistor's maximum allowable operating power?

7.84 A long rectangular plate of AISI 304 stainless steel is initially at 1200 K and is cooled by an array of slot jets (see Figure 7.16). The nozzle width and pitch are $W = 10$ mm and $S = 100$ mm, respectively, and the nozzle-to-plate separation is $H = 200$ mm. The plate thickness and width are $t = 8$ mm and $L = 1$ m, respectively. If air exits the nozzles at a temperature of 400 K and a velocity of 30 m/s, what is the initial cooling rate of the plate?

7.85 Air at 10 m/s and 15°C is used to cool a square hot molded plastic plate 0.5 m to a side having a surface temperature of 140°C. To increase the throughput of the production process, it is proposed to cool the plate using an array of slotted nozzles with width and pitch of 4 mm and 56 mm, respectively, and a nozzle-to-plate separation of 40 mm. The air exits the nozzle at a temperature of 15°C and a velocity of 10 m/s.

(a) Determine the improvement in cooling rate that can be achieved using the slotted nozzle arrangement in lieu of turbulated air at 10 m/s and 15°C in parallel flow over the plate.

(b) Would the heat rates for both arrangements change significantly if the air velocities were increased by a factor of 2?

(c) What is the air mass rate requirement for the slotted nozzle arrangement?

7.86 Consider Problem 7.85 in which the improvement in performance of slot-jet cooling over parallel-flow cooling was demonstrated. Design an optimal round nozzle array, using the same air jet velocity and temperature, 10 m/s and 15°C, respectively, and compare the cooling rates and supply air requirements. Discuss the features associated with each of the three methods relevant to selecting one for this application of cooling the plastic part.

7.87 Consider the plasma spraying process of Problems 5.23 and 7.70. For a nozzle exit diameter of $D = 10$ mm and a substrate radius of $r = 25$ mm, *estimate* the rate of heat transfer by convection, q_{conv}, from the argon plasma to the substrate, if the substrate temperature is maintained at 300 K. Energy transfer to the substrate is also associated with the release of latent heat, q_{lat}, which occurs during solidification of the impacted molten droplets. If the mass rate of droplet impingement is $\dot{m}_p = 0.02$ kg/s · m², estimate the rate of latent heat release.

Packed Beds

7.88 The use of rock pile thermal energy storage systems has been considered for solar energy and industrial process heat applications. A particular system involves a cylindrical container, 2 m long by 1 m in diameter, in which nearly spherical rocks of 0.03-m diameter are packed. The bed has a void space of 0.42, and the density and specific heat of the rock are $\rho = 2300$ kg/m³ and $c_p = 879$ J/kg · K, respectively. Consider conditions for which atmospheric air is supplied to the rock pile at a steady flow rate of 1 kg/s and a temperature of 90°C. The air flows in the axial direction through the container. If the rock is at a temperature of 25°C, what is the total rate of heat transfer from the air to the rock pile?

7.89 A cylindrical container in a high-temperature gas reactor is of length $L = 300$ mm and diameter $D = 100$ mm and houses a packed bed of spherical, uranium oxide fuel pellets. The pellets, which are of diameter $D_p = 10$ mm, are each coated with a layer of graphite, which is of uniform thickness $\delta = 1$ mm. The packed bed has a porosity of $\varepsilon = 0.40$. During steady operation, thermal energy is generated at a uniform volumetric rate of 4×10^7 W/m³ within the uranium oxide, while helium enters the bed at a velocity of $V = 30$ m/s and a temperature of $T_i = 400$ K.

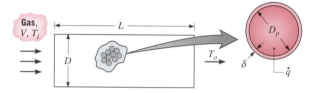

The uranium oxide and graphite each have a thermal conductivity of 2 W/m · K, while properties of the helium may be assumed to be $\rho = 0.089$ kg/m³, $c_p = 5193$ J/kg · K, $k = 0.236$ W/m · K, $\mu = 3 \times 10^{-5}$ kg/s · m, and $Pr = 0.66$.

(a) Using an energy balance of the form $q = \dot{m}c_p(T_o - T_i)$, determine the mean temperature T_o of the helium leaving the packed bed.

(b) What is the maximum temperature of the uranium oxide?

Heat and Mass Transfer

7.90 A turbine blade of length $L = 250$ mm operates at 400 K in atmospheric air at $T_\infty = 800$ K and $V = 100$ m/s. To determine the average heat transfer coefficient over a forward section of area $A_s = 0.05$ m², a mass transfer experiment was performed in which a geometrically similar model of equivalent characteristic length was constructed and the forward region was coated with naphthalene. With the atmospheric air and model temperatures at 300 K, the upstream velocity was adjusted to provide an equivalent Reynolds number.

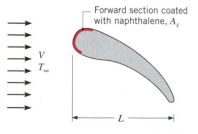

Forward section coated with naphthalene, A_s

The experiment was performed over a 3-h period, and the mass loss due to sublimation was determined to be 0.056 kg. Assuming a correlation of the form $\overline{Nu}_L = C\,Re_L^m\,Pr^{1/3}$ to be appropriate, what is the average convection heat transfer coefficient for the forward section? Naphthalene has a molecular weight of 128.2 kg/kmol and a saturated vapor pressure of 1.308×10^{-4} bar at 300 K.

7.91 Consider mass loss from a smooth wet flat plate due to forced convection at atmospheric pressure. The plate is 0.5 m long and 3 m wide. Dry air at 300 K and a free stream velocity of 35 m/s flows over the surface, which is also at a temperature of 300 K. Estimate the average mass transfer coefficient $\overline{h}_m$ and determine the water vapor mass loss rate (kg/s) from the plate.

7.92 Consider dry, atmospheric air in parallel flow over a 0.5-m long plate whose surface is wetted. The air velocity is 35 m/s, and the air and water are each at a temperature of 300 K.

(a) Estimate the heat loss and evaporation rate per unit width of the plate, q' and n'_A, respectively.

(b) Assuming the air temperature remains at 300 K, generate plots of q' and n'_A for a range of water temperatures from 300 to 350 K, with air velocities of 10, 20, and 35 m/s.

(c) For the air velocities and air temperature of part (b), determine the water temperatures for which the heat loss will be zero.

7.93 A flat plate coated with a volatile substance (species A) is exposed to dry, atmospheric air in parallel flow with $T_\infty = 20°C$ and $u_\infty = 8$ m/s. The plate is maintained at a constant temperature of 134°C by an electrical heating element, and the substance evaporates from the surface. The plate has a width of 0.25 m (normal to the plane of the sketch) and is well insulated on the bottom.

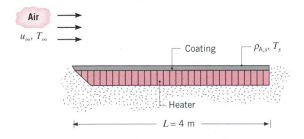

The molecular weight and the latent heat of vaporization of species A are $\mathcal{M}_A = 150$ kg/kmol and $h_{fg} = 5.44 \times 10^6$ J/kg, respectively, and the mass diffusivity is $D_{AB} = 7.75 \times 10^{-7}$ m²/s. If the satu-

rated vapor pressure of the substance is 0.12 atm at 134°C, what is the electrical power required to maintain steady-state conditions?

7.94 Dry air at atmospheric pressure and 350 K, with a free stream velocity of 25 m/s, flows over a smooth, porous plate 1 m long.

(a) Assuming the plate to be saturated with liquid water at 350 K, estimate the mass rate of evaporation per unit width of the plate, n'_A (kg/s · m).

(b) For air and liquid water temperatures of 300, 325, and 350 K, generate plots of n'_A as a function of velocity for the range from 1 to 25 m/s.

7.95 A scheme for dissipating heat from an array of $N = 100$ integrated circuits involves joining the circuits to the bottom of a plate and exposing the top of the plate to a water bath. The water container is of length $L = 100$ mm on a side and is exposed to airflow at its top surface. The flow is turbulated by the protruding lip of the side wall.

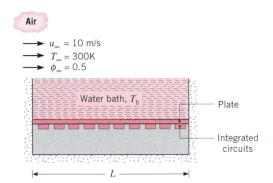

If the sides and bottom of the container are well insulated from the surroundings and heat is uniformly dissipated in each circuit, at which rate may heat be dissipated from each circuit when the water temperature is maintained at $T_b = 350$ K?

7.96 A series of water-filled trays each 222 mm long, experiences an evaporative drying process. Dry air at $T_\infty = 300$ K, flows over the trays with a velocity of 15 m/s, while radiant heaters maintain the surface temperature at $T_s = 330$ K.

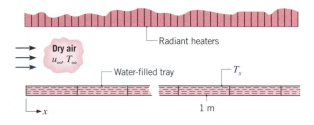

(a) What is the evaporative flux (kg/s · m²) at a distance 1 m from the leading edge?

(b) What is the irradiation (W/m²) that should be supplied to the tray surface at this location to maintain the water temperature at 330 K?

(c) Assuming the water temperature is uniform over the tray at this location, what is the evaporation rate (kg/s · m) from the tray per unit width of the tray?

(d) What irradiation should be applied to each of the first four trays such that the corresponding evaporation rates are identical to that found in part (c).

7.97 Consider the physical system of Problem 7.96 (a series of water-filled trays heated radiatively), but under operating conditions for which each tray is 0.25 m long by 1 m wide and is uniformly irradiated, with $G = 10^4$ W/m². Dry air at $T_\infty = 300$ K continues to flow over the trays at a velocity of 15 m/s.

(a) What is the rate of water loss (kg/s) from the first, third, and fourth trays?

(b) Estimate the temperature of the water in each of the designated trays.

7.98 The apparatus described in Problem 7.40 is used by our students to experimentally determine convection heat and mass transfer coefficients, to confirm the heat–mass analogy, and to compare measured results with predictions based on standard correlations. The velocity, V, of the airstream is measured using a thermistor-based anemometer, and its relative humidity is determined from measurements of the wet and dry bulb temperatures, T_{wb} and T_{db}, respectively. Thermocouples are attached to the *test-plate*, which is covered with a sheet of wet paper in the mass transfer experiments.

(a) *Convection heat transfer coefficient.* Using the data provided in Problem 7.40, determine the heat transfer coefficients for the two velocities, assuming the plate to behave as a spacewise isothermal object. Evaluate the coefficients C and m for a correlation of the form $\overline{Nu}_L = C\, Re^m\, Pr^{-1/3}$. Compare this result with a standard flat-plate correlation. Comment on the *goodness* of the comparison and provide reasons for any differences.

(b) *Convection mass transfer coefficient.* A sheet of water-saturated paper, 133 mm to a side, was used as the test surface and its mass measured at two different times, $m(t)$ and $m\,(t +$

$\Delta t)$. Thermocouples were used to monitor the paper temperature as a function of time, from which the average temperature, $\overline{T}_s$, was determined. The wet and dry bulb temperatures were $T_{wb} = 13°C$ and $T_{db} = 27°C$, and data recorded for two airstream velocities are as follows:

		Water Mass Loss Observations		
V (m/s)	$\overline{T}_s$(°C)	m(t) (g)	m(t + Δt) (g)	Δt (s)
3	15.3	55.62	54.45	475
9	16.0	55.60	54.50	240

Determine the convection mass transfer coefficients for the two flow conditions. Evaluate the coefficients C and m for a correlation of the form $\overline{Sh}_L = C\, Re^m\, Sc^{-1/3}$.

(c) Using the heat–mass analogy, compare the experimental results with each other and against standard correlations. Comment on the *goodness* of the comparison and provide reasons for any differences.

7.99 Dry air at 35°C and a velocity of 20 m/s flows over a wetted plate of length 500 mm and width 150 mm. An imbedded electrical heater supplies power to maintain the plate surface temperature at 20°C.

(a) What is the evaporation rate (kg/h) of water from the plate? What electrical power is required to maintain steady-state conditions?

(b) After a long period of operation, all the water is evaporated from the plate and its surface is dry. For the same free stream conditions and heater power of part (a), estimate the temperature of the plate.

7.100 A graduate student's spouse, after washing a load of diapers, hangs them on a line to dry. Each diaper is 1.0 m long, 0.5 m wide, and initially holds 0.2 kg of water. The relative humidity is 50%, the ambient temperature is 27°C, there is a gentle breeze of 0.5 m/s, and the sky is overcast. The spouse asks the student when the diapers should be dry. What is a good answer?

7.101 A minivan traveling 90 km/h has just passed through a thunderstorm that left a film of water 0.1 mm thick on the top of the van. The top of the van can be assumed to be a flat plate 6 m long. Assume isothermal conditions at 27°C, an ambient air relative humidity of 80%, and turbulent flow over the entire surface. What location on the van top will be the last to dry? What is the water evaporation rate per unit area (kg/s · m²) at the trailing edge of the van top?

7.102 Benzene, a known carcinogen, has been spilled on the laboratory floor and has spread to a length of 2 m. If a film 1 mm deep is formed, how long will it take for the benzene to completely evaporate? Ventilation in the laboratory provides for airflow parallel to the surface at 1 m/s, and the benzene and air are both at 25°C. The mass densities of benzene in the saturated vapor and liquid states are known to be 0.417 and 900 kg/m³, respectively.

7.103 Atmospheric air of 40% relative humidity and temperature $T_\infty = 300$ K is in parallel flow over a series of water-filled trays, with $u_\infty = 12$ m/s.

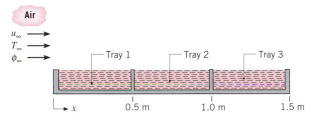

What is the rate at which energy must be supplied to each of the first three trays to maintain the water at 300 K?

7.104 A stream of atmospheric air is used to dry a series of photographic plates that are each of length $L_i = 0.25$ m in the direction of the airflow. The air is dry and at a temperature equal to that of the plates ($T_\infty = T_s = 50$°C). The air speed is $u_\infty = 9.1$ m/s.

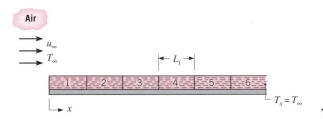

(a) Sketch the variation of the local convection mass transfer coefficient $h_{m,x}$ with distance x from the leading edge. Indicate the specific nature of the x dependence.

(b) Which of the plates will dry the fastest? Calculate the drying rate per meter of width for this plate (kg/s · m).

(c) At what rate would heat have to be supplied to the fastest drying plate to maintain it at $T_s = 50$°C during the drying process?

7.105 Condenser cooling water for a power plant is stored in a cooling pond that is 1000 m long by 500 m wide. However, because of evaporative losses, it is necessary to periodically add "makeup" water to the pond in order to maintain a suitable water level. Assuming isothermal conditions at 27°C for the water and the air, that the free stream air is dry and moving at a velocity of 2 m/s in the direction of the 1000-m pond length, and that the boundary layer on the water surface is everywhere turbulent, determine the amount of makeup water that should be added to the pond daily.

7.106 In a paper mill drying process, a sheet of paper slurry (water–fiber mixture) has a linear velocity of 5 m/s as it is rolled. Radiant heaters maintain a sheet temperature of $T_s = 330$ K, as evaporation occurs to dry, ambient air at 300 K above and below the sheet.

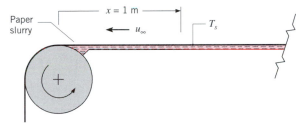

(a) What is the evaporative flux at a distance of $x = 1$ m from the leading edge of the roll? What is the corresponding value of the radiant flux (irradiation, G) that must be supplied to the sheet to maintain its temperature at 330 K? The sheet has an absorptivity of $\alpha = 1$.

(b) To accelerate the drying and paper production processes, the velocity and temperature of the strip are increased to 10 m/s and 340 K, respectively. To maintain a uniform strip temperature, the irradiation G must be varied with x along the strip. For $0 \leq x \leq 1$ m, compute and plot the variations $h_{m,x}(x)$, $N_A''(x)$, and $G(x)$.

7.107 A channel of triangular cross section, which is 25 m long and 1 m deep, is used for the storage of water.

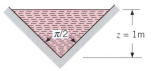

The water and the surrounding air are each at a temperature of 25°C, and the relative humidity of the air is 50%.

(a) If the air moves at a velocity of 5 m/s along the length of the channel, what is the rate of water loss due to evaporation from the surface?

(b) Obtain an expression for the rate at which the water depth would decrease with time due to evaporation. For the above conditions, how long would it take for all the water to evaporate?

7.108 To determine the heat transfer coefficient over a portion of a 50-mm wide flat plate with a corrugated finish, a section was fitted with a naphthalene insert having the properties $\mathcal{M} = 128.16$ kg/kmol and $p_{sat}(300 \text{ K}) = 1.308 \times 10^{-4}$ bar. An experiment was performed in which the mass loss due to sublimation from the insert was determined to be 1.10 g over a 3-h period for which a parallel flow of atmospheric air was maintained with $T_\infty = 300$ K and $u_\infty = 10$ m/s.

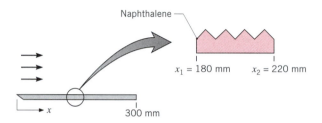

Naphthalene

$x_1 = 180$ mm $x_2 = 220$ mm

x

300 mm

What is the average convection heat transfer coefficient associated with the insert? Contrast this result with that predicted by assuming turbulent flow and using a conventional flat-plate correlation.

7.109 Mass transfer experiments have been conducted on a naphthalene cylinder of 18.4-mm diameter and 88.9-mm length subjected to a cross flow of air in a low-speed wind tunnel. After exposure for 39 min to the airstream at a temperature of 26°C and a velocity of 12 m/s, it was determined that the cylinder mass decreased by 0.35 g. The barometric pressure was recorded at 750.6 mm Hg. The saturation pressure p_{sat} of naphthalene vapor in equilibrium with solid naphthalene is given by the relation $p_{sat} = p \times 10^E$, where $E = 8.67 - (3766/T)$, with T (K) and p (bar) being the temperature and pressure of air. Naphthalene has a molecular weight of 128.16 kg/kmol.

(a) Determine the convection mass transfer coefficient from the experimental observations.

(b) Compare this result with an estimate from an appropriate correlation for the prescribed flow conditions.

7.110 Dry air at 1-atm pressure and a velocity of 15 m/s is to be humidified by passing it in cross flow over

a porous cylinder of diameter $D = 40$ mm, which is saturated with water.

(a) Assuming the water and air to be at 300 K, calculate the mass rate of water evaporated under steady-state conditions from the cylindrical medium per unit length?

(b) How will the evaporation rate change if the air and water are maintained at a higher temperature? Generate a plot for the temperature range 300 to 350 K to illustrate the effect of temperature on the evaporation rate.

7.111 Dry air at 35°C and a velocity of 15 m/s flows over a long cylinder of 20-mm diameter. The cylinder is covered with a thin porous coating saturated with water, and an imbedded electrical heater supplies power to maintain the coating surface temperature at 20°C.

(a) What is the evaporation rate of water from the cylinder per unit length (kg/h · m)? What electrical power per unit length of the cylinder (W/m) is required to maintain steady-state conditions?

(b) After a long period of operation, all the water is evaporated from the coating and its surface is dry. For the same free stream conditions and heater power of part (a), estimate the temperature of the surface.

7.112 Dry air at 20°C and a velocity of 15 m/s flows over a 20-mm diameter rod covered with a thin porous coating that is saturated with water. The rod ($k = 175$ W/m · K) is 250 mm long and its ends are attached to heat sinks maintained at 35°C.

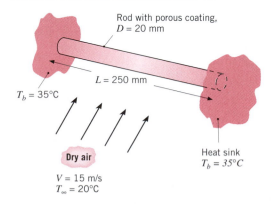

Rod with porous coating,
$D = 20$ mm

$T_b = 35°C$

$L = 250$ mm

Dry air

$V = 15$ m/s
$T_\infty = 20°C$

Heat sink
$T_b = 35°C$

Perform a steady-state, finite-difference analysis of the rod–porous coating system, considering conduction in the rod as well as energy transfer from the surface by convection heat and mass transfer. Use the analysis to estimate the temperature at the midspan of the rod and the evaporation rate from the surface. (*Suggestions:* Use 10 nodes to represent the half-length of the system. Estimate the overall average convection heat transfer coefficient based on an average film temperature for the system, and use the heat–mass transfer analogy to determine the average convection mass transfer coefficient. Validate your code by using it to predict a temperature distribution that agrees with the analytical solution for a fin without evaporation.)

7.113 Approximate the human form as an unclothed vertical cylinder of 0.3-m diameter and 1.75-m length with a surface temperature of 30°C.

(a) Calculate the heat loss in a 10-m/s wind at 20°C.

(b) What is the heat loss if the skin is covered with a thin layer of water at 30°C and the relative humidity of the air is 60%?

7.114 It has been suggested that heat transfer from a surface can be augmented by wetting it with water. As a specific example, consider a horizontal tube that is exposed to a transverse stream of dry air. You may assume that the tube, which is maintained at a temperature $T_s > T_\infty$, is completely wetted on the outside with a thin film of water. Derive an equation to determine the extent of heat transfer enhancement due to wetting. Evaluate the enhancement for $V = 10$ m/s, $D = 10$ mm, $T_s = 320$ K, and $T_\infty = 300$ K.

7.115 In the first stage of a paper drying process, a cylinder of diameter 0.15 m is covered by moisture-soaked paper. The temperature of the paper is maintained at 70°C by imbedded electrical heaters. Dry air at a velocity of 10 m/s and temperature of 20°C flows over the cylinder.

(a) Calculate the required electrical power and the evaporation rate per unit length of the cylinder, q' and n'_A, respectively.

(b) Generate plots of q' and n'_A as a function of the dry air velocity for $5 \leq V \leq 20$ m/s and for paper temperatures of 65, 70, and 75°C.

7.116 Cylindrical dry-bulb and wet-bulb thermometers are installed in a large-diameter duct to obtain the temperature T_∞ and the relative humidity ϕ_∞ of moist air flowing through the duct at a velocity V. The dry-bulb thermometer has a bare glass surface of diameter D_{db} and emissivity ε_g. The wet-bulb thermometer is covered with a thin wick that is saturated with water flowing continuously by capillary action from a bottom reservoir. Its diameter and emissivity are designated as D_{wb} and ε_w. The duct inside surface is at a known temperature T_s, which is less than T_∞. Develop expressions that may be used to obtain T_∞ and ϕ_∞ from knowledge of the dry-bulb and wet-bulb temperatures T_{db} and T_{wb} and the foregoing parameters. Determine T_∞ and ϕ_∞ when $T_{db} = 45°C$, $T_{wb} = 25°C$, $T_s = 35°C$, $p = 1$ atm, $V = 5$ m/s, $D_{db} = 3$ mm, $D_{wb} = 4$ mm, and $\varepsilon_g = \varepsilon_w = 0.95$. As a first approximation, evaluate the dry- and wet-bulb air properties at 45 and 25°C, respectively.

7.117 The thermal pollution problem is associated with discharging warm water from an electrical power plant or from an industrial source to a natural body of water. Methods for alleviating this problem involve cooling the warm water before allowing the discharge to occur. Two such methods, involving wet cooling towers or spray ponds, rely on heat transfer from the warm water in droplet form to the surrounding atmosphere. To develop an understanding of the mechanisms that contribute to this cooling, consider a spherical droplet of diameter D and temperature T, which is moving at a velocity V relative to air at a temperature T_∞ and relative humidity ϕ_∞. The surroundings are characterized by the temperature T_{sur}. Develop expressions for the droplet evaporation and cooling rates. Calculate the evaporation rate (kg/s) and cooling rate (K/s) when $D = 3$ mm, $V = 7$ m/s, $T = 40°C$, $T_\infty = 25°C$, $T_{sur} = 15°C$, and $\phi_\infty = 0.60$. The emissivity of water is $\varepsilon_w = 0.96$.

7.118 A spherical drop of water, 0.5 mm in diameter, is falling at a velocity of 2.15 m/s through dry still air at 1-atm pressure. Estimate the instantaneous rate of evaporation from the drop if the drop surface is at 60°C and the air is at 100°C.

7.119 A spherical droplet of alcohol, 0.5 mm in diameter, is falling freely through quiescent air at a velocity of 1.8 m/s. The concentration of alcohol vapor at the surface of the droplet is 0.0573 kg/m³, and the diffusion coefficient for alcohol in air is 10^{-5} m²/s. Neglecting radiation and assuming steady-state conditions, calculate the surface temperature of the droplet if the ambient air temperature is 300 K. The latent heat of vaporization is 8.42×10^5 J/kg.

7.120 The humidity of air may be controlled by spraying water droplets into an airflow. Consider droplets of diameter $D = 3$ mm in an airflow for which the relative velocity is 5 m/s. If the droplet and air temperatures are 25 and 35°C, respectively, what is the rate of evaporation from a single drop?

7.121 In a home furnace humidification system, water droplets of diameter D are discharged in a direction opposing the motion of warm air emerging from the heater. The air is humidified by evaporation from the droplets, and the excess water is collected on a *splash plate,* from which it is routed to a drain.

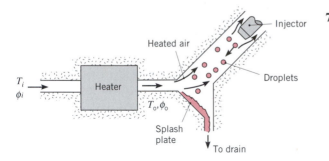

Consider conditions for which air enters the heater at a temperature and relative humidity of 17°C and 70%, respectively, and leaves the heater at a temperature of 47°C. The droplet diameter is 1 mm, and the relative velocity between the droplets and the heated air is 15 m/s. During the time-of-flight, the change in droplet diameter may be neglected and the droplet temperature may be assumed to remain at 47°C. What is the rate of evaporation from a single droplet?

7.122 Evaporation of liquid fuel droplets is often studied in the laboratory by using a porous sphere technique in which the fuel is supplied at a rate just sufficient to maintain a completely wetted surface on the sphere.

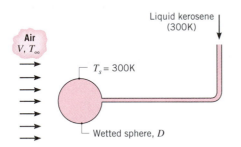

Consider the use of kerosene at 300 K with a porous sphere of 1-mm diameter. At this temperature the kerosene has a saturated vapor density of 0.015 kg/m³ and a latent heat of vaporization of 300 kJ/kg. The mass diffusivity for the vapor–air mixture is 10^{-5} m²/s. If dry, atmospheric air at $V = 15$ m/s and $T_\infty = 300$ K flows over the sphere, what is the minimum mass rate at which kerosene

must be supplied to maintain a wetted surface? For this condition, by how much must T_∞ actually exceed T_s to maintain the wetted surface at 300 K?

7.123 Consider an air conditioning system composed of a bank of tubes arranged normal to air flowing in a duct at a mass rate of $\dot{m}_a$ (kg/s). A coolant flowing through the tubes is able to maintain the surface temperature of the tubes at a constant value of $T_s < T_{a,i}$, where $T_{a,i}$ is the inlet air temperature (upstream of the tube bank). It has been suggested that air cooling may be enhanced if a thin, uniform film of water is maintained on the outer surface of each of the tubes.

(a) Assuming the water film to be at the temperature T_s, develop an expression for the ratio of the amount of cooling that occurs with the water film to the amount of cooling that occurs without the film. The amount of cooling may be defined as $T_{a,i} - T_{a,o}$, where $T_{a,o}$ is the outlet air temperature (downstream of the tube bank). The upstream air may be assumed to be dry, and the driving potentials for convection heat and mass transfer may be approximated as $(T_{a,i} - T_s)$ and $\rho_{A,sat}(T_s)$, respectively. *Note:* The total rate of heat loss from the air may be expressed as $q = \dot{m}_a c_{p,a}(T_{a,i} - T_{a,o})$. Estimate the value of this ratio under conditions for which $T_{a,i} = 35°C$ and $T_s = 10°C$.

(b) Consider a tube bank that is 5 rows deep, with 12 tubes in a row. Each tube is 0.5 m long, with an outside diameter of 8 mm, and a staggered arrangement is used for which $S_T = S_L = 24$ mm. Under conditions for which $\dot{m}_a = 0.5$ kg/s, $V = 3$ m/s, $T_{a,i} = 35°C$, and $T_s = 10°C$, what is the value of $T_{a,o}$ if the tubes are wetted? What is the specific humidity of the air leaving the tube bank?

7.124 In a paper-drying process, the paper moves on a conveyor belt at 0.2 m/s, while dry air from an array of slot jets (Figure 7.16) impinges normal to its surface. The nozzle width and pitch are $W = 10$ mm and $S = 100$ mm, respectively, and the nozzle-to-plate separation is $H = 200$ mm. The wet paper is of width $L = 1$ m and is maintained at 300 K, while the air exits the nozzles at a temperature of 300 K and a velocity of 20 m/s. In kg/s · m², what is the drying rate per unit surface area of the paper?

CHAPTER 8

Internal Flow

*H*aving acquired the means to compute convection transfer rates for external flow, we now consider the convection transfer problem for *internal flow*. Recall that an external flow is one for which boundary layer development on a surface is allowed to continue without external constraints, as for the flat plate of Figure 6.6. In contrast, an internal flow, such as flow in a pipe, is one for which the fluid is *confined* by a surface. Hence the boundary layer is unable to develop without eventually being constrained. The internal flow configuration represents a convenient geometry for heating and cooling fluids used in chemical processing, environmental control, and energy conversion technologies.

We begin by considering velocity (hydrodynamic) effects pertinent to internal flows, focusing on certain unique features of boundary layer development. Thermal boundary layer effects are considered next, and an overall energy balance is applied to determine fluid temperature variations in the flow direction. Finally, correlations for estimating the convection heat transfer coefficient are presented for a variety of internal flow conditions.

8.1
Hydrodynamic Considerations

When considering external flow, it is necessary to ask only whether this flow is laminar or turbulent. However, for an internal flow we must also be concerned with the existence of *entrance* and *fully developed* regions.

8.1.1   Flow Conditions

Consider laminar flow in a circular tube of radius r_o (Figure 8.1), where fluid enters the tube with a uniform velocity. We know that when the fluid makes contact with the surface, viscous effects become important and a boundary layer develops with increasing x. This development occurs at the expense of a shrinking inviscid flow region and concludes with boundary layer merger at the centerline. Following this merger, viscous effects extend over the entire cross

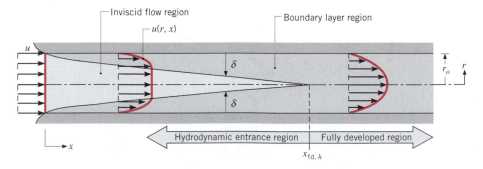

FIGURE 8.1   Laminar, hydrodynamic boundary layer development in a circular tube.

section and the velocity profile no longer changes with increasing x. The flow is then said to be *fully developed,* and the distance from the entrance at which this condition is achieved is termed the *hydrodynamic entry length,* $x_{\text{fd}, h}$. As shown in Figure 8.1, the *fully developed velocity profile* is parabolic for laminar flow in a circular tube. For turbulent flow, the profile is *flatter* due to turbulent mixing in the radial direction.

When dealing with internal flows, it is important to be cognizant of the extent of the entry region, which depends on whether the flow is laminar or turbulent. The Reynolds number for flow in a circular tube is defined as

$$Re_D \equiv \frac{\rho u_m D}{\mu} \tag{8.1}$$

where u_m is the mean fluid velocity over the tube cross section and D is the tube diameter. In a fully developed flow, the critical Reynolds number corresponding to the *onset* of turbulence is

$$Re_{D,c} \approx 2300 \tag{8.2}$$

although much larger Reynolds numbers ($Re_D \approx 10{,}000$) are needed to achieve fully turbulent conditions. The transition to turbulence is likely to begin in the developing boundary layer of the entrance region.

For laminar flow ($Re_D \lesssim 2300$), the hydrodynamic entry length may be obtained from an expression of the form [1]

$$\left(\frac{x_{\text{fd}, h}}{D}\right)_{\text{lam}} \approx 0.05 Re_D \tag{8.3}$$

This expression is based on the presumption that fluid enters the tube from a rounded converging nozzle and is hence characterized by a nearly uniform velocity profile at the entrance (Figure 8.1). Although there is no satisfactory general expression for the entry length in turbulent flow, we know that it is approximately independent of Reynolds number and that, as a first approximation [2],

$$10 \lesssim \left(\frac{x_{\text{fd}, h}}{D}\right)_{\text{turb}} \lesssim 60 \tag{8.4}$$

For the purposes of this text, we shall assume fully developed turbulent flow for $(x/D) > 10$.

8.1.2 The Mean Velocity

Because the velocity varies over the cross section and there is no well-defined free stream, it is necessary to work with a mean velocity u_m when dealing with internal flows. This velocity is defined such that, when multiplied by the fluid density ρ and the cross-sectional area of the tube A_c, it provides the rate of mass flow through the tube. Hence

$$\dot{m} = \rho u_m A_c \tag{8.5}$$

For steady, incompressible flow in a tube of uniform cross-sectional area, $\dot{m}$ and u_m are constants independent of x. From Equations 8.1 and 8.5 it is evident that, for flow in a *circular tube* ($A_c = \pi D^2/4$), the Reynolds number reduces to

$$Re_D = \frac{4\dot{m}}{\pi D \mu} \tag{8.6}$$

Since the mass flow rate may also be expressed as the integral of the mass flux (ρu) over the cross section

$$\dot{m} = \int_{A_c} \rho u(r, x) \, dA_c \tag{8.7}$$

it follows that, for *incompressible* flow in a *circular* tube,

$$u_m = \frac{\int_{A_c} \rho u(r, x) \, dA_c}{\rho A_c} = \frac{2\pi\rho}{\rho\pi r_o^2} \int_0^{r_o} u(r, x)r \, dr = \frac{2}{r_o^2} \int_0^{r_o} u(r, x)r \, dr \tag{8.8}$$

The foregoing expression may be used to determine u_m at any axial location x from knowledge of the velocity profile $u(r)$ at that location.

8.1.3   Velocity Profile in the Fully Developed Region

The form of the velocity profile may readily be determined for the *laminar flow* of an *incompressible, constant property fluid* in the *fully developed region* of a *circular tube*. An important feature of hydrodynamic conditions in the fully developed region is that both the radial velocity component v and the gradient of the axial velocity component ($\partial u/\partial x$) are everywhere zero.

$$v = 0 \qquad \text{and} \qquad \left(\frac{\partial u}{\partial x}\right) = 0 \tag{8.9}$$

Hence the axial velocity component depends only on r, $u(x, r) = u(r)$.

The radial dependence of the axial velocity may be obtained by solving the appropriate form of the x-momentum equation. This form is determined by first recognizing that, for the conditions of Equation 8.9, the net momentum flux is everywhere zero in the fully developed region. Hence the momentum conservation requirement reduces to a simple balance between shear and pressure forces in the flow. For the annular differential element of Figure 8.2, this force balance may be expressed as

$$\tau_r(2\pi r \, dx) - \left\{\tau_r(2\pi r \, dx) + \frac{d}{dr}[\tau_r(2\pi r \, dx)] \, dr\right\}$$

$$+ \, p(2\pi r \, dr) - \left\{p(2\pi r \, dr) + \frac{d}{dx}[p(2\pi r \, dr)] \, dx\right\} = 0$$

which reduces to

$$-\frac{d}{dr}(r\tau_r) = r\frac{dp}{dx} \tag{8.10}$$

With $y = r_o - r$, Newton's law of viscosity, Equation 6.53, assumes the form

$$\tau_r = -\mu\frac{du}{dr} \tag{8.11}$$

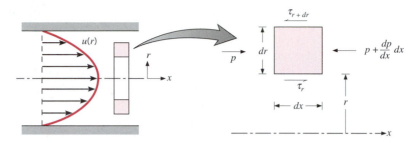

FIGURE 8.2 Force balance on a differential element for laminar, fully developed flow in a circular tube.

and Equation 8.10 becomes

$$\frac{\mu}{r}\frac{d}{dr}\left(r\frac{du}{dr}\right) = \frac{dp}{dx} \tag{8.12}$$

Since the axial pressure gradient is independent of r, Equation 8.12 may be solved by integrating twice to obtain

$$r\frac{du}{dr} = \frac{1}{\mu}\left(\frac{dp}{dx}\right)\frac{r^2}{2} + C_1$$

and

$$u(r) = \frac{1}{\mu}\left(\frac{dp}{dx}\right)\frac{r^2}{4} + C_1 \ln r + C_2$$

The integration constants may be determined by invoking the boundary conditions

$$u(r_o) = 0 \qquad \text{and} \qquad \frac{\partial u}{\partial r}\bigg|_{r=0} = 0$$

which, respectively, impose the requirements of zero slip at the tube surface and radial symmetry about the centerline. It is a simple matter to evaluate the constants, and it follows that

$$u(r) = -\frac{1}{4\mu}\left(\frac{dp}{dx}\right)r_o^2\left[1 - \left(\frac{r}{r_o}\right)^2\right] \tag{8.13}$$

Hence the fully developed velocity profile is *parabolic*. Note that the pressure gradient must always be negative.

The foregoing result may be used to determine the mean velocity of the flow. Substituting Equation 8.13 into Equation 8.8 and integrating, we obtain

$$u_m = -\frac{r_o^2}{8\mu}\frac{dp}{dx} \tag{8.14}$$

Substituting this result into Equation 8.13, the velocity profile is then

$$\frac{u(r)}{u_m} = 2\left[1 - \left(\frac{r}{r_o}\right)^2\right] \tag{8.15}$$

Since u_m can be computed from knowledge of the mass flow rate, Equation 8.14 can be used to determine the pressure gradient.

8.1.4 Pressure Gradient and Friction Factor in Fully Developed Flow

The engineer is frequently interested in the pressure drop needed to sustain an internal flow because this parameter determines pump or fan power requirements. To determine the pressure drop, it is convenient to work with the *Moody* (or Darcy) *friction factor,* which is a dimensionless parameter defined as

$$f \equiv \frac{-(dp/dx)D}{\rho u_m^2/2} \tag{8.16}$$

This quantity is not to be confused with the *friction coefficient,* sometimes called the Fanning friction factor, which is defined as

$$C_f \equiv \frac{\tau_s}{\rho u_m^2/2} \tag{8.17}$$

Since $\tau_s = -\mu(du/dr)_{r=r_o}$, it follows from Equation 8.13 that

$$C_f = \frac{f}{4} \tag{8.18}$$

Substituting Equations 8.1 and 8.14 into 8.16, it follows that, for fully developed laminar flow,

$$f = \frac{64}{Re_D} \tag{8.19}$$

For fully developed turbulent flow, the analysis is much more complicated, and we must ultimately rely on experimental results. Friction factors for a wide Reynolds number range are presented in the *Moody diagram* of Figure 8.3. In addition to depending on Reynolds number, the friction factor is a function of the tube surface condition. It is a minimum for *smooth* surfaces and increases with increasing surface roughness, *e*. Correlations that reasonably approximate the smooth surface condition are of the form

$$f = 0.316Re_D^{-1/4} \qquad Re_D \lesssim 2 \times 10^4 \tag{8.20a}$$

$$f = 0.184Re_D^{-1/5} \qquad Re_D \gtrsim 2 \times 10^4 \tag{8.20b}$$

Alternatively, a single correlation that encompasses a large Reynolds number range has been developed by Petukhov [4] and is of the form

$$f = (0.790 \ln Re_D - 1.64)^{-2} \qquad 3000 \lesssim Re_D \lesssim 5 \times 10^6 \tag{8.21}$$

Note that f, hence dp/dx, is a constant in the fully developed region. From Equation 8.16 the pressure drop $\Delta p = p_1 - p_2$ associated with fully developed flow from the axial position x_1 to x_2 may then be expressed as

$$\Delta p = -\int_{p_1}^{p_2} dp = f\frac{\rho u_m^2}{2D}\int_{x_1}^{x_2} dx = f\frac{\rho u_m^2}{2D}(x_2 - x_1) \tag{8.22a}$$

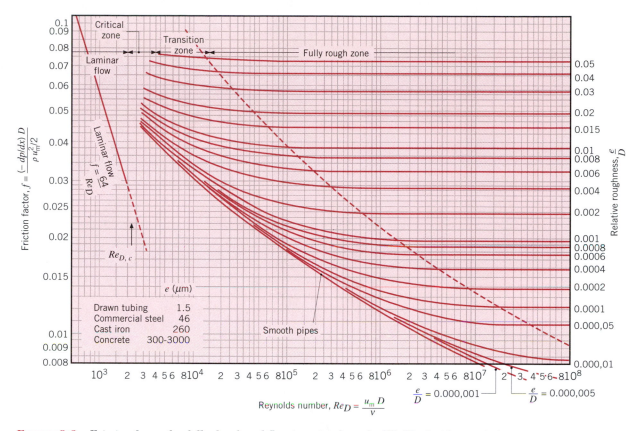

FIGURE 8.3 Friction factor for fully developed flow in a circular tube [3]. Used with permission.

where f is obtained from Figure 8.3 or from Equation 8.19 for laminar flow and from Equation 8.20 or 8.21 for turbulent flow in smooth tubes. The power (W) required to overcome the resistance to flow associated with this pressure drop may be expressed as

$$P = (\Delta p)\dot{\forall} \qquad (8.22b)$$

where the volumetric flow rate $\dot{\forall}$ may, in turn, be expressed as $\dot{\forall} = \dot{m}/\rho$ for an incompressible fluid.

8.2
Thermal Considerations

Having reviewed the fluid mechanics of internal flow, we now consider thermal effects. If fluid enters the tube of Figure 8.4 at a uniform temperature $T(r, 0)$ that is less than the surface temperature, convection heat transfer occurs and a *thermal boundary layer* begins to develop. Moreover, if the tube *surface* condition is fixed by imposing either a uniform temperature (T_s is constant) or a uniform heat flux (q''_s is constant), a *thermally fully developed condition* is eventu-

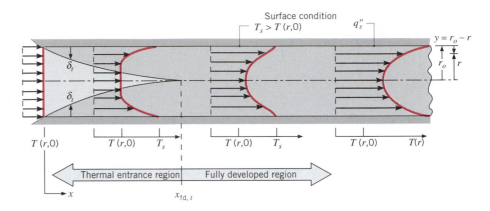

FIGURE 8.4   Thermal boundary layer development in a heated circular tube.

ally reached. The shape of the fully developed temperature profile $T(r, x)$ differs according to whether a uniform surface temperature or heat flux is maintained. For both surface conditions, however, the amount by which fluid temperatures exceed the entrance temperature increases with increasing x.

For laminar flow the *thermal entry length* may be expressed as [2]

$$\left(\frac{x_{fd,t}}{D}\right)_{lam} \approx 0.05 Re_D\, Pr \tag{8.23}$$

Comparing Equations 8.3 and 8.23, it is evident that, if $Pr > 1$, the hydrodynamic boundary layer develops more rapidly than the thermal boundary layer ($x_{fd,h} < x_{fd,t}$), while the inverse is true for $Pr < 1$. For extremely large Prandtl number fluids, such as oils ($Pr \gtrsim 100$), $x_{fd,h}$ is very much smaller than $x_{fd,t}$ and it is reasonable to assume a fully developed velocity profile throughout the thermal entry region. In contrast, for turbulent flow, conditions are nearly independent of Prandtl number, and to a first approximation, we shall assume $(x_{fd,t}/D) = 10$.

Thermal conditions in the fully developed region are characterized by several interesting and useful features. Before we can consider these features (Section 8.2.3), however, it is necessary to introduce the concept of a mean temperature and the appropriate form of Newton's law of cooling.

8.2.1   The Mean Temperature

Just as the absence of a free stream velocity requires use of a mean velocity to describe an internal flow, the absence of a fixed free stream temperature necessitates using a *mean temperature*. The *mean (or bulk) temperature* of the fluid at a given cross section is defined in terms of the *thermal* energy transported by the fluid as it moves past the cross section. The rate at which this transport occurs, $\dot{E}_t$, may be obtained by integrating the product of the mass flux (ρu) and the internal energy per unit mass ($c_v T$) over the cross section. That is,

$$\dot{E}_t = \int_{A_c} \rho u c_v T\, dA_c \tag{8.24}$$

Hence if a mean temperature is defined such that

$$\dot{E}_t \equiv \dot{m}c_v T_m \tag{8.25}$$

we obtain

$$T_m = \frac{\int_{A_c} \rho u c_v T \, dA_c}{\dot{m}c_v} \tag{8.26}$$

For *incompressible flow* in a *circular tube* with constant c_v, it follows from Equations 8.5 and 8.26 that

$$T_m = \frac{2}{u_m r_o^2} \int_0^{r_o} uTr \, dr \tag{8.27}$$

It is important to note that, when multiplied by the mass flow rate and the specific heat, T_m provides the rate at which thermal energy is transported with the fluid as it moves along the tube.

8.2.2 Newton's Law of Cooling

The mean temperature T_m is a convenient reference temperature for internal flows, playing much the same role as the free stream temperature T_∞ for external flows. Accordingly, Newton's law of cooling may be expressed as

$$q_s'' = h(T_s - T_m) \tag{8.28}$$

where h is the *local* convection heat transfer coefficient. However, there is an essential difference between T_m and T_∞. Whereas T_∞ is a constant in the flow direction, T_m must vary in this direction. That is, dT_m/dx is never zero if heat transfer is occurring. The value of T_m increases with x if heat transfer is from the surface to the fluid ($T_s > T_m$); it decreases with x if the opposite is true ($T_s < T_m$).

8.2.3 Fully Developed Conditions

Since the existence of convection heat transfer between the surface and the fluid dictates that the fluid temperature must continue to change with x, one might legitimately question whether fully developed thermal conditions can ever be reached. The situation is certainly different from the hydrodynamic case, for which $(\partial u/\partial x) = 0$ in the fully developed region. In contrast, if there is heat transfer, (dT_m/dx), as well as $(\partial T/\partial x)$ at any radius r, is not zero. Accordingly, the temperature profile $T(r)$ is continuously changing with x, and it would seem that a fully developed condition could never be reached. This apparent contradiction may be reconciled by working with a dimensionless form of the temperature.

Analyses have often been simplified by working with dimensionless temperature differences, as for transient conduction (Chapter 5) and the energy conservation equation (Chapter 6). Introducing a dimensionless temperature difference of the form $(T_s - T)/(T_s - T_m)$, conditions for which this ratio becomes independent of x are known to exist [2]. That is, although the temperature profile $T(r)$ continues to change with x, the *relative* shape of the profile no longer changes and the flow is said to be *thermally fully developed*. The requirement for such a condition is formally stated as

$$\frac{\partial}{\partial x} \left[\frac{T_s(x) - T(r, x)}{T_s(x) - T_m(x)} \right]_{\text{fd, }t} = 0 \tag{8.29}$$

where T_s is the tube surface temperature, T is the local fluid temperature, and T_m is the mean temperature of the fluid over the cross section of the tube.

The condition given by Equation 8.29 is eventually reached in a tube for which there is either a *uniform surface heat flux* (q_s'' is constant) *or a uniform surface temperature* (T_s is constant). These surface conditions arise in many engineering applications. For example, a constant surface heat flux would exist if the tube wall were heated electrically or if the outer surface were uniformly irradiated. In contrast, a constant surface temperature would exist if a phase change (due to boiling or condensation) were occurring at the outer surface. Note that it is impossible to *simultaneously* impose the conditions of constant surface heat flux and constant surface temperature. If q_s'' is constant, T_s must vary with x; conversely, if T_s is constant, q_s'' must vary with x.

Several important features of thermally developed flow may be inferred from Equation 8.29. Since the temperature ratio is independent of x, the derivative of this ratio with respect to r must also be independent of x. Evaluating this derivative at the tube surface (note that T_s and T_m are constants insofar as differentiation with respect to r is concerned), we then obtain

$$\frac{\partial}{\partial r} \left(\frac{T_s - T}{T_s - T_m} \right) \bigg|_{r=r_o} = \frac{-\partial T/\partial r|_{r=r_o}}{T_s - T_m} \neq f(x)$$

Substituting for $\partial T/\partial r$ from Fourier's law, which, from Figure 8.4, is of the form

$$q_s'' = -k \frac{\partial T}{\partial y} \bigg|_{y=0} = k \frac{\partial T}{\partial r} \bigg|_{r=r_o}$$

and for q_s'' from Newton's law of cooling, Equation 8.28, we obtain

$$\frac{h}{k} \neq f(x) \tag{8.30}$$

Hence *in the thermally fully developed flow* of a fluid *with constant properties*, the *local convection coefficient is a constant, independent of x.*

Equation 8.29 is not satisfied in the entrance region, where h varies with x, as shown in Figure 8.5. Because the thermal boundary layer thickness is zero at the tube entrance, the convection coefficient is extremely large at $x = 0$. However, h decays rapidly as the thermal boundary layer develops, until the constant value associated with fully developed conditions is reached.

Additional simplifications are associated with the special case of *uniform surface heat flux*. Since both h and q_s'' are constant in the fully developed region, it follows from Equation 8.28 that

$$\frac{dT_s}{dx} \bigg|_{\text{fd, }t} = \frac{dT_m}{dx} \bigg|_{\text{fd, }t} \qquad q_s'' = \text{constant} \tag{8.31}$$

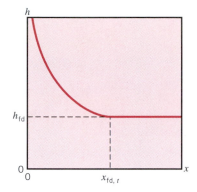

FIGURE 8.5
Axial variation of the convection heat transfer co-efficient for flow in a tube.

If we expand Equation 8.29 and solve for $\partial T/\partial x$, it also follows that

$$\frac{\partial T}{\partial x}\bigg|_{\text{fd},\,t} = \frac{dT_s}{dx}\bigg|_{\text{fd},\,t} - \frac{(T_s - T)}{(T_s - T_m)}\frac{dT_s}{dx}\bigg|_{\text{fd},\,t} + \frac{(T_s - T)}{(T_s - T_m)}\frac{dT_m}{dx}\bigg|_{\text{fd},\,t} \quad (8.32)$$

Substituting from Equation 8.31, we then obtain

$$\frac{\partial T}{\partial x}\bigg|_{\text{fd},\,t} = \frac{dT_m}{dx}\bigg|_{\text{fd},\,t} \qquad q_s'' = \text{constant} \quad (8.33)$$

Hence the axial temperature gradient is independent of the radial location. For the case of *constant surface temperature* ($dT_s/dx = 0$), it also follows from Equation 8.32 that

$$\frac{\partial T}{\partial x}\bigg|_{\text{fd},\,t} = \frac{(T_s - T)}{(T_s - T_m)}\frac{dT_m}{dx}\bigg|_{\text{fd},\,t} \qquad T_s = \text{constant} \quad (8.34)$$

in which case the value of $\partial T/\partial x$ depends on the radial coordinate.

From the foregoing results, it is evident that the mean temperature is a very important variable for internal flows. To describe such flows, its variation with x must be known. This variation may be obtained by applying an *overall energy balance* to the flow.

EXAMPLE 8.1

For flow of a liquid metal through a circular tube, the velocity and temperature profiles at a particular axial location may be approximated as being uniform and parabolic, respectively. That is, $u(r) = C_1$ and $T(r) - T_s = C_2[1 - (r/r_o)^2]$, where C_1 and C_2 are constants. What is the value of the Nusselt number Nu_D at this location?

SOLUTION

Known: Form of the velocity and temperature profiles at a particular axial location for flow in a circular tube.

Find: Nusselt number at the prescribed location.

Schematic:

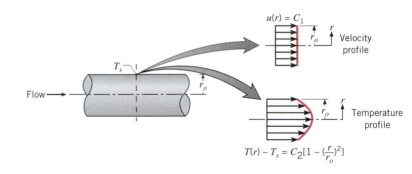

Assumptions: Incompressible, constant property flow.

Analysis: The Nusselt number may be obtained by first determining the convection coefficient, which, from Equation 8.28, is given as

$$h = \frac{q_s''}{T_s - T_m}$$

From Equation 8.27, the mean temperature is

$$T_m = \frac{2}{u_m r_o^2} \int_0^{r_o} uTr \, dr = \frac{2C_1}{u_m r_o^2} \int_0^{r_o} \left\{ T_s + C_2 \left[1 - \left(\frac{r}{r_o} \right)^2 \right] \right\} r \, dr$$

or, since $u_m = C_1$ from Equation 8.8,

$$T_m = \frac{2}{r_o^2} \int_0^{r_o} \left\{ T_s + C_2 \left[1 - \left(\frac{r}{r_o} \right)^2 \right] \right\} r \, dr$$

$$T_m = \frac{2}{r_o^2} \left[T_s \frac{r^2}{2} + C_2 \frac{r^2}{2} - \frac{C_2}{4} \frac{r^4}{r_o^2} \right] \Bigg|_0^{r_o}$$

$$T_m = \frac{2}{r_o^2} \left(T_s \frac{r_o^2}{2} + \frac{C_2}{2} r_o^2 - \frac{C_2}{4} r_o^2 \right) = T_s + \frac{C_2}{2}$$

The heat flux may be obtained from Fourier's law, in which case

$$q_s'' = k \frac{\partial T}{\partial r} \Bigg|_{r=r_o} = -kC_2 2 \frac{r}{r_o^2} \Bigg|_{r=r_o} = -2C_2 \frac{k}{r_o}$$

Hence

$$h = \frac{q_s''}{T_s - T_m} = \frac{-2C_2(k/r_o)}{-C_2/2} = \frac{4k}{r_o}$$

and

$$Nu_D = \frac{hD}{k} = \frac{(4k/r_o) \times 2r_o}{k} = 8 \qquad \triangleleft$$

8.3.1   General Considerations

Because the flow in a tube is completely enclosed, an energy balance may be applied to determine how the mean temperature $T_m(x)$ varies with position along the tube and how the total convection heat transfer q_{conv} is related to the difference in temperatures at the tube inlet and outlet. Consider the tube flow of Figure 8.6. Fluid moves at a constant flow rate $\dot{m}$, and convection heat transfer occurs at the inner surface. Typically, fluid kinetic and potential energy changes, as well as energy transfer by conduction in the axial direction, are negligible. Hence if no shaft work is done by the fluid as it moves through the tube, the only significant effects will be those associated with *thermal energy changes* and with *flow work*. Flow work is performed to move fluid through a control surface [5, 6] and, per unit mass of fluid, may be expressed as the product of the fluid pressure p and specific volume v ($v = 1/\rho$).

Applying conservation of energy, Equation 1.11a, to the differential control volume of Figure 8.6 and recalling the definition of the mean temperature, Equation 8.25, we obtain

$$dq_{conv} + \dot{m}(c_v T_m + pv) - \left[\dot{m}(c_v T_m + pv) + \dot{m}\,\frac{d(c_v T_m + pv)}{dx}\,dx \right] = 0$$

or

$$dq_{conv} = \dot{m}\,d(c_v T_m + pv) \tag{8.35}$$

That is, the rate of convection heat transfer to the fluid must equal the rate at which the fluid thermal energy increases plus the net rate at which work is done in moving the fluid through the control volume. If the fluid is assumed to be an *ideal gas* ($pv = RT_m$, $c_p = c_v + R$) and c_p is assumed to be constant, Equation 8.35 reduces to

$$dq_{conv} = \dot{m} c_p\, dT_m \tag{8.36}$$

This expression may also be used to a good approximation for *incompressible liquids*. In this case $c_v = c_p$, and since v is very small, $d(pv)$ is generally much

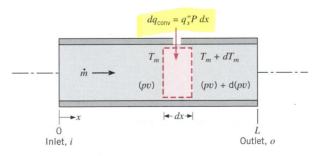

$$dq_{conv} = q_s'' P\, dx$$

FIGURE 8.6   Control volume for internal flow in a tube.

less than $d(c_v T_m)$.[1] Accordingly, Equation 8.36 again follows from Equation 8.35.

A special form of Equation 8.36 relates to conditions for the *entire* tube. In particular, integrating from the tube inlet i to the outlet o, it follows that

$$q_{conv} = \dot{m} c_p (T_{m,o} - T_{m,i}) \qquad (8.37)$$

where q_{conv} is the total tube heat transfer rate. This simple overall energy balance relates three important thermal variables (q_{conv}, $T_{m,o}$, $T_{m,i}$). *It is a general expression that applies irrespective of the nature of the surface thermal or tube flow conditions.*

Equation 8.36 may be cast in a convenient form by expressing the rate of convection heat transfer to the differential element as $dq_{conv} = q_s'' P \, dx$, where P is the surface perimeter ($P = \pi D$ for a circular tube). Substituting from Equation 8.28, it follows that

$$\frac{dT_m}{dx} = \frac{q_s'' P}{\dot{m} c_p} = \frac{P}{\dot{m} c_p} h(T_s - T_m) \qquad (8.38)$$

This expression is an extremely useful result, from which the axial variation of T_m may be determined. If $T_s > T_m$, heat is transferred to the fluid and T_m increases with x; if $T_s < T_m$, the opposite is true.

The manner in which quantities on the right-hand side of Equation 8.38 vary with x should be noted. Although P may vary with x, most commonly it is a constant (a tube of constant cross-sectional area). Hence the quantity ($P/\dot{m} c_p$) is a constant. In the fully developed region, the convection coefficient h is also constant, although it varies with x in the entrance region (Figure 8.5). Finally, although T_s may be constant, T_m must always vary with x (except for the trivial case of no heat transfer, $T_s = T_m$).

The solution to Equation 8.38 for $T_m(x)$ depends on the surface thermal condition. Recall that the two special cases of interest are *constant surface heat flux* and *constant surface temperature*. It is common to find one of these conditions existing to a reasonable approximation.

8.3.2   Constant Surface Heat Flux

For constant surface heat flux we first note that it is a simple matter to determine the total heat transfer rate q_{conv}. Since q_s'' is independent of x, it follows that

$$q_{conv} = q_s''(P \cdot L) \qquad (8.39)$$

This expression could be used with Equation 8.37 to determine the fluid temperature change, $T_{m,o} - T_{m,i}$.

For constant q_s'' it also follows that the right-hand side of Equation 8.38 is a constant independent of x. Hence

$$\frac{dT_m}{dx} = \frac{q_s'' P}{\dot{m} c_p} \neq f(x) \qquad (8.40)$$

---

[1]The only exception arises when the pressure gradient is extremely large. This situation occurs when $\dot{m}$ is very large and/or A_c is very small (see Problem 8.9).

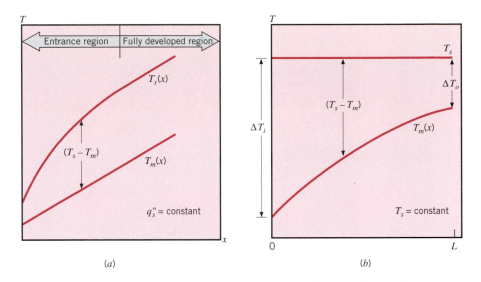

FIGURE 8.7   Axial temperature variations for heat transfer in a tube. (*a*) Constant surface heat flux. (*b*) Constant surface temperature.

Integrating from $x = 0$, it follows that

$$T_m(x) = T_{m,i} + \frac{q_s'' P}{\dot{m} c_p} x \qquad q_s'' = \text{constant} \tag{8.41}$$

Accordingly, the mean temperature varies *linearly* with x along the tube (Figure 8.7a). Moreover, from Equation 8.28 and Figure 8.5 we also expect the temperature difference $(T_s - T_m)$ to vary with x, as shown in Figure 8.7a. This difference is initially small (due to the large value of h at the entrance) but increases with increasing x due to the decrease in h that occurs as the boundary layer develops. However, in the fully developed region we know that h is independent of x. Hence from Equation 8.28 it follows that $(T_s - T_m)$ must also be independent of x in this region.

It should be noted that, if the heat flux is not constant but is, instead, a known function of x, Equation 8.38 may still be integrated to obtain the variation of the mean temperature with x. Similarly, the total heat rate may be obtained from the requirement that $q_{\text{conv}} = \int_0^L q_s''(x) P \, dx$.

EXAMPLE 8.2

A system for heating water from an inlet temperature of $T_{m,i} = 20°C$ to an outlet temperature of $T_{m,o} = 60°C$ involves passing the water through a thick-walled tube having inner and outer diameters of 20 and 40 mm. The outer surface of the tube is well insulated, and electrical heating within the wall provides for a uniform generation rate of $\dot{q} = 10^6$ W/m³.

1. For a water mass flow rate of $\dot{m} = 0.1$ kg/s, how long must the tube be to achieve the desired outlet temperature?

2. If the inner surface temperature of the tube is $T_s = 70°C$ at the outlet, what is the local convection heat transfer coefficient at the outlet?

SOLUTION

Known: Internal flow through thick-walled tube having uniform heat generation.

Find:

1. Length of tube needed to achieve the desired outlet temperature.
2. Local convection coefficient at the outlet.

Schematic:

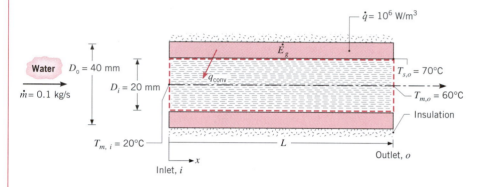

Assumptions:

1. Steady-state conditions.
2. Uniform heat flux.
3. Negligible potential energy, kinetic energy, and flow work changes.
4. Constant properties.
5. Adiabatic outer tube surface.

Properties: Table A.6, water ($\overline{T}_m = 313$ K): $c_p = 4.179$ kJ/kg · K.

Analysis:

1. Since the outer surface of the tube is adiabatic, the rate at which energy is generated within the tube wall must equal the rate at which it is convected to the water.

$$\dot{E}_g = q_{\text{conv}}$$

With

$$\dot{E}_g = \dot{q}\,\frac{\pi}{4}\,(D_o^2 - D_i^2)L$$

it follows from Equation 8.37 that

$$\dot{q}\,\frac{\pi}{4}\,(D_o^2 - D_i^2)L = \dot{m}c_p(T_{m,o} - T_{m,i})$$

or

$$L = \frac{4\dot{m}c_p}{\pi(D_o^2 - D_i^2)\dot{q}} (T_{m,o} - T_{m,i})$$

$$L = \frac{4 \times 0.1 \text{ kg/s} \times 4179 \text{ J/kg} \cdot \text{K}}{\pi(0.04^2 - 0.02^2) \text{ m}^2 \times 10^6 \text{ W/m}^3} (60 - 20)°\text{C} = 17.7 \text{ m} \qquad \triangleleft$$

2. From Newton's law of cooling, Equation 8.28, the local convection coefficient at the tube exit is

$$h_o = \frac{q_s''}{T_{s,o} - T_{m,o}}$$

Assuming that uniform heat generation in the wall provides a constant surface heat flux, with

$$q_s'' = \frac{\dot{E}_g}{\pi D_i L} = \frac{\dot{q}}{4} \frac{D_o^2 - D_i^2}{D_i}$$

$$q_s'' = \frac{10^6 \text{ W/m}^3}{4} \frac{(0.04^2 - 0.02^2) \text{ m}^2}{0.02 \text{ m}} = 1.5 \times 10^4 \text{ W/m}^2$$

it follows that

$$h_o = \frac{1.5 \times 10^4 \text{ W/m}^2}{(70 - 60)°\text{C}} = 1500 \text{ W/m}^2 \cdot \text{K} \qquad \triangleleft$$

Comments:

1. If conditions are fully developed over the entire tube, the local convection coefficient and the temperature difference $(T_s - T_m)$ are independent of x. Hence $h = 1500 \text{ W/m}^2 \cdot \text{K}$ and $(T_s - T_m) = 10°\text{C}$ over the entire tube. The inner surface temperature at the tube inlet is then $T_{s,i} = 30°\text{C}$.

2. The required tube length L could have been computed by applying the expression for $\dot{T}_m(x)$, Equation 8.41, at $x = L$.

8.3.3 Constant Surface Temperature

Results for the total heat transfer rate and the axial distribution of the mean temperature are entirely different for the *constant surface temperature* condition. Defining ΔT as $T_s - T_m$, Equation 8.38 may be expressed as

$$\frac{dT_m}{dx} = -\frac{d(\Delta T)}{dx} = \frac{P}{\dot{m}c_p} h \Delta T$$

Separating variables and integrating from the tube inlet to the outlet,

$$\int_{\Delta T_i}^{\Delta T_o} \frac{d(\Delta T)}{\Delta T} = -\frac{P}{\dot{m}c_p} \int_0^L h \, dx$$

or

$$\ln\frac{\Delta T_o}{\Delta T_i} = -\frac{PL}{\dot{m}c_p}\left(\int_0^L \frac{1}{L}\, h\, dx\right)$$

From the definition of the average convection heat transfer coefficient, Equation 6.5, it follows that

$$\ln\frac{\Delta T_o}{\Delta T_i} = -\frac{PL}{\dot{m}c_p}\,\overline{h}_L \qquad T_s = \text{constant} \qquad (8.42a)$$

where $\overline{h}_L$, or simply $\overline{h}$, is the average value of h for the entire tube. Rearranging,

$$\frac{\Delta T_o}{\Delta T_i} = \frac{T_s - T_{m,o}}{T_s - T_{m,i}} = \exp\left(-\frac{PL}{\dot{m}c_p}\,\overline{h}\right) \qquad T_s = \text{constant} \qquad (8.42b)$$

Had we integrated from the tube inlet to some axial position x within the tube, we would have obtained the similar, but more general, result that

$$\frac{T_s - T_m(x)}{T_s - T_{m,i}} = \exp\left(-\frac{Px}{\dot{m}c_p}\,\overline{h}\right) \qquad T_s = \text{constant} \qquad (8.43)$$

where $\overline{h}$ is now the average value of h from the tube inlet to x. This result tells us that the temperature difference $(T_s - T_m)$ decays *exponentially* with distance along the tube axis. The axial surface and mean temperature distributions are therefore as shown in Figure 8.7*b*.

Determination of an expression for the total heat transfer rate q_{conv} is complicated by the exponential nature of the temperature decay. Expressing Equation 8.37 in the form

$$q_{\text{conv}} = \dot{m}c_p[(T_s - T_{m,i}) - (T_s - T_{m,o})] = \dot{m}c_p(\Delta T_i - \Delta T_o)$$

and substituting for $\dot{m}c_p$ from Equation 8.42a, we obtain

$$q_{\text{conv}} = \overline{h}A_s\Delta T_{\text{lm}} \qquad T_s = \text{constant} \qquad (8.44)$$

where A_s is the tube surface area ($A_s = P \cdot L$) and ΔT_{lm} is the *log mean temperature difference*,

$$\Delta T_{\text{lm}} \equiv \frac{\Delta T_o - \Delta T_i}{\ln\,(\Delta T_o/\Delta T_i)} \qquad (8.45)$$

Equation 8.44 is a form of Newton's law of cooling for the entire tube, and ΔT_{lm} is the appropriate *average* of the temperature difference over the tube length. The logarithmic nature of this average temperature difference [in contrast, e.g., to an *arithmetic mean temperature* difference of the form $\Delta T_{\text{am}} = (\Delta T_i + \Delta T_o)/2$] is due to the exponential nature of the temperature decay.

Before concluding this section, it is important to note that, in many applications, it is the temperature of an *external* fluid, rather than the tube surface temperature, that is fixed (Figure 8.8). In such cases, it is readily shown that the results of this section may still be used if T_s is replaced by T_∞ (the free stream temperature of the external fluid) and $\overline{h}$ is replaced by $\overline{U}$ (the average overall heat transfer coefficient). For such cases, it follows that

$$\frac{\Delta T_o}{\Delta T_i} = \frac{T_\infty - T_{m,o}}{T_\infty - T_{m,i}} = \exp\left(-\frac{\overline{U}A_s}{\dot{m}c_p}\right) \qquad (8.46a)$$

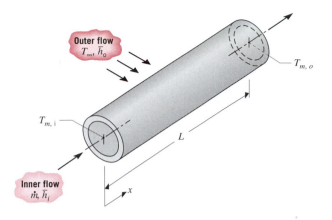

FIGURE 8.8 Heat transfer between fluid flowing over a tube and fluid passing through the tube.

and

$$q = \overline{U}A_s \, \Delta T_{\text{lm}} \qquad (8.47a)$$

The overall heat transfer coefficient is defined in Section 3.3.1, and for this application it would include contributions due to convection at the tube inner and outer surfaces. For a thick-walled tube of small thermal conductivity, it would also include the effect of conduction across the tube wall. Note that the product $\overline{U}A_s$ yields the same result, irrespective of whether it is defined in terms of the inner $(\overline{U}_i A_{s,i})$ or outer $(\overline{U}_o A_{s,o})$ surface areas of the tube (see Equation 3.32). Also note that $(\overline{U}A_s)^{-1}$ is equivalent to the total thermal resistance between the two fluids, in which case Equations 8.46a and 8.47a may be expressed as

$$\frac{\Delta T_o}{\Delta T_i} = \frac{T_\infty - T_{m,o}}{T_\infty - T_{m,i}} = \exp\left(-\frac{1}{\dot{m}c_p R_{\text{tot}}}\right) \qquad (8.46b)$$

and

$$q = \frac{\Delta T_{\text{lm}}}{R_{\text{tot}}} \qquad (8.47b)$$

A common variation of the foregoing conditions is one for which the uniform temperature of an *outer* surface, $T_{s,o}$, rather than the free stream temperature of an external fluid, T_∞, is known. In the foregoing equations, T_∞ is then replaced by $T_{s,o}$, and the total resistance embodies the convection resistance associated with the internal flow, as well as the total resistance due to conduction between the inner surface of the tube and the surface corresponding to $T_{s,o}$.

EXAMPLE 8.3

Steam condensing on the outer surface of a thin-walled circular tube of 50-mm diameter and 6-m length maintains a uniform surface temperature of 100°C. Water flows through the tube at a rate of $\dot{m} = 0.25$ kg/s, and its inlet and outlet temperatures are $T_{m,i} = 15°$C and $T_{m,o} = 57°$C. What is the average convection coefficient associated with the water flow?

| SOLUTION

Known: Flow rate and inlet and outlet temperatures of water flowing through a tube of prescribed dimensions and surface temperature.

Find: Average convection heat transfer coefficient.

Schematic:

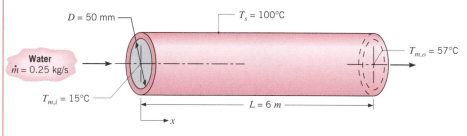

Assumptions:

1. Negligible outer surface convection resistance and tube wall conduction resistance.
2. Negligible kinetic energy, potential energy, and flow work changes.
3. Constant properties.

Properties: Table A.6, water (36°C): $c_p = 4178$ J/kg · K.

Analysis: Combining the energy balance, Equation 8.37, with the rate equation, Equation 8.44, the average convection coefficient is given by

$$\bar{h} = \frac{\dot{m}c_p}{\pi D L} \frac{(T_{m,\,o} - T_{m,\,i})}{\Delta T_{\text{lm}}}$$

From Equation 8.45

$$\Delta T_{\text{lm}} = \frac{(T_s - T_{m,\,o}) - (T_s - T_{m,\,i})}{\ln\,[(T_s - T_{m,\,o})/(T_s - T_{m,\,i})]}$$

$$\Delta T_{\text{lm}} = \frac{(100 - 57) - (100 - 15)}{\ln\,[(100 - 57)/(100 - 15)]} = 61.6°C$$

Hence

$$\bar{h} = \frac{0.25 \text{ kg/s} \times 4178 \text{ J/kg · K}}{\pi \times 0.05 \text{ m} \times 6 \text{ m}} \frac{(57 - 15)°C}{61.6°C}$$

or

$$\bar{h} = 756 \text{ W/m}^2 \cdot \text{K} \qquad \qquad \triangleleft$$

Comments:

1. In this case use of an arithmetic mean temperature difference, $\Delta T_{\text{am}} = T_s - (T_{m,\,i} + T_{m,\,o})/2 = 64°C$, in place of the log mean temperature difference, $\Delta T_{\text{lm}} = 61.6°C$, would have been a reasonable approximation.

2. If conditions were fully developed over the entire tube, the local convection coefficient would be everywhere equal to 756 W/m^2 · K.

8.4
Laminar Flow in Circular Tubes: Thermal Analysis and Convection Correlations

To use many of the foregoing results, the convection coefficients must be known. In this section we outline the manner in which such coefficients may be obtained theoretically for laminar flow in a circular tube. In subsequent sections we consider empirical correlations pertinent to turbulent flow in a circular tube, as well as to flows in tubes of noncircular cross section.

8.4.1 The Fully Developed Region

The problem of laminar flow in a circular tube has been treated theoretically, and the results may be used to determine the convection coefficients. At any point in the tube the boundary layer approximations (Section 6.5) are assumed to be applicable, and for constant properties, the energy equation is

$$u \frac{\partial T}{\partial x} + v \frac{\partial T}{\partial r} = \frac{\alpha}{r} \frac{\partial}{\partial r} \left(r \frac{\partial T}{\partial r} \right) \tag{8.48}$$

This equation, which applies for cylindrical coordinates, is of the same form as the boundary layer equation (6.57), which was developed in rectangular coordinates, except that viscous dissipation has been neglected. The terms on the left-hand side of Equation 8.48 account for net energy transfer by gross fluid motion (advection), and the term on the right accounts for net energy transfer by conduction in the radial direction.

The solution to Equation 8.48 is readily obtained for the *fully developed region*. In this region the velocity boundary layer approximations are satisfied exactly. That is, $v = 0$ and $(\partial u/\partial x) = 0$, in which case the axial velocity component is given by the parabolic profile of Equation 8.15. Moreover, for the case of *constant surface heat flux*, the thermal boundary layer approximation is also satisfied exactly. That is, $(\partial^2 T/\partial x^2) = 0$. Substituting for the axial temperature gradient from Equation 8.33 and for the axial velocity component from Equation 8.15, the energy equation, Equation 8.48, reduces to

$$\frac{1}{r} \frac{d}{dr} \left(r \frac{dT}{dr} \right) = \frac{2u_m}{\alpha} \left(\frac{dT_m}{dx} \right) \left[1 - \left(\frac{r}{r_o} \right)^2 \right] \qquad q_s'' = \text{constant} \tag{8.49}$$

where $(2u_m/\alpha)(dT_m/dx)$ is a constant. Separating variables and integrating twice, we obtain an expression for the radial temperature distribution:

$$T(r) = \frac{2u_m}{\alpha} \left(\frac{dT_m}{dx} \right) \left[\frac{r^2}{4} - \frac{r^4}{16r_o^2} \right] + C_1 \ln r + C_2$$

The constants of integration may be evaluated by applying appropriate boundary conditions. From the requirement that the temperature remain finite at $r = 0$, it follows that $C_1 = 0$. From the requirement that $T(r_o) = T_s$, where T_s varies with x, it also follows that

$$C_2 = T_s - \frac{2u_m}{\alpha}\left(\frac{dT_m}{dx}\right)\left(\frac{3r_o^2}{16}\right)$$

Accordingly, for the fully developed region with constant surface heat flux, the temperature profile is of the form

$$T(r) = T_s - \frac{2u_m r_o^2}{\alpha}\left(\frac{dT_m}{dx}\right)\left[\frac{3}{16} + \frac{1}{16}\left(\frac{r}{r_o}\right)^4 - \frac{1}{4}\left(\frac{r}{r_o}\right)^2\right] \tag{8.50}$$

From knowledge of the temperature profile, all other thermal parameters may be determined. For example, if the velocity and temperature profiles, Equations 8.15 and 8.50, respectively, are substituted into Equation 8.27 and the integration over r is performed, the mean temperature is found to be

$$T_m = T_s - \frac{11}{48}\left(\frac{u_m r_o^2}{\alpha}\right)\left(\frac{dT_m}{dx}\right) \tag{8.51}$$

From Equation 8.40, where $P = \pi D$ and $\dot{m} = \rho u_m(\pi D^2/4)$, we then obtain

$$T_m - T_s = -\frac{11}{48}\frac{q_s'' D}{k} \tag{8.52}$$

Combining Newton's law of cooling, Equation 8.28, and Equation 8.52, it follows that

$$h = \frac{48}{11}\left(\frac{k}{D}\right)$$

or

$$Nu_D \equiv \frac{hD}{k} = 4.36 \qquad q_s'' = \text{constant} \tag{8.53}$$

Hence in a *circular tube* characterized by *uniform surface heat flux* and *laminar, fully developed conditions*, the *Nusselt number is a constant*, independent of Re_D, Pr, and axial location.

For *laminar, fully developed conditions* with a *constant surface temperature*, the velocity boundary layer approximations are again satisfied exactly, and the thermal boundary layer approximation $(\partial^2 T/\partial x^2) \ll (\partial^2 T/\partial r^2)$ is often reasonable. Substituting for the velocity profile from Equation 8.15 and for the axial temperature gradient from Equation 8.34, the energy equation becomes

$$\frac{1}{r}\frac{d}{dr}\left(r\frac{dT}{dr}\right) = \frac{2u_m}{\alpha}\left(\frac{dT_m}{dx}\right)\left[1 - \left(\frac{r}{r_o}\right)^2\right]\frac{T_s - T}{T_s - T_m} \qquad T_s = \text{constant} \tag{8.54}$$

A solution to this equation may be obtained by an iterative procedure, which involves making successive approximations to the temperature profile. The resulting profile is not described by a simple algebraic expression, but the resulting Nusselt number may be shown to be of the form [2]

$$Nu_D = 3.66 \qquad T_s = \text{constant} \tag{8.55}$$

Note that in using Equation 8.53 or 8.55 to determine h, the thermal conductivity should be evaluated at T_m.

EXAMPLE 8.4

One concept used for solar energy collection involves placing a tube at the focal point of a parabolic reflector and passing a fluid through the tube.

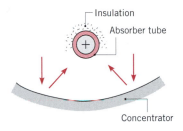

The net effect of this arrangement *may be approximated* as one of creating a condition of uniform heating at the surface of the tube. That is, the resulting heat flux to the fluid q_s'' may be assumed to be a constant along the circumference and axis of the tube. Consider operation with a tube of diameter $D = 60$ mm on a sunny day for which $q_s'' = 2000$ W/m^2.

1. If pressurized water enters the tube at $\dot{m} = 0.01$ kg/s and $T_{m,i} = 20°C$, what tube length L is required to obtain an exit temperature of $80°C$?
2. What is the surface temperature at the outlet of the tube, where fully developed conditions may be assumed to exist?

SOLUTION

Known: Internal flow with uniform surface heat flux.

Find:

1. Length of tube L to achieve required heating.
2. Surface temperature $T_s(L)$ at the outlet section, $x = L$.

Schematic:

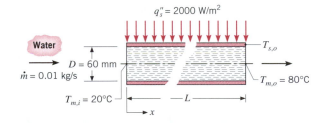

Assumptions:

1. Steady-state conditions.
2. Incompressible flow with constant properties.

3. Negligible kinetic and potential energy and flow work changes.

4. Fully developed conditions at tube outlet.

Properties: Table A.6, water ($\overline{T}_m$ = 323 K): c_p = 4181 J/kg · K. Table A.6, water ($T_{m,o}$ = 353 K): k = 0.670 W/m · K, μ = 352 × 10⁻⁶ N · s/m², Pr = 2.2.

Analysis:

1. For constant surface heat flux, Equation 8.39 may be used with the energy balance, Equation 8.37, to obtain

$$A_s = \pi D L = \frac{\dot{m}c_p(T_{m,o} - T_{m,i})}{q_s''}$$

$$L = \frac{\dot{m}c_p}{\pi D q_s''}(T_{m,o} - T_{m,i})$$

Hence

$$L = \frac{0.01 \text{ kg/s} \times 4181 \text{ J/kg} \cdot \text{K}}{\pi \times 0.060 \text{ m} \times 2000 \text{ W/m}^2}(80 - 20)°\text{C} = 6.65 \text{ m} \qquad \triangleleft$$

2. The surface temperature at the outlet may be obtained from Newton's law of cooling, Equation 8.28, where

$$T_{s,o} = \frac{q_s''}{h} + T_{m,o}$$

To find the local convection coefficient at the tube outlet, the nature of the flow condition must first be established. From Equation 8.6

$$Re_D = \frac{4\dot{m}}{\pi D \mu} = \frac{4 \times 0.01 \text{ kg/s}}{\pi \times 0.060 \text{ m} \times 352 \times 10^{-6} \text{ N} \cdot \text{s/m}^2} = 603$$

Hence the flow is laminar. With the assumption of fully developed conditions, the appropriate heat transfer correlation is then

$$Nu_D = \frac{hD}{k} = 4.36$$

and

$$h = 4.36 \frac{k}{D} = 4.36 \frac{0.670 \text{ W/m} \cdot \text{K}}{0.06 \text{ m}} = 48.7 \text{ W/m}^2 \cdot \text{K}$$

The surface temperature at the tube outlet is then

$$T_{s,o} = \frac{2000 \text{ W/m}^2}{48.7 \text{ W/m}^2 \cdot \text{K}} + 80°\text{C} = 121°\text{C} \qquad \triangleleft$$

Comments: For the conditions given, $(x_{fd}/D) = 0.05 Re_D Pr = 66.3$, while L/D = 110. Hence the assumption of fully developed conditions is justified. Note, however, that with $T_{s,o} > 100°\text{C}$, boiling may occur at the tube surface.

8.4.2   The Entry Region

The solution to the energy equation, Equation 8.48, for the entry region is more difficult to obtain, since velocity and temperature now depend on x as well as r. Even if the radial advection term is neglected, the axial temperature gradient $\partial T/\partial x$ may no longer be simplified through Equation 8.33 or 8.34. However, two different entry length solutions have been obtained. The simplest solution is for the *thermal entry length problem,* and it is based on assuming that thermal conditions develop in the presence of a *fully developed velocity profile.* Such a situation would exist if the location at which heat transfer begins were preceded by an *unheated starting length.* It could also be assumed to a reasonable approximation for large Prandtl number fluids, such as oils. Even in the absence of an unheated starting length, velocity boundary layer development would occur far more rapidly than thermal boundary layer development, and a thermal entry length approximation could be made. In contrast, the *combined* (thermal and velocity) *entry length problem* corresponds to the case for which the temperature and velocity profiles develop simultaneously.

Solutions have been obtained for both entry length conditions [2], and selected results are presented in Figure 8.9. The Nusselt numbers are, in principle, infinite at $x = 0$ and decay to their asymptotic (fully developed) values with increasing x. When plotted against the dimensionless parameter $x/(D\,Re_D\,Pr)$, which is the reciprocal of the *Graetz number* $Gz_D \equiv (D/x)Re_D\,Pr$, the manner in which Nu_D varies with Gz_D^{-1} is independent of Pr for the thermal entry length problem. In contrast, for the combined entry length problem, results depend on the Prandtl number and have been presented for $Pr = 0.7$, which is representative of most gases. At any location within the entry region,

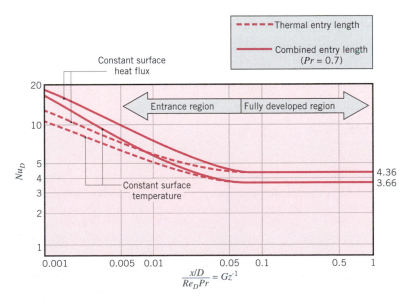

FIGURE 8.9   Local Nusselt number obtained from entry length solutions for laminar flow in a circular tube [2]. Adapted with permission.

Nu_D decreases with increasing Pr and approaches the thermal entry length condition as $Pr \rightarrow \infty$. Note that fully developed conditions are reached for $[(x/D)/Re_D \, Pr] \approx 0.05$.

For the *constant surface temperature* condition, it is desirable to know the *average* convection coefficient for use with Equation 8.44. Kays [7] presents a correlation attributed to Hausen [8], which is of the form

$$\overline{Nu}_D = 3.66 + \frac{0.0668(D/L)Re_D \, Pr}{1 + 0.04[(D/L)Re_D \, Pr]^{2/3}} \tag{8.56}$$

where $\overline{Nu}_D \equiv \overline{h}D/k$. Because this result presumes a thermal entry length, it is not generally applicable. For the combined entry length, a suitable correlation, due to Sieder and Tate [9], is of the form

$$\overline{Nu}_D = 1.86 \left(\frac{Re_D \, Pr}{L/D} \right)^{1/3} \left(\frac{\mu}{\mu_s} \right)^{0.14} \tag{8.57}$$

$$\left[\begin{array}{l} T_s = \text{constant} \\ 0.48 < Pr < 16,700 \\ 0.0044 < \left(\dfrac{\mu}{\mu_s} \right) < 9.75 \end{array} \right]$$

The correlation has been recommended by Whitaker [10] for values of $\{[Re_D \, Pr/(L/D)]^{1/3}(\mu/\mu_s)^{0.14}\} \gtrsim 2$. Below this limit fully developed conditions encompass much of the tube, and Equation 8.55 may be used to a good approximation. All properties appearing in Equations 8.56 and 8.57, except μ_s, should be evaluated at the average value of the mean temperature, $\overline{T}_m \equiv (T_{m,i} + T_{m,o})/2$.

The subject of laminar flow in ducts has been studied extensively, and numerous results are available for a variety of duct cross sections and surface conditions. These results have been compiled in a monograph by Shah and London [11] and in an updated review by Shah and Bhatti [12].

8.5
Convection Correlations: Turbulent Flow in Circular Tubes

Since the analysis of turbulent flow conditions is a good deal more involved, greater emphasis is placed on determining empirical correlations. A classical expression for computing the *local* Nusselt number for *fully developed (hydrodynamically* and *thermally) turbulent flow* in a *smooth circular tube* is due to Colburn [13] and may be obtained from the Chilton–Colburn analogy. Substituting Equation 6.103 into Equation 8.18, the analogy is of the form

$$\frac{C_f}{2} = \frac{f}{8} = St \, Pr^{2/3} = \frac{Nu_D}{Re_D \, Pr} Pr^{2/3} \tag{8.58}$$

Substituting for the friction factor from Equation 8.21, the *Colburn equation* is then

$$Nu_D = 0.023 Re_D^{4/5}\, Pr^{1/3} \tag{8.59}$$

The *Dittus–Boelter equation* [14] is a slightly different and preferred version of the above result and is of the form

$$Nu_D = 0.023 Re_D^{4/5}\, Pr^n \tag{8.60}$$

where $n = 0.4$ for heating ($T_s > T_m$) and 0.3 for cooling ($T_s < T_m$). These equations have been confirmed experimentally for the range of conditions

$$\left[\begin{array}{l} 0.7 \leq Pr \leq 160 \\ Re_D \gtrsim 10{,}000 \\ \dfrac{L}{D} \gtrsim 10 \end{array} \right]$$

The equations should be used only for small to moderate temperature differences, $T_s - T_m$, with all the properties evaluated at T_m. For flows characterized by large property variations, the following equation, due to Sieder and Tate [9], is recommended:

$$Nu_D = 0.027 Re_D^{4/5}\, Pr^{1/3} \left(\frac{\mu}{\mu_s} \right)^{0.14} \tag{8.61}$$

$$\left[\begin{array}{l} 0.7 \leq Pr \leq 16{,}700 \\ Re_D \gtrsim 10{,}000 \\ \dfrac{L}{D} \gtrsim 10 \end{array} \right]$$

where all properties except μ_s are evaluated at T_m. *To a good approximation, the foregoing correlations may be applied for both the uniform surface temperature and heat flux conditions.*

Although Equations 8.60 and 8.61 are easily applied and are certainly satisfactory for the purposes of this text, errors as large as 25% may result from their use. Such errors may be reduced to less than 10% through the use of more recent, but generally more complex, correlations [15]. One correlation, which is widely used and is attributed to Petukhov [4], is of the form

$$Nu_D = \frac{(f/8) Re_D\, Pr}{1.07 + 12.7 (f/8)^{1/2} (Pr^{2/3} - 1)} \tag{8.62}$$

where the friction factor may be obtained from the Moody diagram or, for smooth tubes, from Equation 8.21. The correlation is valid for $0.5 < Pr < 2000$ and $10^4 < Re_D < 5 \times 10^6$. To obtain agreement with data for smaller Reynolds numbers, Gnielinski [16] modified the correlation and proposed an expression of the form

$$Nu_D = \frac{(f/8)(Re_D - 1000) Pr}{1 + 12.7 (f/8)^{1/2} (Pr^{2/3} - 1)} \tag{8.63}$$

where, for smooth tubes, the friction factor is again given by Equation 8.21. The correlation is valid for $0.5 < Pr < 2000$ and $3000 < Re_D < 5 \times 10^6$. In using Equations 8.62 and 8.63, which apply for both uniform surface heat flux and temperature, properties should be evaluated at T_m. If temperature differences are large, additional consideration must be given to variable-property effects and available options are reviewed by Kakac [17].

We note that, unless specifically developed for the transition region ($2300 < Re_D < 10^4$), caution should be exercised when applying a turbulent flow correlation for $Re_D < 10^4$. If the correlation was developed for fully turbulent conditions ($Re_D > 10^4$), it may be used as a first approximation at smaller Reynolds numbers, with the understanding that the convection coefficient will be overpredicted. If a high level of accuracy is desired, the Gnielinski correlation, Equation 8.63, should be used. We also note that Equations 8.59 to 8.63 pertain to smooth tubes. For turbulent flow, the heat transfer coefficient increases with wall roughness, and, as a first approximation, it may be computed by using Equation 8.62 or 8.63 with friction factors obtained from the Moody diagram, Figure 8.3. However, although the general trend is one of increasing h with increasing f, the increase in f is proportionately larger, and when f is approximately four times larger than the corresponding value for a smooth surface, h no longer changes with additional increases in f [18]. Procedures for estimating the effect of wall roughness on convection heat transfer in fully developed turbulent flow are discussed by Bhatti and Shah [15].

Since entry lengths for turbulent flow are typically short, $10 \lesssim (x_{fd}/D) \lesssim 60$, it is often reasonable to assume that the average Nusselt number for the entire tube is equal to the value associated with the fully developed region, $\overline{Nu_D} \approx Nu_{D,\,fd}$. However, for short tubes $\overline{Nu_D}$ will exceed $Nu_{D,\,fd}$ and may be calculated from an expression of the form

$$\frac{\overline{Nu_D}}{Nu_{D,\,fd}} = 1 + \frac{C}{(x/D)^m} \tag{8.64}$$

where C and m depend on the nature of the inlet (e.g., sharp-edged or nozzle) and entry region (thermal or combined), as well as on the Prandtl and Reynolds numbers [2, 15, 19]. Typically, errors of less than 15% are associated with assuming $\overline{Nu_D} = Nu_{D,\,fd}$ for $(L/D) > 60$. When determining $\overline{Nu_D}$, all fluid properties should be evaluated at the arithmetic average of the mean temperature, $\overline{T}_m \equiv (T_{m,i} + T_{m,o})/2$.

Finally, we note that the foregoing correlations do not apply to liquid metals ($3 \times 10^{-3} \lesssim Pr \lesssim 5 \times 10^{-2}$). For fully developed turbulent flow in smooth circular tubes with constant surface heat flux, Skupinski et al. [20] recommend a correlation of the form

$$Nu_D = 4.82 + 0.0185 Pe_D^{0.827} \qquad q_s'' = \text{constant} \tag{8.65}$$
$$\begin{bmatrix} 3.6 \times 10^3 < Re_D < 9.05 \times 10^5 \\ 10^2 < Pe_D < 10^4 \end{bmatrix}$$

Similarly, for constant surface temperature Seban and Shimazaki [21] recommend the following correlation for $Pe_D > 100$:

$$Nu_D = 5.0 + 0.025Pe_D^{0.8} \qquad T_s = \text{constant} \qquad (8.66)$$

Extensive data and additional correlations are available in the literature [22].

EXAMPLE 8.5

Hot air flows with a mass rate of $\dot{m} = 0.050$ kg/s through an uninsulated sheet metal duct of diameter $D = 0.15$ m, which is in the crawlspace of a house. The hot air enters at 103°C and, after a distance of $L = 5$ m, cools to 77°C. The heat transfer coefficient between the duct outer surface and the ambient air at $T_\infty = 0$°C is known to be $h_o = 6$ W/m² · K.

1. Calculate the heat loss (W) from the duct over the length L.
2. Determine the heat flux and the duct surface temperature at $x = L$.

SOLUTION

Known: Hot air flowing in a duct.

Find:

1. Heat loss from the duct over the length L, q (W).
2. Heat flux and surface temperature at $x = L$.

Schematic:

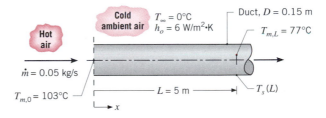

Assumptions:

1. Steady-state conditions.
2. Constant properties.
3. Ideal gas behavior.
4. Negligible kinetic and potential energy changes.
5. Negligible duct wall thermal resistance.
6. Uniform convection coefficient at outer surface of duct.

Properties: Table A.4, air ($\overline{T}_m = 363$ K): $c_p = 1010$ J/kg · K. Table A.4, air ($T_{m,L} = 350$ K): $k = 0.030$ W/m · K, $\mu = 208 \times 10^{-7}$ N · s/m², $Pr = 0.70$.

Analysis:

1. From the energy balance for the entire tube, Equation 8.37,

$$q = \dot{m}c_p(T_{m,L} - T_{m,0})$$

$$q = 0.05 \text{ kg/s} \times 1010 \text{ J/kg} \cdot \text{K} (77 - 103)°\text{C} = -1313 \text{ W} \qquad \triangleleft$$

2. An expression for the heat flux at $x = L$ may be inferred from the resistance network

$$q_s''(L) \longrightarrow \underset{\dfrac{1}{h_x(L)}}{\overset{T_{m,L}}{\circ}\!\!-\!\!\text{\Large www}\!\!-\!\!\overset{T_s(L)}{\circ}} \underset{\dfrac{1}{h_o}}{-\!\!\text{\Large www}\!\!-\!\!\overset{T_\infty}{\circ}}$$

where $h_x(L)$ is the inside convection heat transfer coefficient at $x = L$. Hence

$$q_s''(L) = \frac{T_{m,L} - T_\infty}{[1/h_x(L)] + (1/h_o)}$$

The inside convection coefficient may be obtained from knowledge of the Reynolds number. From Equation 8.6

$$Re_D = \frac{4\dot{m}}{\pi D \mu} = \frac{4 \times 0.05 \text{ kg/s}}{\pi \times 0.15 \text{ m} \times 208 \times 10^{-7} \text{ N} \cdot \text{s/m}^2} = 20{,}404$$

Hence the flow is turbulent. Moreover, with $(L/D) = (5/0.15) = 33.3$, it is reasonable to assume fully developed conditions at $x = L$. Hence from Equation 8.60, with $n = 0.3$,

$$Nu_D = \frac{h_x(L)D}{k} = 0.023 Re_D^{4/5} Pr^{0.3} = 0.023(20{,}404)^{4/5}(0.70)^{0.3} = 57.9$$

$$h_x(L) = Nu_D\frac{k}{D} = 57.9 \frac{0.030 \text{ W/m} \cdot \text{K}}{0.15 \text{ m}} = 11.6 \text{ W/m}^2 \cdot \text{K}$$

Hence

$$q_s''(L) = \frac{(77 - 0)°\text{C}}{[(1/11.6) + (1/6.0)] \text{ m}^2 \cdot \text{K/W}} = 304.5 \text{ W/m}^2$$

Referring back to the network, it also follows that

$$q_s''(L) = \frac{T_{m,L} - T_{s,L}}{1/h_x(L)}$$

in which case

$$T_{s,L} = T_{m,L} - \frac{q_s''(L)}{h_x(L)} = 77°\text{C} - \frac{304.5 \text{ W/m}^2}{11.6 \text{ W/m}^2 \cdot \text{K}} = 50.7°\text{C} \qquad \triangleleft$$

Comments:

1. In using the energy balance of part 1 for the entire tube, properties (in this case, only c_p) are evaluated at $\overline{T}_m = (T_{m,0} + T_{m,L})/2$. However, in using

the correlation for a local heat transfer coefficient, Equation 8.60, properties are evaluated at the local mean temperature, $T_{m,L} = 77°C$.

2. This problem is characterized neither by constant surface temperature nor by constant surface heat flux. It would therefore be erroneous to presume that the total heat loss from the tube is given by $q_s''(L)\pi DL = 717$ W. This result is substantially less than the actual heat loss of 1313 W because $q_s''(x)$ decreases with increasing x. This decrease in $q_s''(x)$ is due to reductions in both $h_x(x)$ and $[T_m(x) - T_\infty]$ with increasing x.

8.6
Convection Correlations: Noncircular Tubes

Although we have thus far restricted our consideration to internal flows of circular cross section, many engineering applications involve convection transport in *noncircular tubes*. At least to a first approximation, however, many of the circular tube results may be applied by using *an effective diameter* as the characteristic length. It is termed the *hydraulic diameter* and is defined as

$$D_h \equiv \frac{4A_c}{P} \tag{8.67}$$

where A_c and P are the *flow* cross-sectional area and the *wetted perimeter*, respectively. It is this diameter that should be used in calculating parameters such as Re_D and Nu_D.

For turbulent flow, which still occurs if $Re_D \gtrsim 2300$, it is reasonable to use the correlations of Section 8.5 for $Pr \gtrsim 0.7$. However, in a noncircular tube the convection coefficients vary around the periphery, approaching zero in the corners. Hence in using a circular tube correlation, the coefficient is presumed to be an average over the perimeter.

For laminar flow, the use of circular tube correlations is less accurate, particularly with cross sections characterized by sharp corners. For such cases the Nusselt number corresponding to fully developed conditions may be obtained from Table 8.1, which is based on solutions of the differential momentum and energy equations for flow through the different duct cross sections. As for the circular tube, results differ according to the surface thermal condition. Nusselt numbers tabulated for a uniform surface heat flux presume a constant flux in the axial (flow) direction, but a constant temperature around the perimeter at any cross section. This condition is typical of highly conductive tube wall materials. Results tabulated for a uniform surface temperature apply when the temperature is constant in both the axial and peripheral directions.

Although the foregoing procedures are generally satisfactory, exceptions do exist. Detailed discussion of heat transfer in noncircular tubes are provided in several sources [11, 12, 23].

TABLE 8.1 Nusselt numbers and friction factors for fully developed laminar flow in tubes of differing cross section

Cross Section	$\dfrac{b}{a}$	$Nu_D \equiv \dfrac{hD_h}{k}$ (Uniform q_s'')	$Nu_D \equiv \dfrac{hD_h}{k}$ (Uniform T_s)	$f\,Re_{D_h}$
●	—	4.36	3.66	64
$a\ \square\ b$	1.0	3.61	2.98	57
$a\ \square\ b$	1.43	3.73	3.08	59
$a\ \square\ b$	2.0	4.12	3.39	62
$a\ \square\ b$	3.0	4.79	3.96	69
$a\ \square\ b$	4.0	5.33	4.44	73
$a\ \square\ b$	8.0	6.49	5.60	82
▭	∞	8.23	7.54	96
△	—	3.11	2.47	53

Used with permission from W. M. Kays and M. E. Crawford, *Convection Heat and Mass Transfer*, McGraw-Hill, New York, 1980.

EXAMPLE 8.6

Consider the finned air heater of Problem 3.114, but operating under symmetrical conditions for which both of the end plates are at an equivalent temperature of $T_o = T_L = T_s = 400$ K. The stack width and depth are $W = 200$ mm and $B = 100$ mm, respectively, and the aluminum fins ($k = 240$ W/m · K) have a thickness of $t = 1$ mm. Consider conditions for which atmospheric air enters the stack at a temperature and velocity of $T_{m,i} = 300$ K and $u_{m,i} = 5$ m/s. If the fin length and pitch are $L = 15$ mm and $S = 3$ mm, respectively, determine the air outlet temperature, $T_{m,o}$, and the heat transfer rate, q, for the stack.

SOLUTION

Known: Dimensions of aluminum, finned-plate heat exchanger. Temperature of exchanger end plates. Inlet air temperature and velocity.

Find: Air outlet temperature and total heat rate.

Schematic:

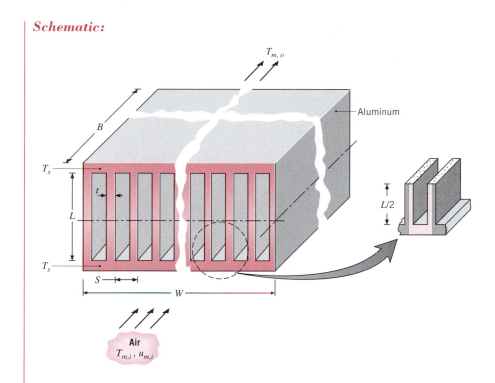

Assumptions:

1. Steady-state conditions.
2. Constant properties.
3. Ideal gas behavior.
4. Negligible kinetic and potential energy changes.
5. Negligible conduction resistance for end plates.
6. Fully developed flow throughout each channel.

Properties: Table A.4, air ($\overline{T}_m \approx 350$ K): $c_p = 1009$ J/kg · K, $\mu = 208.2 \times 10^{-7}$ N · s/m^2, $k = 0.030$ W/m · K. Table A.4 ($T_{m,i} = 300$ K): $\rho = 1.161$ kg/m^3.

Analysis: Heat transfer to the air is enhanced by the longitudinal fins, of which there are $N = W/S \approx 67$. With symmetry existing about its midplane, the stack may be viewed as two equivalent fin arrays, each fin having a length of $L/2 = 7.5$ mm. Hence the heat transfer rate for the stack may be expressed as $q = 2q_t$, where q_t is the total heat rate associated with each of the fin arrays. Modifying Equation 3.98 to account for variations of h and θ_b in the flow direction, it follows that

$$q = 2\overline{h}A_t\eta_o\theta_{b,\text{lm}} \tag{1}$$

where $\overline{h}$ and $\theta_{b,\text{lm}}$ are the average convection coefficient and *log mean temperature difference,* respectively, associated with the longitudinal extent B of the stack. From Equation 3.99, the total surface area of one of the arrays is

$$A_t = NA_f + A_b = N2(L/2)B + (W - Nt)B$$

$$= 67(0.015 \text{ m})0.1 \text{ m} + (0.2 \text{ m} - 67 \times 0.001 \text{ m})0.1 \text{ m}$$

$$= 0.1005 \text{ m}^2 + 0.0133 \text{ m}^2 = 0.1138 \text{ m}^2$$

and from Equation 3.102, the overall efficiency of the array is

$$\eta_o = 1 - \frac{NA_f}{A_t}(1 - \eta_f)$$

Since the plane of symmetry corresponds to an adiabatic surface, Equation 3.89 yields a fin efficiency of

$$\eta_f = \frac{\tanh m(L/2)}{m(L/2)}$$

To evaluate η_f, the average convection coefficient associated with flow in each of the 67 equivalent rectangular channels must be determined. With a channel hydraulic diameter of $D_h = 4A_c/P = 4L(S - t)/2(L + S - t) = 3.53$ mm and a flow rate of $\dot{m}_1 = \rho u_m A_c = 1.161 \text{ kg/m}^3$ (5 m/s)(0.015 × 0.002) m^2 = 1.742×10^{-4} kg/s, the Reynolds number is

$$Re_{D_h} = \frac{\rho u_m D_h}{\mu} = \frac{\dot{m}_1 D_h}{A_c \mu} = \frac{1.742 \times 10^{-4} \text{ kg/s}(0.00353 \text{ m})}{3 \times 10^{-5} \text{ m}^2 \times 208.2 \times 10^{-7} \text{ N} \cdot \text{s/m}^2} = 985$$

Hence flow within each channel is laminar. If a uniform surface temperature is assumed, use of Table 8.1 with an aspect ratio of $b/a = L/(S - t) = 7.5$ yields

$$Nu_D = \frac{hD_h}{k} = 5.46$$

Hence, assuming fully developed flow throughout each channel,

$$\bar{h} \approx \frac{k}{D_h}Nu_D = \frac{0.030 \text{ W/m} \cdot \text{K}}{0.00353 \text{ m}} 5.46 = 46.4 \text{ W/m}^2 \cdot \text{K}$$

With $m \equiv (\bar{h}P/kA_c)^{1/2} = [\bar{h}2(t + B)/k(tB)]^{1/2}$, it follows that

$$m = \left[\frac{46.4 \text{ W/m}^2 \cdot \text{K} \times 2(0.001 + 0.100) \text{ m}}{240 \text{ W/m} \cdot \text{K}(0.001 \times 0.100) \text{ m}^2}\right]^{1/2} = 19.8 \text{ m}^{-1}$$

Hence, with $m(L/2) = 19.8 \text{ m}^{-1} (0.0075 \text{ m}) = 0.148$,

$$\eta_f = \frac{\tanh (0.148)}{0.148} \approx \frac{0.147}{0.148} \approx 0.99$$

and

$$\eta_o \approx 1 - \frac{0.1005 \text{ m}^2}{0.1138 \text{ m}^2}(1 - 0.99) \approx 0.99$$

From Equation 3.103 the thermal resistance associated with each of the arrays is then

$$R_{t,o} = (\eta_o \bar{h} A_t)^{-1} = (0.99 \times 46.4 \text{ W/m}^2 \cdot \text{K} \times 0.1138 \text{ m}^2)^{-1} = 0.191 \text{ K/W}$$

To obtain the *log mean temperature difference*, $\theta_{b, \text{lm}}$, the air outlet temperature, $T_{m, o}$, must be determined. Referring to Equation 8.46b and replacing T_∞ by T_s, the outlet temperature may be obtained from

$$\frac{T_s - T_{m, o}}{T_s - T_{m, i}} = \exp\left(-\frac{1}{\dot{m}c_p R_{\text{tot}}}\right)$$

If this expression is applied to the entire stack, $\dot{m} = 67\dot{m}_1 = 0.01167$ kg/s and $R_{\text{tot}} = R_{t, o}/2 = 0.0955$ K/W, where the total resistance associated with heat transfer from the end plates must include the effect of both fin arrays. Hence

$$T_{m, o} = T_s - (T_s - T_{m, i}) \exp\left(-\frac{1}{\dot{m}c_p R_{\text{tot}}}\right)$$

$$= 400 \text{ K} - 100 \text{ K} \exp\left(-\frac{1}{0.01167 \text{ kg/s} \times 1009 \text{ J/kg} \cdot \text{K} \times 0.0955 \text{ K/W}}\right)$$

$$= 359 \text{ K} \qquad \triangleleft$$

From Equation 8.45, the desired *log mean temperature difference* is then

$$\theta_{b, \text{lm}} = \frac{(T_s - T_{m, o}) - (T_s - T_{m, i})}{\ln\left[(T_s - T_{m, o})/(T_s - T_{m, i})\right]} = \frac{(41 - 100) \text{ K}}{\ln(0.41)} = 66.2 \text{ K}$$

Returning to Equation 1, the heat rate for the entire stack is

$$q = 2\bar{h}A_t\eta_o\theta_{b, \text{lm}} = 2 \times 46.4 \text{ W/m}^2 \cdot \text{K} \times 0.1138 \text{ m}^2 \times 0.99 \times 66.2 \text{ K}$$

$$q = 692 \text{ W} \qquad \triangleleft$$

Comments:

1. With $B/D_h = 0.100$ m/0.00353 m $= 28.3$, the entire channel flow is characterized by developing hydrodynamic and thermal conditions, causing the actual value of $\bar{h}$ to exceed the estimate based on fully developed conditions. The actual heat rate would therefore exceed the estimated value of q.

2. The desired results could have been obtained by considering a unit channel, one-half of which is identified by the shaded region of the inset to the schematic. For the unit channel, Equation 8.46b takes the form

$$\frac{T_s - T_{m, o}}{T_s - T_{m, i}} = \exp\left(-\frac{1}{\dot{m}_1 c_p R_{\text{tot}, 1}}\right)$$

where $R_{\text{tot}, 1} = NR_{\text{tot}} = 67(0.0955 \text{ K/W}) = 6.40$ K/W. Use of this equation also yields $T_{m, o} = 359$ K. The heat rate per channel may be expressed as

$$q_1 = \frac{\theta_{b, \text{lm}}}{R_{\text{tot}, 1}}$$

or

$$q_1 = \dot{m}_1 c_p(T_{m, o} - T_{m, i})$$

which yields values of q_1 equal to 10.34 or 10.37 W, respectively, with the corresponding values of q equal to 693 or 695 W, respectively. Small differences in results are due to round-off error.

3. The heat rate may be increased by increasing the channel length B. Using the foregoing model, values of q and $T_{m,o}$ have been computed for $0.1 \leq B \leq 0.5$ m, and the results are plotted as follows:

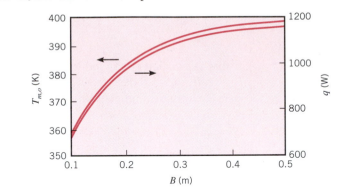

As $T_{m,o}$ approaches T_s with increasing B, the local heat flux decreases and q approaches an upper limit.

8.7
The Concentric Tube Annulus

Many internal flow problems involve heat transfer in a *concentric tube annulus* (Figure 8.10). Fluid passes through the space (annulus) formed by the concentric tubes, and convection heat transfer may occur to or from both the inner and outer tube surfaces. It is possible to independently specify the heat flux or temperature, that is, the thermal condition, at each of these surfaces. In any case the heat flux from each surface may be computed with expressions of the form

$$q_i'' = h_i(T_{s,i} - T_m) \qquad (8.68)$$

$$q_o'' = h_o(T_{s,o} - T_m) \qquad (8.69)$$

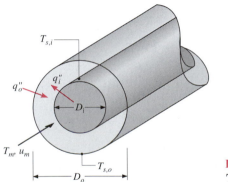

FIGURE 8.10
The concentric tube annulus.

Note that separate convection coefficients are associated with the inner and outer surfaces. The corresponding Nusselt numbers are of the form

$$Nu_i \equiv \frac{h_i D_h}{k} \tag{8.70}$$

$$Nu_o \equiv \frac{h_o D_h}{k} \tag{8.71}$$

where, from Equation 8.67, the hydraulic diameter D_h is

$$D_h = \frac{4(\pi/4)(D_o^2 - D_i^2)}{\pi D_o + \pi D_i} = D_o - D_i \tag{8.72}$$

For the case of fully developed laminar flow with one surface insulated and the other surface at a constant temperature, Nu_i or Nu_o may be obtained from Table 8.2. Note that in such cases we would be interested only in the convection coefficient associated with the isothermal (nonadiabatic) surface.

If uniform heat flux conditions exist at both surfaces, the Nusselt numbers may be computed from expressions of the form

$$Nu_i = \frac{Nu_{ii}}{1 - (q_o''/q_i'')\theta_i^*} \tag{8.73}$$

$$Nu_o = \frac{Nu_{oo}}{1 - (q_i''/q_o'')\theta_o^*} \tag{8.74}$$

The influence coefficients (Nu_{ii}, Nu_{oo}, θ_i^*, and θ_o^*) appearing in these equations may be obtained from Table 8.3. Note that q_i'' and q_o'' may be positive or negative, depending on whether heat transfer is to or from the fluid, respectively. Moreover, situations may arise for which the values of h_i and h_o are negative. Such results, when used with the sign convention implicit in Equations 8.68 and 8.69, reveal the relative magnitudes of T_s and T_m.

For fully developed turbulent flow, the influence coefficients are a function of the Reynolds and Prandtl numbers [23]. However, to a first approximation the inner and outer convection coefficients may be assumed to be equal, and

TABLE 8.2 Nusselt number for fully developed laminar flow in a circular tube annulus with one surface insulated and the other at constant temperature

D_i/D_o	Nu_i	Nu_o
0	—	3.66
0.05	17.46	4.06
0.10	11.56	4.11
0.25	7.37	4.23
0.50	5.74	4.43
1.00	4.86	4.86

Used with permission from W. M. Kays and H. C. Perkins, in W. M. Rohsenow and J. P. Hartnett, Eds., *Handbook of Heat Transfer,* Chap. 7, McGraw-Hill, New York, 1972.

TABLE 8.3 Influence coefficients for fully developed laminar flow in a circular tube annulus with uniform heat flux maintained at both surfaces

D_i/D_o	Nu_{ii}	Nu_{oo}	θ_i^*	θ_o^*
0	—	4.364	∞	0
0.05	17.81	4.792	2.18	0.0294
0.10	11.91	4.834	1.383	0.0562
0.20	8.499	4.833	0.905	0.1041
0.40	6.583	4.979	0.603	0.1823
0.60	5.912	5.099	0.473	0.2455
0.80	5.58	5.24	0.401	0.299
1.00	5.385	5.385	0.346	0.346

Used with permission from W. M. Kays and H. C. Perkins, in W. M. Rohsenow and J. P. Hartnett, Eds., *Handbook of Heat Transfer,* Chap. 7, McGraw-Hill, New York, 1972.

they may be evaluated by using the hydraulic diameter, Equation 8.72, with the Dittus–Boelter equation, Equation 8.60.

8.8
Heat Transfer Enhancement

Several options are available for enhancing heat transfer associated with internal flows. Enhancement may be achieved by increasing the convection coefficient and/or by increasing the convection surface area. For example, h may be increased by introducing surface roughness to enhance turbulence, as, for example,

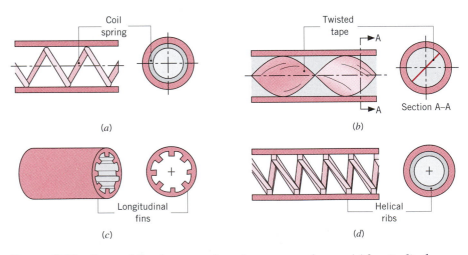

FIGURE 8.11 Internal flow heat transfer enhancement schemes: (*a*) longitudinal section and end view of coil-spring wire insert, (*b*) longitudinal section and cross-sectional view of twisted tape insert, (*c*) cut-away section and end view of longitudinal fins, and (*d*) longitudinal section and end view of helical ribs.

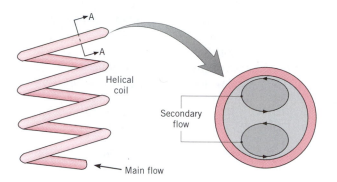

FIGURE 8.12   Schematic of helically coiled tube and secondary flow in enlarged cross-sectional view.

through machining or insertion of a coil-spring wire. The wire insert (Figure 8.11*a*) provides a helical roughness element in contact with the tube inner surface. Alternatively, the convection coefficient may be increased by inducing swirl through insertion of a twisted tape (Figure 8.11*b*). The insert consists of a thin strip that is periodically twisted through 360°. Introduction of a tangential velocity component increases the speed of the flow, particularly near the tube wall. The heat transfer area may be increased by attaching longitudinal fins to the inner surface (Figure 8.11*c*), while both the convection coefficient and area may be increased by using spiral fins or ribs (Figure 8.11*d*). In evaluating any heat transfer enhancement scheme, attention must also be given to the attendant increase in pressure drop and hence fan or pump power requirements. Comprehensive assessments of enhancement options have been published [24–26], and the *Journal of Enhanced Heat Transfer* provides access to recent developments in the field.

By coiling a tube (Figure 8.12), heat transfer may be enhanced without inducing turbulence or additional heat transfer surface area. In this case, centrifugal forces induce a *secondary flow* consisting of a pair of longitudinal vortices that increase the convection coefficient. A comprehensive review of heat transfer in coiled ducts is provided by Shah and Joshi [27].

8.9
Convection Mass Transfer

Mass transfer by convection may also occur for internal flows. For example, a gas may flow through a tube whose surface has been wetted or is sublimable. Evaporation or sublimation will then occur, and a concentration boundary layer will develop. Just as the mean temperature is the appropriate reference temperature for heat transfer, the mean species concentration $\rho_{A,m}$ plays an analogous role for mass transfer. By analogy to Equation 8.27 it follows that, for incompressible flow in a circular tube,

$$\rho_{A,m} = \frac{2}{u_m r_o^2} \int_0^{r_o} u \rho_A r \, dr \tag{8.75}$$

The concentration boundary layer development is characterized by entrance and fully developed regions, and Equation 8.23 may be used (with Pr replaced by Sc) to determine the *concentration entry length* $x_{fd, c}$ for laminar flow. Equation 8.4 may again be used as a first approximation for turbulent flow. Moreover, by analogy to Equation 8.29, for both laminar and turbulent flows, fully developed conditions exist when

$$\frac{\partial}{\partial x} \left[\frac{\rho_{A, s}(x) - \rho_A(r, x)}{\rho_{A, s}(x) - \rho_{A, m}(x)} \right]_{fd, c} = 0 \tag{8.76}$$

The mass flux of species A may be computed from an expression of the form

$$n''_{A, s} = h_m(\rho_{A, s} - \rho_{A, m}) \tag{8.77}$$

where the convection mass transfer coefficient h_m may be obtained from appropriate correlations involving the Sherwood number Sh_D, defined as

$$Sh_D \equiv \frac{h_m D}{D_{AB}} \tag{8.78}$$

Invoking the heat and mass transfer analogy, the specific form of the correlation may be inferred from the foregoing heat transfer results simply by replacing Nu_D with Sh_D and Pr with Sc.

EXAMPLE 8.7

A thin liquid film of ammonia, which has formed on the inner surface of a tube of diameter $D = 10$ mm and length $L = 1$ m, is removed by passing dry air through the tube at a flow rate of 3×10^{-4} kg/s. The tube and the air are at 25°C. What is the average mass transfer convection coefficient?

SOLUTION

Known: Liquid ammonia on the inner surface of a tube is removed by evaporation into an airstream.

Find: Average mass transfer convection coefficient for the tube.

Schematic:

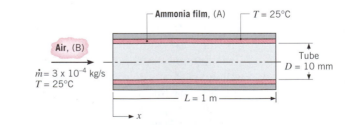

Assumptions:

1. Thin ammonia film with smooth surface.
2. Heat and mass transfer analogy is applicable.

Properties: Table A.4, air (25°C): $\nu = 15.7 \times 10^{-6}$ m²/s, $\mu = 183.6 \times$

10^{-7} N · s/m². Table A.8, ammonia–air (25°C): $D_{AB} = 0.28 \times 10^{-4}$ m²/s, $Sc = (\nu/D_{AB}) = 0.56$.

Analysis: From Equation 8.6

$$Re_D = \frac{4 \times 3 \times 10^{-4} \text{ kg/s}}{\pi \times 0.01 \text{ m} \times 183.6 \times 10^{-7} \text{ N} \cdot \text{s/m}^2} = 2080$$

in which case the flow is laminar. Hence, since a constant ammonia vapor concentration is maintained at the surface of the film, which is analogous to a constant surface temperature, and since

$$\left(\frac{Re_D \, Sc}{L/D} \right)^{1/3} = \left[\frac{(2080)(0.56)}{1/0.01} \right]^{1/3} = 2.27 > 2$$

the mass transfer analog to Equation 8.57 may be used to determine the average mass transfer convection coefficient. It follows that

$$\overline{Sh}_D = 1.86 \left(\frac{Re_D \, Sc}{L/D} \right)^{1/3} = 1.86 \times 2.27 = 4.22$$

Hence

$$\overline{h}_m = \overline{Sh}_D \frac{D_{AB}}{D} = \frac{4.22 \times 0.28 \times 10^{-4} \text{ m}^2/\text{s}}{0.01 \text{ m}} = 0.012 \text{ m/s} \qquad \triangleleft$$

Comments: From Equation 8.23, $x_{fd, c} \approx (0.05 Re_D \, Sc)D = 0.58$ m, and fully developed conditions exist over approximately 40% of the tube length. An assumption of fully developed conditions over the entire tube would provide a value of $\overline{Sh}_D = 3.66$, which is 13% less than the above result.

8.10
Summary

Internal flow is encountered in numerous applications, and it is important to appreciate its unique features. What is the nature of fully developed flow, and how does it differ from flow in the entry region? How does the Prandtl number influence boundary layer development in the entry region? How do thermal conditions in the fluid depend on the surface condition? For example, how do the surface and mean temperatures vary with x for the case of uniform surface heat flux? Or, how do mean temperature and the surface heat flux vary for the case of uniform surface temperature?

You must be able to perform engineering calculations that involve an energy balance and appropriate convection correlations. The methodology involves determining whether the flow is laminar or turbulent and establishing the length of the entry region. After deciding whether you are interested in local conditions (at a particular axial location) or in average conditions (for the entire tube), the convection correlation may be selected and used with the appropriate form of the energy balance to solve the problem. A summary of the correlations is provided in Table 8.4.

Several features that complicate internal flows have not been considered in this chapter. For example, a situation may exist for which there is a prescribed axial variation in T_s or q_s'', rather than uniform surface conditions. Among other things, such a variation would preclude the existence of a fully developed region. There may also exist surface roughness effects, circumferential heat flux or temperature variations, widely varying fluid properties, or transition flow

TABLE 8.4   Summary of convection correlations for flow in a circular tube[a,b,e]

Correlation		Conditions
$f = 64/Re_D$	(8.19)	Laminar, fully developed
$Nu_D = 4.36$	(8.53)	Laminar, fully developed, uniform q_s'', $Pr \gtrsim 0.6$
$Nu_D = 3.66$	(8.55)	Laminar, fully developed, uniform T_s, $Pr \gtrsim 0.6$
$\overline{Nu}_D = 3.66$ $+ \dfrac{0.0668(D/L)Re_D\,Pr}{1 + 0.04[(D/L)Re_D\,Pr]^{2/3}}$	(8.56)	Laminar, thermal entry length ($Pr \gg 1$ or an unheated starting length), uniform T_s
or $\overline{Nu}_D = 1.86\left(\dfrac{Re_D\,Pr}{L/D}\right)^{1/3}\left(\dfrac{\mu}{\mu_s}\right)^{0.14}$	(8.57)	Laminar, combined entry length $\{[Re_D\,Pr/(L/D)]^{1/3}(\mu/\mu_s)^{0.14}\} \gtrsim 2$, uniform T_s, $0.48 < Pr < 16{,}700$, $0.0044 < (\mu/\mu_s) < 9.75$
$f = 0.316Re_D^{-1/4}$	(8.20a)[c]	Turbulent, fully developed, $Re_D \lesssim 2 \times 10^4$
$f = 0.184Re_D^{-1/5}$	(8.20b)[c]	Turbulent, fully developed, $Re_D \gtrsim 2 \times 10^4$
or $f = (0.790 \ln Re_D - 1.64)^{-2}$	(8.21)[c]	Turbulent, fully developed, $3000 \lesssim Re_D \lesssim 5 \times 10^6$
$Nu_D = 0.023Re_D^{4/5}\,Pr^n$	(8.60)[d]	Turbulent, fully developed, $0.6 \le Pr \le 160$, $Re_D \gtrsim 10{,}000$, $(L/D) \gtrsim 10$, $n = 0.4$ for $T_s > T_m$ and $n = 0.3$ for $T_s < T_m$
or $Nu_D = 0.027Re_D^{4/5}\,Pr^{1/3}\left(\dfrac{\mu}{\mu_s}\right)^{0.14}$	(8.61)[d]	Turbulent, fully developed, $0.7 \le Pr \le 16{,}700$, $Re_D \gtrsim 10{,}000$, $L/D \gtrsim 10$
or $Nu_D = \dfrac{(f/8)(Re_D - 1000)Pr}{1 + 12.7(f/8)^{1/2}(Pr^{2/3} - 1)}$	(8.63)[d]	Turbulent, fully developed, $0.5 < Pr < 2000$, $3000 \lesssim Re_D \lesssim 5 \times 10^6$, $(L/D) \gtrsim 10$
$Nu_D = 4.82 + 0.0185(Re_D\,Pr)^{0.827}$	(8.65)	Liquid metals, turbulent, fully developed, uniform q_s'', $3.6 \times 10^3 < Re_D < 9.05 \times 10^5$, $10^2 < Pe_D < 10^4$
$Nu_D = 5.0 + 0.025(Re_D\,Pr)^{0.8}$	(8.66)	Liquid metals, turbulent, fully developed, uniform T_s, $Pe_D > 100$

[a]The mass transfer correlations may be obtained by replacing Nu_D and Pr by Sh_D and Sc, respectively.
[b]Properties in Equations 8.53, 8.55, 8.60, 8.61, 8.63, 8.65, and 8.66 are based on T_m; properties in Equations 8.19, 8.20, and 8.21 are based on $T_f \equiv (T_s + T_m)/2$; properties in Equations 8.56 and 8.57 are based on $\overline{T}_m \equiv (T_{m,i} + T_{m,o})/2$.
[c]Equations 8.20 and 8.21 pertain to smooth tubes. For rough tubes, Equation 8.63 should be used with the results of Figure 8.3.
[d]As a first approximation, Equation 8.60, 8.61, or 8.63 may be used to evaluate the average Nusselt number $\overline{Nu}_D$ over the entire tube length, if $(L/D) \gtrsim 10$. The properties should then be evaluated at the average of the mean temperature, $\overline{T}_m \equiv (T_{m,i} + T_{m,o})/2$.
[e]For tubes of noncircular cross section, $Re_D \equiv D_h u_m/\nu$, $D_h \equiv 4A_c/P$, and $u_m = \dot{m}/\rho A_c$. Results for fully developed laminar flow are provided in Table 8.1. For turbulent flow, Equation 8.60 may be used as a first approximation.

conditions. For a complete discussion of these effects, the literature should be consulted [11, 12, 15, 17, 23].

References

1. Langhaar, H. L., *J. Appl. Mech.,* **64,** A-55, 1942.

2. Kays, W. M., and M. E. Crawford, *Convective Heat and Mass Transfer,* McGraw-Hill, New York, 1980.

3. Moody, L. F., *Trans. ASME,* **66,** 671, 1944.

4. Petukhov, B. S., in T. F. Irvine and J. P. Hartnett, Eds., *Advances in Heat Transfer,* Vol. 6, Academic Press, New York, 1970.

5. Wark, K., *Thermodynamics,* 4th ed., McGraw-Hill, New York, 1983.

6. Bird, R. B., W. E. Stewart, and E. N. Lightfoot, *Transport Phenomena,* Wiley, New York, 1966.

7. Kays, W. M., *Trans. ASME,* **77,** 1265, 1955.

8. Hausen, H., *Z. VDI Beih. Verfahrenstech.,* **4,** 91, 1943.

9. Sieder, E. N., and G. E. Tate, *Ind. Eng. Chem.,* **28,** 1429, 1936.

10. Whitaker, S., *AIChE J.,* **18,** 361, 1972.

11. Shah, R. K., and A. L. London, *Laminar Flow Forced Convection in Ducts,* Academic Press, New York, 1978.

12. Shah, R. K., and M. S. Bhatti, in S. Kakac, R. K. Shah, and W. Aung, Eds., *Handbook of Single-Phase Convective Heat Transfer,* Chap. 3, Wiley-Interscience, New York, 1987.

13. Colburn, A. P., *Trans. AIChE,* **29,** 174, 1933.

14. Dittus, F. W., and L. M. K. Boelter, University of California, Berkeley, Publications on Engineering, Vol. 2, p. 443, 1930.

15. Bhatti, M. S., and R. K. Shah, in S. Kakac, R. K. Shah, and W. Aung, Eds., *Handbook of Single-Phase Convective Heat Transfer,* Chap. 4, Wiley-Interscience, New York, 1987.

16. Gnielinski, V., *Int. Chem. Eng.,* **16,** 359, 1976.

17. Kakac, S., in S. Kakac, R. K. Shah, and W. Aung, Eds., *Handbook of Single-Phase Convective Heat Transfer,* Chap. 18, Wiley-Interscience, New York, 1987.

18. Norris, R. H., in A. E. Bergles and R. L. Webb, Eds., *Augmentation of Convective Heat and Mass Transfer,* ASME, New York, 1970.

19. Molki, M., and E. M. Sparrow, *J. Heat Transfer,* **108,** 482, 1986.

20. Skupinski, E. S., J. Tortel, and L. Vautrey, *Int. J. Heat Mass Transfer,* **8,** 937, 1965.

21. Seban, R. A., and T. T. Shimazaki, *Trans. ASME,* **73,** 803, 1951.

22. Reed, C. B., in S. Kakac, R. K. Shah, and W. Aung, Eds., *Handbook of Single-Phase Convective Heat Transfer,* Chap. 8, Wiley-Interscience, New York, 1987.

23. Kays, W. M., and H. C. Perkins, in W. M. Rohsenow, J. P. Hartnett, and E. N. Ganic, Eds., *Handbook of Heat Transfer, Fundamentals,* Chap. 7, McGraw-Hill, New York, 1985.

24. Bergles, A. E., "Principles of Heat Transfer Augmentation," *Heat Exchangers, Thermal-Hydraulic Fundamentals and Design,* Hemisphere Publishing, New York, 1981, pp. 819–842.

25. Webb, R. L., in S. Kakac, R. K. Shah, and W. Aung, Eds., *Handbook of Single-Phase Convective Heat Transfer,* Chap. 17, Wiley-Interscience, New York, 1987.

26. Webb, R. L., *Principles of Enhanced Heat Transfer,* Wiley, New York, 1993.

27. Shah, R. K., and S. D. Joshi, in *Handbook of Single-Phase Convective Heat Transfer,* Chap. 5, Wiley-Interscience, New York, 1987.

Problems

Hydrodynamic Considerations

8.1 Fully developed conditions are known to exist for water flowing through a 25-mm diameter tube at 0.01 kg/s and 27°C. What is the maximum velocity of the water in the tube? What is the pressure gradient associated with the flow?

8.2 What is the pressure drop associated with water at 27°C flowing with a mean velocity of 0.2 m/s through a 600-m long cast iron pipe of 0.15-m inside diameter?

8.3 Water at 27°C flows with a mean velocity of 1 m/s through a 1-km long pipe of 0.25-m inside diameter.

(a) Determine the pressure drop over the pipe length and the corresponding pump power requirement, if the pipe surface is smooth.

(b) If the pipe is made of cast iron and its surface is clean, determine the pressure drop and pump power requirement.

(c) For the smooth pipe condition, generate a plot of pressure drop and pump power requirement for mean velocities in the range from 0.05 to 1.5 m/s.

8.4 Consider a 25-mm diameter circular tube through which liquid mercury, water, or engine oil at 27°C may flow at a rate of 0.03 kg/s. Determine the velocity, the hydrodynamic entry length, and the thermal entry length for each of the fluids.

8.5 An engine oil cooler consists of a bundle of 64 smooth tubes, each of length $L = 5$ m and diameter $D = 12.7$ mm.

(a) If oil at 300 K and a total flow rate of 8 kg/s is in fully developed flow through the tubes, what is the pressure drop and the pump power requirement?

(b) Compute and plot the pressure drop and pump power requirement as a function of flow rate for $5 \leq \dot{m} \leq 60$ kg/s.

Thermal Entry Length and Energy Balance Considerations

8.6 Compare the thermal and velocity entry lengths for oil, water, and mercury flowing through a 25-mm diameter tube with a mean velocity and temperature of $u_m = 5$ mm/s and $T_m = 27$°C, respectively.

8.7 Velocity and temperature profiles for laminar flow in a tube of radius $r_o = 10$ mm have the form

$$u(r) = 0.1[1 - (r/r_o)^2]$$
$$T(r) = 344.8 + 75.0(r/r_o)^2 - 18.8(r/r_o)^4$$

with units of m/s and K, respectively. Determine the corresponding value of the mean (or bulk) temperature, T_m, at this axial position.

8.8 At a particular axial station, velocity and temperature profiles for laminar flow in a parallel plate channel have the form

$$u(y) = 0.75[1 - (y/y_o)^2]$$
$$T(y) = 5.0 + 95.66(y/y_o)^2 - 47.83(y/y_o)^4$$

with units of m/s and °C, respectively.

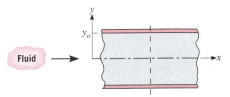

Determine corresponding values of the mean velocity, u_m, and mean (or bulk) temperature, T_m. Plot the velocity and temperature distributions. Do your values of u_m and T_m appear reasonable?

8.9 Water at a flow rate of 2 kg/s enters a long section of pipe with a temperature of 25°C and a pressure of 100 bars. The pipe wall is heated such that 10^5 W are transferred to the water as it flows through the pipe.

(a) If the water leaves the pipe with a pressure of 2 bars, what is its outlet temperature?

(b) What value of the outlet temperature would be obtained if Equation 8.37 were used for the calculation?

8.10 Water enters a tube at 27°C with a flow rate of 450 kg/h. The heat transfer from the tube wall to the fluid is given as q'_s (W/m) $= ax$, where the coefficient a is 20 W/m² and x (m) is the axial distance from the tube entrance.

(a) Beginning with a properly defined differential control volume in the tube, derive an expression for the temperature distribution $T_m(x)$ of the water.

(b) What is the outlet temperature of the water for a heated section 30 m long?

(c) Sketch the mean fluid temperature, $T_m(x)$, and the tube wall temperature, $T_s(x)$, as a function of distance along the tube for fully developed *and* developing flow conditions.

(d) What value of a uniform wall heat flux, q''_s (instead of $q'_s = ax$), would provide the same fluid outlet temperature as that determined in part (b)? For this type of heating, sketch the temperature distributions requested in part (c).

8.11 Consider flow in a circular tube. Within the test section length (between 1 and 2) a constant heat flux q''_s is maintained.

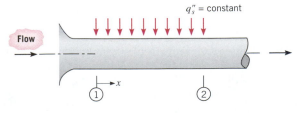

(a) For the two cases identified, sketch, qualitatively, the surface temperature $T_s(x)$ and the fluid mean temperature $T_m(x)$ as a function of distance along the test section x. In case A flow is hydrodynamically and thermally fully developed. In case B flow is not developed.

(b) Assuming that the surface flux q_s'' and the inlet mean temperature $T_{m,1}$ are identical for both cases, will the exit mean temperature $T_{m,2}$ for case A be greater than, equal to, or less than $T_{m,2}$ for case B? Briefly explain why.

8.12 Consider a cylindrical nuclear fuel rod of length L and diameter D that is encased in a concentric tube. Pressurized water flows through the annular region between the rod and the tube at a rate $\dot{m}$, and the outer surface of the tube is well insulated. Heat generation occurs within the fuel rod, and the volumetric generation rate is known to vary sinusoidally with distance along the rod. That is, $\dot{q}(x) = \dot{q}_o \sin(\pi x/L)$, where $\dot{q}_o$ (W/m³) is a constant. A uniform convection coefficient h may be assumed to exist between the surface of the rod and the water.

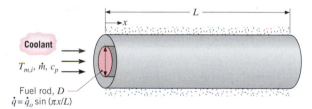

Coolant
$T_{m,i}$, $\dot{m}$, c_p

Fuel rod, D
$\dot{q} = \dot{q}_o \sin(\pi x/L)$

(a) Obtain expressions for the local heat flux $q''(x)$ and the total heat transfer q from the fuel rod to the water.

(b) Obtain an expression for the variation of the mean temperature $T_m(x)$ of the water with distance x along the tube.

(c) Obtain an expression for the variation of the rod surface temperature $T_s(x)$ with distance x along the tube. Develop an expression for the x location at which this temperature is maximized.

8.13 In a particular application involving fluid flow at a rate $\dot{m}$ through a circular tube of length L and diameter D, the surface heat flux is known to have a sinusoidal variation with x, which is of the form $q_s''(x) = q_{s,m}'' \sin(\pi x/L)$. The maximum flux, $q_{s,m}''$, is a known constant, and the fluid enters the tube at a known temperature, $T_{m,i}$. Assuming the convection coefficient to be constant, how do the mean temperature of the fluid and the surface temperature vary with x?

8.14 A flat-plate solar collector is used to heat atmospheric air flowing through a rectangular channel.

The bottom surface of the channel is well insulated, while the top surface is subjected to a uniform heat flux q_o'', which is due to the net effect of solar radiation absorption and heat exchange between the absorber and cover plates.

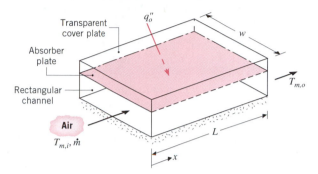

(a) Beginning with an appropriate differential control volume, obtain an equation that could be used to determine the mean air temperature $T_m(x)$ as a function of distance along the channel. Solve this equation to obtain an expression for the mean temperature of the air leaving the collector.

(b) With air inlet conditions of $\dot{m} = 0.1$ kg/s and $T_{m,i} = 40°C$, what is the air outlet temperature if $L = 3$ m, $w = 1$ m, and $q_o'' = 700$ W/m²? The specific heat of air is $c_p = 1008$ J/kg · K.

8.15 Atmospheric air enters the heated section of a circular tube at a flow rate of 0.005 kg/s and a temperature of 20°C. The tube is of diameter $D = 50$ mm, and fully developed conditions with $h = 25$ W/m² · K exist over the entire length of $L = 3$ m.

(a) For the case of uniform surface heat flux at $q_s'' = 1000$ W/m², determine the total heat transfer rate q and the mean temperature of the air leaving the tube $T_{m,o}$. What is the value of the surface temperature at the tube inlet $T_{s,i}$ and outlet $T_{s,o}$? Sketch the axial variation of T_s and T_m. On the same figure, also sketch (qualitatively) the axial variation of T_s and T_m for the more realistic case in which the local convection coefficient varies with x.

(b) If the surface heat flux varies linearly with x, such that q_s'' (W/m²) $= 500x$ (m), what are the values of q, $T_{m,o}$, $T_{s,i}$, and $T_{s,o}$? Sketch the axial variation of T_s and T_m. On the same figure, also sketch (qualitatively) the axial variation of T_s and T_m for the more realistic case in which the local convection coefficient varies with x.

(c) For the two heating conditions of parts (a) and (b), plot the mean fluid and surface temperatures, $T_m(x)$ and $T_s(x)$, respectively, as functions of distance along the tube. What effect will a fourfold increase in the convection coefficient have on the temperature distributions?

(d) For each type of heating process, what heat fluxes are required to achieve an air outlet temperature of 125°C? Plot the temperature distributions.

8.16 Slug flow is an idealized tube flow condition for which the velocity is assumed to be uniform over the entire tube cross section. For the case of laminar slug flow with a uniform surface heat flux, determine the form of the fully developed temperature distribution $T(r)$ and the Nusselt number Nu_D.

8.17 Superimposing a control volume that is differential in x on the tube flow conditions of Figure 8.8, derive Equation 8.46a.

8.18 An experimental nuclear core simulation apparatus consists of a long thin-walled metallic tube of diameter D and length L, which is electrically heated to produce the sinusoidal heat flux distribution

$$q_s''(x) = q_o'' \sin\left(\frac{\pi x}{L}\right)$$

where x is the distance measured from the tube inlet. Fluid at an inlet temperature $T_{m,i}$ flows through the tube at a rate of $\dot{m}$. Assuming the flow is turbulent and fully developed over the entire length of the tube, develop expressions for: (a) the total rate of heat transfer, q, from the tube to the fluid; (b) the fluid outlet temperature, $T_{m,o}$; (c) the axial distribution of the wall temperature, $T_s(x)$; and (d) the magnitude and position of the highest wall temperature. (e) Consider a 40-mm diameter tube of 4-m length with a sinusoidal heat flux distribution for which $q_o'' = 10,000$ W/m². Fluid passing through the tube has a flow rate of 0.025 kg/s, a specific heat of 4180 J/kg · K, an entrance temperature of 25°C, and a convection coefficient of 1000 W/m² · K. Plot the mean fluid and surface temperatures as a function of distance along the tube. Identify important features of the distributions. Explore the effect of ±25% changes in the convection coefficient and the heat flux on the distributions.

Heat Transfer Correlations: Circular Tubes

8.19 Engine oil at a rate of 0.02 kg/s flows through a 3-mm diameter tube 30 m long. The oil has an inlet temperature of 60°C, while the tube wall temperature is maintained at 100°C by steam condensing on its outer surface.

(a) Estimate the average heat transfer coefficient for internal flow of the oil.

(b) Determine the outlet temperature of the oil.

8.20 Engine oil is heated by flowing through a circular tube of diameter $D = 50$ mm and length $L = 25$ m and whose surface is maintained at 150°C.

(a) If the flow rate and inlet temperature of the oil are 0.5 kg/s and 20°C, what is the outlet temperature $T_{m,o}$? What is the total heat transfer rate q for the tube?

(b) For flow rates in the range $0.5 \leq \dot{m} \leq 2.0$ kg/s, compute and plot the variations of $T_{m,o}$ and q with $\dot{m}$. For what flow rate(s) are q and $T_{m,o}$ maximized? Explain your results.

8.21 Engine oil flows through a 25-mm diameter, 10-m long tube at a rate of 0.5 kg/s. The oil enters the tube at 25°C, while the tube surface is maintained at 100°C.

(a) Determine the total heat transfer to the oil and the oil outlet temperature.

(b) Repeat part (a), subject to the assumption of fully developed conditions throughout the tube.

8.22 Engine oil flows through a 25-mm diameter tube at a rate of 0.5 kg/s. The oil enters the tube at a temperature of 25°C, while the tube surface temperature is maintained at 100°C.

(a) Determine the oil outlet temperature for a 5-m and for a 100-m long tube. For each case, compare the log mean temperature difference to the arithmetic mean temperature difference.

(b) For $5 \leq L \leq 100$ m, compute and plot the average Nusselt number $\overline{Nu}_D$ and the oil outlet temperature as a function of L.

8.23 Ethylene glycol flows at 0.01 kg/s through a 3-mm diameter, thin-walled tube. The tube is coiled and submerged in a well-stirred water bath maintained at 25°C. If the fluid enters the tube at 85°C, what heat rate and tube length are required for the fluid to leave at 35°C? Neglect heat transfer enhancement associated with the coiling.

8.24 In the final stages of production, a pharmaceutical is sterilized by heating it from 25 to 75°C as it moves at 0.2 m/s through a straight thin-walled stainless steel tube of 12.7-mm diameter. A uniform heat flux is maintained by an electric resistance heater wrapped around the outer surface of the tube. If the tube is 10 m long, what is the required heat flux? If fluid enters the tube with a fully developed velocity profile and a uniform temperature profile, what is the surface temperature at the tube exit and at a distance of 0.5 m from the entrance? Fluid properties may be approximated at $\rho = 1000$ kg/m^3, $c_p = 4000$ J/kg · K, $\mu = 2 \times 10^{-3}$ kg/s · m, $k = 0.48$ W/m · K, and $Pr = 10$.

8.25 Oil at a temperature of 25°C and a mean velocity of 10 m/s is in hydrodynamically fully developed flow through a circular tube of 5-mm diameter, when it enters a 6-m long heated section. If the surface of the heated section is maintained at 150°C, what is the oil outlet temperature and the total heat rate? Oil properties may be evaluated at an estimated mean temperature of 310 K.

8.26 An electrical power transformer of diameter 300 mm and height 500 mm dissipates 1000 W. It is desired to maintain its surface temperature at 47°C by supplying glycerin at 24°C through thin-walled tubing of 20-mm diameter welded to the lateral surface of the transformer. All the heat dissipated by the transformer is assumed to be transferred to the glycerin.

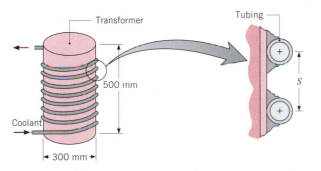

(a) Assuming the maximum allowable temperature rise of the coolant to be 6°C and fully developed flow throughout the tube, determine the required coolant flow rate, the total length of tubing, and the lateral spacing S between turns of the tubing.

(b) For a prescribed tube length of 15 m and a maximum allowable transformer surface temperature of 47°C, compute and plot the maximum allowable transformer power and the outlet temperature of the glycerin as a function of flow rate for $0.05 \le \dot{m} \le 0.25$ kg/s. Account for the fact that the flow is not fully developed.

8.27 Consider a flat-plate solar collector such as that shown in Problem 3.93. Copper tubing of inner diameter $D = 10$ mm and total length $L = 8$ m is soldered to the back of the collector plate, which is maintained at a uniform temperature of $T_p = 70$°C by the solar radiation. The thermal resistance asso-

ciated with conduction in the solder and tube wall may be neglected, as may the effect of the serpentine arrangement on flow in the tubes.

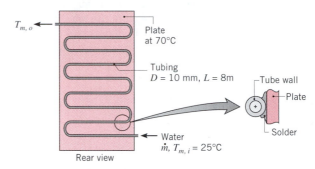

Rear view

(a) If water enters the tube at $T_{m,i} = 25$°C and $\dot{m} = 0.01$ kg/s, what is the exit temperature $T_{m,o}$ and the total heat rate q for the tube?

(b) The water outlet temperature and heat rate depend on the flow rate, which may readily be controlled. Compute and plot $T_{m,o}$ and q as a function of $\dot{m}$ for the range $0.005 \le \dot{m} \le 0.050$ kg/s. For $\dot{m} = 0.005$ and 0.05 kg/s, plot the temperature distribution along the tube.

8.28 The evaporator section of a heat pump is installed in a large tank of water, which is used as a heat source during the winter. As energy is extracted from the water, it begins to freeze, creating an ice/water bath at 0°C, which may be used for air conditioning during the summer. Consider summer cooling conditions for which air is passed through an array of copper tubes, each of inside diameter $D = 50$ mm, submerged in the bath.

(a) If air enters each tube at a mean temperature of $T_{m,i} = 24$°C and a flow rate of $\dot{m} = 0.01$ kg/s, what tube length L is needed to provide an exit temperature of $T_{m,o} = 14$°C? With 10 tubes passing through a tank of total volume $V = 10$ m^3, which initially contains 80% ice by volume, how long would it take to completely melt the ice? The density and latent heat of fusion of ice are 920 kg/m^3 and 3.34×10^5 J/kg, respectively.

(b) The air outlet temperature may be regulated by adjusting the tube mass flow rate. For the tube length determined in part (a), compute and plot $T_{m,o}$ as a function of $\dot{m}$ for $0.005 \le \dot{m} \le 0.05$ kg/s. If the dwelling cooled by this system requires approximately 0.05 kg/s of air at 16°C, what design and operating conditions should be prescribed for the system?

8.29 In a commercial dryer, air is heated from 20 to 50°C by passing it through *thick-walled* copper tubes of length $L = 1$ m and inside diameter $D_i = 0.05$ m. The flow rate per tube is 10^{-3} kg/s, and the air is heated by wrapping each tube with electric resistance heating tape, which provides a uniform heat flux at the *outer surface* of the tube. The tube wall thickness and thermal conductivity are sufficiently large to provide a uniform wall temperature for the prescribed operating conditions. Each heated tube is well insulated from its surroundings.

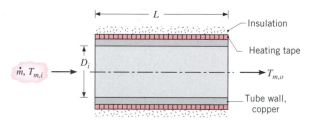

Using air properties of $c_p = 1007$ J/kg · K, $\mu = 188 \times 10^{-7}$ kg/s · m, $k = 0.0269$ W/m · K, and $Pr = 0.71$, evaluate the *average* heat flux at the inner surface of the tube and the tube wall temperature.

8.30 A 25-mm diameter circular tube, whose outer surface is maintained at 100°C, is used to heat water from 30 to 70°C.

(a) For a water flow rate of 1 kg/s, how long must the tube be?

(b) Plot the required tube length as a function of the flow rate for $0.25 \le \dot{m} \le 2$ kg/s.

8.31 Water flowing at 2 kg/s through a 40-mm diameter tube is to be heated from 25 to 75°C by maintaining the tube surface temperature at 100°C.

(a) What is the required tube length for these conditions?

(b) In order to design a water heating system, we wish to consider using tube diameters in the range from 30 to 50 mm. What are the required tube lengths for water flow rates of 1, 2, and 3 kg/s? Represent this design information graphically.

(c) Plot the pressure gradient as a function of tube diameter for the three flow rates. Assume the tube wall is smooth.

8.32 Water flows at 2 kg/s through a 40-mm diameter tube 4 m long. The water enters the tube at 25°C, and the surface temperature is 90°C.

(a) What is the outlet temperature of the water? What is the rate of heat transfer to the water?

(b) By controlling the pressure of the condensation process on the exterior surface of the tube, it is

possible to maintain the surface temperature in the range from 85 to 95°C. Maintaining the outlet temperature found in part (a), plot the required tube length as a function of the surface temperature. All other conditions remain the same.

8.33 A thick-walled, stainless steel (AISI 316) pipe of inside and outside diameters $D_i = 20$ mm and $D_o = 40$ mm is heated electrically to provide a uniform heat generation rate of $\dot{q} = 10^6$ W/m^3. The outer surface of the pipe is insulated, while water flows through the pipe at a rate of $\dot{m} = 0.1$ kg/s.

(a) If the water inlet temperature is $T_{m,i} = 20$°C and the desired outlet temperature is $T_{m,o} = 40$°C, what is the required pipe length?

(b) What are the location and value of the maximum pipe temperature?

8.34 Atmospheric air enters a 10-m long, 150-mm diameter uninsulated heating duct at 60°C and 0.04 kg/s. The duct surface temperature is approximately constant at $T_s = 15$°C.

(a) What are the outlet air temperature, the heat rate q, and pressure drop Δp for these conditions?

(b) To illustrate the tradeoff between heat transfer rate and pressure drop considerations, calculate q and Δp for diameters in the range from 0.1 to 0.2 m. In your analysis, maintain the total surface area, $A_s = \pi D L$, at the value computed for part (a). Plot q, Δp, and L as a function of the duct diameter.

8.35 Liquid mercury at 0.5 kg/s is to be heated from 300 to 400 K by passing it through a 50-mm diameter tube whose surface is maintained at 450 K. Calculate the required tube length by using an appropriate liquid metal convection heat transfer correlation. Compare your result with that which would have been obtained by using a correlation appropriate for $Pr \gtrsim 0.7$.

8.36 The surface of a 50-mm diameter, thin-walled tube is maintained at 100°C. In one case air is in cross flow over the tube with a temperature of 25°C and a velocity of 30 m/s. In another case air is in fully developed flow through the tube with a temperature of 25°C and a mean velocity of 30 m/s. Compare the heat flux from the tube to the air for the two cases.

8.37 Cooling water flows through the 25.4-mm diameter thin-walled tubes of a steam condenser at 1 m/s, and a surface temperature of 350 K is maintained by the condensing steam. The water inlet temperature is 290 K, and the tubes are 5 m long.

(a) What is the water outlet temperature? Evaluate water properties at an assumed average mean temperature, $\overline{T}_m = 300$ K.

(b) Was the assumed value for $\overline{T}_m$ reasonable? If not, repeat the calculation using properties evaluated at a more appropriate temperature.

(c) A range of tube lengths from 4 to 7 m is available to the engineer designing this condenser. Generate a plot to show what coolant mean velocities are possible if the water outlet temperature is to remain at the value found for part (b). All other conditions remain the same.

8.38 The air passage for cooling a gas turbine vane can be approximated as a tube of 3-mm diameter and 75-mm length. The operating temperature of the vane is 650°C, and air enters the tube at 427°C.

(a) For an air flow rate of 0.18 kg/h, calculate the air outlet temperature and the heat removed from the vane.

(b) Generate a plot of the air outlet temperature as a function of flow rate for $0.1 \le \dot{m} \le 0.6$ kg/h. Compare this result with those for vanes having 2- and 4-mm diameter tubes, with all other conditions remaining the same.

8.39 The core of a high-temperature, gas-cooled nuclear reactor has coolant tubes of 20-mm diameter and 780-mm length. Helium enters at 600 K and exits at 1000 K when the flow rate is 8×10^{-3} kg/s per tube.

(a) Determine the uniform tube wall surface temperature for these conditions.

(b) If the coolant gas is air, determine the required flow rate if the heat removal rate and tube wall surface temperature remain the same. What is the outlet temperature of the air?

8.40 Pressurized water in a nuclear reactor enters 25-mm diameter steam generation tubes at 5 m/s, 280°C, and 13.8 MPa. Saturated steam at 4 MPa is generated on the outside of the tubes. Estimate the convection heat transfer coefficient, heat flux, and pressure gradient for the water side of the tube.

8.41 Air at 200 kPa enters a 2-m long, thin-walled tube of 25-mm diameter at 150°C and 6 m/s. Steam at 20 bars condenses on the outer surface.

(a) Determine the outlet temperature and pressure drop of the air, as well as the rate of heat transfer to the air.

(b) Calculate the parameters of part (a) if the pressure of the air is doubled.

8.42 Consider the process by which ice is formed for an indoor rink. A parallel array of cooling tubes is submerged in a shallow layer of water, and a refrigerant (Freon-12) is passed through the tubes.

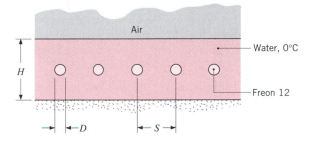

The water layer height is $H = 60$ mm and the tube pitch, diameter, and length are $S = 50$ mm, $D = 12$ mm, and $L = 5$ m, respectively. The temperature and flow rate of refrigerant entering each tube are -33°C and 0.02 kg/s, respectively. The refrigerant remains in the liquid state throughout the tube, and average thermophysical property values may be taken to be $c_p = 900$ J/kg · K, $k = 0.07$ W/m · K, $\mu = 3.5 \times 10^{-4}$ kg/m · s, and $Pr = 4.6$. The density of water is $\rho = 1000$ kg/m³, and its latent heat of fusion is $h_{sf} = 3.34 \times 10^5$ J/kg.

(a) Consider the process whereby the water is taken from the saturated liquid to solid states at the freezing point (0°C). If the tube wall temperature is assumed to be at the freezing point throughout the process, at what temperature does the refrigerant leave the tube? What is the rate of heat transfer to the refrigerant for a single tube length?

(b) For the conditions of part (a), how long would it take to completely freeze the water?

(c) The freezing process may be accelerated by increasing the flow rate of the refrigerant. Compute and plot the rate of heat transfer for a single tube and the refrigerant outlet temperature as a function of flow rate for $0.02 \le \dot{m} \le 0.10$ kg/s. How long would it take to freeze the water for $\dot{m} = 0.10$ kg/s?

8.43 Heated air required for a food-drying process is generated by passing ambient air at 20°C through long, circular tubes ($D = 50$ mm, $L = 5$ m) housed in a steam condenser. Saturated steam at atmospheric pressure condenses on the outer surface of the tubes, maintaining a uniform surface temperature of 100°C.

(a) If an air flow rate of 0.01 kg/s is maintained in each tube, determine the air outlet temperature $T_{m,o}$ and the total heat rate q for the tube.

(b) The air outlet temperature may be controlled by adjusting the tube mass flow rate. Compute and plot $T_{m,o}$ as a function of $\dot{m}$ for $0.005 \leq \dot{m} \leq 0.050$ kg/s. If a particular drying process requires approximately 1 kg/s of air at 75°C, what design and operating conditions should be prescribed for the air heater, subject to the constraint that the tube diameter and length be fixed at 50 mm and 5 m, respectively?

8.44 Consider a horizontal, thin-walled circular tube of diameter $D = 0.025$ m submerged in a container of *n*-octadecane (paraffin), which is used to store thermal energy. As hot water flows through the tube, heat is transferred to the paraffin, converting it from the solid to liquid state at the phase change temperature of $T_\infty = 27.4$°C. The latent heat of fusion and density of paraffin are $h_{sf} = 244$ kJ/kg and $\rho = 770$ kg/m³, respectively, and thermophysical properties of the water may be taken as $c_p = 4.185$ kJ/kg · K, $k = 0.653$ W/m · K, $\mu = 467 \times 10^{-6}$ kg/s · m, and Pr $= 2.99$.

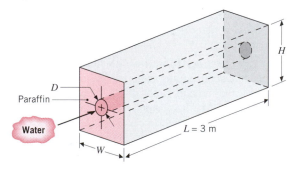

(a) Assuming the tube surface to have a uniform temperature corresponding to that of the phase change, determine the water outlet temperature and total heat transfer rate for a water flow rate of 0.1 kg/s and an inlet temperature of 60°C. If $H = W = 0.25$ m, how long would it take to completely liquefy the paraffin, from an initial state for which all the paraffin is solid and at 27.4°C?

(b) The liquefaction process can be accelerated by increasing the flow rate of the water. Compute and plot the heat rate and outlet temperature as a function of flow rate for $0.1 \leq \dot{m} \leq 0.5$ kg/s. How long would it take to melt the paraffin for $\dot{m} = 0.5$ kg/s?

8.45 A common procedure for cooling a high-performance computer chip involves joining the chip to a heat sink within which circular microchannels are machined. During operation, the chip produces a uniform heat flux q_c'' at its interface with the heat sink, while a liquid coolant (water) is routed through the channels. Consider a square chip and heat sink, each $L \times L$ on a side, with microchannels of diameter D and pitch $S = C_1 D$, where the constant C_1 is greater than unity. Water is supplied at an inlet temperature $T_{m,i}$ and a total mass flow rate $\dot{m}$ (for the entire heat sink).

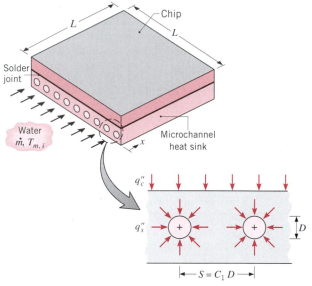

(a) Assuming that q_c'' is dispersed in the heat sink such that a uniform heat flux q_s'' is maintained at the surface of each channel, obtain expressions for the longitudinal distributions of the mean fluid, $T_m(x)$, and surface, $T_s(x)$, temperatures in each channel. Assume laminar, fully developed flow throughout each channel, and express your results in terms of $\dot{m}$, q_c'', C_1, D, and/or L, as well as appropriate thermophysical properties.

(b) For $L = 12$ mm, $D = 1$ mm, $C_1 = 2$, $q_c'' = 20$ W/cm², $\dot{m} = 0.010$ kg/s, and $T_{m,i} = 290$ K, compute and plot the temperature distributions $T_m(x)$ and $T_s(x)$.

(c) A common objective in designing such heat sinks is to maximize q_c'' while maintaining the heat sink at an acceptable temperature. Subject to prescribed values of $L = 12$ mm and $T_{m,i} = 290$ K and the constraint that $T_{s,\max} \leq 50$°C, explore the effect on q_c'' of variations in heat sink design and operating conditions.

8.46 Freon is being transported at 0.1 kg/s through a Teflon tube of inside diameter $D_i = 25$ mm and outside diameter $D_o = 28$ mm, while atmospheric air at $V = 25$ m/s and 300 K is in cross flow over the tube. What is the heat transfer per unit length of tube to Freon at 240 K?

8.47 Oil at 150°C flows *slowly* through a long, thin-walled pipe of 30-mm inner diameter. The pipe is suspended in a room for which the air temperature is 20°C and the convection coefficient at the outer tube surface is 11 W/m² · K. Estimate the heat loss per unit length of tube.

8.48 Exhaust gases from a wire processing oven are discharged into a tall stack, and the gas and stack surface temperatures at the outlet of the stack must be estimated. Knowledge of the outlet gas temperature $T_{m, o}$ is useful for predicting the dispersion of effluents in the thermal plume, while knowledge of the outlet stack surface temperature $T_{s, o}$ indicates whether condensation of the gas products will occur. The thin-walled, cylindrical stack is 0.5 m in diameter and 6.0 m high. The exhaust gas flow rate is 0.5 kg/s, and the inlet temperature is 600°C.

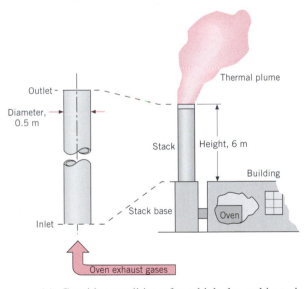

(a) Consider conditions for which the ambient air temperature and wind velocity are 4°C and 5 m/s, respectively. Approximating the thermophysical properties of the gas as those of atmospheric air, estimate the outlet gas and stack surface temperatures for the given conditions.

(b) The gas outlet temperature is sensitive to variations in the ambient air temperature and wind velocity. For $T_\infty = -25°C$, 5°C, and 35°C, compute and plot the gas outlet temperature as a function of wind velocity for $2 \leq V \leq 10$ m/s.

8.49 A hot fluid passes through a thin-walled tube of 10-mm diameter and 1-m length, and a coolant at $T_\infty = 25°C$ is in cross flow over the tube. When the flow rate is $\dot{m} = 18$ kg/h and the inlet temperature is $T_{m, i} = 85°C$, the outlet temperature is $T_{m, o} = 78°C$.

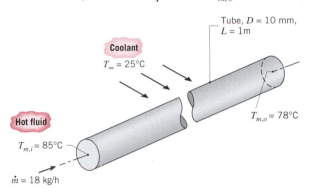

Assuming fully developed flow and thermal conditions in the tube, determine the outlet temperature, $T_{m, o}$, if the flow rate is increased by a factor of 2. That is, $\dot{m} = 36$ kg/h, with all other conditions the same. The thermophysical properties of the hot fluid are $\rho = 1079$ kg/m³, $c_p = 2637$ J/kg · K, $\mu = 0.0034$ N · s/m², and $k = 0.261$ W/m · K.

8.50 Consider a thin-walled tube of 10-mm diameter and 2-m length. Water enters the tube from a large reservoir at $\dot{m} = 0.2$ kg/s and $T_{m, i} = 47°C$.

(a) If the tube surface is maintained at a uniform temperature of 27°C, what is the outlet temperature of the water, $T_{m, o}$? To obtain the properties of water, assume an average mean temperature of $\overline{T}_m = 300$ K.

(b) What is the exit temperature of the water if it is heated by passing air at $T_\infty = 100°C$ and $V = 10$ m/s in cross flow over the tube? The properties of air may be evaluated at an assumed film temperature of $T_f = 350$ K.

(c) In the foregoing calculations, were the assumed values of $\overline{T}_m$ and T_f appropriate? If not, use properly evaluated properties and recompute $T_{m, o}$ for the conditions of part (b).

8.51 Water at a flow rate of $\dot{m} = 0.215$ kg/s is cooled from 70° to 30°C by passing it through a thin-walled tube of diameter $D = 50$ mm and maintaining a coolant at $T_\infty = 15°C$ in cross flow over the tube.

(a) What is the required tube length if the coolant is air and its velocity is $V = 20$ m/s?

(b) What is the tube length if the coolant is water and $V = 2$ m/s?

8.52 Consider a thin-walled, metallic tube of length $L = 1$ m and inside diameter $D_i = 3$ mm. Water enters the tube at $\dot{m} = 0.015$ kg/s and $T_{m,i} = 97°C$.

(a) What is the outlet temperature of the water if the tube surface temperature is maintained at 27°C?

(b) If a 0.5-mm thick layer of insulation of $k = 0.05$ W/m · K is applied to the tube and its outer surface is maintained at 27°C, what is the outlet temperature of the water?

(c) If the outer surface of the insulation is no longer maintained at 27°C but is allowed to exchange heat by free convection with ambient air at 27°C, what is the outlet temperature of the water? The free convection heat transfer coefficient is 5 W/m² · K.

8.53 A thick-walled steel pipe ($k = 60$ W/m · K) carrying hot water is cooled externally by a cross-flow airstream at a velocity of 20 m/s and a temperature of 25°C. The inner and outer diameters of the pipe are $D_i = 20$ mm and $D_o = 25$ mm, respectively. At a certain location along the pipe, the mean temperature of the water is 80°C. Assuming the flow inside the tube is fully developed with a Reynolds number of 20,000, find the heat transfer rate to the airstream per unit pipe length.

8.54 Heat is to be removed from a reaction vessel operating at 75°C by supplying water at 27°C and 0.12 kg/s through a thin-walled tube of 15-mm diameter. The convection coefficient between the tube outer surface and the fluid in the vessel is 3000 W/m² · K.

(a) If the outlet water temperature cannot exceed 47°C, what is the maximum rate of heat transfer from the vessel?

(b) What tube length is required to accomplish the heat transfer rate of part (a)?

8.55 Water enters a thin-walled, 50-mm diameter, 6-m long tube at 0.25 kg/s and 23°C and is heated by hot gases moving in cross flow over the tube with $V = 10$ m/s and $T_\infty = 300°C$.

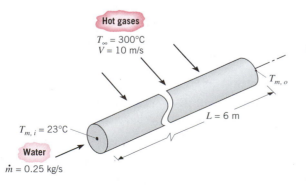

(a) Approximating the thermophysical properties of the hot gases as those of air, estimate the outlet temperature of the water, $T_{m,o}$, the average tube wall temperature, T_s, and the heat rate to the tube, q.

(b) In the operation of this water heater, the hot gas velocity can readily be controlled by changing the fan speed and, of course, the water flow rate can be adjusted. To determine how these parameters affect the water outlet temperature, compute and plot $T_{m,o}$ as a function of gas velocity over the range $10 \leq V \leq 25$ m/s, for water flow rates of 0.20, 0.25, and 0.30 kg/s.

(c) To provide the designer with options in choosing the tube diameter, fix $T_{m,o}$ at 40°C and, for tube diameters of 50, 60, and 70 mm, generate a plot of the water flow rate, $\dot{m}$, as a function of gas velocity for $10 \leq V \leq 25$ m/s.

8.56 A heating contractor must heat 0.2 kg/s of water from 15°C to 35°C using hot gases in cross flow over a thin-walled tube.

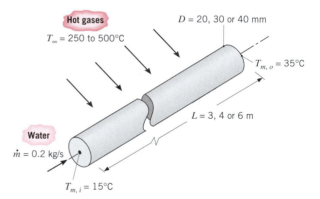

Your assignment is to develop a series of design graphs that can be used to demonstrate acceptable combinations of tube dimensions (D and L) and of hot gas conditions (T_∞ and V) that satisfy this requirement. In your analysis, consider the following parameter ranges: $D = 20$, 30, or 40 mm; $L = 3$, 4, or 6 m; $T_\infty = 250$, 375, or 500°C; and $20 \leq V \leq 40$ m/s.

8.57 A thin-walled tube with a diameter of 6 mm and length of 20 m is used to carry exhaust gas from a smoke stack to the laboratory in a nearby building for analysis. The gas enters the tube at 200°C and with a mass flow rate of 0.03 kg/s. Autumn winds at a temperature of 15°C blow directly across the tube at a velocity of 5 m/s. Assume the thermophysical properties of the exhaust gas are those of air.

(a) Estimate the average heat transfer coefficient for the exhaust gas flowing inside the tube.

(b) Estimate the heat transfer coefficient for the air flowing across the outside of the tube.

(c) Estimate the overall heat transfer coefficient U and the temperature of the exhaust gas when it reaches the laboratory.

8.58 A 50-mm diameter, thin-walled metal pipe covered with a 25-mm thick layer of insulation (0.085 W/m · K) and carrying superheated steam at atmospheric pressure is suspended from the ceiling of a large room. The steam temperature entering the pipe is 120°C, and the air temperature is 20°C. The convection heat transfer coefficient on the outer surface of the covered pipe is 10 W/m² · K. If the velocity of the steam is 10 m/s, at what point along the pipe will the steam begin condensing?

8.59 A thin-walled, uninsulated 0.3-m diameter duct is used to route chilled air at 0.05 kg/s through the attic of a large commercial building. The attic air is at 37°C, and natural circulation provided a convection coefficient of 2 W/m² · K at the outer surface of the duct. If chilled air enters a 15-m long duct at 7°C, what is its exit temperature and the rate of heat gain? Properties of the chilled air may be evaluated at an assumed average temperature of 300 K.

8.60 The problem of heat losses from a fluid moving through a buried pipeline has received considerable attention. Practical applications include the Alaskan pipeline, as well as power plant steam and water distribution lines. Consider a steel pipe of diameter D that is used to transport oil flowing at a rate $\dot{m}_o$ through a cold region. The pipe is covered with a layer of insulation of thickness t and thermal conductivity k_i and is buried in soil to a depth z (distance from the soil surface to the pipe centerline). Each section of pipe is of length L and extends between pumping stations in which the oil is heated to ensure low viscosity and hence low pump power requirements. The temperature of the oil entering the pipe from a pumping station and the temperature of the ground above the pipe are designated as $T_{m,i}$ and T_s, respectively, and are known.

Consider conditions for which the oil (o) properties may be approximated as $\rho_o = 900$ kg/m³, $c_{p,o} = 2000$ J/kg · K, $\nu_o = 8.5 \times 10^{-4}$ m²/s, $k_o = 0.140$ W/m · K, $Pr_o = 10^4$; the oil flow rate is $\dot{m}_o = 500$ kg/s; and the pipe diameter is 1.2 m.

(a) Expressing your results in terms of D, L, z, t, $\dot{m}_o$, $T_{m,i}$, and T_s, as well as the appropriate oil (o), insulation (i), and soil (s) properties, obtain

all the expressions needed to estimate the temperature $T_{m,o}$ of the oil leaving the pipe.

(b) If $T_s = -40°C$, $T_{m,i} = 120°C$, $t = 0.15$ m, $k_i = 0.05$ W/m · K, $k_s = 0.5$ W/m · K, $z = 3$ m, and $L = 100$ km, what is the value of $T_{m,o}$? What is the total rate of heat transfer q from a section of the pipeline?

(c) The operations manager wants to know the tradeoff between the burial depth of the pipe and insulation thickness on the heat loss from the pipe. Develop a graphical representation of this design information.

8.61 In order to maintain pump power requirements per unit flow rate below an acceptable level, operation of the oil pipeline of the preceding problem is subject to the constraint that the oil exit temperature $T_{m,o}$ exceed 110°C. For the values of $T_{m,i}$, T_s, D, t_i, z, L, and k_i prescribed in Problem 8.60, operating parameters that are variable and affect $T_{m,o}$ are the thermal conductivity of the soil and the flow rate of the oil. Depending on soil composition and moisture and the demand for oil, representative variations are $0.25 \leq k_s \leq 1.0$ W/m · K and $250 \leq \dot{m}_o \leq 500$ kg/s. Using the properties prescribed in Problem 8.60, determine the effect of the foregoing variations on $T_{m,o}$ and the total heat rate q. What is the worst case operating condition? If necessary, what adjustments could be made to ensure that $T_{m,o} \geq 110°C$ for the worst case conditions?

8.62 Water flows at 0.25 kg/s through a thin-walled, 40-mm diameter tube 4 m long. The water enters at 30°C and is heated by hot gases moving in cross flow over the tube with $V = 100$ m/s and $T_\infty = 225°C$. Estimate the outlet temperature of the water. The gas properties may be approximated to be those of atmospheric air.

8.63 You are designing an operating room heat exchange device to cool blood (bypassed from a patient) from 40 to 30°C by passing the fluid through a coiled tube sitting in a vat of water–ice mixture. The volumetric flow rate ($\dot{\forall}$) is 10^{-4} m³/min; the tube diameter (D) is 2.5 mm; and $T_{m,i}$ and $T_{m,o}$ represent the inlet and outlet temperatures of the blood. Neglect heat transfer enhancement associated with the coiling.

(a) At what temperature would you evaluate the fluid properties in determining $\bar{h}$ for the entire tube length?

(b) If the properties of blood evaluated at the temperature for part (a) are $\rho = 1000$ kg/m³, $\nu = 7 \times 10^{-7}$ m²/s, $k = 0.5$ W/m · K, and $c_p = 4.0$

kJ/kg · K, what is the Prandtl number for the blood?

(c) Is the blood flow laminar or turbulent?

(d) Neglecting all entrance effects and assuming fully developed conditions, calculate the value of $\bar{h}$ for heat transfer from the blood.

(e) What is the total heat rate loss from the blood as it passes through the tube?

(f) When free convection effects on the outside of the tube are included, the average overall heat transfer coefficient $\bar{U}$ between the blood and the ice–water mixture can be approximated as 300 W/m² · K. Determine the tube length L required to obtain the outlet temperature $T_{m, o}$.

8.64 Pressurized water at $T_{m\ i} = 200°C$ is pumped at $\dot{m} = 2$ kg/s from a power plant to a nearby industrial user through a thin-walled, round pipe of inside diameter $D = 1$ m. The pipe is covered with a layer of insulation of thickness $t = 0.15$ m and thermal conductivity $k = 0.05$ W/m · K. The pipe, which is of length $L = 500$ m, is exposed to a cross flow of air at $T_\infty = -10°C$ and $V = 4$ m/s. Obtain a differential equation that could be used to solve for the variation of the mixed mean temperature of the water $T_m(x)$ with the axial coordinate. As a first approximation, the internal flow may be assumed to be fully developed throughout the pipe. Express your results in terms of $\dot{m}$, V, T_∞, D, t, k, and appropriate water (w) and air (a) properties. Evaluate the heat loss per unit length of the pipe at the inlet. What is the mean temperature of the water at the outlet?

8.65 Water at 290 K and 0.2 kg/s flows through a Teflon tube ($k = 0.35$ W/m · K) of inner and outer radii equal to 10 and 13 mm, respectively. A thin electrical heating tape wrapped around the outer surface of the tube delivers a uniform surface heat flux of 2000 W/m², while a convection coefficient of 25 W/m² · K is maintained on the outer surface of the tape by ambient air at 300 K. What is the fraction of the power dissipated by the tape, which is transferred to the water? What is the outer surface temperature of the Teflon tube?

8.66 The temperature of flue gases flowing through the large stack of a boiler is measured by means of a thermocouple enclosed within a cylindrical tube as shown. The tube axis is oriented normal to the gas flow, and the thermocouple senses a temperature T_t corresponding to that of the tube surface. The gas flow rate and temperature are designated as $\dot{m}_g$ and T_g, respectively, and the gas flow may be assumed to be fully developed. The stack is fabricated from sheet metal that is at a uniform temperature T_s and is exposed to ambient air at T_∞ and large surroundings at T_{sur}. The convection coefficient associated with the outer surface of the duct is designated as h_o, while those associated with the inner surface of the duct and the tube surface are designated as h_i and h_t, respectively. The tube and duct surface emissivities are designated as ε_t and ε_s, respectively.

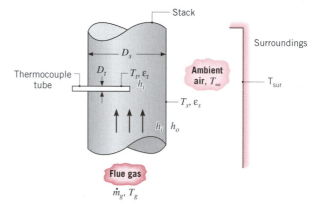

(a) Neglecting conduction losses along the thermocouple tube, develop an analysis that could be used to predict the error $(T_g - T_t)$ in the temperature measurement.

(b) Assuming the flue gas to have the properties of atmospheric air, evaluate the error for $T_t = 300°C$, $D_s = 0.6$ m, $D_t = 10$ mm, $\dot{m}_g = 1$ kg/s, $T_\infty = T_{sur} = 27°C$, $\varepsilon_t = \varepsilon_s = 0.8$, and $h_o = 25$ W/m² · K.

8.67 Four high-powered silicon-controlled rectifiers (SCRs), each dissipating 150 W, are mounted on a water-cooled heat sink. Water at 15°C is supplied at a rate of 4 liters/min. The thermal contact resistance between an SCR and the heat sink is estimated to be 0.1°C/W. Estimate the operating temperature of the SCRs by performing the following series of calculations.

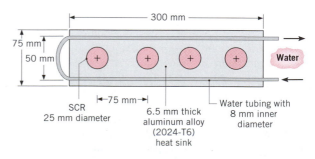

(a) Determine the temperature rise of the water coolant. What is the average mean temperature of the coolant?

(b) Using the proper correlation for forced convection inside a tube, estimate the heat transfer coefficient between the water and the tube wall. What is the approximate temperature rise between the coolant and the tube wall?

(c) Determine the shape factor for the heat sink material by using the two-dimensional graphical flux plotting method. Be sure to recognize symmetry of the system. What is the temperature rise between the tube wall and the location where an SCR contacts the heat sink?

(d) Draw the thermal resistance circuit and identify each of the elements present. Label the temperatures at each node and indicate the operating temperature of the SCRs.

(e) Summarize briefly the assumptions that were made in arriving at the SCR operating temperatures.

8.68 In a biomedical supplies manufacturing process, a requirement exists for a large platen that is to be maintained at $45 \pm 0.25°C$. The proposed design features the attachment of heating tubes to the platen at a relative spacing S. The thick-walled, copper tubes have an inner diameter of $D_i = 8$ mm and are attached to the platen with a high thermal conductivity solder, which provides a contact width of $2D_i$. The heating fluid (ethylene glycol) flows through each tube at a fixed rate of $\dot{m} = 0.06$ kg/s. The platen has a thickness of $w = 25$ mm and is fabricated from a stainless steel with a thermal conductivity of 15 W/m · K.

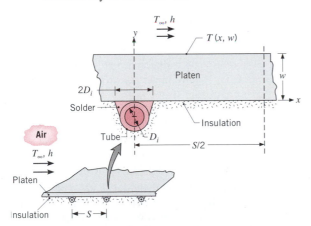

Considering the two-dimensional cross section of the platen shown in the inset, perform an analysis to determine the heating fluid temperature T_m and the tube spacing S required to maintain the surface temperature of the platen, $T(x, w)$, at $45 \pm 0.25°C$, when the ambient temperature is 25°C and the convection coefficient is 100 W/m² · K.

8.69 Ground source heat pumps operate by using a liquid, rather than ambient air, as the heat source (or sink) for winter heating (or summer cooling). The liquid flows in a closed loop through plastic tubing that is buried at a depth for which annual variations in the temperature of the soil are much less than those of the ambient air. For example, at a location such as West Lafayette, Indiana, deep-ground temperatures may remain at approximately 11°C, while annual excursions in the ambient air temperature may range from −25°C to +37°C.

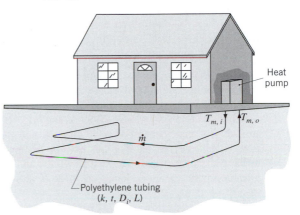

Consider winter conditions for which the liquid is discharged from the heat pump into high-density polyethylene tubing of thickness $t = 8$ mm and thermal conductivity $k = 0.47$ W/m · K. The tubing is routed through soil that maintains a uniform temperature of approximately 10°C at the tube outer surface. The properties of the fluid may be approximated as those of water.

(a) For a tube inner diameter and flow rate of $D_i = 25$ mm and $\dot{m} = 0.03$ kg/s and a fluid inlet temperature of $T_{m,i} = 0°C$, determine the tube outlet temperature (heat pump inlet temperature), $T_{m,o}$, as a function of the tube length L for $10 \leq L \leq 50$ m.

(b) Recommend an appropriate length for the system. How would your recommendation be affected by variations in the liquid flow rate?

8.70 Compare Nusselt number predictions based on the Colburn, Petukhov, and Gnielinski correlations for the fully developed turbulent flow of water ($Pr = 6$) in a smooth circular tube at Reynolds numbers of 4000, 10^4, and 10^5.

8.71 For a sharp-edged inlet and a combined entry region, the average Nusselt number may be computed from Equation 8.64, with $C = 24Re_D^{-0.23}$ and $m = 0.815 - 2.08 \times 10^{-6} Re_D$ [19]. Determine

$\overline{Nu}_D/Nu_{d,\,fd}$ at $x/D = 10$ and 60 for $Re_D = 10^4$ and 10^5.

Noncircular Ducts

8.72 Air at 3×10^{-4} kg/s and 27°C enters a rectangular duct that is 1 m long and 4 mm by 16 mm on a side. A uniform heat flux of 600 W/m² is imposed on the duct surface. What is the temperature of the air and of the duct surface at the outlet?

8.73 Air at 4×10^{-4} kg/s and 27°C enters a triangular duct that is 20 mm on a side and 2 m long. The duct surface is maintained at 100°C. Assuming fully developed flow throughout the duct, determine the air outlet temperature.

8.74 A device that recovers heat from high-temperature combustion products involves passing the combustion gas between parallel plates, each of which is maintained at 350 K by water flow on the opposite surface. The plate separation is 40 mm, and the gas flow is fully developed. The gas may be assumed to have the properties of atmospheric air, and its mean temperature and velocity are 1000 K and 60 m/s, respectively.

(a) What is the heat flux at the plate surface?

(b) If a third plate, 20 mm thick, is suspended midway between the original plates, what is the surface heat flux for the original plates? Assume the temperature and *flow rate* of the gas to be unchanged and radiation effects to be negligible.

8.75 In a gas-fired water heater, heat is transferred from high-temperature products of combustion flowing through long tubes to water moving over the tubes.

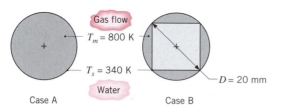

Case A   Case B

(a) Operating the heater in the conventional fashion (case A), gas flow through the tubes is fully developed and unobstructed. If the gas flow rate and temperature are 0.01 kg/s and 800 K, respectively, and the surface temperature is 340 K, what is the rate of heat transfer to the water per unit length of tube? The gas properties may be approximated to be those of air at 1 atm.

(b) It is proposed to increase heater performance by inserting a square rod in the tubes (case B). For the conditions of case A ($\dot{m} = 0.01$ kg/s, $T_m = 800$ K, and $T_s = 340$ K), what is the rate of heat transfer to the water per unit length of tube?

8.76 Air at 1 atm and 285 K enters a 2-m long rectangular duct with cross section 75 mm by 150 mm. The duct is maintained at a constant surface temperature of 400 K, and the air mass flow rate is 0.10 kg/s. Determine the heat transfer rate from the duct to the air and the air outlet temperature.

8.77 A double-wall heat exchanger is used to transfer heat between liquids flowing through semicircular copper tubes. Each tube has a wall thickness of $t = 3$ mm and an inner radius of $r_i = 20$ mm, and good contact is maintained at the plane surfaces by tightly wound straps. The tube outer surfaces are well insulated.

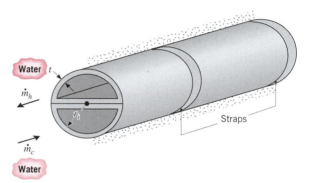

(a) If hot and cold water at mean temperatures of $T_{h,m} = 330$ K and $T_{c,m} = 290$ K flow through the adjoining tubes at $\dot{m}_h = \dot{m}_c = 0.2$ kg/s, what is the rate of heat transfer per unit length of tube? The wall contact resistance is 10^{-5} m² · K/W. Approximate the properties of both the hot and cold water as $\mu = 800 \times 10^{-6}$ kg/s · m, $k = 0.625$ W/m · K, and $Pr = 5.35$. *Hint:* Heat transfer is enhanced by conduction through the semicircular portions of the tube walls, and each portion may be subdivided into two straight fins with adiabatic tips.

(b) Using the thermal model developed for part (a), determine the heat transfer rate per unit length when the fluids are ethylene glycol. Also, what effect will fabricating the exchanger from an aluminum alloy have on the heat rate? Will increasing the thickness of the tube walls have a beneficial effect?

8.78 You have been asked to perform a feasibility study on the design of a blood warmer to be used during the transfusion of blood to a patient. This exchanger is to heat blood taken from the bank at 10°C to 37°C at a flow rate of 200 ml/min. The blood passes through a rectangular cross-section tube, 6.4 mm by 1.6 mm, which is sandwiched between two plates held at a constant temperature of 40°C.

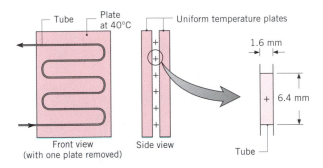

Front view
(with one plate removed)   Side view

(a) Compute the length of the tubing required to achieve the desired outlet conditions at the specified flow rate. Assume the flow is fully developed and the blood has the same properties as water.

(b) Assess your assumptions and indicate whether your analysis over- or underestimates the necessary length.

8.79 A coolant flows through a rectangular channel (*gallery*) within the body of a mold used to form metal injection parts. The *gallery* dimensions are $a = 90$ mm and $b = 9.5$ mm, and the fluid flow rate is 1.3×10^{-3} m³/s. The coolant temperature is 15°C, and the mold wall is at an approximately uniform temperature of 140°C.

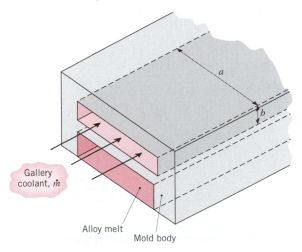

To minimize corrosion damage to the expensive mold, it is customary to use a heat transfer fluid such as ethylene glycol, rather than process water. Compare the convection coefficients of water and ethylene glycol for this application. What is the trade-off between thermal performance and minimizing corrosion?

8.80 A *cold plate* is an active cooling device that is attached to a heat-generating system in order to dissipate the heat while maintaining the system at an acceptable temperature. It is typically fabricated from a material of high thermal conductivity k_{cp}, within which channels are machined and a coolant is passed. Consider a copper cold plate of height H and width W on a side, within which water passes through square channels of width $w = h$. The transverse spacing between channels δ is twice the spacing between the sidewall of an outer channel and the sidewall of the cold plate.

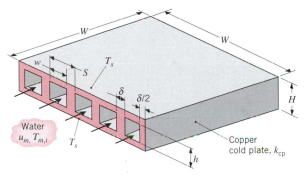

Consider conditions for which *equivalent* heat-generating systems are attached to the top and bottom of the cold plate, maintaining the corresponding surfaces at the same temperature T_s. The mean velocity and inlet temperature of the coolant are u_m and $T_{m,i}$, respectively.

(a) Assuming fully developed turbulent flow throughout each channel, obtain a system of equations that may be used to evaluate the total rate of heat transfer to the cold plate, q, and the outlet temperature of the water, $T_{m,o}$, in terms of the specified parameters.

(b) Consider a cold plate of width $W = 100$ mm and height $H = 10$ mm, with 10 square channels of width $w = 6$ mm and a spacing of $\delta = 4$ mm between channels. Water enters the channels at a temperature of $T_{m,i} = 300$ K and a velocity of $u_m = 2$ m/s. If the top and bottom cold plate surfaces are at $T_s = 360$ K, what is the outlet water temperature and the total rate of heat transfer to the cold plate? The thermal conductivity of the copper is 400 W/m · K, while average properties of the water may be taken to be $\rho = 984$ kg/m^3, $c_p = 4184$ J/kg · K, $\mu = 489 \times 10^{-6}$ N · s/m^2, $k = 0.65$ W/m · K, and $Pr = 3.15$. Is this a good cold plate design? How could its performance be improved?

8.81 The cold plate design of Problem 8.80 has not been optimized with respect to selection of the channel width, and we wish to explore conditions for which the rate of heat transfer may be enhanced. Assume that the width and height of the copper cold plate are fixed at $W = 100$ mm and $H = 10$ mm, while the channel height and spacing between channels are fixed at $h = 6$ mm and $\delta = 4$ mm. The mean velocity and inlet temperature of the water are maintained at $u_m = 2$ m/s and $T_{m,i} = 300$ K, while equivalent heat-generating systems attached to the top and bottom of the cold plate maintain the corresponding surfaces at 360 K. Evaluate the effect of changing the channel width, and hence the number of channels, on the rate of heat transfer to the cold plate. Include consideration of the limiting case for which $w = 96$ mm (one channel).

8.82 An electronic circuit board dissipating 50 W is sandwiched between two ducted, forced-air-cooled heat sinks. The sinks are 150 mm in length and have 24 rectangular passages 6 mm by 25 mm. Atmospheric air at a volumetric flow rate of 0.060 m^3/s and 27°C is drawn through the sinks by a blower. Estimate the operating temperature of the board and the pressure drop across the sinks.

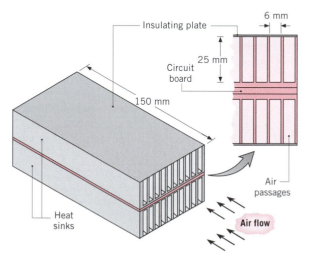

8.83 To slow down large prime movers like locomotives, a process termed dynamic electric braking is used to switch the traction motor to a generator mode in which mechanical power from the drive

wheels is absorbed and used to generate electrical current. As shown in the schematic, the electric power is passed through a resistor grid (a), which consists of an array of metallic blades electrically connected in series (b). The blade material is a high-temperature, high electrical resistivity alloy, and the electrical power is dissipated as heat by internal volumetric generation. To cool the blades, a motor-fan moves high-velocity air through the grid.

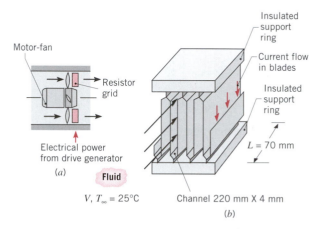

(a) Treating the space between the blades as a rectangular channel of 220-mm × 4-mm cross section and 70-mm length, estimate the heat removal rate per blade if the air stream has an inlet temperature and velocity of 25°C and 50 m/s, respectively, while the blade has an operating temperature of 600°C?

(b) On a locomotive pulling a 10-car train, there may be 2000 of these blades. Based on your result from part (a), how long will it take to slow a train whose total mass is 10^6 kg from a speed of 120 km/h to 50 km/h using dynamic electric braking?

8.84 An extremely effective method of cooling high-power-density silicon chips involves etching microchannels in the back (noncircuit) surface of the chip. The channels are covered with a silicon cap, and cooling is maintained by passing water through the channels.

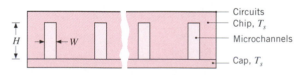

Consider a chip that is 10 mm by 10 mm on a side and in which fifty 10-mm long rectangular microchannels, each of width $W = 50\ \mu m$ and height $H = 200\ \mu m$, have been etched. Consider operating conditions for which water enters each microchannel at a temperature of 290 K and a flow rate of 10^{-4} kg/s, while the chip and cap are at a uniform temperature of 350 K. Assuming fully developed flow in the channel and that all the heat dissipated by the circuits is transferred to the water, determine the water outlet temperature and the chip power dissipation. Water properties may be evaluated at 300 K.

8.85 A novel scheme for dissipating heat from the chips of a multichip array involves machining coolant channels in the ceramic substrate to which the chips are attached. The square chips ($L_c = 5$ mm) are aligned above each of the channels, with longitudinal and transverse pitches of $S_L = S_T = 20$ mm. Water flows through the square cross section ($W = 5$ mm) of each channel with a mean velocity of $u_m = 1$ m/s, and its properties may be approximated as $\rho = 1000$ kg/m^3, $c_p = 4180$ J/kg $\cdot$ K, $\mu = 855 \times 10^{-6}$ kg/s $\cdot$ m, $k = 0.610$ W/m $\cdot$ K, and $Pr = 5.8$. Symmetry in the transverse direction dictates the existence of equivalent conditions for each substrate section of length L_s and width S_T.

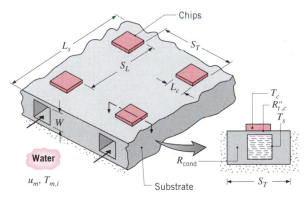

(a) Consider a substrate whose length in the flow direction is $L_s = 200$ mm, thereby providing a total of $N_L = 10$ chips attached in-line above each flow channel. To a good approximation, all the heat dissipated by the chips above a channel may be assumed to be transferred to the water flowing through the channel. If each chip dissipates 5 W, what is the temperature rise of the water passing through the channel?

(b) The chip–substrate contact resistance is $R''_{t,c} = 0.5 \times 10^{-4}$ m$^2 \cdot$ K/W, and the three-dimensional conduction resistance for the $L_s \times S_T$ substrate section is $R_{cond} = 0.120$ K/W. If wa-

ter enters the substrate at 25°C and is in fully developed flow, estimate the temperature T_c of the chips and the temperature T_s of the substrate channel surface.

8.86 Referring to Figure 8.10, consider conditions in an annulus having an outer surface that is insulated ($q''_o = 0$) and a uniform heat flux q''_i at the inner surface. Fully developed, laminar flow may be assumed to exist.

(a) Determine the velocity profile $u(r)$ in the annular region.

(b) Determine the temperature profile $T(r)$ and obtain an expression for the Nusselt number Nu_i associated with the inner surface.

8.87 Consider a concentric tube annulus for which the inner and outer diameters are 25 and 50 mm. Water enters the annular region at 0.04 kg/s and 25°C. If the inner tube wall is heated electrically at a rate (per unit length) of $q' = 4000$ W/m, while the outer tube wall is insulated, how long must the tubes be for the water to achieve an outlet temperature of 85°C? What is the inner tube surface temperature at the outlet, where fully developed conditions may be assumed?

8.88 It is common practice to recover waste heat from an oil- or gas-fired furnace by using the exhaust gases to preheat the combustion air. A device commonly used for this purpose consists of a concentric pipe arrangement for which the exhaust gases are passed through the inner pipe, while the cooler combustion air flows through an annular passage around the pipe.

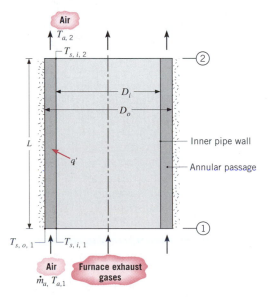

Consider conditions for which there is a uniform heat transfer rate per unit length, $q'_i = 1.25 \times 10^5$ W/m, from the exhaust gases to the pipe inner surface, while air flows through the annular passage at a rate of $\dot{m}_a = 2.1$ kg/s. The thin-walled inner pipe is of diameter $D_i = 2$ m, while the outer pipe, which is well insulated from the surroundings, is of diameter $D_o = 2.05$ m. The air properties may be taken to be $c_p = 1030$ J/kg · K, $\mu = 270 \times 10^{-7}$ N · s/m², $k = 0.041$ W/m · K, and $Pr = 0.68$.

(a) If air enters at $T_{a,1} = 300$ K and $L = 7$ m, what is the air outlet temperature $T_{a,2}$?

(b) If the airflow is fully developed throughout the annular region, what is the temperature of the inner pipe at the inlet ($T_{s,i,1}$) and outlet ($T_{s,i,2}$) sections of the device? What is the outer surface temperature $T_{s,o,1}$ at the inlet?

8.89   An industrial process heat application involves passing ethylene glycol through a concentric tube annulus, which is itself enclosed in a transparent glass tube. Under sunny skies, reflectors are situated such that the blackened outer surface of the annulus is uniformly irradiated to heat the ethylene glycol as it passes through the system. The space between the glass cover and the absorbing surface is evacuated, and the fluid flow is fully developed throughout the annulus. An electrically heated rod is used to heat the fluid during the evening hours, as well as during cloudy days.

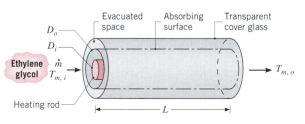

Consider a system for which $D_i = 10$ mm, $D_o = 100$ mm, and $L = 20$ m. The ethylene glycol enters at a rate of $\dot{m} = 0.25$ kg/s and a temperature of $T_{m,i} = 300$ K.

(a) Under sunny skies, there is a net radiation heat flux of $q''_o = 3000$ W/m² uniformly distributed on the outer surface of the annulus, and the heating rod is not operational. What is the outlet fluid temperature $T_{m,o}$? What is the temperature of the absorbing surface at the inlet and outlet?

(b) What is the surface heat flux q''_i that must be provided by the heating rod during the evening to maintain the fluid outlet temperature pre-

dicted in part (a)? What would be the surface temperature of the rod at the inlet and outlet?

8.90   Water at $\dot{m} = 0.02$ kg/s and $T_{m,i} = 20°C$ enters an annular region formed by an inner tube of diameter $D_i = 25$ mm and an outer tube of diameter $D_o = 100$ mm. Saturated steam flows through the inner tube, maintaining its surface at a uniform temperature of $T_{s,i} = 100°C$, while the outer surface of the outer tube is well insulated. If fully developed conditions may be assumed throughout the annulus, how long must the system be to provide an outlet water temperature of 75°C? What is the heat flux from the inner tube at the outlet?

8.91   For the conditions of Problem 8.90, how long must the annulus be if the water flow rate is 0.30 kg/s instead of 0.02 kg/s?

Mass Transfer

8.92   In the processing of very long plastic tubes of 2-mm inside diameter, air flows inside the tubing with a Reynolds number of 1000. The interior layer of the plastic material evaporates into the air under fully developed conditions. Both plastic and air are at 400 K, and the Schmidt number for the mixture of the plastic vapor and air is 2.0. Determine the convection mass transfer coefficient.

8.93   Air at 300 K and a flow rate of 3 kg/h passes upward through a 30-mm tube, as shown in the sketch. A thin film of water, also at 300 K, slowly falls downward on the inner surface of the tube.

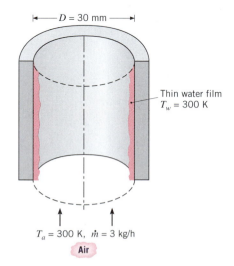

Determine the convection mass transfer coefficient for this situation.

8.94 What is the convection mass transfer coefficient associated with fully developed atmospheric airflow at 27°C and 0.04 kg/s through a 50-mm diameter tube whose surface has been coated with a thin layer of naphthalene? Determine the velocity and concentration entry lengths.

8.95 Air flowing through a tube of 75-mm diameter passes over a 150-mm long roughened section that is constructed from naphthalene having the properties $\mathcal{M} = 128.16$ kg/kmol and $p_{sat}(300\ K) = 1.31 \times 10^{-4}$ bar. The air is at 1 atm and 300 K, and the Reynolds number is $Re_D = 35,000$. In an experiment for which flow was maintained for 3 h, mass loss due to sublimation from the roughened surface was determined to be 0.01 kg. What is the associated convection mass transfer coefficient? What would be the corresponding convection heat transfer coefficient? Contrast these results with those predicted by conventional smooth tube correlations.

8.96 Dry air at 35°C and a velocity of 10 m/s flows over a thin-walled tube of 20-mm diameter and 200-mm length, having a fibrous coating that is water-saturated. To maintain an approximately uniform surface temperature of 27°C, water at a prescribed flow rate and temperature passes through the tube.

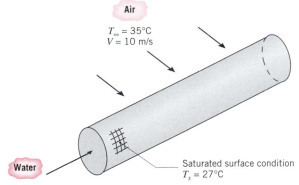

Air
$T_\infty = 35°C$
$V = 10$ m/s

Water
$\dot{m}, T_{m,i}$

Saturated surface condition
$T_s = 27°C$

(a) Considering the heat and mass transfer processes on the external surface of the tube, determine the heat rate from the tube.

(b) For a flow rate of 0.025 kg/s, determine the inlet temperature, $T_{m,i}$, at which water must be supplied to the tube.

8.97 Consider gas flow of mass density ρ and rate $\dot{m}$ through a tube whose inner surface is coated with a liquid or a sublimable solid of uniform vapor density $\rho_{A,s}$. Derive the following expression for vari-

ation of the mean vapor density $\rho_{A,m}$, with distance x from the tube entrance:

$$\frac{\rho_{A,s} - \rho_{A,m}(x)}{\rho_{A,s} - \rho_{A,m,i}} = \exp\left(-\frac{P x \rho}{\dot{m}}\,\bar{h}_m\right)$$

Show that the total rate of vapor transfer for a tube of length L may be expressed as

$$n_A = \bar{h}_m P L \frac{\Delta\rho_{A,o} - \Delta\rho_{A,i}}{\ln(\Delta\rho_{A,o}/\Delta\rho_{A,i})}$$

where $\Delta\rho_A \equiv \rho_{A,s} - \rho_{A,m}$.

8.98 Atmospheric air at 25°C and 3×10^{-4} kg/s flows through a 10-mm diameter, 1-m long circular tube whose inner surface is wetted with a water film. Using results from Problem 8.97, determine the water vapor density at the tube outlet, assuming the inlet air to be dry. What is the rate at which vapor is added to the air?

8.99 Air at 25°C and 1 atm is in fully developed flow at $\dot{m} = 10^{-3}$ kg/s through a 10-mm diameter circular tube whose inner surface is wetted with water. Using results from Problem 8.97, determine the tube length required for the water vapor in the air to reach 99% of saturation. The inlet air is dry.

8.100 A humidifier consists of a bundle of vertical tubes, each of 20-mm diameter, through which dry atmospheric air is in fully developed flow at 10^{-3} kg/s and 298 K. The inner tube surface is wetted with a water film. Using results from Problem 8.97, determine the tube length required for the water vapor to reach 99% of saturation. What is the rate at which energy must be supplied to each tube to maintain its temperature at 298 K?

8.101 A mass transfer operation is preceded by laminar flow of a gaseous species B through a circular tube that is sufficiently long to achieve a fully developed velocity profile. Once the fully developed condition is reached, the gas enters a section of the tube that is wetted with a liquid film (A). The film maintains a uniform vapor density $\rho_{A,s}$ along the tube surface.

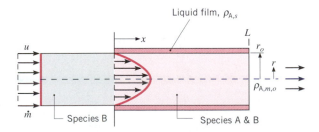

Liquid film, $\rho_{A,s}$

u

L

r_o

r

$\rho_{A,m,o}$

$\dot{m}$

Species B

Species A & B

(a) Write the differential equation and boundary conditions that govern the species A mass density distribution, $\rho_A(x, r)$, for $x > 0$.

(b) What is the heat transfer analog to this problem? From this analog, write an expression for the average Sherwood number associated with mass exchange over the region $0 \leq x \leq L$.

(c) Beginning with application of conservation of species to a differential control volume of ex-

tent $\pi r_o^2 \, dx$, derive an expression that may be used to determine the mean vapor density $\rho_{A,m,o}$ at $x = L$.

(d) Consider conditions for which species B is air at 25°C and 1 atm and the liquid film consists of water, also at 25°C. The flow rate is $\dot{m} = 2.5 \times 10^{-4}$ kg/s, and the tube diameter is $D = 10$ mm. What is the mean vapor density at the tube outlet if $L = 1$ m?

Free Convection

*I*n preceding chapters we considered convection transfer in fluid flows that originate from an *external forcing* condition. For example, fluid motion may be induced by a fan or a pump, or it may result from propulsion of a solid through the fluid. In the presence of a temperature gradient, *forced convection* heat transfer will occur.

Now we consider situations for which there is no *forced* velocity, yet convection currents exist within the fluid. Such situations are referred to as *free* or *natural convection,* and they originate when a *body force* acts on a fluid in which there are *density gradients.* The net effect is a *buoyancy force,* which induces free convection currents. In the most common case, the density gradient is due to a temperature gradient, and the body force is due to the gravitational field.

Since free convection flow velocities are generally much smaller than those associated with forced convection, the corresponding convection transfer rates are also smaller. It is perhaps tempting to therefore attach less significance to free convection processes. This temptation should be resisted. In many systems involving multimode heat transfer effects, free convection provides the largest resistance to heat transfer and therefore plays an important role in the design or performance of the system. Moreover, when it is desirable to minimize heat transfer rates or to minimize operating cost, free convection is often preferred to forced convection.

There are, of course, many applications. Free convection strongly influences heat transfer from pipes and transmission lines, as well as from various electronic devices. It is important in transferring heat from electric baseboard heaters or steam radiators to room air and in dissipating heat from the coil of a refrigeration unit to the surrounding air. It is also relevant to the environmental sciences, where it is responsible for oceanic and atmospheric motions, as well as related heat transfer processes.

9.1
Physical Considerations

In free convection fluid motion is due to buoyancy forces within the fluid, while in forced convection it is externally imposed. *Buoyancy is due to the combined presence of a fluid density gradient* and *a body force that is proportional to density.* In practice, the body force is usually *gravitational,* although it may be a centrifugal force in rotating fluid machinery or a Coriolis force in atmospheric and oceanic rotational motions. There are also several ways in which a mass density gradient may arise in a fluid, but for the most common situation it is due to the presence of a temperature gradient. We know that the density of gases and liquids depends on temperature, generally decreasing (due to fluid expansion) with increasing temperature ($\partial \rho / \partial T < 0$).

In this text we focus on free convection problems in which the density gradient is due to a temperature gradient and the body force is gravitational. However, the presence of a fluid density gradient in a gravitational field does not ensure the existence of free convection currents. Consider the conditions of Figure 9.1. A fluid is enclosed by two large, horizontal plates of different temperature ($T_1 \neq T_2$). In case *a* the temperature of the lower plate exceeds that of

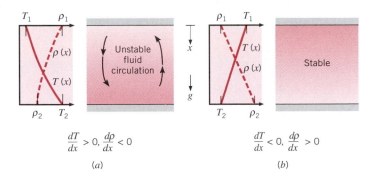

$$\frac{dT}{dx} > 0, \frac{d\rho}{dx} < 0$$

(a)

$$\frac{dT}{dx} < 0, \frac{d\rho}{dx} > 0$$

(b)

FIGURE 9.1 Conditions in a fluid between large horizontal plates at different temperatures. (a) Unstable temperature gradient. (b) Stable temperature gradient.

the upper plate, and the density decreases in the direction of the gravitational force. If the temperature difference exceeds a critical value, conditions are *unstable* and buoyancy forces are able to overcome the retarding influence of viscous forces. The gravitational force on the denser fluid in the upper layers exceeds that acting on the lighter fluid in the lower layers, and the designated circulation pattern will exist. The heavier fluid will descend, being warmed in the process, while the lighter fluid will rise, cooling as it moves. However, this condition does not characterize case b, for which $T_1 > T_2$ and the density no longer decreases in the direction of the gravitational force. Conditions are now *stable,* and there is no bulk fluid motion. In case a heat transfer occurs from the bottom to the top surface by free convection; for case b heat transfer (from top to bottom) occurs by conduction.

Free convection flows may be classified according to whether the flow is bounded by a surface. In the absence of an adjoining surface, *free boundary flows* may occur in the form of a *plume* or a *buoyant jet* (Figure 9.2). A plume is associated with fluid rising from a submerged heated object. Consider the

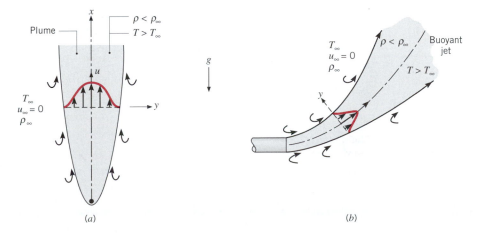

(a)

(b)

FIGURE 9.2 Buoyancy-driven free boundary layer flows in an extensive, quiescent medium. (a) Plume formation above a heated wire. (b) Buoyant jet associated with a heated discharge.

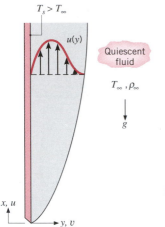

FIGURE 9.3
Boundary layer development on a heated vertical plate.

heated wire of Figure 9.2*a*, which is immersed in an *extensive, quiescent* fluid.[1] Fluid that is heated by the wire rises due to buoyancy forces, entraining fluid from the quiescent region. Although the width of the plume increases with distance from the wire, the plume itself will eventually dissipate as a result of viscous effects and a reduction in the buoyancy force caused by cooling of the fluid in the plume. The distinction between a plume and a buoyant jet is generally made on the basis of the *initial* fluid velocity. This velocity is zero for the plume, but finite for the buoyant jet. Figure 9.2*b* shows a heated fluid being discharged as a horizontal jet into a quiescent medium of lower temperature. The vertical motion that the jet begins to assume is due to the buoyancy force. Such a condition occurs when warm water from the condenser of a central power station is discharged into a reservoir of cooler water. Free boundary flows are discussed in considerable detail by Jaluria [1] and Gebhart et al. [2].

In this text we focus on free convection flows bounded by a surface, and a classic example relates to boundary layer development on a heated vertical plate (Figure 9.3). The plate is immersed in an extensive, quiescent fluid, and with $T_s > T_\infty$ the fluid close to the plate is less dense than fluid that is further removed. Buoyancy forces therefore induce a free convection boundary layer in which the heated fluid rises vertically, entraining fluid from the quiescent region. The resulting velocity distribution is unlike that associated with forced convection boundary layers. In particular, the velocity is zero as $y \rightarrow \infty$, as well as at $y = 0$. A free convection boundary layer also develops if $T_s < T_\infty$. In this case, however, fluid motion is downward.

9.2
The Governing Equations

As for forced convection, the equations that describe momentum and energy transfer in free convection originate from the related conservation principles.

---

[1]An extensive medium is, in principle, an infinite medium. Since a quiescent fluid is one that is otherwise at rest, the velocity of fluid far from the heated wire is zero.

Moreover, the specific processes are much like those that dominate in forced convection. Inertia and viscous forces remain important, as does energy transfer by advection and diffusion. The difference between the two flows is that, in free convection, a major role is played by buoyancy forces. It is such forces that, in fact, sustain the flow.

Consider a laminar boundary layer flow (Figure 9.3) that is driven by buoyancy forces. Assume steady, two-dimensional, constant property conditions in which the gravity force acts in the negative x direction. Also, with one exception, assume the fluid to be incompressible. The exception involves accounting for the effect of variable density in the buoyancy force (the so-called *Boussinesq approximation*), since it is this variation that induces fluid motion. Finally, assume that the boundary layer approximations are valid.

With the foregoing simplifications the x-momentum equation (6.29) reduces to the boundary layer equation (6.55), except that the body force term X is retained. If the only contribution to this force is made by gravity, the body force per unit volume is $X = -\rho g$, where g is the local acceleration due to gravity. The appropriate form of the x-momentum equation is then

$$u\frac{\partial u}{\partial x} + v\frac{\partial u}{\partial y} = -\frac{1}{\rho}\frac{\partial p}{\partial x} - g + \nu\frac{\partial^2 u}{\partial y^2} \tag{9.1}$$

Equation 9.1 may be couched in a more convenient form by first noting that, if there is no body force in the y direction, $(\partial p/\partial y) = 0$ from the y-momentum equation (6.56). Hence the x-pressure gradient at any point *in* the boundary layer must equal the pressure gradient in the quiescent region *outside* the boundary layer. However, in this region $u = 0$ and Equation 9.1 reduces to

$$\frac{\partial p}{\partial x} = -\rho_\infty g \tag{9.2}$$

Substituting Equation 9.2 into 9.1, we obtain the following expression:

$$u\frac{\partial u}{\partial x} + v\frac{\partial u}{\partial y} = \frac{g}{\rho}(\rho_\infty - \rho) + \nu\frac{\partial^2 u}{\partial y^2} \tag{9.3}$$

which must apply at every point in the free convection boundary layer.

The first term on the right-hand side of Equation 9.3 is the buoyancy force, and flow originates because the density ρ is a variable. The origin of this variation may be made more explicit by introducing the *volumetric thermal expansion coefficient*

$$\beta = -\frac{1}{\rho}\left(\frac{\partial \rho}{\partial T}\right)_p \tag{9.4}$$

This *thermodynamic* property of the fluid provides a measure of the amount by which the density changes in response to a change in temperature at constant pressure. If it is expressed in the following approximate form,

$$\beta \approx -\frac{1}{\rho}\frac{\rho_\infty - \rho}{T_\infty - T}$$

it follows that

$$(\rho_\infty - \rho) \approx \rho\beta(T - T_\infty)$$

Substituting into Equation 9.3, the x-momentum equation becomes

$$u\frac{\partial u}{\partial x} + v\frac{\partial u}{\partial y} = g\beta(T - T_\infty) + \nu\frac{\partial^2 u}{\partial y^2} \tag{9.5}$$

where it is now apparent how the buoyancy force, which drives the flow, is related to the temperature difference.

Since buoyancy effects are confined to the momentum equation, the mass and energy conservation equations are unchanged from forced convection. Equations 6.54 and 6.57 may then be used to complete the problem formulation. The set of governing equations is then

$$\frac{\partial u}{\partial x} + \frac{\partial v}{\partial y} = 0 \tag{9.6}$$

$$u\frac{\partial u}{\partial x} + v\frac{\partial u}{\partial y} = g\beta(T - T_\infty) + \nu\frac{\partial^2 u}{\partial y^2} \tag{9.7}$$

$$u\frac{\partial T}{\partial x} + v\frac{\partial T}{\partial y} = \alpha\frac{\partial^2 T}{\partial y^2} \tag{9.8}$$

Note that viscous dissipation has been neglected in the energy equation, (9.8), an assumption that is certainly reasonable for the small velocities associated with free convection. In the mathematical sense the appearance of the buoyancy term in Equation 9.7 complicates matters. No longer may the hydrodynamic problem, given by Equations 9.6 and 9.7, be uncoupled from and solved to the exclusion of the thermal problem, given by Equation 9.8. The solution to the momentum equation depends on knowledge of T, and hence on the solution to the energy equation. Equations 9.6 to 9.8 are therefore strongly coupled and must be solved simultaneously.

Free convection effects obviously depend on the expansion coefficient β. The manner in which β is obtained depends on the fluid. For an ideal gas, $\rho = p/RT$ and

$$\beta = -\frac{1}{\rho}\left(\frac{\partial\rho}{\partial T}\right)_p = \frac{1}{\rho}\frac{p}{RT^2} = \frac{1}{T} \tag{9.9}$$

where T is the *absolute* temperature. For liquids and nonideal gases, β must be obtained from appropriate property tables (Appendix A).

9.3
Similarity Considerations

Let us now consider the dimensionless parameters that govern free convective flow and heat transfer. As for forced convection (Chapter 6), the parameters may be obtained by nondimensionalizing the governing equations. Introducing

$$x^* \equiv \frac{x}{L} \qquad y^* \equiv \frac{y}{L}$$

$$u^* \equiv \frac{u}{u_0} \qquad v^* \equiv \frac{v}{u_0} \qquad T^* \equiv \frac{T - T_\infty}{T_s - T_\infty}$$

where L is a characteristic length and u_0 is an *arbitrary* reference velocity,[2] the *x*-momentum and energy equations (9.7 and 9.8) reduce to

$$u^* \frac{\partial u^*}{\partial x^*} + v^* \frac{\partial u^*}{\partial y^*} = \frac{g\beta(T_s - T_\infty)L}{u_0^2} T^* + \frac{1}{Re_L} \frac{\partial^2 u^*}{\partial y^{*2}} \tag{9.10}$$

$$u^* \frac{\partial T^*}{\partial x^*} + v^* \frac{\partial T^*}{\partial y^*} = \frac{1}{Re_L\, Pr} \frac{\partial^2 T^*}{\partial y^{*2}} \tag{9.11}$$

The dimensionless parameter in the first term on the right-hand side of Equation 9.10 is a direct consequence of the buoyancy force. However, because it is expressed in terms of the unknown reference velocity u_0, it is inconvenient in its present form. It is therefore customary to work with an alternative form that is obtained from multiplying by $Re_L^2 = (u_0 L/\nu)^2$. The result is termed the *Grashof number Gr_L*.

$$Gr_L \equiv \frac{g\beta(T_s - T_\infty)L}{u_0^2} \left(\frac{u_0 L}{\nu} \right)^2 = \frac{g\beta(T_s - T_\infty)L^3}{\nu^2} \tag{9.12}$$

The Grashof number plays the same role in free convection that the Reynolds number plays in forced convection. Recall that the *Reynolds number* provides a measure of the *ratio of the inertial to viscous forces* acting on a fluid element. In contrast, the *Grashof number* indicates the *ratio of* the *buoyancy force to the viscous force* acting on the fluid.

Although Equations 9.10 to 9.12 prompt us to expect heat transfer correlations of the form $Nu_L = f(Re_L, Gr_L, Pr)$, it is important to note that such correlations are pertinent only when forced and free convection effects are comparable. For such cases an external flow is superposed on the buoyancy-driven flow, and there exists a well-defined forced convection velocity. Generally, the combined effects of free and forced convection must be considered when $(Gr_L/Re_L^2) \approx 1$. If the inequality $(Gr_L/Re_L^2) \ll 1$ is satisfied, free convection effects may be neglected and $Nu_L = f(Re_L, Pr)$. Conversely, if $(Gr_L/Re_L^2) \gg 1$, forced convection effects may be neglected and $Nu_L = f(Gr_L, Pr)$. In the strict sense, a free convection flow is one that is induced solely by buoyancy forces, in which case there is no well-defined forced convection velocity and $(Gr_L/Re_L^2) = \infty$.

9.4
Laminar Free Convection on a Vertical Surface

Numerous solutions to the laminar free convection boundary layer equations have been obtained, and a special case that has received much attention involves free convection from an isothermal vertical surface in an extensive qui-

---

[2]Since free stream conditions are quiescent in free convection, there is no logical external reference velocity (V or u_∞), as in forced convection.

escent medium (Figure 9.3). For this geometry Equations 9.6 to 9.8 must be solved subject to boundary conditions of the form[3]

$$y = 0: \qquad u = v = 0 \qquad T = T_s$$

$$y \rightarrow \infty: \qquad u \rightarrow 0 \qquad T \rightarrow T_\infty$$

A similarity solution to the foregoing problem has been obtained by Ostrach [3]. The solution involves transforming variables by introducing a *similarity parameter* of the form

$$\eta \equiv \frac{y}{x}\left(\frac{Gr_x}{4}\right)^{1/4} \tag{9.13}$$

and representing the velocity components in terms of a stream function defined as

$$\psi(x, y) \equiv f(\eta)\left[4\nu\left(\frac{Gr_x}{4}\right)^{1/4}\right] \tag{9.14}$$

With the foregoing definition of the stream function, the *x*-velocity component may be expressed as

$$u = \frac{\partial \psi}{\partial y} = \frac{\partial \psi}{\partial \eta}\frac{\partial \eta}{\partial y} = 4\nu\left(\frac{Gr_x}{4}\right)^{1/4}f'(\eta)\frac{1}{x}\left(\frac{Gr_x}{4}\right)^{1/4}$$

$$= \frac{2\nu}{x}Gr_x^{1/2}f'(\eta) \tag{9.15}$$

where primed quantities indicate differentiation with respect to η. Hence $f'(\eta) \equiv df/d\eta$. Evaluating the *y*-velocity component $v = -\partial\psi/\partial x$ in a similar fashion and introducing the dimensionless temperature

$$T^* \equiv \frac{T - T_\infty}{T_s - T_\infty} \tag{9.16}$$

the three original partial differential equations (9.6 to 9.8) may then be reduced to two ordinary differential equations of the form

$$f''' + 3ff'' - 2(f')^2 + T^* = 0 \tag{9.17}$$

$$T^{*''} + 3Pr\, fT^{*'} = 0 \tag{9.18}$$

where f and T^* are functions of only η and the double and triple primes, respectively, refer to second and third derivatives with respect to η. Note that f is the key dependent variable for the velocity boundary layer and that the continuity equation (9.6) is automatically satisfied through introduction of the stream function.

The transformed boundary conditions required to solve the momentum and energy equations (9.17 and 9.18) are of the form

---

[3]The boundary layer approximations are assumed in using Equations 9.6 to 9.8. However, the approximations are only valid for $(Gr_x\, Pr) \gtrsim 10^4$. Below this value (close to the leading edge), the boundary layer thickness is too large relative to the characteristic length x to ensure the validity of the approximations.

$$\eta = 0: \qquad f = f' = 0 \qquad T^* = 1$$

$$\eta \to \infty: \qquad f' \to 0 \qquad T^* \to 0$$

A numerical solution has been obtained by Ostrach [3], and selected results are shown in Figure 9.4. Note that the *x*-velocity component *u* may readily be obtained from Figure 9.4*a* through the use of Equation 9.15. Note also that, through the definition of the similarity parameter η, Figure 9.4 may be used to obtain values of *u* and *T* for any value of *x* and *y*.

Figure 9.4*b* may also be used to infer the appropriate form of the heat transfer correlation. Using Newton's law of cooling for the local convection coefficient *h*, the local Nusselt number may be expressed as

$$Nu_x = \frac{hx}{k} = \frac{[q_s''/(T_s - T_\infty)]x}{k}$$

Using Fourier's law to obtain q_s'' and expressing the surface temperature gradient in terms of η, Equation 9.13, and T^*, Equation 9.16, it follows that

$$q_s'' = -k \left.\frac{\partial T}{\partial y}\right|_{y=0} = -\frac{k}{x}(T_s - T_\infty)\left(\frac{Gr_x}{4}\right)^{1/4}\left.\frac{dT^*}{d\eta}\right|_{\eta=0}$$

Hence

$$Nu_x = \frac{hx}{k} = -\left(\frac{Gr_x}{4}\right)^{1/4}\left.\frac{dT^*}{d\eta}\right|_{\eta=0} = \left(\frac{Gr_x}{4}\right)^{1/4}g(Pr) \qquad (9.19)$$

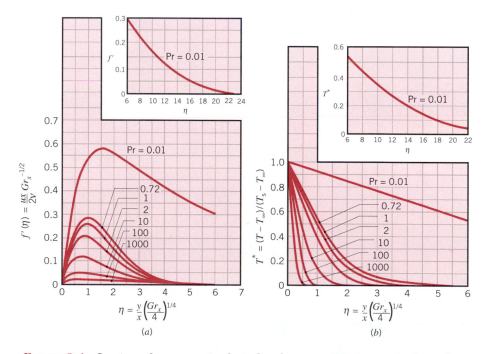

FIGURE 9.4 Laminar, free convection boundary layer conditions on an isothermal, vertical surface. (*a*) Velocity profiles. (*b*) Temperature profiles [3].

which acknowledges that the dimensionless temperature gradient at the surface is a function of the Prandtl number $g(Pr)$. This dependence is evident from Figure 9.4b and has been determined numerically for selected values of Pr [3]. The results have been correlated to within 0.5% by an interpolation formula of the form [4]

$$g(Pr) = \frac{0.75Pr^{1/2}}{(0.609 + 1.221Pr^{1/2} + 1.238Pr)^{1/4}}$$

(9.20)

which applies for $0 \leq Pr \leq \infty$.

Using Equation 9.19 for the local convection coefficient and substituting for the local Grashof number.

$$Gr_x = \frac{g\beta(T_s - T_\infty)x^3}{\nu^2}$$

the average convection coefficient for a surface of length L is then

$$\bar{h} = \frac{1}{L}\int_0^L h\,dx = \frac{k}{L}\left[\frac{g\beta(T_s - T_\infty)}{4\nu^2}\right]^{1/4} g(Pr)\int_0^L \frac{dx}{x^{1/4}}$$

Integrating, it follows that

$$\overline{Nu_L} = \frac{\bar{h}L}{k} = \frac{4}{3}\left(\frac{Gr_L}{4}\right)^{1/4} g(Pr)$$

(9.21)

or substituting from Equation 9.19, with $x = L$,

$$\overline{Nu_L} = \tfrac{4}{3} Nu_L$$

(9.22)

The foregoing results apply irrespective of whether $T_s > T_\infty$ or $T_s < T_\infty$. If $T_s < T_\infty$, conditions are inverted from those of Figure 9.3. The leading edge is at the top of the plate, and positive x is defined in the direction of the gravity force.

9.5
The Effects of Turbulence

It is important to note that free convection boundary layers are not restricted to laminar flow. Free convection flows typically originate from a *thermal instability*. That is, warmer, lighter fluid moves vertically upward relative to cooler, heavier fluid. However, as with forced convection, *hydrodynamic instabilities* may also arise. That is, disturbances in the flow may be amplified, leading to transition from laminar to turbulent flow. This process is shown schematically in Figure 9.5 for a heated vertical plate.

Transition in a free convection boundary layer depends on the relative magnitude of the buoyancy and viscous forces in the fluid. It is customary to correlate its occurrence in terms of the *Rayleigh number,* which is simply the product of the Grashof and Prandtl numbers. For vertical plates the critical Rayleigh number is

$$Ra_{x,c} = Gr_{x,c}\,Pr = \frac{g\beta(T_s - T_\infty)x^3}{\nu\alpha} \approx 10^9$$

(9.23)

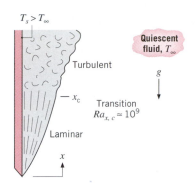

FIGURE 9.5
Free convection boundary layer transition on a vertical plate.

An extensive discussion of stability and transition effects is provided by Gebhart et al. [2].

As in forced convection, transition to turbulence has a strong effect on heat transfer. Hence the results of the foregoing section apply only if $Ra_L \lesssim 10^9$. To obtain appropriate correlations for turbulent flow, emphasis is placed on experimental results.

EXAMPLE 9.1

Consider a 0.25-m long vertical plate that is at 70°C. The plate is suspended in air that is at 25°C. Estimate the boundary layer thickness at the trailing edge of the plate if the air is quiescent. How does this thickness compare with that which would exist if the air were flowing over the plate at a free stream velocity of 5 m/s?

SOLUTION

Known: Vertical plate is in quiescent air at a lower temperature.

Find: Boundary layer thickness at trailing edge. Compare with thickness corresponding to an air speed of 5 m/s.

Schematic:

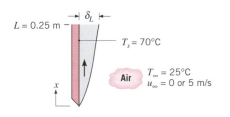

Assumptions:

1. Constant properties.
2. Buoyancy effects negligible when $u_\infty = 5$ m/s.

Properties: Table A.4, air ($T_f = 320.5$ K): $\nu = 17.95 \times 10^{-6}$ m²/s, $Pr = 0.7$, $\beta = T_f^{-1} = 3.12 \times 10^{-3}$ K⁻¹.

Analysis: For the quiescent air, Equation 9.12 gives

$$Gr_L = \frac{g\beta(T_s - T_\infty)L^3}{\nu^2}$$

$$= \frac{9.8 \text{ m/s}^2 \times (3.12 \times 10^{-3} \text{ K}^{-1})(70 - 25)°C(0.25 \text{ m})^3}{(17.95 \times 10^{-6} \text{ m}^2/\text{s})^2} = 6.69 \times 10^7$$

Hence $Ra_L = Gr_L \, Pr = 4.68 \times 10^7$ and, from Equation 9.23, the free convection boundary layer is laminar. The analysis of Section 9.4 is therefore applicable. From the results of Figure 9.4, it follows that, for $Pr = 0.7$, $\eta \approx 6.0$ at the edge of the boundary layer, that is, at $y = \delta$. Hence

$$\delta_L \approx \frac{6L}{(Gr_L/4)^{1/4}} = \frac{6(0.25 \text{ m})}{(1.67 \times 10^7)^{1/4}} = 0.024 \text{ m} \qquad \triangleleft$$

For airflow at $u_\infty = 5$ m/s

$$Re_L = \frac{u_\infty L}{\nu} = \frac{(5 \text{ m/s}) \times 0.25 \text{ m}}{17.95 \times 10^{-6} \text{ m}^2/\text{s}} = 6.97 \times 10^4$$

and the boundary layer is laminar. Hence from Equation 7.19

$$\delta_L \approx \frac{5L}{Re_L^{1/2}} = \frac{5(0.25 \text{ m})}{(6.97 \times 10^4)^{1/2}} = 0.0047 \text{ m} \qquad \triangleleft$$

Comments:

1. Boundary layer thicknesses are typically larger for free convection than for forced convection.

2. $(Gr_L/Re_L^2) = 0.014 \ll 1$, and the assumption of negligible buoyancy effects for $u_\infty = 5$ m/s is justified.

9.6
Empirical Correlations: External Free Convection Flows

In this section we summarize empirical correlations that have been developed for common *immersed* (external flow) geometries. The correlations are suitable for most engineering calculations and are generally of the form

$$\overline{Nu_L} = \frac{\overline{h}L}{k} = C\,Ra_L^n \tag{9.24}$$

where the Rayleigh number,

$$Ra_L = Gr_L\,Pr = \frac{g\beta(T_s - T_\infty)L^3}{\nu\alpha} \tag{9.25}$$

is based on the characteristic length L of the geometry. Typically, $n = \frac{1}{4}$ and $\frac{1}{3}$ for laminar and turbulent flows, respectively. For turbulent flow it then follows that $\bar{h}_L$ is independent of L. Note that all properties are evaluated at the film temperature, $T_f \equiv (T_s + T_\infty)/2$.

9.6.1 The Vertical Plate

Expressions of the form given by Equation 9.24 have been developed for the vertical plate [5–7] and are plotted in Figure 9.6. The coefficient C and the exponent n depend on the Rayleigh number range, and for Rayleigh numbers less than 10^4, the Nusselt number should be obtained directly from the figure.

A correlation that may be applied over the *entire* range of Ra_L has been recommended by Churchill and Chu [8] and is of the form

$$\overline{Nu}_L = \left\{ 0.825 + \frac{0.387 Ra_L^{1/6}}{[1 + (0.492/Pr)^{9/16}]^{8/27}} \right\}^2 \qquad (9.26)$$

Although Equation 9.26 is suitable for most engineering calculations, slightly better accuracy may be obtained for laminar flow by using [8]

$$\overline{Nu}_L = 0.68 + \frac{0.670 Ra_L^{1/4}}{[1 + (0.492/Pr)^{9/16}]^{4/9}} \qquad Ra_L \lesssim 10^9 \qquad (9.27)$$

It is important to recognize that the foregoing results have been obtained for an isothermal plate (constant T_s). If the surface condition is, instead, one of uniform heat flux (constant q_s''), the temperature difference $(T_s - T_\infty)$ will vary with x, increasing from a value of zero at the leading edge. An approximate procedure for determining this variation may be based on results [8, 9] showing

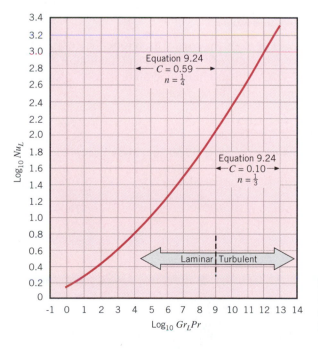

FIGURE 9.6
Nusselt number for free convection heat transfer from a vertical plate [5–7].

that $\overline{Nu}_L$ correlations obtained for the isothermal plate may still be used to an excellent approximation, if $\overline{Nu}_L$ and Ra_L are defined in terms of the temperature difference at the midpoint of the plate, $\Delta T_{L/2} = T_s(L/2) - T_\infty$. Hence, with $\overline{h} \equiv q_s''/\Delta T_{L/2}$, a correlation such as Equation 9.27 could be used (in a trial-and-error solution) to determine $\Delta T_{L/2}$, and hence the midpoint surface temperature $T_s(L/2)$. If it is assumed that $Nu_x \propto Ra_x^{1/4}$ over the entire plate, it follows that

$$\frac{q_s'' x}{k\,\Delta T} \propto \Delta T^{1/4}\, x^{3/4}$$

or

$$\Delta T \propto x^{1/5}$$

Hence the temperature difference at any x is

$$\Delta T_x \approx \frac{x^{1/5}}{(L/2)^{1/5}} \Delta T_{L/2} = 1.15 \left(\frac{x}{L}\right)^{1/5} \Delta T_{L/2} \qquad (9.28)$$

A more detailed discussion of constant heat flux results is provided by Churchill [10].

The foregoing results may also be applied to *vertical* cylinders of height L, if the boundary layer thickness δ is much less than the cylinder diameter D. This condition is known to be satisfied [11] when

$$\frac{D}{L} \gtrsim \frac{35}{Gr_L^{1/4}}$$

Cebeci [12] and Minkowycz and Sparrow [13] present results for slender, vertical cylinders not meeting this condition, where transverse curvature influences boundary layer development and enhances the rate of heat transfer.

EXAMPLE 9.2

A glass-door firescreen, used to reduce exfiltration of room air through a chimney, has a height of 0.71 m and a width of 1.02 m and reaches a temperature of 232°C. If the room temperature is 23°C, estimate the convection heat rate from the fireplace to the room.

SOLUTION

Known: Glass screen situated in fireplace opening.

Find: Heat transfer by convection between screen and room air.

Schematic:

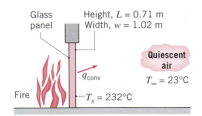

Assumptions:

1. Screen is at a uniform temperature T_s.
2. Room air is quiescent.

Properties:　Table A.4, air ($T_f = 400$ K): $k = 33.8 \times 10^{-3}$ W/m · K, $\nu = 26.4 \times 10^{-6}$ m²/s, $\alpha = 38.3 \times 10^{-6}$ m²/s, $Pr = 0.690$, $\beta = (1/T_f) = 0.0025$ K^{-1}.

Analysis:　The rate of heat transfer by free convection from the panel to the room is given by Newton's law of cooling

$$q = \bar{h} A_s (T_s - T_\infty)$$

where $\bar{h}$ may be obtained from knowledge of the Rayleigh number. Using Equation 9.25,

$$Ra_L = \frac{g\beta(T_s - T_\infty)L^3}{\alpha\nu}$$

$$= \frac{9.8 \text{ m/s}^2 \times 1/400 \text{ K } (232 - 23)°\text{C} \times (0.71 \text{ m})^3}{38.3 \times 10^{-6} \text{ m}^2/\text{s} \times 26.4 \times 10^{-6} \text{ m}^2/\text{s}} = 1.813 \times 10^9$$

and from Equation 9.23 it follows that transition to turbulence occurs on the panel. The appropriate correlation is then given by Equation 9.26

$$\overline{Nu}_L = \left\{ 0.825 + \frac{0.387 Ra_L^{1/6}}{[1 + (0.492/Pr)^{9/16}]^{8/27}} \right\}^2$$

$$\overline{Nu}_L = \left\{ 0.825 + \frac{0.387(1.813 \times 10^9)^{1/6}}{[1 + (0.492/0.690)^{9/16}]^{8/27}} \right\}^2 = 147$$

Hence

$$\bar{h} = \frac{\overline{Nu}_L \cdot k}{L} = \frac{147 \times 33.8 \times 10^{-3} \text{ W/m} \cdot \text{K}}{0.71 \text{ m}} = 7.0 \text{ W/m}^2 \cdot \text{K}$$

and

$$q = 7.0 \text{ W/m}^2 \cdot \text{K } (1.02 \times 0.71) \text{ m}^2 (232 - 23)°\text{C} = 1060 \text{ W} \qquad \triangleleft$$

Comments:

1. If $\bar{h}$ were computed from Equation 9.24, with $C = 0.10$ and $n = \frac{1}{3}$, we would obtain $\bar{h} = 5.8$ W/m² · K and the heat transfer prediction would be approximately 20% lower than the foregoing result. This difference is within the uncertainty normally associated with using such correlations.

2. Radiation heat transfer effects are often significant relative to free convection. Using Equation 1.7 and assuming $\varepsilon = 1.0$ for the glass surface and $T_{\text{sur}} = 23°\text{C}$, the net rate of radiation heat transfer between the glass and the surroundings is

$$q_{\text{rad}} = \varepsilon A_s \sigma(T_s^4 - T_{\text{sur}}^4)$$

$$q_{\text{rad}} = 1(1.02 \times 0.71) \text{ m}^2 \times 5.67 \times 10^{-8} \text{ W/m}^2 \cdot \text{K}^4 (505^4 - 296^4) \text{ K}^4$$

$$q_{\text{rad}} = 2355 \text{ W}$$

Hence in this case radiation heat transfer exceeds free convection heat transfer by more than a factor of 2.

3. The effects of radiation and free convection on heat transfer from the glass depend strongly on its temperature. With $q \propto T_s^4$ for radiation and $q \propto T_s^n$ for free convection, where $1.25 < n < 1.33$, we expect the relative influence of radiation to increase with increasing temperature. This behavior is revealed by computing and plotting the heat rates as a function of temperature for $50 \le T_s \le 250°\text{C}$.

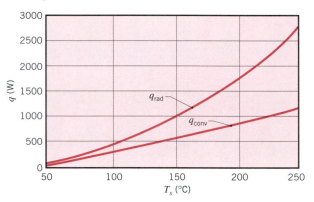

For each value of T_s used to generate the foregoing free convection results, air properties were determined at the corresponding value of T_f.

9.6.2   Inclined and Horizontal Plates

For a vertical plate, heated (or cooled) relative to an ambient fluid, the plate is aligned with the gravitational vector, and the buoyancy force acts exclusively to induce fluid motion in the upward (or downward) direction. However, if the plate is inclined with respect to gravity, the buoyancy force has a component normal, as well as parallel, to the plate surface. With a reduction in the buoyancy force parallel to the surface, there is a reduction in fluid velocities along the plate, and one might expect there to be an attendant reduction in convection heat transfer. Whether, in fact, there is such a reduction depends on whether one is interested in heat transfer from the top or bottom surface of the plate.

As shown in Figure 9.7*a*, if the plate is cooled, the *y* component of the buoyancy force, which is normal to the plate, acts to maintain the descending boundary layer flow in contact with the top surface of the plate. Since the *x* component of the gravitational acceleration is reduced to $g \cos \theta$, fluid velocities along the plate are reduced and there is an attendant reduction in convection heat transfer to the top surface. However, at the bottom surface, the *y* component of the buoyancy force acts to move fluid from the surface, and boundary layer development is interrupted by the discharge of parcels of cool fluid from the surface (Figure 9.7*a*). The resulting flow is three-dimensional, and, as shown by the spanwise (*z*-direction) variations of Figure 9.7*b*, the cool fluid discharged from the bottom surface is continuously replaced by the warmer ambient fluid. The displacement of cool boundary layer fluid by the warmer ambient and the attendant reduction in the thermal boundary layer thickness act

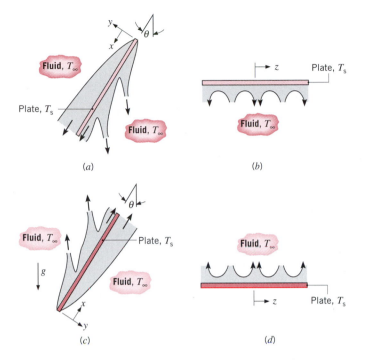

FIGURE 9.7 Buoyancy-driven flows on an inclined plate: (*a*) side view of flows at top and bottom surfaces of a cold plate ($T_s < T_\infty$), (*b*) end view of flow at bottom surface of cold plate, (*c*) side view of flows at top and bottom surfaces of a hot plate ($T_s > T_\infty$), and (*d*) end view of flow at top surface of hot plate.

to increase convection heat transfer to the bottom surface. In fact, heat transfer enhancement due to the three-dimensional flow typically exceeds the reduction associated with the reduced *x* component of *g*, and the combined effect is to increase heat transfer to the bottom surface. Similar trends characterize a heated plate (Figure 9.7*c,d*), and the three-dimensional flow is now associated with the upper surface, from which parcels of warm fluid are discharged. Such flows have been observed by several investigators [14–16].

In an early study of heat transfer from inclined plates, Rich [17] suggested that convection coefficients could be determined from vertical plate correlations, if *g* is replaced by *g* cos θ in computing the plate Rayleigh number. Since then, however, it has been determined that this approach is only satisfactory for the top and bottom surfaces of cooled and heated plates, respectively. It is not appropriate for the top and bottom surfaces of heated and cooled plates, respectively, where the three-dimensionality of the flow has limited the ability to develop generalized correlations. At the top and bottom surfaces of cooled and heated inclined plates, respectively, it is therefore recommended that, for $0 \le \theta \le 60°$, *g* be replaced by *g* cos θ and that Equation 9.26 or 9.27 be used to compute the average Nusselt number. For the opposite surfaces, no recommendations are made and the literature should be consulted [14–16].

If the plate is horizontal, the buoyancy force is exclusively normal to the surface. As for the inclined plate, flow patterns and heat transfer depend strongly on whether the surface is cooled or heated and on whether it is facing upward or downward. For a cold surface facing upward (Figure 9.8*a*) and a hot

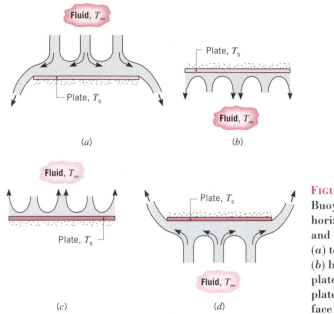

FIGURE 9.8
Buoyancy-driven flows on horizontal cold ($T_s < T_\infty$) and hot ($T_s > T_\infty$) plates: (a) top surface of cold plate, (b) bottom surface of cold plate, (c) top surface of hot plate, and (d) bottom surface of hot plate.

surface facing downward (Figure 9.8d), the tendency of the fluid to descend and ascend, respectively, is impeded by the plate. The flow must move horizontally before it can descend or ascend from the edges of the plate, and convection heat transfer is somewhat ineffective. In contrast, for a cold surface facing downward (Figure 9.8b) and a hot surface facing upward (Figure 9.8c), flow is driven by descending and ascending parcels of fluid, respectively. Conservation of mass dictates that cold (warm) fluid descending (ascending) from a surface be replaced by ascending (descending) warmer (cooler) fluid from the ambient, and heat transfer is much more effective.

Although correlations suggested by McAdams [5] are widely used for horizontal plates, improved accuracy may be obtained by altering the form of the characteristic length on which the correlations are based [18, 19]. In particular with the characteristic length defined as

$$L \equiv \frac{A_s}{P} \tag{9.29}$$

where A_s and P are the plate surface area and perimeter, respectively, recommended correlations for the average Nusselt number are

Upper Surface of Heated Plate or Lower Surface of Cooled Plate:

$$\overline{Nu}_L = 0.54 Ra_L^{1/4} \quad (10^4 \lesssim Ra_L \lesssim 10^7) \tag{9.30}$$

$$\overline{Nu}_L = 0.15 Ra_L^{1/3} \quad (10^7 \lesssim Ra_L \lesssim 10^{11}) \tag{9.31}$$

Lower Surface of Heated Plate or Upper Surface of Cooled Plate:

$$\overline{Nu}_L = 0.27 Ra_L^{1/4} \quad (10^5 \lesssim Ra_L \lesssim 10^{10}) \tag{9.32}$$

EXAMPLE 9.3

Airflow through a long rectangular heating duct that is 0.75 m wide and 0.3 m high maintains the outer duct surface at 45°C. If the duct is uninsulated and exposed to air at 15°C in the crawlspace beneath a home, what is the heat loss from the duct per meter of length?

SOLUTION

Known: Surface temperature of a long rectangular duct.

Find: Heat loss from duct per meter of length.

Schematic:

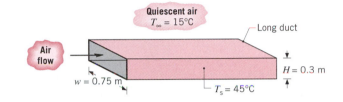

Quiescent air
$T_\infty = 15°C$

Long duct

Air flow

$H = 0.3$ m

$w = 0.75$ m

$T_s = 45°C$

Assumptions:

1. Ambient air is quiescent.
2. Surface radiation effects are negligible.

Properties: Table A.4, air ($T_f = 303$ K): $\nu = 16.2 \times 10^{-6}$ m²/s, $\alpha = 22.9 \times 10^{-6}$ m²/s, $k = 0.0265$ W/m · K, $\beta = 0.0033$ K⁻¹, $Pr = 0.71$

Analysis: Surface heat loss is by free convection from the vertical sides and the horizontal top and bottom. From Equation 9.25

$$Ra_L = \frac{g\beta(T_s - T_\infty)L^3}{\nu\alpha} = \frac{(9.8 \text{ m/s}^2)(0.0033 \text{ K}^{-1})(30 \text{ K}) L^3 \text{ (m}^3)}{(16.2 \times 10^{-6} \text{ m}^2/\text{s})(22.9 \times 10^{-6} \text{ m}^2/\text{s})}$$

$$Ra_L = 2.62 \times 10^9 L^3$$

For the two sides, $L = H = 0.3$ m. Hence $Ra_L = 7.07 \times 10^7$. The free convection boundary layer is therefore laminar, and from Equation 9.27

$$\overline{Nu_L} = 0.68 + \frac{0.670 Ra_L^{1/4}}{[1 + (0.492/Pr)^{9/16}]^{4/9}}$$

The convection coefficient associated with the sides is then

$$\overline{h_s} = \frac{k}{H} \overline{Nu_L}$$

$$\overline{h_s} = \frac{0.0265 \text{ W/m} \cdot \text{K}}{0.3 \text{ m}} \left\{ 0.68 + \frac{0.670(7.07 \times 10^7)^{1/4}}{[1 + (0.492/0.71)^{9/16}]^{4/9}} \right\} = 4.23 \text{ W/m}^2 \cdot \text{K}$$

For the top and bottom, $L = (A_s/P) \approx (w/2) = 0.375$ m. Hence $Ra_L = 1.38 \times 10^8$, and from Equations 9.31 and 9.32, respectively,

$$\bar{h}_t = [k/(w/2)] \times 0.15 Ra_L^{1/3} = \frac{0.0265 \text{ W/m} \cdot \text{K}}{0.375 \text{ m}} \times 0.15 (1.38 \times 10^8)^{1/3}$$

$$= 5.47 \text{ W/m}^2 \cdot \text{K}$$

$$\bar{h}_b = [k/(w/2)] \times 0.27 Ra_L^{1/4} = \frac{0.0265 \text{ W/m} \cdot \text{K}}{0.375 \text{ m}} \times 0.27 (1.38 \times 10^8)^{1/4}$$

$$= 2.07 \text{ W/m}^2 \cdot \text{K}$$

The rate of heat loss per unit length of duct is then

$$q' = 2q'_s + q'_t + q'_b = (2\bar{h}_s \cdot H + \bar{h}_t \cdot w + \bar{h}_b \cdot w)(T_s - T_\infty)$$

$$q' = (2 \times 4.23 \times 0.3 + 5.47 \times 0.75 + 2.07 \times 0.75)(45 - 15) \text{ W/m}$$

$$q' = 246 \text{ W/m} \qquad \qquad \triangleleft$$

Comments:

1. The heat loss may be reduced by insulating the duct. We consider this option for a 25-mm thick layer of blanket insulation ($k = 0.035$ W/m · K) that is wrapped around the duct.

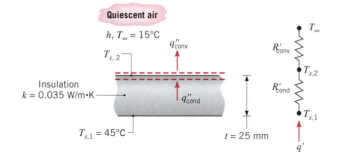

The heat loss at each surface may be expressed as

$$q' = \frac{T_{s,\,1} - T_\infty}{R'_{cond} + R'_{conv}}$$

where R'_{conv} is associated with free convection from the outer surface and hence depends on the unknown temperature $T_{s,\,2}$. This temperature may be determined by applying an energy balance to the outer surface, from which it follows that

$$q''_{cond} = q''_{conv}$$

or

$$\frac{(T_{s,\,1} - T_{s,\,2})}{(t/k)} = \frac{(T_{s,\,2} - T_\infty)}{(1/\bar{h})}$$

Since different convection coefficients are associated with the sides, top, and bottom ($\overline{h}_s$, $\overline{h}_t$, and $\overline{h}_b$), a separate solution to this equation must be obtained for each of the three surfaces. The solutions are iterative, since the properties of air and the convection coefficients depend on T_s. Performing the calculations, we obtain

$$\text{Sides} \quad T_{s,2} = 297.2°C, \quad \overline{h}_s = 3.18 \text{ W/m}^2 \cdot \text{K}$$

$$\text{Top} \quad T_{s,2} = 296.3°C, \quad \overline{h}_t = 3.66 \text{ W/m}^2 \cdot \text{K}$$

$$\text{Bottom} \quad T_{s,2} = 301.5°C, \quad \overline{h}_b = 1.71 \text{ W/m}^2 \cdot \text{K}$$

Neglecting heat loss through the corners of the insulation, the total heat rate per unit length of duct is then

$$q' = 2q'_s + q'_t + q'_b$$

$$q' = \frac{2H(T_{s,1} - T_\infty)}{(t/k) + (1/\overline{h}_s)} + \frac{w(T_{s,1} - T_\infty)}{(t/k) + (1/\overline{h}_t)} + \frac{w(T_{s,1} - T_\infty)}{(t/k) + (1/\overline{h}_b)}$$

which yields

$$q' = (17.5 + 22.8 + 17.3) \text{ W/m} = 57.6 \text{ W/m}$$

The insulation therefore provides a 77% reduction in heat loss to the ambient air by natural convection.

2. Although they have been neglected, radiation losses may still be significant. From Equation 1.7 with ε assumed to be unity and $T_{sur} = 288$ K, $q'_{rad} = 398$ W/m for the uninsulated duct. Inclusion of radiation effects in the energy balance for the insulated duct would reduce the outer surface temperatures, thereby reducing the convection heat rates. With radiation, however, the total heat rate ($q'_{conv} + q'_{rad}$) would increase.

9.6.3 The Long Horizontal Cylinder

This important geometry has been studied extensively, and many existing correlations have been reviewed by Morgan [20]. For an isothermal cylinder, Morgan suggests an expression of the form

$$\overline{Nu}_D = \frac{\overline{h}D}{k} = C\,Ra_D^n \tag{9.33}$$

TABLE 9.1 Constants of Equation 9.33 for free convection on a horizontal circular cylinder [20]

Ra_D	C	n
10^{-10}–10^{-2}	0.675	0.058
10^{-2}–10^{2}	1.02	0.148
10^{2}–10^{4}	0.850	0.188
10^{4}–10^{7}	0.480	0.250
10^{7}–10^{12}	0.125	0.333

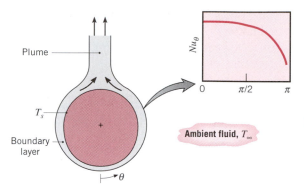

FIGURE 9.9
Boundary layer development
and Nusselt number distribution on a heated horizontal
cylinder.

where C and n are given in Table 9.1 and Ra_D and $\overline{Nu}_D$ are based on the cylinder diameter. In contrast, Churchill and Chu [21] have recommended a single correlation for a wide Rayleigh number range:

$$\overline{Nu}_D = \left\{ 0.60 + \frac{0.387 Ra_D^{1/6}}{[1 + (0.559/Pr)^{9/16}]^{8/27}} \right\}^2 \qquad Ra_D \lesssim 10^{12} \qquad (9.34)$$

The foregoing correlations provide the average Nusselt number over the entire circumference of an isothermal cylinder. As shown in Figure 9.9 for a heated cylinder, local Nusselt numbers are influenced by boundary layer development, which begins at $\theta = 0$ and concludes at $\theta < \pi$ with formation of a plume ascending from the cylinder. If the flow remains laminar over the entire surface, the distribution of the local Nusselt number with θ is characterized by a maximum at $\theta = 0$ and a monotonic decay with increasing θ. This decay would be disrupted at Rayleigh numbers sufficiently large ($Ra_D \gtrsim 10^9$) to permit transition to turbulence within the boundary layer. If the cylinder is cooled relative to the ambient fluid, boundary layer development begins at $\theta = \pi$, the local Nusselt number is a maximum at this location, and the plume descends from the cylinder.

EXAMPLE 9.4

A horizontal, high-pressure steam pipe of 0.1-m outside diameter passes through a large room whose wall and air temperatures are 23°C. The pipe has an outside surface temperature of 165°C and an emissivity of $\varepsilon = 0.85$. Estimate the heat loss from the pipe per unit length.

SOLUTION

Known: Surface temperature of a horizontal steam pipe.

Find: Heat loss from the pipe per unit length q' (W/m).

Schematic:

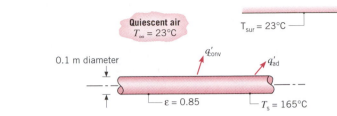

Assumptions:

1. Pipe surface area is small compared to surroundings.
2. Room air is quiescent.

Properties: Table A.4, air ($T_f = 367$ K): $k = 0.0313$ W/m · K, $\nu = 22.8 \times 10^{-6}$ m^2/s, $\alpha = 32.8 \times 10^{-6}$ m^2/s, $Pr = 0.697$, $\beta = 2.725 \times 10^{-3}$ K^{-1}.

Analysis: The total heat loss per unit length of pipe is

$$q' = q'_{conv} + q'_{rad} = \overline{h}\pi D(T_s - T_\infty) + \varepsilon\pi D\sigma(T_s^4 - T_{sur}^4)$$

The convection coefficient may be obtained from Equation 9.34

$$\overline{Nu}_D = \left\{0.60 + \frac{0.387 Ra_D^{1/6}}{[1 + (0.559/Pr)^{9/16}]^{8/27}}\right\}^2$$

where

$$Ra_D = \frac{g\beta(T_s - T_\infty)D^3}{\nu\alpha}$$

$$Ra_D = \frac{9.8 \text{ m/s}^2 \times 2.725 \times 10^{-3} \text{ K}^{-1} (165 - 23)°\text{C} (0.1 \text{ m})^3}{22.8 \times 10^{-6} \text{ m}^2/\text{s} \times 32.8 \times 10^{-6} \text{ m}^2/\text{s}} = 5.073 \times 10^6$$

Hence

$$\overline{Nu}_D = \left\{0.60 + \frac{0.387(5.073 \times 10^6)^{1/6}}{[1 + (0.559/0.697)^{9/16}]^{8/27}}\right\}^2 = 23.3$$

and

$$\overline{h} = \frac{k}{D}\overline{Nu}_D = \frac{0.0313 \text{ W/m} \cdot \text{K}}{0.1 \text{ m}} \times 23.3 = 7.29 \text{ W/m}^2 \cdot \text{K}$$

The total heat loss is then

$$q' = 7.29 \text{ W/m}^2 \cdot \text{K} (\pi \times 0.1 \text{ m})(165 - 23)°\text{C}$$
$$+ 0.85(\pi \times 0.1 \text{ m})5.67 \times 10^{-8} \text{ W/m}^2 \cdot \text{K}^4 (438^4 - 296^4) \text{ K}^4$$
$$q' = (325 + 441) \text{ W/m} = 766 \text{ W/m} \qquad \triangleleft$$

Comments:

1. Equation 9.33 could also be used to estimate the Nusselt number, with the result that $\overline{Nu}_D = 22.8$.

2. To explore the effect of an insulating layer on heat loss from the pipe, we consider a 25-mm thick layer of urethane for which $k = 0.026$ W/m · K and $\varepsilon = 0.85$.

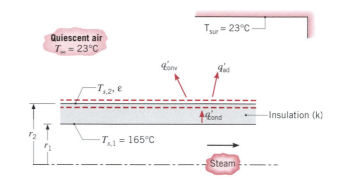

Free convection heat transfer to the ambient air and net radiation transfer to the surroundings depend on the temperature $T_{s,2}$ of the insulation, which may be obtained by performing an energy balance at the outer surface:

$$q'_{cond} = q'_{conv} + q'_{rad}$$

Substituting from Equations 1.7 and 3.27, it follows that

$$\frac{2\pi k(T_{s,1} - T_{s,2})}{\ln(r_2/r_1)} = \bar{h}(2\pi r_2)(T_{s,2} - T_\infty) + \varepsilon 2\pi r_2 \sigma(T_{s,2}^4 - T_{sur}^4)$$

The unknown temperature is determined from an iterative solution for which Equation 9.34 is used to reevaluate the convection coefficient, and hence the properties of air, at each step of the iteration. For the prescribed conditions, the solution yields $T_{s,2} = 35.3°C$, from which it follows that

$$q' = q'_{conv} + q'_{rad}$$

$$q' = 3.71 \text{ W/m}^2 \cdot \text{K}(\pi \times 0.15 \text{ m})(35.3 - 23)°C$$

$$+ 0.85(\pi \times 0.15 \text{ m})(308.3^4 - 296^4) \text{ K}^4$$

$$q' = (21.5 + 30.8) \text{ W/m} = 52.3 \text{ W/m}$$

As expected, the insulation significantly reduces heat loss from the pipe.

9.6.4 Spheres

The following correlation due to Churchill [10] is recommended for spheres in fluids of $Pr \geq 0.7$ and for $Ra_D \lesssim 10^{11}$.

$$\overline{Nu}_D = 2 + \frac{0.589Ra_D^{1/4}}{[1 + (0.469/Pr)^{9/16}]^{4/9}} \tag{9.35}$$

Recommended correlations from this section are summarized in Table 9.2. Results for other immersed geometries and special conditions are presented in comprehensive reviews by Churchill [10] and Raithby and Hollands [22].

TABLE 9.2   Summary of free convection empirical correlations for immersed geometries

Geometry	Recommended Correlation	Restrictions
1. Vertical plates[a]	Equation 9.26	None
2. Inclined plates Cold surface up or hot surface down	Equation 9.26 $g \rightarrow g \cos \theta$	$0 \leq \theta \leq 60°$
3. Horizontal plates (a) Hot surface up or cold surface down	Equation 9.30 Equation 9.31	$10^4 \lesssim Ra_L \lesssim 10^7$ $10^7 \lesssim Ra_L \lesssim 10^{11}$
(b) Cold surface up or hot surface down	Equation 9.32	$10^5 \lesssim Ra_L \lesssim 10^{10}$
4. Horizontal cylinder	Equation 9.34	$Ra_D \lesssim 10^{12}$
5. Sphere	Equation 9.35	$Ra_D \lesssim 10^{11}$ $Pr \geq 0.7$

[a]The correlation may be applied to a vertical cylinder if $(D/L) \gtrsim (35/Gr_L^{1/4})$

9.7
Free Convection within Parallel Plate Channels

A common free convection geometry involves vertical (or inclined) parallel plate channels that are open to the ambient at opposite ends (Figure 9.10). The plates could constitute a fin array used to enhance free convection heat transfer from a base surface to which the fins are attached, or they could constitute an array of circuit boards with heat-dissipating electronic components. Surface thermal conditions may be idealized as being isothermal or isoflux and symmetrical ($T_{s,1} = T_{s,2}$; $q''_{s,1} = q''_{s,2}$) or asymmetrical ($T_{s,1} \neq T_{s,2}$; $q''_{s,1} \neq q''_{s,2}$).

For vertical channels ($\theta = 0$) buoyancy acts exclusively to induce motion in the streamwise (x) direction and, beginning at $x = 0$, boundary layers develop on each surface. For short channels and/or large spacings (small L/S), independent boundary layer development occurs at each surface and conditions correspond to those for an *isolated plate* in an infinite, quiescent medium. For large L/S, however, boundary layers developing on opposing surfaces eventually merge to yield a fully developed condition. If the channel is inclined, there is a component of the buoyancy force normal, as well as parallel, to the streamwise direction, and conditions may strongly be influenced by development of a three-dimensional, secondary flow.

9.7.1   Vertical Channels

Beginning with the benchmark paper by Elenbaas [23], the vertical orientation has been studied extensively for symmetrically and asymmetrically heated plates with isothermal or isoflux surface conditions. For *symmetrically heated, isothermal plates,* Elenbaas obtained the following semiempirical correlation:

$$\overline{Nu}_S = \frac{1}{24} Ra_S \left(\frac{S}{L}\right) \left\{ 1 - \exp\left[-\frac{35}{Ra_S(S/L)} \right] \right\}^{3/4} \qquad (9.36)$$

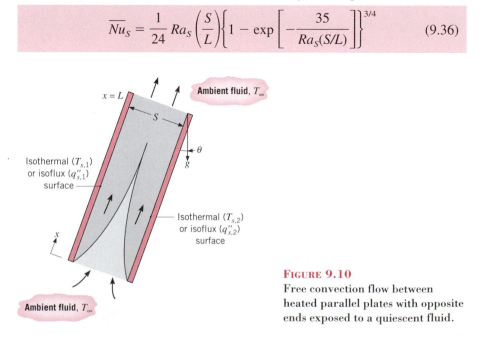

FIGURE 9.10
Free convection flow between heated parallel plates with opposite ends exposed to a quiescent fluid.

where the average Nusselt and Rayleigh numbers are defined as

$$\overline{Nu}_S = \left(\frac{q/A}{T_s - T_\infty} \right) \frac{S}{k} \tag{9.37}$$

and

$$Ra_S = \frac{g\beta(T_s - T_\infty)S^3}{\alpha\nu} \tag{9.38}$$

Knowledge of the average Nusselt number for a plate therefore permits determination of the total heat rate for the plate. In the fully developed limit ($S/L \to 0$), Equation (9.36) reduces to

$$\overline{Nu}_{S(\mathrm{fd})} = \frac{Ra_S(S/L)}{24} \tag{9.39}$$

Retention of the L dependence results from defining $\overline{Nu}_S$ in terms of the fixed inlet (ambient) temperature and not in terms of the fluid mixed-mean temperature, which is not explicitly known. For the common condition corresponding to adjoining isothermal ($T_{s,1}$) and insulated ($q''_{s,2} = 0$) plates, the fully developed limit yields the following expression for the isothermal surface [24]:

$$\overline{Nu}_{S(\mathrm{fd})} = \frac{Ra_S(S/L)}{12} \tag{9.40}$$

For isoflux surfaces, it is more convenient to define a local Nusselt number as

$$Nu_{S,L} = \left(\frac{q''_s}{T_{s,L} - T_\infty} \right) \frac{S}{k} \tag{9.41}$$

and to correlate results in terms of a modified Rayleigh number defined as

$$Ra_S^* = \frac{g\beta q''_s S^4}{k\alpha\nu} \tag{9.42}$$

The subscript L refers to conditions at $x = L$, where the plate temperature is a maximum. For symmetric, isoflux plates the fully developed limit corresponds to [24]

$$Nu_{S,L(\mathrm{fd})} = 0.144[Ra_S^*(S/L)]^{1/2} \tag{9.43}$$

and for asymmetric isoflux conditions with one surface insulated ($q''_{s,2} = 0$) the limit is

$$Nu_{S,L(\mathrm{fd})} = 0.204[Ra_S^*(S/L)]^{1/2} \tag{9.44}$$

Combining the foregoing relations for the fully developed limit with available results for the isolated plate limit, Bar-Cohen and Rohsenow [24] obtained Nusselt number correlations applicable to the complete range of S/L. For isothermal and isoflux conditions, respectively, the correlations are of the form

$$\overline{Nu}_S = \left[\frac{C_1}{(Ra_s\,S/L)^2} + \frac{C_2}{(Ra_s\,S/L)^{1/2}} \right]^{-1/2} \tag{9.45}$$

$$Nu_{S,L} = \left[\frac{C_1}{Ra_S^*\,S/L} + \frac{C_2}{(Ra_S^*\,S/L)^{2/5}} \right]^{-1/2} \tag{9.46}$$

TABLE 9.3   Heat transfer parameters for free convection between vertical parallel plates

Surface Condition	C_1	C_2	S_{opt}	S_{max}/S_{opt}
Symmetric isothermal plates $(T_{s,1} = T_{s,2})$	576	2.87	$2.71(Ra_S/S^3L)^{-1/4}$	1.71
Symmetric isoflux plates $(q''_{s,1} = q''_{s,2})$	48	2.51	$2.12(Ra_S^*/S^4L)^{-1/5}$	4.77
Isothermal/adiabatic plates $(T_{s,1}, q''_{s,2} = 0)$	144	2.87	$2.15(Ra_S/S^3L)^{-1/4}$	1.71
Isoflux/adiabatic plates $(q''_{s,1}, q''_{s,2} = 0)$	24	2.51	$1.69(Ra_S^*/S^4L)^{-1/5}$	4.77

where the constants C_1 and C_2 are given in Table 9.3 for the different surface thermal conditions. In each case the fully developed and isolated plate limits correspond to Ra_S (or Ra_S^*)$S/L \lesssim 10$ and Ra_S (or Ra_S^*)$S/L \gtrsim 100$, respectively.

Bar–Cohen and Rohsenow [24] used the foregoing correlations to infer the optimum plate spacing S_{opt} for maximizing heat transfer from an array of isothermal plates, as well as the spacing S_{max} needed to maximize heat transfer from each plate in the array. Existence of an optimum for the array results from the fact that, although heat transfer from each plate decreases with decreasing S, the number of plates that may be placed in a prescribed volume increases. Hence S_{opt} maximizes heat transfer from the array by yielding a maximum for the product of $\bar{h}$ and the total plate surface area. In contrast, to maximize heat transfer from each plate, S_{max} must be large enough to preclude overlap of adjoining boundary layers, such that the isolated plate limit remains valid over the entire plate. For isoflux plates, the total volumetric heat rate simply increases with decreasing S. However, the need to maintain T_s below prescribed limits precludes reducing S to extremely small values. Hence S_{opt} may be defined as that value of S which yields the maximum volumetric heat dissipation per unit temperature difference, $T_s(L) - T_\infty$. The spacing S_{max} that yields the lowest possible surface temperature for a prescribed heat flux, without regard to volumetric considerations, is again the value of S that precludes boundary layer merger. Values of S_{opt} and S_{max}/S_{opt} are presented in Table 9.3 for plates of negligible thickness.

In using the foregoing correlations, fluid properties are evaluated at average temperatures of $\bar{T} = (T_s + T_\infty)/2$ for isothermal surfaces and $\bar{T} = (T_{s,L} + T_\infty)/2$ for isoflux surfaces.

9.7.2   Inclined Channels

Experiments have been performed by Azevedo and Sparrow [16] for inclined channels in water. Symmetric isothermal plates and isothermal–insulated plates were considered for $0 \leq \theta \leq 45°$ and conditions within the isolated plate limit, $Ra_S(S/L) > 200$. Although three-dimensional secondary flows were observed at the lower plate, when it was heated, data for all experimental conditions were correlated to within $\pm10\%$ by

$$\overline{Nu}_S = 0.645[Ra_S(S/L)]^{1/4} \tag{9.47}$$

Departures of the data from the correlation were most pronounced at large tilt angles with bottom surface heating and were attributed to heat transfer enhancement by the three-dimensional secondary flow. Fluid properties are evaluated at $\overline{T} = (T_s + T_\infty)/2$.

9.8
Empirical Correlations: Enclosures

The foregoing results pertain to free convection between a surface and an extensive fluid medium. However, engineering applications frequently involve heat transfer between surfaces that are at different temperatures and are separated by an *enclosed* fluid. In this section we present correlations that are pertinent to the most common geometries.

9.8.1 Rectangular Cavities

The rectangular cavity (Figure 9.11) has been studied extensively, and comprehensive reviews of both experimental and theoretical results are available [25, 26]. Two of the opposing walls are maintained at different temperatures ($T_1 > T_2$), while the remaining walls are insulated from the surroundings. The tilt angle τ between the heated and cooled surfaces and the horizontal can vary from 0° (*horizontal cavity* with bottom heating) to 90° (*vertical cavity* with sidewall heating) to 180° (*horizontal cavity* with top heating). The heat flux across the cavity, which is expressed as

$$q'' = h(T_1 - T_2) \tag{9.48}$$

can depend strongly on the aspect ratio H/L, as well as the value of τ. For large values of the aspect ratio w/L, its dependence on w/L is small and may be neglected for the purposes of this text.

The horizontal cavity heated from below ($\tau = 0$) has been considered by many investigators. For H/L, $w/L \gg 1$, and Rayleigh numbers less than a critical value of $Ra_{L,c} = 1708$, buoyancy forces cannot overcome the resistance imposed by viscous forces and there is no advection within the cavity. Hence heat transfer from the bottom to the top surface occurs excusively by conduction. Since conditions correspond to one-dimensional conduction through a plane fluid layer, the convection coefficient is $h = k/L$ and $Nu_L = 1$. However, for

$$Ra_L \equiv \frac{g\beta(T_1 - T_2)L^3}{\alpha\nu} > 1708$$

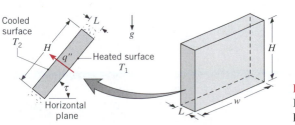

FIGURE 9.11
Free convection in a rectangular cavity.

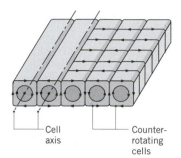

FIGURE 9.12
Longitudinal roll cells characteristic of advection in a horizontal fluid layer heated from below $(1708 < Ra_L \lesssim 5 \times 10^4)$.

Cell axis

Counter-rotating cells

conditions are thermally unstable and there is advection within the cavity. For Rayleigh numbers in the range $1708 < Ra_L \lesssim 5 \times 10^4$, fluid motion consists of regularly spaced roll cells (Figure 9.12), while for larger Rayleigh numbers, the cells break down and the fluid motion is turbulent.

As a first approximation, convection coefficients for the horizontal cavity heated from below may be obtained from the following correlation proposed by Globe and Dropkin [27]:

$$\overline{Nu}_L = \frac{\bar{h}L}{k} = 0.069 Ra_L^{1/3} \, Pr^{0.074} \qquad 3 \times 10^5 < Ra_L < 7 \times 10^9 \quad (9.49)$$

where all properties are evaluated at the average temperature, $\bar{T} \equiv (T_1 + T_2)/2$. The correlation applies for values of L/H sufficiently small to ensure a negligible effect of the sidewalls. More detailed correlations, which apply over a wider range of Ra_L, have been proposed [28, 29]. In concluding the discussion of horizontal cavities, it is noted that, for $\tau = 180°$, heat transfer from the top to bottom surface is exclusively by conduction ($Nu_L = 1$), irrespective of the value of Ra_L.

In the vertical rectangular cavity ($\tau = 90°$), the vertical surfaces are heated and cooled, while the horizontal surfaces are adiabatic. As shown in Figure 9.13, fluid motion is characterized by a recirculating or cellular flow for which fluid ascends along the hot wall and descends along the cold wall. For small Rayleigh numbers, $Ra_L \lesssim 10^3$, the buoyancy-driven flow is weak and heat transfer is primarily by conduction across the fluid. Hence, from Fourier's law, the Nusselt number is again $Nu_L = 1$. With increasing Rayleigh number, the cellular flow intensifies and becomes concentrated in thin boundary layers adjoining the sidewalls. The core becomes nearly stagnant, although additional cells can develop in the corners and the sidewall boundary layers eventually undergo transition to turbulence. For aspect ratios in the range $1 < (H/L) < 10$, the following correlations have been suggested [26]:

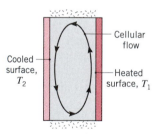

Cellular flow

Cooled surface, T_2

Heated surface, T_1

FIGURE 9.13
Cellular flow in a vertical cavity with different sidewall temperatures.

$$\overline{Nu}_L = 0.22 \left(\frac{Pr}{0.2 + Pr} Ra_L \right)^{0.28} \left(\frac{H}{L} \right)^{-1/4} \qquad (9.50)$$

$$\left[\begin{array}{l} 2 < \dfrac{H}{L} < 10 \\ Pr < 10^5 \\ 10^3 < Ra_L < 10^{10} \end{array} \right]$$

$$\overline{Nu}_L = 0.18 \left(\frac{Pr}{0.2 + Pr} Ra_L \right)^{0.29} \qquad (9.51)$$

$$\left[\begin{array}{c} 1 < \dfrac{H}{L} < 2 \\[2mm] 10^{-3} < Pr < 10^5 \\[2mm] 10^3 < \dfrac{Ra_L\,Pr}{0.2 + Pr} \end{array} \right]$$

while for larger aspect ratios, the following correlations have been proposed [30]:

$$\overline{Nu}_L = 0.42 Ra_L^{1/4}\, Pr^{0.012} \left(\frac{H}{L} \right)^{-0.3} \qquad \left[\begin{array}{c} 10 < \dfrac{H}{L} < 40 \\[2mm] 1 < Pr < 2 \times 10^4 \\[2mm] 10^4 < Ra_L < 10^7 \end{array} \right] \qquad (9.52)$$

$$\overline{Nu}_L = 0.046 Ra_L^{1/3} \qquad \left[\begin{array}{c} 1 < \dfrac{H}{L} < 40 \\[2mm] 1 < Pr < 20 \\[2mm] 10^6 < Ra_L < 10^9 \end{array} \right] \qquad (9.53)$$

Convection coefficients computed from the foregoing expressions are to be used with Equation 9.48. Again, all properties are evaluated at the mean temperature, $(T_1 + T_2)/2$.

In the 1970s, studies of free convection in tilted cavities were stimulated by applications involving flat-plate solar collectors [31–36]. For large aspect ratios, $(H/L) \gtrsim 12$, and tilt angles less than the critical value τ^* given in Table 9.4, the following correlation due to Hollands et al. [36] is in excellent agreement with available data:

$$\overline{Nu}_L = 1 + 1.44 \left[1 - \frac{1708}{Ra_L \cos \tau} \right]^{\cdot} \left[1 - \frac{1708(\sin 1.8\tau)^{1.6}}{Ra_L \cos \tau} \right]$$

$$+ \left[\left(\frac{Ra_L \cos \tau}{5830} \right)^{1/3} - 1 \right]^{\cdot} \qquad \left[\begin{array}{c} \dfrac{H}{L} \gtrsim 12 \\[2mm] 0 < \tau \le \tau^* \end{array} \right] \qquad (9.54)$$

The notation [ ]˙ implies that, if the quantity in brackets is negative, it must be set equal to zero. The implication is that, if the Rayleigh number is less than a critical value $Ra_{L,c} = 1708/\cos \tau$, there is no flow within the cavity. For small

TABLE 9.4   **Critical angle for inclined rectangular cavities**

(H/L)	1	3	6	12	>12
τ^*	25°	53°	60°	67°	70°

aspect ratios Catton [26] suggests that reasonable results may be obtained from a correlation of the form

$$\overline{Nu}_L = \overline{Nu}_L(\tau = 0)\left[\frac{\overline{Nu}_L\,(\tau = 90)}{\overline{Nu}_L\,(\tau = 0)}\right]^{\tau/\tau^*}(\sin\tau^*)^{(\tau/4\tau^*)}\qquad\left[\begin{array}{c}\dfrac{H}{L} \geq 12 \\[1mm] 0 < \tau \leq \tau^*\end{array}\right] \tag{9.55}$$

Beyond the critical tilt angle, the following correlations due to Ayyaswamy and Catton [31] and Arnold et al. [34], respectively, have been recommended [26] for all aspect ratios (H/L):

$$\overline{Nu}_L = \overline{Nu}_L(\tau = 90°)(\sin\tau)^{1/4}\qquad\qquad \tau^* < \tau < 90° \tag{9.56}$$

$$\overline{Nu}_L = 1 + [\overline{Nu}_L(\tau = 90°) - 1]\sin\tau \qquad 90° < \tau < 180° \tag{9.57}$$

9.8.2 Concentric Cylinders

Free convection heat transfer in the annular space between *long,* horizontal concentric cylinders (Figure 9.14) has been considered by Raithby and Hollands [37]. Flow in the annular region is characterized by two cells that are symmetric about the vertical midplane. If the inner cylinder is heated and the outer cylinder is cooled ($T_i > T_o$), fluid ascends and descends along the inner and outer cylinders, respectively. If $T_i < T_o$, the cellular flows are reversed. The heat transfer rate per unit length of cylinder (W/m) may be expressed as

$$q' = \frac{2\pi k_{\text{eff}}}{\ln(D_o/D_i)}\,(T_i - T_o) \tag{9.58}$$

where the *effective thermal conductivity* k_{eff} is the thermal conductivity that a stationary fluid should have to transfer the same amount of heat as the moving fluid. The suggested correlation for k_{eff} is

$$\frac{k_{\text{eff}}}{k} = 0.386\left(\frac{Pr}{0.861 + Pr}\right)^{1/4}(Ra_c^*)^{1/4} \tag{9.59}$$

where

$$Ra_c^* = \frac{[\ln(D_o/D_i)]^4}{L^3(D_i^{-3/5} + D_o^{-3/5})^5}\,Ra_L \tag{9.60}$$

Equation 9.59 may be used for the range $10^2 \leq Ra_c^* \leq 10^7$. For $Ra_c^* < 100$, $k_{\text{eff}} \approx k$. A more detailed correlation, which accounts for cylinder eccentricity effects, has been developed by Kuehn and Goldstein [38].

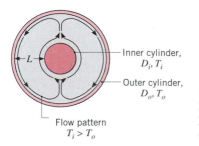

Inner cylinder, D_i, T_i

Outer cylinder, D_o, T_o

Flow pattern $T_i > T_o$

FIGURE 9.14
Free convection flow in the annular space between long, horizontal, concentric cylinders or concentric spheres.

9.8.3   Concentric Spheres

Raithby and Hollands [37] have also considered free convection heat transfer between concentric spheres (Figure 9.14) and express the total heat transfer rate as

$$q = k_{\text{eff}} \pi \left(\frac{D_i D_o}{L} \right) (T_i - T_o) \tag{9.61}$$

The effective thermal conductivity is

$$\frac{k_{\text{eff}}}{k} = 0.74 \left(\frac{Pr}{0.861 + Pr} \right)^{1/4} (Ra_s^*)^{1/4} \tag{9.62}$$

where

$$Ra_s^* = \left[\frac{L}{(D_o D_i)^4} \frac{Ra_L}{(D_i^{-7/5} + D_o^{-7/5})^5} \right] \tag{9.63}$$

The result may be used to a reasonable approximation for $10^2 \leq Ra_s^* \leq 10^4$.

EXAMPLE 9.5

A long tube of 0.1-m diameter is maintained at 120°C by passing steam through its interior. A radiation shield is installed concentric to the tube with an air gap of 10 mm. If the shield is at 35°C, estimate the heat transfer by free convection from the tube per unit length. What is the heat loss if the space between the tube and the shield is filled with glass-fiber blanket insulation?

SOLUTION

Known: Temperatures and diameters of a steam tube and a concentric radiation shield.

Find:

1. Heat loss per unit length of tube.
2. Heat loss if air space is filled with glass-fiber blanket insulation.

Schematic:

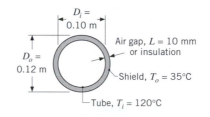

Assumptions:

1. Radiation heat transfer may be neglected.
2. Contact resistance with insulation is negligible.

Properties: Table A.4, air [$T = (T_i + T_o)/2 = 350$ K]: $k = 0.030$ W/m · K, $\nu = 20.92 \times 10^{-6}$ m^2/s, $\alpha = 29.9 \times 10^{-6}$ m^2/s, $Pr = 0.70$, $\beta = 0.00285$ K^{-1}. Table A.3, insulation, glass-fiber ($T \approx 300$ K): $k = 0.038$ W/m · K.

Analysis:

1. From Equation 9.58 the heat loss by free convection is

$$q' = \frac{2\pi k_{\text{eff}}}{\ln (D_o/D_i)} (T_i - T_o)$$

where k_{eff} may be obtained from Equations 9.59 and 9.60. With

$$Ra_L = \frac{g\beta(T_i - T_o)L^3}{\nu\alpha}$$

$$Ra_L = \frac{9.8 \text{ m/s}^2 \times 0.00285 \text{ K}^{-1} \times (120 - 35)°\text{C} \, (0.01 \text{ m})^3}{20.92 \times 10^{-6} \text{ m}^2/\text{s} \times 29.9 \times 10^{-6} \text{ m}^2/\text{s}} = 3795$$

it follows that

$$Ra_c^* = \frac{[\ln (D_o/D_i)]^4 Ra_L}{L^3(D_i^{-3/5} + D_o^{-3/5})^5}$$

$$Ra_c^* = \frac{[\ln (0.12 \text{ m}/0.10 \text{ m})]^4 (3795)}{(0.01 \text{ m})^3 [(0.10 \text{ m})^{-3/5} + (0.12 \text{ m})^{-3/5}]^5} = 171$$

Hence

$$\frac{k_{\text{eff}}}{k} = 0.386 \left(\frac{Pr}{0.861 + Pr} \right)^{1/4} (Ra_c^*)^{1/4}$$

$$\frac{k_{\text{eff}}}{k} = 0.386 \left(\frac{0.70}{0.861 + 0.70} \right)^{1/4} (171)^{1/4} = 1.14$$

The effective thermal conductivity is then

$$k_{\text{eff}} = 1.14(0.030 \text{ W/m} \cdot \text{K}) = 0.0343 \text{ W/m} \cdot \text{K}$$

and the heat loss is

$$q' = \frac{2\pi(0.0343 \text{ W/m} \cdot \text{K})}{\ln (0.12 \text{ m}/0.10 \text{ m})} (120 - 35)°\text{C} = 100 \text{ W/m} \qquad \triangleleft$$

2. With insulation in the space between the tube and the shield, heat loss is by conduction and from Equation 3.27

$$q' = \frac{2\pi k(T_i - T_o)}{\ln (D_o/D_i)}$$

$$q' = \frac{2\pi(0.038 \text{ W/m} \cdot \text{K})}{\ln (0.12 \text{ m}/0.10 \text{ m})} (120 - 35)°\text{C} = 111 \text{ W/m} \qquad \triangleleft$$

Comments: Although there is slightly more heat loss by conduction through the insulation than by free convection across the air space, the total heat loss across the air space may exceed that through the insulation because of the effects of radiation. The heat loss due to radiation may be minimized by using a radiation shield of low emissivity, and the means for calculating the loss will be developed in Chapter 13.

9.9
Combined Free and Forced Convection

In dealing with forced convection (Chapters 6 to 8), we ignored the effects of free convection. This was, of course, an assumption; for, as we now know, free convection is likely when there is an unstable temperature gradient. Similarly, in the preceding sections of this chapter, we assumed that forced convection was negligible. It is now time to acknowledge that situations may arise for which free and forced convection effects are comparable, in which case it is inappropriate to neglect either process. In Section 9.3 we indicated that free convection is negligible if $(Gr_L/Re_L^2) \ll 1$ and that forced convection is negligible if $(Gr_L/Re_L^2) \gg 1$. Hence the *combined free* and *forced* (or *mixed*) *convection* regime is generally one for which $(Gr_L/Re_L^2) \approx 1$.

The effect of buoyancy on heat transfer in a forced flow is strongly influenced by the direction of the buoyancy force relative to that of the flow. Three special cases that have been studied extensively correspond to buoyancy-induced and forced motions having the same direction (*assisting* flow), opposite directions (*opposing* flow), and perpendicular directions (*transverse* flow). Upward and downward forced motions over a heated vertical plate are examples of assisting and opposing flows, respectively. Examples of transverse flow include horizontal motion over a heated cylinder, sphere, or horizontal plate. In assisting and transverse flows, buoyancy acts to enhance the rate of heat transfer associated with pure forced convection; in opposing flows, it acts to decrease this rate.

It has become common practice to correlate mixed convection heat transfer results for external and internal flows by an expression of the form

$$Nu^n = Nu_F^n \pm Nu_N^n \tag{9.64}$$

For the specific geometry of interest, the Nusselt numbers Nu_F and Nu_N are determined from existing correlations for pure forced and natural (free) convection, respectively. The plus sign on the right-hand side of Equation 9.64 applies for assisting and transverse flows, while the minus sign applies for opposing flow. The best correlation of data is often obtained for $n = 3$, although values of $\frac{7}{2}$ and 4 may be better suited for transverse flows involving horizontal plates and cylinders (or spheres), respectively.

Equation 9.64 should be viewed as a first approximation, and any serious treatment of a mixed convection problem should be accompanied by an examination of the open literature. Mixed convection flows received considerable attention in the late 1970s to middle 1980s, and comprehensive literature reviews are available [39–42]. The flows are endowed with a variety of rich and unusual features that can complicate heat transfer predictions. For example, in a horizontal, parallel-plate channel, three-dimensional flows in the form of longitudinal vortices are induced by bottom heating, and the longitudinal variation of the Nusselt number is characterized by a decaying oscillation [43, 44]. Moreover, in channel flows, significant asymmetries may be associated with convection heat transfer at top and bottom surfaces [45]. Finally, we note that, although buoyancy effects can significantly enhance heat transfer for laminar forced convection flows, enhancement is typically negligible if the forced flow is turbulent [46].

9.10
Convection Mass Transfer

Mass density gradients may exist in a fluid as a consequence of spatial variations in the fluid composition. Such variations exist when species transfer is occurring due to a concentration gradient. Accordingly, under the influence of the gravitational field, free convection flows can be induced in species transfer processes. However, discussion of this topic is beyond the scope of this text and is left to more advanced treatments of buoyancy-driven flows [1, 2].

Another aspect of free convection mass transfer involves thermally driven free convection flows that enhance evaporation or sublimation occurring at a surface. For example, evaporation from a horizontal water layer is augmented by the free convection flow that is induced when the temperature of the water exceeds the temperature of the quiescent air above the water. In this case there is *simultaneous* heat and mass transfer by free convection, and a rigorous treatment of the subject must again be left to more advanced texts [1, 2]. However, if species transfer may be assumed to have a negligible influence on the free convection flow, heat transfer correlations, such as those described in the preceding sections, could be used to determine the heat transfer coefficient. From knowledge of this coefficient and use of the heat and mass transfer analogy, Equation 6.92, the convection mass transfer coefficient could then be estimated. The approach is satisfactory if the Prandtl and Schmidt numbers are approximately equal, in which case the Lewis number ($Le = Sc/Pr$) is approximately unity. The approximation ($Le \approx 1$) is usually satisfied in gas–vapor mixtures.

9.11
Summary

We have considered convective flows that originate in part or exclusively from buoyancy forces, and we have introduced the dimensionless parameters needed to characterize such flows. You should be able to discern when free convection effects are important and to quantify the associated heat transfer rates. An assortment of empirical correlations has been provided for this purpose.

References

1. Jaluria, Y., *Natural Convection Heat and Mass Transfer,* Pergamon Press, New York, 1980.

2. Gebhart, B., Y. Jaluria, R. L. Mahajan, and B. Sammakia, *Buoyancy-Induced Flows and Transport,* Hemisphere Publishing, Washington, DC, 1988.

3. Ostrach, S., "An Analysis of Laminar Free Convection Flow and Heat Transfer About a Flat Plate Parallel to the Direction of the Generating Body Force," National Advisory Committee for Aeronautics, Report 1111, 1953.

4. LeFevre, E. J., "Laminar Free Convection from a Vertical Plane Surface," *Proc. Ninth Int. Congr. Appl. Mech.*, Brussels, Vol. 4, 168, 1956.

5. McAdams, W. H., *Heat Transmission,* 3rd ed., McGraw-Hill, New York, 1954, Chap. 7.

6. Warner, C. Y., and V. S. Arpaci, "An Experimental Investigation of Turbulent Natural Convection in Air at Low Pressure for a Vertical Heated Flat Plate," *Int. J. Heat Mass Transfer,* **11,** 397, 1968.

7. Bayley, F. J., "An Analysis of Turbulent Free Convection Heat Transfer," *Proc. Inst. Mech. Eng.,* **169,** 361, 1955.

8. Churchill, S. W., and H. H. S. Chu, "Correlating Equations for Laminar and Turbulent Free Convection from a Vertical Plate," *Int. J. Heat Mass Transfer,* **18,** 1323, 1975.

9. Sparrow, E. M., and J. L. Gregg, "Laminar Free Convection from a Vertical Flat Plate," *Trans. ASME,* **78,** 435, 1956.

10. Churchill, S. W., "Free Convection Around Immersed Bodies," in E. U. Schlünder, Ed.-in-Chief, *Heat Exchange Design Handbook,* Section 2.5.7, Hemisphere Publishing, New York, 1983.

11. Sparrow, E. M., and J. L. Gregg, "Laminar Free Convection Heat Transfer from the Outer Surface of a Vertical Circular Cylinder," *Trans. ASME,* **78,** 1823, 1956.

12. Cebeci, T., "Laminar-Free-Convective Heat Transfer from the Outer Surface of a Vertical Slender Circular Cylinder," *Proc. Fifth Int. Heat Transfer Conf.*, Paper NC1.4, pp. 15–19, 1974.

13. Minkowycz, W. J., and E. M. Sparrow, "Local Nonsimilar Solutions for Natural Convection on a Vertical Cylinder," *J. Heat Transfer,* **96,** 178, 1974.

14. Vliet, G. C., "Natural Convection Local Heat Transfer on Constant-Heat-Flux Inclined Surfaces," *Trans. ASME,* **91C,** 511, 1969.

15. Fujii, T., and H. Imura, "Natural Convection Heat Transfer from a Plate with Arbitrary Inclination," *Int. J. Heat Mass Transfer,* **15,** 755, 1972.

16. Azevedo, L. F. A., and E. M. Sparrow, "Natural Convection in Open-Ended Inclined Channels," *J. Heat Transfer,* **107,** 893, 1985.

17. Rich, B. R., "An Investigation of Heat Transfer from an Inclined Flat Plate in Free Convection," *Trans. ASME,* **75,** 489, 1953.

18. Goldstein, R. J., E. M. Sparrow, and D. C. Jones, "Natural Convection Mass Transfer Adjacent to Horizontal Plates," *Int. J. Heat Mass Transfer,* **16,** 1025, 1973.

19. Lloyd, J. R., and W. R. Moran, "Natural Convection Adjacent to Horizontal Surfaces of Various Planforms," ASME Paper 74-WA/HT-66, 1974.

20. Morgan, V. T., "The Overall Convective Heat Transfer from Smooth Circular Cylinders," in T. F. Irvine and J. P. Hartnett, Eds., *Advances in Heat Transfer,* Vol. 11, Academic Press, New York, 1975, pp. 199–264.

21. Churchill, S. W., and H. H. S. Chu, "Correlating Equations for Laminar and Turbulent Free Convection from a Horizontal Cylinder," *Int. J. Heat Mass Transfer,* **18,** 1049, 1975.

22. Raithby, G. D., and K. G. T. Hollands, in W. M. Rohsenow, J. P. Hartnett, and E. N. Ganic, Eds., *Handbook of Heat Transfer Fundamentals,* Chap. 6, McGraw-Hill, New York, 1985.

23. Elenbaas, W., "Heat Dissipation of Parallel Plates by Free Convection," *Physica,* **9,** 1, 1942.

24. Bar-Cohen, A., and W. M. Rohsenow, "Thermally Optimum Spacing of Vertical Natural Convection Cooled, Parallel Plates," *J. Heat Transfer,* **106,** 116, 1984.

25. Ostrach, S., "Natural Convection in Enclosures," in J. P. Hartnett and T. F. Irvine, Eds., *Advances in Heat Transfer,* Vol. 8, Academic Press, New York, 1972, pp. 161–227.

26. Catton, I., "Natural Convection in Enclosures," *Proc. 6th Int. Heat Transfer Conf.*, Toronto, Canada, 1978, Vol. 6, pp. 13–31.

27. Globe, S., and D. Dropkin, "Natural Convection Heat Transfer in Liquids Confined Between Two Horizontal Plates," *J. Heat Transfer,* **81C,** 24, 1959.

28. Hollands, K. G. T., G. D. Raithby, and L. Konicek, "Correlation Equations for Free Convection Heat Transfer in Horizontal Layers of Air and Water," *Int. J. Heat Mass Transfer,* **18,** 879, 1975.

29. Churchill, S. W., "Free Convection in Layers and Enclosures," in E. U. Schlünder, Ed.-in-Chief, in *Heat Exchanger Design Handbook,* Section 2.5.8, Hemisphere Publishing, New York, 1983.

30. MacGregor, R. K., and A. P. Emery, "Free Convection Through Vertical Plane Layers: Moderate and High Prandtl Number Fluids," *J. Heat Transfer,* **91,** 391, 1969.

31. Ayyaswamy, P. S., and I. Catton, "The Boundary-Layer Regime for Natural Convection in a Differentially Heated, Tilted Rectangular Cavity," *J. Heat Transfer,* **95,** 543, 1973.

32. Catton, I., P. S. Ayyaswamy, and R. M. Clever, "Natural Convection Flow in a Finite, Rectangular Slot Arbitrarily Oriented with Respect to the Gravity Vector," *Int. J. Heat Mass Transfer,* **17,** 173, 1974.

33. Clever, R. M., "Finite Amplitude Longitudinal Convection Rolls in an Inclined Layer," *J. Heat Transfer,* **95,** 407, 1973.

34. Arnold, J. N., I. Catton, and D. K. Edwards, "Experimental Investigation of Natural Convection in Inclined Rectangular Regions of Differing Aspect Ratios," ASME Paper 75-HT-62, 1975.

35. Buchberg, H., I. Catton, and D. K. Edwards, "Natural Convection in Enclosed Spaces—A Review of Application to Solar Energy," *J. Heat Transfer,* **98,** 182, 1976.

36. Hollands, K. G. T., S. E. Unny, G. D. Raithby, and L. Konicek, "Free Convective Heat Transfer Across Inclined Air Layers," *J. Heat Transfer,* **98,** 189, 1976.

37. Raithby, G. D., and K. G. T. Hollands, "A General Method of Obtaining Approximate Solutions to Laminar and Turbulent Free Convection Problems," in T. F. Irvine and J. P. Hartnett, Eds., *Advances in Heat Transfer,* Vol. 11, Academic Press, New York, 1975, pp. 265–315.

38. Kuehn, T. H., and R. J. Goldstein, "Correlating Equations for Natural Convection Heat Transfer Between Horizontal Circular Cylinders," *Int. J. Heat Mass Transfer,* **19,** 1127, 1976.

39. Churchill, S. W., "Combined Free and Force Convection Around Immersed Bodies," in E. U. Schlünder, Ed.-in-Chief, *Heat Exchanger Design Handbook,* Section 2.5.9, Hemisphere Publishing, New York, 1983.

40. Churchill, S. W., "Combined Free and Forced Convection in Channels," in E. U. Schlünder, Ed.-in-Chief, *Heat Exchanger Design Handbook,* Section 2.5.10, Hemisphere Publishing, New York, 1983.

41. Chen, T. S., and B. F. Armaly, in S. Kakac, R. K. Shah, and W. Aung, Eds., *Handbook of Single-Phase Convective Heat Transfer,* Chap. 14, Wiley-Interscience, New York, 1987.

42. Aung, W., in S. Kakac, R. K. Shah, and W. Aung, Eds., *Handbook of Single-Phase Convective Heat Transfer,* Chap. 15, Wiley-Interscience, New York, 1987.

43. Incropera, F. P., A. J. Knox, and J. R. Maughan, "Mixed Convection Flow and Heat Transfer in the Entry Region of a Horizontal Rectangular Duct," *J. Heat Transfer,* **109,** 434, 1987.

44. Maughan, J. R., and F. P. Incropera, "Mixed Convection Heat Transfer for Airflow in a Horizontal and Inclined Channel," *Int. J. Heat Mass Transfer,* **30,** 1307, 1987.

45. Osborne, D. G., and F. P. Incropera, "Laminar, Mixed Convection Heat Transfer for Flow Between Horizontal Parallel Plates with Asymmetric Heating," *Int. J. Heat Mass Transfer,* **28,** 207, 1985.

46. Osborne, D. G., and F. P. Incropera, "Experimental Study of Mixed Convection Heat Transfer for Transitional and Turbulent Flow Between Horizontal, Parallel Plates," *Int. J. Heat Mass Transfer,* **28,** 1337, 1985.

Problems

Properties and General Considerations

9.1 Using the values of density for water in Table A.6, calculate the volumetric thermal expansion coefficient at 300 K from its definition, Equation 9.4, and compare your result with the tabulated value.

9.2 Consider an object having a characteristic length of 0.25 m and a situation for which the temperature difference is 25°C. Using thermophysical properties evaluated at 350 K, calculate the Grashof number for air, hydrogen, water, and ethylene glycol. Assume a pressure of 1 atm.

9.3 Consider an object of characteristic length 0.01 m and a situation for which the temperature difference is 30°C. Evaluating thermophysical properties at the prescribed conditions, determine the Rayleigh number for the following fluids: air (1 atm, 400 K), helium (1 atm, 400 K), glycerin (285 K), and water (310 K).

Vertical Plates

9.4 The heat transfer rate due to free convection from a vertical surface, 1 m high and 0.6 m wide, to quiescent air that is 20 K colder than the surface is known. What is the ratio of the heat transfer rate for that situation to the rate corresponding to a vertical surface, 0.6 m high and 1 m wide, when the quiescent air is 20 K warmer than the surface? Neglect heat transfer by radiation and any influence of temperature on the relevant thermophysical properties of air.

9.5 Consider a large vertical plate with a uniform surface temperature of 130°C suspended in quiescent air at 25°C and atmospheric pressure.

(a) Estimate the boundary layer thickness at a location 0.25 m measured from the lower edge.

(b) What is the maximum velocity in the boundary layer at this location and at what position in the boundary layer does the maximum occur?

(c) Using the similarity solution result, Equation 9.19, determine the heat transfer coefficient 0.25 m from the lower edge.

(d) At what location on the plate measured from the lower edge will the boundary layer become turbulent?

9.6 A number of thin plates are to be cooled by vertically suspending them in a water bath at a temperature of 20°C. If the plates are initially at 54°C and are 0.15 m long, what minimum spacing would prevent interference between their free convection boundary layers?

9.7 A square aluminum plate 5 mm thick and 200 mm on a side is heated while vertically suspended in quiescent air at 40°C. Determine the average heat transfer coefficient for the plate when its temperature is 15°C by two methods: using results from the similarity solution to the boundary layer equations, and using results from an empirical correlation.

9.8 Consider an array of vertical rectangular fins, which is to be used to cool an electronic device mounted in quiescent, atmospheric air at $T_\infty = 27°C$. Each fin has $L = 20$ mm and $H = 150$ mm and operates at an approximately uniform temperature of $T_s = 77°C$.

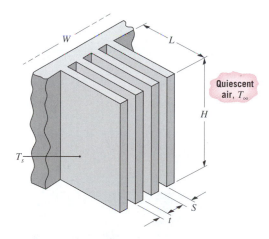

(a) Viewing each fin surface as a vertical plate in an infinite, quiescent medium, briefly describe why there exists an optimum fin spacing S. Using Figure 9.4, estimate the optimum value of S for the prescribed conditions.

(b) For the optimum value of S and a fin thickness of $t = 1.5$ mm, estimate the rate of heat transfer from the fins for an array of width $W = 355$ mm.

9.9 Determine the average convection heat transfer coefficient for the 2.5-m high vertical walls of a home having respective interior air and wall surface temperatures of (a) 20 and 10°C and (b) 27 and 37°C.

9.10 Beginning with the free convection correlation of the form given by Equation 9.24, show that for air at atmospheric pressure and a film temperature of 400 K, the average heat transfer coefficient for a vertical plate can be expressed as

$$\bar{h}_L = 1.40 \left(\frac{\Delta T}{L}\right)^{1/4} \qquad 10^4 < Ra_L < 10^9$$

$$\bar{h}_L = 0.98 \Delta T^{1/3} \qquad 10^9 < Ra_L < 10^{13}$$

9.11 A solid object is to be cooled by submerging it in a quiescent fluid, and the associated free convection coefficient is given by $\bar{h} = C\, \Delta T^{1/4}$, where C is a constant and $\Delta T = T - T_\infty$.

(a) Invoking the lumped capacitance approximation, obtain an expression for the time required for the object to cool from an initial temperature T_i to a final temperature T_f.

(b) Consider a highly polished, 150-mm square aluminum alloy (2024) plate of 5-mm thickness, initially at 225°C, suspended in ambient air at 25°C. Using the appropriate approximate correlation from Problem 9.10, determine the time required for the plate to reach 80°C.

(c) Plot the temperature–time history obtained from part (b) and compare with the results from a lumped capacitance analysis using a constant free convection coefficient, $\bar{h}_o$. Evaluate $\bar{h}_o$ from an appropriate correlation based on an average surface temperature of $\bar{T} = (T_i + T_f)/2$.

9.12 A household oven door of 0.5-m height and 0.7-m width reaches an average surface temperature of 32°C during operation. Estimate the heat loss to the room with ambient air at 22°C. If the door has an emissivity of 1.0 and the surroundings are also at 22°C, comment on the heat loss by free convection relative to that by radiation.

9.13 An aluminum alloy (2024) plate, heated to a uniform temperature of 227°C, is allowed to cool while vertically suspended in a room where the ambient air and surroundings are at 27°C. The plate is 0.3 m square with a thickness of 15 mm and an emissivity of 0.25.

(a) Develop an expression for the time rate of change of the plate temperature, assuming the temperature to be uniform at any time.

(b) Determine the initial rate of cooling (K/s) when the plate temperature is 227°C.

(c) Justify the uniform plate temperature assumption.

(d) Compute and plot the temperature history of the plate from $t = 0$ to the time required to reach a temperature of 30°C. Compute and plot the corresponding variations in the convection and radiation heat transfer rates.

9.14 The plate described in Problem 9.13 has been used in an experiment to determine the free convection heat transfer coefficient. At an instant of time when the plate temperature was 127°C, the time rate of change of this temperature was observed to be -0.0465 K/s. What is the corresponding free convection heat transfer coefficient? Compare this result with an estimate based on a standard empirical correlation.

9.15 The ABC Evening News Report in a news segment on hypothermia research studies at the University of Minnesota claimed that heat loss from the body is 30 times faster in 10°C water than in air at the same temperature. Is that a realistic statement?

9.16 Estimate the rate of heat loss, due to radiation and natural convection, from a person whose clothing surface temperature is 25°C to room surroundings at 20°C. Assume that a person can be approximated as a vertical cylinder of 0.3-m diameter and 1.8-m height. If the surrounding air at 1-atm pressure is replaced by helium at 15-atm pressure (approximating conditions experienced by deep sea divers), calculate the consequent percentage increase in the heat loss.

9.17 A vertical, thin pane of window glass that is 1 m on a side separates quiescent room air at $T_{\infty, i} = 20°C$ from quiescent ambient air at $T_{\infty, o} = -20°C$. The walls of the room and the external surroundings (landscape, buildings, etc.) are also at $T_{sur, i} = 20°C$ and $T_{sur, o} = -20°C$, respectively.

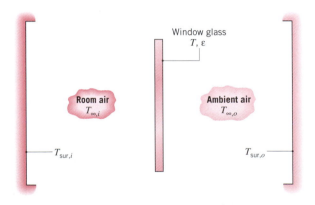

If the glass has an emissivity of $\varepsilon = 1$, what is its temperature T? What is the rate of heat loss through the glass?

9.18 Consider the conditions of Problem 9.17, but now allow for a difference between the inner and outer surface temperatures, $T_{s, i}$ and $T_{s, o}$, of the window. For a glass thickness and thermal conductivity of $t_g = 10$ mm and $k_g = 1.4$ W/m · K, respectively, evaluate $T_{s, i}$ and $T_{s, o}$. What is the heat loss through the window?

9.19 In a study of heat losses from buildings, free convection heat transfer from room air at 305 K to the inner surface of a 2.5-m high wall at 295 K is simulated by performing laboratory experiments using water in a smaller test cell. In the experiments the water and the inner surface of the test cell are maintained at 300 and 290 K, respectively. To achieve similarity between conditions in the room and the test cell, what is the required test cell height? If the average Nusselt number for the wall may be correlated exclusively in terms of the Rayleigh number, what is the ratio of the average convection coefficient for the room wall to the average coefficient for the test cell wall?

9.20 A thin-walled container with a hot process fluid at 50°C is placed in a quiescent, cold water bath at 10°C. Heat transfer at the inner and outer surfaces of the container may be approximated by free convection from a vertical plate.

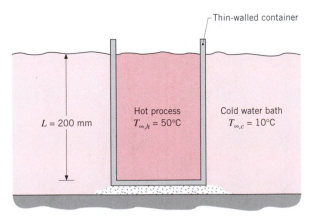

(a) Determine the overall heat transfer coefficient between the hot process fluid and the cold water bath. Assume the properties of the hot process fluid are those of water.

(b) Generate a plot of the overall heat transfer coefficient as a function of the hot process fluid temperature $T_{\infty, h}$ for the range 20 to 60°C, with all other conditions remaining the same.

9.21 A flat, isothermal heating panel, 0.5 m wide by 1 m high, is mounted to one wall of a large room. The surface of the panel has an emissivity of 0.90 and is maintained at 400 K. If the walls and air of the room are at 300 K, what is the net rate at which heat is transferred from the panel to the room?

9.22 A square plate of pure aluminum, 0.5 m on a side and 16 mm thick, is initially at 300°C and is suspended in a large chamber. The walls of the chamber are maintained at 27°C, as is the enclosed air. If the surface emissivity of the plate is 0.25, what is the initial cooling rate? Is it reasonable to assume a uniform plate temperature during the cooling process?

9.23 The vertical rear window of an automobile is of thickness $L = 8$ mm and height $H = 0.5$ m and contains fine-meshed heating wires that can induce nearly uniform volumetric heating, $\dot{q}$ (W/m³).

(a) Consider steady-state conditions for which the interior surface of the window is exposed to quiescent air at 10°C, while the exterior surface is exposed to air at −10°C moving in parallel flow over the surface with a velocity of 20 m/s. Determine the volumetric heating rate needed to maintain the interior window surface at $T_{s,i} = 15$°C.

(b) The interior and exterior window temperatures, $T_{s,i}$ and $T_{s,o}$, depend on the compartment and ambient temperatures, $T_{\infty,i}$ and $T_{\infty,o}$, as well as on the velocity u_∞ of air flowing over the exterior surface and the volumetric heating rate $\dot{q}$. Subject to the constraint that $T_{s,i}$ is to be maintained at 15°C, we wish to develop guidelines for varying the heating rate in response to changes in $T_{\infty,i}$, $T_{\infty,o}$, and/or u_∞. If $T_{\infty,i}$ is maintained at 10°C, how will $\dot{q}$ and $T_{s,o}$ vary with $T_{\infty,o}$ for $-25 \leq T_{\infty,o} \leq 5$°C and $u_\infty = 10, 20,$ and 30 m/s? If a constant vehicle speed is maintained, such that $u_\infty = 30$ m/s, how will $\dot{q}$ and $T_{s,o}$ vary with $T_{\infty,i}$ for $5 \leq T_{\infty,i} \leq 20$°C and $T_{\infty,o} = -25, -10,$ and 5°C?

9.24 Determine the maximum allowable uniform heat flux that may be imposed at a wall heating panel 1 m high, if the maximum temperature is not to exceed 37°C when the ambient air temperature is 25°C.

9.25 The components of a vertical circuit board, 150 mm on a side, dissipate 5 W. The back surface is well insulated and the front surface is exposed to quiescent air at 27°C.

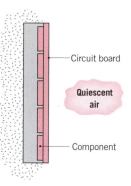

Assuming a uniform surface heat flux, what is the maximum temperature of the board? What is the temperature of the board for an isothermal surface condition?

9.26 A refrigerator door has a height and width of $H = 1$ m and $W = 0.65$ m, respectively, and is situated in a large room for which the air and walls are at $T_\infty = T_{sur} = 25$°C. The door consists of a layer of polystyrene insulation ($k = 0.03$ W/m · K) sandwiched between thin sheets of steel ($\varepsilon = 0.6$) and polypropylene. Under normal operating conditions, the inner surface of the door is maintained at a fixed temperature of $T_{s,i} = 5$°C.

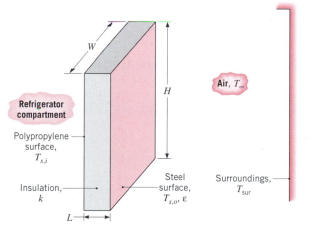

(a) Estimate the heat gain through the door for the worst case condition corresponding to no insulation ($L = 0$).

(b) Compute and plot the heat gain and the outer surface temperature $T_{s,o}$ as a function of insulation thickness for $0 \leq L \leq 25$ mm.

9.27 Air at 3 atm and 100°C is discharged from a compressor into a vertical receiver of 2.5-m height and 0.75-m diameter. Assume that the receiver wall has negligible thermal resistance, is at a uniform temperature, and that heat transfer at its inner and outer surfaces is by free convection from a vertical plate. Neglect radiation exchange and any losses from the top.

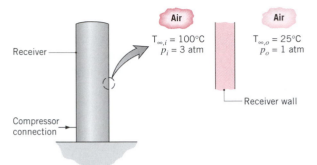

(a) Estimate the receiver wall temperature and the heat transfer to the ambient air at 25°C. To facilitate use of the free convection correlations with appropriate film temperatures, assume that the receiver wall temperature is 60°C.

(b) Were the assumed film temperatures of part (a) reasonable? If not, use an iteration procedure to find consistent values.

(c) Now consider two features of the receiver neglected in the previous analysis: (i) radiation exchange from the exterior surface of emissivity 0.85 to large surroundings, also at 25°C; and (ii) the thermal resistance of a 20-mm thick wall with a thermal conductivity of 0.25 W/m · K. Represent the system by a thermal circuit and estimate the wall temperatures and the heat transfer rate.

9.28 An experimental procedure to determine the free convection coefficient for a vertical plate involves suspending a 200-mm square, 10-mm thick slab of ice in a large enclosure that allows for slight air movement. The ambient air and walls of the enclosure are at a temperature of 27°C.

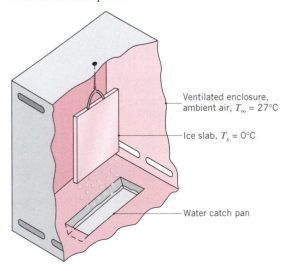

Using a standard convection correlation and considering radiation exchange between the slab and the enclosure, estimate the mass of water that accumulates in the catch pan during a one-hour period. The heat of fusion of ice is $h_{sf} = 333.4$ kJ/kg, and its emissivity is 0.95.

Horizontal and Inclined Plates

9.29 Consider the transformer of Problem 8.26, whose lateral surface is being maintained at 47°C by a forced convection coolant line removing 1000 W. It is desired to explore cooling of the transformer by free convection and radiation, assuming the surface to have an emissivity of 0.80.

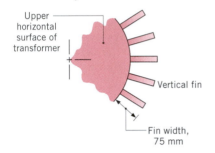

(a) Determine how much power could be removed by free convection and radiation from the lateral *and* the upper horizontal surfaces when the ambient temperature and the surroundings are at 27°C.

(b) Vertical fins, 5 mm thick, 75 mm wide, and 500 mm long, can easily be welded to the lateral surface. What is the heat removal rate by free convection if 30 such fins are attached?

9.30 Airflow through a long, 0.2-m-square air conditioning duct maintains the outer duct surface temperature at 10°C. If the horizontal duct is uninsulated and exposed to air at 35°C in the crawlspace beneath a home, what is the heat gain per unit length of the duct?

9.31 Consider the conditions of Example 9.3, including the effect of adding insulation of thickness t and thermal conductivity $k = 0.035$ W/m · K to the duct. We wish to now include the effect of radiation on the outer surface temperatures and the total heat loss per unit length of duct.

(a) If $T_{s,1} = 45°C$, $t = 25$ mm, $\varepsilon = 1$, and $T_{sur} = 288$ K, what are the temperatures of the side, top, and bottom surfaces? What are the corresponding heat losses per unit length of duct?

(b) For the top surface, compute and plot $T_{s,2}$ and q' as a function of insulation thickness for $0 \leq t \leq 50$ mm. The exposed duct surface ($t = 0$) may also be assumed to have an emissivity of $\varepsilon = 1$.

9.32 An electrical heater in the form of a horizontal disk of 400-mm diameter is used to heat the bottom of a tank filled with engine oil at a temperature of 5°C. Calculate the power required to maintain the heater surface temperature at 70°C.

9.33 Consider a horizontal 6-mm thick, 100-mm long straight fin fabricated from plain carbon steel ($k = 57$ W/m · K, $\varepsilon = 0.5$). The base of the fin is maintained at 150°C, while the quiescent ambient air and the surroundings are at 25°C. Assume the fin tip is adiabatic.

    (a) Estimate the fin heat rate per unit width, q'_f. Use an average fin surface temperature of 125°C to estimate the free convection coefficient and the linearized radiation coefficient. How sensitive is your estimate to the choice of the average fin surface temperature?

    (b) Generate a plot of q'_f as a function of the fin emissivity for $0.05 \leq \varepsilon \leq 0.95$. On the same coordinates, show the fraction of the total heat rate due to radiation exchange.

9.34 A circular grill of diameter 0.25 m and emissivity 0.9 is maintained at a constant surface temperature of 130°C. What electrical power is required when the room air and surroundings are at 24°C?

9.35 Consider Problem 7.32. What is the maximum allowable heat dissipation if the fan malfunctions and the air is quiescent?

9.36 The 4-m by 4-m horizontal roof of an uninsulated aluminum melting furnace is comprised of a 0.08-m thick fireclay brick refractory covered by a 5-mm thick steel (AISI 1010) plate. The refractory surface exposed to the furnace gases is maintained at 1700 K during operation, while the outer surface of the steel is exposed to the air and walls of a large room at 25°C. The emissivity of the steel is $\varepsilon = 0.3$.

    (a) What is the rate of heat loss from the roof?

    (b) If a 20-mm thick layer of alumina–silica insulation (64 kg/m³) is placed between the refractory and the steel, what is the new rate of heat loss from the roof? What is the temperature at the inner surface of the insulation?

    (c) One of the process engineers claims that the temperature at the inner surface of the insulation found in part (b) is too high for safe, long-term operation. What thickness of fireclay brick would reduce this temperature to 1350 K?

9.37 A stereo receiver/amplifier is enclosed in a thin metallic case for which the top, horizontal surface is 0.5 m × 0.5 m on a side. The surface, which is not vented, has an emissivity of $\varepsilon = 0.8$ and is exposed to quiescent ambient air and large surroundings for

which $T_\infty = T_{sur} = 25°C$. For surface temperatures in the range $50 \leq T_s \leq 75°C$, compute and plot the total heat rate from the surface, as well as the contributions due to convection and radiation.

9.38 A 200-mm square, 10-mm thick tile has the thermophysical properties of pyrex ($\varepsilon = 0.80$) and emerges from a curing process at an initial temperature of $T_i = 140°C$. The backside of the tile is insulated while the upper surface is exposed to ambient air and surroundings at 25°C.

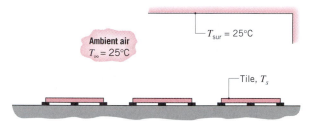

    (a) Estimate the time required for the tile to cool to a final, safe-to-touch temperature of $T_f = 40°C$. Use an average tile surface temperature of $\bar{T} = (T_i + T_f)/2$ to estimate the average free convection coefficient and the linearized radiation coefficient. How sensitive is your estimate to the assumed value for $\bar{T}$?

    (b) Estimate the required cooling time if ambient air is blown in parallel flow over the tile with a velocity of 10 m/s.

9.39 A highly polished aluminum plate of length 0.5 m and width 0.2 m is subjected to an airstream at a temperature of 23°C and a velocity of 10 m/s. Because of upstream conditions, the flow is turbulent over the entire length of the plate. A series of segmented, independently controlled heaters is attached to the lower side of the plate to maintain approximately isothermal conditions over the entire plate. The electrical heater covering the section between the positions $x_1 = 0.2$ m and $x_2 = 0.3$ m is shown in the schematic.

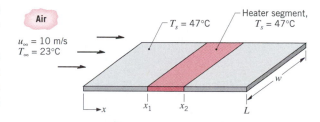

    (a) Estimate the electrical power that must be supplied to the designated heater segment to maintain the plate surface temperature at $T_s = 47°C$.

    (b) If the blower that maintains the airstream velocity over the plate malfunctions, but the power to the heaters remains constant, estimate the surface temperature of the designated segment. Assume that the ambient air is extensive, quiescent, and at 23°C.

9.40 It is desired to estimate the effectiveness of a horizontal, straight fin of rectangular cross section when applied to a surface operating at 45°C in an environment for which the surroundings and ambient air are at 25°C. The fin is to be fabricated from aluminum alloy (2024-T6) with an anodized finish ($\varepsilon = 0.82$) and is 2 mm thick and 100 mm long.

(a) Considering only free convection from the fin surface and estimating an average heat transfer coefficient, determine the effectiveness of the fin.

(b) Estimate the effectiveness of the fin, including the influence of radiation exchange with the surroundings.

(c) Using a numerical method, develop the finite-difference equations and obtain the fin effectiveness. The free convection and radiation exchange modes should be based on local, rather than average, values for the fin.

9.41 Although most wood-burning stoves are equipped with forced convection blowers, certain designs rely exclusively on heat transfer by radiation and natural convection to the surroundings. Consider a stove that forms a cubical enclosure, $L_s = 1$ m on a side, in a large room. The exterior walls of the stove have an emissivity of $\varepsilon = 0.8$ and are at an operating temperature of $T_{s,s} = 500$ K.

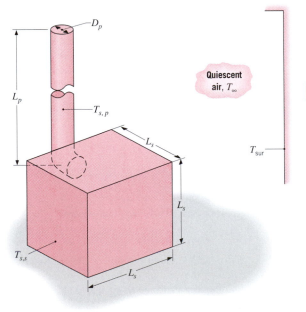

The stove pipe, which may be assumed to be isothermal at an operating temperature of $T_{s,p} = 400$ K, has a diameter of $D_p = 0.25$ m and a height of $L_p = 2$ m, extending from stove to ceiling. The stove is in a large room whose air and walls are at $T_\infty = T_{sur} = 300$ K. Neglecting heat transfer from the small horizontal section of the pipe and radiation exchange between the pipe and stove, estimate the rate at which heat is transferred from the stove and pipe to the surroundings.

9.42 A plate 1 m by 1 m, inclined at an angle of 45°, is exposed to a net radiation heat flux of 300 W/m² at its bottom surface. If the top surface of the plate is well insulated, estimate the temperature the plate reaches when the ambient air is quiescent and at a temperature of 0°C.

Horizontal Cylinders and Spheres

9.43 The heat transfer rate per unit length due to free convection from a horizontal tube is 200 W/m when its surface is maintained at 65°C and the ambient air is at 25°C. Estimate the heat transfer rate per unit length when the tube surface is maintained at 145°C. Neglect heat transfer by radiation and any influence of temperature on the relevant thermophysical properties of air.

9.44 A horizontal rod 5 mm in diameter is immersed in water maintained at 18°C. If the rod surface temperature is 56°C, estimate the free convection heat transfer rate per unit length of the rod.

9.45 A horizontal uninsulated steam pipe passes through a large room whose walls and ambient air are at 300 K. The pipe of 150-mm diameter has an emissivity of 0.85 and an outer surface temperature of 400 K. Calculate the heat loss per unit length from the pipe.

9.46 Consider the conditions of Example 9.4, including the effect of adding insulation of thickness t, thermal conductivity $k = 0.026$ W/m · K, and emissivity $\varepsilon = 0.85$. Compute and plot the surface temperature $T_{s,2}$ and the heat loss q' as a function of insulation thickness for $0 \le t \le 50$ mm. In addition to adding insulation, the heat loss may be reduced by decreasing the surface emissivity of the insulation. For $t = 25$ mm compute and plot $T_{s,2}$ and q' as a function of the emissivity for $0.1 \le \varepsilon \le 1.0$.

9.47 Beverage in cans 150 mm long and 60 mm in diameter is initially at 27°C and is to be cooled by placement in a refrigerator compartment at 4°C. In the interest of maximizing the cooling rate, should the cans be laid horizontally or vertically in the compartment? As a first approximation, neglect heat transfer from the ends.

9.48 Consider Problem 8.44. A more realistic solution would account for the resistance to heat transfer due to free convection in the paraffin during melting. Assuming the tube surface to have a uniform temperature of 55°C and the paraffin to be an infinite, quiescent liquid, determine the convection coefficient associated with the outer surface. Using this result and recognizing that the tube surface temperature is not known, determine the water outlet temperature, the total heat transfer rate, and the time required to completely liquefy the paraffin, for the prescribed conditions. Thermophysical properties associated with the liquid state of the paraffin are $k = 0.15$ W/m · K, $\beta = 8 \times 10^{-4}$ K^{-1}, $\rho = 770$ kg/m^3, $\nu = 5 \times 10^{-6}$ m^2/s, and $\alpha = 8.85 \times 10^{-8}$ m^2/s.

9.49 A horizontal pipe of 70-mm outside diameter passes through a room in which the air and the walls are at 25°C. The outer surface has an emissivity of 0.8 and is maintained at 150°C by steam passing through the pipe. What is the heat loss per unit length of the pipe?

9.50 A horizontal tube of 12.5-mm diameter with an outer surface temperature of 240°C is located in a room with an air temperature of 20°C. Estimate the heat transfer rate per unit length of the tube due to free convection.

9.51 Saturated steam at 4 bars absolute pressure with a mean velocity of 3 m/s flows through a horizontal pipe whose inner and outer diameters are 55 and 65 mm, respectively. The heat transfer coefficient for the steam flow is known to be 11,000 W/m^2 · K.

(a) If the pipe is covered with a 25-mm thick layer of 85% magnesia insulation and is exposed to atmospheric air at 25°C, determine the rate of heat transfer by free convection to the room per unit length of the pipe. If the steam is saturated at the inlet of the pipe, estimate its quality at the outlet of a pipe 30 m long.

(b) Net radiation to the surroundings also contributes to heat loss from the pipe. If the insulation has a surface emissivity of $\varepsilon = 0.8$ and the surroundings are at $T_{sur} = T_{\infty} = 25°C$, what is the rate of heat transfer to the room per unit length of pipe? What is the quality of the outlet flow?

(c) The heat loss may be reduced by increasing the insulation thickness and/or reducing its emissiv-

ity. What is the effect of increasing the insulation thickness to 50 mm if $\varepsilon = 0.8$? Of decreasing the emissivity to 0.2 if the insulation thickness is 25 mm? Of reducing the emissivity to 0.2 and increasing the insulation thickness to 50 mm?

9.52 A horizontal electrical cable of 25-mm diameter has a heat dissipation rate of 30 W/m. If the ambient air temperature is 27°C, estimate the surface temperature of the cable.

9.53 An electric immersion heater, 10 mm in diameter and 300 mm long, is rated at 550 W. If the heater is horizontally positioned in a large tank of water at 20°C, estimate its surface temperature. Estimate the surface temperature if the heater is accidentally operated in air at 20°C.

9.54 The maximum surface temperature of the 20-mm diameter shaft of a motor operating in ambient air at 27°C should not exceed 87°C. Because of power dissipation within the motor housing, it is desirable to reject as much heat as possible through the shaft to the ambient air. In this problem, we will investigate several methods for heat removal.

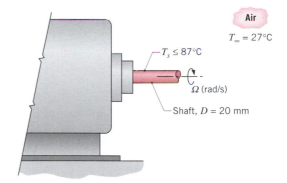

(a) For horizontal rotating cylinders, a suitable correlation for estimating the convection coefficient is of the form

$$\overline{Nu}_D = 0.133Re_D^{2/3} Pr^{1/3}$$
$$(Re_D < 4.3 \times 10^5, \quad 0.7 < Pr < 670)$$

where $Re_D \equiv \Omega D^2/\nu$ and Ω is the rotational velocity (rad/s). Determine the convection coefficient and the maximum heat rate per unit length as a function of rotational speed in the range from 5000 to 15,000 rpm.

(b) Estimate the free convection coefficient and the maximum heat rate per unit length for the stationary shaft. Mixed free and forced convection effects may become significant for

$Re_D < 4.7(Gr_D^3/Pr)^{0.137}$. Are free convection effects important for the range of rotational speeds designated in part (a)?

(c) Assuming the emissivity of the shaft is 0.8 and the surroundings are at the ambient air temperature, is radiation exchange important?

(d) If ambient air is in cross flow over the shaft, what air velocities are required to remove the heat rates determined in part (a)?

9.55 Consider a horizontal pin fin of 6-mm diameter and 60-mm length fabricated from plain carbon steel ($k = 57$ W/m · K, $\varepsilon = 0.5$). The base of the fin is maintained at 150°C, while the quiescent ambient air and the surroundings are at 25°C. Assume the fin tip is adiabatic.

(a) Estimate the fin heat rate, q_f. Use an average fin surface temperature of 125°C in estimating the free convection coefficient and the linearized radiation coefficient. How sensitive is this estimate to your choice of the average fin surface temperature?

(b) Use the finite-difference method of solution to obtain q_f when the convection and radiation coefficients are based on local, rather than average, temperatures for the fin. How does your result compare with the analytical solution of part (a)?

9.56 Water at a temperature of 17°C and a mass rate of 0.012 kg/s is to be used for maintaining a small plate (on which a special sensor is to be mounted) at a fixed temperature. The plate is situated within a hot air environment at a temperature of 235°C. The tube is horizontal and 1 m long. Fabricated from a plastic with a thermal conductivity of 0.05 W/m · K, the tube has an inner diameter $D_i = 1.4$ mm and an outer diameter $D_o = 3.2$ mm.

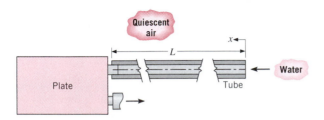

(a) Assuming that the average outer surface temperature of the tube is 120°C, estimate the heat transfer coefficient for free convection between the tube and the ambient air.

(b) Assuming that the flow and thermal conditions within the tube are fully developed, estimate the heat transfer coefficient between the tube and the water.

(c) Determine the overall heat transfer coefficient based on the outer tube area.

(d) Estimate the temperature of the water at the end of the tube ($x = 1$ m).

9.57 Common practice in chemical processing plants is to clad pipe insulation with a durable, thick aluminum foil. The functions of the foil are to confine the batt insulation and to reduce heat transfer by radiation to the surroundings. Because of the presence of chlorine (at chlorine or seaside plants), the aluminum foil surface, which is initially bright, becomes etched with in-service time. Typically, the emissivity might change from 0.12 at installation to 0.36 with extended service. For a 300-mm diameter foil-covered pipe whose surface temperature is 90°C, will this increase in emissivity due to degradation of the foil finish have a significant effect on heat loss from the pipe? Consider two cases with surroundings and ambient air at 25°C: (a) quiescent air and (b) a cross-wind velocity of 10 m/s.

9.58 Consider the electrical heater of Problem 7.45. If the blower were to malfunction, terminating airflow while the heater continued to operate at 1000 W/m, what temperature would the heater assume? How long would it take to come within 10°C of this temperature? Allow for radiation exchange between the heater ($\varepsilon = 0.8$) and the duct walls, which are also at 27°C.

9.59 A computer code is being developed to analyze a 12.5-mm diameter, cylindrical sensor used to determine ambient air temperature. The sensor experiences free convection while positioned horizontally in quiescent air at $T_\infty = 27$°C. For the temperature range from 30 to 80°C, derive an expression for the convection coefficient as a function of only $\Delta T = T_s - T_\infty$, where T_s is the sensor temperature. Evaluate properties at an appropriate film temperature and show what effect this approximation has on the convection coefficient estimate.

9.60 A thin-walled tube of 20-mm diameter passes hot fluid at a mean temperature of 45°C in an experimental flow loop. The tube is mounted horizontally in quiescent air at a temperature of 15°C. To satisfy the stringent temperature control requirements of the experiment, it was decided to wind thin electrical heating tape on the outer surface of the tube to prevent heat loss from the hot fluid to the ambient air.

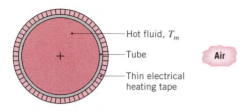

(a) Neglecting radiation heat loss, calculate the heat flux q''_e that must be supplied by the electrical tape to ensure a uniform fluid temperature.

(b) Assuming the emissivity of the tape is 0.95 and the surroundings are also at 15°C, calculate the required heat flux.

(c) The heat loss may be reduced by wrapping the heating tape in a layer of insulation. For 85% magnesia insulation ($k = 0.050$ W/m · K) having a surface emissivity of $\varepsilon = 0.60$, compute and plot the required heat flux q''_e as a function of insulation thickness in the range from 0 to 20 mm. For this range, compute and plot the convection and radiation heat rates per unit tube length as a function of insulation thickness.

9.61 A horizontal insulated pipeline passes through a building whose air and wall temperatures are 27°C. The outer surface of the steel tube containing the process fluid is at 400°C, has a diameter of 168 mm, and is insulated with a layer of calcium silicate having an emissivity of 0.85 and a thermal conductivity of k (W/m · K) = $0.0492 + 0.000137T_a$, where T_a [K] = $(T_i + T_o)/2$ is the average of the insulation inner and outer surface temperatures. Calculate the heat loss from the pipe when it is covered with a 30-mm thick layer of insulation.

9.62 Long stainless steel rods of 50-mm diameter are preheated to a uniform temperature of 1000 K before being suspended from an overhead conveyor for transport to a hot forming operation. The conveyor is in a large room whose walls and air are at 300 K.

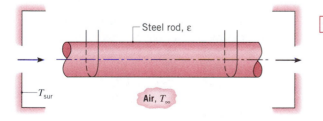

(a) Assuming the linear motion of the rod to have a negligible effect on convection heat transfer from its surface, determine the average convection coefficient at the start of the transport process.

(b) If the surface emissivity of the rod is $\varepsilon = 0.40$, what is the effective radiation heat transfer coefficient at the start of the transport process?

(c) Assuming a constant cumulative (radiation plus convection) heat transfer coefficient corresponding to the results of parts (a) and (b), what is the maximum allowable conveyor transit time, if the centerline temperature of the rod must exceed 900 K for the forming operation? Properties of the steel are $k = 25$ W/m · K and $\alpha = 5.2 \times 10^{-6}$ m²/s.

(d) Heat transfer by convection and radiation are actually decreasing during the transfer operation. Accounting for this reduction, reconsider the conditions of part (c) and obtain a more accurate estimate of the maximum allowable conveyor transit time.

9.63 Consider a long, 30-cm diameter steel cylinder that emerges from a heat treatment process at a temperature of 1000 K. The cylinder is suspended horizontally by cables and allowed to cool by free convection and radiation. The temperature of the ambient air and surrounding walls is 300 K, and the properties of steel are $k = 25$ W/m · K, $\alpha = 7.10 \times 10^{-6}$ m²/s, and $\varepsilon = 0.8$. Assuming the convection and radiation heat transfer coefficients to be constant, calculate the centerline temperature of the cylinder 20 min after it emerges from the heat treatment.

9.64 A horizontal pipe of 0.17-m outside diameter is used to transport steam through a building in which the air and walls are at a temperature of 25°C.

(a) If the pipe has a layer of calcium silicate ($k = 0.072$ W/m · K, $\varepsilon = 0.90$) insulation 25 mm thick and the inner surface of the insulation is at 650 K, what is the temperature of the outer surface and the heat loss per unit length of pipe?

(b) If a polished aluminum foil is applied to the outer surface, what is the outer surface temperature and the heat loss per unit length?

9.65 Hot air flows from a furnace through a 0.15-m diameter, thin-walled steel duct with a velocity of 3 m/s. The duct passes through the crawlspace of a house, and its uninsulated exterior surface is exposed to quiescent air and surroundings at 0°C.

(a) At a location in the duct for which the mean air temperature is 70°C, determine the heat loss per unit duct length and the duct wall temperature. The duct outer surface has an emissivity of 0.5.

(b) If the duct is wrapped with a 25-mm thick layer of 85% magnesia insulation ($k = 0.050$ W/m ·

K) having a surface emissivity of $\varepsilon = 0.60$, what are the duct wall temperature, the outer surface temperature, and the heat loss per unit length?

9.66 A biological fluid moves at a flow rate of $\dot{m} = 0.02$ kg/s through a coiled, thin-walled, 5-mm diameter tube submerged in a large water bath maintained at 50°C. The fluid enters the tube at 25°C.

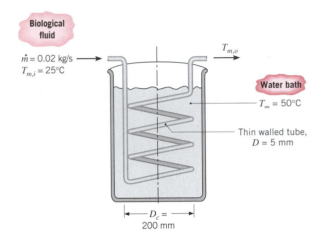

(a) Estimate the length of the tube and the number of coil turns required to provide an exit temperature of $T_{m,o} = 38$°C for the biological fluid. Assume that the water bath is an extensive, quiescent medium, that the coiled tube approximates a horizontal tube, and that the biological fluid has the thermophysical properties of water.

(b) The flow rate through the tube is controlled by a pump that experiences throughput variations of approximately ±10% at any one setting. This condition is of concern to the project engineer because the corresponding variation of the exit temperature of the biological fluid could influence the downstream process. What variation would you expect in $T_{m,o}$ for a ±10% change in $\dot{m}$?

9.67 In the analytical treatment of the fin with uniform cross-sectional area, it was assumed that the convection heat transfer coefficient is constant along the length of the fin. Consider an AISI 316 steel fin of 6-mm diameter and 50-mm length (with insulated tip) operating under conditions for which $T_b = 125$°C, $T_\infty = 27$°C, $T_{sur} = 27$°C, and $\varepsilon = 0.6$.

(a) Estimate average values of the fin heat transfer coefficients for free convection (h_c) and radiation exchange (h_r). Use these values to predict the tip temperature and fin effectiveness.

(b) Use a numerical method of solution to estimate the foregoing parameters when the convection and radiation coefficients are based on local, rather than average, values for the fin.

9.68 A hot fluid at 35°C is to be transported through a tube horizontally positioned in quiescent air at 25°C. Which of the tube shapes, each of equal cross-sectional area, would you use in order to minimize heat losses to the ambient air by free convection?

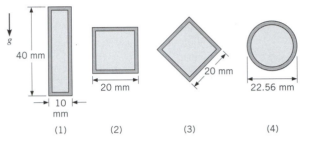

Use the following correlation of Lienhard ("On the Commonality of Equations for Natural Convection from Immersed Bodies," *Int. J. Heat Mass Transfer,* **16,** 2121, 1973) to approximate the laminar convection coefficient for an immersed body on which the boundary layer does not separate from the surface,

$$\overline{Nu}_l = 0.52 Ra_l^{1/4}$$

The characteristic length l is the length of travel of the fluid in the boundary layer across the shape surface. Compare this correlation to that given for a sphere to test its utility.

9.69 Consider a 2-mm diameter sphere immersed in a fluid at 300 K and 1 atm.

(a) If the fluid around the sphere is quiescent and extensive, show that the conduction limit of heat transfer from the sphere can be expressed as $Nu_{D,\text{cond}} = 2$. *Hint:* Begin with the expression for the thermal resistance of a hollow sphere, Equation 3.36, letting $r_2 \rightarrow \infty$, and then expressing the result in terms of the Nusselt number.

(b) Considering free convection, at what surface temperature will the Nusselt number be twice that for the conduction limit? Consider air and water as the fluids.

(c) Considering forced convection, at what velocity will the Nusselt number be twice that for the conduction limit? Consider air and water as the fluids.

9.70 A sphere of 25-mm diameter contains an embedded electrical heater. Calculate the power required to maintain the surface temperature at 94°C when the sphere is exposed to a quiescent medium at 20°C for: (a) air at atmospheric pressure, (b) water, and (c) ethylene glycol.

9.71 Under steady-state operation the surface temperature of a small 20-W incandescent light bulb is 125°C when the temperature of the room air and walls is 25°C. Approximating the bulb as a sphere 40 mm in diameter with a surface emissivity of 0.8, what is the rate of heat transfer from the surface of the bulb to the surroundings?

Parallel Plate Channels

9.72 Consider two long vertical plates maintained at uniform temperatures $T_{s,1} > T_{s,2}$. The plates are open at their ends and are separated by the distance $2L$.

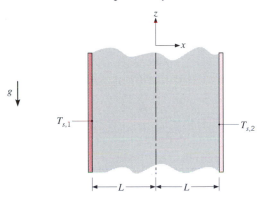

(a) Sketch the velocity distribution in the space between the plates.

(b) Write appropriate forms of the continuity, momentum, and energy equations for laminar flow between the plates.

(c) Evaluate the temperature distribution, and express your result in terms of the mean temperature, $T_m = (T_{s,1} + T_{s,2})/2$.

(d) Estimate the vertical pressure gradient by assuming the density to be a constant ρ_m corresponding to T_m. Substituting from the Boussinesq approximation, obtain the resulting form of the momentum equation.

(e) Determine the velocity distribution.

9.73 Consider the conditions of Problem 9.8, but now view the problem as one involving free convection between vertical, parallel plate channels. What is the optimum fin spacing S? For this spacing and the prescribed values of t and W, what is the rate of heat transfer from the fins?

9.74 A vertical array of circuit boards is immersed in qui-

escent ambient air at $T_\infty = 17°C$. Although the components protrude from their substrates, it is reasonable, as a first approximation, to assume *flat* plates with uniform surface heat flux q_s''. Consider boards of length and width $L = W = 0.4$ m and spacing $S = 25$ mm. If the maximum allowable board temperature is 77°C, what is the maximum allowable power dissipation per board?

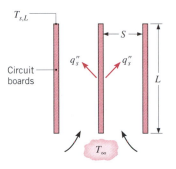

9.75 Consider the forced-air-cooled heat sink arrangement of Problem 8.82 for cooling a circuit board. It is proposed that the insulating plates be removed and the sinks be positioned such that the passages or fins are now vertical.

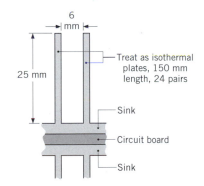

(a) Considering the fins as isothermal at the same temperature as the circuit board, determine how much power can be removed from the circuit board if its temperature is not to exceed 42°C, while the ambient air temperature is 27°C.

(b) Determine the optimum fin spacing for maximizing the heat transfer from the array of fins. Calculate the maximum heat removal rate if the overall size of the heat sinks remains unchanged.

9.76 The front door of a dishwasher of width 580 mm has a vertical air vent that is 500 mm in height with a 20-mm spacing between the inner tub operating at

52°C and an outer plate that is thermally insulated.

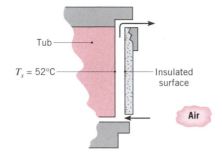

(a) Determine the heat loss from the tube surface when the ambient air is 27°C.

(b) A change in the design of the door provides the opportunity to increase or decrease the 20-mm spacing by 10 mm. What recommendations would you offer with regard to how the change in spacing will alter heat losses?

9.77 A solar collector consists of a parallel plate channel that is connected to a water storage plenum at the bottom and to a heat sink at the top. The channel is inclined $\theta = 30°$ from the vertical and has a transparent cover plate. Solar radiation transmitted through the cover plate and the water maintains the isothermal absorber plate at a temperature $T_s = 67°C$, while water returned to the reservoir from the heat sink is at $T_\infty = 27°C$. The system operates as a *thermosyphon,* for which water flow is driven exclusively by buoyancy forces. The plate spacing and length are $S = 15$ mm and $L = 1.5$ m.

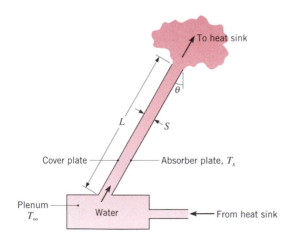

Assuming the cover plate to be adiabatic with respect to convection heat transfer to or from the water, estimate the rate of heat transfer per unit width normal to the flow direction (W/m) from the absorber plate to the water.

Rectangular Cavities

9.78 A building window pane that is 1.2 m high and 0.8 m wide is separated from the ambient air by a storm window of the same height and width. The air space between the two windows is 0.06 m thick. If the building and storm windows are at 20 and −10°C, respectively, what is the rate of heat loss by free convection across the air space?

9.79 The absorber plate and the adjoining cover plate of a flat-plate solar collector are at 70 and 35°C, respectively, and are separated by an air space of 0.05 m. What is the rate of free convection heat transfer per unit surface area between the two plates if they are inclined at an angle of 60° from the horizontal?

9.80 In a solar collector 1 m wide by 3 m long, the glass cover plate at an average temperature of 29°C is spaced 60 mm from the absorber plate at an average temperature of 50°C. Estimate the convection heat loss from the absorber plate to the glass when the collector is positioned horizontally. What is the convection heat loss if the spacing is reduced to 10 mm?

9.81 A rectangular cavity consists of two parallel, 0.5-m-square plates separated by a distance of 50 mm, with the lateral boundaries insulated. The heated plate is maintained at 325 K and the cooled plate at 275 K. Estimate the heat flux between the surfaces for three orientations of the cavity using the notation of Figure 9.7: vertical with $\tau = 90°$, horizontal with $\tau = 0°$, and horizontal with $\tau = 180°$.

9.82 Consider a horizontal flat roof section having the same dimensions as a vertical wall section. For both sections, the surfaces exposed to the air gap are at 18°C (inside) and −10°C (outside).

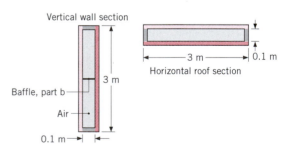

(a) Estimate the ratio of the convection heat rate for the horizontal section to that of the vertical section.

(b) What effect will inserting a baffle at the mid-height of the vertical section have on the convection heat rate for that section?

9.83 The space between the panes of a double-glazed window can be filled with either air or carbon dioxide at atmospheric pressure. The window is 1.5 m high and the spacing between the panes can be varied. Develop an analysis to predict the convection heat transfer rate across the window as a function of pane spacing and determine, under otherwise identical conditions, whether air or carbon dioxide will yield the smaller rate. Illustrate the results of your analysis for two conditions: winter $(-10°C, 20°C)$ and summer $(35°C, 25°C)$.

9.84 The top surface (0.5 m $\times$ 0.5 m) of an oven is 60°C for a particular operating condition when the room air is 23°C. To reduce heat loss from the oven and to minimize burn hazard, it is proposed to create a 50-mm air space by adding a cover plate.

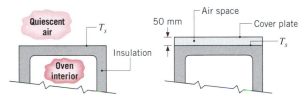

(a) Assuming the same oven surface temperature T_s for both situations, estimate the reduction in the convection heat loss resulting from installation of the cover plate. What is the temperature of the cover plate?

(b) Explore the effect of the cover plate spacing on the convection heat loss and the cover plate temperature for spacings in the range $5 \leq L \leq 50$ mm. Is there an optimum spacing?

9.85 Consider the rectangular cavity of Problem 9.81. Determine the heat flux between the two surfaces for tilt angles of 45° and 75°.

9.86 A solar water heater consists of a flat-plate collector that is coupled to a storage tank. The collector consists of a transparent cover plate and an absorber plate that are separated by an air gap.

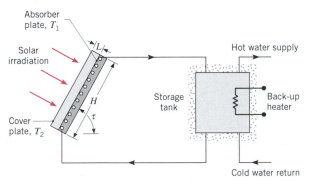

Although much of the solar energy collected by the absorber plate is transferred to a working fluid passing through a coiled tube brazed to the back of the absorber, some of the energy is lost by free convection and net radiation transfer across the air gap. In Chapter 13, we will evaluate the contribution of radiation exchange to this loss. For now, we restrict our attention to the free convection effect.

(a) Consider a collector that is inclined at an angle of $\tau = 60°$ and has dimensions of $H = w = 2$ m on a side, with an air gap of $L = 30$ mm. If the absorber and cover plates are at $T_1 = 70°C$ and $T_2 = 30°C$, respectively, what is the rate of heat transfer by free convection from the absorber plate?

(b) The heat loss by free convection depends on the spacing between the plates. Compute and plot the heat loss as a function of spacing for $5 \leq L \leq 50$ mm. Is there an optimum spacing?

Concentric Cylinders and Spheres

9.87 A solar collector design consists of an inner tube enclosed concentrically in an outer tube that is transparent to solar radiation. The tubes are thin walled with inner and outer diameters of 0.10 and 0.15 m, respectively. The annular space between the tubes is completely enclosed and filled with air at atmospheric pressure. Under operating conditions for which the inner and outer tube surface temperatures are 70 and 30°C, respectively, what is the convective heat loss per meter of tube length across the air space?

9.88 The annulus formed by two concentric, horizontal tubes with inner and outer diameters of 50 and 75 mm is filled with water. If the inner and outer surfaces are maintained at 300 and 350 K, respectively, estimate the convection heat transfer rate per unit length of the tubes.

9.89 The surfaces of two long, horizontal, concentric thin-walled tubes having radii of 100 and 125 mm are maintained at 300 and 400 K, respectively. If the annular space is pressurized with nitrogen at 5 atm, estimate the convection heat transfer rate per unit length of the tubes.

9.90 Liquid nitrogen is stored in a thin-walled spherical vessel of diameter $D_i = 1$ m. The vessel is positioned concentrically within a larger, thin-walled spherical container of diameter $D_o = 1.10$ m, and the intervening cavity is filled with atmospheric helium.

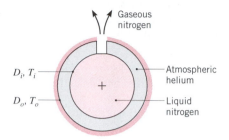

Gaseous nitrogen

D_i, T_i —— Atmospheric helium

D_o, T_o —— Liquid nitrogen

Under normal operating conditions, the inner and outer surface temperatures are $T_i = 77$ K and $T_o = 283$ K. If the latent heat of vaporization of nitrogen is 2×10^5 J/kg, what is the mass rate $\dot{m}$ (kg/s) at which gaseous nitrogen is vented from the system?

9.91 The surfaces of two concentric spheres having radii of 75 and 100 mm are maintained at 325 and 275 K, respectively. If the space between the spheres is filled with air at 3 atm, estimate the convection heat transfer rate.

Mixed Convection

9.92 Water at 35°C with a velocity of 0.05 m/s flows over a horizontal, 50-mm diameter cylinder maintained at a uniform surface temperature of 20°C. Do you anticipate that heat transfer by free convection will be significant? What would be the situation if the fluid were air at atmospheric pressure?

9.93 According to experimental results for parallel airflow over a uniform temperature, heated vertical plate, the effect of free convection on the heat transfer convection coefficient will be 5% when $Gr_L/Re_L^2 = 0.08$. Consider a heated vertical plate 0.3 m long, maintained at a surface temperature of 60°C in atmospheric air at 25°C. What is the minimum vertical velocity required of the airflow such that free convection effects will be less than 5% of the heat transfer rate?

9.94 A vertical array of circuit boards of 150-mm height is to be air cooled such that the board temperature does not exceed 60°C when the ambient temperature is 25°C.

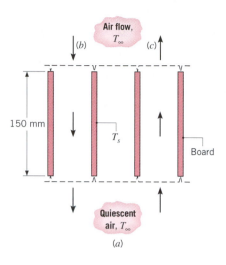

Air flow, T_∞　(b)　　(c)

150 mm

T_s

Board

Quiescent air, T_∞

(a)

Assuming isothermal surface conditions, determine the allowable electrical power dissipation per board for the cooling arrangements:

(a) Free convection only (no forced airflow).

(b) Airflow with a downward velocity of 0.6 m/s.

(c) Airflow with an upward velocity of 0.3 m/s.

(d) Airflow with a velocity (upward or downward) of 5 m/s.

9.95 A horizontal 100-mm diameter pipe passing hot oil is to be used in the design of an industrial water heater. Based on a typical water draw rate, the velocity over the pipe is 0.5 m/s. The hot oil maintains the outer surface temperature at 85°C and the water temperature is 37°C.

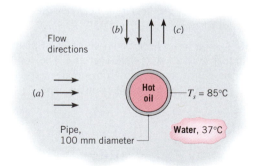

Flow directions　　(b)　(c)

(a)

Pipe, 100 mm diameter

Hot oil　$T_s = 85°C$

Water, 37°C

Investigate the effect of flow direction on the heat rate (W/m) for (a) horizontal, (b) downward, and (c) upward flow.

9.96 Heat-dissipating electrical components are mounted to the inner surface of a long metallic tube. The tube is of 0.10-m diameter and is submerged in a quiescent water bath.

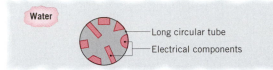

What is the heat dissipation per unit tube length if the tube wall and water temperatures are 350 and 300 K, respectively? What is the heat dissipation per unit length if a forced flow is imposed over the tube, with a cross-flow velocity of 1.0 m/s?

9.97 Square panels (250 mm × 250 mm) with a decorative, highly reflective plastic finish are cured in an oven at 125°C and cooled in quiescent air at 29°C. Quality considerations dictate that the panels remain horizontal and that the cooling rate be controlled. To increase productivity in the plant, it is proposed to replace the batch cooling method with a conveyor system having a velocity of 0.5 m/s.

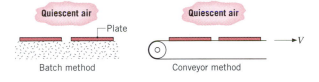

Compare the initial (immediately after leaving the oven) convection heat transfer rates for the two methods.

Mass Transfer

9.98 A garment soaked through with water is hung up to dry in a warm room. The still air is dry and at a temperature of 40°C. The garment may be assumed to have a temperature of 25°C and a characteristic length of 1 m in the vertical direction. Estimate the drying rate per unit width of the garment.

9.99 A 225-mm diameter pan of water maintained at 37°C is exposed to a dry, still atmosphere at 17°C. Estimate the evaporation rate and the total heat transfer rate from the pan.

CHAPTER 10

Boiling and Condensation

*I*n this chapter we focus on convection processes associated with the change in phase of a fluid. In particular, we consider processes that can occur at a solid–liquid interface, namely, *boiling* and *condensation.* For these cases *latent* heat effects associated with the phase change are significant. The change from the liquid to the vapor state due to boiling is sustained by heat transfer from the solid surface; conversely, condensation of a vapor to the liquid state results in heat transfer to the soild surface.

Since they involve fluid motion, boiling and condensation are classified as forms of the convection mode of heat transfer. However, they are characterized by unique features. Because there is a phase change, heat transfer to or from the fluid can occur without influencing the fluid temperature. In fact, through boiling or condensation, large heat tranfer rates may be achieved with small temperature differences. In addition to the *latent heat* h_{fg}, two other parameters are important in characterizing the processes, namely, the *surface tension* σ between the liquid–vapor interface and the *density difference* between the two phases. This difference induces a *buoyancy force,* which is proportional to $g(\rho_l - \rho_v)$. Because of combined latent heat and buoyancy-driven flow effects, boiling and condensation heat transfer coefficients and rates are generally much larger than those characteristic of convection heat transfer without phase change.

Many engineering problems involve boiling and condensation. For example, both processes are essential to all closed-loop power and refrigeration cycles. In a power cycle, pressurized liquid is converted to vapor in a *boiler.* After expansion in a turbine, the vapor is restored to its liquid state in a *condenser,* whereupon it is pumped to the boiler to repeat the cycle. Evaporators, in which the boiling process occurs, and condensers are also essential components in vapor-compression refrigeration cycles. The rational design of such components dictates that the associated phase change processes be well understood.

10.1
Dimensionless Parameters in Boiling and Condensation

In our treatment of boundary layer phenomena (Section 6.6), we nondimensionalized the governing equations to identify relevant dimensionless groups. This approach enhanced our understanding of related physical mechanisms and suggested simplified procedures for generalizing and representing heat transfer results.

Since it is difficult to develop governing equations for boiling and condensation processes, the appropriate dimensionless parameters are obtained by using the Buckingham pi theorem [1]. For either process, the convection coefficient could depend on the difference between the surface and saturation temperatures, $\Delta T = |T_s - T_{\text{sat}}|$, the body force arising from the liquid–vapor density difference, $g(\rho_l - \rho_v)$, the latent heat h_{fg}, the surface tension σ, a characteristic length L, and the thermophysical properties of the liquid or vapor: ρ, c_p, k, μ. That is,

$$h = h[\Delta T, g(\rho_l - \rho_v), h_{fg}, \sigma, L, \rho, c_p, k, \mu] \tag{10.1}$$

Since there are 10 variables in 5 dimensions (m, kg, s, J, K), there will be $(10 - 5) = 5$ pi-groups, which can be expressed in the following forms:

$$\frac{hL}{k} = f\left[\frac{\rho g(\rho_l - \rho_v)L^3}{\mu^2}, \frac{c_p \, \Delta T}{h_{fg}}, \frac{\mu c_p}{k}, \frac{g(\rho_l - \rho_v)L^2}{\sigma}\right] \tag{10.2a}$$

or, defining the dimensionless groups,

$$Nu_L = f\left[\frac{\rho g(\rho_l - \rho_v)L^3}{\mu^2}, Ja, Pr, Bo\right] \tag{10.2b}$$

The Nusselt and Prandtl numbers are familiar from our earlier single-phase convection analyses. The new dimensionless parameters are the Jakob number *Ja*, the Bond number *Bo*, and a nameless parameter that bears a strong resemblance to the Grashof number (see Equation 9.12). This unnamed parameter represents the effect of buoyancy-induced fluid motion on heat transfer. The Jakob number, named after Max Jakob in honor of his early work on phase change phenomena, is the ratio of the maximum sensible energy absorbed by the liquid (vapor) to the latent energy absorbed by the liquid (vapor) during condensation (boiling). In many applications, the sensible energy is much less than the latent energy and *Ja* has a small numerical value. The Bond number is the ratio of the gravitational body force to the surface tension force. In subsequent sections, we will delineate the role of these parameters in boiling and condensation.

10.2
Boiling Modes

When evaporation occurs at a solid–liquid interface, it is termed *boiling*. The process occurs when the temperature of the surface T_s exceeds the saturation temperature T_{sat} corresponding to the liquid pressure. Heat is transferred from the solid surface to the liquid, and the appropriate form of Newton's law of cooling is

$$q_s'' = h(T_s - T_{sat}) = h \, \Delta T_e \tag{10.3}$$

where $\Delta T_e \equiv T_s - T_{sat}$ is termed the *excess temperature*. The process is characterized by the formation of vapor bubbles, which grow and subsequently detach from the surface. Vapor bubble growth and dynamics depend, in a complicated manner, on the excess temperature, the nature of the surface, and thermophysical properties of the fluid, such as its surface tension. In turn, the dynamics of vapor bubble formation affect fluid motion near the surface and therefore strongly influence the heat transfer coefficient.

Boiling may occur under various conditions. For example, in *pool boiling* the liquid is quiescent and its motion near the surface is due to free convection and to mixing induced by bubble growth and detachment. In contrast, for *forced convection boiling,* fluid motion is induced by external means, as well as by free convection and bubble-induced mixing. Boiling may also be classified according to whether it is *subcooled* or *saturated.* In subcooled boiling, the temperature of the liquid is below the saturation temperature and bubbles formed at the surface may condense in the liquid. In contrast, the temperature of the liquid slightly exceeds the saturation temperature in *saturated boiling.* Bubbles formed at the surface are then propelled through the liquid by buoyancy forces, eventually escaping from a free surface.

10.3
Pool Boiling

Saturated pool boiling, as shown in Figure 10.1, has been studied extensively. Although there is a sharp decline in the liquid temperature close to the solid surface, the temperature through most of the liquid remains slightly above saturation. Bubbles generated at the liquid–solid interface therefore rise to and are transported across the liquid–vapor interface. An appreciation for the underlying physical mechanisms may be obtained by examining the *boiling curve.*

10.3.1 The Boiling Curve

Nukiyama [2] was the first to identify different regimes of pool boiling using the apparatus of Figure 10.2. The heat flux from a horizontal *nichrome* wire to saturated water was determined by measuring the current flow I and potential drop E. The wire temperature was determined from knowledge of the manner in which its electrical resistance varied with temperature. This arrangement is referred to as *power-controlled* heating, wherein the wire temperature T_s (hence the excess temperature ΔT_e) is the dependent variable and the power setting (hence the heat flux q''_s) is the independent variable. Following the arrows of the *heating curve* of Figure 10.3, it is evident that as power is applied, the heat flux increases, at first slowly and then very rapidly, with excess temperature.

Nukiyama observed that boiling, as evidenced by the presence of bubbles, did not begin until $\Delta T_e \approx 5°C$. With further increase in power, the heat flux in-

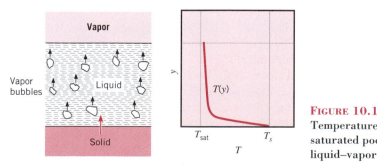

FIGURE 10.1
Temperature distribution in saturated pool boiling with a liquid–vapor interface.

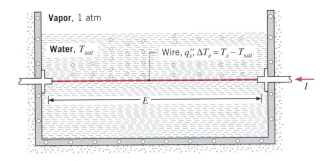

FIGURE 10.2   Nukiyama's power-controlled heating apparatus for demonstrating the boiling curve.

creased to very high levels until *suddenly,* for a value slightly larger than q''_{max}, the wire temperature jumped to the melting point and burnout occurred. However, repeating the experiment with a *platinum* wire having a higher melting point (2045 K vs. 1500 K), Nukiyama was able to maintain heat fluxes above q''_{max} without burnout. When he subsequently reduced the power, the variation of ΔT_e with q''_s followed the *cooling curve* of Figure 10.3. When the heat flux reached the minimum point q''_{min}, a further decrease in power caused the excess temperature to drop abruptly, and the process followed the original heating curve back to the saturation point.

Nukiyama believed that the hysteresis effect of Figure 10.3 was a consequence of the power-controlled method of heating, where ΔT_e is a dependent variable. He also believed that by using a heating process permitting the independent control of ΔT_e, the missing (dashed) portion of the curve could be obtained. His conjecture was subsequently confirmed by Drew and Mueller [3]. By condensing steam inside a tube at different pressures, they were able to control the value of ΔT_e for boiling of a low boiling point organic fluid at the tube outer surface and thereby obtain the missing portion of the boiling curve.

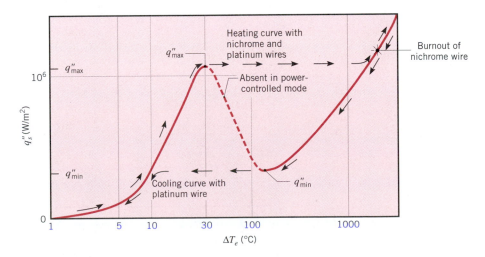

FIGURE 10.3   Nukiyama's boiling curve for saturated water at atmospheric pressure.

10.3.2 Modes of Pool Boiling

An appreciation for the underlying physical mechanisms may be obtained by examining the different modes, or regimes, of pool boiling. These regimes are identified in the boiling curve of Figure 10.4. The specific curve pertains to water at 1 atm, although similar trends characterize the behavior of other fluids. From Equation 10.3 we note that q_s'' depends on the convection coefficient h, as well as on the excess temperature ΔT_e. Different boiling regimes may be delineated according to the value of ΔT_e.

Free Convection Boiling   Free convection boiling is said to exist if $\Delta T_e \leq \Delta T_{e,A}$, where $\Delta T_{e,A} \approx 5°C$. In this regime there is insufficient vapor in contact with the liquid phase to cause boiling at the saturation temperature. As the excess temperature is increased, bubble inception will eventually occur, but below point A (referred to as the *onset of nucleate boiling,* ONB), fluid motion is determined principally by free convection effects. According to whether the flow is laminar or turbulent, h varies as ΔT_e to the $\frac{1}{4}$ or $\frac{1}{3}$ power, respectively, in which case q_s'' varies as ΔT_e to the $\frac{5}{4}$ or $\frac{4}{3}$ power.

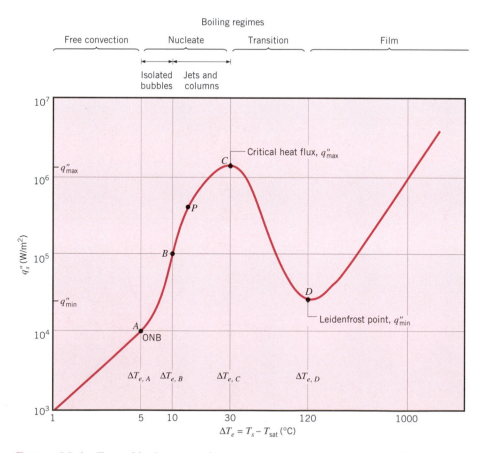

FIGURE 10.4   **Typical boiling curve for water at 1 atm: surface heat flux q_s'' as a function of excess temperature, $\Delta T_e \equiv T_s - T_{\text{sat}}$.**

Nucleate Boiling   Nucleate boiling exists in the range $\Delta T_{e,A} \leq \Delta T_e \leq \Delta T_{e,C}$, where $\Delta T_{e,C} \approx 30°C$. In this range, two different flow regimes may be distinguished. In region *A–B*, *isolated bubbles* form at nucleation sites and separate from the surface, as illustrated in Figure 10.2. This separation induces considerable fluid mixing near the surface, substantially increasing h and q_s''. In this regime most of the heat exchange is through direct transfer from the surface to liquid in motion at the surface, and not through the vapor bubbles rising from the surface. As ΔT_e is increased beyond $\Delta T_{e,B}$, more nucleation sites become active and increased bubble formation causes bubble interference and coalescence. In the region *B–C*, the vapor escapes as *jets* or *columns,* which subsequently merge into slugs of the vapor. This condition is illustrated in Figure 10.5*a*. Interference between the densely populated bubbles inhibits the motion of liquid near the surface. Point *P* of Figure 10.4 corresponds to an inflection in the boiling curve at which the heat transfer coefficient is a maximum. At this point h begins to decrease with increasing ΔT_e, although q_s'', which is the product of h and ΔT_e, continues to increase. This trend results because, for $\Delta T_e > \Delta T_{e,P}$, the relative increase in ΔT_e exceeds the relative reduction in h. At point *C*, however, further increase in ΔT_e is balanced by the reduction in h. The maximum heat flux, $q_{s,C}'' = q_{max}''$, is usually termed the *critical heat flux,* and in water at atmospheric pressure it exceeds 1 MW/m². At the point of this maximum, considerable vapor is being formed, making it difficult for liquid to continuously wet the surface.

Because high heat transfer rates and convection coefficients are associated with small values of the excess temperature, it is desirable to operate many engineering devices in the nucleate boiling regime. The approximate magnitude of the convection coefficient may be inferred by using Equation 10.3 with the boiling curve of Figure 10.4. Dividing q_s'' by ΔT_e, it is evident that convection coefficients in excess of 10^4 W/m² · K are characteristic of this regime. These values are considerably larger than those normally corresponding to convection with no phase change.

Transition Boiling   The region corresponding to $\Delta T_{e,C} \leq \Delta T_e \leq \Delta T_{e,D}$, where $\Delta T_{e,D} \approx 120°C$, is termed *transition boiling, unstable film boiling,* or *partial film boiling.* Bubble formation is now so rapid that a vapor film or blanket begins to form on the surface. At any point on the surface, conditions may oscillate between film and nucleate boiling, but the fraction of the total surface covered by the film increases with increasing ΔT_e. Because the thermal conductivity of the vapor is much less than that of the liquid, h (and q_s'') must decrease with increasing ΔT_e.

Film Boiling   Film boiling exists for $\Delta T_e \geq \Delta T_{e,D}$. At point *D* of the boiling curve, referred to as the *Leidenfrost point,* the heat flux is a minimum, $q_{s,D}'' = q_{min}''$, and the surface is completely covered by a *vapor blanket.* Heat transfer from the surface to the liquid occurs by conduction through the vapor. It was Leidenfrost who in 1756 observed that water droplets supported by the vapor film slowly boil away as they move about a hot surface. As the surface temperature is increased, radiation through the vapor film becomes significant and the heat flux increases with increasing ΔT_e.

Figure 10.5 illustrates the nature of the vapor formation and bubble dynam-

(a)

(b)

(c)

FIGURE 10.5   Boiling of methanol on a horizontal tube. (*a*) Nucleate boiling in the jets and columns regime. (*b*) Transition boiling. (*c*) Film boiling. Photographs courtesy of Professor J. W. Westwater, University of Illinois at Champaign-Urbana.

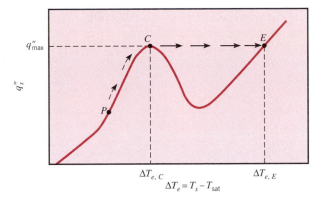

FIGURE 10.6   Onset of the boiling crisis.

ics associated with nucleate boiling, transition boiling, and film boiling. The photographs were obtained for the boiling of methanol on a horizontal tube.

Although the foregoing discussion of the boiling curve assumes that control may be maintained over T_s, it is important to remember the Nukiyama experiment and the many applications that involve controlling q_s'' (e.g., in a nuclear reactor or in an electric resistance heating device). Consider starting at some point P in Figure 10.6 and gradually increasing q_s''. The value of ΔT_e, and hence the value of T_s, will also increase, following the boiling curve to point C. However, any increase in q_s'' beyond this point will induce a sharp departure from the boiling curve in which surface conditions change abruptly from $\Delta T_{e,C}$ to $\Delta T_{e,E} \equiv T_{s,E} - T_{sat}$. Because $T_{s,E}$ may exceed the melting point of the solid, destruction or failure of the system may occur. For this reason point C is often termed the *burnout point* or the *boiling crisis,* and accurate knowledge of the *critical heat flux* (CHF), $q_{s,C}'' \equiv q_{max}''$, is important. We may want to operate a heat transfer surface close to this value, but rarely would we want to exceed it.

10.4
Pool Boiling Correlations

From the shape of the boiling curve and the fact that various physical mechanisms characterize the different regimes, it is no surprise that a multiplicity of heat transfer correlations exist for the boiling process. For the region below $\Delta T_{e,A}$ of the boiling curve (Figure 10.4), appropriate free convection correlations from Chapter 9 can be used to estimate heat transfer coefficients and heat rates. In this section we review some of the more widely used correlations for nucleate and film boiling.

10.4.1   Nucleate Pool Boiling

The analysis of nucleate boiling requires prediction of the number of surface nucleation sites and the rate at which bubbles originate from each site. While mechanisms associated with this boiling regime have been studied extensively, complete and reliable mathematical models have yet to be developed. Yamagata et al. [4] were the first to show the influence of nucleation sites on the heat rate through the relation

$$q_s'' = C \, \Delta T_e^a \, n^b \tag{10.4}$$

where n is the site density (active nucleation sites per unit area) and the exponents are approximately $a = 1.2$ and $b = \frac{1}{3}$. Although C and n vary considerably from one fluid–surface combination to another, it has been found that, for most commercial surfaces, $n \propto \Delta T_e^5$ or ΔT_e^6. Hence, from Equation 10.4, it follows that q_s'' is approximately proportional to ΔT_e^3.

The foregoing dependence of q_s'' on ΔT_e characterizes the first and most useful correlation for nucleate boiling, which was developed by Rohsenow [5],

$$q_s'' = \mu_l h_{fg} \left[\frac{g(\rho_l - \rho_v)}{\sigma} \right]^{1/2} \left(\frac{c_{p,l} \, \Delta T_e}{C_{s,f} h_{fg} \, Pr_l^n} \right)^3 \tag{10.5}$$

where the subscripts l and v, respectively, denote the saturated liquid and vapor states. Appearance of the surface tension σ (N/m) follows from the significant effect this fluid property has on bubble formation and development (see Problem 10.3). The coefficient $C_{s,f}$ and the exponent n depend on the surface–liquid combination, and representative values are presented in Table 10.1. Values for other surface–liquid combinations may be obtained from the literature [6, 7]. Values of the surface tension and the latent heat of vaporization for water are presented in Table A.6 and for selected fluids in Table A.5. Values for other liquids may be obtained from any recent edition of the *Handbook of Chemistry and Physics*. It is evident that the empirical correlation, Equation 10.5, does not explicitly contain the dimensionless parameters identified by Equation 10.2. Note, however, that the Jakob number $Ja = c_{p,l} \, \Delta T_e / h_{fg}$ and the Prandtl number are present and that $q_s'' \propto Ja^3 \, Pr^{-3n}$.

TABLE 10.1 Values of $C_{s,f}$ for Various Surface–Fluid Combinations [5–7]

Surface–Fluid Combination	$C_{s,f}$	n
Water–copper		
Scored	0.0068	1.0
Polished	0.0130	1.0
Water–stainless steel		
Chemically etched	0.0130	1.0
Mechanically polished	0.0130	1.0
Ground and polished	0.0060	1.0
Water–brass	0.0060	1.0
Water–nickel	0.006	1.0
Water–platinum	0.0130	1.0
n-Pentane–copper		
Polished	0.0154	1.7
Lapped	0.0049	1.7
Benzene–chromium	0.101	1.7
Ethyl alcohol–chromium	0.0027	1.7

The Rohsenow correlation applies only for clean surfaces. When it is used to estimate the heat flux, errors can amount to $\pm 100\%$. However, since $\Delta T_e \propto (q_s'')^{1/3}$, this error is reduced by a factor of 3 when the expression is used to estimate ΔT_e from knowledge of q_s''. Also, since $q_s'' \propto h_{fg}^{-2}$ and h_{fg} decreases with increasing saturation pressure (temperature), the nucleate boiling heat flux will increase as the liquid is pressurized.

10.4.2 Critical Heat Flux for Nucleate Pool Boiling

We recognize that the critical heat flux, $q_{s,C}'' = q_{max}''$, represents an important point on the boiling curve. We may wish to operate a boiling process close to this point, but we appreciate the danger of dissipating heat in excess of this amount. Kutateladze [8], through dimensional analysis, and Zuber [9], through a hydrodynamic stability analysis, obtained an expression of the form

$$q_{max}'' = \frac{\pi}{24} h_{fg} \rho_v \left[\frac{\sigma g (\rho_l - \rho_v)}{\rho_v^2} \right]^{1/4} \left(\frac{\rho_l + \rho_v}{\rho_l} \right)^{1/2} \tag{10.6}$$

which, as a first approximation, is independent of surface material and is only weakly dependent on geometry. Replacing the Zuber constant $(\pi/24) = 0.131$ by an experimental value of 0.149 [10] and approximating the last term in parentheses by unity, Equation 10.6 becomes

$$q_{max}'' = 0.149 h_{fg} \rho_v \left[\frac{\sigma g (\rho_l - \rho_v)}{\rho_v^2} \right]^{1/4} \tag{10.7}$$

In principle, since this expression applies to a horizontal heater surface of infinite extent, there is no characteristic length; in practice, however, the expression is applicable if the characteristic length is large compared to the mean bubble diameter parameter D_b (see Problem 10.16). If the ratio of the heater characteristic length to D_b, which is the Bond number, is less than 3, a correction factor must be applied to Equation 10.7. Lienhard and Dhir [11] have developed such correction factors for different geometries including plates, cylinders, spheres, and vertically and horizontally oriented ribbons.

Is important to note that the critical heat flux depends strongly on pressure, mainly through the pressure dependence of surface tension and the heat of vaporization. Cichelli and Bonilla [12] have experimentally demonstrated that the peak flux increases with pressure up to one-third of the critical pressure, after which it falls to zero at the critical pressure.

10.4.3 Minimum Heat Flux

The transition boiling regime is of little practical interest, as it may be obtained only by controlling the surface heater temperature. While no adequate theory has been developed for this regime, conditions can be characterized by periodic, *unstable* contact between the liquid and the heated surface. However, the upper limit of this regime is of interest because it corresponds to formation of a *stable* vapor blanket or film and to a minimum heat flux condition. If the heat flux drops below this minimum, the film will collapse, causing the surface to cool and nucleate boiling to be reestablished.

Zuber [9] used stability theory to derive the following expression for the minimum heat flux, $q''_{s,D} = q''_{\min}$, from a large horizontal plate.

$$q''_{\min} = C\rho_v h_{fg}\left[\frac{g\sigma(\rho_l - \rho_v)}{(\rho_l + \rho_v)^2}\right]^{1/4} \tag{10.8}$$

The constant, $C = 0.09$, has been experimentally determined by Berenson [13]. This result is accurate to approximately 50% for most fluids at moderate pressures but provides poorer estimates at higher pressures [14]. A similar result has been obtained for horizontal cylinders [15].

10.4.4   Film Pool Boiling

At excess temperatures beyond the Leidenfrost point, a continuous vapor film blankets the surface and there is no contact between the liquid phase and the surface. Because conditions in the stable vapor film bear a strong resemblance to those of laminar film condensation (Section 10.7), it is customary to base film boiling correlations on results obtained from condensation theory. One such result, which applies to film boiling on a cylinder or sphere of diameter D, is of the form

$$\overline{Nu}_D = \frac{\overline{h}_{\mathrm{conv}}D}{k_v} = C\left[\frac{g(\rho_l - \rho_v)h'_{fg}D^3}{\nu_v k_v(T_s - T_{\mathrm{sat}})}\right]^{1/4} \tag{10.9}$$

The correlation constant C is 0.62 for horizontal cylinders [16] and 0.67 for spheres [11]. The corrected latent heat h'_{fg} accounts for the sensible energy required to maintain temperatures within the vapor blanket above the saturation temperature. Although it may be approximated as $h'_{fg} = h_{fg} + 0.80c_{p,v}$ $(T_s - T_{\mathrm{sat}})$, it is known to depend weakly on the Prandtl number of the vapor [17]. Vapor properties are evaluated at the film temperature, $T_f = (T_s + T_{\mathrm{sat}})/2$, and the liquid density is evaluated at the saturation temperature.

At elevated surface temperatures ($T_s \gtrsim 300°C$), radiation heat transfer across the vapor film becomes significant. Since radiation acts to increase the film thickness, it is not reasonable to assume that the radiative and convective processes are simply additive. Bromley [16] investigated film boiling from the outer surface of horizontal tubes and suggested calculating the total heat transfer coefficient from a transcendental equation of the form

$$\overline{h}^{4/3} = \overline{h}_{\mathrm{conv}}^{4/3} + \overline{h}_{\mathrm{rad}}\overline{h}^{1/3} \tag{10.10a}$$

If $\overline{h}_{\mathrm{rad}} < \overline{h}_{\mathrm{conv}}$, a simpler form may be used:

$$\overline{h} = \overline{h}_{\mathrm{conv}} + \tfrac{3}{4}\overline{h}_{\mathrm{rad}} \tag{10.10b}$$

The effective radiation coefficient $\overline{h}_{\mathrm{rad}}$ is expressed as

$$\overline{h}_{\mathrm{rad}} = \frac{\varepsilon\sigma(T_s^4 - T_{\mathrm{sat}}^4)}{T_s - T_{\mathrm{sat}}} \tag{10.11}$$

where ε is the emissivity of the solid (Table A.11) and σ is the Stefan–Boltzmann constant.

Note that the analogy between film boiling and film condensation does not hold for small surfaces with high curvature because of the large disparity between vapor and liquid film thicknesses for the two processes. The analogy is also questionable for a vertical surface, although satisfactory predictions have been obtained for limited conditions.

10.4.5 Parametric Effects on Pool Boiling

In this section we briefly consider other parameters that can affect pool boiling, confining our attention to the gravitational field, liquid subcooling, and solid surface conditions.

The influence of the *gravitational field* on boiling must be considered in applications involving space travel and rotating machinery. This influence is evident from appearance of the gravitational acceleration g in the foregoing expressions. Siegel [18], in his review of low gravity effects, confirms that the $g^{1/4}$ dependence in Equations 10.7, 10.8, and 10.9 (for the maximum and minimum heat fluxes and for film boiling) is correct for values of g as low as 0.10 m/s^2. For nucleate boiling, however, evidence indicates that the heat flux is nearly independent of gravity, which is in contrast to the $g^{1/2}$ dependence of Equation 10.5. Above-normal gravitational forces show similar effects, although near the ONB, gravity can influence bubble-induced convection.

If liquid in a pool boiling system is maintained at a temperature that is less than the saturation temperature, the liquid is said to be *subcooled,* where $\Delta T_{\text{sub}} \equiv T_{\text{sat}} - T_l$. In the natural convection regime, the heat flux increases typically as $(T_s - T_l)^{5/4}$ or $(\Delta T_e + \Delta T_{\text{sub}})^{5/4}$. In contrast, for nucleate boiling, the influence of subcooling is considered to be negligible, although the maximum and minimum heat fluxes, q''_{max} and q''_{min}, are known to increase linearly with ΔT_{sub}. For film boiling, the heat flux increases strongly with increasing ΔT_{sub}.

The influence of *surface roughness* (by machining, grooving, scoring or sandblasting) on the maximum and minimum heat fluxes and on film boiling is negligible. However, as demonstrated by Berensen [19], increased surface roughness can cause a large increase in heat flux for the nucleate boiling regime. This influence can be appreciated by considering the Yamagata correlation of Equation 10.4. As Figure 10.7 illustrates, a roughened surface has numerous cavities that serve to trap vapor, providing more and larger sites for bubble growth. It follows that the nucleation site density for a rough surface can be substantially larger than that for a smooth surface. However, under prolonged boiling, the effects of surface roughness generally diminish, indicating

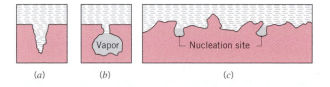

(a)   (b)   (c)

FIGURE 10.7 Formation of nucleation sites. (*a*) Wetted cavity with no trapped vapor. (*b*) Reentrant cavity with trapped vapor. (*c*) Enlarged profile of a roughened surface.

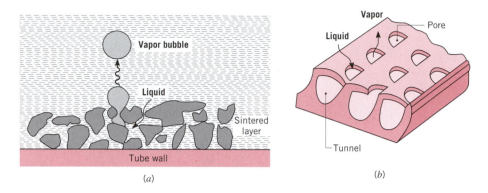

FIGURE 10.8 Typical structured enhancement surfaces for augmentation of nucleate boiling. (*a*) Sintered metallic coating. (*b*) Mechanically formed double-reentrant cavity.

that the new, large sites created by roughening are not stable sources of vapor entrapment.

Special surface arrangments that provide stable *augmentation* (*enhancement*) of nucleate boiling are available commercially and have been reviewed by Webb [20]. *Enhancement surfaces* are of two types: (1) coatings of very porous material formed by sintering, brazing, flame spraying, electrolytic deposition, or foaming, and (2) mechanically machined or formed double-reentrant cavities to ensure continuous vapor trapping (see Figure 10.8). Such surfaces provide for continuous renewal of vapor at the nucleation sites and heat transfer augmentation by more than an order of magnitude. Active augmentation techniques, such as surface wiping–rotation, surface vibration, fluid vibration, and electrostatic fields, have also been reviewed by Bergles [21, 22]. However, because such techniques complicate the boiling system and, in many instances, impair reliability, they have found little practical application.

EXAMPLE 10.1

The bottom of a copper pan, 0.3 m in diameter, is maintained at 118°C by an electric heater. Estimate the power required to boil water in this pan. What is the evaporation rate? Estimate the critical heat flux.

SOLUTION

Known: Water boiling in a copper pan of prescribed surface temperature.

Find:

1. Power required by electric heater to cause boiling.
2. Rate of water evaporation due to boiling.
3. Critical heat flux corresponding to the burnout point.

Schematic:

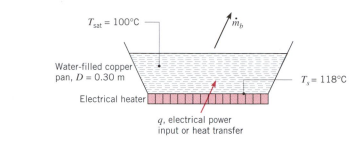

$T_{sat} = 100°C$

$\dot{m}_b$

Water-filled copper pan, $D = 0.30$ m

$T_s = 118°C$

Electrical heater

q, electrical power input or heat transfer

Assumptions:

1. Steady-state conditions.

2. Water exposed to standard atmospheric pressure, 1.01 bar.

3. Water at uniform temperature $T_{sat} = 100°C$.

4. Pan bottom surface of polished copper.

5. Negligible losses from heater to surroundings.

Properties: Table A.6, saturated water, liquid (100°C): $\rho_l = 1/v_f = 957.9$ kg/m³, $c_{p,l} = c_{p,f} = 4.217$ kJ/kg · K, $\mu_l = \mu_f = 279 \times 10^{-6}$ N · s/m², $Pr_l = Pr_f = 1.76$, $h_{fg} = 2257$ kJ/kg, $\sigma = 58.9 \times 10^{-3}$ N/m. Table A.6, saturated water, vapor (100°C): $\rho_v = 1/v_g = 0.5955$ kg/m³.

Analysis:

1. From knowledge of the saturation temperature T_{sat} of water boiling at 1 atm and the temperature of the heated copper surface T_s, the excess temperature ΔT_e is

$$\Delta T_e \equiv T_s - T_{sat} = 118°C - 100°C = 18°C$$

According to the boiling curve of Figure 10.4, nucleate pool boiling will occur and the recommended correlation for estimating the heat transfer rate per unit area of plate surface is given by Equation 10.5.

$$q_s'' = \mu_l h_{fg} \left[\frac{g(\rho_l - \rho_v)}{\sigma} \right]^{1/2} \left(\frac{c_{p,l} \Delta T_e}{C_{s,f} h_{fg} Pr_l^n} \right)^3$$

The values of $C_{s,f}$ and n corresponding to the polished copper surface–water combination are determined from the experimental results of Table 10.1, where $C_{s,f} = 0.0130$ and $n = 1.0$. Substituting numerical values, the boiling heat flux is

$$q_s'' = 279 \times 10^{-6}\, \text{N} \cdot \text{s/m}^2 \times 2257 \times 10^3\, \text{J/kg}$$

$$\times \left[\frac{9.8\, \text{m/s}^2\, (957.9 - 0.5955)\, \text{kg/m}^3}{58.9 \times 10^{-3}\, \text{N/m}} \right]^{1/2}$$

$$\times \left(\frac{4.217 \times 10^3\, \text{J/kg} \cdot \text{K} \times 18°C}{0.0130 \times 2257 \times 10^3\, \text{J/kg} \times 1.76} \right)^3 = 798\, \text{kW/m}^2$$

Hence the boiling heat transfer rate is

$$q_s = q_s'' \times A = q_s'' \times \frac{\pi D^2}{4}$$

$$q_s = 7.89 \times 10^5 \text{ W/m}^2 \times \frac{\pi (0.30 \text{ m})^2}{4} = 55.8 \text{ kW} \qquad \triangleleft$$

2. Under steady-state conditions all heat addition to the pan will result in water evaporation from the pan. Hence

$$q_s = \dot{m}_b h_{fg}$$

where $\dot{m}_b$ is the rate at which water evaporates from the free surface to the room. It follows that

$$\dot{m}_b = \frac{q_s}{h_{fg}} = \frac{5.58 \times 10^4 \text{ W}}{2257 \times 10^3 \text{ J/kg}} = 0.0247 \text{ kg/s} = 89 \text{ kg/h} \qquad \triangleleft$$

3. The critical heat flux for nucleate pool boiling can be estimated from Equation 10.7:

$$q_{\max}'' = 0.149 h_{fg} \rho_v \left[\frac{\sigma g (\rho_l - \rho_v)}{\rho_v^2} \right]^{1/4}$$

Substituting the appropriate numerical values,

$$q_{\max}'' = 0.149 \times 2257 \times 10^3 \text{ J/kg} \times 0.5955 \text{ kg/m}^3$$

$$\times \left[\frac{58.9 \times 10^{-3} \text{ N/m} \times 9.8 \text{ m/s}^2 \, (957.9 - 0.5955) \text{ kg/m}^3}{(0.5955)^2 (\text{kg/m}^3)^2} \right]^{1/4}$$

$$q_{\max}'' = 1.26 \text{ MW/m}^2 \qquad \triangleleft$$

Comments:

1. Note that the critical heat flux $q_{\max}'' = 1.26 \text{ MW/m}^2$ represents the maximum heat flux for boiling water at normal atmospheric pressure. Operation of the heater at $q_s'' = 0.789 \text{ MW/m}^2$ is therefore below the critical condition.

2. Using Equation 10.8, the minimum heat flux at the Leidenfrost point is $q_{\min}'' = 18.9 \text{ kW/m}^2$. Note from Figure 10.4 that, for this condition, $\Delta T_e \approx 120°C$.

EXAMPLE 10.2

A metal-clad heating element of 6-mm diameter and emissivity $\varepsilon = 1$ is horizontally immersed in a water bath. The surface temperature of the metal is 255°C under steady-state boiling conditions. Estimate the power dissipation per unit length of the heater.

SOLUTION

Known: Boiling from outer surface of horizontal cylinder in water.

Find: Power dissipation per unit length for the cylinder, q'_s.

Schematic:

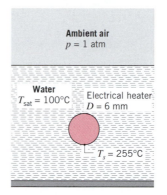

Assumptions:

1. Steady-state conditions.
2. Water exposed to standard atmospheric pressure and at uniform temperature T_{sat}.

Properties: Table A.6, saturated water, liquid (100°C): $\rho_l = 1/v_f = 957.9$ kg/m³, $h_{fg} = 2257$ kJ/kg. Table A.6, saturated water vapor ($T_f \approx 450$ K): $\rho_v = 1/v_g = 4.808$ kg/m³, $c_{p,v} = c_{p,g} = 2.56$ kJ/kg · K, $k_v = k_g = 0.0331$ W/m · K, $\mu_v = \mu_g = 14.85 \times 10^{-6}$ N · s/m².

Analysis: The excess temperature is

$$\Delta T_e = T_s - T_{sat} = 255°C - 100°C = 155°C$$

According to the boiling curve of Figure 10.4, film pool boiling conditions are achieved, in which case heat transfer is due to both convection and radiation. The heat transfer rate follows from Equation 10.3, written on a per unit length basis for a cylindrical surface of diameter D:

$$q'_s = q''_s \pi D = \bar{h} \pi D \, \Delta T_e$$

The heat transfer coefficient $\bar{h}$ is calculated from Equation 10.10a,

$$\bar{h}^{4/3} = \bar{h}^{4/3}_{conv} + \bar{h}_{rad}\bar{h}^{1/3}$$

where the convection and radiation heat transfer coefficients follow from Equations 10.9 and 10.11, respectively. For the convection coefficient:

$$\bar{h}_{conv} = 0.62 \left[\frac{k_v^3 \rho_v (\rho_l - \rho_v) g (h_{fg} + 0.8 c_{p,v} \, \Delta T_e)}{\mu_v D \, \Delta T_e} \right]^{1/4}$$

$$\bar{h}_{conv} = 0.62$$

$$\times \left[\frac{(0.0331)^3 (\text{W/m · K})^3 \times 4.808 \text{ kg/m}^3 (957.9 - 4.808) \text{ kg/m}^3 \times 9.8 \text{ m/s}^2}{1} \right.$$

$$\left. \times \frac{2257 \times 10^3 \text{ J/kg} + 0.8 \times 2.56 \times 10^3 \text{ J/kg · K} \times 155°C}{14.85 \times 10^{-6} \text{ N · s/m}^2 \times 6 \times 10^{-3} \text{ m} \times 155°C} \right]^{1/4}$$

$$\bar{h}_{conv} = 460 \text{ W/m}^2 · \text{K}$$

For the radiation heat transfer coefficient:

$$\bar{h}_{rad} = \frac{\varepsilon\sigma(T_s^4 - T_{sat}^4)}{T_s - T_{sat}}$$

$$\bar{h}_{rad} = \frac{5.67 \times 10^{-8}\ \text{W/m}^2 \cdot \text{K}^4\ (528^4 - 373^4)\ \text{K}^4}{(528 - 373)\ \text{K}} = 21.3\ \text{W/m}^2 \cdot \text{K}$$

Solving Equation 10.10a by trial and error,

$$\bar{h}^{4/3} = 453^{4/3} + 21.3\bar{h}^{1/3}$$

it follows that

$$\bar{h} = 476\ \text{W/m}^2 \cdot \text{K}$$

Hence the heat transfer rate per unit length of heater element is

$$q_s' = 476\ \text{W/m}^2 \cdot \text{K} \times \pi \times 6 \times 10^{-3}\ \text{m} \times 155°\text{C} = 1.39\ \text{kW/m} \qquad \triangleleft$$

Comments: Equation 10.10b is appropriate for estimating $\bar{h}$ from values of $\bar{h}_{conv}$ and $\bar{h}_{rad}$ and gives the identical numerical result.

10.5
Forced Convection Boiling

In *pool boiling* fluid flow is due primarily to the buoyancy-driven motion of bubbles originating from the heated surface. In contrast, for *forced convection boiling,* flow is due to a directed (bulk) motion of the fluid, as well as to buoyancy effects. Conditions depend strongly on geometry, which may involve *external* flow over heated plates and cylinders or *internal* (duct) flow. Internal, forced convection boiling is commonly referred to as *two-phase flow* and is characterized by rapid changes from liquid to vapor in the flow direction.

10.5.1 External Forced Convection Boiling

For external flow over a heated plate, the heat flux can be estimated by standard forced convection correlations up to the inception of boiling. As the temperature of the heated plate is increased, nucleate boiling will occur, causing the heat flux to increase. If vapor generation is not extensive and the liquid is subcooled, Bergles and Rohsenow [7, 23] suggest a method for estimating the total heat flux in terms of components associated with pure forced convection and pool boiling.

Both forced convection and subcooling are known to increase the critical heat flux q_{max}'' for nucleate boiling. Experimental values as high as 35 MW/m² (compared with 1.3 MW/m² for pool boiling of water at 1 atm) have been reported [24]. For a liquid of velocity V moving in cross flow over a cylinder of diameter D, Lienhard and Eichhorn [25] have developed the following expressions for low- and high-velocity flows.

Low Velocity

$$\frac{q''_{\max}}{\rho_v h_{fg} V} = \frac{1}{\pi} \left[1 + \left(\frac{4}{We_D} \right)^{1/3} \right] \qquad (10.12)$$

High Velocity

$$\frac{q''_{\max}}{\rho_v h_{fg} V} = \frac{(\rho_l/\rho_v)^{3/4}}{169\pi} + \frac{(\rho_l/\rho_v)^{1/2}}{19.2\pi \, We_D^{1/3}} \qquad (10.13)$$

The Weber number We_D is the ratio of inertia to surface tension forces and has the form

$$We_D \equiv \frac{\rho_v V^2 D}{\sigma} \qquad (10.14)$$

The high- and low-velocity regions, respectively, are determined by whether the heat flux parameter $q''_{\max}/\rho_v h_{fg} V$ is less than or greater than $[(0.275/\pi) (\rho_l/\rho_v)^{1/2} + 1]$. In most cases, Equations 10.12 and 10.13 correlate $q''_{\max}$ data within 20%.

10.5.2   Two-Phase Flow

Internal forced convection boiling is associated with bubble formation at the inner surface of a heated tube through which a liquid is flowing. Bubble growth and separation are strongly influenced by the flow velocity, and hydrodynamic effects differ significantly from those corresponding to pool boiling. The process is complicated by the existence of different two-phase flow patterns that preclude the development of generalized theories.

Consider flow development in the vertical heated tube of Figure 10.9. Heat transfer to the subcooled liquid that enters the tube is initially by forced convection and may be predicted from the correlations of Chapter 8. However, once boiling is initiated, bubbles that appear at the surface grow and are carried into the mainstream of the liquid. There is a sharp increase in the convection heat transfer coefficient associated with this *bubbly flow regime*. As the volume fraction of the vapor increases, individual bubbles coalesce to form slugs of vapor. This *slug-flow regime* is followed by an *annular-flow* regime in which the liquid forms a film. This film moves along the inner surface, while vapor moves at a larger velocity through the core of the tube. The heat transfer coefficient continues to increase through the *bubbly flow* and much of the *annular-flow* regimes. However, dry spots eventually appear on the inner surface, at which point the convection coefficient begins to decrease. The *transition regime* is characterized by the growth of the dry spots, until the surface is completely dry and all remaining liquid is in the form of droplets appearing in the vapor core. The convection coefficient continues to decrease through this regime. There is little change in this coefficient through the *mist-flow* regime, which persists until all the droplets are converted to vapor. The vapor is then *superheated* by forced convection from the surface.

Discussion of the many correlations that have been developed to quantify the foregoing two-phase flow phenomena is left to the literature [7, 26–29].

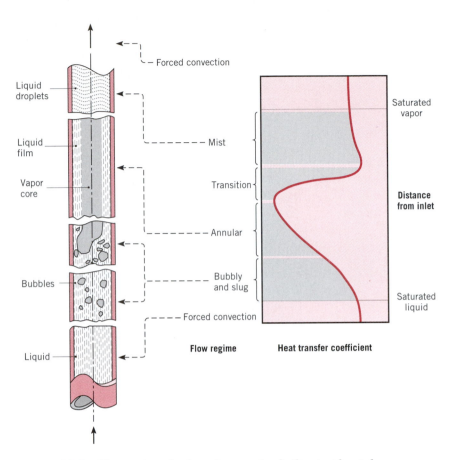

FIGURE 10.9   Flow regimes for forced convection boiling inside a tube.

10.6
Condensation: Physical Mechanisms

Condensation occurs when the temperature of a vapor is reduced below its saturation temperature. In industrial equipment, the process commonly results from contact between the vapor and a cool *surface* (Figures 10.10*a*, *b*). The latent energy of the vapor is released, heat is transferred to the surface, and the condensate is formed. Other common modes are *homogeneous* condensation (Figure 10.10*c*), where vapor condenses out as droplets suspended in a gas phase to form a fog, and *direct contact* condensation (Figure 10.10*d*), which occurs when vapor is brought into contact with a cold liquid. In this chapter we will consider only surface condensation.

As shown in Figures 10.10*a*, *b*, condensation may occur in one of two ways, depending on the condition of the surface. The dominant form of condensation is one in which a liquid film covers the entire condensing surface, and under the action of gravity the film flows continuously from the surface. *Film condensation* is generally characteristic of clean, uncontaminated surfaces. However, if the surface is coated with a substance that inhibits wetting, it is possible to maintain *dropwise condensation*. The drops form in cracks, pits, and

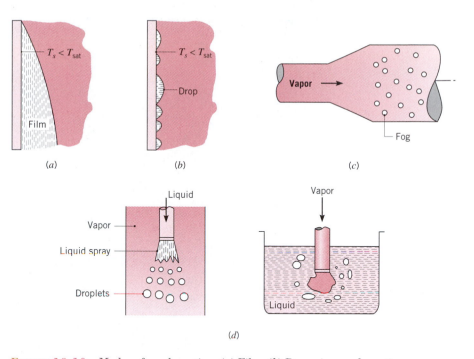

FIGURE 10.10   Modes of condensation. (*a*) Film. (*b*) Dropwise condensation on a surface. (*c*) Homogeneous condensation or fog formation resulting from increased pressure due to expansion. (*d*) Direct contact condensation.

cavities on the surface and may grow and coalesce through condensation. Typically, more than 90% of the surface is covered by drops, ranging from a few micrometers in diameter to agglomerations visible to the naked eye. The droplets flow from the surface due to the action of gravity. Film and dropwise condensation of steam on a vertical copper surface are shown in Figure 10.11. A thin coating of cupric oleate was applied to the left-hand portion of the surface to promote the dropwise condensation. A thermocouple probe of 1-mm diameter extends across the photograph.

Regardless of whether it is in the form of a film or droplets, the condensate provides a resistance to heat transfer between the vapor and the surface. Because this resistance increases with condensate thickness, which increases in the flow direction, it is desirable to use short vertical surfaces or horizontal cylinders in situations involving film condensation. Most condensers therefore consist of horizontal tube bundles through which a liquid coolant flows and around which the vapor to be condensed is circulated. In terms of maintaining high condensation and heat transfer rates, droplet formation is superior to film formation. In dropwise condensation most of the heat transfer is through drops of less than 100-μm diameter, and transfer rates that are more than an order of magnitude larger than those associated with film condensation may be achieved. It is therefore common practice to use surface coatings that inhibit wetting, and hence stimulate dropwise condensation. Silicones, Teflon, and an assortment of waxes and fatty acids are often used for this purpose. However, such coatings gradually lose their effectiveness due to oxidation, fouling, or outright removal, and film condensation eventually occurs.

Although it is desirable to achieve dropwise condensation in industrial ap-

(a) (b)

FIGURE 10.11   Condensation on a vertical surface. (*a*) Dropwise. (*b*) Film. Photograph courtesy of Professor J. W. Westwater, University of Illinois at Champaign-Urbana.

plications, it is often difficult to maintain this condition. For this reason and because the convection coefficients for film condensation are smaller than those for the dropwise case, condenser design calculations are often based on the assumption of film condensation. In the remaining sections of this chapter, we focus on film condensation and mention only briefly the limited results available for dropwise condensation.

10.7
Laminar Film Condensation on a Vertical Plate

As shown in Figure 10.12, there may be several complicating features associated with film condensation. The film originates at the top of the plate and flows downward under the influence of gravity. The thickness δ and the condensate mass flow rate $\dot{m}$ increase with increasing x because of continuous condensation at the liquid–vapor interface, which is at T_{sat}. There is then heat transfer from this interface through the film to the surface, which is maintained at $T_s < T_{\text{sat}}$. In the most general case the vapor may be superheated $(T_{v,\infty} > T_{\text{sat}})$ and may be part of a mixture containing one or more noncondensable gases.

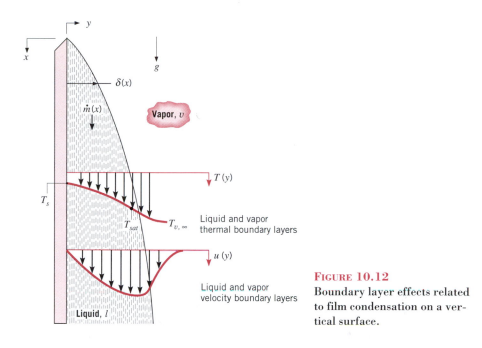

FIGURE 10.12
Boundary layer effects related to film condensation on a vertical surface.

Moreover, there exists a finite shear stress at the liquid–vapor interface, contributing to a velocity gradient in the vapor, as well as in the film [30, 31].

Despite the complexities associated with film condensation, useful results may be obtained by making assumptions that originated with an analysis by Nusselt [32].

1. Laminar flow and constant properties are assumed for the liquid film.

2. The gas is assumed to be a pure vapor and at a uniform temperature equal to T_{sat}. With no temperature gradient in the vapor, heat transfer to the liquid–vapor interface can occur only by condensation at the interface and not by conduction from the vapor.

3. The shear stress at the liquid–vapor interface is assumed to be negligible, in which case $\partial u/\partial y|_{y=\delta} = 0$. With this assumption and the foregoing assumption of a uniform vapor temperature, there is no need to consider the vapor velocity or thermal boundary layers shown in Figure 10.12.

4. Momentum and energy transfers by advection in the condensate film are assumed to be negligible. This assumption is reasonable by virtue of the low velocities associated with the film. It follows that heat transfer across the film occurs only by conduction, in which case the liquid temperature distribution is linear.

Film conditions resulting from the assumptions are shown in Figure 10.13.

From the fourth approximation, momentum advection terms may be neglected, and the x-momentum equation (Chapter 6) may be expressed as

$$\frac{\partial^2 u}{\partial y^2} = \frac{1}{\mu_l}\frac{dp}{dx} - \frac{X}{\mu_l} \tag{10.15}$$

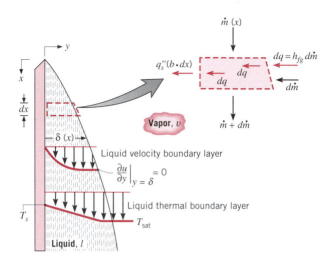

FIGURE 10.13   Boundary layer conditions associated with Nusselt's analysis for a vertical plate of width b.

where the body force X has been retained. The body force within the film is equal to $\rho_l g$, and the pressure gradient may be approximated in terms of conditions outside the film. That is, invoking the boundary layer approximation that $(\partial p/\partial y) \approx 0$, it follows that $(dp/dx) \approx \rho_v g$. The momentum equation may therefore be expressed as

$$\frac{\partial^2 u}{\partial y^2} = -\frac{g}{\mu_l}(\rho_l - \rho_v) \qquad (10.16)$$

Integrating twice and applying boundary conditions of the form $u(0) = 0$ and $\partial u/\partial y|_{y=\delta} = 0$, the velocity profile in the film becomes

$$u(y) = \frac{g(\rho_l - \rho_v)\delta^2}{\mu_l}\left[\frac{y}{\delta} - \frac{1}{2}\left(\frac{y}{\delta}\right)^2\right] \qquad (10.17)$$

From this result the condensate mass flow rate per unit width $\Gamma(x)$ may be obtained in terms of an integral involving the velocity profile:

$$\frac{\dot{m}(x)}{b} = \int_0^{\delta(x)} \rho_l u(y)\, dy \equiv \Gamma(x) \qquad (10.18)$$

Substituting from Equation 10.17, it follows that

$$\Gamma(x) = \frac{g\rho_l(\rho_l - \rho_v)\delta^3}{3\mu_l} \qquad (10.19)$$

The specific variation with x of δ, and hence of Γ, may be obtained by first applying the conservation of energy requirement to the differential element shown in Figure 10.13. At a portion of the liquid–vapor interface of unit width and length dx, the rate of heat transfer into the film, dq, must equal the rate of energy release due to condensation at the interface. Hence

$$dq = h_{fg}\, d\dot{m} \qquad (10.20)$$

Since advection is neglected, it also follows that the rate of heat transfer across the interface must equal the rate of heat transfer to the surface. Hence

$$dq = q_s''(b \cdot dx) \tag{10.21}$$

Since the liquid temperature distribution is linear, Fourier's law may be used to express the surface heat flux as

$$q_s'' = \frac{k_l(T_{sat} - T_s)}{\delta} \tag{10.22}$$

Combining Equations 10.18 and 10.20 to 10.22, we then obtain

$$\frac{d\Gamma}{dx} = \frac{k_l(T_{sat} - T_s)}{\delta h_{fg}} \tag{10.23}$$

Differentiating Equation 10.19, we also obtain

$$\frac{d\Gamma}{dx} = \frac{g\rho_l(\rho_l - \rho_v)\delta^2}{\mu_l} \frac{d\delta}{dx} \tag{10.24}$$

Combining Equations 10.23 and 10.24, it follows that

$$\delta^3 \, d\delta = \frac{k_l\mu_l(T_{sat} - T_s)}{g\rho_l(\rho_l - \rho_v)h_{fg}} \, dx$$

Integrating from $x = 0$, where $\delta = 0$, to any x location of interest on the surface,

$$\delta(x) = \left[\frac{4k_l\mu_l(T_{sat} - T_s)x}{g\rho_l(\rho_l - \rho_v)h_{fg}} \right]^{1/4} \tag{10.25}$$

This result may then be substituted into Equation 10.19 to obtain $\Gamma(x)$.

An improvement to the foregoing result for $\delta(x)$ was made by Nusselt [32] and Rohsenow [33], who showed that, with the inclusion of thermal advection effects, a term is added to the latent heat of vaporization. In lieu of h_{fg}, Rohsenow recommended using a modified latent heat of the form $h_{fg}' = h_{fg} + 0.68c_{p,l}(T_{sat} - T_s)$, or in terms of the Jakob number,

$$h_{fg}' = h_{fg}(1 + 0.68Ja) \tag{10.26}$$

More recently, Sadasivan and Lienhard [17] have shown that the modified latent heat depends weakly on the Prandtl number of the liquid.

The surface heat flux may be expressed as

$$q_s'' = h_x(T_{sat} - T_s) \tag{10.27}$$

Substituting from Equation 10.22, the local convection coefficient is then

$$h_x = \frac{k_l}{\delta} \tag{10.28}$$

or, from Equation 10.25, with h_{fg} replaced by h_{fg}',

$$h_x = \left[\frac{g\rho_l(\rho_l - \rho_v)k_l^3 h_{fg}'}{4\mu_l(T_{sat} - T_s)x} \right]^{1/4} \tag{10.29}$$

Since h_x depends on $x^{-1/4}$, it follows that the average convection coefficient for the entire plate is

$$\overline{h}_L = \frac{1}{L} \int_0^L h_x \, dx = \tfrac{4}{3} h_L$$

or

$$\overline{h}_L = 0.943 \left[\frac{g \rho_l (\rho_l - \rho_v) k_l^3 h'_{fg}}{\mu_l (T_{\text{sat}} - T_s) L} \right]^{1/4} \tag{10.30}$$

The average Nusselt number then has the form

$$\overline{Nu}_L = \frac{\overline{h}_L L}{k_l} = 0.943 \left[\frac{\rho_l g (\rho_l - \rho_v) h'_{fg} L^3}{\mu_l k_l (T_{\text{sat}} - T_s)} \right]^{1/4} \tag{10.31}$$

In using this equation all liquid properties should be evaluated at the film temperature $T_f = (T_{\text{sat}} + T_s)/2$, and h_{fg} should be evaluated at T_{sat}.

A more detailed boundary layer analysis of film condensation on a vertical plate has been performed by Sparrow and Gregg [30]. Their results, confirmed by Chen [34], indicate that errors associated with using Equation 10.31 are less than 3% for $Ja \leq 0.1$ and $1 \leq Pr \leq 100$. Dhir and Lienhard [35] have also shown that Equation 10.31 may be used for inclined plates, if g is replaced by $g \cdot \cos \theta$, where θ is the angle between the vertical and the surface. However, it must be used with caution for large values of θ and does not apply if $\theta = \pi/2$. The expression may be used for condensation on the inner or outer surface of a vertical tube of radius R, if $R \gg \delta$.

The total heat transfer to the surface may be obtained by using Equation 10.30 with the following form of Newton's law of cooling:

$$q = \overline{h}_L A (T_{\text{sat}} - T_s) \tag{10.32}$$

The total condensation rate may then be determined from the relation

$$\dot{m} = \frac{q}{h'_{fg}} = \frac{\overline{h}_L A (T_{\text{sat}} - T_s)}{h'_{fg}} \tag{10.33}$$

Equations 10.32 and 10.33 are generally applicable to any surface geometry, although the form of $\overline{h}_L$ will vary according to geometry and flow conditions.

10.8
Turbulent Film Condensation

As for all previously discussed convection phenomena, turbulent flow conditions may exist in film condensation. Consider the vertical surface of Figure 10.14a. The transition criterion may be expressed in terms of a Reynolds number defined as

$$Re_\delta \equiv \frac{4\Gamma}{\mu_l} \tag{10.34}$$

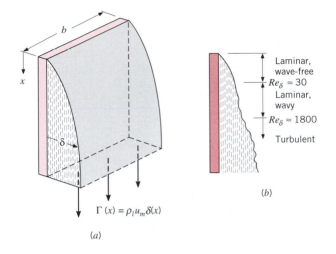

FIGURE 10.14 Film condensation on a vertical plate.
(*a*) Condensate rate for plate of width *b*. (*b*) Flow regimes.

With the condensate mass flow rate given by $\dot{m} = \rho_l u_m b \delta$, the Reynolds number may be expressed as

$$Re_\delta = \frac{4\dot{m}}{\mu_l b} = \frac{4\rho_l u_m \delta}{\mu_l} \tag{10.35}$$

where u_m is the average velocity in the film and δ, the film thickness, is the characteristic length. As in the case of single-phase boundary layers, the Reynolds number is an indicator of flow conditions. As shown in Figure 10.14*b*, for $Re_\delta \lesssim 30$, the film is laminar and wave free. For increased Re_δ, ripples or waves form on the condensate film, and at $Re_\delta \approx 1800$ the transition from laminar to turbulent flow is complete.

For the wave-free laminar region ($Re_\delta \lesssim 30$), Equations 10.34 and 10.19 may be combined to yield

$$Re_\delta = \frac{4g\rho_l(\rho_l - \rho_v)\delta^3}{3\mu_l^2} \tag{10.36}$$

Assuming $\rho_l \gg \rho_v$ and substituting first from Equation 10.25 and then from (10.30), Equation 10.36 can be expressed in terms of a modified Nusselt number:

$$\frac{\bar{h}_L(\nu_l^2/g)^{1/3}}{k_l} = 1.47 Re_\delta^{-1/3} \qquad Re_\delta \lesssim 30 \tag{10.37}$$

In the laminar wavy region, Kutateladze [36] recommends a correlation of the form

$$\frac{\bar{h}_L(\nu_l^2/g)^{1/3}}{k_l} = \frac{Re_\delta}{1.08 Re_\delta^{1.22} - 5.2} \qquad 30 \lesssim Re_\delta \lesssim 1800 \tag{10.38}$$

and for the turbulent region, Labuntsov [37] recommends

$$\frac{\bar{h}_L(\nu_l^2/g)^{1/3}}{k_l} = \frac{Re_\delta}{8750 + 58Pr^{-0.5}(Re_\delta^{0.75} - 253)} \qquad Re_\delta \gtrsim 1800 \tag{10.39}$$

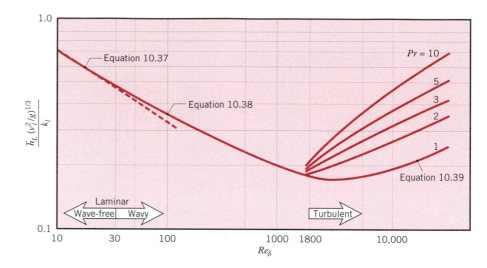

FIGURE 10.15   Modified Nusselt number for condensation on a vertical plate.

Graphical representation of the foregoing correlations is provided in Figure 10.15, and the trends have been verified experimentally by Gregorig et al. [38] for water over the range $1 < Re_\delta < 7200$.

EXAMPLE 10.3

The outer surface of a vertical tube, which is 1 m long and has an outer diameter of 80 mm, is exposed to saturated steam at atmospheric pressure and is maintained at 50°C by the flow of cool water through the tube. What is the rate of heat transfer to the coolant, and what is the rate at which steam is condensed at the surface?

SOLUTION

Known: Dimensions and temperature of a vertical tube experiencing condensation of steam at its outer surface.

Find: Heat transfer and condensation rates.

Schematic:

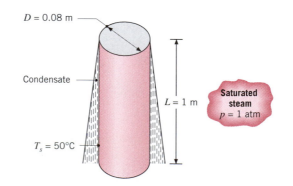

Assumptions: Laminar film condensation on a vertical surface.

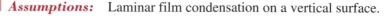

Properties: Table A.6, saturated vapor ($p = 1.0133$ bars): $T_{sat} = 100°C$, $\rho_v = (1/v_g) = 0.596$ kg/m³, $h_{fg} = 2257$ kJ/kg. Table A.6, saturated liquid ($T_f = 75°C$): $\rho_l = (1/v_f) = 975$ kg/m³, $\mu_l = 375 \times 10^{-6}$ N · s/m², $k_l = 0.668$ W/m · K, $c_{p,l} = 4193$ J/kg · K.

Analysis: The heat transfer rate may be determined from Equation 10.32, where $A = \pi D L$. Hence

$$q = \bar{h}_L(\pi D L)(T_{sat} - T_s)$$

Having assumed laminar film condensation on a vertical surface, it follows from Equation 10.30 that

$$\bar{h}_L = 0.943 \left[\frac{g\rho_l(\rho_l - \rho_v)k_l^3 h'_{fg}}{\mu_l(T_{sat} - T_s)L} \right]^{1/4}$$

With

$$Ja = \frac{c_{p,l}(T_{sat} - T_s)}{h_{fg}} = \frac{4193 \text{ J/kg} \cdot \text{K} (100 - 50) \text{ K}}{2257 \text{ kJ/kg}} = 0.0929$$

it follows that

$$h'_{fg} = h_{fg}(1 + 0.68Ja) = 2257 \text{ kJ/kg} (1.0632) = 2400 \text{ kJ/kg}$$

Hence

$\bar{h}_L = 0.943$

$\times [\{9.8 \text{ m/s}^2 \times 975 \text{ kg/m}^3 (975 - 0.596) \text{ kg/m}^3 (0.668 \text{ W/m} \cdot \text{K})^3 2.4$

$\times 10^6 \text{ J/kg}\}/\{375 \times 10^{-6} \text{ kg/s} \cdot \text{m} (100 - 50) \text{ K} \times 1 \text{ m}\}]^{1/4}$

$\bar{h}_L = 4094 \text{ W/m}^2 \cdot \text{K}$

and

$$q = 4094 \text{ W/m}^2 \cdot \text{K} \times \pi \times 0.08 \text{ m} \times 1 \text{ m} (100 - 50) \text{ K} = 51{,}446 \text{ W} \qquad \triangleleft$$

From Equation 10.33 the condensation rate is then

$$\dot{m} = \frac{q}{h'_{fg}} = \frac{51{,}446 \text{ W}}{2.4 \times 10^6 \text{ J/kg}} = 0.0214 \text{ kg/s} \qquad \triangleleft$$

The assumption of laminar film conditions may be checked by calculating Re_δ from Equation 10.35:

$$Re_\delta = \frac{4\dot{m}}{\mu_l b} = \frac{4 \times 0.0214 \text{ kg/s}}{375 \times 10^{-6} \text{ kg/s} \cdot \text{m} \, \pi (0.08 \text{ m})} = 908$$

Since $30 < Re_\delta < 1800$, a significant portion of the condensate is in the wavy-laminar region. Hence the wave-free laminar assumption may be poor.

For the wavy-laminar region, Equation 10.38 may be used with the specified surface dimensions and temperature. Expressing $\bar{h}_L$ in terms of the Reynolds number by combining Equations 10.33 and 10.35,

$$\bar{h}_L = \frac{\dot{m}h'_{fg}}{A(T_{sat} - T_s)} = \frac{Re_\delta(\mu_l b)h'_{fg}}{4A(T_{sat} - T_s)}$$

Equation 10.38 may be expressed as

$$\frac{Re_\delta \mu_l b h'_{fg}}{4A(T_{sat} - T_s)} = \frac{Re_\delta}{1.08 Re_\delta^{1.22} - 5.2} \cdot \frac{k_l}{(v_l^2/g)^{1/3}}$$

which can be solved to obtain a Reynolds number of $Re_\delta = 1173$. It follows from Equations 10.38, 10.32, and 10.33, respectively, that $\bar{h}_L = 5285$ W/m² · K, $q = 66{,}418$ W, and $\dot{m} = 0.0276$ kg/s. Hence estimates of the heat and condensate rates based on the wavy-laminar correlation exceed those based on Nusselt's analysis by 29%.

Note that using Equation 10.25, with the corrected latent heat, the film thickness at the bottom of the tube $\delta(L)$ for the wave-free laminar assumption is

$$\delta(L) = \left[\frac{4k_l\mu_l(T_{\text{sat}} - T_s)L}{g\rho_l(\rho_l - \rho_v)h'_{fg}} \right]^{1/4}$$

$$\delta(L) = \left[\frac{4 \times 0.668 \ \text{W/m} \cdot \text{K} \times 375 \times 10^{-6} \ \text{kg/s} \cdot \text{m} \ (100 - 50) \ \text{K} \times 1 \ \text{m}}{9.8 \ \text{m/s}^2 \times 975 \ \text{kg/m}^3 \ (975 - 0.596) \ \text{kg/m}^3 \times 2.4 \times 10^6 \ \text{J/kg}} \right]^{1/4}$$

$$\delta(L) = 2.18 \times 10^{-4} \ \text{m} = 0.218 \ \text{mm}$$

Hence $\delta(L) \ll (D/2)$, and use of the vertical plate correlation for a vertical cylinder is justified.

Comments: The condensation heat and mass rates may be increased by reducing the temperature of the water flowing through the tube. For $10 \le T_s \le 50°$C, the calculations yield the following variations:

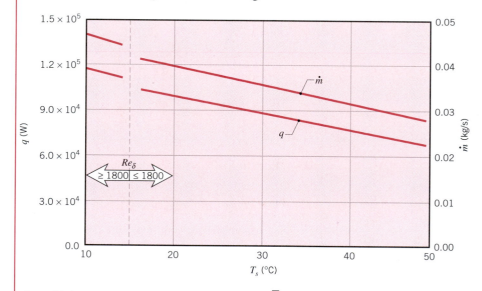

for which $1969 \ge Re_\delta \ge 1173$ and $5155 \le \bar{h}_L \le 5285$ W/m² · K. Due to an increase in the condensation rate, Re_δ increases with decreasing T_s. However, a corresponding increase in the thickness of the film causes a slight reduction in the average convection coefficient. The foregoing calculations were performed by using the wavy-laminar correlation, Equation 10.38, under conditions for which $Re_\delta < 1800$ ($T_s > 15°$C) and the turbulent correlation, Equation 10.39, for $Re_\delta > 1800$. Note, however, that the correlations do not provide equivalent results at $Re_\delta = 1800$. Moreover, there is a narrow Reynolds number range about 1800 for which values of Re_δ computed from Equation 10.38 slightly exceed 1800, while values of Re_δ computed from Equation 10.39 are slightly less than 1800.

10.9
Film Condensation on Radial Systems

The Nusselt analysis may be extended to laminar film condensation on the outer surface of a sphere and a horizontal tube (Figures 10.16*a*, *b*), and the average convection coefficient may be expressed as

$$\bar{h}_D = C \left[\frac{g\rho_l(\rho_l - \rho_v)k_l^3 h'_{fg}}{\mu_l(T_{\text{sat}} - T_s)D} \right]^{1/4} \tag{10.40}$$

where $C = 0.826$ for the sphere [39] and 0.729 for the tube [35].

For a vertical tier of N horizontal tubes, Figure 10.16*c*, the average convection coefficient (over the N tubes) may be expressed as

$$\bar{h}_{D,N} = 0.729 \left[\frac{g\rho_l(\rho_l - \rho_v)k_l^3 h'_{fg}}{N\mu_l(T_{\text{sat}} - T_s)D} \right]^{1/4} \tag{10.41}$$

That is, $\bar{h}_{D,N} = \bar{h}_D N^{-1/4}$, where $\bar{h}_D$ is the heat transfer coefficient for the first (upper) tube. Such an arrangement is often used in condenser design. The reduction in $\bar{h}$ with increasing N may be attributed to an increase in the average film thickness for each successive tube. Equations 10.40 and 10.41 are generally in agreement with or slightly lower than experimental results for pure vapors. Departures may be attributed to ripples in the liquid surface for the single

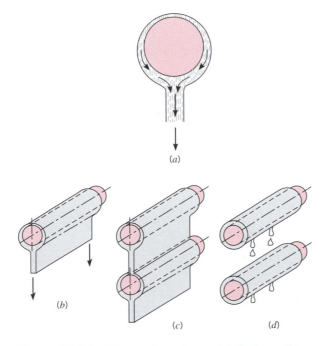

FIGURE 10.16   Film condensation on (*a*) a sphere, (*b*) a single horizontal tube, (*c*) a vertical tier of horizontal tubes with a continuous condensate sheet, and (*d*) with dripping condensate.

horizontal tube. For the tube bank, it is assumed that the condensate falls in a continuous sheet (Figure 10.16c) and two effects are neglected: heat transfer to the condensate sheet between the tubes and momentum gain as the sheet falls freely under gravity. These effects enhance heat transfer, and Chen [40] has accounted for their influence in terms of the Jakob number and the number of tubes. For $Ja < 0.1$, however, heat transfer is enhanced by less than 15%. Despite this correction, experimental results tend to be higher than the predictions. A plausible explanation for the discrepancy is that, rather than flow as a continuous sheet, the condensate drips from tube to tube, as illustrated in Figure 10.16d. Dripping reduces the sheet thickness and promotes turbulence, thereby enhancing heat transfer.

If the length-to-diameter ratio exceeds 1.8 tan θ [41], the foregoing equations may be applied to inclined tubes by replacing g with $g \cos \theta$, where the angle θ is measured from the horizontal position. In the presence of noncondensable gases, however, the convection coefficient will be less than predictions based on the foregoing correlations.

EXAMPLE 10.4

A steam condenser consists of a square array of 400 tubes, each 6 mm in diameter. If the tubes are exposed to saturated steam at a pressure of 0.15 bar and the tube surface temperature is maintained at 25°C, what is the rate at which steam is condensed per unit length of the tubes?

SOLUTION

Known: Configuration and surface temperature of condenser tubes exposed to saturated steam at 0.15 bar.

Find: Condensation rate per unit length of tubes.

Schematic:

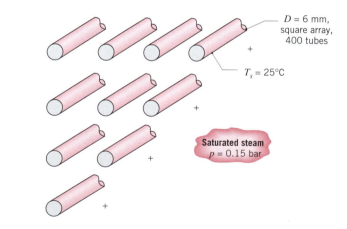

$D = 6$ mm, square array, 400 tubes

$T_s = 25°C$

Saturated steam $p = 0.15$ bar

Assumptions:

1. Negligible concentration of noncondensable gases in the steam.
2. Laminar film condensation on the tubes.

Properties: Table A.6, saturated vapor ($p = 0.15$ bar): $T_{sat} = 327$ K $= 54°C$, $\rho_v = (1/v_g) = 0.098$ kg/m³, $h_{fg} = 2373$ kJ/kg. Table A.6, saturated water ($T_f = 312.5$ K): $\rho_l = (1/v_f) = 992$ kg/m³, $\mu_l = 663 \times 10^{-6}$ N · s/m², $k_l = 0.631$ W/m · K, $c_{p,l} = 4178$ J/kg · K.

Analysis: The *average* condensation rate for a single tube of the array may be obtained from Equation 10.33, where for a unit length of the tube,

$$\dot{m}_1' = \frac{q_1'}{h_{fg}'} = \frac{\overline{h}_{D,N}(\pi D)(T_{sat} - T_s)}{h_{fg}'}$$

From Equation 10.41

$$\overline{h}_{D,N} = 0.729 \left[\frac{g\rho_l(\rho_l - \rho_v)k_l^3 h_{fg}'}{N\mu_l(T_{sat} - T_s)D} \right]^{1/4}$$

or with $N = 20$, $Ja = 0.051$, and $h_{fg}' = 2455$ kJ/kg,

$$\overline{h}_{D,N} = 0.729\{[9.8 \text{ m/s}^2 \times 992 \text{ kg/m}^3 (992 - 0.098) \text{ kg/m}^3$$

$$\times (0.631 \text{ W/m · K})^3 \, 2.455 \times 10^6 \text{ J/kg}]$$

$$\div [20 \times 663 \times 10^{-6} \text{ kg/s · m } (54 - 25) \text{ K} \times 0.006 \text{ m}]\}^{1/4}$$

$$\overline{h}_{D,N} = 5188 \text{ W/m}^2 \cdot \text{K}$$

Hence the average condensation rate for a single tube is

$$\dot{m}_1' = \frac{5188 \text{ W/m}^2 \cdot \text{K}(\pi \times 0.006 \text{ m})(54 - 25) \text{ K}}{2.455 \times 10^6 \text{ J/kg}} = 1.16 \times 10^{-3} \text{ kg/s · m}$$

For the complete array, the condensation rate per unit length is then

$$\dot{m}' = N^2 \dot{m}_1' = 400 \times 1.16 \times 10^{-3} \text{ kg/s · m} = 0.464 \text{ kg/s · m} \quad \triangleleft$$

Comments: Since $Ja < 0.1$, Equation 10.41 provides a reliable estimate of the average heat transfer coefficient.

10.10
Film Condensation in Horizontal Tubes

Condensers used for refrigeration and air-conditioning systems generally involve vapor condensation inside horizontal or vertical tubes. Conditions within the tube are complicated and depend strongly on the velocity of the vapor flowing through the tube. If this velocity is small, condensation occurs in the manner depicted by Figure 10.17a for a horizontal tube. That is, the condensate flow is from the upper portion of the tube to the bottom, from whence it flows in a longitudinal direction with the vapor. For low vapor velocities such that

$$Re_{v,i} = \left(\frac{\rho_v u_{m,v} D}{\mu_v} \right)_i < 35,000$$

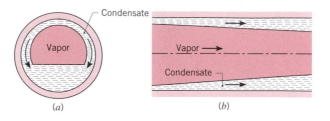

FIGURE 10.17   Film condensation in a horizontal tube. (*a*) Cross section of condensate flow for low vapor velocities. (*b*) Longitudinal section of condensate flow for large vapor velocities.

where *i* refers to the tube inlet, Chato [42] recommends an expression of the form

$$\bar{h}_D = 0.555 \left[\frac{g\rho_l(\rho_l - \rho_v)k_l^3 h'_{fg}}{\mu_l(T_{\text{sat}} - T_s)D} \right]^{1/4} \tag{10.42}$$

where, for this case, the modified latent heat is

$$h'_{fg} \equiv h_{fg} + \tfrac{3}{8}c_{p,\,l}(T_{\text{sat}} - T_s) \tag{10.43}$$

At higher vapor velocities the two-phase flow regime becomes annular (Figure 10.17*b*). The vapor occupies the core of the annulus, diminishing in diameter as the thickness of the outer condensate layer increases in the flow direction. Results for this flow condition are provided by Rohsenow [43].

10.11
Dropwise Condensation

Typically, heat transfer coefficients for dropwise condensation are an order of magnitude larger than those for film condensation. In fact, in heat exchanger applications for which dropwise condensation is promoted, other thermal resistances may be significantly larger than that due to condensation and, therefore, reliable correlations for the condensation process are not needed.

Of the many surface–fluid systems studied [44], most of the data are for steam condensation on well-promoted copper surfaces and are correlated by an expression of the form [45]

$$\bar{h}_{dc} = 51{,}104 + 2044T_{\text{sat}} \qquad 22°C < T_{\text{sat}} < 100°C \tag{10.44}$$

$$\bar{h}_{dc} = 255{,}510 \qquad\qquad 100°C < T_{\text{sat}} \tag{10.45}$$

where the heat transfer coefficient has units of (W/m² · K). The effect of subcooling, $T_{\text{sat}} - T_s$, on $\bar{h}_{dc}$ is small and may be neglected.

The effect of noncondensable vapors in the steam can be very important and has been studied by Shade and Mikic [46]. In addition, if the condensing surface material does not conduct as well as copper or silver, its thermal resistance becomes a factor. Since all the heat is transferred to the drops, which are very small and widely distributed over the surface, heat flow lines within the surface material near the active areas of condensation will *crowd,* inducing a *constriction* resistance. This effect has been studied by Hannemann and Mikic [47].

10.12
Summary

It is apparent that boiling and condensation are complicated processes for which the existence of generalized relations is somewhat limited. This chapter identifies the essential physical features of the processes and presents correlations suitable for approximate engineering calculations. However, a great deal of additional information is available, and much of it has been summarized in several extensive reviews of the subject [7, 14, 24, 26–29, 43, 45, 48–51].

References

1. Fox, R. W., and A. T. McDonald, *Introduction to Fluid Mechanics,* 3rd ed. Wiley, New York, 1985.

2. Nukiyama, S., "The Maximum and Minimum Values of Heat Transmitted from Metal to Boiling Water Under Atmospheric Pressure," *J. Japan Soc. Mech. Eng., 37,* 367, 1934 (Translation: *Int. J. Heat Mass Transfer, 9,* 1419, 1966).

3. Drew, T. B., and C. Mueller, "Boiling," *Trans. AIChE, 33,* 449, 1937.

4. Yamagata, K., F. Kirano, K. Nishiwaka, and H. Matsuoka, "Nucleate Boiling of Water on the Horizontal Heating Surface," *Mem. Fac. Eng. Kyushu, 15,* 98, 1955.

5. Rohsenow, W. M., "A Method of Correlating Heat Transfer Data for Surface Boiling Liquids," *Trans. ASME, 74,* 969, 1952.

6. Vachon, R. I., G. H. Nix, and G. E. Tanger, "Evaluation of Constants for the Rohsenow Pool-Boiling Correlation," *J. Heat Transfer, 90,* 239, 1968.

7. Rohsenow, W. M., "Boiling," in W. M. Rohsenow and J. P. Hartnett, Eds., *Handbook of Heat Transfer,* Chap. 13, McGraw-Hill, New York, 1973.

8. Kutateladze, S. S., "On the Transition to Film Boiling Under Natural Convection," *Kotloturbostroenie, 3,* 10, 1948.

9. Zuber, N., "On the Stability of Boiling Heat Transfer," *Trans. ASME, 80,* 711, 1958.

10. Lienhard, J. H., V. K. Dhir, and D. M. Riherd, "Peak Pool Boiling Heat Flux Measurements on Finite Horizontal Flat Plates," *J. Heat Transfer, 95,* 477, 1973.

11. Lienhard, J. H., and V. K. Dhir, "Extended Hydrodynamic Theory of the Peak and Minimum Pool Boiling Heat Fluxes," National Aeronautics and Space Administration Report NASA-CR-2270, July 1973.

12. Cichelli, M. T., and C. F. Bonilla, "Heat Transfer to Liquids Boiling Under Pressure," *Trans. AIChE, 41,* 755, 1945.

13. Berenson, P. J., "Film Boiling Heat Transfer for a Horizontal Surface," *J. Heat Transfer, 83,* 351, 1961.

14. Hahne, E., and U. Grigull, *Heat Transfer in Boiling,* Hemisphere/Academic Press, New York, 1977.

15. Lienhard, J. H., and P. T. Y. Wong, "The Dominant Unstable Wavelength and Minimum Heat Flux During Film Boiling on a Horizontal Cylinder," *J. Heat Transfer, 86,* 220, 1964.

16. Bromley, L. A., "Heat Transfer in Stable Film Boiling," *Chem. Eng. Prog., 46,* 221, 1950.

17. Sadasivan, P., and J. H. Lienhard, "Sensible Heat Correction in Laminar Film Boiling and Condensation," *J. Heat Transfer, 109,* 45, 1987.

18. Siegel, R., "Effect of Reduced Gravity on Heat Transfer," *Adv. Heat Transfer, 4,* 143, 1967.

19. Berensen, P. J., "Experiments on Pool Boiling Heat Transfer," *Int. J. Heat Mass Transfer, 5,* 985, 1962.

20. Webb, R. L., "The Evolution of Enhanced Surface Geometries for Nucleate Boiling," *Heat Transfer Eng., 2,* 46, 1981, and "Nucleate Boiling on Porous Coated Surfaces," *Heat Transfer Eng., 4,* 71, 1983.

21. Bergles, A. E., "Enhancement of Heat Transfer," *Heat Transfer 1978,* Vol. 6, pp. 89–108, Hemisphere Publishing, New York, 1978.

22. Bergles, A. E., "Augmentation of Boiling and Evaporation," in E. U. Schlünder, Ed.-in-Chief, *Heat Exchanger Design Handbook,* Vol. 2, Chapter 2.7.9, Hemisphere Publishing, New York, 1983.

23. Bergles, A. E., and W. H. Rohsenow, "The Determination of Forced Convection Surface Boiling Heat Transfer," *J. Heat Transfer,* **86,** 365, 1964.

24. van Stralen, S., and R. Cole, *Boiling Phenomena,* McGraw-Hill/Hemisphere, New York, 1979.

25. Lienhard, J. H., and R. Eichhorn, "Peak Boiling Heat Flux on Cylinders in a Cross Flow," *Int. J. Heat Mass Transfer,* **19,** 1135, 1976.

26. Tong, L. S., *Boiling Heat Transfer and Two-Phase Flow,* Wiley, New York, 1965.

27. Collier, J. G., *Convective Boiling and Condensation,* McGraw-Hill, New York, 1972.

28. Griffith, P., "Two-Phase Flow," in W. M. Rohsenow and J. P. Hartnett, Eds., *Handbook of Heat Transfer,* Chap. 14, McGraw-Hill, New York, 1973.

29. Ginoux, J. N., *Two-Phase Flow and Heat Transfer,* McGraw-Hill/Hemisphere, New York, 1978.

30. Sparrow, E. M., and J. L. Gregg, "A Boundary Layer Treatment of Laminar Film Condensation," *J. Heat Transfer,* **81,** 13, 1959.

31. Koh, J. C. Y., E. M. Sparrow, and J. P. Hartnett, "The Two-Phase Boundary Layer in Laminar Film Condensation," *Int. J. Heat Mass Transfer,* **2,** 69, 1961.

32. Nusselt, W., "Die Oberflachenkondensation des Wasserdampfes," *Z. Ver. Deut. Ing.,* **60,** 541, 1916.

33. Rohsenow, W. M., "Heat Transfer and Temperature Distribution in Laminar Film Condensation," *Trans. ASME,* **78,** 1645, 1956.

34. Chen, M. M., "An Analytical Study of Laminar Film Condensation: Part 1—Flat Plate," *J. Heat Transfer,* **83,** 48, 1961.

35. Dhir, V. K., and J. H. Lienhard, "Laminar Film Condensation on Plane and Axisymmetric Bodies in Non-uniform Gravity," *J. Heat Transfer,* **93,** 97, 1971.

36. Kutateladze, S. S., *Fundamentals of Heat Transfer,* Academic Press, New York, 1963.

37. Labuntsov, D. A., "Heat Transfer in Film Condensation of Pure Steam on Vertical Surfaces and Horizontal Tubes," *Teploenergetika,* **4,** 72, 1957.

38. Gregorig, R., J. Kern, and K. Turek, "Improved Correlation of Film Condensation Data Based on a More Rigorous Application of Similarity Parameters," *Wärme Stoffübertrag.,* **7,** 1, 1974.

39. Popiel, Cz. O., and L. Boguslawski, "Heat Transfer by Laminar Film Condensation on Sphere Surfaces," *Int. J. Heat Mass Transfer,* **18,** 1486, 1975.

40. Chen, M. M., "An Analytical Study of Laminar Film Condensation: Part 2—Single and Multiple Horizontal Tubes," *J. Heat Transfer,* **83,** 55, 1961.

41. Selin, G., "Heat Transfer by Condensing Pure Vapours Outside Inclined Tubes," *International Developments in Heat Transfer,* Part 2, International Heat Transfer Conference, University of Colorado, pp. 278–289, ASME, New York, 1961.

42. Chato, J. C., "Laminar Condensation Inside Horizontal and Inclined Tubes," *J. ASHRAE,* **4,** 52, 1962.

43. Rohsenow, W. M., "Film Condensation," in W. M. Rohsenow and J. P. Hartnett, Eds., *Handbook of Heat Transfer,* Chap. 12A, McGraw-Hill, New York, 1973.

44. Tanner, D. W., D. Pope, C. J. Potter, and D. West, "Heat Transfer in Dropwise Condensation of Low Pressures in the Absence and Presence of Non-Condensable Gas," *Int. J. Heat Mass Transfer,* **11,** 181, 1968.

45. Griffith, P., "Dropwise Condensation," in E. U. Schlünder, Ed.-in-Chief, *Heat Exchanger Design Handbook,* Vol. 2, Chap. 2.6.5, Hemisphere Publishing, New York, 1983.

46. Shade, R., and B. Mikic, "The Effects of Non-condensable Gases on Heat Transfer During Dropwise Condensation," Paper 67b presented at the 67th Annual Meeting of the American Institute of Chemical Engineers, Washington, DC, 1974.

47. Hannemann, R., and B. Mikic, "An Experimental Investigation into the Effect of Surface Thermal Conductivity on the Rate of Heat Transfer in Dropwise Condensation," *Int. J. Heat Mass Transfer,* **19,** 1309, 1976.

48. Collier, R., "Pool Boiling," in E. U. Schlünder, Ed.-in-Chief, *Heat Exchanger Design Handbook,* Vol. 2, Chapter 2.7.2, Hemisphere Publishing, New York, 1983.

49. Butterworth, D., "Filmwise Condensation," in D. Butterworth and G. F. Hewitt, Eds., *Two-Phase Flow and Heat Transfer,* Oxford University Press, London, 1977, pp. 426–462.

50. Butterworth, D., "Film Condensation of Pure Vapor," in E. U. Schlünder, Ed.-in-Chief, *Heat Exchanger Design Handbook,* Vol. 2. Chap. 2.6.2, Hemisphere Publishing, New York, 1983.

51. Rose, J. W., "Dropwise Condensation Theory," *Int. J. Heat Mass Transfer,* **24,** 191, 1981.

52. Jakob, M., and G. A. Hawkins, *Elements of Heat Transfer,* 3rd ed., Wiley, New York, 1957.

53. Lienhard, J. H., and V. K. Dhir, "Hydrodynamic Prediction of Peak Boiling Heat Fluxes from Finite Bodies," *J. Heat Transfer,* **95,** 152, 1973.

Problems

General Considerations

10.1 Show that, for water at 1-atm pressure with $T_s - T_{sat}$ = 10°C, the Jakob number is much less than unity. What is the physical significance of this result? Verify that this conclusion applies to other fluids.

10.2 The surface of a horizontal, 20-mm diameter cylinder is maintained at an excess temperature of 5°C in saturated water at 1 atm. Estimate the heat flux using the appropriate free convection correlation and compare your result with the boiling curve of Figure 10.4. For nucleate boiling, estimate the maximum value of the heat transfer coefficient from the boiling curve.

10.3 The role of surface tension in bubble formation can be demonstrated by considering a spherical bubble of pure saturated vapor in *mechanical* and *thermal* equilibrium with its superheated liquid.

(a) Beginning with an appropriate free-body diagram of the bubble, perform a force balance to obtain an expression of the bubble radius,

$$r_b = \frac{2\sigma}{p_{sat} - p_l}$$

where p_{sat} is the pressure of the saturated vapor and p_l is the pressure of the superheated liquid outside the bubble.

(b) On a p–v diagram, represent the bubble and liquid states. Discuss what changes in these conditions will cause the bubble to grow or collapse.

(c) Calculate the bubble size under equilibrium conditions for which the vapor is saturated at 101°C and the liquid pressure corresponds to a saturation temperature of 100°C.

Nucleate Boiling and Critical Heat Flux

10.4 A long, 1-mm diameter wire passes an electrical current dissipating 3150 W/m and reaches a surface temperature of 126°C when submerged in water at 1 atm. What is the boiling heat transfer coefficient? Estimate the value of the correlation coefficient $C_{s, f}$.

10.5 Estimate the nucleate pool boiling heat transfer coefficient for water boiling at atmospheric pressure on the outer surface of a platinum-plated 10-mm

diameter tube maintained 10°C above the saturation temperature.

10.6 Estimate the nucleate pool boiling heat transfer coefficient for water under atmospheric pressure in contact with mechanically polished stainless steel when the excess temperature is 15°C.

10.7 A simple expression to account for the effect of pressure on the nucleate boiling convection coefficient in water (W/m² · K) is [52]

$$h = C(\Delta T_e)^n \left(\frac{p}{p_a}\right)^{0.4}$$

where p and p_a are the system pressure and standard atmospheric pressure, respectively. For a horizontal plate and the range $15 < q_s'' < 235$ kW/m², $C = 5.56$ and $n = 3$. Units of ΔT_e are kelvins. Compare predictions from this expression with the Rohsenow correlation ($C_{s, f} = 0.013$, $n = 1$) for pressures of 2 and 5 bars with $\Delta T_e = 10$°C.

10.8 Calculate the critical heat flux for the following fluids at 1 atm: mercury, ethanol, and refrigerant R-12. Compare these results to the critical heat flux for water at 1 atm.

10.9 The bottom of a copper pan, 150 mm in diameter, is maintained at 115°C by the heating element of an electric range. Estimate the power required to boil the water in this pan. Determine the evaporation rate. What is the ratio of the surface heat flux to the critical heat flux? What pan temperature is required to achieve the critical heat flux?

10.10 Advances in very large scale integration (VLSI) of electronic devices on a chip are often restricted by the ability to cool the chip. For mainframe computers, an array of several hundred chips, each of area 25 mm², may be mounted on a ceramic substrate. A method of cooling the array is by immersion in a low boiling point fluid such as refrigerant R-113. At 1 atm and 321 K, properties of the saturated liquid are $\mu = 5.147 \times 10^{-4}$ N · s/m², $c_p = 983.8$ J/kg · K, and $Pr = 7.183$. Assume values of $C_{s, f} = 0.004$ and $n = 1.7$.

(a) Estimate the power dissipated by a single chip if it is operating at 50% of the critical heat flux. What is the corresponding value of the chip temperature?

(b) Compute and plot the chip temperature as a function of surface heat flux for $0.25 \leq q_s''/q_{max}'' \leq 0.90$.

10.11 Saturated ethylene glycol at 1 atm is heated by a chromium-plated surface which has a diameter of 200 mm and is maintained at 480 K. Estimate the heating power requirement and the rate of evaporation. What fraction is the power requirement of the maximum power associated with the critical heat flux? At 470 K properties of the saturated liquid are $\mu = 0.38 \times 10^{-3}$ N · s/m²; $c_p = 3280$ J/kg · K, and $Pr = 8.7$. The saturated vapor density is $\rho = 1.66$ kg/m³.

10.12 Copper tubes 25 mm in diameter and 0.75 m long are used to boil saturated water at 1 atm.

(a) If the tubes are operated at 75% of the critical heat flux, how many tubes are needed to provide a vapor production rate of 750 kg/h? What is the corresponding tube surface temperature?

(b) Compute and plot the tube surface temperature as a function of heat flux for $0.25 \leq q_s''/q_{max}'' \leq 0.90$. On the same graph, plot the corresponding number of tubes needed to provide the prescribed vapor production rate.

10.13 Estimate the current at which a 1-mm diameter nickel wire will burn out when submerged in water at atmospheric pressure. The electrical resistance of the wire is 0.129 Ω/m.

10.14 Estimate the power (W/m²) required to maintain a brass plate at 115°C while boiling saturated water at 1 atm. What is the power requirement if the water is pressurized to 10 atm? At what fraction of the critical heat flux is the plate operating?

10.15 It has been demonstrated experimentally that the critical heat flux is highly dependent on pressure, primarily through the pressure dependence of the fluid surface tension and latent heat of vaporization. Using Equation 10.7, calculate values of q_{max}'' for water as a function of pressure. Demonstrate that the peak critical heat flux occurs at approximately one-third the critical pressure ($p_c = 221$ bars). Since all common fluids have this characteristic, suggest what coordinates should be used to plot critical heat flux–pressure values to obtain a universal curve.

10.16 In applying dimensional analysis, Kutateladze [8] postulated that the critical heat flux varies with the heat of vaporization, vapor density, surface tension, and a bubble diameter parameter:

$$D_b = \left[\frac{\sigma}{g(\rho_l - \rho_v)} \right]^{1/2}$$

(a) Verify that dimensional analysis would yield the following expression for the critical heat flux:

$$q_{max}'' = Ch_{fg}\rho_v^{1/2}D_b^{-1/2}\sigma^{1/2}$$

(b) The ratio of the characteristic length L of a heater surface (its width or diameter) to the bubble diameter parameter D_b is referred to as the Bond number (Bo) and compares buoyant and capillary forces. The correlation of Equation 10.7 strictly applies when $Bo \geq 3$. Estimate the heater size that will satisfy this requirement for water at 1 atm.

10.17 When the heater surface is not infinite, as is strictly required for use of the critical heat flux correlation of Equation 10.7, more detailed correlations are available in the literature. For the small horizontal cylinder of radius r, Lienhard and Dhir [53] recommend the following expression:

$$q_{max}'' = q_{max, z}''[0.94(Bo)^{-1/4}] \qquad 0.15 \leq Bo \leq 1.2$$

where $q_{max, z}''$ is the critical heat flux as predicted by the Zuber–Kutateladze relation, Equation 10.6, and for Bo, the Bond number, the characteristic length is the radius of the cylinder. Estimate the critical heat flux for horizontal cylinders of 1- and 3-mm diameter in saturated water at 1 atm and compare the result with that for a large horizontal cylinder.

10.18 What is the critical heat flux for boiling water at 1 atm on the surface of the moon, where the gravitational acceleration is one-sixth that of the earth?

10.19 A heater for boiling a saturated liquid consists of two concentric stainless steel tubes packed with dense boron nitride powder. Electrical current is passed through the inner tube, creating uniform volumetric heating $\dot{q}$ (W/m³). The exposed surface of the outer tube is in contact with the liquid and the boiling heat flux is given as

$$q_s'' = C(T_s - T_{sat})^3$$

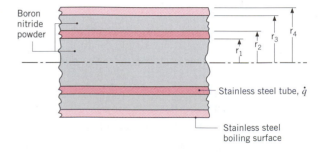

Boron
nitride
powder

Stainless steel tube, $\dot{q}$

Stainless steel
boiling surface

It is feared that under high-power operation the stainless steel tubes would severely oxidize if temperatures exceed $T_{ss,x}$ or that the boron nitride would deteriorate if its temperature exceeds $T_{bn,x}$. Presuming that the saturation temperature of the liquid (T_{sat}) and the boiling surface temperature (T_s) are prescribed, derive expressions for the maximum temperatures in the stainless steel (ss) tubes and in the boron nitride (bn). Express your results in terms of geometric parameters (r_1, r_2, r_3, r_4), thermal conductivities (k_{ss}, k_{bn}), and the boiling parameters (C, T_{sat}, T_s).

10.20 A silicon chip of thickness $L = 2.5$ mm and thermal conductivity $k_s = 135$ W/m · K is cooled by boiling a saturated fluorocarbon liquid ($T_{sat} = 57°C$) on its surface. The electronic circuits on the bottom of the chip produce a uniform heat flux of $q_o'' = 5 \times 10^4$ W/m^2, while the sides of the chip are perfectly insulated.

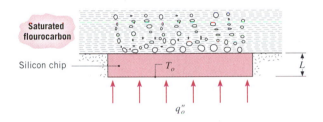

Properties of the saturated fluorocarbon are $c_{p,l} = 1100$ J/kg · K, $h_{fg} = 84{,}400$ J/kg, $\rho_l = 1619.2$ kg/m^3, $\rho_v = 13.4$ kg/m^3, $\sigma = 8.1 \times 10^{-3}$ kg/s^2, $\mu_l = 440 \times 10^{-6}$ kg/m · s, and $Pr_l = 9.01$. In addition, the nucleate boiling constants are $C_{s,f} = 0.005$ and $n = 1.7$.

(a) What is the steady-state temperature T_o at the bottom of the chip? If, during testing of the chip, q_o'' is increased to 90% of the critical heat flux, what is the new steady-state value of T_o?

(b) Compute and plot the chip surface temperatures (top and bottom) as a function of heat flux for $0.20 \leq q_o''/q_{max}'' \leq 0.90$. If the maximum allowable chip temperature is 80°C, what is the maximum allowable value of q_o''?

10.21 A device for performing boiling experiments consists of a copper bar ($k = 400$ W/m · K), which is exposed to a boiling liquid at one end, encapsulates an electrical heater at the other end, and is well insulated from its surroundings at all but the exposed surface. Thermocouples inserted in the bar are used to measure temperatures at distances of $x_1 = 10$ mm and $x_2 = 25$ mm from the surface.

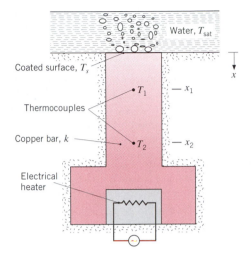

(a) An experiment is performed to determine the boiling characteristics of a special coating applied to the exposed surface. Under steady-state conditions, nucleate boiling is maintained in saturated water at atmospheric pressure and values of $T_1 = 133.7°C$ and $T_2 = 158.6°C$ are recorded. If $n = 1$, what value of the coefficient $C_{s,f}$ is associated with the Rohsenow correlation?

(b) Assuming applicability of the Rohsenow correlation with the value of $C_{s,f}$ determined from part (a), compute and plot the excess temperature ΔT_e as a function of the boiling heat flux for $10^5 \leq q_s'' \leq 10^6$ W/m^2. What are the corresponding values of T_1 and T_2 for $q_s'' = 10^6$ W/m^2? If q_s'' were increased to 1.5×10^6 W/m^2, could the foregoing results be extrapolated to infer the corresponding values of ΔT_e, T_1, and T_2?

Film Boiling

10.22 A small copper sphere, initially at a uniform, elevated temperature $T(0) = T_i$, is suddenly immersed in a large fluid bath maintained at T_{sat}. The initial temperature of the sphere exceeds the Leidenfrost point corresponding to the temperature T_D of Figure 10.4.

(a) Sketch the variation of the average sphere temperature, $\overline{T}(t)$, with time during the quenching process. Indicate on this sketch, the temperatures T_i, T_D, and T_{sat}, as well as the regimes of film, transition, and nucleate boiling and the regime of single-phase convection. Identify key features of the temperature history.

(b) At what time(s) in this cooling process do you expect the surface temperature of the sphere to deviate most from its center temperature? Explain your answer.

10.23 A steel bar, 20 mm in diameter and 200 mm long, with an emissivity of 0.9, is removed from a furnace at 455°C and suddenly submerged in a water bath under atmospheric pressure. Estimate the initial heat transfer rate from the bar.

10.24 Electrical current passes through a horizontal, 2-mm diameter conductor of emissivity 0.5 when immersed in water under atmospheric pressure.

  (a) Estimate the power dissipation per unit length of the conductor required to maintain the surface temperature at 555°C.

  (b) For conductor diameters of 1.5, 2.0, and 2.5 mm, compute and plot the power dissipation per unit length as a function of surface temperature for $250 \leq T_s \leq 650$°C. On a separate figure, plot the percentage contribution of radiation as a function of T_s.

10.25 A polished stainless steel bar of 50-mm diameter and an emissivity of 0.10 is maintained at a surface temperature of 250°C, while horizontally submerged in water at 25°C under atmospheric pressure. Estimate the heat rate per unit length of the bar.

10.26 A cylinder of 120-mm diameter at 1000 K is quenched in saturated water at 1 atm. Describe the quenching process and estimate the maximum heat removal rate per unit length during the process.

10.27 A 1-mm diameter horizontal platinum wire of emissivity $\varepsilon = 0.25$ is operated in saturated water at 1-atm pressure.

  (a) What is the surface heat flux if the surface temperature is $T_s = 800$ K?

  (b) For emissivities of 0.1, 0.25, and 0.95, generate a log–log plot of the heat flux as a function of surface excess temperature, $\Delta T_e \equiv T_s - T_{sat}$, for $150 \leq \Delta T_e \leq 550$ K. Show the critical heat flux and the Leidenfrost point on your plot. Separately, plot the percentage contribution of radiation to the total heat flux for $150 \leq \Delta T_e \leq 550$ K.

10.28 As strip steel leaves the last set of rollers in a hot rolling mill, it is quenched by planar water jets before being coiled. Due to the large plate temperatures, film boiling is achieved shortly downstream of the jet impingement region.

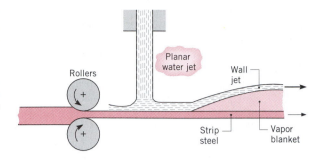

Consider conditions for which the strip steel beneath the vapor blanket is at a temperature of 907 K and has an emissivity of 0.35. Neglecting the effects of the strip and jet motions and assuming convection within the film to be approximated by that associated with a large horizontal cylinder of 1-m diameter, estimate the rate of heat transfer per unit surface area from the strip to the wall jet.

10.29 A copper sphere of 10-mm diameter, initially at a prescribed elevated temperature T_i, is quenched in a saturated (1 atm) water bath. Using the lumped capacitance method, estimate the time for the sphere to cool (a) from $T_i = 130$°C to 110°C and (b) from $T_i = 500$°C to 220°C. Plot the temperature history for each quenching process.

Forced Convection Boiling

10.30 A tube of 2-mm diameter is used to heat saturated water at 1 atm, which is in cross flow over the tube. Calculate and plot the critical heat flux as a function of water velocity over the range 0 to 2 m/s. On your plot identify the pool boiling region and the transition region between the low- and high-velocity ranges.

10.31 Saturated water at 1 atm and velocity 2 m/s flows over a circular heating element of diameter 5 mm. What is the maximum heating rate (W/m) for the element?

10.32 For forced convection local boiling of water inside vertical tubes, the heat transfer coefficient h (W/m^2 · K) can be estimated by the correlation

$$h = 2.54(\Delta T_e)^3 \exp\left(\frac{p}{15.3}\right)$$

with ΔT_e in kelvins and p in bars. Consider water at 4 bars flowing through a vertical tube of 50-mm inner diameter. Local boiling occurs when the tube wall is 15°C above the saturation temperature. Estimate the boiling heat transfer rate per unit length of the tube.

10.33 For forced convection boiling in smooth tubes, the heat flux can be estimated by combining the separate effects of boiling and forced convection. The Rohsenow and Dittus–Boelter correlations may be used to predict nucleate boiling and forced convection effects, with 0.019 replacing 0.023 in the latter expression. Consider water at 1 atm with a mean velocity of 1.5 m/s and a mean temperature of 95°C flowing through a 15-mm diameter brass tube whose surface is maintained at 110°C. Estimate the heat transfer rate per unit length of the tube.

Film Condensation

10.34 Saturated steam at 0.1 bar condenses with a convection coefficient of 6800 W/m² · K on the outside of a brass tube having inner and outer diameters of 16.5 and 19 mm, respectively. The convection coefficient for water flowing inside the tube is 5200 W/m² · K. Estimate the steam condensation rate per unit length of the tube when the mean water temperature is 30°C.

10.35 If advection effects in laminar film condensation on a vertical plate are not neglected, the energy balance, Equation 10.20, on the control volume of Figure 10.13 has the form

$$q_s'' = \frac{d\Gamma}{dx} \left[h_{fg} + \frac{1}{\Gamma} \int_0^\delta \rho_l u c_{p,l} (T_{sat} - T) \, dy \right]$$

Verify this expression and show that the influence of advection for a linear temperature distribution across the condensate film is to replace h_{fg} with $h_{fg}' = h_{fg}(1 + \frac{3}{8} Ja)$ in the Nusselt analysis. The more appropriate expression for h_{fg}' is Equation 10.26, which results from the use of a nonlinear temperature distribution in the film.

10.36 Consider a container exposed to a saturated vapor, T_{sat}, having a cold bottom surface, $T_s < T_{sat}$, and with insulated sidewalls.

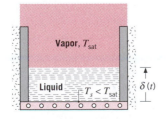

Assuming a linear temperature distribution for the liquid, perform a surface energy balance on the liquid–vapor interface to obtain the following expression for the growth rate of the liquid layer:

$$\delta(t) = \left[\frac{2k_l(T_{sat} - T_s)}{\rho_l h_{fg}} t \right]^{1/2}$$

Calculate the thickness of the liquid layer formed in 1 h for a 200-mm² bottom surface maintained at 80°C and exposed to saturated steam at 1 atm. Compare this result with the condensate formed by a *vertical* plate of the same dimensions for the same period of time.

10.37 Saturated steam at 1 atm is exposed to a vertical plate 1 m high and 0.5 m wide having a uniform surface temperature of 70°C. Estimate the heat transfer rate to the plate and the steam condensation rate.

10.38 Saturated steam at 1 atm condenses on the outer surface of a vertical, 100-mm diameter pipe 1 m long, having a uniform surface temperature of 94°C. Estimate the total condensation rate and the heat transfer rate to the pipe.

10.39 Determine the total condensation rate and the heat transfer rate for Problem 10.38 when the steam is saturated at 1.5 bars.

10.40 Consider the condensation process of Problem 10.38. To maintain the pipe wall at the uniform surface temperature of 94°C, cooling water passes through the steel pipe of inner diameter 92 mm. What water flow rate will result in a 4°C temperature rise of the water between the outlet and inlet of the pipe?

10.41 A vertical plate 500 mm high and 200 mm wide is to be used to condense saturated steam at 1 atm.

  (a) At what surface temperature must the plate be maintained to achieve a condensation rate of $\dot{m} = 25$ kg/h?

  (b) Compute and plot the surface temperature as a function of condensation rate for $15 \leq \dot{m} \leq 50$ kg/h.

  (c) On the same graph and for the same range of $\dot{m}$, plot the surface temperature as a function of condensation rate if the plate is 200 mm high and 500 mm wide.

10.42 Saturated ethylene glycol vapor at 1 atm is exposed to a vertical plate 300 mm high and 100 mm wide having a uniform temperature of 420 K. Estimate the heat transfer rate to the plate and the condensation rate. Approximate the liquid properties as those corresponding to saturated conditions at 373 K (Table A.5).

10.43 A vertical plate 2.5 m high, maintained at a uniform temperature of 54°C, is exposed to saturated steam at atmospheric pressure.

    (a) Estimate the condensation and heat transfer rates per unit width of the plate.

    (b) If the plate height were halved, would turbulent flow still exist?

    (c) For $54 \leq T_s \leq 90°C$, plot the condensation rate as a function of plate temperature for the two plate heights of parts (a) and (b).

10.44 Two configurations are being considered in the design of a condensing system for steam at 1 atm employing a vertical plate maintained at 90°C. The first configuration is a single vertical plate $L \times w$ and the second consists of two vertical plates $(L/2) \times w$, where L and w are the vertical and horizontal dimensions, respectively. Which configuration would you choose?

10.45 Saturated vapor from a chemical process condenses at a slow rate on the inner surface of a vertical, thin-walled cylindrical container of length L and diameter D. The container wall is maintained at a uniform temperature T_s by flowing cold water across its outer surface.

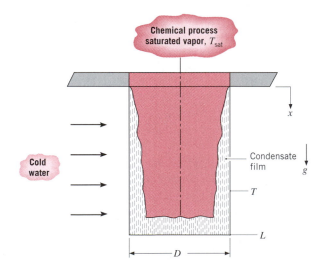

Derive an expression for the time, t_f, required to fill the container with condensate, assuming that

the condensate film is laminar. Express your result in terms of D, L, $(T_{sat} - T_s)$, g, and appropriate fluid properties.

10.46 Determine the total condensation rate and heat transfer rate for the condensation process of Problem 10.38 when the pipe is horizontal.

10.47 Consider laminar film condensation on the outer surface of a tube of length L and diameter D. Obtain an expression for the ratio of the condensation heat transfer rate in the horizontal orientation relative to the vertical orientation.

10.48 An uninsulated, 25-mm diameter pipe with a surface temperature of 15°C passes through a room having an air temperature of 37°C and relative humidity of 75%. Estimate the condensation rate per unit length of the pipe assuming film, rather than dropwise, condensation.

10.49 A horizontal tube of 50-mm diameter, with a surface temperature of 34°C, is exposed to steam at 0.2 bar. Estimate the condensation rate and heat transfer rate per unit length of the tube.

10.50 A horizontal tube 1 m long with a surface temperature of 70°C is used to condense saturated steam at 1 atm.

    (a) What diameter is required to achieve a condensation rate of 125 kg/h?

    (b) Plot the condensation rate as a function of surface temperature for $70 \leq T_s \leq 90°C$ and tube diameters of 125, 150, 175 mm.

10.51 Refrigerant-12 at 1 atm condenses on the outside of a horizontal tube of 10-mm diameter and 1-m length. What surface temperature is required for a condensate rate of 50 kg/h?

10.52 Saturated steam at a pressure of 0.1 bar is condensed over a square array of 100 tubes each of diameter 8 mm.

    (a) If the tube surfaces are maintained at 27°C, estimate the condensation rate per unit tube length.

    (b) Subject to the requirement that the total number of tubes and the tube diameter are fixed at 100 and 8 mm, respectively, what options are available for increasing the condensation rate? Assess these options quantitatively.

10.53 A thin-walled concentric tube heat exchanger of 0.19-m length is to be used to heat deionized water from 40 to 60°C at a flow rate of 5 kg/s. The deionized water flows through the inner tube of 30-mm diameter while saturated steam at 1 atm is supplied to the annulus formed with the outer tube of 60-mm diameter. The thermophysical properties of the deionized water are $\rho = 982.3$ kg/m³, $c_p = 4181$ J/kg · K, $k = 0.643$ W/m · K, $\mu = 548 \times 10^{-6}$ N · s/m², and $Pr = 3.56$. Estimate the convection coefficients for both sides of the tube and determine the inner tube wall outlet temperature. Does condensation provide a fairly uniform inner tube wall temperature equal approximately to the saturation temperature of the steam?

10.54 A technique for cooling a multichip module involves submerging the module in a saturated fluorocarbon liquid. Vapor generated due to boiling at the module surface is condensed on the outer surface of copper tubing suspended in the vapor space above the liquid. The thin-walled tubing is of diameter $D = 10$ mm and is coiled in a horizontal plane. It is cooled by water that enters at 285 K and leaves at 315 K. All the heat dissipated by the chips within the module is transferred from a 100-mm by 100-mm boiling surface, at which the flux is 10^5 W/m², to the fluorocarbon liquid, which is at $T_{sat} = 57°C$. Liquid properties are $k_l = 0.0537$ W/m · K, $c_{p,l} = 1100$ J/kg · K, $h'_{fg} \approx h_{fg} = 84{,}400$ J/kg, $\rho_l = 1619.2$ kg/m³, $\rho_v = 13.4$ kg/m³, $\sigma = 8.1 \times 10^{-3}$ kg/s², $\mu_l = 440 \times 10^{-6}$ kg/m · s, and $Pr_l = 9$.

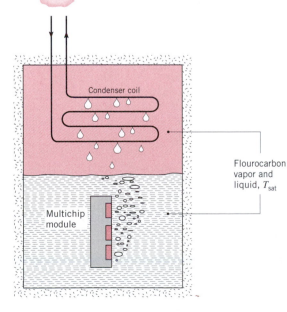

(a) For the prescribed heat dissipation, what is the required condensation rate (kg/s) and water flowrate (kg/s)?

(b) Assuming fully developed flow throughout the tube, determine the tube surface temperature at the coil inlet and outlet.

(c) Assuming a uniform tube surface temperature of $T_s = 53.0°C$, determine the required length of the coil.

10.55 Water at a mean temperature and velocity of 17°C and 2 m/s, respectively, flows through a horizontal brass tube with inner and outer diameters of 28.0 and 30.5 mm, respectively. Estimate the condensation rate per unit tube length of saturated steam at a pressure of 0.15 bar surrounding the tube.

10.56 Determine the rate of condensation on a 100-mm diameter sphere with a surface temperature of 150°C in saturated ethylene glycol vapor at 1 atm. Approximate the liquid properties as those corresponding to saturated conditions at 373 K (Table A.5).

10.57 A 10-mm diameter copper sphere, initially at a uniform temperature of 50°C, is placed in a large container filled with saturated steam at 1 atm. Using the lumped capacitance method, estimate the time required for the sphere to reach an equilibrium condition. How much condensate (kg) was formed during this period?

10.58 Saturated steam at 1.5 bars condenses inside a horizontal, 75-mm diameter pipe whose surface is maintained at 100°C. Assuming low vapor velocities and film condensation, estimate the heat transfer coefficient and the condensation rate per unit length of the pipe.

Dropwise Condensation

10.59 Consider the conditions of Problem 10.48. Estimate the condensation rate for dropwise condensation.

Combined Boiling/Condensation

10.60 A passive technique for cooling heat-dissipating integrated circuits involves submerging the ICs in a low boiling point dielectric fluid. Vapor generated in cooling the circuits is condensed on vertical plates suspended in the vapor cavity above the liquid. The temperature of the plates is maintained below the saturation temperature, and during steady-state operation a balance is established between the rate of heat transfer to the condenser plates and the rate of heat dissipation by the ICs.

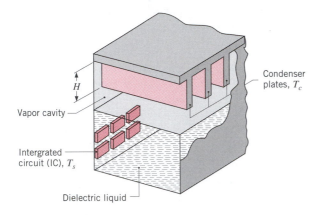

Consider conditions for which the 25-mm^2 surface area of each IC is submerged in a fluorocarbon liquid for which $T_{sat} = 50°C$, $\rho_l = 1700$ kg/m^3, $c_{p,l} = 1005$ J/kg · K, $\mu_l = 6.80 \times 10^{-4}$ kg/s · m, $k_l = 0.062$ W/m · K, $Pr_l = 11.0$, $\sigma = 0.013$ kg/s^2, $h_{fg} = 1.05 \times 10^5$ J/kg, $C_{s,f} = 0.004$, and $n = 1.7$. If the integrated circuits are operated at a surface temperature of $T_s = 75°C$, what is the rate at which heat is dissipated by each circuit? If the condenser plates are of height $H = 50$ mm and are maintained at a temperature of $T_c = 15°C$ by an internal coolant, and laminar film condensation is assumed, how much condenser surface area must be provided to balance the heat generated by 500 integrated circuits?

10.61 A thermosyphon consists of a closed container that absorbs heat along its boiling section and rejects heat along its condensation section. Consider a thermosyphon made from a thin-walled mechanically polished stainless steel cylinder of diameter D. Heat supplied to the thermosyphon boils saturated water at atmospheric pressure on the surfaces of the lower boiling section of length L_b and is then rejected by condensing vapor into a thin film, which falls by gravity along the wall of the condensation section of length L_c back into the boiling section. The two sections are separated by an insulated section of length L_i. The top surface of the condensation section may be treated as being insulated. The thermosyphon dimensions are $D = 20$ mm, $L_b = 20$ mm, $L_c = 40$ mm, and $L_i = 40$ mm.

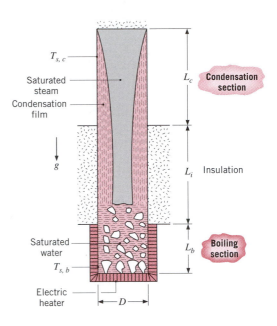

(a) Find the mean surface temperature, $T_{s,b}$, of the boiling surface if the nucleate boiling heat flux is to be maintained at 30% of the critical heat flux.

(b) Find the mean surface temperature, $T_{s,c}$, of the condensation section assuming laminar film condensation.

(c) Find the total condensation flow rate, $\dot{m}$, within the thermosyphon. Explain how you would determine whether the film is laminar, wavy laminar, or turbulent as it falls back into the boiling section.

10.62 A novel scheme for cooling computer chips uses a thermosyphon containing a saturated fluorocarbon. The chip is brazed to the bottom of a cuplike container, within which heat is dissipated by boiling and subsequently transferred to an external coolant (water) via condensation on the inner surface of a thin-walled tube.

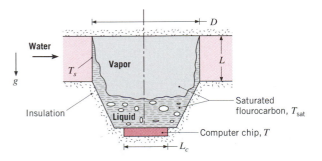

The nucleate boiling constants and the properties of the fluorocarbon are provided in Problem 10.20. In addition, $k_l = 0.054$ W/m · K.

(a) If the chip operates under steady-state conditions and its surface heat flux is maintained at 90% of the critical heat flux, what is its temperature T? What is the total power dissipation if the chip width is $L_c = 20$ mm on a side?

(b) If the tube diameter is $D = 30$ mm and its surface is maintained at $T_s = 25°C$ by the water, what tube length L is required to maintain the designated conditions?

10.63 A condenser–boiler section contains a 2-m × 2-m copper plate operating at a uniform temperature of $T_s = 100°C$ and separating saturated steam, which is condensing, from a saturated liquid-X, which experiences nucleate pool boiling. A portion of the boiling curve for liquid-X is shown below. Both saturated steam and saturated liquid-X are supplied to the system, while water condensate and vapor-X are removed by means not shown in the sketch. At a pressure of 1 bar, fluid-X has a saturation temperature and a latent heat of vaporization of $T_{sat} = 80°C$ and $h_{fg} = 700{,}000$ J/kg, respectively.

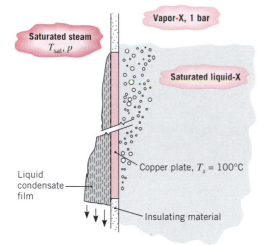

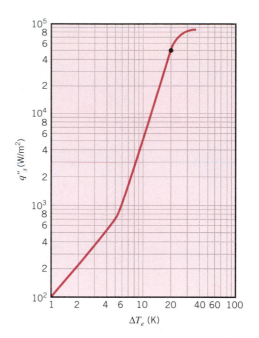

(a) Estimate the rates of evaporation and condensation (kg/s) for the two fluids.

(b) Determine the saturation temperature T_{sat} and pressure p for the steam, assuming that film condensation occurs.

10.64 A thin-walled cylindrical container of diameter D and height L is filled to a height y with a low boiling point liquid (A) at $T_{sat, A}$. The container is located in a large chamber filled with the vapor of a high boiling point fluid (B). Vapor-B condenses into a laminar film on the outer surface of the cylindrical container, extending from the location of the liquid-A free surface. The condensation process sustains nucleate boiling in liquid-A along the container wall according to the relation $q'' = C(T_s - T_{sat})^3$, where C is a known empirical constant.

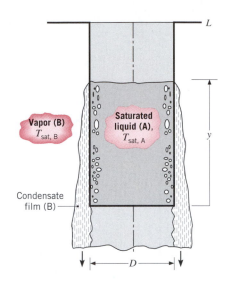

(a) For the portion of the wall covered with the condensate film, derive an equation for the average temperature of the container wall, T_s. Assume that the properties of fluids A and B are known.

(b) At what rate is heat supplied to liquid-A?

(c) Assuming the container is initially filled completely with liquid, that is, $y = L$, derive an expression for the time required to evaporate all the liquid in the container.

CHAPTER 11

Heat Exchangers

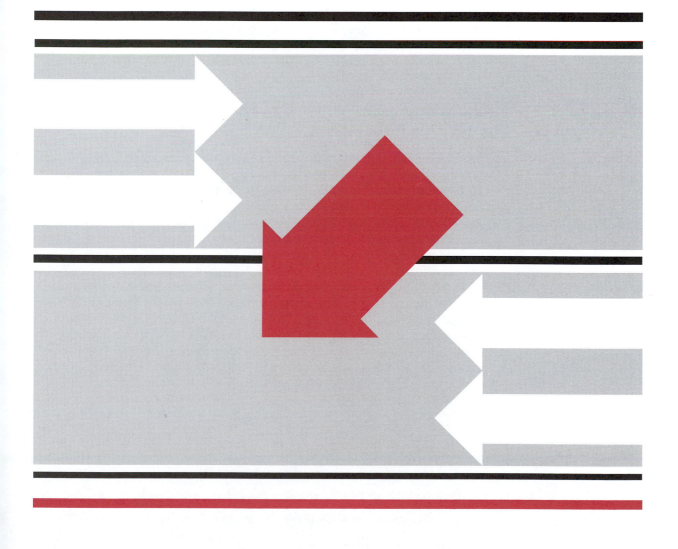

*T*he process of heat exchange between two fluids that are at different temperatures and separated by a solid wall occurs in many engineering applications. The device used to implement this exchange is termed a *heat exchanger,* and specific applications may be found in space heating and air-conditioning, power production, waste heat recovery, and chemical processing. In this chapter we consider the principles of heat transfer needed to design and/or to evaluate the performance of a heat exchanger.

11.1
Heat Exchanger Types

Heat exchangers are typically classified according to *flow arrangement* and *type of construction.* The simplest heat exchanger is one for which the hot and cold fluids move in the same or opposite directions in a *concentric tube* (or *double-pipe*) construction. In the *parallel-flow* arrangement of Figure 11.1*a*, the hot and cold fluids enter at the same end, flow in the same direction, and leave at the same end. In the *counterflow* arrangement of Figure 11.1*b*, the fluids enter at opposite ends, flow in opposite directions, and leave at opposite ends.

Alternatively, the fluids may move in *cross flow* (perpendicular to each other), as shown by the *finned* and *unfinned* tubular heat exchangers of Figure 11.2. The two configurations differ according to whether the fluid moving over the tubes is *unmixed* or *mixed.* In Figure 11.2*a*, the fluid is said to be unmixed because the fins prevent motion in a direction (*y*) that is transverse to the main-flow direction (*x*). In this case the fluid temperature varies with *x* and *y*. In contrast, for the unfinned tube bundle of Figure 11.2*b*, fluid motion, hence mixing, in the transverse direction is possible, and temperature variations are primarily in the main-flow direction. Since the tube flow is unmixed, both fluids are unmixed in the finned exchanger, while one fluid is mixed and the other unmixed in the unfinned exchanger. The nature of the mixing condition can significantly influence heat exchanger performance.

Another common configuration is the *shell-and-tube* heat exchanger [1]. Specific forms differ according to the number of shell-and-tube passes, and the simplest form, which involves single tube and shell *passes,* is shown in Figure 11.3. Baffles are usually installed to increase the convection coefficient of the shell-side fluid by inducing turbulence and a cross-flow velocity component. Baffled heat exchangers with one shell pass and two tube passes and with two shell passes and four tube passes are shown in Figures 11.4*a* and 11.4*b*, respectively.

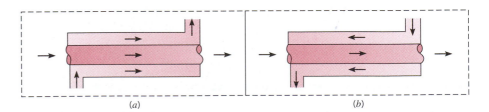

(a)   (b)

FIGURE 11.1   Concentric tube heat exchangers. (*a*) Parallel flow. (*b*) Counterflow.

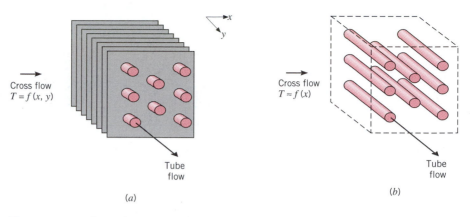

FIGURE 11.2 Cross-flow heat exchangers. (*a*) Finned with both fluids unmixed. (*b*) Unfinned with one fluid mixed and the other unmixed.

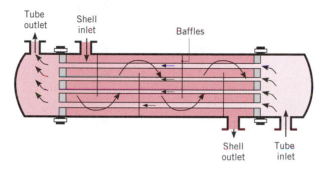

FIGURE 11.3 Shell-and-tube heat exchanger with one shell pass and one tube pass (cross-counterflow mode of operation).

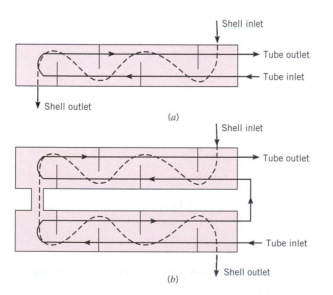

FIGURE 11.4 Shell-and-tube heat exchangers. (*a*) One shell pass and two tube passes. (*b*) Two shell passes and four tube passes.

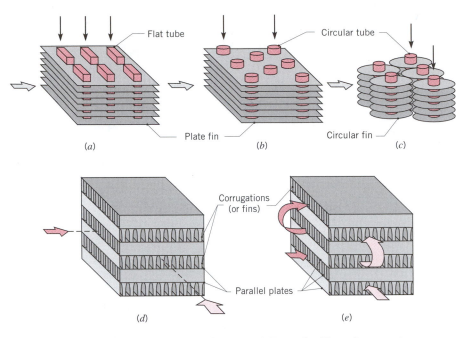

FIGURE 11.5   Compact heat exchanger cores. (*a*) Fin–tube (flat tubes, continuous plate fins). (*b*) Fin–tube (circular tubes, continuous plate fins). (*c*) Fin–tube (circular tubes, circular fins). (*d*) Plate–fin (single pass). (*e*) Plate–fin (multipass).

A special and important class of heat exchangers is used to achieve a very large ($\gtrsim 700$ m^2/m^3) heat transfer surface area per unit volume. Termed *compact heat exchangers,* these devices have dense arrays of finned tubes or plates and are typically used when at least one of the fluids is a gas, and is hence characterized by a small convection coefficient. The tubes may be *flat* or *circular,* as in Figures 11.5*a* and 11.5*b, c,* respectively, and the fins may be *plate* or *circular,* as in Figures 11.5*a, b* and 11.5*c,* respectively. Parallel-plate heat exchangers may be finned or corrugated and may be used in single-pass (Figure 11.5*d*) or multipass (Figure 11.5*e*) modes of operation. Flow passages associated with compact heat exchangers are typically small ($D_h \lesssim 5$ mm), and the flow is usually laminar.

11.2
The Overall Heat Transfer Coefficient

An essential, and often the most uncertain, part of any heat exchanger analysis is determination of the *overall heat transfer coefficient.* Recall from Equation 3.19 that this coefficient is defined in terms of the total thermal resistance to heat transfer between two fluids. In Equations 3.18 and 3.31, the coefficient was determined by accounting for conduction and convection resistances between fluids separated by composite plane and cylindrical walls, respectively. It is important to acknowledge, however, that such results apply only to *clean, unfinned* surfaces.

During normal heat exchanger operation, surfaces are often subject to fouling by fluid impurities, rust formation, or other reactions between the fluid and the wall material. The subsequent deposition of a film or scale on the surface can greatly increase the resistance to heat transfer between the fluids. This effect can be treated by introducing an additional thermal resistance, termed the *fouling factor, R_f.* Its value depends on the operating temperature, fluid velocity, and length of service of the heat exchanger.

In addition, we know that fins are often added to surfaces exposed to either or both fluids and that, by increasing the surface area, they reduce the resistance to convection heat transfer. Accordingly, with inclusion of surface fouling and fin (extended surface) effects, the overall heat transfer coefficient may be expressed as

$$\frac{1}{UA} = \frac{1}{U_c A_c} = \frac{1}{U_h A_h}$$

$$= \frac{1}{(\eta_o h A)_c} + \frac{R''_{f,c}}{(\eta_o A)_c} + R_w + \frac{R''_{f,h}}{(\eta_o A)_h} + \frac{1}{(\eta_o h A)_h} \tag{11.1}$$

where c and h refer to the cold and hot fluids, respectively. Note that calculation of the UA product does not require designation of the hot or cold side ($U_c A_c = U_h A_h$). However, calculation of an overall coefficient depends on whether it is based on the cold or hot side surface area, since $U_c \neq U_h$ if $A_c \neq A_h$. The conduction resistance R_w is obtained from Equation 3.6 for a plane wall or Equation 3.28 for a cylindrical wall. Although representative fouling factors are listed in Table 11.1, the factor is a variable during heat exchanger operation (increasing from zero for a clean surface, as deposits accumulate on the surface). Comprehensive discussions of fouling are provided in References 2 through 4.

The quantity η_o in Equation 11.1 is termed the *overall surface efficiency* or *temperature effectiveness* of a finned surface. It is defined such that, for the hot or cold surface, the heat transfer rate is

$$q = \eta_o h A (T_b - T_\infty) \tag{11.2}$$

where T_b is the base surface temperature (Figure 3.20) and A is the total (fin plus exposed base) surface area. The quantity was introduced in Section 3.6.5, and the following expression was derived:

$$\eta_o = 1 - \frac{A_f}{A}(1 - \eta_f) \tag{11.3}$$

TABLE 11.1   Representative Fouling Factors [1]

Fluid	R''_f (m² · K/W)
Seawater and treated boiler feedwater (below 50°C)	0.0001
Seawater and treated boiler feedwater (above 50°C)	0.0002
River water (below 50°C)	0.0002–0.001
Fuel oil	0.0009
Refrigerating liquids	0.0002
Steam (nonoil bearing)	0.0001

where A_f is the entire fin surface area and η_f is the efficiency of a single fin. To be consistent with the nomenclature commonly used in heat exchanger analysis, the ratio of fin surface area to the total surface area has been expressed as A_f/A. This representation differs from that of Section 3.6.5, where the ratio is expressed as NA_f/A_t, with A_f representing the area of a single fin and A_t the total surface area. If a straight or pin fin of length L (Figure 3.16) is used and an adiabatic tip is assumed, Equations 3.76 and 3.86 yield

$$\eta_f = \frac{\tanh (mL)}{mL} \qquad (11.4)$$

where $m = (2h/kt)^{1/2}$ and t is the fin thickness. For several common fin shapes, the efficiency may be obtained from Table 3.5.

The wall conduction term in Equation 11.1 may often be neglected, since a thin wall of large thermal conductivity is generally used. Also, one of the convection coefficients is often much smaller than the other and hence dominates determination of the overall coefficient. For example, if one of the fluids is a gas and the other is a liquid or a liquid–vapor mixture experiencing boiling or condensation, the gas-side convection coefficient is much smaller. It is in such situations that fins are used to enhance gas-side convection. Representative values of the overall coefficient are summarized in Table 11.2.

For the unfinned, tubular heat exchangers of Figures 11.1 to 11.4, Equation 11.1 reduces to

$$\frac{1}{UA} = \frac{1}{U_i A_i} = \frac{1}{U_o A_o}$$

$$= \frac{1}{h_i A_i} + \frac{R''_{f,i}}{A_i} + \frac{\ln (D_o/D_i)}{2\pi k L} + \frac{R''_{f,o}}{A_o} + \frac{1}{h_o A_o} \qquad (11.5)$$

where subscripts i and o refer to inner and outer tube surfaces ($A_i = \pi D_i L$, $A_o = \pi D_o L$), which may be exposed to either the hot or the cold fluid.

The overall heat transfer coefficient may be determined from knowledge of the hot and cold fluid convection coefficients and fouling factors and from appropriate geometric parameters. For unfinned surfaces, the convection coefficients may be estimated from correlations presented in Chapters 7 and 8. For standard fin configurations, the coefficients may be obtained from results compiled by Kays and London [5].

TABLE 11.2 Representative Values of the Overall Heat Transfer Coefficient

Fluid Combination	U (W/m² · K)
Water to water	850–1700
Water to oil	110–350
Steam condenser (water in tubes)	1000–6000
Ammonia condenser (water in tubes)	800–1400
Alcohol condenser (water in tubes)	250–700
Finned-tube heat exchanger (water in tubes, air in cross flow)	25–50

11.3
Heat Exchanger Analysis: Use of the Log Mean Temperature Difference

To design or to predict the performance of a heat exchanger, it is essential to relate the total heat transfer rate to quantities such as the inlet and outlet fluid temperatures, the overall heat transfer coefficient, and the total surface area for heat transfer. Two such relations may readily be obtained by applying overall energy balances to the hot and cold fluids, as shown in Figure 11.6. In particular, if q is the total rate of heat transfer between the hot and cold fluids and there is negligible heat transfer between the exchanger and its surroundings, as well as negligible potential and kinetic energy changes, application of an energy balance, Equation 1.11a, gives

$$q = \dot{m}_h(i_{h,i} - i_{h,o}) \tag{11.6a}$$

and

$$q = \dot{m}_c(i_{c,o} - i_{c,i}) \tag{11.7a}$$

where i is the fluid enthalpy. The subscripts h and c refer to the hot and cold fluids, whereas i and o designate the fluid inlet and outlet conditions. If the fluids are not undergoing a phase change and constant specific heats are assumed, these expressions reduce to

$$q = \dot{m}_h c_{p,h}(T_{h,i} - T_{h,o}) \tag{11.6b}$$

and

$$q = \dot{m}_c c_{p,c}(T_{c,o} - T_{c,i}) \tag{11.7b}$$

where the temperatures appearing in the expressions refer to the *mean* fluid temperatures at the designated locations. Note that Equations 11.6 and 11.7 are independent of the flow arrangement and heat exchanger type.

Another useful expression may be obtained by relating the total heat transfer rate q to the temperature difference ΔT between the hot and cold fluids, where

$$\Delta T \equiv T_h - T_c \tag{11.8}$$

Such an expression would be an extension of Newton's law of cooling, with the overall heat transfer coefficient U used in place of the single convection coeffi-

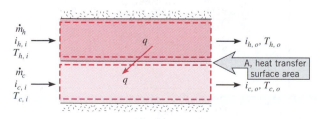

FIGURE 11.6   Overall energy balances for the hot and cold fluids of a two-fluid heat exchanger.

cient *h*. However, since ΔT varies with position in the heat exchanger, it is necessary to work with a rate equation of the form

$$q = UA\,\Delta T_m \qquad (11.9)$$

where ΔT_m is an appropriate *mean* temperature difference. Equation 11.9 may be used with Equations 11.6 and 11.7 to perform a heat exchanger analysis. Before this can be done, however, the specific form of ΔT_m must be established. Consider first the parallel-flow heat exchanger.

11.3.1 The Parallel-Flow Heat Exchanger

The hot and cold fluid temperature distributions associated with a parallel-flow heat exchanger are shown in Figure 11.7. The temperature difference ΔT is initially large but decays rapidly with increasing *x*, approaching zero asymptotically. It is important to note that, for such an exchanger, the outlet temperature of the cold fluid never exceeds that of the hot fluid. In Figure 11.7 the subscripts 1 and 2 designate opposite ends of the heat exchanger. This convention is used for all types of heat exchangers considered. For parallel flow, it follows that $T_{h,i} = T_{h,1}$, $T_{h,o} = T_{h,2}$, $T_{c,i} = T_{c,1}$, and $T_{c,o} = T_{c,2}$.

The form of ΔT_m may be determined by applying an energy balance to differential elements in the hot and cold fluids. Each element is of length *dx* and heat transfer surface area *dA*, as shown in Figure 11.7. The energy balances and the subsequent analysis are subject to the following assumptions.

1. The heat exchanger is insulated from its surroundings, in which case the only heat exchange is between the hot and cold fluids.

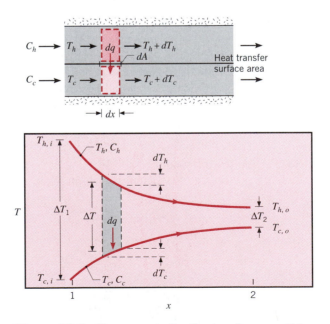

FIGURE 11.7 Temperature distributions for a parallel-flow heat exchanger.

2. Axial conduction along the tubes is negligible.

3. Potential and kinetic energy changes are negligible.

4. The fluid specific heats are constant.

5. The overall heat transfer coefficient is constant.

The specific heats may of course change as a result of temperature variations, and the overall heat transfer coefficient may change because of variations in fluid properties and flow conditions. However, in many applications such variations are not significant, and it is reasonable to work with average values of $c_{p,c}$, $c_{p,h}$, and U for the heat exchanger.

Applying an energy balance to each of the differential elements of Figure 11.7, it follows that

$$dq = -\dot{m}_h c_{p,h}\, dT_h \equiv -C_h\, dT_h \qquad (11.10)$$

and

$$dq = \dot{m}_c c_{p,c}\, dT_c \equiv C_c\, dT_c \qquad (11.11)$$

where C_h and C_c are the hot and cold fluid *heat capacity rates,* respectively. These expressions may be integrated across the heat exchanger to obtain the overall energy balances given by Equations 11.6b and 11.7b. The heat transfer across the surface area dA may also be expressed as

$$dq = U\, \Delta T\, dA \qquad (11.12)$$

where $\Delta T = T_h - T_c$ is the *local* temperature difference between the hot and cold fluids.

To determine the integrated form of Equation 11.12, we begin by substituting Equations 11.10 and 11.11 into the differential form of Equation 11.8

$$d(\Delta T) = dT_h - dT_c$$

to obtain

$$d(\Delta T) = -dq\left(\frac{1}{C_h} + \frac{1}{C_c}\right)$$

Substituting for dq from Equation 11.12 and integrating across the heat exchanger, we obtain

$$\int_1^2 \frac{d(\Delta T)}{\Delta T} = -U\left(\frac{1}{C_h} + \frac{1}{C_c}\right)\int_1^2 dA$$

or

$$\ln\left(\frac{\Delta T_2}{\Delta T_1}\right) = -UA\left(\frac{1}{C_h} + \frac{1}{C_c}\right) \qquad (11.13)$$

Substituting for C_h and C_c from Equations 11.6b and 11.7b, respectively, it follows that

$$\ln\left(\frac{\Delta T_2}{\Delta T_1}\right) = -UA\left(\frac{T_{h,i} - T_{h,o}}{q} + \frac{T_{c,o} - T_{c,i}}{q}\right)$$

$$= -\frac{UA}{q}\left[(T_{h,i} - T_{c,i}) - (T_{h,o} - T_{c,o})\right]$$

Recognizing that, for the parallel-flow heat exchanger of Figure 11.7, $\Delta T_1 = (T_{h,i} - T_{c,i})$ and $\Delta T_2 = (T_{h,o} - T_{c,o})$, we then obtain

$$q = UA \frac{\Delta T_2 - \Delta T_1}{\ln (\Delta T_2/\Delta T_1)}$$

Comparing the above expression with Equation 11.9, we conclude that the appropriate average temperature difference is a *log mean temperature difference*, ΔT_{lm}. Accordingly, we may write

$$q = UA \, \Delta T_{lm} \tag{11.14}$$

where

$$\Delta T_{lm} = \frac{\Delta T_2 - \Delta T_1}{\ln (\Delta T_2/\Delta T_1)} = \frac{\Delta T_1 - \Delta T_2}{\ln (\Delta T_1/\Delta T_2)} \tag{11.15}$$

Remember that, for the *parallel-flow exchanger,*

$$\left[\begin{array}{l} \Delta T_1 \equiv T_{h,1} - T_{c,1} = T_{h,i} - T_{c,i} \\ \Delta T_2 \equiv T_{h,2} - T_{c,2} = T_{h,o} - T_{c,o} \end{array} \right] \tag{11.16}$$

11.3.2   The Counterflow Heat Exchanger

The hot and cold fluid temperature distributions associated with a counterflow heat exchanger are shown in Figure 11.8. In contrast to the parallel-flow exchanger, this configuration provides for heat transfer between the hotter portions of the two fluids at one end, as well as between the colder portions at the

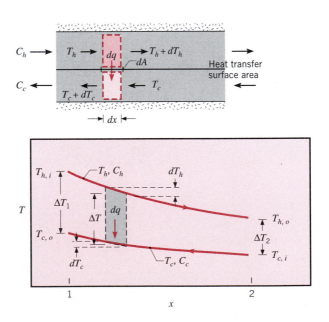

FIGURE 11.8   Temperature distributions for a counterflow heat exchanger.

other. For this reason, the change in the temperature difference, $\Delta T = T_h - T_c$, with respect to x is nowhere as large as it is for the inlet region of the parallel-flow exchanger. Note that the outer temperature of the cold fluid may now exceed the outlet temperature of the hot fluid.

Equations 11.6b and 11.7b apply to any heat exchanger and hence may be used for the counterflow arrangement. Moreover, from an analysis like that performed in Section 11.3.1, it may be shown that Equations 11.14 and 11.15 also apply. However, for the *counterflow exchanger* the endpoint temperature differences must now be defined as

$$\boxed{\begin{aligned} \Delta T_1 &\equiv T_{h,1} - T_{c,1} = T_{h,i} - T_{c,o} \\ \Delta T_2 &\equiv T_{h,2} - T_{c,2} = T_{h,o} - T_{c,i} \end{aligned}} \tag{11.17}$$

Note that, for the same inlet and outlet temperatures, the log mean temperature difference for counterflow exceeds that for parallel flow, $\Delta T_{\mathrm{lm,CF}} > \Delta T_{\mathrm{lm,PF}}$. Hence the surface area required to effect a prescribed heat transfer rate q is smaller for the counterflow than for the parallel-flow arrangement, assuming the same value of U. Also note that $T_{c,o}$ can exceed $T_{h,o}$ for counterflow but not for parallel flow.

11.3.3 Special Operating Conditions

It is useful to note certain special conditions under which heat exchangers may be operated. Figure 11.9a shows temperature distributions for a heat exchanger in which the hot fluid has a heat capacity rate, $C_h \equiv \dot{m}_h c_{p,h}$, which is much larger than that of the cold fluid, $C_c \equiv \dot{m}_c c_{p,c}$. For this case the temperature of the hot fluid remains approximately constant throughout the heat exchanger, while the temperature of the cold fluid increases. The same condition is achieved if the hot fluid is a condensing vapor. Condensation occurs at constant temperature, and, for all practical purposes, $C_h \to \infty$. Conversely, in an evaporator or a boiler (Figure 11.9b), it is the cold fluid that experiences a change in phase and remains at a nearly uniform temperature ($C_c \to \infty$). The same effect is achieved without phase change if $C_h \ll C_c$. Note that, with condensation or evaporation, the heat rate is given by Equation 11.6a or 11.7a. The third special case (Figure 11.9c) involves a counterflow heat exchanger for which the heat

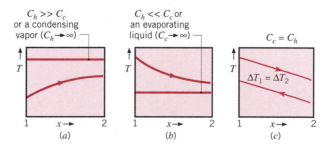

FIGURE 11.9   Special heat exchanger conditions. (*a*) $C_h \gg C_c$ or a condensing vapor. (*b*) An evaporating liquid or $C_h \ll C_c$. (*c*) A counterflow heat exchanger with equivalent fluid heat capacities ($C_h = C_c$).

capacity rates are equal ($C_h = C_c$). The temperature difference ΔT must then be a constant throughout the exchanger, in which case $\Delta T_1 = \Delta T_2 = \Delta T_{lm}$.

11.3.4 Multipass and Cross-Flow Heat Exchangers

Although flow conditions are more complicated in multipass and cross-flow heat exchangers, Equations 11.6, 11.7, 11.14, and 11.15 may still be used if the following modification is made to the log mean temperature difference [6]:

$$\Delta T_{lm} = F \, \Delta T_{lm, \, CF} \tag{11.18}$$

That is, the appropriate form of ΔT_{lm} is obtained by applying a correction factor to the value of ΔT_{lm} that would be computed *under the assumption of counterflow conditions*. Hence from Equation 11.17, $\Delta T_1 = T_{h, \, i} - T_{c, \, o}$ and $\Delta T_2 = T_{h, \, o} - T_{c, \, i}$.

Algebraic expressions for the correction factor F have been developed for various shell-and-tube and cross-flow heat exchanger configurations [1, 6, 7], and the results may be represented graphically. Selected results are shown in Figures 11.10 to 11.13 for common heat exchanger configurations. The notation (T, t) is used to specify the fluid temperatures, with the variable t always assigned to the tube-side fluid. With this convention it does not matter whether the hot fluid or the cold fluid flows through the shell or the tubes. An important implication of Figures 11.10 to 11.13 is that, *if the temperature change of one fluid is negligible,* either P or R is zero and F is 1. *Hence heat exchanger behavior is independent of the specific configuration.* Such would be the case if one of the fluids underwent a phase change.

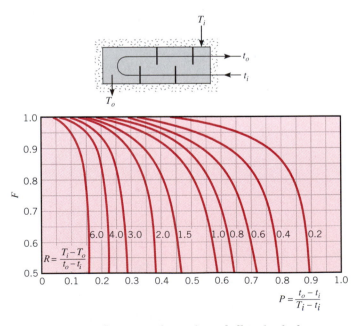

FIGURE 11.10 Correction factor for a shell-and-tube heat exchanger with one shell and any multiple of two tube passes (two, four, etc. tube passes).

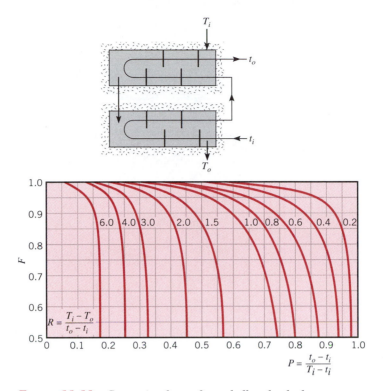

FIGURE 11.11   Correction factor for a shell-and-tube heat exchanger with two shell passes and any multiple of four tube passes (four, eight, etc. tube passes).

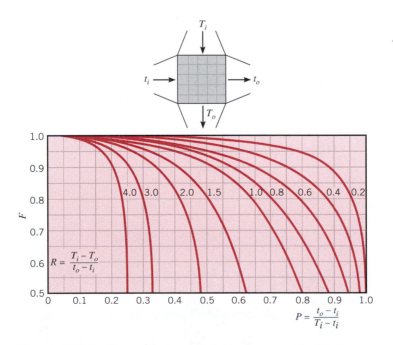

FIGURE 11.12   Correction factor for a single-pass, cross-flow heat exchanger with both fluids unmixed.

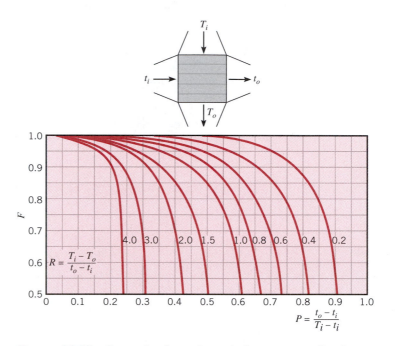

FIGURE 11.13   Correction factor for a single-pass, cross-flow heat exchanger with one fluid mixed and the other unmixed.

EXAMPLE 11.1

A counterflow, concentric tube heat exchanger is used to cool the lubricating oil for a large industrial gas turbine engine. The flow rate of cooling water through the inner tube (D_i = 25 mm) is 0.2 kg/s, while the flow rate of oil through the outer annulus (D_o = 45 mm) is 0.1 kg/s. The oil and water enter at temperatures of 100 and 30°C, respectively. How long must the tube be made if the outlet temperature of the oil is to be 60°C?

SOLUTION

Known: Fluid flow rates and inlet temperatures for a counterflow, concentric tube heat exchanger of prescribed inner and outer diameter.

Find: Tube length to achieve a desired hot fluid outlet temperature.

Schematic:

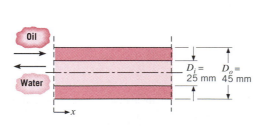

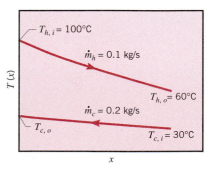

Assumptions:
1. Negligible heat loss to the surroundings.
2. Negligible kinetic and potential energy changes.
3. Constant properties.
4. Negligible tube wall thermal resistance and fouling factors.
5. Fully developed conditions for the water and oil (U independent of x).

Properties: Table A.5, unused engine oil ($\overline{T}_h = 80°C = 353$ K): $c_p = 2131$ J/kg · K, $\mu = 3.25 \times 10^{-2}$ N · s/m², $k = 0.138$ W/m · K. Table A.6, water ($\overline{T}_c \approx 35°C$): $c_p = 4178$ J/kg · K, $\mu = 725 \times 10^{-6}$ N · s/m², $k = 0.625$ W/m · K, $Pr = 4.85$.

Analysis: The required heat transfer rate may be obtained from the overall energy balance for the hot fluid, Equation 11.6b.

$$q = \dot{m}_h c_{p,h}(T_{h,i} - T_{h,o})$$
$$q = 0.1 \text{ kg/s} \times 2131 \text{ J/kg · K } (100 - 60)°C = 8524 \text{ W}$$

Applying Equation 11.7b, the water outlet temperature is

$$T_{c,o} = \frac{q}{\dot{m}_c c_{p,c}} + T_{c,i}$$

$$T_{c,o} = \frac{8524 \text{ W}}{0.2 \text{ kg/s} \times 4178 \text{ J/kg · K}} + 30°C = 40.2°C$$

Accordingly, use of $\overline{T}_c = 35°C$ to evaluate the water properties was a good choice. The required heat exchanger length may now be obtained from Equation 11.14,

$$q = UA \, \Delta T_{\text{lm}}$$

where $A = \pi D_i L$ and from Equations 11.15 and 11.17

$$\Delta T_{\text{lm}} = \frac{(T_{h,i} - T_{c,o}) - (T_{h,o} - T_{c,i})}{\ln [(T_{h,i} - T_{c,o})/(T_{h,o} - T_{c,i})]} = \frac{59.8 - 30}{\ln (59.8/30)} = 43.2°C$$

From Equation 11.5 the overall heat transfer coefficient is

$$U = \frac{1}{(1/h_i) + (1/h_o)}$$

For water flow through the tube,

$$Re_D = \frac{4\dot{m}_c}{\pi D_i \mu} = \frac{4 \times 0.2 \text{ kg/s}}{\pi(0.025 \text{ m})725 \times 10^{-6} \text{ N · s/m}^2} = 14{,}050$$

Accordingly, the flow is turbulent and the convection coefficient may be computed from Equation 8.60

$$Nu_D = 0.023 Re_D^{4/5} \, Pr^{0.4}$$

$$Nu_D = 0.023(14{,}050)^{4/5}(4.85)^{0.4} = 90$$

Hence

$$h_i = Nu_D \frac{k}{D_i} = \frac{90 \times 0.625 \text{ W/m} \cdot \text{K}}{0.025 \text{ m}} = 2250 \text{ W/m}^2 \cdot \text{K}$$

For the flow of oil through the annulus, the hydraulic diameter is, from Equation 8.72, $D_h = D_o - D_i = 0.02$ m, and the Reynolds number is

$$Re_D = \frac{\rho u_m D_h}{\mu} = \frac{\rho(D_o - D_i)}{\mu} \times \frac{\dot{m}_h}{\rho\pi(D_o^2 - D_i^2)/4}$$

$$Re_D = \frac{4\dot{m}_h}{\pi(D_o + D_i)\mu} = \frac{4 \times 0.1 \text{ kg/s}}{\pi(0.045 + 0.025) \text{ m} \times 3.25 \times 10^{-2} \text{ kg/s} \cdot \text{m}} = 56.0$$

The annular flow is therefore laminar. Assuming uniform temperature along the inner surface of the annulus and a perfectly insulated outer surface, the convection coefficient at the inner surface may be obtained from Table 8.2. With $(D_i/D_o) = 0.56$, linear interpolation provides

$$Nu_i = \frac{h_o D_h}{k} = 5.56$$

and

$$h_o = 5.56 \frac{0.138 \text{ W/m} \cdot \text{K}}{0.020 \text{ m}} = 38.4 \text{ W/m}^2 \cdot \text{K}$$

The overall convection coefficient is then

$$U = \frac{1}{(1/2250 \text{ W/m}^2 \cdot \text{K}) + (1/38.4 \text{ W/m}^2 \cdot \text{K})} = 37.8 \text{ W/m}^2 \cdot \text{K}$$

and from the rate equation it follows that

$$L = \frac{q}{U\pi D_i \, \Delta T_{lm}} = \frac{8524 \text{ W}}{37.8 \text{ W/m}^2 \cdot \text{K} \, \pi(0.025 \text{ m})(43.2°C)} = 66.5 \text{ m} \qquad \triangleleft$$

Comments:

1. The hot side convection coefficient controls the rate of heat transfer between the two fluids, and the low value of h_o is responsible for the large value of L. A spiral tube arrangement would be needed.

2. Because $h_i \gg h_o$, the tube wall temperature will follow closely that of the coolant water. Accordingly, the assumption of uniform wall temperature used to obtain h_o is reasonable.

EXAMPLE 11.2

A shell-and-tube heat exchanger must be designed to heat 2.5 kg/s of water from 15 to 85°C. The heating is to be accomplished by passing hot engine oil, which is available at 160°C, through the shell side of the exchanger. The oil is known to provide an average convection coefficient of $h_o = 400$ W/m² · K on the outside of the tubes. Ten tubes pass the water through the shell. Each tube is thin walled, of diameter $D = 25$ mm, and makes eight passes through the shell. If the oil leaves the exchanger at 100°C, what is its flow rate? How long must the tubes be to accomplish the desired heating?

SOLUTION

Known: Fluid inlet and outlet temperatures for a shell-and-tube heat exchanger with 10 tubes making eight passes.

Find:

1. Oil flow rate required to achieve specified outlet temperature.
2. Tube length required to achieve specified water heating.

Schematic:

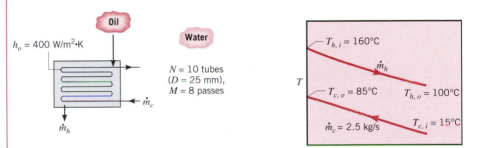

Assumptions:

1. Negligible heat loss to the surroundings and kinetic and potential energy changes.
2. Constant properties.
3. Negligible tube wall thermal resistance and fouling effects.
4. Fully developed water flow in tubes.

Properties: Table A.5, unused engine oil ($\overline{T}_h = 130$°C): $c_p = 2350$ J/kg · K. Table A.6, water ($\overline{T}_c = 50$°C): $c_p = 4181$ J/kg · K, $\mu = 548 \times 10^{-6}$ N · s/m², $k = 0.643$ W/m · K, $Pr = 3.56$.

Analysis:

1. From the overall energy balance, Equation 11.7b, the heat transfer required of the exchanger is

$$q = \dot{m}_c c_{p,c}(T_{c,o} - T_{c,i}) = 2.5 \text{ kg/s} \times 4181 \text{ J/kg} \cdot \text{K} (85 - 15)°\text{C}$$

$$q = 7.317 \times 10^5 \text{ W}$$

Hence, from Equation 11.6b,

$$\dot{m}_h = \frac{q}{c_{p,h}(T_{h,i} - T_{h,o})} = \frac{7.317 \times 10^5 \text{ W}}{2350 \text{ J/kg} \cdot \text{K} \times (160 - 100)^\circ\text{C}} = 5.19 \text{ kg/s} \quad \triangleleft$$

2. The required tube length may be obtained from Equations 11.14 and 11.18, where

$$q = UAF \, \Delta T_{\text{lm, CF}}$$

From Equation 11.5,

$$U = \frac{1}{(1/h_i) + (1/h_o)}$$

where h_i may be obtained by first calculating Re_D. With $\dot{m}_1 \equiv \dot{m}_c/N = 0.25$ kg/s defined as the water flow rate per tube, Equation 8.6 yields

$$Re_D = \frac{4\dot{m}_1}{\pi D \mu} = \frac{4 \times 0.25 \text{ kg/s}}{\pi(0.025 \text{ m})548 \times 10^{-6} \text{ kg/s} \cdot \text{m}} = 23,234$$

Hence the water flow is turbulent, and from Equation 8.60

$$Nu_D = 0.023 Re_D^{4/5} Pr^{0.4} = 0.023(23,234)^{4/5}(3.56)^{0.4} = 119$$

$$h_i = \frac{k}{D} Nu_D = \frac{0.643 \text{ W/m} \cdot \text{K}}{0.025 \text{ m}} 119 = 3061 \text{ W/m}^2 \cdot \text{K}$$

Hence

$$U = \frac{1}{(1/400) + (1/3061)} = 354 \text{ W/m}^2 \cdot \text{K}$$

The correction factor F may be obtained from Figure 11.10, where

$$R = \frac{160 - 100}{85 - 15} = 0.86 \qquad P = \frac{85 - 15}{160 - 15} = 0.48$$

Hence $F \approx 0.87$. From Equations 11.15 and 11.17, it follows that

$$\Delta T_{\text{lm, CF}} = \frac{(T_{h,i} - T_{c,o}) - (T_{h,o} - T_{c,i})}{\ln[(T_{h,i} - T_{c,o})/(T_{h,o} - T_{c,i})]} = \frac{75 - 85}{\ln(75/85)} = 79.9^\circ\text{C}$$

Hence, since $A = N\pi DL$, where $N = 10$ is the number of tubes,

$$L = \frac{q}{UN\pi DF \, \Delta T_{\text{lm, CF}}} = \frac{7.317 \times 10^5 \text{ W}}{354 \text{ W/m}^2 \cdot \text{K} \times 10\pi(0.025 \text{ m})0.87(79.9^\circ\text{C})}$$

$$L = 37.9 \text{ m} \quad \triangleleft$$

Comments:

1. With $(L/D) = 37.9 \text{ m}/0.025 \text{ m} = 1516$, the assumption of fully developed conditions throughout the tube is justified.

2. With eight passes, the shell length is approximately $L/M = 4.7$ m.

11.4
Heat Exchanger Analysis: The Effectiveness–NTU Method

It is a simple matter to use the log mean temperature difference (LMTD) method of heat exchanger analysis when the fluid inlet temperatures are known and the outlet temperatures are specified or readily determined from the energy balance expressions, Equations 11.6b and 11.7b. The value of ΔT_{lm} for the exchanger may then be determined. However, if only the inlet temperatures are known, use of the LMTD method requires an iterative procedure. In such cases it is preferable to use an alternative approach, termed the *effectiveness–NTU* method.

11.4.1 Definitions

To define the *effectiveness of a heat exchanger,* we must first determine the *maximum possible heat transfer rate,* q_{max}, for the exchanger. This heat transfer rate could, in principle, be achieved in a counterflow heat exchanger (Figure 11.8) of infinite length. In such an exchanger, one of the fluids would experience the maximum possible temperature difference, $T_{h,i} - T_{c,i}$. To illustrate this point, consider a situation for which $C_c < C_h$, in which case, from Equations 11.10 and 11.11, $|dT_c| > |dT_h|$. The cold fluid would then experience the larger temperature change, and since $L \to \infty$, it would be heated to the inlet temperature of the hot fluid ($T_{c,o} = T_{h,i}$). Accordingly, from Equation 11.7b,

$$C_c < C_h: \qquad q_{max} = C_c(T_{h,i} - T_{c,i})$$

Similarly, if $C_h < C_c$, the hot fluid would experience the larger temperature change and would be cooled to the inlet temperature of the cold fluid ($T_{h,o} = T_{c,i}$). From Equation 11.6b, we then obtain

$$C_h < C_c: \qquad q_{max} = C_h(T_{h,i} - T_{c,i})$$

From the foregoing results we are then prompted to write the general expression

$$q_{max} = C_{min}(T_{h,i} - T_{c,i}) \tag{11.19}$$

where C_{min} is equal to C_c or C_h, whichever is smaller. For prescribed hot and cold fluid inlet temperatures, Equation 11.19 provides the maximum heat transfer rate that could possibly be delivered by the exchanger. A quick mental exercise should convince the reader that the maximum possible heat transfer rate is *not* equal to $C_{max}(T_{h,i} - T_{c,i})$. If the fluid having the larger heat capacity rate were to experience the maximum possible temperature change, conservation of energy in the form $C_c(T_{c,o} - T_{c,i}) = C_h(T_{h,i} - T_{h,o})$ would require that the other fluid experience yet a larger temperature change. For example, if $C_{max} = C_c$ and one argues that it is possible for $T_{c,o}$ to be equal to $T_{h,i}$, it follows that $(T_{h,i} - T_{h,o}) = (C_c/C_h)(T_{h,i} - T_{c,i})$, in which case $(T_{h,i} - T_{h,o}) > (T_{h,i} - T_{c,i})$. Such a condition is clearly impossible.

It is now logical to define the *effectiveness, ε,* as the ratio of the actual heat transfer rate for a heat exchanger to the maximum possible heat transfer rate:

$$\varepsilon \equiv \frac{q}{q_{max}} \tag{11.20}$$

From Equations 11.6b, 11.7b, and 11.19, it follows that

$$\varepsilon \equiv \frac{C_h(T_{h,\,i} - T_{h,\,o})}{C_{min}(T_{h,\,i} - T_{c,\,i})} \tag{11.21}$$

or

$$\varepsilon = \frac{C_c(T_{c,\,o} - T_{c,\,i})}{C_{min}(T_{h,\,i} - T_{c,\,i})} \tag{11.22}$$

By definition the effectiveness, which is dimensionless, must be in the range $0 \le \varepsilon \le 1$. It is useful because, if ε, $T_{h,i}$, and $T_{c,i}$ are known, the actual heat transfer rate may readily be determined from the expression

$$q = \varepsilon C_{min}(T_{h,\,i} - T_{c,\,i}) \tag{11.23}$$

For any heat exchanger it can be shown that [5]

$$\varepsilon = f\left(\text{NTU}, \frac{C_{min}}{C_{max}}\right) \tag{11.24}$$

where C_{min}/C_{max} is equal to C_c/C_h or C_h/C_c, depending on the relative magnitudes of the hot and cold fluid heat capacity rates. The *number of transfer units* (NTU) is a dimensionless parameter that is widely used for heat exchanger analysis and is defined as

$$\text{NTU} \equiv \frac{UA}{C_{min}} \tag{11.25}$$

11.4.2 Effectiveness–NTU Relations

To determine a specific form of the effectiveness–NTU relation, Equation 11.24, consider a parallel-flow heat exchanger for which $C_{min} = C_h$. From Equation 11.21 we then obtain

$$\varepsilon = \frac{T_{h,\,i} - T_{h,\,o}}{T_{h,\,i} - T_{c,\,i}} \tag{11.26}$$

and from Equations 11.6b and 11.7b it follows that

$$\frac{C_{min}}{C_{max}} = \frac{\dot{m}_h c_{p,\,h}}{\dot{m}_c c_{p,\,c}} = \frac{T_{c,\,o} - T_{c,\,i}}{T_{h,\,i} - T_{h,\,o}} \tag{11.27}$$

Now consider Equation 11.13, which may be expressed as

$$\ln\left(\frac{T_{h,\,o} - T_{c,\,o}}{T_{h,\,i} - T_{c,\,i}}\right) = -\frac{UA}{C_{min}}\left(1 + \frac{C_{min}}{C_{max}}\right)$$

or from Equation 11.25

$$\frac{T_{h,o} - T_{c,o}}{T_{h,i} - T_{c,i}} = \exp\left[-\text{NTU}\left(1 + \frac{C_{\min}}{C_{\max}}\right)\right] \tag{11.28}$$

Rearranging the left-hand side of this expression as

$$\frac{T_{h,o} - T_{c,o}}{T_{h,i} - T_{c,i}} = \frac{T_{h,o} - T_{h,i} + T_{h,i} - T_{c,o}}{T_{h,i} - T_{c,i}}$$

and substituting for $T_{c,o}$ from Equation 11.27, it follows that

$$\frac{T_{h,o} - T_{c,o}}{T_{h,i} - T_{c,i}} = \frac{(T_{h,o} - T_{h,i}) + (T_{h,i} - T_{c,i}) - (C_{\min}/C_{\max})(T_{h,i} - T_{h,o})}{T_{h,i} - T_{c,i}}$$

or from Equation 11.26

$$\frac{T_{h,o} - T_{c,o}}{T_{h,i} - T_{c,i}} = -\varepsilon + 1 - \left(\frac{C_{\min}}{C_{\max}}\right)\varepsilon = 1 - \varepsilon\left(1 + \frac{C_{\min}}{C_{\max}}\right)$$

TABLE 11.3   **Heat Exchanger Effectiveness Relations [5]**

Flow Arrangement	Relation	
Concentric tube		
Parallel flow	$\varepsilon = \dfrac{1 - \exp\left[-\text{NTU}(1 + C_r)\right]}{1 + C_r}$	(11.29a)
Counterflow	$\varepsilon = \dfrac{1 - \exp\left[-\text{NTU}(1 - C_r)\right]}{1 - C_r \exp\left[-\text{NTU}(1 - C_r)\right]} \quad (C_r < 1)$	
	$\varepsilon = \dfrac{\text{NTU}}{1 + \text{NTU}} \quad (C_r = 1)$	(11.30a)
Shell and tube		
One shell pass (2, 4, . . . tube passes)	$\varepsilon_1 = 2\Bigg\{1 + C_r + (1 + C_r^2)^{1/2}$	
	$\quad\quad \times \dfrac{1 + \exp\left[-\text{NTU}(1 + C_r^2)^{1/2}\right]}{1 - \exp\left[-\text{NTU}(1 + C_r^2)^{1/2}\right]}\Bigg\}^{-1}$	(11.31a)
n Shell passes (2n, 4n, . . . tube passes)	$\varepsilon = \left[\left(\dfrac{1 - \varepsilon_1 C_r}{1 - \varepsilon_1}\right)^n - 1\right]\left[\left(\dfrac{1 - \varepsilon_1 C_r}{1 - \varepsilon_1}\right)^n - C_r\right]^{-1}$	(11.32a)
Cross flow (single pass)		
Both fluids unmixed	$\varepsilon = 1 - \exp\left[\left(\dfrac{1}{C_r}\right)(\text{NTU})^{0.22}\left\{\exp\left[-C_r(\text{NTU})^{0.78}\right] - 1\right\}\right]$	(11.33)
$C_{\max}$ (mixed), $C_{\min}$ (unmixed)	$\varepsilon = \left(\dfrac{1}{C_r}\right)(1 - \exp\{-C_r[1 - \exp(-\text{NTU})]\})$	(11.34a)
$C_{\min}$ (mixed), $C_{\max}$ (unmixed)	$\varepsilon = 1 - \exp(-C_r^{-1}\{1 - \exp[-C_r(\text{NTU})]\})$	(11.35a)
All exchangers ($C_r = 0$)	$\varepsilon = 1 - \exp(-\text{NTU})$	(11.36a)

Substituting the above expression into Equation 11.28 and solving for ε, we obtain for the *parallel-flow heat exchanger*

$$\varepsilon = \frac{1 - \exp\{-\text{NTU}[1 + (C_{min}/C_{max})]\}}{1 + (C_{min}/C_{max})} \tag{11.29a}$$

Since precisely the same result may be obtained for $C_{min} = C_c$, Equation 11.29a applies for any parallel-flow heat exchanger, irrespective of whether the minimum heat capacity rate is associated with the hot or cold fluid.

Similar expressions have been developed for a variety of heat exchangers [5], and representative results are summarized in Table 11.3, where C_r is the *heat capacity ratio* $C_r \equiv C_{min}/C_{max}$. In deriving Equation 11.32a, it is assumed that the total NTU is equally distributed between shell passes of the same arrangement, $\text{NTU} = n(\text{NTU})_1$. Hence, when using ε_1 with this expression, NTU is replaced by $(\text{NTU})/n$ in Equation 11.31a. Note that for $C_r = 0$, as in a boiler or condenser, ε is given by Equation 11.36a for *all flow arrangements*. Hence, for this special case, it follows that heat exchanger behavior is independent of flow arrangement. For the cross-flow heat exchanger with both fluids unmixed, Equation 11.33 is exact only for $C_r = 1$. However, it may be used to

TABLE 11.4 Heat Exchanger NTU Relations

Flow Arrangement	Relation	
Concentric tube		
Parallel flow	$\text{NTU} = -\dfrac{\ln[1 - \varepsilon(1 + C_r)]}{1 + C_r}$	(11.29b)
Counterflow	$\text{NTU} = \dfrac{1}{C_r - 1}\ln\left(\dfrac{\varepsilon - 1}{\varepsilon C_r - 1}\right)\quad (C_r < 1)$	
	$\text{NTU} = \dfrac{\varepsilon}{1 - \varepsilon}\qquad (C_r = 1)$	(11.30b)
Shell and tube		
One shell pass (2, 4, . . . tube passes)	$\text{NTU} = -(1 + C_r^2)^{-1/2}\ln\left(\dfrac{E - 1}{E + 1}\right)$	(11.31b)
	$E = \dfrac{2/\varepsilon_1 - (1 + C_r)}{(1 + C_r^2)^{1/2}}$	(11.31c)
n Shell passes (2n, 4n, . . . tube passes)	Use Equations 11.31b and 11.31c with $\varepsilon_1 = \dfrac{F - 1}{F - C_r},\quad F = \left(\dfrac{\varepsilon C_r - 1}{\varepsilon - 1}\right)^{1/n}$	(11.32b,c)
Cross flow (single pass)		
C_{max} (mixed), C_{min} (unmixed)	$\text{NTU} = -\ln\left[1 + \left(\dfrac{1}{C_r}\right)\ln(1 - \varepsilon C_r)\right]$	(11.34b)
C_{min} (mixed), C_{max} (unmixed)	$\text{NTU} = -\left(\dfrac{1}{C_r}\right)\ln[C_r\ln(1 - \varepsilon) + 1]$	(11.35b)
All exchangers ($C_r = 0$)	$\text{NTU} = -\ln(1 - \varepsilon)$	(11.36b)

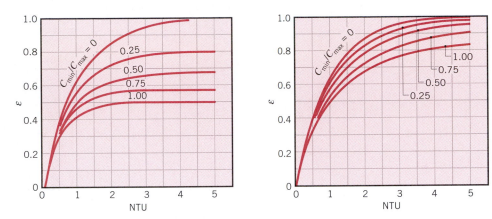

FIGURE 11.14 Effectiveness of a parallel-flow heat exchanger (Equation 11.29).

FIGURE 11.15 Effectiveness of a counterflow heat exchanger (Equation 11.30).

an excellent approximation for all $0 < C_r \leq 1$. For $C_r = 0$, Equation 11.36a must be used.

In heat exchanger design calculations, it is more convenient to work with ε–NTU relations of the form

$$NTU = f\left(\varepsilon, \frac{C_{min}}{C_{max}}\right)$$

Explicit relations for NTU as a function of ε and C_r are provided in Table 11.4. Note that Equation 11.33 may not be manipulated to yield a direct relationship for NTU as a function of ε and C_r. Note also, that in using Equations 11.32b, c with 11.31b, c, it is the NTU *per shell pass* that is computed from Equation

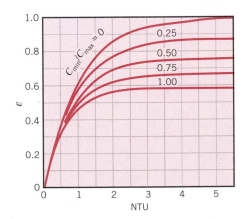

FIGURE 11.16 Effectiveness of a shell-and-tube heat exchanger with one shell and any multiple of two tube passes (two, four, etc. tube passes) (Equation 11.31).

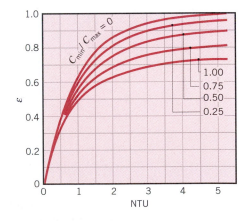

FIGURE 11.17 Effectiveness of a shell-and-tube heat exchanger with two shell passes and any multiple of four tube passes (four, eight, etc. tube passes) (Equation 11.32 with $n = 2$).

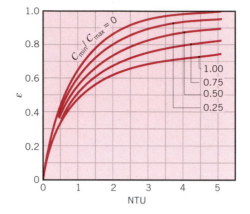

FIGURE 11.18 Effectiveness of a single-pass, cross-flow heat exchanger with both fluids unmixed (Equation 11.33).

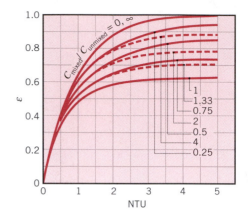

FIGURE 11.19 Effectiveness of a single-pass, cross-flow heat exchanger with one fluid mixed and the other unmixed (Equations 11.34, 11.35).

11.31b. This result is multiplied by n to obtain the NTU for the entire exchanger.

The foregoing expressions are represented graphically in Figures 11.14 to 11.19. For Figure 11.19 the solid curves correspond to C_{min} mixed and C_{max} unmixed, while the dashed curves correspond to C_{min} unmixed and C_{max} mixed. Note that for $C_r = 0$, all heat exchangers have the same effectiveness, which may be computed from Equation 11.36a. Moreover, if NTU $\lesssim 0.25$, all heat exchangers have approximately the same effectiveness, regardless of the value of C_r, and ε may again be computed from Equation 11.36a. More generally, for $C_r > 0$ and NTU $\gtrsim 0.25$, the counterflow exchanger is the most effective. For any exchanger, maximum and minimum values of the effectiveness are associated with $C_r = 0$ and $C_r = 1$, respectively.

EXAMPLE 11.3

Hot exhaust gases, which enter a finned-tube, cross-flow heat exchanger at 300°C and leave at 100°C, are used to heat pressurized water at a flow rate of 1 kg/s from 35 to 125°C. The exhaust gas specific heat is approximately 1000 J/kg · K, and the overall heat transfer coefficient based on the gas-side surface area is $U_h = 100$ W/m² · K. Determine the required gas-side surface area A_h using the NTU method.

SOLUTION

Known: Inlet and outlet temperatures of hot gases and water used in a finned-tube, cross-flow heat exchanger. Water flow rate and gas-side overall heat transfer coefficient.

Find: Required gas-side surface area.

Schematic:

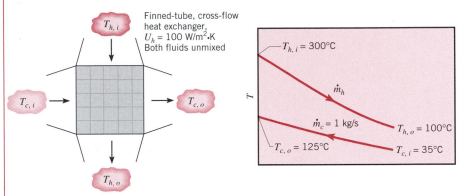

Assumptions:

1. Negligible heat loss to the surroundings and kinetic and potential energy changes.
2. Constant properties.

Properties: Table A.6, water ($\overline{T}_c = 80°C$): $c_{p,c} = 4197$ J/kg · K. Exhaust gases: $c_{p,h} = 1000$ J/kg · K.

Analysis: The required surface area may be obtained from knowledge of the number of transfer units, which, in turn, may be obtained from knowledge of the ratio of heat capacity rates and the effectiveness. To determine the minimum heat capacity rate, we begin by computing

$$C_c = \dot{m}_c c_{p,c} = 1 \text{ kg/s} \times 4197 \text{ J/kg} \cdot \text{K} = 4197 \text{ W/K}$$

Since $\dot{m}_h$ is not specified, C_h is obtained by combining the overall energy balances, Equations 11.6b and 11.7b:

$$C_h = \dot{m}_h c_{p,h} = C_c \frac{T_{c,o} - T_{c,i}}{T_{h,i} - T_{h,o}} = 4197 \frac{125 - 35}{300 - 100} = 1889 \text{ W/K} = C_{\min}$$

From Equation 11.19

$$q_{\max} = C_{\min}(T_{h,i} - T_{c,i}) = 1889 \text{ W/K } (300 - 35)°C = 5.01 \times 10^5 \text{ W}$$

From Equation 11.7b the actual heat transfer rate is

$$q = \dot{m}_c c_{p,c}(T_{c,o} - T_{c,i}) = 1 \text{ kg/s} \times 4197 \text{ J/kg} \cdot \text{K } (125 - 35)°C$$

$$q = 3.77 \times 10^5 \text{ W}$$

Hence from Equation 11.20 the effectiveness is

$$\varepsilon = \frac{q}{q_{\max}} = \frac{3.77 \times 10^5 \text{ W}}{5.01 \times 10^5 \text{ W}} = 0.75$$

With

$$\frac{C_{\min}}{C_{\max}} = \frac{1889}{4197} = 0.45$$

it follows from Figure 11.18 that

$$\text{NTU} = \frac{U_h A_h}{C_{\min}} \approx 2.1$$

or

$$A_h = \frac{2.1(1889 \text{ W/K})}{100 \text{ W/m}^2 \cdot \text{K}} = 39.7 \text{ m}^2 \qquad \triangleleft$$

Comments:

1. The desired heat transfer area may also be determined by using the LMTD method. From Equations 11.14 and 11.18,

$$A_h = \frac{q}{U_h F \, \Delta T_{\text{lm, CF}}}$$

With

$$P = \frac{t_o - t_i}{T_i - t_i} = \frac{125 - 35}{300 - 35} = 0.34 \qquad R = \frac{T_i - T_o}{t_o - t_i} = \frac{300 - 100}{125 - 35} = 2.22$$

it follows from Figure 11.12 that $F \approx 0.87$. From Equations 11.15 and 11.17,

$$\Delta T_{\text{lm, CF}} = \frac{(T_{h,i} - T_{c,o}) - (T_{h,o} - T_{c,i})}{\ln\left[(T_{h,i} - T_{c,o})/(T_{h,o} - T_{c,i})\right]} = 111°\text{C}$$

in which case

$$A_h = \frac{3.77 \times 10^5 \text{ W}}{100 \text{ W/m}^2 \cdot \text{K} \times 0.87 \times 111°\text{C}} = 39.1 \text{ m}^2 \qquad \triangleleft$$

Agreement of the two results is well within the accuracy associated with reading the F and ε charts.

2. With the heat exchanger sized ($A_h = 39.7$ m^2) and placed into operation, its actual performance is subject to uncontrolled variations in the exhaust gas inlet temperature ($200 \leq T_{h,i} \leq 400°$C) and to gradual degradation of the heat exchanger surfaces due to fouling (U_h decreasing from 100 to 60 W/m^2 · K). For a fixed value of $C_{\min} = C_h = 1889$ W/K, the reduction in U_h corresponds to a reduction in the NTU (to NTU ≈ 1.26) and hence to a reduction in the heat exchanger effectiveness, which can be computed from Equation 11.33. The effect of the variations on the water outlet temperature has been computed and is plotted as follows:

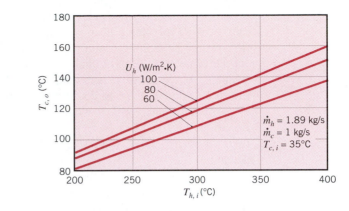

If the intent is to maintain a fixed water outlet temperature of $T_{c,\,o} = 125°C$, adjustments in the flow rates, $\dot{m}_c$ and $\dot{m}_h$, could be made to compensate for the variations. The model equations could be used to determine the adjustments and hence as a basis for designing the requisite *controller*.

11.5
Methodology of a Heat Exchanger Calculation

We have developed two procedures for performing a heat exchanger analysis, the LMTD method and the NTU approach. For any problem both methods may be used to obtain equivalent results. However, depending on the nature of the problem, the NTU approach may be easier to implement.

Clearly, use of the LMTD method, Equations 11.14 and 11.15, is facilitated by knowledge of the hot and cold fluid inlet and outlet temperatures, since ΔT_{lm} may then be readily computed. Problems for which these temperatures are known may be classified as *heat exchanger design problems*. Typically, the fluid inlet temperatures and flow rates, as well as a desired hot or cold fluid outlet temperature, are prescribed. The design problem is then one of selecting an appropriate heat exchanger type and determining the size, that is, the heat transfer surface area A, required to achieve the desired outlet temperature. For example, consider an application for which $\dot{m}_c$, $\dot{m}_h$, $T_{c,\,i}$, and $T_{h,\,i}$ are known, and the objective is to specify a heat exchanger that will provide a desired value of $T_{c,\,o}$. The corresponding values of q and $T_{h,\,o}$ may be computed from the energy balances, Equations 11.7b and 11.6b, respectively, and the value of ΔT_{lm} may be found from its definition, Equation 11.15. Using the rate equation (11.14), it is then a simple matter to determine the required value of A. Of course, the NTU method may also be used to obtain A by first calculating ε and $(C_{\mathrm{min}}/C_{\mathrm{max}})$. The appropriate chart (or equation) may then be used to obtain the NTU value, which in turn may be used to determine A.

Alternatively, the heat exchanger type and size may be known and the objective is to determine the heat transfer rate and the fluid outlet temperatures for prescribed fluid flow rates and inlet temperatures. Although the LMTD method may be used for such a heat exchanger *performance calculation,* the computations would be tedious, requiring iteration. For example, a guess could be made for the value of $T_{c,\,o}$, and Equations 11.7b and 11.6b could be used to determine q and $T_{h,\,o}$, respectively. Knowing all fluid temperatures, ΔT_{lm} could be determined and Equation 11.14 could then be used to again compute the value of q. The original guess for $T_{c,\,o}$ would be correct, if the values of q obtained from Equations 11.7b and 11.14 were in agreement. Such agreement would be fortuitous, however, and it is likely that some iteration on the value of $T_{c,\,o}$ would be needed.

The iterative nature of the above solution could be eliminated by using the NTU method. From knowledge of the heat exchanger type and size and the fluid flow rates, the NTU and $(C_{\min}/C_{\max})$ values may be computed and ε may then be determined from the appropriate chart (or equation). Since $q_{\max}$ may also be computed from Equation 11.19, it is a simple matter to determine the actual heat transfer rate from the requirement that $q = \varepsilon q_{\max}$. Both fluid outlet temperatures may then be determined from Equations 11.6b and 11.7b.

EXAMPLE 11.4

Consider the heat exchanger design of Example 11.3, that is, a finned-tube, cross-flow heat exchanger with a gas-side overall heat transfer coefficient and area of 100 W/m² · K and 40 m², respectively. The water flow rate and inlet temperature remain at 1 kg/s and 35°C. However, a change in operating conditions for the hot gas generator causes the gases to now enter the heat exchanger with a flow rate of 1.5 kg/s and a temperature of 250°C. What is the rate of heat transfer by the exchanger, and what are the gas and water outlet temperatures?

SOLUTION

Known: Hot and cold fluid inlet conditions for a finned-tube, cross-flow heat exchanger of known surface area and overall heat transfer coefficient.

Find: Heat transfer rate and fluid outlet temperatures.

Schematic:

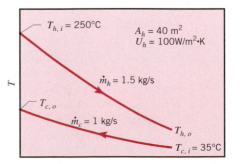

Assumptions:

1. Negligible heat loss to surroundings and kinetic and potential energy changes.

2. Constant properties (unchanged from Example 11.3).

Analysis: The problem may be classified as one requiring a heat exchanger *performance calculation*. Accordingly, it is expedient to base the calculations on the NTU method. The heat capacity rates are

$$C_c = \dot{m}_c c_{p,\, c} = 1 \text{ kg/s} \times 4197 \text{ J/kg} \cdot \text{K} = 4197 \text{ W/K}$$

$$C_h = \dot{m}_h c_{p,\, h} = 1.5 \text{ kg/s} \times 1000 \text{ J/kg} \cdot \text{K} = 1500 \text{ W/K} = C_{\min}$$

in which case

$$\frac{C_{\min}}{C_{\max}} = \frac{1500}{4197} = 0.357$$

The number of transfer units is

$$\text{NTU} = \frac{U_h A_h}{C_{\min}} = \frac{100 \text{ W/m}^2 \cdot \text{K} \times 40 \text{ m}^2}{1500 \text{ W/K}} = 2.67$$

From Figure 11.18 the heat exchanger effectiveness is then $\varepsilon \approx 0.82$, and from Equation 11.19 the maximum possible heat transfer rate is

$$q_{\max} = C_{\min}(T_{h,\, i} - T_{c,\, i}) = 1500 \text{ W/K} (250 - 35)°\text{C} = 3.23 \times 10^5 \text{ W}$$

Accordingly, from the definition of ε, Equation 11.20, the actual heat transfer rate is

$$q = \varepsilon q_{\max} = 0.82 \times 3.23 \times 10^5 \text{ W} = 2.65 \times 10^5 \text{ W} \qquad \triangleleft$$

It is now a simple matter to determine the outlet temperatures from the overall energy balances. From Equation 11.6b

$$T_{h,\, o} = T_{h,\, i} - \frac{q}{\dot{m}_h c_{p,\, h}} = 250°\text{C} - \frac{2.65 \times 10^5 \text{ W}}{1500 \text{ W/K}} = 73.3°\text{C} \qquad \triangleleft$$

and from Equation 11.7b

$$T_{c,\, o} = T_{c,\, i} + \frac{q}{\dot{m}_c c_{p,\, c}} = 35°\text{C} + \frac{2.65 \times 10^5 \text{ W}}{4197 \text{ W/K}} = 98.1°\text{C} \qquad \triangleleft$$

Comments:

1. From Equation 11.33, $\varepsilon = 0.845$, which is in good agreement with the estimate obtained from the charts.

2. The overall heat transfer coefficient has tacitly been assumed to be unaffected by the change in $\dot{m}_h$. In fact, with an approximately 20% reduction in $\dot{m}_h$, there would be a significant, albeit smaller, reduction in U_h.

3. As discussed in Comment 2 of Example 11.3, flow rate adjustments could be made to maintain a fixed water outlet temperature. If, for example, the

outlet temperature must be maintained at $T_{c,\,o} = 125°C$, the water flow rate could be reduced to an amount prescribed by Equation 11.7b. That is,

$$\dot{m}_c = \frac{q}{c_{p,\,c}(T_{c,\,o} - T_{c,\,i})} = \frac{2.65 \times 10^5 \text{ W}}{4197 \text{ J/kg} \cdot \text{K } (125 - 35)°C} = 0.702 \text{ kg/s}$$

The change in flow rate has again been presumed to have a negligible effect on U_h. In this case the assumption is good, since the dominant contribution to U_h is made by the gas-side, and not the water-side, convection coefficient.

EXAMPLE 11.5

The condenser of a large steam power plant is a heat exchanger in which steam is condensed to liquid water. Assume the condenser to be a *shell-and-tube* heat exchanger consisting of a single shell and 30,000 tubes, each executing two passes. The tubes are of thin wall construction with $D = 25$ mm, and steam condenses on their outer surface with an associated convection coefficient of $h_o = 11{,}000 \text{ W/m}^2 \cdot \text{K}$. The heat transfer rate that must be effected by the exchanger is $q = 2 \times 10^9$ W, and this is accomplished by passing cooling water through the tubes at a rate of 3×10^4 kg/s (the flow rate per tube is therefore 1 kg/s). The water enters at 20°C, while the steam condenses at 50°C. What is the temperature of the cooling water emerging from the condenser? What is the required tube length L per pass?

SOLUTION

Known: Heat exchanger consisting of single shell and 30,000 tubes with two passes each.

Find:

1. Outlet temperature of the cooling water.
2. Tube length per pass to achieve required heat transfer.

Schematic:

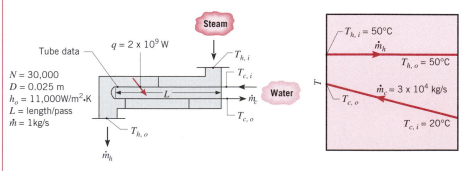

Assumptions:

1. Negligible heat transfer between exchanger and surroundings and negligible kinetic and potential energy changes.

2. Tube internal flow and thermal conditions fully developed.
3. Negligible thermal resistance of tube material and fouling effects.
4. Constant properties.

Properties: Table A.6, water (assume $\overline{T}_c \approx 27°C = 300$ K): $c_p = 4179$ J/kg · K, $\mu = 855 \times 10^{-6}$ N · s/m^2, $k = 0.613$ W/m · K, $Pr = 5.83$.

Analysis:

1. The cooling water outlet temperature may be obtained from the overall energy balance, Equation 11.7b. Accordingly,

$$T_{c,o} = T_{c,i} + \frac{q}{\dot{m}_c c_{p,c}} = 20°C + \frac{2 \times 10^9 \text{ W}}{3 \times 10^4 \text{ kg/s} \times 4179 \text{ J/kg · K}}$$

$$T_{c,o} = 36.0°C \qquad \qquad \triangleleft$$

2. The problem may be classified as one requiring a heat exchanger *design calculation*. Accordingly, either the LMTD or NTU method may be conveniently applied. Using the LMTD method, it follows from Equations 11.14 and 11.18 that

$$q = UAF \, \Delta T_{\text{lm, CF}}$$

where $A = N \times 2L \times \pi D$. From Equation 11.5

$$U = \frac{1}{(1/h_i) + (1/h_o)}$$

where h_i may be estimated from an internal flow correlation. With

$$Re_D = \frac{4\dot{m}}{\pi D \mu} = \frac{4 \times 1 \text{ kg/s}}{\pi (0.025 \text{ m}) 855 \times 10^{-6} \text{ N · s/m}^2} = 59{,}567$$

the flow is turbulent and from Equation 8.60

$$Nu_D = 0.023 Re_D^{4/5} \, Pr^{0.4} = 0.023 (59{,}567)^{0.8} (5.83)^{0.4} = 308$$

Hence

$$h_i = Nu_D \frac{k}{D} = 308 \frac{0.613 \text{ W/m · K}}{0.025 \text{ m}} = 7552 \text{ W/m}^2 · K$$

$$U = \frac{1}{[(1/7552) + (1/11{,}000)]\text{m}^2 · \text{K/W}} = 4478 \text{ W/m}^2 · K$$

From Equations 11.15 and 11.17 the LMTD is

$$\Delta T_{\text{lm, CF}} = \frac{(T_{h,i} - T_{c,o}) - (T_{h,o} - T_{c,i})}{\ln\left[(T_{h,i} - T_{c,o})/(T_{h,o} - T_{c,i})\right]} = \frac{(50 - 36) - (50 - 20)}{\ln(14/30)} = 21°C$$

The correction factor F may be obtained from Figure 11.10, where

$$P = \frac{t_o - t_i}{T_i - t_i} = \frac{(36 - 20)}{(50 - 20)} = 0.53 \qquad R = \frac{T_i - T_o}{t_o - t_i} = \frac{50 - 50}{36 - 20} = 0$$

Accordingly, $F = 1$, and it follows that the tube length per pass is

$$L = \frac{q}{U(N2\pi D)F \, \Delta T_{\mathrm{lm, CF}}}$$

$$L = \frac{2 \times 10^9 \text{ W}}{4478 \text{ W/m}^2 \cdot \text{K} (30{,}000 \times 2\pi \times 0.025 \text{ m}) \times 1 \times 21°\text{C}} = 4.51 \text{ m} \quad \triangleleft$$

Comments:

1. Recognize that L is the tube length per pass, in which case the total tube length is 9.02 m.

2. Using the NTU method, $C_h = C_{\max} = \infty$ and $C_{\min} = \dot{m}_c c_{p, c} = 3 \times 10^4$ (kg/s) $\times 4179$ (J/kg $\cdot$ K) $= 1.25 \times 10^8$ W/K. It follows that $q_{\max} = C_{\min}(T_{h, i} - T_{c, i}) = 3.76 \times 10^9$ W and hence that $\varepsilon = 0.53$. From Figure 11.16 it also follows that NTU ≈ 0.75, from which it may then be shown that $L = 4.46$ m. From Equation 11.36b, note that NTU $= 0.755$.

3. Over time, the performance of the heat exchanger would be degraded by fouling on both the inner and outer tube surfaces. A representative maintenance schedule would call for taking the heat exchanger off-line and cleaning the tubes when fouling factors reached values of $R''_{f, i} = R''_{f, o} = 10^{-4}$ m$^2 \cdot$ K/W. To determine the effect of fouling on performance, the ε–NTU method may be used to calculate the total heat rate as a function of the fouling factor, with $R''_{f, o}$ assumed to equal $R''_{f, i}$. The following results are obtained:

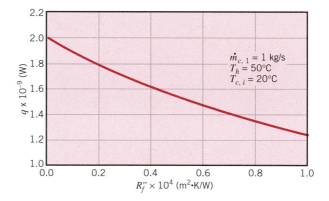

To maintain the requirement of $q = 2 \times 10^9$ W with the maximum allowable fouling and the restriction of $\dot{m}_{c, 1} = 1$ kg/s, the tube length or the number of tubes would have to be increased. Keeping the length per pass at $L = 4.51$ m, $N = 48{,}300$ tubes would be needed to transfer 2×10^9 W for $R''_{f, i} = R''_{f, o} = 10^{-4}$ m$^2 \cdot$ K/W. The corresponding increase in the total flow rate to $\dot{m}_c = N\dot{m}_{c, 1} = 48{,}300$ kg/s would have the beneficial effect of reducing the water outlet temperature to $T_{c, o} = 29.9°$C, thereby ameliorating potentially harmful effects associated with discharge into the environment.

11.6
Compact Heat Exchangers

As discussed in Section 11.1, *compact heat exchangers* are typically used when a large heat transfer surface area per unit volume is desired and at least one of the fluids is a gas. Many different tubular and plate configurations have been considered, where differences are due primarily to fin design and arrangement. Heat transfer and flow characteristics have been determined for specific configurations and are typically presented in the format of Figures 11.20 and 11.21. Heat transfer results are correlated in terms of the Colburn *j* factor $j_H = St \, Pr^{2/3}$ and the Reynolds number, where both the Stanton ($St = h/Gc_p$) and Reynolds ($Re = GD_h/\mu$) numbers are based on the maximum mass velocity

$$G \equiv \rho V_{\max} = \frac{\rho V A_{\mathrm{fr}}}{A_{\mathrm{ff}}} = \frac{\dot{m}}{A_{\mathrm{ff}}} = \frac{\dot{m}}{\sigma A_{\mathrm{fr}}} \tag{11.37}$$

The quantity σ is the ratio of the minimum free-flow area of the finned passages (cross-sectional area perpendicular to flow direction), A_{ff}, to the frontal area, A_{fr}, of the exchanger. Values of σ, D_h (the hydraulic diameter of the flow passage), α (the heat transfer surface area per total heat exchanger volume), A_f/A (the ratio of fin to total heat transfer surface area), and other geometrical parameters are listed for each configuration. The ratio A_f/A is used in Equation 11.3 to evaluate the temperature effectiveness η_o. In a design calculation, α would be used to determine the required heat exchanger volume, after the total

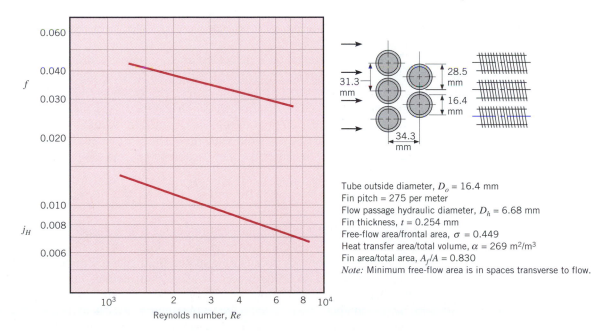

Tube outside diameter, $D_o = 16.4$ mm
Fin pitch = 275 per meter
Flow passage hydraulic diameter, $D_h = 6.68$ mm
Fin thickness, $t = 0.254$ mm
Free-flow area/frontal area, $\sigma = 0.449$
Heat transfer area/total volume, $\alpha = 269$ m²/m³
Fin area/total area, $A_f/A = 0.830$
Note: Minimum free-flow area is in spaces transverse to flow.

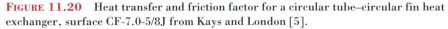

FIGURE 11.20   Heat transfer and friction factor for a circular tube–circular fin heat exchanger, surface CF-7.0-5/8J from Kays and London [5].

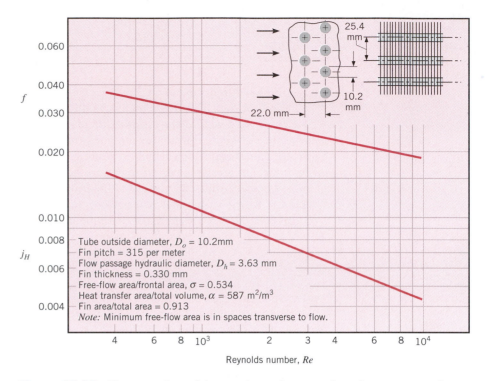

FIGURE 11.21   Heat transfer and friction factor for a circular tube–continuous fin heat exchanger, surface 8.0-3/8T from Kays and London [5].

heat transfer surface area has been found; in a performance calculation it would be used to determine the surface area from knowledge of the heat exchanger volume.

In a compact heat exchanger calculation, empirical information, such as that provided in Figures 11.20 and 11.21, would first be used to determine the average convection coefficient of the finned surfaces. The overall heat transfer coefficient would then be determined, and using the LMTD or ε–NTU method, the heat exchanger design or performance calculations would be performed.

The pressure drop associated with flow across finned-tube banks, such as those of Figures 11.20 and 11.21, may be computed from the expression

$$\Delta p = \frac{G^2 v_i}{2} \left[(1 + \sigma^2) \left(\frac{v_o}{v_i} - 1 \right) + f \frac{A}{A_{\text{ff}}} \frac{v_m}{v_i} \right] \tag{11.38}$$

where v_i and v_o are the fluid inlet and outlet specific volumes and $v_m = (v_i + v_o)/2$. The first term on the right-hand side of Equation 11.38 accounts for the effects of acceleration or deceleration as the fluid passes through the heat exchanger, while the second term accounts for losses due to fluid friction. For a prescribed core configuration, the friction factor is known as a function of Reynolds number, as, for example, from Figures 11.20 and 11.21; and for a prescribed heat exchanger size, the area ratio may be evaluated from the relation $(A/A_{\text{ff}}) = (\alpha V/\sigma A_{\text{fr}})$, where V is the total heat exchanger volume.

The classic work of Kays and London [5] provides Colburn j and friction factor data for many different compact heat exchanger cores, which include flat tube (Figure 11.5a) and plate–fin (Figure 11.5d, e) configurations, as well as other circular tube configurations (Figure 11.5b, c). Other excellent sources of information are provided by References 3, 4, 8, and 9.

EXAMPLE 11.6

Consider a finned-tube, compact heat exchanger having the core configuration of Figure 11.20. The core is fabricated from aluminum, and the tubes have an inside diameter of 13.8 mm. In a waste heat recovery application, water flow through the tubes provides an inside convection coefficient of $h_i = 1500$ W/m$^2 \cdot$ K, while combustion gases at 1 atm and 825 K are in cross flow over the tubes. If the gas flow rate is 1.25 kg/s and the frontal area is 0.20 m^2, what is the gas-side overall heat transfer coefficient? If a water flow rate of 1 kg/s is to be heated from 290 to 370 K, what is the required heat exchanger volume?

SOLUTION

Known: Compact heat exchanger geometry, gas-side flow rate and temperature, and water-side convection coefficient. Water flow rate and inlet and outlet temperatures.

Find: Gas-side overall heat transfer coefficient. Heat exchanger volume.

Schematic:

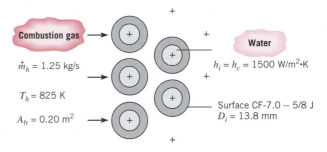

Assumptions:

1. Gas has properties of atmospheric air at an assumed mean temperature of 700 K.
2. Fouling is negligible.

Properties: Table A.1, aluminum ($T \approx 300$ K): $k = 237$ W/m $\cdot$ K. Table A.4, air ($p = 1$ atm, $\overline{T} = 700$ K): $c_p = 1075$ J/kg $\cdot$ K, $\mu = 338.8 \times 10^{-7}$ N $\cdot$ s/m^2, $Pr = 0.695$. Table A.6, water ($\overline{T} = 330$ K): $c_p = 4184$ J/kg $\cdot$ K.

Analysis: Referring to Equation 11.1, the combustion gas and the water are the hot and cold fluids, respectively. Hence, neglecting fouling effects and acknowledging that the tube inner surface is not finned ($\eta_{o,c} = 1$), the overall heat transfer coefficient based on the gas- (hot) side surface area is given by

$$\frac{1}{U_h} = \frac{1}{h_c(A_c/A_h)} + A_h R_w + \frac{1}{\eta_{o,h} h_h}$$

where A_h and A_c are the total gas-side (hot) and water-side (cold) surface areas, respectively. If the fin thickness is assumed to be negligible, it is readily shown that

$$\frac{A_c}{A_h} \approx \frac{D_i}{D_o}\left(1 - \frac{A_{f,h}}{A_h}\right)$$

where $A_{f,h}$ is that portion of the total gas-side area associated with the fins. The approximation is valid to within 10%, and for the heat exchanger core conditions (Figure 11.20)

$$\frac{A_c}{A_h} \approx \frac{13.8}{16.4}(1 - 0.830) = 0.143$$

Obtaining the wall conduction resistance from Equation 3.28, it follows that

$$A_h R_w = \frac{\ln(D_o/D_i)}{2\pi L k/A_h} = \frac{D_i \ln(D_o/D_i)}{2k(A_c/A_h)}$$

Hence

$$A_h R_w = \frac{(0.0138 \text{ m})\ln(16.4/13.8)}{2(237 \text{ W/m} \cdot \text{K})(0.143)} = 3.51 \times 10^{-5} \text{ m}^2 \cdot \text{K/W}$$

The gas-side convection coefficient may be obtained by first using Equation 11.37 to evaluate the mass velocity:

$$G = \frac{\dot{m}}{\sigma A_{\text{fr}}} = \frac{1.25 \text{ kg/s}}{0.449 \times 0.20 \text{ m}^2} = 13.9 \text{ kg/s} \cdot \text{m}^2$$

Hence

$$Re = \frac{13.9 \text{ kg/s} \cdot \text{m}^2 \times 6.68 \times 10^{-3} \text{ m}}{338.8 \times 10^{-7} \text{ kg/s} \cdot \text{m}} = 2740$$

and from Figure 11.20, $j_H \approx 0.0096$. Hence

$$h_h \approx 0.0096 \frac{G c_p}{Pr^{2/3}} = 0.0096 \frac{(13.9 \text{ kg/s} \cdot \text{m}^2)(1075 \text{ J/kg} \cdot \text{K})}{(0.695)^{2/3}}$$

$$= 183 \text{ W/m}^2 \cdot \text{K}$$

To obtain the hot-side temperature effectiveness from Equation 11.3, the fin efficiency must first be determined from Figure 3.19. With $r_{2c} = 14.38$ mm, $r_{2c}/r_1 = 1.75$, $L_c = 6.18$ mm, $A_p = 1.57 \times 10^{-6}$ m^2, and $L_c^{3/2}(h_h/kA_p)^{1/2} = 0.34$, it follows that $\eta_f \approx 0.89$. Hence

$$\eta_{o,h} = 1 - \frac{A_f}{A}(1 - \eta_f) = 1 - 0.830(1 - 0.89) = 0.91$$

We then obtain

$$\frac{1}{U_h} = \left(\frac{1}{1500 \times 0.143} + 3.51 \times 10^{-5} + \frac{1}{0.91 \times 183}\right) m^2 \cdot K/W$$

$$\frac{1}{U_h} = (4.66 \times 10^{-3} + 3.51 \times 10^{-5} + 6.00 \times 10^{-3}) = 0.0107 \; m^2 \cdot K/W$$

or

$$U_h = 93.4 \; W/m^2 \cdot K \qquad \triangleleft$$

With $C_c = \dot{m}_c c_{p,\,c} = 1 \; kg/s \times 4184 \; J/kg \cdot K = 4184 \; W/K$, the heat exchanger must be large enough to transfer heat in the amount

$$q = C_c(T_{c,\,o} - T_{c,\,i}) = 4184 \; W/K \; (370 - 290) \; K = 3.35 \times 10^5 \; W$$

With $C_h = \dot{m}_h c_{p,\,h} = 1.25 \; kg/s \times 1075 \; J/kg \cdot K = 1344 \; W/K$, the minimum heat capacity rate corresponds to the hot fluid and the maximum possible heat transfer rate is

$$q_{max} = C_{min}(T_{h,\,i} - T_{c,\,i}) = 1344 \; W/K \; (825 - 290) \; K = 7.19 \times 10^5 \; W$$

It follows that

$$\varepsilon = \frac{q}{q_{max}} = \frac{3.35 \times 10^5 \; W}{7.19 \times 10^5 \; W} = 0.466$$

Hence, with $(C_{min}/C_{max}) = 0.321$, Figure 11.18 (cross-flow heat exchanger with both fluids unmixed) yields

$$NTU = \frac{U_h A_h}{C_{min}} \approx 0.65$$

The required gas-side heat transfer surface area is then

$$A_h = \frac{0.65 \times 1344 \; W/K}{93.4 \; W/m^2 \cdot K} = 9.35 \; m^2$$

With the gas-side surface area per unit heat exchanger volume corresponding to $\alpha = 269 \; m^2/m^3$ (Figure 11.20), the required heat exchanger volume is

$$V = \frac{A_h}{\alpha} = \frac{9.35 \; m^2}{269 \; m^2/m^3} = 0.0348 \; m^3 \qquad \triangleleft$$

Comments:

1. The effect of the tube wall thermal conduction resistance is negligible, while contributions due to the cold- and hot-side convection resistances are comparable.

2. Knowledge of the heat exchanger volume yields the heat exchanger length in the gas-flow direction, $L = V/A_{fr} = 0.0348 \; m^3/0.20 \; m^2 = 0.174 \; m$, from which the number of tube rows in the flow direction may be determined.

$$N_L \approx \frac{L - D_f}{S_L} + 1 = \frac{(174 - 28.5) \; mm}{34.3 \; mm} + 1 = 5.24 \approx 5$$

3. The temperature of the gas leaving the heat exchanger is

$$T_{h,o} = T_{h,i} - \frac{q}{C_h} = 825 \text{ K} - \frac{3.35 \times 10^5 \text{ W}}{1344 \text{ W/K}} = 576 \text{ K}$$

Hence the assumption of $\overline{T}_h = 700$ K is excellent.

4. From Figure 11.20, the friction factor is $f \approx 0.033$. With $(A/A_{\text{ff}}) = (\alpha V/\sigma A_{\text{fr}}) = (269 \times 0.0348/0.449 \times 0.20) = 104.2$, $v_i(825 \text{ K}) = 2.37$ m³/kg, $v_o(576 \text{ K}) = 1.65$ m³/kg, and $v_m = 2.01$ m³/kg, Equation 11.38 yields a pressure drop of

$$\Delta p = \frac{(13.9 \text{ kg/s} \cdot \text{m}^2)^2(2.37 \text{ m}^3/\text{kg})}{2} [(1 + 0.202)(0.696 - 1)$$

$$+ 0.033 \times 104.2 \times 0.848]$$

$$\Delta p = 584 \text{ kg/s}^2 \cdot \text{m} = 584 \text{ N/m}^2$$

11.7
Summary

Because there are many important applications, heat exchanger research and development has had a long history. Such activity is by no means complete, however, as many talented workers continue to seek ways of improving design and performance. In fact, with heightened concern for energy conservation, there has been a steady and substantial increase in activity. A focal point for this work has been *heat transfer enhancement,* which includes the search for special heat exchanger surfaces through which enhancement may be achieved. In this chapter we have attempted to develop tools that will allow you to perform approximate heat exchanger calculations. More detailed considerations of the subject are available in the literature, including treatment of the uncertainties associated with heat exchanger analysis [3, 4, 8, 10–15].

Although we have restricted ourselves to heat exchangers involving separation of hot and cold fluids by a stationary wall, there are other important options. For example, *evaporative* heat exchangers enable *direct contact* between a liquid and a gas (there is no separating wall), and because of latent energy effects, large heat transfer rates per unit volume are possible. Also, for gas-to-gas heat exchange, use is often made of *regenerators* in which the same space is alternately occupied by the hot and cold gases. In a fixed regenerator such as a packed bed, the hot and cold gases alternately enter a stationary, porous solid. In a rotary regenerator, the porous solid is a rotating wheel, which alternately exposes its surfaces to the continuously flowing hot and cold gases. Detailed descriptions of such heat exchangers are available in the literature [3, 4, 8, 11, 16–19].

References

1. *Standards of the Tubular Exchange Manufacturers Association,* 6th ed., Tubular Exchanger Manufacturers Association, New York, 1978.
2. Chenoweth, J. M., and M. Impagliazzo, Eds., *Fouling in Heat Exchange Equipment,* American Society of Mechanical Engineers Symposium Volume HTD-17, ASME, New York, 1981.
3. Kakac, S., A. E. Bergles, and F. Mayinger, Eds., *Heat Exchangers,* Hemisphere Publishing, New York, 1981.
4. Kakac, S., R. K. Shah, and A. E. Bergles, Eds., *Low Reynolds Number Flow Heat Exchangers,* Hemisphere Publishing, New York, 1983.
5. Kays, W. M., and A. L. London, *Compact Heat Exchangers,* 3rd ed., McGraw-Hill, New York, 1984.
6. Bowman, R. A., A. C. Mueller, and W. M. Nagle, "Mean Temperature Difference in Design," *Trans. ASME,* **62,** 283, 1940.
7. Jakob, M., *Heat Transfer,* Vol. 2, Wiley, New York, 1957.
8. Shah, R. K., C. F. McDonald, and C. P. Howard, Eds., *Compact Heat Exchangers,* American Society of Mechanical Engineers Symposium Volume HTD-10, ASME, New York, 1980.
9. Webb, R. L., "Compact Heat Exchangers," in E. U. Schlünder, Ed., *Heat Exchanger Design Handbook,* Section 3.9, Hemisphere Publishing, New York, 1983.
10. Marner, W. J., A. E. Bergles, and J. M. Chenoweth, "On the Presentation of Performance Data for En-hanced Tubes Used in Shell-and-Tube Heat Exchangers," *Trans. ASME, J. Heat Transfer,* **105,** 358, 1983.
11. Schlünder, E. U., Ed.-in-Chief, *Heat Exchanger Design Handbook,* Vols. 1–5, Hemisphere Publishing, New York, 1983.
12. Webb, R. L., *Principles of Enhanced Heat Transfer,* Wiley, New York, 1994.
13. Andrews, M. J., and L. S. Fletcher, "A Comparative Study of Enhanced Heat Exchanger Technologies," *ASME/JSME Thermal Eng. Conf.,* **4,** 359, 1995.
14. DiGiovanni, M. A., and R. L. Webb, "Uncertainty in Effectiveness–NTU Calculations for Crossflow Heat Exchangers," *Heat Transfer Eng.* **10,** 61, 1989.
15. James, C. A., R. P. Taylor, and B. K. Hodge, "The Application of Uncertainty Analysis to Cross-Flow Heat Exchanger Performance Predictions," *ASME/JSME Thermal Eng. Conf.,* **4,** 337, 1995.
16. Coppage, J. E., and A. L. London, "The Periodic Flow Regenerator: A Summary of Design Theory," *Trans. ASME,* **75,** 779, 1953.
17. Treybal, R. E., *Mass Transfer Operations,* 2nd ed., McGraw-Hill, New York, 1968.
18. Sherwood, T. K., R. L. Pigford, and C. R. Wilkie, *Mass Transfer,* McGraw-Hill, New York, 1975.
19. Schmidt, F. W., and A. J. Willmott, *Thermal Energy Storage and Regeneration,* Hemisphere Publishing, New York, 1981.

Problems

Overall Heat Transfer Coefficient

11.1 In a fire-tube boiler, hot products of combustion flowing through an array of thin-walled tubes are used to boil water flowing over the tubes. At the time of installation, the overall heat transfer coefficient was 400 $W/m^2 \cdot K$. After 1 year of use, the inner and outer tube surfaces are fouled, with corresponding fouling factors of $R''_{f,i} = 0.0015$ and $R''_{f,o} = 0.0005$ $m^2 \cdot K/W$, respectively. Should the boiler be scheduled for cleaning of the tube surfaces?

11.2 A type-302 stainless steel tube of inner and outer diameters $D_i = 22$ mm and $D_o = 27$ mm, respectively, is used in a cross-flow heat exchanger. The fouling factors, R''_f, for the inner and outer surfaces are estimated to be 0.0004 and 0.0002 $m^2 \cdot K/W$, respectively.

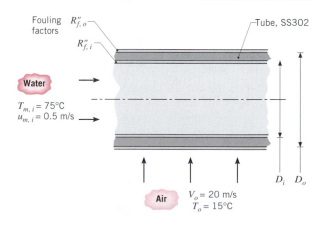

(a) Determine the overall heat transfer coefficient based on the outside area of the tube, U_o. Compare the thermal resistances due to convection, tube wall conduction, and fouling.

(b) Instead of air flowing over the tube, consider a situation for which the cross-flow fluid is water at 15°C with a velocity of $V_o = 1$ m/s. Determine the overall heat transfer coefficient based on the outside area of the tube, U_o. Compare the thermal resistances due to convection, tube wall conduction, and fouling.

(c) For the water–air conditions of part (a) and mean velocities, $u_{m, i}$, of 0.2, 0.5, and 1.0 m/s, plot the overall heat transfer coefficient as a function of the cross-flow velocity for $5 \leq V_o \leq 30$ m/s.

(d) For the water–water conditions of part (b) and cross-flow velocities, V_o, of 1, 3, and 8 m/s, plot the overall heat transfer coefficient as a function of the mean velocity for $0.5 \leq u_{m, i} \leq 2.5$ m/s.

11.3 A copper tube of inner and outer diameters $D_i = 13$ mm and $D_o = 18$ mm, respectively, is used in a shell-and-tube heat exchanger.

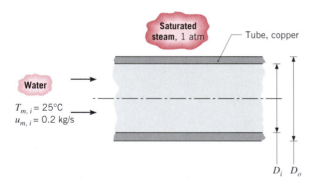

Saturated steam, 1 atm — Tube, copper

Water

$T_{m, i} = 25°C$
$u_{m, i} = 0.2$ kg/s

D_i D_o

(a) Determine the overall heat transfer coefficient based on the outside area of the tube, U_o. Compare the thermal resistances due to convection, tube wall conduction, and condensation.

(b) Plot the overall heat transfer coefficient, U_o, the water-side convection coefficient, h_i, and the steam-side convection coefficient, h_o, as a function of the water flow rate for the range $0.2 \leq \dot{m}_i \leq 0.8$ kg/s.

11.4 A steel tube ($k = 50$ W/m · K) of inner and outer diameters $D_i = 20$ mm and $D_o = 26$ mm, respectively, is used to transfer heat from hot gases flowing over the tube ($h_h = 200$ W/m² · K) to cold water flowing through the tube ($h_c = 8000$ W/m² · K). What is the cold side overall heat transfer coef-

ficient U_c? To enhance heat transfer, 16 straight fins of rectangular profile are installed longitudinally along the outer surface of the tube. The fins are equally spaced around the circumference of the tube, each having a thickness of 2 mm and a length of 15 mm. What is the corresponding overall heat transfer coefficient U_c?

11.5 A heat recovery device involves transferring energy from the hot flue gases passing through an annular region to pressurized water flowing through the inner tube of the annulus. The inner tube has inner and outer diameters of 24 and 30 mm and is connected by eight struts to an insulated outer tube of 60-mm diameter. Each strut is 3 mm thick and is integrally fabricated with the inner tube from carbon steel ($k = 50$ W/m · K).

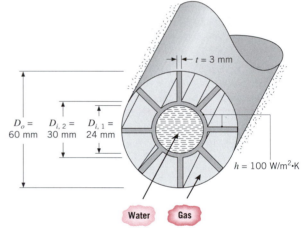

$t = 3$ mm

$D_o = 60$ mm $D_{i, 2} = 30$ mm $D_{i, 1} = 24$ mm

$h = 100$ W/m²·K

Water   Gas

Consider conditions for which water at 300 K flows through the inner tube at 0.161 kg/s while flue gases at 800 K flow through the annulus, maintaining a convection coefficient of 100 W/m² · K on both the struts and the outer surface of the inner tube. What is the rate of heat transfer per unit length of tube from gas to the water?

11.6 A novel design for a condenser consists of a tube of thermal conductivity 200 W/m · K with longitudinal fins snugly fitted into a larger tube. Condensing refrigerant at 45°C flows axially through the inner tube, while water at a flow rate of 0.012 kg/s passes through the six channels around the inner tube. The pertinent diameters are $D_1 = 10$ mm, $D_2 = 14$ mm, and $D_3 = 50$ mm, while the fin thickness is $t = 2$ mm. Assume that the convection coefficient associated with the condensing refrigerant is extremely large.

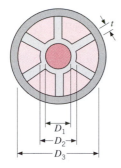

Determine the heat removal rate per unit tube length in a section of the tube for which the water is at 15°C.

11.7 Thin-walled aluminum tubes of diameter $D = 10$ mm are used in the condenser of an air conditioner. Under normal operating conditions, a convection coefficient of $h_i = 5000$ W/m² · K is associated with condensation on the inner surface of the tubes, while a coefficient of $h_o = 100$ W/m² · K is maintained by airflow over the tubes.

(a) What is the overall heat transfer coefficient if the tubes are unfinned?

(b) What is the overall heat transfer coefficient based on the inner surface, U_i, if aluminum annular fins of thickness $t = 1.5$ mm, outer diameter $D_o = 20$ mm, and pitch $S = 3.5$ mm are added to the outer surface? Base your calculations on a 1-m long section of tube. Subject to the requirements that $t \geq 1$ mm and $(S - t) \geq 1.5$ mm, explore the effect of variations in t and S on U_i. What combination of t and S would yield the best heat transfer performance?

11.8 A finned-tube, cross-flow heat exchanger is to use the exhaust of a gas turbine to heat pressurized water. Laboratory measurements are performed on a prototype version of the exchanger, which has a surface area of 10 m², to determine the overall heat transfer coefficient as a function of operating conditions. Measurements made under particular conditions, for which $\dot{m}_h = 2$ kg/s, $T_{h, i} = 325$°C, $\dot{m}_c = 0.5$ kg/s, and $T_{c, i} = 25$°C, reveal a water outlet temperature of $T_{c, o} = 150$°C. What is the overall heat transfer coefficient of the exchanger?

11.9 Water at a rate of 45,500 kg/h is heated from 80 to 150°C in a heat exchanger having two shell passes and eight tube passes with a total surface area of 925 m². Hot exhaust gases having approximately the same thermophysical properties as air enter at 350°C and exit at 175°C. Determine the overall heat transfer coefficient.

Design and Performance Calculations

11.10 A shell-and-tube exchanger (two shells, four tube passes) is used to heat 10,000 kg/h of pressurized water from 35 to 120°C with 5000 kg/h water entering the exchanger at 300°C. If the overall heat transfer coefficient is 1500 W/m² · K, determine the required heat exchanger area.

11.11 Consider the heat exchanger of Problem 11.10. After several years of operation, it is observed that the outlet temperature of the cold water reaches only 95°C rather than the desired 120°C for the same flow rates and inlet temperatures of the fluids. Determine the cumulative (inner and outer surface) fouling factor that is the cause of the poorer performance.

11.12 A counterflow, concentric tube heat exchanger is designed to heat water from 20 to 80°C using hot oil, which is supplied to the annulus at 160°C and discharged at 140°C. The thin-walled inner tube has a diameter of $D_i = 20$ mm, and the overall heat transfer coefficient is 500 W/m² · K. The design condition calls for a total heat transfer rate of 3000 W.

(a) What is the length of the heat exchanger?

(b) After 3 years of operation, performance is degraded by fouling on the water side of the exchanger, and the water outlet temperature is only 65°C for the same fluid flow rates and inlet temperatures. What are the corresponding values of the heat transfer rate, the outlet temperature of the oil, the overall heat transfer coefficient, and the water-side fouling factor, $R''_{f, c}$?

11.13 A thin-walled concentric tube heat exchanger is to be used to cool engine oil from 160 to 60°C, and water, which is available at 25°C, is to be used as the coolant. The oil and water flow rates are each 2 kg/s, and the diameter of the inner tube is 0.5 m. The corresponding value of the overall heat transfer coefficient is 250 W/m² · K. How long must the heat exchanger be to accomplish the desired cooling?

11.14 A concentric tube heat exchanger for cooling lubricating oil is comprised of a thin-walled inner tube of 25-mm diameter carrying water and an outer tube of 45-mm diameter carrying the oil. The exchanger operates in counterflow with an overall heat transfer coefficient of 60 W/m² · K and the tabulated average properties.

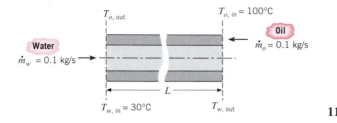

Properties	Water	Oil
ρ (kg/m^3)	1000	800
c_p (J/kg · K)	4200	1900
ν (m^2/s)	7×10^{-7}	1×10^{-5}
k (W/m · K)	0.64	0.134
Pr	4.7	140

(a) If the outlet temperature of the oil is 60°C, determine the total heat transfer and the outlet temperature of the water.

(b) Determine the length required for the heat exchanger.

11.15 Cold water at 20°C and 5000 kg/h is to be heated by hot water supplied at 80°C and 10,000 kg/h. You select from a manufacturer's catalog a shell-and-tube heat exchanger (one shell with two tube passes) having a UA value of 11,600 W/K. Determine the hot water outlet temperature.

11.16 A thin-walled concentric tube heat exchanger of 0.19-m length is to be used to heat deionized water from 40 to 60°C at a flow rate of 5 kg/s. The deionized water flows through the inner tube of 30-mm diameter while hot process water at 95°C flows in the annulus formed with the outer tube of 60-mm diameter. The thermophysical properties of the fluids are:

	Deionized Water	Process Water
ρ (kg/m^3)	982.3	967.1
c_p (J/kg · K)	4181	4197
k (W/m · K)	0.643	0.673
μ (N · s/m^2) $\times 10^6$	548	324
Pr	3.56	2.02

(a) Considering a parallel-flow configuration of the exchanger, determine the minimum flow rate required for the hot process water.

(b) Determine the overall heat transfer coefficient required for the conditions of part (a). Explain

why it is not possible to achieve the conditions prescribed in part (a).

(c) Considering a counterflow configuration, determine the minimum flow rate required for the hot process water. What is the effectiveness of the exchanger for this situation?

11.17 An automobile radiator may be viewed as a cross-flow heat exchanger with both fluids unmixed. Water, which has a flow rate of 0.05 kg/s, enters the radiator at 400 K and is to leave at 330 K. The water is cooled by air that enters at 0.75 kg/s and 300 K.

(a) If the overall heat transfer coefficient is 200 W/m^2 · K, what is the required heat transfer surface area?

(b) A manufacturing engineer claims ridges can be stamped on the finned surface of the exchanger, which could greatly increase the overall heat transfer coefficient. With all other conditions remaining the same and the heat transfer surface area determined from part (a), generate a plot of the air and water outlet temperatures as a function of U for $200 \leq U \leq 400$ W/m^2 · K. What benefits result from increasing the overall convection coefficient for this application?

11.18 Hot air for a large-scale drying operation is to be produced by routing the air over a tube bank (unmixed), while products of combustion are routed through the tubes. The surface area of the cross-flow heat exchanger is $A = 25$ m^2, and for the proposed operating conditions, the manufacturer specifies an overall heat transfer coefficient of $U = 35$ W/m^2 · K. The air and the combustion gases may each be assumed to have a specific heat of $c_p = 1040$ J/kg · K. Consider conditions for which combustion gases flowing at 1 kg/s enter the heat exchanger at 800 K, while air at 5 kg/s has an inlet temperature of 300 K.

(a) What are the air and gas outlet temperatures?

(b) After extended operation, deposits on the inner tube surfaces are expected to provide a fouling resistance of $R_f'' = 0.004$ m^2 · K/W. Should operation be suspended in order to clean the tubes?

(c) The heat exchanger performance may be improved by increasing the surface area and/or the overall heat transfer coefficient. Explore the effect of such changes on the air outlet temperature for $500 \leq UA \leq 2500$ W/K.

11.19 The compartment heater of an automobile exchanges heat between warm radiator fluid and cooler outside air. The flow rate of water is large compared to the air, and the effectiveness, ε, of the heater is known to depend on the flow rate of air according to the relation, $\varepsilon \sim \dot{m}_{air}^{-0.2}$.

(a) If the fan is switched to high and $\dot{m}_{air}$ is doubled, determine the percentage increase in the heat added to the car, if fluid inlet temperatures remain the same.

(b) For the low-speed fan condition, the heater warms outdoor air from 0 to 30°C. When the fan is turned to medium, the airflow rate increases 50% and the heat transfer increases 20%. Find the new outlet temperature.

11.20 A counterflow, twin-tube heat exchanger is made by brazing two circular nickel tubes, each 40 m long, together as shown below. Hot water flows through the smaller tube of 10-mm diameter and air at atmospheric pressure flows through the larger tube of 30-mm diameter. Both tubes have a wall thickness of 2 mm. The thermal contact conductance per unit length of the brazed joint is 100 W/m · K. The mass flow rates of the water and air are 0.04 and 0.12 kg/s, respectively. The inlet temperatures of the water and air are 85 and 23°C, respectively.

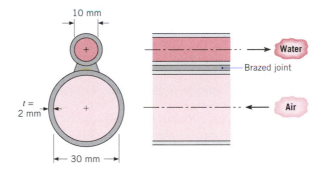

Employ the ε–NTU method to determine the outlet temperature of the air. *Hint:* Account for the effects of circumferential conduction in the walls of the tubes by treating them as extended surfaces.

11.21 A twin-tube, counterflow heat exchanger operates with balanced flow rates of 0.003 kg/s for the hot and cold airstreams. The cold stream enters at 280 K and must be heated to 340 K using hot air at 360 K. The average pressure of the airstreams is 1 atm and the maximum allowable pressure drop for the cold air is 10 kPa. The tube walls may be assumed to act as fins, each with an efficiency of 100%.

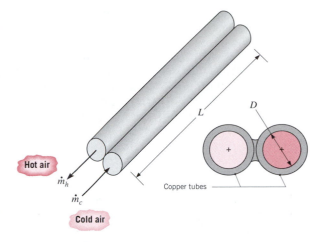

(a) Determine the tube diameter D and length L that satisfy the prescribed heat transfer and pressure drop requirements.

(b) For the diameter D and length L found in part (a), generate plots of the cold stream outlet temperature, the heat transfer rate, and pressure drop as a function of balanced flow rates in the range from 0.002 to 0.004 kg/s. Comment on your results.

11.22 Hot water for an industrial washing operation is produced by recovering heat from the flue gases of a furnace. A cross-flow heat exchanger is used, with the gases passing over the tubes and the water making a single pass through the tubes. The steel tubes ($k = 60$ W/m · K) have inner and outer diameters of $D_i = 15$ mm and $D_o = 20$ mm, while the staggered tube array has longitudinal and transverse pitches of $S_T = S_L = 40$ mm. The plenum in which the array is installed has a width (corresponding to the tube length) of $W = 2$ m and a height (normal to the tube axis) of $H = 1.2$ m. The number of tubes in the transverse plane is therefore $N_T \approx H/S_T = 30$. The gas properties may be approximated as those of atmospheric air, and the convection coefficient associated with water flow in the tubes may be approximated as 3000 W/m² · K.

(a) If 50 kg/s of water are to be heated from 290 to 350 K by 40 kg/s of flue gases entering the exchanger at 700 K, what is the gas outlet temperature and how many tube rows N_L are required?

(b) The water outlet temperature may be controlled by varying the gas flow rate and/or inlet temperature. For the value of N_L determined in part (a) and the prescribed values of H, W, S_T, $\dot{m}_c$, and $T_{c,\,i}$, compute and plot $T_{c,\,o}$ as a function of $\dot{m}_h$ over the range $20 \le \dot{m}_h \le 40$ kg/s for values of $T_{h,\,i} = 500$, 600, and 700 K. Also plot the corresponding variations of $T_{h,\,o}$. If $T_{h,\,o}$ must not drop below 400 K to prevent condensation of corrosive vapors on the heat exchanger surfaces, are there any constraints on $\dot{m}_h$ and $T_{h,\,i}$?

11.23 A single-pass, cross-flow heat exchanger uses hot exhaust gases (mixed) to heat water (unmixed) from 30 to 80°C at a rate of 3 kg/s. The exhaust gases, having thermophysical properties similar to air, enter and exit the exchanger at 225 and 100°C, respectively. If the overall heat transfer coefficient is 200 W/m² · K, estimate the required surface area.

11.24 Consider the fluid conditions and overall heat transfer coefficient of Problem 11.23 for a concentric tube heat exchanger operating in parallel flow. The thin-walled separator tube has a diameter of 100 mm.

(a) Determine the required length for the exchanger.

(b) Assuming water flow inside the separator tube to be fully developed, estimate the convection heat transfer coefficient.

(c) Using the overall coefficient and the inlet temperatures from Problem 11.23, plot the heat transfer rate and fluid outlet temperatures as a function of the tube length for $60 \le L \le 400$ m and the parallel-flow configuration.

(d) If the exchanger were operated in counterflow with the same overall coefficient and inlet temperatures, what would be the reduction in the required length relative to the value found in part (a)?

(e) For the counterflow configuration, plot the effectiveness and fluid outlet temperatures as a function of the tube length for $60 \le L \le 400$ m.

11.25 Saturated steam at 0.14 bar is condensed in a shell-and-tube heat exchanger with one shell pass and two tube passes consisting of 130 brass tubes, each with a length per pass of 2 m. The tubes have inner and outer diameters of 13.4 and 15.9 mm, respectively. Cooling water enters the tubes at 20°C with a mean velocity of 1.25 m/s. The heat transfer coefficient for condensation on the outer surfaces of the tubes is 13,500 W/m² · K.

(a) Determine the overall heat transfer coefficient, the cooling water outlet temperature, and the steam condensation rate.

(b) With all other conditions remaining the same, but accounting for changes in the overall coefficient, plot the cooling water outlet temperature and the steam condensation rate as a function of the water flow rate for $10 \le \dot{m}_c \le 30$ kg/s.

11.26 A feedwater heater that supplies a boiler consists of a shell-and-tube heat exchanger with one shell pass and two tube passes. One hundred thin-walled tubes each have a diameter of 20 mm and a length (per pass) of 2 m. Under normal operating conditions water enters the tubes at 10 kg/s and 290 K and is heated by condensing saturated steam at 1 atm on the outer surface of the tubes. The convection coefficient of the saturated steam is 10,000 W/m² · K.

(a) Determine the water outlet temperature.

(b) With all other conditions remaining the same, but accounting for changes in the overall heat transfer coefficient, plot the water outlet temperature as a function of the water flow rate for $5 \le \dot{m}_c \le 20$ kg/s.

(c) On the plot of part (b), generate two additional curves for the water outlet temperature as a function of flow rate for fouling factors of $R''_f = 0.0002$ and 0.0005 m² · K/W.

11.27 Saturated water vapor leaves a steam turbine at a flow rate of 1.5 kg/s and a pressure of 0.51 bar. The vapor is to be completely condensed to saturated liquid in a shell-and-tube heat exchanger that uses city water as the cold fluid. The water enters the thin-walled tubes at 17°C and is to leave at 57°C. Assuming an overall heat transfer coefficient of 2000 W/m² · K, determine the required heat exchanger surface area and the water flow rate. After extended operation, fouling causes the overall heat transfer coefficient to decrease to 1000 W/m² · K, and to completely condense the vapor, there must be an attendant reduction in the vapor flow rate. For the same water inlet temperature and flow rate, what is the new vapor flow rate required for complete condensation?

11.28 A two-fluid heat exchanger has inlet and outlet temperatures of 65 and 40°C for the hot fluid and 15 and 30°C for the cold fluid. Can you tell whether this exchanger is operating under counterflow or parallel-flow conditions? What is the effectiveness of the exchanger if the cold fluid has the minimum capacity rate?

11.29 Water at 225 kg/h is to be heated from 35 to 95°C by means of a concentric tube heat exchanger. Oil at 225 kg/h and 210°C, with a specific heat of 2095 J/kg · K, is to be used as the hot fluid. If the overall heat transfer coefficient based on the outer diameter of the inner tube is 550 W/m² · K, determine the length of the exchanger if the outer diameter is 100 mm.

11.30 A hot fluid enters a concentric pipe heat exchanger at 150°C and is to be cooled to 100°C by a cold fluid entering at 35°C and heated to 65°C. Would you use parallel-flow or counterflow for the most effective design?

11.31 Consider a *very long,* concentric tube heat exchanger having hot and cold water inlet temperatures of 85 and 15°C. The flow rate of the hot water is twice that of the cold water. Assuming equivalent hot and cold water specific heats, determine the hot water outlet temperature for the following modes of operation: (a) counterflow and (b) parallel flow.

11.32 A plate-type heat exchanger consists of an arrangement of thin, parallel plates that are separated by a distance of $s = 5$ mm and are of length $L = 750$ mm and width $w = 150$ mm. The counterflow arrangement involves passage of hot and cold water streams in opposite directions along the surfaces of each plate. The principles of heat exchanger analysis can be used for this configuration by considering a single plate separating portions of the hot and cold streams, which occupy half the spacing between plates.

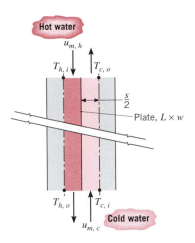

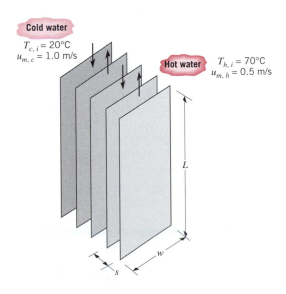

(a) For the prescribed conditions, estimate the overall heat transfer coefficient between the hot and cold fluids, the rate of heat transfer across a single plate, and the fluid outlet temperatures.

(b) For cold water mean velocities of $u_{m, c} = 0.5$, 1.5, and 3.0 m/s, plot the fluid outlet temperatures as a function of the mean velocity of the hot water over the range $0.5 \leq u_{m, h} \leq 3.0$ m/s.

(c) Repeat the analyses of parts (a) and (b) for the parallel-flow arrangement. Compare your results and comment on the relative performance of the two configurations.

11.33 A shell-and-tube heat exchanger is to heat 10,000 kg/h of water from 16 to 84°C by hot engine oil flowing through the shell. The oil makes a single shell pass, entering at 160°C and leaving at 94°C, with an average heat transfer coefficient of 400 W/m² · K. The water flows through 11 brass tubes of 22.9-mm inside diameter and 25.4-mm outside diameter, with each tube making four passes through the shell.

(a) Assuming fully developed flow for the water, determine the required tube length per pass.

(b) For the tube length found in part (a), plot the effectiveness, fluid outlet temperatures, and water-side convection coefficient as a function of the water flow rate for $5000 \leq \dot{m}_c \leq 15,000$ kg/h, with all other conditions remaining the same.

11.34 In a supercomputer, signal propagation delays are reduced by resorting to high-density circuit arrangements which are cooled by immersing them in a special dielectric liquid. The fluid is pumped in a closed loop through the computer and an adjoining shell-and-tube heat exchanger having one shell and two tube passes.

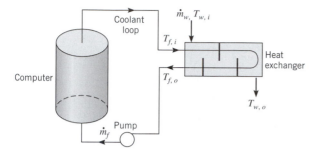

During normal operation, heat generated within the computer is transferred to the dielectric fluid passing through the computer at a flow rate of $\dot{m}_f = 4.81$ kg/s. In turn, the fluid passes through the tubes of the heat exchanger and the heat is transferred to water passing over the tubes. The dielectric fluid may be assumed to have constant properties of $c_p = 1040$ J/kg · K, $\mu = 7.65 \times 10^{-4}$ kg/s · m, $k = 0.058$ W/m · K, and $Pr = 14$. During normal operation, chilled water at a flow rate of $\dot{m}_w = 2.5$ kg/s and an inlet temperature of $T_{w,\,i} = 5°$C passes over the tubes. The water has a specific heat of 4200 J/kg · K and provides an average convection coefficient of 10,000 W/m² · K over the outer surface of the tubes.

(a) If the heat exchanger consists of 72 thin-walled tubes, each of 10-mm diameter, and fully developed flow is assumed to exist within the tubes, what is the convection coefficient associated with flow through the tubes?

(b) If the dielectric fluid enters the heat exchanger at $T_{f,\,i} = 25°$C and is to leave at $T_{f,\,o} = 15°$C, what is the required tube length per pass?

(c) For the exchanger with the tube length per pass determined in part (b), plot the outlet temperature of the dielectric fluid as a function of its flow rate for $4 \leq \dot{m}_f \leq 6$ kg/s. Account for corresponding changes in the overall heat transfer coefficient, but assume all other conditions to remain the same.

(d) The site specialist for the computer facilities is concerned about changes in the performance of the water chiller supplying the cold water ($\dot{m}_w$, $T_{w,\,i}$) and their effect on the outlet temperature $T_{f,\,o}$ of the dielectric fluid. With all other conditions remaining the same, determine the effect of a $\pm10\%$ change in the cold water flow rate on $T_{f,\,o}$.

(e) Repeat the performance analysis of part (d) to determine the effect of a $\pm3°$C change in the water inlet temperature on $T_{f,\,o}$, with all other conditions remaining the same.

11.35 A shell-and-tube heat exchanger consists of 135 thin-walled tubes in a double-pass arrangement, each of 12.5-mm diameter with a total surface area of 47.5 m². Water (the tube-side fluid) enters the heat exchanger at 15°C and 6.5 kg/s and is heated by exhaust gas entering at 200°C and 5 kg/s. The gas may be assumed to have the properties of atmospheric air, and the overall heat transfer coefficient is approximately 200 W/m² · K.

(a) What are the gas and water outlet temperatures?

(b) Assuming fully developed flow, what is the tube-side convection coefficient?

(c) With all other conditions remaining the same, plot the effectiveness and fluid outlet temperatures as a function of the water flow rate over the range from 6 to 12 kg/s.

(d) What gas inlet temperature is required for the exchanger to supply 10 kg/s of hot water at an outlet temperature of 42°C, all other conditions remaining the same? What is the effectiveness for this operating condition?

11.36 An ocean thermal energy conversion system is being proposed for electric power generation. Such a system is based on the standard power cycle for which the working fluid is evaporated, passed through a turbine, and subsequently condensed. The system is to be used in very special locations for which the oceanic water temperature near the surface is approximately 300 K, while the temperature at reasonable depths is approximately 280 K. The warmer water is used as a heat source to evaporate the working fluid, while the colder water is used as a heat sink for condensation of the fluid. Consider a power plant that is to generate 2 MW of electricity at an efficiency (electric power output per heat input) of 3%. The evaporator is a heat exchanger consisting of a single shell with many tubes executing two passes. If the working fluid is evaporated at its phase change temperature of 290 K, with ocean water entering at 300 K and leaving at 292 K, what is the heat exchanger area required for the evaporator? What flow rate must be maintained for the water passing through the evaporator? The overall heat transfer coefficient may be approximated as 1200 W/m² · K.

11.37 Steam condenses on the shell-side of a shell-and-tube (single shell, two tube passes) heat exchanger while coolant passes through N tubes of diameter D_i. This heat exchanger (1) is to be reconfigured by using six times the number of tubes of diameter $0.5D_i$. If the length of the new exchanger (2), the coolant flow rate, and the temperatures of the steam and inlet coolant remain unchanged, by what factor will the total heat transfer rate be altered? Present your result by plotting q_2/q_1 as a function of NTU_1. Assume that coolant flow in the tubes is turbulent and the heat transfer coefficient on the shell side is much larger than on the tube side.

11.38 A shell-and-tube heat exchanger (one shell pass, multiple tube passes) is to be used to condense 2.73 kg/s of saturated steam at 340 K. Condensation occurs on the outer tube surfaces and the corresponding convection coefficient is $h_o = 12,500$ W/m² · K. The condenser will be supplied with cooling water that enters the tubes at 291 K and is to exit the tubes at 303 K. Thin-walled tubes of 19-mm diameter are specified, and the mean velocity of water flow through the tubes is to be maintained at 1.5 m/s.

(a) How many tubes must be used? If the length of the heat exchanger is not to exceed 1.5 m, how many tube passes should be made and what is the required length per pass?

(b) For the heat transfer surface area found in part (a), plot the water outlet temperature and steam condensation rate as a function of the water mean velocity over the range from 0.5 to 3 m/s. Assume all other conditions remain the same, but account for changes in the overall heat transfer coefficient.

(c) Repeat the analysis for the conditions of part (b) with tube diameters of 15 and 25 mm.

11.39 Saturated process steam at 1 atm is condensed in a shell-and-tube heat exchanger (one shell, two tube passes). Cooling water enters the tubes at 15°C with an average velocity of 3.5 m/s. The tubes are thin-walled and made of copper with a diameter of 14 mm and length of 0.5 m. The convective heat transfer coefficient for condensation on the outer surface of the tubes is 21,800 W/m² · K.

(a) Find the number of tubes/pass required to condense 2.3 kg/s of steam.

(b) Find the outlet water temperature.

(c) Find the maximum possible condensation rate that could be achieved with this heat exchanger using the same water flow rate and inlet temperature.

(d) Using the heat transfer surface area found in part (a), plot the water outlet temperature and steam condensation rate for water mean velocities in the range from 1 to 5 m/s. Assume that the shell-side convection coefficient remains unchanged.

11.40 The feedwater heater for a boiler supplies 10,000 kg/h of water at 65°C. The feedwater has an inlet temperature of 20°C and is to be heated in a single-shell, two-tube pass heat exchanger by condensing steam at 1.30 bars. The overall heat transfer coefficient is 2000 W/m² · K. Using both the LMTD and NTU methods, determine the required heat transfer area. What is the steam condensation rate?

11.41 A shell-and-tube heat exchanger with single shell and tube passes (Figure 11.3) is used to cool the oil of a large marine engine. Lake water (the shell-side fluid) enters the heat exchanger at 2 kg/s and 15°C, while the oil enters at 1 kg/s and 140°C. The oil flows through 100 copper tubes, each 500 mm long and having inner and outer diameters of 6 and 8 mm. The shell-side convection coefficient is approximately 500 W/m² · K.

(a) What is the oil outlet temperature?

(b) Over time there is some degradation in the performance of the engine's water pump, as well as an accumulation of deposits on the outer surface of the tubes. To assess the effect of these changes on heat exchanger performance, obtain two plots of the oil outlet temperature as a function of water flow rate in the range from 1 to 2 kg/s. One should correspond to a clean outer surface ($R''_{f,c} = 0$) and the other to surface conditions for which $R''_{f,c} = 0.0003$ m² · K/W.

11.42 A shell-and-tube heat exchanger with one shell pass and 20 tube passes uses hot water on the tube side to heat oil on the shell side. The single copper tube has inner and outer diameters of 20 and 24 mm and a length per pass of 3 m. The water enters at 87°C and 0.2 kg/s and leaves at 27°C. Inlet and outlet temperatures of the oil are 7 and 37°C. What is the average convection coefficient for the tube outer surface?

11.43 The oil in an engine is cooled by air in a cross-flow heat exchanger where both fluids are unmixed. Atmospheric air enters at 30°C and 0.53 kg/s. Oil at 0.026 kg/s enters at 75°C and flows through a tube of 10-mm diameter. Assuming fully developed flow and constant wall heat flux, estimate the oil-side heat transfer coefficient. If the overall convection coefficient is 53 W/m² · K and the total heat transfer area is 1 m², determine the effectiveness. What is the exit temperature of the oil?

11.44 A concentric tube heat exchanger uses water, which is available at 15°C, to cool ethylene glycol from 100 to 60°C. The water and glycol flow rates are each 0.5 kg/s. What are the maximum possible heat transfer rate and effectiveness of the exchanger? Which is preferred, a parallel-flow or counterflow mode of operation?

11.45 Water is used for both fluids (unmixed) flowing through a single-pass, cross-flow heat exchanger. The hot water enters at 90°C and 10,000 kg/h, while the cold water enters at 10°C and 20,000 kg/h. If the effectiveness of the exchanger is 60%, determine the cold water exit temperature.

11.46 A cross-flow heat exchanger consists of a bundle of 32 tubes in a 0.6-m² duct. Hot water at 150°C and a mean velocity of 0.5 m/s enters the tubes having inner and outer diameters of 10.2 and 12.5 mm. Atmospheric air at 10°C enters the exchanger with a volumetric flow rate of 1.0 m³/s. The convection heat transfer coefficient on the tube outer surfaces is 400 W/m² · K. Estimate the fluid outlet temperatures.

11.47 Exhaust gas from a furnace is used to preheat the combustion air supplied to the furnace burners. The gas, which has a flow rate of 15 kg/s and an inlet temperature of 1100 K, passes through a bundle of tubes, while the air, which has a flow rate of 10 kg/s and an inlet temperature of 300 K, is in cross flow over the tubes. The tubes are unfinned, and the overall heat transfer coefficient is 100 W/m² · K. Determine the total tube surface area required to achieve an air outlet temperature of 850 K. The exhaust gas and the air may each be assumed to have a specific heat of 1075 J/kg · K.

11.48 A recuperator is a heat exchanger that heats the air used in a combustion process by extracting energy from the products of combustion (the flue gas). Consider using a single-pass, cross-flow heat exchanger as a recuperator.

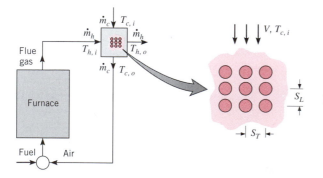

Eighty (80) silicon carbide ceramic tubes ($k = 20$ W/m · K) of inner and outer diameters equal to 55 and 80 mm, respectively, and of length $L = 1.4$ m

are arranged as an aligned tube bank of longitudinal and transverse pitches $S_L = 100$ mm and $S_T = 120$ mm, respectively. Cold air is in cross flow over the tube bank with upstream conditions of $V = 1$ m/s and $T_{c,i} = 300$ K, while hot flue gases of inlet temperature $T_{h,i} = 1400$ K pass through the tubes. The tube outer surface is clean, while the inner surface is characterized by a fouling factor of $R''_f = 2 \times 10^{-4}$ m² · K/W. The air and flue gas flow rates are $\dot{m}_c = 1.0$ kg/s and $\dot{m}_h = 1.05$ kg/s, respectively. As first approximations, (1) evaluate all required air properties at 1 atm and 300 K, (2) assume the flue gas to have the properties of air at 1 atm and 1400 K, and (3) assume the tube wall temperature to be at 800 K for the purpose of treating the effect of variable properties on convection heat transfer.

(a) If there is a 1% fuel savings associated with each 10°C increase in the temperature of the combustion air ($T_{c,o}$) above 300 K, what is the percentage fuel savings for the prescribed conditions?

(b) The performance of the recuperator is strongly influenced by the product of the overall heat transfer coefficient and the total surface area, UA. Compute and plot $T_{c,o}$ and the percentage fuel savings as a function of UA for $300 \leq UA \leq 600$ W/K. Without changing the flow rates, what measures may be taken to increase UA?

11.49 Consider operation of the furnace–recuperator combination of Problem 11.48 under conditions for which chemical energy is converted to thermal energy in the combustor at a rate of $q_{comb} = 2.0 \times 10^6$ W and energy is transferred from the combustion gases to the load in the furnace at a rate of $q_{load} = 1.4 \times 10^6$ W. Assuming equivalent flow rates ($\dot{m}_c = \dot{m}_h = 1.0$ kg/s) and specific heats ($c_{p,c} = c_{p,h} = 1200$ J/kg · K) for the cold air and flue gases in the recuperator, determine $T_{h,i}$, $T_{h,o}$, and $T_{c,o}$ when $T_{c,i} = 300$ K and the recuperator has an effectiveness of $\varepsilon = 0.30$. What value of the effectiveness would be needed to achieve a combustor air inlet temperature of 800 K?

11.50 Saturated steam at 100°C condenses in a shell-and-tube exchanger (single shell, two tube passes) with a surface area of 0.5 m² and overall heat transfer coefficient of 2000 W/m² · K. Water enters at 0.5 kg/s and 15°C. Determine the outlet temperature of the water and the rate of steam condensation.

11.51 Consider a concentric tube heat exchanger characterized by a uniform overall heat transfer coefficient and operating under the following conditions:

	$\dot{m}$ (kg/s)	c_p (J/kg · K)	T_i (°C)	T_o (°C)
Cold fluid	0.125	4200	40	95
Hot fluid	0.125	2100	210	

What is the maximum possible heat transfer rate? What is the heat exchanger effectiveness? Should the heat exchanger be operated in parallel flow or in counterflow? What is the ratio of the required areas for these two flow conditions?

11.52 Consider a concentric tube heat exchanger with hot and cold water inlet temperatures of 200 and 35°C, respectively. The flow rates of the hot and cold fluids are 42 and 84 kg/h, respectively. Assume the overall heat transfer coefficient is 180 W/m² · K.

(a) What is the maximum heat transfer rate that could be achieved for the prescribed inlet conditions?

(b) If the exchanger is operated in counterflow with a heat transfer area of 0.33 m², determine the outlet fluid temperatures.

(c) With all other conditions remaining the same, plot the effectiveness and fluid outlet temperatures as a function of the product UA for the range from 50 to 1000 W/K. As UA becomes very large, what is the asymptotic value for $T_{h,o}$?

(d) What is the largest possible heat transfer rate that could be achieved for the prescribed inlet conditions, if the exchanger is operated in parallel flow and is very long? What is the effectiveness of the exchanger in this configuration?

(e) For operation in parallel flow, plot the effectiveness and fluid outlet temperatures as a function of the product UA for the range from 50 to 1000 W/K. As UA becomes very large, what are the asymptotic values for $T_{h,o}$ and $T_{c,i}$?

11.53 Hot water at 2.5 kg/s and 100°C enters a concentric tube counterflow heat exchanger having a total area of 23 m². Cold water at 20°C enters at 5.0 kg/s, and the overall heat transfer coefficient is 1000 W/m² · K. Determine the total heat transfer rate and the outlet temperatures of the hot and cold fluids.

11.54 Hot exhaust gases are used in a shell-and-tube exchanger to heat 2.5 kg/s of water from 35 to 85°C. The gases, assumed to have the properties of air, enter at 200°C and leave at 93°C. The overall heat transfer coefficient is 180 W/m² · K. Using the effectiveness–NTU method, calculate the area of the heat exchanger.

11.55 In open heart surgery under hypothermic conditions, the patient's blood is cooled before the surgery and rewarmed afterward. It is proposed that a concentric tube, counterflow heat exchanger of length 0.5 m be used for this purpose, with the thin-walled inner tube having a diameter of 55 mm. The specific heat of the blood is 3500 J/kg · K.

(a) If water at $T_{h,i} = 60°C$ and $\dot{m}_h = 0.10$ kg/s is used to heat blood entering the exchanger at $T_{c,i} = 18°C$ and $\dot{m}_c = 0.05$ kg/s, what is the temperature of the blood leaving the exchanger? The overall heat transfer coefficient is 500 W/m² · K.

(b) The surgeon may wish to control the heat rate q and the outlet temperature $T_{c,o}$ of the blood by altering the flow rate and/or inlet temperature of the water during the rewarming process. To assist in the development of an appropriate controller for the prescribed values of $\dot{m}_c$ and $T_{c,i}$, compute and plot q and $T_{c,o}$ as a function of $\dot{m}_h$ for $0.05 \le \dot{m}_h \le 0.20$ kg/s and values of $T_{h,i} = 50, 60,$ and 70°C. Since the dominant influence on the overall heat transfer coefficient is associated with the blood flow conditions, the value of U may be assumed to remain at 500 W/m² · K. Should certain operating conditions be excluded?

11.56 Ethylene glycol and water, at 60 and 10°C, respectively, enter a shell-and-tube heat exchanger for which the total heat transfer area is 15 m². With ethylene glycol and water flow rates of 2 and 5 kg/s, respectively, the overall heat transfer coefficient is 800 W/m² · K.

(a) Determine the rate of heat transfer and the fluid outlet temperatures.

(b) Assuming all other conditions to remain the same, plot the effectiveness and fluid outlet temperatures as a function of the flow rate of ethylene glycol for $0.5 \le \dot{m}_h \le 5$ kg/s.

11.57 A boiler used to generate saturated steam is in the form of an unfinned, cross-flow heat exchanger, with water flowing through the tubes and a high temperature gas in cross flow over the tubes. The gas, which has a specific heat of 1120 J/kg · K and a mass flow rate of 10 kg/s, enters the heat exchanger at 1400 K. The water, which has a flow rate of 3 kg/s, enters as saturated liquid at 450 K and leaves as saturated vapor at the same temperature. If the overall heat transfer coefficient is 50 W/m² · K and there are 500 tubes, each of 0.025-m diameter, what is the required tube length?

11.58 Waste heat from the exhaust gas of an industrial furnace is recovered by mounting a bank of un-finned tubes in the furnace stack. Pressurized water at a flow rate of 0.025 kg/s makes a single pass through *each* of the tubes, while the exhaust gas, which has an upstream velocity of 5.0 m/s, moves in cross flow over the tubes at 2.25 kg/s. The tube bank consists of a square array of 100 thin-walled tubes (10 × 10), each 25 mm in diameter and 4 m long. The tubes are aligned with a transverse pitch of 50 mm. The inlet temperatures of the water and the exhaust gas are 300 and 800 K, respectively. The water flow is fully developed, and the gas properties may be assumed to be those of atmospheric air.

(a) What is the overall heat transfer coefficient?

(b) What are the fluid outlet temperatures?

(c) Operation of the heat exchanger may vary according to the demand for hot water. For the prescribed heat exchanger design and inlet conditions, compute and plot the rate of heat recovery and the fluid outlet temperatures as a function of water flow rate per tube for $0.02 \leq \dot{m}_{c,1} \leq 0.20$ kg/s.

11.59 A heat exchanger consists of a bank of 1200 thin-walled tubes with air in cross flow over the tubes. The tubes are arranged in-line, with 40 longitudinal rows (along the direction of airflow) and 30 transverse rows. The tubes are 0.07 m in diameter and 2 m long, with transverse and longitudinal pitches of 0.14 m. The hot fluid flowing through the tubes consists of saturated steam condensing at 400 K. The convection coefficient of the condensing steam is much larger than that of the air.

(a) If air enters the heat exchanger at $\dot{m}_c = 12$ kg/s, 300 K, and 1 atm, what is its outlet temperature?

(b) The condensation rate may be controlled by varying the airflow rate. Compute and plot the air outlet temperature, the heat rate, and the condensation rate as a function of flow rate for $10 \leq \dot{m}_c \leq 50$ kg/s.

11.60 In analyzing thermodynamic cycles involving heat exchangers, it is useful to express the heat rate in terms of an overall thermal resistance R_t and the inlet temperatures of the hot and cold fluids,

$$q = \frac{(T_{h,i} - T_{c,i})}{R_t}$$

The heat transfer rate can also be expressed in terms of the rate equations,

$$q = UA \, \Delta T_{lm} = \frac{1}{R_{lm}} \Delta T_{lm}$$

(a) Derive a relation for R_{lm}/R_t for a *parallel-flow* heat exchanger in terms of a single dimensionless parameter B, which does not involve any fluid temperatures but only U, A, C_h, C_c (or C_{min}, C_{max}).

(b) Calculate and plot R_{lm}/R_t for values of $B = 0.1$, 1.0, and 5.0. What conclusions can be drawn from the plot?

11.61 Consider the heat exchanger operating under the conditions prescribed in Problem 11.50.

(a) Determine the overall thermal resistance (see Problem 11.60) of the heat exchanger.

(b) Fouling of the tube surfaces is suspected after extended operation. Determine the change in thermal resistance if the fouling factor is presumed to be 0.0002 m² · K/W on the inner and outer surfaces of the tubes.

(c) Plot the thermal resistance of the heat exchanger as a function of the heat capacity rate of the water, assuming all other conditions remain unchanged and no fouling is present. Comment on whether UA will remain constant if the flow rate is changed.

11.62 Cooling water at 17°C and a flow rate of 0.5 kg/s is available to cool the condenser of a Carnot heat engine that must reject 65 kW. Because of the finite thermal resistance (Problem 11.60) of the condenser, the temperature of the Carnot heat reservoir will be greater than 17°C.

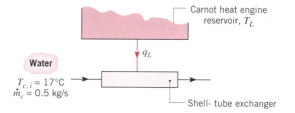

(a) Calculate the reservoir temperature (T_L) if the condenser is a shell-and-tube exchanger (single shell, two tube passes) with a surface area of 0.7 m² and an overall heat transfer coefficient of 1500 W/m² · K.

(b) Is it possible to reduce T_L by changing the flow rate of the water? What are the consequences of doing so?

11.63 An air conditioner operating between indoor and outdoor temperatures of 23 and 43°C, respectively, removes 5 kW from a building. The air conditioner can be modeled as a reversed Carnot heat engine with refrigerant as the working fluid. The efficiency of the motor for the compressor and fan is 80%, and 0.2 kW is required to operate the fan.

(a) Assuming negligible thermal resistances (Problem 11.60) between the refrigerant in the condenser and the outside air and between the refrigerant in the evaporator and the inside air, calculate the power required by the motor.

(b) If the thermal resistances between the refrigerant and the air in the evaporator and condenser sections are the same, 3×10^{-3} K/W, determine the temperature required by the refrigerant in each section. Calculate the power required by the motor.

11.64 In a Rankine power system, 1.5 kg/s of steam leaves the turbine as saturated vapor at 0.51 bar. The steam is condensed to saturated liquid by passing it over the tubes of a shell-and-tube heat exchanger, while liquid water, having an inlet temperature of $T_{c,i} = 280$ K, is passed through the tubes. The condenser contains 100 thin-walled tubes, each of 10-mm diameter, and the total water flow rate through the tubes is 15 kg/s. The average convection coefficient associated with condensation on the outer surface of the tubes may be approximated as $\bar{h}_o = 5000$ W/m² · K. Appropriate property values for the liquid water are $c_p = 4178$ J/kg · K, $\mu = 700 \times 10^{-6}$ kg/s · m, $k = 0.628$ W/m · K, and $Pr = 4.6$.

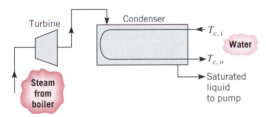

(a) What is the water outlet temperature?

(b) What is the required tube length (per tube)?

(c) After extended use, deposits accumulating on the inner and outer tube surfaces provide a cumulative fouling factor of 0.0003 m² · K/W. For the prescribed inlet conditions and the computed tube length, what mass fraction of the vapor is condensed?

(d) For the tube length computed in part (b) and the fouling factor prescribed in part (c), explore the extent to which the water flow rate and inlet temperature may be varied (within

physically plausible ranges) to improve the condenser performance. Represent your results graphically, and draw appropriate conclusions.

11.65 Consider a Rankine cycle with saturated steam leaving the boiler at a pressure of 2 MPa and a condenser pressure of 10 kPa.

(a) Calculate the thermal efficiency of the ideal Rankine cycle for these operating conditions.

(b) If the net reversible work for the cycle is 0.5 MW, calculate the required flow rate of cooling water supplied to the condenser at 15°C with an allowable temperature rise of 10°C.

(c) Design a shell-and-tube heat exchanger (one-shell, multiple-tube passes) that will meet the heat rate and temperature conditions required of the condenser. Your design should specify the number of tubes and their diameter and length.

11.66 Consider the Rankine cycle of Problem 11.65, which rejects 2.3 MW to the condenser, which is supplied with a cooling water flow rate of 70 kg/s at 15°C.

(a) Calculate UA, a parameter that is indicative of the size of the condenser required for this operating condition.

(b) Consider now the situation where the overall heat transfer coefficient for the condenser, U, is reduced by 10% because of fouling. Determine the reduction in the thermal efficiency of the cycle caused by fouling, assuming that the cooling water flow rate and water temperature remain the same and that the condenser is operated at the same steam pressure.

Compact Heat Exchangers

11.67 Consider the compact heat exchanger conditions of Example 11.6. After extended use, fouling factors of 0.0005 and 0.001 m² · K/W are associated with the water- and gas-side conditions, respectively. What is the gas-side overall heat transfer coefficient?

11.68 Consider the heat exchanger core geometry and frontal area prescribed in Example 11.6. The exchanger must heat 2 kg/s of water from 300 to 350 K, using 1.25 kg/s of combustion gases entering at 700 K. Using the overall heat transfer coefficient determined in the example, find the required heat exchanger volume, assuming single-pass operation. What is the number of tube rows N_L in the longitudinal (gas-flow) direction? If the velocity of water flowing through the tubes is 100 mm/s, what is the number of tube rows N_T in the transverse direction? What is the required tube length?

11.69 A cooling coil consists of a bank of aluminum ($k = 237$ W/m · K) finned tubes having the core configuration of Figure 11.20 and an inner diameter of 13.8 mm. The tubes are installed in a plenum whose square cross section is 0.4 m on a side, thereby providing a frontal area of 0.16 m². Atmospheric air at 1.5 kg/s is in cross flow over the tubes, while saturated refrigerant-12 at 1 atm experiences evaporation in the tubes. If the air enters at 37°C and its exit temperature must not exceed 17°C, what is the minimum allowable number of tube rows in the flow direction? A convection coefficient of 5000 W/m² · K is associated with evaporation in the tubes.

11.70 A cooling coil consists of a bank of aluminum ($k = 237$ W/m · K) finned tubes having the core configuration of Figure 11.20 and an inner diameter of 13.8 mm. The tubes are installed in a plenum whose square cross section is 0.4 m on a side, thereby providing a frontal area of 0.16 m². Atmospheric air at 1.5 kg/s is in cross flow over the tubes, while saturated refrigerant-12 at 1 atm passes through the tubes. There are four rows of tubes in the airflow direction. If the air enters at 37°C, what is its exit temperature? A convection coefficient of 5000 W/m² · K is associated with evaporation in the tubes.

11.71 A steam generator consists of a bank of stainless steel ($k = 15$ W/m · K) tubes having the core configuration of Figure 11.20 and an inner diameter of 13.8 mm. The tubes are installed in a plenum whose square cross section is 0.6 m on a side, thereby providing a frontal area of 0.36 m². Combustion gases, whose properties may be approximated as those of atmospheric air, enter the plenum at 900 K and pass in cross flow over the tubes at 3 kg/s. If saturated water enters the tubes at a pressure of 2.455 bars and a flow rate of 0.5 kg/s, how many tube rows are required to provide saturated steam at the tube outlet? A convection coefficient of 10,000 W/m² · K is associated with boiling in the tubes.

11.72 A steam generator consists of a bank of stainless steel ($k = 15$ W/m · K) tubes having the core configuration of Figure 11.20 and an inner diameter of 13.8 mm. The tubes are installed in a plenum whose square cross section is 0.6 m on a side, thereby providing a frontal area of 0.36 m². Combustion gases, whose properties may be approximated as those of atmospheric air, enter the plenum at 900 K and pass in cross flow over the tubes at 3 kg/s. There are 11 rows of tubes in the gas flow direction. If saturated water at 2.455 bars experiences boiling in the tubes, what is the gas exit temperature? A convection coefficient of 10,000 W/m² · K is associated with boiling in the tubes.

Radiation: Processes and Properties

W e have come to recognize that heat transfer by conduction and convection requires the presence of a temperature gradient in some form of matter. In contrast, heat transfer by *thermal radiation* requires no matter. It is an extremely important process, and in the physical sense it is perhaps the most interesting of the heat transfer modes. It is relevant to many industrial heating, cooling, and drying processes, as well as to energy conversion methods that involve fossil fuel combustion and solar radiation.

In this chapter we consider the means by which thermal radiation is generated, the specific nature of the radiation, and the manner in which the radiation interacts with matter. We give particular attention to radiative interactions at a surface and to the properties that must be introduced to describe these interactions. In Chapter 13 we focus on means for computing radiative exchange between two or more surfaces.

12.1
Fundamental Concepts

Consider a solid that is initially at a higher temperature T_s than that of its surroundings T_{sur}, but around which there exists a vacuum (Figure 12.1). The presence of the vacuum precludes energy loss from the surface of the solid by conduction or convection. However, our intuition tells us that the solid will cool and eventually achieve thermal equilibrium with its surroundings. This cooling is associated with a reduction in the internal energy stored by the solid and is a direct consequence of the *emission* of thermal radiation from the surface. In turn, the surface will intercept and absorb radiation originating from the surroundings. However, if $T_s > T_{sur}$ the *net* heat transfer rate by radiation $q_{rad, net}$ is *from* the surface, and the surface will cool until T_s reaches T_{sur}.

We associate thermal radiation with the rate at which energy is emitted by matter as a result of its finite temperature. At this moment thermal radiation is being emitted by all the matter that surrounds you: by the furniture and walls of the room, if you are indoors, or by the ground, buildings, and the atmosphere and sun if you are outdoors. The mechanism of emission is related to energy released as a result of oscillations or transitions of the many electrons that consti-

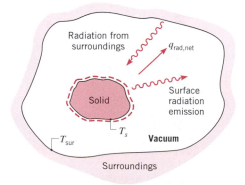

FIGURE 12.1
Radiation cooling of a heated solid.

tute matter. These oscillations are, in turn, sustained by the internal energy, and therefore the temperature, of the matter. Hence we associate the emission of thermal radiation with thermally excited conditions within the matter.

All forms of matter emit radiation. For gases and for semitransparent solids, such as glass and salt crystals at elevated temperatures, emission is a *volumetric phenomenon,* as illustrated in Figure 12.2. That is, radiation emerging from a finite volume of matter is the integrated effect of local emission throughout the volume. However, in this text we concentrate on situations for which radiation is a *surface phenomenon.* In most solids and liquids, radiation emitted from interior molecules is strongly absorbed by adjoining molecules. Accordingly, radiation that is emitted from a solid or a liquid originates from molecules that are within a distance of approximately 1 μm from the exposed surface. It is for this reason that emission from a solid or a liquid into an adjoining gas or a vacuum is viewed as a surface phenomenon.

We know that radiation originates due to emission by matter and that its subsequent transport does not require the presence of any matter. But what is the nature of this transport? One theory views radiation as the propagation of a collection of particles termed *photons* or *quanta.* Alternatively, radiation may be viewed as the propagation of *electromagnetic waves.* In any case we wish to attribute to radiation the standard wave properties of frequency ν and wavelength λ. For radiation propagating in a particular medium, the two properties are related by

$$\lambda = \frac{c}{\nu} \tag{12.1}$$

where c is the speed of light in the medium. For propagation in a vacuum, $c_o = 2.998 \times 10^8$ m/s. The unit of wavelength is commonly the micrometer (μm), where 1 μm $= 10^{-6}$ m.

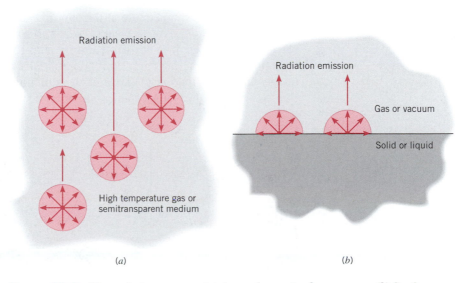

(a)   (b)

FIGURE 12.2   The emission process. (*a*) As a volumetric phenomenon. (*b*) Surface phenomenon.

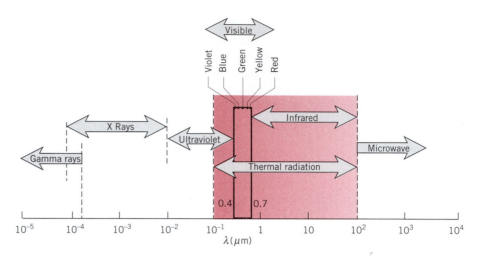

FIGURE 12.3   Spectrum of electromagnetic radiation.

The complete electromagnetic spectrum is delineated in Figure 12.3. The short wavelength gamma rays, X rays, and ultraviolet (UV) radiation are primarily of interest to the high-energy physicist and the nuclear engineer, while the long wavelength microwaves and radio waves are of concern to the electrical engineer. It is the intermediate portion of the spectrum, which extends from approximately 0.1 to 100 μm and includes a portion of the UV and all of the visible and infrared (IR), that is termed *thermal radiation* and is pertinent to heat transfer.

Thermal radiation emitted by a surface encompasses a range of wavelengths. As shown in Figure 12.4*a*, the magnitude of the radiation varies with wavelength, and the term *spectral* is used to refer to the nature of this dependence. Emitted radiation consists of a continuous, nonuniform distribution of *monochromatic* (single-wavelength) components. As we will find, both the magnitude of the radiation at any wavelength and the *spectral distribution* vary with the nature and temperature of the emitting surface.

The spectral nature of thermal radiation is one of two features that complicates its description. The second feature relates to its *directionality*. As shown in Figure 12.4*b*, a surface may emit preferentially in certain directions, creating a *directional distribution* of the emitted radiation. To properly quantify radiation heat transfer, we must be able to treat both spectral and directional effects.

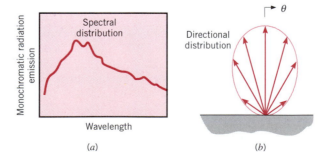

FIGURE 12.4   Radiation emitted by a surface. (*a*) Spectral distribution. (*b*) Directional distribution.

12.2
Radiation Intensity

Radiation emitted by a surface propagates in all possible directions (Figure 12.4*b*), and we are often interested in knowing its directional distribution. Also, radiation incident on a surface may come from different directions, and the manner in which the surface responds to this radiation depends on the direction. Such directional effects may be treated by introducing the concept of *radiation intensity*.

12.2.1 Definitions

Consider emission in a particular direction from an element of area dA_1, as shown in Figure 12.5*a*. This direction may be specified in terms of the zenith and azimuthal angles, θ and ϕ, respectively, of a spherical coordinate system (Figure 12.5*b*). A differentially small surface in space dA_n, through which this radiation passes, subtends a solid angle $d\omega$ when viewed from a point on dA_1. From Figure 12.6*a* we see that the differential plane angle $d\alpha$ is defined by a region between the rays of a circle and is measured as the ratio of the element of arc length dl on the circle to the radius r of the circle. Similarly, from Figure 12.6*b* the differential solid angle $d\omega$ is defined by a region between the rays of a sphere and is measured as the ratio of the element of area dA_n on the sphere to the square of the sphere's radius. Accordingly,

$$d\omega \equiv \frac{dA_n}{r^2} \tag{12.2}$$

The area dA_n is normal to the (θ, ϕ) direction, and as shown in Figure 12.7, it may be represented as $dA_n = r^2 \sin\theta \, d\theta \, d\phi$ for a spherical surface. Accordingly,

$$d\omega = \sin\theta \, d\theta \, d\phi \tag{12.3}$$

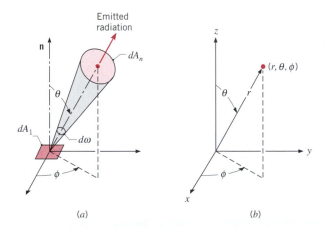

(a)   (b)

FIGURE 12.5   Directional nature of radiation. (*a*) Emission of radiation from a differential area dA_1 into a solid angle $d\omega$ subtended by dA_n at a point on dA_1. (*b*) The spherical coordinate system.

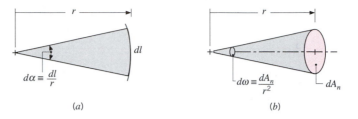

FIGURE 12.6   Definition of (*a*) plane and (*b*) solid angles.

Whereas the plane angle $d\alpha$ has the unit of radians (rad), the unit of the solid angle is the steradian (sr).

   Returning to Figure 12.5*a*, we now consider the rate at which emission from dA_1 passes through dA_n. This quantity may be expressed in terms of the *spectral intensity* $I_{\lambda, e}$ of the emitted radiation. We formally define $I_{\lambda, e}$ as the *rate at which radiant energy is emitted at the wavelength* λ *in the* (θ, ϕ) *direction, per unit area of the emitting surface normal to this direction, per unit solid angle about this direction, and per unit wavelength interval* $d\lambda$ *about* λ. Note that the area used to define the intensity is the component of dA_1 perpendicular to the direction of the radiation. From Figure 12.8 we see that this projected area is equal to $dA_1 \cos \theta$. In effect it is how dA_1 would appear to an observer situated on dA_n. The spectral intensity, which has units of $W/m^2 \cdot sr \cdot \mu m$, is then

$$I_{\lambda, e}(\lambda, \theta, \phi) \equiv \frac{dq}{dA_1 \cos \theta \cdot d\omega \cdot d\lambda} \qquad (12.4)$$

where $(dq/d\lambda) \equiv dq_\lambda$ is the rate at which radiation of wavelength λ leaves dA_1 and passes through dA_n. Rearranging Equation 12.4, it follows that

$$dq_\lambda = I_{\lambda, e}(\lambda, \theta, \phi) \, dA_1 \cos \theta \, d\omega \qquad (12.5)$$

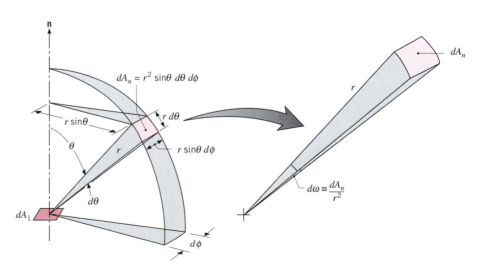

FIGURE 12.7   The solid angle subtended by dA_n at a point on dA_1 in the spherical coordinate system.

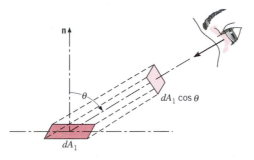

where dq_λ has the units of W/μm. This important expression allows us to compute the rate at which radiation emitted by a surface propagates into the region of space defined by the solid angle $d\omega$ about the (θ, ϕ) direction. However, to compute this rate, the spectral intensity $I_{\lambda, e}$ of the emitted radiation must be known. The manner in which this quantity may be determined is discussed in subsequent sections. Expressing Equation 12.5 per unit area of the emitting surface and substituting from Equation 12.3, the spectral radiation *flux* associated with dA_1 is

$$dq_\lambda'' = I_{\lambda, e}(\lambda, \theta, \phi) \cos \theta \sin \theta \, d\theta \, d\phi \tag{12.6}$$

If the spectral and directional distributions of $I_{\lambda, e}$ are known, that is, $I_{\lambda, e}(\lambda, \theta, \phi)$, the heat flux associated with emission into any finite solid angle or over any finite wavelength interval may be determined by integrating Equation 12.6. For example, the spectral heat flux associated with emission into a hypothetical hemisphere above dA_1, as shown in Figure 12.9, is

$$q_\lambda''(\lambda) = \int_0^{2\pi} \int_0^{\pi/2} I_{\lambda, e}(\lambda, \theta, \phi) \cos \theta \sin \theta \, d\theta \, d\phi \tag{12.7}$$

The solid angle associated with the entire hemisphere may be obtained by integrating Equation 12.3 over the limits $\phi = 0$ to $\phi = 2\pi$ and $\theta = 0$ to $\theta = \pi/2$. Hence

$$\int_h d\omega = \int_0^{2\pi} \int_0^{\pi/2} \sin \theta \, d\theta \, d\phi = 2\pi \int_0^{\pi/2} \sin \theta \, d\theta = 2\pi \text{ sr} \tag{12.8}$$

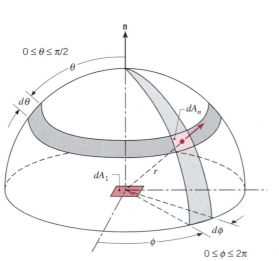

where the subscript h refers to integration over the hemisphere. The total heat flux associated with emission in all directions and at all wavelengths is then

$$q'' = \int_0^\infty q_\lambda''(\lambda)\, d\lambda \tag{12.9}$$

12.2.2   Relation to Emission

The radiation intensity is related to several important radiation fluxes. Recall that emission occurs from any surface that is at a finite temperature. The concept of *emissive power* is introduced to quantify the amount of radiation emitted per unit surface area. The *spectral, hemispherical emissive power* E_λ (W/m² · μm) is defined as the rate at which radiation of wavelength λ is emitted in *all directions* from a surface per unit wavelength $d\lambda$ about λ and per unit surface area. Accordingly, it is related to the spectral intensity of the emitted radiation by an expression of the form

$$E_\lambda(\lambda) = \int_0^{2\pi} \int_0^{\pi/2} I_{\lambda,e}(\lambda, \theta, \phi) \cos\theta \sin\theta\, d\theta\, d\phi \tag{12.10}$$

in which case it is equivalent to the heat flux given by Equation 12.7. Note that E_λ is a flux based on the *actual* surface area, whereas $I_{\lambda,e}$ is based on the *projected* area. The $\cos\theta$ term appearing in the integrand is a consequence of this difference.

The *total, hemispherical emissive power, E* (W/m²), is the rate at which radiation is emitted per unit area at all possible wavelengths and in all possible directions. Accordingly,

$$E = \int_0^\infty E_\lambda(\lambda)\, d\lambda \tag{12.11}$$

or from Equation 12.10

$$E = \int_0^\infty \int_0^{2\pi} \int_0^{\pi/2} I_{\lambda,e}(\lambda, \theta, \phi) \cos\theta \sin\theta\, d\theta\, d\phi\, d\lambda \tag{12.12}$$

Since the term "emissive power" implies emission in all directions, the adjective "hemispherical" is redundant and is often dropped. One then speaks of the *spectral emissive power* E_λ, or the *total emissive power E.*

Although the directional distribution of surface emission varies according to the nature of the surface, there is a special case that provides a reasonable approximation for many surfaces. We speak of a *diffuse emitter* as a surface for which the intensity of the emitted radiation is independent of direction, in which case $I_{\lambda,e}(\lambda, \theta, \phi) = I_{\lambda,e}(\lambda)$. Removing $I_{\lambda,e}$ from the integrand of Equation 12.10 and performing the integration, it follows that

$$E_\lambda(\lambda) = \pi I_{\lambda,e}(\lambda) \tag{12.13}$$

Similarly, from Equation 12.12

$$E = \pi I_e \tag{12.14}$$

where I_e is the *total intensity* of the emitted radiation. Note that the constant appearing in the above expressions is π, not 2π, and has the unit steradians.

EXAMPLE 12.1

A small surface of area $A_1 = 10^{-3}$ m^2 is known to emit diffusely, and from measurements the total intensity associated with emission in the normal direction is $I_n = 7000$ W/m$^2 \cdot$ sr.

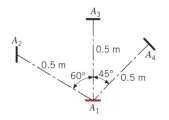

Radiation emitted from the surface is intercepted by three other surfaces of area $A_2 = A_3 = A_4 = 10^{-3}$ m^2, which are 0.5 m from A_1 and are oriented as shown. What is the intensity associated with emission in each of the three directions? What are the solid angles subtended by the three surfaces when viewed from A_1? What is the rate at which radiation emitted by A_1 is intercepted by the three surfaces?

SOLUTION

Known: Normal intensity of diffuse emitter of area A_1 and orientation of three surfaces relative to A_1.

Find:

1. Intensity of emission in each of the three directions.
2. Solid angles subtended by the three surfaces.
3. Rate at which radiation is intercepted by the three surfaces.

Schematic:

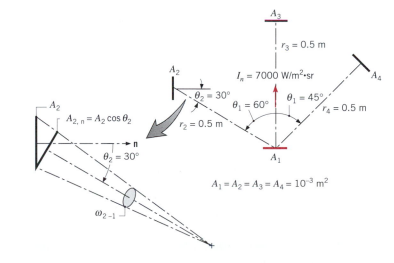

Assumptions:

1. Surface A_1 emits diffusely.
2. $A_1, A_2, A_3,$ and A_4 may be approximated as differential surfaces, $(A_j/r_j^2) \ll 1$.

Analysis:

1. From the definition of a diffuse emitter, we know that the intensity of the emitted radiation is independent of direction. Hence

$$I = 7000 \ \text{W/m}^2 \cdot \text{sr} \qquad \triangleleft$$

for each of the three directions.

2. Treating A_2, A_3, and A_4 as differential surface areas, the solid angles may be computed from Equation 12.2

$$d\omega \equiv \frac{dA_n}{r^2}$$

where dA_n is the projection of the surface normal to the direction of the radiation. Accordingly,

$$dA_{n,j} = dA_j \times \cos \theta_j$$

where θ_j is the angle between the surface normal and the direction of the radiation. The solid angle subtended by surface A_2 with respect to A_1 is then

$$\omega_{2-1} \approx \frac{A_2 \times \cos \theta_2}{r^2} = \frac{10^{-3} \ \text{m}^2 \times \cos 30°}{(0.5 \ \text{m})^2} = 3.46 \times 10^{-3} \ \text{sr} \qquad \triangleleft$$

Similarly,

$$\omega_{3-1} \approx \frac{A_3 \times \cos \theta_3}{r^2} = \frac{10^{-3} \ \text{m}^2 \times \cos 0°}{(0.5 \ \text{m})^2} = 4.00 \times 10^{-3} \ \text{sr} \qquad \triangleleft$$

$$\omega_{4-1} \approx \frac{A_4 \times \cos \theta_4}{r^2} = \frac{10^{-3} \ \text{m}^2 \times \cos 0°}{(0.5 \ \text{m})^2} = 4.00 \times 10^{-3} \ \text{sr} \qquad \triangleleft$$

3. Approximating A_1 as a differential surface, the rate at which radiation is intercepted by each of the three surfaces may be estimated from Equation 12.5, which, for the total radiation, may be expressed as

$$q_{1-j} \approx I \times A_1 \cos \theta_1 \times \omega_{j-1}$$

where θ_1 is the angle between the normal to surface 1 and the direction of the radiation. Hence

$$q_{1-2} \approx 7000 \ \text{W/m}^2 \cdot \text{sr} \ (10^{-3} \ \text{m}^2 \times \cos 60°)3.46 \times 10^{-3} \ \text{sr}$$
$$= 12.1 \times 10^{-3} \ \text{W} \qquad \triangleleft$$

$$q_{1-3} \approx 7000 \ \text{W/m}^2 \cdot \text{sr} \ (10^{-3} \ \text{m}^2 \times \cos 0°)4.00 \times 10^{-3} \ \text{sr}$$
$$= 28.0 \times 10^{-3} \ \text{W} \qquad \triangleleft$$

$$q_{1-4} \approx 7000 \ \text{W/m}^2 \cdot \text{sr} \ (10^{-3} \ \text{m}^2 \times \cos 45°)4.00 \times 10^{-3} \ \text{sr}$$
$$= 19.8 \times 10^{-3} \ \text{W} \qquad \triangleleft$$

Comments:

1. Note the different values of θ_1 for the emitting surface and the values of θ_2, θ_3, and θ_4 for the receiving surfaces.

2. If the surfaces were not small relative to the square of the separation distance, the solid angles and radiation heat transfer rates would have to be obtained by integrating Equations 12.3 and 12.5, respectively.

3. Any spectral component of the radiation rate could also be obtained using these procedures, if the spectral intensity I_λ is known.

4. Even though the intensity of the emitted radiation is independent of direction, the rate at which radiation is intercepted by the three surfaces differs significantly due to differences in the solid angles and projected areas.

12.2.3   Relation to Irradiation

Although we have focused on radiation emitted by a surface, the foregoing concepts may be extended to *incident* radiation (Figure 12.10). Such radiation may originate from emission and reflection occurring at other surfaces and will have spectral and directional distributions determined by the spectral intensity $I_{\lambda,i}(\lambda, \theta, \phi)$. This quantity is defined as the rate at which radiant energy of wavelength λ is incident from the (θ, ϕ) direction, per unit area of the *intercepting surface* normal to this direction, per unit solid angle about this direction, and per unit wavelength interval $d\lambda$ about λ.

The intensity of the incident radiation may be related to an important radiative flux, termed the *irradiation,* which encompasses radiation incident *from all directions.* The *spectral irradiation* G_λ (W/m$^2 \cdot \mu$m) is defined as the rate at which radiation of wavelength λ is incident on a surface, per unit area of the surface and per unit wavelength interval $d\lambda$ about λ. Accordingly,

$$G_\lambda(\lambda) = \int_0^{2\pi} \int_0^{\pi/2} I_{\lambda,i}(\lambda, \theta, \phi) \cos\theta \sin\theta \, d\theta \, d\phi \tag{12.15}$$

where $\sin\theta \, d\theta \, d\phi$ is the unit solid angle. The $\cos\theta$ factor originates because G_λ is a flux based on the actual surface area, whereas $I_{\lambda,i}$ is defined in terms of the projected area. If the *total irradiation* G (W/m^2) represents the rate at which radiation is incident per unit area from all directions and at all wavelengths, it follows that

$$G = \int_0^\infty G_\lambda(\lambda) \, d\lambda \tag{12.16}$$

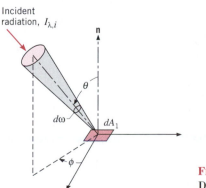

Incident
radiation, $I_{\lambda,i}$

FIGURE 12.10
Directional nature of incident radiation.

or from Equation 12.15

$$G = \int_0^\infty \int_0^{2\pi} \int_0^{\pi/2} I_{\lambda,i}(\lambda, \theta, \phi) \cos \theta \sin \theta \, d\theta \, d\phi \, d\lambda \qquad (12.17)$$

If the incident radiation is *diffuse*, $I_{\lambda,i}$ is independent of θ and ϕ and it follows that

$$G_\lambda(\lambda) = \pi I_{\lambda,i}(\lambda) \qquad (12.18)$$

and

$$G = \pi I_i \qquad (12.19)$$

EXAMPLE 12.2

The spectral distribution of surface irradiation is as follows:

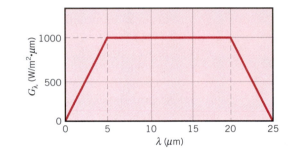

What is the total irradiation?

SOLUTION

Known: Spectral distribution of surface irradiation.

Find: Total irradiation.

Analysis: The total irradiation may be obtained from Equation 12.16

$$G = \int_0^\infty G_\lambda \, d\lambda$$

The integral is readily evaluated by breaking it into parts. That is,

$$G = \int_0^{5 \ \mu m} G_\lambda \, d\lambda + \int_{5 \ \mu m}^{20 \ \mu m} G_\lambda \, d\lambda + \int_{20 \ \mu m}^{25 \ \mu m} G_\lambda \, d\lambda + \int_{25 \ \mu m}^{\infty} G_\lambda \, d\lambda$$

Hence

$$G = \tfrac{1}{2}(1000 \ \text{W/m}^2 \cdot \mu\text{m})(5 - 0) \ \mu\text{m} + (1000 \ \text{W/m}^2 \cdot \mu\text{m})(20 - 5) \ \mu\text{m}$$

$$+ \tfrac{1}{2}(1000 \ \text{W/m}^2 \cdot \mu\text{m})(25 - 20) \ \mu\text{m} + 0$$

$$= (2500 + 15000 + 2500) \ \text{W/m}^2$$

$$G = 20{,}000 \ \text{W/m}^2 \qquad \qquad \triangleleft$$

Comments: Generally, radiation sources do not provide such a regular spectral distribution for the irradiation. However, the procedure of computing the total irradiation from knowledge of the spectral distribution remains the same, although evaluation of the integral is likely to involve more detail.

12.2.4 Relation to Radiosity

The last radiative flux of interest, termed *radiosity,* accounts for *all* the radiant energy leaving a surface. Since this radiation includes the *reflected* portion of the irradiation, as well as direct emission (Figure 12.11), the radiosity is generally different from the emissive power. The *spectral radiosity* J_λ (W/m² · μm) represents the rate at which radiation of wavelength λ leaves a unit area of the surface, per unit wavelength interval $d\lambda$ about λ. Since it accounts for radiation leaving in all directions, it is related to the intensity associated with emission and reflection, $I_{\lambda, e+r}(\lambda, \theta, \phi)$, by an expression of the form

$$J_\lambda(\lambda) = \int_0^{2\pi} \int_0^{\pi/2} I_{\lambda, e+r}(\lambda, \theta, \phi) \cos \theta \sin \theta \, d\theta \, d\phi \qquad (12.20)$$

Hence the *total radiosity J* (W/m²) associated with the entire spectrum is

$$J = \int_0^\infty J_\lambda(\lambda) \, d\lambda \qquad (12.21)$$

or

$$J = \int_0^\infty \int_0^{2\pi} \int_0^{\pi/2} I_{\lambda, e+r}(\lambda, \theta, \phi) \cos \theta \sin \theta \, d\theta \, d\phi \, d\lambda \qquad (12.22)$$

If the surface is both a *diffuse reflector* and a *diffuse emitter,* $I_{\lambda, e+r}$ is independent of θ and ϕ, and it follows that

$$J_\lambda(\lambda) = \pi I_{\lambda, e+r} \qquad (12.23)$$

and

$$J = \pi I_{e+r} \qquad (12.24)$$

Again, note that the radiation flux, in this case the radiosity, is based on the actual surface area, while the intensity is based on the projected area.

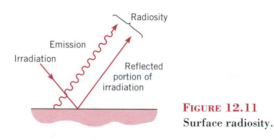

FIGURE 12.11
Surface radiosity.

12.3
Blackbody Radiation

When describing the radiation characteristics of real surfaces, it is useful to introduce the concept of a *blackbody*. The blackbody is an *ideal* surface having the following properties.

1. *A blackbody absorbs all incident radiation, regardless of wavelength and direction.*
2. *For a prescribed temperature and wavelength, no surface can emit more energy than a blackbody.*
3. *Although the radiation emitted by a blackbody is a function of wavelength and temperature, it is independent of direction. That is, the blackbody is a diffuse emitter.*

As the perfect absorber and emitter, the blackbody serves as a *standard* against which the radiative properties of actual surfaces may be compared.

Although closely approximated by some surfaces, it is important to note that no surface has precisely the properties of a blackbody. The closest approximation is achieved by a *cavity* whose inner surface is at a uniform temperature. If radiation enters the cavity through a small aperture (Figure 12.12*a*), it is likely to experience many reflections before reemergence. Hence it is almost entirely absorbed by the cavity, and blackbody behavior is approximated. From thermodynamic principles it may then be argued that radiation leaving the aperture depends only on the surface temperature and corresponds to blackbody emission (Figure 12.12*b*). Since blackbody emission is diffuse, the spectral intensity $I_{\lambda, b}$ of radiation leaving the cavity is independent of direction. Moreover, since the radiation field in the cavity, which is the cumulative effect of emission and reflection from the cavity surface, must be of the same form as the radiation emerging from the aperture, it also follows that a blackbody radiation field exists within the cavity. Accordingly, any small surface in the cavity

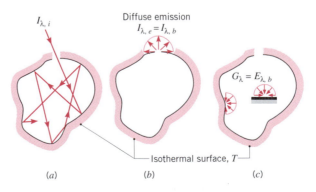

(a)   (b)   (c)

FIGURE 12.12 Characteristics of an isothermal blackbody cavity. (*a*) Complete absorption. (*b*) Diffuse emission from an aperture. (*c*) Diffuse irradiation of interior surfaces.

(Figure 12.12*c*) experiences irradiation for which $G_\lambda = E_{\lambda,b}(\lambda, T)$. This surface is diffusely irradiated, regardless of its orientation. *Blackbody radiation exists within the cavity irrespective of whether the cavity surface is highly reflecting or absorbing.*

12.3.1 The Planck Distribution

The spectral distribution of blackbody emission is well known, having first been determined by Planck [1]. It is of the form

$$I_{\lambda,b}(\lambda, T) = \frac{2hc_o^2}{\lambda^5[\exp(hc_o/\lambda kT) - 1]} \qquad (12.25)$$

where $h = 6.6256 \times 10^{-34}$ J $\cdot$ s and $k = 1.3805 \times 10^{-23}$ J/K are the universal Planck and Boltzmann constants, respectively, $c_o = 2.998 \times 10^8$ m/s is the speed of light in vacuum, and T is the *absolute* temperature of the blackbody (K). Since the blackbody is a diffuse emitter, it follows from Equation 12.13 that its spectral emissive power is of the form

$$E_{\lambda,b}(\lambda, T) = \pi I_{\lambda,b}(\lambda, T) = \frac{C_1}{\lambda^5[\exp(C_2/\lambda T) - 1]} \qquad (12.26)$$

where the first and second radiation constants are $C_1 = 2\pi hc_o^2 = 3.742 \times 10^8$ W $\cdot$ μm^4/m^2 and $C_2 = (hc_o/k) = 1.439 \times 10^4$ μm $\cdot$ K.

Equation 12.26, known as the *Planck distribution,* is plotted in Figure 12.13 for selected temperatures. Several important features should be noted.

1. The emitted radiation varies *continuously* with wavelength.
2. At any wavelength the magnitude of the emitted radiation increases with increasing temperature.
3. The spectral region in which the radiation is concentrated depends on temperature, with *comparatively* more radiation appearing at shorter wavelengths as the temperature increases.
4. A significant fraction of the radiation emitted by the sun, which may be approximated as a blackbody at 5800 K, is in the visible region of the spectrum. In contrast, for $T \lesssim 800$ K, emission is predominantly in the infrared region of the spectrum and is not visible to the eye.

12.3.2 Wien's Displacement Law

From Figure 12.13 we see that the blackbody spectral distribution has a maximum and that the corresponding wavelength λ_{max} depends on temperature. The nature of this dependence may be obtained by differentiating Equation 12.26 with respect to λ and setting the result equal to zero. In so doing, we obtain

$$\lambda_{max}T = C_3 \qquad (12.27)$$

where the third radiation constant is $C_3 = 2897.8$ μm $\cdot$ K.

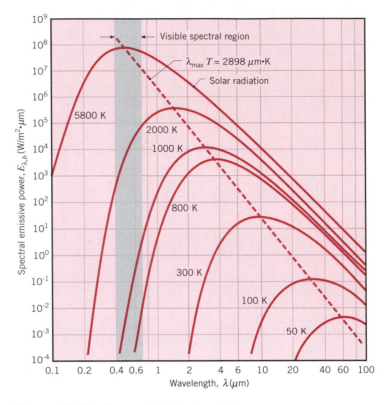

FIGURE 12.13   Spectral blackbody emissive power.

Equation 12.27 is known as *Wien's displacement law,* and the locus of points described by the law is plotted as the dashed line of Figure 12.13. According to this result, the maximum spectral emissive power is displaced to shorter wavelengths with increasing temperature. This emission is in the middle of the visible spectrum ($\lambda \approx 0.50 \ \mu$m) for solar radiation, since the sun emits as a blackbody at approximately 5800 K. For a blackbody at 1000 K, peak emission occurs at 2.90 μm, with some of the emitted radiation appearing visible as red light. With increasing temperature, shorter wavelengths become more prominent, until eventually significant emission occurs over the entire visible spectrum. For example, a tungsten filament lamp operating at 2900 K ($\lambda_{max} = 1 \ \mu$m) emits white light, although most of the emission remains in the IR region.

12.3.3   The Stefan–Boltzmann Law

Substituting the Planck distribution, Equation 12.26, into Equation 12.11, the total emissive power of a blackbody E_b may be expressed as

$$E_b = \int_0^\infty \frac{C_1}{\lambda^5 [\exp{(C_2/\lambda T)} - 1]} \, d\lambda$$

Performing the integration, it may be shown that

$$E_b = \sigma T^4 \tag{12.28}$$

where the *Stefan–Boltzmann* constant, which depends on C_1 and C_2, has the numerical value

$$\sigma = 5.670 \times 10^{-8} \text{ W/m}^2 \cdot \text{K}^4$$

This simple, yet important, result is termed the *Stefan–Boltzmann law.* It enables calculation of the amount of radiation emitted in all directions and over all wavelengths simply from knowledge of the temperature of the blackbody. Because this emission is diffuse, it follows from Equation 12.14 that the total intensity associated with blackbody emission is

$$I_b = \frac{E_b}{\pi} \tag{12.29}$$

12.3.4   Band Emission

It is often necessary to know the fraction of the total emission from a blackbody that is in a certain wavelength interval or *band.* For a prescribed temperature and the interval from 0 to λ, this fraction is determined by the ratio of the shaded section to the total area under the curve of Figure 12.14. Hence

$$F_{(0 \to \lambda)} \equiv \frac{\displaystyle\int_0^{\lambda} E_{\lambda,b}\, d\lambda}{\displaystyle\int_0^{\infty} E_{\lambda,b}\, d\lambda} = \frac{\displaystyle\int_0^{\lambda} E_{\lambda,b}\, d\lambda}{\sigma T^4} = \int_0^{\lambda T} \frac{E_{\lambda,b}}{\sigma T^5}\, d(\lambda T) = f(\lambda T) \tag{12.30}$$

FIGURE 12.14   Radiation emission from a blackbody in the spectral band 0 to λ.

Since the integrand $(E_{\lambda,b}/\sigma T^5)$ is exclusively a function of the wavelength–temperature product λT, the integral of Equation 12.30 may be evaluated to obtain $F_{(0 \to \lambda)}$ as a function of only λT. The results are presented in Table 12.1 and Figure 12.15. They may also be used to obtain the fraction of the radiation between any two wavelengths λ_1 and λ_2, since

$$F_{(\lambda_1 \to \lambda_2)} = \frac{\displaystyle\int_0^{\lambda_2} E_{\lambda,b}\, d\lambda - \int_0^{\lambda_1} E_{\lambda,b}\, d\lambda}{\sigma T^4} = F_{(0 \to \lambda_2)} - F_{(0 \to \lambda_1)} \tag{12.31}$$

Additional blackbody functions are listed in the third and fourth columns of Table 12.1. The third column facilitates calculation of the spectral intensity for a prescribed wavelength and temperature. In lieu of computing this quantity from Equation 12.25, it may be obtained by simply multiplying the tabulated

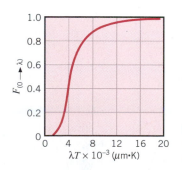

FIGURE 12.15
Fraction of the total blackbody emission in the spectral band from 0 to λ as a function of λT.

TABLE 12.1   Blackbody Radiation Functions[a]

λT (μm · K)	$F_{(0 \to \lambda)}$	$I_{\lambda, b}(\lambda, T)/\sigma T^5$ (μm · K · sr)$^{-1}$	$\dfrac{I_{\lambda, b}(\lambda, T)}{I_{\lambda, b}(\lambda_{max}, T)}$
200	0.000000	0.375034×10^{-27}	0.000000
400	0.000000	0.490335×10^{-13}	0.000000
600	0.000000	0.104046×10^{-8}	0.000014
800	0.000016	0.991126×10^{-7}	0.001372
1,000	0.000321	0.118505×10^{-5}	0.016406
1,200	0.002134	0.523927×10^{-5}	0.072534
1,400	0.007790	0.134411×10^{-4}	0.186082
1,600	0.019718	0.249130	0.344904
1,800	0.039341	0.375568	0.519949
2,000	0.066728	0.493432	0.683123
2,200	0.100888	0.589649×10^{-4}	0.816329
2,400	0.140256	0.658866	0.912155
2,600	0.183120	0.701292	0.970891
2,800	0.227897	0.720239	0.997123
2,898	0.250108	0.722318×10^{-4}	1.000000
3,000	0.273232	0.720254×10^{-4}	0.997143
3,200	0.318102	0.705974	0.977373
3,400	0.361735	0.681544	0.943551
3,600	0.403607	0.650396	0.900429
3,800	0.443382	0.615225×10^{-4}	0.851737
4,000	0.480877	0.578064	0.800291
4,200	0.516014	0.540394	0.748139
4,400	0.548796	0.503253	0.696720
4,600	0.579280	0.467343	0.647004
4,800	0.607559	0.433109	0.599610
5,000	0.633747	0.400813	0.554898
5,200	0.658970	0.370580×10^{-4}	0.513043
5,400	0.680360	0.342445	0.474092
5,600	0.701046	0.316376	0.438002
5,800	0.720158	0.292301	0.404671
6,000	0.737818	0.270121	0.373965
6,200	0.754140	0.249723×10^{-4}	0.345724
6,400	0.769234	0.230985	0.319783
6,600	0.783199	0.213786	0.295973
6,800	0.796129	0.198008	0.274128
7,000	0.808109	0.183534	0.254090
7,200	0.819217	0.170256×10^{-4}	0.235708
7,400	0.829527	0.158073	0.218842
7,600	0.839102	0.146891	0.203360
7,800	0.848005	0.136621	0.189143
8,000	0.856288	0.127185	0.176079
8,500	0.874608	0.106772×10^{-4}	0.147819
9,000	0.890029	0.901463×10^{-5}	0.124801
9,500	0.903085	0.765338	0.105956
10,000	0.914199	0.653279	0.090442
10,500	0.923710	0.560522	0.077600

TABLE 12.1   *Continued*

λT ($\mu m \cdot K$)	$F_{(0 \to \lambda)}$	$I_{\lambda, b}(\lambda, T)/\sigma T^5$ ($\mu m \cdot K \cdot sr)^{-1}$	$\dfrac{I_{\lambda, b}(\lambda, T)}{I_{\lambda, b}(\lambda_{max}, T)}$
11,000	0.931890	0.483321×10^{-5}	0.066913
11,500	0.939959	0.418725	0.057970
12,000	0.945098	0.364394	0.050448
13,000	0.955139	0.279457	0.038689
14,000	0.962898	0.217641	0.030131
15,000	0.969981	0.171866×10^{-5}	0.023794
16,000	0.973814	0.137429	0.019026
18,000	0.980860	0.908240×10^{-6}	0.012574
20,000	0.985602	0.623310	0.008629
25,000	0.992215	0.276474	0.003828
30,000	0.995340	0.140469×10^{-6}	0.001945
40,000	0.997967	0.473891×10^{-7}	0.000656
50,000	0.998953	0.201605	0.000279
75,000	0.999713	0.418597×10^{-8}	0.000058
100,000	0.999905	0.135752	0.000019

[a]The radiation constants used to generate these blackbody functions are:
$C_1 = 3.7420 \times 10^8 \ \mu m^4/m^2$
$C_2 = 1.4388 \times 10^4 \ \mu m \cdot K$
$\sigma = 5.670 \times 10^{-8} \ W/m^2 \cdot K^4$

value of $I_{\lambda, b}/\sigma T^5$ by σT^5. The fourth column is used to obtain a quick estimate of the ratio of the spectral intensity at any wavelength to that at λ_{max}.

EXAMPLE 12.3

Consider a large isothermal enclosure that is maintained at a uniform temperature of 2000 K. Calculate the emissive power of the radiation that emerges from a small aperture on the enclosure surface. What is the wavelength λ_1 below which 10% of the emission is concentrated? What is the wavelength λ_2 above which 10% of the emission is concentrated? Determine the maximum spectral emissive power and the wavelength at which this emission occurs. What is the irradiation incident on a small object placed inside the enclosure?

SOLUTION

Known: Large isothermal enclosure at uniform temperature.

Find:

1. Emissive power of a small aperture on the enclosure.
2. Wavelengths below which and above which 10% of the radiation is concentrated.
3. Spectral emissive power and wavelength associated with maximum emission.
4. Irradiation on a small object inside the enclosure.

Schematic:

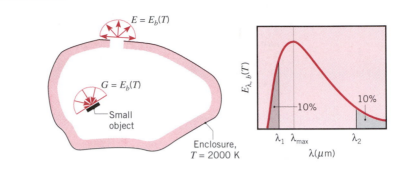

Assumptions: Areas of aperture and object are very small relative to enclosure surface.

Analysis:

1. Emission from the aperture of any isothermal enclosure will have the characteristics of blackbody radiation. Hence, from Equation 12.28,

$$E = E_b(T) = \sigma T^4 = 5.670 \times 10^{-8} \text{ W/m}^2 \cdot \text{K}^4(2000 \text{ K})^4$$

$$E = 9.07 \times 10^5 \text{ W/m}^2 \qquad \triangleleft$$

2. The wavelength λ_1 corresponds to the upper limit of the spectral band $(0 \to \lambda_1)$ containing 10% of the emitted radiation. With $F_{(0 \to \lambda_1)} = 0.10$ it follows from Table 12.1 that $\lambda_1 T \approx 2200 \ \mu\text{m} \cdot \text{K}$. Hence

$$\lambda_1 = 1.1 \ \mu\text{m} \qquad \triangleleft$$

The wavelength λ_2 corresponds to the lower limit of the spectral band $(\lambda_2 \to \infty)$ containing 10% of the emitted radiation. With

$$F_{(\lambda_2 \to \infty)} = 1 - F_{(0 \to \lambda_2)} = 0.1$$

$$F_{(0 \to \lambda_2)} = 0.9$$

it follows from Table 12.1 that $\lambda_2 T = 9382 \ \mu\text{m} \cdot \text{K}$. Hence

$$\lambda_2 = 4.69 \ \mu\text{m} \qquad \triangleleft$$

3. From Wien's law, Equation 12.27, $\lambda_{\max} T = 2898 \ \mu\text{m} \cdot \text{K}$. Hence

$$\lambda_{\max} = 1.45 \ \mu\text{m} \qquad \triangleleft$$

The spectral emissive power associated with this wavelength may be computed from Equation 12.26 or from the third column of Table 12.1. For $\lambda_{\max} T = 2898 \ \mu\text{m} \cdot \text{K}$ it follows from Table 12.1 that

$$I_{\lambda, b}(1.45 \ \mu\text{m}, T) = 0.722 \times 10^{-4} \sigma T^5$$

Hence

$$I_{\lambda, b}(1.45 \ \mu\text{m}, 2000 \text{ K}) = 0.722 \times 10^{-4} \ (1/\mu\text{m} \cdot \text{K} \cdot \text{sr}) \ 5.67$$

$$\times 10^{-8} \text{ W/m}^2 \cdot \text{K}^4(2000 \text{ K})^5$$

$$I_{\lambda, b}(1.45 \ \mu\text{m}, 2000 \text{ K}) = 1.31 \times 10^5 \text{ W/m}^2 \cdot \text{sr} \cdot \mu\text{m}$$

Since the emission is diffuse, it follows from Equation 12.13 that

$$E_{\lambda,b} = \pi I_{\lambda,b} = 4.12 \times 10^5 \text{ W/m}^2 \cdot \mu\text{m} \qquad \triangleleft$$

4. Irradiation of any small object inside the enclosure may be approximated as being equal to emission from a blackbody at the enclosure surface temperature. Hence $G = E_b(T)$, in which case

$$G = 9.07 \times 10^5 \text{ W/m}^2 \qquad \triangleleft$$

EXAMPLE 12.4

A surface emits as a blackbody at 1500 K. What is the rate per unit area (W/m²) at which it emits radiation in directions corresponding to $0° \le \theta \le 60°$ and in the wavelength interval $2 \ \mu\text{m} \le \lambda \le 4 \ \mu\text{m}$?

SOLUTION

Known: Temperature of a surface that emits as a blackbody.

Find: Rate of emission per unit area in directions between $\theta = 0°$ and $60°$ and at wavelengths between $\lambda = 2$ and 4 μm.

Schematic:

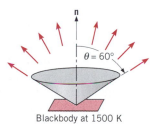

Blackbody at 1500 K

Assumptions: Surface emits as a blackbody.

Analysis: The desired emission may be inferred from Equation 12.12, with the limits of integration restricted as follows:

$$\Delta E = \int_2^4 \int_0^{2\pi} \int_0^{\pi/3} I_{\lambda,b} \cos \theta \sin \theta \ d\theta \ d\phi \ d\lambda$$

or, since a blackbody emits diffusely,

$$\Delta E = \int_2^4 I_{\lambda,b} \left(\int_0^{2\pi} \int_0^{\pi/3} \cos \theta \sin \theta \ d\theta \ d\phi \right) d\lambda$$

$$\Delta E = \int_2^4 I_{\lambda,b} \left(2\pi \left. \frac{\sin^2 \theta}{2} \right|_0^{\pi/3} \right) d\lambda = 0.75 \int_2^4 \pi I_{\lambda,b} \ d\lambda$$

Substituting from Equation 12.13 and multiplying and dividing by E_b, this result may be put in a form that allows for use of Table 12.1 in evaluating the spectral integration. In particular,

$$\Delta E = 0.75 E_b \int_2^4 \frac{E_{\lambda, b}}{E_b} \, d\lambda = 0.75 E_b [F_{(0 \to 4)} - F_{(0 \to 2)}]$$

where from Table 12.1

$$\lambda_1 T = 2 \ \mu m \times 1500 \ K = 3000 \ \mu m \cdot K: \qquad F_{(0 \to 2)} = 0.273$$

$$\lambda_2 T = 4 \ \mu m \times 1500 \ K = 6000 \ \mu m \cdot K: \qquad F_{(0 \to 4)} = 0.738$$

Hence

$$\Delta E = 0.75(0.738 - 0.273) E_b = 0.75(0.465) E_b$$

From Equation 12.28, it then follows that

$$\Delta E = 0.75(0.465)5.67 \times 10^{-8} \ W/m^2 \cdot K^4 \ (1500 \ K)^4 = 10^5 \ W/m^2 \qquad \lhd$$

Comments: The total, hemispherical emissive power is reduced by 25% and 53.5% due to the directional and spectral restrictions, respectively.

12.4
Surface Emission

Having developed the notion of a blackbody to describe *ideal* surface behavior, we may now consider the behavior of *real* surfaces. Recall that the blackbody is an ideal emitter in the sense that no surface can emit more radiation than a blackbody at the same temperature. It is therefore convenient to choose the blackbody as a reference in describing emission from a real surface. A surface radiative property known as the *emissivity*[1] may then be defined as the *ratio* of the radiation emitted by the surface to the radiation emitted by a blackbody at the same temperature.

It is important to acknowledge that, in general, the spectral radiation emitted by a real surface differs from the Planck distribution (Figure 12.16a). Moreover, the directional distribution (Figure 12.16b) may be other than diffuse. Hence the emissivity may assume different values according to whether one is interested in emission at a given wavelength or in a given direction, or in integrated averages over wavelength and direction.

We define the *spectral, directional emissivity* $\varepsilon_{\lambda, \theta}(\lambda, \theta, \phi, T)$ of a surface at the temperature T as the ratio of the intensity of the radiation emitted at the

---

[1]In this text we use the *-ivity*, rather than the *-ance*, ending for material radiative properties (e.g., "emissivity" rather than "emittance"). Although efforts are being made to reserve the -ivity ending for optically smooth, uncontaminated surfaces, no such distinction is made in much of the literature, and none is made in this text.

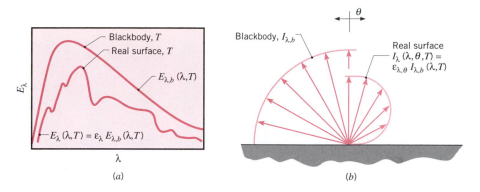

FIGURE 12.16 Comparison of blackbody and real surface emission. (*a*) Spectral distribution. (*b*) Directional distribution.

wavelength λ and in the direction of θ and ϕ to the intensity of the radiation emitted by a blackbody at the same values of T and λ. Hence

$$\varepsilon_{\lambda,\theta}(\lambda, \theta, \phi, T) \equiv \frac{I_{\lambda,e}(\lambda, \theta, \phi, T)}{I_{\lambda,b}(\lambda, T)} \qquad (12.32)$$

Note how the subscripts λ and θ designate interest in a specific wavelength and direction for the emissivity. In contrast, terms appearing within parentheses designate functional dependence on wavelength, direction, and/or temperature. The absence of directional variables in the parentheses of the denominator in Equation 12.32 implies that the intensity is independent of direction, which is, of course, a characteristic of blackbody emission. In like manner a *total, directional emissivity* ε_θ, which represents a spectral average of $\varepsilon_{\lambda,\theta}$, may be defined as

$$\varepsilon_\theta(\theta, \phi, T) \equiv \frac{I_e(\theta, \phi, T)}{I_b(T)} \qquad (12.33)$$

For most engineering calculations, it is desirable to work with surface properties that represent directional averages. A *spectral, hemispherical emissivity* is therefore defined as

$$\varepsilon_\lambda(\lambda, T) \equiv \frac{E_\lambda(\lambda, T)}{E_{\lambda,b}(\lambda, T)} \qquad (12.34)$$

It may be related to the directional emissivity $\varepsilon_{\lambda,\theta}$ by substituting the expression for the spectral emissive power, Equation 12.10, to obtain

$$\varepsilon_\lambda(\lambda, T) = \frac{\int_0^{2\pi} \int_0^{\pi/2} I_{\lambda,e}(\lambda, \theta, \phi, T) \cos\theta \sin\theta \, d\theta \, d\phi}{\int_0^{2\pi} \int_0^{\pi/2} I_{\lambda,b}(\lambda, T) \cos\theta \sin\theta \, d\theta \, d\phi}$$

In contrast to Equation 12.10, the temperature dependence of emission is now acknowledged. From Equation 12.32 and the fact that $I_{\lambda,b}$ is independent of θ

and ϕ, it follows that

$$\varepsilon_\lambda(\lambda, T) = \frac{\displaystyle\int_0^{2\pi}\int_0^{\pi/2} \varepsilon_{\lambda,\theta}(\lambda, \theta, \phi, T) \cos\theta \sin\theta \, d\theta \, d\phi}{\displaystyle\int_0^{2\pi}\int_0^{\pi/2} \cos\theta \sin\theta \, d\theta \, d\phi} \tag{12.35}$$

Assuming $\varepsilon_{\lambda,\theta}$ to be independent of ϕ, which is a reasonable assumption for most surfaces, and evaluating the denominator, we obtain

$$\varepsilon_\lambda(\lambda, T) = 2\int_0^{\pi/2} \varepsilon_{\lambda,\theta}(\lambda, \theta, T) \cos\theta \sin\theta \, d\theta \tag{12.36}$$

The *total, hemispherical emissivity,* which represents an average over all possible directions and wavelengths, is defined as

$$\varepsilon(T) \equiv \frac{E(T)}{E_b(T)} \tag{12.37}$$

Substituting from Equations 12.11 and 12.34, it follows that

$$\varepsilon(T) = \frac{\displaystyle\int_0^\infty \varepsilon_\lambda(\lambda, T) E_{\lambda,b}(\lambda, T) \, d\lambda}{E_b(T)} \tag{12.38}$$

If the emissivities of a surface are known, it is a simple matter to compute its emission characteristics. For example, if $\varepsilon_\lambda(\lambda, T)$ is known, it may be used with Equations 12.26 and 12.34 to compute the spectral emissive power of the surface at any wavelength and temperature. Similarly, if $\varepsilon(T)$ is known, it may be used with Equations 12.28 and 12.37 to compute the total emissive power of the surface at any temperature. Measurements have been performed to determine these properties for many different materials and surface coatings.

The directional emissivity of a *diffuse emitter* is a constant, independent of direction. However, although this condition is often a reasonable *approximation,* all surfaces exhibit some departure from diffuse behavior. Representative variations of ε_θ with θ are shown schematically in Figure 12.17 for conducting and nonconducting materials. For conductors ε_θ is approximately constant over the range $\theta \lesssim 40°$, after which it increases with increasing θ but ultimately decays to zero. In contrast, for nonconductors ε_θ is approximately constant for $\theta \lesssim 70°$, beyond which it decreases sharply with increasing θ. One implication of these variations is that, although there are preferential directions for emission, the hemispherical emissivity ε will not differ markedly from the normal

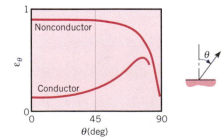

FIGURE 12.17

Representative directional distributions of the total, directional emissivity.

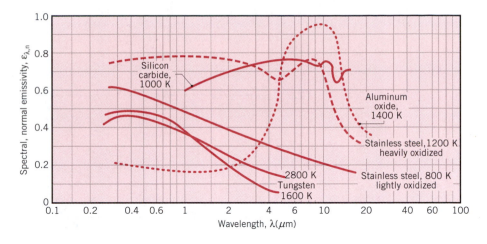

FIGURE 12.18   Spectral dependence of the spectral, normal emissivity $\varepsilon_{\lambda,\,n}$ of selected materials.

emissivity ε_n, corresponding to $\theta = 0$. In fact the ratio rarely falls outside the range $1.0 \le (\varepsilon/\varepsilon_n) \le 1.3$ for conductors and the range of $0.95 \le (\varepsilon/\varepsilon_n) \le 1.0$ for nonconductors. Hence to a reasonable approximation,

$$\varepsilon \approx \varepsilon_n \tag{12.39}$$

Note that, although the foregoing statements were made for the total emissivity, they also apply to spectral components.

Since the spectral distribution of emission from real surfaces departs from the Planck distribution (Figure 12.16a), we do not expect the value of the spectral emissivity ε_λ to be independent of wavelength. Representative spectral distributions of ε_λ are shown in Figure 12.18. The manner in which ε_λ varies with λ depends on whether the solid is a conductor or nonconductor, as well as on the nature of the surface coating.

Representative values of the total, normal emissivity ε_n are plotted in Figures 12.19 and 12.20 and listed in Table A.11. Several generalizations may be made.

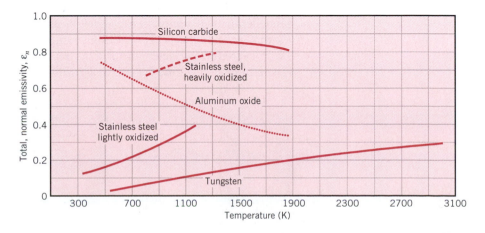

FIGURE 12.19   Temperature dependence of the total, normal emissivity ε_n of selected materials.

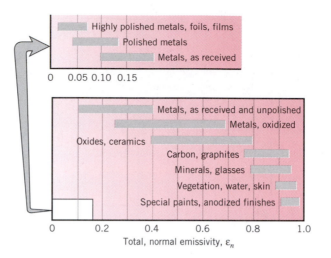

FIGURE 12.20   Representative values of the total, normal emissivity ε_n.

1. The emissivity of metallic surfaces is generally small, achieving values as low as 0.02 for highly polished gold and silver.

2. The presence of oxide layers may significantly increase the emissivity of metallic surfaces. Contrast the value of 0.10 for lightly oxidized stainless steel with the value of approximately 0.50 for the heavily oxidized form.

3. The emissivity of nonconductors is comparatively large, generally exceeding 0.6.

4. The emissivity of conductors increases with increasing temperature; however, depending on the specific material, the emissivity of nonconductors may either increase or decrease with increasing temperature. Note that the variations of ε_n with T shown in Figure 12.19 are consistent with the spectral distributions of $\varepsilon_{\lambda, n}$ shown in Figure 12.18. These trends follow from Equation 12.38. Although the spectral distribution of $\varepsilon_{\lambda, n}$ is approximately independent of temperature, there is proportionately more emission at lower wavelengths with increasing temperature. Hence, if $\varepsilon_{\lambda, n}$ increases with decreasing wavelength for a particular material, ε_n will increase with increasing temperature for that material.

It should be recognized that the emissivity depends strongly on the nature of the surface, which can be influenced by the method of fabrication, thermal cycling, and chemical reaction with its environment. More comprehensive surface emissivity compilations are available in the literature [2–5].

EXAMPLE 12.5

A diffuse surface at 1600 K has the spectral, hemispherical emissivity shown on the next page.

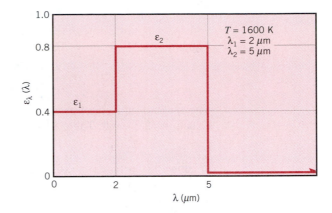

Determine the total, hemispherical emissivity and the total emissive power. At what wavelength will the spectral emissive power be a maximum?

SOLUTION

Known: Spectral, hemispherical emissivity of a diffuse surface at 1600 K.

Find:

1. Total, hemispherical emissivity.
2. Total emissive power.
3. Wavelength at which spectral emissive power will be a maximum.

Assumptions: Surface is a diffuse emitter.

Analysis:

1. The total, hemispherical emissivity is given by Equation 12.38, where the integration may be performed in parts as follows:

$$\varepsilon = \frac{\int_0^\infty \varepsilon_\lambda E_{\lambda, b}\, d\lambda}{E_b} = \frac{\varepsilon_1 \int_0^2 E_{\lambda, b}\, d\lambda}{E_b} + \frac{\varepsilon_2 \int_2^5 E_{\lambda, b}\, d\lambda}{E_b}$$

or

$$\varepsilon = \varepsilon_1 F_{(0 \to 2\ \mu m)} + \varepsilon_2 [F_{(0 \to 5\ \mu m)} - F_{(0 \to 2\ \mu m)}]$$

From Table 12.1 we obtain

$$\lambda_1 T = 2\ \mu m \times 1600\ K = 3200\ \mu m \cdot K: \qquad F_{(0 \to 2\ \mu m)} = 0.318$$
$$\lambda_2 T = 5\ \mu m \times 1600\ K = 8000\ \mu m \cdot K: \qquad F_{(0 \to 5\ \mu m)} = 0.856$$

Hence

$$\varepsilon = 0.4 \times 0.318 + 0.8[0.856 - 0.318] = 0.558 \qquad \triangleleft$$

2. From Equation 12.37 the total emissive power is

$$E = \varepsilon E_b = \varepsilon \sigma T^4$$
$$E = 0.558(5.67 \times 10^{-8}\ W/m^2 \cdot K^4)(1600\ K)^4 = 207\ kW/m^2 \qquad \triangleleft$$

3. If the surface emitted as a blackbody or if its emissivity were a constant, independent of λ, the wavelength corresponding to maximum spectral

emission could be obtained from Wien's law. However, because ε_λ varies with λ, it is not immediately obvious where peak emission occurs. From Equation 12.27 we know that

$$\lambda_{max} = \frac{2898 \ \mu m \cdot K}{1600 \ K} = 1.81 \ \mu m$$

The spectral emissive power at this wavelength may be obtained by using Equation 12.34 with Table 12.1. That is,

$$E_\lambda(\lambda_{max}, T) = \varepsilon_\lambda(\lambda_{max})E_{\lambda, b}(\lambda_{max}, T)$$

or, since the surface is a diffuse emitter,

$$E_\lambda(\lambda_{max}, T) = \pi\varepsilon_\lambda(\lambda_{max})I_{\lambda, b}(\lambda_{max}, T)$$

$$= \pi\epsilon_\lambda(\lambda_{max}) \frac{I_{\lambda, b}(\lambda_{max}, T)}{\sigma T^5} \times \sigma T^5$$

$$E_\lambda(1.81 \ \mu m, 1600 \ K) = \pi \times 0.4 \times 0.722 \times 10^{-4} \ (1/\mu m \cdot K \cdot sr)5.67$$

$$\times 10^{-8} \ W/m^2 \cdot K^4 \times (1600 \ K)^5 = 54 \ kW/m^2 \cdot \mu m$$

Since $\varepsilon_\lambda = 0.4$ from $\lambda = 0$ to $\lambda = 2 \ \mu m$, the foregoing result provides the maximum spectral emissive power for the region $\lambda < 2 \ \mu m$. However, with the change in ε_λ that occurs at $\lambda = 2 \ \mu m$, the value of E_λ at $\lambda = 2 \ \mu m$ may be larger than that for $\lambda = 1.81 \ \mu m$. To determine whether this is, in fact, the case, we compute

$$E_\lambda(\lambda_1, T) = \pi\varepsilon_\lambda(\lambda_1) \frac{I_{\lambda, b}(\lambda_1, T)}{\sigma T^5} \times \sigma T^5$$

where, for $\lambda_1 T = 3200 \ \mu m \cdot K$, $[I_{\lambda, b}(\lambda_1, T)/\sigma T^5] = 0.706 \times 10^{-4} \ (\mu m \cdot K \cdot sr)^{-1}$. Hence

$$E_\lambda(2 \ \mu m, 1600 \ K) = \pi \times 0.80 \times 0.706 \times 10^{-4} \ (1/\mu m \cdot K \cdot sr)5.67$$

$$\times 10^{-8} \ W/m^2 \cdot K^4 \ (1600 \ K)^5$$

$$E_\lambda(2 \ \mu m, 1600 \ K) = 105.5 \ kW/m^2 \cdot \mu m > E_\lambda(1.81 \ \mu m, 1600 \ K)$$

and peak emission occurs at

$$\lambda = \lambda_1 = 2 \ \mu m \qquad\qquad \triangleleft$$

Comments: For the prescribed spectral distribution of ε_λ, the spectral emissive power will vary with wavelength as shown.

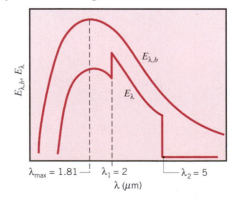

EXAMPLE 12.6

Measurements of the spectral, directional emissivity of a metallic surface at $T = 2000$ K and $\lambda = 1.0$ μm yield a directional distribution that may be approximated as follows:

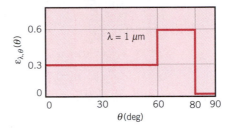

Determine corresponding values of the spectral, normal emissivity; the spectral, hemispherical emissivity; the spectral intensity of radiation emitted in the normal direction; and the spectral emissive power.

SOLUTION

Known: Directional distribution of $\varepsilon_{\lambda, \theta}$ at $\lambda = 1$ μm for a metallic surface at 2000 K.

Find:

1. Spectral, normal emissivity $\varepsilon_{\lambda, n}$ and spectral, hemispherical emissivity ε_λ.
2. Spectral, normal intensity $I_{\lambda, n}$ and spectral emissive power E_λ.

Analysis:

1. From the measurement of $\varepsilon_{\lambda, \theta}$ at $\lambda = 1$ μm, we see that

$$\varepsilon_{\lambda, n} = \varepsilon_{\lambda, \theta}(1 \ \mu\text{m}, 0°) = 0.3 \qquad \triangleleft$$

From Equation 12.36, the spectral, hemispherical emissivity is

$$\varepsilon_\lambda(1 \ \mu\text{m}) = 2 \int_0^{\pi/2} \varepsilon_{\lambda, \theta} \cos \theta \sin \theta \, d\theta$$

or

$$\varepsilon_\lambda(1 \ \mu\text{m}) = 2 \left[0.3 \int_0^{\pi/3} \cos \theta \sin \theta \, d\theta + 0.6 \int_{\pi/3}^{4\pi/9} \cos \theta \sin \theta \, d\theta \right]$$

$$\varepsilon_\lambda(1 \ \mu\text{m}) = 2 \left[0.3 \left. \frac{\sin^2 \theta}{2} \right|_0^{\pi/3} + 0.6 \left. \frac{\sin^2 \theta}{2} \right|_{\pi/3}^{4\pi/9} \right]$$

$$= 2 \left[\frac{0.3}{2} (0.75) + \frac{0.6}{2} (0.97 - 0.75) \right]$$

$$\varepsilon_\lambda(1 \ \mu\text{m}) = 0.36 \qquad \triangleleft$$

2. From Equation 12.32 the spectral intensity of radiation emitted at $\lambda = 1$ μm in the normal direction is

$$I_{\lambda, n}(1 \ \mu\text{m}, 0°, 2000 \text{ K}) = \varepsilon_{\lambda, \theta}(1 \ \mu\text{m}, 0°) \, I_{\lambda, b}(1 \ \mu\text{m}, 2000 \text{ K})$$

where $\varepsilon_{\lambda,\theta}(1\ \mu m, 0°) = 0.3$ and $I_{\lambda,b}(1\ \mu m, 2000\ K)$ may be obtained from Table 12.1. For $\lambda T = 2000\ \mu m \cdot K$, $(I_{\lambda,b}/\sigma T^5) = 0.493 \times 10^{-4}\ (\mu m \cdot K \cdot sr)^{-1}$ and

$$I_{\lambda,b} = 0.493 \times 10^{-4}\ (\mu m \cdot K \cdot sr)^{-1} \times 5.67 \times 10^{-8}\ W/m^2 \cdot K^4(2000\ K)^5$$

$$I_{\lambda,b} = 8.95 \times 10^4\ W/m^2 \cdot \mu m \cdot sr$$

Hence

$$I_{\lambda,n}(1\ \mu m, 0°, 2000\ K) = 0.3 \times 8.95 \times 10^4\ W/m^2 \cdot \mu m \cdot sr$$

$$I_{\lambda,n}(1\ \mu m, 0°, 2000\ K) = 2.69 \times 10^4\ W/m^2 \cdot \mu m \cdot sr \qquad \lhd$$

From Equation 12.34 the spectral emissive power for $\lambda = 1\ \mu m$ and $T = 2000\ K$ is

$$E_{\lambda}(1\ \mu m, 2000\ K) = \varepsilon_{\lambda}(1\ \mu m)E_{\lambda,b}(1\ \mu m, 2000\ K)$$

where

$$E_{\lambda,b}(1\ \mu m, 2000\ K) = \pi I_{\lambda,b}(1\ \mu m, 2000\ K)$$

$$E_{\lambda,b}(1\ \mu m, 2000\ K) = \pi\ sr \times 8.95 \times 10^4\ W/m^2 \cdot \mu m \cdot sr$$

$$= 2.81 \times 10^5\ W/m^2 \cdot \mu m$$

Hence

$$E_{\lambda}(1\ \mu m, 2000\ K) = 0.36 \times 2.81 \times 10^5\ W/m^2 \cdot \mu m$$

or

$$E_{\lambda}(1\ \mu m, 2000\ K) = 1.01 \times 10^5\ W/m^2 \cdot \mu m \qquad \lhd$$

12.5
Surface Absorption, Reflection, and Transmission

In Section 12.2.3 we defined the *spectral irradiation* G_{λ} (W/m² · μm) as the rate at which radiation of wavelength λ is incident on a surface per unit area of the surface and per unit wavelength interval $d\lambda$ about λ. It may be incident from all possible directions, and it may originate from several different sources. The *total irradiation* G (W/m²) encompasses all spectral contributions and may be evaluated from Equation 12.16. In this section we consider the *processes* resulting from interception of this radiation by a solid (or liquid) medium.

In the most general situation the irradiation interacts with a *semitransparent medium,* such as a layer of water or a glass plate. As shown in Figure 12.21 for a spectral component of the irradiation, portions of this radiation may be *reflected, absorbed,* and *transmitted*. From a radiation balance on the medium, it follows that

$$G_{\lambda} = G_{\lambda,\text{ref}} + G_{\lambda,\text{abs}} + G_{\lambda,\text{tr}} \qquad (12.40)$$

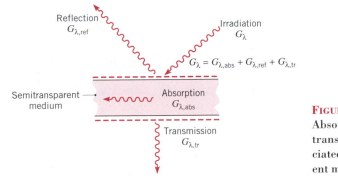

FIGURE 12.21
Absorption, reflection, and transmission processes associated with a semitransparent medium.

In general, determination of these components is complex, depending on the upper and lower surface conditions, the wavelength of the radiation, and the composition and thickness of the medium. Moreover, conditions may be strongly influenced by *volumetric* effects occurring within the medium.

In a simpler situation, which pertains to most engineering applications, the medium is *opaque* to the incident radiation. In this case, $G_{\lambda, tr} = 0$ and the remaining absorption and reflection processes may be treated as *surface phenomena*. That is, they are controlled by processes occurring within a fraction of a micrometer from the irradiated surface. It is therefore appropriate to speak of irradiation being absorbed and reflected *by the surface,* with the relative magnitudes of $G_{\lambda, abs}$ and $G_{\lambda, ref}$ depending on λ and the nature of the surface material. There is no net effect of the reflection process on the medium, while absorption has the effect of increasing the internal thermal energy of the medium.

It is interesting to note that surface absorption and reflection are responsible for our perception of *color*. Unless the surface is at a high temperature $(T_s \gtrsim 1000 \text{ K})$, such that it is *incandescent,* color is in no way due to emission, which is concentrated in the IR region, and is hence imperceptible to the eye. Color is instead due to selective reflection and absorption of the visible portion of the irradiation that is incident from the sun or an artificial source of light. A shirt is "red" because it contains a pigment that preferentially absorbs the blue, green, and yellow components of the incident light. Hence the relative contributions of these components to the reflected light, which is seen, is diminished, and the red component is dominant. Similarly, a leaf is "green" because its cells contain chlorophyll, a pigment that shows strong absorption in the blue and the red and preferential reflection in the green. A surface appears "black" if it absorbs all incident visible radiation, and it is "white" if it reflects this radiation. However, we must be careful how we interpret such *visual* effects. For a prescribed irradiation, the "color" of a surface may not indicate its overall capacity as an absorber or reflector, since much of the irradiation may be in the IR region. A "white" surface such as snow, for example, is highly reflective to visible radiation but strongly absorbs IR radiation, thereby approximating blackbody behavior at long wavelengths.

In Section 12.4 we introduced a property, which we called the emissivity, to characterize the process of surface emission. In the subsections that follow we shall introduce properties to characterize the absorption, reflection, and transmission processes. In general these properties depend on surface material and finish, surface temperature, and the wavelength and direction of the incident radiation.

12.5.1   Absorptivity

The absorptivity is a property that determines the fraction of the irradiation absorbed by a surface. Determination of the property is complicated by the fact that, like emission, it may be characterized by both a directional and spectral dependence. The *spectral, directional absorptivity,* $\alpha_{\lambda,\theta}(\lambda, \theta, \phi)$, of a surface is defined as the fraction of the spectral intensity incident in the direction of θ and ϕ that is absorbed by the surface. Hence

$$\alpha_{\lambda,\theta}(\lambda, \theta, \phi) \equiv \frac{I_{\lambda, i, \text{abs}}(\lambda, \theta, \phi)}{I_{\lambda, i}(\lambda, \theta, \phi)} \tag{12.41}$$

In this expression, we have neglected any dependence of the absorptivity on the surface temperature. Such a dependence is small for most spectral radiative properties.

It is implicit in the foregoing result that surfaces may exhibit selective absorption with respect to the wavelength and direction of the incident radiation. For most engineering calculations, however, it is desirable to work with surface properties that represent directional averages. We therefore define a *spectral, hemispherical absorptivity* $\alpha_{\lambda}(\lambda)$ as

$$\alpha_{\lambda}(\lambda) \equiv \frac{G_{\lambda, \text{abs}}(\lambda)}{G_{\lambda}(\lambda)} \tag{12.42}$$

which, from Equations 12.15 and 12.41, may be expressed as

$$\alpha_{\lambda}(\lambda) = \frac{\int_{0}^{2\pi} \int_{0}^{\pi/2} \alpha_{\lambda,\theta}(\lambda, \theta, \phi) I_{\lambda, i}(\lambda, \theta, \phi) \cos\theta \sin\theta \, d\theta \, d\phi}{\int_{0}^{2\pi} \int_{0}^{\pi/2} I_{\lambda, i}(\lambda, \theta, \phi) \cos\theta \sin\theta \, d\theta \, d\phi} \tag{12.43}$$

Hence α_{λ} depends on the directional distribution of the incident radiation, as well as on the wavelength of the radiation and the nature of the absorbing surface. Note that, if the incident radiation is diffusely distributed and $\alpha_{\lambda,\theta}$ is independent of ϕ, Equation 12.43 reduces to

$$\alpha_{\lambda}(\lambda) = 2 \int_{0}^{\pi/2} \alpha_{\lambda,\theta}(\lambda, \theta) \cos\theta \sin\theta \, d\theta \tag{12.44}$$

The *total, hemispherical absorptivity,* α, represents an integrated average over both direction and wavelength. It is defined as the fraction of the total irradiation absorbed by a surface

$$\alpha \equiv \frac{G_{\text{abs}}}{G} \tag{12.45}$$

and, from Equations 12.16 and 12.42, it may be expressed as

$$\alpha = \frac{\int_{0}^{\infty} \alpha_{\lambda}(\lambda) G_{\lambda}(\lambda) \, d\lambda}{\int_{0}^{\infty} G_{\lambda}(\lambda) \, d\lambda} \tag{12.46}$$

Accordingly, α depends on the spectral distribution of the incident radiation, as well as on its directional distribution and the nature of the absorbing surface. Note that, although α is approximately independent of surface temperature, the same may not be said for the total, hemispherical emissivity, ε. From Equation 12.38 it is evident that this property is strongly temperature dependent.

Because α depends on the spectral distribution of the irradiation, its value for a surface exposed to solar radiation may differ appreciably from its value for the same surface exposed to longer wavelength radiation originating from a source of lower temperature. Since the spectral distribution of solar radiation is nearly proportional to that of emission from a blackbody at 5800 K, it follows from Equation 12.46 that the total absorptivity to solar radiation α_S may be approximated as

$$\alpha_S \approx \frac{\int_0^\infty \alpha_\lambda(\lambda) E_{\lambda, b}(\lambda, 5800 \text{ K}) \, d\lambda}{\int_0^\infty E_{\lambda, b}(\lambda, 5800 \text{ K}) \, d\lambda} \tag{12.47}$$

The integrals appearing in this equation may be evaluated by using the blackbody radiation function $F_{(0 \to \lambda)}$ of Table 12.1.

12.5.2 Reflectivity

The reflectivity is a property that determines the fraction of the incident radiation reflected by a surface. However, its specific definition may take several different forms, because the property is inherently *bidirectional* [6]. That is, in addition to depending on the direction of the incident radiation, it also depends on the direction of the reflected radiation. We shall avoid this complication by working exclusively with a reflectivity that represents an integrated average over the hemisphere associated with the reflected radiation and therefore provides no information concerning the directional distribution of this radiation. Accordingly, the *spectral, directional reflectivity, $\rho_{\lambda, \theta}(\lambda, \theta, \phi)$*, of a surface is defined as the fraction of the spectral intensity incident in the direction of θ and ϕ, which is reflected by the surface. Hence

$$\rho_{\lambda, \theta}(\lambda, \theta, \phi) \equiv \frac{I_{\lambda, i, \text{ref}}(\lambda, \theta, \phi)}{I_{\lambda, i}(\lambda, \theta, \phi)} \tag{12.48}$$

The *spectral, hemispherical reflectivity $\rho_\lambda(\lambda)$* is then defined as the fraction of the spectral irradiation that is reflected by the surface. Accordingly,

$$\rho_\lambda(\lambda) \equiv \frac{G_{\lambda, \text{ref}}(\lambda)}{G_\lambda(\lambda)} \tag{12.49}$$

which is equivalent to

$$\rho_\lambda(\lambda) = \frac{\int_0^{2\pi} \int_0^{\pi/2} \rho_{\lambda, \theta}(\lambda, \theta, \phi) I_{\lambda, i}(\lambda, \theta, \phi) \cos \theta \sin \theta \, d\theta \, d\phi}{\int_0^{2\pi} \int_0^{\pi/2} I_{\lambda, i}(\lambda, \theta, \phi) \cos \theta \sin \theta \, d\theta \, d\phi} \tag{12.50}$$

The *total, hemispherical reflectivity* ρ is then defined as

$$\rho \equiv \frac{G_{\text{ref}}}{G} \tag{12.51}$$

in which case

$$\rho = \frac{\displaystyle\int_0^\infty \rho_\lambda(\lambda) G_\lambda(\lambda)\, d\lambda}{\displaystyle\int_0^\infty G_\lambda(\lambda)\, d\lambda} \tag{12.52}$$

Surfaces may be idealized as *diffuse* or *specular,* according to the manner in which they reflect radiation (Figure 12.22). Diffuse reflection occurs if, regardless of the direction of the incident radiation, the intensity of the reflected radiation is independent of the reflection angle. In contrast, if all the reflection is in the direction of θ_2, which equals the incident angle θ_1, specular reflection is said to occur. Although no surface is perfectly diffuse or specular, the latter condition is more closely approximated by polished, mirrorlike surfaces and the former condition by rough surfaces. The assumption of diffuse reflection is reasonable for most engineering applications.

12.5.3 Transmissivity

Although treatment of the response of a semitransparent material to incident radiation is a complicated problem [6], reasonable results may often be obtained through the use of hemispherical transmissivities defined as

$$\tau_\lambda = \frac{G_{\lambda,\,\text{tr}}(\lambda)}{G_\lambda(\lambda)} \tag{12.53}$$

and

$$\tau = \frac{G_{\text{tr}}}{G} \tag{12.54}$$

The total transmissivity τ is related to the spectral component τ_λ by

$$\tau = \frac{\displaystyle\int_0^\infty G_{\lambda,\,\text{tr}}(\lambda)\, d\lambda}{\displaystyle\int_0^\infty G_\lambda(\lambda)\, d\lambda} = \frac{\displaystyle\int_0^\infty \tau_\lambda(\lambda) G_\lambda(\lambda)\, d\lambda}{\displaystyle\int_0^\infty G_\lambda(\lambda)\, d\lambda} \tag{12.55}$$

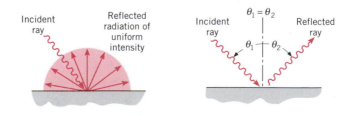

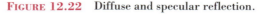

FIGURE 12.22   Diffuse and specular reflection.

12.5.4   Special Considerations

We conclude this section by noting that, from the radiation balance of Equation 12.40 and the foregoing definitions,

$$\rho_\lambda + \alpha_\lambda + \tau_\lambda = 1 \tag{12.56}$$

for a semitransparent medium. For properties that are averaged over the entire spectrum, it also follows that

$$\rho + \alpha + \tau = 1 \tag{12.57}$$

Of course, if the medium is opaque, there is no transmission, and absorption and reflection are surface processes for which

$$\alpha_\lambda + \rho_\lambda = 1 \tag{12.58}$$

and

$$\alpha + \rho = 1 \tag{12.59}$$

Hence knowledge of one property implies knowledge of the other.

Spectral distributions of the normal reflectivity and absorptivity are plotted in Figure 12.23 for selected *opaque* surfaces. A material such as glass or water, which is semitransparent at short wavelengths, becomes opaque at longer wavelengths. This behavior is shown in Figure 12.24, which presents the spectral transmissivity of several common semitransparent materials. Note that the

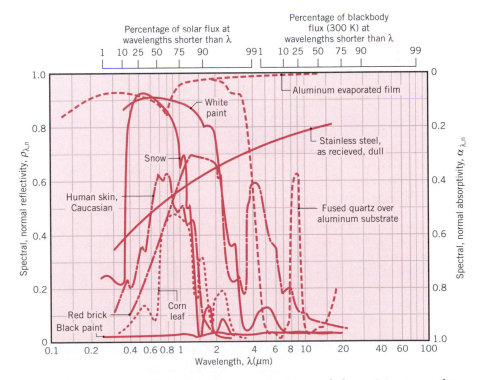

FIGURE 12.23   Spectral dependence of the spectral, normal absorptivity $\alpha_{\lambda,\,n}$ and reflectivity $\rho_{\lambda,\,n}$ of selected materials.

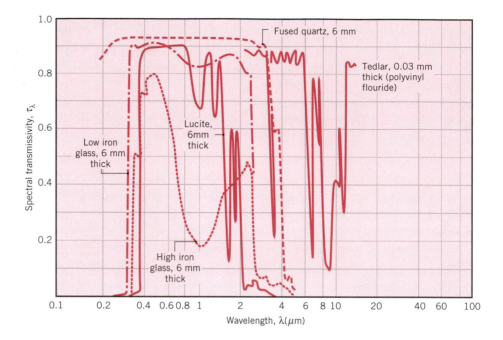

FIGURE 12.24   Spectral dependence of the spectral transmissivities τ_λ of selected materials.

transmissivity of glass is affected by its iron content and that the transmissivity of plastics, such as Tedlar, is greater than that of glass in the IR region. These factors have an important bearing on the selection of cover plate materials for solar collector applications. Values for the total transmissivity to solar radiation of common collector cover plate materials are presented in Table A.12, along with surface solar absorptivities and low-temperature emissivities.

EXAMPLE 12.7

The spectral, hemispherical absorptivity of an opaque surface and the spectral irradiation at the surface are as shown.

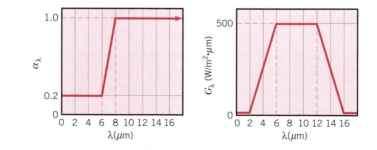

How does the spectral, hemispherical reflectivity vary with wavelength? What is the total, hemispherical absorptivity of the surface? If the surface is initially at 500 K and has a total, hemispherical emissivity of 0.8, how will its temperature change upon exposure to the irradiation?

SOLUTION

Known: Spectral, hemispherical absorptivity and irradiation of a surface. Surface temperature (500 K) and total, hemispherical emissivity (0.8).

Find:

1. Spectral distribution of reflectivity.
2. Total, hemispherical absorptivity.
3. Nature of surface temperature change.

Schematic:

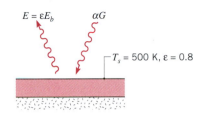

Assumptions:

1. Surface is opaque.
2. Surface convection effects are negligible.
3. Back surface is insulated.

Analysis:

1. From Equation 12.58, $\rho_\lambda = 1 - \alpha_\lambda$. Hence from knowledge of $\alpha_\lambda(\lambda)$, the spectral distribution of ρ_λ is as shown.

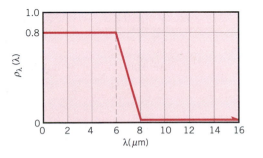

2. From Equations 12.45 and 12.46,

$$\alpha = \frac{G_{\text{abs}}}{G} = \frac{\displaystyle\int_0^\infty \alpha_\lambda G_\lambda \, d\lambda}{\displaystyle\int_0^\infty G_\lambda \, d\lambda}$$

or, subdividing the integral into parts,

$$\alpha = \frac{0.2 \int_2^6 G_\lambda \, d\lambda + 500 \int_6^8 \alpha_\lambda \, d\lambda + 1.0 \int_8^{16} G_\lambda \, d\lambda}{\int_2^6 G_\lambda \, d\lambda + \int_6^{12} G_\lambda \, d\lambda + \int_{12}^{16} G_\lambda \, d\lambda}$$

$$\alpha = \{0.2(\tfrac{1}{2})500 \text{ W/m}^2 \cdot \mu\text{m} \, (6-2) \, \mu\text{m}$$

$$+ \, 500 \text{ W/m}^2 \cdot \mu\text{m} \, [0.2(8-6) \, \mu\text{m} + (1-0.2)(\tfrac{1}{2})(8-6) \, \mu\text{m}]$$

$$+ \, [1 \times 500 \text{ W/m}^2 \cdot \mu\text{m} \, (12-8) \, \mu\text{m}$$

$$+ \, 1(\tfrac{1}{2})500 \text{ W/m}^2 \cdot \mu\text{m} \, (16-12) \, \mu\text{m}]\}$$

$$\div \, [(\tfrac{1}{2})500 \text{ W/m}^2 \cdot \mu\text{m} \, (6-2) \, \mu\text{m} + 500 \text{ W/m}^2 \cdot \mu\text{m} \, (12-6) \, \mu\text{m}$$

$$+ \, (\tfrac{1}{2})500 \text{ W/m}^2 \cdot \mu\text{m} \, (16-12) \, \mu\text{m}]$$

Hence

$$\alpha = \frac{G_{\text{abs}}}{G} = \frac{(200 + 600 + 3000) \text{ W/m}^2}{(1000 + 3000 + 1000) \text{ W/m}^2} = \frac{3800 \text{ W/m}^2}{5000 \text{ W/m}^2} = 0.76 \quad \triangleleft$$

3. Neglecting convection effects, the net heat flux *to* the surface is

$$q''_{\text{net}} = \alpha G - E = \alpha G - \varepsilon \sigma T^4$$

Hence

$$q''_{\text{net}} = 0.76(5000 \text{ W/m}^2) - 0.8 \times 5.67 \times 10^{-8} \text{ W/m}^2 \cdot \text{K}^4 (500 \text{ K})^4$$

$$q''_{\text{net}} = 3800 - 2835 = 965 \text{ W/m}^2$$

Since $q''_{\text{net}} > 0$, the surface temperature will *increase* with time. $\quad \triangleleft$

EXAMPLE 12.8

The cover glass on a flat-plate solar collector has a low iron content, and its spectral transmissivity may be approximated by the following distribution.

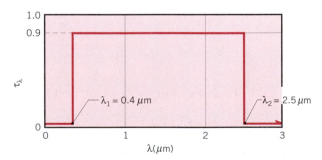

What is the total transmissivity of the cover glass to solar radiation?

SOLUTION

Known: Spectral transmissivity of solar collector cover glass.

Find: Total transmissivity of cover glass to solar radiation.

Assumptions: Spectral distribution of solar irradiation is proportional to that of blackbody emission at 5800 K.

Analysis: From Equation 12.55 the total transmissivity of the cover is

$$\tau = \frac{\int_0^\infty \tau_\lambda G_\lambda \, d\lambda}{\int_0^\infty G_\lambda \, d\lambda}$$

where the irradiation G_λ is due to solar emission. Having assumed that the sun emits as a blackbody at 5800 K, it follows that

$$G_\lambda(\lambda) \propto E_{\lambda, b}(5800 \text{ K})$$

With the proportionality constant canceling from the numerator and denominator of the expression for τ, we obtain

$$\tau = \frac{\int_0^\infty \tau_\lambda E_{\lambda, b}(5800 \text{ K}) \, d\lambda}{\int_0^\infty E_{\lambda, b}(5800 \text{ K}) \, d\lambda}$$

or, for the prescribed spectral distribution of $\tau_\lambda(\lambda)$,

$$\tau = 0.90 \frac{\int_{0.4}^{2.5} E_{\lambda, b}(5800 \text{ K}) \, d\lambda}{E_b(5800 \text{ K})}$$

From Table 12.1

$\lambda_1 = 0.4 \ \mu\text{m}, T = 5800 \text{ K}: \quad \lambda_1 T = 2320 \ \mu\text{m} \cdot \text{K}, F_{(0 \rightarrow \lambda_1)} = 0.1245$

$\lambda_2 = 2.5 \ \mu\text{m}, T = 5800 \text{ K}: \quad \lambda_2 T = 14{,}500 \ \mu\text{m} \cdot \text{K}, F_{(0 \rightarrow \lambda_2)} = 0.9660$

Hence from Equation 12.31

$$\tau = 0.90[F_{(0 \rightarrow \lambda_2)} - F_{(0 \rightarrow \lambda_1)}] = 0.90(0.9660 - 0.1245) = 0.76 \qquad \triangleleft$$

Comments: It is important to recognize that the irradiation at the cover plate is not equal to the emissive power of a blackbody at 5800 K, $G_\lambda \neq E_{\lambda, b}$ (5800 K). It is simply assumed to be proportional to this emissive power, in which case it is assumed to have a spectral distribution of the same form. With G_λ appearing in both the numerator and denominator of the expression for τ, it is then possible to replace G_λ by $E_{\lambda, b}$.

12.6
Kirchhoff's Law

In the foregoing sections we separately considered surface properties associated with emission and absorption. In Sections 12.6 and 12.7 we consider conditions for which these properties are equal.

Consider a large, isothermal enclosure of surface temperature T_s, within which several small bodies are confined (Figure 12.25). Since these bodies are small relative to the enclosure, they have a negligible influence on the radiation field, which is due to the cumulative effect of emission and reflection by the enclosure surface. Recall that, regardless of its radiative properties, such a surface forms a *blackbody cavity.* Accordingly, regardless of its orientation, the irradiation experienced by any body in the cavity is diffuse and equal to emission from a blackbody at T_s.

$$G = E_b(T_s) \tag{12.60}$$

Under steady-state conditions, *thermal equilibrium* must exist between the bodies and the enclosure. Hence $T_1 = T_2 = \cdots = T_s$, and the net rate of energy transfer to each surface must be zero. Applying an energy balance to a control surface about body 1, it follows that

$$\alpha_1 G A_1 - E_1(T_s) A_1 = 0$$

or, from Equation 12.60,

$$\frac{E_1(T_s)}{\alpha_1} = E_b(T_s)$$

Since this result must apply to each of the confined bodies, we then obtain

$$\frac{E_1(T_s)}{\alpha_1} = \frac{E_2(T_s)}{\alpha_2} = \cdots = E_b(T_s) \tag{12.61}$$

This relation is known as *Kirchhoff's law.* A major implication is that, since $\alpha \leq 1$, $E(T_s) \leq E_b(T_s)$. Hence *no real surface can have an emissive power exceeding that of a black surface at the same temperature,* and the notion of the blackbody as an ideal emitter is confirmed.

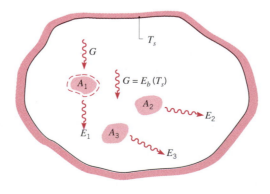

FIGURE 12.25
Radiative exchange in an isothermal enclosure.

From the definition of the total, hemispherical emissivity, Equation 12.37, an alternative form of Kirchhoff's law is

$$\frac{\varepsilon_1}{\alpha_1} = \frac{\varepsilon_2}{\alpha_2} = \cdots = 1$$

Hence, for any surface in the enclosure,

$$\varepsilon = \alpha \tag{12.62}$$

That is, the total, hemispherical emissivity of the surface is equal to its total, hemispherical absorptivity.

We will later find that calculations of radiative exchange between surfaces are greatly simplified if Equation 12.62 may be applied to each of the surfaces. However, the restrictive conditions inherent in its derivation should be remembered. In particular, the surface irradiation has been assumed to correspond to emission from a blackbody at the same temperature as the surface. In Section 12.7 we consider other, less restrictive, conditions for which Equation 12.62 is applicable.

The preceding derivation may be repeated for spectral conditions, and for any surface in the enclosure it follows that

$$\varepsilon_\lambda = \alpha_\lambda \tag{12.63}$$

Conditions associated with the use of Equation 12.63 are less restrictive than those associated with Equation 12.62. In particular, it will be shown that Equation 12.63 is applicable if the irradiation is diffuse *or* if the surface is diffuse. A form of Kirchhoff's law for which there are *no restrictions* involves the spectral, directional properties.

$$\varepsilon_{\lambda, \theta} = \alpha_{\lambda, \theta} \tag{12.64}$$

This equality is always applicable because $\varepsilon_{\lambda, \theta}$ and $\alpha_{\lambda, \theta}$ are *inherent* surface properties. That is, respectively, they are independent of the spectral and directional distributions of the emitted and incident radiation.

More detailed developments of Kirchhoff's law are provided by Planck [1] and Siegel and Howell [6].

12.7
The Gray Surface

In Chapter 13 we will find that the problem of predicting radiant energy exchange between surfaces is greatly simplified if Equation 12.62 may be assumed to apply for each surface. It is therefore important to examine whether this equality may be applied to conditions other than those for which it was derived, namely, irradiation due to emission from a blackbody at the same temperature as the surface.

Accepting the fact that the spectral, directional emissivity and absorptivity are equal under any conditions, Equation 12.64, we begin by considering the conditions associated with using Equation 12.63. According to the definitions of the spectral, hemispherical properties, Equations 12.35 and 12.43, we are really asking under what conditions, if any, the following equality will hold:

$$\varepsilon_\lambda = \frac{\displaystyle\int_0^{2\pi}\int_0^{\pi/2} \varepsilon_{\lambda,\theta}\cos\theta\sin\theta\,d\theta\,d\phi}{\displaystyle\int_0^{2\pi}\int_0^{\pi/2} \cos\theta\sin\theta\,d\theta\,d\phi} \stackrel{?}{=} \frac{\displaystyle\int_0^{2\pi}\int_0^{\pi/2} \alpha_{\lambda,\theta}I_{\lambda,i}\cos\theta\sin\theta\,d\theta\,d\phi}{\displaystyle\int_0^{2\pi}\int_0^{\pi/2} I_{\lambda,i}\cos\theta\sin\theta\,d\theta\,d\phi} = \alpha_\lambda$$

$$(12.65)$$

Since $\varepsilon_{\lambda,\theta} = \alpha_{\lambda,\theta}$, it follows by inspection that Equation 12.63 is applicable if *either* of the following conditions is satisfied:

1. The *irradiation* is *diffuse* ($I_{\lambda,i}$ is independent of θ and ϕ).
2. The *surface* is *diffuse* ($\varepsilon_{\lambda,\theta}$ and $\alpha_{\lambda,\theta}$ are independent of θ and ϕ).

The first condition is a reasonable approximation for many engineering calculations; the second condition is reasonable for many surfaces, particularly for electrically nonconducting materials (Figure 12.17).

    Assuming the existence of either diffuse irradiation or a diffuse surface, we now consider what *additional* conditions must be satisfied for Equation 12.62 to be valid. From Equations 12.38 and 12.46, the equality applies if

$$\varepsilon = \frac{\displaystyle\int_0^\infty \varepsilon_\lambda E_{\lambda,b}(\lambda,T)\,d\lambda}{E_b(T)} \stackrel{?}{=} \frac{\displaystyle\int_0^\infty \alpha_\lambda G_\lambda(\lambda)\,d\lambda}{G} = \alpha \qquad (12.66)$$

Since $\varepsilon_\lambda = \alpha_\lambda$, it follows by inspection that Equation 12.62 applies if *either* of the following conditions is satisfied:

1. The irradiation corresponds to emission from a blackbody at the surface temperature T, in which case $G_\lambda(\lambda) = E_{\lambda,b}(\lambda,T)$ and $G = E_b(T)$.
2. The *surface* is *gray* (α_λ and ε_λ are independent of λ).

Note that the first condition corresponds to the major assumption required for the derivation of Kirchhoff's law (Section 12.6).

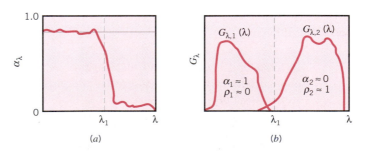

FIGURE 12.26   Spectral distribution of (*a*) the spectral absorptivity of a surface and (*b*) the spectral irradiation at the surface.

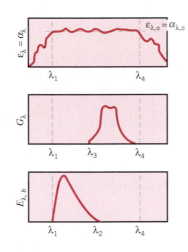
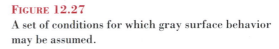

FIGURE 12.27

A set of conditions for which gray surface behavior may be assumed.

Because the total absorptivity of a surface depends on the spectral distribution of the irradiation, it cannot be stated unequivocally that $\alpha = \epsilon$. For example, a particular surface may be highly absorbing to radiation in one spectral region and virtually nonabsorbing in another region (Figure 12.26a). Accordingly, for the two possible irradiation fields $G_{\lambda, 1}(\lambda)$ and $G_{\lambda, 2}(\lambda)$ of Figure 12.26b, the values of α will differ drastically. In contrast, the value of ε is independent of the irradiation. Hence there is *no* basis for stating that α is always equal to ε.

To assume gray surface behavior, hence the validity of Equation 12.62, it is not necessary for α_λ and ε_λ to be independent of λ over the entire spectrum. Practically speaking, a *gray surface* may be defined as *one for which α_λ and ε_λ are independent of λ over the spectral regions of the irradiation and the surface emission.* From Equation 12.66 it is readily shown that gray surface behavior may be assumed for the conditions of Figure 12.27. That is, the irradiation and surface emission are concentrated in a region for which the spectral properties of the surface are approximately constant. Accordingly,

$$\varepsilon = \frac{\varepsilon_{\lambda, o} \int_{\lambda_1}^{\lambda_2} E_{\lambda, b}(\lambda, T)\, d\lambda}{E_b(T)} = \varepsilon_{\lambda, o} \quad \text{and} \quad \alpha = \frac{\alpha_{\lambda, o} \int_{\lambda_3}^{\lambda_4} G_\lambda(\lambda)\, d\lambda}{G} = \alpha_{\lambda, o}$$

in which case $\alpha = \varepsilon = \varepsilon_{\lambda, o}$. However, if the irradiation were in a spectral region corresponding to $\lambda < \lambda_1$ or $\lambda > \lambda_4$, gray surface behavior could not be assumed.

A surface for which $\alpha_{\lambda, \theta}$ and $\varepsilon_{\lambda, \theta}$ are independent of θ and λ is termed a *diffuse, gray surface* (diffuse because of the directional independence and gray because of the wavelength independence). It is a surface for which both Equations 12.62 and 12.63 are satisfied. We assume such surface conditions in many of our subsequent considerations. However, although the assumption of a gray surface is reasonable for many practical applications, some caution should be exercised in its use, particularly if the spectral regions of the irradiation and emission are widely separated.

EXAMPLE 12.9

A diffuse, fire brick wall of temperature $T_s = 500$ K has the spectral emissivity shown and is exposed to a bed of coals at 2000 K.

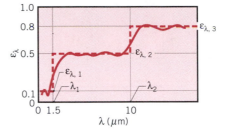

Determine the total, hemispherical emissivity and emissive power of the fire brick wall. What is the total absorptivity of the wall to irradiation resulting from emission by the coals?

SOLUTION

Known: Brick wall of surface temperature $T_s = 500$ K and prescribed $\varepsilon_\lambda(\lambda)$ is exposed to coals at $T_c = 2000$ K.

Find:

1. Total, hemispherical emissivity of the fire brick wall.
2. Total emissive power of the brick wall.
3. Absorptivity of the wall to irradiation from the coals.

Schematic:

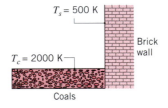

Assumptions:

1. Brick wall is opaque and diffuse.
2. Spectral distribution of irradiation at the brick wall approximates that due to emission from a blackbody at 2000 K.

Analysis:

1. The total, hemispherical emissivity follows from Equation 12.38.

$$\varepsilon(T_s) = \frac{\displaystyle\int_0^\infty \varepsilon_\lambda(\lambda)E_{\lambda,\,b}(\lambda, T_s)\, d\lambda}{E_b(T_s)}$$

Breaking the integral into parts,

$$\varepsilon(T_s) = \varepsilon_{\lambda,1} \frac{\displaystyle\int_0^{\lambda_1} E_{\lambda,b}\, d\lambda}{E_b} + \varepsilon_{\lambda,2} \frac{\displaystyle\int_{\lambda_1}^{\lambda_2} E_{\lambda,b}\, d\lambda}{E_b} + \varepsilon_{\lambda,3} \frac{\displaystyle\int_{\lambda_2}^{\infty} E_{\lambda,b}\, d\lambda}{E_b}$$

and introducing the blackbody functions, it follows that

$$\varepsilon(T_s) = \varepsilon_{\lambda,1} F_{(0\to\lambda_1)} + \varepsilon_{\lambda,2}[F_{(0\to\lambda_2)} - F_{(0\to\lambda_1)}] + \varepsilon_{\lambda,3}[1 - F_{(0\to\lambda_2)}]$$

From Table 12.1

$$\lambda_1 T_s = 1.5\ \mu m \times 500\ K = 750\ \mu m \cdot K: \qquad F_{(0\to\lambda_1)} = 0.000$$

$$\lambda_2 T_s = 10\ \mu m \times 500\ K = 5000\ \mu m \cdot K: \qquad F_{(0\to\lambda_2)} = 0.634$$

Hence

$$\varepsilon(T_s) = 0.1 \times 0 + 0.5 \times 0.634 + 0.8(1 - 0.634) = 0.610 \qquad \triangleleft$$

2. From Equations 12.28 and 12.37, the total emissive power is

$$E(T_s) = \varepsilon(T_s)E_b(T_s) = \varepsilon(T_s)\sigma T_s^4$$

$$E(T_s) = 0.61 \times 5.67 \times 10^{-8}\ W/m^2 \cdot K^4 (500\ K)^4 = 2161\ W/m^2 \qquad \triangleleft$$

3. From Equation 12.46, the total absorptivity of the wall to radiation from the coals is

$$\alpha = \frac{\displaystyle\int_0^{\infty} \alpha_\lambda(\lambda)G_\lambda(\lambda)\, d\lambda}{\displaystyle\int_0^{\infty} G_\lambda(\lambda)\, d\lambda}$$

Since the surface is diffuse, $\alpha_\lambda(\lambda) = \varepsilon_\lambda(\lambda)$. Moreover, since the spectral distribution of the irradiation approximates that due to emission from a blackbody at 2000 K, $G_\lambda(\lambda) \propto E_{\lambda,b}(\lambda, T_c)$. It follows that

$$\alpha = \frac{\displaystyle\int_0^{\infty} \varepsilon_\lambda(\lambda)E_{\lambda,b}(\lambda, T_c)\, d\lambda}{\displaystyle\int_0^{\infty} E_{\lambda,b}(\lambda, T_c)\, d\lambda}$$

Breaking the integral into parts and introducing the blackbody functions, we then obtain

$$\alpha = \varepsilon_{\lambda,1} F_{(0\to\lambda_1)} + \varepsilon_{\lambda,2}[F_{(0\to\lambda_2)} - F_{(0\to\lambda_1)}] + \varepsilon_{\lambda,3}[1 - F_{(0\to\lambda_2)}]$$

From Table 12.1

$$\lambda_1 T_c = 1.5\ \mu m \times 2000\ K = 3000\ \mu m \cdot K: \qquad F_{(0\to\lambda_1)} = 0.273$$

$$\lambda_2 T_c = 10\ \mu m \times 2000\ K = 20{,}000\ \mu m \cdot K: \qquad F_{(0\to\lambda_2)} = 0.986$$

Hence

$$\alpha = 0.1 \times 0.273 + 0.5(0.986 - 0.273) + 0.8(1 - 0.986) = 0.395 \qquad \triangleleft$$

Comments:

1. The emissivity depends on the surface temperature T_s, while the absorptivity depends on the spectral distribution of the irradiation, which depends on the temperature of the source T_c.

2. The surface is not gray, $\alpha \neq \varepsilon$. This result is to be expected. Since emission is associated with $T_s = 500$ K, its spectral maximum occurs at $\lambda_{max} \approx 6$ μm. In contrast, since irradiation is associated with emission from a source at $T_c = 2000$ K, its spectral maximum occurs at $\lambda_{max} \approx 1.5$ μm. Because $\varepsilon_\lambda = \alpha_\lambda$ is not constant over the spectral ranges of both emission and irradiation, $\alpha \neq \varepsilon$. For the prescribed spectral distribution of $\varepsilon_\lambda = \alpha_\lambda$, ε and α decrease with increasing T_s and T_c, respectively, and it is only for $T_s = T_c$ that $\varepsilon = \alpha$. The foregoing expressions for ε and α may be used to determine their equivalent variation with T_s and T_c, and the following result is obtained:

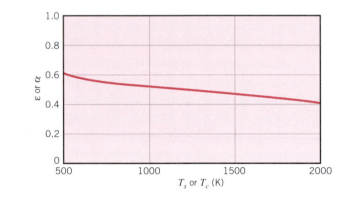

EXAMPLE 12.10

A *small,* solid metallic sphere has an opaque, diffuse coating for which $\alpha_\lambda = 0.8$ for $\lambda \leq 5$ μm and $\alpha_\lambda = 0.1$ for $\lambda > 5$ μm. The sphere, which is initially at a uniform temperature of 300 K, is inserted into a *large* furnace whose walls are at 1200 K. Determine the total, hemispherical absorptivity and emissivity of the coating for the initial condition and for the final, steady-state condition.

SOLUTION

Known: Small metallic sphere with spectrally selective absorptivity, initially at $T_s = 300$ K, is inserted into a large furnace at $T_f = 1200$ K.

Find:

1. Total, hemispherical absorptivity and emissivity of sphere coating for the initial condition.

2. Values of α and ε after sphere has been in furnace a long time.

Schematic:

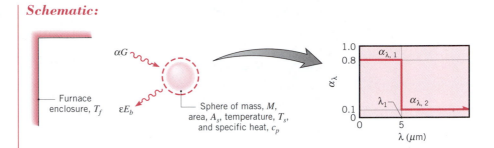

Assumptions:

1. Coating is opaque and diffuse.
2. Since furnace surface is much larger than that of sphere, irradiation approximates emission from a blackbody at T_f.

Analysis:

1. From Equation 12.46 the total, hemispherical absorptivity is

$$\alpha = \frac{\int_0^\infty \alpha_\lambda(\lambda) G_\lambda(\lambda) \, d\lambda}{\int_0^\infty G_\lambda(\lambda) \, d\lambda}$$

or, with $G_\lambda = E_{\lambda, b}(T_f) = E_{\lambda, b}(\lambda, 1200\ \text{K})$,

$$\alpha = \frac{\int_0^\infty \alpha_\lambda(\lambda) E_{\lambda, b}(\lambda, 1200\ \text{K}) \, d\lambda}{E_b(1200\ \text{K})}$$

Hence

$$\alpha = \alpha_{\lambda, 1} \frac{\int_0^{\lambda_1} E_{\lambda, b}(\lambda, 1200\ \text{K}) \, d\lambda}{E_b(1200\ \text{K})} + \alpha_{\lambda, 2} \frac{\int_{\lambda_1}^\infty E_{\lambda, b}(\lambda, 1200\ \text{K}) \, d\lambda}{E_b(1200\ \text{K})}$$

or

$$\alpha = \alpha_{\lambda, 1} F_{(0 \to \lambda_1)} + \alpha_{\lambda, 2}[1 - F_{(0 \to \lambda_1)}]$$

From Table 12.1,

$$\lambda_1 T_f = 5\ \mu\text{m} \times 1200\ \text{K} = 6000\ \mu\text{m} \cdot \text{K}: \qquad F_{(0 \to \lambda_1)} = 0.738$$

Hence

$$\alpha = 0.8 \times 0.738 \times 0.1(1 - 0.738) = 0.62$$

The total, hemispherical emissivity follows from Equation 12.38.

$$\varepsilon(T_s) = \frac{\int_0^\infty \varepsilon_\lambda E_{\lambda, b}(\lambda, T_s) \, d\lambda}{E_b(T_s)}$$

Since the surface is diffuse, $\varepsilon_\lambda = \alpha_\lambda$ and it follows that

$$\varepsilon = \alpha_{\lambda, 1} \frac{\displaystyle\int_0^{\lambda_1} E_{\lambda, b}(\lambda, 300\ \text{K})\, d\lambda}{E_b(300\ \text{K})} + \alpha_{\lambda, 2} \frac{\displaystyle\int_{\lambda_1}^{\infty} E_{\lambda, b}(\lambda, 300\ \text{K})\, d\lambda}{E_b(300\ \text{K})}$$

or,

$$\varepsilon = \alpha_{\lambda, 1} F_{(0 \to \lambda_1)} + \alpha_{\lambda, 2}[1 - F_{(0 \to \lambda_1)}]$$

From Table 12.1,

$$\lambda_1 T_s = 5\ \mu\text{m} \times 300\ \text{K} = 1500\ \mu\text{m} \cdot \text{K}: \qquad F_{(0 \to \lambda_1)} = 0.014$$

Hence

$$\varepsilon = 0.8 \times 0.014 + 0.1(1 - 0.014) = 0.11 \qquad\qquad \triangleleft$$

2. Because the spectral characteristics of the coating and the furnace temperature remain fixed, there is no change in the value of α with increasing time. However, as T_s increases with time, the value of ε will change. After a sufficiently long time, $T_s = T_f$, and $\varepsilon = \alpha(\varepsilon = 0.62)$.

Comments:

1. The equilibrium condition that eventually exists ($T_s = T_f$) corresponds precisely to the condition for which Kirchhoff's law was derived. Hence α must equal ε.

2. Approximating the sphere as a lumped capacitance and neglecting convection, an energy balance for a control volume about the sphere yields

$$\dot{E}_{\text{in}} - \dot{E}_{\text{out}} = \dot{E}_{\text{st}}$$

$$(\alpha G)A_s - (\varepsilon \sigma T_s^4)A_s = Mc_p \frac{dT}{dt}$$

The differential equation could be solved to determine $T(t)$ for $t > 0$, and the variation in ε that occurs with increasing time would have to be included in the solution.

12.8
Environmental Radiation

It would be inappropriate to conclude this chapter without commenting on the radiation that comprises our natural environment. Solar radiation is, of course, essential to all life on earth. Through the process of photosynthesis, it satisfies our need for food, fiber, and fuel. Moreover, through thermal and photovoltaic processes, it has the potential to satisfy much of our demand for space heat, process heat, and electricity.

The sun is a nearly spherical radiation source that is 1.39×10^9 m in diameter and is located 1.50×10^{11} m from the earth. With respect to the magnitude and the spectral and directional dependence of the incident solar radiation, it is necessary to distinguish between conditions at the earth's surface and at the outer edge of the earth's atmosphere. For a horizontal surface outside the earth's atmosphere, solar radiation appears as a beam of nearly *parallel rays* that form an angle θ, the *zenith angle,* relative to the surface normal (Figure 12.28). The *extraterrestrial* solar irradiation $G_{S,o}$ depends on the geographic latitude, as well as the time of day and year. It may be determined from an expression of the form

$$G_{S,o} = S_c \cdot f \cdot \cos \theta \qquad (12.67)$$

where S_c, the *solar constant,* is the flux of solar energy incident on a surface oriented *normal* to the sun's rays, when the earth is at its mean distance from the sun. It is known to have a value of $S_c = 1353 \text{ W/m}^2$. The quantity f is a small correction factor to account for the eccentricity of the earth's orbit about the sun ($0.97 \lesssim f \lesssim 1.03$).

The spectral distribution of solar radiation is significantly different from that associated with emission by engineering surfaces. As shown in Figure 12.29, this distribution approximates that of a blackbody at 5800 K. The radiation is concentrated in the low wavelength region ($0.2 \lesssim \lambda \lesssim 3 \ \mu\text{m}$) of the thermal spectrum, with the peak occurring at approximately 0.50 μm. It is this short wavelength concentration that frequently precludes the assumption of graybody behavior for solar irradiated surfaces, since emission is generally in the spectral region beyond 4 μm and it is unlikely that surface spectral properties would be constant over such a wide spectral range.

As solar radiation passes through the earth's atmosphere, its magnitude and its spectral and directional distributions experience significant change. The change is due to *absorption* and *scattering* of the radiation by the atmospheric constituents. The effect of absorption by the atmospheric gases O_3 (ozone), H_2O, O_2, and CO_2 is shown by the lower curve of Figure 12.29. Absorption by ozone is strong in the UV region, providing considerable attenuation below 0.4 μm and complete attenuation below 0.3 μm. In the visible region there is some absorption by O_3 and O_2; and in the near and far IR regions, absorption is dominated by water vapor. Throughout the solar spectrum, there is also continuous absorption of radiation by the dust and aerosol content of the atmosphere.

Atmospheric scattering provides for *redirection* of the sun's rays and consists of two kinds (Figure 12.30). *Rayleigh* (or *molecular*) *scattering* by the gas

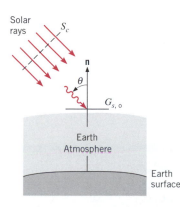

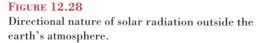

FIGURE 12.28
Directional nature of solar radiation outside the earth's atmosphere.

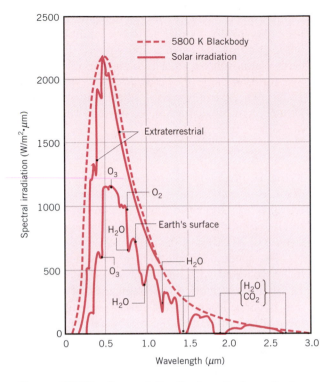

FIGURE 12.29   Spectral distribution of solar radiation.

molecules provides for nearly uniform scattering of radiation in all directions. Hence approximately half of the scattered radiation is redirected to space, while the remaining portion strikes the earth's surface. At any point on this surface, the scattered radiation is incident from all directions. In contrast, *Mie scattering* by the dust and aerosol particles of the atmosphere is concentrated in directions that are close to that of the incident rays. Hence virtually all this radiation strikes the earth's surface in directions close to that of the sun's rays.

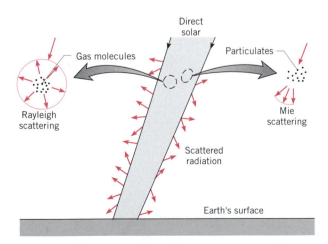

FIGURE 12.30   Scattering of solar radiation in the earth's atmosphere.

The cumulative effect of the scattering processes on the directional distribution of solar radiation striking the earth's surface is shown in Figure 12.31*a*. That portion of the radiation that has penetrated the atmosphere without having been scattered (or absorbed) is in the direction of the zenith angle and is termed the *direct radiation*. The scattered radiation is incident from all directions, although its intensity is largest for directions close to that of the direct radiation. However, because the radiation intensity is often *assumed* to be independent of direction (Figure 12.31*b*), the radiation is termed *diffuse*. The total solar radiation reaching the earth's surface is therefore the sum of direct and diffuse contributions. The diffuse contribution may vary from approximately 10% of the total solar radiation on a clear day to nearly 100% on a totally overcast day.

The foregoing discussion has focused on the *nature* of solar radiation. Heat transfer analyses related to its use are considered in many of the examples and problems of the text. Detailed treatment of solar energy technologies is left to the literature [7–11].

Long wavelength forms of environmental radiation include emission from the earth's surface, as well as emission from certain atmospheric constituents. The emissive power associated with the earth's surface may be computed in the conventional manner. That is,

$$E = \varepsilon \sigma T^4 \tag{12.68}$$

where ε and T are the surface emissivity and temperature, respectively. Emissivities are generally close to unity, and that of water, for example, is approximately 0.97. Since temperatures typically range from 250 to 320 K, emission is concentrated in the spectral region from approximately 4 to 40 μm, with the peak occurring at approximately 10 μm.

Atmospheric emission is largely from the CO_2 and H_2O molecules and is concentrated in the spectral regions from 5 to 8 μm and above 13 μm. Although the spectral distribution of atmospheric emission does not correspond to that of a blackbody, its contribution to irradiation of the earth's surface can be estimated using Equation 12.28. In particular, earth irradiation due to atmospheric emission may be expressed in the form

$$G_{\text{atm}} = \sigma T^4_{\text{sky}} \tag{12.69}$$

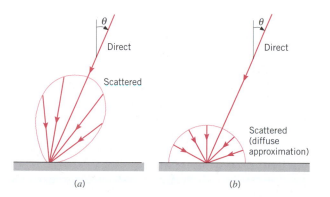

(a)   (b)

FIGURE 12.31   Directional distribution of solar radiation at the earth's surface. (*a*) Actual distribution. (*b*) Diffuse approximation.

TABLE 12.2   Solar Absorptivity α_S and Emissivity
ε of Surfaces Having Spectral Absorptivity
Given in Figure 12.23

Surface	α_S	ε (300 K)	α_S/ε
Evaporated aluminum film	0.09	0.03	3.0
Fused quartz on aluminum film	0.19	0.81	0.24
White paint on metallic substrate	0.21	0.96	0.22
Black paint on metallic substrate	0.97	0.97	1.0
Stainless steel, as received, dull	0.50	0.21	2.4
Red brick	0.63	0.93	0.68
Human skin (Caucasian)	0.62	0.97	0.64
Snow	0.28	0.97	0.29
Corn leaf	0.76	0.97	0.78

where T_{sky} is termed the effective sky temperature. Its value depends on atmospheric conditions, ranging from a low of 230 K under a cold, clear sky to a high of approximately 285 K under warm, cloudy conditions. At night, atmospheric emission is the only source of earth irradiation. When its value is small, as on a cool, clear night, water may freeze even though the air temperature exceeds 273 K.

We close by recalling that values of the spectral properties of a surface at short wavelengths may be appreciably different from values at long wavelengths (Figures 12.18 and 12.23). Since solar radiation is concentrated in the short wavelength region of the spectrum and surface emission is at much longer wavelengths, it follows that many surfaces may not be approximated as gray in their response to solar irradiation. In other words, the solar absorptivity of a surface α_S may differ from its emissivity ε. Values of α_S and the emissivity at moderate temperatures are presented in Table 12.2 for representative surfaces. Note that the ratio α_S/ε is an important engineering parameter. Small values are desired if the surface is intended to reject heat; large values are required if the surface is intended to collect solar energy.

EXAMPLE 12.11

A flat-plate solar collector with no cover plate has a selective absorber surface of emissivity 0.1 and solar absorptivity 0.95. At a given time of day the absorber surface temperature T_s is 120°C when the solar irradiation is 750 W/m², the effective sky temperature is -10°C, and the ambient air temperature T_∞ is 30°C. Assume that the heat transfer convection coefficient for the calm day conditions can be estimated from

$$\overline{h} = 0.22(T_s - T_\infty)^{1/3} \text{ W/m}^2 \cdot \text{K}$$

Calculate the useful heat removal rate (W/m²) from the collector for these conditions. What is the corresponding efficiency of the collector?

SOLUTION

Known: Operating conditions for a flat-plate solar collector.

Find:

1. Useful heat removal rate per unit area, q_u'' (W/m^2).

2. Efficiency η of the collector.

Schematic:

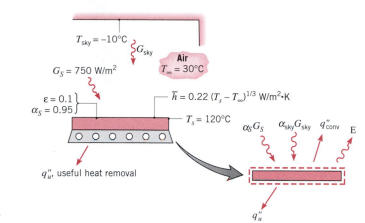

Assumptions:

1. Steady-state conditions.

2. Bottom of collector well insulated.

3. Absorber surface diffuse.

Analysis:

1. Performing an energy balance on the absorber,

$$\dot{E}_{in} - \dot{E}_{out} = 0$$

or, per unit surface area,

$$\alpha_S G_S + \alpha_{sky} G_{sky} - q_{conv}'' - E - q_u'' = 0$$

From Equation 12.69,

$$G_{sky} = \sigma T_{sky}^4$$

Since the sky radiation is concentrated in approximately the same spectral region as that of surface emission, it is reasonable to assume that

$$\alpha_{sky} \approx \varepsilon = 0.1$$

With

$$q_{conv}'' = \overline{h}(T_s - T_\infty) = 0.22(T_s - T_\infty)^{4/3}$$

and

$$E = \varepsilon \sigma T_s^4$$

it follows that

$$q''_u = \alpha_S G_S + \varepsilon \sigma T^4_{\text{sky}} - 0.22(T_s - T_\infty)^{4/3} - \varepsilon \sigma T^4_s$$

$$q''_u = \alpha_S G_S - 0.22(T_s - T_\infty)^{4/3} - \varepsilon \sigma (T^4_s - T^4_{\text{sky}})$$

$$q''_u = 0.95 \times 750 \text{ W/m}^2 - 0.22(120 - 30)^{4/3} \text{ W/m}^2$$

$$- 0.1 \times 5.67 \times 10^{-8} \text{ W/m}^2 \cdot \text{K}^4 (393^4 - 263^4) \text{ K}^4$$

$$q''_u = (712.5 - 88.7 - 108.1) \text{ W/m}^2 = 516 \text{ W/m}^2 \qquad \triangleleft$$

2. The collector efficiency, defined as the fraction of the solar irradiation extracted as useful energy, is then

$$\eta = \frac{q''_u}{G_S} = \frac{516 \text{ W/m}^2}{750 \text{ W/m}^2} = 0.69 \qquad \triangleleft$$

Comments:

1. Since the spectral range of G_{sky} is entirely different from that of G_S, it would be *incorrect* to assume that $\alpha_{\text{sky}} = \alpha_S$.

2. The convection heat transfer coefficient is extremely low ($\bar{h} \approx 1 \text{ W/m}^2 \cdot \text{K}$). With a modest increase to $\bar{h} = 5 \text{ W/m}^2 \cdot \text{K}$, the useful heat flux and the efficiency are reduced to $q''_u = 161 \text{ W/m}^2$ and $\eta = 0.21$. A cover plate can contribute significantly to reducing convection (and radiation) heat loss from the absorber plate.

12.9
Summary

Many new and important ideas have been introduced in this chapter, and at this stage you may well be confused, particularly by the terminology. However, the subject matter has been developed in a systematic fashion, and a careful rereading of the material should make you more comfortable with its application. A glossary has been provided in Table 12.3 to assist you in assimilating the terminology.

You should be able to answer the following questions. What is radiation, and what position does *thermal radiation* occupy in the *electromagnetic spectrum*? What are the physical origins of thermal radiation? What are meant by the terms *irradiation, emissive power,* and *radiosity*? What material property characterizes the ability of a surface to emit thermal radiation? What processes and associated material properties characterize the manner in which a surface responds to irradiation? What is an *opaque* surface? A *semitransparent* surface?

You should recognize the difference between *directional* and *spectral* radiation properties on the one hand and *hemispherical* and *total* properties on the other. Moreover, you should be able to proceed from knowledge of the former

TABLE 12.3 Glossary of Radiative Terms

Term	Definition
Absorption	The process of converting radiation intercepted by matter to internal thermal energy.
Absorptivity	Fraction of the incident radiation absorbed by matter. Equations 12.41, 12.42, and 12.45. Modifiers: *directional, hemispherical, spectral, total.*
Blackbody	The ideal emitter and absorber. Modifier referring to ideal behavior. Denoted by the subscript b.
Diffuse	Modifier referring to the directional independence of the intensity associated with emitted, reflected, or incident radiation.
Directional	Modifier referring to a particular direction. Denoted by the subscript θ.
Directional distribution	Refers to variation with direction.
Emission	The process of radiation production by matter at a finite temperature. Modifiers: *diffuse, blackbody, spectral.*
Emissive power	Rate of radiant energy emitted by a surface in all directions per unit area of the surface, E (W/m^2). Modifiers: *spectral, total, blackbody.*
Emissivity	Ratio of the radiation emitted by a surface to the radiation emitted by a blackbody at the same temperature. Equations 12.32, 12.33, 12.34, and 12.37. Modifiers: *directional, hemispherical, spectral, total.*
Gray surface	A surface for which the spectral absorptivity and the emissivity are independent of wavelength over the spectral regions of surface irradiation and emission.
Hemispherical	Modifier referring to all directions in the space above a surface.
Intensity	Rate of radiant energy propagation in a particular direction, per unit area normal to the direction, per unit solid angle about the direction, I (W/m^2 · sr). Modifier: *spectral.*
Irradiation	Rate at which radiation is incident on a surface from all directions per unit area of the surface, G (W/m^2). Modifiers: *spectral, total, diffuse.*
Kirchhoff's law	Relation between emission and absorption properties for surfaces irradiated by a blackbody at the same temperature. Equations 12.61, 12.62, 12.63, and 12.64.
Planck's law	Spectral distribution of emission from a blackbody. Equations 12.25 and 12.26.
Radiosity	Rate at which radiation leaves a surface due to emission and reflection in all directions per unit area of the surface, J (W/m^2). Modifiers: *spectral, total.*
Reflection	The process of redirection of radiation incident on a surface. Modifiers: *diffuse, specular.*
Reflectivity	Fraction of the incident radiation reflected by matter. Equations 12.48, 12.49, and 12.51. Modifiers: *directional, hemispherical, spectral, total.*
Semitransparent	Refers to a medium in which radiation absorption is a volumetric process.

TABLE 12.3   *Continued*

Term	Definition
Solid angle	Region subtended by an element of area on the surface of a sphere with respect to the center of the sphere, ω (sr). Equations 12.2 and 12.3.
Spectral	Modifier referring to a single-wavelength (monochromatic) component. Denoted by the subscript λ.
Spectral distribution	Refers to variation with wavelength.
Specular	Refers to a surface for which the angle of reflected radiation is equal to the angle of incident radiation.
Stefan–Boltzmann law	Emissive power of a blackbody. Equation 12.28.
Thermal radiation	Electromagnetic energy emitted by matter at a finite temperature and concentrated in the spectral region from approximately 0.1 to 100 μm.
Total	Modifier referring to all wavelengths.
Transmission	The process of thermal radiation passing through matter.
Transmissivity	Fraction of the incident radiation transmitted by matter. Equations 12.53 and 12.54. Modifiers: *hemispherical, spectral, total.*
Wien's law	Locus of the wavelength corresponding to peak emission by a blackbody. Equation 12.27.

to determination of the latter. You should also appreciate the unique role of the *blackbody* in the description of thermal radiation. In what sense is the blackbody *perfect*? Why is it an idealization, and how may it be approximated in practice? What is the *Planck distribution*? How is it altered with increasing surface temperature? What are the *Wien* and *Stefan–Boltzmann* laws? For what purposes may the blackbody emission functions of Table 12.1 be used?

Relations between emissivity and absorptivity are often extremely important in radiative exchange calculations. What is *Kirchhoff's law,* and what restrictive conditions are inherent in its derivation? What is a *gray surface,* and under what conditions might the assumption of a gray surface be particularly suspect? Finally, what are the characteristics of *solar radiation*? In what region of the spectrum is this radiation concentrated, and how is it altered due to passage through the earth's atmosphere? What is the nature of its directional distribution at the earth's surface?

References

1. Planck, M., *The Theory of Heat Radiation,* Dover Publications, New York, 1959.

2. Gubareff, G. G., J. E. Janssen, and R. H. Torberg, *Thermal Radiation Properties Survey,* 2nd ed., Honeywell Research Center, Minneapolis, 1960.

3. Wood, W. D., H. W. Deem, and C. F. Lucks, *Thermal Radiative Properties,* Plenum Press, New York, 1964.

4. Touloukian, Y. S., *Thermophysical Properties of High Temperature Solid Materials,* Macmillan, New York, 1967.

5. Touloukian, Y. S., and D. P. DeWitt, *Thermal Radiative Properties,* Vols. 7, 8, and 9, from *Thermophysical Properties of Matter,* TPRC Data Series, Y. S. Touloukian and C. Y. Ho, Eds., IFI Plenum, New York, 1970–1972.

6. Siegel, R., and J. R. Howell, *Thermal Radiation Heat Transfer,* 2nd ed., McGraw-Hill, New York, 1981.

7. Duffie, J. A., and W. A. Beckman, *Solar Energy Thermal Processes,* Wiley, New York, 1974.

8. Meinel, A. B., and M. P. Meinel, *Applied Solar Energy: An Introduction,* Addison-Wesley, Reading, MA, 1976.

9. Sayigh, A. A. M., Ed., *Solar Energy Engineering,* Academic Press, New York, 1977.

10. Anderson, E. E., *Solar Energy Fundamentals for Designers and Engineers,* Addison-Wesley, Reading, MA, 1982.

11. Howell, J. R., R. B. Bannerot, and G. C. Vliet, *Solar-Thermal Energy Systems, Analysis and Design,* McGraw-Hill, New York, 1982.

Problems

Intensity, Emissive Power, and Irradiation

12.1 What is the irradiation at surfaces A_2, A_3, and A_4 of Example 12.1 due to emission from A_1?

12.2 Consider a small surface of area $A_1 = 10^{-4}$ m^2, which emits diffusely with a total, hemispherical emissive power of $E_1 = 5 \times 10^4$ W/m^2.

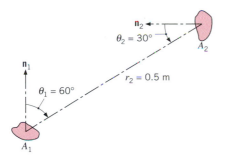

(a) At what rate is this emission intercepted by a small surface of area $A_2 = 5 \times 10^{-4}$ m^2, which is oriented as shown?

(b) What is the irradiation G_2 on A_2?

(c) For zenith angles of $\theta_2 = 0$, 30, and 60°, plot G_2 as a function of the separation distance for $0.25 \leq r_2 \leq 1.0$ m.

12.3 According to its directional distribution, solar radiation incident on the earth's surface may be divided into two components. The *direct* component consists of parallel rays incident at a fixed zenith angle θ, while the *diffuse* component consists of radiation that may be approximated as being diffusely distributed with θ.

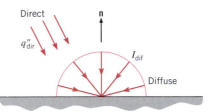

Consider clear sky conditions for which the direct radiation is incident at $\theta = 30°$, with a total flux (based on an area that is normal to the rays) of $q''_{\text{dir}} = 1000$ W/m^2, and the total intensity of the diffuse radiation is $I_{\text{dif}} = 70$ W/m$^2 \cdot$ sr. What is the total solar irradiation at the earth's surface?

12.4 On an overcast day the directional distribution of the solar radiation incident on the earth's surface may be approximated by an expression of the form $I_i = I_n \cos \theta$, where $I_n = 80$ W/m$^2 \cdot$ sr is the total intensity of radiation directed normal to the surface and θ is the zenith angle. What is the solar irradiation at the earth's surface?

12.5 The spectral distribution of the radiation emitted by a diffuse surface may be approximated as follows.

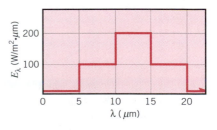

(a) What is the total emissive power?

(b) What is the total intensity of the radiation emitted in the normal direction and at an angle of 30° from the normal?

(c) Determine the fraction of the emissive power leaving the surface in the directions $\pi/4 \leq \theta \leq \pi/2$.

12.6 A furnace with an aperture of 20-mm diameter and emissive power of 3.72×10^5 W/m² is used to calibrate a heat flux gage having a sensitive area of 1.6×10^{-5} m².

(a) At what distance, measured along a normal from the aperture, should the gage be positioned to receive irradiation of 1000 W/m²?

(b) If the gage is tilted off normal by 20°, what will be its irradiation?

(c) For tilt angles of 0, 20, and 60°, plot the gage irradiation as a function of the separation distance for values ranging from 100 to 300 mm.

12.7 Determine the fraction of the total, hemispherical emissive power that leaves a diffuse surface in the directions $\pi/4 \leq \theta \leq \pi/2$ and $0 \leq \phi \leq \pi$.

12.8 What fraction of the total, hemispherical emissive power E (W/m²) leaving a diffuse radiator will be in the directions $0 \leq \theta \leq 30°$?

12.9 Consider a 5-mm square, diffuse surface ΔA_o having a total emissive power of $E_o = 4000$ W/m². The radiation field due to emission into the hemispherical space above the surface is diffuse, thereby providing a uniform intensity $I(\theta, \phi)$. Moreover, if the space is a nonparticipating medium (nonabsorbing, nonscattering, and nonemitting), the intensity is independent of radius for any (θ, ϕ) direction. Hence intensities at any points P_1 and P_2 would be equal.

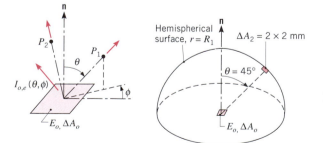

(a) What is the rate at which radiant energy is emitted by ΔA_o, q_{emit}?

(b) What is the intensity $I_{o,e}$ of the radiation field emitted from the surface ΔA_o?

(c) Beginning with Equation 12.10 and presuming knowledge of the intensity $I_{o,e}$, obtain an expression for q_{emit}.

(d) Consider the hemispherical surface located at $r = R_1 = 0.5$ m. Using the conservation of energy requirement, determine the rate at which radiant energy is incident on this surface due to emission from ΔA_o.

(e) Using Equation 12.5, determine the rate at which radiant energy leaving ΔA_o is intercepted by the small area ΔA_2 located in the direction $(45°, \phi)$ on the hemispherical surface. What is the irradiation on ΔA_2?

(f) Repeat part (e) for the location $(0°, \phi)$. Are the irradiations at the two locations equal?

(g) Using Equation 12.15, determine the irradiation G_1 on the hemispherical surface at $r = R_1$.

12.10 During radiant heat treatment of a thin-film material, its shape, which may be hemispherical (a) or spherical (b), is maintained by a relatively low air pressure (as in the case of a rubber balloon). Irradiation on the film is due to emission from a radiant heater of area $A_h = 0.0052$ m², which emits diffusely with an intensity of $I_{e,h} = 169{,}000$ W/m² · sr.

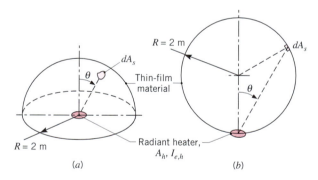

(a) Obtain an expression for the irradiation on the film as a function of the zenith angle θ.

(b) Based on the expressions derived in part (a), which shape provides the more uniform irradiation G and hence provides better quality control for the treatment process?

12.11 In order to initiate a process operation, an infrared motion sensor (radiation detector) is employed to determine the approach of a hot part on a conveyor system. To set the sensor's amplifier discriminator, the engineer needs a relationship between the sensor output signal, S, and the position of the part on the conveyor. The sensor output signal is proportional to the rate at which radiation is incident on the sensor.

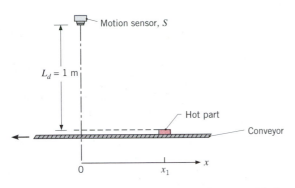

(a) For $L_d = 1$ m, at what location x_1 will the sensor signal S_1 be 75% of the signal corresponding to the position directly beneath the sensor, S_o ($x = 0$)?

(b) For values of $L_d = 0.8$, 1.0, and 1.2 m, plot the signal ratio, S/S_o, versus part position, x, for signal ratios in the range from 1.0 to 0.2. Compare the x-locations for which $S/S_o = 0.75$.

Blackbody Radiation

12.12 Approximations to Planck's law for the spectral emissive power are the Wien and Rayleigh–Jeans spectral distributions, which are useful for the extreme low and high limits of the product λT, respectively.

(a) Show that Planck's spectral distribution will have the form

$$E_{\lambda, b}(\lambda, T) \approx \frac{C_1}{\lambda^5} \exp\left(-\frac{C_2}{\lambda T}\right)$$

when $C_2/\lambda T \gg 1$ and determine the error (compared to Planck's law) for the condition $\lambda T = 2898 \ \mu\text{m} \cdot \text{K}$. This form is known as Wien's law.

(b) Show that the Planck distribution will have the form

$$E_{\lambda, b}(\lambda, T) \approx \frac{C_1}{C_2} \frac{T}{\lambda^4}$$

when $C_2/\lambda T \ll 1$ and determine the error (compared to Planck's law) for the condition $\lambda T = 100{,}000 \ \mu\text{m} \cdot \text{K}$. This form is known as the Rayleigh–Jeans law.

12.13 Isothermal furnaces with small apertures approximating a blackbody are frequently used to calibrate heat flux gages, radiation thermometers, and other radiometric devices. In such applications, it is necessary to control power to the furnace such that the variation of temperature and the spectral intensity of the aperture are within desired limits.

(a) By considering the Planck spectral distribution, Equation 12.26, show that the ratio of the fractional change in the spectral intensity to the fractional change in the temperature of the furnace has the form

$$\frac{dI_\lambda/I_\lambda}{dT/T} = \frac{C_2}{\lambda T} \frac{1}{1 - \exp\left(-C_2/\lambda T\right)}$$

(b) Using this relation, determine the allowable variation in temperature of the furnace operating at 2000 K to ensure that the spectral intensity at 0.65 μm will not vary by more than 0.5%. What is the allowable variation at 10 μm?

12.14 Substitute the Planck distribution, Equation 12.26, into Equation 12.11 and perform the spectral integration to obtain Equation 12.28 for the total emissive power of a blackbody. Show that

$$\sigma = \left(\frac{\pi}{C_2}\right)^4 \frac{C_1}{15}$$

and calculate the value of the Stefan–Boltzmann constant using values of the radiation constants C_1 and C_2.

12.15 A spherical aluminum shell of inside diameter $D = 2$ m is evacuated and is used as a radiation test chamber. If the inner surface is coated with carbon black and maintained at 600 K, what is the irradiation on a small test surface placed in the chamber? If the inner surface were not coated and maintained at 600 K, what would the irradiation be?

12.16 An enclosure has an inside area of 100 m², and its inside surface is black and is maintained at a constant temperature. A small opening in the enclosure has an area of 0.02 m². The radiant power emitted from this opening is 70 W. What is the temperature of the interior enclosure wall? If the interior surface is maintained at this temperature, but is now polished, what will be the value of the radiant power emitted from the opening?

12.17 Assuming the earth's surface is black, estimate its temperature if the sun has an equivalent blackbody temperature of 5800 K. The diameters of the sun and earth are 1.39×10^9 and 1.29×10^7 m, respectively, and the distance between the sun and earth is 1.5×10^{11} m.

12.18 The energy flux associated with solar radiation incident on the outer surface of the earth's atmosphere has been accurately measured and is known to be 1353 W/m². The diameters of the sun and earth are 1.39×10^9 and 1.29×10^7 m, respectively, and the distance between the sun and the earth is 1.5×10^{11} m.

(a) What is the emissive power of the sun?

(b) Approximating the sun's surface as black, what is its temperature?

(c) At what wavelength is the spectral emissive power of the sun a maximum?

(d) Assuming the earth's surface to be black and the sun to be the only source of energy for the earth, estimate the earth's surface temperature.

12.19 A small flat plate is positioned just beyond the earth's atmosphere and is oriented such that the normal to the plate passes through the center of the sun. Refer to Problem 12.18 for pertinent earth–sun dimensions.

(a) What is the solid angle subtended by the sun about a point on the surface of the plate?

(b) Determine the incident intensity, I_i, on the plate using the known value of the solar irradiation above the earth's atmosphere ($G_S = 1353$ W/m²).

(c) Sketch the incident intensity I_i as a function of the zenith angle θ, where θ is measured from the normal to the plate.

12.20 Estimate the wavelength corresponding to maximum emission from each of the following surfaces: the sun, a tungsten filament at 2500 K, a heated metal at 1500 K, human skin at 305 K, and a cryogenically cooled metal surface at 60 K. Estimate the fraction of the solar emission that is in the following spectral regions: the ultraviolet, the visible, and the infrared.

12.21 A 100-W light source consists of a filament that is in the form of a thin rectangular strip, 5 mm long by 2 mm wide, and radiates as a blackbody at 2900 K.

(a) Assuming that the glass bulb transmits all incident visible radiation, what is its efficiency?

(b) Determine the efficiency as a function of filament temperature for the range from 1300 to 3300 K.

12.22 Consider the radiation emerging from a small aperture of a furnace operating at 1000 K. Calcu-

late the emissive power of the aperture. Determine the spectral intensity at 2 μm. What is the ratio of the spectral intensity at 2 μm to the spectral intensity at 6 μm? What fraction of the emissive power is in the spectral range 2 to 6 μm?

12.23 Determine the fraction of the radiation emitted by the sun in the visible region of the spectrum. Plot the percentage of solar emission that is at wavelengths less than λ as a function of λ. On the same coordinates, plot the percentage of emission from a blackbody at 300 K that is at wavelengths less than λ as a function of λ. Compare the plotted results with the upper abscissa scale of Figure 12.23.

12.24 An electrically powered, ring-shaped radiant heating element is maintained at a temperature of $T_h = 3000$ K and is used in a manufacturing process to heat a small part having a surface area of $A_p = 0.007$ m². The surface of the heating element may be assumed to be black.

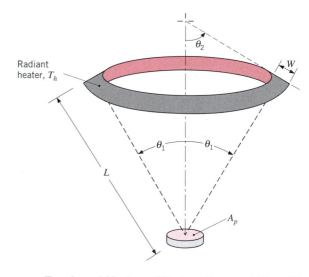

For $\theta_1 = 30°$, $\theta_2 = 60°$, $L = 3$ m, and $W = 30$ mm, what is the rate at which radiant energy emitted by the heater is incident on the part?

Emissivity

12.25 The spectral, hemispherical emissivity of tungsten may be approximated by the distribution depicted below. Consider a cylindrical tungsten filament that is of diameter $D = 0.8$ mm and length $L = 20$ mm. The filament is enclosed in an evacuated bulb and is heated by an electrical current to a steady-state temperature of 2900 K.

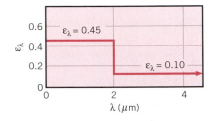

(a) What is the total hemispherical emissivity when the filament temperature is 2900 K?

(b) Assuming the surroundings are at 300 K, what is the initial rate of cooling of the filament when the current is switched off?

(c) Generate a plot of the emissivity as a function of filament temperature for $1300 \leq T \leq 2900$ K.

(d) Estimate the time required for the filament to cool from 2900 to 1300 K.

12.26 For materials A and B, whose spectral hemispherical emissivities vary with wavelength as shown below, how does the total, hemispherical emissivity vary with temperature? Explain briefly.

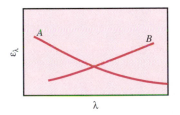

12.27 Various materials are to be evaluated for design of more energy-efficient household cooking pans. Compare the emissive powers of the following materials at 200°C and explain what influence this quantity would have on energy consumption: polished copper, Teflon-coated copper, cleaned stainless steel, Pyrex, and pyroceram.

12.28 The spectral, directional emissivity of a diffuse material at 2000 K has the following distribution:

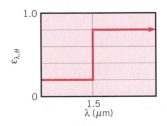

Determine the total, hemispherical emissivity at 2000 K. Determine the emissive power over the spectral range 0.8 to 2.5 μm *and* for the directions $0 \leq \theta \leq 30°$.

12.29 Consider the directionally selective surface having the directional emissivity ε_θ, as shown. Assuming that the surface is isotropic in the ϕ direction, calculate the ratio of the normal emissivity ε_n to the hemispherical emissivity ε_h.

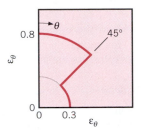

12.30 Consider the metallic surface of Example 12.6. Additional measurements of the spectral, hemispherical emissivity yield a spectral distribution which may be approximated as follows:

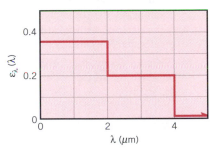

(a) Determine corresponding values of the total, hemispherical emissivity ε and the total emissive power E at 2000 K.

(b) Plot the emissivity as a function of temperature for $500 \leq T \leq 3000$ K. Explain the variation.

12.31 A detector of area $A_2 = 4 \times 10^{-6}$ m^2 is used to measure the total radiation emitted by a surface of area $A_1 = 5 \times 10^{-6}$ m^2 and temperature $T_1 = 1000$ K. When the detector surface views radiation emitted by A_1 in the normal direction ($\theta = 0°$) at a distance of $L = 0.5$ m, it measures a radiant power of 1.155×10^{-6} W.

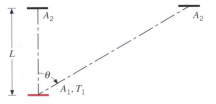

What is the total, normal emissivity of surface 1? When the detector is displaced horizontally, such that $\theta = 60°$, it measures a radiant power of 5.415×10^{-8} W. Is surface 1 a diffuse emitter?

12.32 A sphere is suspended in air in a dark room and maintained at a uniform incandescent temperature. When first viewed with the naked eye, the sphere appears to be brighter around the rim. After several hours, however, it appears to be brighter in the center. Of what type material would you reason the sphere is made? Give plausible reasons for the nonuniformity of brightness of the sphere and for the changing appearance with time.

12.33 A radiation thermometer is a device that responds to a radiant flux within a prescribed spectral interval and is calibrated to indicate the temperature of a blackbody that produces the same flux.

(a) When viewing a surface at an elevated temperature T_s and emissivity less than unity, the thermometer will indicate an apparent temperature referred to as the brightness or spectral radiance temperature T_λ. Will T_λ be greater than, less than, or equal to T_s?

(b) Write an expression for the spectral emissive power of the surface in terms of Wien's spectral distribution (see Problem 12.12) and the spectral emissivity of the surface. Write the equivalent expression using the spectral radiance temperature of the surface and show that

$$\frac{1}{T_s} = \frac{1}{T_\lambda} + \frac{\lambda}{C_2}\ln \varepsilon_\lambda$$

where λ represents the wavelength at which the thermometer operates.

(c) Consider a radiation thermometer that responds to a spectral flux centered about the wavelength 0.65 μm. What temperature will the thermometer indicate when viewing a surface with ε_λ (0.65 μm) = 0.9 and $T_s = 1000$ K? Verify that Wien's spectral distribution is a reasonable approximation to Planck's law for this situation.

12.34 A technique for measuring the total hemispherical emissivity of a coating material involves application of the coating to the surface of a small hollow sphere containing an electric resistance heater. The sphere is placed in a large evacuated enclosure whose walls are cryogenically cooled. When the heater is energized, measurements of the heater power and the surface temperature are used to determine the surface emissivity. Under conditions for which the walls of the enclosure are at 77 K and a heater power of 15 W is needed to maintain a surface temperature of 500 K on a sphere of 40-mm diameter, what is the emissivity of the surface coating?

12.35 Sheet steel emerging from the hot roll section of a steel mill has a temperature of 1200 K, a thickness of $\delta = 3$ mm, and the following distribution for the spectral, hemispherical emissivity.

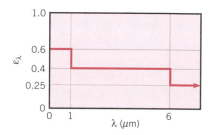

The density and specific heat of the steel are 7900 kg/m^3 and 640 J/kg · K, respectively. What is the total, hemispherical emissivity? Accounting for emission from both sides of the sheet and neglecting conduction, convection, and radiation from the surroundings, determine the initial time rate of change of the sheet temperature $(dT/dt)_i$. As the steel cools, it oxidizes and its total, hemispherical emissivity increases. If this increase may be correlated by an expression of the form $\varepsilon = \varepsilon_{1200}[1200 \text{ K}/T \text{ (K)}]$, how long will it take for the steel to cool from 1200 to 600 K?

12.36 A large body of nonluminous gas at a temperature of 1200 K has emission bands between 2.5 and 3.5 μm and between 5 and 8 μm. The effective emissivity in the first band is 0.8 and in the second 0.6. Determine the emissive power of this gas.

Absorptivity, Reflectivity, and Transmissivity

12.37 A diffusely emitting surface is exposed to a radiant source causing the irradiation on the surface to be 1000 W/m^2. The intensity for emission is 143 W/m^2 · sr and the reflectivity of the surface is 0.8. Determine the emissive power, E (W/m^2), and radiosity, J (W/m^2), for the surface. What is the net heat flux to the surface by the radiation mode?

12.38 An opaque surface with the prescribed spectral, hemispherical reflectivity distribution is subjected to the spectral irradiation shown.

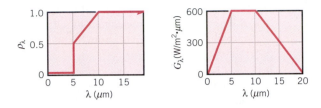

(a) Sketch the spectral, hemispherical absorptivity distribution.

(b) Determine the total irradiation on the surface.

(c) Determine the radiant flux that is absorbed by the surface.

(d) What is the total, hemispherical absorptivity of this surface?

12.39 A small, opaque, diffuse object at $T_s = 400$ K is suspended in a large furnace whose interior walls are at $T_f = 2000$ K. The walls are diffuse and gray and have an emissivity of 0.20. The spectral, hemispherical emissivity for the surface of the small object is given below:

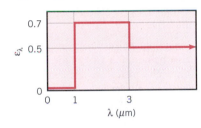

(a) Determine the total emissivity and absorptivity of the surface.

(b) Evaluate the reflected radiant flux and the net radiative flux *to* the surface.

(c) What is the spectral emissive power at $\lambda = 2$ μm?

(d) What is the wavelength $\lambda_{1/2}$ for which one-half of the total radiation emitted by the surface is in the spectral region $\lambda \geq \lambda_{1/2}$?

12.40 The spectral reflectivity distribution for white paint (Figure 12.23) can be approximated by the following stairstep function:

α_λ	0.75	0.15	0.96
λ (μm)	<0.4	0.4–3.0	>3.0

A small flat plate coated with this paint is suspended inside a large enclosure, and its temperature is maintained at 400 K. The surface of the enclosure is maintained at 3000 K and the spectral distribution of its emissivity has the following characteristics:

ε_λ	0.2	0.9
λ (μm)	<2.0	>2.0

(a) Determine the total emissivity, ε, of the enclosure surface.

(b) Determine the total emissivity, ε, and absorptivity, α, of the plate.

12.41 An opaque surface, 2 m by 2 m, is maintained at 400 K and is simultaneously exposed to solar irradiation with $G = 1200$ W/m^2. The surface is diffuse and its spectral absorptivity is $\alpha_\lambda = 0, 0.8, 0$, and 0.9 for $0 \leq \lambda \leq 0.5$ μm, 0.5 μm $< \lambda \leq$ 1 μm, 1 μm $< \lambda \leq 2$ μm, and $\lambda > 2$ μm, respectively. Determine the absorbed irradiation, emissive power, radiosity, and net radiation heat transfer from the surface.

12.42 A diffuse, opaque surface at 700 K has spectral emissivities of $\varepsilon_\lambda = 0$ for $0 \leq \lambda \leq 3$ μm, $\varepsilon_\lambda = 0.5$ for 3 μm $< \lambda \leq 10$ μm, and $\varepsilon_\lambda = 0.9$ for 10 μm $< \lambda < \infty$. A radiant flux of 1000 W/m^2, which is uniformly distributed between 1 and 6 μm, is incident on the surface at an angle of 30° relative to the surface normal.

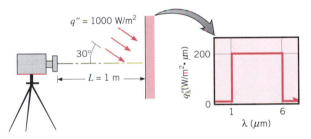

Calculate the total radiant power from a 10^{-4} m^2 area of the surface that reaches a radiation detector positioned along the normal to the area. The aperture of the detector is 10^{-5} m^2, and its distance from the surface is 1 m.

12.43 The spectral, hemispherical absorptivity of an opaque surface is as shown.

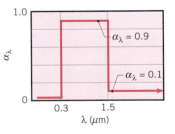

What is the solar absorptivity, α_S? If it is assumed that $\varepsilon_\lambda = \alpha_\lambda$ and that the surface is at a temperature of 340 K, what is its total, hemispherical emissivity?

12.44 The spectral, hemispherical absorptivity of an opaque surface and the spectral distribution of radiation incident on the surface are depicted below.

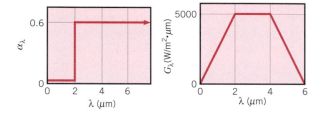

What is the total, hemispherical absorptivity of the surface? If it is assumed that $\varepsilon_\lambda = \alpha_\lambda$ and that the surface is at 1000 K, what is its total, hemispherical emissivity? What is the net radiant heat flux to the surface?

12.45 Consider an opaque, diffuse surface for which the spectral absorptivity and irradiation are as follows:

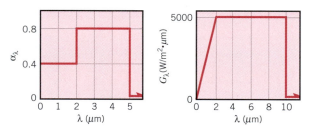

What is the total absorptivity of the surface for the prescribed irradiation? If the surface is at a temperature of 1250 K, what is its emissive power? How will the surface temperature vary with time, for the prescribed conditions?

12.46 The spectral emissivity of an opaque, diffuse surface is as shown.

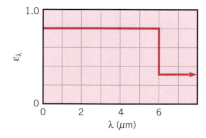

(a) If the surface is maintained at 1000 K, what is the total, hemispherical emissivity?

(b) What is the total, hemispherical absorptivity of the surface when irradiated by large surroundings of emissivity 0.8 and temperature 1500 K?

(c) What is the radiosity of the surface when it is maintained at 1000 K and subjected to the irradiation prescribed in part (b)?

(d) Determine the net radiation flux into the surface for the conditions of part (c).

(e) Plot each of the parameters featured in parts (a)–(d) as a function of the surface temperature for $750 \le T \le 2000$ K.

12.47 An opaque, horizontal flat plate has a top surface area of 3 m², and its edges and lower surface are well insulated. The plate is uniformly irradiated at its top surface at a rate (for the entire plate) of 1300 W. Consider steady-state conditions for which 1000 W of the incident radiation are absorbed, the plate temperature is 500 K, and heat transfer by convection from the surface is 300 W. Determine the irradiation G, emissive power E, radiosity J, absorptivity α, reflectivity ρ, and emissivity ε.

12.48 Square plates freshly sprayed with an epoxy paint must be cured at 140°C for an extended period of time. The plates are located in a large enclosure and heated by a bank of infrared lamps. The top surface of each plate has an emissivity of $\varepsilon = 0.8$ and experiences convection with a ventilation airstream that is at $T_\infty = 27$°C and provides a convection coefficient of $h = 20$ W/m² · K. The irradiation from the enclosure walls is estimated to be $G_{wall} = 450$ W/m², for which the plate absorptivity is $\alpha_{wall} = 0.7$.

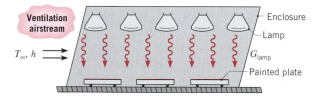

(a) Determine the irradiation that must be provided by the lamps, G_{lamp}. The absorptivity of the plate surface for this irradiation is $\alpha_{lamp} = 0.6$.

(b) For convection coefficients of $h = 15$, 20, and 30 W/m² · K, plot the lamp irradiation, G_{lamp}, as a function of the plate temperature, T_s, for $100 \le T_s \le 300$°C.

(c) For convection coefficients in the range from 10 to 30 W/m² · K and a lamp irradiation of

G_{lamp} = 3000 W/m², plot the airstream temperature T_∞ required to maintain the plate at T_s = 140°C.

12.49 Two small surfaces, A and B, are placed inside an isothermal enclosure at a uniform temperature. The enclosure provides an irradiation of 6300 W/m² to each of the surfaces, and surfaces A and B absorb incident radiation at rates of 5600 and 630 W/m², respectively. Consider conditions after a long time has elapsed.

(a) What are the net heat fluxes for each surface? What are their temperatures?

(b) Determine the absorptivity of each surface.

(c) What are the emissive powers of each surface?

(d) Determine the emissivity of each surface.

12.50 Consider an opaque horizontal plate that is well insulated on its back side. The irradiation on the plate is 2500 W/m², of which 500 W/m² is reflected. The plate is at 227°C and has an emissive power of 1200 W/m². Air at 127°C flows over the plate with a heat transfer convection coefficient of 15 W/m² · K. Determine the emissivity, absorptivity, and radiosity of the plate. What is the net heat transfer rate per unit area?

12.51 A horizontal, opaque surface at a steady-state temperature of 77°C is exposed to an airflow having a free stream temperature of 27°C with a convection heat transfer coefficient of 28 W/m² · K. The emissive power of the surface is 628 W/m², the irradiation is 1380 W/m², and the reflectivity is 0.40. Determine the absorptivity of the surface. Determine the net radiation heat transfer rate for this surface. Is this heat transfer to the surface or from the surface? Determine the combined heat transfer rate for the surface. Is this heat transfer to the surface or from the surface?

12.52 Consider small cylindrical shapes fabricated from materials of different composition and surface finishes as described below. These cylinders, initially at 300 K, are inserted into a large furnace at 600 K. Using tabular values for the emissivity of the materials, indicate which cylinder you expect to heat up faster. Briefly explain your reasoning.

(a) Aluminum: (A) highly polished, film or (B) anodized.

(b) Stainless steel: (A) typical, polished or (B) stably oxidized.

12.53 An apparatus commonly used for measuring the reflectivity of materials is shown. A water-cooled sample, of 30-mm diameter and temperature T_s = 300 K, is mounted flush with the inner surface of a large enclosure. The walls of the enclosure are gray and diffuse with an emissivity of 0.8 and a uniform temperature T_f = 1000 K. A small aperture is located at the bottom of the enclosure to permit sighting of the sample or the enclosure wall. The spectral reflectivity ρ_λ of an opaque, diffuse sample material is as shown. The heat transfer coefficient for convection between the sample and the air within the cavity, which is also at 1000 K, is h = 10 W/m² · K.

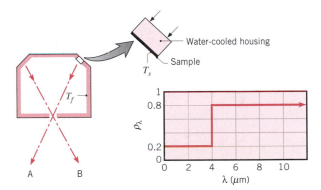

(a) Calculate the absorptivity of the sample.

(b) Calculate the emissivity of the sample.

(c) Determine the heat removal rate (W) by the coolant.

(d) The ratio of the radiation in the A direction to that in the B direction will give the reflectivity of the sample. Briefly explain why this is so.

12.54 A diffuse surface having the following spectral characteristics is maintained at 500 K when situated in a large furnace enclosure whose walls are maintained at 1500 K:

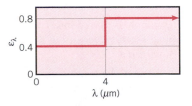

(a) Sketch the spectral distribution of the surface emissive power E_λ and the emissive power $E_{\lambda, b}$ that the surface would have if it were a blackbody.

(b) Neglecting convection effects, what is the net heat flux to the surface for the prescribed conditions?

(c) Plot the net heat flux as a function of the surface temperature for $500 \leq T \leq 1000$ K. On the same coordinates, plot the heat flux for a diffuse, gray surface with total emissivities of 0.4 and 0.8.

(d) For the prescribed spectral distribution of ε_λ, how do the total emissivity and absorptivity of the surface vary with temperature in the range $500 \leq T \leq 1000$ K?

12.55 Consider an opaque, diffuse surface whose spectral reflectivity varies with wavelength as shown. The surface is at 750 K, and irradiation on one side varies with wavelength as shown. The other side of the surface is insulated.

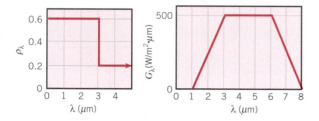

What are the total absorptivity and emissivity of the surface? What is the net radiative heat flux to the surface?

12.56 A very small sample of an opaque surface is initially at 1200 K and has the spectral, hemispherical absorptivity shown.

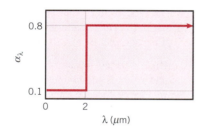

The sample is placed inside a very large enclosure whose walls have an emissivity of 0.2 and are maintained at 2400 K.

(a) What is the total, hemispherical absorptivity of the sample surface?

(b) What is its total, hemispherical emissivity?

(c) What are the values of the absorptivity and emissivity after the sample has been in the enclosure a long time?

(d) For a 10-mm diameter tungsten sphere in an evacuated enclosure, compute and plot the variation of the sample temperature with time, as it is heated from its initial temperature of 1200 K.

12.57 An opaque, gray surface at 27°C is exposed to irradiation of 1000 W/m², and 800 W/m² is reflected. Air at 17°C flows over the surface and the heat transfer convection coefficient is 15 W/m² · K. Determine the net heat flux from the surface.

12.58 Radiation leaves a furnace of inside surface temperature 1500 K through an aperture 20 mm in diameter. A portion of the radiation is intercepted by a detector that is 1 m from the aperture, has a surface area of 10^{-5} m², and is oriented as shown.

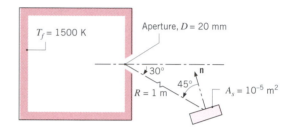

If the aperture is open, what is the rate at which radiation leaving the furnace is intercepted by the detector? If the aperture is covered with a diffuse, semitransparent material of spectral transmissivity $\tau_\lambda = 0.8$ for $\lambda \leq 2$ μm and $\tau_\lambda = 0$ for $\lambda > 2$ μm, what is the rate at which radiation leaving the furnace is intercepted by the detector?

12.59 A thermopile detector of sensitive area 3 mm by 7 mm is to be calibrated using a blackbody laboratory furnace operating at 1500 K. The optical filter placed directly in front of the detector has the spectral transmissivity shown below.

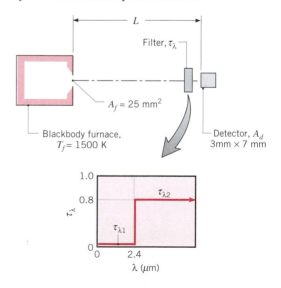

(a) At what distance L from the furnace should the detector be placed such that its irradiation is 50 W/m²?

(b) For blackbody furnace temperatures of 1000, 1500, and 2000 K, plot the irradiation versus L for $100 \leq L \leq 400$ mm.

12.60 A special diffuse glass with prescribed spectral radiative properties is heated in a large oven. The walls of the oven are lined with a diffuse, gray refractory brick having an emissivity of 0.75 and are maintained at $T_w = 1800$ K. Consider conditions for which the glass temperature is $T_g = 750$ K.

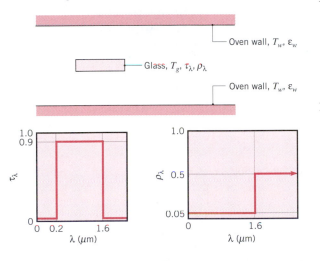

(a) What are the total transmissivity τ, the total reflectivity ρ, and the total emissivity ε of the glass?

(b) What is the net radiative heat flux, $q''_{\text{net, in}}$ (W/m²), to the glass?

(c) For oven wall temperatures of 1500, 1800, and 2000 K, plot $q''_{\text{net, in}}$ as a function of glass temperature for $500 \leq T_g \leq 800$ K.

12.61 A horizontal semitransparent plate is uniformly irradiated from above and below, while air at $T_\infty = 300$ K flows over the top and bottom surfaces, providing a uniform convection heat transfer coefficient of $h = 40$ W/m² · K. The total, hemispherical absorptivity of the plate to the irradiation is 0.40. Under steady-state conditions measurements made with a radiation detector above the top surface indicate a radiosity (which includes transmission, as well as reflection and emission) of $J = 5000$ W/m², while the plate is at a uniform temperature of $T = 350$ K.

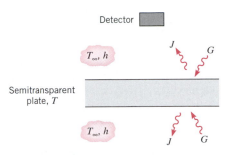

Determine the irradiation G and the total hemispherical emissivity of the plate. Is the plate gray for the prescribed conditions?

12.62 A 20-mm diameter peephole is maintained in the wall of a large furnace whose interior surfaces are at a temperature of 1700 K. What is the heat loss by radiation through an uncovered hole? What is the heat loss by radiation through a fused quartz cover for which $\tau_\lambda = 0.9$ for $\lambda \leq 3.5$ μm and $\tau_\lambda = 0$ for $\lambda > 3.5$ μm?

12.63 The spectral transmissivity of plain and tinted glass can be approximated as follows:

Plain glass: $\tau_\lambda = 0.9$ $0.3 \leq \lambda \leq 2.5$ μm

Tinted glass: $\tau_\lambda = 0.9$ $0.5 \leq \lambda \leq 1.5$ μm

Outside the specified wavelength ranges, the spectral transmissivity is zero for both glasses. Compare the solar energy that could be transmitted through the glasses. With solar irradiation on the glasses, compare the visible radiant energy that could be transmitted.

12.64 Referring to the distribution of the spectral transmissivity of low iron glass (Figure 12.24), describe briefly what is meant by the "greenhouse effect." That is, how does the glass influence energy transfer to and from the contents of a greenhouse?

12.65 The 50-mm peephole of a large furnace operating at 450°C is covered with a material having $\tau = 0.8$ and $\rho = 0$ for irradiation originating from the furnace. The material has an emissivity of 0.8 and is opaque to irradiation from a source at room temperature. The outer surface of the cover is exposed to surroundings and ambient air at 27°C with a convection heat transfer coefficient of 50 W/m² · K. Assuming that convection effects on the inner surface of the cover are negligible, calculate the heat loss by the furnace and the temperature of the cover.

12.66 The spectral absorptivity α_λ and spectral reflectivity ρ_λ for a spectrally selective, diffuse material are as shown.

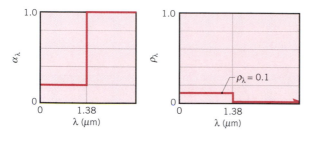

(a) Sketch the spectral transmissivity τ_λ.

(b) If solar irradiation with $G_S = 750$ W/m² and the spectral distribution of a blackbody at 5800 K is incident on this material, determine the fractions of the irradiation that are transmitted, reflected, and absorbed by the material.

(c) If the temperature of this material is 350 K, determine the emissivity ε.

(d) Determine the net heat flux by radiation to the material.

12.67 The window of a large vacuum chamber is fabricated from a material of prescribed spectral characteristics. A collimated beam of radiant energy from a solar simulator is incident on the window and has a flux of 3000 W/m². The inside walls of the chamber, which are large compared to the window area, are maintained at 77 K. The outer surface of the window is subjected to surroundings and room air at 25°C, with a convection heat transfer coefficient of 15 W/m² · K.

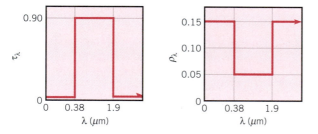

(a) Determine the transmissivity of the window material to radiation from the solar simulator, which approximates the solar spectral distribution.

(b) Assuming that the window is insulated from its chamber mounting arrangement, what steady-state temperature does the window reach?

(c) Calculate the net radiation transfer per unit area of the window to the vacuum chamber wall, excluding the transmitted simulated solar flux.

12.68 A thermocouple whose surface is diffuse and gray with an emissivity of 0.6 indicates a temperature

of 180°C when used to measure the temperature of a gas flowing through a large duct whose walls have an emissivity of 0.85 and a uniform temperature of 450°C.

(a) If the convection heat transfer coefficient between the thermocouple and the gas stream is $\bar{h} = 125$ W/m² · K and there are negligible conduction losses from the thermocouple, determine the temperature of the gas.

(b) Consider a gas temperature of 125°C. Compute and plot the thermocouple *measurement error* as a function of the convection coefficient for $10 \leq \bar{h} \leq 1000$ W/m² · K. What are the implications of your results?

12.69 A thermocouple inserted in a 4-mm diameter stainless steel tube having a diffuse, gray surface with an emissivity of 0.4 is positioned horizontally in a large air-conditioned room whose walls and air temperature are 30 and 20°C, respectively.

(a) What temperature will the thermocouple indicate if the air is quiescent?

(b) Compute and plot the thermocouple *measurement error* as a function of the surface emissivity for $0.1 \leq \varepsilon \leq 1.0$.

12.70 A horizontal, uninsulated pipe of 125-mm diameter passes through a large room having walls and quiescent air at 37 and 25°C, respectively. The pipe is maintained at 125°C; its outer surface is diffuse and gray with an emissivity of 0.8. Calculate the heat loss per unit length from the pipe by convection and radiation.

12.71 A sphere ($k = 185$ W/m · K, $\alpha = 7.25 \times 10^{-5}$ m²/s) of 30-mm diameter whose surface is diffuse and gray with an emissivity of 0.8 is placed in a large oven whose walls are of uniform temperature at 600 K. The temperature of the air in the oven is 400 K, and the convection heat transfer coefficient between the sphere and the oven air is 15 W/m² · K.

(a) Determine the net heat transfer to the sphere when its temperature is 300 K.

(b) What will be the steady-state temperature of the sphere?

(c) How long will it take for the sphere, initially at 300 K, to come within 20 K of the steady-state temperature?

(d) For emissivities of 0.2, 0.4, and 0.8, plot the elapsed time of part (c) as a function of the convection coefficient for $10 \leq h \leq 25$ W/m² · K.

12.72 A vertical flat plate, which is 2 m in height and is insulated on its edges and backside, is suspended in quiescent air at 1 atm and 300 K. The exposed surface is painted with a special diffuse coating having the prescribed absorptivity distribution and is irradiated by solar-simulation lamps that provide a spectral irradiation distribution, G_λ, characteristic of the solar spectrum. Under steady-state conditions, the plate has a temperature of $T_s = 400$ K.

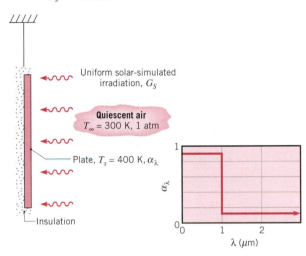

(a) For the prescribed conditions, determine the plate emissivity ε, the plate absorptivity α, the plate irradiation G, and, using an appropriate correlation, the free convection coefficient $\bar{h}$.

(b) If the irradiation G found in part (a) were doubled, what steady-state temperature would the plate reach?

12.73 A manufacturing process involves heating long copper rods, which are coated with a thin film, in a large furnace whose walls are maintained at an elevated temperature T_w. The furnace contains quiescent nitrogen gas at 1-atm pressure and a temperature of $T_\infty = T_w$. The film is a diffuse surface with a spectral emissivity of $\varepsilon_\lambda = 0.9$ for $\lambda \le 2$ μm and $\varepsilon_\lambda = 0.4$ for $\lambda > 2$ μm.

(a) Consider conditions for which a rod of diameter D and initial temperature T_i is inserted in the furnace, such that its axis is horizontal. Assuming validity of the lumped capacitance approximation, derive an equation that could be used to determine the rate of change of the rod temperature at the time of insertion. Express your result in terms of appropriate variables.

(b) If $T_w = T_\infty = 1500$ K, $T_i = 300$ K, and $D = 10$ mm, what is the initial rate of change of the rod temperature? Confirm the validity of the lumped capacitance approximation.

(c) Compute and plot the variation of the rod temperature with time during the heating process.

12.74 A procedure for measuring the thermal conductivity of solids at elevated temperatures involves placement of a sample at the bottom of a large furnace. The sample is of thickness L and is placed in a square container of width W on a side. The sides are well insulated. The walls of the cavity are maintained at T_w, while the bottom surface of the sample is maintained at a much lower temperature T_c by circulating coolant through the sample container. The sample surface is diffuse and gray with an emissivity ε_s. Its temperature T_s is measured optically.

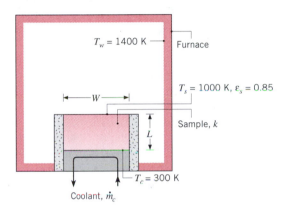

(a) Neglecting convection effects, obtain an expression from which the sample thermal conductivity may be evaluated in terms of measured and known quantities (T_w, T_s, T_c, ε_s, L). The measurements are made under steady-state conditions. If $T_w = 1400$ K, $T_s = 1000$ K, $\varepsilon_s = 0.85$, $L = 0.015$ m, and $T_c = 300$ K, what is the sample thermal conductivity?

(b) If $W = 0.10$ m and the coolant is water with a flow rate of $\dot{m}_c = 0.1$ kg/s, is it reasonable to assume a uniform bottom surface temperature T_c?

12.75 One scheme for extending the operation of gas turbine blades to higher temperatures involves applying a ceramic coating to the surfaces of blades fabricated from a superalloy such as inconel. To assess the reliability of such coatings, an apparatus has been developed for testing samples under laboratory conditions. The sample is placed at the bottom of a large vacuum chamber whose walls are cryogenically cooled and which is equipped with a radiation detector at the top surface. The detector has a surface area of $A_d = 10^{-5}$ m^2, is located a distance of $L_{s-d} = 1$ m from the sample, and views radiation originating from a portion of the ceramic surface having an area of $\Delta A_c = 10^{-4}$ m². An electric heater attached to the bottom of the sample dissipates a uniform heat flux, q_h'', which is transferred upward through the sample. The bottom of the heater and sides of the sample are well insulated.

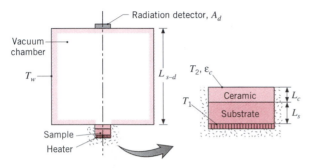

Consider conditions for which a ceramic coating of thickness $L_c = 5$ mm and thermal conductivity $k_c = 60$ W/m · K has been sprayed on a metal substrate of thickness $L_s = 8$ mm and thermal conductivity $k_s = 25$ W/m · K. The opaque surface of the ceramic may be approximated as diffuse and gray, with a total hemispherical emissivity of $\varepsilon_c = 0.8$.

(a) Consider steady-state conditions for which the bottom surface of the substrate is maintained at $T_1 = 1500$ K, while the chamber walls (including the surface of the radiation detector) are maintained at $T_w = 90$ K. Assuming negligible thermal contact resistance at the ceramic–substrate interface, determine the ceramic top surface temperature T_2 and the heat flux q_h''.

(b) For the prescribed conditions, what is the rate at which radiation emitted by the ceramic is intercepted by the detector?

(c) After repeated experiments, numerous cracks develop at the ceramic–substrate interface, creating an interfacial thermal contact resistance. If T_w and q_h'' are maintained at the conditions associated with part (a), will T_1 increase, decrease, or remain the same? Similarly, will T_2 increase, decrease, or remain the same? In each case, justify your answer.

12.76 A small workpiece is placed in a large oven having isothermal walls at $T_f = 1000$ K with an emissivity of $\varepsilon_f = 0.5$. The workpiece experiences convection with moving air at 600 K and a

convection coefficient of $h = 60$ W/m² · K. The surface of the workpiece has a spectrally selective coating for which the emissivity has the following spectral distribution:

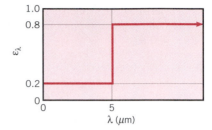

(a) Beginning with identification of all relevant processes for a control surface about the workpiece, perform an energy balance on the workpiece and determine its steady-state temperature, T_s.

(b) Plot the surface temperature T_s as a function of the convection coefficient for $10 \leq h \leq 120$ W/m² · K. On the same plot, show the surface temperature as a function of the convection coefficient for diffuse, gray surfaces with emissivities of 0.2 and 0.8.

12.77 A laser-materials-processing apparatus encloses a sample in the form of a disk of diameter $D = 25$ mm and thickness $w = 1$ mm. The sample has a diffuse surface for which the spectral distribution of the emissivity, $\epsilon_\lambda(\lambda)$, is prescribed. To reduce oxidation, an inert gas stream of temperature $T_\infty = 500$ K and convection coefficient $h = 50$ W/m² · K flows over the sample upper and lower surfaces. The apparatus enclosure is large with isothermal walls at $T_{enc} = 300$ K. To maintain the sample at a suitable operating temperature of $T_s = 2000$ K, a collimated laser beam with an operating wavelength of $\lambda = 0.5$ μm irradiates its upper surface.

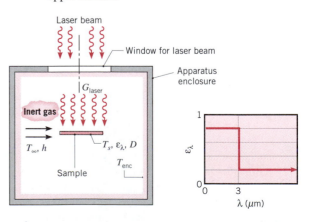

(a) Determine the total emissivity ε of the sample.

(b) Determine the total absorptivity α of the sample for irradiation from the enclosure walls.

(c) Perform an energy balance on the sample and determine the laser irradiation, G_{laser}, required to maintain the sample at $T_s = 2000$ K.

(d) Consider a *cool-down* process, when the laser and the inert gas flow are deactivated. Sketch the total emissivity as a function of the sample temperature, $T_s(t)$, during the process. Identify key features, including the emissivity for the final condition $(t \rightarrow \infty)$.

(e) Estimate the time to cool a sample from its operating condition at $T_s(0) = 2000$ K to a *safe-to-touch* temperature of $T_s(t) = 40°C$. Use the lumped capacitance method and include the effect of convection to the inert gas with $h = 50$ W/m$^2 \cdot$ K and $T_\infty = T_{enc} = 300$ K. The thermophysical properties of the sample material are $\rho = 3900$ kg/m^3, $c_p = 760$ J/kg $\cdot$ K, and $k = 45$ W/m $\cdot$ K.

12.78 A thermograph is a device responding to the radiant power from the scene, which reaches its radiation detector within the spectral region 9–12 μm. The thermograph provides an image of the scene, such as the side of a furnace, from which the surface temperature can be determined.

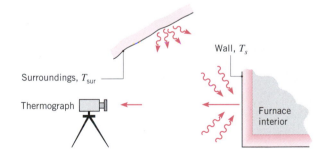

(a) For a black surface at 60°C, determine the emissive power for the spectral region 9–12 μm.

(b) Calculate the radiant power (W) received by the thermograph in the same range (9–12 μm) when viewing, in a normal direction, a small black wall area, 200 mm^2, at $T_s = 60°C$. The solid angle ω subtended by the aperture of the thermograph when viewed from the target is 0.001 sr.

(c) Determine the radiant power (W) received by the thermograph for the same wall area (200 mm^2) and solid angle (0.001 sr) when the wall is a gray, opaque, diffuse material at $T_s = 60°C$ with emissivity 0.7 and the surroundings are black at $T_{sur} = 23°C$.

12.79 A small disk 5 mm in diameter is positioned at the center of an isothermal, hemispherical enclosure. The disk is diffuse and gray with an emissivity of 0.7 and is maintained at 900 K. The hemispherical enclosure, maintained at 300 K, has a radius of 100 mm and an emissivity of 0.85.

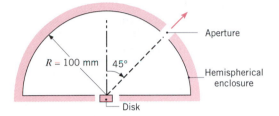

Calculate the radiant power leaving an aperture of diameter 2 mm located on the enclosure as shown.

12.80 A radiation thermometer is a radiometer calibrated to indicate the temperature of a blackbody. A steel billet having a diffuse, gray surface of emissivity 0.8 is heated in a furnace whose walls are at 1500 K. Estimate the temperature of the billet when the radiation thermometer viewing the billet through a small hole in the furnace indicates 1160 K.

12.81 Four diffuse surfaces having the spectral characteristics shown are at 300 K and are exposed to solar radiation.

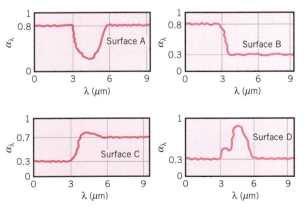

Which of the surfaces may be approximated as being gray?

12.82 A radiation detector has an aperture of area $A_d = 10^{-6}$ m^2 and is positioned at a distance of $r = 1$ m from a surface of area $A_s = 10^{-4}$ m^2. The angle formed by the normal to the detector and the surface normal is $\theta = 30°$.

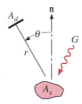

The surface is at 500 K and is opaque, diffuse, and gray with an emissivity of 0.7. If the surface irradiation is 1500 W/m^2, what is the rate at which the detector intercepts radiation from the surface?

12.83 A small anodized aluminum block at 35°C is heated in a large oven whose walls are diffuse and gray with $\varepsilon = 0.85$ and maintained at a uniform temperature of 175°C. The anodized coating is also diffuse and gray with $\varepsilon = 0.92$. A radiation detector views the block through a small opening in the oven and receives the radiant energy from a small area, referred to as the target, A_t, on the block. The target has a diameter of 3 mm, and the detector receives radiation within a solid angle 0.001 sr centered about the normal from the block.

(a) If the radiation detector views a small, but deep, hole drilled into the block, what is the total power (W) received by the detector?

(b) If the radiation detector now views an area on the block surface, what is the total power (W) received by the detector?

12.84 A cylinder of 30-mm diameter and 150-mm length is heated in a large furnace having walls at 1000 K, while air at 400 K is circulating at 3 m/s. Estimate the steady-state cylinder temperature under the following specified conditions.

(a) The cylinder is in cross flow, and its surface is diffuse and gray with an emissivity of 0.5.

(b) The cylinder is in cross flow, but its surface is spectrally selective with $\alpha_\lambda = 0.1$ for $\lambda \leq 3$ μm and $\alpha_\lambda = 0.5$ for $\lambda > 3$ μm.

(c) The cylinder surface is positioned such that the airflow is longitudinal and its surface is diffuse and gray.

(d) For the conditions of part (a), compute and plot the cylinder temperature as a function of the air velocity for $1 \leq V \leq 20$ m/s.

12.85 Consider the diffuse, gray opaque disk A_1, which has a diameter of 10 mm, an emissivity of 0.3, and is at a temperature of 400 K. Coaxial to the disk A_1, there is a black, ring-shaped disk A_2 at 1000 K having the dimensions shown in the sketch. The backside of A_2 is insulated and does not directly irradiate the cryogenically cooled detector disk A_3, which is of diameter 10 mm and is located 2 m from A_1.

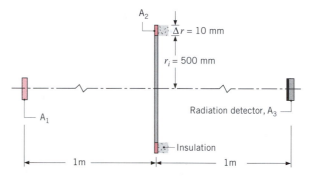

Calculate the rate at which radiation is incident on A_3 due to emission and reflection from A_1.

12.86 Consider a large furnace with opaque, diffuse, gray walls at 3000 K having an emissivity of 0.85. A small, diffuse, spectrally selective object in the furnace is maintained at 300 K.

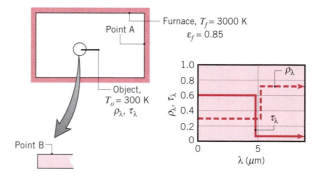

For the specified points on the furnace wall (A) and the object (B), indicate values for ε, α, E, G, and J.

12.87 A small, opaque sample of total hemispherical emissivity $\varepsilon_s = 0.1$ and surface area $A_s = 5$ mm^2 is positioned at the base of a hemispherical enclosure of radius $R = 100$ mm and emissivity $\varepsilon_h = 0.5$ and is viewed by a radiation detector through an opening in the enclosure. The detector has a surface area of $A_d = 2$ mm^2 and is at a distance of $L = 300$ mm from the sample.

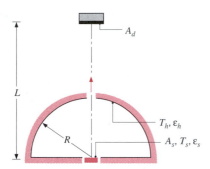

(a) If the sample and hemisphere are diffuse and gray and are maintained at temperatures of $T_s = 400$ K and $T_h = 273$ K, what is the total irradiation on the detector?

(b) Sketch (qualitatively) the spectral distribution of the intensity I_λ of radiation incident on the detector, identifying specific wavelengths at which maxima are likely to occur. Compute the spectral intensity at those wavelengths corresponding to the maxima.

12.88 A radiation detector having a sensitive area of $A_d = 4 \times 10^{-6}$ m^2 is configured to receive radiation from a target area of diameter $D_t = 40$ mm when located a distance of $L_t = 1$ m from the target. For the experimental apparatus shown in the sketch, we wish to determine the emitted radiation from a hot sample of diameter $D_s = 20$ mm. The temperature of the aluminum sample is $T_s = 700$ K and its emissivity is $\varepsilon_s = 0.1$. A ring-shaped cold shield is provided to minimize the effect of radiation from outside the sample area, but within the target area. The sample and the shield are diffuse emitters.

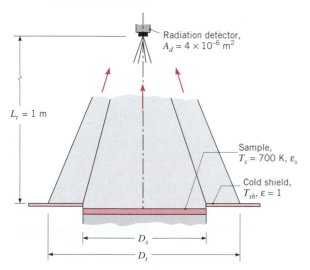

(a) Assuming the shield is black, at what temperature, T_{sh}, should the shield be maintained so that its emitted radiation is 1% of the total radiant power received by the detector?

(b) Subject to the parametric constraint that radiation emitted from the cold shield is 0.5, 1, or 1.5% of the total radiation received by the detector, plot the required cold shield temperature, T_{sh}, as a function of the sample emissivity for $0.05 \le \varepsilon_s \le 0.35$.

12.89 An infrared (IR) scanner (or thermograph) is a radiometer that provides an image of the target scene, indicating the apparent temperature of elements in the scene by a black–white brightness or blue–red color scale. Using a raster scanning approach, radiation originating from an element in the target scene is incident on the radiation detector, which provides a signal proportional to the incident radiant power. The signal sets the image brightness or color scale for the image pixel associated with that element. A scheme is proposed for field calibration of an infrared scanner having a radiation detector with a 3- to 5-μm spectral bandpass. A heated metal plate, which is maintained at 327°C and has four diffuse, gray coatings with different emissivities, is viewed by the IR scanner in surroundings for which $T_{sur} = 87$°C.

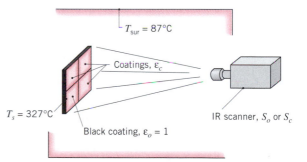

(a) Consider the scanner output when viewing the black coating, $\varepsilon_o = 1$. The radiation reaching the detector is proportional to the product of the blackbody emissive power (or emitted intensity) at the temperature of the surface and the band emission fraction corresponding to the IR scanner spectral bandpass. The proportionality constant is referred to as the responsivity, R (μV · m^2/W). Write an expression for the scanner output signal, S_o, in terms of R, the coating blackbody emissive power, and the appropriate band emission fraction. Assuming $R = 1 \ \mu V \cdot m^2/W$, evaluate $S_o \ (\mu V)$.

(b) Consider the scanner output when viewing one of the coatings for which the emissivity ε_c is less than unity. Radiation from the coating reaches the detector due to emission and the reflection of irradiation from the surroundings. Write an expression for the signal, S_c, in terms of R, the coating blackbody emissive power, the blackbody emissive power of the surroundings, the coating emissivity, and the appropriate band emission fractions. For the diffuse, gray coatings, the reflectivity is $\rho_c = 1 - \varepsilon_c$.

(c) Assuming $R = 1 \ \mu V \cdot m^2/W$, evaluate the scanner signals, $S_c \ (\mu V)$, when viewing panels with emissivities of 0.8, 0.5, and 0.2.

(d) The scanner is calibrated so that the signal S_o (with the black coating) will give a correct scale indication of $T_s = 327°C$. The signals from the other three coatings, S_c, are less than S_o. Hence the scanner will indicate an apparent (blackbody) temperature less than T_s. Estimate the temperatures indicated by the scanner for the three panels of part (c).

12.90 A small billet is heated within a large furnace having isothermal walls at $T_f = 750$ K with a diffuse, gray surface of emissivity $\varepsilon_f = 0.8$. The billet has a temperature of $T_t = 500$ K and a diffuse, gray surface with an emissivity of $\varepsilon_t = 0.9$. A radiation detector of area $A_d = 5.0 \times 10^{-4} \ m^2$ is positioned normal to and at a distance of 0.5 m from the billet. The detector receives radiation from a billet target area of $A_t = 3.0 \times 10^{-6} \ m^2$.

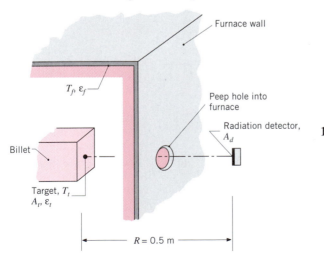

(a) Write a symbolic expression and give the numerical value for each of the following radiation parameters associated with the target surface (t): irradiation on the target, G_t; intensity of the reflected irradiation leaving the target, $I_{t,\text{ref}}$; emissive power of the target, E_t; intensity of the emitted radiation leaving the target, $I_{t,\text{emit}}$; and radiosity of the target, J_t.

(b) The radiation detector is only sensitive in the spectral region beyond 4 μm. Expressing your result in terms of the target reflected and emitted intensities, $I_{t,\text{ref}}$ and $I_{t,\text{emit}}$, respectively, as well as other geometric and radiation parameters, obtain an equation for the rate at which radiation leaves the target in the spectral region $\lambda \geq 4 \ \mu$m and is intercepted by the radiation detector, A_d. Substitute numerical values to obtain $q_{t \rightarrow d}$ for $\lambda \geq 4 \ \mu$m.

12.91 Consider a material that is gray, but directionally selective with $\alpha_\theta(\theta, \phi) = 0.5(1 - \cos \phi)$. Determine the hemispherical absorptivity α when collimated solar flux irradiates the surface of the material in the direction $\theta = 45°$ and $\phi = 0°$. Determine the hemispherical emissivity ε of the material.

12.92 A temperature sensor imbedded in the tip of a small tube having a diffuse, gray surface with an emissivity of 0.8 is centrally positioned within a large air-conditioned room whose walls and air temperature are 30 and 20°C, respectively.

(a) What temperature will the sensor indicate if the convection coefficient between the sensor tube and the air is 5 W/m² · K?

(b) What would be the effect of using a fan to induce airflow over the tube? Plot the sensor temperature as a function of the convection coefficient for $2 \leq h \leq 25$ W/m² · K and values of $\varepsilon = 0.2, 0.5$, and 0.8.

12.93 An instrumentation transmitter pod is a box containing electronic circuitry and a power supply for sending sensor signals to a base receiver for recording. Such a pod is placed on a conveyor system, which passes through a large vacuum brazing furnace as shown in the sketch. The exposed surfaces of the pod have a special diffuse, opaque coating with spectral emissivity as shown.

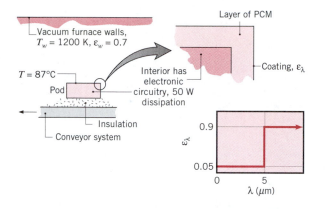

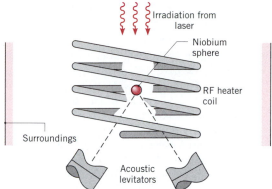

To stabilize the temperature of the pod and prevent overheating of the electronics, the inner surface of the pod is surrounded by a layer of a phase-change material (PCM) having a fusion temperature of 87°C and a heat of fusion of 25 kJ/kg. The pod has an exposed surface area of 0.040 m^2 and the mass of the PCM is 1.6 kg. Furthermore, it is known that the power dissipated by the electronics is 50 W. Consider the situation when the pod enters the furnace at a uniform temperature of 87°C and all the PCM is in the solid state. How long will it take before all the PCM changes to the liquid state?

12.94   To simulate materials processing under the microgravity conditions of space, a niobium sphere of diameter 3 mm is levitated by an acoustical technique in a vacuum chamber. Initially the sphere is at 300 K and is suddenly irradiated with a laser providing an irradiation of 10 W/mm^2 to raise its temperature as quickly as possible to its melting point (2741 K). When the melting point is reached, the laser is switched off and a radio frequency (RF) heater is energized, causing uniform internal volumetric generation $\dot{q}$ to occur within the sphere. Assume that the niobium sphere is isothermal and diffuse and gray with an emissivity of 0.6, and that the chamber walls are at 300 K.

(a) How much time is required to reach the melting point? *Hint:* See Section 5.3 and modify the lumped capacitance analysis accordingly.

(b) How much power must be provided by the RF heater to maintain the sphere at its melting point?

(c) Is the spacewise isothermal assumption realistic for the conditions of parts (a) and (b)?

12.95   A spherical niobium droplet of diameter D = 3 mm is levitated by an acoustical technique in a vacuum chamber having walls that are diffuse and gray with an emissivity of ε_w = 0.8 and a temperature of T_w = 300 K. The niobium surface is diffuse and has the prescribed spectral emissivity distribution.

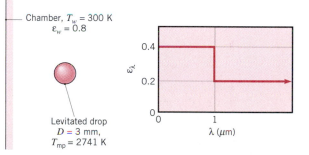

Two heating methods for maintaining the drop at its melting temperature, T_{mp} = 2741 K, are to be investigated.

(a) The effect of applying a radio frequency (RF) field to the drop is to create a uniform internal generation rate, $\dot{q}$ (W/m^3), within the drop. Calculate the value of $\dot{q}$ required to maintain the drop at its melting temperature.

(b) A laser beam, having a diameter larger than that of the drop and operating at 10.6 μm, irradiates the drop. Determine the irradiation, G_{laser} (W/mm^2), required to maintain the drop at its melting temperature. What irradiation would be required if the laser operating wavelength were 0.632 μm?

(c) Estimate the time it would take for the drop to cool to 400 K if the RF field or laser heating were terminated.

12.96 To simulate solidification under the microgravity conditions in space, a 3-mm diameter molten sphere of niobium falls through an evacuated vertical tube of 300-mm diameter and 105-m length whose walls are maintained at 300 K. At one station of this *drop-tube* experimental facility, the sphere has a velocity of 3.5 m/s and its temperature is 3000 K. A fiber optic cable connected to a radiation detector is on the tube wall and has a diameter of 200 μm, which receives radiation within a solid angle defined by the plane angle $\theta_o = 23°$. Assume the niobium is diffuse and gray with an emissivity of 0.6.

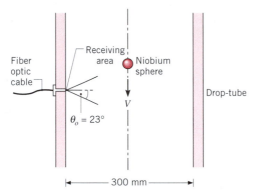

(a) Estimate the average radiant power (W) emitted by the sphere, which is received by the fiber optic as the sphere drops along the centerline of the tube. What is the average radiant power received when the sphere drops very close to the far wall of the drop tube?

(b) Recognizing that the sphere experiences radiative cooling while falling through the drop tube, is it reasonable to assume that the sphere is isothermal during the period of time its radiation is received by the fiber optic?

(c) Estimate the radiant power emitted by the drop-tube wall. Compare this *background* radiation to that from the sphere. Is it important to control carefully the temperature of the drop-tube wall?

12.97 A thin-walled plate separates the interior of a large furnace from surroundings at 300 K. The plate is fabricated from a ceramic material for which diffuse surface behavior may be assumed and the exterior surface is air cooled. With the furnace operating at 2400 K, convection at the interior surface may be neglected.

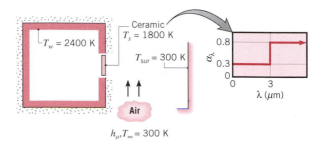

(a) If the temperature of the ceramic plate is not to exceed 1800 K, what is the minimum value of the outside convection coefficient, h_o, that must be maintained by the air-cooling system?

(b) Compute and plot the plate temperature as a function of h_o for $50 \le h_o \le 250$ W/m² · K.

12.98 A thin coating, which is applied to long, cylindrical copper rods of 10-mm diameter, is cured by placing the rods horizontally in a large furnace whose walls are maintained at 1300 K. The furnace is filled with nitrogen gas, which is also at 1300 K and at a pressure of 1 atm. The coating is diffuse, and its spectral emissivity has the distribution shown.

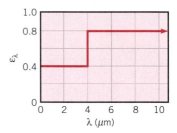

(a) What are the emissivity and absorptivity of the coated rods when their temperature is 300 K?

(b) What is the initial rate of change of their temperature?

(c) What are the emissivity and absorptivity of the coated rods when they reach a steady-state temperature?

(d) Estimate the time required for the rods to reach 1000 K.

12.99 A large combination convection–radiation oven is used to heat treat a small copper cylindrical product of diameter 25 mm and length 0.2 m. The oven walls are at a uniform temperature of 1000 K, and hot air at 750 K is in cross flow over the cylinder with a velocity of 5 m/s. The cylinder surface is opaque and diffuse with the spectral emissivity shown.

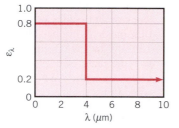

(a) Determine the net rate of heat transfer to the cylinder when it is first placed in the oven at 300 K.

(b) What is the steady-state temperature of the cylinder?

(c) How long will it take for the cylinder to reach a temperature that is within 50°C of its steady-state value?

Environmental Radiation

12.100 Solar irradiation of 1100 W/m² is incident on a large, flat, horizontal metal roof on a day when the wind blowing over the roof causes a convection heat transfer coefficient of 25 W/m² · K. The outside air temperature is 27°C, the metal surface absorptivity for incident solar radiation is 0.60, the metal surface emissivity is 0.20, and the roof is well insulated from below.

(a) Estimate the roof temperature under steady-state conditions.

(b) Explore the effect of changes in the absorptivity, emissivity, and convection coefficient on the steady-state temperature.

12.101 A deep cavity of 50-mm diameter approximates a blackbody and is maintained at 250°C while exposed to solar irradiation of 800 W/m² and surroundings and ambient air at 25°C. A thin window of spectral transmissivity and reflectivity 0.9 and 0, respectively, for the spectral range 0.2 to 4 μm is placed over the cavity opening. In the spectral range beyond 4 μm, the window behaves as an opaque, diffuse, gray body of emissivity 0.95. Assuming that the convection coefficient on the upper surface of the window is 10 W/m² · K, determine the temperature of the window and the power required to maintain the cavity at 250°C.

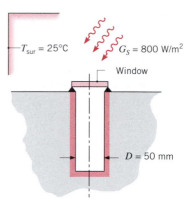

12.102 Consider the evacuated tube solar collector described in Problem 1.53d of Chapter 1. In the interest of maximizing collector efficiency, what spectral radiative characteristics are desired for the outer tube and for the inner tube?

12.103 Solar flux of 900 W/m² is incident on the top side of a plate whose surface has a solar absorptivity of 0.9 and an emissivity of 0.1. The air and surroundings are at 17°C and the convection heat transfer coefficient between the plate and air is 20 W/m² · K. Assuming that the bottom side of the plate is insulated, determine the steady-state temperature of the plate.

12.104 Two plates, one with a black painted surface and the other with a special coating (chemically oxidized copper) are in earth orbit and are exposed to solar radiation. The solar rays make an angle of 30° with the normal to the plate. Estimate the equilibrium temperature of each plate assuming they are diffuse and that the solar flux is 1353 W/m². The spectral absorptivity of the black painted surface can be approximated by $\alpha_\lambda = 0.95$ for $0 \le \lambda \le \infty$ and that of the special coating by $\alpha_\lambda = 0.95$ for $0 \le \lambda < 3$ μm and $\alpha_\lambda = 0.05$ for $\lambda \ge 3$ μm.

12.105 The spectral, hemispherical emissivity distributions for two diffuse panels to be used in a spacecraft are as shown.

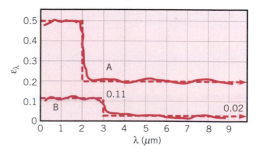

Assuming that the back sides of the panels are insulated and that the panels are oriented normal to the solar flux at 1353 W/m², determine which panel has the higher steady-state temperature.

12.106 An annular fin of thickness t is used as a radiator to dissipate heat for a space power system. The fin is insulated on the bottom and may be exposed to solar irradiation G_S. The fin is coated with a diffuse, spectrally selective material whose spectral reflectivity is specified.

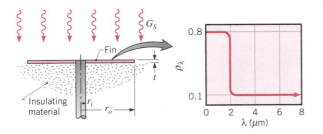

Heat is conducted to the fin through a solid rod of radius r_i, and the exposed upper surface of the fin radiates to free space, which is essentially at absolute zero temperature.

(a) If conduction through the rod maintains a fin base temperature of $T(r_i) = T_b = 400$ K and the fin efficiency is 100%, what is the rate of heat dissipation for a fin of radius $r_o = 0.5$ m? Consider two cases, one for which the radiator is exposed to the sun with $G_S = 1000$ W/m² and the other with no exposure ($G_S = 0$).

(b) In practice, the fin efficiency will be less than 100% and its temperature will decrease with increasing radius. Beginning with an appropriate control volume, derive the differential equation that determines the steady-state, radial temperature distribution in the fin. Specify appropriate boundary conditions.

12.107 The directional absorptivity of a gray surface varies with θ as follows.

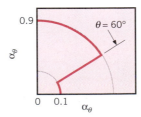

(a) What is the ratio of the normal absorptivity α_n to the hemispherical emissivity of the surface?

(b) Consider a plate with these surface characteristics on both sides in earth orbit. If the solar flux incident on one side of the plate is $q_S'' = 1353$ W/m², what equilibrium temperature will the plate assume if it is oriented normal to the sun's rays? What temperature will it assume if it is oriented at 75° to the sun's rays?

12.108 Two special coatings are available for application to an absorber plate installed below the cover glass described in Example 12.8. Each coating is diffuse and is characterized by the spectral distributions shown.

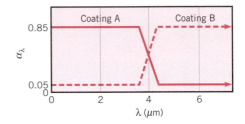

Which coating would you select for the absorber plate? Explain briefly. For the selected coating, what is the rate at which radiation is absorbed per unit area of the absorber plate if the total solar irradiation at the cover glass is $G_S = 1000$ W/m²?

12.109 A flat plate whose surface is gray but has the total, directional emissivity ε_θ shown, is in earth orbit, where the solar flux is 1353 W/m².

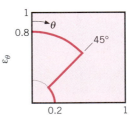

Determine the equilibrium temperatures for the plate when it is oriented normal to the solar flux and at 60° to the solar flux.

12.110 Consider an opaque, gray surface whose directional absorptivity is 0.8 for $0 \le \theta \le 60°$ and 0.1 for $\theta > 60°$. The surface is horizontal and exposed to solar irradiation comprised of direct and diffuse components.

(a) What is the surface absorptivity to direct solar radiation that is incident at an angle of 45° from the normal? What is the absorptivity to diffuse irradiation?

(b) Neglecting convection heat transfer between the surface and the surrounding air, what would be the equilibrium temperature of the surface if the direct and diffuse components of the irradiation were 600 and 100 W/m², respectively? The back side of the surface is insulated.

12.111 A contractor must select a roof covering material from the two diffuse, opaque coatings with $\alpha_\lambda(\lambda)$ as shown. Which of the two coatings would result in a lower roof temperature? Which is preferred for summer use? For winter use? Sketch the spectral distribution of α_λ that would be ideal for summer use. For winter use.

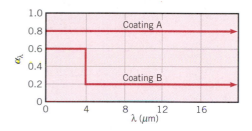

12.112 A radiator on a proposed satellite solar power station must dissipate heat being generated within the satellite by radiating it into space. The radiator surface has a solar absorptivity of 0.5 and an emissivity of 0.95. What is the equilibrium surface temperature when the solar irradiation is 1000 W/m² and the required heat dissipation is 1500 W/m²?

12.113 A diffuse, gray plate of emissivity 0.6 insulated from the ground is exposed to solar irradiation of 1000 W/m². The effective sky temperature is −40°C and the heat transfer coefficient between the plate and the ambient air at 25°C is 6 W/m² · K. The temperature of the plate is 65°C at this particular time.

(a) What is the radiosity of the plate?

(b) What is the net heat flux between the plate and the environment?

(c) In what manner will the temperature of the plate change with respect to time?

12.114 It is not uncommon for the night sky temperature in desert regions to drop to −40°C. If the ambient air temperature is 20°C and the convection coeffi-

cient for still air conditions is approximately 5 W/m² · K, can a shallow pan of water freeze?

12.115 A spherical satellite of diameter D is in orbit about the earth and is coated with a diffuse material for which the spectral absorptivity is $\alpha_\lambda = 0.6$ for $\lambda \leq 3\ \mu m$ and $\alpha_\lambda = 0.3$ for $\lambda > 3\ \mu m$. When it is on the "dark" side of the earth, the satellite sees irradiation from the earth's surface only. The irradiation may be assumed to be incident as parallel rays, and its magnitude is $G_E = 340$ W/m². On the "bright" side of the earth the satellite sees the earth irradiation G_E plus the solar irradiation $G_S = 1353$ W/m². The spectral distribution of radiation from the earth may be approximated as that of a blackbody at 280 K, and the temperature of the satellite may be assumed to remain below 500 K.

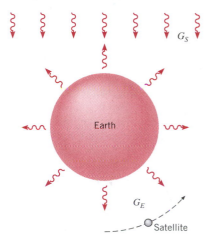

What is the steady-state temperature of the satellite when it is on the dark side of the earth and when it is on the bright side?

12.116 A spherical capsule of 3-m radius is fired from a space platform in earth orbit, such that it travels toward the center of the sun at 16,000 km/s. Assume that the capsule is a lumped capacitance body with a density–specific heat product of 4×10^6 J/m³ · K and that its surface is black.

(a) Derive a differential equation for predicting the capsule temperature as a function of time. Solve this equation to obtain the temperature as a function of time in terms of capsule parameters and its initial temperature T_i.

(b) If the capsule begins its journey at 20°C, predict the position of the capsule relative to the sun at which its destruction temperature, 150°C, is reached.

12.117 A thin sheet of glass is used on the roof of a greenhouse and is irradiated as shown.

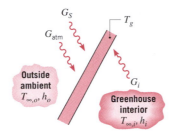

The irradiation comprises the total solar flux G_S, the flux G_{atm} due to atmospheric emission (sky radiation), and the flux G_i due to emission from interior surfaces. The fluxes G_{atm} and G_i are concentrated in the far IR region ($\lambda \gtrsim 8$ μm). The glass may also exchange energy by convection with the outside and inside atmospheres. The glass may be assumed to be totally transparent for $\lambda < 1$ μm ($\tau_\lambda = 1.0$ for $\lambda < 1$ μm) and opaque, with $\alpha_\lambda = 1.0$ for $\lambda \geq 1$ μm.

(a) Assuming steady-state conditions, with all radiative fluxes uniformly distributed over the surfaces and the glass characterized by a uniform temperature T_g, write an appropriate energy balance for a unit area of the glass.

(b) For $T_g = 27°C$, $h_i = 10$ W/m$^2 \cdot$ K, $G_S = 1100$ W/m^2, $T_{\infty,o} = 24°C$, $h_o = 55$ W/m$^2 \cdot$ K, $G_{atm} = 250$ W/m^2, and $G_i = 440$ W/m^2, calculate the temperature of the greenhouse ambient air, $T_{\infty,i}$.

12.118 Components of an electronic package in an orbiting satellite are mounted in a compartment that is well insulated on all but one of its sides. The uninsulated side consists of an isothermal copper plate whose outer surface is exposed to the vacuum of outer space and whose inner surface is attached to the components. The plate dimensions are 1 m by 1 m on a side.

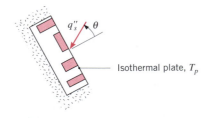

The exposed surface of the plate, which is opaque and diffuse, has a spectral, hemispherical absorp-

tivity of $\alpha_\lambda = 0.2$ for $\lambda \leq 2$ μm and $\alpha_\lambda = 0.8$ for $\lambda > 2$ μm. Consider steady-state conditions for which the plate is exposed to a solar flux of $q_S'' = 1350$ W/m^2, which is incident at an angle of $\theta = 30°$ relative to the surface normal. If the plate temperature is $T_p = 500$ K, how much power is being dissipated by the components?

12.119 Consider a diffuse, opaque 0.25-m^2 plate that is exposed to quiescent air at 27°C and to solar irradiation of 1000 W/m^2 on its top surface and that is well insulated on its bottom surface. The effective sky temperature is $-40°C$ and the spectral, hemispherical emissivity ε_λ is as shown below.

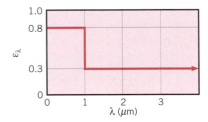

What is the net heat transfer to the plate if its temperature is 127°C?

12.120 A solar furnace consists of an evacuated chamber with transparent windows, through which concentrated solar radiation is passed. Concentration may be achieved by mounting the furnace at the focal point of a large curved reflector that tracks radiation incident directly from the sun. The furnace may be used to evaluate the behavior of materials at elevated temperatures, and we wish to design an experiment to assess the durability of a diffuse, spectrally selective coating for which $\alpha_\lambda = 0.95$ in the range $\lambda \leq 4.5$ μm and $\alpha_\lambda = 0.03$ for $\lambda > 4.5$ μm. The coating is applied to a plate that is suspended in the furnace.

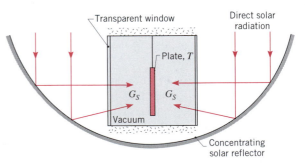

(a) If the experiment is to be operated at a steady-state plate temperature of $T = 2000$ K, how much solar irradiation G_S must be provided? The irradiation may be assumed to be uniformly distributed over the plate surface, and other sources of incident radiation may be neglected.

(b) The solar irradiation may be *tuned* to allow operation over a range of plate temperatures. Compute and plot G_S as a function of temperature for $500 \leq T \leq 3000$ K. Plot the corresponding values of α and ε as a function of T for the designated range.

12.121 In the central receiver concept of solar energy collection, a large number of heliostats (reflectors) provide a concentrated solar flux of $q_S'' = 80,000$ W/m^2 to the receiver, which is positioned at the top of a tower.

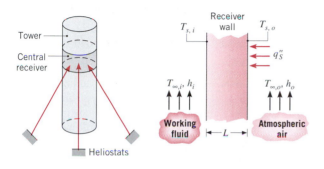

The receiver wall is exposed to the solar flux at its outer surface and to atmospheric air for which $T_{\infty,o} = 300$ K and $h_o = 25$ W/m$^2 \cdot$ K. The outer surface is opaque and diffuse, with a spectral absorptivity of $\alpha_\lambda = 0.9$ for $\lambda < 3$ μm and $\alpha_\lambda = 0.2$ for $\lambda > 3$ μm. The inner surface is exposed to a working fluid (a pressurized liquid) for which $T_{\infty,i} = 700$ K and $h_i = 1000$ W/m$^2 \cdot$ K. The outer surface is also exposed to surroundings for which $T_{sur} = 300$ K. If the wall is fabricated from a high-temperature material for which $k = 15$ W/m $\cdot$ K, what is the minimum thickness L needed to ensure that the outer surface temperature does not exceed $T_{s,o} = 1000$ K? What is the collection efficiency associated with this thickness?

12.122 Consider the central receiver of Problem 12.121 to be a cylindrical shell of outer diameter $D = 7$ m and length $L = 12$ m. The outer surface is opaque and diffuse, with a spectral absorptivity of $\alpha_\lambda = 0.9$ for $\lambda < 3$ μm and $\alpha_\lambda = 0.2$ for $\lambda > 3$ μm. The surface is exposed to *quiescent* ambient air for which $T_\infty = 300$ K.

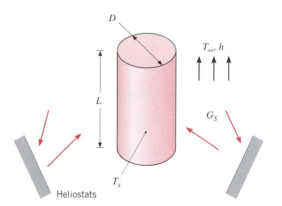

(a) Consider representative operating conditions for which solar irradiation at $G_S = 80,000$ W/m^2 is uniformly distributed over the receiver surface and the surface temperature is $T_s = 800$ K. Determine the rate at which energy is collected by the receiver and the corresponding collector efficiency.

(b) The surface temperature is affected by conditions internal to the receiver. For $G_S = 80,000$ W/m^2, compute and plot the rate of energy collection and the collector efficiency for $600 \leq T_s \leq 1000$ K.

12.123 The flat roof on the refrigeration compartment of a food delivery truck is of length $L = 5$ m and width $W = 2$ m. It is fabricated from thin sheet metal to which a fiberboard insulating material of thickness $t = 25$ mm and thermal conductivity $k = 0.05$ W/m $\cdot$ K is bonded. During normal operation, the truck moves at a velocity of $V = 30$ m/s in air at $T_\infty = 27°$C, with a roof top solar irradiation of $G_S = 900$ W/m^2 and with the interior surface temperature maintained at $T_{s,i} = -13°$C.

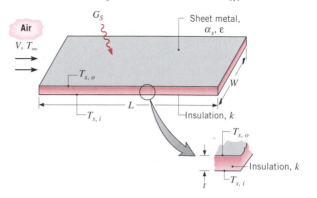

(a) The owner has the option of selecting a roof coating from one of the three paints listed in Table A.12 (Parsons Black, Acrylic White, or

Zinc Oxide White). Which should be chosen and why?

(b) For the preferred paint of part (a), determine the steady-state value of the outer surface temperature $T_{s,o}$. The boundary layer is tripped at the leading edge of the roof, and turbulent flow may be assumed to exist over the entire roof. Properties of the air may be taken to be $\nu = 15 \times 10^{-6}$ m^2/s, $k = 0.026$ W/m · K, and $Pr = 0.71$.

(c) What is the load (W) imposed on the refrigeration system by heat transfer through the roof?

(d) Explore the effect of the truck velocity on the outer surface temperature and the heat load.

12.124 Growers use giant fans to prevent grapes from freezing when the effective sky temperature is low. The grape, which may be viewed as a thin skin of negligible thermal resistance enclosing a volume of sugar water, is exposed to ambient air and is irradiated from the sky above and ground below. Assume the grape to be an isothermal sphere of 15-mm diameter, and assume uniform blackbody irradiation over its top and bottom hemispheres due to emission from the sky and the earth, respectively.

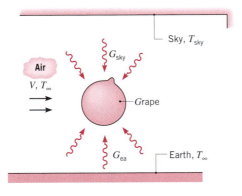

(a) Derive an expression for the rate of change of the grape temperature. Express your result in terms of a convection coefficient and appropriate temperatures and radiative quantities.

(b) Under conditions for which $T_{sky} = 235$ K, $T_\infty = 273$ K, and the fan is off ($V = 0$), determine whether the grapes will freeze. To a good approximation, the skin emissivity is 1 and the grape thermophysical properties are those of sugarless water. However, because of the sugar content, the grape freezes at -5°C.

(c) With all conditions remaining the same, except that the fans are now operating with $V = 1$ m/s, will the grapes freeze?

12.125 A circular metal disk having a diameter of 0.4 m is placed firmly against the ground in a barren horizontal region where the earth is at a temperature of 280 K. The effective sky temperature is also 280 K. The disk is exposed to quiescent ambient air at 300 K and direct solar irradiation of 745 W/m^2. The surface of the disk is diffuse with $\varepsilon_\lambda = 0.9$ for $0 < \lambda < 1$ μm and $\varepsilon_\lambda = 0.2$ for $\lambda > 1$ μm. After some time has elapsed, the disk achieves a uniform, steady-state temperature. The thermal conductivity of the soil is 0.52 W/m · K.

(a) Determine the fraction of the incident solar irradiation that is absorbed.

(b) What is the emissivity of the disk surface?

(c) For a steady-state disk temperature of 340 K, employ a suitable correlation to determine the average free convection heat transfer coefficient at the upper surface of the disk.

(d) Show that a disk temperature of 340 K does indeed yield a steady-state condition for the disk.

12.126 The exposed surface of a power amplifier for an earth satellite receiver of area 130 mm by 130 mm has a diffuse, gray, opaque coating with an emissivity of 0.5. For typical amplifier operating conditions, the surface temperature is 58°C under the following environmental conditions: air temperature, $T_\infty = 27$°C; sky temperature, $T_{sky} = -20$°C; convection coefficient, $h = 15$ W/m^2 · K; and solar irradiation, $G_S = 800$ W/m^2.

(a) For the above conditions, determine the electrical power being generated within the amplifier.

(b) It is desired to reduce the surface temperature by applying one of the diffuse coatings (A, B, C) shown below.

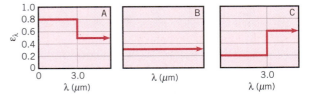

Which coating will result in the coolest surface temperature for the same amplifier operating and environmental conditions?

12.127 Consider a thin opaque, horizontal plate with an electrical heater on its backside. The front side is exposed to ambient air that is at 20°C and provides a convection heat transfer coefficient of 10

W/m² · K, solar irradiation of 600 W/m², and an effective sky temperature of −40°C.

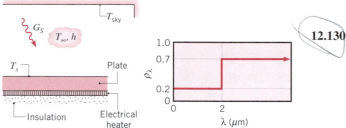

What is the electrical power (W/m²) required to maintain the plate surface temperature at T_s = 60°C if the plate is diffuse and has the designated spectral, hemispherical reflectivity?

Heat and Mass Transfer

12.128 It is known that on clear nights a thin layer of water on the ground will freeze before the air temperature drops below 0°C. Consider such a layer of water on a clear night for which the effective sky temperature is −30°C and the convection heat transfer coefficient due to wind motion is h = 25 W/m² · K. The water may be assumed to have an emissivity of 1.0 and to be insulated from the ground as far as conduction is concerned. Neglecting evaporation, determine the lowest temperature that the air can have without the water freezing? Accounting now for the effect of evaporation, what is the lowest temperature that the air can have without the water freezing? Assume the air to be dry.

12.129 A shallow layer of water is exposed to the natural environment as shown.

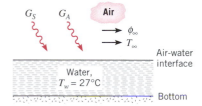

Consider conditions for which the solar and atmospheric irradiations are G_S = 600 W/m² and G_A = 300 W/m², respectively, and the air temperature and relative humidity are T_∞ = 27°C and ϕ_∞ = 0.50, respectively. The reflectivities of the water surface to the solar and atmospheric irradiation are ρ_S = 0.3 and ρ_A = 0, respectively, while

the surface emissivity is ε = 0.97. The convection heat transfer coefficient at the air–water interface is h = 25 W/m² · K. If the water is at 27°C, will this temperature increase or decrease with time?

12.130 A roof-cooling system, which operates by maintaining a thin film of water on the roof surface, may be used to reduce air-conditioning costs or to maintain a cooler environment in nonconditioned buildings. To determine the effectiveness of such a system, consider a sheet metal roof for which the solar absorptivity α_S is 0.50 and the hemispherical emissivity ε is 0.3. Representative conditions correspond to a surface convection coefficient h of 20 W/m² · K, a solar irradiation G_S of 700 W/m², a sky temperature of −10°C, an atmospheric temperature of 30°C, and a relative humidity of 65%. The roof may be assumed to be well insulated from below. Determine the roof surface temperature without the water film. Assuming the film and roof surface temperatures to be equal, determine the surface temperature with the film. The solar absorptivity and the hemispherical emissivity of the film–surface combination are α_S = 0.8 and ε = 0.9, respectively.

12.131 Our students perform a laboratory experiment to determine mass transfer from a wet paper towel experiencing forced convection and irradiation from radiant lamps. For the values of T_∞ and T_{wb} prescribed on the sketch, the towel temperature was found to be T_s = 310 K. In addition, flat-plate correlations yielded average heat and mass transfer convection coefficients of $\bar{h}$ = 28.7 W/m² · K and $\bar{h}_m$ = 0.027 m/s, respectively. The towel has dimensions of 92.5 mm × 92.5 mm and is diffuse and gray with an emissivity of 0.96.

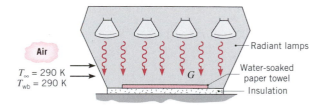

(a) From the foregoing results, determine the vapor densities, $\rho_{A,s}$ and $\rho_{A,\infty}$, the evaporation rate, n_A (kg/s), and the net rate of radiation transfer to the towel, q_{rad} (W).

(b) Using results from part (a) and assuming that the irradiation G is uniform over the towel, determine the emissive power E, the irradiation G, and the radiosity J.

CHAPTER

Radiation Exchange Between Surfaces

*H*aving thus far restricted our attention to radiative processes that occur at a *single surface,* we now consider the problem of radiative exchange between two or more surfaces. This exchange depends strongly on the surface geometries and orientations, as well as on their radiative properties and temperatures. We assume that the surfaces are separated by a *nonparticipating medium.* Since such a medium neither emits, absorbs, nor scatters, it has no effect on the transfer of radiation between surfaces. A vacuum meets these requirements exactly, and most gases meet them to an excellent approximation.

We initially focus on the geometrical features of the radiation exchange problem by developing the notion of a *view factor.* We then consider radiation exchange between *black surfaces* and follow with a treatment of exchange between *diffuse, gray surfaces.* We also consider the problem of radiation transfer in an *enclosure,* a term that describes the region enveloped by a collection of surfaces.

13.1
The View Factor

To compute radiation exchange between any two surfaces, we must first introduce the concept of a *view factor* (also called a *configuration* or *shape factor*).

13.1.1   The View Factor Integral

The view factor F_{ij} is defined as the *fraction of the radiation leaving surface i that is intercepted by surface j.* To develop a general expression for F_{ij}, we consider the arbitrarily oriented surfaces A_i and A_j of Figure 13.1. Elemental areas on each surface, dA_i and dA_j, are connected by a line of length R, which forms the polar angles θ_i and θ_j, respectively, with the surface normals $\mathbf{n}_i$ and $\mathbf{n}_j$. The values of R, θ_i, and θ_j vary with the position of the elemental areas on A_i and A_j.

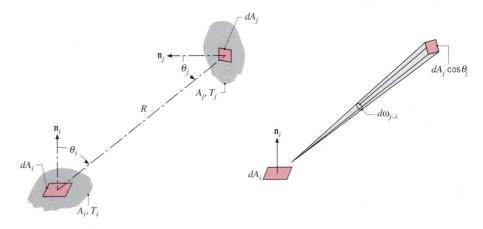

FIGURE 13.1   View factor associated with radiation exchange between elemental surfaces of area dA_i and dA_j.

From the definition of the radiation intensity, Section 12.2.1, and Equation 12.5, the rate at which radiation *leaves* dA_i and is *intercepted* by dA_j may be expressed as

$$dq_{i \to j} = I_i \cos \theta_i \, dA_i \, d\omega_{j-i}$$

where I_i is the intensity of the radiation leaving surface i and $d\omega_{j-i}$ is the solid angle subtended by dA_j when viewed from dA_i. With $d\omega_{j-i} = (\cos \theta_j \, dA_j)/R^2$ from Equation 12.2, it follows that

$$dq_{i \to j} = I_i \frac{\cos \theta_i \cos \theta_j}{R^2} \, dA_i \, dA_j$$

Assuming that surface i *emits* and *reflects diffusely* and substituting from Equation 12.24, we then obtain

$$dq_{i \to j} = J_i \frac{\cos \theta_i \cos \theta_j}{\pi R^2} \, dA_i \, dA_j$$

The total rate at which radiation leaves surface i and is intercepted by j may then be obtained by integrating over the two surfaces. That is,

$$q_{i \to j} = J_i \int_{A_i} \int_{A_j} \frac{\cos \theta_i \cos \theta_j}{\pi R^2} \, dA_i \, dA_j$$

where it is assumed that the radiosity J_i is uniform over the surface A_i. From the definition of the view factor as the fraction of the radiation that leaves A_i and is intercepted by A_j,

$$F_{ij} = \frac{q_{i \to j}}{A_i J_i}$$

it follows that

$$F_{ij} = \frac{1}{A_i} \int_{A_i} \int_{A_j} \frac{\cos \theta_i \cos \theta_j}{\pi R^2} \, dA_i \, dA_j \tag{13.1}$$

Similarly, the view factor F_{ji} is defined as the fraction of the radiation that leaves A_j and is intercepted by A_i. The same development then yields

$$F_{ji} = \frac{1}{A_j} \int_{A_i} \int_{A_j} \frac{\cos \theta_i \cos \theta_j}{\pi R^2} \, dA_i \, dA_j \tag{13.2}$$

Either Equation 13.1 or 13.2 may be used to determine the view factor associated with any two surfaces that are *diffuse emitters* and *reflectors* and have *uniform radiosity*.

13.1.2 View Factor Relations

An important view factor relation is suggested by Equations 13.1 and 13.2. In particular, equating the integrals appearing in these equations, it follows that

$$A_i F_{ij} = A_j F_{ji} \tag{13.3}$$

This expression, termed the *reciprocity relation*, is useful in determining one view factor from knowledge of the other.

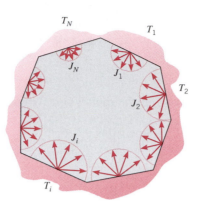

FIGURE 13.2
Radiation exchange in an enclosure.

Another important view factor relation pertains to the surfaces of an *enclosure* (Figure 13.2). From the definition of the view factor, the *summation rule*

$$\sum_{j=1}^{N} F_{ij} = 1 \qquad (13.4)$$

may be applied to each of the N surfaces in the enclosure. This rule follows from the conservation requirement that all radiation leaving surface i must be intercepted by the enclosure surfaces. The term F_{ii} appearing in this summation represents the fraction of the radiation that leaves surface i and is directly intercepted by i. If the surface is concave, it *sees itself* and F_{ii} is nonzero. However, for a plane or convex surface, $F_{ii} = 0$.

To calculate radiation exchange in an enclosure of N surfaces, a total of N^2 view factors is needed. This requirement becomes evident when the view factors are arranged in the matrix form.

$$\begin{bmatrix} F_{11} & F_{12} & \cdots & F_{1N} \\ F_{21} & F_{22} & \cdots & F_{2N} \\ \bullet & \bullet & & \bullet \\ \bullet & \bullet & & \bullet \\ \bullet & \bullet & & \bullet \\ F_{N1} & F_{N2} & \cdots & F_{NN} \end{bmatrix}$$

However, all the view factors need not be calculated *directly*. A total of N view factors may be obtained from the N equations associated with application of the

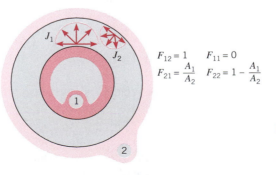

$$F_{12} = 1 \qquad F_{11} = 0$$
$$F_{21} = \frac{A_1}{A_2} \qquad F_{22} = 1 - \frac{A_1}{A_2}$$

FIGURE 13.3
View factors for the enclosure formed by two spheres.

summation rule, Equation 13.4, to each of the surfaces in the enclosure. In addition, $N(N - 1)/2$ view factors may be obtained from the $N(N - 1)/2$ applications of the reciprocity relation, Equation 13.3, which are possible for the enclosure. Accordingly, only $[N^2 - N - N(N - 1)/2] = N(N - 1)/2$ view factors need be determined directly. For example, in a three-surface enclosure this re-

TABLE 13.1 View Factors for Two-Dimensional Geometries [4]

Geometry	Relation

Parallel Plates with Midlines Connected by Perpendicular

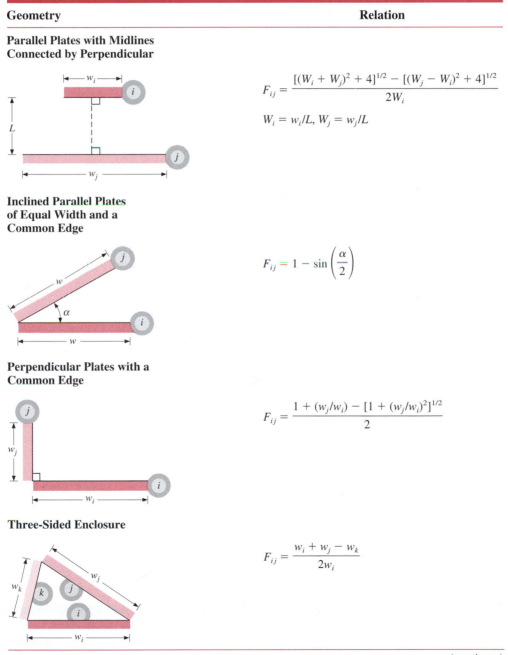

$$F_{ij} = \frac{[(W_i + W_j)^2 + 4]^{1/2} - [(W_j - W_i)^2 + 4]^{1/2}}{2W_i}$$

$$W_i = w_i/L, \quad W_j = w_j/L$$

Inclined Parallel Plates of Equal Width and a Common Edge

$$F_{ij} = 1 - \sin\left(\frac{\alpha}{2}\right)$$

Perpendicular Plates with a Common Edge

$$F_{ij} = \frac{1 + (w_j/w_i) - [1 + (w_j/w_i)^2]^{1/2}}{2}$$

Three-Sided Enclosure

$$F_{ij} = \frac{w_i + w_j - w_k}{2w_i}$$

(continues)

TABLE 13.1   *Continued*

Geometry	Relation

Parallel Cylinders of Different Radii

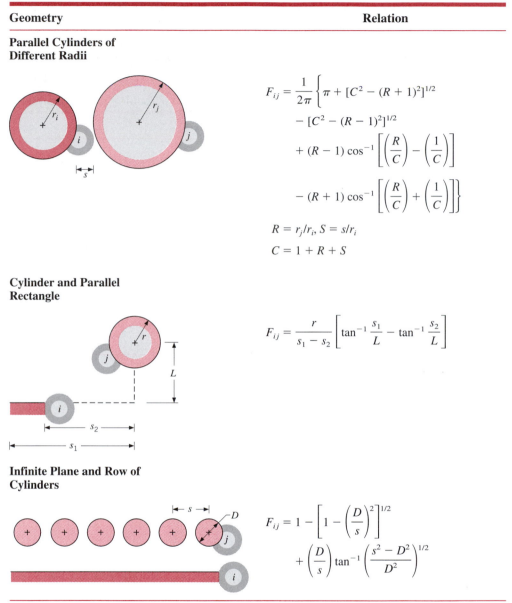

$$F_{ij} = \frac{1}{2\pi} \left\{ \pi + [C^2 - (R + 1)^2]^{1/2} \right.$$
$$- [C^2 - (R - 1)^2]^{1/2}$$
$$+ (R - 1) \cos^{-1} \left[\left(\frac{R}{C} \right) - \left(\frac{1}{C} \right) \right]$$
$$\left. - (R + 1) \cos^{-1} \left[\left(\frac{R}{C} \right) + \left(\frac{1}{C} \right) \right] \right\}$$

$$R = r_j/r_i, \; S = s/r_i$$
$$C = 1 + R + S$$

Cylinder and Parallel Rectangle

$$F_{ij} = \frac{r}{s_1 - s_2} \left[\tan^{-1} \frac{s_1}{L} - \tan^{-1} \frac{s_2}{L} \right]$$

Infinite Plane and Row of Cylinders

$$F_{ij} = 1 - \left[1 - \left(\frac{D}{s} \right)^2 \right]^{1/2}$$
$$+ \left(\frac{D}{s} \right) \tan^{-1} \left(\frac{s^2 - D^2}{D^2} \right)^{1/2}$$

quirement corresponds to only $3(3 - 1)/2 = 3$ view factors. The remaining six view factors may be obtained by solving the six equations that result from use of Equations 13.3 and 13.4.

To illustrate the foregoing procedure, consider a simple, two-surface enclosure involving the spherical surfaces of Figure 13.3. Although the enclosure is characterized by $N^2 = 4$ view factors (F_{11}, F_{12}, F_{21}, F_{22}), only $N(N - 1)/2 = 1$ view factor need be determined directly. In this case such a determination may be made by *inspection*. In particular, since all radiation leaving the inner surface must reach the outer surface, it follows that $F_{12} = 1$. The same may not be

TABLE 13.2 View Factors for Three-Dimensional Geometries [4]

Geometry	Relation

Aligned Parallel Rectangles (Figure 13.4)

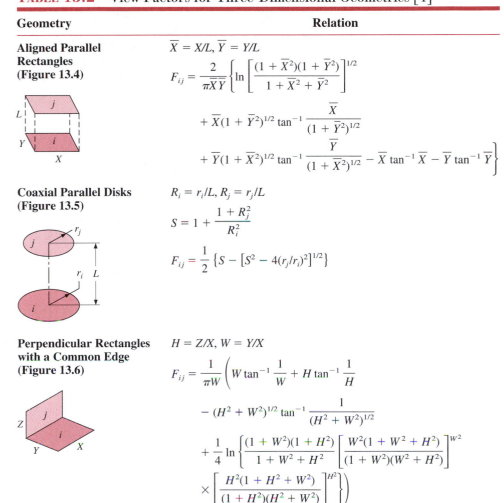

$\overline{X} = X/L, \overline{Y} = Y/L$

$$F_{ij} = \frac{2}{\pi \overline{X}\,\overline{Y}} \left\{ \ln \left[\frac{(1 + \overline{X}^2)(1 + \overline{Y}^2)}{1 + \overline{X}^2 + \overline{Y}^2} \right]^{1/2} \right.$$

$$+ \overline{X}(1 + \overline{Y}^2)^{1/2} \tan^{-1} \frac{\overline{X}}{(1 + \overline{Y}^2)^{1/2}}$$

$$\left. + \overline{Y}(1 + \overline{X}^2)^{1/2} \tan^{-1} \frac{\overline{Y}}{(1 + \overline{X}^2)^{1/2}} - \overline{X}\tan^{-1}\overline{X} - \overline{Y}\tan^{-1}\overline{Y} \right\}$$

Coaxial Parallel Disks (Figure 13.5)

$R_i = r_i/L, R_j = r_j/L$

$$S = 1 + \frac{1 + R_j^2}{R_i^2}$$

$$F_{ij} = \frac{1}{2} \left\{ S - [S^2 - 4(r_j/r_i)^2]^{1/2} \right\}$$

Perpendicular Rectangles with a Common Edge (Figure 13.6)

$H = Z/X, W = Y/X$

$$F_{ij} = \frac{1}{\pi W} \left(W\tan^{-1}\frac{1}{W} + H\tan^{-1}\frac{1}{H} \right.$$

$$- (H^2 + W^2)^{1/2} \tan^{-1} \frac{1}{(H^2 + W^2)^{1/2}}$$

$$+ \frac{1}{4} \ln \left\{ \frac{(1 + W^2)(1 + H^2)}{1 + W^2 + H^2} \left[\frac{W^2(1 + W^2 + H^2)}{(1 + W^2)(W^2 + H^2)} \right]^{W^2} \right.$$

$$\left. \left. \times \left[\frac{H^2(1 + H^2 + W^2)}{(1 + H^2)(H^2 + W^2)} \right]^{H^2} \right\} \right)$$

said of radiation leaving the outer surface, since this surface sees itself. However, from the reciprocity relation, Equation 13.3, we obtain

$$F_{21} = \left(\frac{A_1}{A_2} \right) F_{12} = \left(\frac{A_1}{A_2} \right)$$

From the summation rule, we also obtain

$$F_{11} + F_{12} = 1$$

in which case $F_{11} = 0$, and

$$F_{21} + F_{22} = 1$$

in which case

$$F_{22} = 1 - \left(\frac{A_1}{A_2} \right)$$

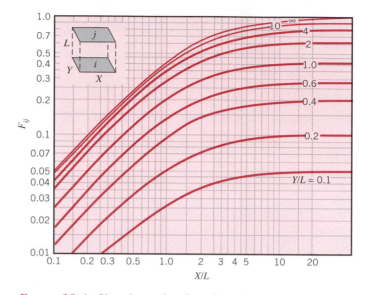

FIGURE 13.4   View factor for aligned parallel rectangles.

For more complicated geometries, the view factor may be determined by solving the double integral of Equation 13.1. Such solutions have been obtained for many different surface arrangements and are available in equation, graphical, and tabular form [1–4]. Results for several common geometries are presented in Tables 13.1 and 13.2 and Figures 13.4 to 13.6. The configurations of Table 13.1 are assumed to be infinitely long (in a direction perpendicular to the page) and are hence two-dimensional. The configurations of Table 13.2 and Figures 13.4 to 13.6 are three-dimensional.

It is useful to note that the results of Figures 13.4 to 13.6 may be used to determine other view factors. For example, the view factor for an end surface

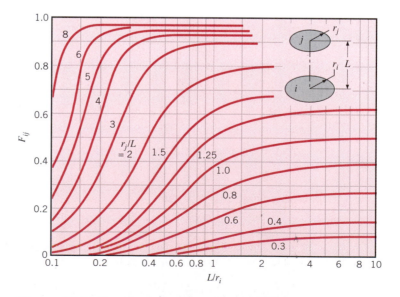

FIGURE 13.5   View factor for coaxial parallel disks.

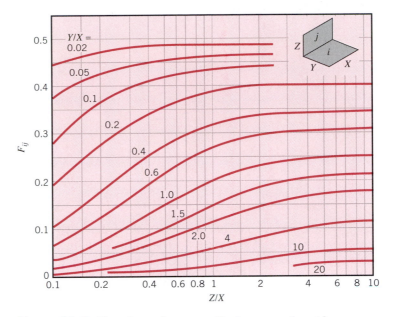

FIGURE 13.6 View factor for perpendicular rectangles with a common edge.

of a cylinder (or a truncated cone) relative to the lateral surface may be obtained by using the results of Figure 13.5 with the summation rule, Equation 13.4. Moreover, Figures 13.4 and 13.6 may be used to obtain other useful results if two additional view factor relations are developed.

The first relation concerns the additive nature of the view factor for a subdivided surface and may be inferred from Figure 13.7. Considering radiation from surface i to surface j, which is divided into n components, it is evident that

$$F_{i(j)} = \sum_{k=1}^{n} F_{ik} \tag{13.5}$$

where the parentheses around a subscript indicate that it is a composite surface, in which case (j) is equivalent to $(1, 2, \ldots, k, \ldots, n)$. This expression simply states that radiation reaching a composite surface is the sum of the radiation reaching its parts. Although it pertains to subdivision of the receiving surface, it may also be used to obtain the second view factor relation, which pertains to subdivision of the originating surface. Multiplying Equation 13.5 by A_i and applying the reciprocity relation, Equation 13.3, to each of the resulting terms, it follows that

$$A_j F_{(j)i} = \sum_{k=1}^{n} A_k F_{ki} \tag{13.6}$$

or

$$F_{(j)i} = \frac{\sum_{k=1}^{n} A_k F_{ki}}{\sum_{k=1}^{n} A_k} \tag{13.7}$$

Equations 13.6 and 13.7 may be applied when the originating surface is composed of several parts.

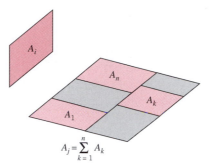

$$A_j = \sum_{k=1}^{n} A_k$$

FIGURE 13.7
Areas used to illustrate view factor relations.

EXAMPLE 13.1

Consider a diffuse circular disk of diameter D and area A_j and a plane diffuse surface of area $A_i \ll A_j$. The surfaces are parallel, and A_i is located at a distance L from the center of A_j. Obtain an expression for the view factor F_{ij}.

SOLUTION

Known: Orientation of small surface relative to large circular disk.

Find: View factor of small surface with respect to disk, F_{ij}.

Schematic:

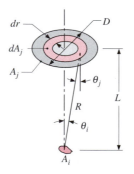

Assumptions:

1. Diffuse surfaces.
2. $A_i \ll A_j$.

Analysis: The desired view factor may be obtained from Equation 13.1.

$$F_{ij} = \frac{1}{A_i} \int_{A_i} \int_{A_j} \frac{\cos \theta_i \cos \theta_j}{\pi R^2} \, dA_i \, dA_j$$

Recognizing that θ_i, θ_j, and R are approximately independent of position on A_i, this expression reduces to

$$F_{ij} = \int_{A_j} \frac{\cos \theta_i \cos \theta_j}{\pi R^2} \, dA_j$$

or, with $\theta_i = \theta_j \equiv \theta$,

$$F_{ij} = \int_{A_j} \frac{\cos^2 \theta}{\pi R^2} \, dA_j$$

With $R^2 = r^2 + L^2$, $\cos \theta = (L/R)$ and $dA_j = 2\pi r \, dr$, it follows that

$$F_{ij} = 2L^2 \int_0^{D/2} \frac{r \, dr}{(r^2 + L^2)^2} = \frac{D^2}{D^2 + 4L^2} \qquad \triangleleft \quad (13.8)$$

Comments: The preceding geometry is one of the simplest cases for which the view factor may be obtained from Equation 13.1. Geometries involving more detailed integrations are considered in the literature [1, 3].

EXAMPLE 13.2

Determine the view factors F_{12} and F_{21} for the following geometries:

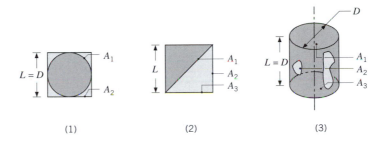

(1) (2) (3)

1. Sphere of diameter D inside a cubical box of length $L = D$.
2. Diagonal partition within a long square duct.
3. End and side of a circular tube of equal length and diameter.

SOLUTION

Known: Surface geometries.

Find: View factors.

Assumptions: Diffuse surfaces with uniform radiosities.

Analysis: The desired view factors may be obtained from inspection, the reciprocity rule, the summation rule, and/or use of the charts.

1. Sphere within a cube:
   By inspection, $\qquad F_{12} = 1$ $\qquad\qquad\qquad\qquad\qquad\qquad\qquad \triangleleft$

   By reciprocity, $\qquad F_{21} = \dfrac{A_1}{A_2} F_{12} = \dfrac{\pi D^2}{6L^2} \times 1 = \dfrac{\pi}{6}$ $\qquad\qquad \triangleleft$

2. Partition within a square duct:

From summation rule, $F_{11} + F_{12} + F_{13} = 1$

where $F_{11} = 0$

By symmetry, $F_{12} = F_{13}$

Hence $F_{12} = 0.50$ ◁

By reciprocity, $F_{21} = \dfrac{A_1}{A_2} F_{12} = \dfrac{\sqrt{2}\,L}{L} \times 0.5 = 0.71$ ◁

3. Circular tube:

From Figure 13.5, with $(r_3/L) = 0.5$ and $(L/r_1) = 2$, $F_{13} \approx 0.17$

From summation rule, $F_{11} + F_{12} + F_{13} = 1$

or, with $F_{11} = 0$, $F_{12} = 1 - F_{13} = 0.83$ ◁

From reciprocity, $F_{21} = \dfrac{A_1}{A_2} F_{12} = \dfrac{\pi D^2/4}{\pi D L} \times 0.83 = 0.21$ ◁

13.2
Blackbody Radiation Exchange

In general, radiation may leave a surface due to both reflection and emission, and on reaching a second surface, experience reflection as well as absorption. However, matters are simplified for surfaces that may be approximated as blackbodies, since there is no reflection. Hence energy only leaves as a result of emission, and all incident radiation is absorbed.

Consider radiation exchange between two black surfaces of arbitrary shape (Figure 13.8). Defining $q_{i \to j}$ as the rate at which radiation *leaves* surface i and is *intercepted* by surface j, it follows that

$$q_{i \to j} = (A_i J_i) F_{ij} \tag{13.9}$$

or, since radiosity equals emissive power for a black surface ($J_i = E_{bi}$),

$$q_{i \to j} = A_i F_{ij} E_{bi} \tag{13.10}$$

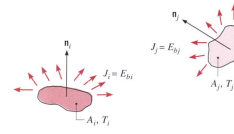

FIGURE 13.8
Radiation transfer between two surfaces that may be approximated as blackbodies.

Similarly,

$$q_{j \to i} = A_j F_{ji} E_{bj} \tag{13.11}$$

The *net radiative exchange* between the two surfaces may then be defined as

$$q_{ij} = q_{i \to j} - q_{j \to i} \tag{13.12}$$

from which it follows that

$$q_{ij} = A_i F_{ij} E_{bi} - A_j F_{ji} E_{bj}$$

or, from Equations 12.28 and 13.3

$$q_{ij} = A_i F_{ij} \sigma (T_i^4 - T_j^4) \tag{13.13}$$

Equation 13.13 provides the *net* rate at which radiation *leaves* surface i as a result of its interaction with j, which is equal to the *net* rate at which j *gains* radiation due to its interaction with i.

The foregoing result may also be used to evaluate the net radiation transfer from any surface in an *enclosure* of black surfaces. With N surfaces maintained at different temperatures, the net transfer of radiation from surface i is due to exchange with the remaining surfaces and may be expressed as

$$q_i = \sum_{j=1}^{N} A_i F_{ij} \sigma (T_i^4 - T_j^4) \tag{13.14}$$

EXAMPLE 13.3

A furnace cavity, which is in the form of a cylinder of 75-mm diameter and 150-mm length, is open at one end to large surroundings that are at 27°C. The sides and bottom may be approximated as blackbodies, are heated electrically, are well insulated, and are maintained at temperatures of 1350 and 1650°C, respectively.

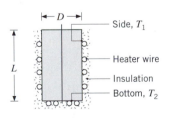

How much power is required to maintain the furnace conditions?

SOLUTION

Known: Surface temperatures of cylindrical furnace.

Find: Power required to maintain prescribed temperatures.

Schematic:

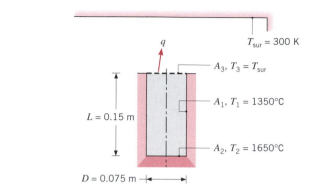

Assumptions:

1. Interior surfaces behave as blackbodies.
2. Heat transfer by convection is negligible.
3. Outer surface of furnace is adiabatic.

Analysis: The power needed to operate the furnace at the prescribed conditions must balance heat losses from the furnace. Subject to the foregoing assumptions, the only heat loss is by radiation through the opening, which may be treated as a hypothetical surface of area A_3. Because the surroundings are large, radiation exchange between the furnace and the surroundings may be treated by approximating the surface as a blackbody at $T_3 = T_{sur}$. The heat loss may then be expressed as

$$q = q_{13} + q_{23}$$

or, from Equation 13.13,

$$q = A_1 F_{13}\sigma(T_1^4 - T_3^4) + A_2 F_{23}\sigma(T_2^4 - T_3^4)$$

From Figure 13.5, it follows that, with $(r_j/L) = (0.0375 \text{ m}/0.15 \text{ m}) = 0.25$ and $(L/r_i) = (0.15 \text{ m}/0.0375 \text{ m}) = 4$, $F_{23} = 0.06$. From the summation rule

$$F_{21} = 1 - F_{23} = 1 - 0.06 = 0.94$$

and from reciprocity

$$F_{12} = \frac{A_2}{A_1} F_{21} = \frac{\pi(0.075 \text{ m})^2/4}{\pi(0.075 \text{ m})(0.15 \text{ m})} \times 0.94 = 0.118$$

Hence, since $F_{13} = F_{12}$ from symmetry,

$$q = (\pi \times 0.075 \text{ m} \times 0.15 \text{ m})0.118 \times 5.67 \times 10^{-8} \text{ W/m}^2 \cdot \text{K}^4$$

$$\times [(1623 \text{ K})^4 - (300 \text{ K})^4] + \left(\frac{\pi}{4}\right)(0.075 \text{ m})^2 \times 0.06$$

$$\times 5.67 \times 10^{-8} \text{ W/m}^2 \cdot \text{K}^4 [(1923 \text{ K})^4 - (300 \text{ K})^4]$$

$$q = 1639 \text{ W} + 205 \text{ W} = 1844 \text{ W} \qquad \triangleleft$$

13.3
Radiation Exchange Between Diffuse, Gray Surfaces in an Enclosure

Although useful to a point, the foregoing results are limited by the assumption of blackbody behavior. The blackbody is, of course, an idealization, which, although closely approximated by some surfaces, is never precisely achieved. A major complication associated with radiation exchange between nonblack surfaces is due to surface reflection. In an enclosure, such as that of Figure 13.9a, radiation may experience multiple reflections off all surfaces, with partial absorption occurring at each.

Analyzing radiation exchange in an enclosure may be simplified by making certain assumptions. Each surface of the enclosure is assumed to be *isothermal* and to be characterized by a *uniform radiosity* and *irradiation. Opaque, diffuse, gray* surface behavior is also assumed, and the medium within the enclosure is taken to be *nonparticipating.* The problem is generally one in which the temperature T_i associated with each of the surfaces is known, and the objective is to determine the *net radiative heat flux q_i''* from each surface.

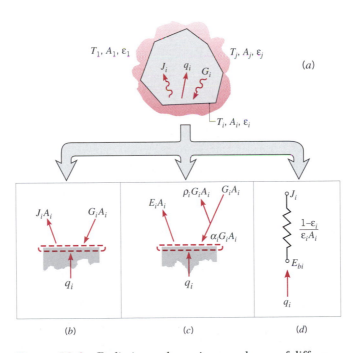

FIGURE 13.9 Radiation exchange in an enclosure of diffuse, gray surfaces with a nonparticipating medium. (*a*) Schematic of the enclosure. (*b*) Radiative balance according to Equation 13.15. (*c*) Radiative balance according to Equation 13.17. (*d*) Network element representing the net radiation transfer from a surface.

13.3.1   Net Radiation Exchange at a Surface

The term q_i, which is the *net* rate at which radiation *leaves* surface i, represents the net effect of radiative interactions occurring at the surface (Figure 13.9b). It is the rate at which energy would have to be transferred to the surface by other means to maintain it at a constant temperature. It is equal to the difference between the surface radiosity and irradiation and may be expressed as

$$q_i = A_i(J_i - G_i) \tag{13.15}$$

From Figure 13.9c and the definition of the radiosity J_i,

$$J_i \equiv E_i + \rho_i G_i \tag{13.16}$$

it is evident that the net radiative transfer from the surface may also be expressed in terms of the surface emissive power and the absorbed irradiation:

$$q_i = A_i(E_i - \alpha_i G_i) \tag{13.17}$$

Substituting from Equation 12.37 and recognizing that $\rho_i = 1 - \alpha_i = 1 - \varepsilon_i$ for an opaque, diffuse, gray surface, the radiosity may also be expressed as

$$J_i = \varepsilon_i E_{bi} + (1 - \varepsilon_i)G_i \tag{13.18}$$

Solving for G_i and substituting into Equation 13.15, it follows that

$$q_i = A_i\left(J_i - \frac{J_i - \varepsilon_i E_{bi}}{1 - \varepsilon_i}\right)$$

or

$$q_i = \frac{E_{bi} - J_i}{(1 - \varepsilon_i)/\varepsilon_i A_i} \tag{13.19}$$

Equation 13.19 provides a convenient representation for the net radiative heat transfer rate from a surface. This transfer, which may be represented by the network element of Figure 13.9d, is associated with the driving potential $(E_{bi} - J_i)$ and a *surface radiative resistance* of the form $(1 - \varepsilon_i)/\varepsilon_i A_i$. Hence if the emissive power that the surface would have if it were black exceeds its radiosity, there is net radiation heat transfer from the surface; if the inverse is true, the net transfer is to the surface.

13.3.2   Radiation Exchange Between Surfaces

To use Equation 13.19, the surface radiosity J_i must be known. To determine this quantity, it is necessary to consider radiation exchange between the surfaces of the enclosure.

The irradiation of surface i can be evaluated from the radiosities of all the surfaces in the enclosure. In particular, from the definition of the view factor, it follows that the total rate at which radiation reaches surface i from all surfaces, including i, is

$$A_i G_i = \sum_{j=1}^{N} F_{ji} A_j J_j$$

or from the reciprocity relation, Equation 13.3,

$$A_i G_i = \sum_{j=1}^{N} A_i F_{ij} J_j$$

Canceling the area A_i and substituting into Equation 13.15 for G_i,

$$q_i = A_i \left(J_i - \sum_{j=1}^{N} F_{ij} J_j \right)$$

or, from the summation rule, Equation 13.4,

$$q_i = A_i \left(\sum_{j=1}^{N} F_{ij} J_i - \sum_{j=1}^{N} F_{ij} J_j \right)$$

Hence

$$q_i = \sum_{j=1}^{N} A_i F_{ij} (J_i - J_j) = \sum_{j=1}^{N} q_{ij} \tag{13.20}$$

This result equates the net rate of radiation transfer from surface i, q_i, to the sum of components q_{ij} related to radiative exchange with the other surfaces. Each component may be represented by a network element for which $(J_i - J_j)$ is the driving potential and $(A_i F_{ij})^{-1}$ is a *space* or *geometrical resistance* (Figure 13.10).

Combining Equations 13.19 and 13.20, we then obtain

$$\frac{E_{bi} - J_i}{(1 - \varepsilon_i)/\varepsilon_i A_i} = \sum_{j=1}^{N} \frac{J_i - J_j}{(A_i F_{ij})^{-1}} \tag{13.21}$$

As shown in Figure 13.10 this expression represents a radiation balance for the radiosity *node* associated with surface i. The rate of radiation transfer (current flow) to i through its surface resistance must equal the rate of radiation transfer

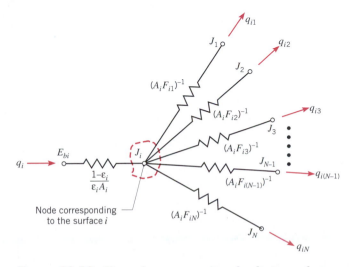

FIGURE 13.10   Network representation of radiative exchange between surface i and the remaining surfaces of an enclosure.

(current flows) from i to all other surfaces through the corresponding geometrical resistances.

Note that Equation 13.21 is especially useful when the surface temperature T_i (hence E_{bi}) is known. Although this situation is typical, it does not always apply. In particular, situations may arise for which the net radiation transfer rate at the surface q_i, rather than the temperature T_i, is known. In such cases the preferred form of the radiation balance is Equation 13.20, rearranged as

$$q_i = \sum_{j=1}^{N} \frac{J_i - J_j}{(A_i F_{ij})^{-1}} \tag{13.22}$$

Use of network representations to solve enclosure radiation problems was first suggested by Oppenheim [5]. The method provides a useful tool for visualizing radiation exchange in the enclosure and, at least for simple enclosures, may be used as the basis for predicting this exchange. However, a more direct approach simply involves working with Equations 13.21 and 13.22. Equation 13.21 is written for each surface at which T_i is known, and Equation 13.22 is written for each surface at which q_i is known. The resulting set of N linear, algebraic equations is solved for the N unknowns, $J_1, J_2, \ldots, J_N$. With knowledge of the J_i, Equation 13.19 may be used to determine the net radiation heat transfer rate q_i at each surface of known T_i or the value of T_i at each surface of known q_i.

For any number N of surfaces in the enclosure, the foregoing problem may readily be solved by iteration or matrix inversion. For each of the N surfaces Equation 13.21 or 13.22 may be rearranged to obtain the following system of N equations:

$$\begin{bmatrix} a_{11}J_1 + \cdots + a_{1i}J_i + \cdots + a_{1N}J_N = C_1 \\ \cdots\cdots\cdots\cdots\cdots\cdots\cdots\cdots\cdots\cdots\cdots\cdots \\ \cdots\cdots\cdots\cdots\cdots\cdots\cdots\cdots\cdots\cdots\cdots\cdots \\ a_{i1}J_1 + \cdots + a_{ii}J_i + \cdots + a_{iN}J_N = C_i \\ \cdots\cdots\cdots\cdots\cdots\cdots\cdots\cdots\cdots\cdots\cdots\cdots \\ \cdots\cdots\cdots\cdots\cdots\cdots\cdots\cdots\cdots\cdots\cdots\cdots \\ a_{N1}J_1 + \cdots + a_{Ni}J_i + \cdots + a_{NN}J_N = C_N \end{bmatrix}$$

where the coefficients a_{ij} and C_i are known quantities. In matrix form these equations may be expressed as

$$[A][J] = [C]$$

where

$$[A] = \begin{bmatrix} a_{11} \cdots a_{1i} \cdots a_{1N} \\ \cdots\cdots\cdots\cdots\cdots \\ \cdots\cdots\cdots\cdots\cdots \\ a_{i1} \cdots a_{ii} \cdots a_{iN} \\ \cdots\cdots\cdots\cdots\cdots \\ \cdots\cdots\cdots\cdots\cdots \\ a_{N1} \cdots a_{Ni} \cdots a_{NN} \end{bmatrix} \qquad [J] = \begin{bmatrix} J_1 \\ . \\ . \\ J_i \\ . \\ . \\ J_N \end{bmatrix} \qquad [C] = \begin{bmatrix} C_1 \\ . \\ . \\ C_i \\ . \\ . \\ C_N \end{bmatrix}$$

Expressing the unknown radiosities as

$$
\begin{bmatrix}
J_1 = b_{11}C_1 + \cdots + b_{1i}C_i + \cdots + b_{1N}C_N \\
\cdots\cdots\cdots\cdots\cdots\cdots\cdots\cdots\cdots\cdots\cdots\cdots \\
\cdots\cdots\cdots\cdots\cdots\cdots\cdots\cdots\cdots\cdots\cdots\cdots \\
J_i = b_{i1}C_1 + \cdots + b_{ii}C_i + \cdots + b_{iN}C_N \\
\cdots\cdots\cdots\cdots\cdots\cdots\cdots\cdots\cdots\cdots\cdots\cdots \\
\cdots\cdots\cdots\cdots\cdots\cdots\cdots\cdots\cdots\cdots\cdots\cdots \\
J_N = b_{N1}C_1 + \cdots + b_{Ni}C_i + \cdots + b_{NN}C_N
\end{bmatrix}
$$

they may be found by obtaining the inverse of $[A], [A]^{-1}$, such that

$$[J] = [A]^{-1}[C]$$

where

$$
[A]^{-1} =
\begin{bmatrix}
b_{11} & \cdots & b_{1i} & \cdots & b_{1N} \\
\cdots & \cdots & \cdots & \cdots & \cdots \\
\cdots & \cdots & \cdots & \cdots & \cdots \\
b_{i1} & \cdots & b_{ii} & \cdots & b_{iN} \\
\cdots & \cdots & \cdots & \cdots & \cdots \\
\cdots & \cdots & \cdots & \cdots & \cdots \\
b_{N1} & \cdots & b_{Ni} & \cdots & b_{NN}
\end{bmatrix}
$$

The foregoing matrix inversion may readily be obtained by using any of numerous computer routines available for this purpose or, for simple problems, from a programmable hand calculator. The system of equations may also be solved by the Gauss–Seidel iteration method described in Chapter 4.

EXAMPLE 13.4

In manufacturing, the special coating on a curved solar absorber surface of area $A_2 = 15 \text{ m}^2$ is cured by exposing it to an infrared heater of width $W = 1$ m. The absorber and heater are each of length $L = 10$ m and are separated by a distance of $H = 1$ m.

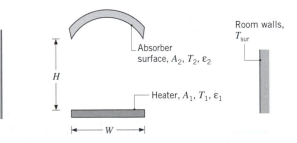

The heater is at $T_1 = 1000$ K and has an emissivity of $\varepsilon_1 = 0.9$, while the absorber is at $T_2 = 600$ K and has an emissivity of $\varepsilon_2 = 0.5$. The system is in a large room whose walls are at 300 K. What is the net rate of heat transfer to the absorber surface?

SOLUTION

Known: A curved, solar absorber surface with a special coating is being cured by use of an infrared heater in a large room.

Find: Net rate of heat transfer to the absorber surface.

Schematic:

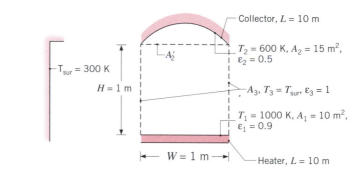

Collector, $L = 10$ m
$T_2 = 600$ K, $A_2 = 15$ m^2, $\varepsilon_2 = 0.5$
A_3, $T_3 = T_{sur}$, $\varepsilon_3 = 1$
$T_1 = 1000$ K, $A_1 = 10$ m^2, $\varepsilon_1 = 0.9$
$T_{sur} = 300$ K
$H = 1$ m
$W = 1$ m
Heater, $L = 10$ m
A'_2

Assumptions:

1. Steady-state conditions exist.

2. Convection effects are negligible.

3. Absorber and heater surfaces are diffuse and gray.

4. The surroundings may be represented by a hypothetical surface A_3, which completes the enclosure and is approximated as a blackbody of temperature $T_3 = 300$ K.

Analysis: The system may be viewed as a three-surface enclosure for which we are interested in obtaining the net rate of radiation transfer to surface 2. From Equation 13.19

$$q_2 = \frac{E_{b2} - J_2}{(1 - \varepsilon_2)/\varepsilon_2 A_2} \tag{1}$$

where all quantities are known except J_2. In most enclosure problems all unknown radiosities must be determined simultaneously, and the matrix inversion procedure is well suited for this purpose. In the present problem, however, matters are simplified by approximation of the hypothetical surface as a blackbody of known temperature. Hence $J_3 = E_{b3}$ is known, and the only unknowns are J_1 and J_2. Since T_1 and T_2 are known, J_1 and J_2 may be obtained by expressing Equation 13.21 for the heater and absorber surfaces. For the absorber surface, it follows that

$$\frac{E_{b2} - J_2}{(1 - \varepsilon_2)/\varepsilon_2 A_2} = \frac{J_2 - J_1}{1/A_2 F_{21}} + \frac{J_2 - J_3}{1/A_2 F_{23}}$$

where $J_3 = E_{b3} = \sigma T_3^4 = 459$ W/m^2 and $E_{b2} = \sigma T_2^4 = 7348$ W/m^2. The quantity $A_2 F_{21}$ may be obtained by recognizing that, from reciprocity,

$$A_2 F_{21} = A_1 F_{12}$$

where F_{12} may be obtained from Figure 13.4. That is, $F_{12} = F_{12'}$, where A'_2 is simply the rectangular base of the absorber surface. Hence with $Y/L = 10/1 = 10$ and $X/L = 1/1 = 1$,

$$F_{12} = 0.39$$

$$F_{21} = \frac{A_1}{A_2} F_{12} = \frac{1 \text{ m} \times 10 \text{ m}}{15 \text{ m}^2} \times 0.39 = 0.26$$

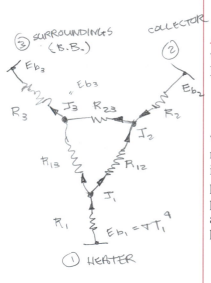

From the summation rule, it also follows that

$$F_{13} = 1 - F_{12} = 1 - 0.39 = 0.61$$

and from reciprocity

$$F_{31} = \frac{A_1}{A_3} F_{13} = \frac{1 \text{ m} \times 10 \text{ m}}{2(10 \times 1) \text{ m}^2} \times 0.61 = 0.305$$

But from symmetry, $F_{31} = F_{32'} = F_{32}$. Hence

$$F_{23} = \frac{A_3}{A_2} F_{32} = \frac{20 \text{ m}^2}{15 \text{ m}^2} \times 0.305 = 0.41$$

Canceling the area A_2, the radiation balance for surface 2 is then

$$\frac{7348 - J_2}{(1 - 0.5)/0.5} = \frac{J_2 - J_1}{1/0.26} + \frac{J_2 - 459}{1/0.41}$$

or

$$7348 - J_2 = 0.26J_2 - 0.26J_1 + 0.41J_2 - 188$$

$$0.26J_1 - 1.67J_2 = -7536 \qquad (2)$$

Expressing Equation 13.21 for the heater surface, it follows that

$$\frac{E_{b1} - J_1}{(1 - \varepsilon_1)/\varepsilon_1 A_1} = \frac{J_1 - J_2}{1/A_1 F_{12}} + \frac{J_1 - J_3}{1/A_1 F_{13}}$$

where $E_{b1} = \sigma T_1^4 = 56{,}700 \text{ W/m}^2$. Hence canceling the area A_1,

$$\frac{56{,}700 - J_1}{(1 - 0.9)/0.9} = \frac{J_1 - J_2}{1/0.39} + \frac{J_1 - 459}{1/0.61}$$

or

$$-10J_1 + 0.39J_2 = -510{,}002 \qquad (3)$$

From Equation 3, we then obtain

$$J_1 = 51{,}000 + 0.039J_2$$

Substituting into Equation 2

$$0.26(51{,}000 + 0.039J_2) - 1.67J_2 = -7536$$

in which case

$$J_2 = 12{,}528 \text{ W/m}^2$$

Substituting into Equation 1, the net heat rate to the absorber is

$$q_2 = \frac{(7348 - 12{,}528) \text{ W/m}^2}{(1 - 0.5)/0.5 \times 15 \text{ m}^2} = -77.7 \text{ kW} \qquad \triangleleft$$

Comments: Recognize the utility of using the hypothetical surfaces, in one case to complete the enclosure with A_3 and in the other case to simplify the evaluation of the shape factor with A_2'.

13.3.3   The Two-Surface Enclosure

The simplest example of an enclosure is one involving two surfaces that exchange radiation only with each other. Such a two-surface enclosure is shown schematically in Figure 13.11a. Since there are only two surfaces, the net rate of radiation transfer *from* surface 1, q_1, must equal the net rate of radiation transfer *to* surface 2, $-q_2$, and both quantities must equal the net rate at which radiation is exchanged between 1 and 2. Accordingly,

$$q_1 = -q_2 = q_{12}$$

The radiation transfer rate may be determined by applying Equation 13.21 to surfaces 1 and 2 and solving the resulting two equations for J_1 and J_2. The results could then be used with Equation 13.19 to determine q_1 (or q_2). However, in this case the desired result is more readily obtained by working with the network representation of the enclosure shown in Figure 13.11b.

From Figure 13.11b we see that the total resistance to radiation exchange between surfaces 1 and 2 is comprised of the two surface resistances and the geometrical resistance. Hence, substituting from Equation 12.28, the net radiation exchange between surfaces may be expressed as

$$q_{12} = q_1 = -q_2 = \frac{\sigma(T_1^4 - T_2^4)}{\dfrac{1 - \varepsilon_1}{\varepsilon_1 A_1} + \dfrac{1}{A_1 F_{12}} + \dfrac{1 - \varepsilon_2}{\varepsilon_2 A_2}} \tag{13.23}$$

The foregoing result may be used for any two diffuse, gray surfaces *that form an enclosure*. Important special cases are summarized in Table 13.3.

13.3.4   Radiation Shields

Radiation shields constructed from low emissivity (high reflectivity) materials can be used to reduce the net radiation transfer between two surfaces. Consider placing a radiation shield, surface 3, between the two large, parallel planes of Figure 13.12a. Without the radiation shield, the net rate of radiation transfer be-

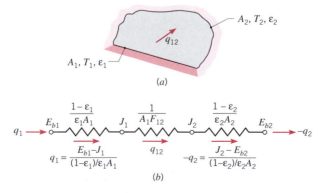

FIGURE 13.11   The two-surface enclosure. (*a*) Schematic. (*b*) Network representation.

TABLE 13.3   Special Diffuse, Gray, Two-Surface Enclosures

Large (Infinite) Parallel Planes

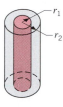

$A_1 = A_2 = A$

$F_{12} = 1$

$$q_{12} = \frac{A\sigma(T_1^4 - T_2^4)}{\dfrac{1}{\varepsilon_1} + \dfrac{1}{\varepsilon_2} - 1} \qquad (13.24)$$

Long (Infinite) Concentric Cylinders

$\dfrac{A_1}{A_2} = \dfrac{r_1}{r_2}$

$F_{12} = 1$

$$q_{12} = \frac{\sigma A_1(T_1^4 - T_2^4)}{\dfrac{1}{\varepsilon_1} + \dfrac{1 - \varepsilon_2}{\varepsilon_2}\left(\dfrac{r_1}{r_2}\right)} \qquad (13.25)$$

Concentric Spheres

$\dfrac{A_1}{A_2} = \dfrac{r_1^2}{r_2^2}$

$F_{12} = 1$

$$q_{12} = \frac{\sigma A_1(T_1^4 - T_2^4)}{\dfrac{1}{\varepsilon_1} + \dfrac{1 - \varepsilon_2}{\varepsilon_2}\left(\dfrac{r_1}{r_2}\right)^2} \qquad (13.26)$$

Small Convex Object in a Large Cavity

$\dfrac{A_1}{A_2} \approx 0$

$F_{12} = 1$

$$q_{12} = \sigma A_1 \varepsilon_1 (T_1^4 - T_2^4) \qquad (13.27)$$

tween surfaces 1 and 2 is given by Equation 13.24. However, with the radiation shield, additional resistances are present, as shown in Figure 13.12*b*, and the heat transfer rate is reduced. Note that the emissivity associated with one side of the shield ($\varepsilon_{3,1}$) may differ from that associated with the opposite side ($\varepsilon_{3,2}$) and the radiosities will always differ. Summing the resistances and recognizing that $F_{13} = F_{32} = 1$, it follows that

$$q_{12} = \frac{A_1\sigma(T_1^4 - T_2^4)}{\dfrac{1}{\varepsilon_1} + \dfrac{1}{\varepsilon_2} + \dfrac{1 - \varepsilon_{3,1}}{\varepsilon_{3,1}} + \dfrac{1 - \varepsilon_{3,2}}{\varepsilon_{3,2}}} \qquad (13.28)$$

Note that the resistances associated with the radiation shield become very large when the emissivities $\varepsilon_{3,1}$ and $\varepsilon_{3,2}$ are very small.

Equation 13.28 may be used to determine the net heat transfer rate if T_1 and T_2 are known. From knowledge of q_{12} and the fact that $q_{12} = q_{13} = q_{32}$, the value of T_3 may then be determined by expressing Equation 13.24 for q_{13} or q_{32}.

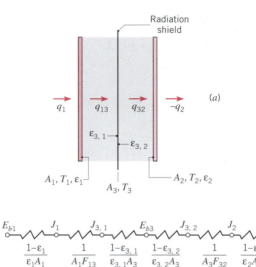

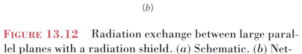

FIGURE 13.12   Radiation exchange between parallel planes with a radiation shield. (*a*) Schematic. (*b*) Network representation.

The foregoing procedure may readily be extended to problems involving multiple radiation shields. In the special case for which all the emissivities are equal, it may be shown that, with N shields,

$$(q_{12})_N = \frac{1}{N+1} (q_{12})_0 \tag{13.29}$$

where $(q_{12})_0$ is the radiation transfer rate with no shields ($N = 0$).

EXAMPLE 13.5

A cryogenic fluid flows through a long tube of 20-mm diameter, the outer surface of which is diffuse and gray with $\varepsilon_1 = 0.02$ and $T_1 = 77$ K. This tube is concentric with a larger tube of 50-mm diameter, the inner surface of which is diffuse and gray with $\varepsilon_2 = 0.05$ and $T_2 = 300$ K. The space between the surfaces is evacuated. Calculate the heat gain by the cryogenic fluid per unit length of tubes. If a thin radiation shield of 35-mm diameter and $\varepsilon_3 = 0.02$ (both sides) is inserted midway between the inner and outer surfaces, calculate the change (percentage) in heat gain per unit length of the tubes.

SOLUTION

Known: Concentric tube arrangement with diffuse, gray surfaces of different emissivities and temperatures.

Find:
1. Heat gain by the cryogenic fluid passing through the inner tube.
2. Percentage change in heat gain with radiation shield inserted midway between inner and outer tubes.

Schematic:

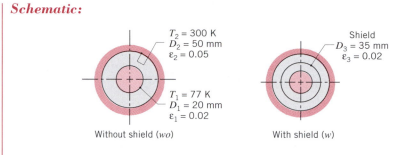

Assumptions:

1. Surfaces are diffuse and gray.
2. Space between tubes is evacuated.
3. Conduction resistance for radiation shield is negligible.
4. Concentric tubes form a two-surface enclosure (end effects are negligible).

Analysis:

1. The network representation of the system without the shield is shown in Figure 13.11, and the desired heat rate may be obtained from Equation 13.25, where

$$q = \frac{\sigma(\pi D_1 L)(T_1^4 - T_2^4)}{\dfrac{1}{\varepsilon_1} + \dfrac{1 - \varepsilon_2}{\varepsilon_2}\left(\dfrac{D_1}{D_2}\right)}$$

Hence

$$q' = \frac{q}{L} = \frac{5.67 \times 10^{-8} \text{ W/m}^2 \cdot \text{K}^4 \,(\pi \times 0.02 \text{ m})[(77 \text{ K})^4 - (300 \text{ K})^4]}{\dfrac{1}{0.02} + \dfrac{1 - 0.05}{0.05}\left(\dfrac{0.02 \text{ m}}{0.05 \text{ m}}\right)}$$

$$q' = -0.50 \text{ W/m} \qquad\qquad\qquad \triangleleft$$

2. The network representation of the system with the shield is shown in Figure 13.12, and the desired heat rate is now

$$q = \frac{E_{b1} - E_{b2}}{R_{\text{tot}}} = \frac{\sigma(T_1^4 - T_2^4)}{R_{\text{tot}}}$$

where

$$R_{\text{tot}} = \frac{1 - \varepsilon_1}{\varepsilon_1(\pi D_1 L)} + \frac{1}{(\pi D_1 L)F_{13}} + 2\left[\frac{1 - \varepsilon_3}{\varepsilon_3(\pi D_3 L)}\right] + \frac{1}{(\pi D_3 L)F_{32}} + \frac{1 - \varepsilon_2}{\varepsilon_2(\pi D_2 L)}$$

or

$$R_{\text{tot}} = \frac{1}{L}\left\{\frac{1 - 0.02}{0.02(\pi \times 0.02 \text{ m})} + \frac{1}{(\pi \times 0.02 \text{ m})1}\right.$$

$$\left. + 2\left[\frac{1 - 0.02}{0.02(\pi \times 0.035 \text{ m})}\right] + \frac{1}{(\pi \times 0.035 \text{ m})1} + \frac{1 - 0.05}{0.05(\pi \times 0.05 \text{ m})}\right\}$$

$$R_{\text{tot}} = \frac{1}{L}(779.9 + 15.9 + 891.3 + 9.1 + 121.0) = \frac{1817}{L}\left(\frac{1}{\text{m}^2}\right)$$

Hence

$$q' = \frac{q}{L} = \frac{5.67 \times 10^{-8} \text{ W/m}^2 \cdot \text{K}^4 \, [(77 \text{ K})^4 - (300 \text{ K})^4]}{1817 \, (1/\text{m})} = -0.25 \text{ W/m} \quad \triangleleft$$

The percentage change in the heat gain is then

$$\frac{q'_w - q'_{wo}}{q'_{wo}} \times 100 = \frac{(-0.25 \text{ W/m}) - (-0.50 \text{ W/m})}{-0.50 \text{ W/m}} \times 100 = -50\%$$

13.3.5   The Reradiating Surface

The assumption of a *reradiating surface* is common to many industrial applications. This idealized surface is characterized by *zero* net radiation transfer ($q_i = 0$). It is closely approached by real surfaces that are well insulated on one side and for which convection effects may be neglected on the opposite (radiating) side. With $q_i = 0$, it follows from Equations 13.15 and 13.19 that $G_i = J_i = E_{bi}$. Hence, if the radiosity of a reradiating surface is known, its temperature is readily determined. In an enclosure, the equilibrium temperature of a reradiating surface is determined by its interaction with the other surfaces, and it is *independent of the emissivity of the reradiating surface*.

A three-surface enclosure, for which the third surface, surface R, is reradiating, is shown in Figure 13.13a, and the corresponding network is shown in Figure 13.13b. Surface R is presumed to be well insulated, and convection effects are assumed to be negligible. Hence, with $q_R = 0$, the net radiation *transfer* from surface 1 must equal the net radiation transfer to surface 2. The network is a simple series–parallel arrangement, and from its analysis it is readily shown that

$$q_1 = -q_2 = \frac{E_{b1} - E_{b2}}{\dfrac{1 - \varepsilon_1}{\varepsilon_1 A_1} + \dfrac{1}{A_1 F_{12} + [(1/A_1 F_{1R}) + (1/A_2 F_{2R})]^{-1}} + \dfrac{1 - \varepsilon_2}{\varepsilon_2 A_2}} \tag{13.30}$$

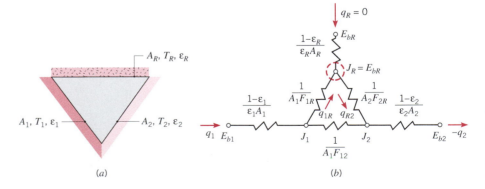

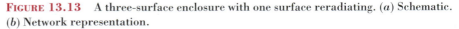

FIGURE 13.13   A three-surface enclosure with one surface reradiating. (a) Schematic. (b) Network representation.

Knowing $q_1 = -q_2$, Equation 13.19 may be applied to surfaces 1 and 2 to determine their radiosities J_1 and J_2. Knowing J_1, J_2, and the geometrical resistances, the radiosity of the reradiating surface J_R may be determined from the radiation balance

$$\frac{J_1 - J_R}{(1/A_1F_{1R})} - \frac{J_R - J_2}{(1/A_2F_{2R})} = 0 \qquad (13.31)$$

The temperature of the reradiating surface may then be determined from the requirement that $\sigma T_R^4 = J_R$.

Note that the general procedure described in Section 13.3.2 may be applied to enclosures with reradiating surfaces. For each such surface, it is appropriate to use Equation 13.22 with $q_i = 0$.

EXAMPLE 13.6

A paint baking oven consists of a long, triangular duct in which a heated surface is maintained at 1200 K and another surface is insulated. Painted panels, which are maintained at 500 K, occupy the third surface. The triangle is of width $W = 1$ m on a side, and the heated and insulated surfaces have an emissivity of 0.8. The emissivity of the panels is 0.4. During steady-state operation, at what rate must energy be supplied to the heated side per unit length of the duct to maintain its temperature at 1200 K? What is the temperature of the insulated surface?

SOLUTION

Known: Surface properties of a long triangular duct that is insulated on one side and heated and cooled on the other sides.

Find:

1. Rate at which heat must be supplied per unit length of duct.
2. Temperature of the insulated surface.

Schematic:

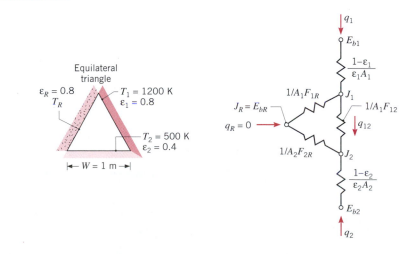

Assumptions:

1. Steady-state conditions exist.
2. All surfaces are opaque, diffuse, gray, and of uniform radiosity.
3. Convection effects are negligible.
4. Surface R is reradiating.
5. End effects are negligible.

Analysis:

1. The system may be modeled as a three-surface enclosure with one surface reradiating. The rate at which energy must be supplied to the heated surface may then be obtained from Equation 13.30:

$$q_1 = \frac{E_{b1} - E_{b2}}{\dfrac{1 - \varepsilon_1}{\varepsilon_1 A_1} + \dfrac{1}{A_1 F_{12} + [(1/A_1 F_{1R}) + (1/A_2 F_{2R})]^{-1}} + \dfrac{1 - \varepsilon_2}{\varepsilon_2 A_2}}$$

From symmetry, $F_{12} = F_{1R} = F_{2R} = 0.5$. Also, $A_1 = A_2 = W \cdot L$, where L is the duct length. Hence

$$q_1' = \frac{q_1}{L} = \frac{5.67 \times 10^{-8} \ \text{W/m}^2 \cdot \text{K}^4 \, (1200^4 - 500^4) \ \text{K}^4}{\dfrac{1 - 0.8}{0.8 \times 1 \ \text{m}} + \dfrac{1}{1 \ \text{m} \times 0.5 + (2 + 2)^{-1} \ \text{m}} + \dfrac{1 - 0.4}{0.4 \times 1 \ \text{m}}}$$

or

$$q_1' = 37 \ \text{kW/m} = -q_2' \qquad \triangleleft$$

2. The temperature of the insulated surface may be obtained from the requirement that $J_R = E_{bR}$, where J_R may be obtained from Equation 13.31. However, to use this expression J_1 and J_2 must be known. Applying the surface energy balance, Equation 13.19, to surfaces 1 and 2, it follows that

$$J_1 = E_{b1} - \frac{1 - \varepsilon_1}{\varepsilon_1 W} q_1' = 5.67 \times 10^{-8} \ \text{W/m}^2 \cdot \text{K}^4 \, (1200 \ \text{K})^4$$

$$- \frac{1 - 0.8}{0.8 \times 1 \ \text{m}} \times 37,000 \ \text{W/m} = 108,323 \ \text{W/m}^2$$

$$J_2 = E_{b2} - \frac{1 - \varepsilon_2}{\varepsilon_2 W} q_2' = 5.67 \times 10^{-8} \ \text{W/m}^2 \cdot \text{K}^4 \, (500 \ \text{K})^4$$

$$- \frac{1 - 0.4}{0.4 \times 1 \ \text{m}} (-37,000 \ \text{W/m}) = 59,043 \ \text{W/m}^2$$

From the energy balance for the reradiating surface, Equation 13.31, it follows that

$$\frac{108,323 - J_R}{\dfrac{1}{W \times L \times 0.5}} - \frac{J_R - 59,043}{\dfrac{1}{W \times L \times 0.5}} = 0$$

Hence

$$J_R = 83,683 \text{ W/m}^2 = E_{bR} = \sigma T_R^4$$

$$T_R = \left(\frac{83,683 \text{ W/m}^2}{5.67 \times 10^{-8} \text{ W/m}^2 \cdot \text{K}^4} \right)^{1/4} = 1102 \text{ K} \qquad \triangleleft$$

Comments:

1. Note that temperature and radiosity discontinuities cannot exist at the corners, and the assumptions of uniform temperature and radiosity are weakest in these regions.

2. The results are independent of the value of ε_R.

3. This problem may also be solved using the matrix inversion method. The solution involves first determining the three unknown radiosities J_1, J_2, and J_R. The governing equations are obtained by writing Equation 13.21 for the two surfaces of known temperature, 1 and 2, and Equation 13.22 for surface R. The three equations are

$$\frac{E_{b1} - J_1}{(1 - \varepsilon_1)/\varepsilon_1 A_1} = \frac{J_1 - J_2}{(A_1 F_{12})^{-1}} + \frac{J_1 - J_R}{(A_1 F_{1R})^{-1}}$$

$$\frac{E_{b2} - J_2}{(1 - \varepsilon_2)/\varepsilon_2 A_2} = \frac{J_2 - J_1}{(A_2 F_{21})^{-1}} + \frac{J_2 - J_R}{(A_2 F_{2R})^{-1}}$$

$$0 = \frac{J_R - J_1}{(A_R F_{R1})^{-1}} + \frac{J_R - J_2}{(A_R F_{R2})^{-1}}$$

Canceling the area A_1, the first equation reduces to

$$\frac{117,573 - J_1}{0.25} = \frac{J_1 - J_2}{2} + \frac{J_1 - J_R}{2}$$

or

$$10J_1 - J_2 - J_R = 940,584 \qquad (1)$$

Similarly, for surface 2,

$$\frac{3544 - J_2}{1.50} = \frac{J_2 - J_1}{2} + \frac{J_2 - J_R}{2}$$

or

$$-J_1 + 3.33J_2 - J_R = 4725 \qquad (2)$$

and for the reradiating surface

$$0 = \frac{J_R - J_1}{2} + \frac{J_R - J_2}{2}$$

or

$$-J_1 - J_2 + 2J_R = 0 \qquad (3)$$

From Equations 1, 2, and 3 the coefficient and right-hand side matrices are

$$\begin{vmatrix} a_{11} & a_{12} & a_{1R} \\ a_{21} & a_{22} & a_{2R} \\ a_{R1} & a_{R2} & a_{RR} \end{vmatrix} = \begin{vmatrix} 10 & -1 & -1 \\ -1 & 3.33 & -1 \\ -1 & -1 & +2 \end{vmatrix}, \quad \begin{vmatrix} C_1 \\ C_2 \\ C_R \end{vmatrix} = \begin{vmatrix} 940{,}584 \\ 4725 \\ 0 \end{vmatrix}$$

Inverting the coefficient matrix, it follows that

$$J_1 = 108{,}328 \text{ W/m}^2 \qquad J_2 = 59{,}018 \text{ W/m}^2 \qquad \text{and} \qquad J_R = 83{,}673 \text{ W/m}^2$$

Recognizing that $J_R = \sigma T_R^4$, it follows that

$$T_R = \left(\frac{J_R}{\sigma} \right)^{1/4} = \left(\frac{83{,}673 \text{ W/m}^2}{5.67 \times 10^{-8} \text{ W/m}^2 \cdot \text{K}^4} \right)^{1/4} = 1102 \text{ K}$$

13.4
Multimode Heat Transfer

Thus far, radiation exchange in an enclosure has been considered under conditions for which conduction and convection could be neglected. However, in many applications, convection and/or conduction are comparable to radiation and must be considered in the heat transfer analysis.

Consider the general surface condition of Figure 13.14*a*. In addition to exchanging energy by radiation with other surfaces of the enclosure, there may be external heat addition to the surface, as, for example, by electric heating, and heat transfer from the surface by convection and conduction. From a surface energy balance, it follows that

$$q_{i,\,ext} = q_{i,\,rad} + q_{i,\,conv} + q_{i,\,cond} \tag{13.32}$$

where $q_{i,\,rad}$, the net rate of radiation transfer from the surface, is determined by standard procedures for an enclosure. Hence, in general, $q_{i,\,rad}$ may be deter-

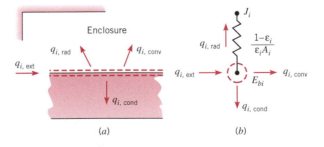

FIGURE 13.14 Multimode heat transfer from a surface in an enclosure. (*a*) Surface energy balance. (*b*) Circuit representation.

mined from Equation 13.19 or 13.20, while for special cases such as a two-surface enclosure and a three-surface enclosure with one reradiating surface, it may be determined from Equations 13.23 and 13.30, respectively. The surface network element of the radiation circuit is modified according to Figure 13.14b, where $q_{i, ext}$, $q_{i, cond}$, and $q_{i, conv}$ represent current flows to or from the surface node. Note, however, that while $q_{i, cond}$ and $q_{i, conv}$ are proportional to temperature differences, $q_{i, rad}$ is proportional to the difference between temperatures raised to the fourth power. Conditions are simplified if the back of the surface is insulated, in which case $q_{i, cond} = 0$. Moreover, if there is no external heating and convection is negligible, the surface is reradiating.

EXAMPLE 13.7

Consider an air heater consisting of a semicircular tube for which the plane surface is maintained at 1000 K and the other surface is well insulated. The tube radius is 20 mm, and both surfaces have an emissivity of 0.8. If atmospheric air flows through the tube at 0.01 kg/s and $T_m = 400$ K, what is the rate at which heat must be supplied per unit length to maintain the plane surface at 1000 K? What is the temperature of the insulated surface?

SOLUTION

Known: Airflow conditions in tubular heater and heater surface conditions.

Find: Rate at which heat must be supplied and temperature of insulated surface.

Schematic:

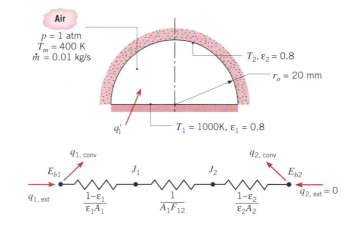

Assumptions:
1. Steady-state conditions.
2. Diffuse, gray surfaces.
3. Negligible tube end effects and axial variations in gas temperature.
4. Fully developed flow.

Properties: Table A.4, air (1 atm, 400 K): $k = 0.0338$ W/m · K, $\mu = 230 \times 10^{-7}$ kg/s · m, $c_p = 1014$ J/kg · K, $Pr = 0.69$.

Analysis: Since the semicircular surface is well insulated and there is no external heat addition, a surface energy balance yields

$$q_{2,\,\text{rad}} = q_{2,\,\text{conv}}$$

Since the tube constitutes a two-surface enclosure, the net radiation transfer to surface 2 may be evaluated from Equation 13.23. Hence

$$\frac{\sigma(T_1^4 - T_2^4)}{\dfrac{1 - \varepsilon_1}{\varepsilon_1 A_1} + \dfrac{1}{A_1 F_{12}} + \dfrac{1 - \varepsilon_2}{\varepsilon_2 A_2}} = hA_2(T_2 - T_m)$$

where the view factor is $F_{12} = 1$ and, per unit length, the surface areas are $A_1 = 2r_o$ and $A_2 = \pi r_o$. With

$$Re_D = \frac{\rho u_m D_h}{\mu} = \frac{\dot{m} D_h}{A_c \mu} = \frac{\dot{m} D_h}{(\pi r_o^2/2)\mu}$$

the hydraulic diameter is

$$D_h = \frac{4A_c}{P} = \frac{2\pi r_o}{\pi + 2} = \frac{0.04\pi \text{ m}}{\pi + 2} = 0.0244 \text{ m}$$

Hence

$$Re_D = \frac{0.01 \text{ kg/s} \times 0.0244 \text{ m}}{(\pi/2)(0.02 \text{ m})^2 \times 230 \times 10^{-7} \text{ kg/s} \cdot \text{m}} = 16,900$$

From the Dittus–Boelter equation,

$$Nu_D = 0.023 Re_D^{4/5} Pr^{0.4}$$

$$Nu_D = 0.023(16,900)^{4/5}(0.69)^{0.4} = 47.8$$

$$h = \frac{k}{D_h} Nu_D = \frac{0.0338 \text{ W/m} \cdot \text{K}}{0.0244 \text{ m}} 47.8 = 66.2 \text{ W/m}^2 \cdot \text{K}$$

Dividing both sides of the energy balance by A_1, it follows that

$$\frac{5.67 \times 10^{-8} \text{ W/m}^2 \cdot \text{K}^4 \, [(1000)^4 - T_2^4] \text{ K}^4}{\dfrac{1 - 0.8}{0.8} + 1 + \dfrac{1 - 0.8}{0.8} \dfrac{2}{\pi}} = 66.2 \frac{\pi}{2} (T_2 - 400) \text{ W/m}^2$$

or

$$5.67 \times 10^{-8} T_2^4 + 146.5 T_2 - 115,313 = 0$$

A trial-and-error solution yields

$$T_2 = 696 \text{ K} \qquad \qquad \triangleleft$$

From an energy balance at the heated surface,

$$q_{1,\,\text{ext}} = q_{1,\,\text{rad}} + q_{1,\,\text{conv}} = q_{2,\,\text{conv}} + q_{1,\,\text{conv}}$$

Hence, on a unit length basis,

$$q'_{1,\,ext} = h\pi r_o(T_2 - T_m) + h2r_o(T_1 - T_m)$$

$$q'_{1,\,ext} = 66.2 \times 0.02[\pi(696 - 400) + 2(1000 - 400)] \text{ W/m}$$

$$q'_{1,\,ext} = (1231 + 1589) \text{ W/m} = 2820 \text{ W/m} \qquad \triangleleft$$

Comments: Applying an energy balance to a differential control volume about the air, it follows that

$$\frac{dT_m}{dx} = \frac{q'_1}{\dot{m}c_p} = \frac{2820 \text{ W/m}}{0.01 \text{ kg/s} (1014 \text{ J/kg} \cdot \text{K})} = 278 \text{ K/m}$$

Hence the air temperature change is significant, and a more representative analysis would subdivide the tube into axial zones and would allow for variations in air and insulated surface temperatures between zones. Moreover, a two-surface analysis of radiation exchange would no longer be appropriate.

13.5
Additional Effects

Although we have developed means for predicting radiation exchange between surfaces, it is important to be cognizant of the inherent limitations. Recall that we have considered *isothermal, opaque, gray* surfaces that *emit* and *reflect diffusely* and that are characterized by *uniform* surface *radiosity* and *irradiation*. For enclosures we have also considered the medium that separates the surfaces to be *nonparticipating;* that is, it neither absorbs nor scatters the surface radiation, and it emits no radiation.

The foregoing conditions and the related equations may often be used to obtain reliable first estimates and, in most cases, highly accurate results for radiation transfer in an enclosure. Sometimes, however, the assumptions are grossly inappropriate and more refined prediction methods are needed. Although beyond the scope of this text, the methods are discussed in more advanced treatments of radiation transfer [3, 6–11].

We have said little about gaseous radiation, having confined our attention to radiation exchange at the surface of an opaque solid or liquid. For *nonpolar* gases, such as O_2 or N_2, such neglect is justified, since the gases do not emit radiation and are essentially transparent to incident thermal radiation. However, the same may not be said for polar molecules, such as CO_2, H_2O (vapor), NH_3, and hydrocarbon gases, which emit and absorb over a wide temperature range. For such gases matters are complicated by the fact that, unlike radiation from a solid or a liquid, which is distributed continuously with wavelength, gaseous radiation is concentrated in specific *wavelength intervals* (called bands). Moreover, gaseous radiation is not a surface phenomenon, but is instead a *volumetric* phenomenon.

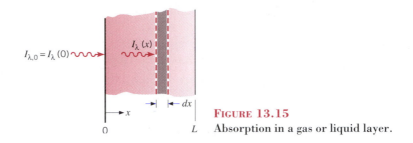

FIGURE 13.15
Absorption in a gas or liquid layer.

13.5.1 Volumetric Absorption

Spectral radiation absorption in a gas (or in a semitransparent liquid or solid) is a function of the absorption coefficient κ_λ (1/m) and the thickness L of the medium (Figure 13.15). If a monochromatic beam of intensity $I_{\lambda,0}$ is incident on the medium, the intensity is reduced due to absorption, and the reduction occurring in an infinitesimal layer of thickness dx may be expressed as

$$dI_\lambda(x) = -\kappa_\lambda I_\lambda(x)\,dx \tag{13.33}$$

Separating variables and integrating over the entire layer, we obtain

$$\int_{I_{\lambda,0}}^{I_{\lambda,L}} \frac{dI_\lambda(x)}{I_\lambda(x)} = -\kappa_\lambda \int_0^L dx$$

where κ_λ is assumed to be independent of x. It follows that

$$\frac{I_{\lambda,L}}{I_{\lambda,0}} = e^{-\kappa_\lambda L} \tag{13.34}$$

This exponential decay, termed *Beer's law,* is a useful tool in approximate radiation analysis. It may, for example, be used to infer the overall spectral absorptivity of the medium. In particular, with the transmissivity defined as

$$\tau_\lambda = \frac{I_{\lambda,L}}{I_{\lambda,0}} = e^{-\kappa_\lambda L} \tag{13.35}$$

the absorptivity is

$$\alpha_\lambda = 1 - \tau_\lambda = 1 - e^{-\kappa_\lambda L} \tag{13.36}$$

If Kirchhoff's law is assumed to be valid, $\alpha_\lambda = \varepsilon_\lambda$, Equation 13.36 also provides the spectral emissivity of the medium.

13.5.2 Gaseous Emission and Absorption

A common engineering calculation is one that requires determination of the radiant heat flux from a gas to an adjoining surface. Despite the complicated spectral and directional effects inherent in such calculations, a simplified procedure may be used. The method was developed by Hottel [12] and involves determining radiation emission from a hemispherical gas mass of temperature T_g to a surface element dA_1, which is located at the center of the hemisphere's base. Emission from the gas per unit area of the surface is expressed as

$$E_g = \varepsilon_g \sigma T_g^4 \tag{13.37}$$

where the gas emissivity ε_g was determined by correlating available data. In particular, ε_g was correlated in terms of the temperature T_g and total pressure p of the gas, the partial pressure p_g of the radiating species, and the radius L of the hemisphere.

Results for the emissivity of water vapor are plotted in Figure 13.16 as a function of the gas temperature, for a total pressure of 1 atm, and for different values of the product of the vapor partial pressure and the hemisphere radius. To evaluate the emissivity for total pressures other than 1 atm, the emissivity from Figure 13.16 must be multiplied by the correction factor C_w from Figure 13.17. Similar results were obtained for carbon dioxide and are presented in Figures 13.18 and 13.19.

The foregoing results apply when water vapor or carbon dioxide appear *separately* in a mixture with other species that are nonradiating. However, the results may readily be extended to situations in which water vapor and carbon dioxide appear *together* in a mixture with other nonradiating gases. In particular, the total gas emissivity may be expressed as

$$\varepsilon_g = \varepsilon_w + \varepsilon_c - \Delta\varepsilon \tag{13.38}$$

where the correction factor $\Delta\varepsilon$ is presented in Figure 13.20 for different values of the gas temperature. This factor accounts for the reduction in emission associated with the mutual absorption of radiation between the two species.

Recall that the foregoing results provide the emissivity of a hemispherical gas mass of radius L radiating to an element of area at the center of its base.

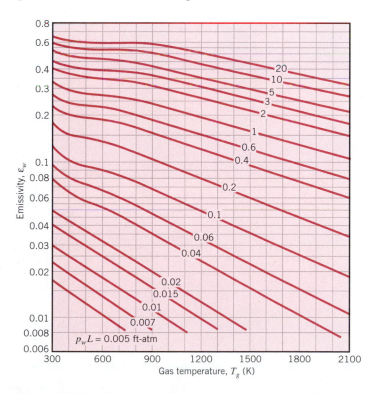

FIGURE 13.16　Emissivity of water vapor in a mixture with nonradiating gases at 1-atm total pressure and of hemispherical shape [12]. Used with permission.

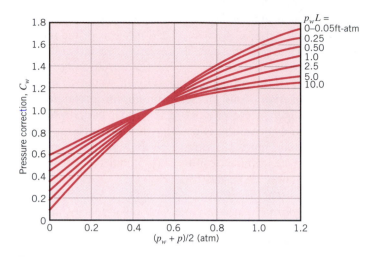

FIGURE 13.17 Correction factor for obtaining water vapor emissivities at pressures other than 1 atm ($\varepsilon_{w,\,p\neq1\text{ atm}} = C_w\varepsilon_{w,\,p=1\text{ atm}}$) [12]. Used with permission.

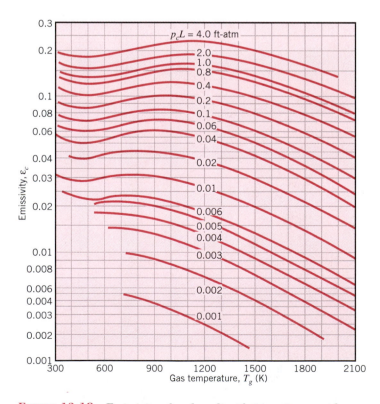

FIGURE 13.18 Emissivity of carbon dioxide in a mixture with nonradiating gases at 1-atm total pressure and of hemispherical shape [12]. Used with permission.

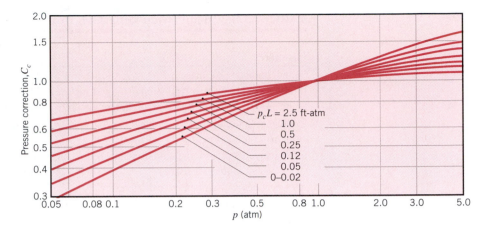

FIGURE 13.19 Correction factor for obtaining carbon dioxide emissivities at pressures other than 1 atm ($\varepsilon_{c,\,p \neq 1\;atm} = C_c \varepsilon_{c,\,p=1\;atm}$) [12]. Used with permission.

However, the results may be extended to other gas geometries by introducing the concept of a *mean beam length, L_e*. The quantity was introduced to correlate, in terms of a single parameter, the dependence of gas emissivity on both the size and the shape of the gas geometry. It may be interpreted as the radius of a hemispherical gas mass whose emissivity is equivalent to that for the geometry of interest. Its value has been determined for numerous gas shapes [12], and representative results are listed in Table 13.4. Replacing L by L_e in Figures 13.16 to 13.20, the emissivity associated with the geometry of interest may then be determined.

Using the results of Table 13.4 with Figures 13.16 to 13.20, it is possible to determine the rate of radiant heat transfer to a surface due to emission from an adjoining gas. This heat rate may be expressed as

$$q = \varepsilon_g A_s \sigma T_g^4 \tag{13.39}$$

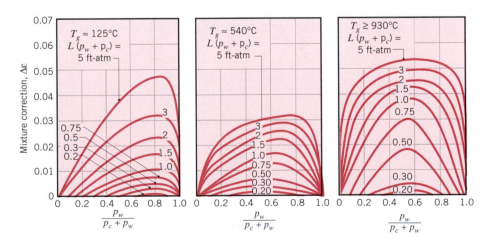

FIGURE 13.20 Correction factor associated with mixtures of water vapor and carbon dioxide [12]. Used with permission.

TABLE 13.4   **Mean Beam Lengths L_e for Various Gas Geometries**

Geometry	Characteristic Length	L_e
Sphere (radiation to surface)	Diameter (D)	$0.65D$
Infinite circular cylinder (radiation to curved surface)	Diameter (D)	$0.95D$
Semi-infinite circular cylinder (radiation to base)	Diameter (D)	$0.65D$
Circular cylinder of equal height and diameter (radiation to entire surface)	Diameter (D)	$0.60D$
Infinite parallel planes (radiation to planes)	Spacing between planes (L)	$1.80L$
Cube (radiation to any surface)	Side (L)	$0.66L$
Arbitrary shape of volume V (radiation to surface of area A)	Volume to area ratio (V/A)	$3.6V/A$

where A_s is the surface area. If the surface is black, it will, of course, absorb all this radiation. A black surface will also emit radiation, and the net rate at which radiation is exchanged between the surface at T_s and the gas at T_g is

$$q_{net} = A_s \sigma (\varepsilon_g T_g^4 - \alpha_g T_s^4) \tag{13.40}$$

For water vapor and carbon dioxide the required gas absorptivity α_g may be evaluated from the emissivity by expressions of the form [12]

Water:

$$\alpha_w = C_w \left(\frac{T_g}{T_s} \right)^{0.45} \times \varepsilon_w \left(T_s, p_w L_e \frac{T_s}{T_g} \right) \tag{13.41}$$

Carbon dioxide:

$$\alpha_c = C_c \left(\frac{T_g}{T_s} \right)^{0.65} \times \varepsilon_c \left(T_s, p_c L_e \frac{T_s}{T_g} \right) \tag{13.42}$$

where ε_w and ε_c are evaluated from Figures 13.16 and 13.18, respectively, and C_w and C_c are evaluated from Figures 13.17 and 13.19, respectively. Note, however, that in using Figures 13.16 and 13.18, T_g is replaced by T_s and $p_w L_e$ or $p_c L_e$ is replaced by $p_w L_e (T_s/T_g)$ or $p_c L_e (T_s/T_g)$, respectively. Note also that, in the presence of both water vapor and carbon dioxide, the total gas absorptivity may be expressed as

$$\alpha_g = \alpha_w + \alpha_c - \Delta\alpha \tag{13.43}$$

where $\Delta\alpha = \Delta\varepsilon$ is obtained from Figure 13.20.

13.6
Summary

In this chapter we focused on the analysis of radiation exchange between the surfaces of an enclosure, and in treating this exchange we introduced the *view factor* concept. Because knowing this geometrical quantity is essential to determining radiation exchange between any two diffuse surfaces, you should be familiar with the means by which it may be determined. You should also be adept at performing radiation calculations for an enclosure of *isothermal, opaque, diffuse,* and *gray* surfaces of *uniform radiosity* and *irradiation.* Moreover, you should be familiar with the results that apply to simple cases such as the two-surface enclosure or the three-surface enclosure with a reradiating surface.

References

1. Hamilton, D. C., and W. R. Morgan, "Radiant Interchange Configuration Factors," National Advisory Committee for Aeronautics, Technical Note 2836, 1952.

2. Eckert, E. R. G., "Radiation: Relations and Properties," in W. M. Rohsenow and J. P. Hartnett, Eds., *Handbook of Heat Transfer,* 2nd ed., McGraw-Hill, New York, 1973.

3. Siegel, R., and J. R. Howell, *Thermal Radiation Heat Transfer,* McGraw-Hill, New York, 1981.

4. Howell, J. R., *A Catalog of Radiation Configuration Factors,* McGraw-Hill, New York, 1982.

5. Oppenheim, A. K., *Trans. ASME,* **65,** 725, 1956.

6. Hottel, H. C., and A. F. Sarofim, *Radiative Transfer,* McGraw-Hill, New York, 1967.

7. Tien, C. L., "Thermal Radiation Properties of Gases," in J. P. Hartnett and T. F. Irvine, Eds., *Advances in Heat Transfer,* Vol. 5, Academic Press, New York, 1968.

8. Sparrow, E. M., "Radiant Interchange Between Surfaces Separated by Nonabsorbing and Nonemitting Media," in W. M. Rohsenow and J. P. Hartnett, Eds., *Handbook of Heat Transfer,* McGraw-Hill, New York, 1973.

9. Dunkle, R. V., "Radiation Exchange in an Enclosure with a Participating Gas," in W. M. Rohsenow and J. P. Hartnett, Eds., *Handbook of Heat Transfer,* McGraw-Hill, New York, 1973.

10. Sparrow, E. M., and R. D. Cess, *Radiation Heat Transfer,* Hemisphere Publishing, New York, 1978.

11. Edwards, D. K., *Radiation Heat Transfer Notes,* Hemisphere Publishing, New York, 1981.

12. Hottel, H. C., "Radiant-Heat Transmission," in W. H. McAdams, Ed., *Heat Transmission,* 3rd ed., McGraw-Hill, New York, 1954.

Problems

View Factors

13.1 Determine F_{12} and F_{21} for the following configurations using the reciprocity theorem and other basic shape factor relations. Do not use tables or charts.

(a) Long duct

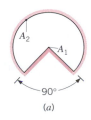

(*a*)

(b) Small sphere of area A_1 under a concentric hemisphere of area $A_2 = 2A_1$

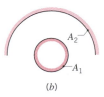

(*b*)

(c) Long duct. What is F_{22} for this case?

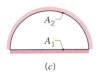

(*c*)

(d) Long inclined plates (point *B* is directly above the center of A_1)

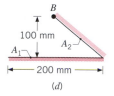

(*d*)

(e) Sphere lying on infinite plane

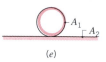

(*e*)

(f) Hemisphere–disk arrangement

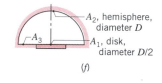

(*f*)

(g) Long, open channel

(*g*)

13.2 Consider the following grooves, each of width *W*, that have been machined from a solid block of material.

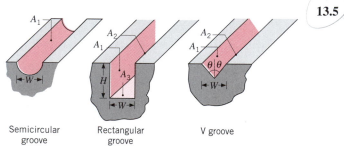

Semicircular groove   Rectangular groove   V groove

(a) For each case obtain an expression for the view factor of the groove with respect to the surroundings outside the groove.

(b) For the V groove, obtain an expression for the view factor F_{12}, where A_1 and A_2 are opposite surfaces.

(c) If $H = 2W$ in the rectangular groove, what is the view factor F_{12}?

13.3 Consider a long cylindrical tube having an inner surface area A_2. Surfaces represented by dashed lines (A_1) in the sketches are inserted into the tube. The curved surfaces (*b*, *c*) are formed by cylindrical segments tangent to one another at their tips.

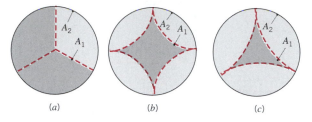

(*a*)   (*b*)   (*c*)

Determine the shape factors F_{12} and F_{21} for each of the arrangements.

13.4 A thin metal hemispherical shell of diameter $D = 0.8$ m is suspended inside a 1.5-m cubical enclosure.

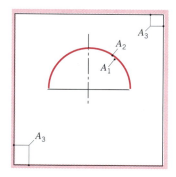

Determine the view factors F_{11}, F_{22}, and F_{33}.

13.5 The "crossed-strings" method of Hottel [12] provides a simple means to calculate view factors between surfaces that are of infinite extent in one direction. For two such surfaces (*a*) with unobstructed views of one another, the view factor is of the form

$$F_{12} = \frac{1}{2w_1} [(ac + bd) - (ad + bc)]$$

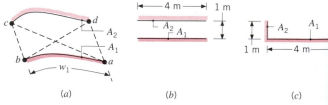

(*a*)   (*b*)   (*c*)

Use this method to evaluate the view factors F_{12} for sketches (*b*) and (*c*). Compare your results with those from the appropriate graphs and analytical expressions.

13.6 Consider the two diffuse surfaces shown in Example 13.1. Beginning with the general definition of $F_{ij} = q''_{i \to j}/J_i$, show that when $L \gg D$, the shape factor for a small area element (i) to a disk (j) is $F_{ij} = D^2/4L^2$. How does this compare with the result of Example 13.1?

13.7 Consider the perpendicular rectangles shown schematically.

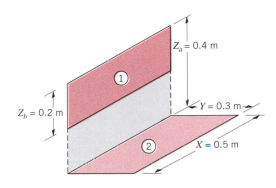

(a) Determine the shape factor F_{12}.

(b) For rectangle widths of $X = 0.5$, 1.5, and 5 m, plot F_{12} as a function of Z_b for $0.05 \leq Z_b \leq 0.4$ m. Compare your results with the view factor obtained from the two-dimensional relation for perpendicular plates with a common edge (Table 13.1).

13.8 The reciprocity relation, the summation rule, and Equations 13.5 to 13.7 can be used to develop view factor relations that allow for applications of Figure 13.4 and/or 13.6 to more complex configurations. Consider the view factor F_{14} for surfaces 1 and 4 of the following geometry. These surfaces are perpendicular but do not share a common edge.

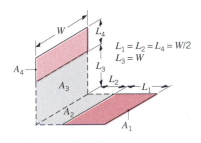

(a) Obtain the following expression for the view factor F_{14}:

$$F_{14} = \frac{1}{A_1}[(A_1 + A_2)F_{(1, 2)(3, 4)} + A_2 F_{23}$$
$$- (A_1 + A_2)F_{(1, 2)3} - A_2 F_{2(3, 4)}]$$

(b) If $L_1 = L_2 = L_4 = (W/2)$ and $L_3 = W$, what is the value of F_{14}?

13.9 Determine the shape factor, F_{12}, for the rectangles shown.

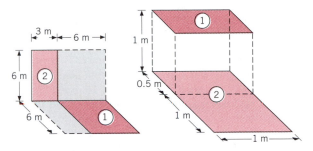

(a) Perpendicular rectangles without a common edge.

(b) Parallel rectangles of unequal areas.

13.10 For a diffuse emitting surface, show that F_ω, the ratio of the radiant power emitted into a solid angle, ω, about the surface normal to the emissive power of the surface, has the form

$$F_\omega = 1 - \cos^2 \theta = \frac{\omega}{2\pi}\left(2 - \frac{\omega}{2\pi}\right)$$

where θ is the plane angle measured from the normal to the surface.

13.11 Consider two diffuse surfaces A_1 and A_2 on the inside of a spherical enclosure of radius R. Using the following methods, derive an expression for the view factor F_{12} in terms of A_2 and R.

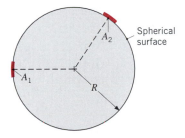

(a) Find F_{12} by beginning with the expression $F_{ij} = q_{i \to j}/A_i J_i$.

(b) Find F_{12} using the view factor integral, Equation 13.1.

13.12 As shown in the sketch, consider the disk A_1 located coaxially 1 m distant, but tilted 30° off the normal, from the ring-shaped disk A_2.

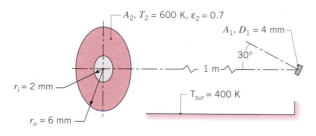

What is the irradiation on A_1 due to radiation from A_2, which is a diffuse, gray surface with an emissivity of 0.7?

13.13   A heat flux gage of 4-mm diameter is positioned normal to and 1 m from the 5-mm diameter aperture of a blackbody furnace at 1000 K. The diffuse, gray cover shield ($\varepsilon = 0.2$) of the furnace has an outer diameter of 100 mm and its temperature is 350 K. The furnace and gage are located in a large room whose walls have an emissivity of 0.8 and are at 300 K.

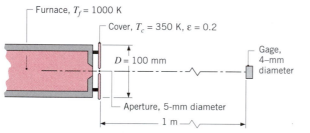

(a) What is the irradiation on the gage, G_g (W/m²), considering only emission from the aperture of the furnace?

(b) What is the irradiation on the gage due to radiation from the cover and aperture?

13.14   Consider the parallel planes of infinite extent normal to the page having opposite edges aligned as shown in the sketch.

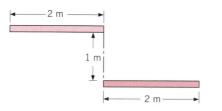

(a) Using appropriate view factor relations and the results for opposing parallel planes, develop an expression for the view factor F_{12}.

(b) Use Hottel's cross-string method described in Problem 13.5 to determine the view factor.

Blackbody Radiation Exchange

13.15   A tubular heater with a black inner surface of uniform temperature $T_s = 1000$ K irradiates a coaxial disk.

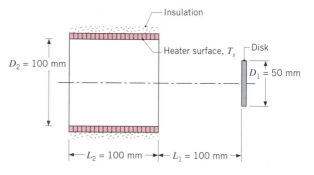

(a) Determine the radiant power from the heater, which is incident on the disk, $q_{s\rightarrow 1}$. What is the irradiation on the disk, G_1?

(b) For disk diameters of $D_1 = 25$, 50, and 100 mm, plot $q_{s\rightarrow 1}$ and G_1 as a function of the separation distance L_1 for $0 \leq L_1 \leq 200$ mm.

13.16   Consider the arrangement of the three black surfaces shown, where A_1 is small compared to A_2 or A_3.

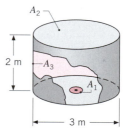

Determine the value of F_{13}. Calculate the net radiation heat transfer from A_1 to A_3 if $A_1 = 0.05$ m², $T_1 = 1000$ K, and $T_3 = 500$ K.

13.17   Two perfectly black, parallel disks 1 m in diameter are separated a distance of 0.25 m. One disk is maintained at 60°C and the other at 20°C. The disks are located in a large room whose walls are 40°C. Assume that the exterior surfaces of the disks (the surfaces that do not face each other) are very well insulated. Determine the net radiation exchange between the disks. Determine the net radiation exchange between the disks and the room.

13.18   A drying oven consists of a long semicircular duct of diameter $D = 1$ m.

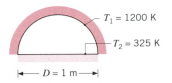

T_1 = 1200 K

T_2 = 325 K

D = 1 m

Materials to be dried cover the base of the oven, while the wall is maintained at 1200 K. What is the drying rate per unit length of the oven (kg/s · m) if a water-coated layer of material is maintained at 325 K during the drying process? Blackbody behavior may be assumed for the water surface and for the oven wall.

13.19 A circular disk of diameter $D_1 = 20$ mm is located at the base of an enclosure that has a cylindrical sidewall and a hemispherical dome. The enclosure is of diameter $D = 0.5$ m, and the height of the cylindrical section is $L = 0.3$ m. The disk and the enclosure surface are black and at temperatures of 1000 and 300 K, respectively.

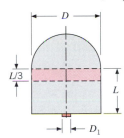

D

$L/3$

L

D_1

What is the net rate of radiation exchange between the disk and the hemispherical dome? What is the net rate of radiation exchange between the disk and the top one-third portion of the cylindrical section?

13.20 An arrangement for curing the surface coating of a panel involves placing the panel under a planar radiation source that is parallel to and midway between the panel. The panel and source may be approximated as blackbodies at temperatures of $T_p = 600$ K and $T_s = 1000$ K, respectively. The panel and source are of widths $W_p = 0.30$ m and $W_s = 0.15$ m, are very long (in the direction perpendicular to the paper), and are separated by a distance of $L = 0.15$ m.

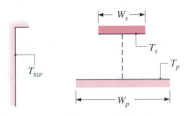

W_s

T_s

T_p

T_{sur}

W_p

If the region between the panel and source is not enclosed, but is exposed to large surroundings at 300 K, what is the rate at which electrical power must be supplied to the radiation source per unit length (perpendicular to the paper)? Convection effects may be neglected. If the region between the panel and source is enclosed by adiabatic sidewalls, what is the electric power requirement per unit length of the radiation source? Once again, convection effects may be neglected.

13.21 Consider coaxial, parallel, black disks separated a distance of 0.20 m. The lower disk of diameter 0.40 m is maintained at 500 K and the surroundings are at 300 K. What temperature will the upper disk of diameter 0.20 m achieve if electrical power of 17.5 W is supplied to the heater on the back side of the disk?

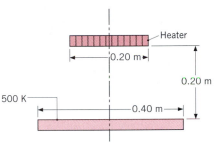

Heater

0.20 m

0.20 m

500 K

0.40 m

13.22 A circular plate of 500-mm diameter is maintained at $T_1 = 600$ K and is positioned coaxial to a conical shape. The back side of the cone is well insulated. The plate and the cone, whose surfaces are black, are located in a large, evacuated enclosure whose walls are at 300 K.

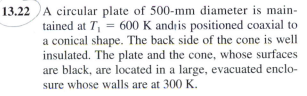

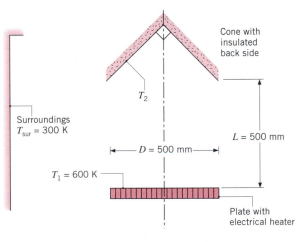

Cone with insulated back side

T_2

Surroundings
T_{sur} = 300 K

D = 500 mm

L = 500 mm

T_1 = 600 K

Plate with electrical heater

(a) What is the temperature of the conical surface, T_2?

(b) What is the electrical power that would be required to maintain the circular plate at 600 K?

13.23 A furnace is constructed in three sections, which include insulated circular (2) and cylindrical (3) sections, as well as an intermediate cylindrical section (1) with imbedded electrical resistance heaters. The overall length and diameter are 200 mm and 100 mm, respectively, and the cylindrical sections are of equal length. The surroundings are at 300 K.

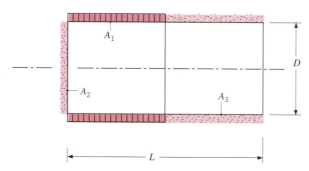

(a) If all the surfaces are black, determine the electrical power, q_1, required to maintain the heated section at 1000 K.

(b) What are the temperatures of the insulated sections, T_2 and T_3?

(c) For $D = 100$ mm, generate a plot of q_1, T_2, and T_3 as functions of the length-to-diameter ratio, with $1 \leq L/D \leq 5$.

13.24 In the arrangement shown, the lower disk has a diameter of 30 mm and a temperature of 500 K. The upper surface, which is at 1000 K, is a ring-shaped disk with inner and outer diameters of 0.15 m and 0.2 m. This upper surface is aligned with and parallel to the lower disk and is separated by a distance of 1 m.

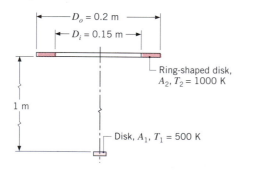

Assuming both surfaces to be blackbodies, calculate their net radiative heat exchange.

13.25 A radiometer views a small target (1) that is being heated by a ring-shaped disk heater (2). The target has an area of $A_1 = 0.0004$ m², a temperature of $T_1 = 500$ K, and a diffuse, gray emissivity of $\varepsilon_1 = 0.8$. The heater operates at $T_2 = 1000$ K and has a black surface. The radiometer views the entire sample area with a solid angle of $\omega = 0.0008$ sr.

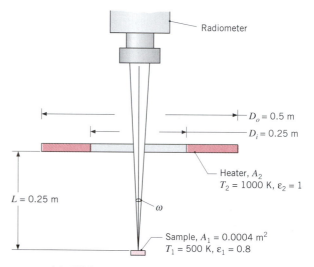

(a) Write an expression for the radiant power leaving the target, which is collected by the radiometer, in terms of the target radiosity J_1 and relevant geometric parameters. Leave in symbolic form.

(b) Write an expression for the target radiosity J_1 in terms of its irradiation, emissive power, and appropriate radiative properties. Leave in symbolic form.

(c) Write an expression for the irradiation on the target, G_1, due to emission from the heater in terms of the heater emissive power, the heater area, and an appropriate view factor. Use this expression to numerically evaluate G_1.

(d) Use the foregoing expressions and results to determine the radiant power collected by the radiometer.

13.26 An electrically heated sample is maintained at a surface temperature of $T_s = 500$ K. The sample coating is diffuse but spectrally selective, with the spectral emissivity distribution shown schematically. The sample is irradiated by a furnace located coaxially at a distance of $L_{sf} = 750$ mm. The furnace has isothermal walls with an emissivity of $\varepsilon_f = 0.7$ and a uniform temperature of $T_f = 3000$ K. A radiation detector of area $A_d = 8 \times 10^{-5}$ m² is positioned at a distance of $L_{sd} = 1.0$ m from the sample along a direction that is 45° from the sample normal. The detector is sensitive to spectral radiant power only in the spectral region from 3 to 5 μm. The sample surface experiences convection with a gas for which $T_\infty = 300$ K and $h = 20$ W/m² · K. The surroundings of the sample mount are large and at a uniform temperature of $T_{sur} = 300$ K.

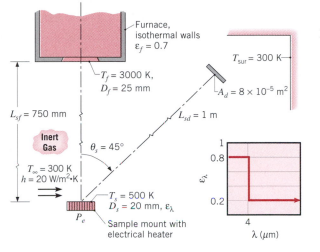

(a) Determine the electrical power, P_e, required to maintain the sample at $T_s = 500$ K.

(b) Considering both emission and reflected irradiation from the sample, determine the radiant power that is incident on the detector within the spectral region from 3 to 5 μm.

13.27 Consider the cylindrical cavity of diameter D and length L having the lateral (A_1) and bottom (A_2) surfaces maintained at temperatures T_1 and T_2, respectively. Assuming that A_1 and A_2 emit as blackbodies, develop an expression for the emissive power of the cavity opening A_3 in terms of T_1, T_2, and the shape factor F_{13}.

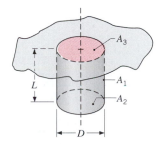

13.28 A cylindrical cavity of diameter D and depth L is machined in a metal block, and conditions are such that the base and side surfaces of the cavity are maintained at $T_1 = 1000$ K and $T_2 = 700$ K, respectively. Approximating the surfaces as

black, determine the emissive power of the cavity if $L = 20$ mm and $D = 10$ mm.

13.29 Two parallel plates 1 m by 1 m, insulated on their back sides and separated by 1 m, may be approximated as blackbodies at 500 and 750 K. The plates are located in a large room whose walls are maintained at 300 K. Determine the net radiative heat transfer from each plate and the net radiative heat transfer to the room walls.

13.30 The arrangement shown is to be used to calibrate a heat flux gage. The gage has a black surface that is 10 mm in diameter and is maintained at 17°C by means of a water-cooled backing plate. The heater, 200 mm in diameter, has a black surface that is maintained at 800 K and is located 0.5 m from the gage. The surroundings and the air are at 27°C and the convection heat transfer coefficient between the gage and the air is 15 W/m² · K.

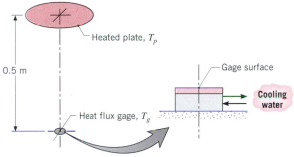

(a) Determine the net radiation exchange between the heater and the gage.

(b) Determine the net transfer of radiation to the gage per unit area of the gage.

(c) What is the net heat transfer rate to the gage per unit area of the gage?

(d) If the gage is constructed according to the description of Problem 3.92, what heat flux will it indicate?

13.31 A long, cylindrical heating element of 20-mm diameter operating at 700 K in vacuum is located 40 mm from an insulated wall.

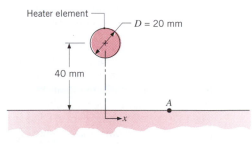

Assuming that the wall thermal conductivity is very low and that both the element and wall are black, estimate the maximum temperature reached by the wall when the surroundings are at 300 K. Estimate the wall temperature at point A, where $x = 40$ mm.

13.32 Water flowing through a large number of long, circular, thin-walled tubes is heated by means of hot parallel plates above and below the tube array. The space between the plates is evacuated, and the plate and tube surfaces may be approximated as blackbodies.

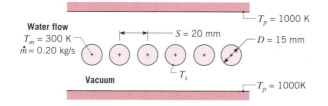

(a) Neglecting axial variations, determine the tube surface temperature, T_s, if water flows through each tube at a mass rate of $\dot{m} = 0.20$ kg/s and a mean temperature of $T_m = 300$ K.

(b) Compute and plot the surface temperature as a function of flow rate for $0.05 \leq \dot{m} \leq 0.25$ kg/s.

13.33 A row of regularly spaced, cylindrical heating elements is used to maintain an insulated furnace wall at 500 K. The opposite wall is at a uniform temperature of 300 K.

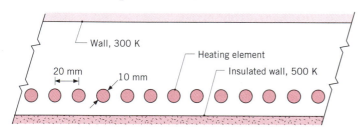

The insulated wall experiences convection with air at 450 K and a convection coefficient of 200 W/m² · K. Assuming the walls and elements are black, estimate the required operating temperature for the elements.

13.34 A manufacturing process calls for heating long copper rods, which are coated with a thin film having $\varepsilon = 1$, by placing them in a large evacuated oven whose surface is maintained at 1650 K. The rods are of 10-mm diameter and are placed in the oven with an initial temperature of 300 K.

(a) What is the initial rate of change of the rod temperature?

(b) How long must the rods remain in the oven to achieve a temperature of 1000 K?

(c) The heating process may be accelerated by routing combustion gases, also at 1650 K, through the oven. For convection coefficients of 10, 100, and 500 W/m² · K, determine the time required for the rods to reach 1000 K.

13.35 Consider the very long, inclined black surfaces (A_1, A_2) maintained at uniform temperatures of $T_1 = 1000$ K and $T_2 = 800$ K.

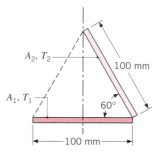

Determine the net radiation exchange between the surfaces per unit length of the surfaces. Consider the configuration when a black surface (A_3), whose back side is insulated, is positioned along the dashed line shown. Calculate the net radiation transfer to surface A_2 per unit length of the surface and determine the temperature of the insulated surface A_3.

13.36 Two plane coaxial disks are separated by a distance $L = 0.20$ m. The lower disk (A_1) is solid with a diameter $D_o = 0.80$ m and a temperature $T_1 = 300$ K. The upper disk (A_2), at temperature $T_2 = 1000$ K, has the same outer diameter but is ring shaped with an inner diameter $D_i = 0.40$ m. Assuming the disks to be blackbodies, calculate the net radiative heat exchange between them.

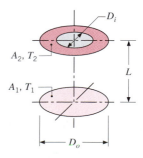

13.37 A circular ice rink 25 m in diameter is enclosed by a hemispherical dome 35 m in diameter. If the ice and dome surfaces may be approximated as blackbodies and are at 0 and 15°C, respectively, what is the net rate of radiative transfer from the dome to the rink?

13.38 A round tube with a diameter of 0.75 m and a length of 0.33 m has an electrical heater wrapped around the outside, and a heavy layer of insulation is wrapped over the tube–heater combination. The tube is open at both ends and is suspended in a very large vacuum chamber whose walls are at 27°C. The inside surface of the tube is black and is maintained at a steady uniform temperature of 127°C.

(a) Determine the electrical power P_e that must be supplied to the heater.

(b) For tube temperatures of 127, 177, and 227°C, plot P_e as a function of the tube length L over a range from 25 to 250 mm.

Enclosures of Diffuse, Gray Surfaces

13.39 Consider two very large parallel plates with diffuse, gray surfaces.

$T_1 = 1000$ K, $\varepsilon_1 = 1$

$T_2 = 500$ K, $\varepsilon_2 = 0.8$

Determine the irradiation and radiosity for the upper plate. What is the radiosity for the lower plate? What is the net radiation exchange between the plates per unit area of the plates?

13.40 A flat-bottomed hole 6 mm in diameter is drilled to a depth of 24 mm in a diffuse, gray material having an emissivity of 0.8 and a uniform temperature of 1000 K.

(a) Determine the radiant power leaving the opening of the cavity.

(b) The effective emissivity ε_e of a cavity is defined as the ratio of the radiant power leaving the cavity to that from a blackbody having the area of the cavity opening and a temperature of the inner surfaces of the cavity. Calculate the effective emissivity of the cavity described above.

(c) If the depth of the hole were increased, would ε_e increase or decrease? What is the limit of ε_e as the depth increases?

13.41 Consider a long V groove 10 mm deep machined on a block that is maintained at 1000 K.

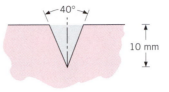

If the groove surfaces are diffuse and gray with an emissivity of 0.6, determine the radiant flux leaving the groove to its surroundings. Also determine the effective emissivity of the cavity, as defined in Problem 13.40.

13.42 Consider a flat-bottomed hole of 10-mm diameter, which is drilled to a depth of 40 mm in a metallic block of emissivity $\varepsilon = 0.7$. If the block is maintained at 1000 K, what is the rate at which radiation leaves the hole? What is the effective, or apparent, emissivity of the hole, as defined in Problem 13.40.

13.43 Consider the cavities formed by a cone, cylinder, and sphere having the same opening size (d) and major dimension (L), as shown in the diagram.

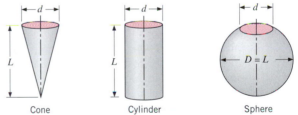

Cone   Cylinder   Sphere

(a) Find the view factor between the inner surface of each cavity and the opening of the cavity.

(b) Find the effective emissivity of each cavity, ε_e, as defined in Problem 13.40, assuming the inner walls are diffuse and gray with an emissivity of ε_w.

(c) For each cavity and wall emissivities of $\varepsilon_w = 0.5, 0.7,$ and 0.9, plot ε_e as a function of the major dimension-to-opening size ratio, L/d, over a range from 1 to 10.

13.44 A very long ($L \gg 1$ m), diffuse, gray, thin-walled tube of 1 m radius is contained inside a very long black duct with a square, 3.2 m $\times$ 3.2 m, cross-section. The top portion of the tube is open, as shown schematically.

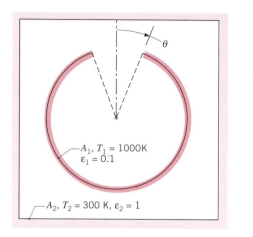

(a) For $\theta = 45°$, determine the net radiant heat transfer rate per unit length of the tube from the opening, q_1/L, and the effective emissivity of the opening, ε_{eff}.

(b) For $0 \leq \theta \leq 180°$, plot q_1/L and ε_{eff} as a function of θ.

13.45 Consider a cylindrical cavity 100 mm in diameter and 50 mm in depth whose walls are diffuse and gray with an emissivity of 0.6 and have a uniform temperature of 1500 K. Assuming the surroundings of the cavity opening are very large and at 300 K, calculate the net radiation transfer from the cavity.

13.46 A long, thin-walled horizontal tube 100 mm in diameter is maintained at 120°C by the passage of steam through its interior. A radiation shield is installed around the tube, providing an air gap of 10 mm between the tube and the shield, and reaches a surface temperature of 35°C. The tube and shield are diffuse, gray surfaces with emissivities of 0.80 and 0.10, respectively. What is the radiant heat transfer from the tube per unit length?

13.47 A very long electrical conductor 10 mm in diameter is concentric with a cooled cylindrical tube 50 mm in diameter whose surface is diffuse and gray with an emissivity of 0.9 and temperature of 27°C. The electrical conductor has a diffuse, gray surface with an emissivity of 0.6 and is dissipating 6.0 W per meter of length. Assuming that the space between the two surfaces is evacuated, calculate the surface temperature of the conductor.

13.48 Under steady-state operation a 50-W incandescent light bulb has a surface temperature of 135°C when the room air is at a temperature of 25°C. If the bulb may be approximated as a 60-mm diameter sphere with a diffuse, gray surface of emissiv-

ity 0.8, what is the radiant heat transfer from the bulb surface to its surroundings?

13.49 An arrangement for direct conversion of thermal energy to electrical power is shown. The inner cylinder of diameter $D_i = 25$ mm is heated internally by a combustion process that brings the ceramic cylinder ($\varepsilon_i = 0.9$) to a surface temperature of $T_i = 1675°C$. The outer cylinder of diameter $D_o = 0.38$ m consists of a semiconductor material ($\varepsilon_o = 0.5$) that converts absorbed incident irradiation to electrical current; the backing material for the semiconductor is a highly conducting metal that is water cooled to 20°C. The converter is assumed to be very long compared to the outer diameter. The space between the two concentric cylinders is evacuated.

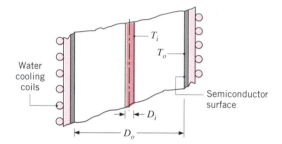

Assuming diffuse, gray surfaces, determine the heat transfer rate per unit area of the outer cylinder. The electrical output of the semiconductor surface is 10% of the absorbed irradiation in the wavelength region 0.6 to 2.0 μm. For the prescribed conditions, determine the power generation in watts per unit of outer surface area.

13.50 Liquid oxygen is stored in a thin-walled, spherical container 0.8 m in diameter, which is enclosed within a second thin-walled, spherical container 1.2 m in diameter. The opaque, diffuse, gray container surfaces have an emissivity of 0.05 and are separated by an evacuated space. If the outer surface is at 280 K and the inner surface is at 95 K, what is the mass rate of oxygen lost due to evaporation? (The latent heat of vaporization of oxygen is 2.13×10^5 J/kg.)

13.51 Consider the three-surface enclosure shown. The lower plate (A_1) is a black disk of 200-mm diameter and is supplied with a heat rate of 10,000 W. The upper plate (A_2), a disk coaxial to A_1, is a diffuse, gray surface with $\varepsilon_2 = 0.8$ and is maintained at $T_2 = 473$ K. The diffuse, gray sides between the plates are perfectly insulated. Assume convection heat transfer is negligible.

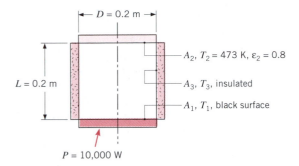

$P = 10,000$ W

Determine the operating temperature of the lower plate T_1 and the temperature of the insulated side T_3.

13.52 Consider the conical cavity of radius r_o and depth L formed in the opaque, diffuse, gray, isothermal material of emissivity ε maintained at temperature T.

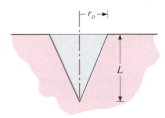

Derive an expression for the radiant power leaving the opening of the cavity in terms of T, r_o, ε, and L.

13.53 Determine the steady-state temperatures of two radiation shields placed in the evacuated space between two infinite planes at temperatures of 600 and 325 K. All the surfaces are diffuse and gray with emissivities of 0.7.

13.54 Coatings applied to long metallic strips are cured by installing the strips along the walls of a long oven of square cross section.

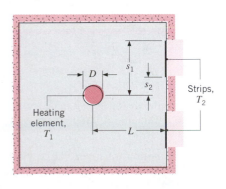

Thermal conditions in the oven are maintained by a long silicon carbide rod (heating element), which is of diameter $D = 20$ mm and is operated at $T_1 = 1700$ K. Each of two strips on one side wall has the same orientation relative to the rod ($s_1 = 60$ mm, $s_2 = 20$ mm, $L = 80$ mm) and is operated at $T_2 = 600$ K. All surfaces are diffuse and gray with $\varepsilon_1 = 0.9$ and $\varepsilon_2 = 0.4$. Assuming the oven to be well insulated at all but the strip surfaces and neglecting convection effects, determine the heater power requirement per unit length (W/m).

13.55 The proposed design for a blackbody simulator consists of a diffuse, gray, circular plate with an emissivity of 0.9 maintained at $T_p = 600$ K and mounted over a well-insulated hemispherical cavity of radius $r_o = 100$ mm. The opening in the circular plate is $r_o/2$.

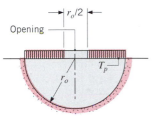

(a) Calculate the radiant power leaving the opening when the surroundings are at 300K.

(b) Calculate the effective emissivity of the cavity, ε_e, which is defined as the ratio of the radiant power leaving the cavity to the rate at which the heated circular plate would emit radiation if it were black.

(c) Determine the surface temperature of the hemispherical cavity, T_{hc}.

(d) For plate emissivities of $\varepsilon_p = 0.5$, 0.7, and 0.9, plot ε_e and T_{hc} as a function of the opening in the circular plate over a range from $r_o/8$ to $r_o/2$. All other conditions remain the same.

13.56 A long, hemicylindrical (1-m radius) shaped furnace used to heat treat sheet metal products is comprised of three zones. The heating zone (1) is constructed from a ceramic plate of emissivity 0.85 and is operated at 1600 K by gas burners. The load zone (2) consists of sheet metal products, assumed to be black surfaces, that are to be maintained at 500 K. The refractory zone (3) is fabricated from insulating bricks having an emissivity of 0.6. Assume steady-state conditions, diffuse, gray surfaces, and negligible convection.

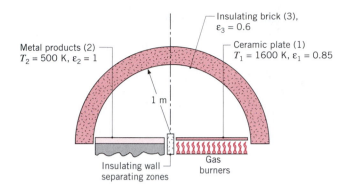

Insulating brick (3), $\varepsilon_3 = 0.6$

Ceramic plate (1) $T_1 = 1600$ K, $\varepsilon_1 = 0.85$

Metal products (2) $T_2 = 500$ K, $\varepsilon_2 = 1$

1 m

Insulating wall separating zones

Gas burners

(a) What is the heat rate per unit length of the furnace (normal to the page) that must be supplied by the gas burners for the prescribed conditions?

(b) What is the temperature of the insulating brick surface for the prescribed conditions?

13.57 The end of a cylindrical liquid cryogenic propellant tank in free space is to be protected from external (solar) radiation by placing a thin metallic shield in front of the tank. Assume the view factor F_{ts} between the tank and the shield is unity; all surfaces are diffuse and gray, and the surroundings are at 0 K.

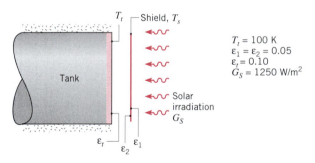

T_t — Shield, T_s

$T_t = 100$ K
$\varepsilon_1 = \varepsilon_2 = 0.05$
$\varepsilon_t = 0.10$
$G_S = 1250$ W/m²

Tank

Solar irradiation G_S

$\varepsilon_t \quad \varepsilon_2 \quad \varepsilon_1$

Find the temperature of the shield T_s and the heat flux (W/m²) to the end of the tank.

13.58 Consider two large, diffuse, gray, parallel surfaces separated by a small distance. If the surface emissivities are 0.8, what emissivity should a thin radiation shield have to reduce the radiation heat transfer rate between the two surfaces by a factor of 10?

13.59 Two concentric spheres of diameters $D_1 = 0.8$ m and $D_2 = 1.2$ m are separated by an air space and have surface temperatures of $T_1 = 400$ K and $T_2 = 300$ K.

(a) If the surfaces are black, what is the net rate of radiation exchange between the spheres?

(b) What is the net rate of radiation exchange between the surfaces if they are diffuse and gray with $\varepsilon_1 = 0.5$ and $\varepsilon_2 = 0.05$?

(c) What is the net rate of radiation exchange if D_2 is increased to 20 m, with $\varepsilon_2 = 0.05$, $\varepsilon_1 = 0.5$, and $D_1 = 0.8$ m? What error would be introduced by assuming blackbody behavior for the outer surface ($\varepsilon_2 = 1$), with all other conditions remaining the same?

(d) For $D_2 = 1.2$ m and emissivities of $\varepsilon_1 = 0.1$, 0.5, and 1.0, compute and plot the net rate of radiation exchange as a function of ε_2 for $0.05 \le \varepsilon_2 \le 1.0$.

13.60 A cryogenic fluid flows through a tube 20 mm in diameter, the outer surface of which is diffuse and gray with an emissivity of 0.02 and temperature of 77 K. This tube is concentric with a larger tube of 50-mm diameter, the inner surface of which is diffuse and gray with an emissivity of 0.05 and temperature of 300 K. The space between the surfaces is evacuated. Determine the heat gain by the cryogenic fluid per unit length of the inner tube. If a thin-walled radiation shield that is diffuse and gray with an emissivity of 0.02 (both sides) is inserted between the inner and outer surfaces, calculate the change (percentage) in heat gain per unit length of the inner tube.

13.61 A diffuse, gray radiation shield of 60-mm diameter and emissivities of $\varepsilon_{2,i} = 0.01$ and $\varepsilon_{2,o} = 0.1$ on the inner and outer surfaces, respectively, is concentric with a long tube transporting a hot process fluid. The tube surface is black with a diameter of 20 mm. The region interior to the shield is evacuated. The exterior surface of the shield is exposed to a large room whose walls are at 17°C and experiences convection with air at 27°C and a convection heat transfer coefficient of 10 W/m² · K.

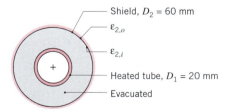

Shield, $D_2 = 60$ mm

$\varepsilon_{2,o}$

$\varepsilon_{2,i}$

Heated tube, $D_1 = 20$ mm

Evacuated

Determine the operating temperature for the inner tube if the shield temperature is maintained at 42°C.

13.62 At the bottom of a very large vacuum chamber whose walls are at 300 K, a black panel 0.1 m in diameter is maintained at 77 K. To reduce the heat gain to this panel, a radiation shield of the same diameter D and an emissivity of 0.05 is placed very close to the panel. Calculate the net heat gain to the panel.

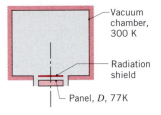

13.63 A small disk of diameter $D_1 = 50$ mm and emissivity $\varepsilon_1 = 0.6$ is maintained at a temperature of $T_1 = 900$ K. The disk is covered with a hemispherical radiation shield of the same diameter and an emissivity of $\varepsilon_2 = 0.02$ (both sides). The disk and cap are located at the bottom of a large evacuated refractory container ($\varepsilon_4 = 0.85$), facing another disk of diameter $D_3 = D_1$, emissivity $\varepsilon_3 = 0.4$, and temperature $T_3 = 400$ K. The view factor F_{23} of the shield with respect to the upper disk is 0.3.

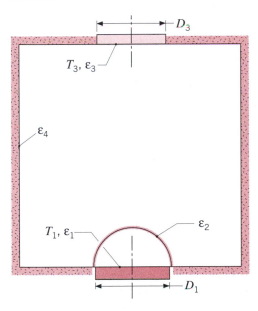

(a) Construct an equivalent thermal circuit for the above system. Label all nodes, resistances, and currents.

(b) Find the net rate of heat transfer between the hot disk and the rest of the system.

13.64 Heat transfer by radiation occurs between two large parallel plates, which are maintained at tem-

peratures $T_1 > T_2$. To reduce the rate of heat transfer between the plates, it is proposed that they be separated by a thin shield that has different emissivities on opposite surfaces. In particular, one surface has the emissivity $\varepsilon_s < 0.5$, while the opposite surface has an emissivity of $2\varepsilon_s$.

(a) How should the shield be oriented to provide the largest reduction in heat transfer between the plates? That is, should the surface of emissivity ε_s or that of emissivity $2\varepsilon_s$ be oriented toward the plate at T_1?

(b) What orientation will result in the larger value of the shield temperature T_s?

13.65 A large slab of meat 0.15 m by 0.30 m in area is taken from a refrigerator at 4°C and is placed on a wire grill 0.15 m above and parallel to a bed of hot coals at 850°C and approximately the same area. Assume that the meat and the coals are essentially black and neglect convection.

(a) What is the initial heat transfer rate between the coals and the meat?

(b) By what percentage would this rate increase if well-insulated sidewalls were placed around the perimeter of the system?

(c) What is the average temperature of the insulated wall?

13.66 The bottom of a steam-producing still of 200-mm diameter is heated by radiation. The heater, maintained at 1000°C and separated 100 mm from the still, has the same diameter as the still bottom. The still bottom and heater surfaces are black.

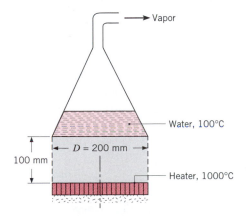

(a) By what factor could the vapor production rate be increased if the cylindrical sides (dashed surface) were insulated rather than open to the surroundings maintained at 27°C?

(b) For heater temperatures of 600, 800, and 1000°C, plot the net rate of radiation heat transfer to the still as a function of the separation distance over the range from 25 to 100 mm. Consider the cylindrical sides to be insulated and all other conditions to remain the same.

13.67 A long cylindrical heater element of diameter $D = 10$ mm, temperature $T_1 = 1500$ K, and emissivity $\varepsilon_1 = 1$ is used in a furnace. The bottom area A_2 is a diffuse, gray surface with $\varepsilon_2 = 0.6$ and is maintained at $T_2 = 500$ K. The side and top walls are fabricated from an insulating, refractory brick that is diffuse and gray with $\varepsilon = 0.9$. The length of the furnace normal to the page is very large compared to the width w and height h.

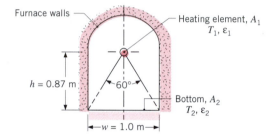

Neglecting convection and treating the furnace walls as isothermal, determine the power per unit length that must be provided to the heating element to maintain steady-state conditions. Calculate the temperature of the furnace wall.

13.68 Consider two aligned, parallel, square planes (0.4 m × 0.4 m) spaced 0.8 m apart and maintained at $T_1 = 500$ K and $T_2 = 800$ K. Calculate the net radiative heat transfer *from surface 1* for the following special conditions:

(a) Both planes are black and the surroundings are at 0 K.

(b) Both planes are black with connecting, reradiating walls.

(c) Both planes are diffuse and gray with $\varepsilon_1 = 0.6$, $\varepsilon_2 = 0.8$, and the surroundings at 0 K.

(d) Both planes are diffuse and gray ($\varepsilon_1 = 0.6$ and $\varepsilon_2 = 0.8$) with connecting, reradiating walls.

13.69 Two parallel, diffuse, gray surfaces, each 1 m by 2 m, face each other and are separated by 1 m. Each surface has an emissivity of 0.2. One surface is at 27°C and the other is at 277°C.

(a) Calculate the net radiation transfer from the hotter surface, if the surroundings are at 0 K.

(b) If well-insulated, diffuse, gray sidewalls are added to this configuration, what will be the net radiation heat transfer rate from the hotter surface?

(c) Review the assumptions you made in performing the calculations and briefly explain what effect they could have on the accuracy of your results.

13.70 Consider a room that is 4 m long by 3 m wide with a floor-to-ceiling distance of 2.5 m. The four walls of the room are well insulated, while the surface of the floor is maintained at a uniform temperature of 30°C by means of electric resistance heaters. Heat loss occurs through the ceiling, which has a surface temperature of 12°C. If all surfaces have an emissivity of 0.9, what is the rate of heat loss by radiation from the room?

13.71 Two parallel, aligned disks, 0.4 m in diameter and separated by 0.1 m, are located in a large room whose walls are maintained at 300 K. One of the disks is maintained at a uniform temperature of 500 K with an emissivity of 0.6, while the back side of the second disk is well insulated. If the disks are diffuse, gray surfaces, determine the temperature of the insulated disk.

13.72 Consider a circular furnace that is 0.3 m long and 0.3 m in diameter. One end (A_1) and the lateral surface (A_2) are black and maintained at $T_1 = 500$ K and $T_2 = 400$ K, respectively. The other end (A_3) is insulated.

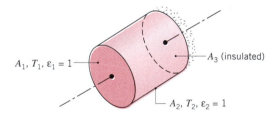

(a) Determine the net radiation heat transfer from each of the surfaces.

(b) Determine the temperature of A_3.

(c) For the tube diameter of 0.3 m, plot T_3 as a function of tube length L over a range from 0.1 to 0.5 m.

13.73 Consider a long duct constructed with diffuse, gray walls 1 m wide.

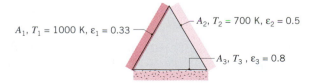

(a) Determine the net radiation transfer from surface A_1 per unit length of the duct.

(b) Determine the temperature of the insulated surface A_3.

(c) What effect would changing the value of ε_3 have on your result? After considering your assumptions, comment on whether you expect your results to be exact.

13.74 The coating on a 1 m by 2 m metallic surface is cured by placing it 0.5 m below an electrically heated surface of equivalent dimensions, and the assembly is exposed to large surroundings at 300 K. The heater is well insulated on its top side and is aligned with the coated surface. Both the heater and coated surfaces may be approximated as blackbodies. During the curing procedure, the heater and coated surfaces are maintained at 700 and 400 K, respectively. Convection effects may be neglected.

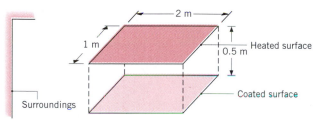

(a) What is the electric power, q_{elec}, required to operate the heater?

(b) What is q_{elec} if the surfaces are connected with reradiating sidewalls?

(c) The facility may be used to cure different surface coatings. With the heater still approximated as a blackbody, compute and plot q_{elec} as a function of the coating's emissivity for values ranging from 0.1 to 1.0. Perform the calculations for both open and reradiating sides.

13.75 A cubical furnace 2 m on a side is used for heat treating steel plate. The top surface of the furnace consists of electrical radiant heaters that have an emissivity of 0.8 and a power input of 1.5×10^5 W. The sidewalls consist of a well-insulated refractory material, while the bottom consists of the steel plate, which has an emissivity of 0.4. Assume diffuse, gray surface behavior for the heater and the plate, and consider conditions for which the plate is at 300 K. What are the corresponding temperatures of the heater surface and the sidewalls?

13.76 An electric furnace consisting of two heater sections, top and bottom, is used to heat treat a coat-

ing that is applied to both surfaces of a thin metal plate inserted midway between the heaters.

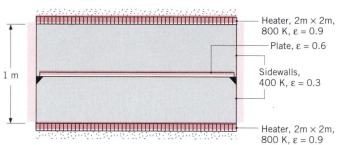

The heaters and the plate are 2 m by 2 m on a side, and each heater is separated from the plate by a distance of 0.5 m. Each heater is well insulated on its back side and has an emissivity of 0.9 at its exposed surface. The plate and the sidewalls have emissivities of 0.6 and 0.3, respectively. Sketch the equivalent radiation network for the system and label all pertinent resistances and potentials. For the prescribed conditions, obtain the required electrical power and the plate temperature.

13.77 A solar collector consists of a long duct through which air is blown; its cross section forms an equilateral triangle 1 m on a side. One side consists of a glass cover of emissivity $\varepsilon_1 = 0.9$, while the other two sides are absorber plates with $\varepsilon_2 = \varepsilon_3 = 1.0$.

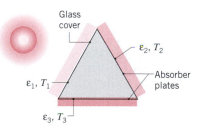

During operation the surface temperatures are known to be $T_1 = 25°C$, $T_2 = 60°C$, and $T_3 = 70°C$. What is the net rate at which radiation is transferred to the cover due to exchange with the absorber plates?

13.78 Consider a circular furnace that is 0.3 m long and 0.3 m in diameter. The two ends have diffuse, gray surfaces that are maintained at 400 and 500 K with emissivities of 0.4 and 0.5, respectively. The lateral surface is also diffuse and gray with an emissivity of 0.8 and a temperature of 800 K. Determine the net radiative heat transfer from each of the surfaces.

13.79 Consider a cylindrical cavity closed at the bottom with an opening in the top surface.

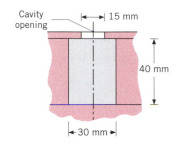

For the following conditions, calculate the rate at which radiation passes through the cavity opening when the surroundings temperature is 0 K. Also determine the effective emissivity of the cavity, ε_e, as defined in Problem 13.40.

(a) All interior surfaces are black at 600 K.

(b) The bottom surface of the cavity is diffuse and gray with an emissivity of 0.6 at 600 K, while all other interior surfaces are reradiating.

(c) All interior surfaces are diffuse and gray with an emissivity of 0.6 and a uniform temperature of 600 K.

(d) For the cavity configurations of parts (b) and (c) and cavity depths of 20, 40, and 80 mm, plot ε_e as a function of the interior surface emissivity over a range from 0.6 to 1.0. All other conditions remain the same.

13.80 A cylindrical furnace for heat treating materials in a spacecraft environment has a 90-mm diameter and an overall length of 180 mm. Heating elements in the 135-mm long section (1) maintain a refractory lining of $\varepsilon_1 = 0.8$ at 800°C. The linings for the bottom (2) and upper (3) sections are made of the same refractory material, but are insulated.

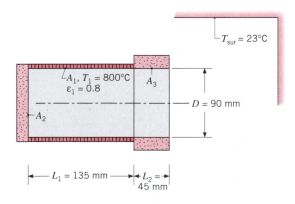

Determine the power required to maintain the furnace operating conditions with the surroundings at 23°C.

13.81 Treat the system of Problem 12.79 as a three-surface enclosure to evaluate the radiant power leaving the 2-mm diameter hole.

13.82 A radiant space heater consists of a *long,* diffuse, gray, cylindrical heating element that is of diameter $D = 30$ mm and is backed by a curved metallic reflector. The surface temperature and emissivity of the heating element are $T_1 = 945$ K and $\varepsilon_1 = 0.8$, respectively, and the temperature of the air and surroundings is 300 K. Convection heat transfer from the element is negligible. The reflector may be assumed to be an isothermal, diffuse, gray surface with an emissivity of $\varepsilon_2 = 0.1$ and an arc length of $S = 0.3$ m. The base of the reflector has a width of $W = 0.15$ m. The steady-state temperature of the reflector plate is $T_2 = 385$ K, and the view factor of the heater with respect to the reflector is $F_{12} = 0.625$.

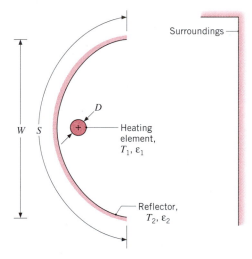

(a) Draw an equivalent circuit to describe radiation heat transfer between the reflector, heating element, and surroundings. Omit that portion of the circuit describing heat transfer from the back of the reflector. Per unit length of the heater, evaluate all the resistances and *known* nodal potentials.

(b) Calculate the electrical power input to the heating element per unit length.

13.83 The enclosure analysis of diffuse, gray surfaces presented in Section 13.3 requires that the irradiation over any surface be uniform. When the geometry of the enclosure precludes this requirement, it is necessary to divide the isothermal surface into zones over which the irradiation is approximately uniform. In the cylindrical cavity shown the irradiations along the lateral surface and on the base are likely to be different.

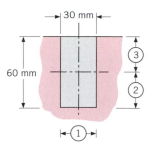

Dividing the cavity into three zones, each having the same surface temperature, perform an enclosure analysis to determine the radiative heat transfer leaving the cavity opening. The surfaces of the cavity are diffuse and gray with an emissivity of 0.2 and with a uniform temperature of 1000 K. Compare your result to the calculation treating the cavity walls (base plus lateral side) as a single surface.

13.84 In manufacturing semiconductors, rapid thermal processing (RTP) is used to rapidly heat a silicon wafer to an elevated temperature in order to induce effects such as ion diffusion, annealing, and oxidation. One type of RTP device consists of a cylindrical enclosure with a coaxially positioned wafer. The top side of the wafer experiences uniform irradiation from a lamp bank, G_{lamp}. The enclosure lateral (A_2) and bottom (A_3) surfaces have a low emissivity $\varepsilon_2 = \varepsilon_3 = 0.07$ and are maintained at 300 K by coolant passages. The wafer (A_1) diameter is $D = 300$ mm and the height of the enclosure is $L = 300$ mm. The aperature (A_4) diameter is $D_a = 30$ mm and provides for optical access to the wafer.

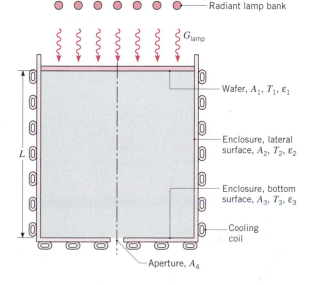

(a) What lamp irradiation, G_{lamp}, is required to maintain the wafer at 1300 K if the wafer emissivity is $\varepsilon_1 = 0.8$? What is the heat removal rate by the cooling coil? Assume there are no heat losses from the top side of the wafer.

(b) If the enclosure were perfectly reflecting, the radiosity of the wafer, J_1, would be equal to its blackbody emissive power, $E_{b,1}$. As such, the radiosity would be independent of the wafer emissivity, thereby minimizing effects due to variation of that property from wafer-to-wafer. For wafer emissivities of $\varepsilon_1 = 0.75$, 0.8, and 0.85, plot the fractional difference, $(E_{b,1} - J_1)/E_{b,1}$, as a function of the aspect ratio for $0.5 \leq L/D \leq 2.5$. How sensitive is this parameter to the enclosure surface emissivity, ε_2?

13.85 Consider the diffuse, gray, four-surface enclosure with all sides equal as shown. The temperatures of three surfaces are specified, while the fourth surface is well insulated and can be treated as a reradiating surface. Determine the temperature of the reradiating surface (4).

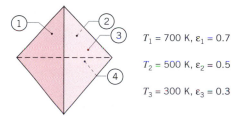

$T_1 = 700$ K, $\varepsilon_1 = 0.7$

$T_2 = 500$ K, $\varepsilon_2 = 0.5$

$T_3 = 300$ K, $\varepsilon_3 = 0.3$

13.86 A room is represented by the following enclosure, where the ceiling (1) has an emissivity of 0.8 and is maintained at 40°C by embedded electrical heating elements. Heaters are also used to maintain the floor (2) of emissivity 0.9 at 50°C. The right wall (3) of emissivity 0.7 reaches a temperature of 15°C on a cold, winter day. The left wall (4) and end walls (5A) and (5B) are very well insulated. To simplify the analysis, treat the two end walls as a single surface (5).

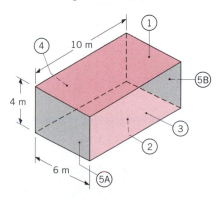

Assuming the surfaces are diffuse and gray, find the net radiation heat transfer from each surface.

Multimode Heat Transfer

13.87   An opaque, diffuse, gray (200 mm by 200 mm) plate with an emissivity of 0.8 is placed over the opening of a furnace and is known to be at 400 K at a certain instant. The bottom of the furnace, having the same dimensions as the plate, is black and operates at 1000 K. The sidewalls of the furnace are well insulated. The top of the plate is exposed to ambient air with a convection coefficient of 25 W/m² · K and to large surroundings. The air and surroundings are each at 300 K.

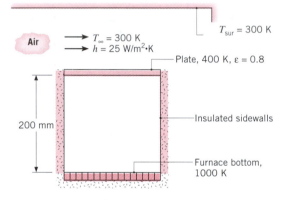

(a) Evaluate the net radiative heat transfer to the bottom surface of the plate.

(b) If the plate has mass and specific heat of 2 kg and 900 J/kg · K, respectively, what will be the change in temperature of the plate with time, dT_p/dt? Assume convection to the bottom surface of the plate to be negligible.

(c) Extending the analysis of part (b), generate a plot of the change in temperature of the plate with time, dT_p/dt, as a function of the plate temperature for $350 \le T_p \le 900$ K and all other conditions remaining the same. What is the steady-state temperature of the plate?

13.88   A tool for processing silicon wafers is housed within a vacuum chamber whose walls are black and maintained by a coolant at $T_{vc} = 300$ K. The thin silicon wafer is mounted close to, but not touching, a chuck, which is electrically heated and maintained at the temperature T_c. The surface of the chuck facing the wafer is black. The wafer temperature is $T_w = 700$ K, and its surface is diffuse and gray with an emissivity of $\varepsilon_w = 0.6$. The function of the grid, a thin metallic foil positioned

coaxial with the wafer and of the same diameter, is to control the power of the ion beam reaching the wafer. The grid surface is black with a temperature of $T_g = 500$ K. The effect of the ion beam striking the wafer is to apply a uniform heat flux of $q''_{ib} = 600$ W/m². The top surface of the wafer is subjected to the flow of a process gas for which $T_\infty = 500$ K and $h = 10$ W/m² · K. Since the gap between the wafer and chuck, δ, is very small, flow of the process gas in this region may be neglected.

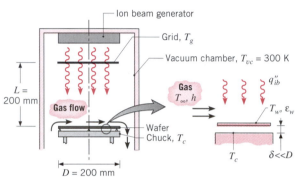

(a) Represent the wafer schematically, showing a control surface and all relevant thermal processes.

(b) Perform an energy balance on the wafer and determine the chuck temperature T_c.

13.89   Boiler tubes exposed to the products of coal combustion in a power plant are subject to fouling by the ash (mineral) content of the combustion gas. The ash forms a solid deposit on the tube outer surface, which reduces heat transfer to a pressurized water/steam mixture flowing through the tubes. Consider a thin-walled boiler tube ($D_t = 0.05$ m) whose surface is maintained at $T_t = 600$ K by the boiling process. Combustion gases flowing over the tube at $T_\infty = 1800$ K provide a convection coefficient of $\bar{h} = 100$ W/m² · K, while radiation from the gas and boiler walls to the tube may be approximated as that originating from large surroundings at $T_{sur} = 1500$ K.

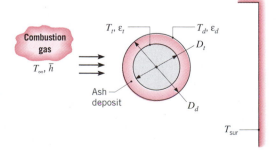

(a) If the tube surface is diffuse and gray, with $\varepsilon_t = 0.8$, and there is *no* ash deposit layer, what is the rate of heat transfer per unit length, q', to the boiler tube?

(b) If a deposit layer of diameter $D_d = 0.06$ m and thermal conductivity $k = 1$ W/m · K forms on the tube, what is the deposit surface temperature, T_d? The deposit is diffuse and gray, with $\varepsilon_d = 0.9$, and T_t, T_∞, $\bar{h}$, and T_{sur} remain unchanged. What is the net rate of heat transfer per unit length, q', to the boiler tube?

(c) Explore the effect of variations in D_d and $\bar{h}$ on q', as well as on relative contributions of convection and radiation to the net heat transfer rate. Represent your results graphically.

13.90 Consider two large parallel, diffuse, gray surfaces. The upper surface is maintained at $T_1 = 400$ K. The lower surface experiences convection (T_∞, h), and its back side is insulated.

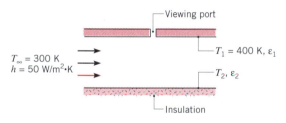

(a) Calculate the temperature of the lower surface, T_2, when $\varepsilon_1 = \varepsilon_2 = 0.5$.

(b) Calculate the radiant flux that leaves the viewing port.

13.91 Coated metallic disks are cured by placing them at the top of a cylindrical furnace whose bottom surface is electrically heated and whose sidewall may be approximated as a reradiating surface. Curing is accomplished by maintaining a disk at $T_2 = 400$ K for an extended period. The electrically heated surface is maintained at $T_1 = 800$ K and is mounted on a ceramic base material of thermal conductivity $k = 20$ W/m · K. The bottom of the base material, as well as the ambient air and large surroundings above the disk, is maintained at a temperature of 300 K. Emissivities of the heater and the disk inner and outer surfaces are $\varepsilon_1 = 0.9$, $\varepsilon_{2,i} = 0.5$, and $\varepsilon_{2,o} = 0.9$, respectively.

Assuming steady-state operation and neglecting convection within the cylindrical cavity, determine the electric power that must be supplied to the heater and the convection coefficient h that must be maintained at the outer surface of the disk in order to satisfy the prescribed conditions.

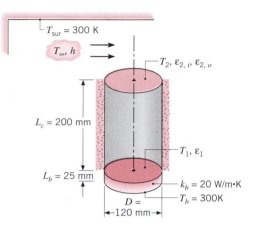

13.92 Electrical conductors, in the form of parallel plates of length $L = 40$ mm, have one edge mounted to a ceramic insulated base and are spaced a distance $w = 10$ mm apart. The plates are exposed to large isothermal surroundings at $T_{sur} = 300$ K. The conductor (1) and ceramic (2) surfaces are diffuse and gray with emissivities of $\varepsilon_1 = 0.8$ and $\varepsilon_2 = 0.6$, respectively. For a prescribed operating current in the conductors, their temperature is $T_1 = 500$ K.

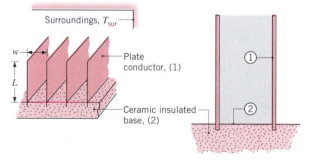

(a) Determine the electrical power dissipated in a conductor plate per unit length, q'_1, considering only radiation exchange. What is the temperature of the insulated base, T_2?

(b) Determine q'_1 and T_2 when the surfaces experience convection with an airstream at 300 K and a convection coefficient of $h = 25$ W/m² · K.

13.93 The spectral absorptivity of a *large* diffuse surface is $\alpha_\lambda = 0.9$ for $\lambda < 1$ μm and $\alpha_\lambda = 0.3$ for $\lambda \geq 1$ μm. The bottom of the surface is well insulated, while the top may be exposed to one of two different conditions.

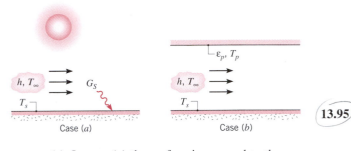

Case (a)   Case (b)

(a) In case (*a*) the surface is exposed to the sun, which provides an irradiation of $G_S = 1200$ W/m² and to an airflow for which $T_\infty = 300$ K. If the surface temperature is $T_s = 320$ K, what is the convection coefficient associated with the airflow?

(b) In case (*b*) the surface is shielded from the sun by a large plate and an airflow is maintained between the plate and the surface. The plate is diffuse and gray with an emissivity of $\varepsilon_p = 0.8$. If $T_\infty = 300$ K and the convection coefficient is equivalent to the result obtained in part (a), what is the plate temperature T_p that is needed to maintain the surface at $T_s = 320$ K?

13.94 Options for thermally shielding the top ceiling of a large furnace include the use of an insulating material of thickness L and thermal conductivity k, case (*a*), or an air space of equivalent thickness formed by installing a steel sheet above the ceiling, case (*b*).

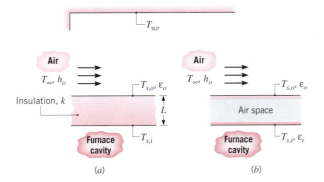

(a)   (b)

(a) Develop mathematical models that could be used to assess which of the two approaches is better. In both cases the interior surface is maintained at the same temperature, $T_{s,i}$, and the ambient air and surroundings are at equivalent temperatures ($T_\infty = T_{sur}$).

(b) If $k = 0.090$ W/m · K, $L = 25$mm, $h_0 = 25$ W/m²·K, the surfaces are diffuse and gray with $\varepsilon_i = \varepsilon_o = 0.50$, $T_{s,i} = 900$ K, and $T_\infty = T_{sur} = 300$ K, what is the outer surface tem-

perature $T_{s,o}$ and the heat loss per unit surface area associated with each option?

(c) For each case, assess the effect of surface radiative properties on the outer surface temperature and the heat loss per unit area for values of $\varepsilon_i = \varepsilon_o$ ranging from 0.1 to 0.9. Plot your results.

13.95 The composite insulation shown, which was described in Chapter 1 (Problem 1.52e), is being considered as a ceiling material.

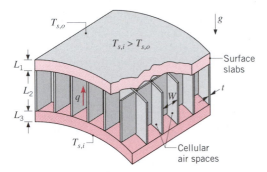

It is proposed that the outer and inner slabs be made from low-density particle board of thicknesses $L_1 = L_3 = 12.5$ mm and that the honeycomb core be constructed from a high-density particle board. The square cells of the core are to have length $L_2 = 50$ mm, width $W = 10$ mm, and wall thickness $t = 2$ mm. The emissivity of both particle boards is approximately 0.85, and the honeycomb cells are filled with air at 1-atm pressure. To assess the effectiveness of the insulation, its total thermal resistance must be evaluated under representative operating conditions for which the bottom (inner) surface temperature is $T_{s,i} = 25°C$ and the top (outer) surface temperature is $T_{s,o} = -10°C$. To assess the effect of free convection in the air space, assume a cell temperature difference of 20°C and evaluate air properties at 7.5°C. To assess the effect of radiation across the air space, assume inner surface temperatures of the outer and inner slabs to be −5 and 15°C, respectively.

13.96 Consider part (b) of Problem 8.75, but now account for the effects of radiation by assuming the rod and tube surfaces to be diffuse and gray and to have an emissivity of 0.8. Determine the temperature of the rod and the rate of heat transfer to the water per unit length of tube.

13.97 Hot coffee is contained in a cylindrical thermos bottle that is of length $L = 0.3$ m and is laying on its side (horizontally).

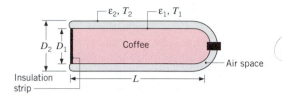

The coffee container consists of a glass flask of diameter $D_1 = 0.07$ m, separated from an aluminum housing of diameter $D_2 = 0.08$ m by air at atmospheric pressure. The outer surface of the flask and the inner surface of the housing are silver coated to provide emissivities of $\varepsilon_1 = \varepsilon_2 = 0.25$. If these surface temperatures are $T_1 = 75°C$ and $T_2 = 35°C$, what is the heat loss from the coffee?

13.98 A vertical air space in the wall of a home is 0.1 m thick and 3 m high. The air separates a brick exterior from a plaster board wall, with each surface having an emissivity of 0.9. Consider conditions for which the temperatures of the brick and plaster surfaces exposed to the air are -10 and $18°C$, respectively. What is the rate of heat loss per unit surface area? What would be the heat loss per unit area if the space were filled with urethane foam?

13.99 To estimate the energy savings associated with using double-paned (thermopane) versus single-paned window construction, consider a window that is oriented vertically and is 0.6 m by 0.6 m on a side. The inner surface of the window is exposed to the air and walls of a room that are maintained at a temperature of T_i, while the outer surface is exposed to ambient air at T_∞ and to surroundings (sky, ground, etc.) having an effective temperature T_{sur}. The convection coefficient associated with the outer surface and the ambient air is designed as h_o. The window is constructed of glass that is opaque to longwave radiation and of known emissivity ε_g. The single-pane window is of thickness t_1, while each pane of the thermopane window is of thickness t_2 with an intervening air space of thickness t_a.

(a) Assuming steady-state, nighttime conditions and negligible temperature gradients within the glass, develop a model that could be used to determine the rate of heat transfer through the window and the glass temperature(s) for both single-paned and thermopane constructions.

(b) Determine the heat transfer rate and the glass temperatures when $T_i = 18°C$, $T_\infty = -20°C$,

$T_{sur} = -30°C$, $h_o = 25$ W/m² · K, $\varepsilon_g = 0.9$, $t_1 = 10$ mm, $t_2 = 5$ mm, and $t_a = 15$ mm.

13.100 The radiation generated in an evacuated 60-W light bulb may be approximated as having the spectral distribution of radiation emitted by a blackbody at 2900 K.

(a) If the radiative properties of the glass envelope are $\tau_\lambda = 1$ and $\alpha_\lambda = 0$ for $0 < \lambda < 1$ μm and $\tau_\lambda = 0$ and $\alpha_\lambda = 1$ for 1 μm $< \lambda$, what fraction of the radiation emitted by the filament is transmitted to the surroundings of the bulb?

(b) Neglecting conduction losses through the socket, what will be the temperature of the glass bulb, which may be approximated as a sphere of diameter $D = 50$ mm, when it is situated in a room whose walls and air are at 20°C?

13.101 A flat-plate solar collector, consisting of an absorber plate and single cover plate, is inclined at an angle of $\tau = 60°$ relative to the horizontal.

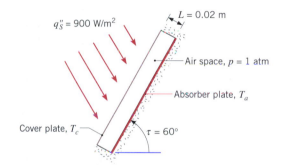

Consider conditions for which the incident solar radiation is collimated at an angle of 60° relative to the horizontal and the solar flux is 900 W/m². The cover plate is perfectly transparent to solar radiation ($\lambda \leq 3$ μm) and is opaque to radiation of larger wavelengths. The cover and absorber plates are diffuse surfaces having the spectral absorptivities shown.

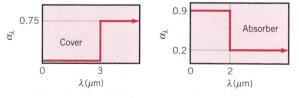

The length and width of the absorber and cover plates are much larger than the plate spacing L. What is the rate at which solar radiation is absorbed per unit area of the absorber plate? With the absorber plate well insulated from below and absorber and cover plate temperatures T_a and T_c of 70 and 27°C, respectively, what is the heat loss per unit area of the absorber plate?

13.102 Consider the flat-plate solar collector of Problem 9.86. The absorber plate has a black coating for which $\varepsilon = 0.96$, and the cover plate has an emissivity of $\varepsilon = 0.92$. With respect to radiation exchange, both plates may be approximated as diffuse, gray surfaces.

(a) For the conditions of Problem 9.86a, what is the rate of heat transfer by free convection from the absorber plate and the net rate of radiation exchange between the plates?

(b) The temperature of the absorber plate varies according to the flow rate of the working fluid routed through the coiled tube. With all other parameters remaining as prescribed, compute and plot the free convection and radiant heat rates as a function of the absorber plate temperature for $50 \le T_1 \le 100°C$.

13.103 The lower side of a 400-mm diameter disk is heated by an electric furnace, while the upper side is exposed to quiescent, ambient air and surroundings at 300 K. The radiant furnace (negligible convection) is of circular construction with the bottom surface ($\varepsilon_1 = 0.6$) and cylindrical side surface ($\varepsilon_2 = 1.0$) maintained at $T_1 = T_2 = 500$ K. The surface of the disk facing the radiant furnace is black ($\varepsilon_{d,1} = 1.0$), while the upper surface has an emissivity of $\varepsilon_{d,2} = 0.8$. Assume the plate and furnace surfaces to be diffuse and gray.

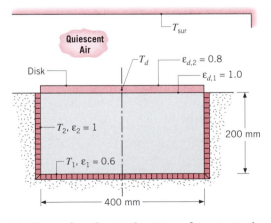

(a) Determine the net heat transfer rate to the disk, $q_{net, d}$, when $T_d = 400$ K.

(b) Plot $q_{net, d}$ as a function of the disk temperature for $300 \le T_d \le 500$ K, with all other

conditions remaining the same. What is the steady-state temperature of the disk?

13.104 The surface of a radiation shield facing a black hot wall at 400 K has a reflectivity of 0.95. Attached to the back side of the shield is a 25-mm thick sheet of insulating material having a thermal conductivity of 0.016 W/m · K. The overall heat transfer coefficient (convection and radiation) at the surface exposed to the ambient air and surroundings at 300 K is 10 W/m² · K.

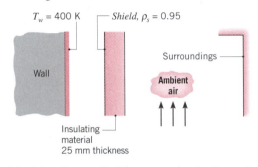

(a) Assuming negligible convection in the region between the wall and the shield, estimate the heat loss per unit area from the hot wall.

(b) Perform a parameter sensitivity analysis on the insulation system, considering the effects of shield reflectivity, ρ_s, and insulation thermal conductivity, k. What influence do these parameters have on the heat loss from the hot wall? What is the effect of an increased overall coefficient on the heat loss? Show the results of your analysis in a graphical format.

13.105 The fire tube of a hot water heater consists of a long circular duct of diameter $D = 0.07$ m and temperature $T_s = 385$ K, through which combustion gases flow at a temperature of $T_{m, g} = 900$ K. To enhance heat transfer from the gas to the tube, a thin partition is inserted along the midplane of the tube. The gases may be assumed to have the thermophysical properties of air and to be radiatively nonparticipating.

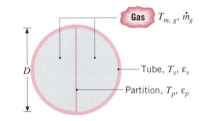

(a) With no partition and a gas flow rate of $\dot{m}_g = 0.05$ kg/s, what is the rate of heat transfer per unit length, q', to the tube?

(b) For a gas flow rate of $\dot{m}_g = 0.05$ kg/s and emissivities of $\varepsilon_s = \varepsilon_p = 0.5$, determine the partition temperature T_p and the total rate of heat transfer q' to the tube.

(c) For $\dot{m}_g = 0.02, 0.05$, and 0.08 kg/s and equivalent emissivities $\varepsilon_p = \varepsilon_s \equiv \varepsilon$, compute and plot T_p and q' as a function of ε for $0.1 \leq \varepsilon \leq 1.0$. For $\dot{m}_g = 0.05$ kg/s and equivalent emissivities, plot the convective and radiative contributions to q' as a function of ε.

13.106 A long uniform rod of 50-mm diameter with a thermal conductivity of 15 W/m · K is heated internally by volumetric energy generation of 20 kW/m³. The rod is positioned coaxially within a larger circular tube of 60-mm diameter whose surface is maintained at 500°C. The annular region between the rod and tube is evacuated, and their surfaces are diffuse and gray with an emissivity of 0.2.

(a) Determine the center and surface temperatures of the rod.

(b) Determine the center and surface temperatures of the rod if atmospheric air occupies the annular space.

(c) For tube diameters of 60, 100, and 1000 mm and for both the evacuated and atmospheric conditions, compute and plot the center and surface temperatures as a function of equivalent surface emissivities in the range from 0.1 to 1.0.

13.107 The cross section of a long circular tube, which is divided into two semicylindrical ducts by a thin wall, is shown below.

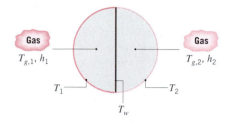

Sides 1 and 2 are maintained at temperatures of $T_1 = 600$ K and $T_2 = 400$ K, respectively, while the mean temperatures of gas flows through ducts 1 and 2 are $T_{g,1} = 571$ K and $T_{g,2} = 449$ K, respectively. The foregoing temperatures are invariant in the axial direction. The gases provide surface convection coefficients of $h_1 = h_2 = 5$ W/m² · K, while all duct surfaces may be approx-

imated as blackbodies ($\varepsilon_1 = \varepsilon_2 = \varepsilon_w = 1$). What is the duct wall temperature, T_w? By performing an energy balance on the gas in side 1, verify that $T_{g,1}$ is, in fact, equal to 571 K.

13.108 A scheme for cooling electronic components involves mounting them to a copper plate that forms one vertical wall of a rectangular cavity containing helium gas at atmospheric pressure.

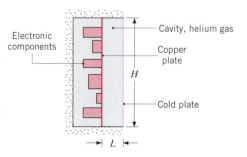

The cavity width and height are $L = 20$ mm and $H = 160$ mm, respectively. The top and bottom surfaces, as well as the back wall of the component compartment, are well insulated. Under steady-state conditions, the copper plate and the adjoining cold plate are maintained at 75 and 15°C, respectively. The plates each have an emissivity of 0.8.

(a) Determine the rate at which heat is being dissipated by the components per unit distance perpendicular to the cross section (W/m).

(b) Assess the effect of varying the plate spacing L on the heat dissipation rate.

13.109 Consider the conditions of Problem 9.90. Accounting for radiation, as well as convection, across the helium-filled cavity, determine the mass rate at which gaseous nitrogen is vented from the system. The cavity surfaces are diffuse and gray with emissivities of $\varepsilon_i = \varepsilon_o = 0.3$. If the cavity is evacuated, how may surface conditions be altered to further reduce the evaporation? Support your recommendation with appropriate calculations.

13.110 A row of regularly spaced, cylindrical heating elements (1) is used to cure a surface coating that is applied to a large panel (2) positioned below the elements. A second large panel (3), whose top surface is well insulated, is positioned above the elements. The elements are black and maintained at $T_1 = 600$ K, while the panel has an emissivity of $\varepsilon_2 = 0.5$ and is maintained at $T_2 = 400$ K. The cavity is filled with a nonparticipating gas and convection heat transfer occurs at surfaces 1 and 2, with $\bar{h}_1 = 10$ W/m$^2 \cdot$ K and $\bar{h}_2 = 2$ W/m$^2 \cdot$ K. (Convection at the insulated panel may be neglected.)

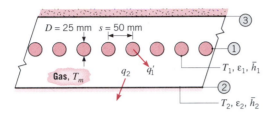

(a) Evaluate the mean gas temperature, T_m.

(b) What is the rate per unit axial length at which electrical energy must be supplied to each element to maintain its prescribed temperature?

(c) What is the rate of heat transfer to a portion of the coated panel that is 1 m wide by 1 m long?

13.111 A special surface coating on a square panel that is 5 m by 5 m on a side is cured by placing the panel directly under a radiant heat source having the same dimensions. The heat source is diffuse and gray and operates with a power input of 75 kW. The top surface of the heater, as well as the bottom surface of the panel, may be assumed to be well insulated, and the arrangement exists in a large room with air and wall temperatures of 25°C. The surface coating is diffuse and gray, with an emissivity of 0.30 and an upper temperature limit of 400 K. Neglecting convection effects, what is the minimum spacing that may be maintained between the heater and the panel to ensure that the panel temperature will not exceed 400 K? Allowing for convection effects at the coated surface of the panel, what is the minimum spacing?

13.112 A long rod heater of diameter $D_1 = 10$ mm and emissivity $\varepsilon_1 = 1.0$ is coaxial with a well-insulated, semicylindrical reflector of diameter $D_2 = 1$ m. A long panel of width $W = 1$ m is aligned with the reflector and is separated from the heater by a distance of $H = 1$ m. The panel is coated with a special paint ($\varepsilon_3 = 0.7$), which is cured by maintaining it at 400 K. The panel is well insulated on its back side, and the entire system is located in a large room where the walls and the atmospheric, quiescent air are at 300 K. Heat transfer by convection may be neglected for the reflector surface.

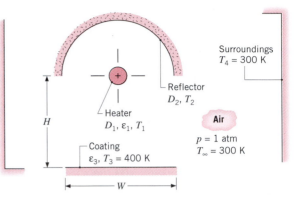

(a) Sketch the equivalent thermal circuit for the system and label all pertinent resistances and potentials.

(b) Expressing your results in terms of appropriate variables, write the system of equations needed to determine the heater and reflector temperatures, T_1 and T_2, respectively. Determine these temperatures for the prescribed conditions.

(c) Determine the rate at which electrical power must be supplied per unit length of the rod heater.

13.113 A radiant heater, which is used for surface treatment processes, consists of a long cylindrical heating element of diameter $D_1 = 0.005$ m and emissivity $\varepsilon_1 = 0.80$. The heater is partially enveloped by a long, thin parabolic reflector whose inner and outer surface emissivities are $\varepsilon_{2i} = 0.10$ and $\varepsilon_{2o} = 0.80$, respectively. Inner and outer surface areas per unit length of the reflector are each $A'_{2i} = A'_{2o} = 0.20$ m, and the average convection coefficient for the combined inner and outer surfaces is $\bar{h}_{2(i, o)} = 2$ W/m$^2 \cdot$ K. The system may be assumed to be in an infinite, quiescent medium of atmospheric air at $T_\infty = 300$ K and to be exposed to large surroundings at $T_{sur} = 300$ K.

(a) Sketch the appropriate radiation circuit, and write expressions for each of the network resistances.

(b) If, under steady-state conditions, electrical power is dissipated in the heater at $P'_1 = 1500$ W/m and the heater surface temperature is $T_1 = 1200$ K, what is the *net* rate at which *radiant* energy is transferred from the heater?

(c) What is the net rate at which radiant energy is transferred from the heater to the surroundings?

(d) What is the temperature, T_2, of the reflector?

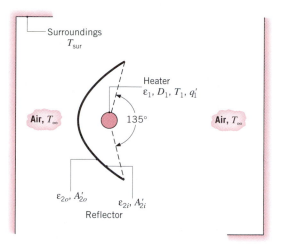

13.114 A steam generator consists of an in-line array of tubes, each of outer diameter $D = 10$ mm and length $L = 1$ m. The longitudinal and transverse pitches are each $S_L = S_T = 20$ mm, while the numbers of longitudinal and transverse rows are $N_L = 20$ and $N_T = 5$. Saturated water (liquid) enters the tubes at a pressure of 2.5 bars, and its flow rate is adjusted to ensure that it leaves the tubes as saturated vapor. Boiling that occurs in the tubes maintains a uniform tube wall temperature of 400 K.

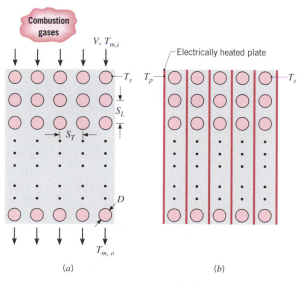

(a) Consider case (a) for which products of combustion *enter* the tube bank with velocity and temperature of $V = 10$ m/s and $T_{m,i} = 1200$ K, respectively. Determine the average gas-side convection coefficient, the gas outlet temperature, and the rate of steam production in kg/s. Properties of the gas may be approxi-

mated to be those of atmospheric air at an average temperature of 900 K.

(b) An alternative steam generator design, case (b), consists of the same tube arrangement, but the gas flow is replaced by an evacuated space with electrically heated plates inserted between each line of tubes. If the plates are maintained at a uniform temperature of $T_p = 1200$ K, what is the rate of steam production? The plate and tube surfaces may be approximated as blackbodies.

(c) Consider conditions for which the plates are installed, as in case (b), and the high-temperature products of combustion flow over the tubes, as in case (a). The plates are no longer electrically heated, but their thermal conductivity is sufficiently large to ensure a uniform plate temperature. Comment on factors that influence the plate temperature and the gas temperature distribution. Contrast (qualitatively) the gas outlet temperature and the steam generation rate with the results of case (a).

13.115 An electrical resistance heating element of diameter d whose surface is black and at a temperature T_s is located in close proximity to a metal plate of prescribed thickness t, thermal conductivity k, and emissivity ε. The plate is exposed to air at T_∞ and surroundings at T_{sur}. The plate and heating element are very long in the direction normal to the page.

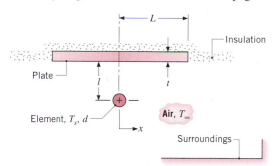

(a) Derive the differential equation and identify the boundary conditions that determine the temperature distribution in the plate.

(b) How would you solve this equation to obtain the temperature distribution?

13.116 The sketch shows a gas-fired radiant tube of the single-ended recuperative (SER) type. A mixture of air and natural gas is injected at the left end of the central tube, and combustion is essentially complete while the gases are still in the inner tube. The products of combustion are exhausted through the annulus. Thermocouple measure-

PAST EXAM PROB

ments at a particular axial location indicate an inner tube wall temperature of 1200 K and a gas temperature of 1050 K in the annulus. The radiant tube is located in a furnace, the walls of which are 950 K, and the furnace atmosphere is quiescent and also at 950 K. The thin-walled tubes are made of silicon carbide with an emissivity of 0.6. The gas mass flow rate is 0.13 kg/s and the gas pressure is 101.5 kPa.

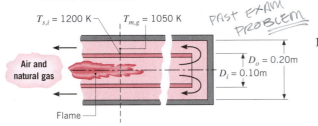

Calculate the temperature of the outer tube at the axial location where thermocouple measurements are being made. Assume the gas to be radiatively nonparticipating and to have the following thermophysical properties: $\rho = 0.32$ kg/m^3, $\nu = 130 \times 10^{-6}$ m^2/s, $k = 70 \times 10^{-3}$ W/m · K, and $Pr = 0.72$.

13.117 A wall-mounted natural gas heater uses combustion on a porous catalytic pad to maintain a ceramic plate of emissivity $\varepsilon_c = 0.95$ at a uniform temperature of $T_c = 1000$ K. The ceramic plate is separated from a glass plate by an air gap of thickness $L = 50$ mm. The surface of the glass is diffuse, and its spectral transmissivity and absorptivity may be approximated as $\tau_\lambda = 0$ and $\alpha_\lambda = 1$ for $0 \le \lambda \le 0.4$ μm, $\tau_\lambda = 1$ and $\alpha_\lambda = 0$ for $0.4 < \lambda \le 1.6$ μm, and $\tau_\lambda = 0$ and $\alpha_\lambda = 0.9$ for $\lambda > 1.6$ μm. The exterior surface of the glass is exposed to quiescent ambient air and large surroundings for which $T_\infty = T_{sur} = 300$ K. The height and width of the heater are $H = W = 2$ m.

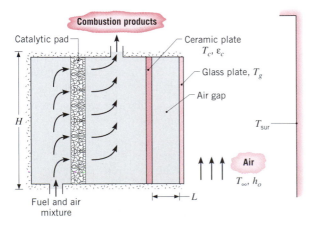

(a) What is the total transmissivity of the glass to irradiation from the ceramic plate? Can the glass be approximated as opaque and gray?

(b) For the prescribed conditions, evaluate the glass temperature, T_g, and the rate of heat transfer from the heater, q_h.

(c) A fan may be used to control the convection coefficient h_o at the exterior surface of the glass. Compute and plot T_g and q_h as a function of h_o for $10 \le h_o \le 100$ W/m^2 · K.

13.118 In recent years there has been considerable interest in developing flat-plate solar collectors for home heating needs. The typical collector configuration consists of an absorbing surface that is thermally insulated on the edges and on the back side. Solar radiation is transmitted through one or more glass *cover plates* before reaching the absorber. The collected energy is removed by circulating water through tubes that are in good thermal contact with the absorber.

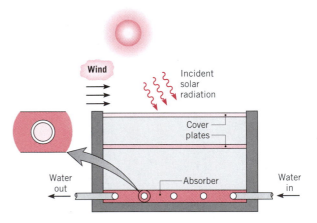

Identify all the energy exchange processes that influence the performance of the collector. Suggest design features that would increase collection efficiency.

13.119 Consider a flat-plate solar collector, such as that shown schematically in Problem 13.118, but assume that it has one, rather than two, cover plates. The following quantities are known:

G_S	the solar irradiation
$\alpha_{cp, S}$	the absorptivity of the cover plate to solar radiation
$\tau_{cp, S}$	the transmissivity of the cover plate to solar radiation
$\alpha_{ap, S}$	the absorptivity of the absorber plate to solar radiation

$\varepsilon_{cp}, \varepsilon_{ap}$	emissivities of the cover and absorber plates, respectively
h_i	air space convection coefficient based on the temperature difference, $T_{ap} - T_{cp}$
h_o	convection coefficient between cover plate and ambient air
T_{ap}	absorber plate temperature
T_∞	ambient air temperature
T_{sky}	effective sky temperature

Assuming the absorber–plate tube assembly to be perfectly insulated from below, develop expressions that could be used to obtain the rate at which useful energy is collected per unit area of the absorber plate surface q_u'' under steady-state conditions.

Gaseous Radiation

13.120 A furnace having a spherical cavity of 0.5-m diameter contains a gas mixture at 1 atm and 1400 K. The mixture consists of CO_2 with a partial pressure of 0.25 atm and nitrogen with a partial pressure of 0.75 atm. If the cavity wall is black, what is the cooling rate needed to maintain its temperature at 500 K?

13.121 A gas turbine combustion chamber may be approximated as a long tube of 0.4-m diameter. The combustion gas is at a pressure and temperature of 1 atm and 1000°C, respectively, while the chamber surface temperature is 500°C. If the combustion gas contains CO_2 and water vapor, each with a mole fraction of 0.15, what is the net radiative heat flux between the gas and the chamber surface, which may be approximated as a blackbody?

13.122 A flue gas at 1-atm total pressure and a temperature of 1400 K contains CO_2 and water vapor at partial pressures of 0.05 and 0.10 atm, respectively. If the gas flows through a long flue of 1-m diameter and 400 K surface temperature, determine the net radiative heat flux from the gas to the surface. Blackbody behavior may be assumed for the surface.

13.123 A furnace consists of two large parallel plates separated by 0.75 m. A gas mixture comprised of O_2, N_2, CO_2, and water vapor, with mole fractions of 0.20, 0.50, 0.15, and 0.15, respectively, flows between the plates at a total pressure of 2 atm and a temperature 1300 K. If the plates may be ap-

proximated as blackbodies and are maintained at 500 K, what is the net radiative heat flux to the plates?

13.124 In an industrial process, products of combustion at a temperature and pressure of 2000 K and 1 atm, respectively, flow through a long, 0.25-m diameter pipe whose inner surface is black. The combustion gas contains CO_2 and water vapor, each at a partial pressure of 0.10 atm. The gas may be assumed to have the thermophysical properties of atmospheric air and to be in fully developed flow with $\dot{m} = 0.25$ kg/s. The pipe is cooled by passing water in cross flow over its outer surface. The upstream velocity and temperature of the water are 0.30 m/s and 300 K, respectively. Determine the pipe wall temperature and heat flux. *Hint:* Emission from the pipe wall may be neglected.

13.125 Waste heat recovery from the exhaust (flue) gas of a melting furnace is accomplished by passing the gas through a vertical metallic tube and introducing saturated water (liquid) at the bottom of an annular region around the tube.

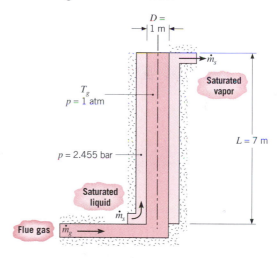

The tube length and inside diameter are 7 and 1 m, respectively, and the tube inner surface is black. The gas in the tube is at atmospheric pressure, with CO_2 and $H_2O(v)$ partial pressures of 0.1 and 0.2 atm, respectively, and its mean temperature may be approximated as $T_g = 1400$ K. The gas flow rate is $\dot{m}_g = 2$ kg/s. If saturated water is introduced at a pressure of 2.455 bars, estimate the water flow rate $\dot{m}_s$ for which there is complete conversion from saturated liquid at the inlet to saturated vapor at the outlet. Thermophysical properties of the gas may be approximated as

$\mu = 530 \times 10^{-7}$ kg/s · m, $k = 0.091$ W/m · K, and $Pr = 0.70$.

Heat and Mass Transfer

13.126 A radiant oven for drying newsprint consists of a long duct ($L = 20$ m) of semicircular cross section. The newsprint moves through the oven on a conveyor belt at a velocity of $V = 0.2$ m/s. The newsprint has a water content of 0.02 kg/m² as it enters the oven and is completely dry as it exits. To assure quality, the newsprint must be maintained at room temperature (300 K) during drying. To aid in maintaining this condition, all system components and the air flowing through the oven have a temperature of 300 K. The inner surface of the semicircular duct, which is of emissivity 0.8 and temperature T_1, provides the radiant heat required to accomplish the drying. The wet surface of the newsprint can be considered to be black. Air entering the oven has a temperature of 300 K and a relative humidity of 20%.

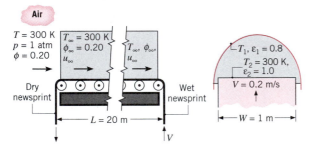

Since the velocity of the air is large, its temperature and relative humidity can be assumed to be

constant over the entire duct length. Calculate the required evaporation rate, air velocity u_∞, and temperature T_1 that will ensure steady-state conditions for the process.

13.127 A grain dryer consists of a long semicircular duct of radius $R = 1$ m. One-half of the base surface consists of an electrically heated plate of emissivity $\varepsilon_p = 0.8$, while the other half supports the grain to be dried, which has an emissivity of $\varepsilon_g = 0.9$. In a batch drying process for which the temperature of the grain is $T_g = 330$ K, 2.50 kg of water are to be removed per meter of duct length over a 1-hour period.

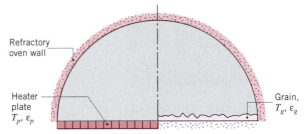

(a) Neglecting convection heat transfer, determine the required temperature T_p of the heater plate.

(b) If the water vapor is swept from the duct by the flow of dry air, what convection mass transfer coefficient h_m must be maintained by the flow?

(c) If the air is at 300 K, is the assumption of negligible convection justified?

Diffusion Mass Transfer

W e have learned that heat is transferred if there is a temperature difference in a medium. Similarly, if there is a difference in the concentration of some chemical species in a mixture, mass transfer *must* occur.

> *Mass transfer is mass in transit as the result of a species concentration difference in a mixture.*

Just as a *temperature gradient* constitutes the *driving potential* for heat transfer, a species *concentration gradient* in a mixture provides the *driving potential* for transport of that species.

It is important to clearly understand the context in which the term *mass transfer* is used. Although mass is certainly transferred whenever there is bulk fluid motion, this is not what we have in mind. For example, we do *not* use the term *mass transfer* to describe the motion of air that is induced by a fan or the motion of water being forced through a pipe. In both cases, there is gross or bulk fluid motion due to mechanical work. We do, however, use the term to describe the relative motion of species in a mixture due to the presence of concentration gradients. One example is the dispersion of oxides of sulfur released from a power plant smoke stack into the environment. Another example is the transfer of water vapor into dry air, as in a home humidifier.

There are *modes* of *mass transfer* that are similar to the conduction and convection modes of heat transfer. In Chapters 6 through 8 we considered mass transfer by convection, which is *analogous* to convection heat transfer; in this chapter we consider mass transfer by diffusion, which is *analogous* to conduction heat transfer.

14.1
Physical Origins and Rate Equations

From the standpoint of physical origins and the governing rate equations, strong analogies exist between heat and mass transfer by diffusion.

14.1.1   Physical Origins

Both conduction heat transfer and mass diffusion are transport processes that originate from molecular activity. Consider a chamber in which two different gas species at the same temperature and pressure are initially separated by a partition. If the partition is removed, both species will be transported by diffusion. Figure 14.1 shows the situation as it might exist shortly after removal of the partition. A higher concentration means more molecules per unit volume, and the concentration of species A (light dots) decreases with increasing x, while the concentration of B increases with x. Since mass diffusion is in the direction of decreasing concentration, there is net transport of species A to the right and of species B to the left. The physical mechanism may be explained by

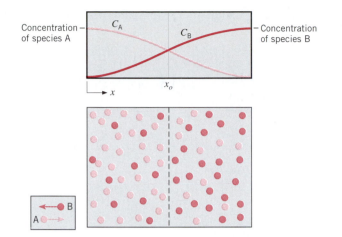

FIGURE 14.1   Mass transfer by diffusion in a binary gas mixture.

considering the imaginary plane shown as a dashed line at x_o. Since molecular motion is random, there is equal probability of any molecule moving to the left or the right. Accordingly, more molecules of species A cross the plane from the left (since this is the side of higher A concentration) than from the right. Similarly, the concentration of B molecules is higher to the right of the plane than to the left, and random motion provides for *net* transfer of species B to the left. Of course, after a sufficient time, uniform concentrations of A and B are achieved, and there is no *net* transport of species A or B across the imaginary plane.

Mass diffusion occurs in liquids and solids, as well as in gases. However, since mass transfer is strongly influenced by molecular spacing, diffusion occurs more readily in gases than in liquids and more readily in liquids than in solids. Examples of diffusion in gases, liquids, and solids, respectively, include nitrous oxide from an automobile exhaust in air, dissolved oxygen in water, and helium in Pyrex.

14.1.2   Mixture Composition

A mixture consists of two or more chemical constituents (*species*), and the amount of any species i may be quantified in terms of its *mass density* ρ_i (kg/m^3) or its *molar concentration* C_i (kmol/m^3). The mass density and molar concentration are related through the species molecular weight, $\mathcal{M}_i$ (kg/kmol), such that

$$\rho_i = \mathcal{M}_i C_i \tag{14.1}$$

With ρ_i representing the mass of species i per unit volume of the mixture, the mixture mass density is

$$\rho = \sum_i \rho_i \tag{14.2}$$

Similarly, the total number of moles per unit volume of the mixture is

$$C = \sum_i C_i \tag{14.3}$$

The amount of species i in a mixture may also be quantified in terms of its *mass fraction*

$$m_i = \frac{\rho_i}{\rho} \tag{14.4}$$

or its *mole fraction*

$$x_i = \frac{C_i}{C} \tag{14.5}$$

From Equations 14.2 and 14.3, it follows that

$$\sum_i m_i = 1 \tag{14.6}$$

and

$$\sum_i x_i = 1 \tag{14.7}$$

For a mixture of ideal gases, the mass density and molar concentration of any constituent are related to the partial pressure of the constituent through the ideal gas law. That is,

$$\rho_i = \frac{p_i}{R_i T} \tag{14.8}$$

and

$$C_i = \frac{p_i}{\mathscr{R} T} \tag{14.9}$$

where R_i is the gas constant for species i and $\mathscr{R}$ is the universal gas constant. Using Equations 14.5 and 14.9 with Dalton's law of partial pressures,

$$p = \sum_i p_i \tag{14.10}$$

it follows that

$$x_i = \frac{C_i}{C} = \frac{p_i}{p} \tag{14.11}$$

14.1.3 Fick's Law of Diffusion

Since the same physical mechanism is associated with heat and mass transfer by diffusion, it is not surprising that the corresponding rate equations are of the same form. The rate equation for mass diffusion is known as *Fick's law,* and for the transfer of species A in a *binary mixture* of A and B, it may be expressed in vector form as

$$\mathbf{j}_A = -\rho D_{AB} \, \nabla m_A \qquad (14.12)$$

or

$$\mathbf{J}_A^* = -CD_{AB} \, \nabla m_A x_A \qquad (14.13)$$

These expressions are analogous to Fourier's law, Equation 2.3. Moreover, just as Fourier's law serves to define one important transport property, the thermal conductivity, Fick's law defines a second important transport property, namely, the *binary diffusion coefficient* or *mass diffusivity, D_{AB}*.

The quantity $\mathbf{j}_A$ (kg/s · m^2) is defined as the mass flux of species A. It is the amount of A that is transferred per unit time and per unit area perpendicular to the direction of transfer, and it is proportional to the mixture mass density, $\rho = \rho_A + \rho_B$ (kg/m^3), and to the gradient in the species mass fraction, $m_A = \rho_A/\rho$. The species flux may also be evaluated on a molar basis, where $\mathbf{J}_A^*$ (kmol/s · m^2) is the molar flux of species A. It is proportional to the total molar concentration of the mixture, $C = C_A + C_B$ (kmol/m^3), and to the gradient in the species mole fraction, $x_A = C_A/C$.[1] The foregoing forms of Fick's law may be simplified when the total mass density ρ or the total molar concentration C is a constant.

14.1.4 Restrictive Conditions

Despite its analogous behavior, mass diffusion is a good deal more complicated than conduction. Complications are associated with two restrictive conditions inherent in Equations 14.12 and 14.13. First, although mass diffusion may result from a temperature gradient, a pressure gradient, or an external force, as well as from a concentration gradient, we are assuming that these additional effects are not present or are negligible. In most problems, this is in fact the case, and the dominant driving potential is the species concentration gradient. This condition is referred to as *ordinary diffusion*. Treatment of the other (higher order) effects is presented by Bird et al. [1–3]. The second restrictive condition is that the fluxes are measured relative to *coordinates that move with the average velocity of the mixture*. If the mass or molar flux of a species is expressed relative to a *fixed set of coordinates,* Equations 14.12 and 14.13 are not generally valid.

To obtain an expression for the mass flux relative to a fixed coordinate system, consider species A in a binary mixture of A and B. The mass flux $\mathbf{n}_A''$ relative to a fixed coordinate system is related to the species absolute velocity $\mathbf{v}_A$ by

$$\mathbf{n}_A'' \equiv \rho_A \mathbf{v}_A \qquad (14.14)$$

A value of $\mathbf{v}_A$ may be associated with any point in the mixture, and it is interpreted as the average velocity of all the A particles in a small volume element

---

[1] Do not confuse x_i, the mole fraction of species i, with the spatial coordinate x. The former variable will always be subscripted with the species designation.

about the point. An average, or aggregate, velocity may also be associated with the particles of species B, in which case

$$\mathbf{n}''_B \equiv \rho_B \mathbf{v}_B \tag{14.15}$$

A *mass-average velocity for the mixture* may then be obtained from the requirement that

$$\rho \mathbf{v} = \mathbf{n}'' = \mathbf{n}''_A + \mathbf{n}''_B = \rho_A \mathbf{v}_A + \rho_B \mathbf{v}_B \tag{14.16}$$

giving

$$\mathbf{v} = m_A \mathbf{v}_A + m_B \mathbf{v}_B \tag{14.17}$$

It is important to note that we have defined the velocities ($\mathbf{v}_A$, $\mathbf{v}_B$, $\mathbf{v}$) and the fluxes ($\mathbf{n}''_A$, $\mathbf{n}''_B$, $\mathbf{n}''$) as *absolute* quantities. That is, they are referred to axes that are fixed in space. The mass-average velocity $\mathbf{v}$ is a useful parameter of the binary mixture, since it need only be multiplied by the total mass density to obtain the total mass flux with respect to fixed axes. Moreover, since the velocities $\mathbf{v}_A$, $\mathbf{v}_B$, and $\mathbf{v}$ are averages associated with *aggregates* of particles, the fluxes $\mathbf{n}''_A$, $\mathbf{n}''_B$, and $\mathbf{n}''$ may be associated with transport due to *bulk* or *gross* motion.

We may now define the *mass flux of species A relative to the mixture mass-average velocity* as

$$\mathbf{j}_A \equiv \rho_A(\mathbf{v}_A - \mathbf{v}) \tag{14.18}$$

Whereas $\mathbf{n}''_A$ is the *absolute flux* of species A, $\mathbf{j}_A$ is the *relative* or *diffusive flux* of the species. It represents the motion of the species relative to the average motion of the mixture. It follows from Equation 14.14 that

$$\mathbf{n}''_A = \mathbf{j}_A + \rho_A \mathbf{v} \tag{14.19}$$

This expression indicates that there are two contributions to the absolute flux of species A: a contribution due to *diffusion* (i.e., due to the motion of A *relative to the mass-average motion* of the mixture) and a contribution due to motion of A *with the mass-average motion* of the mixture. Substituting from Equations 14.12 and 14.16, we obtain

$$\mathbf{n}''_A = -\rho D_{AB} \, \nabla m_A + m_A(\mathbf{n}''_A + \mathbf{n}''_B) \tag{14.20}$$

At this point it is useful to recap what we have done by noting the alternative formulations for the mass flux of species A. The form given by Equation 14.12 determines the transport of A relative to the mixture mass-average velocity, whereas the form given by Equation 14.20 determines the absolute transport of A. If the second term on the right-hand side of Equation 14.20 is not zero, the expression for the *absolute* flux of species A, Equation 14.20, is not analogous to that for the heat flux, Equation 2.3. We will later identify special situations for which the equations become analogous.

The foregoing considerations may be extended to species B. The *mass flux of B relative to the mixture mass-average velocity* (the *diffusive flux*) is

$$\mathbf{j}_B \equiv \rho_B(\mathbf{v}_B - \mathbf{v}) \tag{14.21}$$

where

$$\mathbf{j}_B = -\rho D_{BA} \, \nabla m_B \tag{14.22}$$

It follows from Equations 14.16, 14.18, and 14.21 that the diffusive fluxes in a binary mixture are related by

$$\mathbf{j_A} + \mathbf{j_B} = 0 \qquad (14.23)$$

If Equations 14.12 and 14.22 are substituted into Equation 14.23, and it is recognized that $\nabla m_A = -\nabla m_B$, since $m_A + m_B = 1$ for a binary mixture, it follows that

$$D_{BA} = D_{AB} \qquad (14.24)$$

Hence, as in Equation 14.20, the *absolute* flux of species B may be expressed as

$$\mathbf{n_B''} = -\rho D_{AB} \, \nabla m_B + m_B(\mathbf{n_A''} + \mathbf{n_B''}) \qquad (14.25)$$

Although the foregoing expressions pertain to *mass* fluxes, the same procedures can be used to obtain results on a *molar* basis. The absolute molar fluxes of species A and B may be expressed as

$$\mathbf{N_A''} \equiv C_A \mathbf{v_A} \qquad \text{and} \qquad \mathbf{N_B''} \equiv C_B \mathbf{v_B} \qquad (14.26)$$

and a *molar-average velocity for the mixture,* $\mathbf{v^*}$, is obtained from the requirement that

$$\mathbf{N''} = \mathbf{N_A''} + \mathbf{N_B''} = C\mathbf{v^*} = C_A \mathbf{v_A} + C_B \mathbf{v_B} \qquad (14.27)$$

giving

$$\mathbf{v^*} = x_A \mathbf{v_A} + x_B \mathbf{v_B} \qquad (14.28)$$

The significance of the molar-average velocity is that, when multiplied by the total molar concentration C, it provides the total molar flux $\mathbf{N''}$ with respect to a fixed coordinate system. Equation 14.26 provides the *absolute molar flux* of species A and B. In contrast, the molar flux of A relative to the mixture molar average velocity $\mathbf{J_A^*}$, termed the *diffusive flux,* may be obtained from Equation 14.13 or from the expression

$$\mathbf{J_A^*} \equiv C_A(\mathbf{v_A} - \mathbf{v^*}) \qquad (14.29)$$

To determine an expression similar in form to Equation 14.20, we combine Equations 14.26 and 14.29 to obtain

$$\mathbf{N_A''} = \mathbf{J_A^*} + C_A \mathbf{v^*} \qquad (14.30)$$

or, from Equations 14.13 and 14.27,

$$\mathbf{N_A''} = -CD_{AB} \, \nabla x_A + x_A(\mathbf{N_A''} + \mathbf{N_B''}) \qquad (14.31)$$

Compare the molar fluxes given by Equations 14.13 and 14.31. In the first case the molar flux is relative to the mixture molar-average velocity, and in the second case it is the absolute molar flux. Note also that Equation 14.31 represents the absolute molar flux as the sum of a diffusive flux and a flux due to the bulk motion of the mixture. For the binary mixture, it also follows that

$$\mathbf{J_A^*} + \mathbf{J_B^*} = 0 \qquad (14.32)$$

A *special case* for which the *absolute flux* of a species is *equal* to the *diffusive flux* pertains to what is termed a *stationary* medium. In terms of mass units, it is a medium for which $\mathbf{v} = 0$, in which case $\mathbf{j_A} = \mathbf{n_A''}$. In terms of molar units, it is a medium for which $\mathbf{v^*} = 0$, and hence $\mathbf{J_A^*} = \mathbf{N_A''}$. For this special case, the analogy between heat and mass transfer is complete, since the rate equations have the same physical form regardless of the reference frame.

EXAMPLE 14.1

Gaseous hydrogen is stored at elevated pressure in a rectangular container having steel walls 10 mm thick. The molar concentration of hydrogen in the steel at the inner surface is 1 kmol/m³, while the concentration of hydrogen in the steel at the outer surface is negligible. The binary diffusion coefficient for hydrogen in steel is 0.26×10^{-12} m²/s. What is the molar diffusive flux for hydrogen through the steel?

SOLUTION

Known: Molar concentrations of hydrogen at inner and outer surfaces of a steel wall.

Find: Hydrogen molar diffusive flux.

Schematic:

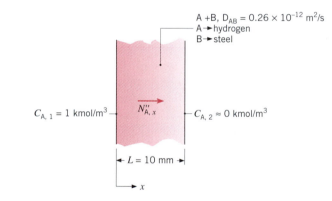

Assumptions:
1. Steady-state, one-dimensional conditions.
2. Molar concentration of hydrogen much less than that of steel ($C_A \ll C_B$); total molar concentration, $C = C_A + C_B$, uniform.
3. No chemical reactions of hydrogen in the steel.

Analysis: Although the molar flux of hydrogen with respect to a fixed reference frame is given by Equation 14.31, the second term on the right-hand side may be neglected to an excellent approximation. In particular $N''_B = 0$, and since $x_A \ll 1$, $(x_A N''_A) \ll N''_A$. Hence Equation 14.31 reduces to

$$N''_{A,x} = -CD_{AB}\frac{dx_A}{dx}$$

In addition, since the total molar concentration is approximately constant,

$$C\frac{dx_A}{dx} = \frac{dC_A}{dx} \qquad \text{and} \qquad N''_{A,x} = -D_{AB}\frac{dC_A}{dx}$$

With steady-state, one-dimensional conditions and no chemical reactions, it also follows from the species conservation requirement that $N''_{A,x}$ is independent of x. Hence dC_A/dx is constant and the species flux may be expressed as

$$N''_{A,x} = -D_{AB} \frac{C_{A,2} - C_{A,1}}{L} = D_{AB} \frac{C_{A,1}}{L}$$

Hence

$$N''_{A,x} = 0.26 \times 10^{-12} \text{ m}^2/\text{s} \frac{1 \text{ kmol/m}^3}{0.01 \text{ m}} = 2.6 \times 10^{-11} \text{ kmol/s} \cdot \text{m}^2 \qquad \triangleleft$$

Comments:

1. It is the existence of the conditions $N''_B = 0$ and $x_A \ll 1$ that will be implied in future use of the *stationary medium approximation.*

2. The mass flux of hydrogen is $n''_{A,x} = \mathcal{M}_A N''_{A,x} = 2 \text{ kg/kmol} \times 2.6 \times 10^{-11}$ kmol/s $\cdot$ m$^2 = 5.2 \times 10^{-11}$ kg/s $\cdot$ m^2.

14.1.5 Mass Diffusion Coefficient

Considerable attention has been given to predicting the mass diffusion coefficient D_{AB} for the binary mixture of two gases, A and B. Assuming ideal gas behavior, kinetic theory may be used to show that

$$D_{AB} \sim p^{-1} T^{3/2}$$

This relation applies for restricted pressure and temperature ranges and is useful for estimating values of the diffusion coefficient at conditions other than those for which data are available. Bird et al. [1–3] provide detailed discussions of available theoretical treatments and comparisons with experiment.

For binary liquid solutions, it is necessary to rely exclusively on experimental measurements. For small concentrations of A (the solute) in B (the solvent), D_{AB} is known to increase with increasing temperature. The mechanism of diffusion of gases, liquids, and solids in solids is extremely complicated and generalized theories are not available. Furthermore, only limited experimental results are available in the literature.

Data for binary diffusion in selected mixtures are presented in Table A.8. Skelland [4] and Ried and Sherwood [5] provide more detailed treatments of this subject.

14.2
Conservation of Species

Just as the first law of thermodynamics (the law of *conservation of energy*) plays an important role in heat transfer analyses, the law of *conservation of*

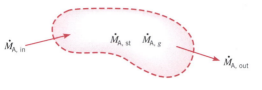

FIGURE 14.2   Conservation of species for a control volume.

species plays an important role in the analysis of mass transfer problems. In this section we consider a general statement of this law, as well as application of the law to species diffusion in a stationary medium.

14.2.1   Conservation of Species for a Control Volume

Just as a general formulation of the energy conservation requirement, Equation 1.11, was expressed for the control volume of Figure 1.7, we may express an analogous species mass conservation requirement for the control volume of Figure 14.2.

> *The rate at which the mass of some species enters a control volume minus the rate at which this species mass leaves the control volume must equal the rate at which the species mass is stored in the control volume.*

For example, any species A may enter and leave the control volume due to both fluid motion and diffusion across the control surface; these processes are *surface* phenomena represented by $\dot{M}_{A, in}$ and $\dot{M}_{A, out}$. The same species A may also be generated, $\dot{M}_{A, g}$, and accumulated or stored, $\dot{M}_{A, st}$, *within* the control volume. The conservation equation may then be expressed on a rate basis as

$$\dot{M}_{A, in} + \dot{M}_{A, g} - \dot{M}_{A, out} = \frac{dM_A}{dt} \equiv \dot{M}_{A, st} \qquad (14.33)$$

Species generation exists when chemical reactions occur in the system. There would, for example, be a net production of species A if a dissociation reaction of the form AB → A + B were occurring.

14.2.2   The Mass Diffusion Equation

The foregoing result may be used to obtain a mass, or species, diffusion equation that is analogous to the heat equation of Chapter 2. We will consider a homogeneous medium that is a binary mixture of species A and B and is *stationary*. That is, the mass-average or molar-average velocity of the mixture is everywhere zero and mass transfer may occur only by diffusion. The resulting equation could be solved for the species concentration distribution, which could, in turn, be used with Fick's law to determine the species diffusion rate at any point in the medium.

    Allowing for concentration gradients in each of the x, y, and z coordinate directions, we first define a differential control volume, $dx \, dy \, dz$, within the

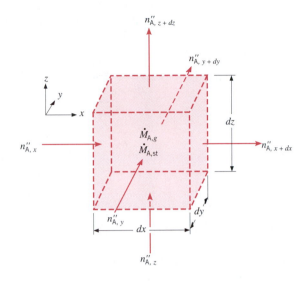

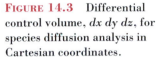

FIGURE 14.3 Differential control volume, $dx\,dy\,dz$, for species diffusion analysis in Cartesian coordinates.

medium (Figure 14.3) and consider the processes that influence the distribution of species A. With the concentration gradients, diffusion must result in the transport of species A through the control surfaces. Moreover, relative to stationary coordinates, the species transport rates at opposite surfaces must be related by

$$n''_{A,\,x+dx}\,dy\,dz = n''_{A,\,x}\,dy\,dz + \frac{\partial[n''_{A,\,x}\,dy\,dz]}{\partial x}\,dx \qquad (14.34a)$$

$$n''_{A,\,y+dy}\,dx\,dz = n''_{A,\,y}\,dx\,dz + \frac{\partial[n''_{A,\,y}\,dx\,dz]}{\partial y}\,dy \qquad (14.34b)$$

$$n''_{A,\,z+dz}\,dx\,dy = n''_{A,\,z}\,dx\,dy + \frac{\partial[n''_{A,\,z}\,dx\,dy]}{\partial z}\,dz \qquad (14.34c)$$

In addition, there may be volumetric (homogeneous) chemical reactions occurring throughout the medium. The rate at which species A is generated within the control volume due to such reactions may be expressed as

$$\dot{M}_{A,\,g} = \dot{n}_A\,dx\,dy\,dz \qquad (14.35)$$

where $\dot{n}_A$ is the rate of increase of the mass of species A per unit volume of the mixture (kg/s · m^3). Finally, these processes may change the mass of species A stored within the control volume, and the rate of change is

$$\dot{M}_{A,\,st} = \frac{\partial \rho_A}{\partial t}\,dx\,dy\,dz \qquad (14.36)$$

With mass inflow rates determined by $n''_{A,\,x}$, $n''_{A,\,y}$, and $n''_{A,\,z}$ and the outflow rates determined by Equations 14.34, Equations 14.34 to 14.36 may be substituted into Equation 14.33 to obtain

$$-\frac{\partial n''_A}{\partial x} - \frac{\partial n''_A}{\partial y} - \frac{\partial n''_A}{\partial z} + \dot{n}_A = \frac{\partial \rho_A}{\partial t}$$

For a *stationary* medium, the mass-average velocity $\mathbf{v}$ is zero, and from Equation 14.19 it follows that $\mathbf{n}''_A = \mathbf{j}_A$. Hence substituting the x, y, and z components of Equation 14.13, we obtain

$$\frac{\partial}{\partial x}\left(\rho D_{AB}\frac{\partial m_A}{\partial x}\right) + \frac{\partial}{\partial y}\left(\rho D_{AB}\frac{\partial m_A}{\partial y}\right) + \frac{\partial}{\partial z}\left(\rho D_{AB}\frac{\partial m_A}{\partial z}\right) + \dot{n}_A = \frac{\partial \rho_A}{\partial t}$$

$$(14.37a)$$

In terms of the molar concentration, a similar derivation yields

$$\frac{\partial}{\partial x}\left(CD_{AB}\frac{\partial x_A}{\partial x}\right) + \frac{\partial}{\partial y}\left(CD_{AB}\frac{\partial x_A}{\partial y}\right) + \frac{\partial}{\partial z}\left(CD_{AB}\frac{\partial x_A}{\partial z}\right) + \dot{N}_A = \frac{\partial C_A}{\partial t}$$

$$(14.38a)$$

In subsequent treatments of species diffusion phenomena, we shall work with simplified versions of the foregoing equations. In particular, if D_{AB} and ρ are constant, Equation 14.37a may be expressed as

$$\frac{\partial^2 \rho_A}{\partial x^2} + \frac{\partial^2 \rho_A}{\partial y^2} + \frac{\partial^2 \rho_A}{\partial z^2} + \frac{\dot{n}_A}{D_{AB}} = \frac{1}{D_{AB}}\frac{\partial \rho_A}{\partial t} \qquad (14.37b)$$

Similarly, if D_{AB} and C are constant, Equation 14.38a may be expressed as

$$\frac{\partial^2 C_A}{\partial x^2} + \frac{\partial^2 C_A}{\partial y^2} + \frac{\partial^2 C_A}{\partial z^2} + \frac{\dot{N}_A}{D_{AB}} = \frac{1}{D_{AB}}\frac{\partial C_A}{\partial t} \qquad (14.38b)$$

The foregoing equations are analogous to the heat equation, Equation 2.15. Hence it follows that, for analogous boundary conditions, the solution to Equation 14.37b for $\rho_A(x, y, z, t)$ or to Equation 14.38b for $C_A(x, y, z, t)$ is of the same form as the solution to Equation 2.15 for $T(x, y, z, t)$.

The foregoing species diffusion equations may also be expressed in cylindrical and spherical coordinates. These alternative forms can be inferred from the analogous expressions for heat transfer, Equations 2.20 and 2.23, and in terms of the molar concentration have the following forms:

Cylindrical coordinates:

$$\frac{1}{r}\frac{\partial}{\partial r}\left(CD_{AB}r\frac{\partial x_A}{\partial r}\right) + \frac{1}{r^2}\frac{\partial}{\partial \phi}\left(CD_{AB}\frac{\partial x_A}{\partial \phi}\right) + \frac{\partial}{\partial z}\left(CD_{AB}\frac{\partial x_A}{\partial z}\right) + \dot{N}_A = \frac{\partial C_A}{\partial t}$$

$$(14.39)$$

Spherical coordinates:

$$\frac{1}{r^2}\frac{\partial}{\partial r}\left(CD_{AB}r^2\frac{\partial x_A}{\partial r}\right) + \frac{1}{r^2\sin^2\theta}\frac{\partial}{\partial \phi}\left(CD_{AB}\frac{\partial x_A}{\partial \phi}\right)$$

$$+ \frac{1}{r^2\sin\theta}\frac{\partial}{\partial \theta}\left(CD_{AB}\sin\theta\frac{\partial x_A}{\partial \theta}\right) + \dot{N}_A = \frac{\partial C_A}{\partial t} \qquad (14.40)$$

Simpler forms are, of course, associated with the absence of chemical reactions ($\dot{n}_A = \dot{N}_A = 0$) and with one-dimensional, steady-state conditions.

14.3
Boundary and Initial Conditions

To determine the species concentration distribution in a stationary medium, it is necessary to solve the appropriate diffusion equation. However, such a solution depends on the physical conditions existing at the *boundaries* of the medium and, if the situation is time dependent, on conditions at an *initial time*. Two kinds of boundary conditions are commonly encountered in species diffusion problems and are *analogous* to the first two heat transfer conditions of Table 2.1. The first condition corresponds to a situation for which the species concentration at the surface is maintained at a constant value, $x_{A,s}$. Expressing this condition on a molar basis for the surface at $x = 0$, we have

$$x_A(0, t) = x_{A,s} \qquad (14.41)$$

The second condition corresponds to existence of a constant species flux $J^*_{A,s}$ at the surface and, using Fick's law, Equation 14.13, may be expressed as

$$-CD_{AB} \frac{\partial x_A}{\partial x}\bigg|_{x=0} = J^*_{A,s} \qquad (14.42)$$

A special case of this condition corresponds to the *impermeable surface* for which $\partial x_A/\partial x|_{x=0} = 0$.

In applying Equation 14.41, we are often interested in knowing the concentration of species A in a liquid or a solid (species B) at the interface with a gas containing species A (Figure 14.4). If species A is only weakly soluble in a *liquid*, species B, *Henry's law* may be used to relate the mole fraction of A *in the liquid* to the partial pressure of A in the gas phase outside the liquid:

$$x_A(0) = \frac{p_A}{H} \qquad (14.43)$$

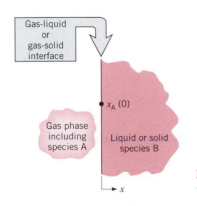

Gas-liquid
or
gas-solid
interface

$x_A(0)$

Gas phase
including
species A

Liquid or solid
species B

x

FIGURE 14.4 Species concentration at the gas–liquid or gas–solid interface.

The coefficient H is known as *Henry's constant*, and values for selected aqueous solutions are listed in Table A.9. Although H depends on temperature, its pressure dependence may generally be neglected for values of p up to 5 bars.

Conditions at a *gas–solid* interface may also be determined if the gas, species A, dissolves in the solid, species B, and a homogeneous solution is formed. In such cases gas transport in the solid is independent of the structure of the solid and may be treated as a diffusion process. In contrast, there are many situations for which the porosity of a solid strongly influences gas transport through the solid. Treatment of such cases is left to more advanced texts [2, 4].

Treating the solid as a uniform substance, the concentration of the gas in the solid at the interface may be obtained through use of a property known as the *solubility, S*. It is defined by the expression

$$C_A(0) = Sp_A \tag{14.44}$$

where p_A is the partial pressure (bars) of the gas adjoining the interface. The molar concentration of A in the solid at the interface, $C_A(0)$, is in units of kilomoles of A per cubic meter of solid, in which case the units of S must be *kilomoles of* A *per cubic meter of solid per bar partial pressure of* A. Values of S for several gas–solid combinations are given in Table A.10.

The foregoing considerations relate to transfer across an interface for which a gas solute differs from the liquid or solid solvent. Another interfacial condition that has numerous applications, but is fundamentally different from those cited above, involves transfer of a species into a gas stream due to evaporation or sublimation from a liquid or solid surface, respectively. In this case, although there are two phases, both involve the same species. The concentration (or partial pressure) of the vapor at the interface may readily be determined by assuming thermodynamic equilibrium. The partial pressure of the vapor will then correspond to saturation at the temperature of the interface and may be determined from standard thermodynamic tables.

EXAMPLE 14.2

Helium gas is stored at 20°C in a spherical container of fused silica (SiO_2), which has a diameter of 0.20 m and a wall thickness of 2 mm. If the container is charged to an initial pressure of 4 bars, what is the rate at which this pressure decreases with time?

SOLUTION

Known: Initial pressure of helium in a spherical, fused silica container of prescribed diameter D and wall thickness L.

Find: The rate of change of the helium pressure, dp_A/dt.

Schematic:

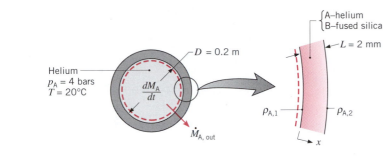

Assumptions:

1. Since $D \gg L$, diffusion may be approximated as being one-dimensional through a plane wall.
2. Quasisteady diffusion (pressure variation is sufficiently slow to permit assuming steady-state conditions for diffusion through the fused silica at any instant).
3. Stationary medium has uniform density ρ.
4. Pressure of helium in outside air is negligible.
5. Helium exhibits ideal gas behavior.

Properties: Table A.8, helium-fused silica (293 K): $D_{AB} = 0.4 \times 10^{-13}$ m²/s. Table A.10, helium-fused silica (293 K): $S = 0.45 \times 10^{-3}$ kmol/m³ · bar.

Analysis: The rate of change of the helium pressure may be obtained by applying the species conservation requirement, Equation 14.33, to a control volume about the helium. It follows that

$$-\dot{M}_{A,\,out} = \dot{M}_{A,\,st}$$

or, since the helium outflow is due to diffusion through the fused silica,

$$\dot{M}_{A,\,out} = n''_{A,\,x} A$$

and the change in mass storage is

$$\dot{M}_{A,\,st} = \frac{dM_A}{dt} = \frac{d(\rho_A V)}{dt}$$

the species balance reduces to

$$-n''_{A,\,x} A = \frac{d(\rho_A V)}{dt}$$

Recognizing that $\rho_A = \mathcal{M}_A C_A$ and applying the ideal gas law

$$C_A = \frac{p_A}{\mathcal{R} T}$$

the species balance becomes

$$\frac{dp_A}{dt} = -\frac{\mathcal{R} T}{\mathcal{M}_A V} A n''_{A,\,x}$$

For a stationary medium the absolute flux of species A through the fused silica is equal to the diffusion flux, $n''_{A,x} = j_{A,x}$, in which case, from Fick's law, Equation 14.12,

$$n''_{A,x} = -\rho D_{AB} \frac{dm_A}{dx} = -D_{AB} \frac{d\rho_A}{dx}$$

or, for the assumed conditions,

$$n''_{A,x} = D_{AB} \frac{\rho_{A,1} - \rho_{A,2}}{L}$$

The species densities $\rho_{A,1}$ and $\rho_{A,2}$ pertain to conditions *within* the fused silica at its inner and outer surfaces, respectively, and may be evaluated from knowledge of the solubility through Equation 14.44. Hence with $\rho_A = \mathcal{M}_A C_A$,

$$\rho_{A,1} = \mathcal{M}_A S p_{A,i} = \mathcal{M}_A S p_A \qquad \text{and} \qquad \rho_{A,2} = \mathcal{M}_A S p_{A,o} = 0$$

where $p_{A,i}$ and $p_{A,o}$ are helium pressures at the inner and outer surfaces, respectively. Hence

$$n''_{A,x} = \frac{D_{AB} \mathcal{M}_A S p_A}{L}$$

and substituting into the species balance it follows that

$$\frac{dp_A}{dt} = -\frac{\mathcal{R} T A D_{AB} S}{L V} p_A$$

or with $A = \pi D^2$ and $V = \pi D^3/6$

$$\frac{dp_A}{dt} = -\frac{6 \mathcal{R} T D_{AB} S}{L D} p_A$$

Substituting numerical values, the rate of change of the pressure is

$$\frac{dp_A}{dt} = [-6(0.08314 \text{ m}^3 \cdot \text{bar/kmol} \cdot \text{K}) \, 293 \text{ K} \, (0.4 \times 10^{-13} \text{ m}^2/\text{s})$$

$$0.45 \times 10^{-3} \text{ kmol/m}^3 \cdot \text{bar} \times 4 \text{ bar}] \div [0.002 \text{ m} \, (0.2 \text{ m})]$$

$$\frac{dp_A}{dt} = -2.63 \times 10^{-8} \text{ bar/s} \qquad \qquad \triangleleft$$

Comments: The foregoing result provides the initial (maximum) leakage rate for the system. The leakage rate decreases as the inside pressure decreases.

14.4
Mass Diffusion Without Homogeneous Chemical Reactions

There are numerous situations for which species transfer occurs under one-dimensional, steady-state conditions. Moreover, there are many special cases for which results are entirely analogous to those obtained in Chapter 3 for heat transfer. In this section we consider some of these special cases, restricting ourselves to situations that do not involve homogeneous chemical reactions. A *homogeneous* reaction is one that occurs *within* the medium. It is a *volumetric phenomenon* whose magnitude may vary from point to point in the medium. Homogeneous reactions involve *species generation* and are therefore analogous to internal sources of heat generation. In contrast, a *heterogeneous* chemical reaction is one that results from contact between the reactants and a surface. It is therefore a *surface phenomenon* that can be treated as a *boundary condition*. Heterogeneous chemical reactions are analogous to the constant surface heat flux condition commonly encountered in heat diffusion.

In Section 14.4.1 we consider species transfer in stationary media for which the species concentrations may be specified at the boundaries. In Section 14.4.2 this treatment is extended to situations characterized by heterogeneous reactions at a surface boundary. Two special cases, both of which involve nonreacting gases, are considered in Sections 14.4.3 and 14.4.4. One situation involves equimolar counterdiffusion in a gas mixture, and the other involves evaporation from a liquid surface in a column. Homogeneous chemical reactions are treated in Section 14.5. For each case, we are interested in determining the species concentration distribution and the corresponding species flux.

14.4.1 Stationary Media with Specified Surface Concentrations

In Section 14.1.4 we defined a *stationary medium* as one for which the molar (or mass) average velocity of the mixture is zero, in which case $\mathbf{N}_A'' = \mathbf{J}_A^*$ (or $\mathbf{n}_A'' = \mathbf{j}_A$). That is, the absolute flux of species A is equivalent to the diffusive flux, and from Equations 14.12 and 14.13 we have

$$\mathbf{n}_A'' = -\rho D_{AB} \nabla m_A \tag{14.45}$$

$$\mathbf{N}_A'' = -C D_{AB} \nabla x_A \tag{14.46}$$

From Equation 14.20 it is evident that Equation 14.45 will apply to a good approximation in nonstationary media for which $m_A \ll 1$ and $\mathbf{n}_B'' \approx 0$. (Recall Example 14.1.) Similarly, from Equation 14.31 we see that Equation 14.46 is a good approximation in media for which $x_A \ll 1$ and $\mathbf{N}_B'' \approx 0$. Such conditions often characterize dilute gas mixtures and liquid solutions (m_A or x_A is small) in which the motion of the solvent is negligible ($\mathbf{n}_B''$ or $\mathbf{N}_B''$ is small). Both these conditions must be satisfied for the species flux due to bulk motion of the mixture to be negligible compared to the diffusive flux. This situation can also occur for species diffusion in a solid, and satisfactory results may be obtained from Equation 14.45 or 14.46.

Consider one-dimensional diffusion of species A through a planar medium of A and B, as shown in Figure 14.5. For steady-state conditions with no homogeneous chemical reactions, the molar form of the species diffusion equation (14.38a) reduces to

$$\frac{d}{dx}\left(C D_{AB} \frac{dx_A}{dx}\right) = 0 \tag{14.47}$$

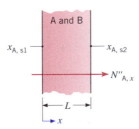

FIGURE 14.5 Mass transfer in a stationary planar medium.

Assuming the total molar concentration and the diffusion coefficient to be constant, Equation 14.47 may be solved and the surface conditions applied to yield

$$x_A(x) = (x_{A,s2} - x_{A,s1})\frac{x}{L} + x_{A,s1} \tag{14.48}$$

From Equation 14.46, it follows that

$$N''_{A,x} = -CD_{AB}\frac{x_{A,s2} - x_{A,s1}}{L} \tag{14.49}$$

Multiplying by the surface area A and substituting for $x_A \equiv C_A/C$, the molar rate is then

$$N_{A,x} = \frac{D_{AB}A}{L}(C_{A,s1} - C_{A,s2}) \tag{14.50}$$

From this expression we can define a resistance to species transfer by diffusion in a planar medium as

$$R_{m,\text{dif}} = \frac{C_{A,s1} - C_{A,s2}}{N_{A,x}} = \frac{L}{D_{AB}A} \tag{14.51}$$

TABLE 14.1 **Summary of Species Diffusion Solutions for Stationary Media with Specified Surface Concentrations**[a]

Geometry	Species Concentration Distribution, $x_A(x)$ or $x_A(r)$	Species Diffusion Resistance, $R_{m,\text{dif}}$
	$x_A(x) = (x_{A,s2} - x_{A,s1})\dfrac{x}{L} + x_{A,s1}$	$R_{m,\text{dif}} = \dfrac{L}{D_{AB}A}$ [b]
	$x_A(r) = \dfrac{x_{A,s1} - x_{A,s2}}{\ln(r_1/r_2)}\ln\left(\dfrac{r}{r_2}\right) + x_{A,s2}$	$R_{m,\text{dif}} = \dfrac{\ln(r_2/r_1)^c}{2\pi L D_{AB}}$
	$x_A(r) = \dfrac{x_{A,s1} - x_{A,s2}}{1/r_1 - 1/r_2}\left(\dfrac{1}{r} - \dfrac{1}{r_2}\right) + x_{A,s2}$	$R_{m,\text{dif}} = \dfrac{1}{4\pi D_{AB}}\left(\dfrac{1}{r_1} - \dfrac{1}{r_2}\right)^c$

[a]Assuming C and D_{AB} are constant.
[b]$N_{A,x} = (C_{A,s1} - C_{A,s2})/R_{m,\text{dif}}$.
[c]$N_{A,r} = (C_{A,s1} - C_{A,s2})/R_{m,\text{dif}}$.

Comparing the foregoing results with those obtained for one-dimensional, steady-state conduction in a plane wall with no generation (Section 3.1), it is evident that a direct analogy exists between heat and mass transfer by diffusion.

The analogy also applies to cylindrical and spherical systems. For one-dimensional, steady diffusion in a cylindrical, nonreacting medium, Equation 14.39 reduces to

$$\frac{d}{dr}\left(rCD_{AB}\frac{dx_A}{dr}\right) = 0 \tag{14.52}$$

Similarly, for a spherical medium,

$$\frac{1}{r^2}\frac{d}{dr}\left(r^2CD_{AB}\frac{dx_A}{dr}\right) = 0 \tag{14.53}$$

Equations 14.52 and 14.53, as well as Equation 14.47, dictate that the molar transfer rate, $N_{A,r}$ or $N_{A,x}$, is constant in the direction of transfer (r or x). Assuming C and D_{AB} to be constant, it is a simple matter to obtain general solutions to Equations 14.52 and 14.53. For prescribed surface species concentrations, the corresponding solutions and diffusion resistances are summarized in Table 14.1.

EXAMPLE 14.3

Hydrogen gas is maintained at 3 bars and 1 bar on opposite sides of a plastic membrane, which is 0.3 mm thick. The temperature is 25°C, and the binary diffusion coefficient of hydrogen in the plastic is 8.7×10^{-8} m²/s. The solubility of hydrogen in the membrane is 1.5×10^{-3} kmol/m³ · bar. What is the mass diffusive flux of hydrogen through the membrane?

SOLUTION

Known: Pressure of hydrogen on opposite sides of a membrane.

Find: The hydrogen mass diffusive flux $n''_{A,x}$ (kg/s · m²).

Schematic:

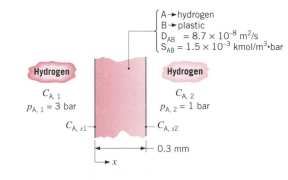

$$\begin{cases} A \rightarrow \text{hydrogen} \\ B \rightarrow \text{plastic} \\ D_{AB} = 8.7 \times 10^{-8} \text{ m}^2/\text{s} \\ S_{AB} = 1.5 \times 10^{-3} \text{ kmol/m}^3\text{·bar} \end{cases}$$

Hydrogen

$C_{A,1}$
$p_{A,1} = 3$ bar

$C_{A,s1}$

Hydrogen

$C_{A,2}$
$p_{A,2} = 1$ bar

$C_{A,s2}$

0.3 mm

x

Assumptions:
1. Steady-state, one-dimensional conditions exist.

2. Membrane is a stationary, nonreacting medium of uniform total molar concentration.

Analysis: For the prescribed conditions, Equation 14.46 reduces to Equation 14.49, which may be expressed as

$$N''_{A,x} = CD_{AB} \frac{x_{A,s1} - x_{A,s2}}{L} = \frac{D_{AB}}{L}(C_{A,s1} - C_{A,s2})$$

The surface molar concentrations of hydrogen may be obtained from Equation 14.44, where

$$C_{A,s1} = 1.5 \times 10^{-3} \text{ kmol/m}^3 \cdot \text{bar} \times 3 \text{ bars} = 4.5 \times 10^{-3} \text{ kmol/m}^3$$

$$C_{A,s2} = 1.5 \times 10^{-3} \text{ kmol/m}^3 \cdot \text{bar} \times 1 \text{ bar} = 1.5 \times 10^{-3} \text{ kmol/m}^3$$

Hence

$$N''_{A,x} = \frac{8.7 \times 10^{-8} \text{ m}^2/\text{s}}{0.3 \times 10^{-3} \text{ m}} (4.5 \times 10^{-3} - 1.5 \times 10^{-3}) \text{ kmol/m}^3$$

$$N''_{A,x} = 8.7 \times 10^{-7} \text{ kmol/s} \cdot \text{m}^2$$

On a mass basis,

$$n''_{A,x} = N''_{A,x} \mathcal{M}_A$$

where the molecular weight of hydrogen is 2 kg/kmol. Hence

$$n''_{A,x} = 8.7 \times 10^{-7} \text{ kmol/s} \cdot \text{m}^2 \times 2 \text{ kg/kmol} = 1.74 \times 10^{-6} \text{ kg/s} \cdot \text{m}^2 \quad \triangleleft$$

Comments: The molar concentrations of hydrogen in the gas phase, $C_{A,1}$ and $C_{A,2}$, differ from the surface concentrations in the membrane and may be calculated from the ideal gas equation of state

$$C_A = \frac{p_A}{\mathcal{R}T}$$

where $\mathcal{R} = 8.314 \times 10^{-2} \text{ m}^3 \cdot \text{bar/kmol} \cdot \text{K}$. It follows that $C_{A,1} = 0.121$ kmol/m^3 and $C_{A,2} = 0.040$ kmol/m^3. Even though $C_{A,s2} < C_{A,2}$, hydrogen transport will occur from the membrane to the gas at $p_{A,2} = 1$ bar. This seemingly anomalous result may be explained by recognizing that the two concentrations are based on *different* volumes; in one case the concentration is per unit volume of the membrane and in the other case it is per unit volume of the adjoining gas phase. For this reason it is not possible to infer the direction of hydrogen transport from a simple comparison of the numerical values of $C_{A,s2}$ and $C_{A,2}$.

14.4.2 Stationary Media with Catalytic Surface Reactions

Many mass transfer problems involve specification of the species flux, rather than the species concentration, at a surface. One such problem relates to the process of catalysis, which involves the use of special surfaces to promote chemical reactions. Often a one-dimensional diffusion analysis may be used to *approximate* the performance of a catalytic reactor.

Consider the system of Figure 14.6. A catalytic surface is placed in a gas stream to promote a heterogeneous chemical reaction involving species A. Assume that the reaction produces species A at a rate $\dot{N}_A''$, which is defined as the molar rate of production per unit surface area of the catalyst. To maintain steady-state conditions, the rate of species transfer from the surface, $N_{A,x}''$, must equal the surface reaction rate:

$$N_{A,x}''(0) = \dot{N}_A'' \tag{14.54}$$

Although bulk motion influences transfer of A through the film, it is reasonable to assume, as a first estimate, that the effect is negligible and that transfer occurs exclusively by diffusion. It is also assumed that species A leaves the surface as a result of one-dimensional transfer through a thin film of thickness L and that no reactions occur within the film itself. The mole fraction of A at $x = L$, $x_{A,L}$, corresponds to conditions in the mainstream of the mixture and is presumed to be known. Representing the remaining species of the mixture as a single species B and assuming the medium to be stationary, Equation 14.38a reduces to

$$\frac{d}{dx}\left(CD_{AB}\frac{dx_A}{dx}\right) = 0 \tag{14.55}$$

where D_{AB} is the binary diffusion coefficient for A in B and B may be a multi-component mixture. Assuming C and D_{AB} to be constant, Equation 14.55 may be solved subject to the conditions that

$$x_A(L) = x_{A,L}$$

and

$$N_{A,x}''(0) = -CD_{AB}\frac{dx_A}{dx}\bigg|_{x=0} = \dot{N}_A'' \tag{14.56}$$

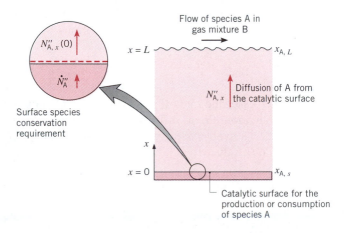

FIGURE 14.6 One-dimensional diffusion with heterogeneous catalysis.

This expression follows from Equation 14.54 and the substitution of Fick's law, Equation 14.46.

For a catalytic surface, the surface reaction rate $\dot{N}''_A$ generally depends on the surface concentration $C_A(0)$. For a *first-order reaction* that results in species consumption at the surface, the reaction rate is of the form

$$\dot{N}''_A = -k''_1 C_A(0) \tag{14.57}$$

where k''_1 (m/s) is the reaction rate constant. Accordingly, the surface boundary condition, Equation 14.56, reduces to

$$-D_{AB} \frac{dx_A}{dx}\bigg|_{x=0} = -k''_1 x_A(0) \tag{14.58}$$

Solving Equation 14.55 subject to the above conditions, it is readily verified that the concentration distribution is linear and of the form

$$\frac{x_A(x)}{x_{A,L}} = \frac{1 + (xk''_1/D_{AB})}{1 + (Lk''_1/D_{AB})} \tag{14.59}$$

At the catalytic surface this result reduces to

$$\frac{x_A(0)}{x_{A,L}} = \frac{1}{1 + (Lk''_1/D_{AB})} \tag{14.60}$$

and the molar flux is

$$N''_A(0) = -CD_{AB} \frac{dx_A}{dx}\bigg|_{x=0} = -k''_1 C x_A(0)$$

or

$$N''_A(0) = -\frac{k''_1 C x_{A,L}}{1 + (Lk''_1/D_{AB})} \tag{14.61}$$

The negative sign implies mass transfer *to* the surface.

Two limiting cases of the foregoing results are of special interest. For the limit $k''_1 \to 0$, $(Lk''_1/D_{AB}) \ll 1$ and Equations 14.60 and 14.61 reduce to

$$\frac{x_{A,s}}{x_{A,L}} \approx 1 \qquad \text{and} \qquad N''_A(0) \approx -k''_1 C x_{A,L}$$

In such cases the rate of reaction is controlled by the reaction rate constant, and the limitation due to diffusion is negligible. The process is said to be *reaction limited*. Conversely, for the limit $k''_1 \to \infty$, $(Lk''_1/D_{AB}) \gg 1$ and Equations 14.60 and 14.61 reduce to

$$x_{A,s} \approx 0 \qquad \text{and} \qquad N''_{A,s} \approx -\frac{CD_{AB} x_{A,L}}{L}$$

In this case the reaction is controlled by the rate of diffusion to the surface, and the process is said to be *diffusion limited*.

14.4.3   Equimolar Counterdiffusion

A special case of the conditions considered in Section 14.4.1 is shown in Figure 14.7. A channel connecting two large reservoirs contains an isothermal, ideal gas mixture of species A and B. The species concentrations are maintained constant in each of the reservoirs, such that $x_{A,0} > x_{AL}$ and $x_{B,0} < x_{B,L}$, while the total pressure $p = p_A + p_B$ is uniform. The species concentration gradients cause the diffusion of A molecules in the direction of increasing x and the diffusion of B molecules in the opposite direction. Under steady-state conditions, diffusion of the species occurs at equal and opposite rates, in which case the total molar flux must be zero relative to stationary coordinates. This requirement is expressed as

$$N''_{A,x} + N''_{B,x} = 0 \tag{14.62}$$

and is necessitated by the fact that, with uniform p and T, the total molar concentration C must also be uniform. This condition may only be maintained if the molar fluxes of A and B are balanced, and the process is termed *equimolar counterdiffusion*. From Equations 14.27 and 14.62, the mixture molar-average velocity $\mathbf{v}_x^*$ is zero, and the mixture is *stationary*. From Equation 14.31 it follows that

$$N''_{A,x} = -CD_{AB}\frac{dx_A}{dx} \tag{14.63}$$

Similarly,

$$N''_{B,x} = -CD_{BA}\frac{dx_B}{dx} \tag{14.64}$$

From Equations 14.24 and 14.62 it is also evident that

$$\frac{dx_A}{dx} = -\frac{dx_B}{dx} \tag{14.65}$$

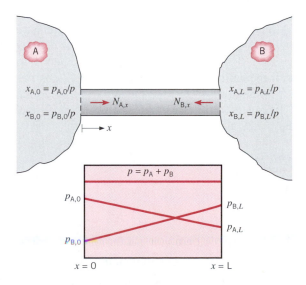

FIGURE 14.7   Equimolar counterdiffusion in a binary, isothermal, ideal gas mixture.

Since the mole fraction of each species is the ratio of its partial pressure to the total pressure, Equation 14.65 yields

$$\frac{dp_A}{dx} = -\frac{dp_B}{dx} \tag{14.66}$$

Since we are dealing with one-dimensional, steady-state conditions involving no homogeneous chemical reactions, the species diffusion rates are independent of x. Assuming D_{AB} to be constant and recalling that C is independent of x, Equation 14.63 may be expressed as

$$N_{A,x} \int_0^L \frac{dx}{A} = -CD_{AB} \int_{x_{A,0}}^{x_{A,L}} dx_A \tag{14.67}$$

If the channel cross-sectional area is uniform, the species A molar diffusion rate is then

$$N_{A,x} = CD_{AB}A \frac{x_{A,0} - x_{A,L}}{L} = D_{AB}A \frac{C_{A,0} - C_{A,L}}{L} \tag{14.68}$$

Alternatively, substituting from the equation of state,

$$p_A = C_A \mathcal{R} T \tag{14.69}$$

we obtain

$$N_{A,x} = \frac{D_{AB}A}{\mathcal{R}T} \frac{p_{A,0} - p_{A,L}}{L} \tag{14.70}$$

Similar results may be obtained for species B. Note that the foregoing results imply that the species mole fraction gradients, and therefore the species pressure gradients, are linear and of opposite sign (Figure 14.7).

EXAMPLE 14.4

To maintain a pressure close to 1 atm, an industrial pipeline containing ammonia gas is vented to ambient air. Venting is achieved by tapping the pipe and inserting a 3-mm diameter tube, which extends for 20 m into the atmosphere. With the entire system operating at 25°C, determine the mass rate of ammonia lost to the atmosphere and the mass rate of contamination of the pipe with air. What are the mole and mass fractions of air in the pipe when the ammonia flow rate is 5 kg/h?

SOLUTION

Known: Conditions in an ammonia pipeline vented to the atmosphere.

Find:

1. Mass rate of ammonia lost through the vent, n_A.
2. Mass rate of air contamination in the pipe, n_B.
3. Mole and mass fractions of air in the pipe.

Schematic:

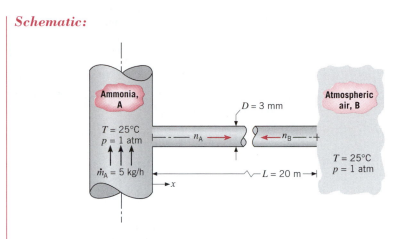

Assumptions:

1. Steady-state, one-dimensional diffusion in the tube.

2. Constant properties.

3. Uniform temperature and total pressure, $p = p_A + p_B$, in the tube.

4. Negligible mole fraction of air in the pipe, $x_{B,0} \ll 1$, and negligible mole fraction of ammonia in the atmosphere, $x_{A,L} \ll 1$.

5. No chemical reactions and ideal gas behavior.

Properties: Table A.8, ammonia–air (298 K): $D_{AB} = 0.28 \times 10^{-4}$ m²/s.

Analysis:

1. The prescribed conditions are precisely those that provide for equimolar counterdiffusion. Hence the mass rate of ammonia lost to the atmosphere may be obtained from Equation 14.70, where $n_A = \mathcal{M}_A N_A$ and

$$N_A = \frac{D_{AB}A}{\mathscr{R}T} \frac{p_{A,0} - p_{A,L}}{L}$$

From assumption 4, $p_{A,0} = p$ and $p_{A,L} = 0$. Hence

$$N_A = \frac{0.28 \times 10^{-4} \text{ m}^2/\text{s} \times \dfrac{\pi}{4}(0.003 \text{ m})^2}{8.205 \times 10^{-2} \text{ m}^3 \cdot \text{atm/kmol} \cdot \text{K} (298 \text{ K})} \times \frac{(1 - 0) \text{ atm}}{20 \text{ m}} \times 3600 \text{ s/h}$$

or

$$N_A = 1.46 \times 10^{-9} \text{ kmol/h}$$

Hence, with a molecular weight for the ammonia of $\mathcal{M}_A = 17$ kg/kmol,

$$n_A = 17 \text{ kg/kmol} \times 1.46 \times 10^{-9} \text{ kmol/h} = 2.48 \times 10^{-8} \text{ kg/h} \qquad \triangleleft$$

2. The mass rate of contamination of the pipe by air is determined from the equimolar diffusion requirement

$$N_B = -N_A = -1.46 \times 10^{-9} \text{ kmol/h}$$

Hence, with the molecular weight of air given by $\mathcal{M}_B = 28.97$ kg/kmol,

$$n_B = \mathcal{M}_B N_B = -28.97 \text{ kg/kmol} \times 1.46 \times 10^{-9} \text{ kmol/h}$$

$$n_B = -4.23 \times 10^{-8} \text{ kg/h} \qquad \triangleleft$$

3. For an ammonia flow rate of $\dot{m}_A = 5$ kg/h, the mass fraction of air in the pipe is

$$m_{B,0} = \frac{-n_B}{\dot{m}_A} = \frac{4.23 \times 10^{-8} \text{ kg/h}}{5 \text{ kg/h}} = 0.85 \times 10^{-8} \qquad \triangleleft$$

With an ammonia molar flow rate in the pipe of $\dot{m}_A/\mathcal{M}_A = (5 \text{ kg/h})/(17$ kg/kmol), the mole fraction of air in the pipe is

$$x_{B,0} = \frac{-N_B}{\dot{m}_A/\mathcal{M}_A} = \frac{1.46 \times 10^{-9} \text{ kmol/h}}{(5 \text{ kg/h})/(17 \text{ kg/kmol})} = 4.96 \times 10^{-9} \qquad \triangleleft$$

Comments: The foregoing results for $m_{B,0}$ and $x_{B,0}$ are based on assumption 4, which presumes that $C \approx C_A$ (for $\rho \approx \rho_A$) in the pipe. The validity of this assumption is confirmed by the small value of $x_{B,0}$ (or $m_{B,0}$).

14.4.4   Evaporation in a Column

Let us now consider diffusion in the binary gas mixture of Figure 14.8. Fixed species concentrations $x_{A,L}$ and $x_{B,L}$ are maintained at the top of a beaker containing a liquid layer of species A, and the system is at constant pressure and temperature. Since equilibrium exists between the vapor and liquid phases at the liquid interface, the vapor concentration corresponds to saturated conditions. With $x_{A,0} > x_{A,L}$, species A *evaporates* from the liquid interface and is transferred upward by diffusion. For steady, one-dimensional conditions with

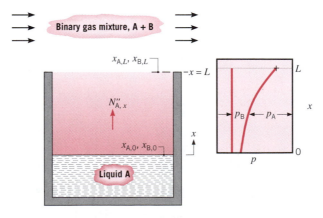

FIGURE 14.8   **Evaporation of liquid A into a binary gas mixture, A + B.**

no chemical reactions, the absolute molar flux of A must be constant through-out the column. Hence

$$\frac{dN''_{A,x}}{dx} = 0 \tag{14.71}$$

Since p and T are constant, it follows that the total molar concentration, $C = C_A + C_B$, is also constant, in which case $x_A + x_B = 1$ throughout the column. Having assumed that $x_{A,0} > x_{A,L}$, we conclude that $x_{B,L} > x_{B,0}$ and therefore that species B must diffuse from the top of the column toward the liquid interface. However, if species B is insoluble in liquid A, steady-state conditions can be maintained only if the downward diffusion of B is bal-anced by an upward bulk motion. That is, the absolute flux of species B must be zero ($N''_{B,x} = 0$). A major implication of this condition is that we are no longer dealing with a stationary medium; hence Equation 14.46 does not apply. However, an appropriate expression for $N''_{A,x}$ may be obtained by sub-stituting the requirement that $N''_{B,x} = 0$ into Equation 14.31, giving

$$N''_{A,x} = -CD_{AB}\frac{dx_A}{dx} + x_A N''_{A,x} \tag{14.72}$$

or, from Equation 14.27,

$$N''_{A,x} = -CD_{AB}\frac{dx_A}{dx} + C_A v^*_x \tag{14.73}$$

From this equation it is evident that the diffusive transport of species A $[-CD_{AB}(dx_A/dx)]$ is augmented by bulk motion ($C_A v^*_x$). Rearranging Equation 14.72, we obtain

$$N''_{A,x} = -\frac{CD_{AB}}{1-x_A}\frac{dx_A}{dx} \tag{14.74}$$

For constant p and T, C and D_{AB} are also constant. Substituting Equation 14.74 into Equation 14.71, we then obtain

$$\frac{d}{dx}\left(\frac{1}{1-x_A}\frac{dx_A}{dx}\right) = 0$$

Integrating twice, we have

$$-\ln(1-x_A) = C_1 x + C_2$$

Applying the conditions $x_A(0) = x_{A,0}$ and $x_A(L) = x_{A,L}$, the constants of inte-gration may be evaluated and the concentration distribution becomes

$$\frac{1-x_A}{1-x_{A,0}} = \left(\frac{1-x_{A,L}}{1-x_{A,0}}\right)^{x/L} \tag{14.75}$$

Since $1 - x_A = x_B$, we also obtain

$$\frac{x_B}{x_{B,0}} = \left(\frac{x_{B,L}}{x_{B,0}}\right)^{x/L} \tag{14.76}$$

To determine the evaporation rate of species A, Equation 14.75 is first used to evaluate the concentration gradient (dx_A/dx). Substituting the result into Equation 14.74, it follows that

$$N''_{A,x} = \frac{CD_{AB}}{L} \ln\left(\frac{1 - x_{A,L}}{1 - x_{A,0}}\right) \tag{14.77}$$

14.5
Mass Diffusion with Homogeneous Chemical Reactions

Just as heat diffusion may be influenced by internal sources of energy, species transfer by diffusion may be influenced by homogeneous chemical reactions. We restrict our attention to stationary media, in which case Equation 14.45 or 14.46 determines the absolute species flux. If we also assume steady, one-dimensional transfer in the x direction and that D_{AB} and C are constant, Equation 14.38b reduces to

$$D_{AB} \frac{d^2 C_A}{dx^2} + \dot{N}_A = 0 \tag{14.78}$$

If there are no homogeneous chemical reactions involving species A, the volumetric species *production* rate, $\dot{N}_A$, is zero. When reactions do occur, they are often of the form

Zero-order reaction:

$$\dot{N}_A = k_0$$

First-order reaction:

$$\dot{N}_A = k_1 C_A$$

That is, the reaction may occur at a constant rate (zero order) or at a rate that is proportional to the local concentration (first order). The units of k_0 and k_1 are kmol/s · m³ and s⁻¹, respectively. If $\dot{N}_A$ is positive, the reaction results in the generation of species A; if it is negative, it results in the consumption of A.

In many applications the species of interest is converted to another form through a first-order chemical reaction, and Equation 14.78 becomes

$$D_{AB} \frac{d^2 C_A}{dx^2} - k_1 C_A = 0 \tag{14.79}$$

This linear, homogeneous differential equation has the general solution

$$C_A(x) = C_1 e^{mx} + C_2 e^{-mx} \tag{14.80}$$

where $m = (k_1/D_{AB})^{1/2}$ and the constants C_1 and C_2 depend on the prescribed boundary conditions. The form of this equation is identical to that characterizing heat conduction in an extended surface.

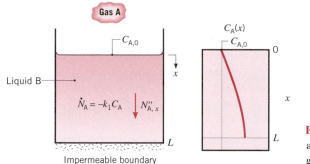

FIGURE 14.9 Diffusion and homogeneous reaction of gas A in liquid B.

Consider the situation illustrated in Figure 14.9. Gas A is soluble in liquid B, where it is transferred by diffusion and experiences a first-order chemical reaction. The solution is dilute, and the concentration of A in the liquid at the interface is a known constant $C_{A,0}$. If the bottom of the container is impermeable to A, the boundary conditions are

$$C_A(0) = C_{A,0} \qquad \text{and} \qquad \left.\frac{dC_A}{dx}\right|_{x=L} = 0$$

Using these boundary conditions with Equation 14.80, it may be shown, after some manipulation, that

$$C_A(x) = C_{A,0}(\cosh mx - \tanh mL \sinh mx) \qquad (14.81)$$

Quantities of special interest are the concentration of A at the bottom and the flux of A across the gas–liquid interface. Applying Equation 14.81 at $x = L$, we obtain

$$C_A(L) = C_{A,0}\frac{(\cosh^2 mL - \sinh^2 mL)}{\cosh mL} = \frac{C_{A,0}}{\cosh mL} \qquad (14.82)$$

Moreover,

$$N''_{A,x}(0) = -D_{AB}\left.\frac{dC_A}{dx}\right|_{x=0}$$

$$= -D_{AB}C_{A,0}\,m(\sinh mx - \tanh mL \cosh mx)\big|_{x=0}$$

or

$$N''_{A,x}(0) = D_{AB}C_{A,0}\,m \tanh mL \qquad (14.83)$$

EXAMPLE 14.5

A solid waste treatment system operates on the principle of aerobic fermentation to decompose organic matter into its basic chemical constituents. Bacteria distributed within the organic matter use oxygen from the atmosphere in the decomposition process. Consider a plane layer of organic matter that has thickness L and is supported by a concrete slab. The top of the layer is exposed to atmospheric air that maintains a fixed molar concentration of oxygen $C_{A,0}$ in the layer (at the exposed surface). The diffusion coefficient D_{AB} of oxygen in the organic matter is known, as is the volumetric rate at which the oxygen is consumed by biochemical reactions. This consumption depends on the local concentration of oxygen and may be expressed as $\dot{N}_A = -k_1 C_A$ (kmol/s · m³). From consideration of a differential control volume within the organic matter, derive a differential equation that could be solved for the local concentration of oxygen $C_A(x)$. Assume one-dimensional, steady-state conditions. Write the general solution to this equation. Applying appropriate boundary conditions, determine the specific form of $C_A(x)$.

SOLUTION

Known: Oxygen transport by diffusion with a homogeneous, first-order chemical reaction.

Find:

1. Differential equation for the concentration distribution $C_A(x)$ and general solution.

2. Appropriate boundary conditions and final solution form.

Schematic:

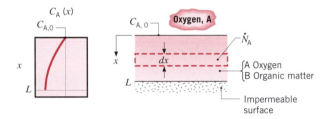

Assumptions:

1. Steady-state, one-dimensional conditions.

2. Stationary medium ($x_A \ll 1$) of uniform concentration and constant properties.

3. Homogeneous chemical reaction.

4. Impermeable bottom.

Analysis:

1. Applying a species mass balance to the differential control volume of the schematic, it follows from the molar form of Equation 14.33 that

$$\dot{N}_{A,\,in} + \dot{N}_{A,\,g} - \dot{N}_{A,\,out} = \dot{N}_{A,\,st}$$

where

$$\dot{N}_{A,\,in} = N_{A,\,x} = -D_{AB}A \,\frac{dC_A}{dx}$$

$$\dot{N}_{A,\,out} = N_{A,\,x} + \frac{dN_{A,\,x}}{dx}\,dx = -D_{AB}A \,\frac{dC_A}{dx} - D_{AB}A \,\frac{d^2C_A}{dx^2}\,dx$$

$$\dot{N}_{A,\,g} = \dot{N}_A A \, dx$$

$$\dot{N}_{A,\,st} = 0$$

Substituting the rate equations into the species balance and dividing by the area A, it follows that

$$D_{AB} \frac{d^2 C_A}{dx^2} - k_1 C_A = 0 \qquad \triangleleft$$

The general solution is

$$C_A(x) = C_1 e^{-mx} + C_2 e^{mx} \qquad \triangleleft$$

where $m \equiv (k_1/D_{AB})^{1/2}$.

2. Appropriate boundary conditions are of the form

$$C_A(0) = C_{A,0} \qquad \triangleleft$$

$$\left. \frac{dC_A}{dx} \right|_{x=L} = 0 \qquad \triangleleft$$

Applying these conditions to the general solution, we obtain

$$C_{A,0} = C_1 + C_2$$

$$\left. \frac{dC_A}{dx} \right|_{x=L} = -mC_1 e^{-mL} + mC_2 e^{mL} = 0$$

Hence from the second condition,

$$C_1 = C_2 e^{2mL}$$

and from the first condition,

$$C_{A,0} = C_2 (e^{2mL} + 1)$$

Hence

$$C_2 = \frac{C_{A,0}}{e^{2mL} + 1} \qquad \text{and} \qquad C_1 = \frac{C_{A,0} e^{2mL}}{e^{2mL} + 1}$$

$$C_A(x) = \frac{C_{A,0}}{e^{2mL} + 1} [e^{m(2L-x)} + e^{mx}] \qquad \triangleleft$$

14.6
Transient Diffusion

Results analogous to those of Chapter 5 may be obtained for the transient diffusion of a dilute species A in a stationary medium. Assuming no homogeneous reactions, constant D_{AB} and C, and one-dimensional transfer in the x direction, Equation 14.38b reduces to

$$D_{AB} \frac{\partial^2 C_A}{\partial x^2} = \frac{\partial C_A}{\partial t} \qquad (14.84)$$

Assuming an initial uniform concentration,

$$C_A(x, 0) = C_{A,i} \tag{14.85}$$

Equation 14.84 may be solved for boundary conditions that depend on the particular geometry and surface conditions. If, for example, the geometry is a plane wall of thickness $2L$ with surface convection, the boundary conditions are

$$\left.\frac{\partial C_A}{\partial x}\right|_{x=0} = 0 \tag{14.86}$$

$$C_A(L, t) = C_{A,s} \tag{14.87}$$

Equation 14.86 describes the symmetry requirement at the midplane. Equation 14.87 represents the surface convection condition if the *mass transfer Biot number*, $Bi_m = h_m L/D_{AB}$, is much larger than unity. In this case the resistance to species transfer by diffusion in the medium is much larger than the resistance to species transfer by convection at the surface. If this situation is taken to the limit of $Bi_m \to \infty$, or $Bi_m^{-1} \to 0$, it follows that the free stream species concentration $C_{A,\infty}$ may be replaced by the surface concentration $C_{A,s}$. Note, however, that $C_{A,s}$ represents the species concentration in the medium, and hence must be determined by using Equation 14.43 or 14.44.

The analogy between heat and mass transfer may conveniently be applied if we nondimensionalize the above equations. Introducing a dimensionless concentration and time, as follows,

$$\gamma^* \equiv \frac{\gamma}{\gamma_i} = \frac{C_A - C_{A,s}}{C_{A,i} - C_{A,s}} \tag{14.88}$$

$$t_m^* \equiv \frac{D_{AB}t}{L^2} \equiv Fo_m \tag{14.89}$$

and substituting into Equation 14.84, we obtain

$$\frac{\partial^2 \gamma^*}{\partial x^{*2}} = \frac{\partial \gamma^*}{\partial Fo_m} \tag{14.90}$$

where $x^* = x/L$. Similarly, the initial and boundary conditions are

$$\gamma^*(x^*, 0) = 1 \tag{14.91}$$

$$\left.\frac{\partial \gamma^*}{\partial x^*}\right|_{x^*=0} = 0 \tag{14.92}$$

and

$$\gamma^*(1, t_m^*) = 0 \tag{14.93}$$

One need only compare Equations 14.90 through 14.93 with Equations 5.34 through 5.36 and Equation 5.37 for the case of $Bi \to \infty$ to confirm the existence of the analogy. Note that for $Bi \to \infty$, Equation 5.37 reduces to $\theta^*(1, t^*) = 0$, which is analogous to Equation 14.93. Hence the two systems of equations must have equivalent solutions.

The correspondence between variables for transient heat and mass diffusion is summarized in Table 14.2. From this correspondence it is possible to

TABLE 14.2 **Correspondence Between Heat and Mass Transfer Variables for Transient Diffusion**

Heat Transfer	Mass Transfer
$\theta^* = \dfrac{T - T_\infty}{T_i - T_\infty}$	$\gamma^* = \dfrac{C_A - C_{A,s}}{C_{A,i} - C_{A,s}}$
$1 - \theta^* = \dfrac{T - T_i}{T_\infty - T_i}$	$1 - \gamma^* = \dfrac{C_A - C_{A,i}}{C_{A,s} - C_{A,i}}$
$Fo = \dfrac{\alpha t}{L^2}$	$Fo_m = \dfrac{D_{AB}t}{L^2}$
$Bi = \dfrac{hL}{k}$	$Bi_m = \dfrac{h_m L}{D_{AB}}$
$\dfrac{x}{2\sqrt{\alpha t}}$	$\dfrac{x}{2\sqrt{D_{AB}t}}$

use many of the previous heat transfer results to solve transient mass diffusion problems. For example, replacing θ_o^* and Fo by γ_o^* and Fo_m, Figure D.1 and Equation 5.41 could be used, for $Bi^{-1} = 0$, to determine the midplane concentration $C_{A,o}$. The remaining Heisler charts and equations may be applied in a similar fashion, along with results obtained for the semi-infinite solid.

EXAMPLE 14.6

A slab of salt (NaCl) of thickness L is used to support a deep layer of water. The salt dissolves in the water, maintaining a fixed mass density $\rho_{A,s}$ (kg/m³) at the water–salt interface. If the salt density in the water is initially zero, how does this density vary with position and time after contact between the solid salt and the water has been made? What is the surface recession rate dL/dt and how does the surface recession vary with time? If the mass density of solid salt is $\rho_A(S) = 2165$ kg/m³ and its density in solution at the surface is $\rho_{A,s} = 380$ kg/m³, by how much will the surface recede after 24 h? The saltwater diffusion coefficient is $D_{AB} = 1.2 \times 10^{-9}$ m²/s.

SOLUTION

Known: Salt from a slab of thickness L is diffusing into a deep layer of water under transient conditions.

Find:

1. Density distribution of salt in the water, $\rho_A(x, t)$.
2. Surface recession rate dL/dt, and surface recession as a function of time.

Schematic:

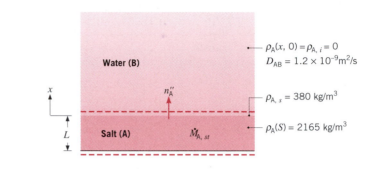

Assumptions:

1. One-dimensional species diffusion in x.
2. No chemical reactions.
3. Solution stationary and semi-infinite.
4. Constant properties, including total density ρ of the solution.

Analysis:

1. With the foregoing assumptions, $n''_A = j_A$, and Equation 14.37b becomes

$$\frac{\partial^2 \rho_A}{\partial x^2} = \frac{1}{D_{AB}} \frac{\partial \rho_A}{\partial t}$$

which is analogous to Equation 5.26. Moreover, the initial condition

$$\rho_A(x, 0) = \rho_{A, i} = 0$$

and the boundary conditions

$$\rho_A(\infty, t) = \rho_{A, i} = 0 \qquad \text{and} \qquad \rho_A(0, t) = \rho_{A, s}$$

are analogous to Equations 5.27, 5.53, and case 1 condition of Figure 5.7. By analogy to Equation 5.57, the density distribution of salt in the water is then

$$\frac{\rho_A(x, t) - \rho_{A, s}}{\rho_{A, i} - \rho_{A, s}} = \text{erf} \left[\frac{x}{2(D_{AB}t)^{1/2}} \right]$$

or

$$\rho_A(x, t) = \rho_{A, s} \left\{ 1 - \text{erf} \left[\frac{x}{2(D_{AB}t)^{1/2}} \right] \right\} \qquad \triangleleft$$

2. Performing a mass balance on a control volume about the slab, it follows from Equation 14.33 that

$$-\dot{M}_{A, \text{out}} = \dot{M}_{A, \text{st}}$$

or, for a unit surface area,

$$-n''_{A, s} = \frac{d[\rho_A(S)L]}{dt}$$

where, by analogy to Equation 5.58, the species mass flux at the surface is

$$n''_{A, s} = \frac{D_{AB}\rho_{A, s}}{(\pi D_{AB}t)^{1/2}} = \left(\frac{D_{AB}}{\pi t}\right)^{1/2}\rho_{A, s}$$

Hence

$$\frac{dL}{dt} = -\left(\frac{D_{AB}}{\pi t}\right)^{1/2}\frac{\rho_{A, s}}{\rho_A(S)}$$

Integrating

$$\int_0^{\Delta L} dL = -\left(\frac{D_{AB}}{\pi}\right)^{1/2}\frac{\rho_{A, s}}{\rho_A(S)}\int_0^t\frac{dt}{t^{1/2}}$$

The surface recession that has occurred after the time t is then

$$\Delta L = -2\frac{\rho_{A, s}}{\rho_A(S)}\left(\frac{D_{AB}t}{\pi}\right)^{1/2} \qquad \triangleleft$$

For the prescribed conditions, it follows that

$$\Delta L = -2\frac{380 \text{ kg/m}^3}{2165 \text{ kg/m}^3}\left(\frac{1.2 \times 10^{-9} \text{ m}^2/\text{s} \times 24 \text{ h} \times 3600 \text{ s/h}}{\pi}\right)^{1/2}$$

$$\Delta L = 2.02 \times 10^{-3} \text{ m} = 2.02 \text{ mm} \qquad \triangleleft$$

Comments:

1. Recognize that D_{AB} is analogous to α through their appearance in the diffusion equations. In contrast, D_{AB} is analogous to k through their appearance in the rate equations (Fick's and Fourier's laws). Note how D_{AB} has been substituted for both α and k in the foregoing implementation of the analogy.

2. These results apply to only a first approximation, since the large salt concentration near the surface precludes the existence of a dilute solution in this region.

References

1. Bird, R. B., *Adv. Chem. Eng.*, **1**, 170, 1956.

2. Bird, R. B., W. E. Stewart, and E. N. Lightfoot, *Transport Phenomena*, Wiley, New York, 1960.

3. Hirschfelder, J. O., C. F. Curtiss, and R. B. Bird, *Molecular Theory of Gases and Liquids*, Wiley, New York, 1954.

4. Skelland, A. H. P., *Diffusional Mass Transfer*, Wiley, New York, 1974.

5. Ried, R. C., and T. K. Sherwood, *The Properties of Gases and Liquids*, McGraw-Hill, New York, 1966.

Problems

Mixture Composition

14.1   Assuming air to be composed exclusively of O_2 and N_2, with their partial pressures in the ratio 0.21:0.79, what are their mass fractions?

14.2   A mixture of CO_2 and N_2 is in a container at 25°C, with each species having a partial pressure of 1 bar. Calculate the molar concentration, the mass density, the mole fraction, and the mass fraction of each species.

14.3   Consider an ideal gas mixture of n species.

  (a)   Derive an equation for determining the mass fraction of species i from knowledge of the mole fraction and the molecular weight of each of the n species. Derive an equation for determining the mole fraction of species i from knowledge of the mass fraction and the molecular weight of each of the n species.

  (b)   In a mixture containing equal mole fractions of O_2, N_2, and CO_2, what is the mass fraction of each species? In a mixture containing equal mass fractions of O_2, N_2, and CO_2, what is the mole fraction of each species?

Fick's Law

14.4   Consider air in a closed, cylindrical container with its axis vertical and with opposite ends maintained at different temperatures. Assume that the total pressure of the air is uniform throughout the container.

  (a)   If the bottom surface is colder than the top surface, what is the nature of conditions within the container? For example, will there be vertical gradients of the species (O_2 and N_2) concentrations? Is there any motion of the air? Does mass transfer occur?

  (b)   What is the nature of conditions within the container if it is inverted (i.e., the warm surface is now at the bottom)?

14.5   Gaseous hydrogen at 10 bars and 27°C is stored in a 100-mm diameter spherical tank having a steel wall 2 mm thick. The molar concentration of hydrogen in the steel is 1.50 kmol/m³ at the inner surface and negligible at the outer surface, while the diffusion coefficient of hydrogen in steel is approximately 0.3×10^{-12} m²/s. What is the initial rate of mass loss of hydrogen by diffusion through

the tank wall? What is the initial rate of pressure drop within the tank?

14.6   A thin plastic membrane is used to separate helium from a gas stream. Under steady-state conditions the concentration of helium in the membrane is known to be 0.02 and 0.005 kmol/m³ at the inner and outer surfaces, respectively. If the membrane is 1 mm thick and the binary diffusion coefficient of helium with respect to the plastic is 10^{-9} m²/s, what is the diffusive flux?

14.7   Estimate values of the mass diffusion coefficient D_{AB} for binary mixtures of the following gases at 350 K and 1 atm: ammonia–air and hydrogen–air.

The Mass Diffusion Equation

14.8   Beginning with a differential control volume, derive the diffusion equation, on a molar basis, for species A in a three-dimensional (Cartesian coordinates), stationary medium, considering species generation with constant properties. Compare your result with Equation 14.38b.

14.9   Consider the radial diffusion of a gaseous species (A) through the wall of a plastic tube (B), and allow for chemical reactions that provide for the depletion of A at a rate $\dot{N}_A$ (kmol/s · m³). Derive a differential equation that governs the molar concentration of species A in the plastic.

14.10   Beginning with a differential control volume, derive the diffusion equation, on a molar basis, for species A in a one-dimensional, spherical, stationary medium, considering species generation. Compare your result with Equation 14.40.

One-Dimensional, Steady Diffusion

14.11   Oxygen gas is maintained at pressures of 2 bars and 1 bar on opposite sides of a rubber membrane that is 0.5 mm thick, and the entire system is at 25°C. What is the molar diffusive flux of O_2 through the membrane? What are the molar concentrations of O_2 on both sides of the membrane (outside the rubber)?

14.12   Insulation degrades (experiences an increase in thermal conductivity) if it is subjected to water vapor condensation. The problem may occur in home insulation during cold periods, when vapor in a humidified room diffuses through the dry wall (plaster board) and condenses in the adjoining insulation. Estimate the mass diffusion rate for a 3 m by 5 m wall, under conditions for which the vapor pressure is 0.03 bar in the room air and 0.0 bar in the insulation. The dry wall is 10 mm thick, and the solubility of water vapor in the wall material is approximately 5×10^{-3} kmol/m^3 · bar. The binary diffusion coefficient for water vapor in the dry wall is approximately 10^{-9} m^2/s.

14.13 A rubber plug that is 20 mm thick and has a surface area of 300 mm^2 is used to contain CO_2 at 25°C and 5 bars in a 10-liter vessel. What is the rate of mass loss of CO_2 from the vessel? What is the reduction in pressure that would be experienced over a 24-h period?

14.14 Helium gas at 25°C and 4 bars is contained in a glass cylinder of 100-mm inside diameter and 5-mm thickness. What is the rate of mass loss per unit length of the cylinder?

14.15 Helium gas at 25°C and 4 bars is stored in a spherical Pyrex container of 200-mm inside diameter and 10-mm thickness. What is the rate of mass loss from the container?

14.16 Hydrogen at a pressure of 2 atm flows within a tube of diameter 40 mm and wall thickness 0.5 mm. The outer surface is exposed to a gas stream for which the hydrogen partial pressure is 0.1 atm. The mass diffusivity and solubility of hydrogen in the tube material are 1.8×10^{-11} m^2/s and 160 kmol/m^3 · atm, respectively. When the system is at 500 K, what is the rate of hydrogen transfer through the tube per unit length (kg/s · m)?

14.17 Nitric oxide (NO) emissions from automobile exhaust can be reduced by using a catalytic converter, and the following reaction occurs at the catalytic surface:

$$\text{NO} + \text{CO} \rightarrow \tfrac{1}{2}\text{N}_2 + \text{CO}_2$$

The concentration of NO is reduced by passing the exhaust gases over the surface, and the rate of reduction at the catalyst is governed by a first-order reaction of the form given by Equation 14.57. As a first approximation it may be assumed that NO reaches the surface by one-dimensional diffusion through a thin gas film of thickness L that adjoins the surface. Referring to Figure 14.6, consider a situation for which the exhaust gas is at 500°C and 1.2 bars and the mole fraction of NO is $x_{A, L} = 0.15$. If $D_{AB} = 10^{-4}$ m^2/s, $k_1'' = 0.05$ m/s, and the film thickness is $L = 1$ mm, what is the mole fraction of NO at the catalytic surface and what is the

NO removal rate for a surface of area $A = 200$ cm^2?

14.18 Pulverized coal pellets, which may be approximated as carbon spheres of radius $r_o = 1$ mm, are burned in a pure oxygen atmosphere at 1450 K and 1 atm. Oxygen is transferred to the particle surface by diffusion, where it is consumed in the reaction $C + O_2 \rightarrow CO_2$. The reaction rate is first order and of the form $\dot{N}_{O_2}'' = -k_1'' C_{O_2}(r_o)$, where $k_1'' = 0.1$ m/s. Neglecting changes in r_o, determine the steady-state O_2 molar consumption rate in kmol/s. At 1450 K, the binary diffusion coefficient for O_2 and CO_2 is 1.71×10^{-4} m^2/s.

14.19 Pulverized coal, which may be approximated as pure carbon spheres of radius $r_o = 1$ mm, is burned in pure oxygen at 1450 K and 1 atm. Oxygen is transferred to the particle surface by diffusion, where it is consumed in a reaction of the form $C + O_2 \rightarrow CO_2$. Assuming the surface reaction rate to be infinite and neglecting the change in r_o, obtain expressions for the radial distributions of the CO_2 and O_2 concentrations. What is the O_2 molar consumption rate?

14.20 To enhance the effective surface, and hence the chemical reaction rate, catalytic surfaces often take the form of porous solids. One such solid may be visualized as consisting of a large number of cylindrical pores, each of diameter D and length L.

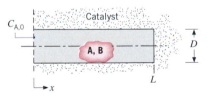

Consider conditions involving a gaseous mixture of A and B for which species A is chemically consumed at the catalytic surface. The reaction is known to be first order, and the rate at which it occurs per unit area of the surface may be expressed as $k_1'' C_A$, where k_1'' (m/s) is the reaction rate constant and C_A (kmol/m^3) is the local molar concentration of species A. Under steady-state conditions, flow over the porous solid is known to maintain a fixed value of the molar concentration $C_{A, 0}$ at the pore mouth. Beginning from fundamentals, obtain the differential equation that governs the variation of C_A with distance x along the pore. Applying appropriate boundary conditions, solve the equation to obtain an expression for $C_A(x)$.

Equimolar Counterdiffusion and Column Evaporation

14.21 Carbon dioxide and nitrogen experience equimolar counterdiffusion in a circular tube whose length and diameter are 1 m and 50 mm, respectively. The system is at a total pressure of 1 atm and a temperature of 25°C. The ends of the tubes are connected to large chambers in which the species concentrations are maintained at fixed values. The partial pressure of CO_2 at one end is 100 mm Hg, while at the other end it is 50 mm Hg. What is the mass transfer rate of CO_2 through the tube?

14.22 Consider evaporation in a column, where the vapor A is transferred through a gas B. Which of the following limiting cases is characterized by the largest evaporation rate? (a) Gas B has unlimited solubility in liquid A. (b) Gas B is completely insoluble in liquid A. What is the ratio of the evaporation rate in part (a) to that in part (b) if the vapor pressure is zero at the top of the column and the saturated vapor pressure is one-tenth of the total pressure?

14.23 An open pan of diameter 0.2 m and height 80 mm (above water at 27°C) is exposed to ambient air at 27°C and 25% relative humidity. Determine the evaporation rate, assuming that only mass diffusion occurs. Determine the evaporation rate, considering bulk motion.

14.24 A technique for growing crystals involves vaporizing the crystal material (A) at one end of a horizontal, cylindrical tube ($x = 0$) and depositing it at the other end ($x = L$). The tube also contains an inert gas B, to which the vapor source and the crystal are impermeable.

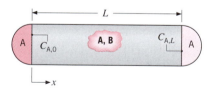

Assuming isothermal conditions, obtain expressions for the vapor molar flux and the spatial distribution of the vapor molar concentration. What is the location of the maximum concentration gradient?

14.25 A spherical droplet of liquid A and radius r_o evaporates into a stagnant layer of gas B. Derive an expression for the evaporation rate of species A in terms of the saturation pressure of species A, $p_A(r_o) = p_{A, sat}$, the partial pressure of species A at an arbitrary radius r, $p_A(r)$, the total pressure p, and other pertinent quantities. Assume the droplet and

the mixture are at a uniform pressure p and temperature T.

14.26 The condenser in a methanol distillation system has a small cooler before the vent pipe. The vapor volume of the cooler is 0.005 m^3, while the vent pipe discharging to the atmosphere at a pressure of 1 bar has a diameter of 35 mm and a length of 0.5 m. The temperature in the cooler and vent is 21°C, and the partial pressure of methanol in the cooler is 100 mm Hg. The binary diffusion coefficient for a methanol–air mixture is 0.13×10^{-4} m^2/s at 273 K, and the molecular weight of methanol is 32.

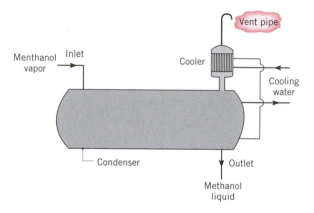

(a) Assuming that air cannot exit the condenser through its inlet and outlet ports, estimate the weekly loss of methanol vapor (kg/week) due to diffusion through the vent pipe to the atmosphere.

(b) Once per hour the heat rate for the distillation process is varied, causing the vapor in the cooler to be expelled. Estimate the additional weekly loss of methanol.

14.27 The presence of a small amount of air may cause a significant reduction in the heat rate to a water-cooled steam condenser surface. For a clean surface with pure steam and the prescribed conditions, the condensate rate is 0.020 $kg/m^2 \cdot s$. With the presence of stagnant air in the steam, the condensate surface temperature drops from 28 to 24°C and the condensate rate is reduced by a factor of 2.

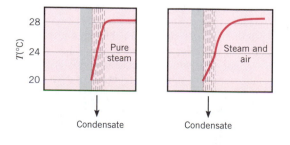

For the air–steam mixture, determine the partial pressure of air as a function of distance from the condensate film.

Homogeneous Chemical Reactions

14.28 As an employee of the Los Angeles Air Quality Commission, you have been asked to develop a model for computing the distribution of NO_2 in the atmosphere. The molar flux of NO_2 at ground level, $N''_{A,0}$, is presumed known. This flux is attributed to automobile and smoke stack emissions. It is also known that the concentration of NO_2 at a distance well above ground level is zero and that NO_2 reacts chemically in the atmosphere. In particular, NO_2 reacts with unburned hydrocarbons (in a process that is activated by sunlight) to produce PAN (peroxyacetylnitrate), the final product of photochemical smog. The reaction is first order, and the local rate at which it occurs may be expressed as $\dot{N}_A = -k_1 C_A$.

(a) Assuming steady-state conditions and a stagnant atmosphere, obtain an expression for the vertical distribution $C_A(x)$ of the molar concentration of NO_2 in the atmosphere.

(b) If an NO_2 partial pressure of $p_A = 2 \times 10^{-6}$ bar is sufficient to cause pulmonary damage, what is the value of the ground level molar flux for which you would issue a smog alert? You may assume an isothermal atmosphere at $T = 300$ K, a reaction coefficient of $k_1 = 0.03$ s^{-1}, and an NO_2–air diffusion coefficient of $D_{AB} = 0.15 \times 10^{-4}$ m^2/s.

14.29 The process of catalysis is usually carried out in a fixed-bed reactor consisting of many porous pellets. The internal porous structure of the pellets is intended to provide a large catalytic surface area per unit volume of the reactor. Typically, the pellets are submerged in a gas stream, and the pellet surface catalyzes a chemical reaction involving one of the species (A) from the gas.

Although the processes that govern the transport of species A within the pellet are extremely complex, reasonable results may be obtained by using Fick's law with an *effective diffusion coefficient.* That is, the molar flux at any radial position within a pellet may be expressed as $J^*_{A,r} = -CD_{eff}(dx_A/dr)$, where x_A is the gas phase mole fraction of species A. Moreover, even though the actual reaction is heterogeneous, it is reasonable to approximate conditions within the pellet in terms of a homogeneous reaction that is a continuous function of radius. The rate

at which species A is consumed per unit volume of the pellet is then expressed as $\dot{N}_A = -k'_1 A_v C_A$, where A_v is the catalytic surface area per unit volume of the pellet. The modified reaction rate constant k'_1 has units of meter per second. It is also reasonable to assume isothermal, constant pressure conditions for the gas in the pellet.

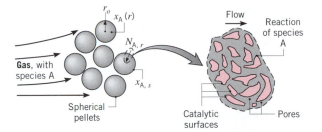

(a) Assuming steady-state conditions in a stationary medium, obtain an expression for the radial distribution of the species mole fraction x_A in a spherical pellet. To obtain this result, you may wish to make use of the transformation $y \equiv rx_A$. Obtain expressions for the total rate at which A is consumed by a pellet and the pellet effectiveness ε, which is defined as the ratio of the actual consumption rate to the consumption rate that would exist if D_{eff} were infinite. If D_{eff} were infinite, dx_A/dr would have to be zero throughout the pellet, in which case x_A would be constant throughout the pellet. Express your results in terms of hyperbolic functions involving r_o, k'_1, D_{eff}, A_v, C (the total molar concentration of gases in the pellet), and $x_{A,s}$ (the mole fraction of A at the pellet outer surface).

(b) A catalytic reactor is used to convert CO, species A, in automobile exhausts to CO_2. The reactor is composed of CuO-coated alumina pellets of 5-mm diameter and $A_v = 10^8$ m^2/m^3. The exhaust gas is at 550°C and 1.2 atm, and the surface mole fraction of A is 0.04. If $D_{eff} = 2 \times 10^{-5}$ m^2/s and $k'_1 = 10^{-3}$ m/s, what is the CO consumption rate and the pellet effectiveness?

14.30 Consider the problem of oxygen transfer from the interior lung cavity, across the lung tissue, to the network of blood vessels on the opposite side. The lung tissue (species B) may be approximated as a plane wall of thickness L. The inhalation process may be assumed to maintain a constant molar concentration $C_A(0)$ of oxygen (species A) in the tissue at its inner surface ($x = 0$), and assimilation of oxygen by the blood may be assumed to maintain a constant molar concentration $C_A(L)$ of oxygen in the tissue at its outer surface ($x = L$). There is oxygen consumption in the tissue due to metabolic processes, and the reaction is zero order, with $N_A = -k_0$. Obtain expressions for the distribution of the oxygen concentration in the tissue and for the rate of assimilation of oxygen by the blood per unit tissue surface area.

14.31 Consider combustion of hydrogen gas in a mixture of hydrogen and oxygen adjacent to the metal wall of a combustion chamber. Combustion occurs at constant temperature and pressure according to the chemical reaction $2H_2 + O_2 \rightarrow 2H_2O$. Measurements under steady-state conditions at a distance of 10 mm from the wall indicate that the molar concentrations of hydrogen, oxygen, and water vapor are 0.10, 0.10, and 0.20 $kmol/m^3$, respectively. The generation rate of water vapor is 0.96×10^{-2} $kmol/m^3 \cdot s$ throughout the region of interest. The binary diffusion coefficient for each of the species (H_2, O_2, and H_2O) in the remaining species is 0.6×10^{-5} m^2/s.

(a) Determine an expression for and make a qualitative plot of C_{H_2} as a function of distance from the wall.

(b) Determine the value of C_{H_2} at the wall.

(c) On the same coordinates used in part (a), sketch curves for the concentrations of oxygen and water vapor.

(d) What is the molar flux of water vapor at $x = 10$ mm?

Transient Diffusion

14.32 In Problem 14.28, NO_2 transport by diffusion in a stagnant atmosphere was considered for steady-state conditions. However, the problem is actually time dependent, and a more realistic approach would account for transient effects. Consider the ground level emission of NO_2 to begin in the early morning (at $t = 0$), when the NO_2 concentration in the atmosphere is everywhere zero. Emission occurs throughout the day at a constant flux $N''_{A,0}$, and the NO_2 again experiences a first-order photochemical reaction in the atmosphere ($\dot{N}_A = -k_1 C_A$).

(a) For a differential element in the atmosphere, derive a differential equation that could be used to determine the molar concentration $C_A(x, t)$. State appropriate initial and boundary conditions.

(b) Obtain an expression for $C_A(x, t)$ under the special condition for which photochemical re-

actions may be neglected. For this condition what are the molar concentrations of NO_2 at ground level and at 100-m elevation 3 h after the start of the emissions, if $N''_{A,0} = 3 \times 10^{-11}$ $kmol/s \cdot m^2$ and $D_{AB} = 0.15 \times 10^{-4}$ m^2/s?

14.33 The presence of CO_2 in solution is essential to the growth of aquatic plant life, with CO_2 used as a reactant in the photosynthesis. Consider a stagnant body of water in which the concentration of CO_2 (ρ_A) is everywhere zero. At time $t = 0$, the water is exposed to a source of CO_2, which maintains the surface ($x = 0$) concentration at a fixed value $\rho_{A,0}$. For time $t > 0$, CO_2 will begin to accumulate in the water, but the accumulation is inhibited by CO_2 consumption due to photosynthesis. The time rate at which this consumption occurs per unit volume is equal to the product of a reaction rate constant k_1 and the local CO_2 concentration $\rho_A(x, t)$.

(a) Write (do not derive) a differential equation that could be used to determine $\rho_A(x, t)$ in the water. What does each term in the equation represent physically?

(b) Write appropriate boundary conditions that could be used to obtain a particular solution, assuming a "deep" body of water. What would be the form of this solution for the special case of negligible CO_2 consumption ($k_1 \approx 0$)?

14.34 A large sheet of material 40 mm thick contains dissolved hydrogen (H_2) having a uniform concentration of 3 $kmol/m^3$. The sheet is exposed to a fluid stream that causes the concentration of the dissolved hydrogen to be reduced suddenly to zero at both surfaces. This surface condition is maintained constant thereafter. If the mass diffusivity of hydrogen is 9×10^{-7} m^2/s, how much time is required to bring the density of dissolved hydrogen to a value of 1.2 kg/m^3 at the center of the sheet?

14.35 A common procedure for increasing the moisture content of air is to bubble it through a column of water. Assume the air bubbles to be spheres of radius $r_o = 1$ mm and to be in thermal equilibrium with the water at 25°C. How long should the bubbles remain in the water to achieve a vapor concentration at the center that is 99% of the maximum possible (saturated) concentration? The air is dry when it enters the water.

14.36 Steel is carburized in a high-temperature process that depends on the transfer of carbon by diffusion. The value of the diffusion coefficient is strongly temperature dependent and may be approximated as D_{c-s} (m^2/s) $\approx 2 \times 10^{-5}$ exp $[-17,000/T$ (K)]. If the process is effected at 1000°C and a carbon mole fraction of 0.02 is maintained at the surface of the steel, how much time is required to elevate the carbon content of the steel from an initial value of 0.1% to a value of 1.0% at a depth of 1 mm?

14.37 A solar pond operates on the principle that heat losses from a shallow layer of water, which acts as a solar absorber, may be minimized by establishing a stable vertical salinity gradient in the water. In practice such a condition may be achieved by applying a layer of pure salt to the bottom and adding an overlying layer of pure water. The salt enters into solution at the bottom and is transferred through the water layer by diffusion, thereby establishing salt-stratified conditions.

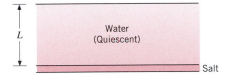

As a first approximation, the total mass density ρ and the diffusion coefficient for salt in water (D_{AB}) may be assumed to be constant, with $D_{AB} = 1.2 \times 10^{-9}$ m^2/s.

(a) If a saturated density of $\rho_{A,s}$ is maintained for salt in solution at the bottom of the water layer of thickness $L = 1$ m, how long will it take for the mass density of salt at the top of the layer to reach 25% of saturation?

(b) In the time required to achieve 25% of saturation at the top of the layer, how much salt is transferred from the bottom into the water per unit surface area (kg/m^2)? The saturation density of salt in solution is $\rho_{A,s} = 380$ kg/m^3.

(c) If the bottom is depleted of salt at the time that the salt density reaches 25% of saturation at the top, what is the final (steady-state) density of the salt at the bottom? What is the final density of the salt at the top?

APPENDIX **A**

Thermophysical Properties of Matter[1]

---

[1] The convention used to present numerical values of the properties is illustrated by this example:

T (K)	$\nu \cdot 10^7$ (m²/s)	$k \cdot 10^3$ (W/m · K)
300	0.349	521

where $\nu = 0.349 \times 10^{-7}$ m²/s and $k = 521 \times 10^{-3} = 0.521$ W/m · K at 300 K.

TABLE A.1   Thermophysical Properties of Selected Metallic Solids[a]

| Composition | Melting Point (K) | Properties at 300 K | | | | Properties at Various Temperatures (K) k (W/m·K)/c_p (J/kg·K) | | | | | | | | | |
		ρ (kg/m³)	c_p (J/kg·K)	k (W/m·K)	$\alpha \cdot 10^6$ (m²/s)	100	200	400	600	800	1000	1200	1500	2000	2500
Aluminum															
Pure	933	2702	903	237	97.1	302	237	240	231	218					
						482	798	949	1033	1146					
Alloy 2024-T6 (4.5% Cu, 1.5% Mg, 0.6% Mn)	775	2770	875	177	73.0	65	163	186	186						
						473	787	925	1042						
Alloy 195, Cast (4.5% Cu)		2790	883	168	68.2			174	185						
								—	—						
Beryllium	1550	1850	1825	200	59.2	990	301	161	126	106	90.8	78.7			
						203	1114	2191	2604	2823	3018	3227	3519		
Bismuth	545	9780	122	7.86	6.59	16.5	9.69	7.04							
						112	120	127							
Boron	2573	2500	1107	27.0	9.76	190	55.5	16.8	10.6	9.60	9.85				
						128	600	1463	1892	2160	2338				
Cadmium	594	8650	231	96.8	48.4	203	99.3	94.7							
						198	222	242							
Chromium	2118	7160	449	93.7	29.1	159	111	90.9	80.7	71.3	65.4	61.9	57.2	49.4	
						192	384	484	542	581	616	682	779	937	
Cobalt	1769	8862	421	99.2	26.6	167	122	85.4	67.4	58.2	52.1	49.3	42.5		
						236	379	450	503	550	628	733	674		
Copper															
Pure	1358	8933	385	401	117	482	413	393	379	366	352	339			
						252	356	397	417	433	451	480			
Commercial bronze (90% Cu, 10% Al)	1293	8800	420	52	14		42	52	59						
							785	460	545						
Phosphor gear bronze (89% Cu, 11% Sn)	1104	8780	355	54	17		41	65	74						
									—						
Cartridge brass (70% Cu, 30% Zn)	1188	8530	380	110	33.9	75	95	137	149						
							360	395	425						
Constantan (55% Cu, 45% Ni)	1493	8920	384	23	6.71	17	19								
						237	362								
Germanium	1211	5360	322	59.9	34.7	232	96.8	43.2	27.3	19.8	17.4	17.4			
						190	290	337	348	357	375	395			

TABLE A.1 *Continued*

Composition	Melting Point (K)	Properties at 300 K				Properties at Various Temperatures (K) k (W/m·K)/c_p (J/kg·K)									
		ρ (kg/m³)	c_p (J/kg·K)	k (W/m·K)	$\alpha \cdot 10^6$ (m²/s)	100	200	400	600	800	1000	1200	1500	2000	2500
Gold	1336	19300	129	317	127	327 109	323 124	311 131	298 135	284 140	270 145	255 155			
Iridium	2720	22500	130	147	50.3	172 90	153 122	144 133	138 138	132 144	126 153	120 161	111 172		
Iron															
Pure	1810	7870	447	80.2	23.1	134 216	94.0 384	69.5 490	54.7 574	43.3 680	32.8 975	28.3 609	32.1 654		
Armco (99.75% pure)		7870	447	72.7	20.7	95.6 215	80.6 384	65.7 490	53.1 574	42.2 680	32.3 975	28.7 609	31.4 654		
Carbon steels															
Plain carbon (Mn ≤ 1%, Si ≤ 0.1%)		7854	434	60.5	17.7			56.7 487	48.0 559	39.2 685	30.0 1169				
AISI 1010		7832	434	63.9	18.8			58.7 487	48.8 559	39.2 685	31.3 1168				
Carbon–silicon (Mn ≤ 1%, 0.1% < Si ≤ 0.6%)		7817	446	51.9	14.9			49.8 501	44.0 582	37.4 699	29.3 971				
Carbon–manganese–silicon (1% < Mn ≤ 1.65%, 0.1% < Si ≤ 0.6%)		8131	434	41.0	11.6			42.2 487	39.7 559	35.0 685	27.6 1090				
Chromium (low) steels															
$\frac{1}{2}$Cr-$\frac{1}{4}$Mo–Si (0.18% C, 0.65% Cr, 0.23% Mo, 0.6% Si)		7822	444	37.7	10.9			38.2 492	36.7 575	33.3 688	26.9 969				
1 Cr-$\frac{1}{2}$Mo (0.16% C, 1% Cr, 0.54% Mo, 0.39% Si)		7858	442	42.3	12.2			42.0 492	39.1 575	34.5 688	27.4 969				
1 Cr–V (0.2% C, 1.02% Cr, 0.15% V)		7836	443	48.9	14.1			46.8 492	42.1 575	36.3 688	28.2 969				

Properties at various temperatures are given as k (W/m·K) / cp (J/kg·K).

Composition	Melting Point (K)	ρ (kg/m³)	cp (J/kg·K)	k (W/m·K)	α·10⁶ (m²/s)	100	200	400	600	800	1000	1200	1500	2000	2500
Stainless steels AISI 302	1670	8055	480	15.1	3.91			17.3 / 512	20.0 / 559	22.8 / 585	25.4 / 606	28.0 / 640	31.7 / 682		
AISI 304		7900	477	14.9	3.95	9.2 / 272	12.6 / 402	16.6 / 515	19.8 / 557	22.6 / 582	25.4 / 611				
AISI 316		8238	468	13.4	3.48			15.2 / 504	18.3 / 550	21.3 / 576	24.2 / 602				
AISI 347		7978	480	14.2	3.71			15.8 / 513	18.9 / 559	21.9 / 585	24.7 / 606				
Lead	601	11340	129	35.3	24.1	39.7 / 118	36.7 / 125	34.0 / 132	31.4 / 142						
Magnesium	923	1740	1024	156	87.6	169 / 649	159 / 934	153 / 1074	149 / 1170	146 / 1267					
Molybdenum	2894	10240	251	138	53.7	179 / 141	143 / 224	134 / 261	126 / 275	118 / 285	112 / 295	105 / 308	98 / 330	90 / 380	86 / 459
Nickel Pure	1728	8900	444	90.7	23.0	164 / 232	107 / 383	80.2 / 485	65.6 / 592	67.6 / 530	71.8 / 562	76.2 / 594	82.6 / 616		
Nichrome (80% Ni, 20% Cr)	1672	8400	420	12	3.4			14 / 480	16 / 525	21 / 545					
Inconel X-750 (73% Ni, 15% Cr, 6.7% Fe)	1665	8510	439	11.7	3.1	8.7 / —	10.3 / 372	13.5 / 473	17.0 / 510	20.5 / 546	24.0 / 626	27.6 / —	33.0 / —		
Niobium	2741	8570	265	53.7	23.6	55.2 / 188	52.6 / 249	55.2 / 274	58.2 / 283	61.3 / 292	64.4 / 301	67.5 / 310	72.1 / 324	79.1 / 347	
Palladium	1827	12020	244	71.8	24.5	76.5 / 168	71.6 / 227	73.6 / 251	79.7 / 261	86.9 / 271	94.2 / 281	102 / 291	110 / 307		
Platinum Pure	2045	21450	133	71.6	25.1	77.5 / 100	72.6 / 125	71.8 / 136	73.2 / 141	75.6 / 146	78.7 / 152	82.6 / 157	89.5 / 165	99.4 / 179	
Alloy 60Pt-40Rh (60% Pt, 40% Rh)	1800	16630	162	47	17.4			52 / —	59 / —	65 / —	69 / —	73 / —	76 / —		
Rhenium	3453	21100	136	47.9	16.7	58.9 / 97	51.0 / 127	46.1 / 139	44.2 / 145	44.1 / 151	44.6 / 156	45.7 / 162	47.8 / 171	51.9 / 186	
Rhodium	2236	12450	243	150	49.6	186 / 147	154 / 220	146 / 253	136 / 274	127 / 293	121 / 311	116 / 327	110 / 349	112 / 376	
Silicon	1685	2330	712	148	89.2	884 / 259	264 / 556	98.9 / 790	61.9 / 867	42.2 / 913	31.2 / 946	25.7 / 967	22.7 / 992		
Silver	1235	10500	235	429	174	444 / 187	430 / 225	425 / 239	412 / 250	396 / 262	379 / 277	361 / 292			
Tantalum	3269	16600	140	57.5	24.7	59.2 / 110	57.5 / 133	57.8 / 144	58.6 / 146	59.4 / 149	60.2 / 152	61.0 / 155	62.2 / 160	64.1 / 172	65.6 / 189
Thorium	2023	11700	118	54.0	39.1	59.8 / 99	54.6 / 112	54.5 / 124	55.8 / 134	56.9 / 145	56.9 / 156	58.7 / 167			
Tin	505	7310	227	66.6	40.1	85.2 / 188	73.3 / 215	62.2 / 243							

TABLE A.1 *Continued*

Composition	Melting Point (K)	ρ (kg/m³)	c_p (J/kg·K)	k (W/m·K)	$\alpha \cdot 10^6$ (m²/s)	100	200	400	600	800	1000	1200	1500	2000	2500
		Properties at 300 K				Properties at Various Temperatures (K) k (W/m · K)/c_p (J/kg · K)									
Titanium	1953	4500	522	21.9	9.32	30.5	24.5	20.4	19.4	19.7	20.7	22.0	24.5		
						300	465	551	591	633	675	620	686		
Tungsten	3660	19300	132	174	68.3	208	186	159	137	125	118	113	107	100	95
						87	122	137	142	145	148	152	157	167	176
Uranium	1406	19070	116	27.6	12.5	21.7	25.1	29.6	34.0	38.8	43.9	49.0			
						94	108	125	146	176	180	161			
Vanadium	2192	6100	489	30.7	10.3	35.8	31.3	31.3	33.3	35.7	38.2	40.8	44.6	50.9	
						258	430	515	540	563	597	645	714	867	
Zinc	693	7140	389	116	41.8	117	118	111	103						
						297	367	402	436						
Zirconium	2125	6570	278	22.7	12.4	33.2	25.2	21.6	20.7	21.6	23.7	26.0	28.8	33.0	
						205	264	300	322	342	362	344	344	344	

[a]Adapted from References 1–7.

TABLE A.2 Thermophysical Properties of Selected Nonmetallic Solids[a]

Composition	Melting Point (K)	Properties at 300 K				Properties at Various Temperatures (K) k (W/m · K)/c_p (J/kg · K)									
		ρ (kg/m³)	c_p (J/kg · K)	k (W/m · K)	$\alpha \cdot 10^6$ (m²/s)	100	200	400	600	800	1000	1200	1500	2000	2500
Aluminum oxide, sapphire	2323	3970	765	46	15.1	450	82	32.4 / 940	18.9 / 1110	13.0 / 1180	10.5 / 1225				
Aluminum oxide, polycrystalline	2323	3970	765	36.0	11.9	133	55	26.4 / 940	15.8 / 1110	10.4 / 1180	7.85 / 1225	6.55	5.66	6.00	
Beryllium oxide	2725	3000	1030	272	88.0			196 / 1350	111 / 1690	70 / 1865	47 / 1975	33 / 2055	21.5 / 2145	15 / 2750	
Boron	2573	2500	1105	27.6	9.99	190	52.5	18.7 / 1490	11.3 / 1880	8.1 / 2135	6.3 / 2350	5.2 / 2555			
Boron fiber epoxy (30% vol) composite	590	2080													
k, ∥ to fibers				2.29		2.10	2.23	2.28							
k, ⊥ to fibers				0.59		0.37	0.49	0.60							
c_p			1122			364	757	1431							
Carbon Amorphous	1500	1950	—	1.60	—	0.67	1.18	1.89	2.19	2.37	2.53	2.84	3.48		
Diamond, type IIa insulator	—	3500	509	2300	—	10000 / 21	4000 / 194	1540 / 853							
Graphite, pyrolytic	2273	2210													
k, ∥ to layers				1950		4970	3230	1390	892	667	534	448	357	262	
k, ⊥ to layers				5.70		16.8	9.23	4.09	2.68	2.01	1.60	1.34	1.08	0.81	
c_p			709			136	411	992	1406	1650	1793	1890	1974	2043	
Graphite fiber epoxy (25% vol) composite	450	1400													
k, heat flow ∥ to fibers				11.1		5.7	8.7	13.0							
k, heat flow ⊥ to fibers				0.87		0.46	0.68	1.1							
c_p			935			337	642	1216							
Pyroceram, Corning 9606	1623	2600	808	3.98	1.89	5.25	4.78	3.64 / 908	3.28 / 1038	3.08 / 1122	2.96 / 1197	2.87 / 1264	2.79 / 1498		

TABLE A.2 *Continued*

Composition	Melting Point (K)	Properties at 300 K				Properties at Various Temperatures (K) k (W/m · K)/c_p (J/kg · K)									
		ρ (kg/m³)	c_p (J/kg · K)	k (W/m · K)	$\alpha \cdot 10^6$ (m²/s)	100	200	400	600	800	1000	1200	1500	2000	2500
Silicon carbide	3100	3160	675	490	230	—	—	—	—	—	87	58	30		
								880	1050	1135	1195	1243	1310		
Silicon dioxide, crystalline (quartz)	1883	2650													
k, ∥ to c axis				10.4		39	16.4	7.6	5.0	4.2					
k, ⊥ to c axis				6.21		20.8	9.5	4.70	3.4	3.1					
c_p			745			—	—	885	1075	1250					
Silicon dioxide, polycrystalline (fused silica)	1883	2220	745	1.38	0.834	0.69	1.14	1.51	1.75	2.17	2.87	4.00			
						—	—	905	1040	1105	1155	1195			
Silicon nitride	2173	2400	691	16.0	9.65	—	—	13.9	11.3	9.88	8.76	8.00	7.16	6.20	
						—	578	778	937	1063	1155	1226	1306	1377	
Sulfur	392	2070	708	0.206	0.141	0.165	0.185								
						403	606								
Thorium dioxide	3573	9110	235	13	6.1			10.2	6.6	4.7	3.68	3.12	2.73	2.5	
								255	274	285	295	303	315	330	
Titanium dioxide, polycrystalline	2133	4157	710	8.4	2.8			7.01	5.02	3.94	3.46	3.28			
								805	880	910	930	945			

^{a}Adapted from References 1, 2, 3, and 6.

TABLE A.3   Thermophysical Properties of Common Materials[a]

Structural Building Materials

	Typical Properties at 300 K		
Description/Composition	**Density, ρ (kg/m³)**	**Thermal Conductivity, k (W/m · K)**	**Specific Heat, c_p (J/kg · K)**
Building Boards			
Asbestos–cement board	1920	0.58	—
Gypsum or plaster board	800	0.17	—
Plywood	545	0.12	1215
Sheathing, regular density	290	0.055	1300
Acoustic tile	290	0.058	1340
Hardboard, siding	640	0.094	1170
Hardboard, high density	1010	0.15	1380
Particle board, low density	590	0.078	1300
Particle board, high density	1000	0.170	1300
Woods			
Hardwoods (oak, maple)	720	0.16	1255
Softwoods (fir, pine)	510	0.12	1380
Masonry Materials			
Cement mortar	1860	0.72	780
Brick, common	1920	0.72	835
Brick, face	2083	1.3	—
Clay tile, hollow			
1 cell deep, 10 cm thick	—	0.52	—
3 cells deep, 30 cm thick	—	0.69	—
Concrete block, 3 oval cores			
Sand/gravel, 20 cm thick	—	1.0	—
Cinder aggregate, 20 cm thick	—	0.67	—
Concrete block, rectangular core			
2 cores, 20 cm thick, 16 kg	—	1.1	—
Same with filled cores	—	0.60	—
Plastering Materials			
Cement plaster, sand aggregate	1860	0.72	—
Gypsum plaster, sand aggregate	1680	0.22	1085
Gypsum plaster, vermiculite aggregate	720	0.25	—

TABLE A.3 *Continued*

Insulating Materials and Systems

Description/Composition	Typical Properties at 300 K		
	Density, ρ (kg/m³)	Thermal Conductivity, k (W/m · K)	Specific Heat, c_p (J/kg · K)
Blanket and Batt			
Glass fiber, paper faced	16	0.046	—
	28	0.038	—
	40	0.035	—
Glass fiber, coated; duct liner	32	0.038	835
Board and Slab			
Cellular glass	145	0.058	1000
Glass fiber, organic bonded	105	0.036	795
Polystyrene, expanded			
Extruded (R-12)	55	0.027	1210
Molded beads	16	0.040	1210
Mineral fiberboard; roofing material	265	0.049	—
Wood, shredded/cemented	350	0.087	1590
Cork	120	0.039	1800
Loose Fill			
Cork, granulated	160	0.045	—
Diatomaceous silica, coarse	350	0.069	—
Powder	400	0.091	—
Diatomaceous silica, fine powder	200	0.052	—
	275	0.061	—
Glass fiber, poured or blown	16	0.043	835
Vermiculite, flakes	80	0.068	835
	160	0.063	1000
Formed/Foamed-in-Place			
Mineral wool granules with asbestos/inorganic binders, sprayed	190	0.046	—
Polyvinyl acetate cork mastic; sprayed or troweled	—	0.100	—
Urethane, two-part mixture; rigid foam	70	0.026	1045
Reflective			
Aluminum foil separating fluffy glass mats; 10–12 layers, evacuated; for cryogenic applications (150 K)	40	0.00016	—
Aluminum foil and glass paper laminate; 75–150 layers; evacuated; for cryogenic application (150 K)	120	0.000017	—
Typical silica powder, evacuated	160	0.0017	—

TABLE A.3 *Continued*

Industrial Insulation

Description/Composition	Maximum Service Temperature (K)	Typical Density (kg/m³)	Typical Thermal Conductivity, k (W/m · K), at Various Temperatures (K)													
			200	215	230	240	255	270	285	300	310	365	420	530	645	750
Blankets																
Blanket, mineral fiber, metal reinforced	920	96–192									0.038	0.046	0.056	0.078		
	815	40–96									0.035	0.045	0.058	0.088		
Blanket, mineral fiber, glass; fine fiber, organic bonded	450	10				0.036		0.040	0.043	0.048	0.052	0.076				
		12				0.035	0.036	0.039	0.042	0.046	0.049	0.069				
		16				0.033	0.035	0.036	0.039	0.042	0.046	0.062				
		24				0.030	0.032	0.033	0.036	0.039	0.040	0.053				
		32				0.029	0.030	0.032	0.033	0.036	0.038	0.048				
		48				0.027	0.029	0.030	0.032	0.033	0.035	0.045				
Blanket, alumina–silica fiber	1530	48												0.071	0.105	0.150
		64												0.059	0.087	0.125
		96												0.052	0.076	0.100
		128												0.049	0.068	0.091
Felt, semirigid; organic bonded	480	50–125	0.023	0.025	0.026	0.027	0.029	0.035	0.036	0.038	0.039	0.051	0.063			
	730	50						0.030	0.032	0.033	0.035	0.051	0.079			
Felt, laminated; no binder	920	120											0.051	0.065	0.087	
Blocks, Boards, and Pipe Insulations																
Asbestos paper, laminated and corrugated																
4-ply	420	190								0.078	0.082	0.098				
6-ply	420	255								0.071	0.074	0.085				
8-ply	420	300								0.068	0.071	0.082				
Magnesia, 85%	590	185									0.051	0.055	0.061	0.075	0.089	
Calcium silicate	920	190									0.055	0.059	0.063		0.104	

TABLE A.3 *Continued*

Industrial Insulation (Continued)

Description/Composition	Maximum Service Temperature (K)	Typical Density (kg/m³)	Typical Thermal Conductivity, k (W/m · K), at Various Temperatures (K)													
			200	215	230	240	255	270	285	300	310	365	420	530	645	750
Cellular glass	700	145			0.046	0.048	0.051	0.052	0.055	0.058	0.062	0.069	0.079			
Diatomaceous	1145	345												0.092	0.098	0.104
silica	1310	385												0.101	0.100	0.115
Polystyrene, rigid																
Extruded (R-12)	350	56	0.023	0.023	0.022	0.023	0.023	0.025	0.026	0.027	0.029					
Extruded (R-12)	350	35	0.023	0.023	0.023	0.025	0.025	0.026	0.027	0.029						
Molded beads	350	16	0.026	0.029	0.030	0.033	0.035	0.036	0.038	0.040						
Rubber, rigid foamed	340	70						0.029	0.030	0.032	0.033					
Insulating Cement Mineral fiber (rock, slag or glass)																
With clay binder	1255	430									0.071	0.079	0.088	0.105	0.123	
With hydraulic setting binder	922	560									0.108	0.115	0.123	0.137		
Loose Fill																
Cellulose, wood or paper pulp	—	45							0.038	0.039	0.042					
Perlite, expanded	—	105	0.036	0.039	0.042	0.043	0.046	0.049	0.051	0.053	0.056					
Vermiculite, expanded	—	122			0.056	0.058	0.061	0.063	0.065	0.068	0.071					
		80			0.049	0.051	0.055	0.058	0.061	0.063	0.066					

TABLE A.3 *Continued*

Other Materials

Description/ Composition	Temperature (K)	Density, ρ (kg/m³)	Thermal Conductivity, k (W/m · K)	Specific Heat, c_p (J/kg · K)
Asphalt	300	2115	0.062	920
Bakelite	300	1300	1.4	1465
Brick, refractory				
Carborundum	872	—	18.5	—
	1672	—	11.0	—
Chrome brick	473	3010	2.3	835
	823		2.5	
	1173		2.0	
Diatomaceous	478	—	0.25	—
silica, fired	1145	—	0.30	
Fire clay, burnt 1600 K	773	2050	1.0	960
	1073	—	1.1	
	1373	—	1.1	
Fire clay, burnt 1725 K	773	2325	1.3	960
	1073		1.4	
	1373		1.4	
Fire clay brick	478	2645	1.0	960
	922		1.5	
	1478		1.8	
Magnesite	478	—	3.8	1130
	922	—	2.8	
	1478		1.9	
Clay	300	1460	1.3	880
Coal, anthracite	300	1350	0.26	1260
Concrete (stone mix)	300	2300	1.4	880
Cotton	300	80	0.06	1300
Foodstuffs				
Banana (75.7% water content)	300	980	0.481	3350
Apple, red (75% water content)	300	840	0.513	3600
Cake, batter	300	720	0.223	—
Cake, fully baked	300	280	0.121	—
Chicken meat, white	198	—	1.60	—
(74.4% water content)	233	—	1.49	
	253		1.35	
	263		1.20	
	273		0.476	
	283		0.480	
	293		0.489	
Glass				
Plate (soda lime)	300	2500	1.4	750
Pyrex	300	2225	1.4	835

TABLE A.3   *Continued*

Other Materials (Continued)

Description/ Composition	Temperature (K)	Density, ρ (kg/m³)	Thermal Conductivity, k (W/m · K)	Specific Heat, c_p (J/kg · K)
Ice	273	920	1.88	2040
	253	—	2.03	1945
Leather (sole)	300	998	0.159	—
Paper	300	930	0.180	1340
Paraffin	300	900	0.240	2890
Rock				
Granite, Barre	300	2630	2.79	775
Limestone, Salem	300	2320	2.15	810
Marble, Halston	300	2680	2.80	830
Quartzite, Sioux	300	2640	5.38	1105
Sandstone, Berea	300	2150	2.90	745
Rubber, vulcanized				
Soft	300	1100	0.13	2010
Hard	300	1190	0.16	—
Sand	300	1515	0.27	800
Soil	300	2050	0.52	1840
Snow	273	110	0.049	—
		500	0.190	—
Teflon	300	2200	0.35	—
	400		0.45	—
Tissue, human				
Skin	300	—	0.37	—
Fat layer (adipose)	300	—	0.2	—
Muscle	300	—	0.41	—
Wood, cross grain				
Balsa	300	140	0.055	—
Cypress	300	465	0.097	—
Fir	300	415	0.11	2720
Oak	300	545	0.17	2385
Yellow pine	300	640	0.15	2805
White pine	300	435	0.11	—
Wood, radial				
Oak	300	545	0.19	2385
Fir	300	420	0.14	2720

[a]Adapted from References 1 and 8–13.

TABLE A.4 Thermophysical Properties
of Gases at Atmospheric Pressure[a]

T (K)	ρ (kg/m³)	c_p (kJ/kg · K)	$\mu \cdot 10^7$ (N · s/m²)	$\nu \cdot 10^6$ (m²/s)	$k \cdot 10^3$ (W/m · K)	$\alpha \cdot 10^6$ (m²/s)	Pr
Air							
100	3.5562	1.032	71.1	2.00	9.34	2.54	0.786
150	2.3364	1.012	103.4	4.426	13.8	5.84	0.758
200	1.7458	1.007	132.5	7.590	18.1	10.3	0.737
250	1.3947	1.006	159.6	11.44	22.3	15.9	0.720
300	1.1614	1.007	184.6	15.89	26.3	22.5	0.707
350	0.9950	1.009	208.2	20.92	30.0	29.9	0.700
400	0.8711	1.014	230.1	26.41	33.8	38.3	0.690
450	0.7740	1.021	250.7	32.39	37.3	47.2	0.686
500	0.6964	1.030	270.1	38.79	40.7	56.7	0.684
550	0.6329	1.040	288.4	45.57	43.9	66.7	0.683
600	0.5804	1.051	305.8	52.69	46.9	76.9	0.685
650	0.5356	1.063	322.5	60.21	49.7	87.3	0.690
700	0.4975	1.075	338.8	68.10	52.4	98.0	0.695
750	0.4643	1.087	354.6	76.37	54.9	109	0.702
800	0.4354	1.099	369.8	84.93	57.3	120	0.709
850	0.4097	1.110	384.3	93.80	59.6	131	0.716
900	0.3868	1.121	398.1	102.9	62.0	143	0.720
950	0.3666	1.131	411.3	112.2	64.3	155	0.723
1000	0.3482	1.141	424.4	121.9	66.7	168	0.726
1100	0.3166	1.159	449.0	141.8	71.5	195	0.728
1200	0.2902	1.175	473.0	162.9	76.3	224	0.728
1300	0.2679	1.189	496.0	185.1	82	238	0.719
1400	0.2488	1.207	530	213	91	303	0.703
1500	0.2322	1.230	557	240	100	350	0.685
1600	0.2177	1.248	584	268	106	390	0.688
1700	0.2049	1.267	611	298	113	435	0.685
1800	0.1935	1.286	637	329	120	482	0.683
1900	0.1833	1.307	663	362	128	534	0.677
2000	0.1741	1.337	689	396	137	589	0.672
2100	0.1658	1.372	715	431	147	646	0.667
2200	0.1582	1.417	740	468	160	714	0.655
2300	0.1513	1.478	766	506	175	783	0.647
2400	0.1448	1.558	792	547	196	869	0.630
2500	0.1389	1.665	818	589	222	960	0.613
3000	0.1135	2.726	955	841	486	1570	0.536
Ammonia (NH_3)							
300	0.6894	2.158	101.5	14.7	24.7	16.6	0.887
320	0.6448	2.170	109	16.9	27.2	19.4	0.870
340	0.6059	2.192	116.5	19.2	29.3	22.1	0.872
360	0.5716	2.221	124	21.7	31.6	24.9	0.872
380	0.5410	2.254	131	24.2	34.0	27.9	0.869

Table A.4 *Continued*

T (K)	ρ (kg/m^3)	c_p (kJ/kg · K)	$\mu \cdot 10^7$ (N · s/m^2)	$\nu \cdot 10^6$ (m^2/s)	$k \cdot 10^3$ (W/m · K)	$\alpha \cdot 10^6$ (m^2/s)	Pr
Ammonia (NH$_3$) *(continued)*							
400	0.5136	2.287	138	26.9	37.0	31.5	0.853
420	0.4888	2.322	145	29.7	40.4	35.6	0.833
440	0.4664	2.357	152.5	32.7	43.5	39.6	0.826
460	0.4460	2.393	159	35.7	46.3	43.4	0.822
480	0.4273	2.430	166.5	39.0	49.2	47.4	0.822
500	0.4101	2.467	173	42.2	52.5	51.9	0.813
520	0.3942	2.504	180	45.7	54.5	55.2	0.827
540	0.3795	2.540	186.5	49.1	57.5	59.7	0.824
560	0.3708	2.577	193	52.0	60.6	63.4	0.827
580	0.3533	2.613	199.5	56.5	63.8	69.1	0.817
Carbon Dioxide (CO$_2$)							
280	1.9022	0.830	140	7.36	15.20	9.63	0.765
300	1.7730	0.851	149	8.40	16.55	11.0	0.766
320	1.6609	0.872	156	9.39	18.05	12.5	0.754
340	1.5618	0.891	165	10.6	19.70	14.2	0.746
360	1.4743	0.908	173	11.7	21.2	15.8	0.741
380	1.3961	0.926	181	13.0	22.75	17.6	0.737
400	1.3257	0.942	190	14.3	24.3	19.5	0.737
450	1.1782	0.981	210	17.8	28.3	24.5	0.728
500	1.0594	1.02	231	21.8	32.5	30.1	0.725
550	0.9625	1.05	251	26.1	36.6	36.2	0.721
600	0.8826	1.08	270	30.6	40.7	42.7	0.717
650	0.8143	1.10	288	35.4	44.5	49.7	0.712
700	0.7564	1.13	305	40.3	48.1	56.3	0.717
750	0.7057	1.15	321	45.5	51.7	63.7	0.714
800	0.6614	1.17	337	51.0	55.1	71.2	0.716
Carbon Monoxide (CO)							
200	1.6888	1.045	127	7.52	17.0	9.63	0.781
220	1.5341	1.044	137	8.93	19.0	11.9	0.753
240	1.4055	1.043	147	10.5	20.6	14.1	0.744
260	1.2967	1.043	157	12.1	22.1	16.3	0.741
280	1.2038	1.042	166	13.8	23.6	18.8	0.733
300	1.1233	1.043	175	15.6	25.0	21.3	0.730
320	1.0529	1.043	184	17.5	26.3	23.9	0.730
340	0.9909	1.044	193	19.5	27.8	26.9	0.725
360	0.9357	1.045	202	21.6	29.1	29.8	0.725
380	0.8864	1.047	210	23.7	30.5	32.9	0.729
400	0.8421	1.049	218	25.9	31.8	36.0	0.719
450	0.7483	1.055	237	31.7	35.0	44.3	0.714
500	0.67352	1.065	254	37.7	38.1	53.1	0.710
550	0.61226	1.076	271	44.3	41.1	62.4	0.710
600	0.56126	1.088	286	51.0	44.0	72.1	0.707

TABLE A.4 *Continued*

T (K)	ρ (kg/m³)	c_p (kJ/kg · K)	$\mu \cdot 10^7$ (N · s/m²)	$\nu \cdot 10^6$ (m²/s)	$k \cdot 10^3$ (W/m · K)	$\alpha \cdot 10^6$ (m²/s)	Pr
Carbon Monoxide (CO) *(continued)0*							
650	0.51806	1.101	301	58.1	47.0	82.4	0.705
700	0.48102	1.114	315	65.5	50.0	93.3	0.702
750	0.44899	1.127	329	73.3	52.8	104	0.702
800	0.42095	1.140	343	81.5	55.5	116	0.705
Helium (He)							
100	0.4871	5.193	96.3	19.8	73.0	28.9	0.686
120	0.4060	5.193	107	26.4	81.9	38.8	0.679
140	0.3481	5.193	118	33.9	90.7	50.2	0.676
160	—	5.193	129	—	99.2	—	—
180	0.2708	5.193	139	51.3	107.2	76.2	0.673
200	—	5.193	150	—	115.1	—	—
220	0.2216	5.193	160	72.2	123.1	107	0.675
240	—	5.193	170	—	130	—	—
260	0.1875	5.193	180	96.0	137	141	0.682
280	—	5.193	190	—	145	—	—
300	0.1625	5.193	199	122	152	180	0.680
350	—	5.193	221	—	170	—	—
400	0.1219	5.193	243	199	187	295	0.675
450	—	5.193	263	—	204	—	—
500	0.09754	5.193	283	290	220	434	0.668
550	—	5.193	—	—	—	—	—
600	—	5.193	320	—	252	—	—
650	—	5.193	332	—	264	—	—
700	0.06969	5.193	350	502	278	768	0.654
750	—	5.193	364	—	291	—	—
800	—	5.193	382	—	304	—	—
900	—	5.193	414	—	330	—	—
1000	0.04879	5.193	446	914	354	1400	0.654
Hydrogen (H₂)							
100	0.24255	11.23	42.1	17.4	67.0	24.6	0.707
150	0.16156	12.60	56.0	34.7	101	49.6	0.699
200	0.12115	13.54	68.1	56.2	131	79.9	0.704
250	0.09693	14.06	78.9	81.4	157	115	0.707
300	0.08078	14.31	89.6	111	183	158	0.701
350	0.06924	14.43	98.8	143	204	204	0.700
400	0.06059	14.48	108.2	179	226	258	0.695
450	0.05386	14.50	117.2	218	247	316	0.689
500	0.04848	14.52	126.4	261	266	378	0.691
550	0.04407	14.53	134.3	305	285	445	0.685

TABLE A.4   *Continued*

T (K)	ρ (kg/m³)	c_p (kJ/kg · K)	$\mu \cdot 10^7$ (N · s/m²)	$\nu \cdot 10^6$ (m²/s)	$k \cdot 10^3$ (W/m · K)	$\alpha \cdot 10^6$ (m²/s)	Pr
Hydrogen (H₂) *(continued)*							
600	0.04040	14.55	142.4	352	305	519	0.678
700	0.03463	14.61	157.8	456	342	676	0.675
800	0.03030	14.70	172.4	569	378	849	0.670
900	0.02694	14.83	186.5	692	412	1030	0.671
1000	0.02424	14.99	201.3	830	448	1230	0.673
1100	0.02204	15.17	213.0	966	488	1460	0.662
1200	0.02020	15.37	226.2	1120	528	1700	0.659
1300	0.01865	15.59	238.5	1279	568	1955	0.655
1400	0.01732	15.81	250.7	1447	610	2230	0.650
1500	0.01616	16.02	262.7	1626	655	2530	0.643
1600	0.0152	16.28	273.7	1801	697	2815	0.639
1700	0.0143	16.58	284.9	1992	742	3130	0.637
1800	0.0135	16.96	296.1	2193	786	3435	0.639
1900	0.0128	17.49	307.2	2400	835	3730	0.643
2000	0.0121	18.25	318.2	2630	878	3975	0.661
Nitrogen (N₂)							
100	3.4388	1.070	68.8	2.00	9.58	2.60	0.768
150	2.2594	1.050	100.6	4.45	13.9	5.86	0.759
200	1.6883	1.043	129.2	7.65	18.3	10.4	0.736
250	1.3488	1.042	154.9	11.48	22.2	15.8	0.727
300	1.1233	1.041	178.2	15.86	25.9	22.1	0.716
350	0.9625	1.042	200.0	20.78	29.3	29.2	0.711
400	0.8425	1.045	220.4	26.16	32.7	37.1	0.704
450	0.7485	1.050	239.6	32.01	35.8	45.6	0.703
500	0.6739	1.056	257.7	38.24	38.9	54.7	0.700
550	0.6124	1.065	274.7	44.86	41.7	63.9	0.702
600	0.5615	1.075	290.8	51.79	44.6	73.9	0.701
700	0.4812	1.098	321.0	66.71	49.9	94.4	0.706
800	0.4211	1.22	349.1	82.90	54.8	116	0.715
900	0.3743	1.146	375.3	100.3	59.7	139	0.721
1000	0.3368	1.167	399.9	118.7	64.7	165	0.721
1100	0.3062	1.187	423.2	138.2	70.0	193	0.718
1200	0.2807	1.204	445.3	158.6	75.8	224	0.707
1300	0.2591	1.219	466.2	179.9	81.0	256	0.701
Oxygen (O₂)							
100	3.945	0.962	76.4	1.94	9.25	2.44	0.796
150	2.585	0.921	114.8	4.44	13.8	5.80	0.766
200	1.930	0.915	147.5	7.64	18.3	10.4	0.737
250	1.542	0.915	178.6	11.58	22.6	16.0	0.723
300	1.284	0.920	207.2	16.14	26.8	22.7	0.711

TABLE A.4 *Continued*

T (K)	ρ (kg/m³)	c_p (kJ/kg · K)	$\mu \cdot 10^7$ (N · s/m²)	$\nu \cdot 10^6$ (m²/s)	$k \cdot 10^3$ (W/m · K)	$\alpha \cdot 10^6$ (m²/s)	Pr
Oxygen (O₂) *(continued)*							
350	1.100	0.929	233.5	21.23	29.6	29.0	0.733
400	0.9620	0.942	258.2	26.84	33.0	36.4	0.737
450	0.8554	0.956	281.4	32.90	36.3	44.4	0.741
500	0.7698	0.972	303.3	39.40	41.2	55.1	0.716
550	0.6998	0.988	324.0	46.30	44.1	63.8	0.726
600	0.6414	1.003	343.7	53.59	47.3	73.5	0.729
700	0.5498	1.031	380.8	69.26	52.8	93.1	0.744
800	0.4810	1.054	415.2	86.32	58.9	116	0.743
900	0.4275	1.074	447.2	104.6	64.9	141	0.740
1000	0.3848	1.090	477.0	124.0	71.0	169	0.733
1100	0.3498	1.103	505.5	144.5	75.8	196	0.736
1200	0.3206	1.115	532.5	166.1	81.9	229	0.725
1300	0.2960	1.125	588.4	188.6	87.1	262	0.721
Water Vapor (Steam)							
380	0.5863	2.060	127.1	21.68	24.6	20.4	1.06
400	0.5542	2.014	134.4	24.25	26.1	23.4	1.04
450	0.4902	1.980	152.5	31.11	29.9	30.8	1.01
500	0.4405	1.985	170.4	38.68	33.9	38.8	0.998
550	0.4005	1.997	188.4	47.04	37.9	47.4	0.993
600	0.3652	2.026	206.7	56.60	42.2	57.0	0.993
650	0.3380	2.056	224.7	66.48	46.4	66.8	0.996
700	0.3140	2.085	242.6	77.26	50.5	77.1	1.00
750	0.2931	2.119	260.4	88.84	54.9	88.4	1.00
800	0.2739	2.152	278.6	101.7	59.2	100	1.01
850	0.2579	2.186	296.9	115.1	63.7	113	1.02

[a]Adapted from References 8, 14, and 15.

TABLE A.5   Thermophysical Properties of Saturated Fluids[a]

Saturated Liquids

T (K)	ρ (kg/m³)	c_p (kJ/kg · K)	$\mu \cdot 10^2$ (N · s/m²)	$\nu \cdot 10^6$ (m²/s)	$k \cdot 10^3$ (W/m · K)	$\alpha \cdot 10^7$ (m²/s)	Pr	$\beta \cdot 10^3$ (K⁻¹)
Engine Oil (Unused)								
273	899.1	1.796	385	4,280	147	0.910	47,000	0.70
280	895.3	1.827	217	2,430	144	0.880	27,500	0.70
290	890.0	1.868	99.9	1,120	145	0.872	12,900	0.70
300	884.1	1.909	48.6	550	145	0.859	6,400	0.70
310	877.9	1.951	25.3	288	145	0.847	3,400	0.70
320	871.8	1.993	14.1	161	143	0.823	1,965	0.70
330	865.8	2.035	8.36	96.6	141	0.800	1,205	0.70
340	859.9	2.076	5.31	61.7	139	0.779	793	0.70
350	853.9	2.118	3.56	41.7	138	0.763	546	0.70
360	847.8	2.161	2.52	29.7	138	0.753	395	0.70
370	841.8	2.206	1.86	22.0	137	0.738	300	0.70
380	836.0	2.250	1.41	16.9	136	0.723	233	0.70
390	830.6	2.294	1.10	13.3	135	0.709	187	0.70
400	825.1	2.337	0.874	10.6	134	0.695	152	0.70
410	818.9	2.381	0.698	8.52	133	0.682	125	0.70
420	812.1	2.427	0.564	6.94	133	0.675	103	0.70
430	806.5	2.471	0.470	5.83	132	0.662	88	0.70
Ethylene Glycol [$C_2H_4(OH)_2$]								
273	1,130.8	2.294	6.51	57.6	242	0.933	617	0.65
280	1,125.8	2.323	4.20	37.3	244	0.933	400	0.65
290	1,118.8	2.368	2.47	22.1	248	0.936	236	0.65
300	1,114.4	2.415	1.57	14.1	252	0.939	151	0.65
310	1,103.7	2.460	1.07	9.65	255	0.939	103	0.65
320	1,096.2	2.505	0.757	6.91	258	0.940	73.5	0.65
330	1,089.5	2.549	0.561	5.15	260	0.936	55.0	0.65
340	1,083.8	2.592	0.431	3.98	261	0.929	42.8	0.65
350	1,079.0	2.637	0.342	3.17	261	0.917	34.6	0.65
360	1,074.0	2.682	0.278	2.59	261	0.906	28.6	0.65
370	1,066.7	2.728	0.228	2.14	262	0.900	23.7	0.65
373	1,058.5	2.742	0.215	2.03	263	0.906	22.4	0.65
Glycerin [$C_3H_5(OH)_3$]								
273	1,276.0	2.261	1,060	8,310	282	0.977	85,000	0.47
280	1,271.9	2.298	534	4,200	284	0.972	43,200	0.47
290	1,265.8	2.367	185	1,460	286	0.955	15,300	0.48
300	1,259.9	2.427	79.9	634	286	0.935	6,780	0.48
310	1,253.9	2.490	35.2	281	286	0.916	3,060	0.49
320	1,247.2	2.564	21.0	168	287	0.897	1,870	0.50

TABLE A.5   *Continued*

Saturated Liquids (Continued)

T (K)	ρ (kg/m³)	c_p (kJ/kg · K)	$\mu \cdot 10^2$ (N · s/m²)	$\nu \cdot 10^6$ (m²/s)	$k \cdot 10^3$ (W/m · K)	$\alpha \cdot 10^7$ (m²/s)	Pr	$\beta \cdot 10^3$ (K⁻¹)
Freon (Refrigerant-12) (CCl_2F_2)								
230	1,528.4	0.8816	0.0457	0.299	68	0.505	5.9	1.85
240	1,498.0	0.8923	0.0385	0.257	69	0.516	5.0	1.90
250	1,469.5	0.9037	0.0354	0.241	70	0.527	4.6	2.00
260	1,439.0	0.9163	0.0322	0.224	73	0.554	4.0	2.10
270	1,407.2	0.9301	0.0304	0.216	73	0.558	3.9	2.25
280	1,374.4	0.9450	0.0283	0.206	73	0.562	3.7	2.35
290	1,340.5	0.9609	0.0265	0.198	73	0.567	3.5	2.55
300	1,305.8	0.9781	0.0254	0.195	72	0.564	3.5	2.75
310	1,268.9	0.9963	0.0244	0.192	69	0.546	3.4	3.05
320	1,228.6	1.0155	0.0233	0.190	68	0.545	3.5	3.5
Mercury (Hg)								
273	13,595	0.1404	0.1688	0.1240	8,180	42.85	0.0290	0.181
300	13,529	0.1393	0.1523	0.1125	8,540	45.30	0.0248	0.181
350	13,407	0.1377	0.1309	0.0976	9,180	49.75	0.0196	0.181
400	13,287	0.1365	0.1171	0.0882	9,800	54.05	0.0163	0.181
450	13,167	0.1357	0.1075	0.0816	10,400	58.10	0.0140	0.181
500	13,048	0.1353	0.1007	0.0771	10,950	61.90	0.0125	0.182
550	12,929	0.1352	0.0953	0.0737	11,450	65.55	0.0112	0.184
600	12,809	0.1355	0.0911	0.0711	11,950	68.80	0.0103	0.187

Saturated Liquid–Vapor, 1 atm[b]

Fluid	T_{sat} (K)	h_{fg} (kJ/kg)	ρ_f (kg/m³)	ρ_g (kg/m³)	$\sigma \cdot 10^3$ (N/m)
Ethanol	351	846	757	1.44	17.7
Ethylene glycol	470	812	1,111[c]	—	32.7
Glycerin	563	974	1,260[c]	—	63.0[c]
Mercury	630	301	12,740	3.90	417
Refrigerant R-12	243	165	1,488	6.32	15.8
Refrigerant R-113	321	147	1,511	7.38	15.9

[a]Adapted from References 15 and 16.
[b]Adapted from References 8, 17, and 18.
[c]Property value corresponding to 300 K.

TABLE A.6   **Thermophysical Properties of Saturated Water**[a]

Temperature, T (K)	Pressure,[b] P (bars)	Specific Volume (m³/kg)		Heat of Vaporization, h_{fg} (kJ/kg)	Specific Heat (kJ/kg·K)		Viscosity (N·s/m²)		Thermal Conductivity (W/m·K)		Prandtl Number		Surface Tension, $\sigma_f \cdot 10^3$ (N/m)	Expansion Coefficient, $\beta_f \cdot 10^6$ (K⁻¹)	Temperature, T (K)
		$v_f \cdot 10^3$	v_g		$c_{p,f}$	$c_{p,g}$	$\mu_f \cdot 10^6$	$\mu_g \cdot 10^6$	$k_f \cdot 10^3$	$k_g \cdot 10^3$	Pr_f	Pr_g			
273.15	0.00611	1.000	206.3	2502	4.217	1.854	1750	8.02	569	18.2	12.99	0.815	75.5	−68.05	273.15
275	0.00697	1.000	181.7	2497	4.211	1.855	1652	8.09	574	18.3	12.22	0.817	75.3	−32.74	275
280	0.00990	1.000	130.4	2485	4.198	1.858	1422	8.29	582	18.6	10.26	0.825	74.8	46.04	280
285	0.01387	1.000	99.4	2473	4.189	1.861	1225	8.49	590	18.9	8.81	0.833	74.3	114.1	285
290	0.01917	1.001	69.7	2461	4.184	1.864	1080	8.69	598	19.3	7.56	0.841	73.7	174.0	290
295	0.02617	1.002	51.94	2449	4.181	1.868	959	8.89	606	19.5	6.62	0.849	72.7	227.5	295
300	0.03531	1.003	39.13	2438	4.179	1.872	855	9.09	613	19.6	5.83	0.857	71.7	276.1	300
305	0.04712	1.005	29.74	2426	4.178	1.877	769	9.29	620	20.1	5.20	0.865	70.9	320.6	305
310	0.06221	1.007	22.93	2414	4.178	1.882	695	9.49	628	20.4	4.62	0.873	70.0	361.9	310
315	0.08132	1.009	17.82	2402	4.179	1.888	631	9.69	634	20.7	4.16	0.883	69.2	400.4	315
320	0.1053	1.011	13.98	2390	4.180	1.895	577	9.89	640	21.0	3.77	0.894	68.3	436.7	320
325	0.1351	1.013	11.06	2378	4.182	1.903	528	10.09	645	21.3	3.42	0.901	67.5	471.2	325
330	0.1719	1.016	8.82	2366	4.184	1.911	489	10.29	650	21.7	3.15	0.908	66.6	504.0	330
335	0.2167	1.018	7.09	2354	4.186	1.920	453	10.49	656	22.0	2.88	0.916	65.8	535.5	335
340	0.2713	1.021	5.74	2342	4.188	1.930	420	10.69	660	22.3	2.66	0.925	64.9	566.0	340
345	0.3372	1.024	4.683	2329	4.191	1.941	389	10.89	668	22.6	2.45	0.933	64.1	595.4	345
350	0.4163	1.027	3.846	2317	4.195	1.954	365	11.09	668	23.0	2.29	0.942	63.2	624.2	350
355	0.5100	1.030	3.180	2304	4.199	1.968	343	11.29	671	23.3	2.14	0.951	62.3	652.3	355
360	0.6209	1.034	2.645	2291	4.203	1.983	324	11.49	674	23.7	2.02	0.960	61.4	697.9	360
365	0.7514	1.038	2.212	2278	4.209	1.999	306	11.69	677	24.1	1.91	0.969	60.5	707.1	365
370	0.9040	1.041	1.861	2265	4.214	2.017	289	11.89	679	24.5	1.80	0.978	59.5	728.7	370
373.15	1.0133	1.044	1.679	2257	4.217	2.029	279	12.02	680	24.8	1.76	0.984	58.9	750.1	373.15
375	1.0815	1.045	1.574	2252	4.220	2.036	274	12.09	681	24.9	1.70	0.987	58.6	761	375
380	1.2869	1.049	1.337	2239	4.226	2.057	260	12.29	683	25.4	1.61	0.999	57.6	788	380
385	1.5233	1.053	1.142	2225	4.232	2.080	248	12.49	685	25.8	1.53	1.004	56.6	814	385

$\rho_{SAT} = 1/v_g$

T															T
390	841	55.6	1.013	1.47	26.3	686	12.69	237	2.104	4.239	2212	0.980	1.058	1.794	390
400	896	53.6	1.033	1.34	27.2	688	13.05	217	2.158	4.256	2183	0.731	1.067	2.455	400
410	952	51.5	1.054	1.24	28.2	688	13.42	200	2.221	4.278	2153	0.553	1.077	3.302	410
420	1010	49.4	1.075	1.16	29.8	688	13.79	185	2.291	4.302	2123	0.425	1.088	4.370	420
430	—	47.2	1.10	1.09	30.4	685	14.14	173	2.369	4.331	2091	0.331	1.099	5.699	430
440		45.1	1.12	1.04	31.7	682	14.50	162	2.46	4.36	2059	0.261	1.110	7.333	440
450		42.9	1.14	0.99	33.1	678	14.85	152	2.56	4.40	2024	0.208	1.123	9.319	450
460		40.7	1.17	0.95	34.6	673	15.19	143	2.68	4.44	1989	0.167	1.137	11.71	460
470		38.5	1.20	0.92	36.3	667	15.54	136	2.79	4.48	1951	0.136	1.152	14.55	470
480		36.2	1.23	0.89	38.1	660	15.88	129	2.94	4.53	1912	0.111	1.167	17.90	480
490	—	33.9	1.25	0.87	40.1	651	16.23	124	3.10	4.59	1870	0.0922	1.184	21.83	490
500	—	31.6	1.28	0.86	42.3	642	16.59	118	3.27	4.66	1825	0.0766	1.203	26.40	500
510	—	29.3	1.31	0.85	44.7	631	16.95	113	3.47	4.74	1779	0.0631	1.222	31.66	510
520	—	26.9	1.35	0.84	47.5	621	17.33	108	3.70	4.84	1730	0.0525	1.244	37.70	520
530	—	24.5	1.39	0.85	50.6	608	17.72	104	3.96	4.95	1679	0.0445	1.268	44.58	530
540	—	22.1	1.43	0.86	54.0	594	18.1	101	4.27	5.08	1622	0.0375	1.294	52.38	540
550	—	19.7	1.47	0.87	58.3	580	18.6	97	4.64	5.24	1564	0.0317	1.323	61.19	550
560	—	17.3	1.52	0.90	63.7	563	19.1	94	5.09	5.43	1499	0.0269	1.355	71.08	560
570	—	15.0	1.59	0.94	76.7	548	19.7	91	5.67	5.68	1429	0.0228	1.392	82.16	570
580	—	12.8	1.68	0.99	76.7	528	20.4	88	6.40	6.00	1353	0.0193	1.433	94.51	580
590	—	10.5	1.84	1.05	84.1	513	21.5	84	7.35	6.41	1274	0.0163	1.482	108.3	590
600	—	8.4	2.15	1.14	92.9	497	22.7	81	8.75	7.00	1176	0.0137	1.541	123.5	600
610	—	6.3	2.60	1.30	103	467	24.1	77	11.1	7.85	1068	0.0115	1.612	137.3	610
620	—	4.5	3.46	1.52	114	444	25.9	72	15.4	9.35	941	0.0094	1.705	159.1	620
625	—	3.5	4.20	1.65	121	430	27.0	70	18.3	10.6	858	0.0085	1.778	169.1	625
630	—	2.6	4.8	2.0	130	412	28.0	67	22.1	12.6	781	0.0075	1.856	179.7	630
635	—	1.5	6.0	2.7	141	392	30.0	64	27.6	16.4	683	0.0066	1.935	190.9	635
640	—	0.8	9.6	4.2	155	367	32.0	59	42	26	560	0.0057	2.075.	202.7	640
645	—	0.1	26	12	178	331	37.0	54	—	90	361	0.0045	2.351	215.2	645
647.3[c]	—	0.0	∞	∞	238	238	45.0	45	∞	∞	0	0.0032	3.170	221.2	647.3[c]

[a] Adapted from Reference 19.
[b] 1 bar = 10^5 N/m^2.
[c] Critical temperature.

TABLE A.7   Thermophysical Properties of Liquid Metals[a]

Composition	Melting Point (K)	T (K)	ρ (kg/m³)	c_p (kJ/kg · K)	$\nu \cdot 10^7$ (m²/s)	k (W/m · K)	$\alpha \cdot 10^5$ (m²/s)	Pr
Bismuth	544	589	10,011	0.1444	1.617	16.4	0.138	0.0142
		811	9,739	0.1545	1.133	15.6	1.035	0.0110
		1033	9,467	0.1645	0.8343	15.6	1.001	0.0083
Lead	600	644	10,540	0.159	2.276	16.1	1.084	0.024
		755	10,412	0.155	1.849	15.6	1.223	0.017
		977	10,140	—	1.347	14.9	—	—
Potassium	337	422	807.3	0.80	4.608	45.0	6.99	0.0066
		700	741.7	0.75	2.397	39.5	7.07	0.0034
		977	674.4	0.75	1.905	33.1	6.55	0.0029
Sodium	371	366	929.1	1.38	7.516	86.2	6.71	0.011
		644	860.2	1.30	3.270	72.3	6.48	0.0051
		977	778.5	1.26	2.285	59.7	6.12	0.0037
NaK, (45%/55%)	292	366	887.4	1.130	6.522	25.6	2.552	0.026
		644	821.7	1.055	2.871	27.5	3.17	0.0091
		977	740.1	1.043	2.174	28.9	3.74	0.0058
NaK, (22%/78%)	262	366	849.0	0.946	5.797	24.4	3.05	0.019
		672	775.3	0.879	2.666	26.7	3.92	0.0068
		1033	690.4	0.883	2.118	—	—	—
PbBi, (44.5%/55.5%)	398	422	10,524	0.147	—	9.05	0.586	—
		644	10,236	0.147	1.496	11.86	0.790	0.189
		922	9,835	—	1.171	—	—	—
Mercury	234			See Table A.5				

[a]Adapted from *Liquid Materials Handbook,* 23rd ed., the Atomic Energy Commission, Department of the Navy, Washington, DC, 1952.

TABLE A.8 Binary Diffusion Coefficients at One Atmosphere[a,b]

Substance A	Substance B	T (K)	D_{AB} (m²/s)
Gases			
NH_3	Air	298	0.28×10^{-4}
H_2O	Air	298	0.26×10^{-4}
CO_2	Air	298	0.16×10^{-4}
H_2	Air	298	0.41×10^{-4}
O_2	Air	298	0.21×10^{-4}
Acetone	Air	273	0.11×10^{-4}
Benzene	Air	298	0.88×10^{-5}
Naphthalene	Air	300	0.62×10^{-5}
Ar	N_2	293	0.19×10^{-4}
H_2	O_2	273	0.70×10^{-4}
H_2	N_2	273	0.68×10^{-4}
H_2	CO_2	273	0.55×10^{-4}
CO_2	N_2	293	0.16×10^{-4}
CO_2	O_2	273	0.14×10^{-4}
O_2	N_2	273	0.18×10^{-4}
Dilute Solutions			
Caffeine	H_2O	298	0.63×10^{-9}
Ethanol	H_2O	298	0.12×10^{-8}
Glucose	H_2O	298	0.69×10^{-9}
Glycerol	H_2O	298	0.94×10^{-9}
Acetone	H_2O	298	0.13×10^{-8}
CO_2	H_2O	298	0.20×10^{-8}
O_2	H_2O	298	0.24×10^{-8}
H_2	H_2O	298	0.63×10^{-8}
N_2	H_2O	298	0.26×10^{-8}
Solids			
O_2	Rubber	298	0.21×10^{-9}
N_2	Rubber	298	0.15×10^{-9}
CO_2	Rubber	298	0.11×10^{-9}
He	SiO_2	293	0.4×10^{-13}
H_2	Fe	293	0.26×10^{-12}
Cd	Cu	293	0.27×10^{-18}
Al	Cu	293	0.13×10^{-33}

[a]Adapted with permission from References 20, 21, and 22.

[b]Assuming ideal gas behavior, the pressure and temperature dependence of the diffusion coefficient for a binary mixture of gases may be estimated from the relation

$$D_{AB} \propto p^{-1} T^{3/2}$$

TABLE A.9   Henry's Constant for Selected Gases in Water at Moderate Pressure[a]

$H = p_{A,i}/x_{A,i}$ **(bars)**

T (K)	NH$_3$	Cl$_2$	H$_2$S	SO$_2$	CO$_2$	CH$_4$	O$_2$	H$_2$
273	21	265	260	165	710	22,880	25,500	58,000
280	23	365	335	210	960	27,800	30,500	61,500
290	26	480	450	315	1300	35,200	37,600	66,500
300	30	615	570	440	1730	42,800	45,700	71,600
310	—	755	700	600	2175	50,000	52,500	76,000
320	—	860	835	800	2650	56,300	56,800	78,600
323	—	890	870	850	2870	58,000	58,000	79,000

[a]Adapted with permission from Reference 23.

TABLE A.10   The Solubility of Selected Gases and Solids[a]

Gas	Solid	T (K)	$S = C_{A,i}/p_{A,i}$ (kmol/m^3 · bar)
O$_2$	Rubber	298	3.12×10^{-3}
N$_2$	Rubber	298	1.56×10^{-3}
CO$_2$	Rubber	298	40.15×10^{-3}
He	SiO$_2$	293	0.45×10^{-3}
H$_2$	Ni	358	9.01×10^{-3}

[a]Adapted with permission from Reference 22.

TABLE A.11 Total, Normal (*n*) or Hemispherical (*h*) Emissivity of Selected Surfaces

Metallic Solids and Their Oxides[a]

Description/Composition		Emissivity, ε_n or ε_h, at Various Temperatures (K)										
		100	200	300	400	600	800	1000	1200	1500	2000	2500
Aluminum												
Highly polished, film	(*h*)	0.02	0.03	0.04	0.05	0.06						
Foil, bright	(*h*)	0.06	0.06	0.07								
Anodized	(*h*)			0.82	0.76							
Chromium												
Polished or plated	(*n*)	0.05	0.07	0.10	0.12	0.14						
Copper												
Highly polished	(*h*)			0.03	0.03	0.04	0.04	0.04				
Stably oxidized	(*h*)					0.50	0.58	0.80				
Gold												
Highly polished or film	(*h*)	0.01	0.02	0.03	0.03	0.04	0.05	0.06				
Foil, bright	(*h*)	0.06	0.07	0.07								
Molybdenum												
Polished	(*h*)					0.06	0.08	0.10	0.12	0.15	0.21	0.26
Shot-blasted, rough	(*h*)					0.25	0.28	0.31	0.35	0.42		
Stably oxidized	(*h*)					0.80	0.82					
Nickel												
Polished	(*h*)					0.09	0.11	0.14	0.17			
Stably oxidized	(*h*)					0.40	0.49	0.57				
Platinum												
Polished	(*h*)						0.10	0.13	0.15	0.18		
Silver												
Polished	(*h*)			0.02	0.02	0.03	0.05	0.08				
Stainless steels												
Typical, polished	(*n*)			0.17	0.17	0.19	0.23	0.30				
Typical, cleaned	(*n*)			0.22	0.22	0.24	0.28	0.35				
Typical, lightly oxidized	(*n*)						0.33	0.40				
Typical, highly oxidized	(*n*)						0.67	0.70	0.76			
AISI 347, stably oxidized	(*n*)					0.87	0.88	0.89	0.90			
Tantalum												
Polished	(*h*)								0.11	0.17	0.23	0.28
Tungsten												
Polished	(*h*)							0.10	0.13	0.18	0.25	0.29

TABLE A.11   *Continued*

Nonmetallic Substances[b]

Description/Composition		Temperature (K)	Emissivity ε
Aluminum oxide	(n)	600	0.69
		1000	0.55
		1500	0.41
Asphalt pavement	(h)	300	0.85–0.93
Building materials			
Asbestos sheet	(h)	300	0.93–0.96
Brick, red	(h)	300	0.93–0.96
Gypsum or plaster board	(h)	300	0.90–0.92
Wood	(h)	300	0.82–0.92
Cloth	(h)	300	0.75–0.90
Concrete	(h)	300	0.88–0.93
Glass, window	(h)	300	0.90–0.95
Ice	(h)	273	0.95–0.98
Paints			
Black (Parsons)	(h)	300	0.98
White, acrylic	(h)	300	0.90
White, zinc oxide	(h)	300	0.92
Paper, white	(h)	300	0.92–0.97
Pyrex	(n)	300	0.82
		600	0.80
		1000	0.71
		1200	0.62
Pyroceram	(n)	300	0.85
		600	0.78
		1000	0.69
		1500	0.57
Refractories (furnace liners)			
Alumina brick	(n)	800	0.40
		1000	0.33
		1400	0.28
		1600	0.33
Magnesia brick	(n)	800	0.45
		1000	0.36
		1400	0.31
		1600	0.40
Kaolin insulating brick	(n)	800	0.70
		1200	0.57
		1400	0.47
		1600	0.53
Sand	(h)	300	0.90
Silicon carbide	(n)	600	0.87
		1000	0.87
		1500	0.85
Skin	(h)	300	0.95
Snow	(h)	273	0.82–0.90

TABLE A.11 *Continued*

Nonmetallic Substances[b]

Description/Composition		Temperature (K)	Emissivity ε
Soil	(h)	300	0.93–0.96
Rocks	(h)	300	0.88–0.95
Teflon	(h)	300	0.85
		400	0.87
		500	0.92
Vegetation	(h)	300	0.92–0.96
Water	(h)	300	0.96

[a]Adapted from Reference 1.
[b]Adapted from References 1, 9, 24, and 25.

TABLE A.12 Solar Radiative Properties for Selected Materials[a]

Description/Composition	α_S	ε^b	α_S/ε	τ_S
Aluminum				
Polished	0.09	0.03	3.0	
Anodized	0.14	0.84	0.17	
Quartz overcoated	0.11	0.37	0.30	
Foil	0.15	0.05	3.0	
Brick, red (Purdue)	0.63	0.93	0.68	
Concrete	0.60	0.88	0.68	
Galvanized sheet metal				
Clean, new	0.65	0.13	5.0	
Oxidized, weathered	0.80	0.28	2.9	
Glass, 3.2-mm thickness				
Float or tempered				0.79
Low iron oxide type				0.88
Metal, plated				
Black sulfide	0.92	0.10	9.2	
Black cobalt oxide	0.93	0.30	3.1	
Black nickel oxide	0.92	0.08	11	
Black chrome	0.87	0.09	9.7	
Mylar, 0.13-mm thickness				0.87
Paints				
Black (Parsons)	0.98	0.98	1.0	
White, acrylic	0.26	0.90	0.29	
White, zinc oxide	0.16	0.93	0.17	
Plexiglas, 3.2-mm thickness				0.90
Snow				
Fine particles, fresh	0.13	0.82	0.16	
Ice granules	0.33	0.89	0.37	
Tedlar, 0.10-mm thickness				0.92
Teflon, 0.13-mm thickness				0.92

[a]Adapted with permission from Reference 25.
[b]The emissivity values in this table correspond to a surface temperature of approximately 300 K.

References

1. Touloukian, Y. S., and C. Y. Ho, Eds., *Thermophysical Properties of Matter,* Vol. 1, *Thermal Conductivity of Metallic Solids;* Vol. 2, *Thermal Conductivity of Nonmetallic Solids;* Vol. 4, *Specific Heat of Metallic Solids;* Vol. 5, *Specific Heat of Nonmetallic Solids;* Vol. 7, *Thermal Radiative Properties of Metallic Solids;* Vol. 8, *Thermal Radiative Properties of Nonmetallic Solids;* Vol. 9, *Thermal Radiative Properties of Coatings,* Plenum Press, New York, 1972.

2. Touloukian, Y. S., and C. Y. Ho, Eds., *Thermophysical Properties of Selected Aerospace Materials,* Part I: Thermal Radiative Properties; Part II: Thermophysical Properties of Seven Materials. Thermophysical and Electronic Properties Information Analysis Center, CINDAS, Purdue University, West Lafayette, IN, 1976.

3. Ho, C. Y., R. W. Powell, and P. E. Liley, *J. Phys. Chem. Ref. Data,* **3,** Supplement 1, 1974.

4. Desai, P. D., T. K. Chu, R. H. Bogaard, M. W. Ackermann, and C. Y. Ho, Part I: Thermophysical Properties of Carbon Steels, Part II: Thermophysical Properties of Low Chromium Steels, Part III: Thermophysical Properties of Nickel Steels, Part IV: Thermophysical Properties of Stainless Steels, CINDAS Special Report, Purdue University, West Lafayette, IN, September 1976.

5. American Society for Metals, *Metals Handbook,* Vol. 1, *Properties and Selection of Metals,* 8th ed., ASM, Metals Park, OH, 1961.

6. Hultgren, R., P. D. Desai, D. T. Hawkins, M. Gleiser, K. K. Kelley, and D. D. Wagman, *Selected Values of the Thermodynamic Properties of the Elements,* American Society of Metals, Metals Park, OH, 1973.

7. Hultgren, R., P. D. Desai, D. T. Hawkins, M. Gleiser, and K. K. Kelley, *Selected Values of the Thermodynamic Properties of Binary Alloys,* American Society of Metals, Metals Park, OH, 1973.

8. American Society of Heating, Refrigerating and Air Conditioning Engineers, *ASHRAE Handbook of Fundamentals,* ASHRAE, New York, 1981.

9. Mallory, J. F., *Thermal Insulation,* Van Nostrand Reinhold, New York, 1969.

10. Hanley, E. J., D. P. DeWitt, and R. E. Taylor, "The Thermal Transport Properties at Normal and Elevated Temperature of Eight Representative Rocks," *Proceedings of the Seventh Symposium on Thermophysical Properties,* American Society of Mechanical Engineers, New York, 1977.

11. Sweat, V. E., "A Miniature Thermal Conductivity Probe for Foods," American Society of Mechanical Engineers, Paper 76-HT-60, August 1976.

12. Kothandaraman, C. P., and S. Subramanyan, *Heat and Mass Transfer Data Book,* Halsted Press/Wiley, New York, 1975.

13. Chapman, A. J., *Heat Transfer,* 4th ed., Macmillan, New York, 1984.

14. Vargaftik, N. B., *Tables of Thermophysical Properties of Liquids and Gases,* 2nd ed., Hemisphere Publishing, New York, 1975.

15. Eckert, E. R. G., and R. M. Drake, *Analysis of Heat and Mass Transfer,* McGraw-Hill, New York, 1972.

16. Vukalovich, M. P., A. I. Ivanov, L. R. Fokin, and A. T. Yakovelev, *Thermophysical Properties of Mercury,* State Committee on Standards, State Service for Standards and Handbook Data, Monograph Series No. 9, Izd. Standartov, Moscow, 1971.

17. Bolz, R. E., and G. L. Tuve, Eds., CRC Handbook of Tables for Applied Engineering Science, 2nd ed., CRC Press, Boca Raton, FL, 1979.

18. Liley, P. E., private communication, School of Mechanical Engineering, Purdue University, West Lafayette, IN, May 1984.

19. Liley, P. E., Steam Tables in SI Units, private communication, School of Mechanical Engineering, Purdue University, West Lafayette, IN, March 1984.

20. Perry, J. H., Ed., *Chemical Engineer's Handbook,* 4th ed., McGraw-Hill, New York, 1963.

21. Geankoplis, C. J., *Mass Transport Phenomena,* Holt, Rinehart & Winston, New York, 1972.

22. Barrer, R. M., *Diffusion In and Through Solids,* Macmillan, New York, 1941.

23. Spalding, D. B., *Convective Mass Transfer,* McGraw-Hill, New York, 1963.

24. Gubareff, G. G., J. E. Janssen, and R. H. Torborg, *Thermal Radiation Properties Survey,* Minneapolis-Honeywell Regulator Company, Minneapolis, MN, 1960.

25. Kreith, F. and J. F. Kreider, *Principles of Solar Energy,* Hemisphere Publishing, New York, 1978.

APPENDIX B

Mathematical Relations and Functions

B.1
Hyperbolic Functions[1]

x	$\sinh x$	$\cosh x$	$\tanh x$	x	$\sinh x$	$\cosh x$	$\tanh x$
0.00	0.0000	1.0000	0.00000	2.00	3.6269	3.7622	0.96403
0.10	0.1002	1.0050	0.09967	2.10	4.0219	4.1443	0.97045
0.20	0.2013	1.0201	0.19738	2.20	4.4571	4.5679	0.97574
0.30	0.3045	1.0453	0.29131	2.30	4.9370	5.0372	0.98010
0.40	0.4108	1.0811	0.37995	2.40	5.4662	5.5569	0.98367
0.50	0.5211	1.1276	0.46212	2.50	6.0502	6.1323	0.98661
0.60	0.6367	1.1855	0.53705	2.60	6.6947	6.7690	0.98903
0.70	0.7586	1.2552	0.60437	2.70	7.4063	7.4735	0.99101
0.80	0.8881	1.3374	0.66404	2.80	8.1919	8.2527	0.99263
0.90	1.0265	1.4331	0.71630	2.90	9.0596	9.1146	0.99396
1.00	1.1752	1.5431	0.76159	3.00	10.018	10.068	0.99505
1.10	1.3356	1.6685	0.80050	3.50	16.543	16.573	0.99818
1.20	1.5095	1.8107	0.83365	4.00	27.290	27.308	0.99933
1.30	1.6984	1.9709	0.86172	4.50	45.003	45.014	0.99975
1.40	1.9043	2.1509	0.88535	5.00	74.203	74.210	0.99991
1.50	2.1293	2.3524	0.90515	6.00	201.71	201.72	0.99999
1.60	2.3756	2.5775	0.92167	7.00	548.32	548.32	1.0000
1.70	2.6456	2.8283	0.93541	8.00	1490.5	1490.5	1.0000
1.80	2.9422	3.1075	0.94681	9.00	4051.5	4051.5	1.0000
1.90	3.2682	3.4177	0.95624	10.000	11013	11013	1.0000

[1]The hyperbolic functions are defined as

$$\sinh x = \tfrac{1}{2}(e^x - e^{-x}) \qquad \cosh x = \tfrac{1}{2}(e^x + e^{-x}) \qquad \tanh x = \frac{e^x - e^{-x}}{e^x + e^{-x}} = \frac{\sinh x}{\cosh x}$$

The derivatives of the hyperbolic functions of the variable u are given as

$$\frac{d}{dx}(\sinh u) = (\cosh u)\frac{du}{dx} \qquad \frac{d}{dx}(\cosh u) = (\sinh u)\frac{du}{dx} \qquad \frac{d}{dx}(\tanh u) = \left(\frac{1}{\cosh^2 u}\right)\frac{du}{dx}$$

B.2
Gaussian Error Function[1]

w	erf w	w	erf w	w	erf w
0.00	0.00000	0.36	0.38933	1.04	0.85865
0.02	0.02256	0.38	0.40901	1.08	0.87333
0.04	0.04511	0.40	0.42839	1.12	0.88679
0.06	0.06762	0.44	0.46622	1.16	0.89910
0.08	0.09008	0.48	0.50275	1.20	0.91031
0.10	0.11246	0.52	0.53790	1.30	0.93401
0.12	0.13476	0.56	0.57162	1.40	0.95228
0.14	0.15695	0.60	0.60386	1.50	0.96611
0.16	0.17901	0.64	0.63459	1.60	0.97635
0.18	0.20094	0.68	0.66378	1.70	0.98379
0.20	0.22270	0.72	0.69143	1.80	0.98909
0.22	0.24430	0.76	0.71754	1.90	0.99279
0.24	0.26570	0.80	0.74210	2.00	0.99532
0.26	0.28690	0.84	0.76514	2.20	0.99814
0.28	0.30788	0.88	0.78669	2.40	0.99931
0.30	0.32863	0.92	0.80677	2.60	0.99976
0.32	0.34913	0.96	0.82542	2.80	0.99992
0.34	0.36936	1.00	0.84270	3.00	0.99998

[1]The Gaussian error function is defined as

$$\text{erf } w = \frac{2}{\sqrt{\pi}} \int_0^w e^{-v^2} \, dv$$

The complementary error function is defined as

$$\text{erfc } w \equiv 1 - \text{erf } w$$

B.3

The First Four Roots of the Transcendental Equation, $\xi_n \tan \xi_n = Bi$, For Transient Conduction in a Plane Wall

$Bi = \dfrac{hL}{k}$	ξ_1	ξ_2	ξ_3	ξ_4
0	0	3.1416	6.2832	9.4248
0.001	0.0316	3.1419	6.2833	9.4249
0.002	0.0447	3.1422	6.2835	9.4250
0.004	0.0632	3.1429	6.2838	9.4252
0.006	0.0774	3.1435	6.2841	9.4254
0.008	0.0893	3.1441	6.2845	9.4256
0.01	0.0998	3.1448	6.2848	9.4258
0.02	0.1410	3.1479	6.2864	9.4269
0.04	0.1987	3.1543	6.2895	9.4290
0.06	0.2425	3.1606	6.2927	9.4311
0.08	0.2791	3.1668	6.2959	9.4333
0.1	0.3111	3.1731	6.2991	9.4354
0.2	0.4328	3.2039	6.3148	9.4459
0.3	0.5218	3.2341	6.3305	9.4565
0.4	0.5932	3.2636	6.3461	9.4670
0.5	0.6533	3.2923	6.3616	9.4775
0.6	0.7051	3.3204	6.3770	9.4879
0.7	0.7506	3.3477	6.3923	9.4983
0.8	0.7910	3.3744	6.4074	9.5087
0.9	0.8274	3.4003	6.4224	9.5190
1.0	0.8603	3.4256	6.4373	9.5293
1.5	0.9882	3.5422	6.5097	9.5801
2.0	1.0769	3.6436	6.5783	9.6296
3.0	1.1925	3.8088	6.7040	9.7240
4.0	1.2646	3.9352	6.8140	9.8119
5.0	1.3138	4.0336	6.9096	9.8928
6.0	1.3496	4.1116	6.9924	9.9667
7.0	1.3766	4.1746	7.0640	10.0339
8.0	1.3978	4.2264	7.1263	10.0949
9.0	1.4149	4.2694	7.1806	10.1502
10.0	1.4289	4.3058	7.2281	10.2003
15.0	1.4729	4.4255	7.3959	10.3898
20.0	1.4961	4.4915	7.4954	10.5117
30.0	1.5202	4.5615	7.6057	10.6543
40.0	1.5325	4.5979	7.6647	10.7334
50.0	1.5400	4.6202	7.7012	10.7832
60.0	1.5451	4.6353	7.7259	10.8172
80.0	1.5514	4.6543	7.7573	10.8606
100.0	1.5552	4.6658	7.7764	10.8871
∞	1.5708	4.7124	7.8540	10.9956

B.4
Bessel Functions of the First Kind

x	$J_0(x)$	$J_1(x)$
0.0	1.0000	0.0000
0.1	0.9975	0.0499
0.2	0.9900	0.0995
0.3	0.9776	0.1483
0.4	0.9604	0.1960
0.5	0.9385	0.2423
0.6	0.9120	0.2867
0.7	0.8812	0.3290
0.8	0.8463	0.3688
0.9	0.8075	0.4059
1.0	0.7652	0.4400
1.1	0.7196	0.4709
1.2	0.6711	0.4983
1.3	0.6201	0.5220
1.4	0.5669	0.5419
1.5	0.5118	0.5579
1.6	0.4554	0.5699
1.7	0.3980	0.5778
1.8	0.3400	0.5815
1.9	0.2818	0.5812
2.0	0.2239	0.5767
2.1	0.1666	0.5683
2.2	0.1104	0.5560
2.3	0.0555	0.5399
2.4	0.0025	0.5202

B.5
Modified Bessel Functions[1] of the First and Second Kinds

x	$e^{-x}I_0(x)$	$e^{-x}I_1(x)$	$e^x K_0(x)$	$e^x K_1(x)$
0.0	1.0000	0.0000	∞	∞
0.2	0.8269	0.0823	2.1407	5.8334
0.4	0.6974	0.1368	1.6627	3.2587
0.6	0.5993	0.1722	1.4167	2.3739
0.8	0.5241	0.1945	1.2582	1.9179
1.0	0.4657	0.2079	1.1445	1.6361
1.2	0.4198	0.2152	1.0575	1.4429
1.4	0.3831	0.2185	0.9881	1.3010
1.6	0.3533	0.2190	0.9309	1.1919
1.8	0.3289	0.2177	0.8828	1.1048
2.0	0.3085	0.2153	0.8416	1.0335
2.2	0.2913	0.2121	0.8056	0.9738
2.4	0.2766	0.2085	0.7740	0.9229
2.6	0.2639	0.2046	0.7459	0.8790
2.8	0.2528	0.2007	0.7206	0.8405
3.0	0.2430	0.1968	0.6978	0.8066
3.2	0.2343	0.1930	0.6770	0.7763
3.4	0.2264	0.1892	0.6579	0.7491
3.6	0.2193	0.1856	0.6404	0.7245
3.8	0.2129	0.1821	0.6243	0.7021
4.0	0.2070	0.1787	0.6093	0.6816
4.2	0.2016	0.1755	0.5953	0.6627
4.4	0.1966	0.1724	0.5823	0.6453
4.6	0.1919	0.1695	0.5701	0.6292
4.8	0.1876	0.1667	0.5586	0.6142
5.0	0.1835	0.1640	0.5478	0.6003
5.2	0.1797	0.1614	0.5376	0.5872
5.4	0.1762	0.1589	0.5279	0.5749
5.6	0.1728	0.1565	0.5188	0.5633
5.8	0.1696	0.1542	0.5101	0.5525
6.0	0.1666	0.1520	0.5019	0.5422
6.4	0.1611	0.1479	0.4865	0.5232
6.8	0.1561	0.1441	0.4724	0.5060
7.2	0.1515	0.1405	0.4595	0.4905
7.6	0.1473	0.1372	0.4476	0.4762
8.0	0.1434	0.1341	0.4366	0.4631
8.4	0.1398	0.1312	0.4264	0.4511
8.8	0.1365	0.1285	0.4168	0.4399
9.2	0.1334	0.1260	0.4079	0.4295
9.6	0.1305	0.1235	0.3995	0.4198
10.0	0.1278	0.1213	0.3916	0.4108

[1] $I_{n+1}(x) = I_{n-1}(x) - (2n/x)I_n(x)$

APPENDIX **C**

Thermal Conditions Associated with Uniform Energy Generation in One-Dimensional, Steady-State Systems

In Section 3.5 the problem of conduction with thermal energy generation is considered for one-dimensional, steady-state conditions. The form of the heat equation differs, according to whether the system is a plane wall, a cylindrical shell, or a spherical shell (Figure C.1). In each case, there are several options for the boundary condition at each surface, and hence a greater number of possibilities for specific forms of the temperature distribution and heat rate (or heat flux).

An alternative to solving the heat equation for each possible combination of boundary conditions involves obtaining a solution by prescribing *boundary conditions of the first kind,* Equation 2.24, at both surfaces and then applying an energy balance to each surface at which the temperature is unknown. For the geometries of Figure C.1, with uniform temperatures $T_{s,1}$ and $T_{s,2}$ prescribed at each surface, solutions to appropriate forms of the heat equation are readily obtained and are summarized in Table C.1. The temperature distributions may be used with Fourier's law to obtain corresponding distributions for the heat flux and heat rate. If $T_{s,1}$ and $T_{s,2}$ are both known for a particular problem, the expressions of Table C.1 provide all that is needed to completely determine related thermal conditions. If $T_{s,1}$ and/or $T_{s,2}$ are not known, the results may still be used with surface energy balances to determine the desired thermal conditions.

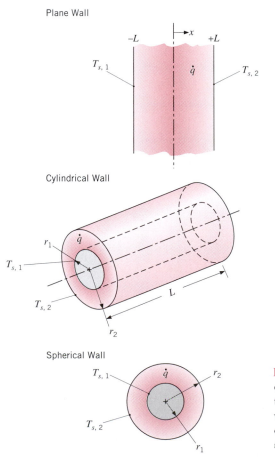

FIGURE C.1   One-dimensional conduction systems with uniform thermal energy generation: a plane wall with asymmetric surface conditions, a cylindrical shell, and a spherical shell.

TABLE C.1 One-Dimensional, Steady-State Solutions to the Heat
Equation for Plane, Cylindrical, and Spherical Walls with
Uniform Generation and Asymmetrical Surface Conditions

Temperature Distribution

Plane Wall
$$T(x) = \frac{\dot{q}L^2}{2k}\left(1 - \frac{x^2}{L^2}\right) + \frac{T_{s,2} - T_{s,1}}{2}\frac{x}{L} + \frac{T_{s,1} + T_{s,2}}{2} \tag{C.1}$$

Cylindrical Wall
$$T(r) = T_{s,2} + \frac{\dot{q}r_2^2}{4k}\left(1 - \frac{r^2}{r_2^2}\right) - \left[\frac{\dot{q}r_2^2}{4k}\left(1 - \frac{r_1^2}{r_2^2}\right) + (T_{s,2} - T_{s,1})\right]\frac{\ln(r_2/r)}{\ln(r_2/r_1)} \tag{C.2}$$

Spherical Wall
$$T(r) = T_{s,2} + \frac{\dot{q}r_2^2}{6k}\left(1 - \frac{r^2}{r_2^2}\right) - \left[\frac{\dot{q}r_2^2}{6k}\left(1 - \frac{r_1^2}{r_2^2}\right) + (T_{s,2} - T_{s,1})\right]\frac{(1/r) - (1/r_2)}{(1/r_1) - (1/r_2)} \tag{C.3}$$

Heat Flux

Plane Wall
$$q''(x) = \dot{q}x - \frac{k}{2L}(T_{s,2} - T_{s,1}) \tag{C.4}$$

Cylindrical Wall
$$q''(r) = \frac{\dot{q}r}{2} - \frac{k\left[\frac{\dot{q}r_2^2}{4k}\left(1 - \frac{r_1^2}{r_2^2}\right) + (T_{s,2} - T_{s,1})\right]}{r\ln(r_2/r_1)} \tag{C.5}$$

Spherical Wall
$$q''(r) = \frac{\dot{q}r}{3} - \frac{k\left[\frac{\dot{q}r_2^2}{6k}\left(1 - \frac{r_1^2}{r_2^2}\right) + (T_{s,2} - T_{s,1})\right]}{r^2[(1/r_1) - (1/r_2)]} \tag{C.6}$$

Heat Rate

Plane Wall
$$q(x) = \left[\dot{q}x - \frac{k}{2L}(T_{s,2} - T_{s,1})\right]A_x \tag{C.7}$$

Cylindrical Wall
$$q(r) = \dot{q}\pi Lr^2 - \frac{2\pi Lk}{\ln(r_2/r_1)} \cdot \left[\frac{\dot{q}r_2^2}{4k}\left(1 - \frac{r_1^2}{r_2^2}\right) + (T_{s,2} - T_{s,1})\right] \tag{C.8}$$

Spherical Wall
$$q(r) = \frac{\dot{q}4\pi r^3}{3} - \frac{4\pi k\left[\frac{\dot{q}r_2^2}{6k}\left(1 - \frac{r_1^2}{r_2^2}\right) + (T_{s,2} - T_{s,1})\right]}{(1/r_1) - (1/r_2)} \tag{C.9}$$

Alternative surface conditions could involve specification of a uniform surface heat flux (*boundary condition of the second kind,* Equation 2.25 or 2.26) or
a convection condition (*boundary condition of the third kind,* Equation 2.27). In
each case, the surface temperature would not be known but could be determined by applying a surface energy balance. The forms that such balances may
take are summarized in Table C.2. Note that, to accommodate situations for
which a surface of interest may adjoin a composite wall in which there is no
generation, the boundary condition of the third kind has been applied by using
the overall heat transfer coefficient U in lieu of the convection coefficient h.

TABLE C.2   Alternative Surface Conditions and Energy Balances for
One-Dimensional, Steady-State Solutions to the Heat Equation for
Plane, Cylindrical, and Spherical Walls with Uniform Generation

Plane Wall

Uniform Surface Heat Flux

$$x = -L: \qquad q''_{s,1} = -\dot{q}L - \frac{k}{2L}(T_{s,2} - T_{s,1}) \tag{C.10}$$

$$x = +L: \qquad q''_{s,2} = \dot{q}L - \frac{k}{2L}(T_{s,2} - T_{s,1}) \tag{C.11}$$

Prescribed Transport Coefficient and Ambient Temperature

$$x = -L: \qquad U_1(T_{\infty,1} - T_{s,1}) = -\dot{q}L - \frac{k}{2L}(T_{s,2} - T_{s,1}) \tag{C.12}$$

$$x = +L: \qquad U_2(T_{s,2} - T_{\infty,2}) = \dot{q}L - \frac{k}{2L}(T_{s,2} - T_{s,1}) \tag{C.13}$$

Cylindrical Wall

Uniform Surface Heat Flux

$$r = r_1: \qquad q''_{s,1} = \frac{\dot{q}r_1}{2} - \frac{k\left[\dfrac{\dot{q}r_2^2}{4k}\left(1 - \dfrac{r_1^2}{r_2^2}\right) + (T_{s,2} - T_{s,1})\right]}{r_1 \ln(r_2/r_1)} \tag{C.14}$$

$$r = r_2: \qquad q''_{s,2} = \frac{\dot{q}r_2}{2} - \frac{k\left[\dfrac{\dot{q}r_2^2}{4k}\left(1 - \dfrac{r_1^2}{r_2^2}\right) + (T_{s,2} - T_{s,1})\right]}{r_2 \ln(r_2/r_1)} \tag{C.15}$$

Prescribed Transport Coefficient and Ambient Temperature

$$r = r_1: \qquad U_1(T_{\infty,1} - T_{s,1}) = \frac{\dot{q}r_1}{2} - \frac{k\left[\dfrac{\dot{q}r_2^2}{4k}\left(1 - \dfrac{r_1^2}{r_2^2}\right) + (T_{s,2} - T_{s,1})\right]}{r_1 \ln(r_2/r_1)} \tag{C.16}$$

$$r = r_2: \qquad U_2(T_{s,2} - T_{\infty,2}) = \frac{\dot{q}r_2}{2} - \frac{k\left[\dfrac{\dot{q}r_2^2}{4k}\left(1 - \dfrac{r_1^2}{r_2^2}\right) + (T_{s,2} - T_{s,1})\right]}{r_2 \ln(r_2/r_1)} \tag{C.17}$$

Spherical Wall

Uniform Surface Heat Flux

$$r = r_1: \qquad q''_{s,1} = \frac{\dot{q}r_1}{3} - \frac{k\left[\dfrac{\dot{q}r_2^2}{6k}\left(1 - \dfrac{r_1^2}{r_2^2}\right) + (T_{s,2} - T_{s,1})\right]}{r_1^2[(1/r_1) - (1/r_2)]} \tag{C.18}$$

$$r = r_2: \qquad q''_{s,2} = \frac{\dot{q}r_2}{3} - \frac{k\left[\dfrac{\dot{q}r_2^2}{6k}\left(1 - \dfrac{r_1^2}{r_2^2}\right) + (T_{s,2} - T_{s,1})\right]}{r_2^2[(1/r_1) - (1/r_2)]} \tag{C.19}$$

TABLE C.2   *Continued*

Prescribed Transport Coefficient and Ambient Temperature

$$r = r_1: \quad U_1(T_{\infty, 1} - T_{s, 1}) = \frac{\dot{q}r_1}{3} - \frac{k\left[\dfrac{\dot{q}r_2^2}{6k}\left(1 - \dfrac{r_1^2}{r_2^2}\right) + (T_{s, 2} - T_{s, 1})\right]}{r_1^2[(1/r_1) - (1/r_2)]} \tag{C.20}$$

$$r = r_2: \quad U_2(T_{s, 2} - T_{\infty, 2}) = \frac{\dot{q}r_2}{3} - \frac{k\left[\dfrac{\dot{q}r_2^2}{6k}\left(1 - \dfrac{r_1^2}{r_2^2}\right) + (T_{s, 2} - T_{s, 1})\right]}{r_2^2[(1/r_1) - (1/r_2)]} \tag{C.21}$$

As an example, consider a plane wall for which a uniform (known) surface temperature $T_{s, 1}$ is prescribed at $x = -L$ and a uniform heat flux $q''_{s, 2}$ is prescribed at $x = +L$. Equation C.11 may be used to evaluate $T_{s, 2}$, and Equations C.1, C.4, and C.7 may then be used to determine the temperature, heat flux, and heat rate distributions, respectively.

Special cases of the foregoing configurations involve a plane wall with one adiabatic surface, a solid cylinder (a circular rod), and a sphere (Figure C.2). Subject to the requirements that $dT/dx|_{x=0} = 0$ and $dT/dr|_{r=0} = 0$, the corresponding forms of the heat equation may be solved to obtain Equations C.22 to C.24 of Table C.3. The solutions are based on prescribing a uniform tempera-

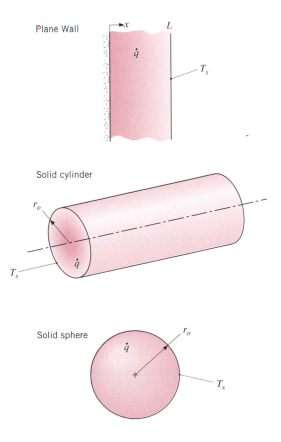

FIGURE C.2   One-dimensional conduction systems with uniform thermal energy generation: a plane wall with one adiabatic surface, a cylindrical rod, and a sphere.

TABLE C.3   One-Dimensional, Steady-State Solutions to the Heat
Equation for Uniform Generation in a Plane Wall with One
Adiabatic Surface, a Solid Cylinder, and a Solid Sphere

Temperature Distribution

Plane Wall
$$T(x) = \frac{\dot{q}L^2}{2k}\left(1 - \frac{x^2}{L^2}\right) + T_s \qquad (C.22)$$

Circular Rod
$$T(r) = \frac{\dot{q}r_o^2}{4k}\left(1 - \frac{r^2}{r_o^2}\right) + T_s \qquad (C.23)$$

Sphere
$$T(r) = \frac{\dot{q}r_o^2}{6k}\left(1 - \frac{r^2}{r_o^2}\right) + T_s \qquad (C.24)$$

Heat Flux

Plane Wall
$$q''(x) = \dot{q}x \qquad (C.25)$$

Circular Rod
$$q''(r) = \frac{\dot{q}r}{2} \qquad (C.26)$$

Sphere
$$q''(r) = \frac{\dot{q}r}{3} \qquad (C.27)$$

Heat Rate

Plane Wall
$$q(x) = \dot{q}xA_x \qquad (C.28)$$

Circular Rod
$$q(r) = \dot{q}\pi Lr^2 \qquad (C.29)$$

Sphere
$$q(r) = \frac{\dot{q}4\pi r^3}{3} \qquad (C.30)$$

ture T_s at $x = L$ and $r = r_o$. Using Fourier's law with the temperature distributions, the heat flux (Equations C.25 to C.27) and heat rate (Equations C.28 to C.30) distributions may also be obtained. If T_s is not known, it may be determined by applying a surface energy balance, appropriate forms of which are summarized in Table C.4.

TABLE C.4 Alternative Surface Conditions and Energy Balances for One-Dimensional, Steady-State Solutions to the Heat Equation for Uniform Generation in a Plane Wall with One Adiabatic Surface, a Solid Cylinder, and a Solid Sphere

Prescribed Transport Coefficient and Ambient Temperature

Plane Wall

$$x = L: \qquad \dot{q}L = U(T_s - T_\infty) \tag{C.31}$$

Circular Rod

$$r = r_o: \qquad \frac{\dot{q}r_o}{2} = U(T_s - T_\infty) \tag{C.32}$$

Sphere

$$r = r_o: \qquad \frac{\dot{q}r_o}{3} = U(T_s - T_\infty) \tag{C.33}$$

Graphical Representation of One-Dimensional, Transient Conduction in the Plane Wall, Long Cylinder, and Sphere

In Sections 5.4 and 5.5, one-term approximations have been developed for transient, one-dimensional conduction in a plane wall (with symmetrical convection conditions) and radial systems (long cylinder and sphere). The results apply for $Fo > 0.2$ and can conveniently be represented in graphical forms that illustrate the functional dependence of the transient temperature distribution on the Biot and Fourier numbers.

Results for the plane wall (Figure 5.6a) are presented in Figures D.1 to D.3. Figure D.1 may be used to obtain the *midplane* temperature of the wall, $T(0, t) \equiv T_o(t)$, at any time during the transient process. If T_o is known for particular values of Fo and Bi, Figure D.2 may be used to determine the corresponding temperature at any location *off the midplane*. Hence Figure D.2 must be used in conjunction with Figure D.1. For example, if one wishes to determine the surface temperature ($x^* = \pm 1$) at some time t, Figure D.1 would first be used to determine T_o at t. Figure D.2 would then be used to determine the surface temperature from knowledge of T_o. The procedure would be inverted if the problem were one of determining the time required for the surface to reach a prescribed temperature.

Graphical results for the energy transferred from a plane wall over the time interval t are presented in Figure D.3. These results were generated from Equation 5.46. The dimensionless energy transfer Q/Q_o is expressed exclusively in terms of Fo and Bi.

Results for the infinite cylinder are presented in Figures D.4 to D.6, and those for the sphere are presented in Figures D.7 to D.9, where the Biot number is defined in terms of the radius r_o.

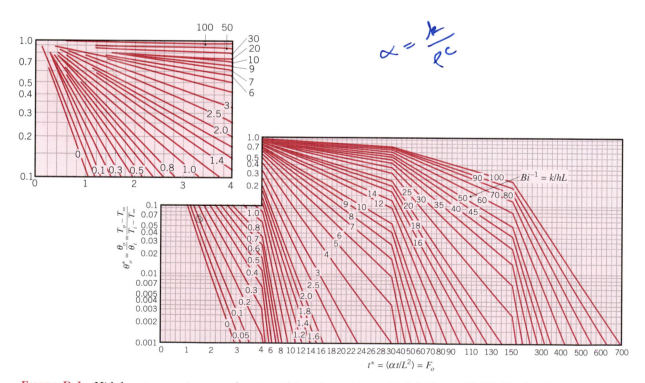

FIGURE D.1   Midplane temperature as a function of time for a plane wall of thickness 2L [1]. Used with permission.

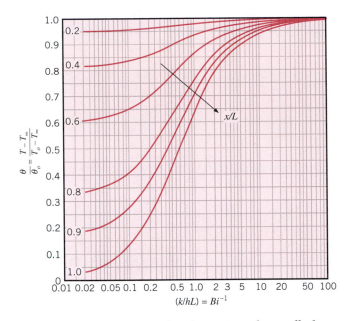

FIGURE D.2 Temperature distribution in a plane wall of thickness $2L$ [1]. Used with permission.

The foregoing charts may also be used to determine the transient response of a plane wall, an infinite cylinder, or sphere subjected to a *sudden change in surface temperature*. For such a condition it is only necessary to replace T_∞ by the prescribed surface temperature T_s and to set Bi^{-1} equal to zero. In so doing, the convection coefficient is tacitly assumed to be infinite, in which case $T_\infty = T_s$.

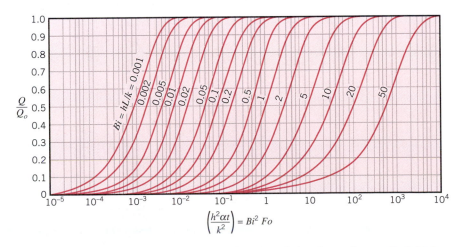

FIGURE D.3 Internal energy change as a function of time for a plane wall of thickness $2L$ [2]. Adapted with permission.

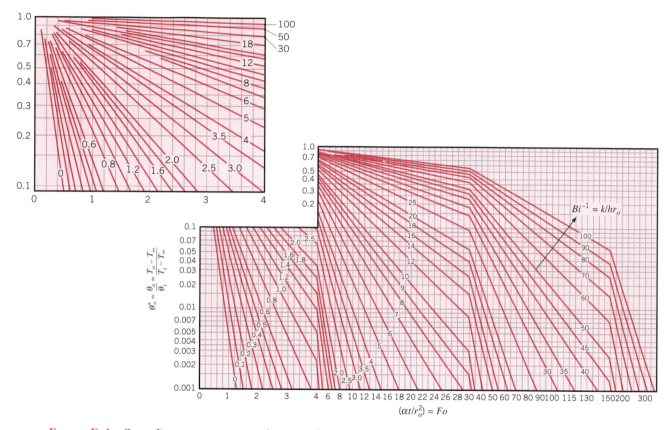

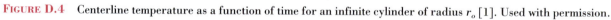

FIGURE D.4   Centerline temperature as a function of time for an infinite cylinder of radius r_o [1]. Used with permission.

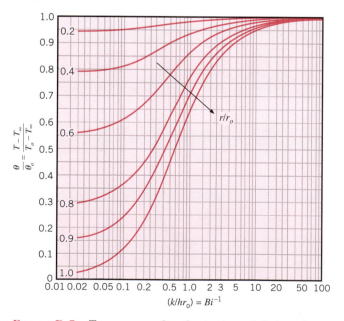

FIGURE D.5   Temperature distribution in an infinite cylinder of radius r_o [1]. Used with permission.

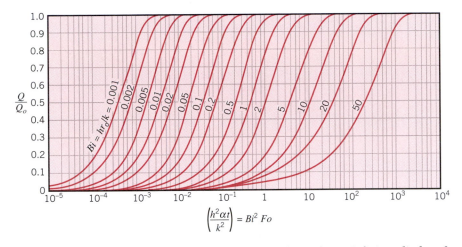

FIGURE D.6 Internal energy change as a function of time for an infinite cylinder of radius r_o [2]. Adapted with permission.

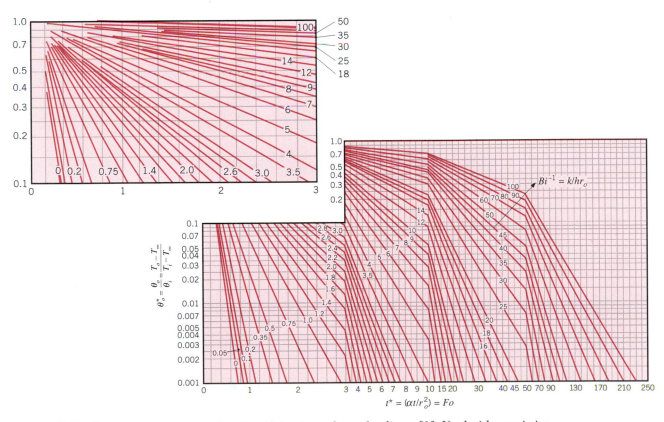

FIGURE D.7 Center temperature as a function of time in a sphere of radius r_o [1]. Used with permission.

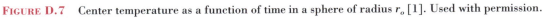

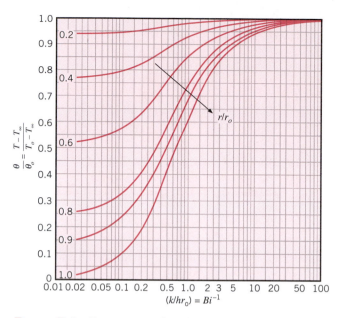

FIGURE D.8   Temperature distribution in a sphere of radius r_o [1]. Used with permission.

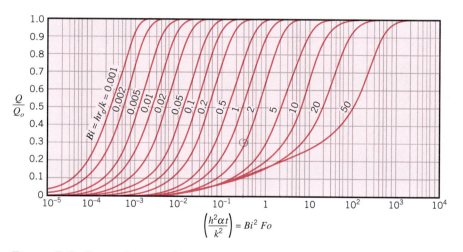
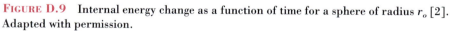

FIGURE D.9   Internal energy change as a function of time for a sphere of radius r_o [2]. Adapted with permission.

References

1. Heisler, M. P., *Trans. ASME,* **69,** 227–236, 1947.
2. Gröber, H., S. Erk, and U. Grigull, *Fundamentals of Heat Transfer,* McGraw-Hill, New York, 1961.

An Integral Laminar Boundary Layer Solution for Parallel Flow over a Flat Plate

An alternative approach to solving the boundary layer equations involves the use of an approximate *integral* method. The approach was originally proposed by von Kármán [1] in 1921 and first applied by Pohlhausen [2]. It is without the mathematical complications inherent in the *exact (similarity)* method of Section 7.2.1; yet it can be used to obtain reasonably accurate results for the key boundary layer parameters (δ, δ_t, δ_c, C_f, h, and h_m). Although the method has been used with some success for a variety of flow conditions, we restrict our attention to parallel flow over a flat plate, subject to the same restrictions enumerated in Section 7.2.1, that is, *incompressible laminar flow* with *constant fluid properties* and *negligible viscous dissipation*.

To use the method, the boundary layer equations, Equations 7.4 to 7.7, must be cast in integral form. These forms are obtained by integrating the equations in the y direction across the boundary layer. For example, integrating Equation 7.4, we obtain

$$\int_0^\delta \frac{\partial u}{\partial x} \, dy + \int_0^\delta \frac{\partial v}{\partial y} \, dy = 0 \tag{E.1}$$

or, since $v = 0$ at $y = 0$,

$$v(y = \delta) = -\int_0^\delta \frac{\partial u}{\partial x} \, dy \tag{E.2}$$

Similarly, from Equation 7.5, we obtain

$$\int_0^\delta u \frac{\partial u}{\partial x} \, dy + \int_0^\delta v \frac{\partial u}{\partial y} \, dy = \nu \int_0^\delta \frac{\partial}{\partial y} \left(\frac{\partial u}{\partial y} \right) dy$$

or, integrating the second term on the left-hand side by parts,

$$\int_0^\delta u \frac{\partial u}{\partial x} \, dy + uv \Big|_0^\delta - \int_0^\delta u \frac{\partial v}{\partial y} \, dy = \nu \frac{\partial u}{\partial y} \Big|_0^\delta$$

Substituting from Equations 7.4 and E.2, we obtain

$$\int_0^\delta u \frac{\partial u}{\partial x} \, dy - u_\infty \int_0^\delta \frac{\partial u}{\partial x} \, dy + \int_0^\delta u \frac{\partial u}{\partial x} \, dy = -\nu \frac{\partial u}{\partial y} \Big|_{y=0}$$

or

$$u_\infty \int_0^\delta \frac{\partial u}{\partial x} \, dy - \int_0^\delta 2u \frac{\partial u}{\partial x} \, dy = \nu \frac{\partial u}{\partial y} \Big|_{y=0}$$

Therefore

$$\int_0^\delta \frac{\partial}{\partial x} (u_\infty \cdot u - u \cdot u) \, dy = \nu \frac{\partial u}{\partial y} \Big|_{y=0}$$

Rearranging, we then obtain

$$\frac{d}{dx} \left[\int_0^\delta (u_\infty - u) u \, dy \right] = \nu \frac{\partial u}{\partial y} \Big|_{y=0} \tag{E.3}$$

Equation E.3 is the integral form of the boundary layer momentum equation. In a similar fashion, the following integral forms of the boundary layer energy and species continuity equations may be obtained:

$$\frac{d}{dx}\left[\int_0^{\delta_t}(T_\infty - T)u\,dy\right] = \alpha\,\frac{\partial T}{\partial y}\bigg|_{y=0} \tag{E.4}$$

$$\frac{d}{dx}\left[\int_0^{\delta_c}(\rho_{A,\infty} - \rho_A)u\,dy\right] = D_{AB}\,\frac{\partial \rho_A}{\partial y}\bigg|_{y=0} \tag{E.5}$$

Equations E.3 to E.5 satisfy the x momentum, the energy, and the species conservation requirements in an *integral* (or *average*) fashion over the entire boundary layer. In contrast, the original conservation equations, (7.5) to (7.7), satisfy the conservation requirements *locally,* that is, at each point in the boundary layer.

The integral equations can be used to obtain *approximate* boundary layer solutions. The procedure involves first *assuming* reasonable functional forms for the unknowns u, T, and ρ_A in terms of the corresponding (*unknown*) boundary layer thicknesses. The assumed forms must satisfy appropriate boundary conditions. Substituting these forms into the integral equations, expressions for the boundary layer thicknesses may be determined and the assumed functional forms may then be completely specified. Although this method is approximate, it frequently leads to accurate results for the surface parameters.

Consider the hydrodynamic boundary layer, for which appropriate boundary conditions are

$$u(y=0) = \frac{\partial u}{\partial y}\bigg|_{y=\delta} = 0 \qquad \text{and} \qquad u(y=\delta) = u_\infty$$

From Equation 7.5 it also follows that, since $u = v = 0$ at $y = 0$,

$$\frac{\partial^2 u}{\partial y^2}\bigg|_{y=0} = 0$$

With the foregoing conditions, we could approximate the velocity profile as a third-degree polynomial of the form

$$\frac{u}{u_\infty} = a_1 + a_2\left(\frac{y}{\delta}\right) + a_3\left(\frac{y}{\delta}\right)^2 + a_4\left(\frac{y}{\delta}\right)^3$$

and apply the conditions to determine the coefficients a_1 to a_4. It is easily verified that $a_1 = a_3 = 0$, $a_2 = \frac{3}{2}$ and $a_4 = -\frac{1}{2}$, in which case

$$\frac{u}{u_\infty} = \frac{3}{2}\frac{y}{\delta} - \frac{1}{2}\left(\frac{y}{\delta}\right)^3 \tag{E.6}$$

The velocity profile is then specified in terms of the unknown boundary layer thickness δ. This unknown may be determined by substituting Equation E.6 into E.3 and integrating over y to obtain

$$\frac{d}{dx}\left(\frac{39}{280}u_\infty^2\,\delta\right) = \frac{3}{2}\frac{\nu u_\infty}{\delta}$$

Separating variables and integrating over x, we obtain

$$\frac{\delta^2}{2} = \frac{140}{13} \frac{\nu x}{u_\infty} + \text{constant}$$

However, since $\delta = 0$ at the leading edge of the plate ($x = 0$), the integration constant must be zero and

$$\delta = 4.64 \left(\frac{\nu x}{u_\infty} \right)^{1/2} = \frac{4.64 x}{Re_x^{1/2}} \tag{E.7}$$

Substituting Equation E.7 into Equation E.6 and evaluating $\tau_s = \mu(\partial u/\partial y)_s$, we also obtain

$$C_{f,x} = \frac{\tau_s}{\rho u_\infty^2/2} = \frac{0.646}{Re_x^{1/2}} \tag{E.8}$$

Despite the approximate nature of the foregoing procedure, Equations E.7 and E.8 compare quite well with results obtained from the exact solution, Equations 7.19 and 7.20.

In a similar fashion one could assume a temperature profile of the form

$$T^* = \frac{T - T_s}{T_\infty - T_s} = b_1 + b_2 \left(\frac{y}{\delta_t} \right) + b_3 \left(\frac{y}{\delta_t} \right)^2 + b_4 \left(\frac{y}{\delta_t} \right)^3$$

and determine the coefficients from the conditions

$$T^*(y = 0) = \left. \frac{\partial T^*}{\partial y} \right|_{y=\delta_t} = 0$$

$$T^*(y = \delta_t) = 1$$

as well as

$$\left. \frac{\partial^2 T^*}{\partial y^2} \right|_{y=0} = 0$$

which is inferred from the energy equation (7.6). We then obtain

$$T^* = \frac{3}{2} \frac{y}{\delta_t} - \frac{1}{2} \left(\frac{y}{\delta_t} \right)^3 \tag{E.9}$$

Substituting Equations E.6 and E.9 into Equation E.4, we obtain, after some manipulation and assuming $Pr \gtrsim 1$,

$$\frac{\delta_t}{\delta} = \frac{Pr^{-1/3}}{1.026} \tag{E.10}$$

This result is in good agreement with that obtained from the exact solution, Equation 7.24. Moreover, the heat transfer coefficient may be then computed from

$$h = \frac{-k \left. \partial T/\partial y \right|_{y=0}}{T_s - T_\infty} = \frac{3}{2} \frac{k}{\delta_t}$$

Substituting from Equations E.7 and E.10, we obtain

$$Nu_x = \frac{hx}{k} = 0.332Re_x^{1/2}\,Pr^{1/3} \tag{E.11}$$

This result agrees precisely with that obtained from the exact solution, Equation 7.23. Using the same procedures, analogous results may be obtained for the concentration boundary layer.

References

1. von Kárman, T., *Z. Angew. Math. Mech.,* **1,** 232, 1921.   2. Pohlhausen, K., *Z. Angew. Math. Mech.,* **1,** 252, 1921.

Index

Conversion Factors

Acceleration	1 m/s^2	$= 4.2520 \times 10^7 \text{ ft/h}^2$
Area	1 m^2	$= 1550.0 \text{ in.}^2$
		$= 10.764 \text{ ft}^2$
Energy	1 J	$= 9.4787 \times 10^{-4} \text{ Btu}$
Force	1 N	$= 0.22481 \text{ lb}_f$
Heat transfer rate	1 W	$= 3.4123 \text{ Btu/h}$
Heat flux	1 W/m^2	$= 0.3171 \text{ Btu/h} \cdot \text{ft}^2$
Heat generation rate	1 W/m^3	$= 0.09665 \text{ Btu/h} \cdot \text{ft}^3$
Heat transfer coefficient	$1 \text{ W/m}^2 \cdot \text{K}$	$= 0.17612 \text{ Btu/h} \cdot \text{ft}^2 \cdot {}^\circ\text{F}$
Kinematic viscosity and diffusivities	$1 \text{ m}^2/\text{s}$	$= 3.875 \times 10^4 \text{ ft}^2/\text{h}$
Latent heat	1 J/kg	$= 4.2995 \times 10^{-4} \text{ Btu/lb}_m$
Length	1 m	$= 39.370 \text{ in.}$
		$= 3.2808 \text{ ft}$
	1 km	$= 0.62137 \text{ mile}$
Mass	1 kg	$= 2.2046 \text{ lb}_m$
Mass density	1 kg/m^3	$= 0.062428 \text{ lb}_m/\text{ft}^3$
Mass flow rate	1 kg/s	$= 7936.6 \text{ lb}_m/\text{h}$
Mass transfer coefficient	1 m/s	$= 1.1811 \times 10^4 \text{ ft/h}$
Pressure and stress[1]	1 N/m^2	$= 0.020886 \text{ lb}_f/\text{ft}^2$
		$= 1.4504 \times 10^{-4} \text{ lb}_f/\text{in.}^2$
		$= 4.015 \times 10^{-3} \text{ in. water}$
		$= 2.953 \times 10^{-4} \text{ in. Hg}$
	$1.0133 \times 10^5 \text{ N/m}^2$	$= 1 \text{ standard atmosphere}$
	$1 \times 10^5 \text{ N/m}^2$	$= 1 \text{ bar}$
Specific heat	$1 \text{ J/kg} \cdot \text{K}$	$= 2.3886 \times 10^{-4} \text{ Btu/lb}_m \cdot {}^\circ\text{F}$
Temperature	K	$= (5/9){}^\circ\text{R}$
		$= (5/9)({}^\circ\text{F} + 459.67)$
		$= {}^\circ\text{C} + 273.15$
Temperature difference	1 K	$= 1{}^\circ\text{C}$
		$= (9/5){}^\circ\text{R} = (9/5){}^\circ\text{F}$
Thermal conductivity	$1 \text{ W/m} \cdot \text{K}$	$= 0.57782 \text{ Btu/h} \cdot \text{ft} \cdot {}^\circ\text{F}$
Thermal resistance	1 K/W	$= 0.52750 {}^\circ\text{F/h} \cdot \text{Btu}$
Viscosity (dynamic)[2]	$1 \text{ N} \cdot \text{s/m}^2$	$= 2419.1 \text{ lb}_m/\text{ft} \cdot \text{h}$
		$= 5.8016 \times 10^{-6} \text{ lb}_f \cdot \text{h/ft}^2$
Volume	1 m^3	$= 6.1023 \times 10^4 \text{ in.}^3$
		$= 35.314 \text{ ft}^3$
		$= 264.17 \text{ gal}$
Volume flow rate	$1 \text{ m}^3/\text{s}$	$= 1.2713 \times 10^5 \text{ ft}^3/\text{h}$
		$= 2.1189 \times 10^3 \text{ ft}^3/\text{min}$
		$= 1.5850 \times 10^4 \text{ gal/min}$

[1] The SI name for the quantity pressure is pascal (Pa) having units N/m^2 or $\text{kg/m} \cdot \text{s}^2$.
[2] Also expressed in equivalent units of $\text{kg/s} \cdot \text{m}$.